Functions and Relations

$f : A \to B$	f is a function with domain (input set) A, and codomain (output set) B
$f : A \to B :: y = f(x)$	The function f from A to B is defined by the formula $f(x)$
$\lfloor x \rfloor$ or floor(x)	The floor function; $=$ the greatest integer k that is less than or equal to x
$\lceil x \rceil$ or ceil(x)	The ceiling function; $=$ the least integer k that is greater than or equal to x
$f(S)$	The image of set S under the function f; $= \{f(s) \mid s \in S\}$
$f^{-1}(S)$	The pre-image of set T under the function f; $= \{x \mid f(x) \in T\}$
$f^{-1} : B \to A$	The inverse function of a (bijective) function $f : A \to B$
$g \circ f : A \to C$	Composition of two functions $f : A \to B_1$, $g : B \to C$ (with $f(A) \subseteq B$)
$i_A : A \to A$	The identity function defined by $i_A(a) = a$, for every $a \in A$
$a \mid b$	a divides b; there exists an integer k such that $b = ak$
$a \equiv b \pmod{m}$	a is congruent to b modulo m; $m \mid (a - b)$
$\overline{x}$	Boolean complement of a Boolean variable x; $\overline{1} = 0, \overline{0} = 1$
f^d	The (Boolean) dual of a the Boolean expression f
$\sum_{i=1}^{n} a_i$	Sigma notation for the sum $a_1 + a_2 + \cdots + a_n$
$n!$	The factorial function; $= 1 \cdot 2 \cdot 3 \cdots n$, if n is a positive integer; $0! = 1$
$\pi(x)$	The number of prime numbers p satisfying $p < x$
$\gcd(a, b)$	The greatest common divisor of a and b
$\text{lcm}(a, b)$	The least common multiple of a and b
$\phi(n)$	Euler's phi-function
$\text{ord}_n(a)$	The order of a relative to n (or order of a mod n)
$f(n) = O(g(n))$	Big-O notation; for some $N \in \mathbb{Z}_+$ and $C > 0$, $f(n) \le Cg(n)$ for every $n \ge N$
$f(n) = \Omega(g(n))$	Big-Omega notation; for some $N \in \mathbb{Z}_+$ and $C > 0$, $f(n) \ge Cg(n)$ for every $n \ge N$
$f(n) = \Theta(g(n))$	Big-Theta notation; $f(n) = O(g(n))$ and $g(n) = O(f(n))$
$f(n) \precsim g(n)$	$f(n)$ is dominated by $g(n)$; $f(n) = O(g(n))$
$f(n) \prec g(n)$	$f(n)$ is strictly dominated by $g(n)$

Number Systems and Matrices

$n \sim [c_K \cdots c_1\ c_0]$ (base b)	The positive integer n has base b representation $[c_K \cdots c_1\ c_0]$
A, B, C, D, E, F	Hexadecimal (base 16) string notation for the integers 10,11,12,14,15, respectively
$\mathbb{Z}_m$	The set of integer mod m; $= \{0, 1, 2, \cdots, m-1\}$
$\mathbb{Z}_m^{\times}$	The set of invertible elements of $\mathbb{Z}_m$
$A = [a_{ij}]$	Matrix notation; a_{ij} is the row i, column j entry of the matrix A
$u \bullet v$	The dot product of two vectors u, v that have the same length
I or I_n	The $(n \times n)$ square identity matrix
A^{-1}	The inverse of the invertible square matrix A

D0022642

Continued inside back cover

DISCRETE STRUCTURES WITH CONTEMPORARY APPLICATIONS

DISCRETE STRUCTURES WITH CONTEMPORARY APPLICATIONS

Alexander Stanoyevitch

California State University–Dominguez Hills

Carson, California, USA

CRC Press
Taylor & Francis Group
Boca Raton   London   New York

CRC Press is an imprint of the
Taylor & Francis Group, an **informa** business

A CHAPMAN & HALL BOOK

MATLAB® is a trademark of The MathWorks, Inc. and is used with permission. The MathWorks does not warrant the accuracy of the text or exercises in this book. This book's use or discussion of MATLAB® software or related products does not constitute endorsement or sponsorship by The MathWorks of a particular pedagogical approach or particular use of the MATLAB® software.

MATLAB® is a registered trademark of The Math Works, Inc. For product information, please contact: The Math Works, Inc., 3 Apple Hull Drive, Natick, MA. Tel. +508-647-7000; Fax: +508-647-7001; E-mail: info@mathworks.com; Web: http://www.mathworks.com

Chapman & Hall/CRC
Taylor & Francis Group
6000 Broken Sound Parkway NW, Suite 300
Boca Raton, FL 33487-2742

© 2011 by Taylor and Francis Group, LLC
Chapman & Hall/CRC is an imprint of Taylor & Francis Group, an Informa business

No claim to original U.S. Government works

Printed in the United States of America on acid-free paper
10 9 8 7 6 5 4 3 2 1

International Standard Book Number: 978-1-4398-1768-1 (Hardback)

Library of Congress Cataloging-in-Publication Data

Stanoyevitch, Alexander.
  Discrete structures with contemporary applications / Alexander Stanoyevitch.
    p. cm.
  Includes bibliographical references and index.
  ISBN 978-1-4398-1768-1 (hardback)
  1. Computer science--Mathematics. 2. Logic, Symbolic and mathematical. 3. Probabilities. I. Title.

  QA76.9.M35S735 2010
  004.01'51--dc22   2010043711

Visit the Taylor & Francis Web site at
http://www.taylorandfrancis.com

and the CRC Press Web site at
http://www.crcpress.com

Contents

Uniform Random Variables, Setting up a Simulation, Generating
Random Permutations and Random Subsets, Expectation and Variance of
a Random Variable, Independence of Random Variables, Linearity of
Expectation, Properties of Variances, Poisson Random Variables,
Exercises, Computer Exercises

Chapter 7: Complexity of Algorithms   449

7.1: Some Algorithms for Searching and Sorting: The Linear Search
Algorithm, The Binary Search Algorithm, The Selection Sort Algorithm,
the Bubble Sort Algorithm, The Quick Sort Algorithm, The Merge Sort
Algorithm, A Randomized Algorithm for Computing Medians, Exercises,
Computer Exercises

7.2: Growth Rates of Functions and the Complexity of Algorithms:
A Brief and Informal Preview, Big-O Notation, Combinations of Big-O
Estimates, Big-Omega and Big-Theta Notation, Complexity of
Algorithms, Optimality of the Merge Sort Algorithm, the Classes P and
NP, Exercises, Computer Exercises

Chapter 8: Graphs, Trees, and Associated Algorithms   495

8.1: Graph Concepts and Properties: Simple Graphs, General Graphs,
Degrees, Regular Graphs, and the Handshaking Theorem, Some
Important Families of Simple Graphs, Bipartite Graphs, Degree
Sequences, Subgraphs, Isomorphism of Simple Graphs, the Complement
of a Simple Graph, Representing Graphs on Computers, Directed Graphs
(Digraphs), Some Graph Models for Optimization Problems, Exercises,
Computer Exercises

8.2: Paths, Connectedness, and Distances in Graphs: Paths, Circuits
and Reachability in Graphs, Paths, Circuits, and Reachability in Digraphs,
Connectedness and Connected Components, Distances and Diameters in
Graphs, Eccentricity, Radius, and Central Vertices, Adjacency Matrices
and Distance Computations in Graphs and Directed Graphs, Edge and
Vertex Cuts in Connected Graphs/Digraphs, Characterization of Bipartite
Graphs Using Cycles, Exercises, Computer Exercises

8.3: Trees: Basic Concepts about Trees, Rooted Trees and Binary Trees,
Models with Rooted Trees, Properties of Rooted Trees, Ordered Tree
Traversal Algorithms, Binary Search Trees, Representing Rooted Trees
on Computers, Exercises, Computer Exercises

        Appendix: Application of Rooted Trees to Data Compression and
                               Coding; Huffman Codes

Preface

The purpose of this book is to provide a modern and comprehensive introduction to the subject of discrete structures. *Discrete structures*, also called *discrete mathematics*, is an exciting and active subject, particularly due to its extreme relevance to both mathematics and to computer science and algorithms. The subject forms a common foundation for rigorous mathematical logical reasoning and proofs, as well as a formal introduction to abstract objects that are essential tools in an assortment of applications and effective computer implementations. Computing skills are now an integral part of most all scientific fields, and students are very enthusiastic about being able to harness the full computing power of these tools. Courses in discrete structures are offered at most all universities and in an increasingly large portion of community colleges as well, and are required by both math and computer science programs.

How the book evolved: Over the past 12 years the author has been regularly teaching the designated one-semester course in discrete structures (to both mathematics and computer science majors) at two universities: the University of Guam and the California State University–Dominguez Hills, at both of which he has been appointed as a professor. In different semesters, he has been working to supplement the core material with exposure to an assortment of the new exciting applications-oriented topics that fall under the heading of "discrete structures," but are not always part of the standard curriculum. This standard curriculum is nonetheless important, and it should be included in any decent book or course on the subject. At the same time, it is also important to develop materials that reflect many of the advances and recent trends and applications of this area. Examples of some of the applications that are extensively treated in this book include: simulation, genetic algorithms, network flows, probabilistic primality tests, public key cryptography, and many applications to coding theory. There have been numerous new developments in this vast subject, and the tasks that can be accomplished by students on their PCs are now very different than what was feasible even just 15 years ago. It is these exciting application areas that really make the topics of discrete structures so interesting and useful in applications, and is the reason that it is a required course for computer science students (and increasingly for mathematics students as well). Some of the topics covered in this book were introduced in various semesters in the author's discrete structures classes, and others were further expounded upon in subsequent special topics classes. Examples include separate upper-level courses in each of the areas of graph theory and algorithms, simulation, genetic algorithms, and in cryptography. Not being able to find books that included many of these new topics or had a good selection of exercises, the author found it necessary to prepare his own notes and exercise sets for many of these classes and topics. His courses in cryptography have resulted in a separate book by the author on this subject that was published (also by CRC Press) shortly before this book. Starting about five years ago, with the aim of expanding his materials into a complete discrete structures textbook, the author began to prepare book materials and exercises for the core topics in discrete structures (i.e., those included in the ACM's recommendations).[1] Thus, the book has been evolving through classroom testing on a number of different courses first with the applications, and then with the core material.

How the book is organized: The resulting chapters are written in an easy-going, yet rigorous, and extremely conscientious style. Each section is replete with clear definitions and theorems; the proofs are carefully explained, and written in a cordial style that students find appealing. There are numerous completely worked-out examples that illustrate key concepts, and figures and tables are employed to help students grasp the more subtle and difficult concepts. The text proper is punctuated with "Exercises for the Reader" that give readers frequent opportunities to assess their understanding of the material. These are meant to be done during a careful reading, and complete solutions of these Exercises for the Reader are provided in an appendix at the end of the book. In addition to these, each section ends with an extensive and well-thought-out set of section exercises that range in difficulty from routine to nontrivial, and sometimes include developments of new topics that complement or extend the material in the section proper. Some groups of the exercises (usually toward the end of a set) introduce and develop new topics and methods. Another appendix supplies answers, and in many cases, solution outlines to most of the odd-numbered exercises. A separate instructor's manual containing solutions to the even-numbered exercises is freely available from CRC Press to all qualified instructors. Many sections have appendices that cover either a new and related topic, or material that is more advanced. These appendices, as well as some of the exercise groups on new material, form an excellent basis for student projects.   In addition to the ordinary exercise sets, most sections contain separate sets of Computer Exercises that are intended to be done with the aid of a computer. Except for a few rare exceptions that are clearly indicated (in Section 6.2, in Chapter 10, and in some of the section appendices), the text proper and ordinary exercises can be done without using computers. This feature makes the book suitable for courses that depend on a variety of computer usage: from none at all to courses where students write their own programs. The computer exercises contain extensive and useful material on how students can learn to write their own programs on any computing platform for most all of the major algorithms that are considered in this book.   The separation of the various sections was done with learning and pedagogy in mind.   One extreme is to have exercises only at the end of the chapters, and this tends to make it difficult for instructors to assign exercises on a daily basis; the other extreme is

---

[1] ACM is the acronym for The Association for Computing Machinery, the premier professional organization for computer science of the United States. It periodically drafts recommendations for discrete structures courses, since the area is so fundamental for computer science.

to split the chapters into many small sections, each one covering a single topic. There is much interaction of concepts in discrete structures, so an intermediate approach of splitting chapters into sections of reasonable sizes has been taken. The sections were conceived to provide an effective separation of learning objectives and quiz units. The large number, scope, organization, and variety of carefully crafted exercises form one of the major strengths of the book.

A few comments about the numbering scheme are in order. All theoretical results—theorems, lemmas, propositions, and corollaries—are numbered sequentially in each chapter. Definitions, examples, labeled equations, figures, tables, algorithms, and exercises for the reader each have their own separate counters. The general index at the end of the book is preceded by separate indices for theoretical results and algorithms.

How technology is incorporated: Although the book highly encourages the use of computing platforms, it can still be used without any computers. The text proper is written for non-computer users, as are all of the ordinary exercises. As a general rule, technical comments and implementation suggestions for computer users are restricted the computer exercise sets located at the end of most sections. A few clearly isolated examples do discuss computing (for example, the simulation material of Section 6.2, and some of the appendices), but these passages may simply be read by non-computer users. The main exception is Chapter 10. By its nature, this chapter makes extensive use of computing, so the exercises and computer exercises have been combined into single sets. Nonetheless, Chapter 10 has been written in a way that will be understandable to all readers, whether or not they are using a computer. Indeed, for non-computer users, it should serve to enlighten them about the synergistic power when computing is effectively combined with discrete structure skills.

The explanations of the algorithms and the computer exercises are platformindependent. Algorithms are explained first in ordinary English, and when appropriate using a natural and easy-to-understand pseudocode that can be readily translated into any computer language (an appendix summarizes the pseudo code). A Web page for the book is being maintained that includes, in particular, sample programs and programming notes for all of the computer material for the MATLAB[2] computing platform (which the author has used for computer implementations of all of the relevant topics in this book). These programs are replete with explanatory comments, and since MATLAB is a very user-friendly language, they should be useful to users of other computing platforms in their efforts to write programs. Rather than print the URL of the Web page, which may change as servers get updated or as (the author's physical or electronic) addresses may change, the easiest way to access this page is through the author's homepage, which can be obtained by a simple Web search of the

---

[2] MATLAB® is a computing platform that is among the top three used by mathematics departments in the world. It is gaining popularity in mathematics and computer science departments because of its wide usage across the spectrum in science and engineering departments.

author's last name. The Web page should also be navigable via the publisher's Web site.

How to get the most out of this book: Although everything in this book is carefully defined from scratch and there are no formal prerequisites, it is tacitly assumed that the reader has passed a course in precalculus. In particular, a course in calculus is not required. Many schools make calculus a prerequisite for the discrete structures course, not so much for the subject matter dependence, but more for the reason that calculus courses give students a good, possibly first, experience with mathematical rigor. There are a few rare and isolated comments and footnotes for the benefit of readers who are familiar with calculus, but these are clearly marked and may be skipped by non-calculus readers. There is more than enough material in this book for two semesters of discrete structures. The chapters are organized in a way that would allow instructors to build many different courses. A chapter/section dependence chart is provided (after this preface) for instructors wishing to plan courses. That being said, readers should not feel constrained from flipping ahead if they wish to read any particular topics of interest. Great care has been taken to make it possible to do this, and to look up and read about any earlier topics when this is necessary.

Acknowledgments: Most of the writing of this book was motivated by the author's teaching of discrete structures classes and related special topics courses over the past 12 years. He would like to express his gratitude to the math departments, computer science departments, and to his colleagues at the University of Guam and at California State University–Dominguez Hills for allowing him to frequently teach such classes, often with new topics and with various levels of technology. He is grateful to many students over the years who have suffered through preliminary drafts of various portions, and have provided much useful feedback that has helped him to make numerous improvements. Two colleagues deserve very special mention: It was delightful to have George Jennings involved in this project. He has done an exemplary job in preparing an instructor solutions manual for the book (instructors should contact CRC Press to obtain a copy), and in the process has read through the entire book and provided numerous useful suggestions for improvement. The author was also highly appreciative to have had his colleague Frank Miles on board. Frank has the precision of a Swiss watch, and he has carefully read through the entire manuscript and provided a plethora of scholarly suggestions and comments. Neighboring colleague Will Murray at California State University–Long Beach has read through significant portions and provided some very useful suggestions. The author thanks Yumi Nishimura for fine work in creating some of the technical drawings, and help with the cover design.[3] The anonymous reviewers have provided some very useful feedback and suggestions that have led to many

---

[3] The cover design is based on a photograph of a green sea turtle taken by the author during a recent dive trip to the beautiful island of Palau. The JPG technology of such digital photographs relies on Huffman codes, which are discussed in Chapter 8.

improvements, including some major restructuring of the book. As with all of his previous books, the author's mother Christa Stanoyevitch has continued to proofread through all of the early drafts (and thus has had the most arduous of all the proofreader jobs), and has helped him to correct many of the initial errors and to make stylistic improvements before the drafts make it to the students. The book's editor, Sunil Nair (who also holds a Ph.D. in mathematics), has been extraordinarily farsighted and helpful in his ideas that have led to significant improvements in the organization and coverage, as well as very nicely and promptly taking care of the numerous issues or concerns that came up during the latter part of the writing and the production process. The author was very pleased to be able to work again on his second CRC book with production editor Tara Nieuwesteeg; she has always been exceptionally hard-working, wise, and accommodating throughout the project. Thanks go also to the copy editor Cindy Gallardo, who has carefully read over the drafts and offered numerous helpful suggestions and corrections. The author takes full responsibility for any errors that remain, and would be grateful to any readers who could direct his attention to any such oversights.

About the Author

Alexander Stanoyevitch completed his doctorate in mathematical analysis at the University of Michigan–Ann Arbor, has held academic positions at the University of Hawaii and the University of Guam, and is presently a professor at California State University–Dominguez Hills. He has published several articles in leading mathematical journals and has been an invited speaker at numerous lectures and conferences in the United States, Europe, and Asia. His research interests include areas of both pure and applied mathematics, and he has taught many upper-level classes to mathematics students as well as computer science students.

Dependency Chart

The following chart should be helpful to readers or instructors aiming to plan courses with this book. Major dependencies are indicated with solid arrows, minor ones with dashed arrows. The dependencies on Section 2.2 (*) pertain only to the material on equivalence relations. The dependence on Section 4.2 by Section 4.3 (**) is only for the material on modular matrices. This material is not used in Section 8.1 (which has a minor dependence on Section 4.3).

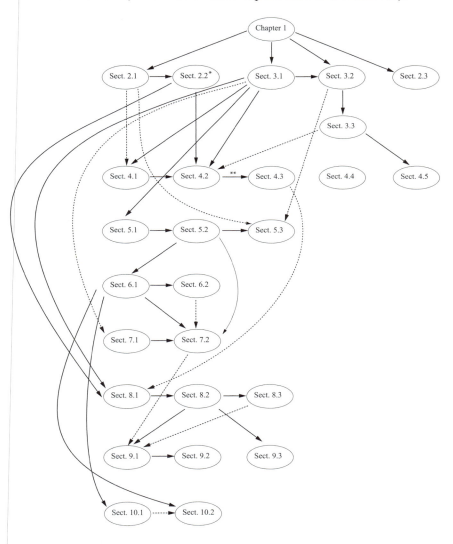

Chapter 1: Logic and Sets

1.1: LOGICAL OPERATORS

Logic is the foundational edifice on which mathematics, computer science, and philosophy all rest. Over the years, protocols and computing languages have evolved and they will continue to change, but the logic governing how they are organized remains the same. The formal subject of **logic** aims to establish a coherent framework in which all scientific thoughts can be expressed, communicated, and synthesized. In this section we will cover the essential principles of logic that will be needed for the reader to understand and write theorems, proofs, algorithms, and programs. The topics covered in this book will put the reader in constant need of logical principles and will continue to reinforce the reader's mastery of logical thinking and inferences. Our first definition introduces the building blocks of logic; although it is nontechnically formulated it will suffice for our purposes.

Statements and Truth Values

DEFINITION 1.1: A **statement** is any declarative sentence or mathematical relation that has a **truth value** of either **true** or **false**.

EXAMPLE 1.1: For each item below, indicate whether it is a statement, and, if possible, indicate the truth value.
(a) Honolulu is the capital of Hawaii.
(b) What is your name?
(c) $-5 < 2$
(d) It is not possible to have $x^3 + y^3 = z^3$, for three nonzero *integers* x, y, and z.

Note: An **integer** is a real number with no decimal part, i.e., a number among the list $0, \pm 1, \pm 2, \cdots$.

SOLUTION: (a) is a true statement. (b) is not a statement. (c) is a true statement, as is (d). Do not worry if you did not know the truth value of (d); it is quite nontrivial.[1]

---

[1] This statement is part of *Fermat's Last Theorem*, named after the French barrister and amateur mathematician Pierre de Fermat, who scribbled his statement in the margin of a book in the 1630s,

Item (d) motivates introducing some general terminology regarding statements in the sciences.

DEFINITION 1.2:   A **theorem** or a **proposition** is a true statement that has been proved.   Theorems are usually of greater significance than propositions.   A **lemma** is a true statement that has been proved, but is usually intended to be used to prove other results (theorems or propositions), rather than being of interest in its own right.   A **corollary** is a true statement that is rather easily seen as a consequence or an interesting special case of a deeper theorem or proposition.   A **conjecture** is a statement that is believed to be true but has not yet been proved.

Note that a conjecture is a statement that has a truth value, even though this truth value is not known (at present).   In the pure scientific fields the unspoken etiquette requires that all (nonconjecture) proclamations of statements need to be backed up with a proof.   This is unfortunately not always the case in some other disciplines.

Negations, Conjunctions, and Disjunctions

New statements can be obtained by combining other statements using **logical connectives** (or **logical operators**); moreover, truth values of such new statements can be inferred from the truth values of the statements from which they are constructed.   Logical inference basically refers to the protocol of writing correct proofs, and thus lies at the foundation of all scientific theories.   We begin by introducing the most basic logical connectives.

DEFINITION 1.3:   Suppose that P and Q represent statements.
(i)   The **negation** of P, denoted $\sim P$, and read as "not P," is another statement whose truth value is the opposite of that of P.   Thus, $\sim P$ is false when P is true and $\sim P$ is true when P is false.
(ii)   The **conjunction** of P and Q, denoted $P \wedge Q$, and read as "P and Q," is another statement that will be true exactly when both P and Q are true (and so will be false in all other cases).
(iii)   The **disjunction** of P and Q, denoted $P \vee Q$, and read as "P or Q,"   is another statement that will be true as long as P is true, or Q is true (or both).

Some comments are in order.   These definitions agree with those of formal (and contractual) English.   In spoken English, the words "not" and "and" are always clear, but there are several standard variations in their wording.   For example, the negation of the statement: "You are rich" is usually not expressed in the formal

---

where he claimed that there was not enough space to contain his "remarkable proof." The complete statement reads that if n is any positive integer greater than two, then the equation $x^n + y^n = z^n$ can have no nonzero integer solutions: x, y, z. This "theorem" remained unproved until the mid 1990s (to be a true statement) by Princeton mathematician Andrew Wiles, who realized a childhood dream. He had spent seven years secretly (in his attic) working feverishly on this problem. See [Kli-00] for a very interesting historical account of this famous theorem.

way: "It is not the case that you are rich," but rather as "you are not rich." Caveat: "You are poor" is not the negation of "You are rich." (Why?) We point out that in this context "but" is a synonym for "and." The word "or," however, is sometimes ambiguous because in certain cases it is not intended that both P and Q can be true. For example, suppose a restaurant waiter asks you: "would you like the soup or the salad that comes with the meal?" If you ask for both, you will either get a snappy retort from the waiter to choose only one, or an unexpected extra charge on your bill. The sciences (and formal English) cannot tolerate such ambiguity, so by default the word "or" has the meaning as in (iii) above. In case one wants to use the more restrictive disjunction of two statements P and Q, one calls it the **exclusive disjunction**, and denotes it as $P \oplus Q$. This is read simply as "P or Q but not both." Any statement that is made using logical connectives and other more basic statements is called a **compound statement**.

Truth Tables

A **truth table** for a compound statement is a complete listing of all of the possible cases of the truth values for the basic statements from which the compound statement is constructed (thought of as *logical variables* of the compound statement) along with the resulting truth values for the compound statement. Tables 1, 2, and 3 below give truth tables for the negation, conjunction, and disjunction logical connectives.

P	$\sim P$
T	F
F	T

P	Q	$P \wedge Q$
T	T	T
T	F	F
F	T	F
F	F	F

P	Q	$P \vee Q$
T	T	T
T	F	T
F	T	T
F	F	F

TABLES 1.1–1.3: Truth tables for a negation, conjunction, and disjunction.

If we had to make a truth table involving three (variable) basic statements, say $(P \wedge Q) \vee R$, the truth table would require eight rows of truth values. In general, a truth table for a compound statement involving n basic statements will contain 2^n rows of truth values. This is an easy consequence of the *multiplication principle* that will be introduced in Section 5.1, but it would behoove readers who have not yet seen this principle to directly convince themselves of this fact.

We next come to the very important logical connective of *implication* that will allow logical inferences. Implications <u>very</u> often get abused (and misunderstood) in spoken English, and furthermore, there are <u>many</u> different ways of expressing them both in written and in spoken language. Consequently it may take some time to get comfortable with them. Before we give the formal definition, we will warm up with the following nontechnical example of an implication:

Conditional Statements (Implications)

Suppose that Professor Saunders tells his student Jimmy: "Jimmy, if you get at least a 90 on the final, then I will give you an A for the course." This is a compound statement of the form: If P, then Q, where P = "Jimmy gets at least a 90 on the final," and Q = "Professor Saunders gives Jimmy an A for the course." As with all compound statements, the truth value of the (whole) implication will be determined by the individual truth values of P and Q. Most all people would agree that the implication (if P then Q) is true when both P and Q are true (Jimmy gets at least a 90 on the final and Prof. S. gives him an A), and false when P is true but Q is false (because if Jimmy gets at least a 90 on the final, but Prof. S. does not give him an A, then Prof. S. has broken his promise). These truth values are indeed logically correct. The remaining cases are often ambiguous and not well understood in spoken language. For example, what would be the truth value of this implication in case P is false, and Q is true? This would mean that Jimmy did not get at least a 90 on the final, but Prof. S. still gave Jimmy an A. Would this make our implication (If P, then Q) true or false? In this case, the implication is true. One helpful way to understand this is to view the implication as a promise or guarantee. As long as the promise is not broken, the implication is true. Thus, if Jimmy did not get at least a 90 on his exam, Prof. S. has no obligation to give Jimmy an A. Whether he does or does not, his promise would not be broken. He may have decided to give Jimmy an A for an assortment of reasons (perhaps the exam was more difficult than Prof. S. had anticipated, and the average score was only 52). In the same fashion, the implication "If P, then Q" is true of both P and Q are false—here, Jimmy fails to get at least a 90 on the exam, and Prof. S. does not give Jimmy an A. Again, Professor Saunders' promise to Jimmy is not broken.

DEFINITION 1.4: Suppose that P and Q represent statements. The **implication** or **conditional statement** $P \rightarrow Q$, which can be read as "If P, then Q," is another statement that will be true in all cases, except when P is true and Q is false. In the implication $P \rightarrow Q$, P is called the **hypothesis**, and Q is called the **conclusion**.

The truth table for the implication is shown in Table 1.4.

P	Q	$P \rightarrow Q$
T	T	T
T	F	F
F	T	T
F	F	T

TABLE 1.4: Truth table for an implication.

Unlike with the other logical connectives that were previously introduced, the truth values of a conditional statement are much less obviously related to spoken English. For example, note that the conditional $P \rightarrow Q$ will <u>always</u> be true if the hypothesis is false. Thus, a nonsensical English statement such as, "If the moon is made of blue cheese, then Germany won World War II," would be a true statement (as far as logic is concerned), since both the hypothesis and conclusion are false (see row four of Table 4).

There are numerous ways of expressing conditional statements both in written and spoken language, especially in the sciences where logical inference is central to the subjects. Here are several common variants of wording the implication "if P, then Q" $(P \rightarrow Q)$:

1. "P is sufficient for Q" (or "P is a sufficient condition for Q").
2. "Q is necessary for P" (or "Q is a necessary condition for P").
3. "P implies Q" (or "Q is implied by P") (or "Q follows from P").
4. "P only if Q."[2]
5. "Q, if P."

In order to be able to understand books, papers, and lectures in any scientific discipline, you must become thoroughly familiar with these different ways of expressing a conditional statement. In particular, one must take care to distinguish between the hypothesis and the conclusion of a conditional; if they are interchanged, you get a different conditional, as we will now see.

Converses and Contrapositives

DEFINITION 1.5: Given an implication $P \rightarrow Q$, the **converse** is the implication $Q \rightarrow P$, obtained by interchanging the hypothesis and conclusion. If, in addition, we also negate both the hypothesis and conclusion, we arrive at the so-called **contrapositive** of the implication $P \rightarrow Q$: $\sim Q \rightarrow \sim P$.

For example, the implication: "if Spot is a Doberman pinscher, then Spot is a dog," is (always) a true statement since a Doberman pinscher is a special breed of dog. The converse, "if Spot is a dog, then Spot is a Doberman pinscher," however, could be false, for example, if Spot were a collie (see line two of Table 1.4). On the other hand, if we form the contrapositive: "if Spot is not a dog, then Spot is not a Doberman pinscher," we get something that makes good sense and is in fact a true statement. This is no coincidence; before we elaborate, we first give the following definition.

---

[2] To help remember this one, you should view "only if" as a synonym for "then."

Logical Equivalence and Biconditionals

DEFINITION 1.6: Two compound statements $\mathcal{A}$, $\mathcal{B}$ are said to be (**logically**) **equivalent** (written as $\mathcal{A} \equiv \mathcal{B}$ or as $\mathcal{A} \Leftrightarrow \mathcal{B}$) if their truth tables are identical.[3] A compound statement that is always true (regardless of the values of its logical variables) is called a **tautology**; if it is always false, it is called a **contradiction**.

EXERCISE FOR THE READER 1.1: (a) Show that an implication $P \rightarrow Q$, and its contrapositive $\sim Q \rightarrow \sim P$, are (logically) equivalent.
(b) Construct a compound statement with one (logical) variable that is a tautology.
(c) Construct a compound statement with two (logical) variables that is a contradiction.

EXAMPLE 1.2: Find a compound statement involving only logical operators from the list $\wedge, \vee, \sim$, which is equivalent to the implication $P \rightarrow Q$.

SOLUTION: Implication (Table 1.4) and disjunction (Table 1.3) have one important property in common: their truth values are true in exactly three out of the four possible cases. Recall that a disjunction is true, except when both variables are false. On the other hand, the implication $P \rightarrow Q$ is true, except when P is true and Q is false, or equivalently, except when $\sim P$ and Q are both false. Putting this all together, we conclude that $P \rightarrow Q$ and $\sim P \vee Q$ are logically equivalent. A truth table can always be used to check any purported equivalence, and any skeptical readers are encouraged to do this. Notice that we did not use conjunction.

EXERCISE FOR THE READER 1.2: (a) Find a compound statement involving only the conjunction ($\wedge$) and negation ($\sim$) operators that is logically equivalent to the disjunction $P \vee Q$.
(b) Find a compound statement involving only conjunctions and negations that is logically equivalent to the implication $P \rightarrow Q$.

Abuses of conditionals in spoken language are rampant. For example, suppose that a parent tells his or her son: "If you don't finish your homework, then you will not be allowed to go out tonight." We know this statement is equivalent to its contrapositive: "If you are allowed to go out tonight, then you will have finished your homework." This latter statement would be true if the son were not allowed to go out tonight even though he finishes his homework ($F \rightarrow T$), but this is certainly not what the parent intended. The parent had really intended that

---

[3] The concept of logical equivalence can be extended to cases where one of $\mathcal{A}$ or $\mathcal{B}$ has more logical variables than the other, if these additional variables are redundant. For example, if $\mathcal{A}$ and $\mathcal{B}$ were tautologies we would certainly have $\mathcal{A} \equiv \mathcal{B}$.

the converse also be true: "If you finish your homework, then you will be allowed to go out tonight."[4] In other words, the parent intended (and the son surely understood) that the truth values of the two parts of this statement should be the same. Thus the intended and interpreted statement is not a conditional, but rather a so-called *biconditional statement*; we enunciate this important logical construction in the following definition.

DEFINITION 1.7: Suppose that P and Q represent statements. The **biconditional statement** $P \leftrightarrow Q$, which can be read as "P if, and only if Q," is another statement that is true whenever P and Q share the same truth values, and false when P and Q have opposite truth values.

The truth table for the biconditional is shown in Table 1.5. It follows that the logical equivalence of two compound statements $\mathscr{A}$, $\mathscr{B}$ ($\mathscr{A} \equiv \mathscr{B}$) can be expressed as saying $\mathscr{A} \leftrightarrow \mathscr{B}$ is a tautology. As pointed out before the definition, biconditionals are often intended and interpreted when a conditional statement is given in spoken language; indeed, you rarely hear the "if and only if" phrase in spoken language. An often used variation (in formal written language) of "P if, and only if Q" is "P is (both) necessary and sufficient for Q."

P	Q	$P \leftrightarrow Q$
T	T	T
T	F	F
F	T	F
F	F	T

TABLE 1.5: Truth table for a biconditional.

Hierarchy of Logical Operators

As with arithmetic operations, there is a convention of a **hierarchy** of logical operations, but parentheses can always overrule any such hierarchy. The precedence is as shown in Table 1.6:

OPERATOR	HIERARCHY
~	highest (do first)
$\wedge$	next highest
$\vee$	third highest
$\rightarrow$	lower
$\leftrightarrow$	lowest (do last)

TABLE 1.6: Hierarchy of logical operators; parentheses can overrule any hierarchy.

---

[4] This is the converse of the contapositive. Since the contrapositive is equivalent to the original statement, its converse is also equivalent to that of the original statement.

The hierarchy conventions of Table 1.6 are standard, and help to avoid unnecessary parentheses. For example, compare the following compound statement with and without redundant parentheses:[5]

$$(P \vee (Q \rightarrow (\sim R))) \rightarrow [(\sim Q) \wedge P] \equiv P \vee (Q \rightarrow \sim R) \rightarrow \sim Q \wedge P.$$

Only one of the five sets of parentheses is actually needed.[6]

We point out that in practice, when one wants to create (by hand) a truth table of all but the simplest compound statements, it is easier if one adds additional columns for portions of the compound statements. It is also good practice to organize the truth values of logical variables as we have been doing. The last variable alternates single truth values as T, F, T, F, ...; the next-to-last variable alternates double truth values as T, T, F, F, T, T, ...; the third-to-last variable then alternates quadruple truth values as: T, T, T, T, F, F, F, F, ..., and so on.

EXAMPLE 1.3: Construct a truth table for the statement: $(P \wedge Q) \rightarrow (Q \rightarrow R)$.

SOLUTION: Two versions of a truth table are shown in Table 1.7; the first one is constructed as explained before the example, and the second is done in a more compact fashion. The basic idea in both is the same: we start by evaluating the logical operators highest in the hierarchy first, eventually working our way to the last operators lowest in the hierarchy, whose truth values determine those of the given compound statement.

P	Q	R	$P \wedge Q$	$Q \rightarrow R$	$(P \wedge Q) \rightarrow (Q \rightarrow R)$
T	T	T	T	T	T
T	T	F	T	F	F
T	F	T	F	T	T
T	F	F	F	T	T
F	T	T	F	T	T
F	T	F	F	F	T
F	F	T	F	T	T
F	F	F	F	T	T

---

[5] As in arithmetic, logical operations of the same hierarchy are performed from left to right.
[6] Unlike in arithmetic, the hierarchy of Table 1.6 is not always so well-known by nonspecialists (of logic); we therefore will not always aim for minimizing parentheses usage in forming compound statements.

P	Q	R	$(P \wedge Q) \rightarrow (Q \rightarrow R)$		
T	T	T	T	T	T
T	T	F	T	F	F
T	F	T	F	T	T
T	F	F	F	T	T
F	T	T	F	T	T
F	T	F	F	T	F
F	F	T	F	T	T
F	F	F	F	T	T

TABLE 1.7: Two versions of truth tables for $(P \wedge Q) \rightarrow (Q \rightarrow R)$. The bottom version is more compact, but in the upper version it is easier to follow the order of truth value evaluations.

Some Useful Logical Equivalences

A logical problem that often arises is to decide whether two compound statements $\mathcal{A}$, $\mathcal{B}$, involving the same logical variables, are logically equivalent. One may, of course, construct truth tables for both to see if they are identical, but it is helpful to know a few common logical equivalences. Some of the more useful logical equivalences are summarized in the following result:

THEOREM 1.1: (*Some Logical Equivalences*) We let P, Q, and R denote any statements, **T** denote any tautology, and **F** denote any contradiction. The following logical equivalences are (always) valid:

PART I: Equivalences involving conjunctions, disjunctions and, negations:
(a) (*Commutativity*) $P \wedge Q \equiv Q \wedge P$, $P \vee Q \equiv Q \vee P$
(b) (*Associativity*) $(P \wedge Q) \wedge R \equiv P \wedge (Q \wedge R)$, $(P \vee Q) \vee R \equiv P \vee (Q \vee R)$
(c) (*Distributivity*)
  $P \wedge (Q \vee R) \equiv (P \wedge Q) \vee (P \wedge R)$, $P \vee (Q \wedge R) \equiv (P \vee Q) \wedge (P \vee R)$
(d) (*De Morgan's Laws*) $\sim (P \vee Q) \equiv \sim P \wedge \sim Q$, $\sim (P \wedge Q) \equiv \sim P \vee \sim Q$ [7]

---

[7] The following observations can help to make the very important De Morgan's laws a bit easier to remember: They can be thought of as distributive laws for negation over disjunctions and conjunctions. The negation gets distributed, but the conjunction gets changed to a disjunction, and vice versa. For the first one: Since a the disjunction $P \vee Q$ is true in three out of the four cases (see Table 1.3); its negation $\sim (P \vee Q)$ is false in three out of four cases, like a conjunction (see Table 1.2). In

(e) (*Double Negation*) $\sim(\sim P) \equiv P$

(f) (*Absorption*) $P \vee (P \wedge Q) \equiv P,\ P \wedge (P \vee Q) \equiv P$

(g) (*Identity Laws*) $P \wedge \mathbf{T} \equiv P,\quad P \vee \mathbf{F} \equiv P$

PART II: Equivalences involving conditionals and biconditionals:

(a) (*Implication as Disjunction*) $P \to Q \equiv \sim P \vee Q$

(b) (*Negation of Implication*) $\sim(P \to Q) \equiv P \wedge \sim Q$

(c) (*Exportation*) $P \to (Q \to R) \equiv (P \wedge Q) \to R$

(d) (*Contrapositive Equivalence*) $P \to Q \equiv \sim Q \to \sim P$

(e) (*Biconditional as Implications*) $P \leftrightarrow Q \equiv (P \to Q) \wedge (Q \to P)$

PART III: Some tautologies and contradictions:

(a) (*Tautologies*) $P \vee \mathbf{T} \equiv \mathbf{T},\ P \vee \sim P \equiv \mathbf{T},\ P \to \mathbf{T} \equiv \mathbf{T},\ \mathbf{F} \to Q \equiv \mathbf{T}$

(b) (*Contradictions*) $P \wedge \mathbf{F} \equiv \mathbf{F},\ P \wedge \sim P \equiv \mathbf{F}$

Proof: Each of these equivalences can be verified using truth tables, and we have already verified some. If they are proved in order, proofs of most parts of this theorem can be completed by combining previously proved equivalences and/or working with definitions. For example, here is such a proof of II(b) $\sim(P \to Q) \equiv P \wedge \sim Q$:

First use II(a): $\sim(P \to Q) \equiv \sim(\sim P \vee Q)$. Next we apply De Morgan's Law I(d) (the first one) to the negation on the right to write $\sim(\sim P \vee Q) \equiv \sim(\sim P) \wedge \sim Q$. Finally double negation I(e), shows the last expression is equivalent to $P \wedge \sim Q$, and putting this all together gives the desired equivalence II(b).

Such proofs are more elegant and valuable than a simple (but rote) pair of truth tables. The exercises will ask the reader to prove more of the above equivalences. The serious reader, however, would be advised to now go through proving each of the above statements, in order, avoiding truth table proofs whenever possible.[8] □

We point out that Part II(b) $\sim(P \to Q) \equiv P \wedge \sim Q$ of Theorem 1.1, provides a practical template for negating an implication in spoken or written English. For example, to negate the implication: "If it rains, then I will go to see a movie," rather than the formal (and highly inelegant) "It is not the case that if it rains, then I will go to see a movie," we could express it as: "It will rain and I will not go to see a movie." Note that this latter statement would always have the opposite truth values of the original implication.

---

words, the formula $\sim(P \vee Q) \equiv \sim P \wedge \sim Q$ can be read as: "the negation of a disjunction is the conjunction of the negations." If we interchange the words disjunction/conjunction in this phrase, we get the wording for the second De Morgan law.

[8] It is fine (and all the better) to recognize which equivalences have previously been proved (and to skip them).

Logical Implication

Recall that a logical equivalence $\mathscr{A} \equiv \mathscr{B}$ (or $\mathscr{A} \Leftrightarrow \mathscr{B}$) means that whenever the statement $\mathscr{A}$ is true, so is $\mathscr{B}$, and whenever $\mathscr{A}$ is false, so is $\mathscr{B}$, i.e., $\mathscr{A}$ and $\mathscr{B}$ have the same truth values in corresponding rows of their truth tables. If we take only the first half of this definition, i.e., whenever the statement $\mathscr{A}$ is true, so is $\mathscr{B}$, we arrive at the important concept of *logical implication*:

DEFINITION 1.8: If $\mathscr{A}$, $\mathscr{B}$ are two compound statements, we say that $\mathscr{A}$ **(logically) implies** $\mathscr{B}$ (written as $\mathscr{A} \Rightarrow \mathscr{B}$) if whenever the statement $\mathscr{A}$ is true, so is $\mathscr{B}$. In other words $\mathscr{A} \rightarrow \mathscr{B}$ is a tautology.

Thus to check an implication $\mathscr{A} \Rightarrow \mathscr{B}$, we need only look at situations (rows of the truth tables) where $\mathscr{A}$ is true. If $\mathscr{B}$ is ever false in such a situation (even just once), then the logical implication $\mathscr{A} \Rightarrow \mathscr{B}$ is invalid; otherwise it is valid.

We explained earlier that a logical equivalence $\mathscr{A} \equiv \mathscr{B}$ is valid if, and only if the compound statement $\mathscr{A} \leftrightarrow \mathscr{B}$ is a tautology. In a similar fashion, it can be easily seen that the logical implication $\mathscr{A} \Rightarrow \mathscr{B}$ is valid if, and only if the compound statement $\mathscr{A} \rightarrow \mathscr{B}$ is a tautology. The following result will summarize a few useful and important logical implications.

THEOREM 1.2: (*Some Logical Implications*) We let P, Q, and R denote any statements, **T** denote any tautology, and **F** denote any contradiction. The following logical implications are (always) valid:

(a) (*Addition*) $P \Rightarrow P \vee Q$

(b) (*Subtraction*) $P \wedge Q \Rightarrow P$

(c) (*Modus Ponens*) $P \wedge (P \rightarrow Q) \Rightarrow Q$

(d) (*Modus Tollens*) $(P \rightarrow Q) \wedge \sim Q \Rightarrow \sim P$

(e) (*Hypothetical Syllogism*) $(P \rightarrow Q) \wedge (Q \rightarrow R) \Rightarrow P \rightarrow R$

(f) (*Disjunctive Syllogism*) $(P \vee Q) \wedge \sim P \Rightarrow Q$

(g) (*Constructive Dilemmas*) $(P \rightarrow Q) \wedge (R \rightarrow S) \Rightarrow [(P \vee R) \rightarrow (Q \vee S)]$

$$(P \rightarrow Q) \wedge (R \rightarrow S) \Rightarrow [(P \wedge R) \rightarrow (Q \wedge S)]$$

Proof: At this point, each of these implications should seem quite reasonable to the reader. Each part can be proved by constructing truth tables for each side, and checking to see that in all cases where the first compound statement ($\mathscr{A}$) is true, the second compound statement ($\mathscr{B}$) is also true. We will try to promote more elegant modes of proof that will encourage a deeper understanding. Below we give proofs of only two parts of the theorem, leaving the rest as exercises.

Part (c): The only situation that could make $P \wedge (P \to Q) \Rightarrow Q$ false is if Q were to be false and $P \wedge (P \to Q)$ were to be true. We will show that this cannot happen.   The truth of the conjunction implies (by subtraction—Part (b) of Theorem 1.2) that both the implication $P \to Q$ and its hypothesis P are true. This forces the conclusion Q of the implication to also be true because otherwise the implication would be false (i.e., it would have the form $\mathbf{T} \to \mathbf{F}$).   This proves that the original implication $P \wedge (P \to Q) \Rightarrow Q$ is therefore valid.[9]

Part (d): We assume that we are in a situation where $(P \to Q) \wedge \sim Q$ is true. The proof can be complete by inferring that $\sim P$ must also be true (i.e., P is false). Since the conjunction $(P \to Q) \wedge \sim Q$ is true, each portion $P \to Q$ and $\sim Q$ must be true (Definition 1.3). By contrapositive equivalence (Theorem 1.1 II(d)), we can infer that $\sim Q \to \sim P$ is also true. Combining this with the fact that $\sim Q$ is true, we infer (Definition 1.3) that the conjunction $\sim Q \wedge (\sim Q \to \sim P)$ is also true.   Finally, we apply modus ponens (Part (c) of the present theorem, which we just proved) to infer that $\sim P$ is also true, as desired. □

Proofs and Counterexamples

New logical equivalences and implications (logical theorems) can be derived and proved using definitions, previously proved theorems, or, as a last resort, truth tables.   Remember, for a logical expression (involving logical variables) to be logically equivalent to or to logically imply another, the corresponding biconditional/implication must be true for <u>any assignment</u> of truth values to the logical variables.   It takes only a <u>single assignment</u> of truth values of the logical variables that would render the corresponding biconditional/implication false (a **counterexample**), to show that a logical equivalence/implication is invalid.

EXERCISE FOR THE READER 1.3:   (a)   Prove or disprove the following equivalence: $(P \vee Q) \vee \sim R \equiv [(P \vee \sim Q) \wedge R] \to P$ using truth tables.

(b)   If the equivalence in Part (a) is true, give another proof of it using other known theorems.

(c) Prove or disprove the following equivalence: $(P \vee Q) \wedge (\sim P \vee R) \equiv Q \vee R$.

Proving or disproving implications or equivalences can be used to assess the validity of written arguments.   One very common type of written argument (an inference) can be represented symbolically as follows:

$\qquad$ Assume that $\mathcal{A}_1, \mathcal{A}_2, \cdots, \mathcal{A}_k$ are all true

---

[9] A more efficient framework to complete this proof would have been by using the method of "proof by contradiction," which will be introduced shortly.

Therefore ($\therefore$) $\mathscr{B}$ is true.

The validity of such an argument can be viewed as the validity of the implication:

$$\mathscr{A}_1 \wedge \mathscr{A}_2 \wedge \quad \cdots \quad \wedge \mathscr{A}_k \Rightarrow \mathscr{B} \qquad\qquad (1)$$

(i.e., $(\mathscr{A}_1 \wedge \mathscr{A}_2 \wedge \quad \cdots \quad \wedge \mathscr{A}_k) \rightarrow \mathscr{B}$ is a tautology.)[10] Logical implications (e.g., from Theorem 1.2) and equivalences (e.g., from Theorem 1.1) can be used to infer the validity of an argument, after the various basic statements are put into symbols. The following example will illustrate this procedure.

EXAMPLE 1.4: Determine whether each of the following arguments is valid:
(a) If Joey is a bungee jumper, then Joey is a fun person. Joey is not a bungee jumper. Therefore, Joey is not a fun person.
(b) If it rains today, Mary will go to a movie. If Mary goes to a movie, she will eat popcorn. Mary did not eat popcorn today. Therefore, it did not rain today.
(c) If Mike goes to the party, then he will either dance with Jane, or not play air hockey with Chuck. If he dances with Jane, then he will come home late. If he does not play air hockey with Chuck, then he will not play cards either. Therefore, if Mike goes to the party, then either he will come home late, or he will not play cards.

SOLUTION: In each part we introduce symbols in the obvious fashion. Let us explain this abstraction just for the first statement of Part (a). We let B represent the statement: "Joey is a bungee jumper," and F represent the statement: "Joey is a fun person."[11] Thus "If Joey is a bungee jumper, then Joey is a fun person" is represented symbolically as $B \rightarrow F$.

Part (a): The argument becomes: $B \rightarrow F$ and $\sim B$ $\therefore$ $\sim F$. If Joey, who we know is not a bungee jumper (i.e., B is false) were to indeed be a fun person (i.e., F is true), then both hypotheses would be satisfied ($B \rightarrow F$ is False $\rightarrow$ True, and $\sim B$ is True), but the conclusion is not satisfied ($\sim F$ is false), so the argument is not a valid one. Invalid arguments are also called **fallacies**.

Part (b): The argument can be symbolized as follows:
$\qquad$ (i) $R \rightarrow M$, (ii) $M \rightarrow P$, (iii) $\sim P$ $\qquad \therefore$ $\sim R$
(We have numbered each of the hypotheses to facilitate the proof that follows.)
By hypothetical syllogism (Theorem 1.2(e)), (i) and (ii) imply (iv) $R \rightarrow P$. Next we can use (iii) and (iv) with modus tollens (Theorem 1.2(d)) to conclude $\sim R$, and we have thus proved the validity of the argument.

---

[10] Any disjunction (or conjunction) of multiple statements can be represented using the usual sigma notation convention; for example, $\bigvee_{k=1}^{n} P_k \equiv P_1 \vee P_2 \vee \cdots \vee P_n$.

[11] For obvious reasons it is not a good idea to use either of the letters T or F to represent a statement; just for this example, however, we gave into the temptation.

Part (c): The argument can be symbolized as follows:

(i) $P \to (D \vee \sim H)$, (ii) $D \to L$, (iii) $\sim H \to \sim C$ $\therefore P \to (L \vee \sim C)$

We can apply the constructive dilemma implication (of Theorem 1.2(g)) to hypotheses (ii) and (iii) to obtain (iv) $(D \vee \sim H) \to (L \vee \sim C)$. We next apply hypothetical syllogism (Theorem 1.2(e)) with (i) and (iv) to obtain the desired conclusion $P \to (L \vee \sim C)$.

A few comments are in order. It is an unfortunate fact of life that (logical) fallacies do in fact occur: in conversations, in more formal scholarly arguments (both with students and academics), and even in published work and speeches. Understanding logical inference will allow you to write bullet-proof arguments and also to detect logical flaws, when they are present, in arguments of others. The above proofs demonstrate the usefulness of being familiar with a few basic theorems in logic. With experience and practice, you will begin to develop an intuition that will aid you in making an initial guess as to whether a given argument is valid. After all, trying to prove an invalid argument will be impossible, as will attempting to find a counterexample for a valid one. More complicated arguments and mathematical proofs can take more time to create or analyze. Admittedly, it is sometimes difficult to decide in which direction one should aim (i.e., either attempt to prove it, or look for a counterexample that will prove it false); but such problems are a fine instance of the art of scientific research and discovery. Truth tables can always be used to prove/disprove any purely logical assertion; in Part (c) of the preceding example, a (hand-drawn) truth table for the corresponding implication would have required 32 lines.

Each of the proofs used in the preceding example (for the valid arguments) was a so-called **direct proof** of (1) $\mathscr{A}_1 \wedge \mathscr{A}_2 \wedge \cdots \wedge \mathscr{A}_k \Rightarrow \mathscr{B}$: we assumed the hypothesis and inferred the conclusion. It is sometimes more convenient to give an **indirect proof**. In this format, we assume the conclusion ($\mathscr{B}$) is false, and attempt to show that one of the hypotheses ($\mathscr{A}_i$) is also false. All this really boils down to is that we are proving instead the contrapositive of (1):

$$\sim \mathscr{B} \Rightarrow \sim (\mathscr{A}_1 \wedge \mathscr{A}_2 \wedge \cdots \wedge \mathscr{A}_k)$$

(Since De Morgan's law tells us that $\sim (\mathscr{A}_1 \wedge \mathscr{A}_2 \wedge \cdots \wedge \mathscr{A}_k) \equiv \sim \mathscr{A}_1 \vee \sim \mathscr{A}_2 \vee \cdots \vee \sim \mathscr{A}_k$, we may show that the contrapositive is true by showing that at least one $\mathscr{A}_i$ is false, when $\mathscr{B}$ is false.)[12]

---

[12] There is one missing detail here: De Morgan's law $\sim (P \wedge Q) \equiv \sim P \vee \sim Q$ "the negation of a conjunction is the disjunction of the negations" actually works for any finite number of logical variables, although the law was stated for just two variables. The proof is rather straightforward, but requires mathematical induction, which we will introduce later on. (Readers familiar with mathematical induction should be able to easily write out the proof now, and are encouraged to do so.) At present, we can write out a proof for any (fixed) finite number of logical variables. For example, here is a proof of the statement with three logical variables: (it uses the corresponding two-variable De Morgan's law twice, along with associativity):

$\sim (P \wedge Q \wedge R) \equiv \sim ([P \wedge Q] \wedge R) \equiv \sim [P \wedge Q] \vee \sim Q \equiv [\sim P \vee \sim Q] \vee \sim Q \equiv \sim P \vee \sim Q \vee \sim R.$

EXERCISE FOR THE READER 1.4:   Use an indirect proof to establish the validity of the following argument:
If the Lakers win both this week's and next week's games, then they will be in the playoffs. As long as the Lakers stay injury-free, then they will win this week's games. However, if the Lakers have an injury or they lose next week's games, then the Heat will be in the playoffs. The Lakers do stay injury-free, and the Heat don't make it to the playoffs.   Also, the Lakers win next week's game. Therefore, the Lakers will play in the playoffs.

In cases where neither a direct nor an indirect proof seems to be working, there is yet another strategy for proving any implication $\mathcal{A} \Rightarrow \mathcal{B}$ that is known as a **proof by contradiction**. This method works as follows: we assume <u>both</u> $\mathcal{A}$ and $\sim \mathcal{B}$ are true (this is the only way that the implication $\mathcal{A} \Rightarrow \mathcal{B}$ could fail to be true) and we proceed to derive a contradiction. Once a contradiction is logically derived, the proof is complete, since it shows that the simultaneous truth of both $\mathcal{A}$ and $\sim \mathcal{B}$ is not possible.   This method is particularly popular for students who might get stuck on a difficult (direct) proof (where they are only assuming $\mathcal{A}$) by providing another assumption ($\sim \mathcal{B}$) with which to work.[13]

EXERCISE FOR THE READER 1.5:   Use the method of proof by contradiction to establish the following implication:
$$[P \to (Q \vee R)] \wedge [\sim R \to P] \wedge \sim Q \Rightarrow R$$

All of our comments regarding logical proofs continue to hold true for general proofs in mathematics and the sciences, where numerous theorems are either equivalences or inferences of the form (1).   We point out that another helpful technique in proving a theorem or analyzing an argument is to separate the proof (or analysis) into cases. This often makes difficult tasks more manageable. A very important example of this is that any logical equivalence $\mathcal{A} \equiv \mathcal{B}$ can be proved by separately proving the corresponding two implications $\mathcal{A} \Rightarrow \mathcal{B}$ and $\mathcal{B} \Rightarrow \mathcal{A}$. In general, proving an implication is an easier task than proving an equivalence, since the equivalences of Theorem 1 can also be used as implications (in either direction)—see Theorem 1.1 II(e). Our next set of examples will further demonstrate the usefulness of breaking things up into cases.

---

[13] In his influential textbook [Roy-88], with which a significant portion of several generations of mathematicians has been trained in their initial graduate analysis courses, Halsey Royden discourages students from resorting to proofs by contradiction, In his words: *"All students are enjoined in the strongest possible terms to eschew proofs by contradiction! There are two reasons for this prohibition: First such proofs are very often fallacious... Second, even when correct, such a proof gives little insight into the connection between $\mathcal{A}$ and $\mathcal{B}$, whereas both direct and indirect proofs construct a chain of argument connecting $\mathcal{A}$ and $\mathcal{B}$."*   By being careful, you can certainly avoid being a reason for his first objection. His second objection does indeed highlight a good point; however, there are some cases where a proof by contraction is the only way to go.

Logical Puzzles

Logical arguments have also been popularized into puzzles, many of which make excellent problems for students of logic to practice their newly learned skills. We close this section with two examples illustrating these sorts of problems; more will be given as exercises. Interested readers can find many more such problems on the Internet, in magazine columns, or in books of **logic puzzles**; logic puzzles frequently appear even on NPR radio shows. Two particularly famous logic puzzle authors are Raymond Smullyan and Martin Gardner. Our first example involves a fictitious island where all of the inhabitants are either liars (they always lie) or truth tellers (they always tell the truth); such liar puzzles actually date back to the ancient Greek philosophers.

EXAMPLE 1.5: Suppose that you are a tourist on an island where all inhabitants are either liars or truth tellers. You meet three inhabitants walking together. After you greet them,

> A says: "We are all liars."
> B says: "No, A is the only liar."
> C says: "The other two are both liars."

Can you determine, who, if any, of these three is telling you the truth?

SOLUTION: We will run through the different possibilities and check for consistency. A could not be a truth teller, for then his statement would be false; therefore A has to be a liar. If B were telling the truth, it would mean that C would also have to be a truth teller, but then C's statement would be false (his statement implies that B is a liar). Therefore, B is also a liar. From this we now know that C's statement is true, so he must be a (and the only) truth teller.

The next example is a famous one for which we present in two slightly different proofs. The first version is a bit less formal and can be understood by people without any background in formal logic. The reader is strongly encouraged to try to answer both questions before peeking at the solution.

EXAMPLE 1.6: (*A Logic Puzzle*) Suppose that you are a prisoner on an island where every resident is either a truth teller or a liar. Truth tellers always tell the truth while liars always lie.   There are two doors (a left one and a right one) and one guard. You know that one of the doors will lead to your freedom but the other will lead to your immediate death, and the guard knows which one is which.
(a) If you were allowed to ask the guard a single yes–no question, what should you ask to help you find the door that would lead to your freedom?
(b) Suppose instead that you were allowed to give the guard one statement and ask him to give you its truth value. What statement would you give the guard to help you to find freedom?

SOLUTION: Part (a): Here is one (very elegant) solution:

If I were to ask you whether the left door would lead to my freedom, what would you say?

That this question will work is simply based on the double negation rule (Theorem 1.1 I(e)). Let's analyze how this question would play out:

Case 1: The guard is a truth teller: This is easy: "Yes" would mean you take the left door, "no" would mean you should take the right door.

Case 2: The guard is a liar: If the left door would lead to your freedom, and you simply asked the guard if it did, he would give you a "no" answer since he is a liar. The question being asked, however, is a bit different. It asks him to tell you what his response would be if you <u>were</u> to simply ask him the former question. Since he would answer "no" to the former question, in order to lie about the actual question being asked, he would need to say "yes." By the same reasoning, a "no" answer would have to mean that the right door would lead to your freedom.

Thus, with this question, it is immaterial whether the guard is a liar or a truth teller, a yes will always mean the left door is the one you should take.

Part (b):   There are many possibilities; we give only one: Consider the following two basic statements:

$$P = \text{You are a liar, and } Q = \text{The left door leads to freedom.}$$

If you ask the guard the truth value of the following statement: $P \oplus Q$, in words:

True or False?   You are a liar or the left door leads to freedom, but not both.

A true response will always mean the left door leads to freedom, while a false response will mean the right door does.   We leave it to the reader to run through the four cases: guard is a truth teller or not, left door leads to freedom or not, to see that this will indeed do the job.

EXERCISE FOR THE READER 1.6: Suppose that you are told of a small town in which there is a barber who shaves every man and only those men who do not shave themselves. Is this possible? (Assume there are no female barbers.)

EXERCISES 1.1:

1.   For each item below, indicate whether it is a statement. For those that are statements, attempt to determine the truth value (without resorting to any resources).
   (a) Marilyn Monroe was born before John F. Kennedy.
   (b) Who assassinated John F. Kennedy?
   (c) $3^8 > 2^9$.
   (d) $x > y^2 - 7$.
   (e) $x > y^2 - 7$, given that $x = 10$ and $y = 5$.
   (f) Ulysses Grant was the 15th president of the United States.

2.   For each item below, indicate whether it is a statement. For those that are statements, attempt to

determine the truth value (without resorting to any resources).
(a) Please take off your shoes.
(b) The capital of Italy is Florence.
(c) There are infinitely many *prime numbers*.
Note: Recall that a **prime number** (or **prime**) is an integer greater than one whose only positive integer divisors are 1 and itself. The first few primes are: 2, 3, 5, 7, 11, 13, 17.

(d) If $x^2 = 25$, then $x = 5$.

(e) If x is a positive real number with $x^2 = 25$, then $x = 5$.

(f) (*Prime Pairs Conjecture*)[14] There are infinitely many **prime pairs**, that is, pairs of prime numbers that have exactly one integer between (e.g., 3, 5;   5, 7;   11, 13;   17, 19;   29, 31).

3.   Determine the truth value for each of the following compound statements.
        (a) The United States has 52 states or Washington DC is not a state.
        (b) The United States has 52 states and Washington DC is not a state.
        (c) If the United States has 52 states, then Washington DC is not a state.
        (d) If Washington DC is not a state, then the United States has 52 states.
        (e) The United States has 52 states, if, and only if Washington DC is not a state.

4.   Determine the truth value for each of the following compound statements.
        (a) Paris is the capital of France or Florence is the capital of Italy.
        (b) Paris is the capital of France and Florence is the capital of Italy.
        (c) If Paris is the capital of France, then Florence is the capital of Italy.
        (d) If Florence is the capital of Italy, then Paris is the capital of France.
        (e) Paris is the capital of France, if, and only if Florence is the capital of Italy.

5.   Determine the truth value for each of the following compound statements. Assume throughout that $x = 2$, $y = -2$, and $z = 10$.
        (a) $x^2 - 5y^3 > z^2$ or $z/(x^2 + y^2) < x$.
        (b) $x^2 - 5y^3 > z^2$ and $z/(x^2 + y^2) < x$.
        (c) If $x^2 - 5y^3 > z^2$, then $z/(x^2 + y^2) < x$.
        (d) If $z/(x^2 + y^2) < x$, then $x^2 - 5y^3 > z^2$.
        (e) $x^2 - 5y^3 > z^2$, if, and only if $z/(x^2 + y^2) < x$.

6.   Determine the truth value for each of the following compound statements. Assume throughout that $x = 3$, $y = 2$, and $z = -5$.
        (a) $x^2 - 5y^3 > z^2$ or $z/(x^2 + y^2) < x$.
        (b) $x^2 - 5y^3 > z^2$ and $z/(x^2 + y^2) < x$.
        (c) If $x^2 - 5y^3 > z^2$, then $z/(x^2 + y^2) < x$.
        (d) If $z/(x^2 + y^2) < x$, then $x^2 - 5y^3 > z^2$.
        (e) $x^2 - 5y^3 > z^2$, if, and only if $z/(x^2 + y^2) < x$.

7.   Suppose that we have five cards, each of which has a positive integer (from $\{1,2,3,\cdots\}$) on one of its sides, and a letter (from $\{A,B,\cdots,Z\}$) on the other side. Suppose that the cards have been laid out on a table and they show (from left to right):

     Card #1: K,   Card #2: 13,   Card #3: A,   Card #4: 6,   Card #5: X

---

[14] The *prime pairs conjecture*, also known as the *twin primes conjecture*, is a very natural conjecture that has been around for centuries, although its exact origin is unknown. As of the writing of this book, the conjecture remains unproved. Its simplicity has helped to make resolving this conjecture (proving or disproving it) a famous problem in mathematics.

For each statement below indicate which cards would need to be turned over to determine the truth value of the statement:

(a) If the letter on a card is a vowel (A, E, I, O, or U), then the number on the other side is greater than 10.

(b) If the letter on a card is a vowel (A, E, I, O, or U), then the number on the other side is less than 20.

(c) If the number on the card is even, then the letter on the other side must be a consonant (not a vowel).

(d) If the letter on a card is a consonant, then the number on the other side is even.

(e) If the letter on a card is a consonant, then the number on the other side is odd.

(f) For the number on a card to be odd, it is necessary that the letter on the opposite side is not a vowel.

(g) A letter on a card is even if, and only if the letter on the other side is a vowel.

8. Repeat the directions and each part of Exercise 7 if (a new set of) cards are now showing:

Card #1: 2,   Card #2: 24,   Card #3: E,   Card #4: Z,   Card #5: 21

9. Write each implication below in the form: "if … then …" and then formulate the converse and the contrapositive.

(a) We will go to a movie only if it rains.

(b) I will go to the party if Yumi will go.

(c) Only if I can beat Norris this weekend, I will enter the tournament.

(d) In order for Tom to make the team, it is necessary for him to be able to run a mile in under six minutes.

(e) An attractive job offer will be sufficient for Carol to move to France.

(f) In order for an infinite series $\sum_{n=1}^{\infty} a_n$ to converge, it is necessary that the terms a_n tend to zero as n tends to infinity.[15]

10. Write each implication below in the form: "if … then …" and then formulate the converse and the contrapositive.

(a) Luis will play football if it is not raining.

(b) I will go to the party only if Yumi is going.

(c) Being able to dunk a basketball is sufficient for being able to join the team.

(d) That he is on the team implies that he can dunk a basketball.

(e) An attractive job offer is necessary for Carol to move to France.

(f) If a prime number p is a factor of a product of two integers ab, then p is a factor of a, or p is a factor of b.

11. Create truth tables for each of the following compound statements. Identify any tautologies or contradictions.

    (a) $\sim P \to P$ (b) $P \wedge \sim (P \to P)$

    (c) $(P \wedge Q) \to P$ (d) $P \vee (Q \to P)$

    (e) $((P \to Q) \leftrightarrow P) \to \sim Q)$ (f) $((P \leftrightarrow Q) \wedge P) \oplus Q$

    (g) $(P \oplus Q) \to (Q \oplus R)$ (h) $P \to (Q \leftrightarrow R)$

    (i) $(P \to Q) \to (Q \vee (R \leftrightarrow \sim P))$ (j) $[P \to (R \wedge Q)] \leftrightarrow [\sim P \to (\sim R \vee Q)]$

12. Create truth tables for each of the following compound statements. Identify any tautologies or contradictions.

    (a) $\sim P \leftrightarrow P$ (b) $\sim P \vee \sim (P \leftrightarrow P)$

    (c) $(P \to Q) \to (Q \to P)$ (d) $(P \leftrightarrow Q) \to (Q \vee P)$

    (e) $(P \leftrightarrow \sim Q) \oplus (P \leftrightarrow Q)$ (f) $((P \to Q) \to P) \to \sim Q$

---

[15] This important (albeit basic) theorem from calculus often gets (logically) misinterpreted.

(g) $(P \rightarrow (P \wedge Q)) \rightarrow (P \wedge Q \wedge R)$ (h) $P \rightarrow [(Q \vee P) \oplus (P \wedge R)]$

(i) $[(P \rightarrow \sim Q) \rightarrow (Q \leftrightarrow R)] \rightarrow \sim Q$ (j) $[R \rightarrow (P \wedge Q)] \leftrightarrow [\sim P \vee (R \rightarrow \sim Q)]$

13.   (a) A statement of the form "P unless Q" is formally equivalent to "P or Q," however, there is sometimes ambiguity in spoken language where at times the exclusive or is intended.   In formal language, which of the following implications is thus equivalent to "P unless Q?"

   (i) $P \rightarrow Q$ (ii) $P \rightarrow \sim Q$ (iii) $\sim P \rightarrow Q$ (iv) $\sim P \rightarrow \sim Q$

   (b) Express the statement: "I will go to the movies unless Diane calls" in the form: "if …, then …".   Then form the converse and contrapositive both in the "if …, then …" form and using the word "unless."

14.   (a) See Exercise 11, and then determine which of the following implications the statement "P unless Q" is equivalent to:

   (i) $Q \rightarrow P$ (ii) $Q \rightarrow \sim P$ (iii) $\sim Q \rightarrow P$ (iv) $\sim Q \rightarrow \sim P$

   (b) Express the statement: "I will spend spring break in Hawaii unless Professor Garnett schedules the midterm the day after the break." in the form: "… only if …".   Then form the converse and contrapositive both in the "… only if …" form and using the word "unless."

15.   (a) Use truth tables (one for the left compound statement and another for the right) to establish De Morgan's law: (Theorem 1.1 Part I(d)) $\sim (P \vee Q) \equiv \sim P \wedge \sim Q$.

   (b) Give another proof of this De Morgan law that is based on directly analyzing and comparing the definitions of the relevant logical operators.   See, for example, the proof of Part (c) of Theorem 1.2 and the solution of Example 1.2 (where the result of Theorem 1.1 Part II(a) is derived).

16.   Repeat both parts of Exercise 15 for De Morgan's Law: $\sim (P \wedge Q) \equiv \sim P \vee \sim Q$.

17.   Repeat both parts of Exercise 15 for each of the following equivalences from Theorem 1.1.
   (i) (*Commutativity of Conjunction*) $P \wedge Q \equiv Q \wedge P$.
   (ii) (*Associativity of Conjunction*) $(P \wedge Q) \wedge R \equiv P \wedge (Q \wedge R)$.
   (iii) (*Double Negation*) $\sim (\sim P) \equiv P$.
   (iv) (*Exportation*) $P \rightarrow (Q \rightarrow R) \equiv (P \wedge Q) \rightarrow R$.

18.   Repeat both parts of Exercise 15 for each of the following equivalences from Theorem 1.1.
   (i) (*Commutativity of Disjunction*) $P \vee Q \equiv Q \vee P$.
   (ii) (*Associativity of Disjunction*) $(P \vee Q) \vee R \equiv P \vee (Q \vee R)$.
   (iii) (*Absorption*) $P \vee (P \wedge Q) \equiv P$.
   (iv) (*Biconditional as Implication*) $P \leftrightarrow Q \equiv (P \rightarrow Q) \wedge (Q \rightarrow P)$.

19.   (a)   Use truth tables to establish the hypothetical syllogism: (Theorem 1.2(e)) $(P \rightarrow Q) \wedge (Q \rightarrow R) \Rightarrow P \rightarrow R$.
   (b) Give another proof of the implication of Part (a) using any earlier implication of Theorem 1.2 (Parts (a) through (d)) and/or any of the equivalences of Theorem 1.1.

20.   Prove each of the following implications using any legitimate method.
   (a) $(P \rightarrow Q) \wedge (R \rightarrow S) \Rightarrow [(P \vee R) \rightarrow (Q \vee S)]$.
   (b) $(P \rightarrow Q) \wedge (R \rightarrow S) \Rightarrow [(P \wedge R) \rightarrow (Q \wedge S)]$.

21.   Repeat both parts of Exercise 19 for the disjunctive syllogism (Theorem 1.2(f)): $(P \vee Q) \wedge \sim P \Rightarrow Q$.

22. Determine whether the following is a tautology: $(P \to Q) \wedge (\sim Q \to R) \to (P \to R)$.

23. Establish the following logical equivalences by using only Theorem 1.1:
    (a) $P \leftrightarrow Q \Leftrightarrow (P \wedge Q) \vee (\sim P \wedge \sim Q)$.
    (b) $(P \vee Q) \vee \sim R \Leftrightarrow [(P \vee \sim Q) \wedge R] \to P$.

24. Express the negations of each of the implications in Exercise 9 in English and without the use of any implication.

25. A radio advertisement states: "If you don't come in to Johnson Toyota this weekend to buy your new Toyota, you will be paying too much for your new car." This exercise will logically analyze the actual and intended meanings of this statement. We let $P =$ "You don't come in to Johnson Toyota this weekend" and $Q =$ "You will be paying too much for your new car." The statement can thus be represented as $P \to Q$.

    (a) Suppose that you did come in to Johnson Toyota, bought a new car and subsequently found out that the competitor Anderson Toyota had been selling the exact same model that you bought for $2,000 less. Would Johnson Toyota's advertised claim be contradicted, i.e., constitute false advertising? Do you think that Johnson Toyota had intended (and radio listeners had interpreted) the possibility of such an occurrence?

    (b) Give examples of situations for the remaining three rows of the truth value for the Johnson Toyota conditional, and compare the intended truth of the statement with the corresponding logical truth value in each case.

    (c) Reword Johnson Toyota's statement into a logically coherent form that correctly conveys their intended (and interpreted) message. How often do you encounter such wording in radio, television, and newspapers?

26. (*A Logic Puzzle*) A car driver was listening to a discussion of the World Cup results on a radio talk show. He heard one speaker say: "Either Italy was first, or Germany was second, but not both." During the next statement, some static cut off one of the words, and all the driver heard was: "Either France was second, or Germany was <static>, but not both." The radio host then said that with this information it was possible to determine the ranking (the host did not know about the static that this driver heard). Nevertheless, the driver was able to determine the top three rankings from what he heard. Which countries came in first, second, and third?
    Suggestion: Separate into the three possible cases for which ranking could have been said under the static. Only one of these three cases leads to a unique permissible ranking.

27. (*A Logic Puzzle*) Three friends, one from France, one from Germany, and one from Italy are driving together when they hear the following comment on the radio about the recent World Cup matches: "The top three countries were Italy, France, and Germany. Either Italy outranked France, or Germany came in first, but not both." After hearing that, the Frenchman said that even though he knew the standing of his country's team, it was not possible for him to determine the ranking. The Italian, who did not listen to what the Frenchman just said, also said it was not possible for him to determine the ranking, even though he knew how his country's team placed. The German, who heard everything but knew nothing about the matches, was able to determine the top three ranking. Which countries came in first, second, and third?
    Suggestion: The radio announcer's comment reduces the number of possible rankings to three. Write them out, then use the Italian and Frenchman's comments to rule out two of these three rankings.

28. (*A Logic Puzzle*) You are a tourist visiting an island where all inhabitants are either liars or truth tellers. You run into two inhabitants who tell you: A: "B would say I lie." B: "This is true." What conclusions can you draw about the types of persons A or B are (liar or truth teller)?

29. (*A Logic Puzzle*) You are a tourist visiting an island where all inhabitants are either liars or truth tellers. You run into two inhabitants who tell you: A: "We are both truth tellers." B: "A is a liar." What conclusions can you draw about the types of persons A or B are (liar or truth teller)?

30.   (*A Logic Puzzle*)   You are a tourist visiting an island where all inhabitants are either liars or truth tellers. You run into four inhabitants who tell you: A: "C and D are both truth tellers.", B: "C is a truth teller, but D is a liar." C: "Neither B nor D is a liar." D: "Neither A nor B is a liar."   What conclusions can you draw about the types of persons A, B, C, D are (liar or truth teller)?

31.   (*A Logic Puzzle*)   In Example 1.6, suppose instead that there were three doors: 1, 2, and 3, of which only one would lead to you (the prisoner) to freedom, while the other two would lead to your death. The guard knows which is which, but you don't know if he is a liar or a truth teller.
(a) Explain why you would not be able to decide the correct door if you were allowed to ask the guard only one true/false question.
(b) Give a strategy for determining the desired door if you were allowed to ask the guard two true/false questions.

32.   (*A Logic Puzzle*)   Four suspects were questioned in a bank robbery, and each made the following two statements: A: "I did not do it, B did." B: "A did not do it. C did." C: "B did not do it, I did." D: C did not do it, A did." A polygraph machine indicated that each person told the truth in just one of the two statements. From this information, determine who did the bank robbery.

NOTE: Exercises 31–35 deal with the concept of a *functionally complete* set of operators. A set of logical operators is called **functionally complete** if any compound statement (involving any of the full set of logical operators introduced in this section: $\{\wedge, \vee, \sim, \rightarrow, \leftrightarrow\}$) can be expressed using only the operators in this set, and perhaps also any number of pairs of parentheses. Certainly the full set of operators $\{\wedge, \vee, \sim, \rightarrow, \leftrightarrow\}$ is a (trivial) example of a functionally complete set of operators. Since any implication can be expressed using only the disjunction and negation operators (Theorem 1.1 II(a)), it follows that we can remove the implication and still have a functionally complete set of operators: $\{\wedge, \vee, \sim, \leftrightarrow\}$.

33.   (*A Functionally Complete Set of Operators*) (a) Show that $\{\wedge, \vee, \sim\}$ is a functionally complete set of operators.
(b) (*Disjunctive Normal Form*)   Given any compound statement with n logical variables: $P_1, P_2, \cdots, P_n$, the corresponding truth table will have 2^n rows in it. For each row of this truth table for which the statement is true, the truth values of the variables will be a certain length n sequence of T's and F's. Form the conjunction $\mathcal{A}$ of $Q_1, Q_2, \cdots, Q_n$, where each Q_i equals either P_i or $\sim P_i$ according as to whether the truth value of P_i (in this particular row of the truth value) is T or F, respectively.   For example, if $n = 4$, and the truth values in a row (where the compound statement is true) are   T, F, F, T, then $\mathcal{A}$ would be $P_1 \wedge \sim P_2 \wedge \sim P_3 \wedge P_4$.

Observe that $\mathcal{A}$ is true only for this single row of the truth table, i.e., when $P_1 = T, P_2 = F$, $P_3 = F$, and $P_4 = T$, and false for all other rows in the truth table. This construction gets repeated, and another such $\mathcal{A}$ gets created for each row in the truth table for which the original statement is true. The disjunction of all of these $\mathcal{A}$'s will do the job. This representation of $\mathcal{A}$ is called the *disjunctive normal form*.
Find the disjunctive normal form of each of the following compound statements:
(i) $P \rightarrow Q$ (ii) $P \vee (Q \rightarrow P)$ (iii) $P \rightarrow (Q \leftrightarrow R)$

34.   (*A Functionally Complete Sets of Operators*) (a) Show that $\{\wedge, \sim\}$ is a functionally complete set of operators.
(b) Show that $\{\vee, \sim\}$ is a functionally complete set of operators.
Suggestion: Use the result of Exercise 33 in conjunction with De Morgan's laws.

35. (*Nonfunctionally Complete Sets of Operators*) (a) Show that $\{\wedge\}$ is not a functionally complete set of operators.

    (b) Show that $\{\vee\}$ is not a functionally complete set of operators.

    (c) Show that $\{\vee, \wedge\}$ is not a functionally complete set of operators.

    Suggestion: Examine (compound) statements with just one logical variable.

36. (*A Single Logical Operator Can Be a Functionally Complete Set of Operators*) We define the **nor** operator on two logical variables P and Q as follows:

    P nor Q (denoted $P \downarrow Q$) is true only when P and Q are both false, and is true in all other cases.

    Thus, the nor operator is just the negation of or (nor = not or), i.e., $P \downarrow Q \equiv \sim (P \vee Q)$.

    This exercise will outline a proof that $\{\downarrow\}$ constitutes a functionally complete set of operators. This is surprising in light of Exercise 35.

    (a) Show that $\sim P \equiv P \downarrow P$.

    (b) Show that $P \vee Q \equiv (P \downarrow Q) \downarrow (P \downarrow Q)$.

    (c) Use the result of a previous exercise to deduce that $\{\downarrow\}$ is a functionally complete set of operators.

37. (*A Single Logical Operator Can Be a Functionally Complete Set of Operators*) The negation of the conjunction operator is called the **nand** (not and) operator; we denote this operator as: $P \uparrow Q$. Thus, $P \uparrow Q \equiv \sim (P \wedge Q)$. Show that $\{\uparrow\}$ is a functionally complete set of operators.

    Suggestion: Try to develop an argument similar to that outlined in Exercise 36 for the nor operator.

COMPUTER EXERCISES 1.1:[16]

1. How do you get your computing platform to test whether a (logical or mathematical) statement is true or false? Perform such a test with the following statements and compare with the actual truth values:

    (a) $3 < 6$

    (b) $-5 \geq 0$

    (c) $x^2 + y^2 = 25$, given that $x = 3$ and $y = 4$.

2. Use your computing platform to compute the truth values of each of the following compound statements, and compare with the actual truth values:

    (a) $(3 < 6) \wedge (-5 \geq 0)$

    (b) $(3 < 6) \rightarrow (-5 \leq 0)$

    (c) $\sim (x\sqrt{y} + 5 \geq 2) \vee [(x^3 + y^3 > z^3) \rightarrow (|x - y| > |z - y|)]$ given that $x = 3$, $y = 4$, and $z = 5$.

    (d) $[(3 < 6) \oplus (-5 \leq 0)] \leftrightarrow [((-3)^2 = 9) \rightarrow (8 \div 4 > 4)]$

NOTE: Implications are often used in computer programming with a syntax such as:[17]

---

[16] In the computer exercises sets it is assumed that you are working with a particular programming language or computing platform. In this first set of computer exercises, in addition to introducing some basic logical implementations, we will introduce the important concepts of loops. Loops will get used in numerous computer implementations of many of the algorithms involving discrete structures (particularly after we introduce recursion in Chapter 3), so it is advisable to gain familiarity with them at an early stage.

```
IF <CONDITION A>   THEN   <DO TASK A>
```

This is not actually a statement, but rather a conditional command. In the course of a program, this command will check the truth value of <CONDITION A>, and if it is true, then <TASK A> will be executed, otherwise, nothing is done. Most computing platforms support more elaborate versions of the above basic syntax. For example a syntax such as:

```
IF <CONDITION A>   THEN   <DO TASK A>,
ELSE IF <CONDITION B>    <DO TASK B>,
ELSE IF <CONDITION C>    <DO TASK C>,
                 ...
         ELSE <DO DEFAULT TASK>
                 ...
```

would work as follows: first check if <CONDITION A> is satisfied, if it is, perform <TASK A>, and bypass the remaining "ELSE" lines. If <CONDITION A> fails, move on to check the next condition in the list (<CONDITION B>), if it is satisfied, perform the corresponding task (<TASK B>), and bypass the remaining "ELSE" lines.   Continue, in this way, moving down the "ELSE IF"'s as necessary. The final "ELSE" line is optional; it allows a default task to be performed in case all of the previously listed conditions had failed.

The tasks can be either single commands, a sequence of commands, or a separate program (a subroutine). For example, suppose that variables $x = 6$, $y = -2$, and $z = 8$ are stored in the current workspace. The following command:[18]

```
IF    z/y > x
      THEN set z = 2*z
      ELSE set x = x - 3 and y = -y
```

would check that the condition $z/y > x$ $(8/-2 > 6)$ is false, and so the "ELSE" task would reset x to be its previous value less 3, or to $x = 3$, and y to the negative of its previously stored value, or $y = 2$.   If this same command were now encountered again, the condition $z/y > x$ $(8/2 > 3)$ would now test true, so the "THEN" task would cause z to be reset to twice its currently stored value (of 8), so $z = 16$.

3.   (a) Check by hand what the values of the variables x, y, and z will be after the following set of tasks are repeatedly run six times with initial variable values: $x = 1$, $y = 1$, and $z = 2$:

```
IF    (z ≥ 2*x + 4) OR   (y < x + 10)
      THEN set x = y, y = z, z = x + y
```

(b) Confirm these hand-computed results by running the same code on your computing platform three times.
(c) The IF condition in the above code is a disjunction of two conditions.   Explain why if this program is repeated a sufficiently large number of times, this disjunction becomes a tautology.

4.   Repeat Parts (a) and (b) of Exercise 3 if the code is modified as follows:

```
IF    (z < x + y) OR   (z/2 ≤ x)
      THEN set x = y, y = z, z = x + y
      ELSE set z = z - 1, x = x + 1
```

---

[17] Different computing platforms will have varying syntaxes and constructions, but the concepts behind the main programming constructions are very similar. We will adopt and adhere to an intuitive *pseudocode*, which will be summarized in Appendix A at the end of the book.

[18] We follow the computer programming custom of allowing a single equals sign to represent the operation of assignment. Thus, for example, if z is stored as 2, the mathematical equation $z = 2*z$ makes no sense; but we take it to mean <u>reassign z to be 2 times its old value</u> (i.e., z will now be stored as 4).

NOTE: One of the important strengths of computing platforms is their ability to perform a task repeatedly, and to update variables in the process. The action of performing such a repeated set of commands is called a **loop**, and we will introduce two sorts of loops. A **for loop** is a loop where the number of iterations is made completely explicit, usually by giving the range (and possibly also the steps) of the counter variable. For example, the following for loop (in our default pseudocode language) would add up the numbers 1, 2, 3, $\cdots$,10.

```
SUM = 0  (initialize the sum)
FOR K = 1 TO 10
    SUM = SUM + K  (update the sum)
END
```

The general syntax of a for loop is as follows:

```
FOR  N = <START> TO <LAST> INCREMENT = <GAP>
    <DO TASKS>
END
```

In the preceding example, we had <START> = 1, <LAST> = 10, and since the increment was (the default value) 1, it did not need to be specified. The <DO TASKS> portion of the loop, called the **body** of the loop, can be any set of commands that will be executed in each iteration of the loop. The END construct is required after the body to tell the platform when the loop is closed. The variable N in the above syntax is called the **counter** of the loop, and it may be included in the body of the loop (as it was in the previous example, where it was denoted by K). The way the loop works is that the counter is first set at its starting value <START>, then the commands in the body <DO TASKS> are executed in order, then the counter gets incremented by the INCREMENT <GAP>, and the commands in the body are again executed. This continues until the counter has reached its <LAST> value.

For example the following for loop could be used to compute the sum $4^2 + 6^2 + 8^2 + \cdots + 26^2$:

```
SUM = 0  (initialize the sum)
FOR K = 4 TO 26  INCREMENT = 2
    SUM = SUM + K^2  (update the sum)
END
DISPLAY(SUM)
```

(The last command after the loop is used to display the final value SUM, i.e., the desired sum.)

In Computer Exercises 5–6, for each part, write a for loop that will compute (as its only output) each of the following finite sums. Run it and give the answer (= the output).

5. (a) $1 + 2 + 3 + 4 + \cdots + 9999$. (b) $500 + 501 + 502 + \cdots + 1500$.
    (c) $500^2 + 501^2 + 502^2 + \cdots + 1500^2$. (d) $500 + 501^2 + 502 + 503^2 + \cdots + 1500$.

    Suggestion: For Part (d), the exponent changes at each iteration. Notice that it toggles between 1 and 2. One way to encode the toggle would be to include an assignment line like: `exponent = 3 - exponent` in the body of the loop. A simpler way would be to use an if branch within the body of the loop.

6. (a) $1 + 2 + 3 + 4 + \cdots + 20000$. (b) $50 + 51 + 52 + \cdots + 5050$.
    (c) $50^3 + 51^3 + 52^3 + \cdots + 2000^3$. (d) $50 + 51^3 + 52 + 53^3 + \cdots + 2000$.

7. (*Compound Interest versus Simple Interest*) (a) Suppose that your family has just discovered a savings account that was started by an ancient ancestor 240 years ago. The account was started with $10 and paid 6.5% interest compounded annually. What is the account balance today?
    (b) Suppose instead that your ancestor had invested the $10 in an account that paid 8% simple interest. This means that each year, the interest earned was 8% of the principal (i.e., 8% of $10,

or 80¢). Without using any loop (just basic arithmetic) what would the account balance of this account be today—240 years later?

8.   (*Retirement Annuities*) (a) Suppose that Jackie sets up a retirement annuity that pays $r = 7.5\%$ interest, compounded annually. Starting when she turns 30 years old, Jackie deposits \$3500 each year, and plans to continue this until her 65th birthday (when she makes her last deposit and plans to retire). How much will Jackie's annuity be worth then?
(b) Repeat Part (a) with r changed to 9%.
(c) Repeat Part (a) with r changed to 12%.

9.   (*Retirement Annuities*) (a) Suppose that Randall sets up a 401(k) retirement annuity that pays $r = 6\%$ interest, compounded annually. When he turns 35 years old, Randall deposits \$3500, and each subsequent year he plans to deposit 5% more than the previous year (due to planned raises and better money management). He continues this until his 65th birthday (when he makes his last deposit and plans to retire). How much will Randall's annuity be worth then?
(b) Repeat Part (a) if the annual deposits are doubled.
(c) Repeat Part (a) if the annual interest rate is doubled.
(d) Repeat Part (a) if he starts the annuity 10 years earlier (when he turns 25), but still continues until his 65th birthday.

In many situations, we do not know in advance how many iterations are needed, but rather we would like the iterations to stop when a certain condition is met. In contrast to a for loop, a while loop is suitable for such a purpose. The general syntax of a **while loop** is shown below:

```
WHILE <CONDITION>
    <DO TASKS>
END
```

Here, `<DO TASKS>` can be any set of commands as in a for loop. When encountered in a program code, the above while loop will function as follows: First the truth value of `<CONDITION>` is checked. If it tests false, the while loop is bypassed with no action having been taken. If it tests true, then all the commands in the body `<DO TASKS>` are performed, and `<CONDITION>` is reevaluated. If it tests false now, the while loop is exited, while if `<CONDITION>` tests true, then `<DO TASKS>` is performed once again, and the program will go back again to check the truth value of `<CONDITION>`. This process continues.[19]

For example the following for loop could be used to compute the sum $4^2 + 6^2 + 8^2 + \cdots$ by continuing to add terms until the sum first equals or exceeds 1000.

```
SUM = 0  (initialize the sum)
TERM = 4 (initialize the term)
WHILE SUM < 1000
    SUM = SUM + K^2  (update the sum)
    TERM = TERM + 2  (increment term)
END
DISPLAY(SUM)
```

Notice one important difference with for loops: In while loops the counter term needed to be initialized and incremented inside the loop. This was done automatically with for loops.

---

[19] **Caution:** Unlike with a for loop, there is always a danger that a while loop can enter into an *infinite loop*. You should learn how to exit from an infinite loop in case your programs accidentally enter into one. One common strategy that works on several platforms is to enter the CTRL-C command.

In Exercises 10–11, for each part, write a while loop that will add up the terms of the given sequence until the sum first exceeds or equals the number M that is given. Your loop should have (exactly) two outputs: both the number of terms that were added up, and the sum of these terms. Run it and give the answers (= the output).

10.   (a) $1+2+3+4+\cdots$; $M = 10,000$.   (b) $500 + 501 + 502 + \cdots$; $M = 1,000,000$.

       (c)   (d)

       $500^2 + 501^2 + 502^2 + \cdots$; $M = 1,000,000$. $500 + 501^2 + 502 + 503^2 \cdots$; $M = 1,000,000$.

11.   (a) $1+2+3+4+\cdots$; $M = 2,000,000$.   (b) $50 + 51 + 52 + \cdots$; $M = 2,000,000$.

       (c)   (d)

       $50^3 + 51^3 + 52^3 + \cdots$; $M = 2,000,000,000$. $50 + 51^3 + 52 + 53^3 + \cdots$; $M = 2,000,000,000$.

12.   (*Compound Interest*) (a)   Suppose that $1000 is invested in a long-term CD that pays 7% interest compounded annually. Write a while loop to determine the number of years that it would take the money to triple (to equal or exceed $3000), and the resulting balance after this number of years.
      (b) Re-do Part (a) if the amount invested is changed to $1 million.
      (c) Re-do Part (a) with the annual interest rate changed to 9%.
      (d) Re-do Part (a) with the annual interest rate changed to 5%.

13.   (*Retirement Annuities*) For each part below, re-do Example 3 with the indicated changes (the changes indicated for each part apply only to that part).
      (a)   Suppose that Michael instead decides to retire when his annuity reaches or exceeds $500,000.
      (b) Suppose that the interest rate is 10% (rather than 9%).
      (c) Suppose that rather than making the same deposit every year, because of regular raises and better money management, Michael's annual deposits grow by 6% each year. So the first year he puts in $5000, the second year $5300, the third year $5618, etc.

14.   (*Retirement Annuities*) Suppose that Jocelyn sets up a 401(k) that pays $r = 10\%$ interest, compounded annually. Starting when she turns 36 years old, Jocelyn deposits $10,000 each year, and plans to continue this until her annuity meets or exceeds $1 million. How many years will it take for this to happen?
      (a)   Suppose instead that Jocelyn decides to retire when her annuity reaches or exceeds $1.5 million.
      (b) Suppose instead that the interest rate is 12% (rather than 10%).
      (c) Suppose that rather than making the same deposit every year, because of regular raises and better money management, Jocelyn's annual deposits grow by $1000 each year. So the first year she puts in $10,000, the second year $11,000, the third year $12,000, etc.

15.   (*Paying off a Loan*) (a)   Suppose that Gomez takes out a loan for $22,000 to buy a new SUV (after trading in his car). The bank charges 6% annual interest on the unpaid balance, compounded monthly (that's 0.5% interest each month). If Gomez pays $300 at the end of each month, how many months will it take him to pay off his loan? What is the amount of his last payment? His last payment will just cover the unpaid balance so will likely be less than $300.
      (b) Suppose that, in writing a while loop to solve this problem, you accidentally typed 220000 (rather than 22000); i.e., you added an extra zero to the loan amount. Explain what would happen. If you don't know what would happen, try it out.

16.   (*Paying off a Loan*) (a)   Suppose that the Jones family borrows $420,000 to buy a house. The bank charges 6% annual interest on the unpaid balance, compounded monthly. If the Jones's pay $3000 at the end of each month, how long will it take them to pay off the loan, and what will the amount of their last payment be? Their last payment will just cover the unpaid balance

so will likely be less than $3000.

(b) Suppose that, in entering a while loop to solve this problem you accidentally typed 4200000 (rather than 420000); i.e., you added an extra zero to the loan amount. Explain what would happen. If you do not know what would happen, try it out.

1.2: LOGICAL QUANTIFIERS

Phrases or mathematical relations that involve unspecified non-logical variables such as: "$x > 2$," or "he is an accountant," are not (logical) statements since each has an unspecified variable that makes it impossible to determine a truth value.[20] Such phrases are called *predicates*, and they can be made into statements by specifying the values of the variable(s). The variable specifications will be accomplished by introducing the so-called *universe* of objects under consideration, and by using two types of *quantifiers*: the *universal quantifier* and the *existential quantifier*. We now proceed with a detailed development of these concepts.

Predicates and Universes

DEFINITION 1.9: A **predicate** is a declarative sentence or mathematical relation involving one or more variables whose values are not specified. As we used letters P, Q, etc., to denote logical statements, we use notations such as $P(x)$, $P(x, y)$, $Q(x, y, z)$, etc., to denote predicates (with one, two, or three variables, etc.). The **universe (of discourse)** (or **domain**) of a predicate is the set of all values under consideration for the variable(s) of the predicate.

In cases where it is clear from the context, the domain of a predicate is sometimes not stated explicitly. Any time the variables of a predicate are replaced by specific values for the variables (in the domain), the predicate becomes a (logical) statement and so will have a truth value. As with statements, predicates can be combined using any of the logical operators of the previous section to form new (compound) predicates.

EXAMPLE 1.7: Consider the following predicates:

$$P(x) \equiv \text{ "} x = \sqrt{x^2} \text{ "}$$

$$Q(x, y) \equiv \text{ "} x + y \leq x^2 + y^2 \text{ "}$$

$$R(x) \equiv \text{ "} x > 0 \text{ "}$$

---

[20] The only variables that were present in Section 1 were logical variables (P, Q, ...). Unlike the variable x in the inequality "x > 2," a logical variable represented an entire logical statement and could only take on two values: True or False.

(a)   If we set the universe of $P(x)$ to be $\mathbb{R} \equiv$ the set of all real numbers, is $P(x)$ always going to be true for any x in this universe?

(b)   If the universe of $R(x)$ is also taken to be $\mathbb{R}$, is the (compound) predicate $R(x) \to P(x)$ always going to be true for any x in this universe?

(c)   If we define the universe of $Q(x, y)$ to allow both x and y to be any real numbers, is $Q(x, y)$ always going to be true for any x, y in $\mathbb{R}$?

(d) Find another universe for $Q(x, y)$, such that $Q(x, y)$ is always true whenever x and y are in the universe.

SOLUTION: Part (a):   If x is any negative number, say $x = -2$, then $\sqrt{x^2} = \sqrt{(-2)^2} = \sqrt{4} = 2$ is positive (since the square root is defined to be positive), so $P(x)$ is false.   Otherwise, if $x \geq 0$, $P(x)$ will be true because the square root and the square are inverse functions.

Part (b):   From what was said in the solution of Part (a), it follows that $R(x) \to P(x)$ must be true for any x in $\mathbb{R}$ (the common universe).   (Remember: The only way for a conditional to be false is if the hypothesis is true but the conclusion is false.   If x is a real number, and $R(x)$ is true, then x is a positive number, so by Part (a)'s solution $P(x)$ must also be true.)

Part (c):   The answer is no.   A simple counterexample is obtained by taking $x = y = 1/2$. ($Q(x, y)$ would then say: "$1 \leq 1/2$.")   Here a more thorough solution: If x is a real number in the interval $0 < x < 1$, then $x > x^2$;   whereas for all other values of x, we have $x \leq x^2$.   (This can be seen most easily by graphing the two functions $y = x$ and $y = x^2$ together.)   It follows that $Q(x, y)$ will be false if both $0 < x < 1$, and $0 < y < 1$, and it will be true if both x and y are outside of these ranges.

Part (d):   From the solution of Part (c), we see that if we take as the universe of $Q(x, y)$ to be all real numbers x, and y that are outside of the interval $(0,1)$, then $Q(x, y)$ will be true for all x and y in this universe.

In scientific discourse and written works, predicates are often asserted to be always true, sometimes true, or never true. The following two logical quantifiers will allow us to extend our logical theory to such assertions.

Universal and Existential Quantifiers

DEFINITION 1.10: (a) (*Universal Quantifier*) For a predicate $P(x)$ with a single variable, the **universal quantification** of $P(x)$, written in symbols as $\forall x\ P(x)$, means that $P(x)$ is (always) true for any x in the universe of discourse.   Thus, the universal quantification $\forall x\ P(x)$ is a (logical) statement. It is often read as "for every x, $P(x)$" or "for all x, $P(x)$."

(b)   (*Existential Quantifier*)   For a predicate $P(x)$ with a single variable, the **existential quantification** of $P(x)$, written in symbols as $\exists x\ P(x)$, is the logical statement meaning that $P(x)$ is true for at least one x in the universe of discourse. The quantification $\exists x\ P(x)$ is often read as "there exists an x, such that $P(x)$" or "there is at least one x, such that $P(x)$."

We will discuss these quantifiers for predicates with several variables a bit later. Our next example gives some quantified statement forms that are common in spoken and written language.

EXAMPLE 1.8: Translate each of the following statements into logical symbols with appropriate quantifiers. Make sure to indicate all universes of discourse.
(a) Someone has forgotten their keys.
(b) All of the marathoners are warmed up and hydrated.
(c) None of the players under six feet tall can dunk a basketball.
(d) To get into this evening's performance, one must have a ticket and be properly attired.

SOLUTION: Part (a): $\exists x\ F(x)$, where $F(x)$ is the predicate "x has forgotten their keys," and the domain of discourse is all people (that were here today).

Part (b): $\forall x\ (W(x) \wedge H(x))$, where $W(x)$ is the predicate "x is warmed up," and $H(x)$ is the predicate "x is hydrated."   The domain of discourse is all marathoners.   Alternatively, we could symbolize the statement as $\forall x\ R(x)$, where $R(x)$ is the predicate "x is warmed up and hydrated."

Part (c): $\forall x\ (\sim D(x))$, where $D(x)$ is the predicate "x can dunk a basketball," and the domain of discourse is all players under six feet tall.

Part (d): $\forall x\ (P(x) \rightarrow [T(x) \wedge A(x)])$, where $P(x)$, $T(x)$, $A(x)$ respectively denote the predicates "x gets into this evening's performance," "x has a ticket," and "x is properly attired."

Negations of Quantifiers

The negation of quantified statements turns out to be quite straightforward. To motivate the general rules that we will soon present, let's begin with the specific task of negating the statement of Part (a) in Example 1.8: "Someone has forgotten their keys." The negation of this statement could formally be written as: "It is not the case that someone has forgotten their keys." In spoken or written English, it would be more common just to say: "Everyone has remembered to take their keys," which less eloquently could be expressed by saying: "Everyone has not forgotten their keys." Notice that the negation of the existential quantification turned out to be a universal quantification of the negation of the predicate. The following equivalence summarizes the general rule:

$$\sim (\exists x \ P(x)) \equiv \forall x \sim P(x) \tag{2}$$

The above example (if read backwards) also demonstrates the following very similar formula for negations of universal quantifications:

$$\sim (\forall x \ P(x)) \equiv \exists x \sim P(x) \tag{3}$$

Notice that equivalence (3) tells us that for a universal quantification of $P(x)$ to be false, it amounts to the existence of a single **counterexample**, that is, the existence of a single value of x in the domain of discourse for which $P(x)$ is false (i.e., $\sim P(x)$). Many conjectures and theorems in mathematics and the sciences are stated in the form of a universal quantification. Thus, only a single counterexample is needed to see that such a statement is false. By contrast, equivalence (2) tells us that for an existential quantification to be false, the predicate must be false for all x in the domain of discourse.

EXERCISE FOR THE READER 1.7: For each of the statements in Parts (b), (c), and (d) of Example 1.8, formulate the negation in symbols, and then translate into proper English. Make sure that your English sentences do not contain any phrases of the form "it is not the case that."

Nested Quantifiers

As promised, we next move on to discuss the quantifiers for predicates with more than one variable. The basic ideas are quite simple. For example, for a predicate $P(x, y)$ with two variables, the universal quantification, written symbolically as $\forall x \forall y \ P(x, y)$ (or $\forall y \forall x \ P(x, y)$), is read as "for all x and for all y, $P(x,y)$." For this to be true, $P(x, y)$ needs to be true for any x and y in the domain of discourse.   Similarly, the existential quantification $\exists x\,\exists y\,P(x,y)$ (or $\exists y\exists x\,P(x,y)$), is read as "there exists an x and y in the domain of discourse such that $P(x,y)$." Note that from the above comments, pairs of universal quantifiers commute, as do pairs of existential quantifiers.[21]   Formulas (2) and (3) can be iterated for any number of variables. Two applications of (3) produce:

$$\sim(\forall x\forall y\,P(x,y)) \equiv \exists x\sim(\forall y\,P(x,y))\equiv\exists x\,\exists y\sim P(x,y).$$

In other words: (just like for single variable quantifiers) for a universal quantification involving several variables to be false, all that is needed is a single counterexample. In the same fashion, four applications of (2) would allow us to obtain the following equivalence:

$$\sim(\exists x\,\exists y\,\exists z\,\exists w\,P(x,y,z,w)) \equiv \forall x\,\forall y\,\forall z\,\forall w\sim P(x,y,z,w).$$

Although the above rules and definitions are quite simple, the reason we put them off until now is that <u>mixed quantifiers</u>, in general, <u>do not commute</u>.   For this reason multiple quantifiers are understood to be **nested**, meaning that they are applied from left to right. Our next example will illustrate this protocol.

EXAMPLE 1.9: (a) Let $P(x,y)$ denote the predicate "$x>y+5$."   Interpret each of the following statements, and determine its truth value.   Then form the negations of each.
(i) $\forall x\exists y\,P(x,y)$ and (ii) $\exists y\forall x\,P(x,y)$.
(Take the domain of discourse to be all pairs of real numbers.)
(b) Suppose that $M(a,b)$ represents the predicate "a has more money than b." Take the universe of discourse to be the set of students at this university. For the following statement, first represent it with symbols, then form the negation in symbols, and finally express this negation in proper English.

Cindy has more money than at least two other students.

(c) What is wrong with the following symbolic nested quantifier statements?

(i) $\forall x\forall y\,M(\text{Cindy},x)$.   (ii) $\forall x\forall y[\exists z\,M(x,z)\rightarrow\exists y\,M(y,x)]$.

SOLUTION: Part (a): (i) For the statement $\forall x\exists y\,P(x,y)$ to be true, it means that for any x, there corresponds at least one y such that $P(x,y)$ is true.   To prove this, we can simply observe that whatever x is given, if we take $y=x-6$, then $P(x,y)$ would become $x>(x-6)+5$ or $x>x-1$, which is certainly (always) true. (Note that y could have been taken to be any real number less than $x-5$, but we

---

[21] This makes it unambiguous to write $\forall x,y$ in place of $\forall x\forall y\equiv\forall y\forall x$, or $\exists x,y$ in place of $\exists x\exists y\equiv\exists y\exists x$. Similar shorthand notations for greater numbers of variables are likewise permissible.

just needed a single y for the truth of the predicate.)   The negation of the statement can be obtained by using both (2) and (3):

$$\sim \forall x \exists y \; P(x, y) \equiv \exists x \sim \exists y \; P(x, y) \equiv \exists x \forall y \sim P(x, y).$$

In words, this states roughly that there exists a single real number x, such that for any real number y, $P(x, y)$ is false. This negation must be false, since the original statement is true. It would behoove the reader to verify directly (by interpreting the nested quantifiers) why this negation is indeed false. (Note that: $\sim P(x, y)$ can be expressed as $x \leq y + 5$.)

(ii) The truth of the statement $\exists y \forall x \; P(x, y)$ would mean that there exists a real number y such that, for any real number x, $P(x, y)$ is true, i.e., $x > y + 5$. But this is clearly not going to be true for any x that satisfies $x \leq y + 5$ (once y has been fixed). Thus the statement is false, so we know its negation:

$$\sim \exists y \forall x \; P(x, y) \equiv \forall y \sim \forall x \; P(x, y) \equiv \forall y \exists x \sim P(x, y).$$

must be true. In English, this negation states that given any real number y, there exists (at least one) real number x such that $x \leq y + 5$. As in Part (a), this latter statement is easily verified to be true.

Part (b): We can put the statement in symbols as follows:

$$\exists x \exists y [x \neq y \wedge M(\text{Cindy}, x) \wedge M(\text{Cindy}, y)].$$

We needed to include the statement $x \neq y$ with the other conjunctions to assure that x and y are not the same people.[22]   In negating this statement, the negation passes through both existential quantifiers and changes them to universal quantifiers:

$$\sim \exists x \exists y [x \neq y \wedge M(\text{Cindy}, x) \wedge M(\text{Cindy}, y)]$$
$$\equiv \forall x \forall y \sim [x \neq y \wedge M(\text{Cindy}, x) \wedge M(\text{Cindy}, y)].$$

Next, we apply De Morgan's law to negate the (triple) conjunction:

$$\sim [x \neq y \wedge M(\text{Cindy}, x) \wedge M(\text{Cindy}, y)]$$
$$\equiv \sim (x \neq y) \vee \sim M(\text{Cindy}, x) \vee \sim M(\text{Cindy}, y).$$

Since $\sim (x \neq y) \equiv (x = y)$, the complete negation now becomes:

$$\forall x \forall y [(x = y) \vee \sim M(\text{Cindy}, x) \vee \sim M(\text{Cindy}, y)].$$

In words, $\sim M(\text{Cindy}, x)$ translates as "Cindy does not have more money than x," or equivalently, "x has at least as much money as Cindy has."[23]   Thus, the complete negation can be expressed in English as: "For any two people, either

---

[22] Some students might write $\exists x \exists y [x \neq y \wedge x \neq \text{Cindy} \wedge y \neq \text{Cindy} \wedge M(\text{Cindy}, x) \wedge M(\text{Cindy}, y)]$. Although this is also correct, the additional two conditions are redundant because $M(\text{Cindy}, \text{Cindy})$ is impossible (Cindy cannot have more money than Cindy). It is best (and most elegant) to avoid such redundancies.

[23] This is not the same as saying "x has more money than Cindy." (Why?).

they are the same, or at least one of them has at least as much money as Cindy has." We can express this a bit more elegantly as: "For any two different people, (at least) one of them must have at least as much money as Cindy has." The reader might wish to contemplate the reasons why this negation statement amounts to saying that "at most one person can have less money than Cindy."

Part (c): (i) The variable y was quantified, but never appears in the predicate that follows. Quantify only those variables that actually appear. (ii) The variable y was quantified twice. This is illegal and makes no sense. Each variable in the predicates should be quantified exactly once.

EXERCISE FOR THE READER 1.8: (a) Use symbols to represent the following statement, where the universe of discourse is the set of all students:

> *Everyone either knows Jimmy or knows someone who knows Jimmy.*

(b) Negate the above statement, first in symbols and then in proper English (without using any phrase of the form: "it is not the case that").

Very common in the sciences are theorems that assert either existence, uniqueness, or both. Our existential quantifier allows us to easily express an existence statement, but uniqueness is a bit more awkward. Consider the following (true) basic mathematical fact:

For any odd positive integer n, the equation $x^n = -1$ has a unique real root. (namely $x = -1$).[24]

We introduce the predicate $P(a, n) =$ "the equation $x^n = -1$ has a as a root," where the domain of discourse is the set of all real numbers for a, and the set of all odd positive integers for n. We can easily represent the fact that $x^n = -1$ has a real root (existence only) symbolically as: $\forall n \exists a\, P(a,n)$. If we wanted to represent the complete existence and uniqueness statement, it would run something like this:

$$\forall n\ \exists a\ [P(a,n) \wedge \forall b\ ((b \neq a) \rightarrow\ \sim P(b,n))].$$

Mathematicians have invented the shorthand notation $\exists!$ for existence and uniqueness. With this new symbol, the theorem can be succinctly represented symbolically as $\forall n \exists! a\, P(a,n)$.

EXERCISES 1.2:

1.   Let $P(x)$ represent the predicate "$x < 0$." Let the domain of discourse be the set of all nonzero real numbers. Write each of the following predicates symbolically.
      (a) Both x and xy are negative.

---

[24] *Proof:* (*Using calculus*) The function $f(x) = x^n$ is strictly increasing and tends to $-\infty$ as x approaches $-\infty$ and tends to ∞ as x approaches ∞. Therefore, by the intermediate value theorem it must cross the line $y = -1$ exactly once.

     (b) Exactly one of x or y is negative.

     (c) If $x < 0$ and $y > 0$, then $xy < 0$.

     (d) If $x < 0$ and $y < 0$, then $xy > 0$.

2.   Let $P(n)$ represent the predicate "n is even," and $Q(n)$ represent the predicate "the square root of n is an integer." Let the domains of discourse be the set of all integers. Write each of the following predicates symbolically. (Do not be concerned about truth values here.)

     (a) Neither n nor m is even.

     (b) At least one of $\sqrt{n}$ or $\sqrt{m}$ is an integer.

     (c) If $\sqrt{n}$ is an integer, then n is even.

     (d) If $\sqrt{nm}$ is an even integer, then at least one of $\sqrt{n}$ or $\sqrt{m}$ must be an integer.

3.   Let $L(a,b)$ represent the predicate "a loves b," and $H(a,b)$ represent the predicate "a hates b." Let the domains of discourse be all human beings. Write each of the following predicates and statements symbolically.

     (a) Jane loves Jimmy.   (b) Jimmy hates Jane.

     (c) Carla and Chuck love each other.   (d) Somebody hates Dan.

     (e) Everybody loves Raymond.   (f) Nobody hates Susan.

4.   Let $C(a)$ represent the predicate "a has taken calculus," and $F(a)$ represent the predicate "a can speak French." Let the domains of discourse be all students in your (or a fictitious) discrete structures class. Write each of the following predicates and statements symbolically.

     (a) Al has taken calculus.   (b) Betty speaks French.

     (c) Cindy has neither taken calculus nor can   (d) Somebody can speak French.
     she speak French.

     (e) Everybody has taken calculus.   (f) Nobody speaks French.

5.   Let $G(a,b)$ represent the predicate "a gossips to b," and $I(a,b)$ represent the predicate "a does not listen to b." Let the domains of discourse be all adults living in a small village. Write each of the following predicates and statements symbolically.

     (a) Helen gossips to Tom.   (b) Tom does not listen to Helen.

     (c) Brad gossips to everyone, but listens to   (d) Esmeralda gossips to no one, but listens
     no one.   to everyone.

     (e) If someone gossips to someone else, then (f) There is someone for whom all of the
     he/she will be gossiped to.   people he/she gossips to do not gossip back
                                     to that person.

6.   Let $T(a,b)$ represent the predicate "a is taller than b," and $H(a,b)$ represent the predicate "a is heavier than b." Let the domains of discourse be all students in a certain discrete structures class. Write each of the following predicates and statements symbolically.

     (a) Joey is taller than Sam.   (b) Linda is heavier than Jane.

     (c) Marty is the tallest person in the class but (d) Linda is the shortest person in the class,
     not the heaviest.   but not the lightest.

     (e) Three people in the class are of the same (f) No two people in the class weigh the
     height.   same.

7.   With $L(a,b)$ and $H(a,b)$ being the predicates of Exercise 3, write symbolically the negations for each part of that exercise, and then translate into proper English. Do not use any phrases of the form "it is not the case that."

8.   With $C(a)$ and $F(a)$ being the predicates of Exercise 4, write symbolically the negations for each part of that exercise, and then translate into proper English. Do not use any phrases of the

form "it is not the case that."

9.   With $G(a,b)$ and $I(a,b)$ being the predicates of Exercise 5, write symbolically the negations for each part of that exercise, and then translate into proper English. Do not use any phrases of the form "it is not the case that."

10.   With $T(a,b)$ and $H(a,b)$ being the predicates of Exercise 6, write symbolically the negations for each part of this exercise, and then translate into proper English. Do not use any phrases of the form "it is not the case that."

11.   Determine the truth value of each statement below if the universe of discourse is the set of all integers.
(a) $\exists n\ (n^2 = n)$ (b) $\forall n\ (n^3 \geq n^2)$
(c) $\forall n \forall m (n + m \leq n^2 + m^2)$ (d) $\forall n \forall m ((n+m)^2 \leq n^2 + m^2)$

12.   Repeat each part of Exercise 11 with the universe of discourse now being changed to the set of all real numbers.

13.   Formulate the negation of each statement of Exercise 11, and indicate the resulting truth values.

14.   Formulate the negation of each statement of Exercise 12, and indicate the resulting truth values.

NOTE: We follow the (unambiguous) scientific usage, by taking the word *sometimes* to mean "at least once." Similarly, we take the word *some* to mean "at least one."

15.   Let $P(a,t)$ represent the predicate "you can please a at time t," where the universe of discourse for a consists of all of your relatives. (i) Write each of the following statements symbolically.
(ii) Formulate the negation of each statement, first symbolically, and then in proper English (without using any phrases of the form "it is not the case that.")
(a) You sometimes can please your wife.
(b) You always please your father.
(c) You can never please your mother-in-law.
(d) You can please some people all of the time.
(e) You can please all of the people some of the time.
(f) You can't please all of the people all of the time.

16.   Let $P(a,b)$ represent the predicate "a loves b," and $Q(a,b)$ represent the predicate "a hates b." Let the domains of discourse be all human beings. (i) Write each of the following statements symbolically. (ii) Formulate the negation of each statement, first symbolically, and then in proper English (without using any phrases of the form "it is not the case that.")
NOTE: Unlike in Exercise 15 where "you" refers to the reader, occurrence of "you" in the parts below refer to any person in the universe, i.e., "you" is in the proverbial sense.
(a) If you love someone, they will not necessarily love you.
(b) There is someone who does not hate anyone.
(c) No one hates everyone.
(d) There is someone who loves at least two people other than him/herself.
(e) There is someone who loves exactly one person other than him/herself.
(f) If you love someone, then you cannot hate him/her.

17.   (i) Translate each statement below involving nested quantifiers into English. (ii) Determine the truth value of each statement. (iii) Formulate the negation of each statement, first symbolically, and then in proper English (without using any phrases of the form "it is not the case that.") Take the universe of discourse to be the set of all real numbers.

(a) $\forall x \forall y \, (xy = yx)$ (b) $\forall x \exists y \, (xy = x + y)$

(c) $\forall x \forall y \exists z \, (xyz = x + y + z)$ (d) $\forall a \forall b \exists x \, (ax + b = 0)$

(e) $\forall a \forall b \forall c \, (a \neq 0 \rightarrow \exists x (ax^2 + bx + c = 0))$ (f) $\forall a \forall b (a \neq 0 \rightarrow \exists x (ax + b = 0))$

18. Re-do Exercise 17, but this time taking the universe of discourse to consist of all integers.

19. Prove each of the following equivalences involving predicated statements.

(a) $\forall x \, P(x) \,\vee\, \forall x \, Q(x) \;\equiv\; \forall x \, (P(x) \vee Q(x))$.

(b) $\exists x \, P(x) \,\vee\, \exists x \, Q(x) \;\equiv\; \exists x \, (P(x) \vee Q(x))$.

(c) $\exists x \, P(x) \,\wedge\, \exists x \, Q(x) \;\equiv\; \exists x \exists y \, (P(x) \wedge Q(y))$.

(d) $\forall x \, P(x) \,\vee\, \exists x \, Q(x) \;\equiv\; \forall x \, \exists y \, (P(x) \vee Q(y))$.

Suggestion: Either work directly with the definitions, or (more easily but with more to write down) for each equivalence separately establish each of the two corresponding implications.

20. (*Negation of Unique Quantification*) Establish the following equivalence for the negation of unique quantification:

$$\sim [\exists! x \, P(x)] \;\equiv\; \sim[\exists x \, P(x)] \,\vee\, \exists y \exists z [(y \neq z) \wedge P(y) \wedge P(z)].$$

Suggestion: Either work directly with the definitions, or (more easily but with more to write down) separately establish each of the two corresponding implications.

21. Determine the truth value of the following statement. Take the universe of discourse to be the set of all real numbers:

$$\forall x \, \forall \, y \exists z \left(\frac{x}{z} + yz = 2x + 2y \right).$$

22. (*Calculus: Definition of a Limit*)[25] Suppose that f is a real-valued function[26] defined (at least) on an open interval I of real numbers containing $x = a$, except possibly at $x = a$. The **limit of** $f(x)$ **as** x **approaches** a, if it exists, is defined to be a number L for which the following condition holds:

$$\forall \varepsilon > 0 \; \exists \delta > 0 \; \forall x \, [\; 0 < |x - a| < \delta \;\; \rightarrow \;\; |f(x) - L| < \varepsilon \;].[27]$$

In case such a limit exists, we denote this as $\lim_{x \to a} f(x) = L$.[28]

(a) Prove (using the above definition) that $\lim_{x \to 0} |x| = 0$.

(b) Prove (using the above definition) that $\lim_{x \to 0} x^3 = 0$.

---

[25] The definition of a limit is one of the cornerstone concepts in calculus; it is fundamental and so does not require any other calculus concepts (only algebra). It is also one of the most sophisticated and difficult concepts for students, and is unfortunately most often presented near the beginning of calculus courses without any logical prelude. Most calculus students tend to not reach a level of understanding this definition; even mathematics majors usually do not come to grips with it until they get to a course in advanced calculus.

[26] Although most people reading this book will likely know what a function is, functions will be formally defined in the next chapter.

[27] We are using the shorthand notation in the above quantifiers to indicate that the universe of discourse for both ε, δ is the set of all positive real numbers, and for x is the set of real numbers. We are also following the standard choice of variables used in nearly all calculus and analysis textbooks; in particular, the Greek letters ε (epsilon) and δ (delta) are standard for the parameters that they represent.

[28] At first glance this notation may appear to be subject to ambiguity (what if two different limits exist?). But in Part (c) of this exercise, it will be shown that if such a limit exists, it must be unique.

(c) Prove that if $\lim_{x \to a} f(x)$ exists, it must be unique.

(d) Using quantifiers, express the negation of the above definition of a limit.

(e) Find an example of a function $f(x)$ that is defined for all real numbers x, but does not have a limit as x approaches 0.

Suggestions: For Part (c), suppose both $L \neq L'$ are both limits of $f(x)$ as x approaches a. Use $\varepsilon = |L - L'|/2$ in both limit definitions along with the *triangle inequality* ($|a-b| \leq |a-c| + |c-b|$). For Part (e), look at (for example) the function defined by: $f(x) = 1$, if $x > 0$, and otherwise $f(x) = 0$.

NOTE: (*Lewis Carroll's Logic Puzzles*) Lewis Carroll[29] helped to bring a better competency in logical reasoning to the general public by creating and publishing a substantial variety of logical puzzles. In one such type of puzzle, Carroll presented a series of assumptions (hypotheses) and asked the reader to write down the ultimate conclusion that could be reached (in proper English). He intentionally made the topics of his puzzles to be rather inane so as to avoid any existing prejudices of the readers. Here is an example of an actual Lewis Carroll puzzle:

Hypotheses: No potatoes of mine, that are new, have been boiled. All my potatoes in this dish are fit to eat. No unboiled potatoes of mine are fit to eat.
Conclusion: ???

FIGURE 1.1: Charles Lutwidge Dodgson (Lewis Carroll) (1832–1898), English mathematician and writer.

Solution: The universe for this problem consists of {my potatoes}. Although each of the assumptions involves a universal quantification, for example the first one can be symbolized as: $\forall x[N(x) \to \sim B(x)]$. It will economize the arguments to convert each of the predicated implications into simple unpredicated implications by changing the wording of the statements to involve a generic element of the universe. More specifically, if we introduce the following logical statements (all involving "my potato"):
N: my potato is new, B: my potato is boiled, D: my potato is in the dish, and E: my potato is fit to eat, then the hypotheses and their contrapositives can be symbolized as follows:

(1) $N \to \sim B$, $B \to \sim N$

(2) $D \to E$, $\sim E \to \sim D$

(3) $E \to B$, $\sim B \to \sim E$

The goal is to string together all three hypotheses (either directly, or using their contrapositives) to obtain a master series of implications that incorporates all hypotheses. We then apply hypothetical syllogism (Theorem 1.2 (e)) to obtain the ultimate conclusion.
Applying (1), followed by the contrapositive of (3), followed by the contrapositive of (2), we obtain the string: $N \to \sim B \to \sim E \to \sim D$. From this we infer (using hypothetical syllogism) the ultimate conclusion: $N \to \sim D$. In English, this would read: If my potato is new, then it is not in the dish.

---

[29] Charles Lutwidge Dodgson was, for most of his career, a mathematics professor at Oxford, though he was also an acclaimed novelist, and is best known for his *Alice's Adventures in Wonderland* (1865), one of the best known and enduring children's books of all time. He wrote numerous mathematical books and papers that he published under his own name, but for his children's novels he used the pen name Lewis Carroll. Carroll followed many of the steps of his father, Charles Dodgson, Sr., who had been a mathematics lecturer at Oxford, until he had to give up his post because he had married his cousin. He subsequently became a reverend. As a student, Carroll excelled in both mathematics and classics to such a degree that he was awarded a lifetime annual fellowship of 25 pounds sterling. This stipend had no requirements for further academic work, but it did require him to take Holy Orders and remain unmarried. Carroll made significant contributions into efficient ways of learning and teaching mathematics.

Putting this back in the original (universal quantification language), it can be worded as: "None of my new potatoes are in the dish," or "All of my potatoes in the dish are not new."

23. (*Lewis Carroll's Logic Puzzles*) For each part below, deduce, and put into spoken but proper English, the ultimate conclusion obtained by using all of the following hypotheses:
(a) All hummingbirds are richly colored. No large birds live on honey. Birds that do not live on honey are dull in color.
(b) Promise breakers are untrustworthy. Wine drinkers are very communicative. A man who keeps his promises is honest. No teetotalers are pawnbrokers. One can always trust a very communicative person.

24. (*Lewis Carroll's Logic Puzzles*) For each part below, follow the instructions of Exercise 23.
(a) Things sold in the street are of no great value. Nothing but rubbish can be had for a song. Eggs of the Great Auk are very valuable. It is only what is sold in the street that is really rubbish.
(b) No one who is going to a party ever fails to brush his hair. No one looks fascinating, if he is untidy. Opium eaters have no self-command. Everyone who has brushed his hair looks fascinating. No one wears white kid gloves, unless he is going to a party. A man is always untidy, if he has no self-command.

25. (*Lewis Carroll's Logic Puzzles*) For each part below, follow the instructions of Exercise 23.
(a) The only books in this library that I do *not* recommend for reading are unhealthy in tone. The bound books are all well written.   All the romances are healthy in tone. I do not recommend that you read any of the unbound books.
(b) The only animals in this house are cats. Every animal is suitable for a pet, that loves to gaze at the moon. When I detest an animal, I avoid it. No animals are carnivorous, unless they prowl at night. No cat fails to kill mice. No animals ever take to me, except what are in this house. Kangaroos are not suitable for pets. None but carnivores kill mice. I detest animals that do not take to me. Animals that prowl at night always love to gaze at the moon.
Suggestion for Part (b): Refer back to Exercise 13 of Section 1.1.

1.3: SETS

Sets and Their Elements

A **set** S is a collection of objects; the objects in a set are called **elements** (or **members**) of the set.[30]   The objects making up a set can be of any sort.   For example, the collection of all students in this class (or any particular class) is a set; we have already spoken of at least two sets of numbers, the set of all real numbers and the set of integers.   Sets are the building blocks of all discrete structures.   In this section we will introduce some fundamental set operations, and present some important examples and concepts relating to sets.   In mathematics, everything can be defined using sets and logic.

---

[30] The definition is informal. Although it will serve all of our purposes, it can lead to subtle *paradoxes* (contradictions); see, for example, Exercise 32. For the formal rigorous definition of a set (and the related axiomatics), the interested reader would do well to consult the book by Enderton ([End-77]).

We write $x \in S$ to denote that object x belongs to the set S, and $x \notin S$ to mean x does not belong to the set S.   For example, if S denotes the set of all states of the US, then Hawaii $\in S$, but Guam $\notin S$. Braces ({ }) are a common notation used to describe a set. Between the braces we either give a listing all of the elements or a description of them. For example, we can write the set of the colors of the flag of the US as {Red, White, Blue}. Listing the elements in a large incongruous set such as {all words in Webster's Dictionary} would not be feasible.   An ellipsis (…) can be used (repeatedly) to replace obvious patterns. For example, the set of all integers can be written as {…, −3, −2, −1, 0, 1, 2, 3, …}. Table 1.8 establishes some convenient notation for several sets of numbers that we will use often.

SET	NAME OF SET	DESCRIPTION OF ELEMENTS
$\mathbb{R}$	The set of all **real numbers**	All numbers on the number line (positive, negative, and zero)
$\mathbb{Z}$	The set of all **integers**	{…, −3, −2, −1, 0, 1, 2, 3, …}
$\mathbb{Q}$	The set of all **rational numbers**	$\{x \in \mathbb{R} \mid x = p/q, \exists\, p,q \in \mathbb{Z}, q \neq 0\}$,[31] or {quotients of integers}.
$\mathbb{N}$ or $\mathbb{Z}_{\geq 0}$	The set of all **natural numbers**	$\{0, 1, 2, …\}$ or $\{a \in \mathbb{Z} \mid a \geq 0\}$
$\mathbb{Z}_{+}$	The set of all **positive integers**	$\{1, 2, 3, …\}$ or $\{a \in \mathbb{Z} \mid a > 0\}$

TABLE 1.8: Some sets of numbers that will arise often. The symbols $\mathbb{R}, \mathbb{Z}, \mathbb{Q}$, and $\mathbb{N}$ have become standard in mathematics literature.   The subscript notations, such as $\mathbb{Z}_{+}$, seem to have been less used in the past, but are fast being adopted due to their intuitive appeal. The subscript notation can be applied to any set of numbers.

We stress that the order in which the elements of a set are listed is immaterial. To decide whether two sets are the same, one needs only to compare elements. Thus, {1, 2, 3, 4} = {2, 1, 4, 3} = {1, 2, 3, 1, 4}, but the duplicate listing as in the last version should be avoided.   We next introduce two operations on pairs of sets that produce new sets. Each of the sets of Table 1.8 are called *infinite sets* since each has a never ending list of elements. Sets like {Red, White, Blue} containing a finite (ending) list of elements are called *finite sets*.

---

[31] Within the bracket notation for set descriptions, the vertical bar (|) is notation for the phrase: "with the property that."   Thus the set of rational numbers is being described as the set of all real numbers that can be expressed as fractions (of integers). Sometimes a colon (:) is used in place of the vertical bar notation.

Unions and Intersections

DEFINITION 1.11: Let A and B be two sets.
(i) The **union** of A and B, denoted $A \cup B$, is the set of all elements that belong to A or to B (or both). In symbols, we can write $A \cup B = \{x \mid x \in A \vee x \in B\}$.
(ii) The **intersection** of A and B, denoted $A \cap B$, is the set of all elements that belong to A and to B. In symbols, we can write $A \cap B = \{x \mid x \in A \wedge x \in B\}$.

EXAMPLE 1.10: (a) The union and intersection of the two sets $A = \{2, 4, 6, 8, 10\}$, and $B = \{3, 6, 9, 12\}$, are $A \cup B = \{2, 3, 4, 6, 8, 9, 10, 12\}$, and $A \cap B = \{6\}$. We point out that a set with just one element in it, like $A \cap B = \{6\}$, is called a *singleton set*.

Venn Diagrams

The union and intersection are the set-theoretical equivalents of the logical disjunction and conjunction operators. In contrast with logical ideas, many concepts about sets can be visualized using so-called **Venn diagrams**. Venn diagrams represent sets by simple geometric shapes (like circles) and use shading to represent any particular set in question. Such diagrams are shown in Figure 1.2 for the union and intersection.

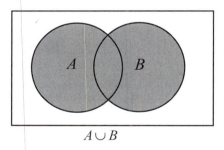 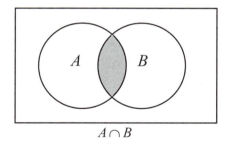

$$A \cup B \qquad\qquad A \cap B$$

FIGURE 1.2: Venn diagrams for the union $(A \cup B)$ and intersection $(A \cap B)$ of two sets A and B.

EXAMPLE 1.11: Let A be the set of all female (undergraduates) at your (or some other) college, and let B denote the set of all mathematics majors at this same college. Describe in plain English the two sets $A \cup B$, and $A \cap B$. Next, suppose that Joey and Amy are mathematics majors at this college, Linda is an economics major also at this college, and Jane is a mathematics major at a different college (say Dartmouth College, assuming you are not at Dartmouth). Indicate the truth value of each set membership statement below and illustrate with a Venn diagram.
(a) Joey $\in A \cup B$
(b) Amy $\in A \cap B$

(c) Linda $\in A \cap B$

(d) Jane $\in A \cup B$

SOLUTION: $A \cup B$ = {students at your college who are either females or math majors (or both)}. $A \cap B$ = {female mathematics majors at your college}.   The Venn diagram in Figure 1.3 (compared with Figure 1.2) justifies the following answers: (a) True, (b) True, (c) False, (d) False.

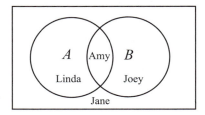

FIGURE 1.3: A Venn diagram for Example 1.11.   The set A represents the set of all female (undergraduate) students at your college, and B represents the set of all mathematics majors at your college.

Subsets and the Empty Set

The next definition introduces a simple yet very important possible relationship between two sets.

DEFINITION 1.12: Let A and B be two sets. We say that A is a **subset** of B, written $A \subseteq B$, if every element of A is also an element of B. The symbols $A \nsubseteq B$ denote that A is not a subset of B. Note that the subset relation does allow the two sets to be equal, i.e., $A \subseteq A$. If $A \subseteq B$ and, moreover, $A \neq B$ (i.e., B has some elements that are not in A), then A is called a **proper subset** of B and we denote this by $A \subset B$.

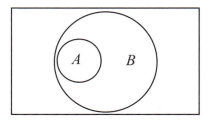

FIGURE 1.4: A Venn diagram illustrating the subset relation $A \subseteq B$.

Figure 1.4 gives a Venn diagram that illustrates the concept of a subset. Among the sets we have considered so far, many subset relations exist. For example, $\mathbb{N} \subseteq \mathbb{Z} \subseteq \mathbb{Q} \subseteq \mathbb{R}$ (these are all proper subset relations). In terms of quantifiers, the definition of $A \subseteq B$ is thus equivalent to the truth of the statement $\forall x(x \in A \rightarrow x \in B)$. Note that $A = B$, if, and only if, $A \subseteq B$ and $B \subseteq A$. Hence, proving a set equality $A = B$ can be accomplished by proving the quantified biconditional $\forall x(x \in A \leftrightarrow x \in B)$. An often convenient way to prove that two sets A and B are the same is to go through two steps to establish the two subset relations: $\forall x(x \in A \rightarrow x \in B)$ and $\forall x(x \in B \rightarrow x \in A)$. The former method can be used only when all of the steps are reversible, otherwise, the (longer) latter proof strategy should be employed. Before we go on, we introduce a very special set that arises very often in practice as well as in theory.

DEFINITION 1.13: The **empty set**, denoted $\varnothing$, is the set that contains no elements, i.e., $\varnothing = \{\ \}$. Two sets A and B for which $A \cap B = \varnothing$ are said to be **disjoint** (or **mutually exclusive**).

Note that for any set A, we (always) have $\varnothing \subseteq A$. The reason for this is simple: the corresponding implication $\forall x(x \in \varnothing \rightarrow x \in A)$, is **vacuously true**, meaning that the hypothesis of the implication is never true (so the implication False $\rightarrow$ (True or False) will automatically be true).

Complements and Differences of Sets

DEFINITION 1.14: In a given context, the **universal set** U is the set of all objects under consideration. In such a context, if A is any set, we define the **complement of** A, denoted $\sim A$, to be the set of all elements of the universal set that are not in A, i.e., $\sim A = \{x \in U \mid x \notin A\}$. Outside the context of universal sets, the notion of *set differences* (*relative complements*) can be formulated as follows: if A and B are two sets, the **set difference** $A \sim B$ is defined to be the set of all elements of A that do not belong to B, i.e., $A \sim B = \{x \in A \mid x \notin B\}$.

In any Venn diagram, the universal set is represented by the bounding rectangle, cf. Figures 1.2–1.4. Venn diagrams for the complement as well as the set difference are given in Figure 1.5.

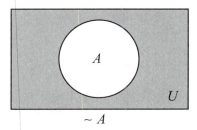

$\sim A$

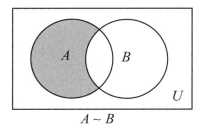

$A \sim B$

FIGURE 1.5: Venn diagrams illustrating the complement $\sim A$ of a set A (left) and the set difference (relative complement) $A \sim B$.

EXAMPLE 1.12: Let the universal set U be the set $\mathbb{R}$ of real numbers. Describe the following sets.

(a) $\{2, 4, 6, 8, 10\} \sim \{3, 6, 9, 12\}$

(b) $\sim \mathbb{Q}$

SOLUTION: Part (a): We simply remove any elements from the first set that appear in the second set (that is, only the number 6), to get $\{2, 4, 8, 10\}$. The universal set is not relevant for this question.

Part (b):   The set $\sim \mathbb{Q}$ consists of all real numbers that are not expressible as fractions. In mathematics, these numbers are called *irrational numbers*. Another (more direct) definition of an irrational number is that it is a real number whose decimal expansion is nonrepeating and nonterminating; see Exercise 27. Despite the fact that there are infinitely many irrational numbers,[32] examples of which include $\sqrt{2}$, π, and e, it is often not an easy task to prove the irrationality of a number.   Exercise 28 contains outlines of proofs of the irrationality of some numbers.[33]

Set Theoretic Identities

EXAMPLE 1.13:   Prove that for any three sets A, B, and C, the following *distributive law* is valid: $A \cap (B \cup C) = (A \cap B) \cup (A \cap C)$.

SOLUTION: We present two different methods:
Method 1: (*Using definitions and logic*)
$x \in A \cap (B \cup C)$

$\quad \Leftrightarrow x \in A \;\wedge\; x \in B \cup C$ (definition of intersection)

$\quad \Leftrightarrow x \in A \;\wedge\; [x \in B \vee x \in C]$ (definition of union)

$\quad \Leftrightarrow [x \in A \;\wedge\; x \in B] \vee [x \in A \;\wedge\; x \in C]$ (distributivity—Theorem 1.1 I(c))

$\quad \Leftrightarrow [x \in A \cap B] \vee [x \in A \cap C]$ (definition of intersection)

$\quad \Leftrightarrow x \in (A \cap B) \cup (A \cap C)$ (definition of union)

---

[32] In fact, there are "more" irrational numbers than there are rational numbers. This has to do with the fact that there are different levels of infinity. The smallest size of infinity is denoted by the symbol $\aleph_0$ (using a Hebrew letter, read as aleph naught), and this is the size of the integers $\mathbb{Z}$. The rational numbers can also be shown to have this same size, however, the irrational numbers and the real numbers have size $\aleph_1$ (aleph one), a strictly larger level of infinity. Interested readers can refer to [End-77] for a nice and rigorous treatment of such topics.
[33] Since we work in a world where everything has finite limitations, outside of pure mathematics it is not feasible (or important) to deal with any subtle properties of irrational numbers.

Method 2: (*Using Venn Diagrams*) It is easily verified that the shaded region in Figure 1.6 represents both sets of the asserted equation. (The figure can be obtained in two steps for each side by graphing the parenthesized sets first.)

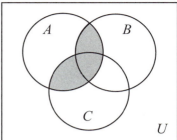

FIGURE 1.6: A Venn diagram for the sets $A \cap (B \cup C)$ and $(A \cap B) \cup (A \cap C)$ of Example 1.13.

EXERCISE FOR THE READER 1.9: (*The Symmetric Difference Operation*) The **symmetric difference** of two sets A and B, denoted $A \Delta B$, is defined to be the set of all elements that belong to either A or B, but not to both A and B.[34] Thus, it easily follows that $A \Delta B = (A \cup B) \sim (A \cap B)$.

(a) Prove that the symmetric difference is commutative: $A \Delta B = B \Delta A$.

(b) Prove that the symmetric difference is associative: $A \Delta (B \Delta C) = (A \Delta B) \Delta C$.

Comparing the two proofs of Example 1.13, many would tend to agree that the elegant Venn diagram proof is preferable and might well ask whether we may altogether avoid logical proofs in set theory. The problem with Venn diagram proofs is that they become very complicated when they involve more than three sets. Their inventor, John Venn (1834–1923, English mathematician), successfully created efficient four-set diagrams using ellipses (see Figure 1.7), but was not satisfied with his diagrams for five sets.

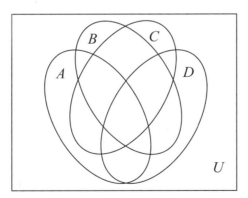

FIGURE 1.7: A generic Venn diagram for four sets A, B, C, and D.

Venn diagrams can actually be drawn for any finite number of sets, but the shapes needed to accomplish this become complicated as the number of sets gets large.

---

[34] This definition can thus be expressed using the exclusive or ($\oplus$) logical operator as follows:
$A \Delta B = \{x \mid (x \in A) \oplus (x \in B)\}$.

What is needed to make such a configuration of shapes function as a Venn diagram for a collection of sets, $A_1, A_2, \cdots, A_n$, is that any intersection of the form $B_1 \cap B_2 \cap \cdots \cap B_n$, where each B_i can either equal the corresponding A_i or its complement $\sim A_i$, must correspond to a unique (single) region partitioned by the shapes.   Since there are two possibilities for each B_i and there are n of them, it follows (from the multiplication principle that will be presented in Chapter 5), that there must be 2^n such regions in a Venn diagram for n sets.   Exercises 29–31 present a systematic method for generating Venn diagrams for any number of sets. Although such diagrams are not so useful for understanding set theory (in cases where the number of sets is large), they have had practical applications in the design of circuit boards.

The following theorem, analogous to Theorem 1.1 for logical equivalences, presents several useful identities involving sets.

THEOREM 1.3: (*Some Set Theoretic Identities*) Let A, B, and C be sets contained in some universal set U. The following identities are then valid:

(a) (*Commutativity*) $A \cup B = B \cup A$, $A \cap B = B \cap A$

(b) (*Associativity*) $(A \cup B) \cup C = A \cup (B \cup C)$, $(A \cap B) \cap C = A \cap (B \cap C)$

(c) (*Distributivity*)
$$A \cap (B \cup C) = (A \cap B) \cup (A \cap C), \ A \cup (B \cap C) = (A \cup B) \cap (A \cup C)$$

(d) (*De Morgan's Laws*) $\sim (A \cup B) = \sim A \cap \sim B$, $\sim (A \cap B) = \sim A \cup \sim B$

(e) (*Double Complementation*) $\sim (\sim A) = A$

(f) (*Absorption*) $A \cup (A \cap B) = A$, $A \cap (A \cup B) = A$

(g) (*Identity Laws*) $A \cap U = A$, $A \cup \varnothing = A$

(h) (*Dominance Laws*) $A \cup U = U$, $A \cap \varnothing = \varnothing$

(i) (*Complement Laws*) $A \cup \sim A = U$, $A \cap \sim A = \varnothing$

The proof of Theorem 1.3 is routine and is left to the reader. All parts can be done using either of the two methods used to establish the distributive law in Example 1.13.

Unions and Intersections of Set Families

The definitions of unions and intersections of sets can be generalized to arbitrary collections of sets. This is done in the following definition.

DEFINITION 1.15: If $\{A_k\}$ $(k \in I)$ is any *family* of sets, indexed using an **index set** I, we define the **union** of this family of sets by:

$$\bigcup_{k \in I} A_k = \{x \in U \mid \exists k \ x \in A_k\}. \tag{4}$$

Similarly, the **intersection** of the sequence is defined by:

$$\bigcap_{k \in I} A_k = \{x \in U \mid \forall k \ x \in A_k\}. \tag{5}$$

For any finite collection of sets, it can be shown that the above definitions are consistent with our previous definition for pairs of sets, done pairwise, in any order. What is needed to prove this is the commutativity and associativity of unions and intersections (Parts (a) and (b) of Theorem 1.3), combined with *mathematical induction*, which will be introduced in Chapter 3.

EXERCISE FOR THE READER 1.10: Use the definition (4) to establish the following generalization of the distributive law in Theorem 1.3 (do not try to prove this with Venn diagram):

$$A \cap (\bigcup_{k \in I} B_k) = \bigcup_{k \in I} (A \cap B_k).$$

In applications, it is often important to consider all possible combinations of elements of a set. This motivates the following definition:

Power Sets

DEFINITION 1.16: For any set A, the **power set** of A, denoted $\mathscr{P}(A)$ (or 2^A) is the set of all subsets of A, i.e., $\mathscr{P}(A) = \{S \mid S \subseteq A\}$.

EXAMPLE 1.14: Find the following power sets:
(a) $\mathscr{P}(\varnothing)$
(b) $\mathscr{P}(\{\text{Red}\})$
(c) $\mathscr{P}(\{\text{Red, White}\})$

SOLUTION: Part (a): The empty set $\varnothing$ has only one subset, namely itself (recall that for <u>any</u> set A, $A \subseteq A$), therefore $\mathscr{P}(\varnothing) = \{\varnothing\}$. We alert the reader to one subtle but important fact: This power set $\{\varnothing\}$ is a singleton set whose only element is the empty set. The set $\varnothing$ is a set with no elements, so $\{\varnothing\} \neq \varnothing$; more accurately $\varnothing \subset \{\varnothing\}$.
Part (b): The only possibilities for a subset of $\{\text{Red}\}$ are $\varnothing$ (if the subset has no elements), and $\{\text{Red}\}$ (if the subset has one element), thus $\mathscr{P}(\{\text{Red}\}) = \{\varnothing, \{\text{Red}\}\}$.

Part (c): We do this part in a way that will soon lead us to an important insight about power sets.   There are two types of subsets of {Red, White}; those that do not contain the element "White" and those that do contain "White."   The first type of subsets are precisely the subsets of the set {Red} that we found in Part (b), namely, the two sets $\varnothing$ and {Red}. Now, any subset of the second type (i.e., that contains "White") must be expressible as one of two sets of the first type with the element "White" adjoined, i.e., $\varnothing \cup$ {White} = {White},   and   {Red} $\cup$ {White} = {Red, White}. We have therefore determined the power set

$\mathscr{P}$ ({Red, White}) = {$\varnothing$, {Red}, {White}, {Red, White}}.

EXERCISE FOR THE READER 1.11:   (a) Determine $\mathscr{P}$ ({Red, White, Blue}). (b) If a set A has n (a finite number of) elements in it, how many elements do you conjecture will be in the corresponding power set $\mathscr{P}$ (A)?

After having completed the preceding exercise for the reader, the reader should have discovered the following important fact (that we will formally prove once we learn mathematical induction in Chapter 3):

FACT: If a set A has a finite number n elements, then its power set $\mathscr{P}$ (A) will have 2^n elements.

Cartesian Products of Sets

We have already observed that, in listing the elements of a set, the order in which we list the elements is irrelevant. The resulting set is determined only by which elements belong to the set. In many applications, the order in which elements in a list appear is important. For example, think of waiting in line at a fast food restaurant. Customers would be very unhappy if cashiers did not always serve first the person in front of the line. The next definition formalizes the concept of an ordered list.

DEFINITION 1.17: A **vector** of length n (or an (**ordered**) **n-tuple**) is an ordered list containing n elements.   Vectors are usually denoted using round (or sometimes square) brackets, for distinction from sets, which use braces.

For example, although the sets {Red, White, Blue} and {White, Blue, Red} are identical, the corresponding length three vectors (Red, White, Blue) and (White, Blue, Red) are different objects. Thus while two sets are equal when the elements are the same, for two vectors $(a_1, a_2, \cdots, a_n)$ and $(b_1, b_2, \cdots, b_n)$ (of the same length) to be equal, we need the corresponding elements in the two lists to be the same, i.e., $a_1 = b_1, a_2 = b_2, \cdots, a_n = b_n$. The next definition shows a general technique for forming large sets of vectors with a common length.

DEFINITION 1.18: (a) Let A and B be sets. The **Cartesian product** of A and B, denoted $A \times B$, is the set of all length two vectors (ordered pairs) (a, b), with $a \in A$ and $b \in B$. In symbols: $A \times B = \{(a,b) \mid (a \in A) \wedge (b \in B)\}$. The sets A and B are called the **factor**s of the Cartesian product.

(b) More generally, if $A_1, A_2, \cdots, A_n$ are sets, their **Cartesian product** is denoted $A_1 \times A_2 \times \cdots \times A_n$ (or $\prod_{k=1}^{n} A_k$) consists of all length n vectors $(a_1, a_2, \cdots, a_n)$, where $a_1 \in A_1, a_2 \in A_2, \cdots$, and $a_n \in A_n$. In symbols:

$$A_1 \times A_2 \times \cdots \times A_n = \{(a_1, a_2, \cdots, a_n) \mid \forall k \, [a_k \in A_k]\}.$$

In case all factors are equal, i.e., $A_1 = A_2 = \cdots = A_n = A$, this Cartesian product is denoted $A \times A \times \cdots \times A$ (with n factors) is denoted by A^n.

EXAMPLE 1.15: (a) Letting $A = \{1, 2\}$, $B = \{1,2,3\}$, and $C = A$, list all elements in the Cartesian products $B \times C$, and $A \times B \times C$.

(b) Describe the vectors in the Cartesian products $\mathbb{R}^2$, and those in $\mathbb{R}^3$.

SOLUTION: Part (a): One way to perform an exhaustive listing (when each of the factors is a finite set) works a bit like an automobile odometer:[35] Rather than explain it in words, the strategy is easily seen with the two examples:

$$B \times C = \{(1,1), (1,2), (2,1), (2,2), (3,1), (3,2)\}.$$

Next, the ordered triples of $A \times B \times C$ can be realized as either starting with a 1 or a 2 (the elements of A) followed by one of the six ordered pairs of $B \times C$ that we just listed:

$$A \times B \times C = \{(1,1,1), (1,1,2), (1,2,1), (1,2,2), (1,3,1), (1,3,2)$$
$$(2,1,1), (2,1,2), (2,2,1), (2,2,2), (2,3,1), (2,3,2)\}.$$

Part (b): The set $\mathbb{R}^2$ consists of all ordered pairs (x, y) of real numbers. These ordered pairs can be identified with points in the two-dimensional plane in analytic geometry, which is the workspace for much of one-dimensional calculus. The set $\mathbb{R}^3$ consists of all ordered triples (x, y, z) of real numbers and can be identified with three-dimensional Euclidean space in which much work in multi-variable calculus is done.

EXERCISE FOR THE READER 1.12: If A, B, and C are sets, is it true that $(A \times B) \times C = A \times B \times C$?

We end this chapter with a brief outline of some of the interesting history of logic and set theory.

Despite their very close conceptual relationships, logic has a much longer history than set theory. Ever since there was spoken language, mankind has been

---

[35] This example would correspond to the Cartesian product D^6 where $D = \{0,1,\cdots,9\}$.

involved in inference and debate. Common standards for reason evolved in all of the major ancient cultures (Chinese, Indian, Western, etc.).

The Historical Development of Logic and Sets

FIGURE 1.8:
Aristotle (384–322 BC),
Greek philosopher.

The first to systematically develop a logical framework that is similar to what is used in contemporary society was the Greek philosopher Aristotle[36] (Figure 1.8) during the third century BC. His contribution had a tremendous and lasting effect on scientific methodology and thought throughout the world, and gave rise to numerous subsequent advances in the sciences. For example, Euclid (325–265 BC, Egyptian mathematician), taking full advantage of the coherent and rigorous logical apparatus created by Aristotle, created his book *The Elements*, which is the most important mathematics book of all time. Aristotle contended that logic was not a science but rather an essential prerequisite that had to be mastered before any serious scientific work could take place—a belief that is still maintained in contemporary scientific investigations (e.g., in this course in discrete structures). Incidentally, Aristotle used the term *analytics* for *logic*, the latter term being coined by Xenocrates (396–314 BC, Greek philosopher).

The symbolism of modern logic was first introduced by English mathematician George Boole[37] (Figure 1.9).

---

[36] Aristotle was born in Macedonia, which then was the northern part of Greece. Macedonia has been an independent country since 1991, after having been part of Yugoslavia since the communist era of Tito's regime. Although Aristotle was orphaned at a young age, his uncle continued to raise him well and saw to it that he got a strong education in the classics. At 17, Aristotle went to attend *Plato's Academy*, where he later became an instructor for 20 years, until Plato's death. His philosophy was at odds to that of Plato, and as such he was not selected to succeed Plato as the head of the academy. This disappointment led to his departure. He subsequently accepted an invitation to serve in the court of King Hermias of Atarneus during which time he wed the ruler's niece.   After three years, Aristotle then moved on to serve for King Philip of Macedonia. He tutored and became good friends with King Philip's son Alexander, who would become Alexander the Great. The latter encouraged Aristotle to open his own school of philosophy in Athens to compete with the views propagated by Plato's Academy. Aristotle founded his Lyceum in 335 BC, where he worked avidly for 13 years until the death of Alexander the Great resulted in anti-Macedonian political turmoil that forced him to flee the city. Aristotle died a year later from a stomach ailment.

[37] George Boole was born into a poor family, but his father had a strong interest in science and mathematics and encouraged his son George in these areas. As a young teenager, Boole developed an exceptional talent for languages.   By age 14, he has mastered both Greek and Latin and he had published a translation of an ancient Greek poem.   He taught himself French and German, which were not offered at his school. When Boole was 16, his father went out of business and he needed to interrupt his education to take a full-time job as an assistant school teacher to support the family. At

This symbolism has been a vital framework for all modern computing languages, and also helped to increase the rigor of logic into the realm of the mathematical sciences. Boole thus paved the way for a very eventful period in the further development of logic as well as the new field of set theory. Boole also invented *Boolean algebra* (which has applications in modern computer design that we will discuss in Section 2.3) and did important work in probability.

FIGURE 1.9:
George Boole
(1815–1864),
English
mathematician.

FIGURE 1.10:
Georg F. L. P.
Cantor (1845–1918),
German/Russian
mathematician.

Boole's work came during an important period in the history of mathematics. Much work in mathematics and physics during this period was based on the framework developed with the invention of calculus. New theories and methods were becoming so sophisticated that Aristotelian logic was no longer a sufficiently rigorous framework. These circumstances and Boole's work triggered the birth of set theory as well as further axiomatic developments in logic and set theory that would continue well into the twentieth century.

Towards the end of the nineteenth century, Georg Cantor[38] (see Figure 1.10) single-handedly developed his innovative new theory of sets on which all

---

age 27, Boole accepted a job to take over a boarding school, and he moved there with his parents and siblings to run the school. During this time, Boole had time to study some mathematics on his own, and his notes on his readings eventually became the basis of his first publication. Boole's mathematical career began to flourish as his papers made extremely favorable impressions on the influential mathematicians of that period. His efforts were rewarded with an appointment as a chair of mathematics at Queens College in Cork (Ireland) when he was 34. His proud father unfortunately passed away the year before. It was in Cork where Boole did his most significant work, a 1854 treatise where he introduced his algebraic formulation of logic. The following year Boole began a happy marriage and had five daughters. One very rainy day when he was 49, Boole got soaked in his two mile walk to give a lecture and fell terribly ill. When he was recovering in bed, Boole's wife believed that the cure of such an illness should resemble the cause, so she threw buckets of water over his bed. Boole unfortunately passed away not long after.

[38] Georg Cantor was born in St. Petersburg, Russia, where his father was a successful businessman. When Georg was 11, the family relocated to Germany because of his father's health, and he remained in Germany for most of his life. After having studied with some of the foremost mathematicians at the time and earning his *habilitation* from the University of Berlin, Cantor obtained an appointment at the University of Halle. Cantor's early work was in number theory, but he became interested in analysis and found that the subject needed a new framework in which to study the many concepts of infinity that came up in the theory. He began to develop this new framework in a series of groundbreaking yet controversial papers. Many influential mathematicians were not ready to accept this new framework and Cantor had to endure much harsh criticism. Such skepticism among leading mathematicians had prevented him from landing a post at a more prestigious university that would have been more commensurate with his achievements. In addition to such professional obstacles, Cantor also had to deal with numerous personal problems, such as having lost his father at an early age, having lost his youngest son, and having had documented problems with serious mental illnesses. It is remarkable that, despite all of these hindrances, he was able to make such significant and lasting contributions to mathematics. David Hilbert (1862–1943), one of the most respected and powerful (German)

mathematical fields could be based and that attempted to explain many of the notions of infinity that the contemporary theoreticians often avoided.

Cantor's set theory caused a great deal of controversy and was criticized by many since it involved many nonconstructable objects.   This caused him much personal anguish and professional difficulties in getting his papers published. Nonetheless, his papers were well-received by other influential mathematicians, and journals and his work significantly changed the course of mathematics. Cantor also wrote papers in philosophical journals, successfully adding set theory (like logic) as a research area in philosophy as well as in mathematics.   Cantor's major works were done during the latter part of the nineteenth century and they included numerous extensions and ramifications of his theory, including the associated concepts of infinity (cardinalities of sets). After his ideas had been generally accepted by the worldwide mathematics community, some paradoxes began to appear with his theory. The most notable and simplest one is *Russell's Paradox* (see Exercise 32) that was discovered by English logician Bertrand Russell[39] in 1901. Such paradoxes arise only in situations where the sets under question get unrealistically general and, in particular, will never surface in the discrete mathematical topics covered in this book.

FIGURE 1.11: Bertrand A. W. Russell (1872–1970), English philosopher.

The controversies created by the discovery of paradoxes in Cantor's set theory shook the mathematical community, since the validity of much new and important mathematical research that was based on it was now threatened. Careful studies were made on some of the axioms (both explicit and tacit) that were assumed by Cantor's theory. One such axiom is the so-called *Axiom of Choice*, which states that given any family $\{A_k\}_{k \in I}$ of mutually exclusive nonempty sets, there exists a set A which has exactly one element from each of the sets in the family (i.e., $\forall k \exists ! x \in A_k \cap A$). In less formal language, the Axiom of Choice simply says that

---

mathematicians ever, summarized Cantor's work lavishly as follows: "...*the finest product of mathematical genius and one of the supreme achievements of purely intellectual human activity.*"

[39] Bertrand Russell was one the foremost logicians of the twentieth century. Although best known in mathematical circles for his set theoretical paradox, he wrote numerous influential works on logic and set theory and, in particular, on the formulation of all of mathematics in purely set-theoretical concepts. Russell began his career as a professor at Trinity College in Cambridge, but was forced to relinquish his position after a conviction for anti-war activities. He had to spend six months in prison but he made impressively good use of this time by writing his important book *Mathematical Philosophy* (1919). Russell moved to the United States and worked at the City College in New York. He was quite passionate about his political views, and joined forces with Albert Einstein to protest nuclear proliferation. He became founding president of anti-nuclear protest group and was again imprisoned for these anti-American activities. Russell was married four times and was known to have been involved in numerous affairs.   His writing was outstanding and much of it was accessible to the general public. He was awarded the Nobel Prize for Literature in 1950.

an infinite number of arbitrary choices can be made. It was an unsolved problem for quite some time whether the Axiom of Choice followed from (or contradicted) the remaining set theory axioms. The situation has been gradually resolved during the course of the 20th century, most notably by the works of logicians Kurt Gödel (1906–1978, Austrian mathematician) and Stanford mathematician Paul Cohen (1934–2007). Gödel showed in 1940 that the Axiom of Choice was *consistent* with the contemporary axioms of set theory. In 1963, Cohen showed the Axiom of Choice to be independent of these axioms. Cohen was awarded the prestigious *Fields Medal* for this important contribution to logic. With the exception of some logicians, nearly all research mathematicians accept the Axiom of Choice in their work, since it seems like a quite reasonable axiom and it allows one to prove more theorems. Logicians have developed two different set theories: with and without the Axiom of Choice. Gödel's and Cohen's work allow both theories to live in peaceful coexistence.

EXERCISES 1.3:

1. Let $U = \{0,1, 2, 3, 4, 5, 6, 7, 8, 9\}$, $A = \{0, 2, 4, 6, 8\}$, and $B = \{0, 3, 6, 9\}$. Find each of the following sets:
   (a) $A \cup B$
   (b) $A \cap B$
   (c) $\sim A$
   (d) $A \sim B$
   (e) $B \sim A$
   (f) $B \cap \sim A$

2. Let $U = \{a, b, c, d, e, f, g, h, i, j\}$, $A = \{a, e, i\}$, and $B = \{a, b, c, d, i, j\}$. Find each of the following sets:
   (a) $A \cup B$
   (b) $A \cap B$
   (c) $\sim A$
   (d) $A \sim B$
   (e) $B \sim A$
   (f) $B \cap \sim A$

3. Let $U = \{a, b, c, d, e, f, g, h, i, j\}$, $A = \{a, e, i\}$, $B = \{a, b, c, d, i, j\}$, and $C = \{a, d, e, j\}$. Find each of the following sets:
   (a) $A \cup B \cup C$
   (b) $A \cap B \cap C$
   (c) $\sim (A \cup B)$
   (d) $\sim A \cap \sim B$
   (e) $\sim (A \cup B \cup C)$
   (f) $\sim A \cap \sim B \cap \sim C$

4. Let $U = \{0,1, 2, 3, 4, 5, 6, 7, 8, 9\}$, $A = \{0, 2, 4, 6, 8\}$, $B = \{0, 3, 6, 9\}$, and $C = \{7, 8, 9\}$. Find each of the following sets:
   (a) $A \cup B \cup C$
   (b) $A \cap B \cap C$
   (c) $\sim (A \cup B)$
   (d) $\sim A \cap \sim B$
   (e) $\sim (A \cup B \cup C)$
   (f) $\sim A \cap \sim B \cap \sim C$

5. Either list the members of, or describe with words, each of the following sets:
   (a) $\{n \in \mathbb{Z} \mid \exists k \in \mathbb{Z}[n = 2k]\}$
   (b) $\{x \in \mathbb{R} \mid x^2 = 2\}$
   (c) $\{n \in \mathbb{Z} \mid n^2 = 2\}$
   (d) $\{n \in \mathbb{Z} \mid 2n > n^2\}$

6. Either list the members of, or describe with words (or interval notation), each of the following sets:
   (a) $\{n \in \mathbb{Z} \mid \exists k \in \mathbb{Z}[n = 2k + 1]\}$
   (b) $\{x \in \mathbb{R} \mid 0 < x^2 < 1\}$
   (c) $\{n \in \mathbb{Z} \mid n^2 = \pm 2n\}$
   (d) $\{a \in \mathbb{Q} \mid 25a^4 = 4\}$

7. Let U = {all college students}, F = {all female college students}, M = {all male college students}, and B = {all basketball playing college students}. Describe in plain English each of the following sets.
    (a) $M \cap B$ (b) $M \cup F$
    (c) $F \sim B$ (d) $\sim M \cap \sim B$

8. Let U = {all California universities}, P = {all private California universities}, and E = {all California universities with engineering programs}. Describe in plain English each of the following sets.
    (a) $\sim P$ (b) $P \cap E$
    (c) $P \sim E$ (d) $P \cup E$

9. Let U = {all college students}, R = {all college students who like R&B music}, C = {all college students who like classical music}, and J = {all college students who like jazz music}. Use set-theoretic notation to describe the following sets:
    (a) {all college students who don't like classical music}
    (b) {all college students who like R&B but not jazz}
    (c) {all college students who like neither Jazz nor classical music}
    (d) {all college students who like neither classical, jazz, nor R&B}

10. A utility company uses the following sets to classify credit of its customers. A = {customers who always have paid their bills on time}, B = {customers whose bills are always more than $200 per month}, L = {customers who have maintained an account for over two years}, S = {customers have had an account for less than nine months}. Use set-theoretic notation to describe the following sets:
    (a) {customers who have had their account for at most two years}
    (b) {customers who have always paid their bills on time but have had an account for less than nine months}
    (c) {customers whose bills are not more than $200 per month or have had an active account for less than nine months}
    (d) {customers who have not both always paid their bills on time and had an account for more than two years, but whose bills sometimes do not exceed $200 per month}

11. For each set-theoretic identity below, do the following: (i) Prove the identity with Venn diagrams. (ii) Give an analytical (logical) proof of the identity.
    (a) $(A \cup B) \cup C = A \cup (B \cup C)$ (*Associativity*)
    (b) $\sim (A \cup B) = \sim A \cap \sim B$ (*De Morgan's Law*)
    (c) $A \cup (A \cap B) = A$ (*Absorption*)
    (d) $\sim (A \cup B \cup C) = \sim A \cap \sim B \cap \sim C$ (*De Morgan's Law*)

12. For each set-theoretic identity below, do the following: (i) Prove the identity with Venn diagrams. (ii) Give an analytical (logical) proof of the identity.
    (a) $(A \cap B) \cap C = A \cap (B \cap C)$ (*Associativity*)
    (b) $A \cup (B \cap C) = (A \cup C) \cap (B \cup C)$ (*Distributivity*)
    (c) $\sim (A \cap B) = \sim A \cup \sim B$ (*De Morgan's Law*)
    (d) $\sim (A \cap B \cap C) = \sim A \cup \sim B \cup \sim C$ (*De Morgan's Law*)

13. Each part below gives a pair of sets. (i) Using Venn diagrams, for each pair, decide what is the strongest general subset relation between them: are the sets always equal?—if not, is one set always a subset of the other? Or does no such relation exist (in general)? (ii) In cases where you have a general subset or set equality relation, provide a logical proof of your assertion. (iii) Unless you have a set equality, provide counterexample(s) to show that one (or both) subset relationships can fail to be true. (This will be possible if you have done item (i) correctly.)
    (a) $A \cap B$, A

(b) $A \cap B \cap C, \ A \cap B$

(c) $(A \sim B) \sim C, \ A \sim C$

(d) $(A \sim B) \sim C, \ (A \sim C) \sim (B \sim C)$

14. Repeat the directions of Exercise 13 for the following pairs of sets.
    (a) $A, \ A \cup B$
    (b) $(A \cap B) \sim C, \ (A \sim C) \cap B$
    (c) $(A \sim B) \sim C, \ A \sim C$
    (d) $(A \sim B) \sim C, \ (A \sim C) \sim (B \sim C)$

15. Below are some statements about sets. For each one, indicate if it is always true, sometimes true, or never true. In the first and third cases, provide explanations (or proofs), while if the statement is sometimes true, provide examples of cases that will show it can be both true and false.
    (a) If $A \subseteq B$, then $\sim B \subseteq \sim A$.
    (b) If $A \Delta C = B \Delta C$, then $A = B$.
    (c) If $\{A_k\}_{k \in \mathbb{Z}_+}$ and $\{B_k\}_{k \in \mathbb{Z}_+}$ are two families of sets with $\forall k \ A_k \subseteq B_k$ then $\bigcup_k A_k \subseteq \bigcup_k B_k$.

16. Repeat the directions of Exercise 15 for the following statements about sets.
    (a) If $A \subseteq B$, $B \nsubseteq C$, then $A \nsubseteq C$.
    (b) If $A \subseteq B$, then $A \cap B = A$.
    (c) $A \Delta (B \cap C) = (A \cap B) \Delta (A \cap C)$.

17. (a) Use Venn diagrams to prove that for two sets $A, B \subseteq U$, we have $\sim (A \ \Delta \ B) = \sim A \ \Delta \ B$.
    (b) Prove the result of Part (a) using only logic and the definitions (without resorting to Venn diagrams). This part should amplify the reader's appreciation for Venn diagrams.

18. For any positive integer k, we define
    $$A_k = \{a \in \mathbb{Z} \mid a \ge k\} = \{k, k+1, k+2, \cdots\}.$$
    Describe the following sets:
    (a) $\bigcup_k A_k$.   (b) $\bigcap_{k \ge 1} A_k$.
    (c) $\bigcup_{k=5}^{20} A_k$.   (d) $\bigcap_{k=5}^{20} A_k$.

19. Determine the truth value of each of the following statements involving the empty set.
    (a) $\varnothing \subseteq \varnothing$.   (b) $\varnothing \in \varnothing$.
    (c) $\varnothing \subset \varnothing$.   (d) $\varnothing \in \{\varnothing\}$.
    (e) $\varnothing \subset \{\varnothing\}$.   (f) $\varnothing \in \{\{\varnothing\}\}$.

20. Given the set $A = \{\varnothing, \{\varnothing\}, \{\varnothing, \{\varnothing\}\}\}$, determine the truth value of each of the following statements.
    (a) $\varnothing \subseteq A$.   (b) $\varnothing \in A$.
    (c) $\{\{\varnothing\}\} \subseteq A$.   (d) $\mathscr{P}(A)$ consists of 16 elements.

21. Consider the following set of colors:
    $$C = \{\text{Red, White, Blue, Green, Yellow, Orange, Brown, Purple}\}.$$
    (a) How many subsets does C have?
    (b) How many proper subsets does C have?
    (c) How many subsets does C have that contain the colors Yellow and Purple?
    (d) How many subsets does C have that do not contain any of the colors Red, White, and Blue?
    (e) How many nonempty subsets does C have that do not contain any of the colors Red, White, and Blue?

22.   Suppose that A and B are two sets with $\mathscr{P}(A) \subseteq \mathscr{P}(B)$. From this information can we
.   conclude that $A \subseteq B$? Justify, if you answer yes, otherwise provide a counterexample.

23.   Consider the following two sets of colors: $A = \{\text{Red, White, Blue}\}$, $B = \{\text{Green, Yellow}\}$.
        (a) Find $A \times B$.
        (b) Find $B \times A$.
        (c) Find B^3.
        (d) How many elements are in the set $\mathscr{P}(A^2)$? Give three elements in this set.

24.   Consider the following two sets: $A = \{1, 2, 3, 4\}$, $B = \{7, 8\}$.
        (a) Find $A \times B$.
        (b) Find $B \times A$.
        (c) Find B^3.
        (d) How many elements are in the set $\mathscr{P}(A^2)$? Give three elements in this set.

25.   Below are some statements about sets. For each one, indicate if it is always true, sometimes
        true, or never true. In the first and third cases, provide explanations (or proofs), while if the
        statement is sometimes true, provide examples of cases that will show it can be both true and
        false.
        (a) If $A \times B = \varnothing$, then $A = \varnothing$ and $B = \varnothing$.
        (b) If A, B, and C, are three sets with $A \in B$, and $B \in C$, then $A \in C$.

26.   Repeat the directions of Exercise 25 for the following statement about sets.
        (a) If $A = \varnothing$ or $B = \varnothing$, then $A \times B = \varnothing$.
        (b) If A and B are sets, then $\mathscr{P}(A) \times \mathscr{P}(B) \subseteq \mathscr{P}(A \times B)$.

27.   (*Rational Numbers—Decimal Form*) The two parts below will show that rational numbers are
        exactly those real numbers whose decimal forms either terminate or repeat. Equivalently, it
        shows that irrational numbers are precisely those real numbers whose decimal expansions do not
        terminate or repeat.[40]
        (a) Prove that the decimal form representation of any rational number must either terminate
        (like $1/4 = 0.25$) or repeat (like $7/12 = 0.58\overline{3}$).
        (b) Prove that any real number whose decimal expansion either terminates or repeats is a
        rational number.
        Suggestions: For Part (a), one needs to think about the division algorithm that one learned in
        grade school for putting a fraction into decimal form (long division). (For example, to convert
        $7/12$, one performs the long division $12\overline{)7}$, in which 12 is called the *divisor*.) Each iteration of
        a long division produces a (single) digit in the decimal expansion (that goes above the division
        bar). Each nonzero digit encountered needs to be multiplied by the divisor and subtracted from
        accumulated number (at the bottom of the division symbols) to produce a remainder. The
        remainder will always be nonnegative integer less than the divisor. If a zero remainder is ever
        encountered, the division (and hence the decimal) terminates. Otherwise, since there are only a
        finite number of choices for the remainder, a remainder must occur twice. The first
        reoccurrence of a remainder determines the repeating segment of digits.
        For Part (b), use the strategy suggested by the following example. Consider the real number,
        $12.42\overline{875}$. Isolate the repeating portion $x = 0.00\overline{875}$. Note that $1000x = 8.75 + x$, so $999x =$

---

[40] We point out that the decimal representation of a real number is unique, provided that we do not use
any infinite strings of nines; these should be rounded up. To see, for example, that $0.\overline{9} = 1$, is quite
simple. *Proof:* Set $x = 0.\overline{9}$. Then $10x = 9.\overline{9} = 9 + x$. This implies $9x = 9$, or $x = 1$.

8.75 = 875/100. Therefore, $x = 875/99900$, and so $12.42\overline{875} = 1242/100 + 875/99900$ can now be expressed as a quotient of integers (get a common denominator).

28. (*Irrational Numbers—Examples*) (a) Write down an example of an irrational number in decimal form.

(b) Prove that $\sqrt{2}$ is irrational.

(c) Prove that $\sqrt{k}$ is irrational whenever k is a positive integer that is not a perfect square.

(d) Prove that $\sqrt[3]{2}$ is irrational.

(e) Prove that there exist two irrational numbers a and b such that a^b is rational.

Suggestion: For Part (a), use the result of Exercise 20, with, say, the number:
$$0.101001000100001000001\cdots$$
(the digits are only zeros and ones, but between successive ones, the blocks of zero increase by one). For Part (b), use a proof by contradiction. Assume that $\sqrt{2} = p/q$, where the fraction on the right is in lowest terms, i.e., p and q have no common factors. Squaring both sides leads us to $2q^2 = p^2$. From this we see that 2 must be a factor of p. But then 4 must be a factor of p^2. We can thus cancel a factor of 2 from both sides of $2q^2 = p^2$, and still have a factor of 2 left on the right. This means that 2 must be a factor of q^2, and so must also be a factor of q. We have reached a contradiction (p and q have a common factor of 2) to the assumption that p/q was in lowest terms. These ideas used some basic facts about prime factorization of integers; this topic will be discussed in Chapter 3. Here is a sketch of a slick proof for Part (e). Consider $A = \sqrt{2}$. We know from Part (b) that A is irrational. Case 1: $A^A = \sqrt{2}^{\sqrt{2}}$ is rational. We can take $a = b = A$. Case 2: $\sqrt{2}^{\sqrt{2}}$ is irrational. We can take $a = \sqrt{2}^{\sqrt{2}}$ and $b = \sqrt{2}$ (so $a^b = \sqrt{2}^{\sqrt{2}\sqrt{2}} = 2$).[41]

NOTE: The following three exercises will study the construction of Venn diagrams for finite families of sets. The text of the section showed examples of Venn diagrams with two, three, and four sets. Cambridge mathematician Anthony W. F. Edwards has developed a procedure for (inductively) constructing Venn diagrams for any finite number of sets. We outline this procedure here; for further details we encourage the interested reader to examine the original references [Edw-89a], and [Edw-89b]. Interestingly, this seems to have become the most general procedure for creating Venn diagrams since Venn's original work [Ven-80] was done over 100 years earlier. It will be helpful to define exactly what we mean by a *Venn diagram*. A *Venn diagram* is a collection $\mathscr{C} = \{C_1, C_2, \cdots, C_n\}$ of simple closed curves (e.g., circles, squares, or any closed nonself-intersecting curves)[42] drawn inside a square (representing the universal set) with the following properties:

(i) The region formed by any intersection of the form $X_1 \cap X_2 \cap \cdots \cap X_n$, where each X_i is either $\text{int}(C_i)$ or $\text{ext}(C_i)$, is a nonempty connected region, and

(ii) There are only a finite number of intersection points between any pair of curves in $\mathscr{C}$.

---

[41] Note that this existence proof is *nonconstructive*. The proof was completed without actually knowing which of the two cases holds, so it does not provide an explicit example.

[42] We informally define a simple closed curve as a curve in the plane that is closed and never crosses over itself (e.g., ellipses are examples of simple closed curves, whereas a figure eight curve is not since it crosses itself, and, an S-shaped curve is not since it not closed). One important (and intuitively obvious) property of any a simple closed curve C in the plane is that it will separate the plane into two portions: the interior $\text{int}(C)$ (the bounded portion) and the exterior $\text{ext}(C)$ (the unbounded portion). Rigorous definitions and derivations of properties of simple closed curves can be found in any decent book on topology (e.g. [Mun-75]) or complex analysis (e.g., [Ahl-79]).

In set theory, the function of a Venn diagram with n curves is to represent all possible subsets generated by taking intersections from a collection of n sets and/or their complements. Such sets can then be visualized as conglomerates of "pieces" of the Venn diagram. From this it is easy to show (using mathematical induction that will be introduced in Chapter 3) that the number of regions in a Venn diagram for n sets is 2^n. Edwards' construction of Venn diagrams is inductive. He started with a one-set Venn diagram (Figure 1.12a) where the corresponding curve C_1 is simply the left rectangular half of the (square) universal set. The next iteration adds a second curve C_2 that is the lower half rectangle. The third generation adds a centrally centered circle C_3.

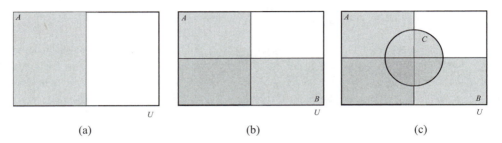

FIGURE 1.12: The first three iterations of Edwards' inductive construction of general Venn diagrams. (a) (left) One set A is enclosed by the left (half) rectangle. (b) (middle) A new set B is added, enclosed by the lower (half) rectangle. (c) (right) A new set C is added, enclosed by a circle.

At each new iteration, the idea is to add a new simple closed curve whose interior will meet all of the existing portions in one nonempty connected region. Figure 1.13 shows the next three iterations of Edwards' construction. Note that each new curve added must cut each existing Venn region into two regions (inside and outside the new set).

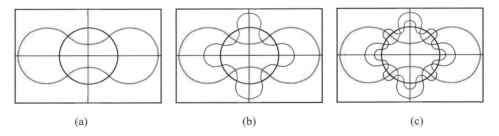

FIGURE 1.13: The iterations four through six of Edwards' inductive construction of general Venn diagrams. (a) (left) For four sets. (b) (middle) For five sets. (c) (right) For six sets.

29.   (a) Perform the next iteration of Edwards' Venn diagram construction for seven sets by adding a new curve to Figure 1.13(c).
       (b) Create a Venn diagram for four sets using four rectangles of the same dimensions.

30.   (a) Is the diagram in Figure 1.14(a) Venn diagram for four sets? Explain your answer.
       (b) Is the diagram in Figure 1.14(b) Venn diagram for four sets? Explain your answer.

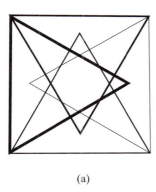   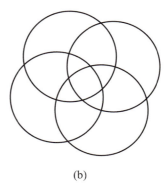

                    (a)   (b)

FIGURE 1.14: (a) A diagram for four sets made up of four triangles (drawn with different line widths). (b) A diagram for four sets made up of four circles.

31. (a) Show that it is not possible to create a Venn diagram for four sets using four circles of the same radius.
     (b) Is it possible to create a Venn diagram for four sets using four circles (of different radii)?
     (c) Create a Venn diagram for five sets using five ellipses.

32. (*Russell's Paradox*)   Consider the set R consisting of all sets that are not members of themselves. In symbols, $R = \{S \mid S \notin S\}$. This "set" seems to be well defined using the naïve set theory developed in this section. However, its very definition, as Russell found, leads to a contradiction. Obtain a contradiction by considering the question of whether $R \in R$ or $R \notin R$.

COMPUTER EXERCISES 1.3:

NOTE: Finite sets can be stored as objects on most computing platforms. Depending on your particular platform, how a set gets stored might depend on the nature of the elements. If the elements are numbers, the set is probably stored as a list or vector, where the order does not matter (nor should any duplicated entries). If the elements are text strings, such as in the set {Red, White, Blue}, then the storage will require a different type of data structure. In the computer exercises that follow, the objective will be to learn how to store both types of finite sets on your platform and to learn how to perform the set operations. We point out that the size of a set being stored on a given computer platform is limited by the available memory. Also, there are different grades of efficiency for storing given sets. At this juncture, we will be working with small sets so memory and efficiency issues will not be relevant. These issues are important for many practical problems, however, and we will revisit these topics as needs arise.

In Exercises 1–4 below, in addition to learning how to store the sets given, you should also find out how to perform (at least) the following set operations: union, intersection, and (relative) complement.

1.   Store the sets of (regular) Exercise 1 of this section on your computing platform, and perform the set operations asked for in Parts (a) through (f) of that exercise.

2.   Store the sets of (regular) Exercise 2 of this section on your computing platform, and perform the set operations asked for in Parts (a) through (f) of that exercise.

3.   Store the sets of (regular) Exercise 3 of this section on your computing platform, and perform the set operations asked for in Parts (a) through (f) of that exercise.

4.   Store the sets of (regular) Exercise 4 of this section on your computing platform, and perform the set operations asked for in Parts (a) through (f) of that exercise.

5. Find out how your computing platform can test whether a number is an element of a set of numbers.[43]   Similarly, find out how you can test the subset $(A \subseteq B)$ and equality $(A = B)$ relationships between two sets. Next, using the sets A, B, C, and U of (regular) Exercise 4, run through and test each of the identities in Parts (a) through (d) of (regular) Exercise 11.

6. Read Exercise 5 and then again using the sets A, B, C, and U of (regular) Exercise 4, run through and test each of the identities in Parts (a) through (d) of (regular) Exercise 12.

---

[43] Try to avoid the obvious brute force approach of using a for loop to run through the elements of the set, one-by-one, checking whether the given element equals the set element. Look for a more elegant approach, or better yet, a built-in function that will do the job.

Chapter 2: Relations and Functions, Boolean Algebra, and Circuit Design

In the preceding chapter we established the language and concepts of logic, and then moved on to develop the most fundamental of all discrete structures, namely sets. All other discrete structures can be completely described in terms of sets, using the language of logic. This chapter has two main themes: relations and Boolean algebra. Relations are simply subsets of Cartesian products of sets. The most important example of a relation is a function. Although most students have been dealing with functions since their pre-high school mathematics courses, the proper and rigorous definition of a function will probably be new to many, and may even seem a bit strange. But it is a versatile concept that will be prevalent in nearly all of the developments throughout the rest of the book. In Section 2.1, we will study some general concepts about relations, and then give an in-depth coverage of functions. Section 2.2 will make detailed excursions into two other very important types of relations: equivalence relations, and (partial) orders. These latter relations will prove extremely useful for comparisons between numerous sorts of discrete structures that we will be encountering.

Boolean algebra is a special sort of algebra in which all variables are logical variables that can take on only one of only two values: 0 (for false) and 1 (for true). The theory was first introduced in the mid-nineteenth century by George Boole to formalize the subject of logic into a rigorous mathematical framework. Subsequently, in the 1930s, Claude Shannon showed how Boolean algebra could be applied to the design of general electronic circuits, made up of certain circuit elements that correspond to basic Boolean (logical) operations. Section 2.3 will develop Boolean algebra and techniques for designing circuits. A beautiful application of simplifying Boolean expressions will be to the creation simplified circuits that possess a given functionality. Techniques for such simplifications will be provided.

2.1: RELATIONS AND FUNCTIONS

The notion of a *function* is fundamental to all of mathematics and the applied sciences. Students learn about functions, with varying degrees of sophistication, throughout middle and high school. Functions are special cases of more primitive objects called *relations*, which allow for very versatile descriptions of relationships between two sets, both abstract and concrete. Both functions and relations will be investigated and elaborated on in this section, and they will be described completely using the set theory of the preceding chapter. We begin with the simple definition of a relation.

Binary Relations

DEFINITION 2.1: Suppose that A and B are two sets.   A (**binary**) **relation** R **from** A **to** B is simply a subset of $A \times B$, i.e., $R \subseteq A \times B$. In other words, a relation is simply a set of ordered pairs of elements (a, b), where $a \in A$, and $b \in B$. We employ the notation aRb to denote $(a, b) \in R$, and we say that a is related to b by R. We use the notation $a \not\!R b$ to indicate $(a, b) \notin R$. In case $B = A$ (i.e., a relation from A to A), we simply call R a **relation on** A.

The adjective "binary" in the preceding definition is only to distinguish this type of relation from a more general n-ary relation, which is a subset of a Cartesian product of n sets. (Thus a binary relation is a 2-ary relation.) Since nearly all of the relations that will be needed in this book are binary, this adjective will often be omitted.

EXAMPLE 2.1: (a) Let A denote the set of all 50 states in the US and B denote the set of all professional sports teams in the NBA, NFL, or NHL (basketball, football, or hockey). We define the relation R from A to B to consist of all pairs ("state", "team") where "state" represents one of the 50 states in A, and "team" represents any team in B whose home is in "state." Here are examples and non-examples for the relation R: Michigan R Pistons,   Pennsylvania R Steelers, Florida R Lightning, Kentucky $\not\!R$ Bulls, Colorado $\not\!R$ Cowboys, Illinois $\not\!R$ Kings.   Readers who are not sports fans need not worry about justifying the truth value of these six statements; those who are may wish to change the states of the last three to turn them into positive examples.

(b)   Consider the relation R on $\mathbb{R}$ consisting of all ordered pairs (x, y) for which $0 \le y \le x^2$.   The reader should verify the truth of the following statements regarding this relation: $-2R3$, $2 \not\!R 5$, $\forall x \forall y < 0[x \not\!R y]$. If we identify the Cartesian product $\mathbb{R}^2$ with the plane, the relation R can be efficiently visualized by its graph shown in Figure 2.1.

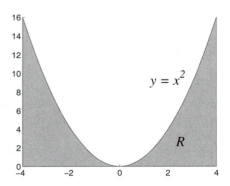

FIGURE 2.1: Graph of the relation of Example 2.1(b) (shaded region).

Depending on the size and the types of the sets involved, there are many different ways to represent relations, including simply listing all elements, graphs as in Figure 2.1, charts, and diagrams.

EXERCISE FOR THE READER 2.1: Consider the following relation R on the set $\{1,2,3,4,5,6\}$: $R = \{(a,b) \mid 2b < a \text{ or } b > 3a\}$. Draw a chart and make a table that will each represent this relation. Also, determine the truth values of these statements: $2R5$, and $6 \not{R} 6$.

Functions

We next move on to discuss functions, which are the most important and useful relations. We will say more about some other important types of relations in the next section. It is often useful to associate with every element of one set, a <u>unique</u> element from another set. For example, the assignment to each subject in a medical experiment to his or her weight is a function from the set of subjects to the set of (positive) numbers.

DEFINITION 2.2: Let A and B be two sets. A **function** (or **mapping**) f **from** A **to** B, denoted $f : A \rightarrow B$, is a relation from A to B in which every element $a \in A$ appears exactly once as the first component of an ordered pair in the relation. For $a \in A$, the corresponding unique $b \in B$, for which $(a,b) \in f$, is called the **image** of a under f, and we write this as $b = f(a)$. The set A is called the **domain** of the function f, and the set B is the **co-domain** of f.[1] The **range** of f, denoted $f(A)$, is the set of all images of elements of A under f, i.e.,

$$f(A) = \{b \in B \mid \exists a \in A[b = f(a)]\}.$$

It is helpful to visualize a function with a diagram such as the one shown in Figure 2.2.

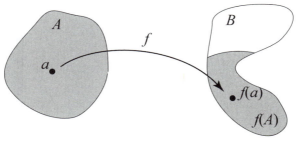

FIGURE 2.2: Schematic diagram of a function $f : A \rightarrow B$ with the image $f(a)$ of an element a in the domain A, and the range $f(A)$ shown (shaded).

---

[1] There is, unfortunately, not much of a consensus on the terminology for what we have called the co-domain. Indeed, in many mathematics texts, there is no term for this set.

Functions can be specified in many ways. First the domain and co-domain sets should be specified, and then the function is defined using formulas, rules, graphs, or algorithms. Functions are often programmed into computers, and in this context it is useful to consider elements $a \in A$ as **inputs** of the function, and the images $f(a)$ as the corresponding **outputs**. The next example will provide a small sampling of functions that will help us to introduce some general attributes that functions can possess.

EXAMPLE 2.2: Below are some examples of functions.
(a) If C = {countries of the world}, and D = {cities of the world}, a function cap : $C \to D$ is specified by the rule: cap(a Country) = its capital city. For example, cap(France) = Paris, and cap(Japan) = Tokyo.
(b) Of the two relations on the set {1, 2, 3} shown in Figure 2.3, only the second is a function.

FIGURE 2.3: Two relations on the set {1, 2, 3}. (a) (left) Is not a function since the image of 2 is not unique. (b) (right) Is a function.

(c) The formula $f(x) = 1/(x^2 - 1)$, defines a function $f : \{x \in \mathbb{R} \mid x \neq \pm 1\} \to \mathbb{R}$. In earlier mathematics courses, students are often led to believe that such formulas alone are the functions. In such courses, the co-domain is often taken by default to be $\mathbb{R}$, and the domain is taken to be the largest subset of $\mathbb{R}$ for which the formula makes sense. Note that if we tried to evaluate $f(\pm 1)$, we would get $1/0$, which is undefined. Any other real number has an image under f, for example, $f(1/2) = -4/3$.

(d) Three useful functions with domain $\mathbb{R}$ and co-domain $\mathbb{Z}$ are the **round**, **floor**, and **ceiling** functions, which are defined as follows:

$\lceil x \rceil$ = the **ceiling** of x = the least integer k that is greater than or equal to x.

round(x) = the integer that is closest to x. [2]

$\lfloor x \rfloor$ = the **floor** of x = the greatest integer k that is less than or equal to x.

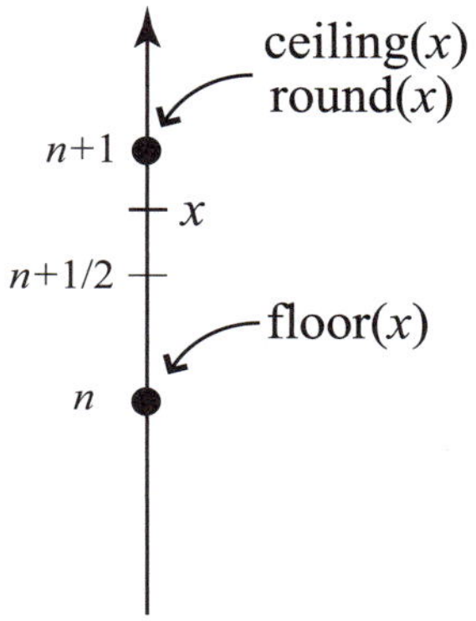

FIGURE 2.4: Picture illustrating the floor, round, and ceiling functions.

For example, $\lceil 2.1 \rceil = 3$, $\lfloor 2.1 \rfloor = 2$, round(2.1) = 2, and $\lfloor -8.7 \rfloor = -9$. Note also that if n is an integer then $\lfloor n \rfloor = \text{round}(n) = \lceil n \rceil = n$.

Function Images and Pre-Images

With a concept of such versatility as that of a function, it should be expected that there will be many definitions and theorems about functions. We move on to some important definitions and properties that a function might possess. The following definition gives two useful ways that a function induces natural relations between subsets of its domain and co-domain.

DEFINITION 2.3: Suppose that $f : A \to B$ is a function. If S is a subset of the domain A, we define the **image** of S under f, denoted $f(S)$, to be the set of all images of elements of S under f, i.e., $f(S) = \{b \in B \mid \exists s \in S[b = f(s)]\}$.

If T is a subset of the co-domain B, we define the **pre-image** of T under f, denoted $f^{-1}(T)$, to be the set of all elements of S whose images under f belong to T, i.e., $f^{-1}(T) = \{a \in A \mid f(a) \in T\}$.

---

[2] For the round function we need to be more specific: In case x lies halfway between two integers, since the image of a function must be unique. We follow the usual convention of rounding upward (e.g., round(3.5) = 4). There does not seem to be a commonly adopted abbreviation for the round function. The ceiling function is sometimes referred to as the *least integer function*, and the floor function is sometimes called the *greatest integer function*.

Informally, the image $f(S)$ of a set S of inputs is the set of all possible outputs under f, with inputs taken from S. Similarly, the pre-image $f^{-1}(T)$ of a set of T in the co-domain of f is the set of all possible inputs whose outputs under f lie in T. As a very simple example, consider the function $f : \{1,2,3\} \to \{a,b,c\}$, defined by $f(1) = f(2) = a$, and $f(3) = c$. If $S = \{1,2\}$, then $f(S) = \{a\}$; while if $T = \{b\}$, $f^{-1}(T) = \emptyset$.

In case the graph of a function f is available, an image $f(S)$ can be visualized by using the graph to "project" the set S (placed one the x-axis) to the y-axis; Figure 2.5(a) illustrates this for the image $f([1,3)) = [1,9)$ for the function $f : \mathbb{R} \to \mathbb{R} :: f(x) = x^2$.

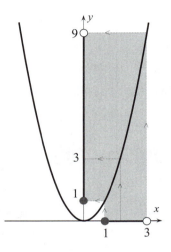 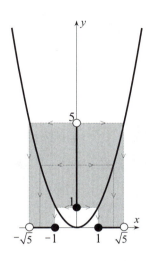

FIGURE 2.5: Graphical illustration of an image and a pre-image of the function $f : \mathbb{R} \to \mathbb{R} :: f(x) = x^2$.

(a) (left) The image $f([1,3)) = [1,9)$.

(b) (right) The pre-image $f^{-1}([1,5)) = (-\sqrt{5},-1] \cup [1,\sqrt{5})$.

Similarly, a pre-image $f^{-1}(T)$ can be visualized by using the graph to "project" the set T (now placed on the y-axis) to the x-axis; Figure 2.5(b) illustrates this for the pre-image $f^{-1}([1,5)) = (-\sqrt{5},-1] \cup [1,\sqrt{5})$, for the function $f : \mathbb{R} \to \mathbb{R} :: f(x) = x^2$.

EXAMPLE 2.3: (a) For the function "cap" of Part(a) of Example 2.2, cap(North America) = {Washington DC, Ottawa}, since North America is the set of the two countries US and Canada.

(b) For the function of Figure 2.3(b), the pre-image of {1} is the set {1, 2}.

(c) For the function with formula $f(x) = 1/(x^2 - 1)$ of Example 2.2(c), the pre-image $f^{-1}(2)$ consists of all solutions to the equation $f(x) = 2$: $1/(x^2 - 1) = 2 \Leftrightarrow x^2 - 1 = 1/2 \Leftrightarrow x^2 = 3/2 \Leftrightarrow x = \pm\sqrt{3/2}$. Thus, $f^{-1}(2) = \{\pm\sqrt{3/2}\}$.

(d) We give a few examples of images and pre-images for the functions $\lceil x \rceil$, $\lfloor x \rfloor$, and round(x). To facilitate the pre-image notation, we temporarily denote $\lceil x \rceil$ by ceiling(x), and $\lfloor x \rfloor$ by floor(x).[3]

$$\text{ceiling}([0,1)) = \text{round}([0,1)) = \{0,1\}, \quad \text{floor}([0,1)) = \{0\},$$

$$\text{ceiling}^{-1}(\{5\}) = (4,5], \quad \text{round}^{-1}(\{0,1\}) = [-0.5,1.5), \quad \text{floor}^{-1}(\{3\}) = [3,4).$$

For general functions, the pre-image operation behaves quite nicely in that it preserves unions, intersections, differences, and inclusions of sets. Of these, the image operation preserves only unions. These facts are made more precise in the following exercise for the reader and in the exercises.

EXERCISE FOR THE READER 2.2: Suppose that $f : A \to B$ is a function.

(a) Prove that if $B_1, B_2 \subseteq B$, then $f^{-1}(B_1 \cap B_2) = f^{-1}(B_1) \cap f^{-1}(B_2)$.

(b) Provide a counterexample to show that the equation corresponding to that in Part (a) for images, i.e., $A_1, A_2 \subseteq A \Rightarrow f(A_1 \cap A_2) = f(A_1) \cap f(A_2)$, can fail to be true.

(c) Can the false statement in Part (b) be salvaged into a true statement by replacing the equality by an inclusion? Either prove such a result or give a counterexample to show the inclusion can fail.

One-to-One, Onto, and Bijective Functions

There are two very important properties that a function can possess: One of these is that the range coincides with the co-domain; the other states that different inputs will always result in different outputs. The next definition will formalize these concepts.

DEFINITION 2.4: Suppose that $f : A \to B$ is a function.

(a) We say f is **injective**, or **one-to-one**, or an **injection**, if the following condition holds: $\forall x, y \in A, \quad f(x) = f(y) \Rightarrow x = y$.

(b) We say f is **surjective**, or **onto**, or a **surjection**, if $f(A) = B$.

(c) We say f is **bijective**, or a **bijection**, if it is both one-to-one and onto.

---

[3] We employ here (and elsewhere in this book) the standard notation for intervals in $\mathbb{R}$: If $a < b$ are real numbers, the open interval $\{x \in \mathbb{R} \mid a < x < b\}$ is denoted (a, b). The closed interval $\{x \in \mathbb{R} \mid a \le x \le b\}$ is denoted by $[a, b]$. Half closed intervals are defined by combining these notations, e.g., $[a, b) = \{x \in \mathbb{R} \mid a \le x < b\}$. Infinite intervals can be represented by letting $a = -\infty$, or $b = \infty$, e.g., $(-\infty, b] = \{x \in \mathbb{R} \mid x \le b\}$.

Note that the (equivalent) contrapositive of the implication in (a) is $x \neq y \implies f(x) \neq f(y)$. Informally, a function is one-to-one if different inputs always have different outputs. If the graph of f is available, then f will be one-to-one if its graph satisfies the *horizontal line test*: no horizontal line can intersect the graph more than once. (Why?) A function is onto if everything in the co-domain is an output (for some input in the domain).

EXAMPLE 2.4: For each of the functions of Example 2.2, determine which are one-to-one, and which are onto.

SOLUTION: Part (a): The function "cap" of Part (a) is certainly one-to-one (different countries cannot have the same capital city) but not onto (not all cities of the world are capitals).
Part (b): The function of Figure 2.3(b) is neither one-to-one (since 1 has two different pre-images), nor onto (since 3 is not an image).
Part (c): The function is neither one-to-one (e.g., since $f(-1) = f(1)$) nor onto (since 0 is not in the range of f).
Part (d): The three functions round, floor, and ceiling are all certainly onto, but none are one-to-one, since each can map an entire interval to a single output number.

EXERCISE FOR THE READER 2.3: Determine whether each of the following functions is one-to-one and/or onto.
(a) $g : \mathbb{Z} \to \mathbb{Z} :: g(k) = 101 - k$ (b) $h : \mathbb{Z} \to \mathbb{Z} :: h(k) = k^3$

(c) $f : \mathbb{R} \to \mathbb{R} :: f(x) = x^2$ (d) $F : [0, \infty) \to [0, \infty) :: F(x) = x^2$

EXERCISE FOR THE READER 2.4: For any finite set $A = \{1, 2, \cdots, n\}$ (n is a positive integer) and any function $f : A \to A$, show that: f is one-to-one $\Leftrightarrow f$ is onto $\Leftrightarrow f$ is bijective.

Inverse Functions

A function $f : A \to B$ that is both one-to-one and onto (i.e., a bijection) is very special, since the input to output process can be reversed.

DEFINITION 2.5: If $f : A \to B$ is a bijection, then a function $f^{-1} : B \to A$ can be defined by the rule: $f^{-1}(b) = a$

if, and only if $f(a) = b$. This function is called the **inverse** (**function**) of f.

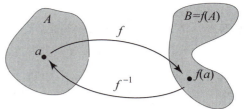

FIGURE 2.6: Illustration of the inverse function $f^{-1} : B \to A$ of a bijection $f : A \to B$.

The concept of an inverse function is illustrated in Figure 2.6. Using the pre-image notation of Definition 2.3, $f^{-1}(b)$ is the unique element of the set $f^{-1}(\{b\})$. Although the latter sets $f^{-1}(\{b\})$ are always defined, if f is not a bijection, these sets will sometimes either be empty (if f is not onto), or contain more than one element (if f is not one-to-one). Notice that if f is a bijection, so must be its inverse function f^{-1}, and that $(f^{-1})^{-1} = f$.

EXAMPLE 2.5: If A is any finite set, say $A = \{1, 2, \cdots, n\}$, then a bijection from A to A is usually called a *permutation* of the set A. For example, when $n = 6$, a particular permutation $\sigma : \{1, 2, \cdots, 6\} \to \{1, 2, \cdots, 6\}$ is represented in *tabular form* by:

$$\sigma : \begin{pmatrix} 1 & 2 & 3 & 4 & 5 & 6 \\ 4 & 6 & 2 & 1 & 5 & 3 \end{pmatrix}.$$

(The numbers in the second row represent the images of the corresponding numbers in the first row, e.g., $\sigma(1) = 4$.) This function is clearly both one-to-one and onto, and hence it is a permutation. The inverse function is given by (as the reader should verify):

$$\sigma^{-1} : \begin{pmatrix} 1 & 2 & 3 & 4 & 5 & 6 \\ 4 & 3 & 6 & 1 & 5 & 2 \end{pmatrix}.$$

The exercises will examine more properties of permutations.

In case a bijection is given by a formula $y = f(x)$, a formula for the inverse function can sometimes be obtained by solving the formula for the independent variable (here x) in terms of the dependent variable (here y). For example, for the bijection $g : \mathbb{Z} \to \mathbb{Z} :: g(k) = 101 - k$ of Exercise for the Reader 2.3, the formula $y = 101 - k$, is easily solved for the independent variable: $k = 101 - y$. This gives the way to reverse the process: for a given integer y, $k = 101 - y$ will be the unique integer that satisfies $g(k) = y$. This gives the formula $g^{-1}(y) = 101 - y$ for the inverse function $g^{-1} : \mathbb{Z} \to \mathbb{Z}$. Note that in this case, the inverse function is really the same function as the original function, i.e., $g = g^{-1}$ (it does not matter that we used different letters for the independent/dependent variables).

DEFINITION 2.6: If a function $f : A \rightarrow B_1$ has as its range a subset of the domain of another function $g : B \rightarrow C$ (i.e., $f(A) \subseteq B$) then these functions can be combined in a natural way to from a new function $g \circ f : A \rightarrow C$ called the **composition** of f and g, defined by $(g \circ f)(a) = g(f(a))$.

The composition is illustrated in Figure 2.7.

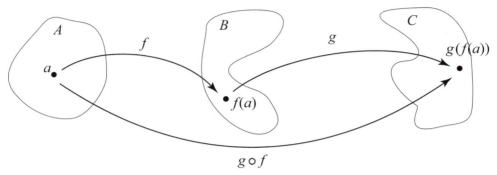

FIGURE 2.7: The composition $g \circ f : A \rightarrow C$ of two functions $f : A \rightarrow B_1$ and $g : B \rightarrow C$, where $f(A) \subseteq B$.

EXAMPLE 2.6: (a) Let $f : \mathbb{R} \rightarrow \mathbb{R} :: f(x) = 3x + 5$, and $g : \mathbb{R} \rightarrow \mathbb{R} :: g(x) = x^2$. Obtain formulas for the two composition functions $f \circ g$, and $g \circ f$.
(b) For any set A, we define the *identity function* for A, to be the function $i_A : A \rightarrow A$ given by the formula $i_A(a) = a$. Show that if $f : A \rightarrow B$ is a bijection, then $f^{-1} \circ f = i_A$, and $f \circ f^{-1} = i_B$.

SOLUTION: Part (a): Using the definition of composition, we obtain:

$(f \circ g)(x) = f(g(x)) = f(x^2) = 3x^2 + 5$

$(g \circ f)(x) = g(f(x)) = g(3x + 5) = (3x + 5)^2$ or $9x^2 + 30x + 25$

Notice that $f \circ g \neq g \circ f$. Compositions are usually not commutative.
Part (b): For each $a \in A$, the image $f(a)$ will be some element b of B. By definition of the inverse function, we have $f^{-1}(b) = a$. Hence, $(f^{-1} \circ f)(a) = f^{-1}(f(a)) = f^{-1}(b) = a = i_A(a)$. This proves that $f^{-1} \circ f = i_A$. Since $(f^{-1})^{-1} = f$, the second equality follows from the first.

EXERCISE FOR THE READER 2.5:   (a) Let $\sigma, \tau : \{1, 2, \cdots, 6\} \rightarrow \{1, 2, \cdots, 6\}$ be the permutations specified by:

$$\sigma: \begin{pmatrix} 1 & 2 & 3 & 4 & 5 & 6 \\ 4 & 6 & 2 & 1 & 5 & 3 \end{pmatrix} \text{ and } \tau: \begin{pmatrix} 1 & 2 & 3 & 4 & 5 & 6 \\ 3 & 4 & 5 & 6 & 1 & 2 \end{pmatrix}.$$

Find similar representations for the two compositions: $\sigma \circ \tau$ and $\tau \circ \sigma$.

(b) Let $F : \mathbb{Z} \times \mathbb{Z} \to \mathbb{Z} :: F(n,m) = 3n - 4m^3$, and $g : \mathbb{R} \to \mathbb{R} :: g(x) = 2x - 5$. Find a formula for the composition function $g \circ F : \mathbb{Z} \times \mathbb{Z} \to \mathbb{R}$. Is this function one-to-one?

EXERCISES 2.1:

1. For each relation on the set $A = \{-2, -1, 1, 2\}$ described below do the following: (i) List all ordered pairs in the relation. (ii) Create a table describing the relation. (iii) Draw a diagram describing the relation.

   (a) $R = \{(a,b) \in A^2 \mid a = b\}$.   (b) $R = \{(a,b) \in A^2 \mid 2a = b\}$.

   (c) $R = \{(a,b) \in A^2 \mid a \text{ is a factor of } b\}$.   (d) $R = \{(a,b) \in A^2 \mid a^2 = b^2\}$.

   Suggestion: For items (ii) and (iii) look at the solution to Exercise for the Reader 2.1.

2. Repeat all parts of Exercise 1, but change the set A to $\{1, 2, 3, 4\}$.

3. Draw a graph for each of the following relations on $\mathbb{R}$.

   (a) $R = \{(x,y) \in \mathbb{R}^2 \mid 5x = y\}$.   (b) $R = \{(x,y) \in \mathbb{R}^2 \mid x^2 > y^2\}$.

   (c) $R = \{(x,y) \in \mathbb{R}^2 \mid x^2 > |y|\}$.   (d) $R = \{(x,y) \in \mathbb{R}^2 \mid y \geq \tan x, \ |y| \leq 1,$

   $\qquad\qquad\qquad\qquad\qquad\qquad\qquad\qquad\qquad\qquad \text{and } 0 \leq x \leq \pi/4\}$.

4. Draw a graph for each of the following relations on $\mathbb{R}$.

   (a) $R = \{(x,y) \in \mathbb{R}^2 \mid x^2 + y^2 \leq 4\}$.   (b) $R = \{(x,y) \in \mathbb{R}^2 \mid x < \sin(y)\}$.

   (c) $R = \{(x,y) \in \mathbb{R}^2 \mid |x| \leq |y|\}$.   (d) $R = \{(x,y) \in \mathbb{R}^2 \mid |x| + |y| \leq 1\}$.

5. Consider the following relations on $\mathbb{R}$:

   $$R_1 = \{(a,b) \in \mathbb{R}^2 : a > b\}, \ R_2 = \{(a,b) \in \mathbb{R}^2 : a = b\}, \ R_3 = \{(a,b) \in \mathbb{R}^2 : a < b^2\}.$$

   Describe each of the following relations in set-theoretic notation (as we did the R_i 's above) and/or using a graph:

   (a) $R_1 \cup R_2$.   (b) $R_1 \Delta R_2$.   (c) $\sim (R_1 \cup R_3)$.

   (d) $R_1 \cap R_3$.   (e) $R_3 \sim R_1$.   (f) $R_1 \sim R_3$.

6. Consider the following relations on $\mathbb{R}$:

   $$R_1 = \{(a,b) \in \mathbb{R}^2 : a \leq b\}, \ R_2 = \{(a,b) \in \mathbb{R}^2 : a \geq b\}, \ R_3 = \{(a,b) \in \mathbb{R}^2 : a < |b|\}.$$

   Describe each of the following relations in set-theoretic notation (as we did the R_i 's above) and/or using a graph:

   (a) $R_1 \cup R_2$.   (b) $R_1 \Delta R_2$.   (c) $R_1 \cup R_3$.

   (d) $R_1 \cup R_2 \cup R_3$.   (e) $R_3 \sim R_1$.   (f) $R_1 \sim R_3$.

7. Find the largest possible domain D (as a subset of the indicated set) and the corresponding range of each of the following functions:

   (a) $f : D \to \mathbb{R} :: f(x) = \sqrt{x}, \quad D \subseteq \mathbb{R}.$

   (b) $h : D \to \mathbb{R} :: h(x) = 1/\sqrt{1-x}, \quad D \subseteq \mathbb{R}.$

   (c) $g : D \to \mathbb{Z} :: g((n,m)) = n + 2m, \quad D \subseteq \mathbb{Z}^2.$

   (d) $\pi_1 : D \to S :: \pi_1((s,t)) = s, \quad D \subseteq S \times T.$ (π_1 is called a *projection*; here S and T can be any sets.)

8. Find the largest possible domain D (as a subset of the indicated set) and the corresponding range of each of the following functions:

   (a) $f : D \to \mathbb{R} :: f(x) = 2x, \quad D \subseteq \mathbb{R}.$

   (b) $g : D \to \mathbb{Z} :: g(x) = 2x, \quad D \subseteq \mathbb{Z}.$

   (c) $k : D \to \mathbb{R} :: k(x) = \dfrac{x^2 - 2}{x^2 - 4}, \quad D \subseteq \mathbb{R}.$

   (d) $h : D \to \mathbb{R} :: g((n,m)) = \sqrt{n - 2m}, \quad D \subseteq \mathbb{Z}^2.$

9. For each formula below, explain why it does not represent a function $f : \mathbb{R} \to \mathbb{R}$. Then, if possible, find a *maximal* subset $D \subseteq \mathbb{R}$ for which the formula does represent a function $f : D \to \mathbb{R}$. "Maximal" here means that for any set D' that properly includes D as a subset (i.e., $D \subset D'$), the formula will not represent a function $f : D' \to \mathbb{R}$.

   (a) $f(x) = \sqrt{x+4}.$ (b) $f(x) = \dfrac{x+1}{x-1}.$ (c) $f(x) = \ln(5-x).$ (d) $f(x) = \pm\sqrt{x^2 - 1}.$

10. Repeat the directions of Exercise 9 for each of the following formulas:

   (a) $f(x) = \dfrac{2}{\lfloor x \rfloor}.$ (b) $f(x) = \dfrac{1}{\lceil 3x \rceil}.$ (c) $f(x) = \lceil \pm x \rceil.$ (d) $f(x) = \lfloor \pm 10x \rfloor.$

11. For each of the functions given below do the following: (i) Determine whether it is one-to-one. (ii) Determine whether it is onto. (iii) Find the indicated image and pre-image.

   (a) $f : \mathbb{R} \to \mathbb{R} :: f(x) = e^x; \quad f((-\infty, 0]), \quad f^{-1}((-\infty, 0]).$

   (b) $g : \mathbb{R} \to \mathbb{R}_+ :: g(x) = e^x; \quad g((-\infty, 0]), \quad g^{-1}((0,1]).$

   (c) $F : \mathbb{R} \to \mathbb{R} :: F(x) = -2x + 5; \quad F([-2,2]), \quad F^{-1}((-1,1]).$

   (d) $G : \mathbb{R} \to \mathbb{R} :: G(x) = x^3; \quad G([-2,2]), \quad G^{-1}((-64,-1]).$

   (e) $C : \mathbb{R} \to [-1,1] :: C(x) = \cos(x); \quad C([0,\pi)), \quad C^{-1}(\{1\}).$

   (f) $h : \mathbb{Z} \times \mathbb{Z} \to \mathbb{Z} :: h((n,m)) = n^2 + m^2, \quad h(\{(n,m) \in \mathbb{Z}^2 \mid n = -m\}), \quad h^{-1}(\{0, \pm 1, \pm 2\}).$

   (g) $k : \mathbb{Z} \times \mathbb{Z}_+ \to \mathbb{Q} :: k((n,m)) = n/m, \quad g(\mathbb{Z} \times \{1\}), \quad g^{-1}(\{0, \pm 1, \pm 2\}).$

12. For each of the functions given below do the following: (i) Determine whether it is one-to-one. (ii) Determine whether it is onto. (iii) Find the indicated image and pre-image.

   (a) $f : \mathbb{R} \to \mathbb{R} :: f(x) = x^2 - 4x - 6; \quad f((-\infty, 0]), \quad f^{-1}((-\infty, 0]).$

   (b) $g : (-\infty, 2] \to [-10, \infty) :: g(x) = x^2 - 4x - 6; \quad g((-\infty, 0]), \quad g^{-1}((-1,1]).$

   (c) $F : [-2, \infty) \to \mathbb{R} :: F(x) = -\sqrt{x+2}; \quad F([-2,2]), \quad F^{-1}((-3,3]).$

   (d) $G : \mathbb{R} \to [-1,1] :: G(x) = \cos(x); \quad G([0,\pi)), \quad G^{-1}(\{1\}).$

   (e) $L : (0, \infty) \to \mathbb{R} :: L(x) = \ln(x); \quad L([e, e^3)), \quad L^{-1}(\{1, 2, 3\}).$

   (f) $h : \mathbb{Z} \times \mathbb{Z} \to \mathbb{Z} :: h((n,m)) = n - m, \quad h(\{(n,m) \in \mathbb{Z}^2 \mid n > m\}), \quad h^{-1}(\{0, \pm 1, \pm 2\}).$

(g) $k : \mathbb{Z} \to \mathbb{Z} :: k(n) = n^2 + n$, $h(\mathbb{Z}_+)$, $h^{-1}(\{0, \pm 1, \pm 2\})$.

13. For each of the functions given below do the following: Determine whether it is bijection. For those that are not bijections, is it possible to change the co-domain of the function so as to make it into a bijection? For those that are bijections, or can be made into bijections as indicated, find a formula for the inverse function.

(a) $f : \mathbb{R} \to \mathbb{R} :: f(x) = 3x + 5$.

(b) $g : \mathbb{R} \to \mathbb{R} :: g(x) = e^x$.

(c) $h : \mathbb{Z} \to \mathbb{Z} :: h(x) = 3x + 5$.

(d) $k : \mathbb{R} \to \mathbb{R} :: k(x) = x / (x^2 + 1)$.

14. For each of the functions given below do the following: Determine whether it is bijection. For those that are not bijections, is it possible to change the co-domain of the function so as to make it into a bijection? For those that are bijections, or can be made into bijections as indicated, find a formula for the inverse function.

(a) $f : \mathbb{R} \to \mathbb{R} :: f(x) = 2x + 1$.

(b) $g : \mathbb{R} \to \mathbb{R} :: g(x) = \cos(x)$.

(c) $h : \mathbb{Z} \to \mathbb{Z} :: h(x) = 2x + 1$.

(d) $k : [0, \infty) \to \mathbb{R} :: k(x) = \sqrt{x}$.

15. For each part below, two functions f and g are given. (i) Find a formula for the composition function $g \circ f$. (ii) Find the indicated values.

(a) $f : \mathbb{R} \to \mathbb{R} :: f(x) = \cos(x)$, $g : \mathbb{R} \to \mathbb{R} :: g(x) = 2x - 1$; $(g \circ f)(\pi)$, $(g \circ f)(-4\pi)$.

(b) $f : \mathbb{R} \to \mathbb{R} :: f(x) = x^2$, $g : [0, \infty) \to \mathbb{R} :: g(x) = \sqrt{x}$; $(g \circ f)(3)$, $(g \circ f)(-4)$.

16. For each part below, two functions f and g are given. (i) Find a formula for the composition function $g \circ f$. (ii) Find the indicated values.

(a) $f : \mathbb{R} \to \mathbb{R} :: f(x) = x^2 + x + 1$, $g = f$; $(g \circ f)(3)$, $(g \circ f)(-4)$.

(b) $f : \mathbb{R} \to \mathbb{R} :: f(x) = 2x + 1$, $g : \mathbb{R} \to \mathbb{Z} :: g(x) = \lfloor x \rfloor$; $(g \circ f)(3)$, $(g \circ f)(-\pi)$.

17. Let $\sigma, \tau, \omega : \{1, 2, \cdots, 7\} \to \{1, 2, \cdots, 7\}$ be the permutations specified in tabular forms by:

$$\sigma : \begin{pmatrix} 1 & 2 & 3 & 4 & 5 & 6 & 7 \\ 4 & 6 & 2 & 3 & 5 & 7 & 1 \end{pmatrix}, \quad \tau : \begin{pmatrix} 1 & 2 & 3 & 4 & 5 & 6 & 7 \\ 3 & 4 & 5 & 6 & 7 & 1 & 2 \end{pmatrix}, \quad \text{and} \quad \omega : \begin{pmatrix} 1 & 2 & 3 & 4 & 5 & 6 & 7 \\ 5 & 1 & 3 & 6 & 7 & 4 & 2 \end{pmatrix}.$$

Compute the tabular form representations for the following related permutations:

(a) $\sigma \circ \tau$. (b) $(\sigma \circ \tau) \circ \omega$. (c) $\tau \circ \omega$. (d) $\sigma \circ (\tau \circ \omega)$.

After you complete this exercise, read Exercise 28.

18. Repeat the directions of Exercise 13 for each of the following permutations.

(a) σ^{-1}. (b) τ^{-1}. (c) $(\sigma \circ \tau)^{-1}$. (d) $\tau^{-1} \circ \sigma^{-1}$.

After you complete this exercise, read Exercise 29.

19. For each statement below, decide whether it is true or false. Prove the true statements and provide counterexamples for those that are false. Throughout we assume that $f : A \to B$ and $g : B \to C$ are functions.

(a) If f and g are both one-to-one, then so is $g \circ f$.

(b) If $g \circ f$ is one-to-one then so is f.

(c) If $g \circ f$ is one-to-one then so is g.

20. For each statement below, decide whether it is true or false. Prove the true statements and

provide counterexamples for those that are false. Throughout we assume that $f : A \rightarrow B$ and $g : B \rightarrow C$ are functions.

(a) If f and g are both onto, then so is $g \circ f$.

(b) If $g \circ f$ is onto then so is f.

(c) If $g \circ f$ is onto then so is g.

21. For each statement below, decide whether it is true or false. Prove the true statements and provide counterexamples for those that are false.

(a) $\forall x \in \mathbb{R}, \ \lfloor 2x \rfloor = 2 \lfloor x \rfloor.$ (b) $\forall x \in \mathbb{R}, \ \lfloor -x \rfloor = -\lfloor x \rfloor.$ (c) $\forall x \in \mathbb{R}, \ \lfloor -x \rfloor = -\lceil x \rceil.$

22. For each statement below, decide whether it is true or false. Prove the true statements and provide counterexamples for those that are false.

(a) $\forall x \in \mathbb{R} \sim \mathbb{Z}, \ \lceil x \rceil = \lfloor x + 1 \rfloor.$ (b) $\forall x \in \mathbb{R}, \ \lfloor \sqrt{x} \rfloor = \lfloor \sqrt{\lfloor x \rfloor} \rfloor.$

(c) $\forall x, y \in \mathbb{R}, \ \lceil x + y \rceil \leq \lceil x \rceil + \lceil y \rceil.$ (d) $\forall n \in \mathbb{Z}_+, \ \lceil n/2 \rceil = \lfloor (n-1)/2 \rfloor + 1.$

23. For each of the equations given below, determine the set of all real numbers x for which the equation is valid.

(a) $\lfloor x + 2 \rfloor = \lfloor x \rfloor + 2.$ (b) $\lceil \lfloor x \rfloor \rceil = \lfloor x \rfloor.$ (c) $1/\lceil x \rceil = \lceil 1/x \rceil.$

24. For each of the equations given below, determine the set of all real numbers x for which the equation is valid.

(a) $\lceil x - 1 \rceil = \lceil x \rceil - 1.$ (b) $\lfloor 2x \rfloor = \lfloor x \rfloor + \lfloor x + 1/2 \rfloor.$ (c) $\lceil 2x \rceil = \lceil x \rceil + \lceil x + 1/2 \rceil.$

25. Suppose that $f : A \rightarrow B$ is a function, and that $A_k \subseteq A$ and $B_k \subseteq B$ for any index k.

(a) Prove that if $B_1 \subseteq B_2$, then $f^{-1}(B_1) \subseteq f^{-1}(B_2)$.

(b) Prove that if $A_1 \subseteq A_2$, then $f(A_1) \subseteq f(A_2)$.

(c) Prove that $f^{-1}(\cap_k B_k) = \cap_k f^{-1}(B_k)$.

Note: We have already seen in Exercise for the Reader 2.2 that intersections are not in general preserved under images of functions, so there can be no identity corresponding to that of Part (c) for images.

26. (*Complements are Preserved Under Pre-Images*) Suppose that $f : A \rightarrow B$ is a function, and that $A_k \subseteq A$ and $B_k \subseteq B$ for $k = 1, 2$.

(a) Prove that $f^{-1}(B_1 \sim B_2) = f^{-1}(B_1) \sim f^{-1}(B_2)$.

(b) Construct a counterexample to show that the corresponding identity for images: $f(A_1 \sim A_2) = f^{-1}(A_1) \sim f^{-1}(A_2)$, is false.

(c) Can the false statement in Part (b) be salvaged into a true statement by replacing the equality by an inclusion? Either prove such a result or give a counterexample to show the inclusion can fail.

27. (*Unions are Preserved Under Images and Pre-Images*) Suppose that $f : A \rightarrow B$ is a function, and that $A_k \subseteq A$ and $B_k \subseteq B$ for any index k.

(a) Prove that $f^{-1}(B_1 \cup B_2) = f^{-1}(B_1) \cup f^{-1}(B_2)$.

(b) Prove that $f(A_1 \cup A_2) = f(A_1) \cup f(A_2)$.

(c) Prove that $f^{-1}(\cup_k B_k) = \cup_k f^{-1}(B_k)$.

(d) Prove that $f(\cup_k A_k) = \cup_k f(A_k)$.

28. (*Composition of Functions is Associative*)   Suppose that $f : A \to B$, $g : B \to C$, and $h : C \to D$ are functions.   Prove that $h \circ (g \circ f) = (h \circ g) \circ f$.

29. (*An Inverse Function Criterion and a Formula for Inverses of Compositions*) (a) Suppose that $f : A \to B$, and $F : B \to A$ are functions.   Recall that in Example 2.6, it was shown that if $F = f^{-1}$, then we have $F \circ f = i_A$, and $f \circ F = i_B$, where $i_A : A \to A :: i_A(a) = a$ is the identity function on A and i_B is the identity function on B.   Prove the converse: If $f : A \to B$, and $F : B \to A$ are functions that satisfy $F \circ f = i_A$, and $f \circ F = i_B$, then f is a bijection and $f^{-1} = F$.

    (b) Suppose that $f : A \to B$, and $g : B \to C$ are bijections.   Prove that $(g \circ f)^{-1} = f^{-1} \circ g^{-1}$.   In words: "the inverse of a composition is the composition of the inverses in the reverse order."

30. Sketch graphs of the following functions $\mathbb{R} \to \mathbb{R}$. :

       (a) $y = \lceil x \rceil$.   (b) $y = \lfloor x \rfloor$.   (c) $y = \lceil x \rceil - \lfloor x \rfloor$.   (d) $y = 2\lceil x \rceil + \lfloor x \rfloor$.

31. Suppose that $a < b$ are real numbers.

    (a) Prove that the interval $[a, b]$ contains exactly $\lfloor b \rfloor - \lceil a \rceil + 1$ integers.

    (b) Prove that the interval $[a, b)$ contains exactly $\lceil b \rceil - \lceil a \rceil$ integers.

    (c) Prove that the interval $(a, b]$ contains exactly $\lfloor b \rfloor - \lfloor a \rfloor$ integers.

    (d) Prove that the interval $(a, b]$ contains exactly $\lceil b \rceil - \lfloor a \rfloor - 1$ integers.

32. (*Cycles and Orbits of a Permutation*) Suppose that we have a permutation of the set $\{1, 2, \cdots, n\}$: $\sigma : \{1, 2, \cdots, n\} \to \{1, 2, \cdots, n\}$.   We have introduced the tabular form representation of σ:

$$\sigma : \begin{pmatrix} 1 & 2 & \cdots & n-1 & n \\ \sigma(1) & \sigma(2) & \cdots & \sigma(n-1) & \sigma(n) \end{pmatrix}.$$

In this exercise it will be convenient to use the power notation for compositions of σ with itself.   For a positive integer k, we define σ^k to be $\underbrace{\sigma \circ \sigma \circ \cdots \circ \sigma}_{k \text{ times}}$.   Thus, $\sigma^1 = \sigma$, $\sigma^2 = \sigma \circ \sigma$, $\sigma^3 = \sigma \circ \sigma \circ \sigma$, etc.   Since composition of functions is associative (see Exercise 28), the order in which the compositions are done has no effect on the resulting permutation.   Since compositions of bijections are bijections, each σ^k is a permutation of the set $\{1, 2, \cdots, n\}$.

(a) For any element i of the set $\{1, 2, \cdots, n\}$, the list of elements: $i, \sigma(i), \sigma^2(i), \sigma^3(i), \cdots$ must eventually have a duplication (since the elements all lie in the same finite set $\{1, 2, \cdots, n\}$).   Let $\sigma^k(i)$ be the first such duplicated element (thus $i, \sigma(i), \sigma^2(i), \sigma^3(i), \cdots, \sigma^{k-1}(i)$ are all distinct).   Show that $\sigma^k(i) = i$.   The exponent k is called the *order* of i with respect to the permutation σ.   The set of the k corresponding distinct elements $\{i, \sigma(i), \sigma^2(i), \sigma^3(i), \cdots, \sigma^{k-1}(i)\}$ is called the *orbit* of i with respect to the permutation σ.   The action of the permutation σ on this orbit is represented by the following mapping diagram:

$$i \to \sigma(i) \to \sigma^2(i) \to \sigma^3(i) \to \cdots \to \sigma^{k-1}(i) \to i.$$

In general, given any ordered list $(a_1 \quad a_2 \quad \cdots \quad a_{k-1} \quad a_k)$ of distinct elements in $\{1, 2, \cdots, n\}$, a mapping τ on $\{1, 2, \cdots, n\}$ is defined by using the induced mapping diagram

$$a_1 \to a_2 \to a_3 \to \cdots \to a_{k-1} \to a_k \to a_1.$$

to define τ on elements of the set $\{a_1, a_2, \cdots, a_{k-1}, a_k\}$, and taking τ to be the identity mapping when applied to all other elements of $\{1, 2, \cdots, n\}$. Such a permutation τ is called a **cycle**, and is written as the ordered list used to define it, i.e., $\tau = (a_1, a_2, \cdots, a_{k-1}, a_k)$. Thus, each orbit of a permutation gives rise to an associated cycle. Show also that if $\tau = (a_1, a_2, \cdots, a_{k-1}, a_k)$ is a cycle of length k, then $\tau^k = identity$, but $\tau^j \neq identity$ for any smaller positive integer j. Because of this latter property, a cycle of length k is also said to have order k.

(b) Explain why a cycle of a permutation is itself a permutation on $\{1, 2, \cdots, n\}$. If two elements i and j belong to the same orbit of σ, then their corresponding cycles represent the same mapping diagram (with wraparound notation) so their cycles are considered the same. Thus, the cycles (1 5 2 4) and (2 4 1 5) are considered identical.

(c) A *fixed point* of a permutation σ is an element $i \in \{1, 2, \cdots, n\}$ such that $\sigma(i) = i$. Thus the orbit of a fixed point i is the singleton $\{i\}$ and the corresponding cycle is (i).   If $\tau = (a_1, a_2, \cdots, a_{k-1}, a_k)$ is a cycle associated (to an orbit of) a permutation σ, then show that the permutation $\sigma \circ \tau^{-1}$ fixes all elements in $\{a_1, a_2, \cdots, a_{k-1}, a_k\}$.

(d) (*Cycle Decomposition of a Permutation*) Show that any permutation σ of $\{1, 2, \cdots, n\}$ can be uniquely represented as a composition of its cycles. Since the cycles act on disjoint orbits, the order in which they are written has no effect. In this representation, called the *cycle decomposition* of σ, order one cycles (if any) are often omitted since they correspond to fixed points and have no effect.   For example, the cycle decomposition of the permutation $\sigma : \begin{pmatrix} 1 & 2 & 3 & 4 & 5 & 6 \\ 4 & 6 & 2 & 1 & 5 & 3 \end{pmatrix}$ is (1 4)(2 6 3).

Suggestion: For Part (a), assume (for a contradiction) that $\sigma^k(i) = \sigma^j(i)$, where $1 \leq j < k$. Apply σ^{-1} to this equation j times to produce $\sigma^{k-j}(i) = i$, a contradiction.

33. (*Cycles and Orbits of a Permutation, continued*) (a) Establish the following simple formula for inverses of cycles: $(a_1 \quad a_2 \quad \cdots \quad a_{k-1} \quad a_k)^{-1} = (a_k \quad a_{k-1} \quad \cdots \quad a_2 \quad a_1)$.

(b) Obtain the cycle decomposition for the permutations $\sigma, \tau, \omega : \{1, 2, \cdots, 7\} \rightarrow \{1, 2, \cdots, 7\}$ of Exercise 17, as well as for their inverses.

2.2: EQUIVALENCE RELATIONS AND PARTIAL ORDERINGS

We now return to the setting of general relations. The next definition describes a very important category of relations on a given set.

Equivalence Relations

DEFINITION 2.7: A relation R on a set A is called an **equivalence relation** if it satisfies the following three properties:

(i) (*Reflexivity*) $\forall a \in A \quad a R a$.

(ii) (*Symmetry*) $\forall a,b \in A$ $\quad a\,R\,b \Rightarrow b\,R\,a$.

(iii) (*Transitivity*) $\forall a,b,c \in A$ $\quad [a\,R\,b$ and $b\,R\,c] \Rightarrow a\,R\,c$.

In the context of equivalence relations, alternative notations such as $a \sim_R b$, $a \sim b$, $a \equiv b$, $a \approx b$, etc., are often preferred over $a\,R\,b$.

EXAMPLE 2.7: Check each of the three axioms in Definition 2.7 for each relation below:

(a) The equality relation "$x = y$" on the set of real numbers $\mathbb{R}$.

(b) The relation "$x \geq y$" on $\mathbb{R}$.

(c) The relation $a \equiv b \,(\mathrm{mod}\ 12)$ on the integers $\mathbb{Z}$, defined by $a \equiv b \,(\mathrm{mod}\ 12)$ if, and only if 12 is a factor of the difference $a - b$.

(d) The relation $S \times S$ on an arbitrary set S.

(e) The relation on the set A of all adults in the world defined by $a \sim b$ if, and only if a loves b.

SOLUTION: Part (a): The equality relation on <u>any</u> set is easily seen to satisfy all three axioms of an equivalence relation. It is the smallest equivalence relation on any set (because of the reflexivity axiom).

Part (b): Let x, y, and z be real numbers. Certainly $x \geq x$, so we have reflexivity. Note, however, that $2 \geq 1$ is true but $1 \geq 2$ is false, so this counterexample shows the relation is not symmetric. If $x \geq y$ and $y \geq z$ then we have $x \geq z$, so we have transitivity. Thus the greater than or equal to relation fails to be an equivalence relation on $\mathbb{R}$, since it is not symmetric.

Part (c): Let a, b, and c be integers. Since 12 is a factor of $0 = a - a$ (i.e., $0 = 12 \cdot 0$), we conclude that $a \equiv a \,(\mathrm{mod}\ 12)$ and the relation is reflexive. Symmetry is likewise easy to verify: if $a \equiv b \,(\mathrm{mod}\ 12)$, then 12 is a factor of $a - b$, but then 12 will also be a factor of $-(a-b) = b-a$, so $b \equiv a \,(\mathrm{mod}\ 12)$. To show transitivity, we assume $a \equiv b \,(\mathrm{mod}\ 12)$, and $b \equiv c \,(\mathrm{mod}\ 12)$. This means that 12 is a factor of both $a - b$, and $b - c$. It follows that we can write $a-b = 12e$, and $b-c = 12f$, for some integers e, and f. Adding these two equations produces $a-b+b-c = 12e+12f$, or $a-c = 12(e+f)$, which shows 12 is a factor of $a - c$, hence $a \equiv c \,(\mathrm{mod}\ 12)$. We have proved that this relation, called **congruence modulo** 12, is an equivalence relation on the set of integers.

Part (d): The relation $S \times S$ contains all possible ordered pairs of elements taken from the set S and thus will trivially satisfy all three of the axioms for an equivalence relation. This relation is the largest equivalence relation on a given set S.

Part (e): We leave it to the reader to explain why this loving relation can fail to satisfy each of the axioms for an equivalence relation. For example, transitivity can fail as follows: a man (a) and a woman (b) love one another, the woman and her father (c) love one another, but the father does not at all respect (or love) the man (*a*) and forbids marriage (so $c \sim b$, and $b \sim a$, but $c \not\sim a$).   This is one possible scenario for why couples may need to elope to get married.

EXERCISE FOR THE READER 2.6:   Suppose that R_1 and R_2 are both equivalence relations on a set *S*.
(a) Will the intersection $R_1 \cap R_2$ also be an equivalence relation on *S*?
(b) Will the union $R_1 \cup R_2$ also be an equivalence relation on *S*?

In Part (c) of Example 2.7, the number 12 was not very special. Indeed, we could replace 12 with any positive integer $m > 1$, and the above proof would easily translate to show that the resulting relation is an equivalence relation (Exercise 17).   We enunciate these important examples in the next definition.

Congruence Modulo a Positive Integer

DEFINITION 2.8:   Let $m > 1$ be a positive integer.   Then the relation of **congruence modulo** *m* on the integers $\mathbb{Z}$ is defined by

$$a \equiv b \,(\text{mod } m) \iff m \text{ is a factor of } b - a.$$

As pointed out above, this is an equivalence relation on $\mathbb{Z}$.

Equivalence Classes and Their Representatives

Any equivalence relation on a set *S* naturally breaks the set up into pairwise disjoint subsets called *equivalence classes*. We define these precisely as follows:

DEFINITION 2.9: Let *R* be an equivalence relation on a set *S*.   For any element $a \in S$, the **equivalence class** of *a* is denoted $[a]_R$ or just $[a]$ if the equivalence relation *R* is clear from the context, and consists of all elements of *S* that are related to *a* under the equivalence relation *R*.   In symbols:

$$[a]_R = [a] = \{s \in S \mid s \sim_R a\}.$$

Any element $b \in [a]$ is called a **representative** for the equivalence class $[a]$.

EXAMPLE 2.8: (a) What are the equivalence classes of the congruence modulo 12 equivalence relation (Example 2.7(c) and Definition 2.8)?
(b) What does it mean for two integers *a, b* to be congruent mod 2? What are the equivalence classes mod 2?

SOLUTION: Part (a): Any equivalence class $[a]$ under this equivalence relation consists of all integers *k* such that 12 is a factor of $a - k$, i.e., $a - k = 12n$, or

$k = a - 12n$ for some integer n. Thus (replacing n with $-n$) the equivalence class $[a]$ is simply the set of all integers expressible as a added to some (integer) multiple of 12: $[a] = \{a + 12n : n \in \mathbb{Z}\}$. Any two of the following 12 equivalence classes below are disjoint:

$$[0] = \{\cdots, -24, -12, 0, 12, 24, \cdots\},$$
$$[1] = \{\cdots, -23, -11, 1, 13, 25, \cdots\},$$
$$[2] = \{\cdots, -22, -10, 2, 14, 26, \cdots\},$$
$$\cdots$$
$$[11] = \{\cdots, -13, -1, 11, 23, 35, \cdots\},$$

and they clearly comprise all of the equivalence classes. This can be seen in two ways: First, starting with the middle column (with zero on top) of the array of numbers on the right, as we go down and then move down successive columns, we obtain the list of all natural numbers. If we move to the left (going up successive columns) we similarly obtain a listing of all negative integers. The fact that these equivalence classes cover all of the integers also follows from the fact that any integer a may be expressed as $a = 12n + r$, where $r \in \{0, 1, \cdots, 11\}$. In this representation of a, n is called the *quotient*, and r is called the *remainder* in the division of a by 12. The algorithm for performing such an integer division is usually taught in grade school (at least for positive integers) and will be presented in the next chapter. From $a = 12n + r$, it follows that $[a] = [r]$.

Part (b): The relation $a \equiv b \pmod 2$ means (by Definition 2.8) that 2 is a factor of $a - b$, i.e., that we may write $a - b = 2k$, or $a = b + 2k$, for some integer k. In words: a equals b plus an even number. This simply means that a and b have the same parity: i.e., they are either both even integers, or both odd integers. Since the argument is clearly reversible, we may conclude that there are only two equivalence classes mod 2: $[1] = \{\text{odd integers}\}$, and $[2] = \{\text{even integers}\}$.

The results of the above example are nicely generalized into the following theorem and proposition.

THEOREM 2.1: If R is an equivalence relation on a set S, then any two equivalence classes $[a]$ and $[b]$ are either equal (as subsets of S), or disjoint. Also, the union of all of the equivalence classes equals S.

Proof: Let $a, b \in S$.

Case 1: $a \sim b$. In this case we claim that $[a] = [b]$. To prove this, let $c \in [a]$. This means (by Definition 2.9) $c \sim a$. But since (by the Case 1 assumption) $a \sim b$, transitivity allows us to conclude that $c \sim b$, and (again by Definition 2.9) this gives us that $c \in [b]$. This proves that $[a] \subseteq [b]$. A similar argument (Exercise 18) shows that $[b] \subseteq [a]$, and therefore $[a] = [b]$.

Case 2: $a \not\sim b$. Here we will show that $[a] \cap [b] = \varnothing$. Indeed, suppose there were a common element in this intersection: $c \in [a] \cap [b]$.   Then by the definitions of membership in each of these equivalence classes, we would have $c \sim a$, and $c \sim b$.   Applying symmetry to the first of these relations produces $a \sim c$. But from the latter two relations, transitivity would then imply that $a \sim b$ —a contradiction to the Case 2 assumption. We thus have proved that $[a] \cap [b] = \varnothing$. The first statement of the theorem is completely proved.   As for the second statement, since each equivalence class is a subset of S, we need only show that every element $a \in S$ is covered by some equivalence class.   But clearly any element is in its own equivalence class: $a \in [a]$. $\square$

For a simple example, the relation of the set S of all automobiles defined by $a \sim b$ if, and only if (car) a has the same number of (engine) cylinders as (car) b, is easily seen to be an equivalence relation.   If a is a certain four cylinder Toyota sedan, then $[a]$ consists of all four cylinder cars.   A more important example of Theorem 2.1 is furnished by congruence modulo m relations on the integers.

The proof of the following proposition can be modeled after the solution to Examples 2.7(c) and 2.9, and is left as the next exercise for the reader.

PROPOSITION 2.2: If $m > 1$ is a positive integer, then the relation on $\mathbb{Z}$ of congruence modulo m, i.e., $a \equiv b \pmod{m}$, if, and only if m is a factor of $a - b$, is an equivalence relation.   There are precisely m equivalence classes: $[0], [1], [2], \cdots, [m-1]$.

EXERCISE FOR THE READER 2.7: Prove Proposition 2.2.

Strings

Discrete mathematical structures often involve the concept of strings, which basically are juxtapositions of characters belonging to some alphabet.

DEFINITION 2.10: An **alphabet** $\mathscr{A}$ is any set of formal symbols. A **string** s in the alphabet $\mathscr{A}$ is any formal word (ordered finite list) made up entirely of symbols from $\mathscr{A}$. The **length** of a string is the number of symbols it has (counting repetitions).

For example, *binary* (or *bit*) *strings*, are strings from the alphabet $\{0,1\}$, or finite words made up of 0's and 1's.   Thus, 0011111 and 101010001 are bit strings length 7 and 9, respectively, and JQALMZZT is a length 8 string from the alphabet $\{A, B, \cdots, Z\}$.

EXERCISE FOR THE READER 2.8: Let S denote the set of all bit strings of length 5. Define a relation on S, by $s \sim s' \Leftrightarrow$ the total number of 1's in s and s' are the same. Show that this is an equivalence relation on S. Write down all elements of the equivalence class [01101].

Any function naturally induces an equivalence relation on its domain. This useful construction is described in the following proposition.

PROPOSITION 2.3: Suppose that $f : A \to B$ is a function. Use f to define a relation on the domain A of f as follows: $a \sim a' \Leftrightarrow f(a) = f(a')$. This defines an equivalence relation on A, the equivalence classes of which are the so-called *level sets* of the function: $f^{-1}(\{b\})$, for any b in the range of f.

The proof of this result is left as the next exercise for the reader. The result can be used to easily identify many equivalence relations that are defined analytically. For example, the relation on $\mathbb{R}^2$ defined by $(x,y) \sim (x',y')$ $\Leftrightarrow$ $x^2 - y^2 = (x')^2 - (y')^2$ is an equivalence relation by Proposition 2.3 with $f : \mathbb{R}^2 \to \mathbb{R}$ defined by $f(x,y) = x^2 - y^2$. The equivalence classes are the planar sets $x^2 - y^2 = c$, for each (constant) real number c. These sets are hyperbolas, except when $c = 0$, when the set is a pair of intersecting lines.

EXERCISE FOR THE READER 2.9: Prove Proposition 2.3.

Partial Order(ings)

We now move on to a discussion of another useful type of relation, that of a partial ordering.

DEFINITION 2.11: A relation R on a set A is called a **partial order(ing)** if it satisfies the following three properties:
(i) *(Reflexivity)* $\forall a \in A \;\; a R a$.
(ii) *(Antisymmetry)* $\forall a,b \in A \;\; [a R b \text{ and } b R a] \Rightarrow a = b$.
(iii) *(Transitivity)* $\forall a,b,c \in A \;\; [a R b \text{ and } b R c] \Rightarrow a R c$.
In the context of partial orderings, the notation $a \preceq b$ is often preferred over $a R b$. A set A along with a partial ordering $\preceq$ is called a **partially ordered set**, or a **poset**.

The three axioms for a partial ordering are similar to those for an equivalence relation: both are reflexive and transitive, but equivalence relations are symmetric, whereas partial orderings are antisymmetric. This seemingly small difference turns out to be quite major, as will become evident when we contrast the examples below of posets with some of the previous examples of equivalence relations.

EXAMPLE 2.9: Show that each of the following is a poset:
(a) The set of real numbers $\mathbb{R}$ with the "less than or equal to" relation: $x \leq y$.

(b) For any set A, the power set $\mathscr{P}(A) = 2^A$, with the subset relation: $S \preceq T \Leftrightarrow S \subseteq T$.

SOLUTION: Part (a):   Let x, y, and $z \in \mathbb{R}$.   Certainly $x \leq x$, so we have reflexivity.   Also, if both $x \leq y$ and $y \leq x$, then we must have $x = y$, so the relation is antisymmetric.   Finally, if both $x \leq y$ and $y \leq z$, then $x \leq z$, so the relation is transitive, and we have established that we have a poset.

Part (b):   Let S, T, and $W \in \mathscr{P}(A)$ (i.e., S, T, and W are subsets of A).   Since $S \subseteq S$, we have reflexivity.   Also, since $S \subseteq T$, and $T \subseteq S$ are equivalent to the set equality $S = T$, we have antisymmetry.   Finally, if $S \subseteq T$, and $T \subseteq W$ then we must have (by definition of subsets) that $S \subseteq W$, so we have transitivity.   Thus we have shown that the subset relation is a partial ordering.

The natural ordering $x \leq y$ (from Part (a) of the preceding example) on $\mathbb{R}$ has the additional property that any two real numbers x and y are *comparable*:   either $x \leq y$ or $y \leq x$.   Such partial orderings are given a special name:

DEFINITION 2.12: A partial ordering on a set A under which any two elements are comparable is called a **total** (or **linear**) **ordering**.

For contrast, we point out that the partial ordering of Part (b) of Example 2.9 is not a total ordering.

EXERCISE FOR THE READER 2.10: (*Lexicographic Order*)   Suppose that we have two posets: $(A, \preceq_A)$ and $(B, \preceq_B)$.   Define a relation on the Cartesian product $C = A \times B$, as follows:

$(a, b) \preceq (a', b') \Leftrightarrow$ Either (i) $a \neq a'$ and $a \preceq_A a'$ or (ii) $a = a'$ and $b \preceq_B b'$.

This relation is called the *lexicographic order* (or the *dictionary order*) on $C = A \times B$.   Show that this is a partial order.

Hasse Diagrams

It is often convenient to display a partial ordering on a finite set by means of a so-called *Hasse*[4] *diagram*.

DEFINITION 2.13: A **Hasse diagram** for a finite poset $(A, \preceq)$ is a two-dimensional diagram consisting of the elements of A (represented by points) along with some line segments connecting pairs of points $a \preceq a'$, drawn in such a way that the following conditions are met: (i) If $a \preceq a'$, then a is drawn below a'. (ii) A line segment is drawn between distinct pairs of points $a \preceq a'$, only when there is not a third distinct point a'' satisfying $a \preceq a''$, and $a'' \preceq a'$.

FIGURE 2.8: Helmut Hasse (1898–1979), German mathematician

The requirements for Hasse diagrams provide flexibility, but the order relations and structure of a poset are often easily read from a nicely constructed Hasse diagram. The purpose of condition (ii) is to eliminate redundant order relations that can be easily obtained using transitivity. Conversely, any Hasse diagram contructed using the conditions stipulated in the above definition gives rise to a poset. The ordering on the set elements is obtained first by making the poset reflexive, (i.e., including all relations of the form $a \preceq a$), and then by using the Hasse diagram and transitivity to include all relations $a \preceq a'$, where the element a lies below the element a' and is connected to it by a sequence of line segments of the Hasse diagram.

EXAMPLE 2.10: Draw Hasse diagrams for the following two posets:
(a) The subset partial order on $\mathscr{P}(\{1,2,3\})$: $S \preceq T \Leftrightarrow S \subseteq T$. (See Example 2.9(b).)

---

[4] Helmut Hasse was a German mathematician who grew up in the city of Kassel. He moved to Berlin during his high school ("gymnasium" in Germany) years. From high school, Hasse joined the German Navy, and was stationed in Kiel, where he attended some lectures of the renowned mathematician Otto Toeplitz. After a short time of service in the Navy, Hasse studied higher mathematics at the University of Göttingen (in the most prestigious mathematics department in Germany at the time), where, in 1921, he wrote an important doctoral dissertation in the area of abstract algebra. He subsequently obtained appointments in the cities of Kiel, Halle, and Marburg, making major contributions to algebra and number theory. In 1933, when the Nazis came to power, the effects on academia in Germany were drastic. At the University of Göttingen, 18 mathematicians left or were summarily dismissed for being Jewish. Hasse accepted a position there, and throughout the Nazi years, he maintained relationships with many respected Jewish former colleagues. He worked hard to keep the Nazis from interfering with academic research. Yet, he also publicly supported many of Hitler's nationalistic policies. After the defeat of the Nazis in 1945, Hasse was forced by the occupation forces to give up his professorship in Göttingen. He then found a suitable position in Berlin, and in 1950 accepted an appointment at the University of Hamburg, where he remained until his retirement in 1966.

(b) The lexicographic order on $\{1,2\} \times \{1,2,3\}$, where each of the two factor sets is given the less than or equal to ordering. (See Example 2.9(a)[5]).

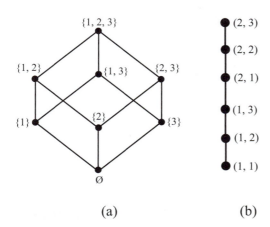

(a)   (b)

SOLUTION: Corresponding Hasse diagrams are shown in Figure 2.9. The latter Hasse diagram is typical for linear orders.

FIGURE 2.9: Hasse diagrams: (a) (left) for the subset partial order of Example 2.10(a), and (b) (right) for the lexicographic order of Example 2.10(b).

Poset Isomorphisms

Bijections that map one poset to another and that preserve the underlying order structure of the posets lead to a natural equivalence relation among posets: Two posets are considered to be the same if the elements in the underlying set of one can be relabeled to correspond to those of the second set in such a way that the resulting ordering of the first poset coincides with that of the latter. The concepts are made precise in the following definition.

DEFINITION 2.14: A function f from a poset $(P, \preceq_P)$ to a poset $(Q, \preceq_Q)$ is said to be **order-preserving** if: $x \preceq_P x' \implies f(x) \preceq_Q f(x')$ for all $x, x' \in P$. In case f is also a bijection and the inverse function is also order-preserving, then f is called a **poset isomorphism**, and the posets $(P, \preceq_P)$ and $(Q, \preceq_Q)$ are said to be **isomorphic**.

EXERCISE FOR THE READER 2.11: (*Isomorphism Classes of Posets*) Show that the relation $(P, \preceq_P) \sim (Q, \preceq_Q)$ if, and only if $(P, \preceq_P)$ and $(Q, \preceq_Q)$ are isomorphic is an equivalence relation on the set of all posets. The resulting equivalence classes are known as *isomorphism classes* of posets.

---

[5] Although the partial ordering axioms are easily verified here, the fact that these two sets are posets directly follows from the useful fact that if $(A, \preceq)$ is any poset and $B \subseteq A$, then the induced relation on B defined by $b \preceq_B b' \Leftrightarrow b \preceq b'$ (i.e., simply restrict the partial order of A to pairs of elements of B) makes $(B, \preceq)$ into a poset. The verification of this fact is routine and is left as an exercise at the end of this section.

In particular, finite isomorphic posets must have underlying sets of the same size, but the preservation of order says much more. In terms of Hasse diagrams, two posets are isomorphic if, and only if the same Hasse diagram can be used for both posets, after only a relabeling of the set elements. Thus, Hasse diagrams can be helpful in comparing posets or finding different poset isomorphism classes.

EXAMPLE 2.11: (a) How many different isomorphism classes are there for posets containing two elements? (b) How about for posets with three elements?

SOLUTION: We will use Hasse diagrams to not only count, but also to display all equivalence classes of posets. This scheme is most straightforward if we separate Hasse diagrams according to the number of levels.

Part (a): With only two elements, there can be at most two levels. With one level, there is always only the isomorphism class that only includes the minimal comparison relations required by reflexivity. The corresponding Hasse diagram is shown on the left in Figure 2.10. With two levels and only two elements,

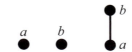

FIGURE 2.10: Hasse diagrams for the two isomorphism classes for two-element posets.

the only Hasse diagram corresponds to the linear order shown on the right in Figure 2.10.

Part (b): For three-element sets, we can similarly run through the Hasse diagrams by doing them one level at a time. As always, there will only be single Hasse diagrams with one level and with (the maximum number) three levels. The corresponding Hasse diagrams, as well as those for the three possible isomorphism classes with two levels, are shown in Figure 2.11.

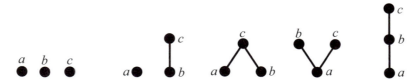

FIGURE 2.11: Hasse diagrams for the five isomorphism classes of three-element posets.

We close this section with an exercise for the reader that will introduce an interesting partial order corresponding to divisibility of integers.

EXERCISE FOR THE READER 2.12 (*The Divisibility Relation*): Let $A \subset \mathbb{Z}_+$. Define a relation on A, by $a \preceq b \iff a$ is a factor of b, i.e., $b = ka$ for some integer k (not necessarily belonging to A). In number theory (and in Chapter 3) this condition is referred to by saying a *divides* b (or a *divides into* b), and is written as $a \mid b$. Show that this *divisibility relation* makes A into a poset. Draw a Hasse diagram for this relation in case A is the set of all nonnegative divisors of 60: $A = \{1, 2, 3, 4, 5, 6, 10, 12, 20, 30, 60\}$.

EXERCISES 2.2:

In Exercises 1–4, for each part, determine whether the given relation is an equivalence relation. For those that are not, which of the axioms (i) reflexivity, (ii) symmetry, and/or (iii) transitivity fail?

1. (a) The relation $a R b \Leftrightarrow a - b \le 0$ on $\mathbb{R}$.
   (b) The relation $a R b \Leftrightarrow a - b < 0$ on $\mathbb{R}$.
   (c) The relation $a R b \Leftrightarrow a$ and b have the same weight (rounded to the nearest pound) on the set of all people.
   (d) The relation $a R b \Leftrightarrow a$ and b have a common friend on the set of all people.

2. (a) The relation $a R b \Leftrightarrow a \ne b$ on $\mathbb{R}$.
   (b) The relation $a R b \Leftrightarrow |a| = |b|$ on $\mathbb{R}$.
   (c) The relation $a R b \Leftrightarrow a$ and b live within five miles of one another on the set of all residents of New York.
   (d) The relation $a R b \Leftrightarrow a$ and b have the same number of siblings on the set of all people.

3. (a) The relation $a R b \Leftrightarrow ab > 0$ on $\mathbb{R} \sim \{0\}$.
   (b) The relation $a R b \Leftrightarrow ab \le 0$ on $\mathbb{R}$.
   (c) The relation $a R b \Leftrightarrow a$ and b have a common ancestor on the set of all people.
   (d) The relation $a R b \Leftrightarrow a$ and b attended the same high school on the set of all US citizens with high school diplomas.

4. (a) The relation $a R b \Leftrightarrow |a| - |b|$ is even on $\mathbb{Z}$.
   (b) The relation $a R b \Leftrightarrow ab$ is the square of an integer on $\mathbb{Z}$.
   (c) The relation $a R b \Leftrightarrow a$ and b have a common parent on the set of all people.
   (d) The relation $a R b \Leftrightarrow a$ and b share a common boundary on the set of all of the 48 contiguous United States.

In Exercises 5–6, for each part, a relation on a set is given, along with an element x of the set. Show that the relation is an equivalence relation, and then describe the equivalence class $[x]$.

5. (a) The relation $a \sim b \Leftrightarrow \sin(a) = \sin(b)$ on $\mathbb{R}$; $x = \pi$.
   (b) The relation $a \sim b \Leftrightarrow a - b$ is even on $\mathbb{Z}$; $x = 11$.
   (c) The relation $s \sim s' \Leftrightarrow$ the number of 1's of $s \equiv$ the number of 1's of s' (mod 2) on the set of all length four bit strings; $x = 1011$.
   (d) The relation $a R b \Leftrightarrow$ the zip codes of a and b are the same on the set of all physical street addresses in the United States; $x =$ 1600 Pennsylvania Ave. NW, Washington, DC 20500.

6. (a) The relation $a \sim b \Leftrightarrow \lceil a \rceil = \lceil b \rceil$ on $\mathbb{R}$; $x = \pi$.
   (b) The relation $R = \{(1,1), (2,2), (3,3), (4,4), (1,2), (2,1)\}$ on $\{1,2,3,4\}$; $x = 2$.
   (c) The relation $S \sim T \Leftrightarrow S \cap \{1,2,3\} = T \cap \{1,2,3\}$ on $\mathscr{P}(\{1,2,3,4,5,6\})$; $x = \{2,3,4\}$.
   (d) The relation $a R b \Leftrightarrow$ the names of a and b begin with the same letter on the set of all of the 48 contiguous United States; $x =$ New Hampshire.

7. (a) Show that the relation $R = \{(n,m) \in \mathbb{Z}^2 : n^2 = m^2\}$ on $\mathbb{Z}$ is an equivalence relation by

verifying each of the axioms, and determine the equivalence classes.
(b) Use Proposition 2.3 to show that the relation of Part (a) is an equivalence relation.

8. Show that the relation on the Cartesian plane defined by $(x, y) \sim (x', y') \Leftrightarrow x = x'$ is an equivalence relation. What are the equivalence classes?

9. For each relation below, determine whether it is (i) reflexive, (ii) symmetric, and (iii) transitive. For those that satisfy all three axioms (and hence are equivalence relations), describe the equivalence classes.

   (a) The relation $R = \{(n, m) \in \mathbb{Z}^2 : |n - m| \le 1\}$ on $\mathbb{Z}$.

   (b) The relation $xRy \Leftrightarrow xy \ge y/x$, on the positive real numbers $\mathbb{R}_+$.

   (c) The relation $R = \{[(n, m), (n', m')] \in \mathbb{Z}^2 \times \mathbb{Z}^2 : |n| + |m| = |n'| + |m'|\}$ on $\mathbb{Z}^2$ (points in the plane with both coordinates being integers).

10. For each relation below, determine whether it is (i) reflexive, (ii) symmetric, and (iii) transitive. For those that satisfy all three axioms (and hence are equivalence relations), describe the equivalence classes.

   (a) The relation $R = \{(n, m) \in \mathbb{Z}^2 : n - m \text{ is odd}\}$ on $\mathbb{Z}$.

   (b) The relation $R = \{a, b\} \times \{a, b\} \cup \{(c, c)\} \cup \{d, f\} \times \{d, f\} \cup \{(e, e)\}$ on the set $\{a, b, c, d, e, f\}$; $x = f$.

   (c) The relation $R = \{(n, m) \in \mathbb{Z}^2 : 12 \text{ is a factor of } n^2 - m^2\}$ on $\mathbb{Z}$.

11. Let $S = \{1, 2, 3\}$.
   (a) Is there a relation on S that is reflexive and symmetric, but not transitive?
   (b) Is there a relation on S that is reflexive and transitive, but not symmetric?
   (c) Is there a relation on S that is symmetric and transitive, but not reflexive?

12. Repeat each part of Exercise 11 with S changed to $\{1, 2\}$.

13. Define a relation $\sim$ on $\mathbb{Z}^2$ by $(n, m) \sim (n', m') \Leftrightarrow$ Either (i) $m = m' = 0$, or (ii) $n/m = n'/m'$ (and both m and m' are nonzero). Show that this is an equivalence relation, and describe the equivalence classes.

14. Let $\mathscr{S}$ be the set of all nonempty subsets of $\mathbb{Z}$. Thus, $\mathscr{S} = \{A \subseteq \mathbb{Z} : A \ne \varnothing\}$. Define a relation R on $\mathscr{S}$ by $(A, B) \in R \Leftrightarrow A \cap B \ne \varnothing$. In words, two nonempty subsets of integers are related if, and only if, they have nonempty intersection. Determine whether this relation is (i) reflexive, (ii) symmetric, and (iii) transitive.

15. (*Using a Partition to Define an Equivalence Relation*) Let $\mathscr{P}$ be any partition of a set A into nonempty disjoint subsets. This means that $\mathscr{P}$ consists of nonempty subsets of A such that (i) if $S, T \in \mathscr{P}$ with $S \ne T$, then $S \cap T = \varnothing$, and also (ii) $\bigcup_{S \in \mathscr{P}} S = A$.

   (a) Show that the relation $\sim$ (or $\sim_{\mathscr{P}}$) on A defined by $a \sim a' \Leftrightarrow$ both a and a' lie in the same subset of $\mathscr{P}$, is an equivalence relation by directly verifying the three axioms.
   (b) Show that the equivalence classes consist precisely of the elements of the partition $\mathscr{P}$.
   (c) Use Proposition 2.3 to give another verification of the assertion of Part (a).
   (d) Show that any equivalence relation on A is obtainable by the method of Part (a).

16. (a) How many different equivalence relations are there on the set $\{1, 2\}$?
   (b) How many different equivalence relations are there on the set $\{1, 2, 3, 4\}$?
   Suggestion: Use the results of the preceding exercise.

17. Let $m > 1$ be an integer.   Show that the relation of *congruence modulo m*, defined by:

$$a \equiv b \pmod{m} \iff \exists k \in \mathbb{Z} \ [b - a = mk],$$

    is an equivalence relation on $\mathbb{Z}$.
    Suggestion: Mimic what was done in Example 2.7 for the special case $m = 12$.

18. Complete the details of Case 1 in the proof of Theorem 2.1.

In Exercises 19–20, for each part, determine whether the given relation is a partial ordering. For those that are not, which of the axioms (i) reflexivity, (ii) antisymmetry, and/or (iii) transitivity fail?   For those that are partial orderings, are they total orderings?

19. (a) The relation $a R b \iff a \ge b$ on $\mathbb{R}$.
    (b) The relation $a R b \iff a > b$ on $\mathbb{R}$.
    (c) The relation $a R b \iff$ the weight of a is less than or equal to that of b (both rounded to the nearest pound), on the set of all people.
    (d) The relation $s R s' \iff$ the total number of 1's of s is at most equal to the total number of 1's of s', on the set of all finite length bit strings.

20. (a) The relation $a R b \iff a^2 < b^2$ on $\mathbb{R}$.
    (b) The relation $(n, m) R (n', m') \iff nm \le n'm'$ on $\mathbb{Z} \times \mathbb{Z}$.
    (c) The relation $a R b \iff a$ does not divide into b $(a \nmid b)$ on $\mathbb{Z}_+$. (C.f., Exercise for the Reader 2.12.)
    (d) The relation $s R s' \iff$ the length of the longest string of consecutive 1's of s is at most equal to the length of the longest string of consecutive 1's of s', on the set of all finite length bit strings.

21. For each relation below, verify that it is a partial order, and construct a Hasse diagram.
    (a) The relation $R = \{(a,a),(b,b),(c,c),(d,d),(b,d),(c,a)\}$ on $A = \{a,b,c,d\}$.
    (b) The subset relation on the set of nonempty proper subsets of $\{1,2,3\}$.
    (c) The divisibility relation $a \mid b$ on the set of divisors of 20: $\{1,2,4,5,10,20\}$.
    (d) The divisibility relation $a \mid b$ on the set of $\{1,2,3,4,5,6\}$.

22. For each relation below, verify that it is a partial order, and construct a Hasse diagram.
    (a) The relation $R = \{(a,a),(b,b),(c,c),(d,d),(e,e),(b,c),(c,e),(d,e)\}$ on $A = \{a,b,c,d,e\}$.
    (b) The subset relation on the set of all subsets of $\{1,2,3,4\}$ with an even number of elements.
    (c) The divisibility relation $a \mid b$ on the set of divisors of 36: $\{1,2,3,6,9,12,18,36\}$.
    (d) The divisibility relation $a \mid b$ on the set of $\{3,4,6,7,8,9,10,11,12\}$.

23. Define a relation on the plane $\mathbb{R}^2$ by $(x, y) \preceq (x', y') \iff x \le x'$ and $y \le y'$. Show that this is a partial order, and explain whether it is a linear order.

24. Suppose that $(A, \preceq_A)$ and $(B, \preceq_B)$ are linearly ordered posets. Define a relation on the Cartesian product $C = A \times B$, as follows: $(a,b) \preceq (a',b') \iff a \preceq_A a'$, and $b \preceq_B b'$. Show that this defines a partial order on C. Is it always a linear order?

25. (*Induced Partial Order on a Subset of a Poset*) Suppose that $(A, \preceq)$ is a poset,   and that

$A' \subseteq A$. Define a relation $\preceq'$ on A' by $a_1' \preceq' a_2' \Leftrightarrow a_1' \preceq a_2'$. Show that $(A', \preceq')$ is a poset. Show also that if $\preceq$ is a linear order on A, then $\preceq'$ will be a linear order on A'.

26. Show that if a relation on a set A is both an equivalence relation and a partial order, then this relation must be the equality relation.

27. Suppose that R_1 and R_2 are partial orders on the same set A.

(a) Will the intersection $R_1 \cap R_2$ necessarily be a partial order on A?

(b) Will the union $R_1 \cup R_2$ necessarily be a partial order on A?

28. Find all order preserving functions between the indicated posets:
(a) $f : (\{1,2\}, \leq) \rightarrow (\{1,2\}, \leq)$.
(b) $f : (\{1,2\}, \leq) \rightarrow (\{1,2,3\}, \leq)$.
(c) $f : (\{1,2,3,4\}, |) \rightarrow (\{1,2\}, \leq)$.

29. Find all order preserving functions between the indicated posets:
(a) $f : (\{1,2\}, =) \rightarrow (\{1,2\}, \leq)$.
(b) $f : (\{1,2,3\}, \leq) \rightarrow (\{1,2\}, \leq)$.
(c) $f : (\{1,2\}, \leq) \rightarrow (\{1,2,3,4\}, |)$.

30. Use Hasse diagrams to identify all of the 16 isomorphism classes of posets with four elements.

31. Use Hasse diagrams to identify all of the isomorphism classes of posets with five elements that require four levels in a Hasse diagram.

NOTE: In any poset $(A, \preceq)$, it is natural to define the notions of maximal, minimal, greatest, and least elements. The definitions are as follows: throughout, we assume $a, x \in A$.

The element a is a **minimal element** of A, if $\forall x \in A \, [x \neq a \Rightarrow x \npreceq a]$ (i.e., no element lies below a).

The element a is a **maximal element** of A, if $\forall x \in A \, [x \neq a \Rightarrow a \npreceq x]$ (i.e., no element lies above a).

The element a is a **least element** of A, if $\forall x \in A \, [a \preceq x]$ (i.e., a lies below all other elements).

The element a is a **greatest element** of A, if $\forall x \in A \, [x \preceq a]$ (i.e., a lies above all other elements).

Any of these elements is easily identified from Hasse diagrams (using the parenthetical interpretations given above). For example, in the one-level poset $\{(a,a), (b,b), (c,c)\}$ (first one on the left in Figure 2.11), each of the three elements of the set a, b, and c is a minimal element and maximal element, but the poset has neither a least element nor a greatest element. Also, the two-level poset $\{(a,a), (b,b), (c,c), (a,b), (a,c)\}$ (with the V-shaped Hasse diagram appearing second from the right in Figure 2.11), has only one minimal element (a) that is also the least element. Elements b and c are both maximal elements but the poset has no greatest element. The next few exercises will give examples and further developments of these concepts.

32. (a) Show that a least element in any poset is necessarily a minimal element.
(b) Show that a least element in any poset, if exists, is the unique minimal element.
(c) Show that a poset need not possess any minimal elements, but if the poset is finite, then such an element will always exist.

33. (a) Show that a greatest element in any poset is necessarily a maximal element.
(b) Show that a greatest element in any poset, if exists, is the unique maximal element.
(c) Show that a poset need not possess any maximal elements, but if the poset is finite, then such an element will always exist.

34.   Find all minimal, maximal, least, and greatest elements of the posets of Exercise 22.

35.   Find all minimal, maximal, least, and greatest elements of the posets of Exercise 21.

36.   Find all minimal, maximal, least, and greatest elements of the posets of Example 2.10.

37.   Find all minimal, maximal, least, and greatest elements of the poset of Exercise for the Reader 2.12.

38.   For each statement below, decide whether it is true or false. Prove those statements that are true and provide counterexamples for those that are false. Throughout we assume that $(A, \preceq_A)$ and $(B, \preceq_B)$ are posets, and that the Cartesian product $C = A \times B$ is endowed with the lexicographic order (see Exercise for the Reader 2.10).
(a) If $(A, \preceq_A)$ and $(B, \preceq_B)$ both have a minimal element, then so does C.
(b) If $(A, \preceq_A)$ and $(B, \preceq_B)$ both have a least element, then so does C.
(c) If $(A, \preceq_A)$ has a minimal element, then so does C.
(d) If $(B, \preceq_B)$ has a least element, then C has a minimal element.

39.   For each statement below, decide whether it is true or false. Prove those statements that are true and provide counterexamples for those that are false. Throughout we assume that $(A, \preceq_A)$ and $(B, \preceq_B)$ are posets, and that the Cartesian product $C = A \times B$ is endowed with the lexicographic order.
(a) The posets $(A, \preceq_A)$ and $(B, \preceq_B)$ both have a maximal element, if, and only if, C has a greatest element.
(b) The posets $(A, \preceq_A)$ and $(B, \preceq_B)$ both have a greatest element, if, and only if, C has a maximal element.

2.3: BOOLEAN ALGEBRA AND CIRCUIT DESIGN

Designing electronic circuits to perform specified tasks has been an important problem since scientists first learned how to generate electricity for commercial and industrial use. As the tasks get more complicated and the capacity to produce more circuits on smaller circuit board chips and wafers continues to increase, the efficiency of such circuit designs remains crucial. The functionality of a circuit is determined by its construction, consisting of a series of inputs, internal components, and corresponding output(s). Circuits need to be designed so that they will produce specified outputs for certain combinations of inputs. The internal components can be broken down into a collection of logical gates, which are analogous to the logical operators that we learned about in Section 1.1. This important

FIGURE 2.12: Claude E. Shannon (1916–2001), American applied mathematician

systematization of circuits was developed in the 1930s by the American applied mathematician Claude Shannon[6] (Figure 2.12); it rests on the foundation of Boolean algebras, a theory discovered in the mid-1800s by English mathematician George Boole. After introducing the concepts and symbolism for modern circuit design, we will study some techniques for reducing the complexity of circuit designs that can perform specified functions.

Boolean Operations, Variables, and Functions

An electric conducting wire can have current running through it or not. These two states are represented mathematically by the bit values:

$$1 = \text{“on” (or “true”) and } 0 = \text{“off” (or “false”)}$$

On the set $B = \{0, 1\}$, we define the complementation, addition, multiplication operations as follows:

Boolean complementation: $\overline{0} = 1; \ \overline{1} = 0.$

Boolean multiplication: $0 \cdot 0 = 0 \cdot 1 = 1 \cdot 0 = 0; \ 1 \cdot 1 = 1.$

Boolean addition: $0 + 0 = 0; \ 0 + 1 = 1 + 0 = 1 + 1 = 1.$

Without parentheses, the hierarchy of these Boolean operations is as follows: first do complements, next do multiplications, and last do additions. For example, the expression $1 + \overline{(0 \cdot 1)} + \overline{0} \cdot 1$ would be evaluated as: $1 + \overline{(0)} + 1 \cdot 1 = 1 + 1 + 1 \cdot 1 = 1 + 1 + 1 = 1$. In the notation of Section 1.1, the Boolean complement, sum, and product simply correspond to the logical negation ($\sim$), conjunction ($\wedge$), and disjunction ($\vee$) operators, respectively.

DEFINITION 2.15: A **Boolean variable** x is a variable that assumes values only from the set $B = \{0, 1\}$. A **Boolean function of n variables** is a function $f : B^n \rightarrow B$.

---

[6] Claude Shannon grew up in Michigan. He earned a bachelor's degree with a double major in mathematics and electrical engineering from the University of Michigan-Ann Arbor. His landmark discovery of an effective symbolism for electric circuits actually came from his master's thesis at MIT: *A symbolic analysis of relay and switching circuits*. This thesis had a tremendous impact on industry by changing circuit design from an art to a science. Shannon went on to earn a doctorate at MIT, and continued to make valuable contributions to the electronics and communications fields during his career working at Bell Labs, where his laboratory office ceiling was adorned with a rainbow of gowns from universities that awarded him honorary doctorates. He developed a secure cryptosystem that was used by Roosevelt and Churchill for transoceanic communications during WWII. His work in this area motivated the development of the field of coding theory, for which he is considered the founder. Coding theory studies what are called error correcting codes, which are used in everything from CDs to routine data transmissions. We have Shannon to thank, for example, when a scratched music CD will still play perfectly well.

A Boolean function in n variables is usually represented by an expression in the Boolean variables $x_1, x_2, \cdots, x_n$ that represents the output $f(x_1, x_2, \cdots, x_n)$, where the expression is made up entirely using some or all of the three Boolean operations (complementation, addition, multiplication) and the Boolean variables $x_1, x_2, \cdots, x_n$. A Boolean function in n variables is analogous to a logical expression in n logical variables, and thus can also be represented by a table of values, analogous to a truth table.   For electrical and computer applications, Boolean functions are more often thought of as switching functions.   In applications, switching functions need to be created to produce certain prescribed outputs (0/1s, or "off/on's") for given input values.   Simplicity of design of such switching functions translates to finding a simplest expression that represents a given switching requirement.

EXAMPLE 2.12:   (a) For any Boolean variable x, the properties of Boolean addition (in particular, the facts that $1 + 1 = 1$ and $0 + 0 = 0$) give us the (Boolean) identity $x + x = x$. Similarly, the properties of Boolean complementation give us the (Boolean) identity $\overline{\overline{x}} = x.$ Such identities can be verified by comparing tables of Boolean values (similar to the truth tables of Section 1.1). Alternatively, these identities are analogous to (and, indeed, they directly follow from) the corresponding identities from propositional logic: $P \vee P \equiv P$ and $\sim (\sim P) \equiv P.$ Some useful Boolean identities are summarized in Theorem 2.4 below.

(b) Consider the Boolean function $f : B^3 \to B :: f(x, y, z) = xy\overline{z} + x\overline{y} + \overline{x}.$ Using the rules of Boolean arithmetic given above, values of this function are easily computed. For example,

$$f(1,1,0) = 1 \cdot 1 \cdot \overline{0} + 1 \cdot \overline{1} + \overline{1} = 1 \cdot 1 \cdot 1 + 1 \cdot 0 + 0 = 1 + 0 + 0 = 1.$$

The reader should verify the rest of the values of this function that are given in Table 2.1.

x	y	z	$xy\overline{z}$	$x\overline{y}$	$\overline{x}$	$f(x, y, z) = xy\overline{z} + x\overline{y} + \overline{x}$
1	1	1	0	0	0	0
1	1	0	1	0	0	1
1	0	1	0	1	0	1
1	0	0	0	1	0	1
0	1	1	0	0	1	1
0	1	0	0	0	1	1
0	0	1	0	0	1	1
0	0	0	0	0	1	1

TABLE 2.1: A table of values for the Boolean function $f(x, y, z) = xy\overline{z} + x\overline{y} + \overline{x}.$

Boolean Algebra Identities

THEOREM 2.4: (*Some Boolean Algebra Identities*) Suppose that x, y, and z are Boolean variables. The following identities are (always) valid:
(a) (*Commutativity*) $x + y = y + x$, $\quad xy = yx$
(b) (*Associativity*) $x + (y + z) = (x + y) + z$, $\quad x(yz) = (xy)z$
(c) (*Distributivity*) $x(y + z) = xy + xz$, $\quad x + yz = (x + y)(x + z)$
(d) (*De Morgan's Laws*) $\overline{x + y} = \overline{x}\,\overline{y}$, $\quad \overline{xy} = \overline{x} + \overline{y}$
(e) (*Double Complementation*) $\overline{\overline{x}} = x$
(f) (*Absorption*) $x + xy = x$, $\quad x(x + y) = x$
(g) (*Identity Laws*) $x \cdot 1 = x$, $\quad x + 0 = x$
(h) (*Idempotent Laws*) $x + x = x$, $\quad xx = x$
(i) (*Unit/Zero Identity*) $x + \overline{x} = 1$, $\quad x\overline{x} = 0$
(j) (*Dominance Laws*) $x + 1 = 1$, $\quad x \cdot 0 = 0$

Proof: Parts (a) through (f) follow directly from the corresponding identities of Part I of Theorem 1.1. (Simply change all occurrences of the logical variables P, Q, and R, to the Boolean variables x, y, and z, respectively, and change the logical operators $(\sim, \vee, \wedge)$ to their Boolean counterparts $(\overline{}, +, \cdot)$. The remaining parts all involve just a single Boolean variable, and are easily verified by constructing (two-row) tables of values. These verifications are left as an exercise. □

As was the case for logical equivalences (the propositional calculus counterpart for Boolean algebra identities), these identities may be combined to obtain or prove new identities, but such proofs can be tricky. Keep in mind that any Boolean identity (or non-identity) can always be checked by constructing tables of values (just like any purported logical equivalence can be verified with truth tables). The next exercise for the reader contrasts three different approaches for proving a Boolean identity.

EXERCISE FOR THE READER 2.13: Use each of the following three approaches to establish the so-called *consensus law*:

$$xy + \overline{x}z + yz = xy + \overline{x}z.$$

(a) By constructing tables of values for each side.
(b) By setting each side separately equal to zero, and "solving" for the Boolean variables.
(c) By using some of the Boolean algebra identities of Theorem 2.4 to convert the left side of the identity to the right side. (NOTE: This will take several steps.)

Sums, Products, and Complements of Boolean Functions

Since Boolean arithmetic may be applied to the values of any Boolean function, we may combine such functions in the natural ways indicated in the following definition.

DEFINITION 2.16: Suppose that f and g are Boolean functions in n variables. Then we can define three new Boolean functions in n variables: the **complement**, $\overline{f}$, the **sum** $f + g$, and the **product** $f \cdot g$, by:

$$\overline{f}(x_1, x_2, \cdots, x_n) = \overline{f(x_1, x_2, \cdots, x_n)},$$
$$(f + g)(x_1, x_2, \cdots, x_n) = f(x_1, x_2, \cdots, x_n) + g(x_1, x_2, \cdots, x_n), \text{ and}$$
$$(f \cdot g)(x_1, x_2, \cdots, x_n) = f(x_1, x_2, \cdots, x_n) \cdot g(x_1, x_2, \cdots, x_n).$$

The Boolean functions f and g are said to be **equivalent** if the functions agree on every possible input, i.e.,

$$\forall (x_1, x_2, \cdots, x_n) \in B^n \ [f(x_1, x_2, \cdots, x_n) = g(x_1, x_2, \cdots, x_n)].$$

Equivalent functions thus represent the same object as a function $B^n \to B$, but with possibly different looking Boolean formulas that define them. It is desirable to find a simplest possible formula that represents a given function, for this will make the design of the circuit or switching function more efficient. The next example illustrates such a simplification that can be achieved using some of the Boolean algebra identities of Theorem 2.4.

EXAMPLE 2.13: Use the Boolean algebra identities of Theorem 2.4 to simplify the formula representing the following Boolean function in four variables:

$$f(x, y, z, w) = \overline{x}yzw + xyzw + y\overline{z}w.$$

SOLUTION: First we use the distributive law (Theorem 2.4(c)) to write:

$$\overline{x}yzw + xyzw = (\overline{x} + x)yzw.$$

The unit identity $x + \overline{x} = 1$ and then the identity law (with commutativity) $1 \cdot x = x$, allow us to rewrite the above expression as yzw. Thus, we can simplify

$$f(x, y, z, w) = yzw + y\overline{z}w.$$

The same procedure that got us this far allows us to further simplify the above (equivalent) expression for f as:

$$f(x, y, z, w) = yw.$$

Later in this section we will develop more systematic procedures for accomplishing such tasks. At this point we will present a method for constructing a formula for a Boolean function of n variables by starting with a table giving all of the required output values. To facilitate such a construction, the following terminology will be convenient:

Sums of Products Expansions (Disjunctive Normal Form)

DEFINITION 2.17: A **literal** is a Boolean variable or a complement of a Boolean variable. A **minterm** in n Boolean variables $x_1, x_2, \cdots, x_n$ is a product of n literals $y_1 y_2 \cdots y_n$, where the ith literal in the product, y_i, equals either x_i or $\overline{x}_i$.

If a table of values is constructed for a minterm in n Boolean variables, only one of the 2^n rows of the table will have the value 1 (hence the reason for the terminology: min[imum]term), namely, the row in which the values of the $x_i = 1$ if, and only if, $y_i = x_i$. By the properties of Boolean addition, it easily follows that any Boolean function in n variables is equivalent to the sum of the minterms formed using the rows of the function's table that correspond to the function having value 1. This sum of minterms representation of a Boolean function is known as the **disjunctive normal form**, or, more descriptively, the **sum-of-products expansion**.[7]

EXAMPLE 2.14: Construct the sum-of-products expansion for the function of f of Example 2.12(b).

SOLUTION: *Method 1:* (*Work with the table of values.*) The only minterm (of the eight possible minterms) that the sum-of-products expansion will not contain is xyz (see Table 2.1). Summing the remaining minterms produces the following sum-of-products expansion for f:[8]

$$f(x, y, z) = xy\overline{z} + x\overline{y}z + x\overline{y}\,\overline{z} + \overline{x}yz + \overline{x}y\overline{z} + \overline{x}\,\overline{y}z + \overline{x}\,\overline{y}\,\overline{z}.$$

Method 2: (*Work directly with the formula for the function.*) This method begins with the formula that was given: $f(x, y, z) = xy\overline{z} + x\overline{y} + \overline{x}$, expands it into separate products, if necessary (here it is not), and uses Boolean arithmetic rules (from

---

[7] See Exercise 24 of Section 1.1 for a development of the corresponding concept in the terminology of logic.

[8] We point out the following caveat regarding the bar notation: Since $\overline{x}\,\overline{y} \neq \overline{xy}$ (the left expression is the logical "not x and not y" and the right expression is "not (x and y)," one should take care (when writing or reading) to distinguish between a single bar over a product of two Boolean variables versus two separate bars over each Boolean variable.

Theorem 2.4) to build the products up so that each includes literals from all of the variables.   The first of the three terms of f is fine.   We build up the second as follows: $x\overline{y} = x\overline{y} \cdot 1 = x\overline{y}(z + \overline{z}) = x\overline{y}z + x\overline{y}\,\overline{z}$ (we have used parts c, g, and i of Theorem 2.4).   Similarly, $\overline{x} = \overline{x}(y + \overline{y})(z + \overline{z}) = (\overline{x}y + \overline{x}\,\overline{y})(z + \overline{z}) = \overline{x}yz + \overline{x}\,\overline{y}z + \overline{x}y\overline{z} + \overline{x}\,\overline{y}\,\overline{z}$.   Summing these $1 + 2 + 4$ products produces the same representation that was obtained in Method 1.

EXERCISE FOR THE READER 2.14:   Find the sum-of-products decomposition for the following Boolean function: $f(x, y, z, w) = \overline{x\overline{y}z} + \overline{z}w$.

Duality

We make one important observation about the Boolean arithmetic rules appearing in Theorem 2.4.   Each of these rules (with the exception of the double complementation rule of Part (e)) comes in a pair.   In any of these pairs, if we take one of the identities, and change any sum to a product, any product to a sum, any 0 to a 1, and any 1 to a 0, then we will arrive at the other identity of the pair.   (The reader should check this.)   This is not just a coincidence—in general, when this process is applied to any Boolean algebra identity, the result will be another Boolean algebra identity.   We state this important fact after recording the following definition that is stated informally.   The formal definition requires a concept called structural induction, and will be presented in the Appendix to Section 3.2.   The definition below will be sufficient for the purposes of this section, but for more advanced applications (and for writing programs) the structural induction-based version will be indispensable.

DEFINITION 2.18 (*Informal*):   If f is a Boolean expression, then the **dual** of f, denoted f^d, is the Boolean expression that is obtained by changing all Boolean sums to Boolean products, all Boolean products to Boolean sums, all 0's to 1's, and all 1's to 0's.   For example, the dual of the Boolean expression $x\overline{y} + z$ is $[x\overline{y} + z]^d = (x + \overline{y})z$, also $[(0 + x)(y + 1)]^d = 1 \cdot x + y \cdot 0$.

EXERCISE FOR THE READER 2.15:   (a)   Find the duals of the Boolean expressions $(x + 1)(yz)$, $x + \overline{y} + z + 0$.

(b)   Obtain a formula for the dual of the Boolean function f of Exercise for the Reader 2.14.

What makes this concept useful is that if we take the duals of two different expressions that represent the same Boolean function, then these duals will also represent the same Boolean function (i.e., they will be equivalent).

PROPOSITION 2.5: (*The Duality Principle*) Suppose that f and g are equivalent Boolean functions in n variables, then the duals f^d and g^d are also equivalent Boolean functions, i.e.,

$$\forall (x_1, x_2, \cdots, x_n) \in B^n \ [f^d(x_1, x_2, \cdots, x_n) = g^d(x_1, x_2, \cdots, x_n)].$$

The proof of Proposition 2.5 requires structural induction, and will be presented in the appendix to Section 3.2. The duality principle is useful in obtaining new identities from old. One nice example is obtained by taking the dual of the sum-of-products (disjunctive normal form) of a function. When we take the dual of this, we get a **product-of-sums** (also known as the **conjunctive normal form**) representation of the dual function. The next example illustrates this procedure.

EXAMPLE 2.15: Use the duality principle on the sum-of-products representation of the Boolean function $f(x, y, z) = xy\overline{z} + x\overline{y} + \overline{x}$ that was obtained in Example 2.14 to obtain the product-of-sums representation of the dual $f^d(x, y, z)$.

SOLUTION: Taking duals of the identity obtained in Example 2.14:

$$f(x, y, z) = xy\overline{z} + x\overline{y}z + x\overline{y}\,\overline{z} + \overline{x}yz + \overline{x}y\overline{z} + \overline{x}\,\overline{y}z + \overline{x}\,\overline{y}\,\overline{z},$$

produces the desired representation of the dual of f: $f^d(x, y, z) =$

$$(x + y + \overline{z})(x + \overline{y} + z)(x + \overline{y} + \overline{z})(\overline{x} + y + z)(\overline{x} + y + \overline{z})(\overline{x} + \overline{y} + z)(\overline{x} + \overline{y} + \overline{z}).$$

The product-of-sums representation can also be obtained directly in a procedure that is similar to (in fact "dual to") that which was introduced for obtaining the sum-of-products representation, see Exercise 33.

EXERCISE FOR THE READER 2.16: Apply the duality principle to the consensus law Boolean identity of Exercise for the Reader 2.13: $xy + \overline{x}y + yz = xy + \overline{x}z$, to obtain another Boolean identity (that is also known as a consensus law).

Logic Gates and Circuit Designs

In the design of electrical and computer circuits, the basic components correspond to the Boolean operators of complementation, conjunction, and disjunction. These basic components are called **logic gates**, and are called respectively, **inverters**, **AND gates**, and **OR gates**. The IEEE (Institute of Electrical & Electronics Engineers, Inc.) has established the symbols shown in Figure 2.13 to schematically represent these logic gates.

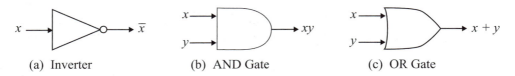

(a) Inverter   (b) AND Gate   (c) OR Gate

FIGURE 2.13: The basic logic gates of Boolean circuit design: (a) (left) The inverter, (b) (middle) The AND gate, and (c) (right) The OR gate.

Because of associativity and commutativity, the AND and OR gates may take any number of inputs (in any order) to produce (unambiguous) outputs.   From these basic logic gates, more intricate circuits can be designed that will correspond to any given Boolean function.   For a given Boolean expression representing a Boolean function, there is some flexibility in how the corresponding circuit can be drawn (by putting together the basic logic gates), we give an example to illustrate two of the more commonly used schemes.

EXAMPLE 2.16:   Draw a circuit design for the Boolean function represented by the expression $xyz + x(\overline{y} + \overline{z})$.

SOLUTION: The gates needed to put together a circuit for this Boolean expression share some common inputs.   Such common inputs can either be branched off from single input lines for each variable, or input lines can be drawn separately for each initial gate.   Circuit diagrams corresponding to these two possible approaches are illustrated in Figure 2.14.

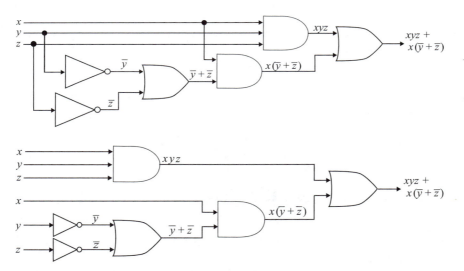

FIGURE 2.14: Two circuit drawings for the Boolean expression of Example 2.16: (a) (top) Using single input lines for each input variable. (b) (bottom) Using separate input lines for each occurrence of each input variable.

EXERCISE FOR THE READER 2.17: Draw a circuit design for the Boolean function represented by the expression $\overline{(xy\overline{z})}(x+\overline{y})$.

EXAMPLE 2.17: Design a circuit that will control a light using three (standard light) switches in such a way that whenever one of the switches is flipped the status of the light will change (either from "on" to "off," or from "off" to "on").

SOLUTION: Each switch can be thought of as an inputted Boolean variable. We let the value 0 correspond to a switch being in the down position, and 1 correspond to the switch being in the up position. We let the variables x, y, and z correspond to the three switch inputs, and $f(x,y,z)$ denote the corresponding Boolean function, where the output will be 1 if the light is "on," and 0 if the light is "off." Any particular state of the switches can be assigned to produce either a 0 or a 1 output, and from this all other states (and thus the Boolean function) will be completely determined. We arbitrarily assign the value $f(0,0,0)=0$ (i.e., when all switches are down, we stipulate that the light is "off"). From this state, if exactly one of the switches is flipped up, the light turns "on," so $f(1,0,0)=f(0,1,0)=f(0,0,1)=1$. Next, if exactly two switches are up, the light will be off, i.e., $f(1,1,0)=f(1,0,1)=f(0,1,1)=0$. Finally, with all three switches up, the light will be on again: $f(1,1,1)=1$. The resulting sum-of-products expansion for this Boolean (switching) function is thus:

$$f(x,y,z) = x\overline{y}\,\overline{z} + \overline{x}y\overline{z} + \overline{x}\,\overline{y}z + xyz.$$

A corresponding circuit diagram for this function is shown in Figure 2.15.

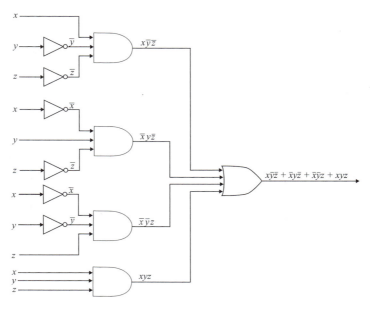

FIGURE 2.15: Diagram for the switching circuit of Example 2.17.

Karnaugh Maps

We next turn our attention to the very interesting and practical problem of simplifying circuit designs. Since we have learned how to create a diagram of a circuit directly from the corresponding Boolean expression, the problem amounts to finding simple Boolean expressions that will represent a given Boolean function (i.e., that will perform the specified tasks). We have performed such a simplification in Example 2.13 using the Boolean identities of Theorem 2.4, but this method was not very systematic, and is difficult to apply in general. The general problem of simplifying a given Boolean expression/circuit design has been rigorously proved to be among the most difficult problems in discrete mathematics. More precisely, in the language of Chapter 7, this problem has been shown to belong to the class of *NP-complete problems*, a class that includes many other important problems that we will see throughout the rest of the text (for example, the problem of factoring a positive integer into primes, which will be discussed in the next chapter). There are several general methods for simplifying Boolean expressions, but we will present a very visual method that was discovered by American applied mathematician and physicist Maurice Karnaugh.[9] This method is a general one, but in practice it gets very complicated to apply in case there are more than six Boolean variables. It can be programmed, however, to work for more complex problems. The method basically consists of creating a so-called **Karnaugh map**, which is a table constructed from the outputs of the Boolean function. This Karnaugh map is then examined for the possibility of combining terms in a similar fashion as was done in the solution of Example 2.13 where pairs of products that differ in exactly one literal may be combined into a shorter product, e.g., $xyz + x\overline{y}z = x(y + \overline{y})z = x \cdot 1 \cdot z = xz$. The difference is that such "blocks" of terms are easily located and combined in efficient ways directly from the diagram (without any Boolean algebra). The method is most easily described using examples. We begin with the case of two Boolean variables.

EXAMPLE 2.18: We use Karnaugh maps to create a simple Boolean expression for the Boolean function f presented in Table 2.2.

SOLUTION: For any Boolean function in two variables x and y, there can be four possible minterms in the sum-of-products representation. The corresponding Karnaugh map will be a two-by-two table where the rows correspond to the literals in one

x	y	$f(x, y)$
1	1	0
1	0	1
0	1	1
0	0	1

TABLE 2.2: The Boolean function of Example 2.18.

---

[9] Maurice Karnaugh (1924–) grew up in New York City. A year after earning his Ph.D. from Yale in physics in 1952, he published his now famous geometric method for simplifying electric circuits [Kar-53], while he was employed as a researcher at Bell Labs. In 1966 he moved on to a similar position at IBM, and in 1980 he accepted a professorship at the Polytechnic Institute of New York, where he remained until his retirement in 1999. He was the founding governor of the International Council for Computer Communication.

variable (i.e., either x or $\overline{x}$) and the columns to the literals in the other variable (i.e., either y or $\overline{y}$). Each of the four squares in the table of a Karnaugh map corresponds uniquely to one of the literals in the obvious fashion, as shown in Figure 2.16(a). The Karnaugh map for the Boolean function is then constructed by placing 1's in the squares that correspond to the minterms that are present in the sum-of-products representation of the function. (The remaining cells are left blank.) Since the sum-of-products representation of f is (from Table 2.2), $f(x, y) = x\overline{y} + \overline{x}y + \overline{x}\,\overline{y}$, the Karnaugh map for $f(x,y)$ is as shown in Figure 2.16(b).

FIGURE 2.16: (a) (left) The correspondence of minterms with cells for a general Karnaugh map for a Boolean function of two variables. (b) (right) The specific Karnaugh map for the Boolean function presented in Table 2.2.

We say that two cells in a general Karnaugh map are **adjacent** if their corresponding minterms differ by exactly one literal. Thus, for example, the two cells in the first row, corresponding to the literals xy and $x\overline{y}$ are adjacent, as are the cells in the second row, or in either of the two columns. Using the Boolean identities of Theorem 2.4, sums of minterms in rectangular **blocks** of adjacent cells can always be combined into a single product of a smaller number of literals. For example, the sum of the minterms in the first row block can be combined as $xy + x\overline{y} = x(y + \overline{y}) = x \cdot 1 = x$. Similarly, the sum of the minterms in the block consisting of the second row would simplify as $\overline{x}y + \overline{x}\,\overline{y} = \overline{x}$. (The reader should check this.) For the block of all four cells, using the just mentioned simplifications, we could take it one step further to obtain:

$$xy + x\overline{y} + \overline{x}y + \overline{x}\,\overline{y} = x + \overline{x} = 1.$$

We used the unit identity (Theorem 2.4(i)) once more. This corresponds to the fact that the Boolean function with all possible minterms present is simply the constant function 1.

In general, to complete the Karnaugh map procedure, one searches for the largest possible (rectangular) blocks consisting of 2, 4, 8, 16, ..., adjacent cells, circling them off, until all of the 1s have been covered. In selecting each new grouping of cells to circle off, previously circled cells (with 1's) may be included in the new circling, but the largest possible number of uncircled 1's should be enclosed. There may be more than one choice for a new group of cells to circle; this is fine, but we point out that such situations will generally lead to different final simplified Boolean expressions.[10]   We then replace each of these blocks with the corresponding simplified Boolean expression, and the sum of all of these will be the simplified expression that is

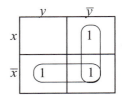

FIGURE 2.17: Completed Karnaugh map for the function of Table 2.2.

equivalent to the original Boolean function. For the Karnaugh map at hand, there are two blocks needed to cover the 1s; these are indicated in Figure 2.17. The two blocks correspond to the simplified expressions $\overline{x}$ and $\overline{y}$, so the Boolean function f is equivalent to the expression $\overline{x} + \overline{y}$.[11]

EXERCISE FOR THE READER 2.18: (a) The Boolean function of Table 2.2 is simply the complement of the AND gate, i.e., $f(x,y) = \overline{xy}$ and is often called the **NAND gate**.   Use Theorem 2.4 to reproduce the identity $\overline{xy} = \overline{x} + \overline{y}$ that follows from the solution of Example 2.18.

(b) Write down the Karnaugh map for the Boolean function presented in Table 2.3, and find the resulting representation of this function that is produced by the Karnaugh map method.

x	y	$f(x,y)$
1	1	1
1	0	0
0	1	0
0	0	1

TABLE 2.3:   The Boolean function of Exercise for the Reader 2.18.

We now turn to an example that will demonstrate how to use Karnaugh maps to find simplified representations of a Boolean function with three variables.   The possible minterms for a Boolean function of x, y, and z, come from the four minterms in the two variables x and y, with each multiplied by either the literal z or the literal $\overline{z}$.[12] Thus, a Karnaugh map for a function of three variables will require 8 cells. One way to do this is to use a 2-row 4-column table where the first and second rows correspond to minterms having x and $\overline{x}$, respectively, and the four columns correspond to the four possible minterm combinations in the variables y and z (ordered so that neighboring minterms differ

---

[10] Thus, a "simplest" equivalent Boolean expression is generally not a unique object.

[11] In case any readers might be concerned that the minterm $\overline{xy}$ was added twice, this is not a problem since in Boolean algebra we always have the idempotent law: $z + z = z$, for any Boolean variable z (Theorem 2.4(h)).

[12] We have basically shown that whenever we add a new variable, the number of possible minterms doubles. Thus, for example, the number of possible minterms in four Boolean variables is 16, in five Boolean variables is 32, etc.   It follows that the number of minterms for a Boolean function with n variables will be 2^n, and this must be the number of cells in the Karnaugh map. Technically the above argument requires mathematical induction, which we will introduce in the next chapter.

by just one literal, so their cells will be adjacent). The general structure of such a Karnaugh map is shown in Figure 2.18.

	yz	$y\bar{z}$	$\bar{y}\bar{z}$	$\bar{y}z$
x	xyz	$xy\bar{z}$	$x\bar{y}\bar{z}$	$x\bar{y}z$
$\bar{x}$	$\bar{x}yz$	$\bar{x}y\bar{z}$	$\bar{x}\bar{y}\bar{z}$	$\bar{x}\bar{y}z$

FIGURE 2.18: The correspondence of minterms with cells for a general Karnaugh map with three Boolean variables.

One thing that is new here is that adjacent cells need not be horizontally or vertically touching one another. Either cell on the left edge of the Karnaugh map is adjacent to the cell in the same row on the right edge (since the minterms differ only in the y-literal). Blocks of adjacent cells (for which the sums of the corresponding minterms can be combined to simpler expressions) can take on different forms than before, but in any Karnaugh map, the number of cells in any block must always be a power of two: 1, 2, 4, 8, … . Some such blocks, with the simplified expressions they represent are shown in Figure 2.19.

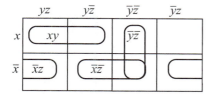
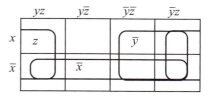

FIGURE 2.19: Sample blocks for the three variable Karnaugh map: (a) (left) 2-cell blocks demonstrating the Boolean identities: $xy = xyz + xy\bar{z}$, $\bar{y}\bar{z} = x\bar{y}\bar{z} + \bar{x}\bar{y}\bar{z}$, $\bar{x}z = \bar{x}yz + \bar{x}\bar{y}z$, $\bar{x}\bar{z} = \bar{x}y\bar{z} + \bar{x}\bar{y}\bar{z}$. (b) (right) 4-cell blocks demonstrating the Boolean identities: $z = xyz + x\bar{y}z + \bar{x}yz + \bar{x}\bar{y}z$, $\bar{y} = x\bar{y}\bar{z} + x\bar{y}z + \bar{x}\bar{y}\bar{z} + \bar{x}\bar{y}z$, $\bar{x} = \bar{x}yz + \bar{x}y\bar{z} + \bar{x}\bar{y}\bar{z} + \bar{x}\bar{y}z$.

The reader should verify each of the Boolean identities demonstrated in Figure 2.19 (using Theorem 2.4). Any 1-cell block is simply a single minterm, whereas the (only) 8-cell block (for three variable Karnaugh maps) contains all minterms so represents the constant Boolean function 1. It is perhaps helpful to "roll" the Karnaugh map of Figure 2.18 into a cylinder by joining the vertical edges. With this visualization, all adjacent cells really do lie next to one another.

EXAMPLE 2.19: Use Karnaugh maps to find a simplified Boolean expression for the Boolean function $f(x, y, z) = x + z(\overline{x\bar{y}})$. Compare the number of logical gates for the circuit corresponding to the original expression with that for the simplified expression that you find.

SOLUTION: We first compute a table of values using the given expression for f:

x	y	z	$\overline{xy}$	$z\overline{(xy)}$	$f(x,y,z) = x + z\overline{(xy)}$
1	1	1	0	1	1
1	1	0	0	0	1
1	0	1	1	0	1
1	0	0	1	0	1
0	1	1	0	1	1
0	1	0	0	0	0
0	0	1	0	1	1
0	0	0	0	0	0

TABLE 2.4: A table of values for the Boolean function $f(x,y,z) = x + z\overline{(xy)}$.

The Karnaugh map is shown in Figure 2.20. The procedure for identifying blocks starts with the largest possible blocks first, then goes down in size, as necessary until all of the 1-cells are covered. In this case, two 4-cell blocks did the job, and the corresponding simplified Boolean expression gives us the

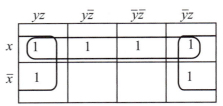

FIGURE 2.20: Completed Karnaugh map for the Boolean function of Table 2.4.

representation $f(x,y,z) = x + z$. With this simplified representation, the corresponding circuit design requires just a single logical gate, whereas with the original representation, the circuit would require 13 logical gates (13 to process the 6 minterms, and a final OR gate).

EXERCISE FOR THE READER 2.19: Apply the Karnaugh map method to the Boolean expression for the three-way switching circuit that was found in Example 2.17. Compare the number of logical gates for the circuit corresponding to the original expression versus that for the simplified expression that you find.

We next give a brief explanation for the case of four variable Karnaugh maps. Since there are 16 possible minterms, such Karnaugh maps will require 16 cells. In general with an even number of variables, the Karnaugh maps are taken to be square grids, and with an odd number of variables, rectangular grids with twice as many columns as rows are used (see Figure 2.20 for the case of three variables). As always, minterms are placed so that cells that are horizontal or vertical neighbors will be adjacent cells. Such a minterm correspondence for four variable Karnaugh maps is shown in Figure 2.21. Notice that here, in addition to the horizontal and vertical neighbor adjacencies, and the left–right edge adjacencies (as in the three variable Karnaugh maps), we also have that each cell in the top row is adjacent to the corresponding cell in the bottom row.

	yz	$y\bar{z}$	$\bar{y}\bar{z}$	$\bar{y}z$
wx	$wxyz$	$wxy\bar{z}$	$wx\bar{y}\bar{z}$	$wx\bar{y}z$
$w\bar{x}$	$w\bar{x}yz$	$w\bar{x}y\bar{z}$	$w\bar{x}\bar{y}\bar{z}$	$w\bar{x}\bar{y}z$
$\bar{w}\bar{x}$	$\bar{w}\bar{x}yz$	$\bar{w}\bar{x}y\bar{z}$	$\bar{w}\bar{x}\bar{y}\bar{z}$	$\bar{w}\bar{x}\bar{y}z$
$\bar{w}x$	$\bar{w}xyz$	$\bar{w}xy\bar{z}$	$\bar{w}x\bar{y}\bar{z}$	$\bar{w}x\bar{y}z$

FIGURE 2.21: The correspondence of minterms with cells for a general Karnaugh map with four Boolean variables.

If we were to first "roll" the Karnaugh map of Figure 2.21 into a cylinder by joining the vertical edges (as we suggested with three variable Karnaugh maps) and then bend the cylinder into a doughnut so that the top and bottom edges are joined at adjacent cells, then all adjacent cells really would lie next to one another.[13] Some adjacency blocks with the simplified expressions they represent are shown in Figure 2.22.

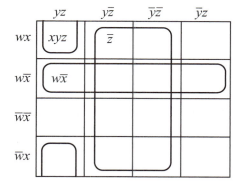
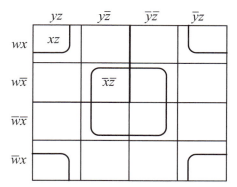

FIGURE 2.22: Sample blocks for the four variable Karnaugh map along with the Boolean expressions that the sum of their minterms represent.

The reader is encouraged to write down and verify (using Theorem 2.4) each of the five Boolean identities represented by Figure 2.22.

Our next example will illustrate another concept in circuit design known informally as **don't care conditions**. These arise when certain combinations of inputs in a Boolean function can never occur, and thus, for such inputs, any convenient outputs can be assigned. For Karnaugh maps, cells with don't care conditions are usually marked with an "X," and they can be helpful in forming blocks of adjacent cells that need to cover all of the 1-cells.

---

[13] For five or more Boolean variables, it is no longer possible to morph the Karnaugh map into a three-dimensional surface where all adjacent cells really are neighboring cells (one would need more than two dimensions to do this). This is part of the reason that it is more difficult to recognize blocks of adjacent cells in Karnaugh maps as the number of Boolean variables increases.

EXAMPLE 2.20: Binary (base 2) representation of integers will be discussed further in Chapter 4, but for now, we point out that the (single) decimal digits $0, 1, 2, \cdots, 9$ are represented by length 4 bit strings as follows: $0 = 0000$, $1 = 0001$, $2 = 0010$, $3 = 0011$, $4 = 0100$, $5 = 0101$, $6 = 0110$, $7 = 0111$, $8 = 1000$, $9 = 1001$. The remaining six bit strings: 1010, 1011, 1100, 1101, 1110, 1111 are thus left unused. Letting w, x, y, and z correspond to the binary digits in the first, second, third, and fourth places, respectively, we consider the Boolean function f that assigns 1 to each binary string corresponding to a digit that is either even or equal to 9 (i.e., 0, 2, 4, 6, or 8, 9). Use Karnaugh maps to find a simplified expression to represent this function, taking advantage of the don't care conditions.

SOLUTION: The completed Karnaugh map is shown in Figure 2.23. The don't care cells have been assigned to be 1s inside, so that the indicated pair of eight-cell blocks could be formed. This gives the simplified Boolean expression $w + \bar{z}$ that will match the required Boolean outputs.

EXERCISE FOR THE READER 2.20: (a) Use Karnaugh maps to find two different simplified expressions for the Boolean function with the formula:

$$f(w, x, y, z) = \overline{w + \bar{x}y + \overline{w}\,\overline{x}\,\overline{y}z},$$

but with don't care conditions on each of the following inputs: $wx\bar{y}\bar{z}$, $w\bar{x}\bar{y}\bar{z}$, $w\bar{x}\bar{y}z$, $\overline{w}\overline{x}y\bar{z}$.

FIGURE 2.23: Completed Karnaugh map for the Boolean function of Example 2.20 with the six don't care conditions marked by Xs.

(b) Are these two expressions equivalent? If not, explain why this is possible.

EXERCISES 2.3:

1. Evaluate each of the following Boolean expressions:
   (a) $(0 + \bar{1}) \cdot 1$.

   (b) $\overline{(0 + 1)} \cdot 1$.

   (c) $\overline{\overline{(0 + 1)} \cdot \overline{(1 + 1)}}$.

   (d) $0 + \bar{0} + \bar{\bar{0}}$.

2. Evaluate each of the following Boolean expressions:
   (a) $(0 + \bar{1}) \cdot 1 + 0 \cdot \bar{1}$.

   (b) $(1 + 1) \cdot (1 + 1)$.

   (c) $\overline{\overline{(0 + 1 \cdot 0)} + 1 \cdot \bar{1}}$.

   (d) $0 \cdot \bar{0} \cdot \bar{\bar{0}}$.

3. Construct tables of values for each of the following Boolean functions:
   (a) $f(x, y) = x\bar{y} + \bar{x}y$.

   (b) $g(x, y) = \overline{x\bar{y}} + \overline{\bar{x}y}$.

   (c) $h(x, y) = \overline{x\bar{y} + \bar{x}y}$.

   (d) $k(x, y) = 1 + x\bar{y} + \bar{x}y$.

4. Construct tables of values for each of the following Boolean functions:

(a) $f(x,y) = (x+\bar{y})(\bar{x}+y)$.

(b) $g(x,y) = \overline{(x+\bar{y})}\,\overline{(\bar{x}+y)}$.

(c) $h(x,y) = \overline{(x+\bar{y})(\bar{x}+y)}$.

(d) $k(x,y) = \overline{(x+\bar{y})}(\bar{x}+y)$.

5. Construct tables of values for each of the following Boolean functions:

(a) $f(x,y,z) = x\bar{y} + \bar{x}y$.

(b) $g(x,y,z) = x\bar{y}z + \overline{\bar{x}y\bar{z}}$.

(c) $h(x,y,z) = \overline{(x+\bar{y}+z)}(\bar{x}+y\bar{z})$.

(d) $k(w,x,y,z) = w + \bar{y} + xz$.

6. Construct tables of values for each of the following Boolean functions:

(a) $f(x,y,z) = (x+\bar{y})(\bar{x}+y)$.

(b) $g(w,x,y,z) = xyzw + \overline{wy\bar{z}}$.

(c) $h(w,x,y,z) = \overline{(x+y+z)}(wz)$.

(d) $k(w,x,y,z) = w+x+y+z$.

7. Use each of the following three approaches to establish the following Boolean identity:
$$x + \overline{(x\bar{z})}y = x + y.$$
(a) By constructing tables of values for each side.
(b) By setting each side separately equal to zero (or one, whichever is more convenient), and "solving" for the Boolean variables.
(c) By using some of the Boolean algebra identities of Theorem 2.4 to convert the left side of the identity to the right side. (NOTE: This will take several steps.)

8. Repeat each of the three methods of Exercise 7 to establish the following Boolean identity: $(x+y)(y+z)(x+z) = xy + yz + xz$.

9. Repeat each of the first two methods of Exercise 7 (i.e., Parts (a) and (b)) to establish the following Boolean identity: $x\bar{y} + y\bar{z} + \bar{x}z = \bar{x}y + \bar{y}z + x\bar{z}$.

10. Repeat each of the first two methods of Exercise 7 (i.e., Parts (a) and (b)) to establish the following Boolean identity: $\bar{x} + \bar{y}(x+\bar{z}) = \bar{x}yz + \bar{x}y\bar{z} + \bar{x}\,\bar{y}z + \bar{x}\,\bar{y}\bar{z} + x\bar{y}z + x\bar{y}\bar{z}$.

11. (a) Let f, g, h, and k be the Boolean functions of Exercise 5. Find Boolean expressions for each of the following Boolean functions. Next, use Boolean algebra identities from Theorem 2.4 to simplify these expressions.

(i) $(f \cdot g)(x,y,z)$.

(ii) $(f+\bar{h})(x,y,z)$.

(b) Find the sum-of-products representations for each of the Boolean functions of Exercise 5.
(c) Find expressions for the duals of each of the Boolean functions of Exercise 5.

12. (a) Let f, g, h, and k be the Boolean functions of Exercise 6. Find Boolean expressions for each of the following Boolean functions. Next, use Boolean algebra identities from Theorem 2.4 to simplify these expressions.

(i) $(f \cdot g)(x,y,z)$.

(ii) $(f+\bar{h})(x,y,z)$.

(b) Find the sum-of-products representations for each of the Boolean functions of Exercise 6.
(c) Find expressions for the duals of each of the Boolean functions of Exercise 6.

13. (a) Find the dual Boolean identity of Exercise 7.
(b) Give a separate verification by constructing tables of values for each side.
(c) Give another verification by setting each side separately equal to zero (or one, whichever is more convenient), and "solving" for the Boolean variables.
(d) Prove this identity using the Boolean algebra identities of Theorem 2.4 to convert the left side of the identity to the right side.

14. Repeat the directions of Exercise 13 for the dual Boolean identity to that of Exercise 8.

15.   Write down the Boolean expression that is represented by the circuit diagram shown in Figure 2.24(a).

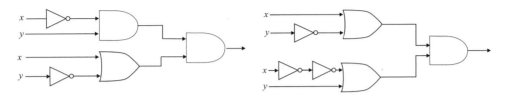

FIGURE 2.24: (a) (left) Circuit diagram for Exercise 15. (b) (right) Circuit diagram for Exercise 16.

16.   Write down the Boolean expression that is represented by the circuit diagram shown in Figure 2.24(b).

17.   Write down the Boolean expression that is represented by the circuit diagram shown in Figure 2.25(a).

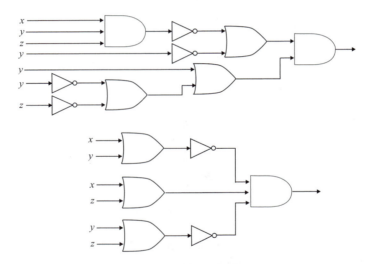

FIGURE 2.25: (a) (top) Circuit diagram for Exercise 17. (b) (bottom) Circuit diagram for Exercise 18.

18.   Write down the Boolean expression that is represented by the circuit diagram shown in Figure 2.25(b).

19.   Draw circuit designs for the Boolean functions represented by the following expressions.

   (a) $\overline{x}y + \overline{y}$ (b) $x\overline{y}\overline{\overline{z}}(xy + \overline{z})$ (c) $\overline{(x\overline{y} + \overline{w}z(1 + y))}$

20.   Draw circuit designs for the Boolean functions represented by the following expressions.

   (a) $(x + y)(\overline{x} + \overline{y})$ (b) $(x + y + z)(\overline{x}\,\overline{y} + \overline{y}\,\overline{z})xyz$ (c) $w + x\overline{y} + wx\overline{z}\overline{(w + x)}$

21.   Draw a circuit design for the Boolean function that has three inputs corresponding to the votes of the individuals of a three-person committee (1 = yes, 0 = no) on whether or not to pass a motion. The output should be 1 if the motion passes by a majority of at least two votes, and 0 otherwise, indicating that the motion failed to pass.

22. Draw a circuit design for the Boolean function that has five inputs corresponding to the votes of the individuals of a five-person committee (1 = yes, 0 = no) on whether or not to pass a motion. The output should be 1 if the motion passes by a majority of at least three votes, and 0 otherwise, indicating that the motion failed to pass.

23. Draw a circuit design for the Boolean function that has four inputs corresponding to the votes of three board members (w, x, and y) and a chairperson (z) (1 = yes, 0 = no) on whether or not to pass a motion. The chairperson's vote counts as two board member votes. The output should be 1 if the motion passes by a majority of at least three votes, and 0 otherwise, indicating that the motion failed to pass.

24. Draw a circuit design for the Boolean function that has five inputs corresponding to the votes of four board members (u, w, x, and y) and a chairperson (z) (1 = yes, 0 = no) on whether or not to pass a motion. The chairperson's vote is only used to break a 2/2 tie. The output should be 1 if the motion passes by a majority, and 0 otherwise, indicating that the motion failed to pass.

25. For each of the two-variable Boolean functions given in Exercise 3, complete the Karnaugh map, and write down the resulting simplified expression.

26. For each of the two-variable Boolean functions given in Exercise 4, complete the Karnaugh map, and write down the resulting simplified expression.

27. For each of the Boolean functions given in Exercise 5, complete the Karnaugh map, and write down the resulting simplified expression.

28. For each of the Boolean functions given in Exercise 6, complete the Karnaugh map, and write down the resulting simplified expression.

In the Exercises 29 and 30, complete the Karnaugh map for the indicated circuit diagram and draw a new circuit using the resulting simplified expression that results from the Karnaugh method.

29. (a) The circuit shown in Figure 2.24(a).
    (b) The circuit shown in Figure 2.25(a).

30. (a) The circuit shown in Figure 2.24(b).
    (b) The circuit shown in Figure 2.25(b).

In the Exercises 31 and 32, apply the Karnaugh map method to determine simplified expressions for the given Boolean functions with the indicated don't care conditions.

31. (a) The Boolean function g of Exercise 5(b), but with don't care conditions on $\overline{x}\overline{y}z$, $x\overline{y}\,\overline{z}$, $\overline{x}\overline{y}\overline{z}$, and $\overline{x}yz$.

    (b) The Boolean function given by the expression of Exercise 19(c) with don't care conditions on $wx\overline{y}z$, $wx\overline{y}\overline{z}$, $\overline{w}\overline{x}\,\overline{y}\,\overline{z}$, $\overline{w}\overline{x}\,\overline{y}z$, and $\overline{w}\overline{x}yz$.

    (c) The Boolean function given by the expression of Exercise 19(d) with don't care conditions on $wxyz$, $\overline{w}\overline{x}yz$, and $\overline{w}\overline{x}y\overline{z}$

32. (a) The Boolean function g of Exercise 6(b), but with don't care conditions on $wx\overline{y}\,\overline{z}$, $w\overline{x}\,\overline{y}z$, $wx\overline{y}\overline{z}$, and $\overline{w}\overline{x}\,\overline{y}\overline{z}$.

    (b) The Boolean function given by the expression of Exercise 20(b) with don't care conditions on xyz, $x\overline{y}z$, $x\overline{y}\,\overline{z}$, and $\overline{x}\,\overline{y}\,\overline{z}$.

    (c) The Boolean function given by the expression of Exercise 20(c) with don't care conditions on $wxyz$, $w\overline{x}y\overline{z}$, $w\overline{x}yz$, $wx\overline{y}\overline{z}$, $\overline{w}x\overline{y}\overline{z}$, and $\overline{w}\overline{x}\,\overline{y}z$.

33.   (*Conjunctive Normal Form of a Boolean Function*) Throughout this exercise, we assume that we are dealing with Boolean functions in the Boolean variables $x_1, x_2, \cdots, x_n$.

(a) We call a sum of literals $y_1 + y_2 + \cdots + y_n$ in which the ith literal y_i, equals either x_i or $\bar{x}_i$, a *maxterm*. Show that a maxterm has the value 0 at exactly one (of the 2^n) combinations of the Boolean variables.

(b) Show that any Boolean function in the variables $x_1, x_2, \cdots, x_n$ can be uniquely expressed as a product of maxterms. This representation is called the *conjunctive normal form*, or the *product-of-sums* representation.

Suggestion: For the existence part in (b), take one maxterm for each row of the table of values where the function equals 0.

34.   Using the procedure suggested in Exercise 33, find the product-of-sums representation for each of the Boolean functions of Exercise 6.

35.   Using the procedure suggested in Exercise 33, find the product-of-sums representation for each of the Boolean functions of Exercise 5.

36.   (a) In a Karnaugh map for a Boolean function with five input variables, how many cells would be adjacent to a given cell?

(b) Draw a generic Karnaugh map with the input variables u, w, x, y, z. (Label the rows and columns but not the individual cells.) Pick any cell and mark all of its adjacent cells.

(c) In your Karnaugh map of Part (b), draw in some different sorts of 4-blocks of adjacent cells (perhaps using different markings/colors).

37.   (a) In a Karnaugh map for a Boolean function with six input variables, how many cells would be adjacent to a given cell?

(b) Draw a generic Karnaugh map with the input variables u, v, w, x, y, z. (Label the rows and columns but not the individual cells.) Pick any cell and mark all of its adjacent cells.

(c) In your Karnaugh map of Part (b), draw a black-filled circle in the upper-left corner square of your figure in Part (b), and then shade in black all neighboring squares. Next, pick some other square in the interior all of whose adjacent squares are disjoint from the ones that you have just shaded in black. Draw in a gray-filled circle in this new square and shade all of its adjacent squares in gray.

38.   Write proofs of the following Boolean algebra identities that use the identities of Theorem 2.4 to translate the left side of the identity to the right side.

(a) The identity $x\bar{y} + y\bar{z} + \bar{x}z = \bar{x}y + \bar{y}z + x\bar{z}$ of Exercise 9.

(b) The identity $\bar{x} + \bar{y}(x + \bar{z}) = \bar{x}yz + \bar{x}y\bar{z} + \bar{x}\,\bar{y}z + \bar{x}\,\bar{y}\,\bar{z} + x\bar{y}z + x\bar{y}\bar{z}$ of Exercise 10.

Chapter 3: The Integers, Induction, and Recursion

We introduced the set $\mathbb{Z} = \{0, \pm 1, \pm 2, \cdots\}$ of the integers in Chapter 1. Most discrete mathematical models can be formulated in terms of the integers rather than the real numbers. This is important not just because of the simplicity of not needing to work with decimals, but also because with integer arithmetic, we can avoid round-off errors that always occur in computer applications of real numbers.

One of the basic (and rather obvious) properties of the positive integers is that any nonempty subset must have a smallest element. This property has a very important consequence known as the principle of mathematical induction. This principle is an important tool that is often used throughout mathematics to prove theorems. We will develop mathematical induction in Section 3.1. Section 3.2 will focus on recursive definitions of objects corresponding to a sequence of positive integers where only the first object is defined explicitly, and the rest are defined in terms of those corresponding to smaller integers.

The (positive) integers go back to antiquity, and there are many seemingly simple questions about them that are extremely difficult. In Section 3.3 we will develop some fundamental yet very useful properties of the integers. Mathematics has a whole branch, called number theory, that is devoted to the study of the integers. One basic property of positive integers (greater than 1) is that they can be uniquely factored into primes. The best known algorithms today (and the consensus among experts in the field is that future algorithms as well) work very slowly. Mathematicians have turned such difficulties into advantageous schemes for sending secret codes. This subject, known as cryptography, is the most important application of number theory, and is assured to be of high importance in the foreseeable future. The number theory that we study here will pave the way for cryptography that we will develop in Chapter 6.

3.1: MATHEMATICAL INDUCTION

The set of positive integers $\mathbb{Z}_+$ has a property known as the *well-ordering principle*, which states that any nonempty subset has a smallest element.[1] This

---

[1] This simple property is possessed, for example, by neither the set of positive real numbers, nor the set of positive rational numbers. For example, although the set of (all) positive integers has a smallest

simple property gives rise to a very important tool in mathematics known as (*the principle of*) *mathematical induction*.   This principle can be often be used when one needs to prove that an infinite sequence of statements: $S(1)$, $S(2)$, $\cdots$, are all true; it is explained in the following theorem:

The Principle of Mathematical Induction: Basic Form

THEOREM 3.1:   (*The Principle of Mathematical Induction–Basic Form*) Suppose that $S(n)$ represents a statement, depending on the parameter *n*, whenever *n* is a positive integer. Suppose that we can prove:

1. *The Basis Step*: $S(1)$ is true, and
2. *The Inductive Step*:   If $S(k)$ is true, then $S(k + 1)$ is true, whenever k is a positive integer.

Then it follows that $S(n)$ is true for all $n \in \mathbb{Z}_{+}$.

Proof: Consider the set A of positive integers n having the property that $S(n)$ is false. We need to show that $A = \varnothing$. Indeed, if $A \neq \varnothing$, then by the well-ordering principle, A must have a smallest element, call it a_0. By the basis step, $a_0 \neq 1$, so $a_0 > 1$. This means that $k \equiv a_0 - 1 \notin A$; in other words, $S(k)$ is true. But then by the inductive step, $S(k + 1)$ must also be true, and this means that $a_0 (= k + 1) \notin A$—a contradiction! □

The principle of mathematical induction can be thought of as an infinite domino effect: Visualize the statements $S(n)$ as a sequence of dominoes, lined up consecutively so that when one falls, the next will also fall; see Figure 3.1.

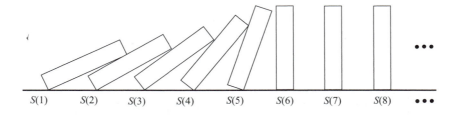

FIGURE 3.1:   The principle of mathematical induction is represented by an infinite sequence of dominoes. If we know that the first domino falls ($S(1)$ is true), and we know that the kth domino falling causes the $(k + 1)$st domino to fall ($S(k)$ is true implies $S(k + 1)$ is true) then all of the dominoes will fall (i.e., $S(n)$ is true for all n).

---

element (1), if x is any positive real or rational number, then so is $x/2$, so the set of positive real numbers and the set of positive rational numbers cannot have smallest elements. The well-ordering principle is either taken as an axiom of the positive integers, or is proved using other axioms. The interested reader may consult Enderton's book [End-77] for a complete development.

In doing the inductive step of a proof by induction, we assume that $S(k)$ is true for some positive integer k, and use this fact to prove that $S(k + 1)$ is also true. It is important to point out that we are only using the assumption that $S(k)$ is true to show that the truth of $S(k + 1)$ will follow (i.e., we are proving an implication). In an induction proof, the assumption that $S(k)$ is true (in the inductive step) is referred to as the *inductive hypothesis*. This is not the same as simply assuming that $S(k)$ is true for all k (this is what we are trying to prove). The principle of mathematical induction can be concisely expressed using quantifiers as follows: $[S(1) \wedge \forall k \ (S(k) \rightarrow S(k+1)] \Rightarrow \forall n \ S(n)$. We proceed with an example that aims to show how to write proofs by induction.

EXAMPLE 3.1: Use mathematical induction to prove the following summation formula for any positive integer n:[2]

$$\sum_{i=1}^{n} i \ (\equiv 1 + 2 + \cdots + n) = \frac{n(n+1)}{2}. \tag{1}$$

SOLUTION: For each $n \in \mathbb{Z}_+$, we let $S(n)$ denote the statement contained in equation (1). We need to go through the two steps of (a proof by) mathematical induction:

1. *Basis Step*: We must show directly that $S(1)$ is true. Substituting $n = 1$ into equation (1) above gives the equation: $\sum_{i=1}^{1} i (\equiv 1) = \frac{1(1+1)}{2}$. Since both sides clearly equal 1, the basis step is complete.

2. *Inductive Step*: Let $k \in \mathbb{Z}_+$ be arbitrary. We assume that $S(k)$ is true, i.e., that $\sum_{i=1}^{k} i = \frac{k(k+1)}{2}$. Using this assumption, our goal is to show that $S(k + 1)$ is also true, i.e., that $\sum_{i=1}^{k+1} i = \frac{(k+1)(k+2)}{2}$ (this equation came from (1) by substituting $k+1$ for n). Since $\sum_{i=1}^{k+1} i = \sum_{i=1}^{k} i + (k+1)$, the inductive hypothesis allows us to write $\sum_{i=1}^{k+1} i = \frac{k(k+1)}{2} + (k+1)$. The right side of the latter can be simplified

---

[2] From now on, we freely use *sigma notation* to represent summations. Sigma notation works as follows: If a_i are numbers (depending on the integer index i), and $r < s$ are integers, then $\sum_{i=r}^{s} a_i$ represents the sum $a_r + a_{r+1} + \cdots + a_s$.

using basic algebra as $\dfrac{k(k+1)}{2}+(k+1)=\dfrac{k(k+1)}{2}+\dfrac{2(k+1)}{2}=\dfrac{(k+1)}{2}[k+2]$. We

have thus arrived at the desired equation for $S(k+1)$: $\displaystyle\sum_{i=1}^{k+1}i=\dfrac{(k+1)(k+2)}{2}$. □

EXERCISE FOR THE READER 3.1: Use mathematical induction to prove the following summation formula for any positive integer n:

$$\sum_{i=1}^{n}i^2\ (\equiv 1^2+2^2+\cdots+n^2)=\frac{n(n+1)(2n+1)}{6}. \tag{2}$$

The reader is encouraged to study the mechanics of the above proof(s) carefully since all induction proofs share the same basic structure. Some comments are in order. There are other ways to prove such particular formulas,[3] but induction is ideally suited for many proofs such as these, and they tend to work rather mechanically once one has mastered the method. Notice that the inductive proof does not allow for the discovery of the formula, but rather serves only to prove the formula (provided it is true). The discovery of possible formulas, or more general relations is often done by performing experiments. Our next example will give a simple illustration of such experimentation. Formulas such as (1) and (2) can be derived using matrix methods (see p. 202 of [Sta-04]). Before moving on to our next example, we first elaborate on the point that it is a simple matter to extend mathematical induction to situations where the first proposition does not have to be indexed by 1.

The Principle of Mathematical Induction: General Form

COROLLARY 3.2: (*The Principle of Mathematical Induction–General Form*) Suppose that $S(n)$ represents a statement, depending on the parameter n, whenever n is an integer in the set $\{n\in\mathbb{Z}:n\geq N\}$, where N is some (fixed) integer. Suppose that we can prove:

---

[3] Here is a proof of identity (1) that was discovered by the illustrious German mathematician Carl Friedrich Gauss (1777–1855): Let S denote the sum $1+2+\cdots+n$. Writing this in the reverse order $S=n+(n-1)+\cdots+2+1$, and adding corresponding terms with the first expression for S, gives $2S=(n+1)+(n+1)+\cdots+(n+1)=n(n+1)$. Dividing the left and right sides by 2 gives (1). The amazing thing about this elegant discovery by Gauss is that he made it while he was in second grade. His teacher was trying to find something to occupy this precocious child for a while so she asked him to add up all of the numbers from 1 through 100. Two minutes later he came back and shocked his teacher by giving the correct answer.

1. *The Basis Step*: $S(N)$ is true, and
2. *The Inductive Step*: If $S(k)$ is true, then $S(k + 1)$ is true, whenever k is an integer in the set $\{n \in \mathbb{Z} : n \geq N\}$.

Then it follows that $S(n)$ is true for all $n \in \mathbb{Z}$ with $n \geq N$.

Proof: For $n \in \mathbb{Z}_+$, define the statement $T(n)$ to be the statement $S(n + N - 1)$. If we translate the above two steps for the S-statements into corresponding steps for the T-statements, we get precisely the two steps of Theorem 3.1 (for an inductive proof of the the the T-statements). Thus, from mathematical induction, the two steps would imply that $T(n)$ for each $n \in \mathbb{Z}_+$, and this translates into $S(n)$ being true for all $n \in \mathbb{Z}$ with $n \geq N$. □

EXAMPLE 3.2: Which of the two terms is eventually larger for positive integers n: 2^n or n^2?

SOLUTION: The table on the right shows a comparison of these two expressions for a few positive integers. It seems to appear that $2^n > n^2$, whenever $n \geq 5$. Let's use mathematical induction (in the general form of Corollary 3.2) to attempt to prove this conjecture. In the notation of the corollary, $S(n)$ denotes the statement $2^n > n^2$, and $N = 5$. In symbols, our goal is to prove that $\forall n \geq N[S(n)]$.

n	2^n	n^2
1	2	1
2	4	4
3	8	9
4	16	16
5	32	25
6	64	36
7	128	49
8	256	64

1. *Basis Step*: In the table, we have already verified $S(5)$, namely that $2^5 > 5^2$.

2. *Inductive Step*: Let k denote any integer that is at least 5. We must show that the implication $S(k) \to S(k+1)$ is valid. So we assume that $S(k)$ is true, i.e., that $2^k > k^2$, and we must try to use this hypothesis to deduce the validity of $S(k+1)$, i.e., that $2^{k+1} > (k+1)^2$. In order to make use of the inductive hypothesis, we break both sides of this latter inequality into pieces that are more related to the terms in the former inequality. First note that $2^{k+1} = 2 \cdot 2^k = 2^k + 2^k$, and that $(k+1)^2 = k^2 + 2k + 1$. Since we know (the induction hypothesis) that $2^k > k^2$, it is sufficient to delete these these terms from the respective sides of the inequality we are trying to show $(2^k + 2^k > k^2 + 2k + 1)$ and prove the resulting inequality, i.e., that $2^k > 2k + 1$. Put differently: If we can show that $2^k > 2k + 1$, then it will follow that $2^{k+1} > (k+1)^2$. But this task is easy if we once again invoke the inductive hypothesis and use the fact that $k \geq 5$: $2^k > k^2 \geq 5k = 2k + 3k > 2k + 1$. □

EXERCISE FOR THE READER 3.2: Obtain a formula for the sum of the first n odd positive integers: $1+3+5+\cdots+(2n-1)$, and then prove your formula by mathematical induction.

EXERCISE FOR THE READER 3.3: Find the flaw in the following "proof" by mathematical induction that all horses have the same color. We will show that any group of n horses must have the same color, where n can be any positive integer. Basis step: Any single horse ($n=1$) must have the same color as any horse in this single horse group. Inductive step: Assume that any group of k horses has the same color. Using this assumption, we must prove that a group of $k+1$ horses all have the same color. Take any horse in such a group, call it X. The rest of the horses must all have the same color, by the inductive hypothesis; say this color is brown. Now exchange X for another horse Y in the brown group. The group of k horses (with X but without Y) must all have the same color, so X too must be brown, and thus all horses in our $k+1$ sized group have the same color.

We next use induction to prove the following fact that was asserted in Section 1.3:

PROPOSITION 3.3: If a set A has a finite number n elements, then its power set $\mathscr{P}(A)$ (i.e., the set of all subsets of A) will have 2^n elements.

Proof: The proof will be done by induction on n, the number of elements of the set A.
1. *Basis Step*: $n=1$. A set with one element has the form $A=\{a\}$. Such a set has exactly two subsets: $\mathscr{P}(A)=\{\varnothing,\{a\}\}$, so indeed $\mathscr{P}(A)$ has $2=2^1=2^n$ elements.
2. *Inductive Step*: We assume that a set with k (= positive integer) elements will (always) have 2^k subsets. Our task is to show that a set with $k+1$ elements will have 2^{k+1} subsets. Let $A=\{a_1,a_2,\cdots,a_{k+1}\}$ be such a set, and let B be the subset of A with the last element removed: $B=\{a_1,a_2,\cdots,a_k\}$. Since B is a set with k elements, the inductive hypothesis tells us that B has 2^k subsets. Now, any subset of A must fall into one of two categories: either it does not contain a_{k+1}, and is thus a subset of B, or it does contain a_{k+1}, and thus can be written (uniquely) as $S\cup\{a_{k+1}\}$, where S is some subset of B. Since there are 2^k different subsets of each of these two forms, it follows that there are a grand total of $2^k+2^k=2\cdot2^k$ $=2^{k+1}$ subsets of A. This completes the inductive step, and hence the proof. $\square$

EXERCISE FOR THE READER 3.4: (a) Use mathematical induction to generalize the associative rules for unions and intersections of sets from Theorem 1.3(b): $(A\cup B)\cup C=A\cup(B\cup C)$, $(A\cap B)\cap C=A\cap(B\cap C)$, to unions and intersections of any finite collection of sets. It follows that in taking unions or intersections of any finite number of sets, parentheses are not needed.

(b) Use mathematical induction to establish the following extension of De Morgan's law (Theorem 1.3(d)) to finite collections of sets: $\sim(A_1 \cup A_2 \cup \cdots \cup A_n) = \sim A_1 \cap \sim A_2 \cap \cdots \cap \sim A_n$.

All of the proofs by induction presented thus far have involved establishing an equality or inequality. Mathematical induction is not limited to such analytic proofs; in our next example we prove a nonobvious geometric result.

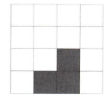

EXAMPLE 3.3: Prove that for any positive integer n, if a single square is removed from a $2^n \times 2^n$ grid of squares, then the resulting configuration can be partitioned (or tiled) by L-shaped tiles. By an L-shaped tile, we mean a union of three different squares that share a common vertex; see Figure 3.2. For a collection of L-shaped tiles to partition the configuration, we mean that different tiles are disjoint (expect for boundary edge intersections) and their union covers exactly the grid without the deleted square.

FIGURE 3.2: An L-shaped tile for Example 3.3.

SOLUTION: Our proof will be accomplished by induction on n. We let $S(n)$ denote the statement that any $2^n \times 2^n$ grid of squares with a single square removed can be partitioned (or tiled) by L-shaped tiles.

1. *Basis Step*: For $n = 1$, the proposition states that any 2×2 grid with one square removed can by partitioned by L-shaped regions. But this is easy since such a configuration <u>is</u> an L-shaped tile; see Figure 3.3.

FIGURE 3.3: The four possible configurations in the basis step for Example 3.3.

2. *Inductive Step*: In order to show that $S(k) \to S(k+1)$, we will assume that any $2^k \times 2^k$ grid with one square deleted can be partitioned into L-shaped tiles, and use this assumption to prove that any $2^{k+1} \times 2^{k+1}$ grid with a single square removed can also be partitioned into L-shaped tiles. A simple way to make this work is to first partition the $2^{k+1} \times 2^{k+1}$ grid into four $2^k \times 2^k$ grids (in the obvious and only way possible). Now, the removed square will lie in exactly one of these four smaller grids. We then insert an L-shaped tile with common vertex at the center

of the main grid (the common vertex of the four subgrids), that contains a square from each of the three other subgrids (which do not have the deleted square); see Figure 3.4. With this being done, each of the four $2^k \times 2^k$ grids now has exactly one square either deleted or covered by the single L-shaped region that was just deployed. It follows from the inductive hypothesis that each of these resulting $2^k \times 2^k$ configurations can be partitioned into L-shaped regions. These four partitions, along with the first L-shaped region that was separately deployed thus form a partition of the original grid with the single deleted square. The proof is complete. □

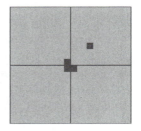

FIGURE 3.4: The inductive step for Example 3.3.

In practice, when one is going through the inductive step $(S(k) \to S(k+1))$ in a mathematical induction proof, it can sometimes happen that the single hypothesis $S(k)$ is not by itself adequate to deduce $S(k+1)$, but rather one needs some previous statement(s) $S(j)$ ($j \le k$). In other instances, these extra hypotheses, although not necessary, can make the inductive step easier. The good news is that these extra assumptions can be used freely, and this fact follows from the general form of mathematical induction. We formulate this more flexible version of mathematical induction in the following corollary.

Strong Mathematical Induction

COROLLARY 3.4: (*Strong Mathematical Induction*) Suppose that $S(n)$ represents a statement, depending on the parameter n, whenever n is an integer in the set $\{n \in \mathbb{Z} : n \ge N\}$, where N is some (fixed) integer. Suppose that we can prove:

1. *The Basis Step*: $S(N)$ is true, and
2. *The Inductive Step*: Whenever k is an integer in the set $\{n \in \mathbb{Z} : n \ge N\}$, we have $[S(N) \wedge S(N+1) \wedge \cdots \wedge S(k)] \to S(k+1)$.

Then it follows that $S(n)$ is true for all $n \in \mathbb{Z}$ with $n \ge N$.

Proof: This follows directly from Corollary 3.2, if we apply it to the modified statements $\tilde{S}(n) \equiv [S(N) \wedge S(N+1) \wedge \cdots \wedge S(n)]$.

Note that $\tilde{S}(n+1) = \tilde{S}(n) \wedge S(n+1)$. □

EXAMPLE 3.4: Suppose that we have an unlimited supply of three and five cent postage stamps. Show that for any (integer) postage value of at least eight cents, we can always find a combination of three and five cent stamps that will add up exactly to the desired value.

SOLUTION: We let $S(n)$ denote the statement that a postage value of n cents can be achieved using only three cent or five cent stamps. Our goal is to show that $\forall n \geq 8[S(n)]$. We prove this using strong mathematical induction.

1. *Basis Step*: To show $S(8)$ is true we simply observe that one five cent stamp and one three cent stamp will add up to give eight cents of postage.

2. *Inductive Step*: We let k be an integer greater than seven, and we assume that any postage value from eight cents through k cents can be achieved using five cent or three cents stamps. Our goal is to show that the postage value $(k + 1)$ cents can be so expressed. We may assume that $k > 9$, since if k were 8, then $k + 1 = 9$ cents could easily be expressed with 3 three cent stamps, while if k were 9, $k + 1 = 10$ cents and we could use 2 five cent stamps. By the strong induction hypothesis (since $k - 2 \geq 8$), $(k - 2)$ cents may be expressed using some combination of three cent and five cent stamps. If we add a three cent stamp to this combination, then we will have a way to express $(k + 1)$ cents using three cent or five cent stamps. □

The result of the above example could have been proved using ordinary induction (this will be the next exercise for the reader), but the additional hypotheses provided through strong induction gave us more flexibility. In practice, when writing induction proofs, it is not necessary to explicitly state whether one is using strong or ordinary induction; one may use the additional hypotheses whenever convenient.

EXERCISE FOR THE READER 3.5: Prove the result of Example 3.4 using ordinary mathematical induction (Corollary 3.2).

EXERCISE FOR THE READER 3.6: (*A Theorem for Chocoholics*) Chocolate bars are usually rectangular in shape and can be broken into squares. For the purpose of this problem, we thus assume that a chocolate bar is a rectangular grid of squares, and that any permissible "break" is made along one of the horizontal or vertical grid lines (so will not destroy any of the squares of the grid). Prove that to break a chocolate bar that consists of n squares into these n squares, it will always take $n - 1$ breaks.

Finite Geometric Series

We end this section with a useful formula involving finite sums of terms that are in a *geometric progression*, meaning that successive terms will always have the same ratio. If we denote this ratio by r, the terms of a geometric progression thus take the form $a, ar, ar^2, ar^3, \cdots$.

PROPOSITION 3.5: (*Finite Geometric Series*) If a and r are real numbers with $r \neq 0, 1$, and n is a nonnegative integer, then

$$\sum_{i=0}^{n} ar^i \ (\equiv a + ar + ar^2 + \cdots + ar^n) = \frac{ar^{n+1} - a}{r - 1} = a\left[\frac{r^{n+1} - 1}{r - 1}\right]. \tag{3}$$

NOTE: Although the formula is not valid in case r equals 0, or 1, in these cases the sum is trivial. When $r = 0$, the sum just has a single (nonzero) term a, and when $r = 1$, all of the $n+1$ terms equal a, so the sum is just $(n+1)a$.

Proof: We proceed by induction on n.

1. *Basis Step*: If $n = 0$, the right side of (3) equals $\dfrac{ar - a}{r - 1} = \dfrac{a(r-1)}{r-1} = a$, which is the (single-term sum making up the) left side.

2. *Inductive Step*: We let $k \geq 0$ be an integer, and assume the validity of (3) when $n = k$. The task is to show the validity in case $n = k + 1$:

$$\sum_{i=0}^{k+1} ar^i = \sum_{i=0}^{k} ar^i + ar^{k+1}$$

$$= \frac{ar^{k+1} - a}{r - 1} + ar^{k+1} \quad \text{(using the inductive hypothesis)}$$

$$= \frac{ar^{k+1} - a}{r - 1} + \frac{ar^{k+1}(r-1)}{r-1} = \frac{\cancel{ar^{k+1}} - a + ar^{k+2} - \cancel{ar^{k+1}}}{r - 1}$$

$$= \frac{ar^{k+2} - a}{r - 1}. \quad \square$$

EXERCISES 3.1:

1. (a) Use induction to prove that $\sum_{i=1}^{n} 2^i = 2^{n+1} - 2$, for any positive integer n.

   (b) Derive the identity of Part (a) using the geometric series identity (3).

   Suggestion: For Part (b), rewrite $\sum_{i=1}^{n} 2^i$ as (by substituting $i = j + 1$) $\sum_{j=0}^{n-1} 2^{j+1} = \sum_{j=0}^{n-1} 2 \cdot 2^j$.

2. (a) Use induction to prove that $\sum_{i=1}^{n} 3^i = (3^{n+1} - 3)/2$, for any positive integer n.

   (b) Derive the identity of Part (a) using the geometric series identity (3). (Use the suggestion of Exercise 1 with 2 changed to 3.)

3. Prove that for any positive integer n, we have $3 + 7 + 11 + \cdots + (4n - 1) = n(2n + 1)$.

4. Prove that for any positive integer n, we have:
$$1^2 + 3^2 + 5^2 + \cdots + (2n+1)^2 = (n+1)(2n+1)(2n+3)/3.$$

5. If n is a positive integer, and $a_1, a_2, \cdots, a_n$ are real numbers, prove the following *general triangle inequality*: $|a_1 + a_2 + \cdots + a_n| \leq |a_1| + |a_2| + \cdots + |a_n|$.

   Note: This generalizes the so-called *triangle inequality* for pairs of real numbers: $|x + y| \leq |x| + |y|$.

6. If n is a positive integer, and $x > -1$ is a real number, prove that $(1+x)^n \geq 1 + nx$.

7. Use induction to establish the following summation formula $\sum_{i=1}^{n} i^3 = [n(n+1)/2]^2$ $(n \in \mathbb{Z}_+)$.

8. Establish the following inequality: $1 + \dfrac{1}{4} + \dfrac{1}{9} + \cdots + \dfrac{1}{n^2} \leq 2 - \dfrac{1}{n}$ $(n \in \mathbb{Z}_+)$.

9. Establish the following identity: $1 \cdot 2 + 2 \cdot 3 + \cdots + n(n+1) = \dfrac{n(n+1)(n+2)}{3}$ $(n \in \mathbb{Z}_+)$.

10. (a) Use induction to prove the following identity: $\dfrac{1}{1 \cdot 2} + \dfrac{1}{2 \cdot 3} + \cdots + \dfrac{1}{n(n+1)} = \dfrac{n}{n+1}$ $(n \in \mathbb{Z}_+)$.
    (b) Give a noninductive proof of the identity in Part (a).
    Suggestion: For Part (b), use the algebraic identity $\dfrac{1}{i} - \dfrac{1}{i+1} = \dfrac{1}{i(i+1)}$.

11. Use induction to establish the following summation formula $\sum_{i=1}^{n} (-1)^i i^2 = (-1)^n [n(n+1)/2]$, for any positive integer n.

12. (a) Use induction to establish the identity $\left(1 - \dfrac{1}{2}\right) \cdot \left(1 - \dfrac{1}{3}\right) \cdots \cdots \left(1 - \dfrac{1}{n+1}\right) = \dfrac{1}{n+1}$, for any positive integer n.
    (b) Give a noninductive proof of the identity of Part (a).

13. (a) What are the postages that can be achieved by combining an unlimited number of two cent and/or five cent stamps?
    (b) Prove that the answer you found in Part (a) is correct using mathematical induction.

14. (a) What are the postages that can be achieved by combining an unlimited number of four cent and/or five cent stamps?
    (b) Prove that the answer you found in Part (a) is correct using mathematical induction.

15. (*Finite Geometric Series*) Give a noninductive proof of the identity (3) of Proposition 3.5.
    Note: The identities of Exercises 1 and 2 are special cases of this (3).
    Suggestion: Expand and simplify the expression $(1 - r)(1 + r + r^2 + \cdots + r^n)$.

16. (*An Interesting Fractional Identity for Positive Integers*) Use induction to prove that any positive integer n can be expressed as the following sum:
$$n = \sum_{\{a_1, a_2, \cdots, a_k\} \subseteq \{1, 2, \cdots, n\}} \frac{1}{a_1 \cdot a_2 \cdots \cdots a_k},$$
    where the sum on the right is taken over all nonempty subsets $\{a_1, a_2, \cdots, a_k\} \subseteq \{1, 2, \cdots, n\}$.

17. Establish the following trigonometric identity:
$$\sqrt{2 + \sqrt{2 + \sqrt{2 + \cdots + \sqrt{2}}}} = 2\cos\left(\frac{\pi}{2^{n+1}}\right) \quad (n \in \mathbb{Z}_+),$$
    where there are n 2s in the expression on the left side.

18. Establish the following trigonometric identity:

$$\sum_{j=1}^{n} \sin(j\theta) = \frac{\sin((n+1)\theta/2)\sin(n\theta/2)}{\sin(\theta/2)} \quad (n \in \mathbb{Z}_+),$$

provided that $\sin(\theta/2) \neq 0$.

19. (a) Use mathematical induction to establish the validity of the following formula, whenever n is

a positive integer: $\sum_{i=1}^{n} i^3 = \left(\sum_{i=1}^{n} i\right)^2$.

(b) Give a direct proof of the formula in Part (a) using formula (1) (of Example 3.1), and the result of Exercise 7.
Suggestion: For Part (a), formula (1) of Example 3.1 can be of use.

20. Use mathematical induction to establish the following extension of De Morgan's law (Theorem 1.3(d)) to finite collections of sets: $\sim (A_1 \cap A_2 \cap \cdots \cap A_n) = \sim A_1 \cup \sim A_2 \cup \cdots \cup \sim A_n$.

21. Use mathematical induction to establish the following extension of the distributive law (Theorem 1.3(c)) to finite collections of sets:
$$A \cap (B_1 \cup B_2 \cup \cdots \cup B_n) = (A \cap B_1) \cup (A \cap B_2) \cup \cdots \cup (A \cap B_n).$$

22. (*Associativity for Conjunctions of Finite Collections of Logical Statements*) Use mathematical induction to prove that in taking conjunctions of a finite collection of three or more logical statements, the order in which the conjunctions are taken (i.e., the way that the parentheses are grouped) does not matter. This generalizes the associative law $(P \wedge Q) \wedge R \equiv P \wedge (Q \wedge R)$ of Theorem 1.1(b), and means that a conjunction such as $P_1 \wedge P_2 \wedge \cdots \wedge P_n$ makes sense without any parentheses being needed.
Suggestion: In the inductive step, treat the innermost conjunction as a separate statement and apply the inductive hypothesis.

23. (*Associativity for Disjunctions of Finite Collections of Logical Statements*) Use mathematical induction to prove the analogue of Exercise 22 for disjunctions.

24. Prove that with an unlimited supply of 10 cent and 13 cent stamps, any postage greater than 107 cents can be achieved using 10 cent and/or 13 cent stamps.

25. How many regions does the plane get partitioned into by n lines, all of which intersect at the same point? Prove your result with mathematical induction.

26. Prove that any n distinct lines drawn in the plane such no two are parallel and no three intersect at the same point will always partition the plane into $(n^2 + n + 2)/2$ regions.

27. Prove that for any positive integer $n \geq 2$, an $n \times 6$ grid of squares can always be partitioned into L-shaped tiles (see Example 3.3).

28. How many different ways are there to tile a $2 \times n$ grid of squares using 2×1 or 1×2 (domino) shaped tiles? Prove your answer using mathematical induction.

29. How many handshakes will there be in a meeting of n diplomats if each diplomat shakes the hand of every other diplomat (exactly once)? Prove your answer using mathematical induction.

30. (*Geometry*) Recall that a polygon is a region in the plane whose boundary consists of line

segments. A polygon with n edges is called an n-gon (thus an n-gon will have n vertices, which are the points where adjacent edges meet). 3-gons are just triangles, 4-gons are quadrilaterals (of which rectangles are special cases), 5-gons are pentagons, etc. Use induction to prove that the sum of the interior angles of any n-gon is $(n-2) \cdot 180°$.

NOTE: The *Jordan curve theorem* states that any (continuous) closed curve C in the plane separates the plane into exactly two components, an *inside* and an *outside*, the latter containing all far away points. Informally this means that any two points on the inside can be joined by drawing a path without lifing your pencil and without crossing C, similarly for any two points on the outside. But no point in the inside can be joined to a point on the outside in this fashion. Most people think this result is rather obvious when they learn about it. But it is not an easy theorem to prove, even in the polygonal case. The paper [Tve-80] proves the Jordan curve theorem by first proving it for polygonal curves, and then approximating general closed curves with polygons. Readers may assume the Jordan curve theorem (in the polygonal case) when doing this exercise.

31. Use induction to establish the identity $\left(1-\dfrac{1}{\sqrt{2}}\right) \cdot \left(1-\dfrac{1}{\sqrt{3}}\right) \cdots \left(1-\dfrac{1}{\sqrt{n}}\right) < \dfrac{2}{n^2}$, for any positive integer n.

    Suggestion: First (use induction to) prove the inequality $n \geq \sqrt{n+1}$, whenever $n \geq 2$ is an integer.

32. Use induction to establish the identity $1 + \dfrac{1}{\sqrt{2}} + \dfrac{1}{\sqrt{3}} + \cdots + \dfrac{1}{\sqrt{n}} < 2\sqrt{n}$, for any positive integer n.

33. Use induction to establish the inequality $\left(1+\dfrac{1}{1^3}\right) \cdot \left(1+\dfrac{1}{2^3}\right) \cdot \left(1+\dfrac{1}{3^3}\right) \cdots \left(1+\dfrac{1}{n^3}\right) < 3$, for any positive integer n.

    Suggestion: The above inequality is not so amenable to an inductive proof, but the stronger inequality $\left(1+\dfrac{1}{1^3}\right) \cdot \left(1+\dfrac{1}{2^3}\right) \cdot \left(1+\dfrac{1}{3^3}\right) \cdots \left(1+\dfrac{1}{n^3}\right) < 3 - \dfrac{1}{n}$, turns out to be easier to deal with.

34. (*Convex Functions*) If $I = (a, b)$ is an open interval in the set of real numbers $\mathbb{R}$, a function $f : I \to \mathbb{R}$ is called *convex* if for each real number $\lambda \in [0,1]$, and any two numbers $x, y \in I$, we have $f(\lambda x + (1-\lambda)y) \leq \lambda f(x) + (1-\lambda)f(y)$.

    (a) Which of these functions is convex (with $I = \mathbb{R}$): $f(x) = e^x$, $f(x) = x^2$, $f(x) = -x^2$?

    (b) Prove that for a convex function, $f : I \to \mathbb{R}$, if n is a positive integer, $x_1, x_2, \cdots, x_n \in I$, and $\lambda_1, \lambda_2, \cdots, \lambda_n$ are nonnegative real numbers that add up to 1 $(\sum_{i=1}^{n} \lambda_i = 1)$, then we have $f(\sum_{i=1}^{n} \lambda_i x_i) \leq \sum_{i=1}^{n} \lambda_i f(x_i)$.

35. (*The Arithmetic–Geometric Mean Inequality*) For a set of positive real numbers $a_1, a_2, \cdots, a_n$ $(n \in \mathbb{Z}_+)$, their *arithmetic mean* is simply their average: $(a_1 + a_2 + \cdots + a_n)/n$, and their *geometric mean* is defined to be $(a_1 \cdot a_2 \cdots a_n)^{1/n}$. It turns out that the geometric mean can never be larger than the arithmetic mean. This inequality is known as the *arithmetic-geometric mean inequality*: $(a_1 \cdot a_2 \cdots a_n)^{1/n} \leq (a_1 + a_2 + \cdots + a_n)/n$. Prove this inequality.

    Suggestion: First use mathematical induction to prove the inequality in the case that n is a power of two: $n = 2^m$ (i.e., use induction on the exponent m). After this, to deal with the case where n is not a power of 2, let $N = 2^m$ denote the first power of 2 that exceeds n. For a given set of real numbers $a_1, a_2, \cdots, a_n$, let $\alpha \equiv (a_1 + a_2 + \cdots + a_n)/n$, set $a_{n+1} = a_{n+2} = \cdots a_N = \alpha$, and deduce the desired inequality involving $a_1, a_2, \cdots, a_n$, by applying the arithmetic-geometric mean inequality to the larger set of $N = 2^m$ numbers $a_1, a_2, \cdots, a_N$.

36.   (*Hall's Marriage Theorem*)[4]   Suppose that we have a set of n boys and a set of n girls. Each girl has a certain list of boys that she likes and would find acceptable for marriage. The marriage problem is to marry off all of the boys and girls (in pairs, of course) in such a way that each girl likes her future husband. In order for this to be possible, it is clearly necessary that any set of r girls $(r \le n)$ must collectively like at least r different boys. Here is a more precise formulation: For each girl g, let $M(g)$ be the set of boys she likes. The condition says that given r different girls: $g_1, g_2, \cdots, g_r$, the union $\bigcup_{i=1}^{r} M(g_i)$ has cardinality at least r. This latter condition is called *Hall's condition*. Hall's theorem is that the converse implication holds, i.e., if Hall's condition holds, then each of the girls can be married to a boy that she likes. Prove this.
Suggestion: Proceed by induction on n. For the inductive step (with $k + 1$ boys and girls), split off into two cases: *Case 1*: Any set of r girls $(1 \le r \le k)$ collectively likes at least $r + 1$ boys. Here, simply marry off one of the girls to a boy she likes; the remaining k boys and girls will still satisfy Hall's condition. *Case 2*: There is a set of r girls $(1 \le r \le k)$ that collectively likes exactly r boys. Reason that the complementary set of $k + 1 - r$ girls can be paired off with the complementary set of boys.

3.2: RECURSION

Infinite Sequences

Much of the material in this section both concerns and can be motivated by (infinite) *sequences*.

DEFINITION 3.1: An **(infinite) sequence** is simply a function from a set of integers of the form $\{n \in \mathbb{Z} : n \ge N\}$ into some set S. Usually $N = 0$ or $N = 1$. In mathematical notation, the values of a particular sequence are usually denoted using subscripts, rather than by using functional notation, but in computer programming, the functional notation is usually adhered to (since subscripts are not natural for single-line computer codes). Thus, if a particular sequence is denoted by the letter a, it can be denoted mathematically by $\{a_n\}_{n=N}^{\infty}$, where $a_n = a(n)$.

---

[4] Philip Hall (1904–1982) was an English mathematician who spent most of his career as a distinguished professor at Cambridge. He is famous for his numerous important discoveries in the field of group theory. He was also quite a linguist and, during World War II, he worked for the British intelligence agency to help them break codes of the enemy nations.

Recursion and Recursively Defined Sequences

The term **recursion** is a general one that pertains to the construction of sequences of numbers or other objects. Up to now, all such sequential objects and functions that we have dealt with were **explicit**, in that they (or their values) were constructed (or solved for) directly. Often such explicit constructions are not available; the construction of the object or sequence value corresponding to a certain index n will depend on objects (or sequence values) with lower-valued indices. We begin with a very simple example of a sequence that can be defined both explicitly and recursively.

EXAMPLE 3.5: The sequence $\{a_n\}_{n=1}^{\infty}$ of odd positive integers: $1, 3, 5, \cdots$ has a simple explicit representation scheme: $a_n = 2n - 1$, as is easily verified. The sequence can also be described using the following recursive definition:

$$\begin{cases} 1.\ (\textit{Basis Step})\ \ a_1 = 1. \\ 2.\ (\textit{Recursive Step})\ \text{For } n \geq 1,\ a_{n+1} = a_n + 2. \end{cases}$$

By the principle of mathematical induction, it follows that this recursive definition defines a_n, for any positive integer n. Recursive definitions are less efficient to work with than explicit ones. For example, if we wanted to compute a_5 using the explicit formula, we would simply substitute $n = 5$ into the formula to get $a_5 = 2 \cdot 5 - 1 = 9$. If we instead had to use the recursive definition, we would first need to find a_1, a_2, a_3, and, a_4 (in this order). Here are all of the steps: First the basis step tells us that $a_1 = 1$. The recursive step always tells us how to get the next value of the sequence from the value that we know. So it first tells us that (put $n = 1$); $a_2 = a_{1+1} = a_1 + 2$, so since we know $a_1 = 1$, we conclude that $a_2 = 1 + 2 = 3$. In the same fashion, now that we know a_2, we can use the recursive step (by putting $n = 2$) to get $a_3 = a_{2+1} = a_2 + 2 = 3 + 2 = 5$. From our newly acquired knowledge (a_3), the recursive step can get us the next term: $a_4 = a_{3+1} = a_3 + 2 = 5 + 2 = 7$. Finally, one last application of the recursive step gets us to the term that we wanted: $a_5 = a_{4+1} = a_4 + 2 = 7 + 2 = 9$ (whew!). We point out that the recursive step could be equivalently expressed as: For $n \geq 2$, $a_n = a_{n-1} + 2$.

From the above example, it is already clear that explicit definitions are much more efficient and desirable than recursive ones, but it is unfortunately not always possible to convert recursive definitions to explicit ones. Recursive definitions are often much more simple to formulate and are very natural in many applications. Recursive definitions often arise naturally in applications and problems where one needs to count numbers of possibilities in a system of size n. Often it is difficult to count directly, but much simpler to make such count in terms of the corresponding counts for smaller systems.

EXAMPLE 3.6: (*The Towers of Hanoi*) A puzzle game known as the *Towers of Hanoi* was quite popular in Europe during the nineteenth century. As with many such interesting puzzles (like the *Rubik's Cube* puzzle of the twentieth century), there are strong mathematical underpinnings that lead to solutions. In this example we will describe the Towers of Hanoi puzzle, and use recursion to solve it.

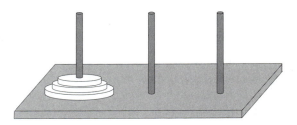

FIGURE 3.5: Illustration of the Towers of Hanoi puzzle with three disks.

The puzzle consists of three pegs on a board and a set of different sized disks. The disks are initially placed on the first peg where larger disks always lie below smaller disks; Figure 3.5 shows this initial configuration when there are three disks. The objective of the game is to transfer all of the disks to the third peg so that smaller pegs always lie above larger ones, and that this be done in the smallest number of "moves." A "move" consists of moving a single disk from one peg to another peg with the restriction that a larger disk cannot be moved on top of a smaller disk. The puzzle originally came with a total of 64 different sized disks (not all had to be used in each game) and a story of monks working in a tower of a temple in Hanoi on the original 64-disk puzzle with the disks made of gold. The legend states that once the monks have completed the game (of transferring the 64 disks to the third peg following allowable moves), the world will come to an end.

For each positive integer n, we let h_n denote the minimum number of moves needed to solve the Towers of Hanoi puzzle with n (different sized) disks. We will derive a recursive formula for the sequence $\{h_n\}_{n=1}^{\infty}$. Clearly $h_1 = 1$, since a 1-disk game can be solved in a single move of the single disk. Now assume that $n > 1$. We will derive an expression for h_n in terms of h_{n-1}. In order to solve the n-disk problem, we will need to get the largest disk to the bottom of the third peg in as few moves as possible. The only way to do this is to move the remaining $(n-1)$ disks, which lie above the largest disk to the second peg. This task is essentially equivalent to the $(n-1)$-problem, except that the intermediate second peg now plays the role of the third peg, and the largest nth disk is irrelevant (since it cannot get moved to the third peg until the disks above it have been moved to the second peg). After these h_{n-1} moves have been completed, the largest peg is then moved to the (bottom of the) third peg, and then, in the same fashion, h_{n-1} moves will be required to transfer the remaining disks to the third peg. This argument proves that $h_n = 2h_{n-1} + 1$. We have thus established the following recursive definition of the Towers of Hanoi puzzle sequence:

$$\begin{cases} 1. \ (\textit{Basis Step}) \ \ h_1 = 1. \\ 2. \ (\textit{Recursive Step}) \ \text{For } n \geq 2, \ h_n = 2h_{n-1} + 1. \end{cases}$$

This recursive definition is easy to convert to an explicit one. One general method that sometimes works is to repeatedly apply the recursive step to the nth term until all instances of the sequence have disappeared. Here is how this would work for the above sequence:

$$\begin{aligned} h_n = 2h_{n-1} + 1 &= 2(2h_{n-2} + 1) + 1 = 2^2 h_{n-2} + 2^1 + 1 \\ &= 2^2(2h_{n-3} + 1) + 2^1 + 1 = 2^3 h_{n-3} + 2^2 + 2^1 + 1 \\ &= 2^3(2h_{n-4} + 1) + 2^2 + 2^1 + 1 = 2^4 h_{n-4} + 2^3 + 2^2 + 2^1 + 1 \\ &\vdots \\ &= 2^{n-2}(2h_{n-(n-1)} + 1) + 2^{n-3} + 2^{n-4} + \cdots + 1 \underset{\substack{\text{since} \\ h_{n-(n-1)} = h_1 = 1}}{=} 2^{n-1} + 2^{n-2} + \cdots + 1. \end{aligned}$$

We have thus obtained an explicit expression for h_n as a finite geometric series. The identity (3) (of Proposition 3.5) lets us express this sum as (using these parameters in (3) $a \to 1, r \to 2, n \to n-1$): $h_n = (2^n - 1)/(2 - 1) = 2^n - 1$.[5] In particular, going back to the legend of the Vietnamese monks, it would take $2^{64} - 1$ moves before the world came to an end. Even if they arranged to make one move per second 24/7, this would still take over half a trillion years to complete! This works out to over 100 times the current age of our solar system, so our world will be safe for quite some time, at least from this form of destruction.

We point out such clean and simple derivations of recursive and explicit solutions, as we obtained for the Towers of Hanoi puzzle, are not always so feasible. Such an example is the simple variation of the Towers of Hanoi puzzle where a fourth peg is added, but all else stays the same. This variation is known as *Reve's puzzle*; it was introduced in 1907, but no one has yet found an explicit solution for it, and not even a recursive solution. Even Donald Knuth, one of the most prominent and influential computer scientists of his era, is reported to have said about Reve's puzzle that "I doubt if anyone will ever resolve this … it is truly difficult."

---

[5] This formula could have also been discovered by doing some experiments directly from the recursion formula: since $h_1 = 1, h_2 = 3, h_3 = 7, h_4 = 15,$ one should eventually recognize this sequence coincides with $2^1 - 1, 2^2 - 1, 2^3 - 1, 2^4 - 1, \cdots$. Then (an easy) induction proof is needed to prove the formula $h_n = 2^n - 1$. The method presented in the text, although (in this case) longer, is more systematic and reliable.

EXERCISE FOR THE READER 3.7: Suppose Rose gets a government job with a starting annual salary of \$32,500. Each year her salary increases by 3.5% of the previous year's salary (to adjust for inflation) with an additional \$1000 added.
(a) Represent Rose's salary as a recursively defined sequence $\{r_n\}$.

(b) Obtain an explicit formula for r_n.

Although the recursive step of the above two examples used only the immediately preceding term to define the next term, in general, it is possible for the next term to depend on any number of preceding terms. Our next example gives a famous recursively defined sequence in which the recursive step uses the two previous terms to define the next term in the sequence.

The Fibonacci Sequence

EXAMPLE 3.7: (*The Fibonacci Sequence*)[6] The **Fibonacci sequence** $\{f_n\}_{n=1}^{\infty}$, is defined recursively as follows:

$$\begin{cases} 1.\ (\textit{Basis Steps})\ \ f_1 = 1,\ \ f_2 = 1, \\ 2.\ (\textit{Recursive Step})\ \text{For}\ n \geq 3,\ f_n = f_{n-1} + f_{n-2}. \end{cases}$$

Since the formula in the recursive step requires the two preceding values of the sequence, it is necessary that the first two sequence values are specified in order for the process to work. It follows from (strong) mathematical induction that the above recursion scheme does indeed define the infinite sequence. We use the recursive step to find the next few values of the Fibonacci sequence:

$$f_3 = f_2 + f_1 = 1 + 1 = 2,$$
$$f_4 = f_3 + f_2 = 2 + 1 = 3,$$
$$f_5 = f_4 + f_3 = 3 + 2 = 5,$$
$$f_6 = f_5 + f_4 = 5 + 3 = 8,$$
$$f_7 = f_6 + f_5 = 8 + 5 = 13.$$

In writing computer programs for recursively defined sequences, one natural scheme is to use a for loop within the program that repeats the recursion formula. Many computing platforms have a very useful feature that allows a program to call on itself. The former type of program, or in general, any that repeatedly computes

---

[6] Fibonacci (Italian for Son of Boncaccio) was the nickname of the Italian mathematician Leonardo Pisano (1180–1250). Fibonacci spent his childhood in and was educated in Algeria, where his father worked as a diplomat. He later moved back to Italy and became a prominent mathematician who wrote several important books (and this was in the era before the press was invented, so all copies had to be handwritten). He is credited with bringing the Arabic (decimal) number system to Europe. Fibonacci invented his namesake sequence in the solution of a problem in population growth. The sequence has numerous applications and there is even a journal (*The Fibonacci Quarterly*) that is dedicated to applications of this important sequence.

and updates everything internally within a single run of the program is usually called an **iterative program**. A program of the latter type that calls on itself with smaller sized inputs is called a **recursive program**. The next example will contrast these two types of programs by considering the problem of writing a program for computing terms of the Fibonacci sequence. Although we will see that recursive programs are often much more natural and easier to write, their performance can sometimes be much slower than iterative programs. Performance issues of programs are very important (correctness of a program is assumed), and will be considered in detail in Chapter 7.

EXAMPLE 3.8: (a) Write an iterative pseudocode program, call it `y = fibonacci(n)`, whose input `n` is a positive integer and whose output `y` is the *n*th term of the Fibonacci sequence, where the program is <u>not</u> allowed to call on itself. (b) Write a recursive pseudocode program, call it `y = fibonacciv2(n)`, whose input `n` is a positive integer and whose output `y` is the *n*th term of the Fibonacci sequence, where the program does call on itself.

SOLUTION: Part (a): Since we only care about the *n*th term, although we will need to compute previous terms, these can be overwritten in the course of the program. The following pseudocode will perform the desired task.

```
PROGRAM:   y = fibonacci(n)
IF n < 3, THEN OUTPUT y = 1, STOP
ELSE    fn_oneb4 = 1, fn_twob4 = 1
        FOR k = 3 TO n
            y = fn_oneb4 + fn_twob4   (kth term of Fibonacci sequence)
            fn_twob4 = fn_oneb4, fn_oneb4 = y  (update previous terms for
                                                             next iteration)
        END k FOR
END IF
OUTPUT y
```

Part (b): If the program is allowed to call on itself, the program can be further simplified:

```
PROGRAM:   y = fibonacciv2(n)
IF n < 3, THEN set y = 1,   ENDIF
ELSE   y = fibonacciv2(n-1) + fibonacciv2(n-2)
OUTPUT y
```

The reader is advised to carefully compare these two programs and understand how they function. Although the recursive program of Part (b) is simpler than the iterative one, the former is much less efficient, as is demonstrated in Figure 3.6, which compares the number of function calls and information flows that the programs would need to go through to evaluate f_{10}. Each of the dots having (at least) two arrows going to it represents a program evaluation (and hence an addition). Notice also that the arrows on the recursive algorithm go both ways: the recursive algorithm calls on itself (in other words, it subcontracts some of the jobs that it needs to get done), and then takes back the information when it is available.   One reason that the recursive algorithm takes so much more work is that information needs to get reconstructed from scratch each time it needs to be accessed, so there are numerous duplicated computations, whereas the iterative algorithm uses each intermediate Fibonacci value exactly twice.   The reader is encouraged to write both programs into their computing language of choice and compare performances. The iterative version will be able to deal quickly with inputs of any needed size (10s of digits, for example) whereas the recursive one will be left hanging at even moderately sized inputs of say $n = 50$.   A more quantitative comparison of these two programs is the subject of the next exercise for the reader.

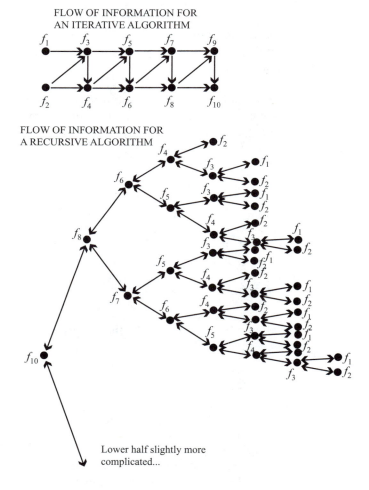

FIGURE 3.6:   Information flow diagrams for an instance of the iterative Fibonacci program (top) and the recursive version (bottom) of Example 3.8.

Although it is not at all obvious, it is possible to find an explicit formula for the Fibonacci sequence. The formula is as follows:

$$f_n = \frac{1}{\sqrt{5}}\left(\frac{1+\sqrt{5}}{2}\right)^n - \frac{1}{\sqrt{5}}\left(\frac{1-\sqrt{5}}{2}\right)^n. \tag{4}$$

(The reader is encouraged to verify that the above values of the Fibonacci sequence check with this formula.) We will show a general procedure later in this section that will allow us to obtain explicit formulas for general classes of recursively defined sequences that include, in particular, the Fibonacci sequence. Even with such results, however, we stress that not all recursively defined sequences have an explicit formula, or when there is one, it may be difficult or not feasible to find it.

EXERCISE FOR THE READER 3.8: (*Quantitative performance comparison: iterative algorithm versus recursive algorithm*)
(a) How many mathematical operations are required for the iterative program `fibonnacci` of Example 3.8 to compute the output when a positive integer n is inputted? (Your answer should be explicit, and will depend on n.)
(b) How many mathematical operations are required for the recursive program `fibonnacciRec` of Example 3.8 to compute the output when a positive integer n is inputted? Either derive and explicit formula or a recursive formula for this sequence of numbers.
(c) With the same number of mathematical operations that `fibonnacciRec` would require to compute the output when $n = 30$, how large an input value n could `fibonnacci` deal with?
Suggestion: The answer to Part (a) can easily be deduced by examination of (the top part of) Figure 3.6. For Part (b), do some hand calculations with small values of n to find a candidate formula, and prove it by induction.

EXERCISE FOR THE READER 3.9: (*The factorial sequence*) The **factorial** of any nonnegative integer n, denoted $n!$, is defined recursively by:

$$\begin{cases} 1. \ (\textit{Basis Step}) \quad 0! = 1. \\ 2. \ (\textit{Recursive Step}) \ \text{For } n \geq 1, \ n! = n(n-1)!. \end{cases}$$

Compute $1!$, $2!$, $3!$, $4!$, and $5!$. Then find an explicit formula for $n!$ (when $n > 0$), and use mathematical induction to prove that it is correct.

EXERCISE FOR THE READER 3.10: (a) Use the following recursive definition of a sequence $\{a_n\}_{n=1}^{\infty}$, to compute a_6.

$$\begin{cases} 1. \ (\textit{Basis Step}) \ a_1 = 1, \ a_2 = 3, \ a_3 = 2. \\ 2. \ (\textit{Recursive Step}) \ \text{For } n \geq 4, \ a_n = a_{n-1} + a_{n-2}a_{n-3}. \end{cases}$$

(b) Write a computer program that will compute the nth term of this sequence, and run it to (try) to compute a_9 and a_{15}.

In general, a sequence can be recursively defined where the recursive step defines the "next" term in terms of previously occurring elements of the sequence that go back any number of (k) steps. For this to work, the basis steps will need to define the first k elements of the sequence. We now formulate this general definition.

Recursive Sequences of Higher Degree

DEFINITION 3.2: If N is an integer, and k is a nonnegative integer, a definition of the form:

$$\begin{cases} 1.\ (\textit{Basis Step})\ a_N = \alpha_1,\ \ a_{N+1} = \alpha_2, \cdots\ a_{N+k-1} = \alpha_k, \\ 2.\ (\textit{Recursive Step})\ \text{For } n \geq N,\ a_{n+k} = f(a_n, a_{n+1}, \cdots, a_{n+k}), \end{cases}$$

where $f(a_n, a_{n+1}, \cdots, a_{n+k})$ is some function of the k variables $a_n, a_{n+1}, \cdots, a_{n+k}$, will define an infinite sequence $\{a_n\}_{n=N}^{\infty}$.   Such a sequence is said to be **recursively defined**, and to have **degree** k.

It follows from the principle of (strong) mathematical induction that the above two steps define a_n for all $n \in \mathbb{Z},\ n \geq N$.   (This routine verification is left as an exercise at the end of the section.)   The reader should also notice that if the values $\alpha_1, \alpha_2, \cdots, \alpha_k$ are viewed as parameters, the definition will define a sequence regardless of what values are substituted for these parameters in the basis step (as long as the functional values $f(a_n, a_{n+1}, \cdots, a_{n+k})$ that are needed in the recursive step are always defined).

As in Example 3.8, it is always possible to create an iterative program to compute values of a recursively defined formula (even when there is no explicit formula). This is unfortunately not true for more general recursive algorithms, which we define next.

DEFINITION 3.3:   An **algorithm** is a sequence of steps (or a procedure) for determining the output (answer) for a given problem starting with the inputs. An algorithm is called **recursive** if it starts off by showing as separate case(s) how to solve the given problem with the "smallest" inputs (*the Basis Step(s)*), and then shows how to solve a general case in terms of repeated applications of solving the problem with smaller inputs (*the Recursive Step*).

EXERCISE FOR THE READER 3.11: (a) Write a recursive program, $y =$ `EvenRec(n)`, that will input a positive integer `n`, and will output the *n*th even positive integer. Thus, for example, the output of `EvenRec(3)`, should be 6.
(b) Write an iterative version `EvenIter(n)` of the program of Part (a).
(c) Compare the number of arithmetic operations that each of the above two programs require with an input value of `n`.

We will discuss algorithms in greater detail in Chapter 5. But we give here an example of a recursive algorithm that is used to compute the determinant of a square matrix (i.e., a square array of numbers). Matrices will be discussed in greater generality in the next chapter, but the following example is self-contained. Below are square matrices of sizes $1 \times 1, 2 \times 2,$ and $3 \times 3,$ respectively:

$$[2], \quad \begin{bmatrix} -1 & 4 \\ 3 & 6 \end{bmatrix}, \quad \begin{bmatrix} 8 & 2 & -2 \\ 4 & 0 & 6 \\ -2 & 6 & 5 \end{bmatrix}.$$

EXAMPLE 3.9: In linear algebra the **determinant** of a square $n \times n$ matrix A, denoted **det(A)**, is a certain number associated with the matrix that can be determined using the following recursive algorithm called *cofactor expansion along the first row.*
1. *Basis Step: Case n* = 1: If $A = [a]$ (1×1 matrix), det(A) = a.

Case n = 2: If $A = \begin{bmatrix} a & b \\ c & d \end{bmatrix}$, det($A$) = $ad - bc$, that is, just the product of the

main diagonal entries (top left to bottom right) less the product of the off diagonal entries (top right to bottom left).
2. *Recursive Step*: $n > 2$:   We follow the standard practice of using double subscripts for the entries of $n \times n$ matrix[7]

$$A = \begin{bmatrix} a_{11} & a_{12} & \cdots & a_{1n} \\ a_{21} & a_{22} & \cdots & a_{2n} \\ \vdots & \vdots & \ddots & \\ a_{n1} & a_{n2} & & a_{nn} \end{bmatrix}.$$

For any entry a_{ij} of the $n \times n$ matrix A, we define the corresponding **submatrix** A_{ij} to be the $(n-1) \times (n-1)$ matrix obtained from A by deleting the row and column of A that contain the entry a_{ij} . Thus, for example, in case $n = 3$, the matrix A_{13} is the 2×2 matrix obtained from A by deleting the row and column

---

[7] One can find the formal definition of the determinant in books on linear algebra such as [HoKu-71].
See [HoKu-71] for a proof that the above recursive definition is equivalent to the formal definition.

determined by a_{13} (so the first row, and the third column). This operation is shown below.

$$A_{13} = \begin{bmatrix} a_{11} & a_{12} & a_{13} \\ a_{21} & a_{22} & a_{23} \\ a_{31} & a_{32} & a_{33} \end{bmatrix} = \begin{bmatrix} a_{21} & a_{22} \\ a_{31} & a_{32} \end{bmatrix}.$$

The recursive formula for the determinant of A is now given by:

$$\det(A) = a_{11}\det(A_{11}) - a_{12}\det(A_{12}) + a_{13}\det(A_{13}) - \cdots + (-1)^{n+1}a_{1n}\det(A_{1n}). \qquad (5)$$

(Note that the signs alternate as we move across the entries of the top row of A.) It is proved in linear algebra books (e.g., see [HoKu-71]) that one could instead take the corresponding (cofactor) expansion along any row or column of A, using the following rule to choose the alternating signs: The sign of the term containing $\det(A_{ij})$ is $(-1)^{i+j}$.

Thus the determinant of any $n \times n$ matrix (with $n > 2$) can be computed by repeatedly applying the recursion step (with (5)) until we have a combination that involves only 2×2 determinants, which can be handled directly with the basis step. For example, we may compute

$$\det\begin{bmatrix} 8 & 2 & -2 \\ 4 & 0 & 6 \\ -2 & 6 & 5 \end{bmatrix} = 8\det\begin{bmatrix} 0 & 6 \\ 6 & 5 \end{bmatrix} - 2\det\begin{bmatrix} 4 & 6 \\ -2 & 5 \end{bmatrix} + (-2)\det\begin{bmatrix} 4 & 0 \\ -2 & 6 \end{bmatrix}$$
$$= 8(0\cdot5 - 6\cdot6) - 2(4\cdot5 - 6\cdot(-2)) - 2(4\cdot6 - 0\cdot(-2))$$
$$= -288 - 64 - 48$$
$$= -400.$$

To write a program for computing the determinant of a square matrix based on this cofactor expansion formula, only a recursive algorithm is possible. The pseudocode for such a program is as follows:[8]

```
PROGRAM:   y = DetCofactor(A)
y=0;   (initialize y)
n = size(A)   %record the size of the square matrix A
IF n = 1, set y = A(1,1), OUTPUT y
  ELSEIF n = 2, set y = A(1,1)*A(2,2) - A(1,2)*A(2,1), OUTPUT y
  ELSE
    FOR i= 1 TO n
    SubA = (matrix formed from A by deleting ith column and first row)
    y=y+(-1)^(i+1)*A(1,i)*DetCofactor(SubA);
    END <i FOR>
    OUTPUT y
END <IF>
```

---

[8] Within this program, the submatrix "SubA" represents A_{1i}. How this matrix is formed from A will depend on how your particular computing platform handles matrix array operations.

It is straightforward to show that the number of arithmetic operations (additions, subtractions, and multiplications) required by this algorithm to compute the determinant of an $n \times n$ matrix exceeds $n \cdot (n-1) \cdot (n-2) \cdots 2 \cdot 1 (\equiv n!)$. (see Exercise 22). Thus, for a 30×30 matrix, cofactor expansion would require more than 2.65×10^{32} arithmetic operations. Even with a computer that could perform a billion arithmetic operations per second, since there are roughly 3×10^7 seconds per year, such a determinant would require over 10^{15} years to compute. Thus, the cofactor expansion algorithm, which is often taught in linear algebra courses, is horribly inefficient as a practical method for computing determinants of larger matrices. Other numerical algorithms have been developed that can easily compute determinants of, say, a 500×500 matrix in a split second. See Chapter 7 of [Sta-04], or, for a more comprehensive treatment, see [GoVL-84].

Explicit Solution Methods for Linear Recursion Formulas

We close this section with some results about a class of recursively defined sequences for which explicit formulas can be determined. For convenience of notation (only), we assume that the sequences start at $n = 0$.

DEFINITION 3.4: If k is a nonnegative integer, a sequence defined by:

$$\begin{cases} 1. \ (\textit{Basis Steps}) \ a_0 = \alpha_0, \ a_1 = \alpha_1, \cdots a_{k-1} = \alpha_{k-1}, \\ 2. \ (\textit{Recursive Step}) \ \text{For } n \geq k, \ a_n = c_1 a_{n-1} + c_2 a_{n-2} + \cdots + c_k a_{n-k} + f(n), \end{cases}$$

where $c_1, c_2, \cdots, c_k$ are real (number) constants with $c_k \neq 0$, and $f(n)$ is some function of the variable n is called a **linear recurrence relation of degree k with constant coefficients**. In case $f(n) \equiv 0$, we say additionally that the recurrence relation is **homogeneous**.[9]

We first deal with the homogeneous case. The key to solving such recurrence relations is to look for solutions of the form $a_n = r^n$, for some real number r. Indeed, substituting this expression into the homogeneous recurrence $a_n = c_1 a_{n-1} + c_2 a_{n-2} + \cdots + c_k a_{n-k}$ yields $r^n = c_1 r^{n-1} + c_2 r^{n-2} + \cdots + c_k r^{n-k}$. Dividing both sides by r^{n-k}, and moving all the terms to the left gives us:

$$r^k - c_1 r^{k-1} - c_2 r^{k-2} - \cdots - c_k = 0.$$

---

[9] For readers who have studied differential equations, the theory of constant coefficient recurrence relations closely parallels that of linear constant coefficient ordinary differential equations. A recurrence relation is, after all, just a discrete version of a differential equation.

This equation is called the **characteristic equation** for the recurrence relation, and the roots of this equation are called the **characteristic roots** of the recurrence relation.   In summary, $a_n = r^n$ solves the recurrence if, and only if r is a characteristic root. From these solutions, we will show we can obtain the solution to the original recurrence. We first look at the case in which $k = 2$. (This will, in particular, cover the Fibonacci recurrence relation of Example 3.7.)   This termporary restriction on k is so that the main idea of the proof will not be cluttered with excess notation. Extending the result to general values of k will then be a simple matter.

THEOREM 3.6: (*Solving second order linear constant coefficient homogeneous recurrence relations*) Suppose that $c_1, c_2 \in \mathbb{R}$, with $c_2 \neq 0$. Let λ_1, λ_2 be the two characteristic roots of the characteristic equation $r^2 - c_1 r - c_2 = 0$. The general solution $\{a_n\}_{n=0}^{\infty}$ of the corresponding recurrence equation $a_n = c_1 a_{n-1} + c_2 a_{n-2}$ is given as follows:

Case 1: The characteristic roots are distinct $\lambda_1 \neq \lambda_2$: $a_n = x_1 \lambda_1^n + x_2 \lambda_2^n$, where x_1, x_2 are constants.

Case 2: The characteristic roots are the same $\lambda_1 = \lambda_2 = \lambda$: $a_n = x_1 \lambda^n + x_2 n \lambda^n$, where x_1, x_2 are constants.

Proof: We prove only Case 1; the proof of Case 2 uses similar ideas and is left as an exercise for the reader.

Step 1:   We first need to show that any sequence of the indicated form $a_n = x_1 \lambda_1^n + x_2 \lambda_2^n$ will indeed solve the recurrence $a_n = c_1 a_{n-1} + c_2 a_{n-2}$. Substituting the former into the right side of the recurrence, and using the fact that λ_1, λ_2 are characteristic roots (so, for example $c_1 \lambda_1 + c_2 = \lambda_1^2$), we obtain:

$$
\begin{aligned}
c_1 a_{n-1} + c_2 a_{n-2} &= c_1 (x_1 \lambda_1^{n-1} + x_2 \lambda_2^{n-1}) + c_2 (x_1 \lambda_1^{n-2} + x_2 \lambda_2^{n-2}) \\
&= x_1 \lambda_1^{n-2} (c_1 \lambda_1 + c_2) + x_2 \lambda_2^{n-2} (c_1 \lambda_2 + c_2) \\
&= x_1 \lambda_1^{n-2} (\lambda_1^2) + x_2 \lambda_2^{n-2} (\lambda_2^2) \\
&= x_1 \lambda_1^n + x_2 \lambda_2^n \\
&= a_n.
\end{aligned}
$$

Step 2:   We next need to show that any solution of the recurrence $a_n = c_1 a_{n-1} + c_2 a_{n-2}$ must be expressible in the indicated form $a_n = x_1 \lambda_1^n + x_2 \lambda_2^n$. So assume that $\{a_n\}_{n=0}^{\infty}$ solves the recurrence. In order for this to work, we must first of all have that $a_0 = x_1 + x_2$, and $a_1 = x_1 \lambda_1 + x_2 \lambda_2$. Using algebra, we can solve these two equations simultaneously for x_1 and x_2, to obtain: $x_1 = (a_1 - a_0 \lambda_2)/(\lambda_1 - \lambda_2)$, and $x_1 = (a_0 \lambda_1 - a_1)/(\lambda_1 - \lambda_2)$. The reader should check this. It is only at this part of the proof where we are using the Case 1 assumption that $\lambda_1 \neq \lambda_2$. Since the given sequence $\{a_n\}_{n=0}^{\infty}$ and $x_1 \lambda_1^n + x_2 \lambda_2^n$ both agree at the initial conditions ($n = 0, 1$), and both satisfy the recurrence relation, it follows (by induction) that $a_n = x_1 \lambda_1^n + x_2 \lambda_2^n$ for all n. $\square$

EXAMPLE 3.10: (*The Fibonacci sequence, cont.*) Use Theorem 2.6 to deduce the explicit formula for the Fibonacci sequence

$$\begin{cases} 1. \ (Basis\ Steps) \ f_1 = 1, \ f_2 = 1, \\ 2. \ (Recursive\ Step)\ \text{For } n \geq 3, \ f_n = f_{n-1} + f_{n-2}, \end{cases}$$

that was introduced in Example 3.7.

SOLUTION: The   characteristic   equation   for   the   recursive   formula $f_n = f_{n-1} + f_{n-2}$ is $r^2 - r - 1 = 0$.   Using the quadratic formula, we see that there are two distinct characteristic roots: $(1 \pm \sqrt{5})/2$.   It follows from Theorem 2.6 (Case 1) that the general solution of the recurrence is

$$f_n = x_1 \left(\frac{1+\sqrt{5}}{2} \right)^{n-1} + x_2 \left(\frac{1-\sqrt{5}}{2} \right)^{n-1},$$

where x_1, x_2 can be any real constants.[10] The basis steps $f_1 = 1$, $f_2 = 1$ force the conditions $1 = x_1 + x_2$, and $1 = x_1(1+\sqrt{5})/2 + x_2(1-\sqrt{5})/2$.   These two equations can be simultaneously solved for x_1, x_2 to give $x_1 = (1+\sqrt{5})/2\sqrt{5}$, and $x_2 = (\sqrt{5}-1)/2\sqrt{5}$, thus leading to the aforementioned explicit representation (4) of the Fibonacci sequence:

$$f_n = \frac{1}{\sqrt{5}} \left(\frac{1+\sqrt{5}}{2} \right)^n - \frac{1}{\sqrt{5}} \left(\frac{1-\sqrt{5}}{2} \right)^n.$$

EXERCISE FOR THE READER 3.12: Solve explicitly the following recurrence relation:

$$\begin{cases} 1. \ (Basis\ Steps) \ a_0 = 1, \ a_1 = 0, \\ 2. \ (Recursive\ Step)\ \text{For } n \geq 2, \ a_n = 4a_{n-1} - 4a_{n-2}. \end{cases}$$

For (linear constant coefficient) recurrence relations of general order k, we let the distinct roots be denoted by $\lambda_1, \lambda_2, \cdots, \lambda_s$.   Each of these roots λ_i has a corresponding algebraic multiplicity $m_i > 0$ (this is simply the largest exponent m for which $(r - \lambda_i)^m$ is a factor of the characteristic polynomial).   These algebraic multiplicities must add up to n, the degree of the characteristic polynomial.   Note that unless each of these multiplicities equals 1, there will be multiple roots. Using this notation, the following general theorem includes the result of Theorem

---

[10] Since the recursion begins at $n = 1$, it will simplify the algebra to use the exponent $n - 1$ rather than n in this general solution form. This useful idea is justified since the missing fixed powers can be absorbed in the arbitrary constants.

3.6 as a special case. The proof of this result involves similar ideas to what we have used with second order recurrence relations, but it is quite long, so we omit it here and refer the interested reader to Chapter 7 of [Bru-04]

THEOREM 3.7: (*Solving general linear order constant coefficient homogeneous recurrence relations*)   Suppose that $c_1, c_2, \cdots, c_k \in \mathbb{R}$, with $c_k \neq 0$. Let $\lambda_1, \lambda_2, \cdots, \lambda_s$ be the distinct characteristic roots of the characteristic equation $r^k - c_1 r^{k-1} - c_2 r^{k-2} - \cdots - c_k = 0$, and $m_1, m_2, \cdots, m_s \in \mathbb{Z}_+$ be their corresponding multiplicities. The general solution $\{a_n\}_{n=0}^{\infty}$ of the corresponding recurrence equation $a_n = c_1 a_{n-1} + c_2 a_{n-2} + \cdots + c_k a_{n-k}$ is given as follows:

$$a_n = (x_{1,0} + x_{1,1}n + \cdots + x_{1,m_1-1}n^{m_1-1})\lambda_1^n + (x_{2,0} + x_{2,1}n + \cdots + x_{2,m_2-1}n^{m_2-1})\lambda_2^n +$$
$$\cdots + (x_{s,0} + x_{s,1}n + \cdots + x_{s,m_s-1}n^{m_s-1})\lambda_s^n,$$

where the $x_{i,j}$ are constants.

Notice from the formula in Theorem 3.7 that each characteristic root λ_i of multiplicity m_i gives rise to a term in the solution that has the form $P_i(n)\lambda^n$, where $P_i(n)$ is a polynomial of degree $m_i - 1$.

COMPUTATIONAL NOTE: For recurrence formulas of order $k > 2$, unless the characteristic polynomial can be readily factored, the roots (and their multiplicities) will need to be found either numerically or by using some symbolic software package. The theory works fine in cases of complex roots, but our examples will involve only real roots. Also, for higher order recurrence relations, to determine the unknown coefficients $x_{i,j}$ of Theorem 3.7, it is best to use matrix methods to solve the linear systems.

For linear constant coefficient recurrence relations that are inhomogeneous, the general solution can be obtained by finding one particular solution to the recurrence and adding to it the general solution of the corresponding homogeneous recurrence relation (the latter can be found using Theorem 3.7).   This simple result is explained in our next theorem.

THEOREM 3.8: (*Solving general inhomogeneous linear constant coefficient recurrence relations*)   If $\{a_n^p\}_{n=0}^{\infty}$ is a particular solution of the inhomogeneous recurrence relation $a_n = c_1 a_{n-1} + c_2 a_{n-2} + \cdots + c_k a_{n-k} + f(n)$, and $\{a_n^h\}_{n=0}^{\infty}$ is the general solution of the corresponding homogeneous recurrence relation $a_n = c_1 a_{n-1} + c_2 a_{n-2} + \cdots + c_k a_{n-k}$, then the general solution of the inhomogeneous recurrence relation $a_n = c_1 a_{n-1} + c_2 a_{n-2} + \cdots + c_k a_{n-k} + f(n)$ is expressible as $\{a_n^p + a_n^h\}_{n=0}^{\infty}$.

Thus, the only new task needed to solve an inhomogeneous (linear constant coefficient) recurrence relation is to find (just a single) particular solution to it. In general, particular solutions will take on a similar form to the inhomogeneity function $f(n)$. The following result gives the form of $\{a_n^p\}_{n=0}^{\infty}$ in the common case where $f(n)$ is a polynomial and/or exponential.

THEOREM 3.9: (*Particular solutions of inhomogeneous linea constant coefficient recurrence relations with polynomial or exponential inhomogeneity*) In the above notation for a recurrence relation $a_n = c_1 a_{n-1} + c_2 a_{n-2} + \cdots + c_k a_{n-k}$ $+ f(n)$, if the inhomogeneity $f(n)$ has the form $f(n) = (b_s n^s + b_{s-1} n^{s-1} + \cdots + b_1 n + b_0) \lambda^n$, where b_i, λ are real numbers, then a particular solution $\{a_n^p\}_{n=0}^{\infty}$ of the recurrence can be found to be of the following form:

Case 1: λ is not a characteristic root. $a_n^p = (d_s n^s + d_{s-1} n^{s-1} + \cdots + d_1 n + d_0) \lambda^n$.

Case 2: λ is a characteristic root of multiplicity m.
$$a_n^p = n^m (d_s n^s + d_{s-1} n^{s-1} + \cdots + d_1 n + d_0) \lambda^n.$$

We caution the reader to be aware that when 1 is a characteristic root we will be in Case 2 of Theorem 3.9. The proofs of the above two theorems are left as exercises (with copious hints). We give an example.

EXAMPLE 3.11: Find an explicit form solution for the following recursively defined sequence.
$$\begin{cases} 1. \ (Basis\ Step) \ \ a_0 = 1, \\ 2. \ (Recursive\ Step) \ \text{For } n \geq 2, \ a_n = 4a_{n-1} + n + 1. \end{cases}$$

SOLUTION: The associated homogeneous recurrence relation is $a_n = 4a_{n-1}$. The characteristic polynomial is $r - 4 = 0$, which has a single characteristic root $r = 4$. By Theorem 3.7, the general solution of this homogeneous recurrence is $a_n^h = x \cdot 4^n$, where x is an arbitrary constant. Since the inhomogeneity $f(n) = n + 1$ is a first degree polynomial, it follows from Theorem 3.9 that a particular solution of the inhomogeneous recurrence $a_n = 4a_{n-1} + n + 1$ will have the form $a_n^p = dn + e$, where d and e are constants. Substituting the latter for a_n into the former recurrence results in the equation:
$$dn + e = 4[d(n-1) + e] + n + 1 = 4dn - 4d + 4e + n + 1.$$

Equating coefficients of n gives: $d = 4d + 1 \implies d = -1/3$.

Equating constant coefficients gives: $e = -4d + 4e + 1 = 4e + 7/3 \implies e = -7/9$.

Thus, we get $a_n^p = -n/3 - 7/9$, and so by Theorem 3.8, it follows that the general solution of the original recurrence is $a_n^h + a_n^p = x \cdot 4^n - n/3 - 7/9$. Substituting the initial condition $a_0 = 1$ (of the basis step) into this general solution gives $1 = x \cdot 4^0 - 0/3 - 7/9$, or $1 = x - 7/9$, which determines $x = 16/9$. Thus, the explicit solution of the original recurrence is given by $a_n = (16/9) \cdot 4^n - n/3 - 7/9$.

EXERCISE FOR THE READER 3.13: (a) Explicitly solve the following recurrence relation:

$$\begin{cases} 1. \ (Basis\ Step)\ \ a_0 = 1,\ \ a_1 = 0,\ \ a_2 = 1, \\ 2. \ (Recursive\ Step)\ \text{For } n \geq 3,\ a_n = a_{n-1} + a_{n-2} - a_{n-3}. \end{cases}$$

(b) Compute a_{10} first using the recurrence relation, and then using the explicit formula that you obtained in Part (a).

(c) Repeat Part (a) following recurrence relation:

$$\begin{cases} a_0 = 1,\ \ a_1 = 2 \\ a_n = 4a_{n-1} - 4a_{n-2} + 2^n n \ (n \geq 2). \end{cases}$$

(d) Compute a_6 first using the recurrence relation, and then using the explicit formula that you obtained in Part (c).

EXERCISES 3.2:

1. Find the first five terms of each of the following recursively defined sequences:

   (a) $\begin{cases} 1. \ (Basis\ Step)\ \ a_0 = 1 \\ 2. \ (Recursive\ Step)\ \text{For } n \geq 1,\ a_n = 2a_{n-1}. \end{cases}$

   (b) $\begin{cases} 1. \ (Basis\ Step)\ \ a_1 = 1,\ a_2 = -1 \\ 2. \ (Recursive\ Step)\ \text{For } n \geq 3,\ a_n = 2a_{n-1} + a_{n-2}. \end{cases}$

   (c) $\begin{cases} 1. \ (Basis\ Step)\ \ a_1 = 1,\ a_2 = 2,\ a_3 = 2 \\ 2. \ (Recursive\ Step)\ \text{For } n \geq 4,\ a_n = 2a_{n-1} + a_{n-2} + 2a_{n-3}. \end{cases}$

2. Find the first five terms of each of the following recursively defined sequences:

   (a) $\begin{cases} 1. \ (Basis\ Step)\ \ a_{-2} = 2 \\ 2. \ (Recursive\ Step)\ \text{For } n \geq -1,\ a_n = 3a_{n-1}. \end{cases}$

   (b) $\begin{cases} 1. \ (Basis\ Step)\ \ a_0 = 0,\ a_1 = 1 \\ 2. \ (Recursive\ Step)\ \text{For } n \geq 2,\ a_n = a_{n-1} - 2a_{n-2}. \end{cases}$

   (c) $\begin{cases} 1. \ (Basis\ Step)\ \ a_0 = 1,\ a_1 = 1,\ a_2 = -1 \\ 2. \ (Recursive\ Step)\ \text{For } n \geq 3,\ a_n = 3a_{n-1} + 2a_{n-2} + a_{n-3}. \end{cases}$

3. Find the first five terms of each of the following recursively defined sequences:

   (a) $\begin{cases} 1. \ (Basis\ Step)\ \ a_0 = 1 \\ 2. \ (Recursive\ Step)\ \text{For } n \geq 1,\ a_n = 2a_{n-1} + 1. \end{cases}$

   (b) $\begin{cases} 1. \ (Basis\ Step)\ \ a_1 = 1,\ a_2 = -1 \\ 2. \ (Recursive\ Step)\ \text{For } n \geq 3,\ a_n = 2a_{n-1} + a_{n-2}^2. \end{cases}$

(c) $\begin{cases} 1.\ (\textit{Basis Step}) \quad a_1 = 1,\ a_2 = 2,\ a_3 = 2 \\ 2.\ (\textit{Recursive Step})\ \text{For } n \geq 4,\ a_n = 2a_{n-1} + a_{n-2} + 2a_{n-3} + 2(n-2)^2. \end{cases}$

4. Find the first five terms of each of the following recursively defined sequences:

   (a) $\begin{cases} 1.\ (\textit{Basis Step}) \quad a_{-2} = 2, \\ 2.\ (\textit{Recursive Step})\ \text{For } n \geq -1,\ a_n = 3a_{n-1} + n. \end{cases}$

   (b) $\begin{cases} 1.\ (\textit{Basis Step}) \quad a_0 = 0,\ a_1 = 1 \\ 2.\ (\textit{Recursive Step})\ \text{For } n \geq 2,\ a_n = na_{n-1} - a_{n-2}^2. \end{cases}$

   (c) $\begin{cases} 1.\ (\textit{Basis Step}) \quad a_0 = 1,\ a_1 = 1,\ a_2 = -1 \\ 2.\ (\textit{Recursive Step})\ \text{For } n \geq 3,\ a_n = 3a_{n-1} + 2a_{n-2} + a_{n-3} - 2^n. \end{cases}$

5. Give recursive definitions for each of the following explicitly defined sequences $\{a_n\}_{n \in \mathbb{Z}_+}$:

   (a) $a_n = 2$ (b) $a_n = 2n$ (c) $a_n = 2^n$ (d) $a_n = 2/n$

6. Give recursive definitions for each of the following explicitly defined sequences $\{a_n\}_{n \in \mathbb{Z}_+}$:

   (a) $a_n = 1 + 2 + \cdots n$ (b) $a_n = 2n + 3$ (c) $a_n = \begin{cases} 1, & \text{if } n \text{ is even} \\ -1, & \text{if } n \text{ is odd} \end{cases}$ (d) $a_n = \sqrt{n}$

7. Give recursive definitions for sequences $\{a_n\}_{n \in \mathbb{Z}_+}$ whose first few terms are as specified:

   (a) $11, 13, 15, 17, 19, \cdots$ (b) $2, 5, 8, 11, 14, \cdots$

   (c) $5, 6, 8, 11, 15, 20, \cdots$ (d) $6, -12, 36, -144, 720, -4320, \cdots$

   Suggestion: Look for any noticeable patterns; some patterns might emerge by looking at the sequence of consecutive differences of terms $a_{n+1} - a_n$, or the quotients a_{n+1}/a_n.

8. Give recursive definitions for sequences $\{a_n\}_{n \in \mathbb{Z}_+}$ whose first few terms are as specified:

   (a) $1, -2, 3, -4, 5, \cdots$ (b) $2, 5, 11, 20, 32, 47, \cdots$

   (c) $3, 6, 18, 72, 360, 2160, \cdots$ (d) $1, 3, 7, 15, 31, \cdots$

   Suggestion: Look for any noticeable patterns; some patterns might emerge by looking at the sequence of consecutive differences of terms $a_{n+1} - a_n$, or the quotients a_{n+1} / a_n.

9. Find an explicit formula for each of the following recursively defined sequences without resorting to Theorems 3.6–3.9.

   (a) $\begin{cases} 1.\ (\textit{Basis Step}) \quad a_1 = 1 \\ 2.\ (\textit{Recursive Step})\ \text{For } n \geq 2,\ a_n = a_{n-1} + 3. \end{cases}$

   (b) $\begin{cases} 1.\ (\textit{Basis Step}) \quad a_1 = 1 \\ 2.\ (\textit{Recursive Step})\ \text{For } n \geq 2,\ a_n = a_{n-1} + n. \end{cases}$

   (c) $\begin{cases} 1.\ (\textit{Basis Step}) \quad a_1 = 1 \\ 2.\ (\textit{Recursive Step})\ \text{For } n \geq 2,\ a_n = 1.1a_{n-1} + 1. \end{cases}$

10. Find an explicit formula for each of the following recursively defined sequences without resorting to Theorems 3.6–3.9.

    (a) $\begin{cases} 1.\ (\textit{Basis Step}) \quad a_1 = 1 \\ 2.\ (\textit{Recursive Step})\ \text{For } n \geq 2,\ a_n = a_{n-1} + 3. \end{cases}$

    (b) $\begin{cases} 1.\ (\textit{Basis Step}) \quad a_1 = 1 \\ 2.\ (\textit{Recursive Step})\ \text{For } n \geq 2,\ a_n = a_{n-1} + n. \end{cases}$

(c) $\begin{cases} 1.\ (Basis\ Step)\ a_1 = 1 \\ 2.\ (Recursive\ Step)\ \text{For } n \geq 2,\ a_n = 1.1a_{n-1} + 1. \end{cases}$

11.   (*Finance: Future Value Annuities*) A future value annuity is a savings plan in which regular periodic payments are made into an interest-bearing account over a number of periods, and interest is paid (added into the account) each period. Examples of these annuities are supplemental retirement plans (like 401(k) and 403(b) plans, where the federal government encourages contributions by allowing them to be made directly from an individual's gross income before it is taxed. This problem deals with an ordinary annuity, where the interest rate and periodic payments are constant and made at the end of each period. The assumptions are as follows: Payments equaling *PMT* are made at the end of k equally spaced periods per year. The annuity pays an annual rate of 100r% interest on the account balance compounded each period. Thus, for example, the values $PMT = 500$, $k = 12$, and $r = .075$ would correspond to an annuity that pays 7.5% interest every year compounded monthly (so 0.625% interest per month) and in which $500 is added at the end of each month. Let $A(n)$ be the account balance of the annuity at the end of the nth period.

(a) Establish the following recursive formula for $A(n)$:

$\begin{cases} 1.\ (Basis\ Step)\ A(1) = PMT \\ 2.\ (Recursive\ Step)\ \text{For } n \geq 1,\ A(n+1) = (1 + r/k)A(n) + PMT. \end{cases}$

(b) Establish the following explicit formula for $A(n)$:

$$A(n) = PMT \cdot \frac{(1 + r/k)^n - 1}{r/k}.$$

(c) With the values $PMT = 500$, $k = 12$, and $r = .075$, use both formulas to find the account balance at the end of three months, and compare answers.

(d) With the values in Part (c), use the explicit formula to compute the account balance after 40 years. How much of this amount is interest?

12.   (*Finance: Present Value Annuities*) A present value annuity is a financial instrument in which regular periodic payments are made toward paying off a loan, with interest on the unpaid balance being added to the account balance each month. Examples of these annuities are home mortgages and car loans. This problem deals with an ordinary annuity, where the interest rate and periodic payments are constant and made at the end of each period. The assumptions are as follows: A loan of value PV is to be paid off with regular payments equaling *PMT* that are made at the end of k equally spaced periods per year. The annuity charges an annual rate of 100r% interest on the unpaid account balance compounded each period. Thus, for example, the values $PV = 20,000$, $PMT = 500$, $k = 12$, and $r = .075$ would correspond to an annuity that pays 7.5% interest every year compounded monthly (so 0.625% interest per month) and in which $500 is paid at the end of each month until the account balance is paid off (the last payment might be less than the regular payment in order to exactly pay off the remaining balance). Let $A(n)$ be the account balance of the annuity at the end of the nth period.

(a) Establish the following recursive formula for $A(n)$:

$\begin{cases} 1.\ (Basis\ Step)\ A(0) = PV \\ 2.\ (Recursive\ Step)\ \text{For } n \geq 0,\ A(n+1) = \begin{cases} (1 + r/k)A(n) - PMT, & \text{if } (1 + r/k)A(n) \geq PMT \\ 0, & \text{otherwise} \end{cases} \end{cases}$

Explain also that as soon as $A(n)$ reaches a balance of zero, it will remain at zero on after (the first point that this happens corresponds to the moment that the loan is completely paid off so that the loan is essentially inactive after this point).

(b) Establish the following explicit formula for $A(n)$ that is valid while $(1 + r/k)A(n) \geq PMT$ (i.e, before the last payment is made):

$$A(n+1) = (1 + r/k)^{n+1} \cdot PV - \frac{(1 + r/k)^{n+1} - 1}{r/k} \cdot PMT.$$

(c) Explain why the last payment will equal PMT (i.e., all payments are equal) if the following condition holds:

$$PV = PMT \cdot \frac{1-(1+r/k)^{-n}}{r/k},$$

for some positive integer n (which is the number of payments needed to pay off the loan).

(d) Use the formula in Part (c) to determine the amount of the monthly payment to pay off a $25,000 car loan in 72 months, if the car loan charges 8% annual interest compounded monthly. Then use the formulas in Parts (a) and (b) to compute the unpaid balance after three months, and compare anwers. When the loan is paid off, how much interest will have been paid?

13. Suppose that a person can climb either one or two stairs at a time and must climb a staircase with n stairs, where n is a positive integer. Let a_n denote the number of ways that this person can climb n stairs. For example, if $n = 3$, there are three ways: (i) one stair at a time, (ii) two stairs, then one, and (iii) one stair, then two. Establish a recursive formula along with initial conditions for a_n, and use these to compute a_{10}.

14. Suppose that a tall person can climb one, two, or three stairs at a time and must climb a staircase with n stairs, where n is a positive integer. Let a_n denote the number of ways that this person can climb n stairs. For example, if $n = 3$, there are four ways: (i) one stair at a time, (ii) two stairs, then one, (iii) one stair, then two, or (iv) all three stairs at once. Establish a recursive formula along with initial conditions for a_n, and use these to compute a_{10}.

15. A pair of baby rabbits of opposite sexes is let loose on an island. Baby rabbits reach maturity to become adult rabbits after two months, and at the end of every month starting with the third month, adult rabbits give birth to another pair of baby rabbits of opposite sexes. This cycle repeats, with adults continuing to reproduce pairs of baby rabbits, and baby rabbits becoming adults after two months and beginning to reproduce monthly at the end of their third month.

(a) Assuming that the rabbits never die, find a recursive scheme for the computation of the sequence $\{a_n\}_{n=1}^{\infty}$ whose nth term gives the number of pairs of rabbits n months after the first pair of rabbits is released.

(b) Use use your scheme to find the first seven terms of the sequence.

(c) Derive an explicit formula for a_n.

Suggestion: Consider separate sequences as follows:

α_n = the number of rabbits of age 0 months at the end of month n,

β_n = the number of rabbits of age 1 month at the end of month n,

γ_n = the number of rabbits of age at least 2 months at the end of month n.

16. Redo Exercise 15, but this time assume each month the mature rabbit pairs each give birth to two pairs of rabbits.

17. Redo Exercise 15(a)(b), but this time assume that after giving birth to their second pair of baby rabbits at the end of four months, the adult rabbits immediately die.

18. Redo Exercise 16(a)(b), but this time assume that after giving birth to their third pair of baby rabbits at the end of five months, the adult rabbits immediately die.

19. Suppose that in the Towers of Hanoi puzzle (Example 3.6), the pegs, lined up from left to right, are numbered by 1, 2, and 3, respectively, where the disks are originally placed on peg 1, and need to be moved to peg 3. If rules are changed so that disks can only be moved between adjacent pegs (i.e., peg 1 to/from peg 2, or peg 2 to/from peg 3), let k_n denote the resulting minimum number of moves needed to solve this new puzzle with n (different sized) disks.

(a) Compute $\mathscr{B}_n$ k_1, k_2.

(b) Find a recurrence relation for k_n.

(c) Obtain an explicit formula for k_n.

(d) Use both formulas in Parts (b) and (c) to compute k_7.

20. For an integer n greater than 1, we let p_n denote the number of different ways that n cents of postage can be created using (an unlimited supply) of two cent and/or three cent stamps, where the stamps are assumed to be arranged in a single line (from left to right) and order matters.

(a) Compute p_2, p_3, p_4, p_5.

(b) Find a recurrence relation along with initial conditions for p_n.

(c) Compute p_{10}.

21. Repeat Exercise 18, with the additional assumption that we now also have an unlimited supply of five cent stamps, and p_n now denotes the number of different ways to line up n cents of postage using two, three, and/or five cent stamps.

22. For an integer n greater than 3, we let b_n denote the number of binary strings of length n that contain three consecutive 0's.

(a) Compute b_3, b_4, b_5.

(b) Find a recurrence relation along with initial conditions for b_n.

(c) Compute b_7.

23. For an integer n greater than 3, we let b_n denote the number of binary strings of length n that do not contain three consecutive 0's.

(a) Compute b_3, b_4, b_5.

(b) Find a recurrence relation along with initial conditions for b_n.

(c) Compute b_{10}.

24. Let a_n (for $n > 1$ a positive integer) denote the number of multiplications plus the number of additions that are needed to compute the determinant of a general $n \times n$ matrix using the cofactor expansion algorithm of Example 3.9.

(a) Obtain a recurrence relation along with initial conditions for the sequence $\{a_n\}_{n=2}^{\infty}$.

(b) Use mathematical induction to prove that $a_n \geq n! \ (\equiv n \cdot (n-1) \cdots 2 \cdot 1)$.

25. Obtain a recursive definition for the *odd factorial function* of a nonnegative integer n. This function, denoted as OddFact(n), is defined to be 1 if $n = 0$, and if n is a positive integer, it is defined to be the product of all odd positive integers less than or equal to n: $1 \cdot 3 \cdot 5 \cdots$. Thus, OddFact(1) = OddFact(2) = 1, OddFact(3) = OddFact(4) = $1 \cdot 3 = 3$, OddFact(5) = OddFact(6) = $1 \cdot 3 \cdot 5$ = 15, and so on.

26. Obtain a recursive definition for the *even factorial function* of a nonnegative integer n. This function, denoted as EvenFact(n), is defined to be 1 if $n = 0$, and if n is a positive integer, it is defined to be the product of all even positive integers less than or equal to n: $2 \cdot 4 \cdot 6 \cdots$. Thus, EvenFact(1) = 1 EvenFact(2) = EvenFact(3) = 2, EvenFact(4) = EvenFact(5) = $2 \cdot 4 = 8$, and so on.

27. Given a set of n real numbers: $x_1, x_2, \cdots, x_n$ their *maximum* $\max(x_1, x_2, \cdots, x_n)$ may be directly computed with $n-1$ comparisons with the following simple algorithm:

    Step 1: Initialize Max $= x_1$

    Step 2:
    FOR $j = 2$ TO n

         IF $x_j > $ Max

            Update Max $= x_j$

         END IF
    END FOR
    Output Max
    (a) Obtain a recursive algorithm that calls on itself for this max function.
    (b) With an input of n real numbers, how many comparisons would your recursive algorithm of Part (b) require?

28. (a) Formulate a direct algorithm that will compute the *minimum* $\min(x_1, x_2, \cdots, x_n)$ of a set of n real numbers: $x_1, x_2, \cdots, x_n$ that is similar to the algorithm given in the preceding exercise for finding the maximum.
    (b) Obtain a recursive algorithm that calls on itself for this min function.
    (c) With an input of n real numbers, how many comparisons would your recursive algorithm of Part (b) require?

29. Apply either Theorem 3.6 or Theorem 3.7 to find explicit formulas for each of the following recursively defined sequences.

    (a) $\begin{cases} a_1 = -5 \\ a_n = 2a_{n-1} \ (n \ge 2) \end{cases}$ (b) $\begin{cases} a_1 = 2, \ a_2 = 1 \\ a_n = 2a_{n-1} + a_{n-2} \ (n \ge 3) \end{cases}$

    (c) $\begin{cases} a_1 = 0, \ a_2 = 2 \\ a_n = a_{n-1} + a_{n-2} \ (n \ge 3) \end{cases}$ (d) $\begin{cases} a_7 = 5, \ a_8 = 13 \\ a_n = 6a_{n-1} - 5a_{n-2} \ (n \ge 9) \end{cases}$

30. Apply either Theorem 3.6 or Theorem 3.7 to find explicit formulas for each of the following recursively defined sequences.

    (a) $\begin{cases} a_1 = 2 \\ a_n = -3a_{n-1} \ (n \ge 2) \end{cases}$ (b) $\begin{cases} a_0 = 8, \ a_1 = -6 \\ a_n = 9a_{n-2} \ (n \ge 2) \end{cases}$

    (c) $\begin{cases} a_1 = 1, \ a_2 = 1 \\ a_n = 2a_{n-1} + 2a_{n-2} \ (n \ge 3) \end{cases}$ (d) $\begin{cases} a_{10} = 0, \ a_{11} = -1 \\ a_n = 2a_{n-1} + 3a_{n-2} \ (n \ge 11) \end{cases}$

31. Apply either Theorem 3.6 or Theorem 3.7 to find explicit formulas for each of the following recursively defined sequences.

    (a) $\begin{cases} a_0 = 3, \ a_1 = 1 \\ a_n = 4a_{n-1} - 4a_{n-2} \ (n \ge 2) \end{cases}$ (b) $\begin{cases} a_2 = 17, \ a_3 = 21 \\ a_n = 2a_{n-1} - a_{n-2} \ (n \ge 4) \end{cases}$

32. Apply either Theorem 3.6 or Theorem 3.7 to find explicit formulas for each of the following recursively defined sequences.

    (a) $\begin{cases} a_1 = -2, \ a_2 = 4, \\ a_n = -2a_{n-1} - a_{n-2} \ (n \ge 3) \end{cases}$ (b) $\begin{cases} a_6 = 5, \ a_7 = 35 \\ a_n = 10a_{n-1} - 25a_{n-2} \ (n \ge 8) \end{cases}$

33. Find the general solution to each of the following recurrences.

(a) $a_n = 8a_{n-2} - 16a_{n-4}$ (b) $a_n = 10a_{n-2} + 9a_{n-4}$

(c) $a_n = 3a_{n-1} - 3a_{n-2} + a_{n-3}$ (d) $a_n = 8a_{n-1} + 24a_{n-2} - 32a_{n-3} + 16a_{n-4}$

34.   Find the general solution to each of the following recurrences.

(a) $a_n = 2a_{n-2} + a_{n-4}$ (b) $a_n = 2a_{n-1} + 4a_{n-2} - 8a_{n-3}$

(c) $a_n = -15a_{n-1} - 75a_{n-2} - 125a_{n-3}$ (d) $a_n = -12a_{n-1} - 54a_{n-2} - 108a_{n-3} - 81a_{n-4}$

35.   Make use of Theorems 3.8 and 3.9 to find explicit formulas for each of the following recursively
       defined sequences:

(a) $\begin{cases} a_2 = 19 \\ a_n = 2a_{n-1} + 3^n \ (n \geq 3) \end{cases}$ (b) $\begin{cases} a_4 = 32 \\ a_n = 2a_{n-1} + 2^n \ (n \geq 4) \end{cases}$

(c) $\begin{cases} a_0 = 0, \ a_1 = 1 \\ a_n = 6a_{n-1} - 5a_{n-2} + n2^n \ (n \geq 2) \end{cases}$ (d) $\begin{cases} a_2 = 1, \ a_3 = 6 \\ a_n = 4a_{n-1} - 4a_{n-2} + 2^{n-1} \ (n \geq 4) \end{cases}$

36.   Make use of Theorems 3.8 and 3.9 to find explicit formulas for each of the following recursively
       defined sequences:

(a) $\begin{cases} a_1 = 2 \\ a_n = -3a_{n-1} + 2n^2 \ (n \geq 2) \end{cases}$ (b) $\begin{cases} a_1 = 2 \\ a_n = -3a_{n-1} + 2n^2 3^n \ (n \geq 2) \end{cases}$

(c) $\begin{cases} a_{10} = 0, \ a_{11} = 1 \\ a_n = 2a_{n-1} + 3a_{n-2} + n5^n \ (n \geq 11) \end{cases}$ (d) $\begin{cases} a_6 = 1, \ a_7 = 3 \\ a_n = 10a_{n-1} - 25a_{n-2} - (n+3)5^{n+3} \ (n \geq 8) \end{cases}$

37.   Prove Theorem 3.8.
       Suggestion: Show that the difference of two particular solutions of the inhomogeneous
       recurrence is a solution of the corresponding homogeneous recurrence.

38.   Prove Theorem 3.9.

APPENDIX TO SECTION 3.2: RECURSIVE DEFINITIONS AND STRUCTURAL INDUCTION

Yet another use of recursion is in the definitions of sets and other structures. In such constructions, the
basis step will specify that the set contains some certain elements (the seeds of the recursive process) or
give some properties of the structure, and the recursive step will show how to obtain new elements of
the set from elements that were previously obtained (or new properties of the structure), either by the
basis step or from previous applications of the recursive step. This sort of definition is useful in
computer algorithms as well as to prove properties of objects in the set. If we are trying to prove a
property of a recursively defined set, it suffices to prove it for the basis step elements, and then to prove
that if it holds for any elements, then it will also hold for any other elements obtained from the former
using the recursive step. This latter method proof technique is sometimes called **structural induction**,
but it is really just ordinary mathematical induction (of the last section) in disguise. There will only be
a few places in this book where we need to use structural induction; we will use it here to tie up a loose
end that we left in the (optional) Section 2.3 on Boolean Algebra. Structural induction is a tool for
proving statements about recursively defined objects. The way that it works is that we first prove the
validity of the statement for the objects defined in the basis step (this is the *basis step* of structural
induction), and then we show that the statement is valid for any object that is obtainable from a single
application of the inductive step, from an object that is assumed to satisfy the statement (this is the
inductive step of the structural induction). It then follows from mathematical induction that the

statement is valid for all objects (obtainable from the basis elements by any finite number of applications of the recursive step of the recursive definition). We begin with some simple examples and then move on to give a proof of Proposition 2.5. We begin with a very simple example.

EXAMPLE 3.12: Let S be the set recursively defined with the following two steps:

1. *Basis Step*: $2 \in S$
2. *Recursive Step*: If $x, y \in S$, then $x + y, x - y \in S$

(*Exclusion Clause*: No other objects are contained in S.) The exclusion clause is usually not stated (it is a default assumption).

Thus, from a single application of the recursive step, we get that $2 + 2 = 4 \in S$, and $2 - 2 = 0 \in S$. Knowing now that $\{0, 2, 4\} \subseteq S$, with other second applications of the recursive step using any possible combination of these three numbers for x and y, leads us the expanded collection of elements that we know are contained in S: $\{-4, -2, 0, 2, 4, 6, 8\}$. It is appearing that S might consist precisely of the set of even integers $2\mathbb{Z} = \{0, \pm 2, \pm 4, \pm 6, \cdots\}$. We make use of structural induction to prove this fact.

Proof that $S = 2\mathbb{Z}$:

Step 1: We show $S \subseteq 2\mathbb{Z}$: 1. Basis Step: The only basis element is 2, which is certainly contained in $2\mathbb{Z}$. 2. (Structural) Inductive Step: Assume that x, y, have been obtained from a finite number of applications of the basis step, and these element have been shown to be in $2\mathbb{Z}$. We must show that $x \pm y \in 2\mathbb{Z}$. But this follows from the simple fact that a sum/difference of even integers is an even integer.

Step 1: We show $2\mathbb{Z} \subseteq S$: Here we will use ordinary mathematical induction to prove that $\pm 2n \in S$ for all nonnegative integers n. 1. Basis Step: The fact that $0 \in S$ was observed above (using a single application of the inductive step of the recursive definition of S). 2. Inductive Step: Assume that $\pm 2n \in S$, where n is some nonnegative integer. We need to show that $\pm 2(n+1) \in S$. The inductive hypothesis gives us that $x = \pm 2n \in S$, and it was pointed out above that (through a single application of the recursive step in the recursive definition of S) $y = 2 \in S$. Thus, by the recursive step in the definition of S, we may conclude that S contains $x \pm y = \pm 2n \pm 2 = \pm 2(n+1)$, as desired. $\square$

Concepts about binary (and more general) strings, can usually be defined recursively. This makes it possible to write computer programs that can compute and analyze such concepts, and it makes it possible to use structural induction to prove related facts. Our next example will use structural induction to give a formal proof a useful, albeit not surprising, property of binary strings.

EXAMPLE 3.13: The reversal of a binary string is simply the sting obtained by reversing the order of the binary digits. For example, if s is the binary string 00011, then its **reversal**, $R(s)$, is the string 11000. Although intuitively clear (to human beings), this informal definition cannot be programmed into a computer, and is not amenable to formal proofs. Instead, we define the concept using the following recursive definition:

1. *Basis Step*: $R(\varnothing) = \varnothing$, where $\varnothing$ denotes the empty binary string.
2. *Recursive Step*: If s is a binary string, and x is a binary digit (i.e., $x \in \{0,1\}$), then
   $R(sx) = xR(s)$.

This definition can be used to program the reversal operation on strings into a computer program.[11] We use structural induction to prove the following fact about the reversal operation:

CLAIM: If s, s' are binary strings, then $R(ss') = R(s')R(s)$.

---

[11] The details will depend on the particular platform being used, and make use of an appropriate data structure for strings.

Proof by Structural Induction on the latter string s': Throughout the proof we fix an arbitrary binary string s. 1. Basis Step: We must show the claim is true for $s' = \emptyset$. In this case we have $R(s\emptyset) = R(s) = \emptyset R(s) = R(\emptyset)R(s)$, as required. 2. (Structural) Inductive Step: We assume that the claim is true for (s and) a string s' that has been obtained from the basis step using a finite number of applications of the inductive step of the recursive definition. Let x be a binary digit. We must show that the claim is true for (s and) the string $s'x$. By the recursive step of the definition of reversal, we have $R(ss'x) = R((ss')x) = xR(ss')$. By the inductive step of this structural induction, we know that $R(ss') = R(s')R(s)$. Combining these last two equations yields $R(ss'x) = xR(s')R(s)$, while another application of the inductive step of structural induction gives $xR(s') = R(s'x)$, and the desired equality $R(ss'x) = R(s'x)R(s)$ now follows. □

The reader should convince him/herself that all that was done in Example 3.13 would work equally well if rather than binary strings, we considered reversals of strings in any alphabet. Our next example of structural induction will be to prove the duality principle (Proposition 2.5) for Boolean expressions. In order to accomplish this, we will need a better (recursive) definition for the informal Definition 2.8 for the dual of a Boolean expression:

DEFINTION 2.8 (*Revised*): The dual f^d of a Boolean expression f, is defined as follows:

1. *Basis Step*: $(0)^d = 1$, $(1)^d = 0$, and $x^d = x$, whenever x is a Boolean variable.

2. *Recursive Step*: If f_1, f_2 are Boolean expressions, then $(f_1 + f_2)^d = f_1^d \cdot f_2^d$, $\left(\overline{f_1}\right)^d = \overline{f_1^d}$, and $(f_1 \cdot f_2)^d = f_1^d + f_2^d$.

This definition can be used to program the dual operation into a computer platform, and to prove properties of it. As promised, we are now ready to give a proof of the duality principle (Proposition 2.5), which states that if f and g are equivalent Boolean expressions, then so are their duals f^d and g^d.

Proof of Proposition 2.5: The result will directly follow from the following identity that expresses the dual of a Boolean expression in terms of the expression:

$$f^d(x_1, x_2, \cdots, x_n) = \overline{f(\overline{x_1}, \overline{x_2}, \cdots, \overline{x_n})}, \tag{6}$$

whenever $f = f(x_1, x_2, \cdots, x_n)$ is a Boolean expression in the Boolean variables $x_1, x_2, \cdots, x_n$. It thus suffices to prove this identity for all Boolean expressions f in n Boolean variables, and we proceed to establish this fact using structural induction. We note that any Boolean expression in any subset of the variables $x_1, x_2, \cdots, x_n$ can trivially be viewed as a Boolean expression in the full set of variables $x_1, x_2, \cdots, x_n$ (simply treat the additional variables as redundant when doing function evaluations).

1. Basis Step: When $f = 0$ or $f = 1$, the identity follows from the definition of Boolean complementation: $\overline{0} = 1$ and $\overline{1} = 0$. When $f = x$ (a Boolean variable), the identity follows from the double complementation law $\overline{\overline{x}} = x$ (Theorem 2.4(e)).

2. (Structural) Inductive Step: We assume that (6) holds for two Boolean expressions f_1, f_2 in the Boolean variables $x_1, x_2, \cdots, x_n$. Our task is to show that it also holds for each of the Boolean expressions: (i) $f_1 + f_2$, (ii) $\overline{f_1}$, and (iii) $f_1 \cdot f_2$. We go through each of these verifications in turn.

(i):

$$(f_1 + f_2)^d (x_1, x_2, \cdots, x_n) = f_1^{\,d}(x_1, x_2, \cdots, x_n) \cdot f_2^{\,d}(x_1, x_2, \cdots, x_n) \quad \text{(recursive step of the definition of dual)}$$
$$= \overline{f_1(\overline{x_1}, \overline{x_2}, \cdots, \overline{x_n})} \cdot \overline{f_2(\overline{x_1}, \overline{x_2}, \cdots, \overline{x_n})} \quad \text{(structural inductive hypothesis)}$$
$$= \overline{(f_1 + f_2)(\overline{x_1}, \overline{x_2}, \cdots, \overline{x_n})} \quad \text{(De Morgan law—Theorem 2.4(d))}.$$

(ii)

$$(\overline{f_1})^d (x_1, x_2, \cdots, x_n) = \overline{f_1^{\,d}(x_1, x_2, \cdots, x_n)} \quad \text{(recursive step of the definition of dual)}$$
$$= \overline{\overline{f_1(\overline{x_1}, \overline{x_2}, \cdots, \overline{x_n})}} \quad \text{(structural inductive hypothesis)}.$$

(iii)

$$(f_1 \cdot f_2)^d (x_1, x_2, \cdots, x_n) = f_1^{\,d}(x_1, x_2, \cdots, x_n) + f_2^{\,d}(x_1, x_2, \cdots, x_n) \quad \text{(recursive step of the definition of dual)}$$
$$= \overline{f_1(\overline{x_1}, \overline{x_2}, \cdots, \overline{x_n})} + \overline{f_2(\overline{x_1}, \overline{x_2}, \cdots, \overline{x_n})} \quad \text{(structural inductive hypothesis)}$$
$$= \overline{(f_1 \cdot f_2)(\overline{x_1}, \overline{x_2}, \cdots, \overline{x_n})} \quad \text{(De Morgan law—Theorem 2.4(d))}.$$

The proof of Proposition 2.5 is now complete. □

EXERCISES FOR APPENDIX TO SECTION 3.2:

1. Let S be the set recursively defined with the following two steps:
    1. *Basis Step*: $3 \in S$
    2. *Recursive Step*: If $x, y \in S$, then $x + y \in S$
    Give a nonrecursive description of the set S.

2. Let S be the set recursively defined with the following two steps:
    1. *Basis Step*: $3 \in S$
    2. *Recursive Step*: If $x, y \in S$, then $x - y \in S$
    Give a nonrecursive description of the set S.

3. Let S be the set recursively defined with the following two steps:
    1. *Basis Step*: $3, 5 \in S$
    2. *Recursive Step*: If $x, y \in S$, then $x \cdot y \in S$
    Give a nonrecursive description of the set S.

4. Let S be the set recursively defined with the following two steps:
    1. *Basis Step*: $5, 16 \in S$
    2. *Recursive Step*: If $x, y \in S$, then $x + y, x - y \in S$
    Give a nonrecursive description of the set S.

5. Let S be the set recursively defined with the following two steps:
    1. *Basis Step*: $3, 5 \in S$
    2. *Recursive Step*: If $x, y \in S$, then $x + y, x \cdot y \in S$
    (a) Find all elements of S that are expressible by a single application of the recursive step.
    (b) Find all elements of S that are expressible by a two applications of the recursive step.

6. Give a recursive definition of each of the following sets of integers:
    (a) The set of odd integers.

(b) The set of negative integers.

(c) The set of integers congruent to 3 mod 5.

7. Give a recursive definition of each of the following sets of integers:

    (a) The set of even integers.

    (b) The set of positive integers greater than 6.

    (c) The set of positive integers congruent to 7 mod 12.

NOTE: The following concepts regarding strings, although quite intuitive, require recursion to be formally defined:

The **length of a string** in an alphabet $\mathscr{A}$ is defined recursively as follows:

1. *Basis Step*: length($\varnothing$) = 0, where $\varnothing$ denotes the empty binary string.

2. *Recursive Step*: If STR is a string, and $x \in \mathscr{A}$ then length(STRx) = length(STR) + 1.

The **concatenation STR1 · STR2 of two strings** STR1, STR2, in an alphabet $\mathscr{A}$ is defined recursively as follows:

1. *Basis Step*: $\varnothing \cdot x = x$, where $x \in \mathscr{A}$ and $\varnothing$ denotes the empty binary string.

2. *Recursive Step*: If STR1, STR2 are strings in the alphabet $\mathscr{A}$, and $x \in \mathscr{A}$ then STR1 · (STR2 x) = (STR1 · STR2) x.

The next exercise asks for a recursive definition of an even more basic concept:

8. Give a recursive definition of the set of all strings in an alphabet $\mathscr{A}$.

9. Use structural induction to prove that length(STR1 · STR2) = length(STR1) + length(STR2), for any two strings STR1, STR2, in an alphabet $\mathscr{A}$.

10. The set S of fully bracketed arithmetic expressions in a variable x is the set of strings in the alphabet $\mathscr{A} = \{x, 0, 1, +, -, \cdot, [,]\}$ defined recursively as follows:

    1. *Basis Step*: $x, 0, 1 \in S$.

    2. *Recursive Step*: If STR1, STR2 $\in S$, then $[\text{STR1} + \text{STR2}] \in S$, $[\text{STR1} \cdot \text{STR2}] \in S$, and $[-\text{STR1}] \in S$.

    Use structural induction to prove that every fully bracketed arithmetic expression (in some variable x) has an equal number of left and right brackets.

11. Let S be the set of pairs of integers (i.e., $S \subseteq \mathbb{Z}^2$) recursively defined with the following two steps:

    1. *Basis Step*: $(2,3) \in S$.

    2. *Recursive Step*: If $(a,b) \in S$, then $(b, 3b - 2a) \in S$.

    (a) Find all elements of S that are expressible by 1, 2, 3, or 4 applications of the recursive step.

    (b) Use structural induction to prove that all elements of S have the form $(2^k + 1, 2^{k+1} + 1)$, where k is a nonnegative integer.

COMPUTER EXERCISES 3.2:

NOTE: If your computing platform is a floating point arithmetic system, it may allow you up to only 15 or so significant digits of accuracy. Symbolic systems allow for much greater precision, being able to handle 100s or 1000s of significant digits. Some platforms allow the user to choose if they wish to work in floating point or symbolic arithmetic, but will work in floating point arithmetic by default since operations are faster and usually sufficiently accurate for general purposes. If you have access to only a floating point system, you should keep these limitations in mind when you do computer calculations with large integers. Some particular questions below may need to be skipped or modified so the numbers are of a manageable size.   See Section 4.4 for more on the important differences between floating point and symbolic computing systems.

1. (*Empirical Comparison of the Iterative versus Recursive Programs for the Fibonacci Sequence*) Write programs (in your computing platform) `y = fibonacci(n)`, `y = fibonacciv2(n)`, corresponding to the programs of the same names/syntax that were presented in Example 3.8. Run each program with inputs n = 2, 4, 6, 8, ... until it takes more than 2 minutes to execute. Record the runtimes and compare them with the complexity analysis that was presented in Example 3.8.

2. (*Empirical Comparison of the Iterative versus Recursive Programs*) Write programs (in your computing platform) corresponding to the iterative and recursive programs of Exercise for the Reader 3.11.   Run each program with inputs n = 2, 4, 6, 8, ... until it takes more than 2 minutes to execute. Record the runtimes and comment on the relative complexity of the two algorithms.

3. (*Empirical Examination of the Cofactor Expansion Program*) Write a program (in your computing platform) `y = DetCofactor(A)`, corresponding to the program of the same name/syntax that was presented in Example 3.9. Run the program with on some $n \times n$ matrices whose entries are taken (preferably "randomly") in the range {0, 1, 2, ..., 9}, with $n = 2, 4, 6, 8,$ ... until it takes more than 2 minutes to execute. Record the runtimes and compare them with the complexity analysis that was presented in Example 3.9.
    Note: Although the topic of random integer generation will be formally discussed in Chapter 6, most computing platforms have utilities (often under the name `rand`) that at each call will generate a "pseudorandom" real number in the range (0,1). If you can find such a utility, use the composition floor(`rand`) to generate a random integer in the range {0, 1, 2, ..., 9}.

4. (*Empirical Comparison of the Iterative versus Recursive Programs for Finding the Maximum in a List of Numbers*) Write programs (in your computing platform) `Max = IterativeMax(vec)`, and `Max = RecursiveMax(vec)` each inputting a vector (i.e., an ordered list) `vec` of real numbers, and outputting the maximum number `Max` in the list. The first program is the iterative algorithm given in (ordinary) Exercise 27, and the latter is the recursive algorithm of Part (a) of that exercise.   Run the program with on some length n vectors whose entries are taken (preferably "randomly") to be real numbers in the range (0,n), with n = 10, 100, 1000, etc., until either n reaches 1 million or the computation takes more than two minutes. Record the runtimes and compare with the answer to Part (b) of Exercise 27.
    Note: See the note to the previous exercise. If you can find such a `rand` utility, generate the entries of the length n vector as `n*rand`.

5. (*Recursive Algorithm for Towers of Hanoi Puzzle Solutions*) (a) Write a recursive program that will delineate the moves needed to solve the Towers of Hanoi puzzle (based on the solution of Example 3.6) with n disks. The syntax should be: `Moves = TowerOfHanoiSol(n)`, where the input n is the number of disks, and the output `Moves` is a 3 column matrix whose rows indicate the sequence of moves (in order) that will solve the Towers of Hanoi puzzle with n disks. Each row of `Moves` will be a vector of three integers: (d, pStart, pEnd), where the first entry d indicates the disk that will be moved, the second entry pStart the peg from where the disk will be moved, and the third entry pEnd the peg to where the disk will be moved. For definiteness assume that the smallest disk has index d = 1, the next smallest has index d = 2, and so forth.
    (b) Run your program of Part (a) with n = 1, 2, 3, and 4, and verify the correctness of the ouputs.

6. (*Recursive Algorithm for a Geometric Tiling*) Recall that in Example 3.3 of Section 3.1, it was proved by induction that any $2^n \times 2^n$ grid of squares with a single square removed can be tiled by 3-square L-shaped tiles. The proof also provided a recursive scheme to achieve such a tiling.
    (a) (*Non-Graphical Interface Implementation*) Identify the grid as a $2^n \times 2^n$ matrix G whose rows and columns are indexed from 1 to 2^n.   Identify each 3-square L-shaped tile as a triple, (i, j, α), where the first two indices give the row and column index of the central square (so

$1 \le i, j \le 2^n$), and $\alpha \in \{1, 2, 3, 4\}$ indicates the orientation of the tile (once the central square of the tile has its location specified, there are four possible ways to lay the tile in a grid). Note that when the central square is on a border, some values of α are infeasible. Write a recursive program `Tiles = ThreeSquareLTiling(n, DelSq)`, whose first input is a positive integer n, whose second input `DelSq` is the vector of indices (i,j) of a deleted square in the $2^n \times 2^n$ matrix (representing a grid) G, and whose output `Tiles` is a three-column matrix whose rows represent a tiling of G with the prescribed single deleted square.

(b) Run your program of Part (a) with n = 1, 2, and 3, each with two different deleted squares and graphically verify the correctness of the ouputs.

(c) (*Graphical Interface Implementation–For Readers Who Have Experience with Computer Graphics*) Create a graphically enhanced version of the program of Part (a). The inputs and outputs should be the same, but the program should also produce a graphic of the grid with the tiles indicated in some easy to read fashion.

(d) Run your program of Part (c) with n = 1, 2, and 3, each with two different deleted squares and graphically verify the correctness of the ouputs.

Suggestion: For Part (a), as the algorithm progresses, let the matrix G representing the grid have a 0 entry if the corresponding square is not yet covered by a tile, and a 1 entry if it is covered or is the deleted square. One simple way to implement Part (c) would be to indicate each tile by a skeletal L drawn within the tile in heavy line font, and to have the grid lines in much lighter dotted line font.

3.3: SOME TOPICS IN ELEMENTARY NUMBER THEORY

Divisibility

One of the most fundamental concepts of the integers is that of divisibility, which is first learned in grade school. Here is the formal definition:

DEFINITION 3.5: Suppose that a and b are integers with $a \ne 0$. We say that a **divides** b (written $a \mid b$) if there is an integer c such that $b = ac$. This can also be expressed by saying: a is a **factor** of b, or b is a **multiple** of a. If a does not divide b, we write $a \nmid b$.

Here are some simple examples: $3 \mid 6$, since $6 = 3 \cdot 2$. Also, $-5 \mid 15$, since $15 = (-5) \cdot (-3)$. But $8 \nmid 20$, because 20/8 is not an integer. Notice also that for any nonzero integer a, we have $a \mid a$ (since $a = a \cdot 1$), $a \mid 0$ (since $0 = a \cdot 0$). The above definition can be expressed concisely using the logical notation of Chapter 1: $a \mid b \Leftrightarrow \exists c[b = ac]$. (The domain of discourse here is $\mathbb{Z}$.) The following theorem contains some basic yet very useful properties of divisibility.

THEOREM 3.10: Let a, b, and c be integers.

(a) (*Divisibility is Transitive*) If $a \mid b$ and $b \mid c$, then $a \mid c$.

(b) If $a \mid b$ and $a \mid c$, then $a \mid (bx + cy)$ for any integers x and y.

Proof: Part (a): Since $a \mid b$ we can write $b = ae$ for some integer e. Similarly, since $b \mid c$, we can write $c = bf$ for some integer f. Substituting the former into the latter gives $c = (ae)f = a(ef)$. Since ef is an integer, we conclude that $a \mid c$.

Part (b): The hypotheses allow us to write $b = ae$ and $c = af$, for some integers e and f. Substituting these gives us $bx + cy = aex + afy = a(ex + fy)$. Since $ex + fy$ is an integer, we conclude that $a \mid (bx + cy)$. $\square$

Primes

DEFINITION 3.6: An integer $p > 1$ is called **prime** if the only positive factors of p are 1 and itself. An integer $a > 1$ that is not prime is called **composite**.

The first few primes are 2, 3, 5, 7, 11, 13, 17, 19, 23, 29, 31, ... Prime numbers are the building blocks of the integers because any integer greater than 1 can always be uniquely factored into primes. This is the so-called *fundamental theorem of arithmetic*. We will state this important theorem here and will give its proof shortly.

THEOREM 3.11: (*Fundamental Theorem of Arithmetic*) Every positive integer $a > 1$ can be uniquely expressed as the product of primes. In other words, there exist unique prime numbers: $p_1 < p_2 < \cdots < p_n$, and corresponding positive exponents $\alpha_1, \alpha_2, \cdots, \alpha_n \in \mathbb{Z}_+$ such that $a = p_1^{\alpha_1} p_2^{\alpha_2} \cdots p_n^{\alpha_n}$.[12]

In general, it becomes difficult to verify whether or not a positive integer is prime, if the integer is large. If a positive integer a has a nontrivial factorization $a = bc$, with $b, c > 1$, then one of b or c must be $\leq \sqrt{a}$ (otherwise we would have the contradiction $a = bc > \sqrt{a}\sqrt{a} = a$). This means that to check if a given positive integer a is prime, we need to look only for (prime) factors that are at most equal to $\sqrt{a}$. But testing primality and, more generally, determining the prime factorization of large positive integers can take an inordinate amount of time, even with the best computers and algorithms.[13]

---

[12] This important theorem is the reason why the number 1 is not considered to be prime. If 1 were prime, we would no longer have unique factorization; for example $18 = 2 \cdot 3^2 = 1^3 \cdot 2 \cdot 3^2 = 1^{12} \cdot 2 \cdot 3^2$, etc.

[13] To illustrate this, we point out that RSA Security (a high-tech cryptographic security company) had offered a number of public challenges on their company Web site. One of these offered a $100,000 prize to the first person to factor a certain 304 digit number (larger prizes were available for factoring larger integers). This particular challenge had remained open for several years. Such challenges actually benefit the company by helping to test the security of some of their secret codes (that rest on the infeasibility of being able to factor such large or even larger integers) against potential hackers. We will discuss such topics in greater detail in Section 4.5.

EXAMPLE 3.14: Find the prime factorizations of each of the following integers: (a) 847, (b) 4808, (c) 6177.

SOLUTION: (a) Using the basic principle mentioned above, we begin checking, in order, for prime factors of 847 (knowing that we can stop after we check primes up to $\sqrt{847} = 29.1033$). Certainly $2 \nmid 847$ (since the latter is odd), also since $847/3 = 282\ 1/3 \notin \mathbb{Z}$, we know that $3 \nmid 847$. Since 847 does not end in 0 or 5, $5 \nmid 847$. But $847/7 = 121$, and we are now reduced to looking for prime factors of 121, so we can stop when we get to $\sqrt{121} = 11$. But you probably already knew that $11^2 = 121$. A diagram as is shown on the right is often used when such factorizations are done by hand. The resulting prime factorization is thus $847 = 7 \cdot 11^2$.

$$\begin{array}{r} 11 \\ \overline{11|121} \\ \overline{7|847} \end{array}$$

(b) and (c): Going through the same procedure, the corresponding prime factorizations are $4808 = 2^3 \cdot 601$, and $6177 = 3 \cdot 29 \cdot 71$.

One natural question arises: how many primes are there (infinitely many, or does the list eventually end)? This question was resolved a very long time ago by Euclid—the Greek mathematician who lived 325 BC–265 BC and is most famous for his timeless geometry book: *The Elements*—who proved that there are infinitely many primes. Euclid's elegant proof uses the fundamental theorem of arithmetic.[14]

THEOREM 3.12: (*Euclid*) There are infinitely many primes.

Proof: Suppose the assertion were false. Then the list of all primes would be finite: $p_1 < p_2 < \cdots < p_M$. Consider the integer $N = p_1 \cdot p_2 \cdots p_M + 1$. By the fundamental theorem of arithmetic, N can be factored (uniquely) into primes. Let p_i be (any) one of the prime factors of N. Then, since $p_i \mid p_1 \cdot p_2 \cdots p_M$, and $p_i \mid N$, it follows from Theorem 3.10(b) that $p_i \mid (N - p_1 \cdot p_2 \cdots p_M)$, i.e., $p_i \mid 1$. But this is a contradiction since no prime can divide 1. □

The Prime Number Theorem

Euclid's proof that there are infinitely many primes, although very elegant, does not tell us much about how the primes are distributed among the positive integers. A more informative but much deeper theorem is known as *the prime number*

---

[14] By contrast, the problem of whether there are infinitely many *prime pairs* has not yet been resolved. A prime pair consists of two primes whose difference is two, for example: 3 and 5, 5 and 7, 11 and 13, and 17 and 19, are the first few prime pairs.

theorem; it gives a very sharp estimate of the number of primes less than any given number x.

THEOREM 3.13: (*The Prime Number Theorem*) If $\pi(x)$ denotes the number of prime numbers p satisfying $p < x$, then we have

$$\pi(x) \sim \frac{x}{\ln x},$$

(in words: $\pi(x)$ is asymptotic to the ratio $x / \ln x$), meaning that the ratio $\pi(x)/(x / \ln x) \to 1$ as $x \to \infty$.

The prime number theorem was first proved independently in 1896 by Jacques Hadamard (1865–1963, French mathematician) and Charles Jean Gustave Nicolas Baron de la Vallée Poussin (1866–1962, Belgian mathematician). Their proof was quite sophisticated and used complex analysis. Elementary (but difficult) proofs have later been found and the interested reader may refer to the books by Hardy and Wright [HaWr-80] or Nagell [Nag-01]. We will forgo giving a proof here. The prime number theorem shows that the integers are quite densely populated by primes. For example, if one were to attempt to factor the 304 digit number n_{364} referenced in the preceding footnote by checking for prime factors up to $\sqrt{n_{364}}$, since $\sqrt{n_{364}} > 10^{152}$, this would possibly entail checking up to $\pi(10^{152}) \sim 10^{152} / \ln(10^{152}) \approx 2.86 \times 10^{149}$ divisions. Even if all the computers in the world could be programmed to work together on this and if they each could check 1 trillion divisions per second, this would take many millions of life spans of our universe. There are better algorithms for factoring into primes, but the prime factorization problem is a centuries old problem that is computationally very hard. An efficient factoring algorithm does not exist, and experts believe that one never will. These guaranteed difficulties are at the heart of many effective cryptographic applications of prime numbers.

The prime number theorem has many practical applications. For example, suppose that we are looking to find a prime number with 50 digits. Since 10^{50} is the smallest 51-digit number, and 10^{49} is the smallest 5-digit number, it follows that there are $\pi(10^{50}) - \pi(10^{49})$ 50-digit prime numbers. The prime number theorem estimates this number to be $10^{50} / \ln(10^{50}) - 10^{49} / \ln(10^{49}) \approx 7.7996 \times 10^{47}$. Since there are $10^{50} - 10^{49} = 9 \cdot 10^{49}$ 50-digit numbers, it follows that if we were to randomly select odd 50-digit numbers, the chances would be $7.7996 \times 10^{47} / 4.5 \times 10^{49} \approx 1/58$ that we would select a prime number.

Greatest Common Divisors, Relatively Prime Integers

DEFINITION 3.7:   Suppose that a and b are integers not both equal to zero. The **greatest common divisor** of a and b, denoted $\gcd(a,b)$, is the largest integer d that divides both a and b.   We say that a and b are **relatively prime** if $\gcd(a,b) = 1$.

For a simple example, since the common factors of 12 and 20 are 1, 2, and 4, we have $\gcd(12,20) = 4$. Similarly, since the only common (positive) factor of 8 and 15 is 1, $\gcd(8,15) = 1$, 8 and 15 are relatively prime. For integers of moderate size that can be readily factored into primes, the greatest common divisor can be easily read off from the prime factorizations—simply take all common prime factors, and use the minimum exponent of each prime. It is routine to verify that this product of common prime powers is the desired gcd (see Exercise for the Reader 3.15). This method is illustrated in the following example.

EXAMPLE 3.15: Find $\gcd(50,165)$, and $\gcd(1960,10800)$.

SOLUTION: The prime factorizations of the first pair of numbers are $50 = 2 \cdot 5^2$ and $165 = 3 \cdot 5 \cdot 11$,   therefore $\gcd(50, 165) = 5$.   Similarly, after computing the prime factorizations $1960 = 2^3 \cdot 5 \cdot 7^2$ and $10800 = 2^4 \cdot 3^3 \cdot 5^2$, we conclude that $\gcd(1960, 10800) = 2^3 \cdot 5 = 40$.

EXERCISE FOR THE READER 3.14: (a) Find the prime factorizations of 16000 and of 42757.
(b) Compute $\gcd(100, 76)$, $\gcd(16000, 960)$.

EXERCISE FOR THE READER 3.15: For a pair of nonzero integers a and b, the **least common multiple** of a and b, denoted $\mathrm{lcm}(a,b)$, is the smallest integer m that is divisible by both a and b.
(a) Find $\mathrm{lcm}(12, 28)$, and $\mathrm{lcm}(100, 76)$.
(b) Show that if $p_1 < p_2 < \cdots < p_n$ are the distinct primes appearing in the prime factorizations of either a or b, if $a = p_1^{\alpha_1} p_2^{\alpha_2} \cdots p_n^{\alpha_n}$, and if $b = p_1^{\beta_1} p_2^{\beta_2} \cdots p_n^{\beta_n}$, then $\mathrm{lcm}(a,b) = p_1^{\mu_1} p_2^{\mu_2} \cdots p_n^{\mu_n}$, where $\mu_i = \max(\alpha_i, \beta_i)$, and $\gcd(a,b) = p_1^{\sigma_1} p_2^{\sigma_2} \cdots p_n^{\sigma_n}$, where $\sigma_i = \min(\alpha_i, \beta_i)$.
(c) Show that $\mathrm{lcm}(a,b) \cdot \gcd(a,b) = ab$.

For pairs of large integers, the above procedure for finding greatest common divisors is very slow (because there is no known fast algorithm for prime factorization); a much more efficient method circumvents the need to factor. This simple yet useful method is called *the Euclidean algorithm*, and is also due to

Euclid.   We first formalize the procedure of dividing one integer by another nonzero integer.

The Division Algorithm

PROPOSITION 3.14:   (*The Division Algorithm*) If a is an integer and d is any positive integer, then there exist unique integers q and r, satisfying $0 \le r < d$, such that $a = dq + r$.   Here, a is called the **dividend**, d is called the **divisor**, q is called the **quotient**, and r is called the **remainder**.

Finding q and r is really just the "long division" problem $a \div d$ that one learns about in grade school, but Exercise for the Reader 3.16 will show how to quickly compute q and r, if one is using a calculator (or computer). The uniqueness proof of Proposition 3.14 is routine, and is left as an exercise. Although the proposition is not really an algorithm, the terminology is nonetheless standard in number theory, so we will adhere to it. In the language of modular arithmetic that we will introduce later, the result of the division algorithm can be expressed as $a \equiv r$ (mod d). Given the integers a and d, the dividend d is most easily expressed in terms of the "floor" function. Recall (from Section 2.2) that the floor function inputs any real number x and outputs floor$(x) = \lfloor x \rfloor =$ the largest integer that is less than or equal to x.

EXERCISE FOR THE READER 3.16: (a) Show that if the division algorithm is applied to an integer division $a \div d$, where (as usual) $d > 0$, then the quotient and remainder are given as follows: $q = \lfloor a / d \rfloor$ and $r = a - qd$.
(b) Use Part (a) to find the quotient and remainder when the division algorithm is applied to the following integer divisions: (i) $123 \div 5$, (ii) $-874 \div 15$.

The Euclidean Algorithm

The Euclidean algorithm consists of repeatedly applying the division algorithm. It is based on the following simple property:

PROPOSITION 3.15: If a, d, q, and r are as in the division algorithm, then $\gcd(a, d) = \gcd(d, r)$.

Proof: From the equation $a = dq + r$, and Theorem 3.10(b), we see that if $e \mid r$ and $e \mid d$, then $e \mid a$. If we rewrite the equation as $r = a - dq$, then by the same token we get that if $e \mid a$ and $e \mid d$, then $e \mid r$. We have proved that the set of all common divisors of r and d equals the set of all common divisors of a and d, from which the result of the theorem directly follows. □

For the pair of integers 100, 76, let us observe what happens when we repeatedly apply the division algorithm by dividing all new remainders into the previous divisors:

$$100 = 1 \cdot 76 + 24$$
$$76 = 3 \cdot 24 + 4$$
$$24 = 6 \cdot 4 + 0$$

From Proposition 3.15, we see that $\gcd(100,76) = \gcd(76,24) = \gcd(24,4) = \gcd(4,0) = 4$. In general, this procedure will always stop since the sequence of remainders is strictly decreasing (by the division algorithm, the new remainder must be less than the previous one because the previous remainder has become the divisor). The last nonzero remainder will be the gcd of the two starting integers. This procedure is the Euclidean algorithm. We now make a formal statement of it:

ALGORITHM 3.1: (*The Euclidean Algorithm*) Input: A pair of integers a and b, not both equal to zero. Output: The greatest common divisor, $\gcd(a,b)$.

Since $\gcd(\pm a, \pm b) = \gcd(a,b)$, we may assume that $a \geq b$ (if not, switch a and b), and that $b > 0$ (if $b = 0$, $\gcd(a,b) = a$). Apply the division algorithm to write $a = q_1 b + r_1$. If $r_1 = 0$, then $\gcd(a,b) = b$, otherwise continue by dividing successive divisors by successive remainders until a zero remainder is reached:

$$b \ = q_2 r_1 + r_2, \qquad 0 \leq r_2 < r_1$$
$$r_1 \ = q_2 r_2 + r_3, \qquad 0 \leq r_3 < r_2$$
$$\cdots$$
$$r_{n-2} = q_{n-1} r_{n-1} + r_n, \ 0 \leq r_n < r_{n-1}$$
$$r_{n-1} = q_n r_n + 0$$

The last nonzero remainder r_n is $\gcd(a,b)$.

Since the sequence of successive remainders is strictly decreasing: $b > r_1 > r_2 > \cdots > r_n > 0$, the algorithm must eventually terminate (in at most b steps). Since Proposition 3.15 implies that $\gcd(a,b) = \gcd(b,r_1) = \gcd(r_1,r_2) = \gcd(r_2,r_3) = \cdots = \gcd(r_{n-1},r_n) = \gcd(r_n,0) = r_n$, it follows that $\gcd(a,b) = r_n$, the last nonzero remainder that is encountered in this process.

We summarize how the division algorithm data are used in going from one round to the next:

remainder → divisor → dividend → not used

In practice, the Euclidean algorithm is a very efficient method for computing greatest common divisors, and it does not require any factorizations. As we will soon discover, it can also be used to solve other interesting problems, and it is readily translated into computer programs.

EXERCISE FOR THE READER 3.17: Use the Euclidean algorithm to compute gcd(65,91) and gcd(1665,910).

One very useful consequence of the Euclidean algorithm can be previewed by looking at the preceding example where we used it to find gcd(100,76) = 4. If we start with the last equation where this gcd (= 4) appeared as the remainder, and work our way up, we will be able to express 4 in the form of $100x + 76y$ for some integers x and y, i.e., as an *integer combination* of 100 and 76 (the integers that we wanted to find the gcd of). Here are the steps: we start with $76 = 3 \cdot 24 + 4$ and isolate the gcd (= 4) to write it as an integer combination of the two previous remainders: $4 = 3 \cdot 24 - 1 \cdot 76$. We then use the next equation up $100 = 1 \cdot 76 + 24$, solve it for 24 and substitute the result into what we had just before obtained: $4 = 3 \cdot 24 - 1 \cdot 76 = 3 \cdot (100 - 1 \cdot 76) - 1 \cdot 76 = 3 \cdot 100 - 4 \cdot 76$. The following theorem contains the general result.

THEOREM 3.16: Suppose that a and b are integers not both equal to zero, and let $d = \gcd(a,b)$. Then there exist integers x and y such that $d = ax + by$. In the special case in which a and b are relatively prime, we can write $1 = ax + by$.

Proof: The theorem is trivial in case either a or b is zero, so we assume that both are nonzero. The proof is a constructive one in that it provides an algorithm for finding such an x and y. If we set $r_0 = b$ and $r_{-1} = a$, then the Euclidean algorithm consists of $n + 1$ applications of the division algorithm, and these can all be expressed as $r_{i-1} = q_i r_i + r_{i+1}$ $(i = 0, 2, \cdots, n)$. Each of these is then rewritten to be solved for the last remainder: $r_{i+1} = q_i r_i - r_{i-1}$ $(i = 0, 2, \cdots, n)$. Since $d = r_n$, we will start with the second-to-last equation $(i = n - 1)$, and rewrite it as: $d = x_n r_{n-2} + y_n r_{n-1}$. Thus d is expressed as an integer combination of r_{n-2} and r_{n-1}. If we substitute the next equation up $(i = n - 2)$ $r_{n-1} = r_{n-3} - q_{n-2} r_{n-2}$ into our expression for d, we arrive at: $d = x_n r_{n-2} + y_n r_{n-1} = x_n r_{n-2} + y_n (r_{n-3} - q_{n-2} r_{n-2}) = y_n r_{n-3} + (x_n - q_{n-2}) r_{n-2} = x_{n-1} r_{n-3} + y_{n-1} r_{n-2}$. We continue this process of successively moving up the list of division algorithm equations and substituting them into our existing integer combination of d. At the kth step $(i = n - k)$, we will have obtained an expression for d as an integer combination $x_{n-k+1} r_{n-k-1} + y_{n-k+1} r_{n-k}$. At the final step $(k = n; i = 0)$, we will have $d = x_1 r_{-1} + y_1 r_0 = x_1 a + y_1 b$, as desired. $\square$

Although the proof was a bit technical, the idea is simple enough, and the whole scheme is nicely amenable to translate into a computer program. In Section 4.2 we will provide a very efficient implementation of this algorithm (that can be directly translated into a computer program).

EXERCISE FOR THE READER 3.18:   (a) Use the procedure described in the proof of Theorem 3.16 to express gcd(65,91) as an integer combination of 91 and 65. Similarly, express gcd(1665,910) as an integer combination of 1665 and 910. (b) Explain why the integers x and y in Theorem 3.16 are not unique.

Aside from its practical applications, Theorem 3.16 turns out to be very useful for obtaining new theoretical results. We demonstate this by using it to prove the following result, which, in turn, will allow us to prove the fundamental theorem of arithmetic.

LEMMA 3.17: (*Euclid's Lemma*) (a) Suppose that p is a prime, and that a and b are integers. If $p \mid ab$, then either $p \mid a$ or $p \mid b$.
(b) Suppose that p is a prime, and that $a_1, a_2, \cdots, a_n$ are integers. If $p \mid a_1 a_2 \cdots a_n$, then p must divide at least one of the factors $a_1, a_2, \cdots, a_n$.

We point out that the assumption that p is prime in the above lemma is crucial. For example, $6 \mid 2 \cdot 3$, but 6 divides neither 2 nor 3.

Proof: (a) Assuming that $p \mid ab$, if also $p \mid a$, we are done, so assume that $p \nmid a$. We need to show that $p \mid b$. Since p is a prime, and $p \nmid a$, it follows that gcd(a,p) = 1. Theorem 3.16 thus allows us to write $1 = ax + py$, for some integers x and y. We multiply both sides of this equation by b to obtain $b = (ab)x + pyb$. But since $p \mid ab$, and certainly $p \mid p$, it follows from Theorem 3.10(b) that $p \mid b$, as desired.
(b) We can achieve the proof of Part (b) by using the just proved special case of Part (a) (when $n = 2$) to repeatedly chip away at it: Assuming that $p \mid a_1(a_2 \cdots a_n)$, Part (a) tells us that either $p \mid a_1$, in which case we are done, or we get that $p \mid a_2 \cdots a_n$, which involves one less factor. Applying Part (a) again to this smaller case $p \mid a_2(a_3 \cdots a_n)$, we find that either $p \mid a_2$, in which case we are done, or we get that $p \mid a_3 \cdots a_n$. If we continue this process, we will either be done, or arrive at a division involving the final two factors $p \mid a_{n-1} a_n$, to which one final application of Part (a) will complete the proof. □

We are now nicely poised to prove the fundamental theorem of arithmetic:

Proof of the Fundamental Theorem of Arithmetic (Theorem 3.11):
Part 1: (*Existence*) Suppose that there were positive integers greater than one that could not be expressed as a product of primes. Let n be the smallest such integer.

Since n cannot be prime (because a single prime is a product of primes), it must be composite, so we can write $n = ab$, where a and b are smaller integers with $1 < a, b < n$. But since n was chosen to be the smallest integer that cannot be written as a product of primes, both a and b must be expressible as a product of primes. Since $n = ab$, we can multiply prime factorizations of a and b to obtain a prime factorization of n. With this contradiction, the existence proof is complete.

Part 2: (*Uniqueness*) Suppose that a positive integer n had two different prime factorizations:

$$n = p_1^{\alpha_1} p_2^{\alpha_2} \cdots p_k^{\alpha_k} = q_1^{\beta_1} q_2^{\beta_2} \cdots q_\ell^{\beta_\ell},$$

where $p_1 < p_2 < \cdots < p_k$ and $q_1 < q_2 < \cdots < q_\ell$ are primes, and $\alpha_1, \alpha_2, \cdots, \alpha_k$ and $\beta_1, \beta_2, \cdots, \beta_\ell$ are positive exponents. If there are any primes among the p's and q's that are common, they can be divided through (canceled) on each side of the equations so that the lists $p_1 < p_2 < \cdots < p_k$ and $q_1 < q_2 < \cdots < q_\ell$ can be assumed to have no primes in common, and we assume that this is indeed the case. Now, since $p_1 \mid p_1^{\alpha_1} p_2^{\alpha_2} \cdots p_k^{\alpha_k} = q_1^{\beta_1} q_2^{\beta_2} \cdots q_\ell^{\beta_\ell}$, it follows from Euclid's lemma (Lemma 3.17(b)) that $p_1 \mid q_j$ for some index j. But since p_1 and q_j are both primes, it follows that $p_1 = q_j$, which contradicts the assumption that the p's and q's have no primes in common. This completes the uniqueness proof. $\square$

Our next example demonstrates a useful application of Euclid's lemma to showing square roots of nonsquare integers are irrational.

EXAMPLE 3.16: Prove that $\sqrt{2}$ is irrational.

SOLUTION: We will prove this by the method of contradiction. If $\sqrt{2}$ were rational, this would mean that it could be written as a fraction of integers a/b, which we may assume is in lowest terms (i.e., a and b have no common prime factors). This means that $b\sqrt{2} = a$. Squaring both sides gives that $2b^2 = a^2$. Since $2 \mid 2b^2$, we get that $2 \mid a^2$, so by Lemma 3.17 (with $a = b$), we may conclude that 2 is a factor of a. If we write $a = 2a'$, the equation $2b^2 = a^2$ can be expressed as $2b^2 = 4(a')^2$ or $b^2 = 2(a')^2$. But once again by Lemma 3.17, this would give that 2 is a factor of b. We thus have a contradiction, since a and b were shown to have a common prime factor, so the result is established.

EXERCISE FOR THE READER 3.19: Show that for any positive integer n that is not a perfect square (i.e., n cannot be expressed in the form $n = k^2$ for some integer k), $\sqrt{n}$ is irrational.

Congruent Substitutions in Modular Arithmetic

We next move on to discuss some properties of congruences; we will further elaborate on this subject in Section 4.2.   Recall (from Section 2.2) that for any positive integer $m > 1$, the relation $a \equiv b \pmod{m}$ if, and only if, $m \mid (a - b)$, is an equivalence relation on $\mathbb{Z}$.   Also, the equivalence class $[a]$ of a given integer a consists of all integers in the set $\{a + km : k \in \mathbb{Z}\}$, and the (disjoint) equivalence classes are: $[0]$, $[1]$, $[2]$, …, $[m - 1]$.   These equivalence relations are very important in number theory and, in particular, in cryptographic applications, as we will see in Section 4.5.   We first give a theorem that basically says that congruences can be treated as ordinary equations when multiplying or adding things to both sides.   The numbers being added or multiplied to both sides do not have to be the same, but only congruent (mod m).   For example, if we start with the (true) equation $3 \equiv 15 \pmod{12}$, then we can multiply both sides by different integers that are congruent mod 12 and still get a valid equation.   For example, since 2 and 38 are congruent mod 12 (since $12 \mid (2 - 38)$), we can multiply both sides of the original congruence by either of these integers (which represent the same equivalence class) and the result will be true.   For example, $2 \cdot 3 \equiv 38 \cdot 15 \pmod{12}$.   This is easily checked since $(2 \cdot 3 - 38 \cdot 15)/12 = -564/12 = -47$ is an integer.   The general results are contained in the following theorem.

THEOREM 3.18:   (*Validity of Congruent Substitutions in Modular Arithmetic*) Suppose that m is a positive integer, and that a, b, a', b' are integers with $a \equiv a' \pmod{m}$ and $b \equiv b' \pmod{m}$.   The following congruences are then valid:

(a) $a + b \equiv a' + b' \pmod{m}$

(b) $-a \equiv -a' \pmod{m}$

(c) $a \cdot b \equiv a' \cdot b' \pmod{m}$

(d) $a - b \equiv a' - b' \pmod{m}$

(e) $a^k \equiv (a')^k \pmod{m}$, for any positive integer k.

The proof of this result is left as the next exercise for the reader.

EXERCISE FOR THE READER 3.20:   Prove Theorem 3.18.

Let us first relish some ramifications of Theorem 3.18.   In the motivating example, we previewed some of these consequences when we replaced each integer with its remainder mod 12.   Remainders are often convenient replacements, but the theorem tells us that we are free to use any replacements that we find convenient. As another example, consider the problem of computing (mod 12) the power 47^{129}.   If we computed this integer directly, it would have nearly 500 digits!   But since $47 \equiv 11 \pmod{12}$ (its remainder), Part (e) of the proposition tells us we could instead compute 11^{129}, and will get the same answer (mod 12).   This number still has about 300 digits, but if we notice that $11 \equiv -1 (\mathrm{mod}\, 12)$, the proposition would tell us that we could simply compute $(-1)^{129}$, which we immediately see (by hand) is -1 (as is any odd power of -1). So, we may conclude that $47^{129} \equiv -1 \equiv 11 (\mathrm{mod}\, 12)$.

A core component in some of the public key cryptosystems that we will develop in Section 4.5 is the ability to efficiently compute high powers in modular arithmethic. The next example previews two useful techniques that will be very useful for such purposes.

EXAMPLE 3.17: Compute 2^{1452} mod 19.

SOLUTION: A horribly inefficient way to do this would be to first compute 2^{1452} directly, using integer arithmetic, and then convert it to a representative, modulo 19, in the set $\{0, 1, 2, \cdots, 18\}$. This number would be so large that it would overflow on many computer systems. We now present two much more efficient methods that are based on Theorem 3.18.[15]

Method 1: (*Squaring Method*) We begin with $2^2 \equiv 4 \pmod{19}$, and continue to square both sides until the exponents exceed at least half of the desired exponent:

$$
\begin{aligned}
2^4 &\equiv 4^2 \equiv 16, \\
2^8 &\equiv 16^2 \equiv 256 \equiv 9, \\
2^{16} &\equiv 9^2 \equiv 81 \equiv 5, \\
2^{32} &\equiv 6, \\
2^{64} &\equiv 17, \\
2^{128} &\equiv 4, \\
2^{256} &\equiv 16, \\
2^{512} &\equiv 9, \\
2^{1024} &\equiv 5.
\end{aligned}
$$

We will now be able to use the above powers to compute the desired power of 2 (mod 19). This is because $1452 = 1024 + 256 + 128 + 32 + 8 + 4$, as the reader can easily check. It follows that (in light of Theorem 3.18), we may compute

$$2^{1452} = 5 \cdot 16 \cdot 4 \cdot 6 \cdot 9 \cdot 16 \equiv 11 \pmod{19}.$$

---

[15] Computing Note: This should serve as a caution to students not to take for granted, or to rely too much on the capabilities of computing platforms. Depending on what type of system you are working on: floating point or symbolic, a floating point system has accuracy to only about 16 digits. For example, the integer 13^{20} has 23 digits, so its computation on a floating point system would be accurate to only the first 15 or so of these digits. Thus, if we took the remainder of this computation, say mod 29, we would probably get the wrong answer. Symbolic systems have much greater accuracy, but usually work more slowly, and even these have limitations in the sizes of the numbers they can deal with. More on these topics will be addressed in Section 4.4.

Some comments are in order. The way that we came up with the decomposition of 1452 as a sum of powers of two was simple. We started with the largest power of two that was less than or equal to half the exponent. We then continued to go down the list of decreasing powers of two, adding them whenever the cumulative sum would not exceed the desired exponent. We will show in Section 4.1 that the resulting sum will always add up to the desired exponent. The method described is a general one that can be used to compute any power a^e (mod m). The resulting algorithm (fast modular exponentiation) will be more efficiently implemented in Section 4.2, but for now we can use it as was done above. The procedure will require at most $2\log_2(e)$ multiplications mod m, so that we will never have to deal with any number larger than $(m-1)^2$.

Method 2: At first glance, this method will appear to use a lucky trick. But we will soon give a theorem to show that this trick can be easily replicated in general situations. We note that $2^{18} \equiv 1$ (mod 19). If we apply the division algorithm to the integer division of 1452 by 18, we get $1452 = 80 \cdot 18 + 12$. It follows (in light of Theorem 3.18) that $2^{1452} \equiv (2^{18})^{80} \cdot 2^{12} \equiv 1^{80} \cdot 11 \equiv 11$ (mod 19). This was even less work than Method 1.

Fermat's Little Theorem

In general, the same trick used in the second method of the above example can be used to compute any power a^e (mod m), provided that we can find a special exponent s (less than m) such that $a^s \equiv 1$ (mod m). We will show that such an exponent always exists (the same will work for any a), as long as a and m are relatively prime, and show how to find it. We first deal with the case in which the modulus m is prime (as in the above example when m was 19). The following classical theorem of Pierre de Fermat (1601–1665, French barrister and mathematician)[16] is surprisingly simple:

THEOREM 3.19: (*Fermat's Little Theorem*) Suppose that p is a prime and a is an integer that is not a multiple of p, then $a^{p-1} \equiv 1$ (mod p).

Proof: We consider the set of nonzero equivalence classes of the integers modulo p: $A = \{[1],[2],\cdots,[p-1]\}$, and consider the function $f : A \to A$, defined by $f([x]) = [ax]$. By Theorem 3.18, this definition will give the same equivalence class output no matter which representative we use of $[x]$, so it is a well-defined function on equivalence classes. But we still need to check that the images will never be $[0]$ (i.e., so the co-domain of the function can be taken to be A. Indeed, if $[ax] = [0]$, this would mean that $p \mid ax$. But then Lemma 3.17 would imply that

---

[16] More biographical information on Fermat will be given in Section 6.1, which deals with probability.

either $p \mid a$ or $p \mid x$. Both of these options are not possible since $[a] \neq [0]$ and since $[x] \neq [0]$.

We next will show that f is one-to-one. Suppose that $f([x]) = f([y])$. This means that $[ax] = [ay]$ or $ax \equiv ay \pmod{p}$. By definition, this means that $p \mid (ax - ay)$, or $p \mid a(x - y)$. Lemma 3.17 then tells us that either $p \mid a$ or $p \mid (x - y)$. Since we know the former is false, the latter must hold, which means that $x \equiv y \pmod{p}$ or $[x] = [y]$, so f is one-to-one.

Since f is a one-to-one function of the set A to itself, it follows that the images of f: $f([1]), f([2]), \cdots, f([p-1]) = [a], [2a], \cdots, [(p-1)a]$ are simply a relisting of the elements of A: $[1], [2], \cdots, [p-1]$, in perhaps a different order. It follows (again from Theorem 3.18) that if we multiply representatives from the equivalence classes from each of these lists, we will get the same answer $(\bmod\, p)$:

$$1 \cdot 2 \cdots \cdot (p-1) \equiv a \cdot 2a \cdots \cdot (p-1)a \equiv a^{p-1}(1 \cdot 2 \cdots \cdot (p-1)) \pmod{p}.$$

This equation implies that $p \mid [a^{p-1}(1 \cdot 2 \cdots \cdot (p-1)) - 1 \cdot 2 \cdots \cdot (p-1)]$, or $p \mid [(a^{p-1} - 1)(1 \cdot 2 \cdots \cdot (p-1))]$, and Lemma 3.17 tells us that p must divide one of the factors on the right. The only possibility is that $p \mid (a^{p-1} - 1)$, so that $a^{p-1} \equiv 1 \pmod{p}$, as we wished to prove. $\square$

In light of Example 3.17, it is now easy to see how Fermat's little theorem can help us to efficiently raise any integer a to any power e modulo a prime p; we simply make use of the "magic" exponent $p-1$ and use the division algorithm to write $e = q(p-1) + r$, where $0 \leq r < p-1$. It then follows that $a^e \equiv a^r \pmod{p}$.

EXERCISE FOR THE READER 3.21: Compute 18^{802} mod 29, using each of the two methods shown in Example 3.17 (in Method 2 use Fermat's little theorem).

Our next theorem, which is due to Leonhard Euler[17] will generalize Fermat's little theorem to work for any modulus m. The analogue of the "magic" exponent p is determined by the integer function of the following definition:

DEFINTION 3.8: Euler's phi function is the function $\phi : \mathbb{Z}_+ \to \mathbb{Z}_+$, defined by $\phi(n) = $ the number of integers in the set $\{1, 2, \cdots, n\}$ that are relatively prime to n.

FIGURE 3.7: Leonhard Euler (1707–1783), Swiss mathematician.

In case $n = p$ is a prime number, then each of 1, 2, $\cdots$, $p - 1$, is relatively prime to p, so that $\phi(p) = p - 1$. The numbers in the set $\{1, 2, \cdots, 10\}$ that are relatively prime to 10 are 1, 3, 7, 9, so $\phi(10) = 4$. The following result makes it easy to compute $\phi(n)$ for any positive integer n, provided that we have the prime factorization of n.

PROPOSITION 3.20: If $n = p_1^{\alpha_1} p_2^{\alpha_2} \cdots p_k^{\alpha_k}$, where $p_1, p_2, \cdots, p_k$ are distinct primes, and $\alpha_1, \alpha_2, \cdots, \alpha_k \in \mathbb{Z}_+$, then

$$\phi(n) = (p_1 - 1) p_1^{\alpha_1 - 1} (p_2 - 1) p_2^{\alpha_2 - 1} \cdots (p_k - 1) p_k^{\alpha_k - 1}.$$

For example, since the prime factorization of 10 is $2 \cdot 5$, Proposition 3.20 tells us that $\phi(10) = (2 - 1) \cdot 2^0 \cdot (5 - 1) \cdot 5^0 = 4$, as we had shown earlier. For another example, since the prime factorization of 378 is $2 \cdot 3^3 \cdot 7$, the proposition tells us that $\phi(378) = \phi(2 \cdot 3^3 \cdot 7) = (2 - 1) \cdot 2^0 \cdot (3 - 1) \cdot 3^2 \cdot (7 - 1) \cdot 7^0 = 108$. A proof of Proposition 3.20 will be outlined in the exercises; see Exercises 41–43.

---

[17] Leonhard Euler (pronounced "Oiler") entered into mathematics during one of its most exciting eras; calculus had recently been invented and the field was transforming with numerous consequences and applications. Euler's life was nothing short of phenomenal. Educated in Switzerland, he was first appointed as a professor at age 19 at the renowned St. Petersburg University in Russia, and six years later he was appointed to the Berlin Academy and became its leader. His published works were significant, and touched on practically all of the fields of mathematics. He was the most prolific mathematician ever, even during the last 17 years of his life when he was completely blind (in fact, this was perhaps his most productive period). His papers were assembled into a collected works compendium that filled over 100 encyclopedia-sized tomes! His mental skills remained remarkably acute throughout his life. At age 70, for example, he could recite an entire novel, as well as tell you the first and last sentences on each page, and he once settled an argument between two students whose answers differed in the fifteenth decimal place, by a fast computation in his head. Euler had 13 children, and he told stories about having made some of his most seminal mathematical discoveries as he was holding one child on his lap while others were playing at his feet.

EXERCISE FOR THE READER 3.22: Compute the following values of Euler's phi function: $\phi(15), \phi(20), \phi(208), \phi(2208), \phi(6624)$.

Euler's Theorem

Notice that when $n = p$ is prime, the formula in Proposition 3.20 gives $\phi(p) = p - 1$, which was the "magic" exponent in Fermat's little theorem. Euler generalized Fermat's little theorem to work for any modulus, with his phi function continuing to play the role of the "magic" exponent.

THEOREM 3.21: (*Euler's Theorem*) Suppose that a and m are relatively prime positive integers with $m > 1$, then $a^{\phi(m)} \equiv 1 \pmod{m}$.

EXERCISE FOR THE READER 3.23: Prove Euler's theorem.
Suggestion: Mimic the proof of Fermat's Little Theorem, but now take the set A to be the set of all positive integers less than m that are relatively prime to m.

Euler's theorem greatly expands the situations for which Method 2 of Example 3.17 can be used for modular exponentiation.

EXAMPLE 3.18: (a) Compute $18^{2551} \bmod 25$.
(b) Find the last (one's) digit of the integer 13^{2017}.
Note: The integer in Part (b) has 2246 digits.

SOLUTION: Part (a): Since $\gcd(18,25) = 1$ and $\phi(25) = \phi(5^2) = (5-1) \cdot 5^1 = 20$, Euler's theorem tells us that $18^{20} \equiv 1 \pmod{25}$. Using the division algorithm for the integer division of 2551 by 20 gives $2551 = 127 \cdot 20 + 11$, and consequently $18^{2551} \equiv (18^{20})^{127} \cdot 18^{11} \equiv 1 \cdot 7 \equiv 7 \pmod{25}$. Note that the only calculation really needed was $18^{11} \equiv 7 \pmod{25}$. This could be done in a few steps as follows: Since $18^2 \equiv 324 \equiv -1 \pmod{25}$, we get $18^{10} \equiv (18^2)^5 \equiv (-1)^5 \equiv -1 \pmod{25}$, from which $18^{11} \equiv 18^{10} \cdot 18 \equiv -18 \equiv 7 \pmod{25}$.
Part (b): Finding the one's digit of any number is the same as the answer we would get by converting it to an integer modulo 10. Thus, we wish to find 13^{2017} (mod 10). Since $\gcd(13,10) = 1$ and $\phi(10) = 4$, Euler's theorem tells us that $13^4 \equiv 1 \pmod{10}$. Applying the division algorithm to the given exponent divided by 4 gives $2017 = 504 \cdot 4 + 1$, hence $13^{2017} \equiv (13^4)^{504} \cdot 13^1 \equiv 1 \cdot 13 \equiv 3 \pmod{10}$.

EXERCISE FOR THE READER 3.24: (a) Compute $7^{8486} \bmod 24$.
(b) Find the last three digits of the integer 13^{2017}.

Despite the speed and apparent magic with which Euler's theorem helps us to perform such modular exponentiations, there is one serious drawback: Evaluating $\phi(n)$ requires the prime factorization of n.   As we had mentioned, this is a very hard problem for which no efficient algorithm exists (and according to the general consensus will never exist).   Consequently, the first method that was used in Example 3.17 actually works more quickly.   We will formalize this method in Section 4.2 (see Algorithm 4.6), after we introduce the concept of binary expansions in Section 4.1.   Modular exponentiation has many applications, for example, it is at the core of the so-called El Gamal cryptosystem that will be studied in Section 4.5.

Orders and Primitive Roots

We end this section with another related and interesting concept, that of a *primitive root*. We first need a definition.

DEFINITION 3.9:   For integers $1 \le a < n$, with a and n relatively prime, we define the **order of a relative to n** (or **order of a mod n**) to be the smallest positive exponent k for which $a^k \equiv 1 \pmod{n}$. We write this as $k = \mathrm{ord}_n(a)$. [18]

We know from Euler's theorem, that $\mathrm{ord}_n(a) \le \phi(n)$.   The next propostion is often useful to help further narrow down the possibilities.

PROPOSTION 3.22:   If a and $n > 1$ are relatively prime positive integers, then $\mathrm{ord}_n(a) \mid \phi(n)$.

Proposition 3.22 is not difficult to prove if one uses the division algorithm in conjunction with Euler's theorem; see Exercise 48.   The following example illustrates the different orders of elements in two small moduli.

EXAMPLE 3.19: (a) Compute the orders of all positive integers less than (and relatively prime to) $n = 7$.
(b) Do the same for $n = 8$.

SOLUTION: Tables 3.1 and 3.2 below illustrate all of the powers (up to the $\phi(n)$th).

---

[18] This notation for orders is due to C. F. Gauss.

TABLE 3.1: The powers (mod 7) of integers relatively prime to $n = 7$; in each row, the order is at the top of the column containing the shaded box.

a^k	$k = 1$	$k = 2$	$k = 3$	$k = 4$	$k = 5$	$k = 6$
$a = 1$	1	1	1	1	1	1
$a = 2$	2	4	1	2	4	1
$a = 3$	3	2	6	4	5	1
$a = 4$	4	2	1	4	2	1
$a = 5$	5	4	6	2	3	1
$a = 6$	6	1	6	1	6	1

TABLE 3.2: The powers (mod 8) of integers relatively prime to $n = 8$; in each row, the order is at the top of the column containing the shaded box.

a^k	$k = 1$	$k = 2$	$k = 3$	$k = 4$
$a = 1$	1	1	1	1
$a = 3$	3	1	3	1
$a = 5$	5	1	5	1
$a = 7$	7	1	7	1

Note that in regards to Table 3.1, Euler's theorem tells us (since $\phi(7) = 6$) that $a^6 \equiv 1 \pmod 7$, for each a relatively prime to 7, while in the context of Table 3.2, it says that $a^4 \equiv 1 \pmod 8$, for each a relatively prime to 8 (since $\phi(8) = 4$). The tables are more revealing. Table 3.1 shows that only two elements have (maximum) order equal to Euler's exponent 6, while Table 3.2 shows that none of the elements have order equal to Euler's exponent 4. This motivates the following definition.

DEFINITION 3.10: We say that a positive integer g that is less than n and relatively prime to n is a **primitive root** mod n if the order of g is $\phi(n)$.

From Tables 3.1 and 3.2, we see that there are two primitive roots mod 7, namely 3 and 5, but that there are no primitive roots mod 8. Notice also that the powers of the primitive roots cycle through all of the $\phi(n)$ positive integers that are less than and relatively prime to n. This is true in general (Exercise 31). Primitive roots will be useful in a certain type of cryptographic system (the El Gamal system) that we will study in Chapter 7, so it is fortunate that we have the following result that completely describes when primitive roots exist, and, when they do, how many there are.

THEOREM 3.23: (*Existence and Number of Primitive Roots*)
    (1) Suppose that n is a positive integer greater than 1. Primitive roots exist mod n if, and only if n is of the form, 2, 4, p^s, or $2p^s$, where p is an odd prime. In this case there are exactly $\phi(\phi(n))$ primitive roots mod n.

(2) If g is a primitive root mod n, and $j < \phi(n)$ is a positive integer that is relatively prime to $\phi(n)$, then $g^j \pmod{n}$ is also a primitive root mod n, and all primitive roots are obtainable in this way.

The proof of Theorem 3.23(1) is rather lengthy, although not particularly difficult, so we will omit it. Interested readers may refer to any good book on number theory, such as [Ros-05]. For a proof of (2), see Exercise 48 (and some of the exercises leading up to it). This theorem is quite practical: (1) tells us whether primitive roots exist and if so how many there are, while (2) shows that once we find one primitive root, it is easy to get all of them.

EXAMPLE 3.20: For which of the following moduli n do primitive roots exist? In cases where primitive roots exist, how many will there be (mod n)?

(a) $n = 20$ (b) $n = 59$ (c) $n = 30$ (d) $n = 1250$

SOLUTION: The prime factorizations of these integers are: $20 = 2^2 \cdot 5$, 59 (prime), $30 = 2 \cdot 3 \cdot 5$, and $1250 = 2 \cdot 5^4$. Only those of Parts (b) and (d) satisfy the conditions of Theorem 3.23, and thus have primitive roots. The corresponding number of primitive roots are given by Euler's phi function (by Theorem 3.23), so that (using Proposition 3.20) 59 has $\phi(\phi(59)) = \phi(58) = \phi(2 \cdot 29) = 28$ primitive roots, and 1250 has $\phi(\phi(1250)) = \phi(\phi(2 \cdot 5^4)) = \phi(500) = \phi(2^2 \cdot 5^3) = 200$ primitive roots.

Once they are known to exist, primitive roots can easily be found by some simple computer procedures. Making use of the result of Propositon 3.22, we could simply run through the integers that are greater than 1, less than n, and relatively prime to n, and check their orders until one is reached that has order $\phi(n)$. This will be a primitive root of n. Such computations will proceed much faster with the fast exponentiation program that we give in Section 4.2. In our cryptographic applications in Section 4.5, we will need to do this when $n = p$ is prime.

EXERCISE FOR THE READER 3.25:
(a) How many primitive roots does $n = 334$ have?
(b) What is the smallest primitive root?

EXERCISES 3.3:

1.   Determine whether each of the statements below is true or false.
        (a) $9 \mid 128$ (b) $3 \mid 111$ (c) $13 \mid 5271$
        (d) $\forall a \in \mathbb{Z} \; [-1 \mid a]$ (e) $\forall a, b \in \mathbb{Z} \; [a \mid ab]$ (f) $0 \mid 12$

2.   Determine whether each of the statements below is true or false.
     (a) $7 \mid -49$ (b) $2 \nmid 111$ (c) $17 \mid 5271$
     (d) $\forall a \in \mathbb{Z} \, [1 \mid a]$ (e) $\forall a \in \mathbb{Z} \, [a \mid a]$ (f) $12 \mid 0$

3.   If a and b are integers such that $a \mid b$ and $b \mid a$, show that $a = \pm b$.

4.   If a, b and c are positive integers such that $a \mid b$, show that $ac \mid bc$.

5.   Which of the following integers is prime?
     (a) 67   (b) 91   (c) 893
     (d) 8671   (e) 6581   (f) 148,877

6.   Which of the following integers is prime?
     (a) 83   (b) 97   (c) 893
     (d) 1229   (e) 46,189   (f) 12,499

7.   Find the prime factorization of each of the following positive integers.
     (a) 24   (b) 88   (c) 675
     (d) 6400   (e) 74,529   (f) 183,495,637

8.   Find the prime factorization of each of the following positive integers.
     (a) 52   (b) 96   (c) 512
     (d) 4725   (e) 130,321   (f) 7,817,095

9.   (a) Use the prime number theorem to estimate the number of primes that are less than 1 billion.
     (b) Use the prime number theorem to estimate the number of primes that lie between 1 billion and 10 billion.
     (c) Use the prime number theorem to estimate the number of primes that lie between 1 billion and 1 trillion.

10. (a) Use the prime number theorem to estimate the number of primes that are less than 1 thousand.
     (b) Use the prime number theorem to estimate the number of primes that lie between 1 thousand and 10 thousand.
     (c) Use the prime number theorem to estimate the number of primes that lie between 1 thousand and 1 million.

11. Find the quotient and remainder when the division algorithm is applied to each of the following integer divisions.
     (a) $67 \div 2$ (b) $108 \div 5$ (c) $-77 \div 2$
     (d) $882 \div 13$ (e) $1228 \div 25$ (f) $-1582 \div 36$

12. Find the quotient and remainder when the division algorithm is applied to each of the following integer divisions.
     (a) $67 \div 3$ (b) $180 \div 5$ (c) $-90 \div 13$
     (d) $-564 \div 14$ (e) $1268 \div 42$ (f) $-8888 \div 25$

13. Using prime factorizations, compute the indicated greatest common divisors or least common multiples.
     (a) gcd(12,36)   (b) lcm(20,25)   (c) gcd(100,56)
     (d) gcd(560,1400)   (e) lcm(120,50)   (f) gcd(121275,5788125)

14. Using prime factorizations, compute the indicated greatest common divisors or least common multiples.
     (a) gcd(15,40)   (b) lcm(15,40)   (c) gcd(136,86)
     (d) gcd(1925,1568)   (e) lcm(150,350)   (f) gcd(256500,109395)

15.   Use the Euclidean algorithm to compute each of the quantities in Exercise 11. For those that are lcm's, first find the gcd (using the Euclidean algorithm), and then use the formula of Exercise for the Reader 3.15(c) to get the lcm.

16.   Use the Euclidean algorithm to compute each of the quantities in Exercise 12. For those that are lcm's, first find the gcd (using the Euclidean algorithm), and then use the formula of Exercise for the Reader 3.15(c) to get the lcm.

17.   For each pair of integers a, b, that are given, use the Euclidean algorithm (as explained in the proof of Theorem 3.16 and the paragraph before the theorem) to determine integers x and y such that $\gcd(a,b) = ax + by$.

   (a) 12, 36   (b) 100, 56
   (c) 560, 1400   (d) 121275, 5788125

18.   For each pair of integers a, b, that are given, use the Euclidean algorithm (as explained in the proof of Theorem 3.16 and the paragraph before the theorem) to determine integers x and y such that $\gcd(a,b) = ax + by$.

   (a) 15, 40   (b) 136, 86
   (c) 1925, 1568   (d) 256500, 109395

19.   Determine whether each of the statements below is true or false.

   (a) $43 \equiv 1 \ (\mathrm{mod}\, 2)$ (b) $-43 \equiv 3 \ (\mathrm{mod}\, 4)$ (c) $488 \equiv 10 \ (\mathrm{mod}\, 12)$
   (d) $2205 \equiv 45 \ (\mathrm{mod}\, 360)$ (e) $-443 \equiv 18 \ (\mathrm{mod}\, 22)$ (f) $7^7 \equiv 5^8 \ (\mathrm{mod}\, 371)$

20.   Determine whether each of the statements below is true or false.

   (a) $43 \equiv 2 \ (\mathrm{mod}\, 3)$ (b) $-43 \equiv 3 \ (\mathrm{mod}\, 6)$ (c) $2207 \equiv 11 \ (\mathrm{mod}\, 12)$
   (d) $11340 \equiv 90 \ (\mathrm{mod}\, 360)$ (e) $-2444 \equiv -446 \ (\mathrm{mod}\, 666)$ (f) $3^6 \equiv 5^2 \ (\mathrm{mod}\, 44)$

21.   Perform the following operations in mod 24 arithmetic.

   (a) $18 + 20$ (b) $5 - 21$ (c) $8 \cdot 8$ (d) $2^8 - 3^8$ (e) 21^{223}

22.   Perform the following operations in mod 53 arithmetic.

   (a) $39 + 47$ (b) $25 - 36$ (c) $18 \cdot 35$ (d) $12^5 - 19^4$ (e) 33^{100}

23.   Compute each of the indicated values of Euler's phi function:

   (a) $\phi(60)$ (b) $\phi(248)$ (c) $\phi(1224)$ (d) $\phi(9900)$

24.   Compute each of the indicated values of Euler's phi function:

   (a) $\phi(50)$ (b) $\phi(360)$ (c) $\phi(987)$ (d) $\phi(10,000)$

25.   (a) Show that if n is an even positive integer then $\phi(2n) = 2\phi(n)$.
      (b) Show that if n is an odd positive integer then $\phi(2n) = \phi(n)$.

26.   (a) Show that if n is a positive integer with $n \equiv 0 (\mathrm{mod}\, 3)$, then $\phi(3n) = 3\phi(n)$.
      (b) Show that if n is a positive integer with $n \not\equiv 0 (\mathrm{mod}\, 3)$, then $\phi(3n) = 2\phi(n)$.

27.   Find all positive integer solutions (if any) of the following equations:

   (a) $\phi(n) = 1$ (b) $\phi(n) = 4$ (c) $\phi(n) = 5$ (d) $\phi(n) = 12$

28. Find all positive integer solutions (if any) of the following equations:

   (a) $\phi(n) = 2$ (b) $\phi(n) = 3$ (c) $\phi(n) = 6$ (d) $\phi(n) = 14$

29. Use Euler's theorem to compute each of the following modular exponentiations. Write each answer as an integer in $\{1, 2, \cdots, m-1\}$, if you are working mod m.

   (a) 2^{1256} (mod 15)   (b) 7^{3945} (mod 20)

   (c) $2^{22,970}$ (mod 25)   (d) $8^{32,149}$ (mod 35)

30. Compute each of the indicated powers, working in modular arithmetic as specified. Write each answer as an integer in $\{1, 2, \cdots, m-1\}$, if you are working mod m.

   (a) 3^{1256} (mod 8)   (b) 12^{3945} (mod 25)

   (c) $3^{22,970}$ (mod 40)   (d) $13^{32,149}$ (mod 15)

31. (a) As in the solution of Example 3.19, create a table of all modular powers of the modular integers a that are relatively prime with $n = 6$ (up to the $\phi(n)th$). Use the table to identify the orders of each of these modular integers as well as any primitive roots.
   (b) Repeat Part (a) for $n = 12$.

32. (a) As in the solution of Example 3.19, create a table of all modular powers of the modular integers a that are relatively prime with $n = 10$ (up to the $\phi(n)th$). Use the table to identify the orders of each of these modular integers as well as any primitive roots.
   (b) Repeat Part (a) for $n = 11$.

33. Compute each of the following orders, if they exist.
   (a) $\text{ord}_{10}(3)$ (b) $\text{ord}_{21}(6)$ (c) $\text{ord}_{304}(21)$

34. Compute each of the following orders, if they exist.
   (a) $\text{ord}_{11}(5)$ (b) $\text{ord}_{17}(2)$ (c) $\text{ord}_{427}(21)$

35. For each of the following integers n, do the following:
   (i) Use Theorem 3.23 to determine whether there are any primitive roots mod n, and if so how many there are.
   (ii) If there are primitive roots, find one.

   (a) $n = 12$ (b) $n = 13$ (c) $n = 14$

36. For each of the following integers n, do the following:
   (i) Use Theorem 3.23 to determine whether there are any primitive roots mod n, and if so how many there are.
   (ii) If there are primitive roots, find one.

   (a) $n = 16$ (b) $n = 17$ (c) $n = 18$

37. For each of the following integers n, do the following:
   (i) Determine whether there are any primitive roots mod n, and if so how many there are.
   (ii) If there are primitive roots, find one. For Parts (a) through (c) give the smallest primitive root. For Parts (e) through (f), give any primitive root.
   (iii) If there are primitive roots, use the one you found in (ii) to construct another.

   (a) $n = 25$ (b) $n = 39$ (c) $n = 31$
   (d) $n = 50$ (e) $n = 52$ (f) $n = 961$

38. For each of the following integers n, do the following:

(i) Determine whether there are any primitive roots mod n, and if so how many there are.
(ii) If there are primitive roots, find one.   For Parts (a) through (c) give the smallest primitive root.   For Parts (e) through (f), give any primitive root.
(iii) If there are primitive roots, use the one you found in (ii) to construct another.

(a) $n = 17$ (b) $n = 81$ (c) $n = 323$
(d) $n = 289$ (e) $n = 4913$ (f) $n = 162$

39. For each of the following divisibility statements, either prove it or give a counterexample.
Assume throughout that all variables represent integers.
(a) If $a \mid b$ and $a \mid (b+1)$, then $a = \pm 1$.

(b) If n is even, then $4 \mid n^2$.

(c) If a and b are both even or both odd, then $a^2 - b^2$ is even.

40. For each of the following divisibility statements, either prove it or give a counterexample.
Assume throughout that all variables represent integers.
(a) If $a \mid b$ and $c \mid d$, then $ac \mid bd$.

(b) If $a \mid b$ and $a \mid c$, then $a \mid \gcd(b,c)$.

(c) If n is an integer, then $3 \mid n^3 - n$.

41. For each of the following statements, either prove it or give a counterexample.   Assume
throughout that all variables represent integers, unless otherwise specified.
(a) $4 \mid [a(a+1)(a+2)]$.

(b) $4 \mid (a^4 - a^2)$.

(c) If $n > 1$ is an odd integer, then $4^n - 3$ is prime.

42. For each of the following statements, either prove it or give a counterexample.   Assume
throughout that all variables represent integers, unless otherwise specified.
(a) If ab is a multiple of 4, then either a is a multiple of 4 or b is a multiple of 4.

(b) If a is odd, then $4 \mid (a^2 - 1)$.

(c) If a is odd, then $8 \mid (a^2 - 1)$.

43. (a) Prove the following identity: $\gcd(ab, ac) = a \gcd(b,c)$, whenever a, b, and c are positive
integers.

(b) Prove that if b and c are positive integers, with $d = \gcd(b,c)$, then $\dfrac{b/d}{c/d}$ will be the lowest-

terms representation of the fraction $\dfrac{b}{c}$.

(c) Explain how the Euclidean algorithm can be used to create algorithm for obtaining the

lowest-terms representation of any fraction $\dfrac{b}{c}$ of positive integers.   Apply your algorithm to

obtain lowest-terms representation of the fraction $\dfrac{1474}{39,463}$.

44. Prove the following variation of Euclid's lemma:
If $a \mid bc$ and $\gcd(a,b) = 1$, then $a \mid c$.
Suggestion: Use Theorem 3.16.

45. Prove that if $n \in \mathbb{Z}_+$, and $2^n - 1$ is prime, then n must be prime.
Suggestion: Use the following (easily verified) algebraic factorization identity:

$$x^{ab} - 1 = (x^a - 1)(x^{a(b-1)} + x^{a(b-2)} + \cdots + x^a + 1).$$

HISTORICAL NOTE: The converse of Exercise 45 is false, for example, $2^{11} - 1 = 2047 = 23 \cdot 89$. Prime numbers of the form $2^n - 1$ are called *Mersenne primes*, after Marin Mersenne (1588–1648, French priest and scholar). Mersenne boldly stated (without proof) in the preface of his book *Cogitata Physico-Mathematica* (1644) that $2^n - 1$ is prime for whenever $n = 2, 3, 5, 7, 13, 17, 19, 31, 67, 127$, and 257, and composite for all other values of n. The last of these numbers has 77 digits, so the claim would have been very difficult to verify at the time, when all calculations had to be done by hand. Mersenne's conjecture initiated much activity in the area, and it was not until over a century later in 1783, that someone discovered that Mersenne had missed one: $2^{61} - 1$ was proved to be a Mersenne prime. Taking advantage of their unique form, specialized and very efficient algorithms have been developed to check whether $2^n - 1$ is prime, and because of these algorithms, the largest known primes are Mersenne primes. Since the mid-1990s there has been an open project: *The Great Internet Mersenne Prime Search* (GIMPS—http://www.mersenne.org/) that has scientists across the world attempting to break new records. GIMPS offers $3000 for each new Mersenne prime that is discovered, and their website provides free specialized software programs that can help. In 2009, the first (Mersenne) prime that broke the 10 million digit benchmark was discovered on a UCLA computer set up (using GIMPS software) by Edson Smith. This prime, $2^{37,156,667} - 1$, has 12,978,189 digits, and garnered a long-standing $100,000 prize, offered by an anonymous donor for being the first to find a prime number with at least 10 million digits. The discovery of the 37th known Mersenne prime was made by Roland Clarkson, who worked independently, and at the time was a 19 year-old student at California State University–Dominguez Hills. There are still many theoretical questions that remain unanswered regarding Mersenne primes. For example, it is not known whether there are infinitely many Mersenne primes.

46. Suppose that a and $n > 1$ are relatively prime positive integers. Prove each of the following. (Note that Part (b) is Proposition 3.22.)

    (a) If k is a positive integer with $a^k \equiv 1 \pmod{n}$, then $\text{ord}_n(a) \mid k$.

    (b) $\text{ord}_n(a) \mid \phi(n)$.

    (c) If i and j are nonnegative integers, then $a^i \equiv a^j \pmod{n}$, if, and only if $i \equiv j \pmod{\text{ord}_n(a)}$.

    Suggestion: For Part (a) use the division algorithm.

47. Show that if a is a primitive root modulo a positive integer n, then the powers $a, a^2, a^3, \cdots,$ $a^{\phi(n)}$ are all different, and hence this list of powers is congruent to the set of all positive integers less than n that are relatively prime to n.
    Suggestion: Assume there is a repeated term. Use Part (c) of Exercise 46.

48. (a) (*Order of Powers Formula*) Prove that if a, j, and $n > 1$ are positive integers, with a relatively prime to n, then

    $$\text{ord}_n(a^j) = \frac{\text{ord}_n(a)}{\gcd(j, \text{ord}_n(a))}.$$

    (b) Show that if a is a primitive root modulo a positive integer n, and j is a positive integer that is relatively prime to $\phi(n)$, then a^j is also a primitive root modulo n, i.e., prove Part (2) of Theorem 3.22.
    Suggestion: The results of the previous two exercises will be useful for Part (a); Part (b) follows easily from Part (a).

49. (a) Verify that $g = 3$ is a primitive root of 223.
    (b) How many integers mod 223 have order 6? If such elements exist, find one.

(c) How many integers mod 223 have order 74? If such elements exist, find one.

(d) How many integers mod 223 have order 10? If such elements exist, find one.

Suggestion: For Parts (b), (c), and (d), use the result of Exercise 48(a).

50.   (a) Verify that $g = 3$ is a primitive root of 566.

(b) How many integers mod 556 have order 12? If such elements exist, find one.

(c) How many integers mod 556 have order 6? If such elements exist, find one.

(d) How many integers mod 223 have order 94? If such elements exist, find one.

Suggestion: For Parts (b), (c), and (d), use the result of Exercise 48(a).

51.   Prove that if a is any integer and n is any nonnegative integer then a and a^{4n+1} have the same last digit.

52.   Suppose that $a > 1$ is an integer, and that $k > \ell$ are positive integers.

(a) Let r be the remainder of the division $k \div \ell$, i.e, $r = k - \mathrm{floor}(k / \ell)$. Prove that the remainder of the division $(a^k - 1) \div (a^\ell - 1)$ is $a^r - 1$.

(b) Prove that $\gcd(a^k - 1, a^\ell - 1) = a^{\gcd(k, \ell)} - 1$.

Suggestion for Part (b): Apply the Euclidean algorithm in conjuction with the result of Part (a).

COMPUTER EXERCISES 3.3:

NOTE: If your computing platform is a floating point arithmetic system, it may only allow you up to 15 or so significant digits of accuracy. Symbolic systems allow for much greater precision, being able to handle 100s or 1000s of significant digits. Some platforms allow the user to choose if they wish to work in floating point or symbolic arithmetic, but will work in floating point arithmetic by default since operations are faster and usually sufficiently accurate for general purposes. If you only have access to a floating point system, you should keep these limitations in mind when you do computer calculations with large integers. Some particular questions below may need to be skipped or modified so the numbers are of a manageable size. See Section 4.4 for more on the important differences between floating point and symbolic computing systems.

1.   (*Program for Check if a Positive Integer is Prime*) (a) Write a program $y = \mathtt{PrimeCheck(n)}$ that inputs an integer $n > 1$, and outputs an integer y that is 1 if n is a prime number and 0 if it is not. The method used should be a brute force check to see whether n has any positive integer factor k, checking all values of k (if necessary) up to $\mathrm{floor}(\sqrt{n})$.

(b) Run your program with each of the following input values: $n = 30, 31, 487, 8893,$ 987654323, 131317171919.

(c) Assuming that your computing platform can perform 1 billion divisions per second (and assuming that the rest of the program in Part (a) takes negligible time), what is the largest number of digits an inputted integer n could have so that the program could be guaranteed to execute in less than 1 minute?

(d) Under the assumption of Part (c), how long could it take for the program in Part (a) to check whether a 100 digit integer is prime?

Note: Of course, the program could execute very fast if a small prime factor is found quickly. A prime input would always take the most time since the full range of k values would need to be checked. There are some efficiency enhancements that we could incorporate into the above program: For example, after it is checked that 2 is not a factor, we need only check odd integers after 2. Such a fix could cut the runtimes essentially in half, but there are more efficient (and sophisticated) prime checking algorithms than such brute-force methods. For more on this interesting area, we refer the reader to [BaSh-96].

2. (*Program for Prime Factorization of Positive Integers*) (a) Write a program `FactorList = PrimeFactors(n)` that inputs an integer n > 1, and outputs a vector `FactorList` that lists all of the prime factors, from smallest to largest, and with repetitions for multiple factors. For example, since the prime factorization of 24 is $2^3 \cdot 3$, the output of `PrimeFactors(24)` should be the vector [2 2 2 3]. Starting with k = 2, and running k through successive integers, the method should check to see whether n has k as a factor. If it does, then the k should be appended to the `FactorList` output vector. We then replace n with `n/k`, and continue to check whether k divides into (the new) n until it no longer does, and then we move on to update k to k + 1. With this scheme, only prime factors will be found. (Why?) Also, we may stop as soon as k reaches the current value of floor($\sqrt{n}$).

   (b) Run your program with each of the following input values: n = 30, 31, 487, 8893 987654323, 131317171919.

   Note: Although there are faster algorithms, the factorization problem of this computer exercise is more untractable than the primality checking algorithm of the preceding exercise. In the language of Chapter 7, a polynomial time algorithm has been found for the primality checking problem, but it is strongly believed that there can exist no such algorithm for the prime factorization problem. This latter fact is the basis for the success of the widely used RSA cryptosystem that we will study in Section 4.5. See also the note of the preceding exercise, and the reference given there.

3. (*Program for Finding the Next Prime*) (a) Write a program `p = NextPrime(n)` that inputs an integer n > 1, and outputs the smallest prime number p ≥ n.

   (b) Run your program with each of the following input values: n = 8, 30, 32, 487, 8899 987654321, 131317171919.

   Suggestion: Call on the program of Computer Exercise 1.

4. (*Program for the Division Algorithm*) (a) Write a program `(q,r) = DivAlg(a,d)` that inputs an integer a = the dividend, and a positive integer d = the divisor. The output will be the (unique) integers q = the quotient, and r = the remainder, satisfying a = dq + r, with 0 ≤ r < a (as guaranteed by Proposition 3.14).

   (b) Run your program with each of the following pairs of input values: (a, d) = (5, 2), (501, 13), (1848, 18), (123456, 321).

   Suggestion: Use the idea of Exercise for the Reader 3.16.

NOTE: (**mod** *Function on Computing Platforms*) Most computing platforms have some sort of remainder or modular integer converter; this will be particularly useful for modular arithmetic computations. The next exercise will ask you to either find such a function on your platform or (if you cannot find one or there is none) to write a (simple) program for one.

5. (a) Either find a program on your computing platform that performs as follows, or write one of you own: Write a program with syntax: `b = mod(a,m)` that will take as inputs an integer a, and an integer m > 0 (the modulus). The output should be a nonnegative integer b, with $0 \le b < m$, that is congruent to a (mod m). In other words, b should be the remainder when a is divided by m using the division algorithm.

   (b) Use this function to redo the hand calculations that were asked in (ordinary) Exercises 21 and 22.

   Suggestion: (For Part (b)) In case your system works in floating point arithmetic (or if you are not sure), in Parts (e) of (ordinary) Exercises 21 and 21, directly evaluating the very large ordinary integers as inputs in the mod function would result in inaccurate results. Instead, for example, to evaluate 21^{223} (mod 24), start by iteratively computing b1 = `mod(21^2,m)`, b2 = `mod(b1^2,m)` $\equiv 21^4 = 21^{2^2}$, b3 = `mod(b2^2,m)` $\equiv 21^8 = 21^{2^3}$, b4 = `mod(b3^2,m)` $\equiv 21^{16} = 21^{2^4}$, etc., until we get to the largest such exponent less than (or equal to) the desired exponent.   Then multiply out some of the appropriate intermediate results (using the `mod` function) to obtain the desired power. This idea can be streamlined into a very fast and effective modular exponentiation algorithm, and this will be done in Section 4.2.

6.   (*Program for the Euclidean Algorithm*) (a)   Write a program `d = EuclidAlg(a,b)` that inputs two positive integers `a` and `b`, and outputs `d =` gcd(a,b), computed using the Euclidean algorithm (Algorithm 3.1).
      (b) Check your program with the results of Exercise for the Reader 3.17, and then run it on each of the following pairs of input values: `(a,b)` = (525, 223), (12364, 9867), (1234567890, 0987654321), (13131717191919, 191917171313).

7.   (*Program for the Euler Phi Function*) (a) Write a program `y = EulerPhi(n)` that inputs an integer n > 1, and outputs y = $\phi(n)$.
      (b) Check your program with the results of (ordinary) Exercise 23, and then use it to compute $\phi(n)$ for each of the following values of n: 18,365, 222,651, 1,847,773, 22,991,877.
      Suggestion: Call on the program of Computer Exercise 2 for factoring integers and use the formula for $\phi(n)$ given in Proposition 3.20.   Since this program relies on factorization, it is not intended to be used with large integers (of say 10 digits or more).

8.   (*Program for Computing Orders*) (a)   Write a program `k = Order(a,n)` that inputs two relatively prime positive integers a, and n, with a < n, and outputs k = $\text{ord}_n(a)$.
      (b) Check your program with the results of (ordinary) Exercise 33, and then use it to compute these two orders: $\text{ord}_{1807}(3)$, $\text{ord}_{10543}(54)$.
      Suggestion: This can either be done with a brute-force approach (of computing successive powers of a until we reach 1), or using Proposition 3.22 (and calling on the program of Computer Exercise 1). In either case, the programs are not suitable for large integers.

9.   (*Program for Finding Primitive Roots*) (a)   Write a program `pRoot = SmallestPrimitiveRoot(n)` that inputs a positive integer n, that is of the form listed in Theorem 3.23, and that outputs the smallest primitive root of n (guaranteed to exist by Theorem 3.23).   It is fine to call on the program `Order` of the preceding exercise.
      (b) Check your program with the results of (ordinary) Exercise 37(a)–(c), and then run it on the following values of n: 8893, 17786, 123457, 246914.
      (c) Write a related program `PRoots = AllPrimitiveRoots(n)` that inputs a positive integer n, that is of the form listed in Theorem 3.23, and that outputs a vector `PRoots` of all of the $\phi(\phi(n))$ primitive roots of n (see Theorem 3.23).
      (d) Run your program of Part (c) on each of the inputs of Part (b).
      Suggestion: For Part (c), make use of (ordinary) Exercise 49.

APPENDIX TO SECTION 3.3: PROBABILISTIC PRIMALITY TESTS[19]

Number theory has gained tremendous interest since the advent of public key cryptography. Common to many of the most widely used public key cryptosystems is the need to generate very large prime numbers. As pointed out in the section proper, the prime number theorem provides us with good

---

[19] This appendix uses some very basic and intuitive concepts about probability and randomness. More details on these concepts can be found in Chapter 6.

estimates on the probability that a randomly generated number of a certain size is prime. Since this probability is quite high (i.e., the primes are quite densely populated in the positive integers), one possible scheme would be to randomly generate integers of a certain size, and perform a primality test on each one until we obtain a prime. The tests that we study here are very fast and widely used, but they actually can only prove that a number is composite. If they fail to do this for a given inputted number, the conclusion is that the number is *probably prime*. The probabilities can be made as close to 100% as one wishes, and these tests are extremely effective for large numbers. There are more elaborate primality tests (both probabilistic and deterministic) that can actually prove that the number being tested is prime. Such tests work much more slowly than the ones we introduce here. To learn more about this interesting topic, we refer the interested reader to the book [Wil-98].

Here is an outline of a typical scheme for constructing a prime of a prescribed size: Say that we need a 300 digit prime number. We would randomly generate the first digit from the set $\{1, 2, ..., 9\}$ (to assure the number will have 300 digits), the last digit gets randomly selected from the odd digits $\{1, 3, 5, 7, 9\}$ (to assure the number is odd), and then randomly assign the remaining 298 digits. By the prime number theorem (Theorem 3.13) the density of primes among integers of such size is approximately $1/\ln(10^{300}) \approx 1/691$. Since we are checking only odd integers, this doubles the density of primes to be approximately 1/345. Thus, on average, it will take about 345 randomly generated odd 300 digit numbers before we hit on a prime. On each attempt we use one of the suitable fast primality tests to screen out composite numbers. The probable primes that remain could be tested with one of the more expensive primality tests to certify primality. The analogy is similar to the public health problem of screening people for certain chronic illnesses. Usually a quick inexpensive test is given to the mass population, and for those who test positive, a more accurate (and expensive and slower) test is given to see if they really are infected.

One basic reason that primality checks can be done faster than factoring is that integers can be proved to be composite without actually producing a factorization. This phenomenon will be seen in our first simple primality test, which is based on the contrapositive of Fermat's little theorem (Theorem 3.19): If p is a prime and $1 < a < p-1$, then $a^{p-1} \equiv 1 \pmod{p}$.[20] The contrapositive can be formulated as:

CONTRAPOSITIVE OF FERMAT'S LITTLE THEOREM: If n is a positive integer and $a^{n-1} \not\equiv 1 \pmod{n}$, for some number a, $1 < a < n-1$, then n is composite.

As a simple example, we can compute (using fast modular exponentiation as in Example 3.17) that $2^{1002} \equiv 990 \pmod{1003}$, so by the contrapositive of Fermat's little theorem (with $a = 2$) this proves that 1003 is composite. This is a nice example to demonstrate how it is possible to determine that an integer is composite without actually factoring it. Since it is not clear how to choose an appropriate a to attempt to use this criterion to prove a given integer n is composite, it is best to simply make random choices, and to apply the test a certain number k times. If one of these k trials results in $a^{n-1} \not\equiv 1 \pmod{n}$, then the test has proved n to be composite, and a corresponding base a for which $a^{n-1} \not\equiv 1 \pmod{n}$, is called a **witness** to the fact that n is composite. If each of the k trials resulted in $a^{n-1} \equiv 1 \pmod{n}$, then n is declared as probably prime. Here is a formal summary of this randomized algorithm.

ALGORITHM 3.2: (*Randomized Fermat Primality Test*):
Inputs: An integer $n > 3$, and a positive integer k.
Output: Either a declaration that n is composite, along with a witness integer a that satisfies $a^{n-1} \not\equiv 1 \pmod{n}$ (and thus proves that n is composite), or a declaration that n is probably prime.

Step 1: Initialize iteration counter: $i = 0$.

---

[20] We intentionally omitted the values $a = 1$, and $a = p - 1$, since, even if p were composite (but an odd number greater than 1), these values would always equal 1 when raised to the (even) power $p - 1$.

Step 2: Randomly choose an integer a, $1 < a < n-1$, compute (using fast modular exponentiation) $a^{n-1} (\bmod\, n)$, and update the iteration counter $i \to i+1$.

Step 3: If $a^{n-1} \not\equiv 1 (\bmod\, n)$, then declare n is composite, and a as a witness to this fact, and exit the algorithm. Otherwise, go back to Step 2, unless $i = k$, in which case we declare n is probably prime, and exit the algorithm.

We point out that even though the above algorithm and the others which follow are called primality tests, they are not capable of proving that a number is prime; they can only prove compositeness. One drawback of the Fermat primality test is that it does not come with any guarantee on the probability that any declared probable prime really is prime. (Our next primality test will come with such a guarantee, however.)

EXAMPLE 3.21: Apply the Fermat primality test (Algorithm 3.2) with $k = 4$ to the following odd integers n: (a) $n = 409$ (b) $n = 721$

SOLUTION:
Part (a): $n = 409$

Step 1: Initialize the trial counter $i = 0$.

Step 2: Randomly generate a base: $a = 238$. We use Algorithm 6.5 to compute the modular power $a^{n-1} \equiv 238^{408} \equiv 1 (\bmod\, 409)$. So 409 has passed the Fermat primality test with this value of a, since i is now 1, we repeat Step 2:

Step 2: (2nd repetition) Randomly generate a base: $a = 222$. We use fast modular exponentiation to compute the modular power $a^{n-1} \equiv 222^{408} \equiv 1 (\bmod\, 409)$. So 409 has passed the Fermat primality test with this value of a, since i is now 2, we repeat Step 2:

Step 2: (3rd repetition) Randomly generate a base: $a = 356$. We use fast modular exponentiation to compute the modular power $a^{n-1} \equiv 356^{408} \equiv 1 (\bmod\, 409)$. So 409 has passed the Fermat primality test with this value of a, since i is now 3, we repeat Step 2:

Step 2: (4th repetition) Randomly generate a base: $a = 109$. We use fast modular exponentiation to compute the modular power $a^{n-1} \equiv 109^{408} \equiv 1 (\bmod\, 409)$. So 409 has passed the Fermat primality test with this value of a, and since i is now 4, this was the final iteration of Step 2.

Step 3: Declare 409 as probably prime. (The reader may check that 409 is indeed prime so the test worked.)

Part (b): $n = 721$

Step 1: Initialize the trial counter $i = 0$.
Step 2: Randomly generate a base: $a = 230$. We use fast modular exponentiation to compute the modular power $a^{n-1} \equiv 230^{720} \equiv 484 (\bmod\, 721)$.

Step 3: Since $a^{n-1} \not\equiv 1 (\bmod\, n)$, Fermat's test has proved that $n = 721$ is composite with witness $a = 230$. The reader may check this by factoring $721 = 7 \cdot 103$.

Fermat's test worked very well in the above example. For Part (b), compositeness was detected in just one (random) try. Indeed, a computer calculation can quickly verify that 684 out of the 718 possible bases in the range $1 < a < 721 - 1$ would have worked as witnesses to show that 721 is composite.

Carmichael Numbers

Usually if a number n is not prime, it will succumb to Fermat's primality test for most all values of a. Although very rare, there are extreme exceptions for which Fermat's test will detect compositeness only if the base a that is selected is an actual factor of n. (Thus the Fermat test for detecting compositeness for one of these numbers would be about as effective as trying to factor the number by randomly guessing at factors.)

DEFINITION 3.11: A composite number $n > 1$ is called a **Carmichael number** if $a^{n-1} \equiv 1 \pmod{n}$, for each integer a that is relatively prime to n.

These numbers are named after the American mathematician Robert D. Carmichael (1879–1967), who introduced these numbers and some of their interesting properties. Carmichael conjectured in 1912 that there were infinitely many Carmichael numbers, but this fact was not proved until 80 years later (see [AlGrPo-92]). The first three Carmichael numbers are 561, 1105, and 1729, and there are just 2163 Carmichael numbers that are less than 25 billion.

The Miller–Rabin Test

Apart from the existence of Carmichael numbers, another drawback of the Fermat primality test is that no matter how many iterations are used, it comes with no guaranteed confidence level of the probability that any probable primes produced will actually be prime. With a bit more work, a much more effective probabilistic primality test, known as the *Miller–Rabin primality test*,[21] can be developed that transcends both of these weaknesses. There will be a very quantitative performance guarantee, and there will be no analogue of Carmichael numbers for the Miller–Rabin test. As with Fermat's test, this one will depend on Fermat's little theorem, but it will hinge on the following two additional results, the second of which is a refined version of the contrapositive of Fermat's little theorem.

LEMMA 3.24: (*Square Roots of One mod p*) If p is an odd prime then $\sqrt{1} \equiv \pm 1 \pmod{p}$, i.e., modulo an odd prime, 1 has exactly two square roots, namely ± 1.

Proof: Modulo any integer n, $(\pm 1)^2 \equiv 1 \pmod{n}$, so ± 1 are always (modular) square roots of 1. We need to show that modulo an odd prime p, there are no others. Indeed, if x is a square root of 1 (mod p), then $x^2 \equiv 1 \Rightarrow x^2 - 1 \equiv 0 \Rightarrow (x+1)(x-1) \equiv 0 \pmod{p} \Rightarrow p \mid (x+1)(x-1)$. By Euclid's lemma (Lemma 3.17) this, in turn, implies that either $p \mid (x+1)$ or $p \mid (x-1)$, i.e., $x \equiv -1$ or $x \equiv 1 \pmod{p}$, as asserted. □

The test will follow from the contrapositive of the following result, just as the Fermat primality test came from the contrapositive of Fermat's little theorem.

PROPOSITION 3.25: Suppose that p is an odd prime, and $1 < a < p - 1$. Write $p - 1 = 2^f m$, where m is an odd integer. Then either $a^m \equiv 1 \pmod{p}$ or $a^{2^j m} \equiv -1 \pmod{p}$, for some j, $0 \le j < f$.

Proof: Fermat's little theorem (Theorem 3.19) tells us that $a^{p-1} \equiv 1 \pmod{p}$. Since $a^{(p-1)/2} \equiv a^{2^{f-1} m}$ is a square root of $a^{p-1} \pmod{p}$, Lemma 3.24 allows us to conclude that $a^{2^{f-1} m} \equiv \pm 1 \pmod{p}$. If

---

[21] The non-probabilistic version of this test was discovered by Gary L. Miller in 1976 [Mil-76]. It came with an effectiveness guarantee, but the drawback was that this guarantee rested upon an unproved conjecture in number theory (the generalized Riemann hypothesis). In 1980, Michael O. Rabin ([Rab-80]) converted this algorithm into a probabilistic one that (importantly) came with the following unconditional guarantee: If the test finds a number to be composite, it is guaranteed to be composite. If the test finds the number to be "probably prime," it has at least a 75% chance of being prime. This guarantee can be improved to attain as high a percentage as one wishes, simply by (independently) iterating the test. Moreover, it has been found that the Miller–Rabin algorithm does much better, on average, than the conservative 75% guarantee level, see [DaLaPo-93].

$a^{2^{f-1}m} \equiv -1(\mathrm{mod}\,p)$, then the second assertion holds. Otherwise, we can continue to take square roots in this fashion, either verifying the second assertion of the proposition, or else getting all the way to $a^m \equiv 1(\mathrm{mod}\,p)$, which is the first assertion of the proposition. □

The contrapositive of Proposition 3.25, can be formulated as follows:

CONTRAPOSITIVE OF PROPOSITION 3.25: If $n > 1$ is an odd integer, with $n - 1 = 2^f m$, where m is an odd integer, and we can find an integer a, with $1 < a < n - 1$, such that $a^m \not\equiv 1(\mathrm{mod}\,n)$ and $a^{2^j m} \not\equiv -1(\mathrm{mod}\,n)$, for all j, $0 \le j < f$, then n must be composite.

As with the Fermat test, there does not seem to be a good (deterministic) method for choosing such an a to prove compositeness, so a random choice is most effective. The following implementation of the Miller–Rabin test is set up to minimize its complexity. It relies on the simple fact that if $a^{2^j m} \equiv 1(\mathrm{mod}\,n)$, for some nonnegative integer j, then $a^{2^\ell m} \equiv 1(\mathrm{mod}\,n)$, for each $\ell \ge j$.

ALGORITHM 3.3: (*The Miller–Rabin Primality Test*):
Inputs: An odd integer $n > 3$, suspected to be prime, and a positive integer k.
Output: Either a declaration that n is composite, along with a witness integer a that violates the contrapositive of Proposition 3.25 (and thus proves that n is composite), or a declaration that n is probably prime.

Step 1: First express $n - 1$ as $2^f m$, where m is an odd integer. Initialize the main iteration counter: $i = 0$.

Step 2: Randomly choose an integer a, $1 < a < n - 1$. Calculate (using fast modular exponentiation) $A_0 \equiv a^m(\mathrm{mod}\,n)$. If $A_0 \equiv \pm 1(\mathrm{mod}\,n)$, update main iteration counter $i \to i+1$, and move on to Step 3, otherwise enter into the following for loop:

FOR $j = 1$ TO $f - 2$

    Compute $A_j \equiv A_{j-1}^2(\mathrm{mod}\,n)$ (this is $a^{2^j m}(\mathrm{mod}\,n)$)

    IF $A_j \equiv 1 \ (\mathrm{mod}\,n)$

        Declare n as composite, output the witness a, and EXIT algorithm
    ELSE IF $A_j \equiv -1 \ (\mathrm{mod}\,n)$

        Update main iteration counter $i \to i+1$, and move on to Step 3
    END IF
END FOR

Compute $A_{f-1} \equiv A_{f-2}^2(\mathrm{mod}\,n)$ (this is $a^{(n-1)/2}(\mathrm{mod}\,n)$)

IF $A_{f-1} \not\equiv -1(\mathrm{mod}\,n)$

    Declare n as composite, output the witness a, and EXIT program
ELSE Update the main iteration counter $i \to i+1$.
END IF

Step 3: If i equals k, declare n is probably prime, and exit the algorithm. Otherwise, go back to Step 2.

An integer a that proves n is composite in the Miller–Rabin primality test is called a **witness** for the compositeness of n. Observe that if the Miller–Rabin test declares n to be probably prime by using a certain integer a, then $a^{2^j m} \equiv \pm 1 (\mod n)$, for some j, $0 \le j < k$, and from this it follows (by repeated squaring) that $a^{n-1} \equiv a^{2^k m} \equiv 1 (\mod n)$, so that the Fermat test would have also declared n to be probably prime. Thus the Miller–Rabin test is at least as effective as the Fermat test; in fact, it is more effective—the exercises at the end of this appendix will examine this phenomenon. But what is even more important and useful is that the Miller–Rabin algorithm comes with the following performance guarantee:

THEOREM 3.26: (*Performance Guarantee for the Miller–Rabin Primality Test*) If $n \ge 3$ is an odd composite number, then at most $(n-1)/4$ of the numbers in the set $\{1, 2, \cdots, n-1\}$ that are relatively prime to n will not serve as witnesses to the compositeness of n in the Miller–Rabin primality test. Thus, the probability that the Miller–Rabin test of Algorithm 3.3 with k independent iterations declares n to be probably prime is at most $(1/4)^k$.

So, for example, if we perform $k = 20$ iterations, and the Miller–Rabin test has declared that n is probably prime, the probability that this is incorrect is smaller than $(1/4)^{20} = 9.0949... \times 10^{-13}$, or less than 1 in 1 trillion! Of course, just like with the Fermat test, if a compositeness conclusion is produced, the result is correct with probability 100%. Such a test can be performed very quickly on most computing platforms to produce primes of several hundred digits; the failure rate is so low that the results are reliable for most practical purposes.[22] The contemporary French number theorist Henri Cohen has referred to such probable primes as **industrial-grade primes**.

EXERCISES AND COMPUTER EXERCISES FOR THE APPENDIX TO SECTION 3.3:

NOTE: As mentioned in chapter text, if your computing platform is a floating point arithmetic system, it may allow you only up to 15 or so significant digits of accuracy. Symbolic systems allow for much greater precision, being able to handle 100s of significant digits. Some platforms allow the user to choose if they wish to work in floating point or symbolic arithmetic. If you are working on such a dual capability platform, you may wish to create two separate programs (for those that might work with large integers), an ordinary version, and a symbolic version (perhaps attaching a `Sym` suffix to the names of those of the latter type). In case you do not have access to a symbolic system, some particular questions below may need to be skipped or modified so the numbers are of a manageable size.

1. Apply the Fermat primality test (Algorithm 3.2) with $k = 4$ to the following odd integers:

   (a) $n = 527$ (b) $n = 523$ (c) $n = 943$
   (d) $n = 5963$ (e) $n = 11,303$ (f) $n = 1811$

2. Apply the Fermat primality test (Algorithm 3.2) with $k = 4$ to the following odd integers:

   (a) $n = 449$ (b) $n = 629$ (c) $n = 1147$
   (d) $n = 4559$ (e) $n = 8893$ (f) $n = 9727$

---

[22] The reader will have the opportunity to experiment with such examples in exercises. One important fact to point out is that unless one is working on a symbolic computing platform, the default floating point arithmetic systems on most computing platforms will be able to deal accurately only with integers up to 15 digits or so. Thus, in order to effectively implement the algorithms in this chapter with larger integers, one will need to make sure that their computing platform has symbolic functionality. (Most computing platforms have this capability, but not necessarily as a default mode, so some modifications in syntax might be required.) All of the applets for this book are designed to have symbolic functionality.

3. Apply the Miller–Rabin test (Algorithm 3.3) with $k = 4$ to each of the odd integers given in Exercise 1.

4. Apply the Miller–Rabin test (Algorithm 3.3) with $k = 4$ to each of the odd integers given in Exercise 2.

5. (*Fermat's Primality Test*) (a) Write a program with syntax y = FermatTest(n, k), that inputs an integer n > 3, and will apply the Fermat primality test k times. The second input variable k is optional (default value is 1).   The output y will be 1 if at least one of the k tests has found n to be composite, and 0 in case all of the tests were inconclusive (meaning, informally, they all found n to be probably prime).   Set your program up so that in cases that it proves *n* is composite, it should also output the witness *a* that resulted in the composite conclusion.
    (b) Apply your program to the following integers *n*, using $k = 10$: $n = 215$, 841, 1931, 3973, 22879.   Check the outputs by factoring each of these integers (in the usual way or using a built-in utility); also, for each corresponding witness *a*, compute $a^{n-1}(\bmod\, n)$.

    (c)   Use your program to attempt to prove compositness of each of the following (composite) integers: 3668963, 154915253, 6271549451, 6732432725687, 52322940983667496651.
    (d) The following two large prime numbers give rise to a famous factoring challenge by the RSA Cryptosystems company:

    p = 1634733645809253848443133883865090859841783670033092312181110852389333100
          10450815121218167511579
    q = 1900871281664822113126851573935413975471896789968515493666638539088027103
          8021044989571912614655571

    Their product $n = pq$, is known as RSA-640.[23] Apply your program of Part (a) to the RSA-640 (174 digit) number using $k = 1$, and repeat this 10 times.
    Note: As with any probabilistic algorithm, results will vary. When the author did Part (d), in all ten trials, the (single application of) Fermat's primality test proved that RSA-640 was prime very quickly (in a couple of seconds).   By contrast, recall that it took over four years before the international challenge to factor RSA-640 was finally met. This fact alone gives good evidence that primality checking is a much easier problem than factoring.

6. (*The Miller–Rabin Primality Test*) (a) Write a program with syntax

    MillerRabinTest(n, k),

    that inputs an odd integer n > 3, and will apply the Miller–Rabin primality test k times. The second input variable k is optional (default value is 1).   The program will indicate that n was found to be composite, if at least one of the k tests has found n to be composite, and that n is probably prime, in case all of the tests were inconclusive (meaning, informally, they all found n to be probably prime).   Set your program up so that in cases that it proves *n* is composite, it should also output the witness *a*, and corresponding exponent parameters [j m], that resulted in the composite conclusion.
    (b) through (d): Apply your program to redo each of the corresponding parts of Computer Exercise 5.

7. (*Comparision of the Miller–Rabin and the Fermat Primality Tests*) As it was explained above, the Miller–Rabin primality test is at least as effective as the Fermat primality test in detecting compositeness. Construct an example that will show the Miller–Rabin test to be stronger than the Fermat test. More specifically, find an example of a composite integer *n* and a corresponding Miller–Rabin witness *a*, for which the Fermat primality test would find *n* to be probably prime (with this same witness *a*).
    Suggestion: Apply both tests to some randomly generated numbers with a large number of digits.

---

[23] The name RSA-640 refers to size of the binary representation. RSA-640 has 640 binary digits, or 174 decimal digits.

8.   (*Comparision of the Miller–Rabin and the Fermat Primality Tests*) For this exercise, it will be
     convenient to modify the programs of Computer Exercises 5, and 6 into basic versions with sytax
     `y = FermatTestBasic(n, k)`, and `y = MillerRabinTestBasic(n, k)`, that
     work exactly in the same fashion as the corresponding programs in those exercises, but which
     only output the value of `y` (1 if n is found composite, and 0 if n is found probably prime by the
     test); so all other output is suppressed.
     (a) There are exactly 560 primes p, in the range $5000 < p < 10,000$.   Use the Fermat primality
     test with $k = 1$ to run through all odd integers in this range. How many possible primes were
     encountered? Repeat this using $k = 4$, and then 10.
     (b) Repeat Part (a) using the Miller–Rabin primality test.
     (c) There are exactly 9590 primes in the range $3 < p < 100,000$.   Use the Fermat primality test
     with $k = 1$ to run through all odd integers in this range. How many possible primes were
     encountered? Repeat this using $k = 4$, and then 10.
     (d)   Repeat Part (a) using the Miller–Rabin primality test.
     Note: As with any probabilistic algorithm, results will vary.

9.   (*Pseudoprime Generating Program*) (a) Write a program with syntax

           `pProb = MillerRabinPrimeGenerator(nDigits, tol)`,

     that inputs an integer `nDigits > 3`, as well as a real number `tol`, $0 < \text{tol} < 1$. The program
     will ouput an integer `pProb` that is of size `nDigits`, such that the probability that it is not
     prime is at most equal to the second input variable `tol` (for tolerance). The program proceeds
     as follows: A random odd integer n of size `nDigits` is generated (the first digit is a nonzero
     digit, the last digit is odd, and the remaining inside digits have no restriction). The program
     `MillerRabinTest(n, T)` of the preceding exercise is applied, where the second input T is
     chosen to be $\lfloor -(1/2)\log_2(\text{tol}) \rfloor$.[24] This is repeated until a such randomly generated integer n
     is encountered for which this primality test declares as probably prime. This n will be the ouput
     `pProb`.
     (a) Run this program with `nDigits` = 7, and `tol` = .1, and then with `tol` = .01.
     (b) (If you have a factoring utility on your computing platform), run this program 200 times with
     `nbits` = 7, and `tol` = .1, and check the veracity of the 200 probable primes that were
     produced. The proportion of composites that were declared probable primes should be less than
     10% (the tolerance goal), and show that the error bound was a conservative one.
     (c) Use your program to produce a 30 digit and a 60 digit probable prime using `tol` = .1.

---

[24] From what was mentioned above, this value for T is the smallest number of iterations of the Miller–
Rabin test to assure that the probability that a nonprime is produced is less than `tol`.

Chapter 4: Number Systems

Most of the numbers in this book are real numbers, that is, numbers on the number line. It is quite often convenient to restrict our attention to certain subsets of the real numbers such as the (positive) integers $(\mathbb{Z}_+)$ $\mathbb{Z}$, the rational numbers $\mathbb{Q}$, etc. Since any real number has an infinite amount of decimal precision, computers are not able to deal with general real numbers or unlimited amounts of significant figures. Moreover, because of the electric circuitry on which computers, their programs, and software are built, computers are unable to internally store real numbers per se, so what is needed is some sort of conversion process that converts real numbers into their associated computer numbers. Computers often work with so-called *binary* (base 2) expansions, or expansions with respect to some other power of two. In Section 4.1, we discuss how to represent integers with respect to an arbitrary base b. This will be followed in the final Section 4.4 to show how computers represent arbitrary real numbers as so-called floating point numbers.

In Chapter 2, we introduced the important equivalence relation of congruence modulo m, where $m > 1$ is a positive integer. We explored some useful properties of this relation in Section 3.3. In Section 4.2, we will give some more properties of the arithmetic of the integers mod m. The resulting arithmetic is very useful for an assortment of discrete structures applications; in particular, modular arithmetic will be a vital component of many of the very powerful and widely used public key cryptosystems, which will be the topic of Section 4.5. Another important data structure is that of a matrix, which is simply a rectangular spreadsheet of numbers. We will discuss some properties of matrices in Section 4.3. Matrices will be very useful for an assortment of applications that we will discuss in later chapters.

4.1: REPRESENTATIONS OF INTEGERS IN DIFFERENT BASES

We begin this section on some very common ground. Across contemporary societies, languages vary, but all have adopted base 10 arithmetic with Arabic numerals. For example, if you take a taxi ride in any country in which you cannot speak the language, and afterwards give the driver a pad of paper and pull out your wallet (or purse); he or she will understand your request and write down a number that you will understand is the fee. We can view any positive integer as a series in powers of 10. Here is a simple example of how this works:

$$12,307 = 1 \times 10^4 + 2 \times 10^3 + 3 \times 10^2 + (0 \times 10^1) + 7 \times 10^0.$$

This seems so natural to us since we have been working in base 10 arithmetic our whole lives. It is interesting to note that we could have also built our number system using any integer $b > 1$ for the base. This fact is summarized in the following theorem.

Representation of Integers in a Base b

THEOREM 4.1: (*Base b Representations of Integers*) Let $b \geq 2$ (the **base**) be an integer. Any positive integer n can be uniquely represented in the form:

$$n = c_K b^K + c_{K-1} b^{K-1} + \cdots + c_2 b^2 + c_1 b + c_0 = \sum_{k=0}^{K} c_k b^k, \tag{1}$$

where K, and $c_i \, (0 \leq i \leq K)$ are nonnegative integers with $c_i \in \{0, 1, \cdots, b-1\}$, and $c_K > 0$. The expansion in (1) is called the **base b expansion of n**, and is abbreviated as:

$$n \sim [c_K \, c_{K-1} \cdots c_2 \, c_1 \, c_0] \text{ (base } b).$$

Thus, in the base b expansion, each digit corresponds to (and gets multiplied by) a power of b. In the notation of this theorem, we may write $12307 \sim [12307]$ (base 10). Since $11 = 2^3 + 2^1 + 2^0 = 1 \times 2^3 + 0 \times 2^2 + 1 \times 2^1 + 1 \times 2^0$, we may write $11 \sim [1011]$ (base 2). Note that exactly b digits are available to use in the base b expansion of an integer. We will prove Theorem 4.1 a bit later; for now we give a general algorithm that explains how to obtain such expansions. (Later, when we prove the existence portion of this theorem, we will need only to explain why this algorithm works.) The reverse process is quite simple: to convert a base b expansion to its integer equivalent, we simply sum the corresponding series (1).

ALGORITHM 4.1: (*Construction of the base b expansion of a positive integer n*):
Step 1: Put $R = n$ (initialize remainder).
Step 2: Let k be the largest nonnegative integer such that $b^k \leq R$. Next, let c_k be the largest positive integer such that $c_k b^k \leq R$. Update $R \to R - c_k b^k$.
Step 3: If $R = 0$, go to Step 4, otherwise, return to Step 2.
Step 4: Let K be the first (i.e., the largest) value of k found in Step 2. The base b expansion of n will be $n \sim [c_K \, c_{K-1} \cdots c_2 \, c_1 \, c_0]$ (base b), where any unassigned coefficients (i.e., skipped in Step 2) are taken as zeros.

The next example should help to familiarize the reader with the use of Algorithm 4.1 to perform conversions among different bases and integer equivalents. We point out that in each instance of Step 2, once k is determined, the corresponding coefficient c_k is simply the quotient (in the integer division algorithm from Section 3.3) when R is divided by b^k, which is just floor$(c_k/b^k) = \lfloor c_k/b^k \rfloor$.

EXAMPLE 4.1: (a) Find the integer equivalents for each of the following expansions: [10011011] (base 2), [1234567] (base 8), and [22222] (base 3).
(b) Use Algorithm 4.1 to find the base 2 expansion of 69, the base 8 expansion of 225, and the base 16 expansion of 729.
SOLUTION: Part (a): To translate any base b expansion into its integer equivalent, we simply substitute the expansion coefficients into formula (1), and then evaluate the sum. The expansion [10011011] (base 2) has eight place holders (digits), so the leading (highest) power of the base 2 will be 2^7.[1] The corresponding series expansion (1) of this number is thus:

$$1 \times 2^7 + 0 \times 2^6 + 0 \times 2^5 + 1 \times 2^4 + 1 \times 2^3 + 0 \times 2^2 + 1 \times 2^1 + 1 \times 2^0$$
$$= 128 + 32 + 16 + 8 + 2 + 1 = 187.$$

Similarly, working in base 8, we obtain

$$[1234567] \sim 1 \cdot 8^6 + 2 \cdot 8^5 + 3 \cdot 8^4 + 4 \cdot 8^3 + 5 \cdot 8^2 + 6 \cdot 8 + 7 \cdot 1 = 342,391,$$

while working in base 3, we have $[22222] \sim 2 \cdot (3^4 + 3^3 + 3^2 + 3^1 + 1) = 242$.

Part (b): In each case we use Algorithm 4.1:

(i) In Step 1, we simply initialize the remainder as the number to be converted: $R = 69$. In the first application of Step 2, we find the largest power of 2 not exceeding $R = 69$ is $2^6 = 64$, so we put $(K = 6)$ and $c_6 = 1$. R is now updated to $69 - 64 = 5$, and Step 2 is repeated. Now, $k = 2$ is the largest exponent such that $2^k \leq 5(= R)$, so we set $(c_5 = c_4 = c_3 = 0$, and) $c_2 = 1$, and update R to be $5 - 4 = 1$. In the next (and last) iteration of Step 2, $k = 0$ is the largest exponent for which $2^k \leq 1(= R)$, so we determine the final digits of the expansion $c_1 = 0$, $c_0 = 1$. We now arrive at Step 4, to produce the resulting binary expansion: $69 \sim [1000101]$ (base 2).

(ii) In Step 1, we initialize the remainder $R = 225$. Since $K = 2$ is the largest power of 8 not exceeding $R = 225$, we determine the coefficient $c_2 = \lfloor 225/64 \rfloor = 3$, and we update $R = 225 - 3 \cdot 8^2 = 33$. In the next iteration of Step 2, we get $k = 1$, and $c_1 = \lfloor 33/8 \rfloor = 4$, so we update $R = 33 - 4 \cdot 8 = 1$. One final application of Step 2 gives the remaining coefficient $c_0 = 1$, and we then move on to Step 4 of Algorithm 4.1 to obtain the base 8 expansion $225 \sim [341]$ (base 8).

(iii) In Step 1, we initialize the remainder $R = 729$. The largest power of 16 not exceeding 729 is $16^2 = 256$. Hence $K = 2$, $c_2 = \lfloor 729/256 \rfloor = 2$, and we update R

---

[1] Always one less than the number of digits, since the powers of b descend to the 0th power.

to be $729 - 2 \cdot 256 = 217$. In the next application of Step 2, $k = 1$, $c_1 = \lfloor 217 / 16 \rfloor = 13$, and R becomes $217 - 13 \cdot 16 = 9$. The final iteration of Step 2 gives $c_0 = 9$, and hence $729 \sim [2 \; 13 \; 9]$ (base 16).

This algorithm is nicely amenable to be coded into computer programs, and this will be done in the computer exercises at the end of the section. We give one useful fact here that will often be helpful in implementing the above algorithm with the aid of either a scientific calculator or computer.

COMPUTING NOTE: Step 2 of the above algorithm involves finding the largest integer with the property that $b^k \le R$. A brute-force approach would be to simply start off with $k = 0$, and continue computing increasing powers of 2 until the result equals or exceeds R: $b^0 = 1$, $b^1 = b$, $b^2, b^3, \cdots.$ [2]   Although this would work, it is not very efficient. Since most calculators (and computers) have built-in logarithm functions, it is much faster to use one of these when we compute this largest integer k. Indeed, for each base b, the associated logarithm function, denoted $y = \log_b(x)$ associates for each positive real number x (in the domain) the real number y with the property that $b^y = x$. The output of this function will generally not be an integer, even when the input is, but since we are seeking the largest integer k such that $b^k \le x$, we can simply take $k = \mathrm{floor}(\log_b(x))$. Finally, since any calculator or computing platform will only have a handful of built-in log functions, the good news is that we only need one. Suppose that we have some logarithm function, denoted as $\mathrm{LOG}(x)$ on our computing tool. The so-called *change of base formula* allows us to express any log in terms of our log as follows:

$$\log_b(x) = \mathrm{LOG}(x) / \mathrm{LOG}(b).$$

In summary, the determination of k in Step 2 of Algorithm 4.1 can be accomplished in a single shot with the formula:

$$k = \mathrm{floor}(\mathrm{LOG}(R) / \mathrm{LOG}(b)). \tag{2}$$

For example with $b = 2$ and $R = 136{,}825$, the largest integer k such that $2^k \le 136{,}825$ is given by $k = \mathrm{floor}(\mathrm{LOG}(136{,}825) / \mathrm{LOG}(2)) = \mathrm{floor}(17.062...)$ $= 17$.

Hex(adecimal) and Binary Expansions

The most often used bases are given special terminology and notation; some of these will be given in the following definition:

DEFINITION 4.1:   As we have mentioned, base 2 expansions are also known as **binary expansions** (and are made up of *bit strings*). Expansions in base 3, 8,

---

[2] On a computer, this would be implemented with a while loop.

and 16 are known, respectively, as **ternary**, **octal**, and **hexadecimal**. In hexadecimal expansions, the double digits 10, 11, 12, 13, 14, and 15 are customarily replaced with the letters, A, B, C, D, E, and F, respectively, so that each element in the representing expansion string will be a single character. Thus, from the above example, we may write 729 ~ [2D9] (base 16), which can also be written as 729 ~ [2D9] (hex). This convention allows hexadecimal expansions to be viewed as strings rather than vectors (without it, the string [125] could mean either [12 5] = [C5], or [1 2 5]).

In applications, the most important bases will be decimal, binary, and hex(adecimal). For reference, Table 4.1 converts the hex digits into these other bases. The reader is advised to verify the entries of this table using (1). In the binary expansions, we have made all of the strings be of equal length four, so for example the binary expansion for 2, which is 10 $(= 1 \cdot 2^1 + 0 \cdot 2^0)$ is expressed as 0010 $(= 0 \cdot 2^3 + 0 \cdot 2^2 + 1 \cdot 2^1 + 0 \cdot 2^0)$. The hex format is thus the least cumbersome. For example, a dizzying 64 bit string (i.e., a binary string of length 64), in hex notation would have a much smaller length of 16.

TABLE 4.1: Conversions between hexadecimal, decimal, and binary digits.

Hex	Decimal	Binary
0	0	0000
1	1	0001
2	2	0010
3	3	0011
4	4	0100
5	5	0101
6	6	0110
7	7	0111
8	8	1000
9	9	1001
A	10	1010
B	11	1011
C	12	1100
D	13	1101
E	14	1110
F	15	1111

EXERCISE FOR THE READER 4.1: (a) Find the integer equivalents for each of the following expansions: [1101001111] (binary), [777] (octal), and [123ABC] (hexadecimal).
(b) Use Algorithm 4.1 to find the binary expansion of 122, the base 32 expansion of 9675, and the hexadecimal expansion of 52,396.

Since computing platforms compute internally using binary (or a base that is some other power of two), larger sets of alphabets need to get translated into appropriate strings that can be processed by a computer. The following example provides some typical schemes for such conversions.

EXAMPLE 4.2: (*Conversions between English Plaintexts and Strings of Digits*)
(a) Develop a corresponding natural scheme for translating the 26 English letters into bit strings (binary strings) starting with the representations: A ~ 0 (with a prefix of an appropriate number of zeros), and use this scheme to translate the English plaintext 'retreat' into its corresponding bit string.
(b) Develop a similar scheme for translating English letters into hexadecimal strings, and use this scheme to translate the English plaintext 'retreat' into its corresponding bit string.

SOLUTION: Part (a): Since the (non-case-sensitive) English alphabet has 26 characters, we would need to use length 5 bit strings to be able to represent all of them. There are $2^5 = 32$ such strings, so we would have 6 left unused. Length four bit strings would not be adequate since there are only $2^4 = 16$ of them—see Table 4.1. Table 4.2 shows such a scheme, along with integer and hex string equivalents.

TABLE 4.2: Correspondence of the English alphabet using the integers mod 26 ($\mathbb{Z}_{26}$), length five binary strings, and length 2 hexadecimal strings.

ENGLISH	A	B	C	D	E	F	G	H	I	J	K	L	M
INTEGER	0	1	2	3	4	5	6	7	8	9	10	11	12
BINARY	00000	00001	00010	00011	00100	00101	00110	00111	01000	01001	01010	01011	01100
HEX	00	01	02	03	04	05	06	07	08	09	0A	0B	0C

ENGLISH	N	O	P	Q	R	S	T	U	V	W	X	Y	Z
INTEGER	13	14	15	16	17	18	19	20	21	22	23	24	25
BINARY	01101	01110	01111	10000	10001	10010	10011	10100	10101	10110	10111	11000	11001
HEX	0D	0E	0F	10	11	12	13	14	15	16	17	18	19

Using this table, the English string 'retreat' would be represented by the following bit string: (100001 00100 10011 10001 00100 00000 10011)

$$\texttt{10000100100100111000100100000010011.}$$

Part (b): Since there are only 16 single digit hexadecimal strings, but $16^2 = 256$ to digit hexadecimal schemes, we would need to use length two hexadecimal strings to accommodate all 26 English letters. The natural correspondence is shown in Table 4.2, which leads us to the following representation for the English plaintext 'retreat': (11 04 13 11 04 00 13) 11041311040013.

COMPUTING NOTE: Most computing platforms have built-in conversion utilities for all of the 256 ASCII (*American Standard Code for Information Interchange*) characters, which consist of the standard English letters, digits, some common foreign letters, special symbols, and control characters. We caution the reader that the correspondence that we have set up in Table 4.2 for a single set of 26 (non-case-sensitive) English letters is different from the ASCII correspondence. For example, in ASCII, the digits 0–9 have (decimal) integer equivalents 48–57, the lower case English letters a–z have integer equivalents 97–122, and the upper case English letters A–Z have integer equivalents 65–90. Thus the complete ASCII alphabet of 256 symbols has a 2 digit hexadecimal representation, and an 8 digit binary representation.

We will next give a proof of Theorem 4.1 in which we will explain why Algorithm 4.1 always works.

Proof of Theorem 4.1: Fix a base b (an integer greater than 1), and let n be a positive integer. We must show two things: (i) n has a base b expansion (existence), and (ii) any two base b expansions for n must be the same (uniqueness).

Part (i): (*Existence*) Existence of a base b expansion of n will be accomplished by showing that Algorithm 4.1 performs correctly. The iterated Step 2 will clearly construct a decreasing sequence of integers $K = k_1 > k_2 > \cdots > k_t \geq 0$, along with a corresponding sequence of coefficients $c_{k_i} = c(k_i)$ that are nonnegative integers less than b. Keeping track of all of the terms that get subtracted from n, we see that each k_i is the largest exponent such that

$$b^{k_i} \leq n - c(k_1)b^{k_1} - c(k_2)b^{k_2} - \cdots - c(k_{i-1})b^{k_{i-1}}.$$

(The right sides are the updated values of R in the algorithm.) We need to check that when all of the coefficients multiplied by their respective powers of b are subtracted from n, we are left with zero, i.e.,

$$n - c(k_1)b^{k_1} - c(k_2)b^{k_2} - \cdots - c(k_t)b^{k_t} = 0.$$

This would prove that R is eventually zero, so Step 3 eventually moves to Step 4 (also, this equation is equivalent to (1) and corresponds to the base b expansion produced by the algorithm). Let us temporarily denote the left side of the above equation by S. By the way the coefficient $c(k_t)$ is chosen, we must have $0 \leq S < b^{k_t}$. Since the process has ended, there can be no other nonnegative power of b that is less than or equal to S. This means that $S < b^0 = 1$, so indeed $S = 0$, as we needed to show.

Part (ii): (*Uniqueness*) Suppose that we have two base b expansions of n:

$$c_K b^K + c_{K-1}b^{K-1} + \cdots + c_2 b^2 + c_1 b + c_0 = n = d_L b^L + d_{L-1}b^{L-1} + \cdots + d_2 b^2 + d_1 b + d_0,$$

where $c_i, d_j \in \{0, 1, 2, \cdots, b-1\}$, and $c_K, d_L > 0$. From the formula for the sum of a finite geometric series (Formula (3) of Proposition 3.5), we note that

$$c_{K-1}b^{K-1} + \cdots + c_2 b^2 + c_1 b + c_0 \leq (b-1)b^{K-1} + \cdots + (b-1)b^2 + (b-1)b + (b-1)$$
$$= \frac{(b-1)(b^K - 1)}{(b-1)} = b^K - 1 < b^K.$$

Thus, no matter what the lower power coefficients are, the leading coefficient (of the highest power of b) on both sides of the equations determines entirely which side is larger, and the only way for both sides to be equal is that $K = L$, and $c_K = d_L$. If we subtract the corresponding equal terms from both sides of the equation, we are left with

$$c_{K-1}b^{K-1} + c_{K-2}b^{K-2} + \cdots + c_2b^2 + c_1b + c_0 = d_{K-1}b^{K-1} + d_{K-2}b^{K-2} + \cdots + d_2b^2 + d_1b + d_0.$$

Repeating the same argument, we will successively obtain that $c_{K-1} = d_{K-1}$, $c_{K-2} = d_{K-2}, \cdots, c_1 = d_1$, and $c_0 = d_0$. Uniqueness of base b expansions is thus established. $\square$

Addition Algorithm with Base b Expansions

The way computer and software systems work to perform arithmetic operations is as follows: First they input the numbers in the usual decimal format (the human interface), then the numbers get converted to computer numbers (in some base b that is a power of 2), then the computer performs the arithmetic operation(s) directly with these computer numbers, and finally the answer gets converted back to decimal format. We have already discussed how such translations work. As it turns out, the decimal arithmetic algorithms that one learns in grade school easily generalize to base b algorithms. For example, after having first learned how to add single digits (like 6 + 8), we then learned how to add two multi-digit positive integers in grade school by stacking the numbers vertically (so corresponding digits are directly above/below each other), and then, starting from the right, we add successive digits and moving any *carries* (when single digit additions yield a sum larger than 9) over to the next digits on the left. When adding two numbers, any carry must equal 1 (since the most that two single decimal digits could add up to is 18). In order to avoid too much formality, we will assume that we are able to add, subtract, and multiply single digits in any base (perhaps with carries or borrows), just as in grade school arithmetic. This can be accomplished by converting to ordinary integer arithmetic. For example, to perform the hexadecimal subtraction E – B, we convert to the corresponding integer subtraction (see Table 4.1) $14 - 11 = 3$, whose decimal answer is the same as its hexadecimal form. To perform the single digit hexadecimal multiplication $E \cdot B$, we first perform the corresponding integer multiplication $14 \cdot 11 = 154$. Since the result exceeds 15 (the largest for a single hexadecimal digit), we apply the division algorithm to this result with divisor 16: $154 = 9 \cdot 16 + 10 = 9 \cdot 16 + A$, to find that the hexadecimal result is [9A], or A with a carry of 9.

We will next give an analogous algorithm that will show how to add two integers $c = [c_{K-1} \cdots c_2\ c_1\ c_0]$ and $d = [d_{K-1} \cdots d_2\ d_1\ d_0]$ of the same base b.[3] We begin by adding the right-most digits in base b: $[c_0] + [d_0] = [\text{Car}_0\ s_0]$, where Car_0 (the "carry") can be either 0 or 1. In ordinary integer language, this corresponds to the

---

[3] We may always arrange things so that the two base b numbers being added have the same number of (base b) digits by padding shorter numbers with additional zeros on the left.

equation $c_0 + d_0 = b \cdot \text{Car}_0 + s_0$.[4] We then continue this process, each time moving one digit to the left and adding the existing carry to the sum. The next step would be to add second-to-right digits with Car_0: $[c_1] + [d_1] + [\text{Car}_0] = [\text{Car}_1 \ s_1]$. Note that since in ordinary integer arithmetic this corresponds to the equation $c_1 + d_1 + \text{Car}_0 = b \cdot \text{Car}_1 + s_1$, it follows that $\text{Car}_1 = \text{floor}([c_1 + d_1 + \text{Car}_0]/b)$. We also point out the new carry, Car_1, is either 0 or 1, because $0 \le c_1 + d_1 + \text{Car}_0 \le (b-1) + (b-1) + 1 \le 2b - 1$. This process continues as we move through the digits, with the carries always being either 0 or 1. After we have moved through the K digits, to obtain $s_0, s_1, \cdots, s_{K-1}$, we take s_K to be the value of the last carry Car_{K-1}. It thus follows that $[c_{K-1} \cdots c_2 \ c_1 \ c_0] + [d_{K-1} \cdots d_2 \ d_1 \ d_0] = [s_K s_{K-1} \cdots s_2 \ s_1 \ s_0]$, in base b. This process is summarized in the following algorithm in which we do not bother storing the individual carries, since they may be discarded (overwritten) once they have been used.

ALGORITHM 4.2: (*Addition of Two Base b Integers*):[5]
Assume that we have two K-digit base b integers of the form:

$$c = [c_{K-1} \cdots c_2 \ c_1 \ c_0] \text{ and } d = [d_{K-1} \cdots d_2 \ d_1 \ d_0].$$

This algorithm will compute the base b expansion $[s_K s_{K-1} \cdots s_2 \ s_1 \ s_0]$ of the sum $c + d$.

Step 1: Put Car = 0 (*Initialize carry*)

Step 2: FOR $i = 0$ TO $K - 1$
   NewCar = floor($[c_i + d_i + \text{Car}]/b$) (*Compute new carry*)
   SET $s_i = c_i + d_i + \text{Car} - b \cdot \text{NewCar}$
   Update Car = NewCar
END i FOR

Step 3: SET $s_K = \text{Car}$

Step 4: Form $[s_K s_{K-1} \cdots s_2 \ s_1 \ s_0]$, the base b expansion of $c + d$.

This algorithm may be carried out with a similar notation to what is taught in elementary school for addition of numbers by hand. We will use this notation in the following example.

---

[4] In other words, if we apply the division algorithm to the division of $c_0 + d_0$ by b, Car_0 will be the quotient and s_0 will be the remainder.

[5] This algorithm subsumes additions of single digit base b numbers, along with some very restricted floor operations involving divisions of small integers by b. Of course, in the design of any computer arithmetic system, such simple operations would need to be programmed in before Algorithm 4.2 could function correctly.

EXAMPLE 4.3: Use Algorithm 4.2 to perform each of the indicated additions, and then translate the operation to an ordinary (base 10) integer addition.
(a) The binary (base 2) addition: $[11011] + [10110]$.
(b) The octal (base 8) addition: $[744] + [552]$.

SOLUTION: Part (a): The process, which is illustrated on the right, proceeds as follows: In Step 2, the first value of NewCar is floor($[1 + 0 + 0]/2) = 0$, so $s_0 = 1 + 0 + 0 - 0 = 1$.

```
 1  1  1  1
 1  1  0  1  1
    1  0  1  1  0
 1  1  0  0  0  1
```

Moving into the $i = 1$ iteration of Step 2, the carry (Car) is 0, and so NewCar becomes floor($[1 + 1 + 0]/2) = 1$, and $s_1 = 1 + 1 + 0 - 2 = 0$. Moving into the $i = 2$ iteration of Step 2, the carry (Car) is now 1, and we have NewCar = floor($[0 + 1 + 1]/2) = 1$, and so, $s_2 = 0 + 1 + 1 - 2 = 0$. In the $i = 3$ iteration, we have Car = 1, NewCar = floor($[1 + 0 + 1]/2) = 1$, and $s_3 = 1 + 0 + 1 - 2 = 0$. In the final $i = 4$ iteration, we have Car = 1, NewCar = floor($[1 + 1 + 1]/2) = 1$, and $s_4 = 1 + 1 + 1 - 2 = 1$. In Step 3, the final carry Car = 1 gets transferred to $s_5 = 1$. Thus we obtain $[11011] + [10110] = [110001]$ (base 2). This corresponds to the integer addition: $27 + 22 = 49$.

Part (b): The process, which is illustrated on the right, proceeds as follows: In Step 2, the first value of NewCar is floor($[4 + 2 + 0]/8) = 0$, so $s_0 = 4 + 2 + 0 - 0 = 6$. Moving into the $i = 1$

```
 1  1
 7  4  4
 5  5  2
 1  5  1  6
```

iteration of Step 2, the carry (Car) is 0, and so NewCar becomes floor($[4 + 5 + 0]/8) = 1$, and $s_1 = 4 + 5 + 0 - 8 = 1$. Moving to the final ($i = 2$) iteration of Step 2, the carry (Car) is now 1, and we have NewCar = floor($[7 + 5 + 1]/8) = 1$, and so, $s_2 = 7 + 5 + 1 - 8 = 5$. In Step 3, the final carry Car = 1 gets transferred to $s_3 = 1$. Thus we obtain $[744] + [552] = [1516]$ (base 8). This corresponds to the integer addition: $484 + 362 = 846$.

EXERCISE FOR THE READER 4.2: Use Algorithm 4.2 to perform each of the indicated additions, and then translate the operation to an ordinary (base 10) integer addition.
(a) The binary addition: $[101111] + [001111]$.
(b) The hexadecimal addition: $[7D4E] + [1AA2]$.

Subtraction Algorithm with Base b Expansions

Just as was done for addition, an algorithm for subtraction in terms of base b expansions will be fashioned after the usual method that is taught in grade school for subtracting integers by hand. We first explain how the algorithm works, and then formally summarize the steps.

Suppose that we wish to perform a subtraction $c - d$ of two integers $c = [c_{K-1} \cdots c_2 \ c_1 \ c_0]$ and $d = [d_{K-1} \cdots d_2 \ d_1 \ d_0]$ expanded in the same base b, with $c > d$. We begin by subtracting the right-most digits in base b: $c_0 - d_0 = \text{Borr}_0 \cdot b + s_0$, where Borr_0 (the "borrow") can be either 0, if $c_0 - d_0 \geq 0$, or -1. The borrow is what is needed to take from the next digit of c to ensure that the current digit subtraction will have a nonnegative result. In ordinary integer language, this corresponds to the equation $c_0 - d_0 = b \cdot \text{Borr}_0 + s_0$.[6] We then continue this process, each time moving one digit to the left and adding the existing borrow to the difference of the new digits. The next step would be to subtract the second-to-right digits: $c_1 - d_1 + \text{Borr}_0 = b \cdot \text{Borr}_1 + s_1$. Note that since in ordinary integer arithmetic this corresponds to the equation $c_1 - d_1 + \text{Borr}_0 = b \cdot \text{Borr}_1 + s_1$, it follows that $\text{Borr}_1 = \text{floor}([c_1 - d_1 + \text{Borr}_0]/b)$. We also point out that the new borrow Borr_1, is again either 0 or -1, because $-b \leq c_1 - d_1 + \text{Borr}_0 < b$. This process continues as we move through the digits, with the borrows always being either 0 or -1. After we have moved through the K digits, to obtain $s_0, s_1, \cdots, s_{K-1}$, we note that because of the assumption that $c > d$, the last borrow Borr_{K-1} will always be 0. It thus follows that $[c_{K-1} \cdots c_2 \ c_1 \ c_0] - [d_{K-1} \cdots d_2 \ d_1 \ d_0] = [s_{K-1} \cdots s_2 \ s_1 \ s_0]$, in base b. This process is summarized in the following algorithm in which we do not bother storing the individual borrows, since they may be discarded (overwritten) once they have been used.

ALGORITHM 4.3: (*Subtraction of Two Base b Integers*):[7]
Assume that we have two K-digit base b integers of the form:

$$c = [c_{K-1} \cdots c_2 \ c_1 \ c_0] \text{ and } d = [d_{K-1} \cdots d_2 \ d_1 \ d_0],$$

with $c > d$. This algorithm will compute the base b expansion $[s_{K-1} \cdots s_2 \ s_1 \ s_0]$ of the difference $c - d$.

Step 1: Put Borr = 0 (*Initialize borrow*)

Step 2: FOR $i = 0$ TO $K - 1$
  NewBorr = floor$([c_i - d_i + \text{Borr}]/b)$ (*Compute new borrow*)
  SET $s_i = c_i - d_i + \text{Borr} - b \cdot \text{NewBorr}$
  UPDATE Borr = NewBorr
END i FOR

---

[6] If we apply the division algorithm to the division of $c_0 - d_0$ by b, Borr_0 will be the quotient, and s_0 will be the remainder. In case $c_0 - d_0 \geq 0$, since, $c_0 - d_0 < b$, we will have $\text{Borr}_0 = 0$; while in case $c_0 - d_0 < 0$, we will have $\text{Borr}_0 = -1$.

[7] This algorithm subsumes subtractions of single digit base b numbers, along with some very restricted floor operations involving divisions of small integers by b. Of course, in the design of any computer arithmetic system, such simple operations would need to be programmed in before Algorithm 4.3 could function correctly.

Step 3: Form $[s_{K-1} \cdots s_2 \; s_1 \; s_0]$, the base b expansion of $c - d$.

This algorithm may be carried out with a similar notation to what is taught in elementary school for subtraction of numbers by hand. We will use this notation in the following example.

EXAMPLE 4.4: Use Algorithm 4.3 to perform the subtraction $[6FAA] - [4FED]$ in hexadecimal arithmetic.

SOLUTION: The process, which is illustrated on the right, proceeds as follows: Step 1 always initializes the borrow to be 0: Borr = 0. In Step 2, the first value of NewBorr is floor($[A - D + 0]/16$) = floor($[10 - 13 + 0]/16$) = -1, so $s_0 = 10 - 13 + 0 - 16 \cdot (-1) = 26 - 13 = 13 = $ D.

$$
\begin{array}{cccc}
-1 & -1 & -1 & \\
6 & F & A & A \\
4 & F & E & D \\
\hline
1 & F & B & D \\
\end{array}
$$

Moving into the $i = 1$ iteration of Step 2, the borrow (Borr) is -1, and so NewBorr becomes floor($[A - E - 1]/16$) = floor($[10 - 14 - 1]/16$) = -1, and so

$$s_1 = c_1 - d_1 + \text{Borr} - 16 \cdot \text{NewBorr} = 10 - 14 - 1 - 16 \cdot (-1) = 26 - 15 = 11 = \text{B}.$$

Moving into the $i = 2$ iteration of Step 2, the borrow (Borr) is now -1, and we have NewBorr = floor($[F - F - 1]/16$) = floor($[15 - 15 - 1]/16$) = -1, and so,

$$s_1 = c_2 - d_2 + \text{Borr} - 16 \cdot \text{NewBorr} = 15 - 15 - 1 - 16 \cdot (-1) = 16 - 1 = 15 = \text{F}.$$

In the final $i = 3$ iteration of Step 2, the borrow (Borr) is still -1, NewBorr = floor($[6 - 4 - 1]/16$) = 0, and

$$s_3 = c_3 - d_3 + \text{Borr} - 16 \cdot \text{NewBorr} = 6 - 4 - 1 - 16 \cdot (0) = 1.$$

We have thus obtained $[6FAA] - [4FED] = [1FBD]$ (base 16). This corresponds to the integer subtraction:

$$[6 \cdot 16^3 + 15 \cdot 16^2 + 10 \cdot 16^1 + 10 \cdot 16^0] - [4 \cdot 16^3 + 15 \cdot 16^2 + 14 \cdot 16^1 + 13 \cdot 16^0]$$
$$= 1 \cdot 16^3 + 15 \cdot 16^2 + 11 \cdot 16^1 + 13 \cdot 16^0,$$

or $28,586 - 20,461 = 8125$.

EXERCISE FOR THE READER 4.3: Use Algorithm 4.3 to perform each of the indicated subtractions, and then translate the operation to an ordinary (base 10) integer subtraction.
(a) The binary subtraction: $[101101] - [001111]$.
(b) The hexadecimal subtraction: $[7D4E] - [1AA2]$.

Multiplication Algorithm in Base b Expansions

The general algorithm for multiplying two numbers using their base b expansions will follow easily from the following two simple observations:

1. If k is a positive integer, the base b expansion of b^k is $[1\ 0\ 0\cdots 0]$, where there

are k zeros following the 1.

2. If we multiply a base b expansion $[c_{K-1}\cdots c_2\ c_1\ c_0]$ by $b^k \sim [1\ 0\ 0\cdots 0]$, the former expansion gets shifted to the left k places, with a string of k zeros appended on the right: $b^k \cdot [c_{K-1}\cdots c_2\ c_1\ c_0] = [c_{K-1}\cdots c_2\ c_1\ c_0\ \underbrace{0\ 0\cdots 0}_{k\ zeros}]$.

Since the algorithm for multiplying two integers $c = [c_{K-1}\cdots c_2\ c_1\ c_0]$ and $d = [d_{K-1}\cdots d_2\ d_1\ d_0]$ of the same base b is a direct generalization of the grade school multiplication algorithm (in base 10), we motivate it with a simple grade school hand multiplication of 126×63.

The algorithm is usually performed schematically, as shown on the right. We line the two numbers up at the right margin, and begin by multiplying the last digit of the lower number, by each of the digits of the upper number (starting from the right and going left). Each time we get a product of digits larger than 9 (= $b - 1$), we will have a carry that we move to the next digit (to

```
        1   3/1
        1   2   6
    ×       6   3
    ———————————————
        3   7   8
    7   5   6
    ———————————————
    7   9   3   8
```

add to the product). When we get through multiplying all of the top number's digits by the last digit of the bottom, we move to the second-to-last digit of the bottom number and repeat this, but we shift the answers that we get one digit to the left, and put them in a second row. The number of such shifted rows that we get will equal the number of digits of the lower number. Note that each row can give rise to different carries; previous carries can be crossed off when we are done using them. We simply included the single pair of multiple carries with a slash symbol. We then need to add these rows using base b addition.[8]

The role of the carries is quite similar to their role in base b addition (Algorithm 4.2); the reason for the shifting of rows can be seen by the distributive law and the shifting property 2 above as follows:

$$cd = c[d_0 \cdot b^0 + d_1 \cdot b^1 + \cdots d_{K-1} \cdot b^{K-1}]$$
$$= (cd_0) \cdot b^0 + (cd_1) \cdot b^1 + \cdots (cd_{K-1}) \cdot b^{K-1}.$$

Each term in the latter sum is thus a single digit of the second factor, multiplied by (all the digits of) the first factor, multiplied by a power of the base. This latter multiplication has the effect of shifting all of the digits of the multiplication to the

---

[8] That is, using Algorithm 4.2. Technically, when there are more than two rows, we would need to use Algorithm 4.2 repeatedly, adding two numbers at a time.

left (by the shifting property 2), with the number of slots shifted equaling the power of the base. The general procedure is summarized in the following algorithm.

ALGORITHM 4.4: (*Multiplication of Two Base b Integers*):
Assume that we have two K-digit base b integers of the form:[9]

$$c = [c_{L-1} \cdots c_2 \; c_1 \; c_0] \text{ and } d = [d_{K-1} \cdots d_2 \; d_1 \; d_0].$$

This algorithm will compute the base b expansion of the product $c \cdot d$.

Step 1: SET $P = 0$ (*Initialize product; will consist of a sum of terms*)

Step 2: FOR $i = 0$ TO $K - 1$ (*i will be the index of a digit of d*)
SET $p_0 = p_1 = \cdots = p_{i-1} = 0$ (*this is the shifting corresponding to* d_i)
SET Car $= 0$ (Initialize carry)

FOR $j = 0$ TO $L - 1$ (*j will be the index of a digit of c*)
    NewCar = floor($[c_j \cdot d_i + \text{Car}] / b$) (*Create new carry*)

    SET $p_{i+j} = c_j \cdot d_i + \text{Car} - b \cdot \text{NewCar}$

    UPDATE Car = NewCar (*Update carry*)
END j FOR

SET $p_{i+L} = \text{Car}$
UPDATE $P \rightarrow P + [p_{i+L} \cdots p_2 \; p_1 \; p_0]$. (*Using Algorithm 4.2*)
END i FOR

Step 3: Form $P = [p_{K+L} \; p_{K+L-1} \cdots p_2 \; p_1 \; p_0]$, the base b expansion of $c \cdot d$.

We point out that (in case $b = 10$) this algorithm deviates only slightly from the elementary school algorithm in that the updating of the cumulative sum (P) is done after each d-digit multiplication, rather than all at once at the end.[10]

EXAMPLE 4.5:   Use Algorithm 4.4 to perform each of the indicated multiplications, and then translate each   to an ordinary (base 10) integer multiplication.
(a) The binary multiplication: $[1101] \times [110]$.
(b) The hexadecimal multiplication: $[2A4] \times [12E]$.

---

[9] In contrast with Algorithm 4.2, no advantage is gained by assuming the base b expansions of the numbers being multiplied have the same length.
[10] This more properly fits with the use of Algorithm 4.2, since the latter was developed for adding two (base b) numbers.

SOLUTION:[11] Part (a): In binary arithmetic, carries will never arise in the digit multiplication process since the largest that a product of digits can be is 1 (a single binary digit). The process is outlined in the diagram on the right. In Step 2 of Algorithm 4.4, all of the values of Car and NewCar are zero and the formula $s_{i+j} = c_j \cdot d_i + \text{Car} - b \cdot \text{NewCar}$ reduces to $s_{i+j} = c_j \cdot d_i$.

$$
\begin{array}{rcccc}
 & 1 & 1 & 0 & 1 \\
\times & 1 & 1 & 0 \\
\hline
0 & 0 & 0 & 0 \\
1 & 1 & 0 & 1 \\
1 & 1 & 0 & 1 \\
\hline
1 & 0 & 0 & 1 & 1 & 1 & 0
\end{array}
$$

It follows that in Step 2, for each digit multiplication d_j the result will be either a string of zeros (which we can ignore) in case $d_j = 0$ (like d_0), or simply a shifted copy of the binary string for c (as happens for d_1, d_2). Thus, the result will be $[0000] + [11010] + [110100] = [1001110]$. In the language of ordinary (base 10) integers, this result corresponds to the multiplication $13 \times 6 = 78$.

Part (b): The process is outlined in the diagram on the right, and the reader might wish to simply verify this diagram rather than read though the following details. In Step 2, starting with $i = 0$ (corresponding to the digit E of the latter number), we need to multiply each of the digits of the top number by E = 14. The first (right-most) digit multiplication is $4 \cdot \text{E}$ $= 4 \cdot 14 = 56 = 3 \cdot 16 + 8$, so $p_0 = 8$, and the first carry

$$
\begin{array}{ccccc}
 & 2/ & 8/1 & 3/ \\
 & 2 & \text{A} & 4 \\
\times & 1 & 2 & \text{E} \\
\hline
 2 & 4 & \text{F} & 8 \\
 5 & 4 & 8 \\
2 & \text{A} & 4 \\
\hline
3 & 1 & \text{D} & 7 & 8
\end{array}
$$

Car = 3. Next, with $j = 1$, we multiply $\text{A} \cdot \text{E} + \text{Car} = 10 \cdot 14 + 3 = 143 = 8 \cdot 16 + 15$, so $p_1 = 15 = \text{F}$, and the next carry is Car = 8. Moving to the last digit, where $j = 2$, the multiplication (and carry) is $2 \cdot \text{E} + \text{Car} = 2 \cdot 14 + 8 = 36 = 2 \cdot 16 + 4$, so $p_2 = 4$, and the last carry (2) is the value of p_3. The cumulative sum P is now updated from its previous value (0) to the new value [24F8] (being added to it). Next we move to $i = 1$, corresponding to the middle digit 2 of the latter number. Here we pad ($i = 1$ digit) $p_0 = 0$. The first multiplication $4 \cdot 2 = 8$ has no carry (Car = 0), and gives $p_1 = 8$. Next, with $j = 1$, we multiply $\text{A} \cdot 2 + \text{Car} = 10 \cdot 2 + 0 = 20 = 1 \cdot 16 + 4$, to obtain $p_2 = 4$ and Car = 1. Moving to the last digit with $j = 2$, the multiplication (and carry) is $2 \cdot 2 + \text{Car} = 2 \cdot 2 + 1 = 5$, so $p_3 = 5$, and there is no final carry to import to p_4. The cumulative sum is now updated to $P = [24\text{F}8] + [5480] = [7978]$ (base 16; using Algorithm 4.2). The last ($i = 2$) iteration is simple, since it corresponds to multiplying by the digit one, and the resulting product is simply the top number with two zeros padded at the right. Hence, the final answer of the multiplication is obtained by updating $P = [7978] + [2\text{A}400] = [31\text{D}78]$. The resulting

---

[11] To become more comfortable with such multiplications, readers are encouraged to contruct their own single digit multiplication tables, say for binary and hex multiplication. For binary, the table would be two by two, while for hex it would be sixteen by sixteen, and include all single digit hex products such as $\text{F} \times \text{B} = \text{A}5$ (which corresponds to the integer multiplication $15 \times 11 = 165 = 10 \cdot 16 + 5 (= \text{A}5)$).

multiplication $[2A4] \times [12E] = [31D78]$ (base 16) translates to $676 \times 302 = = 204,152$ (base 10) in ordinary integer language.

EXERCISE FOR THE READER 4.4: Use Algorithm 4.4 to perform each of the indicated multiplications, and then translate the operation to an ordinary (base 10) integer multiplication.
(a) The binary multiplication: $[1111] \times [1111]$.
(b) The base 7 multiplication: $[262] \times [520]$.

EXERCISES 4.1:

1.   Count from 0 to 25 in each of the following number systems:
     (a) binary (base 2)   (b) octal (base 8)   (c) hexadecimal (base 16)

2.   Count from 0 to 25 in each of the following number systems:
     (a) base 3   (b) base 5   (c) base 11

3.   Count from 100 to 125 in each of the following number systems:
     (a) binary (base 2)   (b) octal (base 8)   (c) hexadecimal (base 16)

4.   Count from 100 to 125 in each of the following number systems:
     (a) base 3   (b) base 5   (c) base 11

5.   Convert each of the following expansions to an (ordinary) decimal integer:
     (a) [101010] (binary)   (b) [3123123] (base 4)   (c) [ABCDEF] (hex)
     (d) [22333] (base 4)   (e) [12345AAAA] (hex)   (f) [9876412] (base 11)

6.   Convert each of the following expansions to an (ordinary) decimal integer.
     (a) [111000111] (base 2)   (b) [70073] (base 8)   (c) [123ABC] (hex)
     (d) [244343] (base 5)   (e) [CAB12FF] (hex)   (f) [1122334455] (base 6)

7.   Convert each of the following decimal (base 10) integers to (i) binary (base 2) form, (ii) octal (base 8) form, and (iii) hexadecimal (base 16) form:
     (a) 66   (b) 237   (c) 1925
     (d) 12,587   (e) 28,000   (f) 150,269

8.   Convert each of the following decimal (base 10) integers to (i) binary (base 2) form, (ii) octal (base 8) form, and (iii) hexadecimal (base 16) form:
     (a) 87   (b) 126   (c) 8000
     (d) 12,347   (e) 77,895   (f) 186,000

9.   Convert each of the following decimal (base 10) integers to (i) base 3 form, (ii) base 9 form, and (iii) base 27 form:
     (a) 66   (b) 237   (c) 1925
     (d) 12,587   (e) 28,000   (f) 150,269

10. Convert each of the following decimal (base 10) integers to (i) base 3 form, (ii) base 9 form, and (iii) base 27 form:
     (a) 87   (b) 126   (c) 8000
     (d) 12,347   (e) 77,895   (f) 186,000

11. (a) Using Table 2.2, convert the following English plaintexts into binary notation, and into hexadecimal notation:

    (i) agent   (ii) met   (iii) liaison

    (b) Using Table 4.2, convert the following strings (which are either binary or hexadecimal) into its corresponding English plaintext.

    (i) 00111001000101101111   (ii) 0A040B0B0411   (iii) 0D0E16

12. (a) Using Table 4.2, convert the following English plaintexts into binary notation, and into hexadecimal notation:

    (i) take   (ii) cover   (iii) intown

    (b) Using Table 4.2, convert the following strings (which are either binary or hexadecimal) into its corresponding English plaintext.

    (i) 13070417   (ii) 00111000001010100100   (iii) 1604000F0E0D12

13. Perform the following additions in the indicated bases; then translate each into decimal (ordinary) integer language as a check.

    (a) [1100] + [1111] in binary arithmetic.

    (b) [5551] + [3333] in octal arithmetic.

    (c) [AACC] + [9998] in hexadecimal arithmetic.

    (d) [22 8] + [2 10] in base 25.

14. Perform the following additions in the indicated bases; then translate each into decimal (ordinary) integer language as a check.

    (a) [1111] + [1010] in binary arithmetic.

    (b) [6767] + [3277] in octal arithmetic.

    (c) [FACE] + [1AA2] in hexadecimal arithmetic.

    (d) [22 18] + [22 13] in base 23.

15. Perform the following additions in the indicated bases; then translate each into decimal (ordinary) integer language as a check.

    (a) [110101100] + [10111011] in binary arithmetic.

    (b) [55544471] + [333322] in octal arithmetic.

    (c) [AABBCC] + [99988FF] in hexadecimal arithmetic.

    (d) [22 18 9 8] + [2 10 22 13] in base 25.

16. Perform the following additions in the indicated bases; then translate each into decimal (ordinary) integer language as a check.

    (a) [101100111000] + [10011101100] in binary arithmetic.

    (b) [5678765] + [1357531] in base 9 arithmetic.

    (c) [11AAFF] + [92929292] in hexadecimal arithmetic.

    (d) [22 18 13 8] + [2 20 29 13] in base 30.

17. Perform the following subtractions in the indicated bases; then translate each into decimal (ordinary) integer language as a check.

    (a) [1100] – [1011] in binary arithmetic.

    (b) [7211] – [1127] in octal arithmetic.

    (c) [AA22CC] – [9988FF] in hexadecimal arithmetic.

    (d) [22 18 9 8] – [2 10 22 13] in base 25.

18. Perform the following subtractions in the indicated bases; then translate each into decimal (ordinary) integer language as a check.

    (a) [1101] – [11] in binary arithmetic.

    (b) [5674235] – [1357538] in base 9 arithmetic.

    (c) [92929292] – [11AAFF] in hexadecimal arithmetic.

    (d) [22 18 13 8] – [2 20 29 13] in base 30.

19. Perform the following multiplications in the indicated bases; then translate each into decimal (ordinary) integer language as a check.

    (a) [110] × [111] in binary arithmetic.

(b) [555] × [33] in octal arithmetic.

(c) [ACC] × [999] in hexadecimal arithmetic.

(d) [22 8] × [2 10] in base 25.

20.   Perform the following multiplications in the indicated bases; then translate each into decimal (ordinary) integer language as a check.

(a) [1111] × [100] in binary arithmetic.

(b) [676] × [377] in octal arithmetic.

(c) [CAB] × [1A9] in hexadecimal arithmetic.

(d) [22 18] × [22 13] in base 23.

21.   Perform the following multiplications in the indicated bases; then translate each into decimal (ordinary) integer language as a check.

(a) [11001] × [11111] in binary arithmetic.

(b) [5544] × [3333] in octal arithmetic.

(c) [AACC] × [99FF] in hexadecimal arithmetic.

(d) [22 18 9] × [2 22 13] in base 25.

22.   Perform the following multiplications in the indicated bases; then translate each into decimal (ordinary) integer language as a check.

(a) [10110] × [101] in binary arithmetic.

(b) [56787] × [31] in base 9 arithmetic.

(c) [AAFF2] × [92] in hexadecimal arithmetic.

(d) [22 18 13 8] × [2 20] in base 30.

23.   Develop a (simple) algorithm for directly converting a binary expansion into its corresponding hexadecimal expansion (without converting to decimal expansions as an intermediate step).

24.   Develop a (simple) algorithm for directly converting a binary expansion into its corresponding octal expansion (without converting to decimal expansions as an intermediate step).

25.   Show that when Algorithm 4.2 is used to add two base b expansions, each with a total of n digits, it will always require between $2n$ and $3n$ single digit (or carry digit) additions.

26.   Formulate a simple criterion to determine which of two base b expansions (represents an integer that) is greater than or equal to the other.

27.   Explain why any integer weight between 0 and $2^n - 1$ (inclusive) can be determined exactly using a balance scale if we have at our disposal exactly one each of the following weights: $\{1, 2, 2^2, \cdots, 2^{n-1}\}$.

28.   Show that it would not be possible to determine any integer weight between 0 and $2^n - 1$ (inclusive) on a balance scale if we had at our disposal any set of $n - 1$ weights.
Note: Exercise 29 gives a set of n weights where all such weights could be determined.

NOTE: (*The Two's Complement Representation Scheme*) A common scheme that computers use to internally store integers is the so-called **two's complement representation** scheme, which works as follows: Given a positive integer n, any integer a in the range $-2^{n-1} \le a < 2^{n-1}$ is represented by a length n bit string $a \sim [b_{n-1} \, b_{n-2} \, \cdots b_1 \, b_0]$, where the left-most bit stores the sign of a: $b_{n-1} = 0$ if a is a nonnegative integer, and $b_{n-1} = 1$ if a is negative. In case a is nonnegative, the remaining bits $[b_{n-2} \, \cdots b_1 \, b_0]$ are just those of the binary expansion of a, whereas if a is negative, $[b_{n-2} \, \cdots b_1 \, b_0]$ is the binary expansion of $2^{n-1} - |x|$. Thus, for example, if we use length $n = 5$ bit strings, the two's complement representation of 8 would be [01000], and the two's complement representation of -5 would be (since $2^{5-1} - |-5| = 16 - 5 = 11$) [11011].

29. (a) Using $n = 6$ bits, find the two's complement representation of each of the following integers:

   (i) 17   (ii) -22 (iii) -32

   (b) Determine the integers that have the following 6-bit two's complement representations:

   (i) [110011]   (ii) [001100]   (iii) [011111]

30. (a) Using $n = 6$ bits, find the two's complement representation of each of the following integers:

   (i) -2 (ii) -17 (iii) 25

   (b) Determine the integers that have the following 6-bit two's complement representations:

   (i) [011011]   (ii) [101100]   (iii) [110001]

31. (a) Find a simple relationship between the two's complement representation of an integer in the range $-2^{n-1} \le a < 2^{n-1}$, and the remainder when a is divided by 2^{n-1}.

   (b) Write out a simple algorithm that inputs an integer n and an integer a within the range $-2^{n-1} \le a < 2^{n-1}$, and outputs the two's complement representation of a. Apply your algorithm to Exercise 29(a).

   (c) Write out a simple algorithm that inputs a two's complement representation vector $[b_{n-1} \ b_{n-2} \cdots b_1 \ b_0]$, and outputs the integer that it represents. Apply your algorithm to Exercise 29(b).

32. (a) Write out a simple algorithm that inputs two length n vectors $[c_{n-1} \ c_{n-2} \cdots c_1 \ c_0]$, and $[d_{n-1} \ d_{n-2} \cdots d_1 \ d_0]$, that are two's complement representations of a pair of integers, c and d, in the range $-2^{n-1} \le c, d < 2^{n-1}$, and outputs a length $n + 1$ vector $[s_n \ s_{n-1} \cdots s_1 \ s_0]$, that gives the two's complement representation of sum $c + d$.

   (b) Apply your algorithm to evaluate the following integer sums: $2 + 7$, $(-22) + 16$, and $(-22) + (-20)$.

33. Suppose that we are working in a very small computing platform whose (hexadecimal) word length is just 3. Outline a scheme by which this system could be used to perform the addition [8FF8] + [9BA2] (hexadecimal). Then check the result using ordinary integer arithmetic.

34. Suppose that we are working in a very small computing platform whose (hexadecimal) word length is just 4. Outline a scheme by which this system could be used to perform the addition [A8FF8] + [29BA2] (hexadecimal). Then check the result using ordinary integer arithmetic.

35. (*Complexity of Addition of Binary Sequences*) Suppose that $[c_{n-1} \ c_{n-2} \cdots c_1 \ c_0]$ and $[d_{n-1} \ d_{n-2} \cdots d_1 \ d_0]$ are two length n bit strings. Show that the number of bit operations to use either Algorithm 4.2 to perform the corresponding addition, or Algorithm 4.3 to perform the corresponding subtraction (if possible) is at most Cn, where C is a universal constant (in particular, C does not depend on n).

36. (*Complexity of Multiplication of Binary Sequences*) Suppose that $[c_{n-1} \ c_{n-2} \cdots c_1 \ c_0]$ and $[d_{n-1} \ d_{n-2} \cdots d_1 \ d_0]$ are two length n bit strings. Show that the number of bit operations to use Algorithm 4.3 to perform the corresponding multiplication is Cn^2, where C is a universal constant (in particular, C does not depend on n).

COMPUTER EXERCISES 4.1:

NOTE: There are two natural *data structures* for storing base b expansions: either as vectors or as strings. Strings are only appropriate in case the base b "digits" are single characters: This includes the cases $b \leq 10$ and $b = 16$ (due to our special hexadecimal notation). Vectors tend to be more amenable to writing programs, but strings display more efficiently. The reader should contemplate both possibilities on his/her particular computing platform and decide which of these options (or perhaps another option) would be most suitable for the exercises and applications of this section.

1. Write a program n = bin2int(v) that will take as input a vector v for a binary expansion (zeros and/or ones), and will output its corresponding equivalent decimal integer n. Perform, by hand, the corresponding conversions for the binary strings [1011], [11111], and [1011110], and run your program on these inputs (debug, as necessary).

2. Write a program n = oct2int(v) that will take as input a vector v for an octal (base 8) expansion, and will output its corresponding equivalent decimal integer n. Perform, by hand, the corresponding conversions for the octal expansions [5027], [23456], and [7031410], and run your program on these inputs (debug, as necessary).

3. Write a program n = hex2int(v) that will take as input a vector v for a hexadecimal (base 16) expansion, and will output its corresponding equivalent decimal integer n. Perform, by hand, the corresponding conversions for the hexadecimal expansions [8B7], [1AEE], and [AAAA6], and run your program on these inputs (debug, as necessary).
   Note: The reader may wish to intead use vectors of integers in the range 0–15 for hexadecimal sequences.

4. Write a program n = base92int(v) that will take as input a vector v viewed as a base 9 expansion, and will output its corresponding equivalent decimal integer n. Perform, by hand, the corresponding conversions for the base 9 expansions [847], [1444], and [65626], and run your program on these inputs (debug, as necessary).

5. Write a program n = baseb2int(v,b) that will take two inputs: a vector v for a base b expansion, and, b an integer greater than 1 (for the base of the expansion). The output will be the corresponding equivalent decimal integer n. Run your program on each of the expansions of ordinary Exercise 5 (debug, as necessary).

6. Write a program v = int2bin(n) that will take as input a nonnegative integer n, and will output its binary expansion vector v using Algorithm 4.1. Perform, by hand, the corresponding conversions to binary expansions for n = 8, 107, 327, and 12,557, and run your program on these inputs (debug, as necessary).

7. Write a program v = int2oct(n) that will take as input a nonnegative integer n, and will output its octal (base 8) expansion vector v using Algorithm 4.1. Perform, by hand, the corresponding conversions to binary expansions for n = 8, 107, 327, and 12,557, and run your program on these inputs (debug, as necessary).

8. Write a program v = int2hex(n) that will take as input a nonnegative integer n, and will output its hexadecimal (base 16) expansion vector v using Algorithm 4.1. Perform, by hand, the corresponding conversions to binary expansions for n = 8, 107, 327, and 12,557, and run your program on these inputs (debug, as necessary).
   Note: Read Computer Exercise 3.

9. Write a program v = int2baseb(n, b) that will take as inputs a nonnegative integer n and an integer b (the base) greater than one. The output will be the base b expansion vector v of the

integer n, determined by using Algorithm 4.1. Perform, by hand, the corresponding conversions for n = 8, 107, 327, and 12,557, with each of the bases 2, 8, 16, and run your program on these (14) inputs (debug, as necessary).

10. Write a program `w = bin_add(u,v)` that will take as inputs two vectors u and v representing binary (base 2) expansions, and will output the vector w representing the binary expansion of the sum u + v, computed using Algorithm 4.2. Perform, by hand, the binary additions [101] + [111], [110110] + [1010111], and run your program for these additions (debug, as necessary).

11. Write a program `w = hex_add(u,v)` that will take as inputs two vectors u and v representing hexadecimal (base 16) expansions (using digits from 0 to 15), and will output the vector w representing the hexadecimal expansion of the sum u + v, computed using Algorithm 4.2. Perform, by hand, the hexadecimal additions [C42] + [A1A], [86B4D] + [76A0C], and run your program for these additions (debug, as necessary).
Note: The reader may wish to instead use vectors of integers in the range 0–15 for hexadecimal sequences.

12. Write a program `w = baseb_add(u,v,b)` that will take as inputs two vectors u and v representing base b expansions, and a third input b (the base) being an integer greater than 1. The output will be the vector w representing the base b expansion of the sum u + v, computed using Algorithm 4.2. Run your program on the base *b* subtractions of ordinary Exercise 15 (debug, as necessary).

13. Write a program `w = bin_sub(u,v)` that will take as inputs two vectors u and v representing binary (base 2) expansions where u ≥ v, and will output the vector w representing the binary expansion of the difference u − v, computed using Algorithm 4.3. Perform, by hand, the binary subtractions [101] − [011], [110110] − [101011], and run your program on them (debug, as necessary).

14. Write a program `w = hex_sub(u,v)` that will take as inputs two vectors u and v representing hexadecimal (base 16) expansions where u ≥ v, and will output the vector w representing the hexadecimal expansion of the difference u − v, computed using Algorithm 4.3. Perform, by hand, the hexadecimal subtractions [A42] − [A1A], [86A4D] + [76C8B], and run your program on them (debug, as necessary).
Note: The reader may wish to use instead vectors of integers in the range 0–15 for hexadecimal sequences.

15. Write a program `w = baseb_sub(u,v,b)` that will take as inputs two vectors u and v representing base b expansions where u ≥ v, and a third input b (the base) being an integer greater than 1. The output will be the vector w representing the base b expansion of the difference u − v, computed using Algorithm 4.3. Run your program on the base *b* subtractions of ordinary Exercise 17 (debug, as necessary).

16. Write a program `w = bin_mult(u,v)` that will take as inputs two vectors u and v representing binary (base 2) expansions, and will output the vector w representing the binary expansion of the product u × v, computed using Algorithm 4.4. Perform, by hand, the binary multiplications [101] × [111], [110110] × [1010111], and run your program for these multiplications (debug, as necessary).

17. Write a program `w = hex_mult(u,v)` that will take as inputs two vectors u and v representing hexadecimal (base 16) expansions (using digits from 0 to 15), and will output the vector w representing the hexadecimal expansion of the product u × v, computed using Algorithm 4.4. Perform, by hand, the hexadecimal multiplications [C42] × [A1A], [86B4D] × [76A0C], and run your program for these multiplications (debug, as necessary).
Note: The reader may wish to instead use vectors of integers in the range 0–15 for hexadecimal sequences.

18.   Write a program $w = \texttt{baseb\_mult(u,v,b)}$ that will take as inputs two vectors $\texttt{u}$ and $\texttt{v}$ representing base b expansions, and a third input b (the base) being an integer greater than 1. The output will be the vector $\texttt{w}$ representing the base b expansion of the product $\texttt{u} \times \texttt{v}$, computed using Algorithm 4.4.   Run your program on the base b multiplications of ordinary Exercises 19, 21 (debug, as necessary).

4.2: MODULAR ARITHMETIC AND CONGRUENCES

The integers, like the real numbers, form an infinite set. Section 3.3 touched on some of their very rich and interesting number theory. For each integer $m > 1$, we can form a finite set of integers $\mathbb{Z}_m = \{0, 1, 2, \cdots, m-1\}$, which, when endowed with its own special multiplication and addition operations, becomes a very important number system known as the *integers modulo m*. These number systems are important and inherent in numerous discrete structures, and in this section we will develop some of their more important features.

We first briefly recall a few facts about the congruence relations that were introduced in Section 2.2 and further elaborated on in Section 3.3. For more details, the reader might wish to review the relevant portions in these sections. Recall that we proved in Section 2.2 that if $m > 1$, then congruence mod m is an equivalence relation on the set of all integers $\mathbb{Z}$, and the resulting equivalence classes are $[0], [1], [2], \cdots, [m-1]$ (Proposition 2.2). Each such equivalence class $[j]$ represents those infinitely many integers that are congruent to j (mod m): $[j] = \{j + km : k = 0, \pm1, \pm2, \cdots\}$. The equivalence classes partition the integers into disjoint subsets (this is a general property of equivalence relations).   From Theorem 3.18, we know that addition and multiplication of equivalence classes by

$$[j] + [k] = [j+k] \text{ and } [j] \cdot [k] = [j \cdot k]$$

are well-defined operations that do not depend on the particular elements (representatives) of the equivalence classes.   Because of this property, we may represent each of these equivalence classes by any single representative.   An effective analogy for this modular arithmetic is clockwork arithmetic, about which we all have extensive experience. This corresponds to the special case of $m = 12$. If we work only with integer hours, then, no matter how many hours are added to a certain time, the answer will always land between 1 and 12.   For example, if we start working at 9 and work for 8 hours, the time that we finish work is $9 + 8 = 17 \equiv 5$ (mod 12).   If we ever had to multiply times, this would work in the same way; for example if we multiplied 8 by 7 we would get $8 \cdot 7 = 56 \equiv 8$ (mod 12). In mod 12 arithmetic, the only difference is that we always take representatives from the set $\mathbb{Z}_{12} = \{0, 1, \cdots, 11\}$ (so 12 o'clock would be replaced by 0 o'clock).   In general modular arithmetic (mod m), we always take representatives from the set

$\{0,\ 1,\ \cdots, m-1\}$, and when endowed with the corresponding addition and multiplication operations, this set is as denoted $\mathbb{Z}_m$. Note that since we always take designated representatives, we do not need to use the equivalence class notation $[j]$ for an integer mod m. But never forget that any integer modulo m really should be thought of as an infinite set of all of the integers that are congruent to it (mod m).

Modular Integer Systems

FIGURE 4.1: Carl Friedrich Gauss (1777–1855), German mathematician.

Among his numerous contributions to mathematics and science, the illustrious mathematician Carl Friedrich Gauss (Figure 4.1)[12] developed the extremely useful number-theoretic concepts of congruences and modular arithmetic. These concepts lead to an infinite supply of abstract number systems that have turned out to play a pivitol role in an assortment of applications. We recall from Section 2.2 the important definition of congruence, and restate it in terms of the divisibility symbol of Section 3.3:

Let m be a positive integer. We say that two integers a and b are **congruent mod(ulo) m**, and denote this as $a \equiv b \pmod{m}$, if $m \mid (a-b)$. The number m is called the **modulus** of the congruency. If $m \nmid (a-b)$, we say that a and b are **incongruent mod m**, and write this as $a \not\equiv b \pmod{m}$.

---

[12] Carl F. Gauss is widely considered to be the greatest mathematician who ever lived. His potential was discovered early, and his mathematical aptitude was astounding. While he was in second grade, his teacher, needing to keep him occupied for a while, asked him to perform the addition of the first 100 integers: $S = 1 + 2 + \ldots + 100$. Two minutes later, Gauss gave the teacher the answer. He did it by rewriting the sum in the reverse order $S = 100 + 99 + \ldots + 1$, adding vertically to the original to get $2S = 101 + 101 + \ldots + 101 = 100 \cdot 101$, so $S = 50 \cdot 101 = 5050$. This idea yields a general proof of an important mathematical series identity. Apart from his numerous groundbreaking contributions to mathematics, Gauss did significant work in physics and astronomy, as well as in other sciences. His brilliant ideas came to him so rapidly, that he had a file cabinet full of them waiting to be written up for formal publication. He would often receive visits from other internationally prominent mathematicians who would proudly share with Gauss recent discoveries, and very often Gauss would simply reach into his file cabinet to pull out his ideas on the topic that frequently eclipsed those of the visitor. For many years until the inception of the Euro, Germany honored Gauss by placing his image on the very common 10 Deutsche Mark banknote (the value was approximately US $5). Figure 4.1 is an image of this banknote, with a drawing of Gauss's important normal (bell-shaped) curve, a cornerstone of statistics.

EXAMPLE 4.6: (*Two Familiar Modulii*) (a) Notice that $15 \equiv 3 \pmod{12}$, since $15 - 3 = 12$; similarly, since $27 - 3 = 24 = 2 \cdot 12$, we have $27 \equiv 3 \pmod{12}$. The reader can similarly check that $3 \equiv -9 \equiv -21 \equiv -33 \cdots \pmod{12}$. Congruences mod 12 can be visualized by means of a traditional (as opposed to a digital) clock; see Figure 4.2. Two times are congruent in the clock if one can be made into the other by turning the (hour) hand of the clock a complete number of revolutions either clockwise (corresponding to adding 12), or counterclockwise (corresponding to subtracting 12).

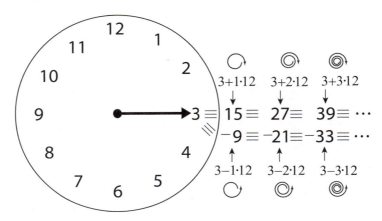

FIGURE 4.2: Congruence modulo 12 is like clockwork; two integers are congruent mod 12 if one can be obtained from the other by adding or subtracting a multiple of 12, corresponding to the hour hand making an integral number of revolutions around the clock.

(b) Anyone who has studied angles or trigonometry will already be familiar with 360 as a modulus, since $360°$ corresponds to a complete revolution angle (so adding any multiple of it results in the same angle as wherever we started). Thus the angular equalities: $-90° = 270° = 630° = \cdots$ correspond to the congruences $-90 \equiv 270 \equiv 630 \equiv \cdots \pmod{360}$. To see that $-90 \equiv 630 \pmod{360}$, for example, we note that $-90 - 630 = -720 = -2 \cdot 360$.

If $a \equiv b \pmod{m}$, we may rewrite the condition $m \mid (a - b)$ as $a - b = km$ (for some integer k), which we express as $a = b + km$. We summarize this alternate formulation:

$$a \equiv b \pmod{m} \quad \Leftrightarrow \quad a = b + km, \text{ for some } k \in \mathbb{Z} \qquad (3)$$

This formula is illustrated in Figure 4.2, showing how to get all of the integers that are congruent to 3 (mod 12).

EXERCISE FOR THE READER 4.5: Show that $a \equiv b \pmod 2$ if, and only if a and b have the **same parity**, i.e., a and b are both even or both odd. Describe the congruence classes mod 2.

DEFINTION 4.2: If m is a positive integer, the set of **integers modulo m**, denoted by $\mathbb{Z}_m$, is the set of possible remainders when dividing by m:

$$\mathbb{Z}_m = \{0, 1, 2, \cdots, m-1\}.$$

We define the arithmetic operations of addition, subtraction, multiplication, and exponentiation on $\mathbb{Z}_m$ by performing the corresponding arithmetic operations on the integers, and converting to an element of $\mathbb{Z}_m$.[13]

Thus, to add or multiply a pair of integers modulo m, we could first perform the corresponding usual operation on the integers, and take the remainder of this number when it is divided by m (using the division algorithm). Theorem 3.18 assures us that the results of such operations will always be consistent. Our next example will look at addition and multiplication tables for $\mathbb{Z}_m$, for two small values of m.

EXAMPLE 4.7: Create addition and multiplication tables for $\mathbb{Z}_5$ and $\mathbb{Z}_6$. Do you notice any similarities or differences in the corresponding tables?

SOLUTION: In Tables 4.3 and 4.4, we construct addition and multiplication tables for $\mathbb{Z}_5$, and Tables 4.5 and 4.6 give the corresponding tables for $\mathbb{Z}_6$.

+	0	1	2	3	4
0	0	1	2	3	4
1	1	2	3	4	0
2	2	3	4	0	1
3	3	4	0	1	2
4	4	0	1	2	3

×	0	1	2	3	4
0	0	0	0	0	0
1	0	1	2	3	4
2	0	2	4	1	3
3	0	3	1	4	2
4	0	4	3	2	1

TABLES 4.3 and 4.4: Addition and multiplication tables for $\mathbb{Z}_5$.

---

[13] With its addition and multiplication operations, the system $\mathbb{Z}_m$ of integers modulo m inherits most all of the nice properties of arithmetic that the system $\mathbb{Z}$ of integers possess, such as commutativity of addition and multiplication: $a+b = b+c$ and $ab = ba$, associativity of addition and multiplication: $(a+b)+c = a+(b+c)$ and $(ab)c = a(bc)$, and the distributive law: $a(b+c) = ab+ac$. The modular integers are examples of what are called *commutative rings* in abstract algebra.

+	0	1	2	3	4	5
0	0	1	2	3	4	5
1	1	2	3	4	5	0
2	2	3	4	5	0	1
3	3	4	5	0	1	2
4	4	5	0	1	2	3
5	5	0	1	2	3	4

×	0	1	2	3	4	5
0	0	0	0	0	0	0
1	0	1	2	3	4	5
2	0	2	4	0	2	4
3	0	3	0	3	0	3
4	0	4	2	0	4	2
5	0	5	4	3	2	1

TABLES 4.5 and 4.6: Addition and multiplication tables for $\mathbb{Z}_6$.

EXERCISE FOR THE READER 4.6: Perform the following operations in $\mathbb{Z}_{12}$: $11+8$, $5 \cdot 8$, 11^2. Is there an element $b \in \mathbb{Z}_{12}$ such that $5b = 1$ in $\mathbb{Z}_{12}$?

The addition tables for $\mathbb{Z}_5$ and $\mathbb{Z}_6$ are quite similar in structure. The row for 0 is simply a copy of the row of the second numbers (upper column) corresponding to the fact that 0 is the *additive identity*: $0 + a = a$. The remaining rows are simply cyclic shifts of the first row. Each time we shift to the left by 1 (with wrap-around) from the previous row, corresponding to the next higher number being added. This simple structure is common to addition tables for any $\mathbb{Z}_m$. There is a stark difference, though, in the multiplication tables. Notice that each nonzero row (or column) of the multiplication table for $\mathbb{Z}_5$ contains each of the elements of $\mathbb{Z}_5$ (as is the case for the addition tables), but this is not the case for the multiplication table for $\mathbb{Z}_6$. For example in $\mathbb{Z}_6$, we can get 0 by multiplying the two nonzero numbers 2 and 3. This sort of problem generally occurs in any $\mathbb{Z}_m$ when m is composite, but never occurs in $\mathbb{Z}_p$ for a prime modulus p (Why?). Furthermore, when p is prime, any nonzero row in the multiplication table of $\mathbb{Z}_p$ will always contain all of the elements of $\mathbb{Z}_p$. In order to elaborate on these concepts, we will first need a definition.

Modular Inverses

DEFINITION 4.3: For any $a \in \mathbb{Z}_m$, we say that a is **invertible** (or has an inverse) if there exists another element $a^{-1} \in \mathbb{Z}_m$ such that $a \cdot a^{-1} = a^{-1} \cdot a = 1$. The element a^{-1}, if it exists, is called the (**multiplicative**) **inverse** of a.

For example, if we have the multiplication table available, it can be determined if an element $a \in \mathbb{Z}_m$ has an inverse simply by checking if the row for this element in the table has a 1 in it. For example, from Table 4.4 (multiplication table for $\mathbb{Z}_5$), we see that in $\mathbb{Z}_5$ we have $1^{-1} = 1$, $2^{-1} = 3$, $3^{-1} = 2$, and $4^{-1} = 4$. (In any $\mathbb{Z}_m$, the element 1 will always be its own inverse, and the element 0 never will have an inverse—why?) From Table 4.6, we see that in $\mathbb{Z}_6$, $1^{-1} = 1$, $5^{-1} = 5$, and no other elements have inverses. We point out that since multiplication in $\mathbb{Z}_m$ is commutative (i.e., $ab = ba$, $\forall a, b \in \mathbb{Z}_m$),[14] only one of the two conditions for inverses $(a \cdot a^{-1} = 1$ or $a^{-1} \cdot a = 1)$ needs to be checked. Also, there can only be one inverse of any element $a \in \mathbb{Z}_m$. (*Proof:* Suppose that both b and $a^{-1} \in \mathbb{Z}_m$ were inverses of $a \in \mathbb{Z}_m$. Then $b = b \cdot 1 = b \cdot (a \cdot a^{-1}) = (b \cdot a) \cdot a^{-1} = 1 \cdot a^{-1} = a^{-1}$.□) Inverses are important for an assortment of reasons some of which we will discuss shortly. The following result provides a simple criterion to determine whether an element $a \in \mathbb{Z}_m$ has an inverse.

PROPOSITION 4.2: (*Inverses in* $\mathbb{Z}_m$) An element $a \in \mathbb{Z}_m$ is invertible precisely when $\gcd(a, m) = 1$, i.e., (in the language of Section 3.3) when it is relatively prime to m. Moreover, the inverse a^{-1} can be obtained from the integer equation $1 = ax + my$ (that follows from Theorem 3.16), as $a^{-1} \equiv x \pmod{m}$. In particular, all nonzero elements of $\mathbb{Z}_m$ are invertible precisely when $m = p$ is prime.

Proof: If $\gcd(a, m) = 1$, then by Theorem 3.16, there exist integers x and y such that $1 = ax + my$. If we rewrite this as $1 - ax = my$, we see that $m \mid (1 - ax)$, and so $ax \equiv 1 \pmod{m}$. This implies (resetting x to be its representative between 1 and $m - 1$) that $x = a^{-1}$ in $\mathbb{Z}_m$. Conversely, if a has an inverse a^{-1} in $\mathbb{Z}_m$, then $a \cdot a^{-1} \equiv 1 \pmod{m}$, or $m \mid (1 - a \cdot a^{-1})$. It follows (from the definition of divisibility) that $1 - a \cdot a^{-1} = mk$, or $mk + a \cdot a^{-1} = 1$, for some $k \in \mathbb{Z}$. From this latter equation, it readily follows that $\gcd(a, m) = 1$, because any (prime) factor of both a and m, would also (by Theorem 3.10(b)) necessarily have to be a (prime) factor of 1 (which has no prime factors). The last statement easily follows from the criterion since any prime number p is relative prime to all positive integers that are less than p, i.e., to all nonzero elements in $\mathbb{Z}_p$. □

Fast Modular Exponentiation

With the thorough development of binary expansions of Section 4.1 behind us, we are now well poised to formalize an algorithm for modular exponentiation that was

---

[14] It is an easy consequence of Theorem 3.18 that all of the algebraic identity properties of the integers are inherited by each $\mathbb{Z}_m$. Apart from commutativity, this includes associativity and the distributive laws.

promised in Section 3.3. The following algorithm is extremely efficient, and will be very useful in our work with public key cryptography of Section 4.5.

ALGORITHM 4.5: (*Fast Modular Exponentiation*):
Input: An integer base c, an integer exponent x, and an integer modulus $m > 1$.
Output: A nonnegative integer $a < m$ that satisfies $a \equiv c^x \pmod{m}$.

Step 1: Use Algorithm 4.1 to create the binary expansion of the exponent x:
$$x \sim [d_K \ d_{K-1} \ \cdots d_1 \ d_0] \ \text{(base 2)}.$$
Step 2: (*Repeatedly square the number c (mod m) as we run through the binary digits d_k of x, including the result in the cumulative product only when $d_k = 1$.*)
SET $a = 1$ (*Initialize cumulative product a*)
SET $s = c \pmod{m}$ (*Initialize squaring*)
FOR $k = 0$ TO K
  IF $d_k = 1$
    UPDATE $a \rightarrow a \times s \pmod{m}$
  END IF
  UPDATE $s \rightarrow s^2 \pmod{m}$ (*Squaring need not be done when k = K*)
END k FOR

Step 3: Output: a

NOTE: At each step of the algorithm, it is assumed that the numbers are reduced mod m.

The proof that this algorithm works follows simply from writing out the binary expansion of the exponent and using the laws of exponents:
$$c^x \equiv c^{\sum_{k=0}^{K} d_k 2^k} \equiv c^{d_0 2^0} c^{d_1 2^1} \cdots c^{d_K 2^K} \pmod{m},$$
and since $c^{2^{k+1}} = (c^{2^k})^2$. All of the successive squares need to be computed (mod m), but only those for which the corresponding $d_k = 1$ make an appearance in the above (cumulative) expression for b^x. Algorithm 4.5 turns out to be a very fast and efficient scheme for modular exponentiation, just requiring an amount of work roughly equal to performing K successive squarings mod m, where K is the number of binary digits in the exponent x.

EXAMPLE 4.8: Use the fast exponentiation algorithm to compute the following modular power: $2^{825} \pmod{173}$.

SOLUTION: We first apply Algorithm 4.1 to obtain the binary expansion $825 \sim [1100111001]$ (base 2). After initializing $a = 1$ and $s = 2$, the iterations of Step 2 produce the following updates:

$k = 0$: Since $d_k = 1$, we update $a \rightarrow a \times s = 2$, and then update the square $s \rightarrow s^2 = 2^2 = 4$. .

$k = 1$: Since $d_k = 0$, we only update the square $s \rightarrow s^2 = 4^2 = 16$.

$k = 2$: Since $d_k = 0$, we only update the square $s \rightarrow s^2 = 16^2 = 256$
$$\equiv 83 (\text{mod} 173).$$

$k = 3$: Since $d_k = 1$, we update $a \rightarrow a \times s = 2 \cdot 83 = 166$ and then update the square $s \rightarrow s^2 = 83^2 = 6889 \equiv 142 (\text{mod} 173)$.

$k = 4$: Since $d_k = 1$, we update $a \rightarrow a \times s = 166 \cdot 142 = 23{,}572 \equiv 44$. and then update the square $s \rightarrow s^2 = 142^2 = 20{,}164 \equiv 96 (\text{mod} 173)$.

$k = 5$: Since $d_k = 1$, we update $a \rightarrow a \times s = 44 \cdot 96 = 4224 \equiv 72$. and then update the square $s \rightarrow s^2 = 96^2 = 9216 \equiv 47 (\text{mod} 173)$.

$k = 6$: Since $d_k = 0$, we only update the square $s \rightarrow s^2 = 47^2 = 2209$
$$\equiv 133 (\text{mod} 173).$$

$k = 7$: Since $d_k = 0$, we only update the square $s \rightarrow s^2 = 133^2 = 17689$
$$\equiv 43 (\text{mod} 173).$$

$k = 8$: Since $d_k = 1$, we update $a \rightarrow a \times s = 72 \cdot 43 = 3096 \equiv 155$, and then update the square $s \rightarrow s^2 = 43^2 = 1849 \equiv 119 (\text{mod} 173)$.

$k = 9$: Since $d_k = 1$, we update $a \rightarrow a \times s = 155 \cdot 119 = 18445 \equiv 107$.

The algorithm thus tells us that $2^{825} \equiv 107 (\text{mod} 173)$.

COMPUTING NOTE: On any computing platform that operates in standard floating point arithmetic with binary word length of about 50, the simple approach of evaluating 2^{825} and taking its remainder (mod 173), would lead to inaccurate results since the number is too large to be accurately stored; Section 4.4 will elaborate on such issues. The algorithm works around this issue since at each iteration of Step 2, the numbers that arise are less than 172^2, and then immediately converted to an integer (mod 173). Even if one is using a computing platform with symbolic capabilities, the above algorithm is a much more efficient method.

EXERCISE FOR THE READER 4.7: Use the fast exponentiation algorithm to compute the following modular powers: $289^{225} (\text{mod} 311)$.

Congruences

One important task relating to modular arithmetic concerns *solving congruences*. For example, a simple congruence such as $x+5 \equiv 2 \pmod{12}$ can easily be solved $3x+2 \equiv 1 \pmod 5$. by subtracting 5 from both sides (just like in basic algebra): $x \equiv 2-5 \equiv -3 \equiv 9 \pmod{12}$. From Proposition 3.9, we can always add or multiply both sides of a congruence by any number (or equivalent representative of that number, mod m). Dividing both sides of a congruence is more difficult; in general, to divide both sides of a congruence by a number a, a needs to be invertible in $\mathbb{Z}_m$, i.e., (from Proposition 4.2) we must have $\gcd(a,m)=1$. Furthermore, once it is ascertained that you can divide both sides by a, you will actually be multiplying both sides by a^{-1}, rather than dividing (with real numbers) by a. This (unlike with multiplication and addition) is where modular arithmetic and ordinary arithmetic are very different. This process is illustrated in the following example.

EXAMPLE 4.9: Solve each of the following congruences for x:

(a) $3x+2 \equiv 1 \pmod 5$ (b) $5x-4 \equiv 4 \pmod 6$.

SOLUTION: Part (a): From Table 4.4 (or by simple trial and error multiplying each of the four nonzero elements of $\mathbb{Z}_5$ until we get 1), we see that $3^{-1} = 2$ in $\mathbb{Z}_5$. Thus, when we need to divide the congruence by 3, we will by multiplying both sides by 2: $\quad 3x+2 \equiv 1 \Rightarrow 3x \equiv 1-2 \equiv -1 \equiv 4 \Rightarrow x = 2 \cdot 4 \equiv 8 \equiv 3 \pmod 5$. This answer is easily checked in the original congruence: $3 \cdot 3 + 2 \equiv 11 \equiv 1 \pmod 5$. Part (b): In $\mathbb{Z}_6$, 5 is invertible since it is relatively prime to 6, so we will be able to perform the necessary division by 5 to solve this congruence. From Table 4.6 (or by simple trial and error), $5^{-1} = 5$, so when we need to divide this congruence by 5, we will multiply both sides by (its inverse) 5: $5x-4 \equiv 4 \Rightarrow 5x \equiv 4+4 \equiv 8 \equiv 2 \Rightarrow x = 5 \cdot 2 \equiv 10 \equiv 4 \pmod 6$. Once again, this answer is easily checked.

The Extended Euclidean Algorithm

We will discuss more general congruences shortly, but we first provide an algorithm for computing the inverse of an invertible integer a mod m. From Proposition 4.2, we know that a must satisfy $\gcd(a,m) = 1$, which means that there exist integers x and y, such that $1 = ax + my$. As shown in the proof of Proposition 4.2, a^{-1} can be taken to be the integer $x \pmod m$ in this equation. We explained in Section 3.3 how these integers x and y can be computed by working backwards through the intermediate steps of the Euclidean algorithm. The algorithm below is an extended version of the Euclidean algorithm that works in a more organized fashion to output the numbers x and y, along with the greatest common divisor d. The algorithm operates on ordered lists (vectors) with three components. The components of such an ordered list V will be denoted (in order) as $V(1)$, $V(2)$, $V(3)$. Thus, for example if $V = [2, 4, 6]$, then $V(1) = 2$, $V(2) = 4$, $V(3) = 6$. The algorithm will also be multiplying ordered lists by numbers, and this is done by multiplying each of the components by the number, for example: $5[2, 4, 6] = [10, 20, 30]$.

ALGORITHM 4.6: (*The Extended Euclidean Algorithm*):
Input: A pair of positive integers a and b, with $a \geq b$.
Output: Three integers $d = \gcd(a, b)$, x, and y, that satisfy the equation $d = ax + by$, along with d.

Step 1: SET $U = [a, 1, 0]$, $V = [b, 0, 1]$ (*Initialize recordkeeping vectors*)

Step 2: WHILE $V(1) > 0$
(*Tasks below will be repeated while first component of V is positive*)
$\qquad W = U - \text{floor}(U(1)/V(1))V$
$\qquad$ UPDATE $U = V$
$\qquad$ UPDATE $V = W$
END WHILE[15]

Step 3: OUTPUT $d = U(1)$, $x = U(2)$, $y = U(3)$

The ordinary Euclidean algorithm (Algorithm 3.1) had the same inputs, but outputted only d. It is not so obvious that this algorithm actually does what claimed. We will explain it and prove that it does indeed work, after the following illustrative example.

EXAMPLE 4.10: (a) Use Algorithm 4.6 to compute $d = \gcd(148, 75)$, and integers x and y such that $d = 148x + 75y$.

(b) If it exists, compute $75^{-1} \bmod 148$.

SOLUTION: Part (a): *Step 1*: We initialize $U = [148, 1, 0]$, $V = [75, 0, 1]$
Step 2: Since $V(1) = 75 > 0$
We set: $W = U - \text{floor}(U(1)/V(1))V = [148, 1, 0] - \lfloor 148/75 \rfloor [75, 0, 1] = [148, 1, 0] - 1 \cdot [75, 0, 1] = [73, 1, -1]$
We update $U = V = [75, 0, 1]$ and $V = W = [73, 1, -1]$

Since $V(1) = 73 > 0$ we repeat this with the updates:

---

[15] Recall that the operations after the "WHILE" instruction and before its "END" are to be executed repeatedly until the condition indicated after the "WHILE," in this case $V(1) > 0$, fails to be valid.

$W = U - \text{floor}(U(1)/V(1))V = [75, 0, 1] - \lfloor 75/73 \rfloor [73, 1, -1] = [75, 0, 1] - 1 \cdot [73, 1, -1] = [2, -1, 2].$

$U = V = [73, 1, -1]$ and $V = W = [2, -1, 2].$

Since $V(1) = 2 > 0$ we again repeat these updates:

$W = U - \text{floor}(U(1)/V(1))V = [73, 1, -1] - \lfloor 73/2 \rfloor [2, -1, 2] = [73, 1, -1] - 36 \cdot [2, -1, 2] = [1, 37, -73].$

$U = V = [2, -1, 1]$ and $V = W = [1, 37, -73].$

Since $V(1) = 1 > 0$ we need one final updating:

$W = U - \text{floor}(U(1)/V(1))V = [2, -1, 1] - \lfloor 2/1 \rfloor [1, 37, -73] = [2, -1, 1] - 2 \cdot [1, 37, -73] = [0, -75, 148]$

$U = V = [1, 37, -73]$ and $V = W = [0, -75, 148].$

Step 3: Output: $d = U(1) = 1$, $x = U(2) = 37$, $y = U(3) = -73$

The resulting relationship is easily checked: $1 = 37 \cdot 148 - 73 \cdot 75.$

Part (b): The result of Part (a) tells us that gcd(148, 75) = 1, so from Proposition 4.2 and the equation $1 = 37 \cdot 148 - 73 \cdot 75$, and since $-73 \equiv 75 \pmod{148}$, we get that $75^{-1} = 75$ in $\mathbb{Z}_{148}$. Thus 75 is its own inverse mod 148; this is not such a common occurrence.

EXERCISE FOR THE READER 4.8:   (a)   Use Algorithm   4.6 to compute $d = \gcd(1155, 862)$, and integers x and y such that $d = 1155x + 862y$.
(b) If it exists, compute the 862^{-1} in $\mathbb{Z}_{1162}$.

Algorithm 4.6 is really just the Euclidean algorithm in disguise, with some additional recordkeeping (hence the three-element vectors). The proof below will use the notation of the Euclidean algorithm (Algorithm 3.1), so it might be helpful for the reader to review this algorithm before reading this proof.

Proof that the outputs d, x, and y of Algorithm 4.6 satisfy $d = gcd(a,b)$ and $d = ax + by$: We first point out that throughout the algorithm, any of the length three vectors $Z = U$, V, or W always corresponds to a valid equation:

$$Z(1) = a \cdot Z(2) + b \cdot Z(3), \text{ where } Z = [Z(1), Z(2), Z(3)].$$

To see this, note first that it is clearly true for the initial vectors $U = [a, 1, 0]$ and $V = [b, 0, 1]$.   (For example, for $Z = U$, the equation becomes $a = a \cdot 1 + b \cdot 0$.)
All other vectors created or updated in the algorithm are either taken to be a previously constructed vector, or (in the case of a W vector) taken as a vector of the form $U + \alpha V$, where α is an integer. It suffices to show if the vectors U and V both correspond to a valid equation with the above scheme, then so will the vector

$U + \alpha V$. Indeed, from the corresponding equations for U and V: $U(1) = a \cdot U(2) + b \cdot U(3)$, $V(1) = a \cdot V(2) + b \cdot V(3)$, if we add α times the second to the first, we get: $U(1) + \alpha V(1) = a \cdot [U(2) + \alpha V(2)] + b \cdot [U(3) + \alpha V(3)]$, which is the (valid) equation corresponding to the vector $U + \alpha V$. With this being done, it now suffices to show that the algorithm eventually terminates, and when it does, we have (the final value of) $U(1) = \gcd(a,b)$. As in the proof of Theorem 3.7, if we set $r_0 = b$ and $r_{-1} = a$, the Euclidean algorithm can be expressed as successive applications of the division algorithm, where each one defines the next element of the remainder sequence: $r_{i-1} = q_i r_i + r_{i+1}$ ($i = 0, 1, 2, \cdots, n$). Recall that the sequence of remainders is strictly decreasing and the final nonzero remainder (r_n) is $\gcd(a,b)$. If we look at the first component of the recursive formula of Algorithm 4.6, i.e., $W(1) = U(1) - \mathrm{floor}(U(1)/V(1)) \cdot V(1)$, we see that $W(1)$ is simply the remainder when the division algorithm is applied to the integer division of $U(1)$ by $V(1)$. Since $U(1)$ starts off at a, $V(1)$ starts off at b, and at each iteration, $U(1)$ is updated to $V(1)$ and $V(1)$ to (the new remainder) $W(1)$, we see that at the ith iteration of Algorithm 4.6, $W(1)$ is exactly the value of the new remainder in the ith iteration of the Euclidean algorithm. It follows that the values of $U(1)$ are strictly decreasing integers (so the algorithm terminates) whose last nonzero value is $\gcd(a,b)$, as claimed. $\square$

Solving Linear Congruences

We have completely described an efficient method for solving any linear congruence:

$$ax + b \equiv c \pmod{m}, \tag{4}$$

whenever $\gcd(a,m) = 1$, in which case there is always a unique solution. Since the first step of subtracting b from both sides (in modular arithmetic) is always easy, the heart of solving such a congruence is the (modular) division step, so we really can focus attention on the simpler equation (obtained by setting $b = 0$ in (4)):

$$ax \equiv c \pmod{m}. \tag{5}$$

We complete our analysis of (5) by moving to the remaining situation where $\gcd(a,m) = d > 1$. In order for a solution to exist, we must have $d \mid c$. (*Proof:* For any solution x, we would have $m \mid ax - c$, so since $d \mid m$, we get also that $d \mid ax - c$, and since $d \mid a$, it follows that $d \mid ax - (ax - c) = c$. $\square$) In case $d \mid c$, it turns out that the congruence (5) will always have d distinct solutions (mod m).

ALGORITHM 4.7: (*Procedure for solving* $ax \equiv c \pmod{m}$ *in case* $d = \gcd(a,m)$ > 1 *and* $d \mid c$): (Recall that if $d \nmid c$ there are no solutions.)

Step 1: Solve the modified congruence $(a/d)y \equiv (c/d) \pmod{m/d}$ as explained earlier in this section. This is possible, and there will be a unique solution y_0, since $\gcd(a/d, m/d) = 1$.

Step 2: The d solutions of the original congruence are $y_0, y_0 + m/d, y_0 + 2m/d,$ $\cdots, y_0 + (d-1)m/d \pmod{m}$.

Before we explain why this algorithm works, we give an example to illustrate its use.

EXAMPLE 4.11: Find all solutions of the following congruences:

(a) $2x \equiv 7 \pmod{10}$ (b) $6x \equiv 12 \pmod{21}$

SOLUTION: Part (a): Since $d = \gcd(2,10) = 2$ does not divide 7, there is no solution.

Part (b): Since $d = \gcd(6,21) = 3$ does divide 12, there will be 3 $(= d)$ distinct solutions (mod 21). We use Algorithm 4.7 to find them. The modified congruence from Step 1 is $(6/3)y \equiv (12/3) \pmod{21/3}$, or $2y \equiv 4 \pmod{7}$, which has the (unique) solution $y_0 = 2 \pmod{7}$.[16] Step 2 now gives us the set of two solutions of the original congruence: $\{2, 2 + 21/3, 2 + 2 \cdot 21/3\} = \{2, 9, 16\}$. These are easily checked to satisfy the original congruence.

EXERCISE FOR THE READER 4.9: Find all solutions of the following congruences:

(a) $123x \equiv 12 \pmod{456}$ (b) $15x + 4 \equiv 20 \pmod{25}$.

We now explain why Algorithm 4.7 does its job.

Proof that Algorithm 4.7 correctly finds all solutions of the indicated congruence: Since the d solutions indicated by the algorithm are distinct integers mod m, there are two things we need to do: (*i*) we must show that the d solutions indicated by the algorithm actually solve the original congruence, and (*ii*) there are no other solutions (mod m).

---

[16] Up to now, our method for solving the congruence $2y \equiv 4 \pmod{7}$ would be to first find that the inverse of 2 (mod 7) is 4 (since $2 \cdot 4 = 8 \equiv 1 \pmod 7$), and then multiply both sides of the congruence to obtain $y \equiv 4 \cdot 4 \equiv 16 \equiv 2 \pmod 7$. Whenever a congruence can be solved by ordinary integer arithmetic, the resulting solution will also be a valid one for the congruence. This is because if two real numbers are equal, then they will be congruent modulo any m. However, this method should not be applied directly to any congruence $ax \equiv b \pmod m$, where $d > 1$ and $d = \gcd(a,b) \mid c$, because it will only give one of the d solutions: For example, $5x \equiv 15 \pmod{25}$ has four solutions, but $x \equiv 3 \pmod{25}$ has only one!

Part (*i*): The fact that $(a/d)y_0 \equiv (c/d) \pmod{m/d}$ means that $(m/d) \mid [(a/d)y_0 - (c/d)]$. This divisibility relation implies that $m \mid [ay_0 - c]$, which is equivalent to the congruence $ay_0 \equiv c \pmod{m}$. Now, since $d \mid a$, for any integer i, we have $a(y_0 + im/d) \equiv ay_0 + (a/d)im \equiv c + 0 \equiv c \pmod{m}$, so $y_0 + im/d$ solves the indicated congruence. In particular, so do the d indicated solutions.

Part (*ii*): We first observe that there can be no other solutions of the original congruence of the form $y_0 + im/d$ $(i \in \mathbb{Z})$ $\pmod{m}$, other than the d solutions indicated by the algorithm. This is because for any integer i, if r is the remainder when i is divided by d, $y_0 + im/d \equiv y_0 + rm/d \pmod{m}$, so $y_0 + im/d$ is one of the solutions produced by the algorithm. It remains to show that there can be no solutions other than these of the original congruence. Indeed, suppose that there was a solution z_0, $az_0 \equiv c \pmod{m}$, which is not of this form. Therefore, there is a unique integer i_0, such that $y_0 + i_0 m/d < z_0 < y_0 + (i_0 + 1)m/d$. If we rewrite this double inequality as $i_0 m/d < z_0 - y_0 < (i_0 + 1)m/d$, it is clear that $z_0 \not\equiv y_0 \pmod{m/d}$. However, the argument in Part (i) shows that since $az_0 \equiv c \pmod{m}$, we have $(a/d)z_0 \equiv c/d \pmod{m/d}$, and this contradicts the fact that y_0 was the unique solution of this latter congruence. $\square$

We have thus completely described how to solve a single linear congruence of form (5) (or (4)). We summarize the procedure:

Summary of Procedure for Solving the Single Linear Congruence (4):
$$ax + b \equiv c \pmod{m}.$$

Step 1: Subtract b from both sides to obtain the equation $ax \equiv c - b \pmod{m}$.

Step 2: First compute $d = \gcd(a, m)$.

> **Case 1:** $d = 1$. (*Unique Solution*) Use the extended Euclidean Algorithm 4.6 to compute integers e and f such that $1 = ae + mf$, to obtain $a^{-1} \equiv e \pmod{m}$. The unique solution of (5) is given by $x \equiv a^{-1} \cdot (c - b) \pmod{m}$.

> **Case 2:** $d > 1$ and $d \nmid c$. (*No Solution*) There are no solutions of the congruence (5) $\pmod{m}$.

> **Case 3:** $d > 1$ and $d \mid c$. (*Multiple Solutions*) Use the Extended Euclidean Algorithm 4.6 to compute integers e' and f' such that $1 = (a/d)e' + (m/d)f'$, to obtain $(a/d)^{-1} \equiv e' \pmod{m/d}$. Use this to find the unique solution of the modified congruence $(a/d)y \equiv ([c - b]/d)$: $y_0 = (a/d)^{-1} \cdot ([c - b]/d) \pmod{m/d}$. The d

solutions of the original congruence are $y_0, y_0 + m/d, y_0 + 2m/d,$ $\cdots, y_0 + (d-1)m/d \pmod{m}$.

EXERCISE FOR THE READER 4.10:   Find all solutions of the following congruences:
(a) $6x + 2 \equiv 5 \pmod 9$.   (b) $6x + 2 \equiv 3 \pmod 9$.   (c) $5x + 2 \equiv 3 \pmod 9$.

The Chinese Remainder Theorem

In many applications, including some in cryptography that we will see in Section 4.5, it is necessary to simultaneously solve a system of linear congruences of different moduli:[17]

$$\begin{cases} a_1 x \equiv c_1 \pmod{m_1} \\ a_2 x \equiv c_2 \pmod{m_2} \\ \;\vdots \\ a_k x \equiv c_k \pmod{m_k}. \end{cases} \tag{6}$$

More precisely, we would like to know when we find an integer x, that solves each of the congruences in (6).   Furthermore, in cases where such a simultaneous integer solution exists, we would like to classify all of the solutions.

Puzzles that can be modeled by simultaneous congruences such as (6) have appeared in various ancient mathematical documents, including those from the Greeks (dating back to the first century AD), the Chinese (dating to back to the third century AD), and to the Hindus (dating back to the seventh century AD). The following is an example of such an ancient Hindu puzzle.

EXAMPLE 4.12: Determine a system of simultaneous congruences that models the following puzzle:

> (*A Hindu puzzle from the Seventh Century AD*) While a woman is on her way to the market, a horse steps on her basket and crushes all her eggs. The rider agrees to pay for the damage and asks how many eggs she had. She does not recall the exact number, but she knows that when she had taken them out two at a time, there was one egg left. The same thing happened when she removed them three, four, five, and six at a time, but when she took them out seven at a time, they all came out. What was the smallest number of eggs she could have had?

---

[17] Since the reduction from a more general linear congruence $ax + b \equiv c \pmod{m}$ is simple, it suffices to assume that our linear congruences are in the form $ax \equiv c \pmod{m}$.

SOLUTION: Letting x denote the (unknown) number of eggs that were in the woman's basket, the problem tells us that x must solve each of the following congruences:

$$\begin{cases} x \equiv 1 \ (\text{mod } 2) \\ x \equiv 1 \ (\text{mod } 3) \\ x \equiv 1 \ (\text{mod } 4) \\ x \equiv 1 \ (\text{mod } 5) \\ x \equiv 1 \ (\text{mod } 6) \\ x \equiv 0 \ (\text{mod } 7) \end{cases} \tag{7}$$

The problem seeks the smallest positive solution of the system (7). We point out one very simple observation about congruences that will sometimes help to simplify such systems.

PROPOSITION 4.3: Suppose that m_1, m_2 are positive integers with $m_1 \mid m_2$. Any solution of a linear congruence $ax \equiv c \ (\text{mod } m_2)$ will also be a solution of the same congruence $(\text{mod } m_1)$.

Proof: By definition, x solves the first congruence means that $m_2 \mid (ax - c)$. Since we are assuming that $m_1 \mid m_2$, it follows by transitivity of divisibility (Theorem 3.10(a)) that $m_1 \mid (ax - c)$, which means that $ax \equiv c \ (\text{mod } m_1)$. $\square$

If we apply this proposition to the system (7), since both 2 and 3 divide 6, the congruences $x \equiv 1 \ (\text{mod } 2)$, $x \equiv 1 \ (\text{mod } 3)$ are redundant consequences of the congruence $x \equiv 1 \ (\text{mod } 6)$. Thus, they can be safely removed from the system to produce the following simpler, but equivalent system:[18]

$$\begin{cases} x \equiv 1 \ (\text{mod } 4) \\ x \equiv 1 \ (\text{mod } 5) \\ x \equiv 1 \ (\text{mod } 6) \\ x \equiv 0 \ (\text{mod } 7) \end{cases} \tag{8}$$

We will return to the system (8) and the Hindu puzzle momentarily, but we first consider the problem of solving the general system (6). First of all, it is clear that in order for a simultaneous solution to exist, each individual congruence must have a solution, and from our development for single linear congruences, this means that we must have $d_i \mid c_i \ (1 \le i \le k)$, where $d_i = \gcd(a_i, m_i)$. With these

---

[18] An *equivalent* system of equations is one that has the same solution set as the original system.

conditions being satisfied, in light of Algorithm 4.7, (6) can be reduced to the simpler system:

$$\begin{cases} x \equiv b_1 \pmod{n_1} \\ x \equiv b_2 \pmod{n_2} \\ \vdots \\ x \equiv b_k \pmod{n_k}, \end{cases} \tag{9}$$

where $n_i = m_i / d_i$ and $b_i = (a_i / d_i)^{-1}(c_i / d_i) \pmod{n_i}$. The following theorem shows that (9) always has a solution in case the moduli are *pairwise relatively prime*, i.e., $\gcd(n_i, n_j) = 1$, whenever $i \neq j$ $(1 \leq i, j \leq k)$, and it includes a uniqueness statement. This theorem has been found to date back to a Chinese mathematics book that was published in 1247 AD, by the Chinese mathematician Qin Jiushao (1202–1261),[19] and has come to be known as the *Chinese remainder theorem*. We will give a constructive (algorithmic) proof of the existence of the solution, and thus provide a practical method of solving the system (9) (and hence also the system (6)).

THEOREM 4.4: (*The Chinese Remainder Theorem*) Suppose that $n_1, n_2, \cdots, n_k > 1$ are pairwise relatively prime integers. Then for any integers $b_1, b_2, \cdots, b_k$, the system of congruences:

$$\begin{cases} x \equiv b_1 \pmod{n_1} \\ x \equiv b_2 \pmod{n_2} \\ \vdots \\ x \equiv b_k \pmod{n_k}, \end{cases} \tag{9}$$

has a simultaneous integer solution x that is unique modulo $N = n_1 n_2 \cdots n_k$.

As the following proof contains an algorithm, we will defer giving an example until after proving the theorem.

---

[19] Qin Jiushao (also transliterated as Ch'in Chiu-Shao) was a strong and influential ancient mathematician whose principle scientific contributions were published in his 1247 book *Shushu Jiuzhang* (*Mathematical Treatise in Nine Sections*). The first chapter of his book contained the development and proof of (what is now called) the Chinese remainder theorem. The book also included analyses of higher order equations that modeled certain interesting applied problems such as the following (from Chapter 2 of his book): Determine the height of rainfall on level ground, given that it reached a height h in a cylindrical vessel with circular top and bottom having respective radii $a > b$. Aside from his mathematics, Qin had quite an interesting military and government career. In his youth, he fought the armies of Ghengis Khan. He was notorious for his corruption and manipulations of his government posts, from which he amassed a tremendous amount of wealth. His book was actually written during a hiatus from work when he returned to his hometown to mourn the death of his mother.

Proof: Part (a): *Existence*: (*Constructive Proof*) For each index i $(1 \le i \le k)$, since $\gcd(N / n_i, n_i) = 1$, there is a (unique) solution e_i of the congruence $e_i(N / n_i) \equiv 1 \pmod{n_i}$. We claim that

$$x = \sum_{i=1}^{k} b_i e_i (N / n_i) \tag{10}$$

is a simultaneous solution of (9). Indeed, for each index j, we have $\pmod{n_j}$ $N / n_i \equiv 0$, unless $j = i$, and this forces all terms other than the jth term in the sum of (10) to be zero $\pmod{n_j}$. Thus $x \equiv b_j e_j (N / n_j) \equiv b_j \cdot 1 \equiv b_j \pmod{n_j}$, as desired.

Part (b): *Uniqueness*: Suppose that x' is another simultaneous solution of (9). It follows that for each index i, $x \equiv b_i \equiv x' \pmod{n_i}$, so that $n_i \mid (x - x')$. Since the n_i's are pairwise relatively prime, it follows (from the fundamental theorem of arithmetic) that their product N also divides $x - x'$, i.e., $x \equiv x' \pmod N$. $\square$

EXAMPLE 4.13: Solve the following system of congruences:

$$\begin{cases} x \equiv 2 \pmod 3 \\ x \equiv 3 \pmod 5 \\ x \equiv 6 \pmod{14}. \end{cases}$$

SOLUTION: Since the moduli are pairwise relatively prime, (10) (in the proof of the Chinese remainder theorem) provides us with a scheme for obtaining a simultaneous solution. We first set $N = 3 \cdot 5 \cdot 14 = 210$. With $b_1, b_2, b_3 = 2, 3, 6$, and $n_1, n_2, n_3 = 3, 5, 14$, in order to use equation (10), we must first determine e_1, e_2, e_3, by their defining equations: $e_i(N / n_i) \equiv 1 \pmod{n_i}$.

For e_1: $e_1 \cdot 70 \equiv 1 \pmod 3 \Leftrightarrow e_1 \cdot 1 \equiv 1 \pmod 3 \Leftrightarrow e_1 \equiv 1 \pmod 3$.

For e_2: $e_2 \cdot 42 \equiv 1 \pmod 5 \Leftrightarrow e_2 \cdot 2 \equiv 1 \pmod 5 \Leftrightarrow e_2 \equiv 3 \pmod 5$. (Since $2^{-1} = 3$ $\pmod 5$.)

For e_3: $e_3 \cdot 15 \equiv 1 \pmod{14} \Leftrightarrow e_3 \cdot 1 \equiv 1 \pmod{14} \Leftrightarrow e_3 \equiv 1 \pmod{14}$.

Now we have all that we need to apply (10) to get a desired solution:

$$x = \sum_{i=1}^{3} b_i e_i (N / n_i) = 2 \cdot 1 \cdot (70) + 3 \cdot 3 \cdot (42) + 6 \cdot 1 \cdot (15) = 608 \equiv 188 \pmod{210}.$$

Thus 188 is the smallest positive integer solution of the original system.

EXERCISE FOR THE READER 4.11: Determine the general solution of the following system of congruences:

$$\begin{cases} x \equiv 0(\text{mod } 2) \\ x \equiv 2(\text{mod } 5) \\ 3x \equiv 4(\text{mod } 7). \end{cases}$$

Although we cannot apply the Chinese remainder theorem to the system (8) of the Hindu problem, Part (b) of the following proposition, which generalizes Proposition 4.3, will allow us to convert the system (8) into a form to which the theorem is applicable.

PROPOSITION 4.5: (a) Any set of divisibility relations of the form: $m_1 \mid b$, $m_2 \mid b, \cdots, m_k \mid b$ is equivalent to the single divisibility relation $\text{lcm}(m_1, m_2, \cdots, m_k) \mid b$.

(b) Any system of congruences of the form

$$\begin{cases} ax \equiv c(\text{mod } m_1) \\ ax \equiv c(\text{mod } m_2) \\ \vdots \\ ax \equiv c(\text{mod } m_k), \end{cases}$$

(i.e., the same congruence under different moduli) is equivalent to the single congruence:

$$ax \equiv c \ (\text{mod } \text{lcm}(m_1, m_2, \cdots, m_k)).$$

NOTE: Unlike in the Chinese remainder theorem, the moduli need not be pairwise relatively prime in this result.

The proof is similar to that of Proposition 4.3, and is left to the following exercise for the reader.

EXERCISE FOR THE READER 4.12: Prove Proposition 4.5.

EXERCISE FOR THE READER 4.13: What is the answer to the ancient Hindu problem of Example 4.12?

The proof that this algorithm works follows simply from writing out the binary expansion of the exponent and using the laws of exponents:

$$b^x \equiv b^{\sum_{k=0}^{K} d_k 2^k} \equiv \prod_{k=0}^{K} b^{d_k 2^k} \ (\text{mod } m),$$

and since $b^{2^{k+1}} = (b^{2^k})^2$. All of the successive squares need to be computed (mod m), but only those for which the corresponding $d_k = 1$ make an appearance in the above (cumulative) expression for b^x. The reader is encouraged to re-do Example 3.17, in the context of the above algorithm (and thus reconcile the previewed

Method 1 of that example's solution with the present algorithm). Algorithm 4.5 is considered a very fast and efficient scheme for modular exponentiation, just requiring an amount of work roughly equal to performing K successive squarings mod m, where K is the number of binary digits in the exponent x.

Pseudo-Random Numbers: The Linear Congruential Method

Congruences have numerous applications, and we will encounter several of these later in this book as well as in the exercise set at the end of this section. We end this section by describing a method that uses congruences to generate *pseudo-random numbers*. *Random numbers* are needed in many computer applications where simulating random outcomes is required (such applications will be developed in the area of simulations in Chapter 6 and in the area of stochastic algorithms of artificial intelligence in Chapter 10). Since computers need algorithms to perform tasks, it is theoretically not possible to produce completely random numbers by a computer program (such true randomness can occur only in nature). Nevertheless, successful algorithms have been developed to generate numbers that mimic random numbers for most all practical purposes (and such sequences of numbers have been statistically verified to be essentially random numbers). Such computer-generated random numbers are called **pseudo-random numbers**. Linear congruences provide a simple and commonly used method, the so-called *linear congruential method*, to generate pseudo-random numbers. This method is explained in the following definition.

DEFINITION 4.4: The **linear congruential method** for generating pseudo-random numbers depends on the following four integer parameters: a **modulus** m > 1, a **multiplier** a, $1 < a < m$, an **increment** c, $0 \le c < m$, and the **seed** x_0, $0 \le x_0 < m$. The following recursive congruence formula is then used to generate the corresponding sequence $\{x_n\}$ of integers modulo m:

$$x_{n+1} \equiv ax_n + c \ (\text{mod } m). \tag{11}$$

The parameters should be chosen so that the values are readily computed (using a computer), and for any initial seed, a large number of sequence elements can be generated before repetition begins. For this to work well, the modulus should be a large prime number. One such choice that is used in several computing platforms is: $m = 2^{31} - 1$, $a = 7^5$, and $c = 0$. It can be shown that with any seed, the sequence will go through $m - 1$ numbers before repetition begins.

We illustrate the method with a relatively small example. With the parameters m = 17389, $a = 505$, $c = 12$, and $x_0 = 1237$, the first four elements of $\{x_n\}$ are:

$x_1 = 505x_0 + 123 = 624808 \equiv 16193 \ (\text{mod } 17389),$

$x_2 = 505x_1 + 123 = 8177588 \equiv 4758 \pmod{17389},$

$x_3 = 505x_2 + 123 = 2402913 \equiv 3231 \pmod{17389},$

$x_4 = 505x_3 + 123 = 1631778 \equiv 14601 \pmod{17389}.$

The linear congruential method is used by many computing platforms in their random number generators. Computer random number generators often produce random "real" numbers in the interval [0, 1]. The way this works with the above method is (using a large value for the modulus *m*) to convert the outputs x_n to numbers between 0 and 1, by simply dividing each by *m*.

EXERCISES 4.2:

1.   Determine whether each of the statements below is true or false.

  (a) $43 \equiv 1 \pmod 2$ (b) $-43 \equiv 3 \pmod 4$ (c) $488 \equiv 10 \pmod{12}$

  (d) $2205 \equiv 45 \pmod{360}$ (e) $-443 \equiv 18 \pmod{22}$ (f) $7^7 \equiv 5^8 \pmod{371}$

2.   Determine whether each of the statements below is true or false.

  (a) $43 \equiv 2 \pmod 3$ (b) $-43 \equiv 3 \pmod 6$ (c) $2207 \equiv 11 \pmod{12}$

  (d) $11340 \equiv 90 \pmod{360}$ (e) $-2444 \equiv -446 \pmod{666}$ (f) $3^6 \equiv 5^2 \pmod{44}$

3.   Perform the following operations in $\mathbb{Z}_{24}$:

  (a) $18 + 20$ (b) $5 - 21$ (c) $8 \cdot 8$ (d) $2^8 - 3^8$ (e) 21^{223}

4.   Perform the following operations in $\mathbb{Z}_{53}$:

  (a) $39 + 47$ (b) $25 - 36$ (c) $18 \cdot 35$ (d) $12^5 - 19^4$ (e) 33^{100}

5.   Create addition and multiplication tables for (a) $\mathbb{Z}_2$ and (b) $\mathbb{Z}_4$. (See Example 4.11.)   By looking at the multiplication tables, determine all invertible elements.

6.   Create addition and multiplication tables for (a) $\mathbb{Z}_3$ and (b) $\mathbb{Z}_9$. (See Example 4.11.)   By looking at the multiplication tables, determine all invertible elements.

7.   Find all invertible elements of $\mathbb{Z}_8$, and for each one find the inverse.

8.   Find all invertible elements of $\mathbb{Z}_{10}$, and for each one find the inverse.

9.   Find all invertible elements of $\mathbb{Z}_7$, and for each one find the inverse.

10.   Find all invertible elements of $\mathbb{Z}_{11}$, and for each one find the inverse.

11. Use the fast modular exponentiation algorithm (Algorithm 4.5) to perform the following modular exponentiations:

    (a) 2^{58} (mod 5) (b) 7^{394} (mod 17)

    (c) 24^{1422} (mod 29) (d) 177^{998} (mod 223)

12. Use the fast modular exponentiation algorithm (Algorithm 4.5) to perform the following modular exponentiations:

    (a) 3^{97} (mod 5) (b) 12^{117} (mod 17)

    (c) 28^{213} (mod 43) (d) 275^{884} (mod 307)

13. Solve each of the following congruences working in mod 8.

    (a) $3x \equiv 5$ (b) $7x + 2 \equiv 3$ (c) $5x - 2 \equiv 2$

14. Solve each of the following congruences working in mod 10.

    (a) $3x \equiv 5$ (b) $7x + 2 \equiv 3$ (c) $9x - 8 \equiv 7$

15. For each integer below, use the extended Euclidean algorithm (Algorithm 4.6) together with Proposition 4.2 to find the inverse in $\mathbb{Z}_{388}$, if the inverse exists.

    (a) 3 (b) 55 (c) 149 (d) 97

16. For each integer below, use the extended Euclidean algorithm (Algorithm 4.6) together with Proposition 4.2 to find the inverse in $\mathbb{Z}_{299}$, if the inverse exists.

    (a) 2 (b) 52 (c) 80 (d) 199

17. For each integer below, use the extended Euclidean algorithm (Algorithm 4.6) together with Proposition 4.2 to find the inverse in $\mathbb{Z}_{1353}$, if the inverse exists.

    (a) 2 (b) 44 (c) 886 (d) 350

18. For each integer below, use the extended Euclidean algorithm (Algorithm 4.6) together with Proposition 4.2 to find the inverse in $\mathbb{Z}_{2555}$, if the inverse exists.

    (a) 2 (b) 74 (c) 98 (d) 1972

19. Find all solutions for each of the following congruences:

    (a) $3x \equiv 59 \pmod{388}$ (b) $149x \equiv 225 \pmod{388}$

    (c) $2x \equiv 1225 \pmod{1353}$ (d) $886x \equiv 35 \pmod{1353}$

20. Find all solutions for each of the following congruences:

    (a) $2x \equiv 59 \pmod{299}$ (b) $199x \equiv 99 \pmod{299}$

    (c) $2x \equiv 847 \pmod{2555}$ (d) $1972x \equiv 363 \pmod{2555}$

21. Find all solutions for each of the following congruences:

    (a) $3x \equiv 6 \pmod{18}$ (b) $15x \equiv 21 \pmod{51}$

    (c) $8x \equiv 12 \pmod{28}$ (d) $8x \equiv 6 \pmod{28}$

22. Find all solutions for each of the following congruences:

    (a) $2x \equiv 6 \pmod{16}$ (b) $6x \equiv 16 \pmod{27}$

    (c) $14x \equiv 21 \pmod{88}$ (d) $25x \equiv 55 \pmod{95}$

23. Find all solutions for each of the following congruences:

   (a) $6x \equiv 28 \pmod{776}$ (b) $15x \equiv 21 \pmod{1940}$

   (c) $596x \equiv 900 \pmod{1552}$ (d) $3544x \equiv 900 \pmod{5412}$

24. Find all solutions for each of the following congruences:

   (a) $8x \equiv 16 \pmod{1196}$ (b) $400x \equiv 125 \pmod{1495}$

   (c) $1393x \equiv 175 \pmod{2093}$ (d) $17748x \equiv 6642 \pmod{22995}$

25. Find all solutions for each of the following systems of congruences:

   (a) $\begin{cases} x \equiv 3 \pmod 5 \\ x \equiv 4 \pmod 7 \end{cases}$ (b) $\begin{cases} x \equiv 2 \pmod 3 \\ x \equiv 1 \pmod 5 \\ x \equiv 3 \pmod{11} \end{cases}$ (c) $\begin{cases} x \equiv 2 \pmod 6 \\ x \equiv 1 \pmod 5 \\ x \equiv 3 \pmod 7 \\ x \equiv 1 \pmod{13} \end{cases}$

26. Find all solutions for each of the following systems of congruences:

   (a) $\begin{cases} x \equiv 3 \pmod 4 \\ x \equiv 4 \pmod 5 \end{cases}$ (b) $\begin{cases} x \equiv 0 \pmod 2 \\ x \equiv 1 \pmod 5 \\ x \equiv 6 \pmod 9 \end{cases}$ (c) $\begin{cases} x \equiv 1 \pmod 4 \\ x \equiv 2 \pmod 5 \\ x \equiv 3 \pmod 9 \\ x \equiv 1 \pmod{11} \end{cases}$

27. A group of 15 pirates has just looted a stash of identical and very valuable gold coins. They plan to equally divide them the next morning. During the night, one pirate, who does not trust the others, gets up to divide the coins into 15 equal parts, finds there are 8 remaining, so takes these 8 with him. But he was unknowingly followed by another pirate who then kills him. This second pirate then divides the remaining coins by 14, finds there are 11 left, and stashes these 11 (plus the 8 he got from the first pirate) under the hull of the boat, but in the process falls overboard and drowns. The next morning the remaining 13 pirates divide the coins by 13 and find there are 5 left. What is the smallest number of coins that could have originally been present?

28. Suppose that a certain computer server has less than 1 GB (1 billion bytes) of memory, and any time it runs jobs, it allocates an equal number of bytes of memory to each job, and leaves any remaining bytes unused. Suppose that when it runs 95 jobs, it has 86 unused bytes, when it runs 98 jobs, it has 13 unused bytes, when it runs 99 jobs, it has 46 unused bytes, and when it runs 101 jobs, all bytes get allocated. Determine the exact size of the computer's memory.

NOTE: Exercises 29–31 give some applications of congruences to coding theory. **Coding theory** is an area of applied mathematics involved with the efficient transportation of information that is prone to transmission errors (for example, either through incorrectly entering the data on the senders end, or through a noisy channel). The basic idea is to transmit some additional (redundant) information that can be used to detect and sometimes even correct errors. The exercises below deal with error detecting codes in ISBN (*International Standard Book Number*) numbers, and in credit card numbers. Error correcting codes have built-in mechanisms that can actually correct errors. An example is with audio CDs, which can play fine with scratches on the playing surface, or even if one were to drill a 2.5 mm hole through the playing surface. Coding theory also makes use of other areas of mathematics such as linear algebra and abstract algebra. For a nice general introduction, see, for example [LiXi-04]. A more comprehensive treatment in error correcting codes can be found in [Moo-05].

29. (*Application of Congruences: ISBN Error Detecting Codes*) Since the 1970s, to facilitate inventory control and the ordering/selling of books, most all books published have an attached unique ISBN number. For over 30 years the same system had been in widespread use, and as of 2007, the 10 digit ISBNs were replaced by 13 digit ISBNs. This exercise will discuss 13 digit ISBNs, and the next one will look at the previously used 10 digit system. To distinguish, we refer to the systems as ISBN-13 and ISBN-10. An ISBN-13 consists of five blocks of digits.

For example, the ISBN-13 of the author's book, *Introduction to Numerical Ordinary and Partial Differential Equations Using MATLAB*, is 978-0-471-69738-1. The first block always consists of three digits. Most books presently in print use 978 in this field (for US ISBN Agency). The second group is a single digit encoding the country or language of the publisher (0 indicates English), the third group of digits may range from two to seven digits and indicates the publisher (471 indicates John Wiley & Sons), the fourth block of digits has length eight less the number in the preceding field, and indicates the publisher's assigned number for the particular book. (Thus, larger publishers will be assigned smaller publisher codes to allow for larger capacities in the book field (up to 6 digits, or 1 million books). The fifth and final group is a single **check digit** from 0 to 9. If the digits (in order) of an ISBN-13 number are $x_1 x_2 \cdots x_{13}$, then the check digit x_{13} is determined by the equation

$$x_{13} \equiv 10 - (x_1 + 3x_2 + x_3 + 3x_4 + \cdots + x_{11} + 3x_{12}) \pmod{10}. \qquad (12)$$

For example, the right-hand side of (12) for the above-mentioned ISBN-13 number works out to be $10 - (9 + 3 \cdot 7 + 8 + 3 \cdot 0 + 4 + 3 \cdot 7 + 1 + 3 \cdot 6 + 9 + 3 \cdot 7 + 3 + 3 \cdot 8) = -129 \equiv 1 \pmod{10}$. This indeed coincides with the last check digit $x_{13} = 1$. This check system (12) was designed to detect any single error of the following most common ones occurring in typing an ISBN number: mistyping one of the digits, or switching two adjacent digits. This is important, since with a single error, the resulting incorrect ISBN could correspond to a totally different book and would otherwise go unnoticed.

(a) Each of the following is the first 12 digits of an ISBN-13 number. Find the 13th check digit for each: 978055215169?, 978082482223?, 978006123400.

(b) Show that if a valid ISBN-13 number has exactly one mistyped digit, then equation (12) will fail to hold, i.e., the error was detected.

(c) Show that if a valid ISBN-13 number has two different adjacent digits that were typed in the wrong order, then equation (12) may fail to detect this error.

(d) Give an example of an incorrectly typed ISBN-13 number, along with two corresponding valid ISBN-13 numbers that differ from the former in exactly one digit. This shows that although the ISBN-13 system can detect common errors, it cannot correct them.

Suggestion: For Part (b): If $x_1 x_2 \cdots x_{13}$ was incorrectly typed as $y_1 y_2 \cdots y_{13}$, where each $y_i = x_i$, with a single exception $y_j \neq x_j$, assume that both ISBNs checked with (12). Then, by subtracting the corresponding sides of the two equations, we would be left with either $y_j \equiv x_j$, or $3y_j \equiv 3x_j \pmod{10}$. But since 3 is invertible (mod 10), we could multiply both sides of the latter equation by the inverse to obtain $y_j \equiv x_j \pmod{10}$, which forces $y_j = x_j$ — a contradiction! A similar argument works for Part (c).

30. (*Application of Congruences: ISBN Error Detecting Codes*) The reader should first read Exercise 43 for some general background. From 1970 to 2007, the ISBN-10 system was used for published books. Books published prior to 2007 needed to have their ISBN-10 numbers converted to ISBN-13. For example, the ISBN-10 number of the book, *Introduction to Numerical Ordinary and Partial Differential Equations Using MATLAB*, is 0-471-69738-9. The first three blocks correspond to the second through fourth blocks of the ISBN-13 numbers. The fourth and final block is a single **check digit** that is either a digit from 0 to 9, or the letter X (corresponding to 10). If the digits (in order) of an ISBN-10 number are $x_1 x_2 \cdots x_{10}$, then the check digit x_{10} is determined by the equation

$$x_{10} \equiv x_1 + 2x_2 + 3x_3 + 4x_4 + \cdots + 9x_9 \pmod{11}. \qquad (13)$$

For example, the right-hand side of (13) for the above-mentioned ISBN-10 number works out to be $0 + 2 \cdot 4 + 3 \cdot 7 + 4 \cdot 1 + 5 \cdot 6 + 6 \cdot 9 + 7 \cdot 7 + 8 \cdot 3 + 9 \cdot 8 = 262 \equiv 9 \pmod{11}$. This indeed checks with the last check digit $x_{10} = 9$. This check system (13) was designed to detect any single error of the following two types: mistyping one of the digits, or switching <u>any</u> two unequal digits. Thus, the ISBN-10 system can detect more general errors than the ISBN-13 system.

(a) Each of the following is the first 9 digits of an ISBN-10 number. Find the 10th check digit for each: 951020387?, 082482223?, 013014400?

(b) Show that if a valid ISBN-10 number has exactly one mistyped digit, then equation (13) will fail to hold, i.e., the error was detected.

(c) Show that if a valid ISBN-10 number has two different digits that were switched, then equation (13) will fail to hold, i.e., the error was detected.

(d) Give an example of an incorrectly typed ISBN-10 number along with two corresponding valid ISBN-10 numbers that differ from the former in exactly one digit. This shows that although the ISBN-10 system can detect common errors, it cannot correct them.

Suggestion: See the suggestion for Exercise 29 for ideas for Parts (b) and (c).

Note: Comparing the error detecting capabilities of ISBN-10 versus ISBN-13, we see that although the latter system has a larger capacity, the former is better at detecting permuation errors (c.f., Part (c) of this and the preceding exercise).

31. (*Application of Congruences: Credit Card Error Detecting Codes*) Credit cards use similar coding systems that include identifying information as well as a check digit. This is how web sites can often immediately inform you if the number you keyed in is not a valid credit card number. In this exercise we will explore the system that VISA cards use. VISA card numbers either contain 13 or 16 digits, and the first digit is always 4, indicating it is a VISA card (Master Cards always begin with 5). The 2nd through the 6th digits identify the bank (that issued the VISA card), and the 7th through the second-to-last digit give the account number. The final digit is the check digit. If the digits (in order) of a 16 digit VISA card number are $x_1 x_2 \cdots x_{16}$, then the check digit x_{16} is determined by the equation

$$x_{16} \equiv -[2x_1 + x_2 + 2x_3 + x_4 + 2x_5 \cdots + x_{14} + 2x_{15}] - r \pmod{10}, \qquad (14)$$

where r is the number of terms in the bracketed expression that are greater than or equal to 10. For example, in the VISA card number 4784 5580 0246 1888, the bracket expression in (14) is $[2 \cdot 4 + 7 + 2 \cdot 8 + 4 + 2 \cdot 5 + 5 + 2 \cdot 8 + 0 + 2 \cdot 0 + 2 + 2 \cdot 4 + 6 + 2 \cdot 1 + 8 + 2 \cdot 8] = [8 + 7 + 16 + 4 + 10 + 5 + 16 + 0 + 0 + 2 + 8 + 6 + 2 + 8 + 16]$ has $r = 4$ two digit terms and equals 108, thus, the right-hand side of (14) equals $-108 - 4 \equiv 8 \pmod{10}$, which coincides (as it should) with the (last) check digit x_{16}.

(a) Which of the following are valid 16 digit VISA card numbers? For those that are not, explain why: 4238 1678 1139 5207, 5602 8333 5495 1777, 4671 8899 3663 1942.

(b) Show that if a valid VISA number has exactly one mistyped digit, then equation (14) will fail to hold, i.e., the error was detected.

(c) Show that if a valid VISA number has two different adjacent digits that were typed in the wrong order, then equation (14) will fail to hold, i.e., the error was detected. Can an error still be detected if the digits are not adjacent?

(d) Give an example of an incorrectly typed VISA card number along with two corresponding valid VISA card numbers that differ from the former in exactly one digit. This shows that although the VISA card system can detect common errors, it cannot correct them.

(e) Suppose that a 16 digit VISA card number was correctly sent in but one of the digits printed out illegibly. Is it always possible to recover the missing digit? Explain your answer.

Suggestion: See the suggestion for Exercise 29 for ideas for Parts (b) and (c).

32. (*Application of Congruences: Divisibility Criteria*) (a) Prove that for any positive integer n, $3 | n$ if, and only if 3 divides the sum of the digits of n. Equivalently, if we write $n = \sum_{k=0}^{D} d_k \cdot 10^k$, where $0 \le d_k \le 9$ (the digits), $3 | n \Leftrightarrow n = \sum_{k=0}^{D} d_k$.

(b) With the notation of Part (a), prove that $4 | n \Leftrightarrow 4 | (d_0 + 10 d_1)$.

Suggestion: For Part (a), show that $n \equiv \sum_{k=0}^{D} d_k \pmod 3$. For Part (b) use the fact that $4 | 10^2$.

33. (*Application of Congruences: Divisibility Criteria*) (a) Prove that for any positive integer $n = \sum_{k=0}^{D} d_k \cdot 10^k$, $11 \mid n$ if, and only if $11 \mid (d_0 + d_2 + d_4 + \cdots - d_1 - d_3 - d_5 - \cdots)$. For example, $11 \mid 930391$, since $11 \mid (9 + 0 + 9 - 3 - 3 - 1)$.
    (b) With the notation of Part (a), prove that
    $$7 \mid n \iff 7 \mid (d_0 + 10 d_1 + 100 d_2 - d_3 - 10 d_4 - 100 d_5 + d_6 + 10 d_7 + 100 d_8 - \cdots).$$
    For example, $7 \mid 4001006002$, since $7 \mid (4 - 1 + 6 - 2) = (4 + 10 \cdot 0 + 100 \cdot 0 - 1 - 10 \cdot 0 - 100 \cdot 0 + 6 + 10 \cdot 0 + 100 \cdot 0 - 2)$.

34. Prove that for any odd modulus $m > 2$, we have $\sum_{k=1}^{m-1} k \equiv 0 \pmod{m}$.

35. Prove that for any even modulus $m > 1$, we have $\sum_{k=1}^{m-1} k \equiv m/2 \pmod{m}$.

36. In the notation of our (constructive) proof of the Chinese remainder theorem (Theorem 4.4) prove that $\sum_{i=1}^{k-1} e_i (N/n_i) \equiv 1 \pmod{N}$.

37. Suppose that a is an integer and n and m are relatively prime integers, both greater than one.
    (a) If $x \equiv a \pmod{m}$ and $x \equiv a \pmod{n}$, show that $x \equiv a \pmod{mn}$.
    (b) Does the result of Part (a) remain valid without the assumption that m and n are relatively prime? Either prove that it is or provide a counterexample.

NOTE: An integer a is said to have a **square root modulo** m ($m > 1$ an integer) if the equation $x^2 \equiv a \pmod{m}$ has at least one solution. Any solution is called a **square root of a mod m**. Exercises 38–41 will explore some situations where it can be determined if a certain number a has a square root modulo a certain m, and also the problem of determining how many square roots a has, once it is known to have at least one.

38. (*Square Roots Modulo a Prime*) Let p be an odd prime (positive) integer.
    (a) Prove that the integers that have square roots mod p are precisely those in the set $\{0^2, 1^2, 2^2, \cdots, [(p-1)/2]^2\} \pmod{p}$.
    (b) Show that the elements listed in the set of Part (a) are all different mod p.
    (c) Show that if $p > 2$, then each of the $(p-1)/2$ <u>nonzero</u> elements listed in the set of Part (a), has exactly two (distinct) square roots mod p.
    (d) Find all the numbers in $\mathbb{Z}_{11}$ that have square roots, and for each one, find all of its square roots (mod 11).
    (e) Repeat the instructions of Part (d) for $\mathbb{Z}_{13}$.
    (f) What happens to the results of Parts (a), (b), (c) in case $p = 2$?
    Suggestion: For Part (a), if $z > (p-1)/2$, show that $w = p - z$ is $\leq (p-1)/2$, and $z^2 \equiv w^2 \pmod{p}$. For Part (b), suppose that $0 \leq w < z \leq (p-1)/2$, but that $w^2 \equiv z^2 \pmod{p}$. This means that $p \mid z^2 - w^2 = (z + w)(z - w)$. Use Euclid's lemma (Lemma 3.17) to obtain a contradiction. For Part (c), if z is a square root, then so is $-z (\equiv p - z) \pmod{p}$.

39. (*Square Roots Modulo a Product of Distinct Primes*) Let $p < q$ both be odd primes.
    (a) Show that the equation $x^2 \equiv a \pmod{pq}$ is equivalent to the system $\begin{cases} x^2 \equiv a \pmod{p} \\ x^2 \equiv a \pmod{q} \end{cases}$.
    Indicate a scheme for finding all square roots of an integer a mod pq.
    (b) Using Exercise 38 and the result of Part (a), obtain a result for the existence of square roots modulo a product of distinct odd primes.

(c) Find (i) all (if any) square roots of 9 mod 35, and then (ii) compute $\sqrt{51}$ (mod 493).

(d) How would the result found in Parts (a) and (b) change in case $p = 2$?

(e) Find the following square roots (if they exist): (i) $\sqrt{11}$ (mod 26), and (ii) $\sqrt{68}$ (mod 86).

Suggestion: Make use of the Chinese remainder theorem.

40. (*Square Roots Modulo a Prime-Efficient Algorithms*) Exercise 38 completely explained the existence and number of square roots that an integer can have modulo a prime, however, the resulting method for extracting square roots was not very much faster than a brute-force search. The following proposition gives a fast method for finding all square roots of any integer modulo a prime p in the case that $p \equiv 3 \pmod{4}$.

Proposition 4.6: Assume that p is a prime number that is congruent to 3 (mod 4), and let $x \neq 0$ be any integer (mod p). Then either x or $-x$, but not both, will have two square roots (mod p), and these square roots are given by $\pm w$, where $w = x^{(p+1)/4} \pmod{p}$.

For a proof as well as another theorem covering the remaining case where $p \equiv 1 \pmod{4}$, we refer to [Coh-93]. Here we look only at the practical applications of Proposition 4.6.[20]
Use Proposition 4.6 to determine all square roots of the following:

   (i) $\sqrt{7}$ (mod 59), (ii) $\sqrt{142}$ (mod 607), (iii) $\sqrt{10}$ (mod 2143).

41. (*More Square Roots Modulo a Product of Distinct Primes*) Let $p \neq q$ both be prime (positive) integers. The technique for extracting square roots mod pq given in Exercise 61 can be speeded up if one of the two primes is congruent to 3 (mod 4), since Proposition 4.6 of Exercise 40 can then be applied. Use these ideas to find all of the following square roots:

   (i) $\sqrt{5}$ (mod 413), (ii) $\sqrt{32}$ (mod 22459), (iii) $\sqrt{34}$ (mod 23573).

42. (*Wilson's Theorem*) **Wilson's theorem** states that
If p is a prime, then $(p-1)! \equiv -1 \pmod{p}$.

(a) Show that the converse of Wilson's theorem is true, i.e., show that if $n > 1$ an integer that satisfies $(n-1)! \equiv -1 \pmod{n}$, then n must be prime.

(b) Use Wilson's theorem to prove that $p \mid 2(p-3)! + 1$, for any odd prime number p.

COMPUTER EXERCISES 4.2:

NOTE: We point out here that certain computing platforms are accurate to only 16 or so significant digits (in the default configuration) since they operate in so-called *floating point arithmetic*. Other platforms can handle arbitrarily large amounts of precision, but usually at a greater cost of computation time. These concepts will be further examined in Section 4.4, but for now, unless you are aware of how to manipulate more than 16 digits of accuracy on your system, you should restrict your computer applications to this limit.

Most computing platforms have some sort of remainder or modular integer converter; this will be particularly useful for modular arithmetic computations. Recall that Computer Exercise 5 of Section 3.3 asked you to either find such a function on your platform or (if you cannot find one or there is none) to write a (simple) program for one. The reader should do this before beginning to work on the following set.

---

[20] Proposition 4.6 can be efficiently implemented using fast modular exponentiation (Algorithm 4.5).

1. (*Brute-Force Program for Finding Modular Inverses*) (a) Write a program with syntax: `ainv = modinv_bf(a, m)` that will take as inputs a modular integer a, and an integer m > 1 (the corresponding modulus). If a has an inverse (mod m), the output `ainv` will be the corresponding (unique) inverse, if a has no inverse, then there will be no output for `ainv`, but only a message to the effect that "there is no inverse mod m." The programming should be done by brute-force, i.e., checking through all of the integers b mod m, and multiplying these by a to see if we get 1 (mod m). If this ever happens, b will be the inverse, so we can output `ainv` as b, if not, then a has no inverse. You may save time in this search by using the necessary condition gcd(a,m) = 1, and gcd(b, m) = 1 for elements to be invertible (or be inverses).
   (b) Use this program to re-do Exercises 15 through 18.
   (c) Use this program to determine the following inverses, if they exist: 1335^{-1} (mod 39467) and 87451^{-1} (mod 139467).

2. (*Program for the Extended Euclidean Algorithm*) (a) Write a program with syntax: `outVec = ExtEucAlg(a,m)` that takes two inputs: a and b, which are positive integers such that a ≥ b. The output will consist of a length three vector `outVec` that has three integer components, d, x, and y, where d = gcd(a,b), and x and y satisfy $d = ax + by$ (as in the Algorithm 4.6). The program should follow Algorithm 4.6.
   (b) Use your program to check the results of Example 4.10 (a) and of Exercise for the Reader 4.8.
   (c) For each pair of integers a and b that we list here, use your program to compute gcd(a,b), and two integers x, and y such that $d = ax + by$. (i) $a = 8359$, $b = 4962$, (ii) $a = 95{,}243$, $b = 24{,}138$.
   (c) Use your program to solve the equation $88243x + 16947y = 1$, for integers x and y (or to determine that such a solution does not exist).

3. (*Program for Finding Inverses with the Extended Euclidean Algorithm*) (a) Write a program with syntax: `ainv = modinv(a,b)` that has the same syntax, inputs and outputs as the program of Computer Exercise 1, but now the programming should use the extended Euclidean algorithm (Algorithm 4.6). Alternatively, your program can directly call on the one of Computer Exercise 2, if you have done that one.
   (b) Use this program to re-do Exercises 9 through 12.
   (c) Use this program to determine the following inverses, if they exist: 1335^{-1} (mod 39467), and 87451^{-1} (mod 139467).
   (d) Compare the performance times of `modinv` and `modinv_bf` (of Computer Exercise 1) for the following pairs of inputs: a = 967, and (i) $m = 10{,}001$, (ii) $m = 100{,}001$, (iii) $m = 1{,}000{,}001$, (iv) $m = 10{,}000{,}001$.

4. (*Program for Solving General Congruences of the Form* $ax \equiv c$ *(mod m)*) (a) Write a program with syntax: `SolVec = LinCongSolver(a,c,m)`, that will take as inputs the parameters a, c, and m that determine a single linear congruence $ax \equiv c$ (mod m), and will output a vector `SolVec` (possibly empty, if there are no solutions) containing all of the solutions of this congruence (mod m). In case gcd(a,m) > 1, the program should follow Algorithm 4.7. The program can be made simpler if it calls on the program `modinv` of Computer Exercise 3.
   (b) Use your program to check the results of Example 4.11 and of Exercise for the Reader 4.9.
   (c) Use this program to re-do Exercises 13 and 14.
   (d) Use this program to re-do Exercises 19 and 20.

5. (*Program to Check ISBN-13 Numbers*) (a) Write a program with syntax: `check = ISBN13(vec)`, where the input is a <u>vector</u> vec of the 13 digits representing an ISBN-13 number (read Exercise 29 for the necessary background). For example, the ISBN-13 number 978-0-471-69738-1 would be inputted as the vector [9 7 8 0 4 7 1 6 9 7 3 8 1] (dashes are left out). The output check will be a string of text: either "Valid ISBN-13 number" if the vector

entered does indeed correspond to a valid ISBN number (i.e., satisfies (12)) or "Invalid ISBN-13 number."

(b) Use this program in conjunction with a for loop to re-do Part (a) of Exercise 29.

6. (*Program to Check VISA Credit Card Numbers*) (a) Write a program with syntax: `check = VISA16(vec)`, where the input is a <u>vector</u> `vec` of the 16 digits representing a 16 digit VISA card number (read Exercise 31 for the necessary background). For example, the VISA card number 4784 5580 0246 1888 would be inputted as the vector [4 7 8 4 5 5 8 0 0 2 4 6 1 8 8 8]. The output `check` will be a string of text: either "Valid VISA card number" if vector entered does indeed correspond to a valid ISBN number (i.e., satisfies the congruence (14)), or "Invalid VISA card number."

(b) Use this program in conjunction with a for loop to re-do Part (a) of Exercise 31.

7. (*Program for Fast Modular Exponentiation*) (a) Write a program with syntax: `a = ModularExponentiation(b,x,m)` that will take as inputs a modular integer $b \not\equiv 0 \pmod{m}$, a positive integer exponent x and a corresponding modulus m > 1. The output a will be the unique nonnegative integer < m that satisfies $a \equiv b^x \pmod{m}$. The mechanics of the program should follow Algorithm 4.5.

(b) Use this program to re-do Exercise 11.

(c) Use this program to compute the following:

(i) $2^{1234567890} \pmod{169}$ and (ii) $12^{123456789012} \pmod{1865}$.

8. (*Brute-force Program for Finding Square Roots Modulo a Prime p*) Before doing this problem, you need to read Exercise 38. (a) Write a program with syntax: `sqrtvec = modsqrt_bf(a, p)` that will take as inputs a modular integer $a \not\equiv 0 \pmod{p}$, and an odd <u>prime</u> integer p > 2 (the corresponding modulus). If a has a square root (mod p), the output `sqrtvec` will be the length two vector containing the two square roots of a, if a has no square roots, then there will be no output for `sqrtvec`, but only a message to the effect that "there are no square roots mod p." The programming should be done by brute-force as suggested in Exercise 38, i.e., checking through all of the integers $\{0^2, 1^2, 2^2, \cdots, [(p-1)/2]^2\}$, to see if one of them is $\equiv a \pmod{p}$. If this happens, say $b^2 \equiv a \pmod{p}$, then the two square roots of a will be b and $-b \pmod{p}$, and they should be outputted into the vector `sqrtvec`.

(b) Use this program to re-do Parts (d) and (e) of Exercise 38.

(c) Use this program to re-do Parts (c) and (d) of Exercise 39.

(d) Use this program to re-do Part (b) of Exercise 40.

9. (*An Efficient Program for Finding Square Roots Modulo a Prime $p \equiv 3 \pmod{4}$*) Before doing this problem, you need to read Exercise 40. (a) Write a program with syntax: `sqrtvec = modsqrt_p3m4(a, p)` that will take as inputs a modular integer $a \not\equiv 0 \pmod{p}$, and <u>prime</u> integer $p \equiv 3 \pmod{4}$ (the corresponding modulus). If a has a square root (mod p), the output `sqrtvec` will be the length two vector containing the two square roots of a, if a has no square roots, then there will be no output for `sqrtvec`, but only a message to the effect that "there are no square roots mod p." The programming should be done using Proposition 4.6 (within Exercise 37).

(b) Use this program to re-do Part (b) of Exercise 40.

(c) For each of the following odd primes p, which are congruent to 3 (mod 4), compare the performance of your program `modsqrt_p3m4` with that for the brute-force version, of the preceding computer exercise in extracting $\sqrt{(p-1)/2} : \pmod{p}$, with p = 103, 1019, 10007, 100003, 1000003, 10000019. Simply compare the runtimes on the particular machine you are working on. Abort any computation if the runtime exceeds five minutes.

Suggestion: You should also make use of Algorithm 4.5. This not only increases the efficiency of your program, but if your computing platform works in floating point arithmetic (see the note at the beginning of this computer exercise section), it can prevent inaccurate results when intermediate computations involve numbers with too many significant digits.

4.3: MATRICES

We have already had several encounters with vectors, which are simply ordered lists of numbers. A more general and very natural data structure in computing is that of a **matrix** (plural: **matrices**), which is simply a rectangular array of numbers, see Figure 4.3. Matrices can be thought of simply as data spreadsheets, and these should certainly be familiar to anyone having basic computer experience. A matrix with n rows and m columns is said to be an $n \times m$ matrix, and these two numbers determine the **dimensions** or the **size** of A. The **element** (or **entry**) a_{ij} of the matrix A is located in the ith row and the jth column (see Figure 4.3),

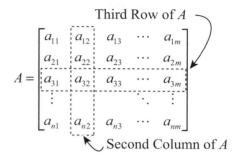

FIGURE 4.3: The anatomy of a matrix with n rows and m columns. The (3,2) entry that lies in the third row and the second column is denoted a_{32}.

and is called the **(i, j) entry** of A. By default, we will consider the entries of a matrix to be real numbers, but as needs arise we will want to allow matrix elements to belong to other number systems.[21] For example, later in this chapter, we will be dealing with matrices of modular integers. For shorthand, a matrix A is often denoted by

$$A = [a_{ij}],$$

where the dimensions of A are either known from the context or are unimportant. An $n \times 1$ matrix has only one column, and is called a **column matrix**. Similarly, a $1 \times m$ matrix is called a **row matrix**. Row and column matrices are simply ordered lists of numbers, otherwise known as **vectors**.[22] An $n \times n$ matrix has an equal number of rows and columns and is called a **square matrix**.

---

[21] In computer applications, the elements of a spreadsheet matrix may be any sort of objects, not necessarily belonging to any number system (for example elements could be English letters or words). In order to define the matrix operations of addition, subtraction, and multiplication, we will need the matrix elements to belong to a number system in which such operations may be performed.

[22] Often when we are dealing with vectors, one is concerned only that the list of numbers is ordered, and it is unimportant whether the vector is written as a row vector or a column vector. Also, vectors are sometimes written with commas inserted between elements, e.g., [2 0 3] = [2, 0, 3].

Matrix Addition, Subtraction, and Scalar Multiplication

In order to define the matrix operations of addition, subtraction, and multiplication, we will need the matrix elements to belong to a number system in which such operations may be performed.

DEFINITION 4.5:   **Addition/subtraction of matrices** $A = [a_{ij}]$ and $B = [b_{ij}]$, of <u>the same size</u> is defined *componentwise*, i.e., simply add/subtract corresponding entries. Similarly, if α is a number,[23] then the **scalar multiplication** αA is the matrix obtained from A by multiplying each of its entries by α. These matrix operations are summarized by the following formulas:

$$[a_{ij}] \pm [b_{ij}] = [a_{ij} \pm b_{ij}], \quad \alpha[a_{ij}] = [\alpha a_{ij}].$$

Multiplication of matrices, on the other hand, is not done componentwise. To define this concept, we first need to introduce the **dot product** $u \bullet v$ of two vectors $u = [u_i]$, and $v = [v_i]$ of the same length. If the two vectors have length n, then this dot product $u \bullet v$ of the two vectors is the <u>number</u> (or scalar) defined by the formula

$$u \bullet v = [u_i] \bullet [v_i] = u_1 v_1 + u_2 v_2 + \cdots + u_n v_n = \sum_{i=1}^{n} u_i v_i.$$

For this definition, we do not require that u and v have the same dimensions, only that they have the same length. For example, it still applies if u is a row matrix and v is a column matrix.

EXAMPLE 4.14: Given the matrices and vectors:

$$A = \begin{bmatrix} 2 & -4 \\ 1 & 6 \end{bmatrix}, \; B = \begin{bmatrix} 1 & 0 & 9 \\ -4 & -2 & 4 \end{bmatrix}, \; C = \begin{bmatrix} 3 & 7 \\ 5 & 5 \end{bmatrix}, \; u = [6, -3, 12], \; v = [4, 2, -2],$$

compute (if possible) each of the following:

(a) $A + B$,
(b) $A - 2C$, and
(c) $u \bullet v$.

SOLUTION:
Part (a):   Since A and B are different-sized matrices (A is 2×2, B is 2×3), their sum is undefined.

---

[23] When dealing with vectors or matrices, numbers are sometimes referred to as *scalars*.

Part (b): $A - 2C = \begin{bmatrix} 2 & -4 \\ 1 & 6 \end{bmatrix} - 2 \cdot \begin{bmatrix} 3 & 7 \\ 5 & 5 \end{bmatrix} = \begin{bmatrix} 2 & -4 \\ 1 & 6 \end{bmatrix} - \begin{bmatrix} 6 & 14 \\ 10 & 10 \end{bmatrix} = \begin{bmatrix} -4 & -18 \\ -9 & -4 \end{bmatrix}.$

Part (c): $u \bullet v = [6, -3, 12] \bullet [4, 2, -2] = 6 \cdot 4 - 3 \cdot 2 + 12 \cdot (-2) = 24 - 6 - 24 = -6.$

Matrix Multiplication

DEFINITION 4.6:   Suppose that $A = [a_{ij}]$ is an $n \times m$ matrix and $B = [b_{ij}]$ is an $m \times r$ matrix (i.e., the number of columns of A is the number of rows of B).   Then the **matrix product** $C = AB$ is defined as the matrix of dimensions $n \times r$, whose (i, j) entry c_{ij} is simply the dot product of the ith row of A with the jth column of B, i.e.,

$$c_{ij} = a_{i1}b_{1j} + a_{i2}b_{2j} + a_{i3}b_{3j} + \cdots + a_{im}b_{mj} = \sum_{k=1}^{m} a_{ik}b_{kj}.$$

This procedure is illustrated in Figure 4.4.

$$\begin{bmatrix} a_{11} & a_{12} & a_{13} & \cdots & a_{1m} \\ a_{21} & a_{22} & a_{23} & \cdots & a_{2m} \\ a_{31} & a_{32} & a_{33} & \cdots & a_{3m} \\ \vdots & & & \ddots & \vdots \\ a_{n1} & a_{n2} & a_{n3} & \cdots & a_{nm} \end{bmatrix} \begin{bmatrix} b_{11} & b_{12} & b_{13} & \cdots & b_{1m} \\ b_{21} & b_{22} & b_{23} & \cdots & b_{2m} \\ b_{31} & b_{32} & b_{33} & \cdots & b_{3m} \\ \vdots & & & \ddots & \vdots \\ b_{m1} & b_{m2} & b_{m3} & \cdots & b_{mr} \end{bmatrix} = \begin{bmatrix} c_{11} & c_{12} & c_{13} & \cdots & c_{1r} \\ c_{21} & c_{22} & c_{23} & \cdots & c_{2r} \\ c_{31} & c_{32} & c_{33} & \cdots & c_{3r} \\ \vdots & & & \ddots & \vdots \\ c_{n1} & c_{n2} & c_{n3} & \cdots & c_{nr} \end{bmatrix}$$
$$\qquad n \times m \qquad\qquad\qquad m \times r \qquad\qquad\qquad n \times r$$

FIGURE 4.4: Illustration of the computation of an element of the product matrix $AB = C$: To get the (3,2) entry c_{32} of C, take the dot product of the 3rd row of A (the first matrix) with the 2nd column of B (the second matrix).

At first glance, this definition of matrix multiplication probably seems very unnatural, especially compared with the componentwise definitions of matrix addition and subtraction.   The reason for this definition is that it gives matrices a very rich arithmetic structure that results in a powerful theory and manifold applications.

EXAMPLE 4.15: (*Matrix multiplication is not commutative*) Given the matrices:

$$A = \begin{bmatrix} 2 & -4 \\ 1 & 6 \end{bmatrix}, \quad B = \begin{bmatrix} 1 & 0 & 9 \\ -4 & -2 & 4 \end{bmatrix}, \quad C = \begin{bmatrix} 3 & 7 \\ 5 & 5 \end{bmatrix},$$

compute (if possible) each of the following:

(a) *AB* and *BA*
(b) *AC* and *CA*

SOLUTION:

Part (a): $AB = \begin{bmatrix} 2 & -4 \\ 1 & 6 \end{bmatrix} \cdot \begin{bmatrix} 1 & 0 & 9 \\ -4 & -2 & 4 \end{bmatrix} = \begin{bmatrix} 18 & 8 & 2 \\ -23 & -12 & 33 \end{bmatrix}.$

Each of the six entries of the 2×3 product matrix on the right was computed as the appropriate dot product. For example the boxed (2,3) entry equals the dot product of the second row of *A*, with the third column of *B* (both boxed), and was computed as $1 \cdot 9 + 6 \cdot 4 = 9 + 24 = 33$.

The matrix *BA* is undefined since the inside dimensions of the two matrices do not match.

Part (b): Since both *A* and *C* are square matrices of the same size, both products may be computed:

$$AC = \begin{bmatrix} 2 & -4 \\ 1 & 6 \end{bmatrix}\begin{bmatrix} 3 & 7 \\ 5 & 5 \end{bmatrix} = \begin{bmatrix} -14 & -6 \\ 33 & 37 \end{bmatrix}, \quad CA = \begin{bmatrix} 3 & 7 \\ 5 & 5 \end{bmatrix}\begin{bmatrix} 2 & -4 \\ 1 & 6 \end{bmatrix} = \begin{bmatrix} 13 & 30 \\ 15 & 10 \end{bmatrix}.$$

This example shows that <u>matrix multiplication is not commutative</u>: In Part (a) we found that $AB \neq BA$, since the latter matrix was not defined. In Part (b), even though both products are defined, they were not equal (in fact, each of the four corresponding entries was different). Despite this drawback, matrix arithmetic does possess many of the properties that we know are true for real numbers.

EXERCISE FOR THE READER 4.14: (*Preview of the fact that matrix multiplication is associative*) Given the matrices:

$$A = \begin{bmatrix} 2 & -4 \\ 1 & 6 \end{bmatrix}, \quad B = \begin{bmatrix} 8 & 0 \\ -4 & 1 \end{bmatrix}, \quad C = \begin{bmatrix} 3 & 7 \\ 5 & 5 \end{bmatrix},$$

compute the two matrix products (*AB*)*C* and *A*(*BC*).

The associativity property (*AB*)*C* = *A*(*BC*) that was witnessed in the preceding exercise for the reader turns out to be generally true, provided the sizes of the constituent matrices are such that the multiplications are defined. This and several other useful properties of matrix arithmetic are collected in the following:

Matrix Arithmetic

PROPOSITION 4.7: (*Some Properties of Matrix Arithmetic*) Suppose that *A*, *B*, and *C* are matrices, and that α is a scalar. The following identities hold, where we assume in each part that the sizes of A, B, and C are such that the matrices on both sides of the identity are defined.

(a) (*Commutativity of Addition*) $A + B = B + A$.

(b) (*Associativity*) $(A+B)+C = A+(B+C)$, $(AB)C = A(BC)$.

(c) (*Distributive Laws*) $A(B+C) = AB+AC$, $(A+B)C = AB+AC$,
$$\alpha(A+B) = \alpha A + \alpha B.$$

(d) $\alpha(AB) = (\alpha A)B = A(\alpha B)$.

Proof: Part (a) and the first identity of Part (b) involve only matrix addition, and so these identities are direct consequences of the corresponding identities for real numbers (one entry at a time). To prove that matrix multiplication is associative (i.e., the second identity of Part (b)), we let $[f_{ij}] = (AB)C$ and $[g_{ij}] = A(BC)$. Since these two matrices have the same size, we need to show that the corresponding entries are equal, i.e., we fix a pair of indices i and j, and aim to show that $f_{ij} = g_{ij}$. For definiteness, let the sizes of A, B, and C be $n \times m$, $m \times r$, and $r \times s$, respectively. By definition, f_{ij} is the dot product of the ith row of AB with the jth column of C. Since the kth entry in the ith row of AB is $\sum_{\ell=1}^{m} a_{i\ell} b_{\ell k}$, it follows that $f_{ij} = \sum_{k=1}^{r} \left(\sum_{\ell=1}^{m} a_{i\ell} b_{\ell k} \right) c_{kj} = \sum_{\ell=1}^{m} a_{i\ell} \left(\sum_{k=1}^{r} b_{\ell k} c_{kj} \right)$. (The commutative and associative properties for numbers allowed us to switch the order of summation.) The parenthesized sum on the right is simply the ℓth entry in the jth column of BC. Hence, $\sum_{\ell=1}^{n} a_{i\ell} \left(\sum_{k=1}^{r} b_{\ell k} c_{kr} \right) = g_{ij}$, so we have proved $f_{ij} = g_{ij}$. The proofs of the identities of Parts (c) and (d) are easier than this one and are left to the next exercise for the reader and to the exercises. $\square$

EXERCISE FOR THE READER 4.15: Prove the identities $A(B+C) = AB + AC$ and $\alpha(A+B) = \alpha A + \alpha B$ from Part (c) of Proposition 4.7.

It quickly becomes impractical to multiply (or even add) large matrices without a computer. Indeed, using Definition 4.6 to multiply an $n \times m$ matrix by an $m \times r$ matrix, each of the nr terms requires computing a dot product involving m multiplications and $m-1$ additions, giving a grand total of $nr(2m-1)$ mathematical operations. If all the dimensions are the same: $n = m = r$, this becomes $n^2(2n-1)$, which is asymptotic to $2n^3$,[24] or a constant times n^3. Since matrices have so many applications, the expense of multiplying them has been a bottleneck for many problems. Many mathematicians and computer scientists thought that this price of matrix multiplication could not be lowered until 1969, when a German mathematician Volker Strassen [Str-69] found a faster

---

[24] Saying that two infinite sequences a_n and b_n are *asymptotic* means that $a_n / b_n \to 1$, as $n \to \infty$.

algorithm that multiplies a pair of $n \times n$ matrices using at most a constant times $n^{2.81}$ arithmetic operations.   After this landmark discovery scientists have been working hard to find ever more efficient algorithms for *fast matrix multiplication*. The fastest method known at present was discovered in 1990 by Don Coppersmith and Shmuel Winograd [CoWi-90], and brings the number of arithmetic operations needed to multiply two $n \times n$ matrices down to a constant times $n^{2.376}$.   This method relies on increasingly complex generalizations of Strassen's idea, and on ideas from the area of mathematics known as group theory.   Researchers believe that it is likely that an algorithm can be found that will be able to multiply a pair of $n \times n$ matrices in using only a constant times n^2.   This would, of course, be the holy grail of all efficiencies for this problem since having to compute the n^2 entries will take at least n^2 of operations.   The constants in these complexity estimates, as well as the overall intricacy of the algorithm can make such algorithms practical only when the matrices involved are sufficiently large.   The reader can find a nice survey article on this very interesting topic in [Rob-05]. The exercises will provide opportunities for the reader to examine Strassen's algorithm.

Definition of an Invertible (Square) Matrix

Thus far, we have learned how to add, subtract, and multiply matrices. It is natural to next consider the question whether matrices can be divided. With real numbers, we are allowed to divide by any nonzero number. In modular arithmetic, we learned that division is possible as long as the divisor is relatively prime to the modulus. In each of these cases, the division by a number a can be viewed as multiplying by the inverse element a^{-1}, that (if it exists) satisfies $a \cdot a^{-1} = a^{-1} \cdot a = 1$.   For example, in the real numbers, the division $7 \div 2$ can be viewed as the multiplication $7 \cdot 2^{-1} = 7 \cdot (1/2) = 3.5$.   A similar situation exists for matrices.   In order to define the inverse of a matrix, one needs to define an analogue for the number 1 for matrices; this will be the so-called *identity matrix* that is defined as follows:

DEFINITION 4.7:   For any positive integer n, the $n \times n$ **identity matrix**, denoted as I_n, or just I (when the size is clear from the context or unimportant), is the $n \times n$ matrix $[\delta_{ij}]$, where $\delta_{ij} = 1$, if $i = j$, and $\delta_{ij} = 0$, if $i \neq j$.

Below are the 1×1, 2×2, and 3×3 identity matrices.

$$I_1 = [1], \ I_2 = \begin{bmatrix} 1 & 0 \\ 0 & 1 \end{bmatrix}, \ I_3 = \begin{bmatrix} 1 & 0 & 0 \\ 0 & 1 & 0 \\ 0 & 0 & 1 \end{bmatrix}.$$

The identity matrix is so named because it behaves like the number 1 does for numbers, i.e., when it gets multiplied by a matrix, it does not change the matrix:

$$AI = A = IA. \tag{15}$$

EXERCISE FOR THE READER 4.16: Prove (15).

The definition of an *invertible matrix* is easily translated from that of an invertible real number by simply changing numbers to matrices and 1 to I:

DEFINITION 4.8: A square matrix A is said to be **invertible**, if there exists another matrix A^{-1} of the same size that, when multiplied by A on either side, gives the identity matrix:

$$AA^{-1} = A^{-1}A = I. \tag{16}$$

The matrix A^{-1} (if it exists) is unique (*Proof:* If B is another such matrix, then using (15), (16), and associativity, we obtain that $B = BI = B(AA^{-1}) = (BA)A^{-1} = IA^{-1} = A^{-1}$. □), and is called the **inverse of A**.

EXAMPLE 4.16: Let $A = \begin{bmatrix} 3 & 5 \\ 4 & 7 \end{bmatrix}$, $B = \begin{bmatrix} 0 & 0 \\ 3 & 6 \end{bmatrix}$, $C = \begin{bmatrix} 7 & -5 \\ -4 & 3 \end{bmatrix}$.

(a) Show that $A^{-1} = C$.
(b) Show that the matrix B is not invertible.

SOLUTION: Part (a): The two multiplications:

$$AC = \begin{bmatrix} 3 & 5 \\ 4 & 7 \end{bmatrix}\begin{bmatrix} 7 & -5 \\ -4 & 3 \end{bmatrix} = \begin{bmatrix} 1 & 0 \\ 0 & 1 \end{bmatrix} \quad \text{and} \quad CA = \begin{bmatrix} 7 & -5 \\ -4 & 3 \end{bmatrix}\begin{bmatrix} 3 & 5 \\ 4 & 7 \end{bmatrix} = \begin{bmatrix} 1 & 0 \\ 0 & 1 \end{bmatrix}, \quad \text{show}$$

that $AC = CA = I$, which means that $A^{-1} = C$ (and also that $C^{-1} = A$).

Part (b): If $M = \begin{bmatrix} a & b \\ c & d \end{bmatrix}$ is any 2×2 matrix, and we compute the product BM, we

notice that: $BM = \begin{bmatrix} 0 & 0 \\ 3 & 6 \end{bmatrix}\begin{bmatrix} a & b \\ c & d \end{bmatrix} = \begin{bmatrix} 0 & 0 \\ * & * \end{bmatrix}$ (where the asterisks denote numbers).

The point is that no matter what the matrix M is, the product BM will inherit the property that the first row of B is all zeros. Thus it is impossible to have $BM = I$, so that B cannot be invertible.

EXERCISE FOR THE READER 4.17: (a) Generalize the idea given in the solution of Part (b) of the preceding example to show that any square matrix A with a row of zeros cannot be invertible.
(b) Show, likewise, that if a square matrix A has a column of zeros, then it cannot be invertible.

The Determinant of a Square Matrix

As with real numbers and with modular arithmetic, it is sometimes important to know whether a given square matrix is invertible. Every square matrix A has a number, called its *determinant*, denoted det(A), associated with it. Whereas a real number is invertible if, and only if it is nonzero, it turns out that a square matrix is invertible, if, and only if the determinant is nonzero (i.e., if the determinant is an invertible real number). Many computing platforms have built-in functions that will compute determinants of square matrices (and inverses of invertible matrices). The following recursive algorithm that was first introduced in Example 3.9 gives a method for computing determinants.

ALGORITHM   4.8:   (*Cofactor   Expansion   Algorithm   for   Computing   Determinants*)

Input: A square $n \times n$ matrix A.

Output: The determinant of A: det(A), *cofactor expansion along the first row.*

Case n = 1: If $A = [a]$ (1×1 matrix), det(A) = a.

Case n = 2: If $A = \begin{bmatrix} a & b \\ c & d \end{bmatrix}$, det($A$) = $ad - bc$, that is, just the product of the main diagonal entries (top left to bottom right) less the product of the off diagonal entries (top right to bottom left).

Case n > 2: For larger sized matrices the following recursive formula for the determinant of A is given by:

$$\det(A) = a_{11} \det(A_{11}) - a_{12} \det(A_{12}) + a_{13} \det(A_{13}) - \cdots + (-1)^{n+1} a_{1n} \det(A_{1n}), \qquad (17)$$

(Note that the signs alternate as we move across the entries of the top row of A.)[25]

where, for any entry a_{ij} of the $n \times n$ matrix A, the corresponding *submatrix* A_{ij} is the $(n-1) \times (n-1)$ matrix obtained from A by deleting the row and column of A that contain the entry a_{ij} .

---

[25] It is proved in linear algebra books (e.g., see [HoKu-71]) that one could instead take the corresponding (cofactor) expansion along any row or column of A, using the following rule to choose the alternating signs: The sign of det(A_{ij}) is $(-1)^{i+j}$. See Proposition 4.6 in the exercise set for a more complete statement of this result.

EXERCISE FOR THE READER 4.18: Use the cofactor expansion algorithm (Algorithm 4.8) to compute the determinant of the matrix $\begin{bmatrix} 5 & -6 & 9 \\ -12 & 2 & 7 \\ 2 & 3 & -7 \end{bmatrix}$.

We now formally state the important relation that was mentioned above between the determinant and the invertibility of a matrix.

THEOREM 4.8: (*On the Invertibility of Square Matrices*) A square matrix A is invertible if, and only if its determinant $\det(A)$ is nonzero.

We will not prove this result here (since it would require the formal definition of determinants), but the interested reader may refer to a decent book on linear algebra (e.g., see [HoKu-71]) .

Inverses of 2×2 Matrices

We have given an example of the inverse of a matrix, but so far have given no clue as to how inverses of invertible matrices can be determined. The following result includes a simple formula for the inverse of a 2×2 matrix; later in this section we give a more general formula that will provide a way to find inverses of any invertible matrix.

THEOREM 4.9: (*On the Invertibility of 2×2 Matrices*)

For a 2×2 matrix $A = \begin{bmatrix} a & b \\ c & d \end{bmatrix}$, with nonzero determinant $\det(A) = ad - bc$, the

inverse matrix is given by $A^{-1} = \dfrac{1}{\det(A)} \begin{bmatrix} d & -b \\ -c & a \end{bmatrix}$.

Proof: We need only check that the matrix given by the formula satisfies equation (2) (definition of an inverse matrix). We temporarily denote the matrix defined by the formula in the theorem as B. We will show that $BA = I$. The proof that $AB = I$ is similar.

$$BA = \frac{1}{\det(A)} \begin{bmatrix} d & -b \\ -c & a \end{bmatrix} \begin{bmatrix} a & b \\ c & d \end{bmatrix} = \frac{1}{\det(A)} \begin{bmatrix} ad - bc & db - bd \\ -ca + ac & -cb + ad \end{bmatrix}$$

$$= \frac{1}{\det(A)} \begin{bmatrix} \det(A) & 0 \\ 0 & \det(A) \end{bmatrix} = I. \;\square$$

For example, the matrix $A = \begin{bmatrix} 2 & 3 \\ 2 & 2 \end{bmatrix}$ has determinant $\det(A) = 2 \cdot 2 - 3 \cdot 2 = -2$, so

this matrix has an inverse given by $A^{-1} = \dfrac{1}{-2} \begin{bmatrix} 2 & -3 \\ -2 & 2 \end{bmatrix} = \begin{bmatrix} 1 & -3/2 \\ -1 & 1 \end{bmatrix}$.

EXERCISE FOR THE READER 4.19: Use Theorem 4.9 to compute the inverses of the following matrices, if they exist: $M = \begin{bmatrix} 2 & 6 \\ 3 & -9 \end{bmatrix}$, $N = \begin{bmatrix} 2 & 6 \\ 3 & 9 \end{bmatrix}$.

Two final important yet basic concepts about matrices that we wish to mention are included in the following two definitions.

The Transpose of a Matrix

DEFINITION 4.9: If $A = [a_{ij}]$ is an $n \times m$ matrix, then the **transpose** of A, denoted A', is the $m \times n$ matrix whose rows are the columns of A, and whose columns are the rows of A (in order). Put differently, $A' = [b_{ij}]$, where $b_{ij} = a_{ji}$ $(1 \le i \le m, 1 \le j \le n)$. A square matrix A is called **symmetric** if $A = A'$.

EXAMPLE 4.17:   Find the transposes of the matrices $\begin{bmatrix} 1 & 2 & 1 \\ 4 & 6 & 9 \end{bmatrix}$ and $\begin{bmatrix} 6 & 5 \\ 5 & 1 \end{bmatrix}$.

SOLUTION: Changing rows to columns, the transpose of the first matrix is $\begin{bmatrix} 1 & 4 \\ 2 & 6 \\ 1 & 9 \end{bmatrix}$, while the transpose of the second matrix is itself; thus the latter matrix is symmetric.

For a square matrix A, nonnegative **powers of A** can be defined recursively in the usual fashion: $A^0 = I$, $A^{n+1} = A^n \cdot A$ $(n \ge 0)$.

Modular Integer Matrices

In some applications, it is useful to consider matrices whose entries are stipulated to lie in some number system other than the real numbers, which have been used in all matrices considered up to this point. The next example illustrates how some of the preceding concepts play out when the elements of matrices are integers mod

m (and the corresponding arithmetic is done (mod m)). We refer to such matrices as **modular integer matrices**.

DEFINITION 4.10: For a fixed integer $m > 1$, we consider matrices whose entries are integers mod m, i.e., are elements in $\mathbb{Z}_m$. Such a matrix is called a **modular integer matrix**. All of the previously defined matrix operations are extended to such matrices with the proviso that all integer calculations are performed (mod m).

EXAMPLE 4.18: If we are considering matrices mod 5, then here is how we would add and multiply the two matrices $A = \begin{bmatrix} 4 & 2 \\ 1 & 3 \end{bmatrix}$ and $B = \begin{bmatrix} 3 & 0 \\ 4 & 3 \end{bmatrix}$:

$$A + B = \begin{bmatrix} 4+3 & 2+0 \\ 1+4 & 3+3 \end{bmatrix} = \begin{bmatrix} 7 & 2 \\ 5 & 6 \end{bmatrix} \equiv \begin{bmatrix} 2 & 2 \\ 0 & 1 \end{bmatrix} \text{ (mod 5)},$$

$$AB = \begin{bmatrix} 4 & 2 \\ 1 & 3 \end{bmatrix} \begin{bmatrix} 3 & 0 \\ 4 & 3 \end{bmatrix} = \begin{bmatrix} 20 & 6 \\ 15 & 9 \end{bmatrix} \equiv \begin{bmatrix} 0 & 1 \\ 0 & 4 \end{bmatrix} \text{ (mod 5)}.$$

Most all of the properties of matrix arithmetic that we have seen carry over very nicely to modular matrices. For example, each of the identities of Proposition 4.7 is valid for modular integer matrices. The proof is simple because the identities hold when real number arithmetic is used to compute the entries. Corresponding entries will remain equal when we take their remainders mod m. Determinants of square modular integer matrices can be defined using the same definition as with ordinary matrices. But more importantly for our purposes, all of the alternative definitions, such as the cofactor expansion that was given in Algorithm 4.8, will also be valid. It is delightfully surprising that both of Theorems 4.8 and 4.9 on invertibility of matrices have the following very natural analogue for modular matrices.

THEOREM 4.10: (*On the Invertibility of Square Modular Integer Matrices*)
(1) A square modular integer matrix A is invertible (mod m) if, and only if its determinant det(A) is relatively prime to the modulus m, i.e., gcd(det(A), m) = 1.

(2) For a 2×2 modular integer matrix $A = \begin{bmatrix} a & b \\ c & d \end{bmatrix}$, with determinant det($A$) = $ad - bc$ relatively prime to m, the inverse matrix is given by

$A^{-1} = \det(A)^{-1} \begin{bmatrix} d & -b \\ -c & a \end{bmatrix}$ (mod m), where $\det(A)^{-1}$ is the inverse of det(A) (mod m).

Note that from Proposition 4.2, the condition that gcd(det(A), m) = 1 is equivalent to det(A) having an inverse (mod m). Since a real number is invertible precisely when it is nonzero, Part (1) of Theorem 4.10 is really a direct translation of Part

(1) of Theorem 4.8, from the language of real number entries into the language of modular integer entries.

For the proof Part (1) of Theorem 4.10, we refer to the books by Koblitz [Kob-94] and Stinson [Sti-05]. The proof of Part (2) is accomplished in the same way as for ordinary matrices in our proof of Theorem 4.8.

EXAMPLE 4.19: If it exists, find the inverse of the matrix $A = \begin{bmatrix} 3 & 2 \\ 2 & 3 \end{bmatrix}$.

(a) (mod 10) and (b) (mod 12).

SOLUTION: Part (a): $\det(A) = 3 \cdot 3 - 2 \cdot 2 = 5$ is not relatively prime to 10, therefore by Theorem 4.10, A has no inverse (mod 10).

Part (b): Since gcd($\det(A)$, 12) = gcd(5, 12) = 1, and since $5^{-1} = 5$ (mod 12) (either by trial and error or the extended division algorithm), it follows from Theorem 4.10 (2) that

$$A^{-1} = 5^{-1} \begin{bmatrix} 3 & -2 \\ -2 & 3 \end{bmatrix} \equiv 5 \cdot \begin{bmatrix} 3 & 10 \\ 10 & 3 \end{bmatrix} \equiv \begin{bmatrix} 15 & 50 \\ 50 & 15 \end{bmatrix} \equiv \begin{bmatrix} 3 & 2 \\ 2 & 3 \end{bmatrix} \text{ (mod 12)}.$$

The reader may wish to check that $AA^{-1} = I$. Coincidentally, here we have that $A^{-1} = A$. We also point out that this inverse is totally different from the real number version (obtained from Theorem 4.9(2)), which has fractions in all of its entries.

EXERCISE FOR THE READER 4.20: Consider the following matrices of integers: $A = \begin{bmatrix} 2 & 7 \\ 4 & 1 \end{bmatrix}, B = \begin{bmatrix} 1 & 2 \\ 9 & 8 \end{bmatrix}$.

(a) Working mod 3, compute the following related matrices, if they are defined: $A + B$, AB, B^{-1}.

(b) Repeat Part (a), but this time working mod 10.

The Classical Adjoint (for Matrix Inversions)

The formulas of Part (2) in both Theorems 4.9 and 4.10 are special cases of the so-called *classical adjoint formula* for the inverse of an invertible square matrix. To state this formula, it will be convenient to first introduce the following definition.

DEFINITION 4.11: Let $A = [a_{ij}]$ be an $n \times n$ matrix. As in Algorithm 4.8, for each pair of indices i, j, we let A_{ij} denote the $(n-1) \times (n-1)$ submatrix obtained from A by deleting the row and column containing a_{ij}. The **classical adjoint** of A is defined to be the matrix

$$\text{adj}(A) = \begin{bmatrix} +\det(A_{11}) & -\det(A_{12}) & \cdots & (-1)^{1+n}\det(A_{1n}) \\ -\det(A_{21}) & +\det(A_{22}) & \cdots & (-1)^{2+n}\det(A_{2n}) \\ \vdots & & \ddots & \\ (-1)^{n+1}\det(A_{n1}) & (-1)^{n+2}\det(A_{n2}) & \cdots & +\det(A_{nn}) \end{bmatrix}'$$

$$= \left[(-1)^{i+j}\det(A_{ji}) \right].$$

(Note the transpose operation in the upper right matrix.)

PROPOSITION 4.11: (*Classical Adjoint Formula for the Inverse of an $n \times n$ Matrix*) If $A = [a_{ij}]$ is an invertible $n \times n$ (real-valued or modular integer) matrix, then $A^{-1} = \det(A)^{-1} \cdot \text{adj}(A)$.

For an $n \times n$ matrix, this formula requires the computation of an $n \times n$ determinant, and n^2 $(n-1) \times (n-1)$ determinants, so it is practical only for small-sized matrices.[26] Since we will need to invert only matrices up to size 3×3 in this book, it will be sufficient for our purposes and can be used in conjunction with the cofactor expansion method for computing determinants of Algorithm 4.8. The proof of this formula easily follows from a slightly more general version of the cofactor expansion (see Algorithm 4.8 and its footnote), and will be outlined in the exercises. The proof and the formula are valid for modular integer matrices as well as ordinary matrices, provided (Theorem 4.10) that $\det(A)$ is relatively prime to the modulus. Our next example involves the inversion of a 3×3 integer modular matrix using Proposition 4.11.

EXAMPLE 4.20: Find the inverse of the integer modular matrix $A = \begin{bmatrix} 7 & 5 & 2 \\ 0 & 6 & 4 \\ 8 & 2 & 5 \end{bmatrix}$ (mod 9), if it exists.

SOLUTION: Using the cofactor expansion formula (17), we find $\det(A) = 434 \equiv 2 \pmod 9$. (Alternatively, all intermediate calculations could have

---

[26] More efficient techniques for computing determinants of square matrices and inverses of invertible matrices can be found in any decent book on elementary linear algebra (such techniques are adaptations of Gaussian elimination); see, for example, [Poo-05] or [KoHi-99]; a more theoretically comprehensive text is the classical reference [HoKu-71].

been performed (mod 9).) Since $\gcd(2, 9) = 1$, it follows that A^{-1} exists (mod 9). The classical adjoint matrix is

$$\text{adj}(A) = \begin{bmatrix} \det\begin{pmatrix} 6 & 4 \\ 2 & 5 \end{pmatrix} & -\det\begin{pmatrix} 5 & 2 \\ 2 & 5 \end{pmatrix} & \det\begin{pmatrix} 5 & 2 \\ 6 & 4 \end{pmatrix} \\ -\det\begin{pmatrix} 0 & 4 \\ 8 & 5 \end{pmatrix} & \det\begin{pmatrix} 7 & 2 \\ 8 & 5 \end{pmatrix} & -\det\begin{pmatrix} 7 & 2 \\ 0 & 4 \end{pmatrix} \\ \det\begin{pmatrix} 0 & 6 \\ 8 & 2 \end{pmatrix} & -\det\begin{pmatrix} 7 & 5 \\ 8 & 2 \end{pmatrix} & \det\begin{pmatrix} 7 & 5 \\ 0 & 6 \end{pmatrix} \end{bmatrix} = \begin{bmatrix} 22 & -21 & 8 \\ 32 & 19 & -28 \\ -48 & 26 & 42 \end{bmatrix},$$

so that $\text{adj}(A) = \begin{bmatrix} 4 & 6 & 8 \\ 5 & 1 & 8 \\ 6 & 8 & 6 \end{bmatrix}$.[27]   Since $2 \cdot 5 = 10 \equiv 1 (\text{mod} 9)$, we have $2^{-1} \equiv 5$

(mod 9), so (by Proposition 4.11),

$$A^{-1} = \det(A)^{-1} \cdot \text{adj}(A) = 5 \cdot \begin{bmatrix} 4 & 6 & 8 \\ 5 & 1 & 8 \\ 6 & 8 & 6 \end{bmatrix} \equiv \begin{bmatrix} 2 & 3 & 4 \\ 7 & 5 & 4 \\ 3 & 4 & 3 \end{bmatrix} \;(\text{mod} 9).$$

(Note that in the final modular multiplication by 5, all but the two odd entries could have simply been divided by 2, rather than multiplied by its inverse.)   The reader may wish to check that $AA^{-1} = I = A^{-1}A$.

The above procedure is easily programmed, and this will be tasked in the computer exercises at the end of this section.   But many computing platforms already have built-in programs (that are more efficient than Algorithm 4.8) for computing determinants. The following note shows how to take advantage of such a feature to render an efficient and easy way of obtaining modular integer matrix inverses, whenever they exist.

COMPUTING NOTE: As a consequence of Proposition 4.11, if you are working on a computing platform that already has programs for computing (ordinary) matrix inverses, any modular matrix inverse (if it exists) can easily be obtained from the corresponding (ordinary) matrix inverse by the following formula:

$$A^{-1}(\text{mod } n) \equiv (\det(A)^{-1}(\text{mod } n)) \cdot (\det(A)A^{-1}), \tag{18}$$

---

[27] Again, we could have been working (mod 9) throughout all intermediate calculations, and would have obtained the same result.   By the way, when viewed as an ordinary (real-valued) matrix, Proposition 4.11 tells us that the inverse matrix would be the prereduced (mod 9) adjoint divided by the prereduced determinant (434).

where A^{-1} denotes the (ordinary) matrix inverse, $A^{-1} (\mathrm{mod}\, n)$ and $\det(A)^{-1} (\mathrm{mod}\, n)$ denote inverses (mod n), and the formula is valid whenever the modular inverses exist. See Computer Exercise 11 at the end of this section for more on how to implement (18). Moreover, if your system has a built-in determinant function (for computing determinants of real-valued matrices), it should be faster to compute determinants of modular matrices by directly using this function and then converting the answer to a mod n integer (by applying the mod function as: $\mathrm{mod}(\det(A),\, n))$.[28]

EXERCISE FOR THE READER 4.21: If it exists, find the inverse of the matrix

$$A = \begin{bmatrix} 1 & 1 & 3 \\ 8 & 4 & 5 \\ 5 & 0 & 1 \end{bmatrix} \text{ (a) (mod 15) and (b) (mod 16)}.$$

EXERCISE FOR THE READER 4.22: Show that the formula of Theorem 4.9 and of Part (2) of Theorems 4.10 are both special cases of the classical adjoint inverse formula of Proposition 4.11.

Application of Modular Matrices: The Hill Cryptosystem

Cryptography is the science of protecting data and communications. One of its main components involves communicating messages or information between designated parties by changing the appearance of the messages (or data) in ways that aim to make it extremely difficult or impossible for other parties to eavesdrop on or interfere with the transmission. This task is accomplished by means of a *cryptosystem*, which, roughly speaking, is any algorithm that can be used to encode a (say English) *plaintext* string (which is readily understood by all parties) into a corresponding *ciphertext* string (which is meant to be not understandable and is what will be transmitted over unsecure channels). An effective cryptosystem should have the property that it would be extremely difficult (ideally

---

[28] Even on a fast computer, the cofactor expansion method starts to get impractically slow when the size of the matrices exceeds about 9×9. Faster determinant and matrix inverse methods that are usually built into computing platforms are based on Gaussian elimination (a technique in linear algebra). These can routinely invert and compute determinants of matrices exceeding sizes of 500×500. In any case, since we will not be needing to invert (or compute inverses of) matrices larger than 9×9 (even in the computer exercises), the methods and algorithms in this section will be perfectly adequate for our purposes. There is one other issue concerning *round-off errors* that results in most standard (floating point arithmetic) computing systems rendering the results unreliable if the number of significant digits is too large (usually about 15, which coincidentally occurs when we try to compute determinants of mod 26 matrices of size larger than 9×9). We will discuss such concepts in greater detail in the next section.

impossible) for any unintended recipient of the ciphertext to decode it back into the plaintext. Figure 4.5 illustrates the idea of a cryptosystem, where a sender (Alice) encrypts her plaintext before sending it to the intended recipient (Bob). The intended recipient will have the knowledge to decode the ciphertext back into the plaintext.

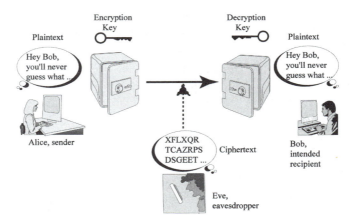

FIGURE 4.5: A basic reference illustration for a cryptosystem. Alice, the sender, wishes to send Bob, the intended recipient, a confidential message. On Alice's end, the message gets encrypted before it is sent to Bob, who, as the designated recipient, will be able to decrypt the message. Eve, the eavesdropper (a hacker) tries to intercept this message, but will not have the key to decode it.[29]

We will develop a cryptosystem that was invented in the early twentieth century known as the *Hill cryptosytem*.[30] The Hill cryptosystem has an important historical significance since it marked the beginning of a revolution in cryptography where cryptosystems were reliant on mathematical sophistication. It is based on modular matrix multiplication by invertible matrices. The Hill system is no longer in widespread use today because of its vulnerability to attacks. Section 4.5 will discuss some much more powerful cryptosystems that continue to be in widespread use. Readers who wish to learn more about the subject of cryptography may wish to consult [Sta-10].

---

[29] The names in our picture have become folk tradition in cryptography circles.

[30] This system is named after its creator, Lester S. Hill (1891–1961), an American mathematician who developed his cryptosystem in a couple of published papers in 1929 and 1931. During this time, since computers had not yet been invented, the calculations involving matrix multiplications would have been extremely tedious. To make his idea more marketable, Hill went further and patented a purely mechanical machine to perform the encryptions and decryptions. Since Hill's cryptosystem depends on a particular invertible matrix (key) being used, and his machine only worked for a single matrix, he advocated use of a so-called *involutory matrix*, that is, a matrix A which is its own inverse: $A = A^{-1}$. (Example 4.18 exhibited such a matrix.) Hill's cryptosystem, along with his mechanical devices were apparently used for radio call signals during World War II.

DEFINITION 4.12: (*The Hill cryptosystem*) The key in a **Hill cryptosystem** is an invertible **encoding matrix** A of integers mod 26. The *block length r* of the cryptosystem will be the number of rows (columns) of the matrix A. Given a plaintext message, we first translate it into a string of integers mod 26, using, for example the natural scheme of Table 4.2 (0 $\leftrightarrow$ a, 1 $\leftrightarrow$ b, $\cdots$, 25 $\leftrightarrow$ z). Thus, we may represent the plaintext message as a vector $[p_1 \, p_2 \cdots p_\ell]$, where each p_i is the element of $\mathbb{Z}_{26}$ representing the corresponding English letter, and ℓ is the length of the plaintext message. We then arrange the elements of this message into an r rowed matrix P by vertically stacking the plaintext elements p_i in order:

$$P = \begin{bmatrix} p_1 & p_{r+1} & \cdots \\ p_2 & p_{r+2} & \\ \vdots & \vdots & \ddots & \vdots \\ p_r & p_{2r} & \cdots \end{bmatrix}$$

Unless ℓ is a multiple of r, the final column of P will need to be padded with some additional elements. For the purposes of illustration and definiteness, we could use 13's (corresponding to a final string of n's). The encryption is performed by modular matrix multiplication,

$$C \equiv AP, \quad C \equiv \begin{bmatrix} c_1 & c_{r+1} & \cdots \\ c_2 & c_{r+2} & \\ \vdots & \vdots & \ddots & \vdots \\ c_r & c_{2r} & \cdots \end{bmatrix} \pmod{26}, \tag{19}$$

where the cipher text vector $[c_1 \, c_2 \cdots c_\ell]$, is read off in the same order as P was formed, from the matrix product C of (5). When translated back to English characters, the final string of n's (if any) would be ignored. Because the matrix A was chosen to be invertible, the decryption is performed simply by multiplying the encrypted matrix C by A^{-1}. (*Proof that this works:* $A^{-1}C = A^{-1}(AP) = (A^{-1}A)P = IP = P$. $\square$)

EXAMPLE 4.21: Use the Hill cryptosystem with encoding matrix $A = \begin{bmatrix} 1 & 2 \\ 1 & 3 \end{bmatrix}$ to encrypt the message "code blue alert." Note that $\det(A) = 1 \equiv 1 \pmod{26}$, so the matrix A is invertible (mod 26), and is thus a legitimate encoding matrix.

SOLUTION: The $\mathbb{Z}_{26}$ vector corresponding to the plaintext is (see Table 4.2)

$$[2 \ 14 \ 3 \ 4 \ 1 \ 11 \ 20 \ 4 \ 0 \ 11 \ 4 \ 17 \ 19].$$

Forming a 2 rowed matrix (filling in the final entry with a 13) and multiplying by the encryption matrix gives the corresponding matrix of ciphertext elements:

$$C \equiv AP \equiv \begin{bmatrix} 1 & 2 \\ 1 & 3 \end{bmatrix} \begin{bmatrix} 2 & 3 & 1 & 20 & 0 & 4 & 19 \\ 14 & 4 & 11 & 4 & 11 & 17 & 13 \end{bmatrix} \equiv \begin{bmatrix} 4 & 11 & 23 & 2 & 22 & 12 & 19 \\ 18 & 15 & 8 & 6 & 7 & 3 & 6 \end{bmatrix}$$

(mod 26). Translating the resulting $\mathbb{Z}_{26}$ ciphertext vector back into English letters gives the final ciphertext message:

plaintext:	c	o	d	e	b	l	u	e	a	l	e	r	t	n
plaintext in $\mathbb{Z}_{26}$:	2	14	3	4	1	11	20	4	0	11	4	17	19	13
ciphertext in $\mathbb{Z}_{26}$:	4	18	11	15	23	8	2	6	22	7	12	3	19	6
ciphertext:	E	S	L	P	X	I	C	G	W	H	M	D	T	G

Notice that the three instances of the plaintext letter e (shaded) were encrypted into three different letters in the ciphertext.

For the decryption process, the modular inverse of the encryption matrix needs to be computed. We presented methods for computing such modular matrix inverses (both by hand and by computer) earlier in this chapter. We recall that a $\mathbb{Z}_{26}$ modular matrix will be invertible, if, and only if its determinant is relatively prime to 26.

EXERCISE FOR THE READER 4.23: The following ciphertext was encrypted using the Hill cryptosystem using the encoding matrix $A = \begin{bmatrix} 1 & 1 & 0 \\ 1 & 0 & 1 \\ 1 & 1 & 1 \end{bmatrix}$:

TARIDWXGXWNUANFHHU. Decode this message.

One natural question arises: How does one go about creating an encoding matrix of a desired size for the Hill cryptosystem? One effective method is to randomly fill in the entries of a square matrix of specified size with integers (mod 26), and check to see whether the determininant is relatively prime to 26. If it is, we have produced a suitable encoding matrix. Otherwise we repeat this process until we create one. The computer exercises will illustrate this method.

As was demonstrated in the previous example, the Hill cryptosystem is a *polyalphabetic* cryptosystem, meaning that plaintext letters may encrypt to different ciphertext characters in different instances. Polyalphabetic cryptosystems were a huge breakthrough over the simple substitution cryptosystems that were used for centuries and succumb to frequency analysis attacks. Since changing one letter in the plaintext in the Hill system affects each of the letters in the corresponding ciphertext length r block, a frequency analysis approach would need to analyze blocks governed by the size of the encryption matrix. For block sizes greater than three, the number of r-letter sequences is at least 26^4, or about half a millon, so that frequency attacks are rendered unfeasible. If, however, the hacker possesses some information other than just a collection of ciphertexts, the Hill cryptosystem may succumb to an attack. The common scheme for such attacks is illustrated in the following example.

EXAMPLE 4.22: Suppose that the Hill cryptosystem is being used and it is known that the plaintext message "stand by for further instructions," was encrypted into "WTANVBOPGZRIJPTKZGBKTTQGZOARAH." Later on, the following ciphertext was intercepted: "TAMDAQVADEWPWDTKAUTCVE OMKMESXZQIWZIIRBOKTMADIKQNVSMMGCNZ." Decode this message.

SOLUTION: The first step is to determine the size of the encryption matrix A. Since we know that the two ciphertext messages that were intercepted have lengths 30 and 56, it follows that the block length r (A is an $r \times r$ matrix) must divide $\gcd(30,56) = 2$, so the key A must be a 2×2 matrix, which we write as $A = \begin{bmatrix} a & b \\ c & d \end{bmatrix}$. We got a bit lucky here that the gcd was a prime number, so the block length was completely determined. In cases where it is not, we would proceed in the same fashion as we do below, using trial and error on the possible block length sizes.

The basic idea of of computing A is simple. The first four letters of the known plaintext/ciphertext correspondence stan $\rightarrow$ WTAN are related by the matrix equation:

$$A \begin{bmatrix} 18 & 0 \\ 19 & 13 \end{bmatrix} \equiv \begin{bmatrix} 22 & 0 \\ 19 & 13 \end{bmatrix} (\bmod 26).$$

Now, if the matrix that is being multiplied by the unknown key A is invertible (mod 26), then we could simply right multiply this equation by its inverse to compute A. Since the determinant of this matrix is $18 \cdot 13 \equiv 0 \pmod{26}$, it is not invertible, so we cannot directly find A in this way. Nonetheless, the above matrix equation does allow us to narrow down the possibilities for A. Rather than going this route, it is simpler to just try another two pairs of plaintext/ciphertext correspondence. For example, the next two pairs give us dbyf $\rightarrow$ VBOP, leading us to the matrix equation:

$$A \begin{bmatrix} 3 & 24 \\ 1 & 5 \end{bmatrix} \equiv \begin{bmatrix} 21 & 14 \\ 1 & 15 \end{bmatrix} (\bmod 26).$$

The matrix $M = \begin{bmatrix} 3 & 24 \\ 1 & 5 \end{bmatrix}$ has determinant equal to 17 (mod 26), which is relatively prime to 26, so that we may now invert this matrix and use it to compute A:

$$M^{-1} \equiv 17^{-1} \begin{bmatrix} 5 & -24 \\ -1 & 3 \end{bmatrix} \equiv 23 \begin{bmatrix} 5 & 2 \\ 25 & 3 \end{bmatrix} \equiv \begin{bmatrix} 11 & 20 \\ 3 & 17 \end{bmatrix} (\bmod 26), \text{ so}$$

$$AM \equiv \begin{bmatrix} 21 & 14 \\ 1 & 15 \end{bmatrix} \Rightarrow A \equiv \begin{bmatrix} 21 & 14 \\ 1 & 15 \end{bmatrix} \cdot M^{-1} \equiv \begin{bmatrix} 21 & 14 \\ 1 & 15 \end{bmatrix} \cdot \begin{bmatrix} 11 & 20 \\ 3 & 17 \end{bmatrix} \equiv \begin{bmatrix} 13 & 8 \\ 4 & 15 \end{bmatrix} \pmod{26} \cdot$$

Now that we have the key, the rest is straightforward: To perform the required decryption, we first compute the inverse of the key (since $\det(A) \equiv 7 \pmod{26}$),

$$A^{-1} \equiv \begin{bmatrix} 13 & 8 \\ 4 & 15 \end{bmatrix}^{-1} \equiv 7^{-1} \begin{bmatrix} 15 & -8 \\ -4 & 13 \end{bmatrix} \equiv 15 \begin{bmatrix} 15 & -8 \\ -4 & 13 \end{bmatrix} \equiv \begin{bmatrix} 17 & 10 \\ 18 & 13 \end{bmatrix} \pmod{26}.$$

Converting the ciphertext to a two-rowed matrix of integers using Table 4.2, left multiplying by the key inverse, and then converting back to text produces the plaintext message:

      Leave at once to the safe house your location has been compromised.

(A final n was discarded, and spaces have been inserted.) We recommend that computations such as those carried out in this example be done on a computer.

EXERCISES 4.3:

1.   Given the matrices $A = \begin{bmatrix} 2 & -4 \\ 6 & 5 \end{bmatrix}$, $B = \begin{bmatrix} 8 & 4 \\ 3 & -2 \end{bmatrix}$, $C = \begin{bmatrix} 9 & -4 & 3 \\ -2 & 0 & 7 \\ -5 & -6 & 6 \end{bmatrix}$, $D = \begin{bmatrix} 2 & 6 \\ 7 & 4 \\ 5 & 6 \end{bmatrix}$, $E = \begin{bmatrix} 1 & 3 & 2 \\ 0 & 4 & 2 \\ 8 & 7 & 9 \end{bmatrix}$,

   do the following:

     (a) Determine the size of D.   (b) Find the (1,3) entry, c_{13} of the matrix C.

     (c) Find $A + B$.   (d) Find $E - C$.

     (e) Find $C + D$.   (f) Find $E + 2C$.

2.   Given the matrices $A = \begin{bmatrix} 3 & 5 \\ 9 & 7 \end{bmatrix}$, $B = \begin{bmatrix} -4 & 7 \\ -7 & 5 \end{bmatrix}$, $C = \begin{bmatrix} 4 \\ 0 \\ 9 \end{bmatrix}$, $D = \begin{bmatrix} 8 & 0 & 1 \\ -5 & 5 & 2 \\ -5 & -9 & 9 \end{bmatrix}$, $E = \begin{bmatrix} 6 & 5 & -2 \\ -4 & 7 & 7 \\ -9 & 5 & -6 \end{bmatrix}$,

   do the following:

     (a) Determine the size of C.   (b) Find the (2,1) entry, d_{21} of the matrix D.

     (c) Find $A + B$.   (d) Find $C + D$.

     (e) Find $E + D$.   (f) Find $2E - 3D$.

3.   With the matrices A, B, C, D, and E as in Exercise 1, find each of the following matrices (if they exist):

     (a) AB (b) BA (c) A^2 (d) CD (e) DE (f) EC

4.   With the matrices A, B, C, D, and E as in Exercise 2, find each of the following matrices (if they exist):

     (a) AB (b) BA (c) A^2 (d) CD (e) DE (f) EC

5.   Find the transposes of the matrices A, C, and D of Exercise 1.

6.   Find the transposes of the matrices A, C, and E of Exercise 2.

7. Find the inverses of the matrices A and B of Exercise 1 (if they exist).

8. Find the inverses of the matrices A and B of Exercise 2 (if they exist).

9. (*Chain Matrix Multiplication*) One might expect, since matrix multiplication is associative (Proposition 4.7(b)), that in computing a matrix product ABC, it does not matter if we compute AB first or BC first. In practice though, for large sized matrices it can make significant differences, as you will see in this exercise. Suppose that A has size 100×2, B has size 2×100, and C has size 100×2. Compare the total number of multiplications and additions required to compute $(AB)C$ and $A(BC)$.

10. (*Chain Matrix Multiplication*) (First read Exercise 9.) Suppose that A has size 100×1, B has size 1×1, C has size 1×100, and D has size 100×1.

    Compare the total number of multiplications and additions required to compute $((AB)C)D$ and $A(B(CD))$.

11. Given the matrices $A = \begin{bmatrix} 2 & 1 \\ 3 & 5 \end{bmatrix}$, $B = \begin{bmatrix} 7 & 4 \\ 0 & 3 \end{bmatrix}$, find each of the following matrices (if they are defined):

    (a) $A + B$ (mod 8)   (b) $A - B$ (mod 8)

    (c) $A + 2B$ (mod 8)   (d) $3A - B$ (mod 8)

12. Repeat each part of Exercise 11, but this time work mod 9.

13. Re-do Exercise 3, where the matrices are viewed as modular integer matrices (a) (mod 5) and (b) (mod 10). Then compare these answers to the answers obtained in Exercise 3 after the entries of each answer of Exercise 3 is converted into the appropriate modular integers.

14. Re-do Exercise 3, where the matrices are viewed as modular integer matrices (a) (mod 7) and (b) (mod 12). Then compare these answers to the answers obtained in Exercise 3 after the entries of each answer of Exercise 3 is converted into the appropriate modular integers.

15. Find the determinants and the inverses of the matrices A and B of Exercise 11 (mod 9), if they exist.

16. Find the determinants and the inverses of the matrices A and B of Exercise 11 (mod 10), if they exist.

17. Let $A = \begin{bmatrix} 9 & 1 & 5 \\ 9 & 5 & 6 \\ 8 & 6 & 7 \end{bmatrix}$. Do the following:

    (a) Compute $\det(A)$.   (b) Compute $\det(A)$ (mod 21).   (c) Compute $\det(A)$ (mod 23).

    (d) Compute A^{-1}.   (e) Compute A^{-1} (mod 21).   (f) Compute A^{-1} (mod 23).

18. Let $A = \begin{bmatrix} 2 & 5 & 0 \\ 6 & 0 & 7 \\ 5 & 11 & 7 \end{bmatrix}$. Do the following:

    (a) Compute $\det(A)$.   (b) Compute $\det(A)$ (mod 21).   (c) Compute $\det(A)$ (mod 23).

    (d) Compute A^{-1}.   (e) Compute A^{-1} (mod 21).   (f) Compute A^{-1} (mod 23).

19. (a) Use the Hill cryptosystem with encryption matrix $A = \begin{bmatrix} 5 & 2 \\ 11 & 5 \end{bmatrix}$ to encrypt each of the

following plaintext strings:

   (i) the shipment will arrive at noon   (ii) lay low until friday
   (iii) always use the back door   (iv) the phone is bugged

   (b) Decrypt each of the following ciphertexts that were encrypted using the Hill encryption matrix of Part (a).

   (i)   YFYCOUZEEPRTICADFJOCHS   (ii) RBGAGUDSRTSQKNOXILANQVBBWWTCKE
   (iii) KMUKKHLLFKIODBRO   (iv) NSOXEOTCLLQBIBEPQVMSQWFVMRQVQR

20. (a) Use the Hill cryptosystem with encryption matrix $A = \begin{bmatrix} 5 & 6 \\ -2 & -3 \end{bmatrix}$ to encrypt each of the

following plaintext strings:

   (i) two minutes until alarm sounds   (ii) spread out your team
   (iii) reconnaissance is on schedule   (iv) this hotel is safe

   (b) Decrypt each of the following ciphertexts that were encrypted using the Hill encryption matrix of Part (a).

   (i)   KIUFBWYBLOBJSIKVUNZUSYRB   (ii) QGXVMUBJHAAEOIANOXHAHCOXUI
   (iii) KEHGHRHAFOVZQZCIFROG   (iv) ICWUYFJOAUZGWUEYSOGYNI

21. (a) Use the Hill cryptosystem with encryption matrix $A = \begin{bmatrix} 1 & 1 & 0 \\ 1 & 0 & 1 \\ 1 & 1 & 1 \end{bmatrix}$ to encrypt each of the

strings of plaintext of Exercise 19, Part (a).

   (b) Decrypt each of the following ciphertexts that were encrypted using the Hill encryption matrix of Part (a).

   (i) MFFPZHZSM   (ii) PSAALAXGKXWAPZHKRZLYC   (iii) AXEGNOOOSHUMTMXGF
   TLLWTARIDWXGXWTAJEW   (iv) GNRFVWBGZKZDVWZLJJHMPPEGNRGUFLPHWLLY

22. (a) Use the Hill cryptosystem with encryption matrix $A = \begin{bmatrix} 5 & 2 & 9 \\ -4 & 2 & 3 \\ 3 & -1 & -5 \end{bmatrix}$ to encrypt each of the

strings of plaintext of Exercise 20, Part (a).

   (b) Decrypt each of the following ciphertexts that were encrypted using the Hill encryption matrix of Part (a).

   (i) URMSCSWQKUBY   (ii) MFKBLJTNDPFVHJMZVWWPINDW   (iii) HRZCCLNWYHXMO
   OTBLLLQDRHKOWJPYZLEYWYAFOI (iv) RRVRRKBFBCSVDIFVCCPCEHBPUUPLZYQMH

23. In each part, a ciphertext is given, along with a portion of the plaintext. The Hill cryptosystem was used for each encryption, and the size of the encrypting matrix is also provided. Indicate whether the key (the encoding matrix) can be uniquely determined mathematically from the given information. Explain why this is possible or impossible, and in cases where it is possible, perform the decryption.

   (a) A 2×2 encoding matrix was used.
   Ciphertext: G R P Z B V L T Z C N V I R G C Q Z C I J Q M H Z N U C
   Plaintext: * * * * * * * * t w * * * * e o n b * * * * * * * * * *

(b) A 2×2 encoding matrix was used.

Ciphertext: I G H O E S Z W J F P I T W B M L L P O F R X J O R T JZI

Plaintext: * * * r e o u * * * * * * e r * * * * e l * * * * * * * * *e

(c) A 3×3 encoding matrix was used.

Ciphertext: S W M L Q L B U C I D V Q P J Y V Y J G K F K Y Z L S

Plaintext: * * * * * * r y l * * * * * * w i n * * * r o v * * *

24.   Repeat the directions of Exercise 23 for each of the following:

(a) A 2×2 encoding matrix was used.

Ciphertext: M T A S W M T G S O J F D B N N A O K S K T U P L Z

Plaintext: * * s s * * * * * * * * a t * * * * * * * * * r s *

(b) A 2×2 encoding matrix was used.

Ciphertext: E G S O I T T I W P M B H E I E Z F T S P C M O L W J MXW

Plaintext: a w * * * * * * * * * * * * * n b e * * * * * o v * **

(c) A 3×3 encoding matrix was used.

Ciphertext: N T B E A H L H K Y N A A V C X R S A A R Y C J N K H

Plaintext: * * * m i l * * * n d o * * * r s i * * * * * * * *

25.   (*A Chosen Ciphertext Attack Against the Hill Cryptosystem*) Suppose that we know that a Hill cryptosystem is being used with an encoding matrix A of size $n \times n$. Suppose further that we have access to the encoder and are able to enter n plaintext strings, each having length n. What would be a good choice for these n plaintexts that would allow us to completely read off the encoding matrix A from the resulting ciphertexts?

26.   (*Problems with a Noninvertible Encoding Matrix for the Hill Cryptosystem*) Give an example of a 2×2 matrix A of mod 26 modular integers that is not invertible (mod 26), and two different digraphs (plaintext pairs) that would encrypt under A to the same ciphertext under the Hill cryptosystem. (Of course, A would not be a legitimate encryption matrix since it is not invertible.)

27.   Prove the following matrix distributive law from Proposition 4.7(c): $(A + B)C = AC + BC$.

28.   Prove Proposition 4.7(d): $\alpha(AB) = (\alpha A)B = A(\alpha B)$.

29.   (*The Transpose of the Sum is the Sum of the Transposes*) Prove that if A and B are matrices of the same size, then $(A + B)' = A' + B'$.

30.   (*The Transpose of the Product is the Reverse Order Product of the Transposes*)

(a) Prove that if A and B are matrices, then $(AB)' = B'A'$, whenever the left side is defined.

(b) Extend the result in Part (a) to general matrix products: $(A_1 A_2 \cdots A_t)' = A_t' A_{t-1}' \cdots A_1'$, whenever the left side is defined.

31.   (a) Prove that if A and B are invertible matrices, then so is AB, and $(AB)^{-1} = B^{-1}A^{-1}$.

(b) Prove that if A is an invertible matrix and t is a positive integer, then A^t is also invertible and $(A^t)^{-1} = (A^{-1})^t$.

32.   If A is an invertible matrix, prove that its transpose A' is also invertible. What is $(A')^{-1}$?

NOTE: The following two exercises will provide a proof of the classical adjoint formula for the inverse of a matrix (Propositon 4.11). This proof will rely on a generalization of the cofactor expansion formula for determinants that was given in formula (17). That formula corresponds to a cofactor expansion along the first row, the general version below allows for an analogous expansion along any row or column of a square matrix A. As in the section, for an $n \times n$ matrix A and indices i, and j, A_{ij} denotes the $(n-1) \times (n-1)$ submatrix obtained from A by deleting the ith row and jth column.

PROPOSITION 4.12: (*General Cofactor Expansions for Determinants of Square Matrices*) Suppose that $A = [a_{ij}]$ is a square matrix. For any indices i, j $(1 \leq i, j \leq n)$ we have the following expansions for the determinant of A:

(a) (*Cofactor Expansion Along the ith Row*)

$$\det(A) = (-1)^{i+1} a_{i1} \det(A_{i1}) + (-1)^{i+2} a_{i2} \det(A_{i2}) + \cdots + (-1)^{i+n} a_{in} \det(A_{in}).$$

(b) (*Cofactor Expansion Along the jth Column*)

$$\det(A) = (-1)^{1+j} a_{1j} \det(A_{1j}) + (-1)^{2+j} a_{2j} \det(A_{2j}) + \cdots + (-1)^{n+j} a_{nj} \det(A_{nj}).$$

For a proof the interested reader is referred to [HoKu-71].

33.   Prove that if A is a square matrix that has either two identical columns or two identical rows, then $\det(A) = 0$.
Suggestion: Show that A cannot have an inverse as follows: if the i_1 th row and the i_2 th row of A were the same, and if B is any matrix (that can be multiplied by A), then the i_1 th row and the i_2 row of AB must be the same. So there is no way that we could ever have $AB = I$. A similar argument works if A has two identical columns. Now apply Theorem 4.8.

34.   Prove Proposition 4.11 (on the classical adjoint formula for the inverse of a matrix).
Suggestion: Use Proposition 4.12 along with the result of Exercise 33.

35.   (*Block Matrix Multiplication*) Suppose first that A and B are $n \times n$ matrices, where n is even.
(a) Show that if $C = AB$, and if these three matrices are partitioned in the natural way as:

$$\begin{bmatrix} C_{11} & C_{12} \\ C_{21} & C_{22} \end{bmatrix} = \begin{bmatrix} A_{11} & A_{12} \\ A_{21} & A_{22} \end{bmatrix} \begin{bmatrix} B_{11} & B_{12} \\ B_{21} & B_{22} \end{bmatrix},$$

(where each of the submatrices has size $(n/2) \times (n/2)$), then we have:

$$\begin{aligned} C_{11} &= A_{11}B_{11} + A_{12}B_{21} & C_{12} &= A_{11}B_{12} + A_{12}B_{22} \\ C_{21} &= A_{21}B_{11} + A_{22}B_{21} & C_{22} &= A_{21}B_{12} + A_{22}B_{22} \end{aligned}.$$

In other words, this block matrix multiplication behaves exactly like a 2×2 (ordinary) matrix multiplication. This technique is known as **block matrix multiplication**, and the smaller submatrices are known as the **blocks**.
(b) Develop a similar version of the block matrix multiplication method of Part (a) that uses a 3×3 array of block matrices in case n is a multiple of 3.
(c) The general formulation of block matrix multiplication does not require square matrix blocks. This part will examine what is needed for the most general formulation. Suppose that the matrices A and B are partitioned into blocks as follows:

$$A = \begin{bmatrix} A_{11} & A_{12} & \cdots & A_{1m} \\ A_{21} & A_{22} & \cdots & A_{2m} \\ \vdots & \vdots & \ddots & \vdots \\ A_{n1} & A_{n2} & \cdots & A_{nm} \end{bmatrix}, \quad B = \begin{bmatrix} B_{11} & B_{12} & \cdots & B_{1r} \\ B_{21} & B_{22} & \cdots & B_{2r} \\ \vdots & \vdots & \ddots & \vdots \\ B_{m1} & B_{m2} & \cdots & B_{mr} \end{bmatrix},$$

where the number of columns of each block A_{ik} equals the number of rows of each block B_{kj}. Find a corresponding decomposition into blocks of the product matrix $C = AB$, and write down a formula for each block of the product C_{ij} in terms of the blocks of A and B, so that the resulting method generalizes those of Parts (a) and (b).

Note: Recall (as explained in this section) that the number of multiplications for an $n \times n$ matrix multiplication is n^3. The above block matrix multiplication scheme in Part (a) involves eight $(n/2) \times (n/2)$ matrix multiplications, and thus a total of $8(n/2)^3 = n^3$ matrix multiplications; so we did not save any time with this scheme.

36. (*Strassen's Method for Fast Matrix Multiplication*) (First read Part (a) of Exercise 35.) Suppose first that A and B are $n \times n$ matrices, where n is even. In Part (a) of Exercise 35, the block matrix multiplication method worked out to computing eight $(n/2) \times (n/2)$ matrix multiplications, and this worked out to exactly the same number of real number multiplications as ordinary matrix multiplication. Strassen [Str-69] discovered a very clever way to reorganize the arithmetic so as to reduce the problem to computing only the following <u>seven</u> $(n/2) \times (n/2)$ matrix multiplications:

$$P_1 = (A_{11} + A_{22})(B_{11} + B_{22}) \qquad P_5 = (A_{11} + A_{12})B_{22}$$
$$P_2 = (A_{21} + A_{22})B_{11} \qquad P_6 = (A_{21} - A_{11})(B_{11} + B_{12})$$
$$P_3 = A_{11}(B_{12} - B_{22}) \qquad P_7 = (A_{12} - A_{22})(B_{21} + B_{22})$$
$$P_4 = A_{22}(B_{21} - B_{11})$$

(a) Show that the blocks of the product matrix $C = AB$, as defined in the decomposition of Part (a) of Exercise 31, are given in terms of the above products by:

$$C_{11} = P_1 + P_4 - P_5 + P_7 \qquad C_{12} = P_3 + P_5$$
$$C_{21} = P_2 + P_4 \qquad C_{22} = P_1 - P_2 + P_3 + P_6$$

This form of block matrix multiplication is called **Strassen's method**.

(b) Show that if A and B are $n \times n$ matrices where $n = 2^K$, and if we recursively apply Strassen's method to perform the matrix multiplication AB until all of the block submatrices are of size 1×1 (i.e., after K recursions), then the total number of (single number) multiplications and additions required is less than 7^{K+1}. Explain how this fact can be translated into an upper bound for the total number of arithmetic operations being less than a constant times $n^{\log_2 7} \approx n^{2.81}$.

(c) Show more precisely that the number of single number arithmetic operations in recursively using Strassen's formula as indicated in Part (b) is $7 \cdot 7^K - 6 \cdot 4^K$.

Suggestion: For Part (a), verify that each of the asserted blocks agree with the basic block formulas in Part (a) of Exercise 31. Use the properties of matrix multiplication, and be careful to recall that it is not commutative. For Part (c) use mathematical induction.

NOTE: (*Zero-One Matrices*) A **zero-one matrix** is simply a matrix each of whose entries is either zero or one. When endowed with the "special" operations that are described in the following definition, zero-one matrices have some applications to graph theory. These definitions will be based on the Boolean arithmetic operators:

If x and y are Boolean variables (this simply means that x and y equal either 0 or 1), we define

$$x \wedge y = \begin{cases} 1, \text{ if } x = y = 1 \\ 0, \text{ otherwise} \end{cases}, \quad x \vee y = \begin{cases} 0, \text{ if } x = y = 0 \\ 1, \text{ otherwise} \end{cases}. \text{ }_{31}$$

DEFINITION 4.13: Suppose that $A = [a_{ij}]$ and $B = [b_{ij}]$ are zero-one matrices. If A and B have the same size, the **join** of A and B, denoted $A \vee B$ is the zero-one matrix whose (i, j) entry is given by $a_{ij} \vee b_{ij}$. The **meet** of A and B, denoted $A \wedge B$ is the zero-one matrix whose (i, j) entry is given by $a_{ij} \wedge b_{ij}$. If A has size $n \times m$ and B has size $m \times r$, then the Boolean product of A and B, denoted $A \odot B$, is the $n \times r$ matrix whose (i, j) entry is $(a_{i1} \wedge b_{1j}) \vee (a_{i2} \wedge b_{2j}) \vee \cdots \vee (a_{im} \wedge b_{mj})$. If A is a square zero-one matrix, Boolean powers of A are defined recursively by: $A^{[0]} = I$, $A^{[n+1]} = A^{[n]} \odot A$.

Exercises 37–40 will investigate this arithmetic.

37. Given the zero-one matrices $A = \begin{bmatrix} 1 & 1 \\ 0 & 1 \end{bmatrix}$, $B = \begin{bmatrix} 0 & 1 \\ 1 & 0 \end{bmatrix}$, $C = \begin{bmatrix} 1 & 1 & 0 \\ 0 & 0 & 1 \\ 0 & 1 & 1 \end{bmatrix}$, $D = \begin{bmatrix} 1 & 1 & 1 \\ 1 & 0 & 1 \\ 0 & 0 & 1 \end{bmatrix}$, compute

    each of the following

    (a) $A \wedge B$ (b) $A \vee B$ (c) $C \wedge D$
    (d) $C \vee D$ (e) $A \odot B$ and $B \odot A$ (f) $C \odot D$

38. Given the zero-one matrices $A = \begin{bmatrix} 1 & 1 \\ 1 & 1 \end{bmatrix}$, $B = \begin{bmatrix} 1 & 1 \\ 1 & 0 \end{bmatrix}$, $C = \begin{bmatrix} 0 & 1 & 1 \\ 0 & 1 & 0 \\ 0 & 1 & 1 \end{bmatrix}$, $D = \begin{bmatrix} 1 & 1 & 1 \\ 1 & 0 & 0 \\ 1 & 0 & 0 \end{bmatrix}$, compute

    each of the following

    (a) $A \wedge B$ (b) $A \vee B$ (c) $C \wedge D$
    (d) $C \vee D$ (e) $A \odot B$ and $B \odot A$ (f) $C \odot D$

39. Prove the following identities for zero-one matrix arithmetic (valid whenever either side is defined):
    (a) (*Commutativity of Join and Meet*) $A \vee B = B \vee A$, $A \wedge B = B \wedge A$.
    (b) (*Associativity of Join and Meet*) $(A \vee B) \vee C = A \vee (B \vee C)$, $(A \wedge B) \wedge C = A \wedge (B \wedge C)$.

40. Prove the following identities for zero-one matrix arithmetic (valid whenever either side is defined):
    (a) (*Distributive Laws*) $A \vee (B \wedge C) = (A \vee B) \wedge (A \vee C)$, $A \wedge (B \vee C) = (A \wedge B) \vee (A \wedge C)$.
    (b) (*Associativity of Boolean Product*) $(A \odot B) \odot C = A \odot (B \odot C)$.

---

[31] It is not necessary to have studied Boolean arithmetic (Section 2.3) to understand this topic. For those who have, these two operations are simply the Boolean product and Boolean sum, respectively (see Section 2.3). For those who have not, if we view 0 as false and 1 as true, these two operators also correspond to the logical AND and OR operators, respectively.

COMPUTER EXERCISES 4.3:

1. (*Program for Matrix Multiplication*) (a) Even if your computing platform has a built-in function for this, write your own program with syntax `C = MatrixMult(A,B)`, that inputs two matrices `A` and `B`, and if the matrix product `C = AB` is defined, the program will output this product. If the matrix multiplication is not possible, the program will output an error message of the form "Error: the inside dimensions must agree in order for matrix multiplication to be possible."
    (b) Use your program to re-do Exercise 3.
    (c) Use your program to re-do Exercise 4.

2. Consider the matrices $A = \begin{bmatrix} 1 & 1 \\ 1 & 1 \end{bmatrix}$ and $B = \begin{bmatrix} 1 & 0 \\ 1 & 1 \end{bmatrix}$.

    (a) Compute $A^2, A^3, A^4, A^{10}, A^{40}$. Do you notice any patterns?

    (b) Compute $B^2, B^3, B^4, B^{10}, B^{40}$. Do you notice any patterns?

3. (*Roots of Matrices*)
    (a) Can you find a 2×2 matrix A such that $A \neq I$ and $A^2 = I$?

    (b) Can you find a 2×2 matrix S such that $S^2 = A$, where A is the matrix of Computer Exercise 2?

    (c) Can you find a 2×2 matrix T such that $T^2 = B$, where B is the matrix of Computer Exercise 2?

    (d) Can you find a 3×3 matrix $U \neq I$ such that $U^3 = I$?
    Suggestion: Try to do these by hand first; if you get stuck, do some computer searches.

4. (*Program for Inverting a 2×2 matrix*)
    (a) Write a program with syntax `Ainv = MatInv2(A)`, that inputs a 2×2 matrix `A`, computes its inverse (if it exists) `Ainv`, using Theorem 4.9, and outputs the result. If the determinant is zero, the program will output an error message of the form "Error: the matrix is not invertible."
    (b) Use your program to re-do Exercise 7.
    (c) Use your program to re-do Exercise 8.

5. (*Program for Modular Integer Matrix Addition and Scalar Multiplication*)
    (a) Write a program with syntax `C = ModMatAdd(A,B,m)`, that inputs two same-sized modular integer matrices `A` and `B`, along with a modulus `m` (an integer > 1), and outputs the sum `C = A + B` (mod m).
    (b) Write a program with syntax `C = ModMatScalMult(a,A,m)`, whose inputs are: a, a modular integer, `A`, a modular integer matrix, and `m`, a modulus (an integer > 1), and outputs the matrix `C = aA` (mod m).
    (c) Use your program to re-do Exercise 11.
    (d) Use your program to re-do Exercise 12.

6. (a) Write a program with syntax `C = ModMatrixMult(A,B,m)`, that inputs two modular integer matrices `A` and `B`, along with a modulus `m` (an integer > 1), and outputs the matrix product `C = AB` (mod m). If the matrix multiplication is not possible, the program will output an error message of the form "Error: the inside dimensions must agree in order for matrix multiplication to be possible."
    (b) Use your program to re-do Exercise 13.
    (c) Use your program to re-do Exercise 14.

7. (*Program for Inverting a 2×2 Modular Integer Matrix*)
   (a) Write a program with syntax `Ainv = ModMatInv2(A,m)`, that inputs a 2×2 matrix modular integer A, and a modulus m (an integer > 1), computes its inverse (if it exists) Ainv, using Proposition 4.10, and outputs the result. If the determinant is not relatively prime to the modulus, the program will output an error message of the form "Error: the matrix is not invertible mod *m*."
   (b) Use your program to re-do the inverse questions of Exercise 15.
   (c) Use your program to re-do the inverse questions of Exercise 16.
   Suggestion: It will be helpful to call on one of the programs of the computer exercises from the previous section for computing inverses of modular integers.

8. (*Program for Inverting a 3×3 modular integer matrix*)
   (a) Write a program with syntax `Ainv = ModMatInv3(A,m)`, that inputs a 3×3 matrix of modular integers A, and a modulus m (an integer > 1), computes its inverse (if it exists) Ainv, using Proposition 4.11, and outputs the result. If the determinant is not relatively prime to the modulus, the program will output an error message of the form "Error: the matrix is not invertible mod *m*."
   (b) Use your program to re-do Parts (e) and (f) of Exercise 17.
   (c) Use your program to re-do Parts (e) and (f) of Exercise 18.

9. (*Program for Computing Determinants of Modular Matrices Using Cofactor Expansion*)
   (a) Write a program with syntax `detA = CofactorModDet(A,m)`, that inputs an $n \times n$ matrix of modular integers A, and a modulus m (an integer > 1), computes its inverse (if it exists) Ainv, using Algorithm 4.8 (cofactor expansion along the first row), and outputs the result.
   (b) Use your program to re-do the inverse questions of Exercises 15 and 16.
   (c) Use your program to re-do Parts (b) and (c) of Exercises 17 and 18.

10. (*Program for Computing Inverses of Modular Matrices Using Cofactor Expansion and Classical Adjoints*)
    (a) Write a program with syntax, `invA = CofactorModMatrixInv(A,m)`, that inputs an $n \times n$ matrix of modular integers A, and a modulus m (an integer > 1), computes its inverse (if it exists) Ainv, using Algorithm 4.8 (cofactor expansion along the first row) and the classical adjoint formula for matrix inverses (Proposition 4.11), and outputs the result. Alternatively, you may wish to directly call on the program of Computer Exercise 9.
    (b) Use your program to re-do the inverse questions of Exercises 15 and 16.
    (c) Use your program to re-do Parts (e) and (f) of Exercises 17 and 18.

11. (*Program for Computing Modular Matrix Inverses on Platforms with Ordinary Inverse Matrix Programs*) As mentioned in the text, computing classical adjoints and determinants using cofactor expansions turns out to be not very efficient. Many computing platforms have built-in programs for numerically computing matrix inverses and determinants. This exercise assumes the reader is working on such a platform.
    (a) Write a program with syntax, `Ainv = ModMatInv(A,m)`, that inputs an $m \times m$ matrix of modular integers A, and a modulus m (an integer > 1), and outputs its modular integer inverse (if it exists) Ainv. The computation should be done using formula (18), and your platform's numerical programs for (ordinary) determinants and (ordinary) matrix inverses. If the determinant is not relatively prime to the modulus, the program will output an error message of the form "Error: the matrix is not invertible mod *m*."
    (b) Use your program to re-do the inverse questions of Exercise 15 and 16.
    (c) Use your program to re-do Parts (e) and (f) of Exercises 17 and 18.
    Suggestion: Suppose that your platform's numerical programs for computing the determinant and inverse of a square matrix *A* have syntax det(A) and inv(A), respectively. To avoid any roundoff errors that might come up in your platform's numerical functions, results that are known to be integer values should be rounded off; thus (18) should be implemented in fashion such as:
    `ModMatInv(A, m) = modinv(round(det(A)), m)*round(det(A)*inv(A))`

where the `modinv` program is that of Computer Exercise 3 of Section 4.2.

12. (*Program for Hill Cryptosystem*) Write a program `StrOut = HillCrypt(str,A)` that inputs a string `str` of plaintext in lower-case English letters and an invertible encryption matrix A (mod 26), representing the key. The output `StrOut` should be the corresponding ciphertext (in upper-case letters) after the Hill cryptosystem with encryption matrix A is applied to the plaintext.
    (a) Use your program to re-do the encryptions of (ordinary) Exercise 19(a).
    (b) Use your program to re-do the encryptions of (ordinary) Exercise 21(a).
    (c) Use your program to re-do the decryptions of (ordinary) Exercise 19(b).
    Note: In Part (c), you will need to change your ciphertexts to lower-case and compute the inverse of the encryption matrix (mod 26).

13. (*Program for Decryption of the Hill Cryptosystem*) Write a program `strOut = HillDeCrypt(STR, A)` that inputs a string `STR` of ciphertext in upper-case English letters, and an invertible encryption matrix A (mod 26), representing the key. The output `strOut` should be the corresponding plaintext (in lower-case letters) before the Hill cryptosystem with encryption matrix A was applied.
    (a) Use your program to re-do the decryptions of (ordinary) Exercise 19(b).
    (b) Use your program to re-do the decryptions of (ordinary) Exercise 21(b).
    Note: This program can be written very simply if it uses the corresponding encryption program of the preceding exercise. The only thing needed will be to compute (inside the program) the inverse of the encryption matrix (mod 26), and this can be done by calling on one of the programs for computing modular inverses of Computer Exercises 10 or 11.

14. (*Known Plaintext Attack on the Hill Cryptosystem*) For each part, a corresponding pair of a known plaintext and ciphertext are given, along with an additional ciphertext. The encryption was performed using a Hill cryptosystem. Determine the encryption matrix, and decrypt the additional ciphertext.
    (a) plaintext: `gotothecourtyardatthreeoclock`, ciphertext: `EWEJBACEUONYSQV WXRBAOBWMJNQCBL`; additional ciphertext: `IUHRLIUGDFUFKRRFBAJZDHLTSVBA PRRFZC`
    (b) plaintext: `proceedtothehamburgmaintrainstationplatformthree`, ciphertext: `BOGUWIHYALIQXVOVOVKIQWWJVHKTSLOJMCSJDITPMTOYVDRG`; additional ciphertext: `LIQNJIIDXKOKSHCWSCKUDIRMPJPILYTBGGKYFAZSQJ`
    (c) plaintext: `threemillionisinthesuitcase`, ciphertext: `NACVRYPNKJNCPWRF XBWAALEXJCPL`; additional ciphertext: `TLOYDFGRFFWLETWCRKUXHSDDOYCYPWRF WXHTZVGNUNJDYWELXOWTZDAX`
    (d) plaintext: `thediamondsareinasafedepositboxatthecentralbankinkalam azoo`, ciphertext: `JGGIAGJJVAOXKKKNHBDQCALBFYFKSCYBAFPCFFHELPXYOGIMKYSH PBYANYJB`; additional ciphertext: `ACDQVKDJXCRCRWWHYJUYVTVUFHPBZDVFFBHHFSS TYBYMHATMXLBEXDEHEWWE`

15. (*Program for Generating Encoding Matrices for the Hill Cryptosystem*) (a) Write a program having the following syntax: `A = HillEncodingMatrixGenerator(n)`, that inputs a positive integer n > 1 and outputs an n×n matrix A of integers that is invertible (mod 26), and thus may be used as a key for the Hill cryptosystem. The program should be based on the random generation trial and error method indicated in the text: Generate an n×n matrix of integers in the range 0 through 25, and check whether the determinant is relatively prime to 26. If it is, we have an invertible Hill matrix, otherwise repeat this process until we obtain one.
    (b) Use your program to generate a 3×3 Hill encryption matrix; check that the outputted matrix is indeed invertible.
    (c) Use your program to generate a 5×5 Hill encryption matrix; check that the outputted matrix is indeed invertible.
    (d) Use the encryption matrix that you obtained in Part (c) in the Hill cryptosystem to encrypt the plaintext message: "gotothecourtyardatthreeoclock." Then decrypt the resulting ciphertext.

16. (*Estimating the Probability that a Randomly Generated Matrix is Invertible*)
(a) How many modular integer 2×2 matrices are there (mod 26)?
(b) Estimate the probability that a randomly generated modular integer 2×2 matrix will be invertible (mod 26) (and thus be a legitimate encoding matrix for the Hill cryptosystem) as follows: First initiate a counter: `count = 0`; then start a for loop to run through 10,000 trials of the following experiment: randomly generate a modular integer 2×2 matrix of integers in the range 0 through 25, and check whether its determinant is relatively prime to 26. If it is (so the matrix is invertible) then add one to the counter: `count = count + 1`. After the for loop has executed, estimate the desired probability as the final value of the counter divided by the number of trials (i.e., `count/10,000`).
(c) Exactly how many 2×2 modular integer matrices are invertible (mod 26)? Use this answer to compute the exact probability of the quantity that was estimated in Part (b).
Suggestion: Part (a) is a simple counting problem; use the multiplication principle. For Part (c), simply use a nested for loop to run through all of the matrices (mod 26) along with a counter to count the number that are invertible (mod 26). The exact probability will be the total number of invertible matrices divided by the total number of matrices (mod 26).

17. (*Estimating the Probability that a Randomly Generated Matrix is Invertible*)
Repeat Parts (a) and (b) of Computer Exercise 16, but this time for 3×3 matrices (mod 26).
Note: There are too many matrices here to be able to do Part (c) of Computer Exercise 16.

18. (*Programming Strassen's Algorithm for Fast Matrix Multiplication*)
(a) Write a recursive program with syntax `C = Strassen(A,B)`, that inputs two $n \times n$ matrices A and B, where n is a power of 2. The output will be the product matrix $C = AB$, computed by recursively applying Strassen's algorithm as explained in Part (b) of Exercise 36.
(b) Randomly generate some pairs of $n \times n$ matrices A and B for each of the following values of n: 4, 8, 16, 64, 512.
(c) Compare the runtimes for computing the matrix products AB for each of the pairs of matrices generated in Part (b) with (i) the program you wrote in Part (a), (ii) the `MatrixMult` program that you wrote in Computer Exercise 1, and (iii) your platform's built-in matrix multiplying program (if available). Comment on the results.
Note: Implementations and performance results of Strassen's algorithm for Part (a) will vary and will depend on the particular platform that you are using.

19. (*Programs for Zero-One Matrix Calculations*)
(a) Write a program with syntax `C = ZeroOneJoin(A,B)`, that inputs two zero-one matrices of the same size: A and B, and outputs their join $C = A \vee B$ (see Definition 4.9).
(b) Write a program with syntax `C = ZeroOneMeet(A,B)`, that inputs two zero-one matrices of the same size: A and B, and outputs their meet $C = A \wedge B$ (see Definition 4.9).
(c) Write a program with syntax `C = BooleanProduct(A,B)`, that inputs two zero-one matrices A and B, and outputs the Boolean product $C = A \odot B$. If this operation is not possible, the program will output an error message of the form "Error: the inside dimensions must agree in order for the Boolean product to be defined."
(d) Use your program to re-do Exercise 25.
(e) Use your program to re-do Exercise 26.

4.4: FLOATING POINT ARITHMETIC

In (pure) mathematics courses, we learn all about the **exact arithmetic** of real numbers. This arithmetic has unlimited precision, in which any real number can be written with an infinite decimal expansion. When computers are used to do mathematics, it is not possible to store and manipulate infinite numbers of digits. In fact, the decimal expansion of just a single irrational number, like

$$e = 2.718281828459045235360287471352662497757247096369995957...,$$

that goes on forever without termination or repetition, has more information in it than any or all of the computers in the world could possibly store (now or at any time in the future). This simple fact is easily seen since any computer can have only a finite amount of memory and thus cannot store an infinite amount of information. Because of this harsh reality, computer arithmetic systems need to conform to some sort of compromise. This is done in one of two ways: Using either a **floating point arithmetic system,** where a fixed number of significant digits is used for all numbers and computations, or a **symbolic arithmetic system,** where numbers are stored as abstract symbols that are manipulated, combined, and simplified according to various mathematical axioms, identities, and theorems. Both systems will have their advantages and disadvantages, and some platforms are able to work with both. Floating point arithmetic has much less baggage and thus typically performs faster than symbolic systems. It is the arithmetic system most often used in scientific and engineering computations.

The easiest way to understand a floating point arithmetic system is to use a notation similar to the familiar scientific notation. Any nonzero real number x can be written in its **decimal** (or **base 10**) **expansion** as follows:

$$x = \pm .d_1 d_2 \cdots d_s d_{s+1} \cdots \times 10^e, \tag{20}$$

where e is an integer and, in general, the sequence of digits $d_i = 0, 1, 2, \cdots,$ or 9 (but $d_1 \neq 0$) may be infinite.[32] This representation of a real number is unique, provided that any decimal expansion containing an infinite string of nines is identified with its rounded up representation (see Exercise 19). In a **floating point arithmetic system** with s **significant digits,** the number x of (20) is identified with the number shown in (15) below that corresponds to using only the first s digits in the decimal expansion of x:[33]

$$x = \pm .d_1 d_2 \cdots d_s \times 10^e. \tag{21}$$

Since the numbers in (21) still constitute a infinite set (so still too many for a computer to be able to deal with), limits are also placed on the size of the exponent

---

[32] Scientific notation differs from expansion (20) in that the former requires that a single nonzero digit appear before the decimal. Expansions can be formed that use other bases than 10, e.g., binary (base 2) or hexadecimal (base 16), and all of these are similar to the integer systems developed in Section 4.1. Internal computer calculations are done in a base that is a power of two, and the results are translated to decimal form when displayed. The concepts of floating point arithmetic, however, are most easily grasped (and well represented) in the decimal setting, so we will restrict our discussion of floating point arithmetic to base 10.

[33] Notice that (21) came from (20) by simply "chopping" all digits after the sth digit. This type of floating point arithmetic conversion system is known as *chopped arithmetic.* All floating point arithmetic systems work in either chopped arithmetic (as above) or in *rounded arithmetic,* where the last digit d_s is obtained by first looking at the subsequent digit d_{s+1} and rounding d_s up (in the usual way) when $d_{s+1} \geq 5$. Throughout this section (and book) we work in chopped arithmetic.

$e: m \le e \le M$.   Most floating point arithmetic systems conform to the so-called *IEEE*[34] *double precision standard*, which corresponds approximately (when translated from its binary arithmetic system) to the parameters $s = 15$, $m = -324$, and $M = 308$.   Some computing systems may display more than the s significant digits in which they are working.   In such a case, any digits past the first s digits must be viewed as unreliable "white noise."   In a floating point arithmetic system, the **unit roundoff** u (also known as the **machine epsilon**) is the smallest positive number that can be added to 1 so that the result is distinguishable from 1 in the system.

EXAMPLE 4.23:   Write a simple program in pseudocode to estimate the unit roundoff/machine epsilon u in your (floating point) computing platform.

SOLUTION: The pseudocoded program below will simply add smaller and smaller (negative) powers 10^{-k} of 10 to 1 until the sum $1 + 10^{-k} = 1$ (in floating point arithmetic).

```
k = 1  (initialize exponent)
WHILE 1 + 10^(-(k+1))  ~= 1
     update k = k + 1
END WHILE
OUTPUT  k
```

The output for an IEEE (double precision floating point) system for this code would be 15, corresponding to the fact that there are 15 or 16 significant digits for the system.   This will mean that the actual machine epsilon will be somewhere between 10^{-k} and $10^{-k}/10 = 10^{-(k+1)}$ (the next smaller power of 10), where k is the final value of k in the loop, i.e., the machine epsilon u will satisfy $10^{-16} < u \le 10^{-15}$.   We stress this important point once again: As far as the floating point system is concerned, 1 and $1 + 10^{-16} = 1.0000000000000001$ are indistinguishable!

Due to the parameters m and M of a floating point arithmetic system, the smallest number that can be recognized in an IEEE standard system is about 10^{-324}, and the largest is about 10^{308}.   A natural question to ask is what will happen if a number arises that is smaller than this minimum or larger than this maximum. We will do an example in a typical IEEE system and then explain what has happened.

```
INPUT: s = 10^(-100), t = 1/s   OUTPUT: s = 1.0000e-100, t = 1.0000e+100
INPUT: a = s^4  OUTPUT: a = 0
INPUT: b = t^4  OUTPUT: b= Infinity
```

---

[34] The *IEEE* (*Institute of Electrical and Electronics Engineers*) is a nonprofit, technical professional association of more than 350,000 individual members in 150 countries.

The number s^4, which equals 10^{-400} (in exact arithmetic), is too small to be recognizable by the system (although $s = 10^{-100}$ was fine). Such a computation is said to have **underflowed**. The system automatically (and without any warning) set the underflow to equal zero. This can sometimes lead to problems so the users must be aware if and when this can happen in the course of a series of computations. In contrast, the number t^4, which equals 10^{400} (in exact arithmetic), is too large to be recognizable by the system, and the computation is said to have **overflowed**. The system set the overflow equal to `Infinity` (∞). From this point on, the usual rules of infinity will apply to the variable b and can lead to discrepancies with exact arithmetic. For example, in exact arithmetic, we certainly know that $ab = 10^{-400} \cdot 10^{400} = 1$, whereas if we try to compute this quantity in the above session, the computation becomes $0 \cdot \infty$, an indeterminate form. The system would give you some sort of error message:

INPUT: c = a*b OUTPUT: c = Not a Number

Thus we see that overflows and underflows can lead to undesirable and inaccurate results. For example, in probability (a topic to be studied in Chapter 6), some problems often involve very large numbers, and one must be careful to avoid such under and overflows. A specific example might be to find the probability that among 180 randomly chosen people (whose birthdays are assumed to be random) none of their birthdays falls on Christmas. This probability equals $364^{180} / 365^{180}$ (this will be apparent to the reader after he/she has studied Section 6.1, but for now please take it on faith). What would happen if we used a floating point system to compute this quantity (in decimal form)? If we computed it as:

INPUT: 364^180/365^180

we would get as output "Not a Number," because both the numerator and denominator overflowed to ∞, so the result was the indeterminate form ∞ / ∞. In this case, the difficulty is easily circumvented by computing this quantity using the mathematically equivalent form:

INPUT: (364/365)^180 OUTPUT: 0.6103

(Interpretation: With 180 randomly chosen people, there is a 61.03% chance that no one in the group has a birthday on Christmas.)

A more insidious problem with floating point arithmetic is the accumulation of roundoff errors that occur each time numbers are translated into floating point numbers with the approximately ($s =$) 15 significant digits. Such errors can crop up in a variety of ways. Estimating and controlling such accumulated roundoff errors forms the central core of the vast subject of numerical analysis, but we sometimes need to be aware of them in discrete structure problems and algorithms. The following example will illustrate some of these problems in the more illustrative setting of 4-digit chopped arithmetic. All computations can be modeled using a computer platform (or hand-held calculator). The way the arithmetic works is that every inputted number is chopped to have four significant digits (as in form (21) with $s = 4$). Each time an arithmetic operation is performed, the answer is again chopped to four significant digits before the next computation takes place. Since we have already illustrated the problems with over- and underflows, in this example, and for the rest of the section we make the following assumption:

Unless stated otherwise we assume that m and M (the magnitude limits for the exponent e of a floating point number in (21) are sufficiently large (in magnitude) so that under- and overflows will not occur.

NOTATION: In order to make clear comparisons between exact (ordinary) arithmetic and floating point arithmetic, we will denote the floating point number corresponding to a real number x by **fl(x)**. Thus, if x is given by (20), then fl(x) is given by (21). For brevity (and slightly abusing notation), we extend this notation for single floating point operations, thus, for example, fl($x + y$) will be used to denote fl(fl(x) + fl(y)). Indeed, when we work in floating point arithmetic, any real number must first be converted to its floating point representative before any (floating point) mathematical operation can be performed.

EXAMPLE 4.24: (*Exact Versus Floating Point Arithmetic*) Perform the following operations in 4-digit chopped arithmetic. Compare the answers with the exact arithmetic answers.

(a) Let $x = 6000$, $y = 1/x$, and $z = 1 + y$. Compute $w = (z - 1)x^2$.

(b) Let $x = 1$, $y = 0.0005$. Compute $(x + y) + y$, and $x + (y + y)$.

SOLUTION: Part (a): In exact arithmetic, we have $w = (z - 1)x^2 = (1 + 1/x - 1)x^2$ $= (1/x)x^2 = x = 6000$.

In floating point arithmetic: Let us proceed step-by-step: First, fl(x) = 6000 and fl(y) = 0.0001666.[35] Then fl(z) = 1 (=1.000), so that fl(w) = $(1-1) \cdot 36,000,000 =$ $0 \cdot 36,000,000 = 0$! A totally different answer.

Part (b): In exact arithmetic, both quantities equal the same number 1.001, because addition is associative.

In floating point arithmetic, fl(x) = 1, fl(y) = 0.0005, but fl($x + y$) = 1, and so fl(($x + y$) + y) = fl(1 + .0005) = 1. However, fl($y + y$) = 0.001, and so fl($x + (y + y)$) = fl(1 + 0.001) = 1.001.

The above example shows that we must be careful not to take for granted many of the nice properties of exact arithmetic when we work in computing platforms that employ floating point arithmetic. In particular, we enunciate the lesson learned in Part (b) of this example:

---

[35] In the notation of (21), we would write fl(x) = $.6000 \times 10^4$, and fl(y) = $.1666 \times 10^{-3}$, but we may use ordinary decimal notation to represent floating point numbers.

CAUTION: In floating point arithmetic, *addition need not be associative*, i.e., for three numbers, x, y, and z, it is possible that $fl(fl(x+y)+z) \neq fl(x+fl(y+z))$.

Notice that the correct answer to Part (b) was obtained by adding up the two smaller numbers first. The basic reason for this is that when we added the small number (y) to the larger number (x), all of the significant digits of the smaller number got chopped. However, when we added the two smaller numbers first, the sum got large enough to be noticed when added to x. The general principle is as follows:

A GENERAL PRINCIPLE OF FLOATING POINT ARITHMETIC: When numerically computing a sum of floating point numbers, it is best to add up the smallest numbers first.

For a more rigorous explanation of this principle as well as related developments in floating point arithmetic, the reader is referred to Chapter 5 of [Sta-04] or [Sta-05].

EXERCISE FOR THE READER 4.24: Perform the following operations in 2-digit chopped arithmetic and compare with the results in exact arithmetic.
(a) 1.22×8.64
(b) $9 + 1 + 0.7 + 0.4 + 0.87 + 0.72 + 0.06 + 0.05$ added from left to right, and then from right to left.

EXERCISE FOR THE READER 4.25: (a) In floating point arithmetic, is multiplication commutative? In other words, is the identity $fl(x) \cdot fl(y) = fl(y) \cdot fl(x)$ valid for all numbers x and y? Either explain why it is true, or provide a counterexample.
(b) In floating point arithmetic, is multiplication associative? In other words, is the identity $fl(fl(x \cdot y) \cdot z) = fl(x \cdot fl(y \cdot z))$ valid for all numbers x, y, and z? Either explain why it is true, or provide a counterexample.

In most floating point computing platforms, integers are stored and manipulated in their own (nondecimal arithmetic). As long as fractions do not enter into the computations, all numbers stay as integers, and no decimal "noise" will enter into the computations. In discrete structures, many problems can be restricted to integer calculations, and this can greatly reduce the need to worry about roundoff errors. But we once again caution the reader that in, say, the standard IEEE double precision standard system, any significant figures after the 15th or 16th digit will be either ignored or misrepresented by the system. Thus in such a system, the numbers 12345678901234560000 and 12345678901234569999 would be considered equal, even though they differ by 9999! In public-key cryptography (the topic of the next section), large integer calculations are frequent, so such a floating point arithmetic system would be inadequate. But, for most other applications in this text, the faster floating point arithmetic system will be sufficient. One needs simply to keep aware of the cautions that we are bringing up in this section, and the corresponding limitations that these may place on any associated programs that we wish to write.

We close this section with another rather telling example that should serve to alert the reader of another simple trap that can sometimes arise in floating point arithmetic calculations. The lesson to be learned is that when one needs to use a floating point system to test whether two quantities are related by equality or inequality, it is usually a good idea to use some safety margins of error.

EXAMPLE 4.25: (*Testing Equality and Inequalities in a Floating Point System*) Often in the writing of a program to implement an algorithm, one needs to test whether two quantities a and b, are equal ($a = b$) or related by an inequality (say $a > b$). When at least one of these quantities is obtained through mathematical operations on (floating point) numbers, errors can be introduced that can distort such equality or inequality relationships. As an example, from simple algebra, we know that $(\sqrt{2} + \sqrt{3})^2 = (\sqrt{2})^2 + 2 \cdot \sqrt{2} \cdot \sqrt{3} + (\sqrt{3})^2 = 5 + 2\sqrt{6}$. However, observe what happens if we test (on any IEEE double precision floating point system) such a relation:

INPUT: set a = (2^0.5 + 3^0.5)^2,   b = 5 + 2*6^.5
        a <= b
OUTPUT: FALSE

This seems to contradict intuition. The computer told us that two numbers a and b (that we knew were the same) did not even satisfy the weaker relation $a \leq b$. What has actually happened is that in computing the floating point representative numbers for a and b, round off errors have accumulated to result in two numbers that are no longer the same. To circumvent such problems, it usually helps to add in a "safety cushion" to the relation. If the number of operations is not too large, this can usually be taken to equal to Mu, where M is the approximate magnitude of the fixed quantity (on one side of the relation), and u is the machine epsilon.   For example if b represents the fixed quantity, we could take $M = |b|$.[36] Most floating point systems have a built-in constant for u, usually denoted as something like MACHEPS. If we modify the above test using this "safety cushion," we indeed get the expected result:

INPUT: set a = (2^0.5 + 3^0.5)^2,   b = 5 + 2*6^.5
        a <= b + |b|*MACHEPS
OUTPUT: TRUE

Again, when working in integer arithmetic (as long as the number of significant digits is not too large, as explained earlier), the problem in the above example will never occur. But there are situations (in discrete structures) where one needs to deal with non-integers, so the lesson of the above example should be remembered. For problems involving very large integers or very fine decimal comparisons, a symbolic arithmetic system should be used.

---

[36] This assignment is fine as long as $b \neq 0$.

EXERCISES 4.4:

NOTE: Throughout this exercise set, when we refer to floating point arithmetic, we will be referring to the chopped arithmetic explained in the text involving the representations (21) $x = \pm .d_1 d_2 \cdots d_s \times 10^e$, where we will usually specify s (the number of digits) and sometimes specify the bounds m, and M on e ($m \leq e \leq M$).

1. In each part, a real number x given, along with the floating point parameter s (the number of digits). Write down the corresponding floating point representative $fl(x)$ exactly in the form of

    (21) $x = \pm .d_1 d_2 \cdots d_s \times 10^e$.

    (a) $x = 12,956, \quad s = 3$ (b) $x = 0.0012345678, \ s = 5$
    (c) $x = 12.0345, \quad s = 3$ (d) $x = 0.000542, \quad s = 5$

2. Repeat the directions of Exercise 1 for each part below.

    (a) $x = 572.5, \quad s = 2$ (b) $x = 0.01010101, \ s = 4$
    (c) $x = 814.099, \quad s = 4$ (d) $x = \pi, \quad s = 5$

3. Perform the following computations in 2-digit floating point arithmetic.

    (a) $1100 + 13$ (b) $\pi^2 + e^2$
    (c) $(111/210)^2$ (d) $111^2 / 210^2$

4. Perform the following computations in 2-digit floating point arithmetic.

    (a) $8759 / 990$ (b) $\cos(\pi)$
    (c) $(22 + 33)^2$ (d) $22^2 + 2 \cdot 22 \cdot 33 + 33^2$

    Note: In Part (b), π needs to be converted into a floating point number before the cosine is evaluated.

5. Repeat each part of Exercise 3, but this time working in 3-digit floating point arithmetic.

6. Repeat each part of Exercise 4, but this time working in 3-digit floating point arithmetic.

7. Perform the following computations in 4-digit floating point arithmetic.

    (a) $11,066,000 / 0.000626$ (b) $1 + 11 + 111 + 1111 + 11111 + 111111$
    (c) $111111 + 11111 + 1111 + 111 + 11 + 1$ (d) $2 \cdot 21 \cdot 211 \cdot 2111 \cdot 21111$

8. Perform the following computations in 4-digit floating point arithmetic.

    (a) $11,066,000 \times 0.000626$ (b) $11 \times 111 \times 1111 \times 11111 \times 111111$
    (c) $111111 \times 11111 \times 1111 \times 111 \times 11$ (d) $23.5 + \sqrt{2122^2 - 41 \times 5 \times 313}$

9. You might recall from calculus that the infinite series $\sum_{n=1}^{\infty} 1/n = 1 + 1/2 + 1/3 + \cdots$ diverges to infinity.
    (a) Compute this infinite sum in 2-digit floating point arithmetic adding the numbers in left to right order. (This sum will turn out to be a finite sum.)
    (b) Explain why for any positive integer s, this infinite sum will converge to a finite number in s-digit floating point arithmetic.

10.   You might recall from calculus that the infinite series $\sum_{n=1}^{\infty} 1/n^2 = 1 + 1/2^2 + 1/3^2 + \cdots$
       converges to $\pi^2/6$.
       (a) Compute this infinite sum in 2-digit floating point arithmetic (adding the numbers in left to
       right order).
       (b) Repeat Part (a), but this time in 3-digit floating point arithmetic.

11.   Suppose that $S > s$ are positive integers, and $a_1, a_2, \cdots, a_N$ are positive numbers. Explain that
       when we compute the sum $a_1, a_2, \cdots, a_N$ (adding the numbers in left to right order) in S-digit
       floating point arithmetic, the sum (when viewed as a real number) will always be at least as large
       as the corresponding sum computed in s-digit floating point arithmetic.

12.   Draw rough sketches of the graphs (on the real number line) of the set of all floating point
       numbers with the following parameters:

       (a) $s = 1, m = -1, M = 0$ (b) $s = 1, m = 0, M = 1$ (c) $s = 2, m = 0, M = 1$

13.   If x_1 and x_2 are numbers, it is elementary that the average $\overline{x} = (x_1 + x_2)/2$ satisfies
       $\min(x_1, x_2) \leq \overline{x} \leq \max(x_1, x_2)$. Explain why the analogue of this double inequality need not be
       valid in floating point arithmetic and back up your explanation with specific examples.

14.   Does the distributive law $x(y + z) = xy + xz$ remain valid in floating point arithmetic? Either
       prove the (floating point arithmetic) identity $\mathrm{fl}(x \cdot \mathrm{fl}(y + z)) = \mathrm{fl}(\mathrm{fl}(xy) + \mathrm{fl}(xz))$, or provide a
       counterexample.

15.   In our solution of Exercise for the Reader 4.25(b), we provided a counterexample to demonstrate
       that the associative law of multiplication need not hold in floating point arithmetic, i.e., we found
       three numbers x, y, and z, along with corresponding parameters s, m, and M (for the floating
       point arithmetic system, for which $\mathrm{fl}(\mathrm{fl}(x \cdot y) \cdot z) \neq \mathrm{fl}(x \cdot \mathrm{fl}(y \cdot z))$. Our counterexample involved
       an overflow. What if we stipulate that no overflows take place; will multiplication then be
       associative in floating point arithmetic?

16.   Write down all of the floating point numbers in the system that is governed by (21)
       $x = \pm.d_1 d_2 \cdots d_s \times 10^e$, with parameters $s = 1$, $m = -1$, and $M = 1$. The number 0 is not of the
       form (21) (since it requires that $d_1 \neq 0$), but should always be included in any floating point
       number system.

17.   We know that in exact arithmetic, any linear equation $ax = b$, with $a \neq 0$, always has exactly one
       solution, namely $x = -b/a$. Give an example of such an equation along with parameters s, m,
       and M, for a floating point arithmetic system governed by (21), in which the equation does not
       have a unique solution, i.e., it either has no solution or more than one solution.

18.   (a) Working in floating point (chopped) arithmetic with parameters $s = 5$, $m = -6$, and $M = 6$,
       show that the quadratic equation $x^2 = 100$ has exactly two floating point solutions, and these
       are the same as the two solutions in exact arithmetic.
       (b) Continuing to work in the floating point arithmetic of Part (a), and with the usual convention
       of underflows being converted to zero, show that the quadratic equation $x^2 = 0$ has more than
       10 solutions in floating point arithmetic.
       (c) In the floating point arithmetic of Part (a), is it possible to have a quadratic equation $x^2 = a$,
       where $a > 0$, with more than two floating point solutions? Either provide an example that
       demonstrates this can happen, or prove that it cannot.

19. As pointed out in the text, any nonzero real number x can be expressed in the form (20)
$x = \pm .d_1 d_2 \cdots d_s d_{s+1} \cdots \times 10^e$.

    (a) Prove that if this expansion contains an infinite string of 9's, i.e., if for some positive index j, we have $d_j = d_{j+1} = d_{j+2} = \cdots 9$, then x has an equivalent representation obtained by changing all of these digits to zero and rounding up d_{j-1} by 1. Here, if $j = 1$, this would mean that
$x = \pm .10 \cdots 00 \cdots \times 10^{e+1}$.

    (b) Prove that apart from the equivalences of Part (a), the representation (20) of nonzero real numbers is unique.

    Suggestion for Part (a): Although this can be proved using infinite series, here is an elementary approach: Note that $a = 0.3333 \cdots = 1/3$. Therefore $3a = 0.9999 \cdots = 3(1/3) = 1$. Thus we have proved that $0.9999 \cdots = 1$. Obtain the general result by multiplying this equation by a suitable power of 10.

20. Prove that for any nonzero real number x, the exponent e in the representation of x by (20):
$x = \pm .d_1 d_2 \cdots d_s d_{s+1} \cdots \times 10^e$, is given by $e = 1 + \lfloor \log(|x|) \rfloor = 1 + \text{floor}(\log(|x|))$, provided that (20) does not have $d_1 = 1, d_i = 0$ $(i = 2, 3, \cdots)$. In this latter case, it is clear that since x is a power of 10, $e = 1 + \lfloor \log(|x|) \rfloor$. Here, log denotes the common (base) 10 log. Note that this latter phenomenon can be avoided; see Exercise 19.

21. (*An Algorithm to Construct the Floating Point Representative of a Real Number*)
    (a) Given a real number x, prove that the following algorithm correctly produces its floating point representative fl(x) in the form of (21) $x = \pm .d_1 d_2 \cdots d_s \times 10^e$.

    Step 1: If $x = 0$, set fl(x) = 0, Stop.

    Step 2: If $x \neq 0$, obtain the exponent e in the representation (20) $x = \pm .d_1 d_2 \cdots d_s d_{s+1} \cdots \times 10^e$, using the formulas $e = 1 + \lfloor \log(|x|) \rfloor = 1 + \text{floor}(\log(|x|))$, in case $d_1 = 1, d_i = 0$ $(i = 2, 3, \cdots)$, of Exercise 20.

    Step 3: fl(x) = $10^{-s} \text{floor}(10^{s-e} x) \times 10^e$.

    (b) Use the algorithm in Part (a) to re-do each of the four floating point conversions of (ordinary) Exercise 1.

COMPUTER EXERCISES 4.4:

NOTE: Throughout this computer exercise set, when we refer to floating point arithmetic, we will be referring to the chopped arithmetic explained in the text involving the representations (21) $x = \pm .d_1 d_2 \cdots d_s \times 10^e$, where we will usually specify s (the number of digits) and sometimes specify the bounds m, and M on e $(m \leq e \leq M)$.

1. (*Program for Computing the Floating Point Representation* fl(x) *of a Real Number x.*)
    (a) Write a program with syntax [dVec, exp] = real2float(x,s) that inputs a real number x and a parameter s for a floating point arithmetic, and whose output dVec will either be the integer zero (if x = 0) or an s-digit integer representing the s-significant digits $[d_1 d_2 \cdots d_s]$ for x. The second output exp is an integer representing the exponent e in the representation (21) fl(x) = $\pm .d_1 d_2 \cdots d_s \times 10^e$.

    (b) Apply this algorithm to re-do Exercise 1.
    (c) Use the program to re-do Example 4.16.

    Suggestion: For Part (a), follow steps 1 and 2 of the algorithm of (ordinary) Exercise 21, and take $\text{dVec} = \text{floor}(10^{s-e}x)$.

Note: This program can work only when s is less than the number of significant digits of the accuracy of the system that you are using. Later in this computer exercise set we will develop a vector based system of floating point arithmetic that can work with any number s of significant digits.

2. (*A Simple Program for Adding Floating Point Numbers*) (a) Write a program with syntax `[dSum, expSum] = FloatAdd(x,y,s)` that inputs two real numbers x and y, and a parameter s for a floating point arithmetic, and whose outputs `dSum` and `expSum` are the parameters (as in Computer Exercise 1) for the sum $x + y$ in s-digit floating point arithmetic. (b) Use this program to re-do Exercise 3(a), and then use it (iteratively) to re-do Exercise for the Reader 4.12(b) and Exercise 7(b) and (c).
Suggestion: For Part (a), use the program of Computer Exercise 1.

3. (*A Simple Program for Multiplying Floating Point Numbers*) (a) Write a program with syntax `[dProd, expProd] = FloatMult(x,y,s)` that inputs two real numbers x and y, and a parameter s for a floating point arithmetic, and whose outputs `dProd` and `expProd` are the parameters (as in Computer Exercise 1) for the product $x*y$ in s-digit floating point arithmetic. (b) Use this program to re-do Exercise for the Reader 4.12(a) and Exercise 8(a)(b)(c).
Suggestion: For Part (a), use the program of Computer Exercise 1.

4. (*A Simple Program for Dividing Floating Point Numbers*) (a) Write a program with syntax `[dQuot, expQuot] = FloatDiv(x,y,s)` that inputs two real numbers x and y, and a parameter s for a floating point arithmetic, and whose outputs `dQuot` and `expQuot` are the parameters (as in Computer Exercise 1) for the quotient x/y in s-digit floating point arithmetic. In case $\text{fl}(y) = 0$ an error message should be produced since division by zero is not defined.
 (b) Use this program to re-do Exercises 4(a) and 7(a).
Suggestion: For Part (a), use the program of Computer Exercise 1.

5. Use the program of Computer Exercise 2 in conjunction with a while loop to re-do Exercise 9(a). To compute each fraction $1/n$ that is to be summed, you may either use the program of Computer Exercise 1 two times, or make use the program of Computer Exercise 4.

6. Use the program of Computer Exercise 2 in conjunction with a while loop to re-do Exercise 10. To compute each fraction $1/n^2$ that is to be summed, you may either use the program of Computer Exercise 1 several times, or make use of the programs of Computer Exercises 3, and 4.

NOTE: The rest of the computer exercises here will be concerned with developing programs that will perform basic floating point arithmetic operations using <u>any</u> number s of significant digits. This was not possible with the simple approach in the preceding exercises that simply relied on the computing platform's arithmetic system and chopped off the digits that were not needed. Here is a summary of the method. First, a floating point number of form (21) $x = \pm.d_1 d_2 \cdots d_s \times 10^e$ will be stored as a vector of digits

$$\pm.d_1 d_2 \cdots d_s \times 10^e \sim [\text{sign}, e, d_1, d_2, \cdots, d_s], \tag{22}$$

where sign $= -1$, 0, or 1, corresponds to the sign of x (sign $= 0$ simply means that the floating point number is zero, and the remaining digits of the vector are irrelevant), and the remaining digits $e, d_1, d_2, \cdots, d_s$ have the same meaning as in (15). We call the vector on the right-hand side of (22) as the **representing vector** of the floating point number x. All arithmetic operations thus need to be carried out on these vectors of digits, and will be analogous to the base 10 arithmetic operations that were developed in Section 4.1.

7. (a) Write a simple program with syntax `flvec = FloatConverter(vec_init, s)` that inputs an initial component `vec_init = [sign, e, d_1, d_2, ···, d_K]` string of a vector `flvec = [sign, e, d_1, d_2, ···, d_s]` $(K \le s)$ that represents the floating point number $\pm .d_1 d_2 \cdots d_K 00 \cdots 0 \times 10^e$ (whose expansion ends in a string of 0's), and whose output is the full representing vector (with the string of zeros appended).
(b) Use your program to create representing vectors for $x = 0.75$, with $s = 10$, and with $s = 40$.

8. (*An Unlimited Precision Program for Adding Positive Floating Point Numbers*) (a) Write a program that will add two positive floating point numbers in terms of their representing vectors. The syntax will be as follows:
     `vecflsum = FloatPosAdd_HighPrecision(vecx,vecy,s)`,
   where the first two inputs `vecx` and `vecy` are the representing vectors of the two floating point numbers x and y to be added, assumed positive, and each with the same number s (the third input variable) of significant digits.
   (b) Use this program in conjunction with a for loop to (exactly) compute the sum, $1 + 11 + 111 + 1111 + \cdots + 11 \cdots 11$, where the last term has 30 digits. Check that you get the same answer if you perform the addition in the reverse order (if you take s sufficiently large). If your computing platform has access to a symbolic arithmetic system, use this to check your answer. Compare the answer with the resulting answers obtained by using the default floating point arithmetic system of your computing platform. (If your platform is symbolic, compare the answers with those obtained with the same program, but with s taken to be 15.)
   Suggestion for Part (a): Suppose that `vecx = [sgx, ex, x1, x2, ..., xs]`, and `vecy = [sgy, ey, y1, y2, ..., ys]`. Separate into three cases: Case 1: `ex = ey`. Use Algorithm 4.2 to add the two base 10 integers `[x1, x2, ..., xs]` and `[y1, y2, ..., ys]`. The result will be a length s + 1 vector `[z1, z2, ..., zs zsplus]`. Case 2: `ex > ey`. Let `diff = ex - ey`. If `diff ≥ s`, then `vecflsum = vecx`. Otherwise, use Algorithm 4.2 to add the two base 10 integers `[x1, x2, ..., xs]` and `[0, 0,...,0, y1, y2, ..., yK]`, where K = `s - diff`. Case 3: `ex < ey`. This case is dealt with in a similar fashion as was done in the preceding case (simply reverse the roles of x and y).

9. (*An Unlimited Precision Program for Multiplying Floating Point Numbers*) (a) Write a program that will multiply two floating point numbers in terms of their representing vectors. The syntax will be as follows: `vecflprod = FloatMult_HighPrecision(vecx,vecy,s)`, where the first two inputs `vecx` and `vecy` are the representing vectors of the two floating point numbers x and y to be added, assumed positive, and each with the same number s (the third input variable) of significant digits.
   (b) Use this program in conjunction with a for loop to (exactly) compute the product $1 \times 11 \times 111 \times 1111 \times \cdots \times 11 \cdots 11$, where the last factor has 30 digits. Check that you get the same answer if you perform the multiplication in the reverse order (if you take s sufficiently large). If your computing platform has access to a symbolic arithmetic system, use this to check your answer. Compare the answer with the resulting answers obtained by using the default floating point arithmetic system of your computing platform. (If your platform is symbolic, compare the answers with those obtained with the same program, but with s taken to be 15.)
   (c) Use the program of Part (a) in conjunction with a for loop (and a large enough value of s) to exactly compute $25! \equiv 25 \cdot 24 \cdot 23 \cdots 3 \cdot 2 \cdot 1$.
   Suggestions for Part (a): Suppose that `vecx = [1, ex, x1, x2, ..., xs]` and `vecy = [1, ey, y1, y2, ..., ys]`. If either `sgx` or `sgy` is zero (i.e., if either factor is zero), set `vecflsum = [0,0,0,..., 0]`. Otherwise, use Algorithm 4.3 to multiply the two base 10 integers `[x1, x2, ..., xs]` and `[y1, y2, ..., ys]`. The result will be a length t = 2s + 1 vector `[z1, z2, ..., zs, ...,zt]`. (This corresponds to the vector $[p_{K+L} \ p_{K+L-1} \cdots p_2 \ p_1 \ p_0]$ of the algorithm.) Take `vecflprod` to be `[sgx*sgy, ex+ey, z1, z2, ..., zs]`.

10.   (*A Program for Comparing Magnitudes of Floating Point Numbers*) (a) Write a program that has the following syntax: comp = FloatCompare(vecx,vecy,s), where the first two inputs vecx and vecy are the representing vectors of the two floating point numbers x and y, each with the same number s (the third input variable) of significant digits. The output will be 1 if $|x| > |y|$, 0 if $|x| = |y|$, and -1, if $|x| < |y|$.
      (b) Use this program to check the relative magnitudes of the following pairs of numbers: $x = 10$, $y = -10$, $x = 1234566890$, $y = 1234567890$.

11.   (*An Unlimited Precision Program for Adding General Floating Point Numbers*) (a) Write a program that will add two floating point numbers in terms of their representing vectors. The syntax will be as follows: vecflsum = FloatAdd_HighPrecision(vecx,vecy,s), where the first two inputs vecx and vecy are the representing vectors of the two floating point numbers x and y to be added, and each with the same number s (the third input variable) of significant digits.
      (b) Use this program in conjunction with a for loop to (exactly) compute the sum, $1 - 12 + 123 - 1234 + \cdots - 123456789012\cdots90$, where the last term has 30 digits. Check that you get the same answer if you perform the addition in the reverse order (if you take s sufficiently large). If your computing platform has access to a symbolic arithmetic system, use this to check your answer. Compare the answer with the resulting answers obtained by using the default floating point arithmetic system of your computing platform. (If your platform is symbolic, compare the answers with those obtained with the same program, but with s taken to be 15.)
      Suggestion for Part (a): First take care of the trivial cases where one of x or y is zero. In case both x and y have the same sign, the addition can easily be done using the program of Computer Exercise 8 (or, alternatively, by following the suggestions given for that computer exercise). For the final case in which x and y have opposite signs, first use the program of Computer Exercise 10 to determine which of x or y has the larger magnitude (if the magnitudes are the same the sum will be zero). The problem is now reduced to subtracting one positive floating point number from a larger floating point number. To do this, follow the suggestions given for Computer Exercise 8 to "line up" the digits. Then perform (if necessary) the base 10 subtraction of the digits using the same (base 10) algorithm that is taught in elementary school for subtracting a positive integer from a larger integer.

4.5: PUBLIC KEY CRYPTOGRAPHY

This section will apply the ideas of modular arithmetic of Section 4.2 and some of the number theory of Section 3.3 to a remarkable revolution in cryptography known as **public key cryptography** or **asymmetric key cryptography**, that occurred in the 1970s.   The discovery was first published in a groundbreaking 1976 paper by American cryptographers Whit Diffie (1944–) and Martin Hellman (1945–) [DiHe-76].   Although Diffie and Hellman did not provide a complete practical implementation of a public key cryptosystem, they provided an important key exchange protocol (the *Diffie–Hellman key exchange*) by which two remote parties can establish a secure key using public (insecure) channels. Inspired by the Diffie Hellman paper, and the need for a practical implementation of their conceptual cryptosystem, MIT scientists Ronald Rivest, Adi Shamir, and Leonard Adleman introduced their RSA cryptosystem.[37]   This has turned out to be one of

---

[37] It was in their RSA paper [RiShAd-78], that the characters "Alice" and "Bob" were introduced as permanent fixtures into the cryptography saga.

the most important and widely used public key cryptosystems. We have already developed the mathematical ingredients necessary to completely describe some of the most important contemporary public key cryptosystems. In addition to describing the Diffie–Hellman protocol and RSA, this section will describe the El Gamal cryptosystem and the knapsack cryptosystems of Hellman and Merkle. We will limit ourselves to developing the Diffie–Hellman key exchange, which started the revolution, and three well-known public key cryptosystems. The section will close with a brief account of some of the political ramifications of these powerful cryptosystems, which are considered weapons by many national governments (including that of the USA). Readers interested in learning more about the exciting subject of cryptography in general, or about some further applications of public key cryptography, such as to providing digital signatures and nonrepudiation, may consult the author's book [Sta-10].

An Informal Analogy for a Public Key Cryptosystem

The reader should glance back to the end of Section 4.3 where some basic concepts about cryptography were described. Up to the 1970s, it was generally taken as an axiom of cryptography that all cryptosystems had to be *symmetric*, meaning that knowing the key to "lock up" (i.e., encrypt) a plaintext message, was essentially equivalent to knowing the key needed to "un lock" (i.e., decrypt) a ciphertext message back into the plaintext. With the Hill cryptosystem introduced in Section 4.3, the key needed to encrypt was the encoding matrix and the key needed to decrypt was its inverse (easily obtained from the matrix). This adds a serious inconvenience to using such a cryptosystem: the key must be securely exchanged before the system can be implemented.

A good analogy between a symmetric key cryptosystem (like Hill cryptosystem) and a public key cryptosystem is the following: Imagine a mathematically primitive world where the only way to encrypt a message would be to lock it in a case. If Alice, who resides in California, wants to send Bob, who lives in Maryland, a confidential message using a symmetric key cryptosystem, she would need to arrange ahead of time with Bob to have a matching set of keys to a certain padlock. She could then place her message in a steel case, lock the case with her padlock, and mail it to Bob. Since Bob, the intended recipient, has Alice's key, he (and only he) should be able to unlock the case and read the message. The main drawback is the need to exchange keys before the communication can take place. The need for a common key was thought for many years to be an essential ingredient in cryptography. Now consider the following change of protocol: Alice and Bob each have their own padlocks and keys, but have not exchanged any common keys. Alice puts her confidential message to Bob in a steel case, and locks the case with her padlock. She then mails the locked case to Bob. When Bob receives the case, he cannot open it because he does not have the key to Alice's lock. But he locks it up a second time with his own padlock, and mails the double locked case back to Alice. Alice then removes her lock, and mails the case back to Bob who is now the only one able to open the case and read Alice's message. Although the second protocol was more time consuming (and this would not be relevant if the mailings were done electronically), its ramifications are quite striking!

NOTE: <u>With one minor exception, throughout this section, plaintexts and ciphertexts will be modular integers in some base.</u> As explained in Section 4.1, it is a routine matter to convert such integers into strings of English letters (or more general ASCII characters by, e.g., first converting into binary strings), so we will not devote further attention to such matters here.

NOTATION: Since most of the cryptosystems in this section will involve modular arithmetic, it will be convenient for us to adopt the following notation for the set of **invertible elements** in $\mathbb{Z}_n$:

$$\mathbb{Z}_n^\times = \{\text{invertible elements of } \mathbb{Z}_n\}. \tag{23}$$

By Proposition 4.2 (and the definition of the Euler phi function), $\mathbb{Z}_n^\times$ consists of the $\phi(n)$ positive integers that are less than n and relatively prime to n. In particular, if p is prime, then $\mathbb{Z}_p^\times = \{1, 2, \cdots, p-1\}$.

The Quest for Secure Electronic Key Exchange

The above analogy kept the early innovators Diffie and Hellman captivated with the problem of developing a feasible implementation.   The reason that this method cannot be readily applied stems from a simple observation about composition of functions. In mathematical terms, Alice would have her own cryptosystem (corresponding to her padlock in the above analogy), along with a corresponding encryption and decryption functions: $E_A, D_A = E_A^{-1}$. Likewise, Bob would have his own encryption and decryption functions: $E_B, D_B = E_B^{-1}$.   Now Alice locking her plaintext P and sending it to Bob would correspond to Alice sending Bob $E_A(P)$. Bob applying his padlock to this would,

FIGURE 4.6: American cryptographers Martin Hellman (1945–) (middle), and Whit Diffie (1944–) (right), pictured with American cryptographer Ralph Merkle (1952–).[38]

---

[38] The author acknowledges Chuck Painter/Stanford News Service for permission to include this photograph. Merkle and Hellman later collaborated to develop one of the first public key cryptosystems; it will be discussed later in this chapter.

in turn, produce $E_B \circ E_A(P)$. Next, when he sends this back to Alice, Alice will not be able to remove her padlock by applying her decryption function; she would only get $D_A \circ E_B \circ E_A(P)$. The general rule is that the inverse of a composition of functions is the composition of the inverses in the reverse order (see Exercise 29 of Section 2.2)—informally, *"first on, first off."* Despite these seemingly insurmountable mathematical difficulties and discouragement from many colleagues and other cryptographers, Diffie and Hellman thought to the future of a networked society and the absolute need for such communications.[39] They continued their efforts and eventually were able to achieve the important public key distribution algorithm. It was not a complete cryptosystem (although one would be developed shortly thereafter, based on their innovation), but it basically allowed secure cryptographic communications to take place without the old-fashioned hindrance of having to manually agree upon and exchange a common key. Prior to this breakthrough, if say a bank and a client needed to establish a secure communication channel (for exchanges of large sums of money), a courier or some other confidential means would have been needed to first exchange keys for a symmetric key cryptosystem.

One-Way Functions

The following definition of a one-way function, although somewhat informal, lies at the core of any symmetric key system. The basic idea is that a one-way function can be given out publicly so that anyone can use it, but it would be an intractable problem for anyone to compute the inverse function.

DEFINITION 4.14: A **one-way function** (also known as a **trapdoor function**) is any bijective function $f : D \to R$, such that any of its values $f(x)$ is easy to compute, but whose inverse function values $f^{-1}(y)$ are computationally intractable to compute.

The trapdoor terminology stems from the fact that with some additional information (the private key), the inverse function will be easy to compute. For

---

[39] Bailey Whitfield (Whit) Diffie went straight from earning his BS degree (1965) in mathematics at MIT to a job at the MITRE Corporation where he became very interested in cryptography. This interest motivated him to accept a position four years later at Stanford's artificial intelligence laboratory. Martin E. Hellman earned his BS, MS, and PhD degrees in electrical engineering from New York University. After completing post-doctoral positions at IBM and MIT, in 1971 he moved on to take an academic position at Stanford, where he met Diffie. Both received numerous accolades for their pioneering work, including an honorary doctorate for Diffie from the *Swiss Federal Institute of Technology*. Hellman remained at Stanford until his retirement, where he had an illustrious career with continuous strong research activity and as an award-winning teacher. Diffie worked for most of the rest of his career in industry, and currently serves as a Vice President and Chief Security Officer at Sun Microsystems.

example, if $\mathbf{P}_N = \{p \mid p \geq N, \text{ and } p \text{ is prime}\}$, then the function f with domain being the set of finite length vectors of primes in $\mathbf{P}_N$ and codomain being all finite products of such primes, and which is defined by $f([p_1, p_2, \cdots, p_k]) = p_1 \cdot p_2 \cdots p_k$, is a one-way function if N is sufficiently large. Indeed, as we have pointed out earlier, the problem of factoring integers with more than 300 digits (according to present standards) is intractable, whereas the inverse problem of multiplying numbers even with millions of digits is easy (on a computer). A trapdoor into factoring a certain output $f([p_1, p_2, \cdots, p_k])$ of the above function might be knowing some of the factors, thereby reducing the factoring problem to one for an integer of tractable size (say 150). As computer technology increases, the value of N needed to assure that the above function is one-way will increase, but it is generally believed that with such modifications, this function should remain a one-way function in perpetuity.

We next explain the Diffie–Hellman key exchange system. Its elegance and simplicity are striking, but the reader should keep in mind that generations of cryptographers had been convinced that such a system could not possibly exist. They published their discovery in a landmark 1976 paper [DiHe-76] that completely revolutionized the field. As pointed out earlier, this is not a complete cryptosystem, but it can be used for two parties (Alice and Bob) to create a mutual symmetric key over a public channel (to which Eve has access). Its security rests on the difficulty of computing discrete logarithms, as does the related El Gamal complete cryptosystem that we will introduce later. The algorithm requires a large prime p and a corresponding primitive root g. Algorithms for generating such parameters were provided in the appendix of Section 3.3.[40]

Review of the Discrete Logarithm Problem

The security of the Diffie–Hellman key exchange, as well as that of the El Gamal cryptosystem that will be introduced later in this chapter, both rely on the difficulty of computing *discrete logarithms*. Recall (from Section 3.3) that a primitive root modulo a prime number p is a positive integer $g < p$, whose order (mod p) equals $p - 1$ (i.e., is as large as possible). This means that $k = p - 1$ is the smallest positive integer for which $g^k \equiv 1 \pmod{p}$. There always exist primitive roots modulo any prime, and the powers of any primitive root: $g, g^2, g^3, \cdots, g^{p-1} \equiv 1$ are all distinct and consist precisely of the integers mod p that are relatively prime to p, i.e., $\mathbb{Z}_p^{\times}$.

---

[40] As in Section 3.3, our illustrative examples will use primes of smaller size so the concepts can be more easily understood. The computer implementation material at the end of this chapter will provide opportunities for readers to use this and other public key algorithms from this chapter with realistically large primes (provided that they have access to a computing platform with symbolic functionality).

DEFINITION 4.15: (*Discrete Logarithms*) Given a prime number p, and a primitive root g (mod p), the **discrete logarithm (mod p) with base g** is the inverse function of the modular exponentiation function

$$E_g : \mathbb{Z}_p^{\times} \to \mathbb{Z}_p^{\times} :: E_g(\ell) \equiv g^{\ell} \pmod{p}.$$

Since g is a primitive root, this modular exponential function is both one-to-one and onto, so has an inverse function $L_g : \mathbb{Z}_p^{\times} \to \mathbb{Z}_p^{\times}$ which is the corresponding discrete logarithm function. Although it would be more proper to include the modulus p in these notations (e.g., to use $E_{g,p}$ and $L_{g,p}$ instead), the prime p will usually be clear from the context and this will allow us to avoid such more cumbersome symbols. Computing any value $L_g(a)$, where we are given $a \in \mathbb{Z}_p^{\times}$, is called an instance of the **discrete logarithm problem**.

Unlike its inverse function E_g, the discrete logarithm function L_g has no explicit formula. More importantly for cryptographic applications, although there are fast algorithms for evaluating E_g (using fast modular exponentiation—Algorithm 4.5), evaluating discrete logarithms is extremely difficult, and it is believed (like the factoring problem on which RSA's security rests) that there cannot exist an efficient algorithm for evaluating discrete logarithms.

EXAMPLE 4.26: The number $p = 53$ is a prime number, so it has $\phi(\phi(53)) = \phi(52) = 26$ different primitive roots (Theorem 2.23(1)) (mod p). Thus, half of all the integers (mod 53) serve as primitive roots. Very often $g = 2$ and $g = 3$ are primitive roots, but in any case, we can find one by computing orders of successive integers (which is easily done on a computer, using Algorithm 4.5). To determine the smallest primitive roots, we could simply run through all integers greater than 1 and compute their orders until we find one whose order equals $\phi(p) = p - 1$, i.e., $\text{ord}_{53}(g) = 52$. We begin testing with the integer 2. By Proposition 3.22, $\text{ord}_{52}(2) \mid 52 = 2^2 \cdot 13$, so the only possibilities for $\text{ord}_{52}(2)$ are 2, 4, 13, 26, and 52, and we need only show that the first four options are not possible by showing that $2^2, 2^4, 2^{13}, 2^{26} \not\equiv 1 \pmod{53}$. Indeed:

$$2^2 \equiv 4,$$
$$2^4 \equiv (2^2) \equiv 16,$$
$$2^{13} \equiv (2^4)^2 \cdot 2^4 \cdot 2 \equiv 16^2 \cdot 16 \cdot 2 \equiv 44 \cdot 32 \equiv 30,$$
$$2^{26} \equiv (2^{13})^2 \equiv 30^2 \equiv 52 (\equiv -1) \pmod{53}.$$

Thus $g = 2$ is the smallest primitive root of 52.

Using fast exponentiation (Algorithm 4.5), we can quickly compute any power of g (mod 53), but the inverse problem of computing a discrete logarithm is much more tedious. For example, if we wanted to find $\ell = L_g(5)$, i.e., to solve the equation $g^{\ell} \equiv 5 \pmod{53}$, the only obvious way would be a brute-force approach of computing successive powers of g until we get 5 (mod 53).   Of course, with such a relatively small prime, a computer could quickly get us the answer $\ell = L_g(5) = 47$, but for a very large value of p, say of size 10^{200},   this brute-force scheme would require, on average, $(p-1)/2$ fast exponential evaluations, or more than 10^{199} of them, way too many to make this approach feasible.   By contrast, the inverse function would involve only a single fast exponentiation.

The Diffie–Hellman Key Exchange

ALGORITHM 4.9: (*The Diffie–Hellman Key Exchange Protocol*):

Purpose: Alice and Bob need to create a secret key between them using a public (insecure) communication channel.
Requirements: A large prime number p, and a primitive root g (mod p); both g and p can be made public.
NOTE: All modular powers in this algorithm should be computed with fast modular exponentiation (Algorithm 4.5).

Step 1: Alice and Bob select integers a and b, respectively (preferably randomly), with $1 \le a, b \le p - 2$. Each keeps their exponent number secret.

Step 2:   Alice computes $A \equiv g^a \pmod{p}$, and sends this to Bob, while Bob computes $B \equiv g^b \pmod{p}$, and sends this to Alice.   These numbers A and B may be sent over public channels.
Step 3:   Alice and Bob each obtain the common key K by raising the number they received from the other to their secret exponent (mod p).   Indeed,

$$B^a \equiv (g^b)^a \equiv g^{ba} \equiv g^{ab} \equiv (g^a)^b \equiv A^b \pmod{p}.$$

EXAMPLE 4.27:   Using the primitive root $g = 2$ of the prime $p = 53$ (see Example 4.26), suppose that Alice chooses her secret exponent to be $a = 22$, and Bob chooses his to be $b = 47$. Determine the resulting Diffie–Hellman key.

SOLUTION: Bob   (computes   and)   sends   Alice   the   number $B = g^b \equiv 2^{47} \equiv 5 \pmod{53}$,   and Alice sends Bob the number $A = g^a \equiv 2^{22}$ $\equiv 43 \pmod{53}$.   On her end, Alice computes the common (secret) key as $K = B^a \equiv 5^{22} \equiv 29 \pmod{53}$, and Bob computes it as $K = A^b \equiv 43^{47} \equiv 29 \pmod{53}$. Notice that the resulting common key $K = 29$ is now a shared secret between Alice and Bob, and it was never sent over the (insecure) channel.

COMPUTING NOTE: In practice, a large-sized prime p would be needed to assure a sufficiently large keyspace (in order to meet the security needs of the particular symmetric key cipher for which the key is going to be used). As we learned in Section 3.3, finding a primitive root can be a difficult mathematical problem in general, particularly if a factorization of $p - 1$ is not available. The task is much more manageable if p is of a special form such as $p = 2q + 1$, where q is another prime.[41] Another approach is to simply search for an integer g whose order (mod p) is sufficiently large. A common key will be created by the Diffie–Hellman algorithm, regardless of what integer is used for g (primitive root or not); but higher order g's result in added security.

EXERCISE FOR THE READER 4.26: (a) Determine the smallest primitive root g of the prime $p = 79$.
(b) Using the prime p and the primitive root g of Part (a), suppose that Alice chooses her secret exponent to be $a = 51$, and Bob chooses his to be $b = 33$. Determine the resulting Diffie–Hellman key.

Suppose that Eve intercepts the agreed upon parameters p and g, as well as the numbers A and B that were sent. How could she determine the resulting Diffie–Hellman key K of Algorithm 4.9? If she knew either Bob's secret exponent b, or Alice's a, she could easily obtain K with a singular modular exponentiation, but there is no other known way to do this. So Eve would be faced with the problem of either determining a from the knowledge that $g^a \equiv A(\bmod p)$, or determining b from the equation $g^b \equiv B(\bmod p)$. But this is just the discrete logarithm problem that is a very difficult problem for which it is believed that there can exist no efficient algorithm. Up to now, this is the only known attack on the Diffie–Hellman key exchange. It is an unproved conjecture that a successful attack on the Diffie–Hellman key exchange would translate into an efficient scheme for computing discrete logarithms. Of course, in the small scale setting of Example 4.27, this could be easily be done by a brute-force check of all possible modular exponents. But if instead p had 500 digits, this procedure would probably take about 10^{500} attempts. This is much too large even if all the computers in the world were to be able work together on it for billions of billions of years!

---

[41] The reason is that $\phi(p) = p - 1 = 2q$, and thus the order of any element in $\mathbb{Z}_p^\times$ can only be 1, 2, q, or $2q$ (in which case it is a primitive root). Prime numbers q for which $2q + 1$ is also prime are called *Sophie Germain primes*. Many Sophie Germain primes have been discovered, so there is no shortage of them for cryptographic applications. In 2007, the largest known Sophie Germain prime had 51,910 digits. It is conjectured, but has not yet been proved, that there are infinitely many Sophie Germain primes; they are named after the French mathematician Marie-Sophie Germain (1776–1831).

The Quest for a Complete Public Key Cryptosystem

Soon after the revelation of Diffie and Hellman's groundbreaking discovery, cryptographers worked hard to create a full public key cryptosystem, for which we adopt the following definition:

DEFINITION 4.16: A **public key cryptosystem** requires that any participating party have two keys: A **public key**, which is available to everyone, inside the system or not, and corresponds to a one-way encryption function, and a **private key**, which is kept secret, and corresponds to the decryption function. If Alice needs to send Bob a private message, she simply encrypts the plaintext P to Bob using his (readily available) public key E_B. When Bob receives the resulting ciphertext $C = E_B(P)$, only he has the needed decryption function from his private key $D_B = E_B^{-1}$, so he applies it to the ciphertext to recover the plaintext, i.e.,

$$D_B(C) = D_B(E_B(P)) = E_B^{-1}(E_B(P)) = P.$$

The security of such a system relies on the security of the one-way functions corresponding to the public keys, i.e., it should be essentially impossible to determine (within say, 200 years) the corresponding private key from any public key. In order for this to be done, the problem of this determination should boil down to an intractable mathematical problem, just as breaking the Diffie–Hellman key exchange depended on the discrete logarithm problem. Table 4.7 shows the intractable mathematical problems on which each of the public key cryptosystems that we will present are based.

TABLE 4.7: The intractable mathematical problems which form the basis of the security of the public key cryptosystems that will be developed in this section.

Public Key Cryptosystem	Mathematical Problem that Provides Security
RSA	Factoring Integers
El Gamal	The Discrete Logarithm Problem
Knapsack	The Knapsack Problem[42]

The RSA Cryptosystem

The first publicly announced discovery of a complete and secure public key cryptosystem was made by a team of three researchers at MIT: Ronald Rivest,

---

[42] The knapsack problem will be discussed later in this section.

Adi Shamir, and Leonard Adleman,[43]   see Figure 4.7.   Rivest was extremely intrigued by the 1976 Diffie–Hellman paper, which suggested that such a cryptosystem could be contructed.   He recruited the other two to join him in a quest for building the first public key cryptosystem.   The difficulty was in designing a true one-way function that could be incorporated into such a system, and the team was perfectly suited for the task.   Rivest and Shamir would come up with new ideas for implementation schemes with candidate one-way functions, and pass them on to Adleman, who had the strongest mathematical background of the three.   Adleman would routinely find flaws in the systems or one-way functions, which sent Rivest and Shamir back to look for new ideas.   The ideas being presented by Rivest and Shamir were primarily based on number theoretic problems, which was Adleman's area of expertise.

FIGURE 4.7: Adi Shamir (1952–) Israeli cryptographer, Ronald Rivest (1947–) American cryptographer, and Leonard Adleman (1945–) American mathematician and cryptographer

This back-and-forth went on for a about year, with Rivest and Shamir never losing hope, while gradually hovering in towards their cherished holy grail, and Adleman helping to keep them from wasting time going down the wrong roads.   One April evening in 1977, after the three had celebrated at a holiday party given by an MIT student, Rivest was returning to his home and a new idea for a public key cryptosystem sparked into his mind.   He was so excited and confident about this idea that he did not sleep that night, rather, he worked straight until dawn ironing out the details which seemed to be coming together very nicely.   By morning, Rivest had a draft paper written up that he apprehensively gave to Adleman.   This time, however, Adleman could find no flaws and was able to certify the system as a legitimate public key cryptosystem. Adleman's only request was that Rivest change the ordering of the names that he had listed on the draft joint paper, to list his name last rather than using

---

[43] Shamir was born in Tel Aviv. He obtained his computer science PhD at the Weizmann Institute where he later returned as a faculty member (joint in mathematics and computer science—the most prevalent subject areas for cryptographers). He was a visiting professor at MIT from 1977–1980, when he helped to create RSA. He has made many other important contributions to cryptography and has been awarded several very prestigious prizes for this work. Rivest was born in New York. He earned his PhD in computer science from Stanford in 1974, and subsequently earned a professorship at MIT, which he still retains. He founded the company *RSA Data Security*. He has authored a major textbook on algorithms, and created several important cryptographic tools (apart from RSA).
Adleman was born in California. In 1976 he earned his PhD in electrical engineering and computer science from UC Berkeley. His first position was as a mathematics professor at MIT. His research has grown to include several areas of biology, and he is now a chaired professor at USC.

alphabetical order.[44]   This change was made, and their system was coined as the **RSA cryptosystem**.   Its security is based on the difficulty of the factorization problem, and it remains one of the most successful and powerful public key cryptosystems.   A year later they showed how to incorporate *digital signatures* that allow receivers to *authenticate* that a certain message or document did actually come from the person who allegedly sent it.   It also makes it impossible for the sender to deny having sent a message that has the sender's digital signature (*nonrepudiation*).

We now describe the RSA cryptosystem:

ALGORITHM 4.10: (*The RSA Public Key Cryptosystem*):
Purpose: Alice needs to send Bob a private message, but they have not yet met to exchange any cryptosystem keys.
Requirements: Bob will first need two different prime numbers, $p \neq q$, that he should keep secret. He multiplies these primes to obtain an integer $n = pq$, which will be made public. He selects (preferably randomly) an integer d that is relatively prime to $\phi(n) = (p-1)(q-1)$.[45]   He then computes (using the extended Euclidean Algorithm 2.2) the inverse e of d $(\mathrm{mod}\,\phi(n))$.
Plaintext, Ciphertext Spaces: $\mathcal{P} = \mathcal{C} = \mathbb{Z}_n$, where $n = pq$, with p and q prime numbers.
Keyspace: $\mathcal{K} = \{(p,q,d,e): de \equiv 1\ (\mathrm{mod}\,\phi(n)),\ \text{where}\ n = pq\}$. From a system key $\kappa = (p,q,d,e)$, Bob's private key is (n, d), and the corresponding public key is (n, e). The parameter e is called the **encryption exponent**, and the parameter d is called the **decryption exponent**.
Encryption Scheme: $e_\kappa:\mathbb{Z}_n \to \mathbb{Z}_n$, defined by $e_\kappa(P) \equiv P^e\ (\mathrm{mod}\,n)$.
Decryption Scheme: $d_\kappa:\mathbb{Z}_n \to \mathbb{Z}_n$, defined by $d_\kappa(C) \equiv C^d\ (\mathrm{mod}\,n)$.

Both encryption and decryption can be performed efficiently using the squaring algorithm for fast modular exponentiation (Algorithm 4.5).   Although the RSA algorithm was set up to accept only single integers as plaintexts, there are natural and effective ways to expand the algorithm into a block cipher (see Exercise 27). Before showing that the decryption functions above are the inverses of the corresponding encryption functions, we give a simple example of RSA.   As explained earlier, the primes used in the chapter examples are too small to be secure, but chosen so that the concepts are more easily illustrated.   The computer

---

[44] In mathematical publications, it is customary to list the author's names in alphabetical order. This is most certainly not the case in some of the other sciences, such as biology, where being the "first author" of a paper carries significantly more weight.   Practices in the other academic fields vary between these two extremes.
[45] This evaluation of $\phi(pq)$ easily follows from Proposition 3.20.

implementation material will provide the reader with opportunities to work with much larger keys.

EXAMPLE 4.28: Suppose that Bob adopts the RSA cryptosystem with primes $p = 37$ and $q = 41$. He chooses the (public key) encryption exponent to be $e = 49$.
(a) Show that Bob's choice of encryption exponent is legitimate, and find his corresponding (private key) decryption exponent d.
(b) Suppose that Alice encrypts the plaintext message $P = 44$ using the RSA cryptosystem with Bob's public key $(n, e) = (1517, 49)$. What is the resulting ciphertext that would be sent to Bob?
(c) Go through the decryption process that would need to get done at Bob's end, using his private key (n, d) with the decryption exponent that was determined in Part (a).

SOLUTION: Part (a): Since $\phi(n) = (p-1)(q-1) = 36 \cdot 40 = 1440$, we may compute that $\gcd(e, \phi(n)) = \gcd(49, 1440) = 1$, so $e = 49$ is indeed a legitimate encryption exponent. We use the extended Euclidean Algorithm 2.2 to compute $d \equiv e^{-1} \equiv 529$, which is the decryption exponent.
Part (b): The ciphertext is computed as $C \equiv 44^{49} \equiv 1069 \pmod{1517}$.[46]
Part (c): Bob's decryption system would need to raise the ciphertext $c = 1069$ to the decryption exponent $d = 529 \pmod{n = 1517}$: $1069^{529} \equiv 44 \pmod{1517}$—the original plaintext message!

PROPOSITION 4.13: The decryption functions listed in Algorithm 4.10 work, i.e., $d_\kappa(e_\kappa(m)) \equiv m \pmod{n}$, for any $m \in \mathbb{Z}_n$.

Proof: Since $de \equiv 1 \pmod{\phi(n)}$, we can write $de = \ell\phi(n) + 1$, for some $\ell \in \mathbb{Z}_+$. By Proposition 3.20, we have $\phi(n) = \phi(pq) = (p-1)(q-1)$. Also, by Fermat's little theorem (Theorem 3.19), it follows that $d_\kappa(e_\kappa(m)) \equiv (m^e)^d \equiv m^{\ell\phi(n)+1} \equiv (m^{(p-1)})^{(q-1)\ell} m \equiv 1^{(q-1)\ell} m \equiv m \pmod{p}$, provided that $\gcd(m, p) = 1$. But if this latter condition does not hold, this would mean that $p \mid m$, so that $d_\kappa(e_\kappa(m)) \equiv$

---

[46] We remind the reader of the problems of working with large integers on computing platforms (as we have dicussed in Sections 3.3, 4.2, and 4.5). The number 44^{49} has 81 digits and could be easily computed (exactly) on a symbolic computing platform. However, for users of a floating point system, it should not be computed directly since the answer would not be accurate. Using Algorithm 4.5 would circumvent this problem. The same comments will apply to the calculation needed in Part (b), where the power, if computed first without modular arithmetic, would have over 1600 digits. Working with integers having more than 16 digits or so would require a symbolic computing platform for all calculations. Thus, with the exception of a few (clearly marked) exercises, all of the material in this section will involve values of n with at most eight digits (so the product of two numbers would have at most 16 digits, before reducing mod n). This will be sufficient for presenting and understanding the concepts of RSA.

$(m^e)^d \equiv 0 \equiv m(\mathrm{mod}\, p)$. We have thus shown that $d_\kappa(e_\kappa(m)) \equiv m(\mathrm{mod}\, p)$, for any $m \in \mathbb{Z}_n$. The same argument shows that $d_\kappa(e_\kappa(m)) \equiv m(\mathrm{mod}\, q)$, for any $m \in \mathbb{Z}_n$. It follows from the Chinese remainder theorem (Theorem 4.4) (or more directly since p and q are different primes that both divide $d_\kappa(e_\kappa(m)) - m$, that $d_\kappa(e_\kappa(m)) \equiv m(\mathrm{mod}\, n)$, for any $m \in \mathbb{Z}_n$, as was needed to show. $\square$

The above proof breaks down if $p = q$ (Why?), and thus shows one reason why the RSA cryptosystem requires $p \neq q$. Besides, if n were chosen to be a perfect square, it would be easy to factor just by taking its square root. Care should also be taken to assure that p and q are not too close, since in this case also n can be easily factored (see Exercise 28). From the proof of Proposition 4.13, we extract the following corollary that will have important consequences regarding the security of RSA systems.

COROLLARY 4.14: If d' is any positive integer that satisfies $d'e \equiv 1$ (mod $\mathrm{lcm}(p-1, q-1)$), then d' could also be used as a decryption exponent for the RSA algorithm.

As an extreme case where this corollary can show a weakness in an RSA system, consider the two Mersenne primes $p = 2^{521} + 1$ and $q = 2^{607} + 1$. Ostensibly, the RSA decryption exponent d is found modulo $(p-1)(q-1) = 2^{1128} \approx 10^{340}$. Corollary 4.14, however, says that the search can be restricted modulo $\mathrm{lcm}(p-1, q-1) = 2^{607} \approx 10^{183}$, a significantly less complex problem. One way to avoid this sort of weakness in RSA algorithms is to use primes p and q such that $p-1$, $q-1$ do not have a lot of small factors. There are some other recommendations that can be made regarding effective choices for the two primes p and q, in order to enhance the security of RSA against factorization of the modulus attacks; for more details see [Mol-03] or [MeOoVa-96].

We next summarize a few implementation issues.

In order to use RSA, one first needs to generate or obtain two sufficiently large primes $p \neq q$. The appendix to Section 3.3 dealt with this problem and presented some practical algorithms. Note that the primes used to construct $n = pq$ will be much smaller, and thus much more tractable in size than n. The size of the modulus n for an RSA system is determined by the current state-of-the-art methods for factoring. For example, if a 300 digit modulus is required, then we will need two primes with at least 150 digits each. Once n is determined, an encrypting exponent e can be randomly generated from the integers less than $\phi(n)$ and relatively prime to it (i.e., from $\mathbb{Z}_{\phi(n)}^\times$). The decryption exponent is then the inverse of e (mod $\phi(n)$), and can be found using the extended Euclidean algorithm (Algorithm 4.6).

EXERCISE FOR THE READER 4.27: Suppose that it is desired to create an RSA cryptosystem using the two primes $p = 67$ and $q = 37$.

(a) Determine the smallest encryption exponent e that is greater than 1000, and could be used for such an encryption exponent, and use it to encrypt the plaintext message $P = 2012$.

(b) What is the decryption exponent d for the RSA of Part (a)? Decrypt the ciphertext message $C = 2095$.

The El Gamal Cryptosystem

We are now nicely prepared to discuss the *El Gamal cryptosystem*.[47] It is a bit more complicated than the RSA system, but it is still quite elementary, and serves as a nice extension of the original Diffie–Hellman key exchange into a full cryptosystem. Moreover, unlike the RSA system, El Gamal has natural extensions to more sophisticated cryptosystems (e.g., elliptic curve cryptosystems).

ALGORITHM 4.11: (*The El Gamal Public Key Cryptosystem*):

Purpose: Alice needs to send Bob a private message, but they have not yet met to exchange any cryptosystem keys.

Requirements: Bob will first need a (large) prime number p, and a corresponding primitive root g (mod p). These parameters will be made public. He chooses a secret exponent b (preferably randomly) in the range $0 < b < p-1$. Alice will need to also choose her own private exponent a in the range $0 < a < p-1$.

Plaintext, Ciphertext Spaces: $\mathscr{P} = \mathbb{Z}_p$, $\mathscr{C} = \{(x, y) : x, y \in \mathbb{Z}_p\}$, where p is a prime number. Thus the ciphertexts consist of ordered pairs of mod p integers (i.e., length two vectors of mod p integers).

Keyspace: $\mathscr{K} = \{(p, g, b) : g$ is a primitive root (mod p), $0 < b < p-1\}$. An element $\kappa = (p, g, b)$ is (Bob's) **private key**, and it will contain all the information needed to encrypt or decrypt a message. The corresponding (Bob's) **public key** is (p, g, B), where $B \equiv g^b \pmod{p}$.

Encryption Scheme: (Alice) randomly selects an integer a (her private key), with $0 < a < p-1$, and she computes $A \equiv g^a \pmod{p}$. She then computes

---

[47] This system was developed in 1985, along with an associated digital signature scheme, by Taher El Gamal (1955–), an Egyptian-American cryptographer. Elgamal did his undergraduate work in Cairo, and then studied at Stanford where he earned a PhD in computer science. His digital signature algorithm was adopted by the NIST as the *Digital Signature Standard* (*DSS*). Elgamal founded his own security company (*Securify*), and before that he worked as a senior scientist at *Netscape* and at *RSA Security*.

$C \equiv B^a P (\bmod\, p),$ [48] where P is the plaintext message.   The ciphertext will be the ordered pair (A, C). In summary $e_\kappa : \mathbb{Z}_p \to \mathbb{Z}_p^2 :: e_\kappa(P) \equiv (A, C)(\bmod\, p).$

Decryption Scheme: $d_\kappa : e_\kappa(\mathbb{Z}_p) \to \mathbb{Z}_p :: d_\kappa((A, C)) \equiv A^{p-1-b} C (\bmod\, p).$

Note that the C, the second component of the ciphertext, is simply the plaintext message multiplied by B^a, which is the common Diffie–Hellman key between Alice and Bob, so that $B^a \equiv A^b (\bmod\, p)$.   Thus, Bob can decrypt by simply multiplying C by the inverse of the Diffie–Hellman key.   The formula that was given amounts to the same operation, but avoids having to compute a modular inverse (so is a bit quicker).   Here is a proof of this fact:

$$A^{p-1-b}C \equiv g^{a(p-1-b)}A^b P \equiv (g^{(p-1)})^a (g^a)^{-b} A^b P \equiv 1^a\, A^{-b} A^b P \equiv P (\bmod\, p).$$

Notice that since $0 < b < p-1$, the **decryption exponent** $p-1-b$ also lies in the same (nonnegative) range.

As with RSA, both El Gamal encryption and decryption can be performed efficiently using the squaring algorithm for fast modular exponentiation (Algorithm 4.5).

EXAMPLE 4.29:   Using the primitive root $g = 2$ of the prime $p = 53$ from Example 4.26, suppose that Alice chooses her secret exponent to be $a = 22$, and Bob chooses his to be $b = 47$.
(a) Using the El Gamal system, what would be the ciphertext when Alice encrypts the message $P = 44$ to send to Bob?
(b) Perform the El Gamal decryption process that would need to be done at Bob's end to decrypt Alice's message.

SOLUTION: Part (a): As in Example 4.27, $A = 43$, $B = 5$, so that Alice can compute $C \equiv B^a P \equiv 29 \cdot 44 \equiv 4 (\bmod\, 53)$.   Thus, the entire ciphertext would be $(A, C) = (43, 4)$.

Part (b): The decryption exponent is $p - 1 - b = 5$ and $A^5 \equiv 43^5 \equiv 11 (\bmod\, 53)$, and so $d_\kappa((A, C)) \equiv A^{p-1-b} C \equiv 11 \cdot 4 \equiv 44 (\bmod\, 53)$, as expected.

EXERCISE FOR THE READER 4.28:   (a) Determine the smallest primitive root g greater than 700 for the prime $p = 1231$.
(b) Suppose that Alice and Bob will be using the El Gamal cryptosystem with these parameters and that Alice chooses her secret exponent to be $a = 212$, and Bob chooses his to be $b = 954$. What would be the ciphertext when Alice encrypts the message $P = 44$ to send to Bob?

---

[48] Note that a, b, A, and B, are exactly as in the Diffie–Hellman key exchange, so a and b are Alice's and Bob's private keys, and A and B are Alice and Bob's public keys, respectively. Furthermore, $B^a \equiv A^b$ is the Diffie–Hellman key.

(c) Perform the El Gamal decryption process that would need to be done at Bob's end to decrypt Alice's message.

Many of the guidelines for choosing "safe" primes that make $p - 1$ resistant to factorizations for use in RSA cryptosystems apply equally well to El Gamal cryptosystems. In particular, it is important to use primes p for which $p - 1$ does not have a lot of small prime factors. See our previous comments and the references cited for more details on this.

The last public key cryptosystem that we will present has its security based on the following difficult number theoretic problem, which does not involve prime numbers.

Knapsack Problems

DEFINITION 4.17: Given a vector of distinct positive integers $[a_1 \ a_2 \ a_3 \ \cdots a_n]$, the **object weights**, and a subcollection of these numbers: $a_{i1}, a_{i2}, \cdots, a_{ik}$ (thought of as the weights of objects selected to be put into a knapsack), it is easy to compute the sum $s = a_{i1} + a_{i2} + \cdots + a_{ik}$ (the total weight of the objects put in the knapsack). The **knapsack problem** is the opposite: given a positive integer s, determine, if possible, a subcollection $a_{i1}, a_{i2}, \cdots, a_{ik}$ having total weight s. (In other words, if we know the total weight of the objects that were put into the knapsack, the problem is to determine the objects.) Such a subcollection of objects is called a **solution** to the given instance of the knapsack problem.

Instances of the knapsack problem need not have any solution, and they may have multiple solutions.

EXAMPLE 4.30: Find all solutions to the knapsack problem with the following parameters: $[a_1, \ a_2, \ a_3, \ a_4, a_5, a_6] = [3, 4, 6, 8, 10, 12]$, and $s = 28$.

SOLUTION: By inspection we find that there are two different solutions to this knapsack problem:

$$a_3 + a_5 + a_6 = 6 + 10 + 12 = 28 \text{ and } a_2 + a_3 + a_4 + a_5 = 4 + 6 + 8 + 10 = 28.$$

Although the above example was small enough to do by hand, note that the number of possible knapsack collections that would have needed to be considered using a brute-force approach would be $2^6 = 64$ (this follows the multiplication principle that will be introduced in Section 5.1, since for each of the six objects we have two choices: include it or do not include it). In general, the knapsack problem is an *NP complete problem*, which puts it in the same class as the prime factorization problem as being among the most difficult discrete problems for which polynomial time solutions are strongly believed not to exist. Since both problems are NP complete, if a polynomial time algorithm for the knapsack problem is ever found, then it could be translated into a polynomial time algorithm for prime factorization (and vice versa)—see [GaJo-79] for more details.

It will be convenient to reformulate the knapsack problem as follows: We introduce a length n binary vector $[x_1 \ x_2 \ x_3 \ \cdots x_n]$. If we use the interpretation that object i (with weight) a_i is included in the knapsack (solution) if, and only if $x_i = 1$, then the total weight of the included objects is simply the dot product $[x_1 \ x_2 \ x_3 \ \cdots x_n] \cdot [a_1 \ a_2 \ a_3 \ \cdots a_n] = x_1 a_1 + x_2 a_2 + x_3 a_3 + \cdots + x_n a_n$. So a solution of the knapsack problem is a bit vector $[x_1 \ x_2 \ x_3 \ \cdots x_n]$ for which the above dot product equals s. In the previous example, the two solutions correspond to the binary vectors $[x_1 \ x_2 \ x_3 \ x_4 \ x_5 \ x_6] = [0,0,1,0,1,1]$ and $[0,1,1,1,1,0]$.

We will need the vector of object weights to have some additional properties when we create knapsack-based cryptosystems. Although general knapsack problems are intractable, certain classes can be solved very quickly. One such class is described in the following definition:

DEFINITION 4.18: An object weight vector $[a_1 \ a_2 \ a_3 \ \cdots a_n]$ in a knapsack problem is said to be **superincreasing** if each object weight is greater than the sum of all of the preceding object weights: $a_i > a_1 + a_2 + \cdots + a_{i-1}$, for $i = 2, 3, \ldots, n$.

Any knapsack problem with superincreasing weights will have at most one solution, and it can be solved very quickly. We give an example that will motivate the general algorithm, and then present the algorithm.

EXAMPLE 4.31: (a) Check that the sequence $[a_1, \ a_2, \ a_3, \ a_4, a_5, a_6] = [1,2,4,9,20,48]$ is superincreasing.
(b) Find all solutions to the knapsack problem with object weights specified by the sequence of Part (a), and $s = 27$.

SOLUTION:
Part (a): $a_2 = 2 > 1 = a_1, a_3 = 4 > 3 = a_1 + a_2, a_4 = 9 > 7 = a_1 + a_2 + a_3, \quad a_5 = 20 > 16 = a_1 + a_2 + a_3 + a_4, a_6 = 48 > 46 = a_1 + a_2 + a_3 + a_4 + a_5$. This shows that $[a_1, \ a_2, \ a_3, \ a_4, a_5, \ a_6]$ is superincreasing.
Part (b): We seek a binary vector $[x_1 \ x_2 \ x_3 \ x_4 \ x_5 \ x_6]$ whose dot product with $[a_1, \ a_2, \ a_3, a_4, a_5, \ a_6]$ is $s = 27$, i.e., $x_1 + 2x_2 + 4x_3 + 9x_4 + 20x_5 + 48x_6 = 27$. First it is clear that we must have $x_6 = 0$ (the last weight is too heavy). Next, it is easy to see that $x_5 = 1$. The reason is that if we left out the fifth weight of 20

(which fits), our total weight could add up to at most $a_1 + a_2 + a_3 + a_4$ which is less than the fifth weight. Now the problem is reduced to solving the smaller binary vector equation $x_1 + 2x_2 + 4x_3 + 9x_4 = 7$. By the same token, we see that $x_4 = 0$ and $x_3 = 1$, and the problem is again reduced to solving $x_1 + 2x_2 = 3$, which forces $x_1 = x_2 = 1$. We have thus found the unique solution: $1 + 2 + 4 + 20 = 27$.

In general, the approach that was used in Part (b) of the above example can be used to prove (see Exercise 29) that any knapsack problem with superincreasing weights can be solved with the following simple approach: Go through all weights, starting with the heaviest and going down. Each time a weight is considered, add it to the knapsack if it will fit, otherwise move on. After we have finished with the last weight, if the total weight of the knapsack equals s, the unique solution has been found, otherwise there is no solution. We record this general result, and follow it with a formal statement of this algorithm.

PROPOSITION 4.15: A knapsack problem with superincreasing weights can have at most one solution, and the following is a fast algorithm for solving any such problem.

ALGORITHM 4.12: (*Solving a Knapsack Problem with Superincreasing Weights*):

Input: A superincreasing weight vector $[a_1 \ a_2 \ a_3 \ \cdots a_n]$, and a positive integer s.

Ouput: Either a binary vector $[x_1 \ x_2 \ x_3 \ \cdots x_n]$ that specifies the unique solution of the knapsack problem with inputted parameters, or a message that there is no solution.

Step 1: Initialize $S = s$, Index $= n$

Step 2: IF $S \geq a_{\text{Index}}$,

   SET $x_{\text{Index}} = 1$, and UPDATE $S = S - a_{\text{Index}}$

   ELSE

   SET $x_{\text{Index}} = 0$

   END IF

   UPDATE Index = Index $- 1$

Step 3: If Index > 0, go back to Step 2, otherwise go to Step 4.

Step 4: IF S equals 0

   OUTPUT solution vector x

   ELSE

   OUTPUT message: No Solution

   END IF

EXERCISE FOR THE READER 4.29:   (a)   Check that the sequence $[a_1, a_2, a_3, a_4, a_5, a_6] = [3,5,9,18,36,100]$ is superincreasing.

(b) Find all solutions to the knapsack problem with object weight specified by the sequence of Part (a) and $s = 27$.

EXERCISE FOR THE READER 4.30:   (a) Suppose that $[a_1 \ a_2 \ a_3 \ \cdots a_n] = [1 \ 2 \ 4 \ \cdots 2^n]$. Show that this sequence is superincreasing, and describe the set of all positive integers s for which the corresponding knapsack problem will have a solution.
(b) Show that any superincreasing sequence $[a_1 \ a_2 \ a_3 \ \cdots a_n]$ must satisfy $a_i \geq 2^{i-1}$ for each i. Thus, the superincreasing sequences of Part (a) are the smallest superincreasing sequences.

The Merkle–Hellman Knapsack Cryptosystem

In the same year that RSA was published, another public key cryptosystem was introduced by Ralph Merkle and Martin Hellman [MeHe-78]. Such systems are known as *knapsack cryptosystems*, and their security rests on the difficulty of corresponding knapsack problems. There are many different special sorts of knapsack problems that give rise to an assortment of knapsack cryptosystems. In this development, we will restrict our attention to a single prototypical knapsack problem and its associated cryptosystem. This knapsack cryptosystem is the original one in the Merkle–Hellman paper. For a more complete treatment of knapsack problems, along with their complexity and applications, we refer the reader to [KePfPi-04].

The Merkle–Hellman cryptosystem is based on the clever idea that the secret decryption key will convert an intractable general knapsack problem into one with superincreasing object weights that can therefore be easily solved. As with any public key cryptosystem, the essential ingredient is an appropriate one-way function. The following proposition provides the one that will be used.

PROPOSITION 4.16: (*One-Way Functions for Knapsack Cryptosystems*) Given a superincreasing object weight vector $[a_1 \ a_2 \ a_3 \ \cdots a_n]$ for a knapsack problem, we define an associated function $f_a : \{$length n binary vectors$\} \rightarrow \mathbb{Z}_+$ by $f_a(X) = x_1 a_1 + x_2 a_2 + x_3 a_3 + \cdots + x_n a_n$, where $X = [x_1 \ x_2 \ x_3 \ \cdots x_n]$. Choose a modulus $m > a_1 + a_2 + a_3 + \cdots + a_n$. Next, randomly choose a multiplier w, $1 < w < m$, that is relatively prime to m. We use these parameters to define a new vector of object weights $[b_1 \ b_2 \ b_3 \ \cdots b_n]$, where, for each index i, b_i is the smallest nonnegative integer that is congruent to $wa_i (\mathrm{mod}\, m)$.

(a) Both f_a and f_b are injective.[49] Also, if their codomains are restricted to their ranges (so both become bijections), the function f_a is never a one-way function, but f_b will typically be a one-way function.

(b) (*Trap door for* f_b) Given a knapsack problem with object weights $[b_1 \; b_2 \; b_3 \; \cdots b_n]$, and parameter $s > 0$, a solution, which corresponds to a solution of $f_b(X) = s$, can be found (using the number w—the **trap door**) as the corresponding solution of $f_a(X) = s'$, where w^{-1} is the inverse of w (mod m), and s' is the least positive integer congruent to $w^{-1} \cdot s$ (mod m).

Proof: Part (a): Since $f_a(X) = s$, if, and only if $X = [x_1 \; x_2 \; x_3 \; \cdots x_n]$ solves the knapsack problem with parameters $[a_1 \; a_2 \; a_3 \; \cdots a_n]$ and s, the fact that f is injective follows from the uniqueness of solutions to knapsack problems with superincreasing weight sequences. If we have $f_b(X) = f_b(X')$ for two binary vectors $X = [x_1 \; x_2 \; x_3 \; \cdots x_n]$ and $X' = [x_1' \; x_2' \; x_3' \; \cdots x_n']$, this means that

$$\sum_{i=1}^{n} b_i x_i = \sum_{i=1}^{n} b_i x_i',$$ which in turn implies $\sum_{i=1}^{n} w^{-1} b_i x_i \equiv \sum_{i=1}^{n} w^{-1} b_i x_i' \pmod{m}$. But

since $w^{-1} b_i \equiv a_i$, the latter congruence can be rewritten as

$$\sum_{i=1}^{n} a_i x_i \equiv \sum_{i=1}^{n} a_i x_i' \pmod{m}.$$ By choice of m, each of these sums is a nonnegative

integer less than m; thus the sums must be equal, so that $f_a(X) = f_a(X')$. But since f_a is one-to-one, we conclude that $X = X'$, and the proof that f_b is one-to-one is thus complete. Since solving $f_a(X) = s$ is equivalent to solving a superincreasing knapsack problem, which is easy to solve using Algorithm 4.12. The function f_a is not a one-way function. Solving $f_b(X) = s$ is equivalent to a knapsack problem with a weight sequence $[b_1 \; b_2 \; b_3 \; \cdots b_n]$ where each of the weights is obtained from the a_i's by modular multiplication by w (mod m). The sequence typically will no longer be superincreasing, so that the knapsack problem will be a general one which typically will be difficult to solve.

Part (b): If we need to solve $f_b(X) = s$, and we know w, we can proceed as

follows: Since $f_b(X) = s$ is equivalent to $\sum_{i=1}^{n} b_i x_i = s$, which imples $\sum_{i=1}^{n} b_i x_i \equiv s$

(mod m), we can (first compute w^{-1}) multiply both sides of this congruence by

w^{-1} to obtain: $\sum_{i=1}^{n} w^{-1} b_i x_i \equiv w^{-1} s \Rightarrow \sum_{i=1}^{n} a_i x_i \equiv w^{-1} s \triangleq s' \pmod{m}$. Since the sum on

---

[49] The function f_b is defined in the same fashion as f_a. (The sequences need not be superincreasing for this definition.)

the left is a nonnegative integer less than m (by choice of m), the last congruence is actually an equality $\sum_{i=1}^{n} a_i x_i = s'$. This equation can be uniquely solved using Algorithm 4.12, thus determining the vector X. $\quad\square$

We are now nicely poised to explain the Merkle–Hellman knapsack cryptosystem.

ALGORITHM 4.13: (*Merkle–Hellman Knapsack Cryptosystem*):
Purpose: Alice needs to send Bob a private message, but they have not yet met to exchange any cryptosystem keys.
Requirements: Bob will first need a (large) integer n, a superincreasing sequence $[a_1 \ a_2 \ a_3 \ \cdots a_n]$, and an integer m larger than $a_1 + a_2 + a_3 + \cdots + a_n$. He then chooses (preferably randomly) a positive integer w in the range $1 < w < m$ that is relatively prime to m.
Plaintext Space: $\mathscr{P} = \{$length n binary vectors$\}$
Ciphertext Space: $\mathscr{C} = \mathbb{Z}_{\geq 0}$ (the nonnegative integers)
Keyspace:

$\mathscr{K} = \{\ w, [a_1 \ a_2 \ a_3 \ \cdots a_n], m : [a_1 \ a_2 \ a_3 \ \cdots a_n]$ is superincreasing, $m > \sum_{i=1}^{n} a_i$, and $1 < w < m$ with $\gcd(w,m) = 1\}$
From a system key $\{w, [a_1 \ a_2 \ a_3 \ \cdots a_n], m\}$, which is Bob's **private key**, the corresponding **public key** is $\{[b_1 \ b_2 \ b_3 \ \cdots b_n]\}$, where b_i is the smallest nonnegative integer that is congruent to $wa_i \pmod{m}$, for each index i.
Encryption Scheme: $e_\kappa : \{$length n binary vectors$\} \to \mathbb{Z}_{\geq 0}$, defined by
$e_\kappa([x_1 \ x_2 \ x_3 \ \cdots x_n]) = x_1 b_1 + x_2 b_2 + x_3 b_3 + \cdots + x_n b_n = s$.
Decryption Scheme: $d_\kappa : \text{Range}(e_\kappa) \to \{$length n binary vectors$\}$, defined by $d_\kappa(s) =$ the (unique) solution of the superincreasing knapsack problem with weight vector $[a_1 \ a_2 \ a_3 \ \cdots a_n]$ and knapsack weight s', the least positive integer congruent to $w^{-1} \cdot s \pmod{m}$. This can be quickly found using Algorithm 4.12.

Notice that the encryption function is the function f_b of Proposition 4.16. As with the rest of the section examples, we use an artificially small keyspace for ease of illustration. The computer exercises will allow readers to experiment with larger parameters.

EXAMPLE 4.32: Suppose that Bob uses the Merkle–Hellman knapsack cryptosystem with superincreasing sequence $[a_1 \ a_2 \ a_3 \ a_4 \ a_5 \ a_6] = [1,2,4,9, 20,48]$, $m = 101$ (which is greater than Σa_i), and $w = 38$ (which is relatively prime to m).
(a) What is Bob's public key?

(b) If Alice uses Bob's public key to encrypt the plaintext 011101, determine the resulting ciphertext.

(c) Perform the decryption process that would need to be done when Bob receives the ciphertext of Part (b).

SOLUTION: Part (a): Working mod 101, we have $w \cdot [a_1 \ a_2 \ a_3 \ a_4 \ a_5 \ a_6] \equiv 38 \cdot [1, 2, 4, 9, 20, 48] \equiv [38, 76, 51, 39, 53, 6]$. Thus, this latter vector is Bob's public key $[b_1 \ b_2 \ b_3 \ \cdots b_n]$.

Part (b): The ciphertext is $f_b([0,1,1,1,0,1]) = x_1 b_1 + x_2 b_2 + x_3 b_3 + x_4 b_4 + x_5 b_5 + x_6 b_6 = 0 \cdot 38 + 1 \cdot 76 + 1 \cdot 51 + 1 \cdot 39 + 0 \cdot 53 + 1 \cdot 6 = 172 = s$.

Part (c): Using the extended Euclidean algorithm (Algorithm 4.6) we compute $w^{-1} \equiv 8$. (This need be computed only once, and could be supplied with the rest of Bob's private key.) Since $w^{-1} \cdot s \equiv 8 \cdot 172 \equiv 63$, the plaintext will be the solution of the superincreasing knapsack problem with weight vector $[a_1 \ a_2 \ a_3 \ a_4 \ a_5 \ a_6]$, and knapsack weight $s' = 63$. Algorithm 4.12 quickly produces the original plaintext.

EXERCISE FOR THE READER 4.31: Suppose that Bob uses the Merkle–Hellman knapsack cryptosystem with superincreasing sequence $[a_1 \ a_2 \ a_3 \ a_4 \ a_5 \ a_6] = [3, 5, 9, 18, 36, 100]$, $m = 201$ (which is greater than Σa_i), and $w = 77$ (which is relatively prime to m).

(a) What is Bob's public key?

(b) If Alice uses Bob's public key to encrypt the plaintext 111000, determine the resulting ciphertext.

(c) Perform the decryption process that would need to be done when Bob receives the ciphertext of Part (b).

In 1982, Adi Shamir [Sha-82] discovered a polynomial time algorithm that can effectively crack the Merkle–Hellman cryptosystem. His attack was based on the fact that an adversary need not exactly determine the private key parameters m and w, but any other pair m', w' that will render $(w')^{-1} \cdot [b_1 \ b_2 \ b_3 \ \cdots b_n] \pmod{m'}$; a superincreasing sequence will work to decrypt ciphertexts. A stronger knapsack cryptosystem (the *Chor–Rivest cryptosystem*) was subsequently developed; see [Mol-03], but this system met its demise in 2001 by an attack discovered by Serge Vaudenay [Vau-01]. Although promising new knapsack cryptosystems have been developed, much confidence in them has been lost, due to the relatively short life of the above two much heralded versions, so they are presently not so often used in practice.

Government Controls on Cryptography

With the emergence of simple yet highly secure public key cryptosystems, RSA in particular, in tandem with the development of personal wireless and electronic communications, national governments have been struggling to adapt their national security policies. Governments have always striven to remain one step ahead of cryptographers, and to a great extent this has been possible until the advent of public key cryptography. The US government, for example, has strict regulations on what sorts of cryptographic technology can be sold or given, and to which countries, as well as the permissible key sizes. Violations of these rules (even people who freely distribute cryptosystems on Web sites) are considered acts of treason, and are subject to imprisonment. While on one hand, many people feel that it is their right to be able to use cryptography to protect their secrets, the flip side of this coin is that cryptography may be used as a weapon. Terrorist groups have taken advantage of simple public key cryptosystems for secure communications, and many of these systems have remained unbreakable.

With sufficiently large keysizes, for example, a properly implemented RSA cryptosystem appears to be essentially unbreakable. This state of affairs could change only in two ways:

1. If a polynomial time algorithm for factoring primes gets discovered, or
2. If a new revolution occurs in computing hardware.

Most experts believe that both of these possibilities are extremely remote. We have already discussed that a polynomial-time algorithm for factoring primes, if it was ever found, would translate into polynomial-time algorithms for most all of the intractable problems in discrete structures. After hundreds of years of research on developing the best algorithms for such problems, researchers have all but conceded total defeat to the problem of finding such an algorithm.

According to *Moore's law*, computing speeds on the latest computers double approximately every 18 months. This was first observed in the 1960s by Gordon Moore (1929–), one of the founders of the Intel Corp. At present, it takes about an hour (on a network of computers) for an exhaustive search attack on a 56 bit key. It would require $512 - 56 = 456$ powers of 2 more time for such an attack on a 512 bit key. Moore's law's estimated waiting time for computers to be able to achieve the same performance on a 512 bit keysearch as they do today with a 56 bit key search would therefore be $456 \cdot 1.5 = 684$ years! One idea that originated in the early 1980's was a suggestion by Nobel Physics Prize Laureate Richard Feynman that it might be possible to build a new type of **quantum computer** that is designed on quantum mechanical principles, which would operate under different physical axioms than present computers. In 1994, it was shown by AT&T Labs researcher Peter Shor that if such a computer could be built, then a polynomial time prime factorization algorithm could be implemented on it. Interested readers may refer to [KaLaMo-07] for more information on quantum computing.

Although the general public is not privy to the latest cryptographic technologies in use by large government organizations (such as the NSA), much evidence suggests that governments are unable to decrypt strong RSA encryptions. For example, so-called **tempest devices** are extremely sensitive electromagnetic detectors that can be used to intercept keystrokes on a computer. This would allow a government to park a tempest-enabled van outside a suspect's house/office to pick up any plaintexts before they are encrypted. Buildings can be fitted using a special insulation procedure that protects against tempest devices, but any company or individual in the US who has this insulation must first obtain a license from the federal government.

EXERCISES 4.5:

1. Suppose that Alice and Bob wish to create a secret key between them using the Diffie–Hellman protocol. They select the prime $p = 773$ and corresponding primitive root $g = 2$. Alice takes $a = 333$ as her secret key and Bob takes his to be $b = 603$.
   (a) Verify that g is indeed a primitive root mod p.
   (b) Compute the number A that Alice (publicly) sends Bob and the number B that Bob sends Alice.
   (c) Compute the shared secret Diffie–Hellman key for Alice and Bob in two different ways, as would be done on Alice's end and on Bob's end.

2. Suppose that Alice and Bob wish to create a secret key between them using the Diffie–Hellman protocol. They select the prime $p = 821$ and corresponding primitive root $g = 2$. Alice takes $a = 404$ as her secret key and Bob takes his to be $b = 769$.
   (a) Verify that g is indeed a primitive root mod p.
   (b) Compute the number A that Alice (publicly) sends Bob and the number B that Bob sends Alice.
   (c) Compute the shared secret Diffie–Hellman key for Alice and Bob in two different ways, as would be done on Alice's end and on Bob's end.

3. Suppose that Alice and Bob wish to create a secret key between them using the Diffie–Hellman protocol. They select the prime $p = 1553$ and use the smallest odd primitive root $g > 300$. Alice takes $a = 1333$ as her secret key and Bob takes his to be $b = 807$.
   (a) Determine the primitive root g that they use.
   (b) Compute the number A that Alice (publicly) sends Bob and the number B that Bob sends Alice.
   (c) Compute the shared secret Diffie–Hellman key for Alice and Bob in two different ways, as would be done on Alice's end and on Bob's end.

4. Suppose that Alice and Bob wish to create a secret key between them using the Diffie–Hellman protocol. They select the prime $p = 2267$ and use the smallest primitive root $g > 2000$. Alice takes $a = 1197$ as her secret key and Bob takes his to be $b = 62$.
   (a) Determine the primitive root g that they use.
   (b) Compute the number A that Alice (publicly) sends Bob and the number B that Bob sends Alice.
   (c) Compute the shared secret Diffie–Hellman key for Alice and Bob in two different ways, as would be done on Alice's end and on Bob's end.

5. (a) Use the RSA algorithm with (Bob's) public key $(n, e) = (6887, 143)$ to encrypt the plaintext message $P = 1234$.

(b) Go through the resulting decryption process that would need to get done on the recipient's (Bob's) end to decode the ciphertext that was   produced in Part (a) using the private key $(n,d) = (6887, 47)$.

6.   (a) Use the RSA algorithm with (Bob's) public key $(n,e) = (7493, 229)$ to encrypt the plaintext message $P = 125$.
      (b) Go through the resulting decryption process that would need to get done on the recipient's (Bob's) end to decode the ciphertext that was   produced in Part (a) using the private key $(n,d) = (7493, 4021)$.

7.   (a) Use the RSA algorithm with (Bob's) public key $(n,e) = (69353, 4321)$ to encrypt the plaintext message $P = 12345$.
      (b) Go through the resulting decryption process that would need to get done on the recipient's

      (Bob's) end to decode the ciphertext that was   produced in Part (a) using the private key $\kappa = (n,d) = (69353, 29401)$.

8.   (a)   Use the RSA algorithm with (Bob's) public key $(n,e) = (66277, 4321)$ to encrypt the plaintext message $P = 12345$.
      (b) Go through the resulting decryption process that would need to get done on the recipient's (Bob's) end to decode the ciphertext that was   produced in Part (a) using the private key $\kappa = (n,d) = (66277, 25301)$.

9.   Suppose that Bob adopts the RSA cryptosystem with primes $p = 37$ and $q = 67$. He chooses the (public key) encryption exponent to be $e = 169$.
      (a) Show that Bob's choice of encryption exponent is legitimate, and find his corresponding (private key) decryption exponent d.
      (b) Suppose that Alice encrypts the plaintext message $P = $ 1234 using the RSA cryptosystem with Bob's public key $(n,e) = (2479, 169)$. What is the resulting ciphertext that would be sent to Bob?
      (c) Go through the decryption process that would need to get done at Bob's end, using his private key (n,d) with the decryption exponent that was determined in Part (a).

10.   Suppose that Bob adopts the RSA cryptosystem with primes $p = 43$ and $q = 73$. He chooses the (public key) encryption exponent to be $e = 1195$.
      (a) Show that Bob's choice of encryption exponent is legitimate, and find his corresponding (private key) decryption exponent d.
      (b) Suppose that Alice encrypts the plaintext message $P = $ 1234 using the RSA cryptosystem with Bob's public key $(n,e) = (3139, 1195)$. What is the resulting ciphertext that would be sent to Bob?
      (c) Go through the decryption process that would need to get done at Bob's end, using his private key (n,d) with the decryption exponent that was determined in Part (a).

11.   Suppose that Alice and Bob decide to communicate with an El Gamal cryptosystem using the prime p, corresponding primitive root g, and individual keys a, and b as given in Exercise 1.
      (a) Compute the ciphertext in this system if Alice sends Bob the message $P = 321$.
      (b) Perform the El Gamal decryption process that would need to get done at Bob's end to decrypt Alice's message.

12.   Suppose that Alice and Bob decide to communicate with an El Gamal cryptosystem using the prime p, corresponding primitive root g, and individual keys a and b as given in Exercise 2.
      (a) Compute the ciphertext in this system if Alice sends Bob the message $P = 321$.
      (b) Perform the El Gamal decryption process that would need to get done at Bob's end to decrypt Alice's message.

13. Suppose that Alice and Bob decide to communicate with an El Gamal cryptosystem using the prime $p = 6469$, and individual keys $a = 2256$ and $b = 4127$, and using the smallest primitive root g of p that satisfies $g > 5050$.

    (a) Determine the primitive root g.
    (b) Compute the ciphertext in this system if Alice sends Bob the message $P = 4321$.

14. Suppose that Alice and Bob decide to communicate with an El Gamal cryptosystem using the prime $p = 8263$, and individual keys $a = 856$ and $b = 3127$, and using the smallest primitive root g of p that satisfies $g > 1700$.

    (a) Determine the primitive root g.
    (b) Compute the ciphertext in this system if Alice sends Bob the message $P = 4321$.
    (c) Perform the El Gamal decryption process that would need to get done at Bob's end to decrypt Alice's message.

15. For each of these Merkle–Hellman knapsack cryptosystem keys:

    (i) $[3,5,9,18,36,100]$, $m = 175$, $w = 88$
    (ii) $[18,36,100, 184, 360, 750]$, $m = 1450$, $w = 371$
    (iii) $[5,9,18,34,72,144]$, $m = 286$, $w = 205$

    do the following:
    (a) Verify that the key is a legitimate Merkle–Hellman knapsack cryptosystem key.
    (b) Determine the corresponding public key.
    (c) Use the public key of Part (a) to encrypt the plaintext message $P = 101010$.
    (d) Perform the decryption process that would need to be done on the receiving end for each of the ciphertexts of Part (c).

16. For each of these Merkle–Hellman knapsack cryptosystem keys:

    (i) $[1,2,6,10,25,55]$, $m = 101$, $w = 77$
    (ii) $[2,6,10,25,55,205]$, $m = 310$, $w = 161$
    (iii) $[3,13,23,43,83,173]$, $m = 339$, $w = 220$

    do the following:
    (a) Verify that the key is a legitimate Merkle–Hellman knapsack cryptosystem key.
    (b) Determine the corresponding public key.
    (c) Use the public key of Part (a) to encrypt the plaintext message $P = 101010$.
    (d) Perform the decryption process that would need to be done on the receiving end for each of the ciphertexts of Part (c).

17. Suppose that Bob uses the Merkle–Hellman knapsack cryptosystem with superincreasing sequence $[a_1\ a_2\ a_3\ a_4\ a_5\ a_6\ a_7\ a_8\ a_9] = [3,5,9,18,36,100,184,360,750]$, $m = 1499$ (which is greater than Σa_i), and $w = 365$ (which is relatively prime to m).

    (a) What is Bob's public key?
    (b) If Alice uses Bob's public key to encrypt each of the following plaintexts

    (i) 111000111   (ii) 101010101   (iii) 110011001

    determine the resulting ciphertexts.
    (c) Perform the decryption process that would need to be done when Bob receives each of the ciphertexts of Part (b).

18. Suppose that Bob uses the Merkle–Hellman knapsack cryptosystem with superincreasing sequence $[a_1\ a_2\ a_3\ a_4\ a_5\ a_6\ a_7\ a_8\ a_9] = [1,2,6,10,25,55,105,205,505]$, $m = 999$ (which is greater than Σa_i), and $w = 334$ (which is relatively prime to m).

    (a) What is Bob's public key?
    (b) If Alice uses Bob's public key to encrypt each of the following plaintexts

    (i) 111000111   (ii) 101010101   (iii) 110011001

    determine the resulting ciphertexts.

(c) Perform the decryption process that would need to be done when Bob receives each of the ciphertexts of Part (b).

19. (*A Common Modulus Attack on RSA*) (a) Suppose that in a certain RSA cryptosystem, Alice needs to send the same message P to two individuals, Bob and Ben, whose public keys are e and $\tilde{e}$, respectively. Suppose that these two encryption exponents are relatively prime. If Eve intercepts both of the corresponding ciphertexts $C \equiv P^e$ and $\tilde{C} \equiv P^{\tilde{e}} \pmod{n}$, show how she will be able to decrypt the plaintext message P. (b) Demonstrate your technique by decrypting the two common modulus ciphertexts (of the same plaintext message) $C = 2254$ and $\tilde{C} = 1902$ that were created using the public keys $e = 143$ and $\tilde{e} = 2209$, respectively, with the common RSA modulus $n = 6887$. (c) Demonstrate your technique by decrypting the two common modulus ciphertexts (of the same plaintext message) $C = 747126$ and $\tilde{C} = 189255$ that were created using the public keys $e = 31547$ and $\tilde{e} = 6251$, respectively, with the common RSA modulus $n = 976901$.

20. Explain whether it would add any additional security to an RSA system if a certain user required that any incoming messages be encrypted twice using different encryption exponents e, $\tilde{e}$, both relatively prime to $\phi(n)$. More precisely, this user's public key would consist of the RSA modulus n, along with the two RSA encryption exponents e, $\tilde{e}$, with the instructions to first encrypt with exponent e, and then encrypt the result with the exponent $\tilde{e}$.

21. Consider the following cryptosystem: Fix a large prime number p. The plaintext and ciphertext spaces are both $\mathbb{Z}_p$. For a given element $e \in \mathbb{Z}_{p-1}^{\times}$, the associated encryption function is defined by $E_e : \mathbb{Z}_p \to \mathbb{Z}_p :: E_e(P) \equiv P^e \pmod{p}$.
    (a) Describe the corresponding decryption function, and find it explicitly in case $p = 1009$ and $e = 275$. In the latter specific setting, decrypt the ciphertext $C = 777$.
    (b) How does the security of such a cryptosystem compare with that of RSA with a comparably sized modulus? Could this system be considered a public key cryptosystem?

22. (*Broadcast Attack on RSA*) Suppose that Alice broadcasts the same message P to a group of users. Each user is free to choose his/her own RSA modulus and encryption exponent. If in the group, there are e users that share the same small encryption exponent e, and whose RSA modulii are pairwise relatively prime, then Eve will be able to use the corresponding e ciphertexts along with the Chinese remainder theorem to recover the plaintext message m. For example, suppose that three recipients with RSA pairwise relatively prime modulii n_1, n_2, and n_3 all use the encryption exponent $e = 3$. This means that the corresponding ciphertexts satisfy $C \equiv P^3 \pmod{n_1}$, $C' \equiv P^3 \pmod{n_2}$, and $\tilde{C} \equiv P^3 \pmod{n_3}$. Eve can use the (constructive proof of the) Chinese remainder theorem (Theorem 4.4) to find an integer D that simultaneously solves all three congruences, and from this it follows that $D \equiv P^3 \pmod{n_1 n_2 n_3}$. Since $P^3 < n_1 n_2 n_3$, this last congruence is actually valid in regular arithmetic, i.e., $D = P^3$, so the plaintext message can be decoded by taking the cube root: $P = \sqrt[3]{D}$.
    (a) Apply the broadcast attack to decrypt the plaintext message m, if it was broadcast to a network of users, including three that share the same encryption exponent $e = 3$, and with modulii $n_1 = 1207$, $n_2 = 2407$, and $n_3 = 3649$. The corresponding three ciphertexts were $C_1 = 494$, $C_2 = 113$, and $C_3 = 2372$.
    (b) Give an example of a broadcast attack on RSA with $e = 4$, if possible.

23. (*A Chosen Ciphertext Attack on RSA*) Assume that Alice has used RSA to send Bob a certain plaintext message P, and that Eve has intercepted the ciphertext C. Eve really wants to know the

plaintext message P. Assume that Bob, playing a risky game with Eve, offers to decrypt for her any other single ciphertext $\tilde{C} \neq C$. Show how by judiciously choosing $\tilde{C}$, Eve will be able to deduce the original plaintext P. We assume, as usual, that Eve knows the public key (n, e). **Note:** This does not suggest Eve will be able to determine Bob's secret key d.

Suggestion: Let x be any element of $\mathbb{Z}_n^{\times}$ such that $Cx^e \not\equiv C (\mathrm{mod}\, n)$, and take $\tilde{C} \equiv Cx^e (\mathrm{mod}\, n)$.

24. (*Cracking the El Gamal Cryptosystem is Equivalent to Cracking the Diffie–Hellman Key Exchange*) Prove the following two assertions:
    (a) If one is able to determine the plaintext P from any given ciphertext C of a given El Gamal cryptosystem with knowledge only of the public key (without knowing the private key), then one will also be able to determine Diffie–Hellman keys from knowing only the transmitted information (A or B in Algorithms 4.9 and 4.12) and the public key. It is assumed that the publicly known prime p, and corresponding primitive root g are the same for both algorithms.
    (b) Prove the converse of the statement in Part (a).

    Sketch of Proof for Part (a): Assume that we know Bob's transmission B ($\equiv g^b (\mathrm{mod}\, p)$), and Alice's transmission A ($\equiv g^a (\mathrm{mod}\, p)$). Take $C = 1$ in Algorithm 4.11, and decrypt the El Gamal ciphertext (A, C) to get $P \equiv 1 \cdot A^{-b} \equiv g^{-ab} (\mathrm{mod}\, p)$, so the Diffie–Hellman key ($= g^{ab} (\mathrm{mod}\, p)$) will be $P^{-1} (\mathrm{mod}\, p)$.

25. (*Properties of Discrete Logarithm Functions*) Establish each of the following "laws of discrete logarithms" that are direct analogues to the laws for ordinary logarithms that one learns about in (pre-)calculus courses. Throughout, g is a primitive root for some prime integer p, and $L_g : \mathbb{Z}_p^{\times} \to \mathbb{Z}_p^{\times}$ is the corresponding discrete logarithm function (as in Definition 4.15).
    (a) $L_g(ab) \equiv L_g(a) + L_g(b) \;(\mathrm{mod}\, p - 1)$.
    (b) $L_g(a^{-1}) = -L_g(a) \;(\mathrm{mod}\, p - 1)$.
    (c) $L_g(a^k) = kL_g(a) \;(\mathrm{mod}\, p - 1)$.
    (d) If h is another primitive root mod p, then $L_h(a) \equiv L_h(g) \cdot L_g(a) \;(\mathrm{mod}\, p - 1)$.

26. (*A Known Plaintext Attack on El Gamal*) Assume that Alice has sent Bob two messages P_1, and P_2, resulting in ciphertext (second components) C_1 and C_2, respectively, using the El Gamal system, and that she used the same secret key a to encrypt both messages. Show how if Eve finds out P_1, and had learned that it is the plaintext for C_1, then she will be able to determine P_2.
    Note: Such an attack can be prevented by (randomly) generating a new secret key for each message sent.

27. (*Formulation of RSA into a Block Cryptosystem*) Suppose that we wish to send messages in a certain alphabet that has K symbols in it. We first use $\mathbb{Z}_K$ to represent this alphabet. For example, with the ordinary (lower-case) English alphabet, Table 4.2 shows a natural representation using $\mathbb{Z}_{26}$. We assume that the RSA modulus n being used is several times larger than K. (This will usually be automatic in any real-life RSA system, since n will have to be very large to assure security.) We let $r = \mathrm{floor}(\log_K(n))$; this will be the *block length* of plaintext messages in the alphabet $\mathbb{Z}_K$ that will be transmitted. The way that an r-block of letters, represented by a length r vector $V = [m_1 \; m_2 \; \cdots \; m_r]$ with elements $m_i \in \mathbb{Z}_K$ is transmitted is that we view V as a base K representation of an integer, and convert V to an ordinary integer as in Section 4.1:

$$M = [m_1 \ m_2 \ \cdots \ m_r] \quad \rightarrow \quad m = \sum_{j=1}^{r} K^{r-j} m_j.$$

(Recall that this is a one-to-one correspondence, and the representing integer m will always lie in the range $0 \le m \le \sum_{j=1}^{r} K^{r-j}(K-1) = K^r - 1 < n.$) We then use the (ordinary) RSA system (with modulus n) to encrypt the integer m into an integer c. This integer c is converted into its corresponding base K representation, which will have length at most $r + 1$.

$$c = \sum_{j=0}^{r} K^{r-j} c_j \rightarrow C = [c_0 \ c_1 \ c_2 \ \cdots \ c_r].$$

The ciphertext is this length $r + 1$ vector C (padded with zeros on the left, if necessary) of elements of $\mathbb{Z}_K$.

(a) Use the RSA system of Exercise 1 to transmit the length 8 plaintext block [0 0 1 1 1 1 0 1] (in the binary alphabet) into a corresponding length 9 ciphertext vector.

(b) Use the RSA system of Exercise 3 to transmit the message "clearout" (in a 26 lower-case English letter alphabet) using the correspondence of Table 4.2) into corresponding ciphertext blocks.

(c) With the block version of RSA of Part (a), decrypt the ciphertext block: [0 1 1 0 1 0 1 0].

(d) With the block version of RSA of Part (b), decrypt the ciphertext blocks: [FQYL][FGOU] [CXET].

28. (*An Elementary Factoring Method*) It was mentioned in the text that the RSA prime factors p and q should not be chosen too close together since then $n = pq$ would be easy to factor. The following simple factoring method shows why this is so.

(a) If $n = pq$ and $p > q$, show that $q < \sqrt{n} < p$.

(b) (*For readers who have studied calculus*) Show that the quantity $\frac{1}{2}(p+q)$, thought of a function of the real variable p $(q = n/p)$, strictly increases from $\sqrt{n}$ to $(n+1)/2$ as p runs from $\sqrt{n}$ to n.

(c) Show that $n = \left(\dfrac{p+q}{2}\right)^2 - \left(\dfrac{p-q}{2}\right)^2$.

(d) If p and q are close together, then by Part (b), the integer $x = \frac{1}{2}(p+q)$ will be (greater than but) close to $\sqrt{n}$, and the integer $y = \frac{1}{2}(p-q)$ will be small. Use Part (c) to obtain the factorization $n = (x+y)(x-y)$.

NOTE: In practice, since $x^2 - n = y^2$, if p and q are close together, then the factorization of Part (d) can be found by checking integer values of x larger than $\sqrt{n}$ until $x^2 - n$ is a perfect square. Then this perfect square will be y^2 and we will have $n = (x+y)(x-y)$.

(e) Use the above factoring method to factor the RSA modulus $n = 3301453$.

29. Prove Proposition 4.15.

COMPUTER EXERCISES 4.5:

NOTE: Although some of the cryptographic systems in this chapter are very simple to program using algorithms from previous sections (like the fast exponentiation Algorithm 6.5), but we will nonetheless aim for the creation of a complete set of programs. As mentioned in chapter text, if your computing platform is a floating point arithmetic system, it may only allow you up to 15 or so significant digits of accuracy. Symbolic systems allow for much greater precision, being able to handle hundreds of significant digits. Some platforms allow the user to choose if they wish to work in floating point or symbolic arithmetic. If you are working on such a dual capability system, you may wish to create two separate programs (for those that might work with large integers), an ordinary version and a symbolic version (perhaps attaching a `Sym` suffix to the names of those of the latter type). In case you do not have access to a symbolic system, some particular questions below may need to be skipped or modified so the numbers are of a manageable size.

1. (*Program for the Diffie–Hellman Key Exchange*) (a) Write a program with syntax:

    K = DiffieHellmanKey(p,g,B,a)

    that will produce the Diffie–Hellman key (using Algorithm 4.9) given the (public) inputs: p = the prime number, g = the primitive root (mod p), B =Bob's public key, and a = Alice's private key. The output, K, is a mod p integer representing the resulting Diffie–Hellman key.
    (b) Run your program with the parameters of Example 4.27, and confirm the results.
    (c) Run your program on the parameters of Example 4.27, but with Bob's and Alice's parameters switched (so you should run: DiffieHellmanKey(p,g,A,b)), and check to see that the result is the same as was obtained in Part (b).

2. Suppose that Alice and Bob wish to create a secret key between them using the Diffie–Hellman protocol. They select the prime p = 2425967623052370772757633156976982469681, and corresponding primative root $g = 3$. Suppose that Alice takes a = 8866446688 as her secret key, and Bob takes his to be b = 196819691970.
    (a) Apply the Miller–Rabin test (of the Appendix to Section 3.3) to certify the primality of p so that the probability of an incorrect certification is less than 1 in 1 trillion. Then check that that g is the smallest primitive root mod p.
    (b) Compute the number A that Alice (publicly) sends Bob and the number B that Bob sends Alice.
    (c) Compute the shared secret Diffie–Hellman key for Alice and Bob in two different ways, as would be done on Alice's end and on Bob's end.
    Answers:
    (a) $\operatorname{ord}_p(2) = (p-1)/2$.
    (b) A = 1749037741156461472834806859126142492373, and
    B = 1519836956974222839962608060419998836078.
    (c) DH Key = 836565563359907232787680145037578098830.

3. (*Program for the RSA Encyrption*) (a) Write a program with syntax:

    C = RSAEncrypt(P,e,n)

    that will perform the RSA encryption (using Algorithm 4.10) given the inputs: P = the plaintext (an integer mod n), n = the RSA modulus, and e = the (public) encryption exponent. The output, C, is a mod n integer representing the ciphertext.
    (b) Run your program with the parameters of Example 4.28(b), and confirm the results.
    (c) Suppose that Bob adopts the RSA cryptosystem with primes p = 153,817 and q = 1,542,689, and public key encryption exponent e = 202,404,606. If Alice uses this system to send Bob the plaintext message P = 888,999,000, apply your program of Part (a) to determine the ciphertext.

4. (*Program for the RSA Decryption*) (a) Write a program with syntax:

    P = RSADecrypt(C,d,n)

that will perform the RSA decryption (using Algorithm 4.10) given the inputs: C = the ciphertext (an integer mod n),   n = the RSA modulus, and d = the (private) decryption exponent. The output, P, is a mod n integer representing the plaintext.
(b) Run your program with the parameters of Example 4.28(c), and confirm the results.
(c) For the RSA system described in Computer Exercise 3(c), compute Bob's corresponding decryption exponent d. Apply your program of Part (a) to decrypt Alice's ciphertext (produced in Computer Exercise 3(c)).

5.   (a) Verify that the integer $n = 21463366383055728841$ factors as $n = pq$, with $p = 5915587277$ and $q = 3628273133$.
(b) Verify that the (Bob's) exponent $e = 22347$ is relatively prime to $\phi(n)$, and so is an admissible RSA encryption exponent. Use the RSA algorithm with modulus n as in Part (a), and encryption exponent e to encrypt the plaintext message $P = 123456789$.
(c) Determine the (Bob's) corresponding decryption exponent d, and then go through the resulting decryption process that would need to get done on the recipient's (Bob's) end to decode the ciphertext that was produced in Part (b).
Answers: (b) $c = 127300030462912338$ (c) $d = 18670353019854889267$

6.   (a) Verify that the integer $n = 345236661798421353184887569396010441044 81$ factors as $n = pq$, with $p = 71755440315342536873$ and $q = 48112959837082048697$.
(b) Verify that the (Bob's) exponent $e = 8877665544332211$ is relatively prime to $\phi(n)$, and so is an admissible RSA encryption exponent. Use the RSA algorithm with modulus n as in Part (a), and encryption exponent e to encrypt the plaintext message $P = 1234567890987654$.
(c) Determine the (Bob's) corresponding decryption exponent d, and then go through the resulting decryption process that would need to get done on the recipient's (Bob's) end to decode the ciphertext that was produced in Part (b).
Answers:
(b) $c = 178923608668085431633426701601012925713 4$
(c) $d = 301492479245913500806753797929958328063 5$

7.   (*An Industrial-Grade RSA System*) The following two prime numbers give rise to a rather special RSA cryptosystem:

p = 16347336458092538484431338838650908598417836700330923121811108523893331001045081512121181675115 79

q = 1900871281664822113126851573935413975471896789968515493666638539088027103802104498957191261465571

Their product $n = pq$ is known as RSA-640 that was introduced in the last chapter. The RSA corporation put out a number of public challenges (with monetary awards) to factor some very large RSA moduli. RSA no longer offers such challenges, but RSA-640 was the last of these factoring challenges that they put out in 2001 that got factored. The factorization was announced in 2005 by a team of scientists and it took 30 2.2 GHz-Opteron-CPU years (in over five months of calendar time).
Note: Rather then entering p and q by hand into your computer, it would be much quicker (and more pleasant) to navigate to the appropriate RSA Web page, and copy and paste these integers into your computer.
(a) If you have a built-in factoring utility on your platform, factor p and q (to check that they are prime); then factor $p - 1$. You will find that "factoring" p and q should go quite a bit quicker (since these really amounted to primality checks). **Caution:** Do not try to factor n or even $q - 1$ (using your built-in factoring utility).
(b) Verify that the (Bob's) exponent $e = 12345678910111213141516171819$ is relatively prime to $\phi(n)$, and so is an admissible RSA encryption exponent. Use the RSA algorithm with modulus $n = pq$, and encryption exponent e to encrypt the plaintext message $P = 112233445566778899009988776655544332211$.
(c) Determine the (Bob's) corresponding decryption exponent d, and then go through the resulting decryption process that would need to get done on the recipient's (Bob's) end to decode the ciphertext that was produced in Part (b).

Answers:

(a) $p - 1 = 2^2 \cdot 3 \cdot 53 \cdot 470419654972168112208103587 07 \cdot 295964577748188802772167 \cdot$
$1554503367019 \cdot 6706111 \cdot 10987 \cdot 55057 \cdot 7129.$

(b) $c = 131537329878824712495795027040738628028002439712122949572589255092108999$
$0223014363655486752537354908334458799119962908664763119873266777469874179 43255$
$203789613021840199762 59$

(c) $d = 134001668668994239120753362505352565327756949565457343285686874309095239$
$7320150413520426159362370409347608762664068906816929230248530625919382362 09281$
$6519876175056170399006 19$

8.   (*Program for the El Gamal Encryption*) (a) Write a program with syntax:
$$A, C = \texttt{ElGamalEncrypt(P,a,B,p,g)}$$
that will perform the El Gamal encryption (using Algorithm 4.11) given the inputs: P = the plaintext (an integer mod p), p = the El Gamal modulus (a prime), g = a corresponding primitive root that has been adopted, a = the (private) El Gamal encryption exponent of the sender, and B = the (public) El Gamal key of the receiving party. The outputs, A, C, are mod p integers representing the ciphertext.
(b) Run your program with the parameters of Example 4.29(a), and confirm the results.
(c) Determine the smallest primitive root g of the prime p of Computer Exercise 7. If Alice and Bob choose their corresponding El Gamal private keys to be $d_A = 111,222,333$,   and $d_B = 444,666,888$, use your program from Part (a) to determine the ciphertext that Alice would send to Bob as the encryption of the plaintext message $P = 333,555,777$.

9.   (*Program for the El Gamal Decryption*) (a) Write a program with syntax:
$$P = \texttt{ElGamalDecrypt(A,C,b,p)}$$
that will perform the El Gamal decryption (using Algorithm 4.11) given the inputs: A, C,   mod p integers representing the ciphertext, p = the El Gamal modulus (a prime), b = the (private) El Gamal encryption exponent of the receiver.   The outputs A and C are   mod p integers representing the ciphertext.
(b) Run your program with the parameters of Example 4.29(a), and confirm the results.
(c)   Use your program of Part (a) to decrypt the ciphertext that was produced in Computer Exercise 7(c).

10.   (*Brute-Force Program for General Knapsack Problem*) (a) Write a program with syntax:
$$\texttt{Bestx BestValue = KnapsackBruteFindAllBest(Weights, s)}$$
that will perform a brute-force search for a knapsack problem and collect all solution vectors corresponding to the best solution.   The inputs are: `Weights`, the vector of object weights (positive integers), and s = the knapsack capacity. The outputs will be a matrix `Bestx` and a scalar `BestValue` that have the following meaning: `BestValue` will be the greatest possible knapsack weight that is less than or equal to s, and `Bestx` will be a matrix of all corresponding binary vectors that describe the different knapsack configurations. Each row of this matrix is a binary vector of the same size as `Weights`, and corresponds to such a configuration.
(b) Run your program with the parameters of Example 4.30, and confirm the results.
(c) Randomly generate some knapsack problems with increasing numbers of weights, and run your program of Part (a) on them until it takes more than 5 seconds to execute. What is the size of the problem for which this first occurs (results will vary)?
Suggestion for Part (c):   For the weight vectors, randomly generate individual weights in a specified range (say between 1 and 40); it is fine if some weights are the same. For the value of s, randomly generate some values between SumWgts/2 and SumWgts, where SumWgts is the sum of all the weights. Run two or three randomly generated datasets for each size vector, starting with $n = 8$ weights.

11.   (*Brute-Force Program for General Knapsack Problem*) (a) Write a program with syntax:
$$\texttt{Bestx BestValue = KnapsackBruteFindAllBest(Weights, s)}$$

that will perform a brute-force search for a knapsack problem and collect all solution vectors corresponding to the best solution. The inputs are: `Weights`, the vector of object weights (positive integers), and `s` = the knapsack capacity. The outputs will be a matrix `Bestx` and a scalar `BestValue` that have the following meaning: `BestValue` will be the greatest possible knapsack weight that is less than or equal to `s`, and `Bestx` will be a matrix of all corresponding binary vectors that describe the different knapsack configurations. Each row of this matrix is a binary vector of the same size as `Weights`, and corresponds to such a configuration.

(b) Run your program with the parameters of Example 4.30, and confirm the results.

(c) Randomly generate some knapsack problems with increasing numbers of weights, and run your program of Part (a) on them until it takes more than 5 seconds to execute. What is the size of the problem for which this first occurs (results will vary)?

Suggestion for Part (c): For the weight vectors, randomly generate individual weights in a specified range (say between 1 and 40); it is fine if some weights are the same. For the value of `s`, randomly generate some values between SumWgts/2 and SumWgts, where SumWgts is the sum of all the weights. Run two or three randomly generated data sets for each size vector, starting with $n = 8$ weights.

12. (*Efficient Program for Superincreasing Knapsack Problem*) (a) Write a program with syntax:

$$x = \mathtt{KnapsackSuperIncreasing(Weights, s)}$$

that will use Algorithm 4.12 to solve a superincreasing knapsack problem. The inputs are: `Weights`, the vector of superincreasing object weights (positive integers), and `s` = the knapsack capacity. The output will be a binary vector `x` that solves the problem (if a solution exists). If there is no solution, the program should print an error message to this effect.

(b) Run your program with the parameters of Exericse for the Reader 4.29, and confirm the results.

(c) Randomly generate some <u>superincreasing</u> knapsack problems with increasing numbers of weights, and run your program of Part (a) on them until it takes more than 5 seconds to execute. What is the size of the problem for which this first occurs (results will vary)?

13. (*Program for Merkle–Hellman Encryption*) (a) Write a program with syntax:

$$C = \mathtt{MerkleHellmanEncrypt(PublicWeights, xPlaintext)}$$

that will use Algorithm 4.13 to perform encryptions using the Merkle–Hellman knapsack cryptosystem. The inputs are: `PublicWeights`, the public key vector object weights (positive integers), and `xPlaintext` = a binary vector representing the plaintext. The output is a nonnegative integer `C` that is the corresponding ciphertext.

(b) Run your program with the parameters of Example 4.32(b) and confirm the results.

(c) Run your program on the encryptions of Ordinary Exercise 17.

14. (*Program for Merkle–Hellman Decryption*) (a) Write a program with syntax:

$$\mathtt{xPlaintext = MerkleHellmanDecrypt(PrivateWeights, m, w, C)}$$

that will use Algorithm 4.13 to perform decryptions using the Merkle–Hellman knapsack cryptosystem. The inputs are: `PrivateWeights`, the private key vector superincreasing object weights (positive integers), `m`, `w`, the private key modulus and multiplier, and `C` = an integer representing the ciphertext. The output is the binary vector `xPlaintext` representing the corresponding plaintext.

(b) Run your program with the parameters of Example 4.32(c), and confirm the results.

(c) Run your program on the decryptions of Ordinary Exercise 17.

Chapter 5: Counting Techniques, Combinatorics, and Generating Functions

In discrete mathematics problems, one often needs to know how many (or approximately how many) objects belong to a certain set. Such counting problems may be interesting in their own right, or constitute an integral part of the solution to another problem. Many important and difficult problems in probability (Chapter 6) amount to counting problems. The counting techniques that we learn in this chapter will also be useful in assessing and comparing the speeds of algorithms, where it is required to get rough estimates of how many logical/arithmetical operations need to be performed in the execution of an algorithm (usually as a function of the input size). This latter application of counting methods is known as the *complexity theory of algorithms*, and will be studied in Chapter 7. After introducing an assortment of useful counting methods in Sections 5.1 and 5.2, Section 5.3 will discuss the theory of *generating functions*. Every sequence has a (unique) generating function. Generating functions are a powerful tool that often allow difficult or seemingly intractable counting problems to be translated into much simpler questions by translating a combinatorial question into a corresponding question about an appropriately formulated generating function.

5.1: FUNDAMENTAL PRINCIPLES OF COUNTING

The subject of sophisticated counting methods has evolved into an important branch of mathematics called *combinatorics*. In this and the next section we will introduce some of the central ideas of combinatorics that frequently arise in discrete structures. Since students often have difficulty remembering how and when to apply some of these methods, we will motivate several key principles by examples. Keeping a collection of such examples in mind for comparisons and contrasts will help the reader in deciding what principles are applicable in order to solve various problems. We first introduce some notation for the number of elements in a finite set.

NOTATION: If S is any finite set, the symbol $|S|$, which can be read as the **cardinality** of S, denotes the number of elements in the set S.

The Multiplication Principle

EXAMPLE 5.1:   (*Motivating example for the multiplication principle*)   Arlo packs three shirts, two ties, and three pairs of pants for a business trip. How many different outfits can Arlo put together during this trip? Assume that an outfit consists of one choice each of a shirt, tie, and a pair of pants, and that any differences in the choices lead to different outfits.

SOLUTION: One approach is to represent the sequence of choices by a so-called *tree diagram*;[1]   such a diagram is shown in Figure 5.1. Notice that each outfit corresponds to a unique sequence of choices of a shirt (from S1, S2, and S3), tie (from T1 and T2), and pants (from P1, P2, and P3), and this in turn corresponds to a unique path down the tree, which is completely determined by where it lands on the bottom. Notice that at the end of each stage, total number of choices is the number from the previous stage multiplied by the number of choices available at the current stage. Thus we have shown that Arlo is able to put together a total of $3 \cdot 2 \cdot 3 = 18$ outfits.

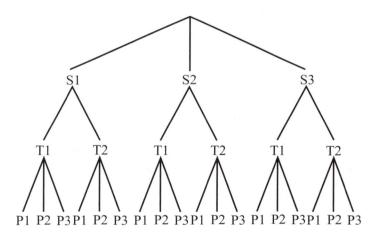

FIGURE 5.1: A tree diagram for the counting problem of Example 5.1. To put together an outfit, we start at the top (root) of the tree, and first choose one of three shirts {S1, S2, S3}, next we choose a tie from {T1, T2}, and finally we choose a pair of pants from {P1, P2, P3}. Each path from top to bottom represents a different permissible outfit, and no other outfits can be put together.

Using a set theoretic approach, the outfits can be viewed to correspond to elements in the Cartesian product $S \times T \times P$, where S denotes the set of all shirts, $S = \{$S1,

---

[1] We will give a more formal development of trees (tree diagrams) in Chapter 8. For now we treat the concept as an intuitive one.

S2, S3}, and similarly for T and P. For example, the ordered triple (S2, T1, P1) corresponds to the outfit with the second shirt, the first tie, and the first pair of pants. Since from the following simple set-theoretic proposition it follows that $|S \times T \times P| = |S| \cdot |T| \cdot |P|$, we have another way to count the different outfits.

PROPOSITION 5.1: If $S_1, S_2, \cdots, S_k$ are finite sets, then the cardinality of their Cartesian product $S_1 \times S_2 \times \cdots \times S_k$ (see Section 1.3) is given by:

$$|S_1 \times S_2 \times \cdots \times S_k| = |S_1| \cdot |S_2| \cdots |S_k|.$$

Proof: We use induction on k.

1. *Basis Step*: The identity simply states that $|S_1| = |S_1|$.

2. *Inductive Step*: Assuming that $k \geq 1$ and the identity is true for k factors, we need to show it is valid for $k + 1$ factors. Write out the elements of the last factor S_{k+1} as $\{a_1, a_2, \cdots, a_N\}$, so that $N = |S_{k+1}|$. Now the elements of $S_1 \times S_2 \times \cdots \times S_k \times S_{k+1}$ fall into N disjoint subsets, $T_1, T_2, \cdots, T_N$, depending on their last entry (of the S_{k+1} factor). Therefore, for $1 \leq j \leq N$, we can write

$$T_j = \{(s_1, s_2, \cdots, s_k, a_j) \mid \forall_{i, 1 \leq i \leq k} [s_i \in S_i]\}.$$

The elements of each of these T_j's are thus in one-to-one correspondence with the elements of $S_1 \times S_2 \times \cdots \times S_k$, so by the inductive hypothesis, we have for each index j, that $|T_j| = |S_1 \times S_2 \times \cdots \times S_k| = |S_1| \cdot |S_2| \cdots |S_k|$. Since there are $N = |S_{k+1}|$ T_j's, and they are disjoint, it follows that $|S_1 \times S_2 \times \cdots \times S_k \times S_{k+1}| = |T_1 \cup T_2 \cup \cdots \cup T_N| = N \cdot |S_1| \cdot |S_2| \cdots |S_k| = |S_1| \cdot |S_2| \cdots |S_{k+1}|$. □

Either of the two methods of the above example works easily to establish the following more general principle:

THE MULTIPLICATION PRINCIPLE: Suppose that a sequence of choices is to be made and that there are m_1 options for the first choice, m_2 options for the second choice, and so on, up to the kth choice. If these choices can be combined freely, then the total number of possible outcomes for the whole set of choices is $m_1 \cdot m_2 \cdots m_k$.

The multiplication principle is extremely useful. To apply it to a counting problem, one must be able to recast the problem at hand into a sequence of unrestricted choices. The following examples will demonstrate this technique.

EXAMPLE 5.2: A standard Hawaii license plate consists of a group of three letters followed by a group of three digits; see Figure 5.2.

(a) How many (standard) Hawaii license plates can the state produce?

(b) If on the island of Maui, the first letter of the plate must be "M," how many (standard) Maui plates can be produced?

FIGURE 5.2: A standard Hawaii license plate.

SOLUTION: Part (a): We view creating a Hawaii plate as making a sequence of six unrestricted choices; for each letter slot we have 26 choices, while for the digit slots we have 10 choices. Hence, by the multiplication principle, the total number of Hawaii plates will be $26 \cdot 26 \cdot 26 \cdot 10 \cdot 10 \cdot 10 = 17,576,000$.

Part (b): Since the first letter is already specified, making a Maui plate can be viewed as a sequence of five choices, with the total number being $26^2 \cdot 10^3 = 676,000$.

EXAMPLE 5.3: A three-member committee is to be formed from the US Senate, which has 100 members (2 from each state). The committee will have a chairperson, a vice-chair, and a spokesperson.

(a) How many different such committees can be formed?

(b) How many if Senator A must be on it?

(c) How many if Senators B and C will serve together or not at all?

SOLUTION: Part (a): We break up the formation of the committee into the following sequence of three choices: first choose a chair (100 senators to choose from), next, after a chair has been chosen, choose a vice-chair (from the 99 senators remaining), finally, from the 98 senators remaining, we choose the spokesperson.[2] The multiplication principle tells us that there can be a total of $100 \cdot 99 \cdot 98 = 970,200$ such committees.

Part (b): We give two different methods:

Method 1: (*Separate into disjoint cases first*) We have learned early on in the last chapter that problems can often be reduced to simpler ones using cases. For the problem at hand there are three natural cases: A serves as chair, vice-chair, or spokesperson. These three cases are disjoint (no matter how the rest of the committee is formed). (Why?) Using the multiplication principle to fill the remaining slots, by disjointness, we may add up the results to get the answer to Part (b): $1 \cdot 99 \cdot 98 + 99 \cdot 1 \cdot 98 + 99 \cdot 98 \cdot 1 = 3 \cdot 99 \cdot 98 = 29,106$ (the factor 1 in each of the three terms represents that there is only one choice for the corresponding slot, since Senator A will occupy that slot in each case).

---

[2] Of course, this method of choosing a committee has no bearing on the process of how the Senate might actually put together such a committee (usually by nominations and voting); we cast the task as a sequence of choices solely as a mathematical device to solve the counting problem.

Method 2: (*Use the multiplication principle directly*):

$$3 \quad \cdot \quad 99 \quad \cdot \quad 99$$

places to put   choices to fill   choices to fill
Senator A   first remaining   second remaining
slot   slot

Part (c): Separating into the two natural cases: (i) neither B nor C serves, and (ii) both B and C serve (which give rise to disjoint sets of committees) seems like the only way to go here. Each of the two cases is amenable to the multiplication principle:

$$\underbrace{98 \quad \cdot \quad 97 \quad \cdot \quad 96}_{\text{neither B nor C serve}} + \underbrace{3 \quad \cdot \quad 2 \quad \cdot \quad 98}_{\text{both B and C serve}} = 913{,}164.$$

choices for   choices for   choices for the   positions   positions   choices for the
the chair   the vice-chair   spokesperson   for B   for C   remaining position

EXERCISE FOR THE READER 5.1: A professional basketball team is arranging a publicity photograph with 5 players taken from its active list of 12 players.
(a) If the players are to be lined up in a row, how many different photograph arrangements are possible?
(b) How many such arrangements are possible if players K and S refuse to appear in the lineup together?

The multiplication principle has both practical and theoretical utility. We use it next to give a proof of an important fact that was mentioned in Section 1.3 (recall that we also gave a proof of this result in Section 3.1 using mathematical induction, see Proposition 3.3).

PROPOSITION 5.2: A set S with a finite number n of elements has 2^n subsets.

Proof: We list the elements of the S as $\{a_1, a_2, \cdots, a_n\}$. We can view the formation of a subset $B \subseteq S$ as a sequence of n choices, the ith choice being whether to include the element a_i in the subset B. Since each of these n steps has two choices (i.e., either $a_i \in B$ or $a_i \notin B$), it follows from the multiplication principle that there are a total of $\underbrace{2 \cdot 2 \cdots \cdots 2}_{n \text{ factors}} = 2^n$ subsets of S. $\square$

The Complement Principle

Another simple yet often useful rule is a consequence of the basic fact that for any subset $S \subseteq U$ (the universal set), U is the disjoint union of S and its complement $\sim S$. If U is a finite set, this implies that $|U| = |S| + |\sim S| \Rightarrow |S| = |U| - |\sim S|$. We reiterate this in words:

THE COMPLEMENT PRINCIPLE: Suppose the universal set is finite. The number of elements in a set equals the number of elements in the (finite) universal set, less the number of elements that are not in the set.

EXAMPLE 5.4: For security reasons, a university's finance office requires students to create a six-character password to log into their accounts. Passwords must contain at least one digit and at least one letter.
(a) How many passwords are possible if the protocol is not case-sensitive?
(b) What if the protocol is case-sensitive?

SOLUTION: Let D denote the set of all six character strings that contain at least one digit, and L the set of all six character strings that contain at least one letter. We wish to count the number of passwords in the set $D \cap L$. The sets D and L are difficult to count directly, but their complements are easy. For example, $\sim D$ is the set of all six character passwords that contain no digits, and therefore consist only of letters. By the multiplication principle, the number of such passwords is 26^6 for Part (a) and 52^6 for Part (b). In the same fashion, $|\sim L| = 10^6$ (for both Parts (a) and (b)). Also, letting S denote the (universal) set of all six character passwords, the multiplication principle gives that $|S| = 36^6$ for Part (a) and $|S| = 62^6$ for Part (b).

The complement principle and then De Morgan's law allow us to write

$$|D \cap L| = |S| - |\sim (D \cap L)| = |S| - |\sim D \cup \sim L|.$$

Now, (fortunately) the sets $\sim D$ and $\sim L$ are disjoint, so $|\sim D \cup \sim L| = |\sim D| + |\sim L|$. We now have all the information we need to answer the questions:

Part (a): $|S| - (|\sim D| + |\sim L|) = 36^6 - 26^6 - 10^6 \approx 1.86687 \times 10^9$.
Part (b): $|S| - (|\sim D| + |\sim L|) = 62^6 - 52^6 - 10^6 \approx 3.70286 \times 10^{10}$.
(Certainly either protocol should be sufficient to accommodate any university.)

The Inclusion-Exclusion Principle

When counting elements in unions of sets that are disjoint, one simply can add up the numbers of elements of the individual sets. In cases of nondisjoint sets, one needs to be more careful.

THE INCLUSION-EXCLUSION PRINCIPLE: (a) (*For two sets*) If A and B are finite sets, then $|A \cup B| = |A| + |B| - |A \cap B|$.
(b) (*For three sets*) If A, B, and C are finite sets, then
$|A \cup B \cup C| = |A| + |B| + |C| - |A \cap B| - |A \cap C| - |B \cap C| + |A \cap B \cap C|$.
(c) (*General case*) If $A_1, A_2, \cdots, A_n$ is a collection of finite sets, then

$$|A_1 \cup A_2 \cup \cdots \cup A_n| = \sum_{i=1}^{n} |A_i| - \sum_{i_1 < i_2} |A_{i_1} \cap A_{i_2}| + \cdots$$

$$+ (-1)^{a+1} \sum_{i_1 < i_2 < \cdots < i_a} |A_{i_1} \cap A_{i_2} \cap \cdots \cap A_{i_a}| + \cdots \qquad (1)$$

$$+ (-1)^{n+1} |A_1 \cap A_2 \cap \cdots \cap A_n|.$$

In this identity, we start off by adding the numbers of elements in each set, then subtract the numbers of elements in each possible intersection of two of the sets, then add the numbers of elements in all possible intersections of three of the sets, and so on.

We point out that in the special case in which the sets are pairwise disjoint, i.e. $\forall i \neq j \, [A_i \cap A_j = \varnothing]$, all of the intersection cardinalities are 0s, so formula (1) simply becomes $|A_1 \cup A_2 \cup \cdots \cup A_n| = |A_1| + |A_2| + \cdots + |A_n|$. We have already used this simple formula when we broke counting arguments into disjoint cases (see the solution of Example 5.3(b), for example). We caution the reader to be extremely careful to resist using this tempting formula, unless he/she is absolutely certain that the needed pairwise disjointness requirement is satisfied.

Proof: We give different proofs for each Part (a) and (b). We prove Part (a) directly and analytically, whereas for Part (b), we use Venn diagrams and go at it sequentially. The reader might wish to supply the alternative proof for each part.

Part (a): We can express A as the disjoint union of $A \sim B$ and $A \cap B$. Consequently, $|A| = |A \sim B| + |A \cap B|$. In the same fashion, $|B| = |B \sim A| + |A \cap B|$. But $A \cup B$ is the disjoint union of the three sets $A \sim B$, $B \sim A$, and $A \cap B$. This yields $|A \cup B| = |A \sim B| + |B \sim A| + |A \cap B|$. Comparing these three equations produces the desired result.

Part (b): We use $|A| + |B| + |C|$ as our naïve "first approximation" to $|A \cup B \cup C|$. If we draw a Venn diagram to see how many times each constituent portion is counted, we arrive at the picture in Figure 5.3(a), where we put an integer in each portion to indicate how many times it was counted in $|A| + |B| + |C|$. Our goal is to have each portion counted exactly once. The two set intersection portions are all counted twice (except for the three-set portion in the middle), so as our next approximation, we subtract off the counts of all two-set intersections: $|A| + |B| + |C| - |A \cap B| - |A \cap C| - |B \cap C|$. This second approximation gives the modified counts shown in Figure 5.3(b). All of the portions of the Venn diagram are fine, except for the central portion, corresponding to where all three sets intersect. The count for this portion is now zero. This is easy to compensate for—we simply need to add $|A \cap B \cap C|$ to get our final approximation, which is the asserted inclusion-exclusion formula for three sets. Figure 5.3(c) shows that we now have the desired counts (=1) on all portions of the Venn diagram.

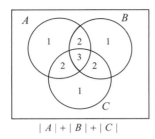
$| A | + | B | + | C |$

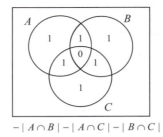
$- | A \cap B | - | A \cap C | - | B \cap C |$

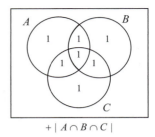
$+ | A \cap B \cap C |$

FIGURE 5.3: Proof of the inclusion-exclusion principle for three sets, in three steps. (a) (left) the initial approximation for $| A \cup B \cup C |$ counts some regions twice and one region three times, (b) (middle) the compensated approximation counts all regions once except for the central region, (c) (right) the final approximation counts all regions correctly.

Part (c): Equation (1) can be proved by mathematical induction (Exercise 31), but a more elegant proof can be given with the material developed in the next section (Exercise for the Reader 5.10). ☐

EXAMPLE 5.5: A university mathematics department has 75 applied mathematics majors, 50 pure mathematics majors, and 32 mathematics education majors. Of these students, there are 17 who list both pure and applied math as their majors, 13 and 11 who list math education and pure, or applied math, respectively, and finally there are five triple majors. How many math majors are there? How many of these are only majoring in applied math?

SOLUTION: The reader is encouraged to draw a Venn diagram for this problem. With the obvious notation, we are given that $| A | = 75$, $| P | = 50$, $| E | = 32$, $| A \cap P \cap E | = 5$, $| A \cap P | = 17$, $| E \cap P | = 13$, and $| E \cap A | = 11$. The inclusion-exclusion principle for three sets now gives us the total number of math majors $| A \cup P \cup E |$ to be $75 + 50 + 32 - (17 + 13 + 11) + 5 = 121$. To get the number of students who are majoring only in applied mathematics ($A \sim (P \cup E)$), we subtract from the total number 75 of applied math majors those who are double majors with pure math ($| A \cap P | - | A \cap P \sim E | = 17 - 5 =$) 12, those who are double majors with math education ($| A \cap E | - | A \cap \sim P \cap E | = 11 - 5 =$) 6, and the five who are triple majors ($A \cap P \cap E$). This gives us $75 - 12 - 6 - 5 = 52$ students who are single majors in applied mathematics.

EXAMPLE 5.6: How many positive integers less than 2009 are divisible by none of 3, 4, or 10?

SOLUTION: We first introduce some convenient notation. For each positive integer n, we let D_n denote the set of all positive integers (less than 2009) that are divisible by n. Thus, for example, $D_4 = \{4, 8, 12, \cdots, 2008\}$. Clearly $| D_n | = \lfloor 2008/n \rfloor$. Also, since an integer a is divisible by both n and m if, and only if a is divisible by the least common multiple of n and m $(= \text{lcm}(n,m))$, we may conclude that $D_n \cap D_m = D_{\text{lcm}(n,m)}$, and also $D_n \cap D_m \cap D_k = D_{\text{lcm}(n,m,k)}$.

The problem is to find $|\sim D_3 \cap \sim D_4 \cap \sim D_{10}|$. By De Morgan's law, this number is the same as $|\sim (D_3 \cup D_4 \cup D_{10})|$, and by the complement principle, this number equals $2008 - |D_3 \cup D_4 \cup D_{10}|$. From the facts mentioned above and the inclusion-exclusion principle, we may now easily perform the needed computations. We first find the three single set counts:

$$|D_3| = \lfloor 2008/3 \rfloor = 669, \quad |D_4| = \lfloor 2008/4 \rfloor = 502, \quad |D_{10}| = \lfloor 2008/10 \rfloor = 200.$$

Next we compute the three double-set counts,

$$|D_3 \cap D_4| = |D_{12}| = \lfloor 2008/12 \rfloor = 167, \quad |D_3 \cap D_{10}| = |D_{30}| = \lfloor 2008/30 \rfloor = 66,$$
$$|D_4 \cap D_{10}| = |D_{20}| = \lfloor 2008/20 \rfloor = 100,$$

and, finally, the single triple-set count:

$$|D_3 \cap D_4 \cap D_{10}| = |D_{60}| = \lfloor 2008/60 \rfloor = 33.$$

Invoking the inclusion-exclusion principle allows us now to arrive at the answer:

$$2008 - |D_3 \cup D_4 \cup D_{10}| = 2008 - (669 + 502 + 200) + (167 + 66 + 100) - 33 = 937.$$

This result can be easily verified using a simply programmed computer loop, and readers are encouraged to perform such a check.

EXERCISE FOR THE READER 5.2: How many positive integers less than 3601 are divisible by at least one of 2, 3, 5, or 11?

EXERCISE FOR THE READER 5.3: How many Hawaiian license plates (see Example 5.2) do not contain any of the strings, "CIA," "FBI," or "GOD?"

The Pigeonhole Principle

Our next principle, known as the *pigeonhole principle*, will probably seem so intuitively obvious that it is hardly worth mentioning. Along with its generalization, the pigeonhole principle turns out to be quite a useful tool.

THE PIGEONHOLE PRINCIPLE: If there are more than k pigeons placed into k pigeonholes, then there must be at least one pigeonhole with more than one pigeon occupying it (see Figure 5.4).

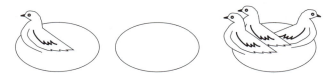

FIGURE 5.4: Illustration of the pigeonhole principle with $k = 3$ pigeonholes: if there are more pigeons (here four) roosting in the pigeonholes than there are pigeonholes, then at least one pigeonhole must have more than one pigeon.

The proof of the pigeonhole principle is an easy proof by contradiction. If every pigeonhole has at most one pigeon in it, then the total number of pigeons would have to be less than or equal to the number of pigeonholes, which is k, a contradiction to the fact that there were supposed to be more pigeons than pigeonholes.

We first give an example of some rather basic consequences of the pigeonhole principle and then proceed to give some more surprising applications.

EXAMPLE 5.7: (a) If a company has 1000 employees, it must have at least two employees who share the same birthday. This follows from the pigeonhole principle with the $k = 366$ possible birthdays being the pigeonholes, and the 1000 ($> k$) employees serving as the pigeons. (We really needed only 367 employees for this to be true.)

(b) If Joey works at Vitali's restaurant 20 evenings in March and Vivian works there 12 evenings in March then they must share at least one common evening of work. This follows from the pigeonhole principle with the pigeonholes being the $31 - 20 = 11$ days that Joey does not work and the pigeons being the 12 days that Vivian works. If there were no overlap, one of the pigeonholes would have two pigeons, i.e., this would mean that Vivian was working twice in the same evening. This is clearly impossible, so Joey and Vivian must indeed share a shift.

EXAMPLE 5.8: If seven points are randomly selected on (the circumference of) a circle of radius 1, show that at least two of these points will lie at a distance of less than 1 from each other.

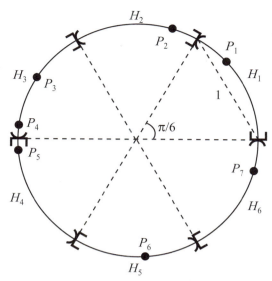

FIGURE 5.5: The six pigeonholes (H_i) and seven (random) pigeons (P_i) for Example 5.8.

SOLUTION: We cut the circle into six adjacent arcs: $H_1, H_2, \cdots, H_6$ —these will serve as the pigeonholes; see Figure 5.5. As shown in the figure, each of these arcs has distance between its endpoints equal to one. But since each arc contains only one of the endpoints, it follows that any two points in such an arc H_i will be separated by a distance less than one. We let the pigeons be the seven random points on the circle. By the pigeonhole principle, at least one of the arcs must contain at least two of the randomly selected points. (In Figure 5.5, H_3 contains P_3 and P_4.)

EXERCISE FOR THE READER 5.4: (a) Show that if five points are randomly selected inside or on an equilateral triangle of side length one, then there will be two of these points whose distance between is not more than 1/2. (b) Show that if the five points are selected to be inside the triangle, then there will be two whose distance is less than 1/2.

EXERCISE FOR THE READER 5.5: Show that if we take any set of 51 integers from the set $\{1, 2, \cdots, 100\}$, then one of the integers in this set must divide some other integer in this set.

EXERCISE FOR THE READER 5.6: Suppose that n is a positive integer. Show that in any set of $n + 1$ integers none of which is divisible by n, there must exist two integers whose difference is divisible by n.

The following application of the pigeonhole principle comes from a paper coauthored by the illustrious Hungarian mathematician Paul Erdös,[3] see [ErSz-35].

PROPOSITION 5.3: Let n be a positive integer. Every sequence of $N = n^2 + 1$ distinct real numbers $(a_1, a_2, \cdots, a_N)$ contains a subsequence of length $n + 1$ that is either increasing or decreasing.

FIGURE 5.6: Paul Erdös (1913–1996), Hungarian mathematician

The proposition implies, for example, that if a brigade of 101 soldiers are standing in a lineup, then it is always possible to find 11 to take a step forward so that their heights will be nonincreasing or nondecreasing, as we go from left to right.[4]

Proof: We proceed by the method of contradiction. Assume that no such subsequences exist for a given sequence $(a_1, a_2, \cdots, a_N)$. We will apply the pigeonhole principle by letting the N terms of the sequence be the pigeons. To each pigeon a_j, we associate an ordered pair of positive integers (I_j, D_j), where I_j is the length of the longest increasing subsequence starting at a_j, and D_j is the length of the longest decreasing subsequence starting at a_j. For example, if $n = 2$ (so $N = 5$) and $(a_1, a_2, \cdots, a_N) = (5, 3, 7, 6, 8)$, then $I_2 = 3$ (corresponding to the

---

[3] Paul Erdös (pronounced "air-dish") was born in Hungary shortly before the outbreak of World War I. His parents were both mathematics teachers. Erdös's two elder sisters had perished to scarlet fever only a few days before his birth, so his parents were particularly protective of their last child. His parents were non-practicing Jews, and this led to numerous difficulties for the family. Erdös' mathematics focused on problem solving rather than general theoretical developments, and he was one of the greatest problem solvers of all time, publishing over 1500 papers in his lifetime, mostly in the areas of combinatorics and number theory. He led a simple life that allowed him to focus almost exclusively on mathematics. Although he had been offered many permanent decent positions that his friends encouraged him to accept, he preferred to live out of his suitcase, to travel around the world, and meet other mathematicians with whom to work. Additionally he was very modest and noncompetitive. For example, he had independently discovered a very elegant proof of the *prime number theorem* with (Princeton mathematician) Atle Selberg. Although both had agreed to publish their papers back-to-back in the same journal, the latter jumped ahead and won the prestigious *Fields Medal* for it. Erdös spent very little of the money that he earned (from prizes, lectures, and temporary contracts), instead, he used it to put up prizes to encourage work on difficult problems. He had such a wide array of collaborators that the concept of a mathematician's *Erdös number* came into being. Erdös's Erdös number is 0. All of his coauthors have Erdös number 1. Others who have written a joint paper with someone with Erdös number 1 have number 2, and so on. If there is no chain of coauthorships connecting someone with Erdös, then that person's Erdös number is said to be infinite.

[4] In case some heights are the same, we (artificially) perturb them by very small numbers, so as to make them all different. For example, if we are measuring heights only to the nearest half inch, and if we had six men who were 75.5 inches tall, we would put these six numbers to be 75.5000, 75.5001, …, 75.5005. Once the proposition is applied, we could convert the heights back to their original numbers, and then still have a sequence that is nondecreasing/nonincreasing.

increasing subsequence $(3,6,8)$) and $D_2 = 1$ (corresponding to the decreasing subsequence (3)). These ordered pairs (I_j, D_j) will serve as our pigeonholes. In light of our (contradiction seeking) hypothesis, we must have $I_j, D_j \leq n$ for every index j. Thus, by the multiplication principle, there are at most n^2 pigeonholes, and since we have more pigeons, the pigeonhole principle implies that there must be two different terms a_j, a_k with $j < k$, and with $I_j = I_k$ and $D_j = D_k$. We will separate into two cases to arrive at contradictions.

Case 1: $a_j < a_k$ Here we can juxtapose a_j to the beginning of an increasing subsequence of length $I_j = I_k$ that starts at a_k to get an increasing subsequence of length $I_j + 1$ —a contradiction!

Case 2: $a_j > a_k$ Here we can juxtapose a_j to the beginning of a decreasing subsequence of length $D_j = D_k$ that starts at a_k to get a decreasing subsequence of length $D_j + 1$—a contradiction!

Since we have arrived at a contradiction in all of the possible cases, the proof of the theorem is complete. $\square$

The Generalized Pigeonhole Principle

The pigeonhole principle guarantees that as long as there is at least one more pigeon than there are pigeonholes, then (at least) one pigeonhole has double occupancy. In case there are a lot more pigeons than pigeonholes, the following generalization of the pigeonhole principle allows us to draw more accurate conclusions.

THE GENERALIZED PIGEONHOLE PRINCIPLE: If there are N pigeons placed into k pigeonholes, then there must be at least one pigeonhole with at least $\lceil N/k \rceil$ pigeons occupying it.

Proof: Just as for the first version of the pigeonhole principle, we will proceed by the method of contradiction. Suppose that all of the pigeonholes contained fewer than $\lceil N/k \rceil$ pigeons. Then the total number of pigeons contained in all of the pigeonholes could be at most $k \cdot (\lceil N/k \rceil - 1)$, but this number is less than $k \cdot ([N/k + 1] - 1) = N$, which is a contradiction. $\square$

EXAMPLE 5.9: If we apply the generalized pigeonhole principle to example 5.7(a), (the company that had 1000 employees), we would arrive at the stronger conclusion that there must be at least 3 ($= \lceil 1000/367 \rceil$) employees who share the same birthday. If there were more than 1101 employees, we could say that at least four employees share the same birthday.

EXERCISES 5.1:

1.   A standard California license plate consists of a single digit, followed by three letters, followed by three digits.
      (a) How many standard California license plates can be made?
      (b) How many can be made if no letters or digits can be used twice?

2.   (a) A menu special at a restaurant offers three courses: appetizer, main course, and dessert. For the appetizer, one can choose either house salad, Caesar salad, or soup of the day. For the main course, the choices are either prime rib, chicken Marengo, or sautéed shrimp. The dessert choices are chocolate mousse, mixed fresh fruit platter, or ice cream cake. How many different menus can be created if one must choose one item from each course?
      (b) Repeat Part (a) with an additional cheese course with choices of either: camembert, garlic herb cheese or goat cheese is added along with a beverage choice of beer, iced tea, cola, or milk.

3.   A man is deciding among four restaurants: Italian, Thai, Chinese, or steakhouse, and then four after-dinner activities: a movie, dancing, bowling, or a basketball game on which to take his date. How many different dates can he put together if he includes both dinner and one after-dinner activity?

4.   In a trip from San Diego to Seattle, suppose that we are considering three routes from San Diego to Los Angeles, four routes from Los Angeles to San Francisco, and five routes from San Francisco to Seattle. How many different trips could we plan from San Diego to Seattle that go through Los Angeles and San Francisco?

5.   A restaurant manager is trying to assign five workers, Andy, Beth, Charlie, Doris, and Earl to five different jobs for the evening.
      (a) If all of these workers can do any of these jobs, how many job assignments are possible?
      (b) How about if Andy, Beth, and Earl cannot do the first two jobs?
      (c) How about if Andy, Beth, and Earl cannot do the last job?

6.   How many positive integers are less than 8000 that have no repeated digits and no occurrences of the digits 2, 4, or 8?

7.   An international student club has 12 members: 3 Chinese, 2 Vietnamese, 1 French, 3 Germans, 2 Japanese, and 1 Australian.
      (a) The group elects a president, vice president, and treasurer. In how many ways can this be done?
      (b) Same question as (a) but with the requirement that at least one of the elected be Chinese.
      (c) Same question as (a) but with the restriction that the Germans and Japanese refuse to serve together.
      (d) Same question as (a) but with the requirement that two members A and B will either serve together or not at all.

8.   (a) How many functions are there from $\{1, 2, 3, 4, 5\}$ to $\{1, 2, 3, 4, 5, 6, 7\}$?
      (b) How many of the functions f in Part (a) satisfy $f(i) \in \{1, 2\}$ for $i = 1, 2, 3$?
      (c) How many of the functions f in Part (a) are one-to-one functions?

9.   (a) How many functions are there from $\{1, 2, 3, 4, 5, 6, 7\}$ to $\{1, 2, 3, 4, 5\}$?
      (b) How many of the functions f in Part (a) satisfy $f(i) \in \{1, 2\}$ for $i = 1, 2, 3$?
      (c) How many of the functions f in Part (a) are one-to-one functions?

10.   Suppose that n and m are positive integers.
      (a) How many functions are there from $\{1, 2, \ldots, n\}$ to $\{1, 2, \ldots, m\}$?

(b) How many of the functions f in Part (a) satisfy $f(i) \in \{1, 2\}$ for $i = 1, 2, 3$? (Assume that $n > 2$ and $m > 1$.)

(c) How many of the functions f in Part (a) are one-to-one functions?

11. A restaurant's lunch menu has three courses:
    Cheeses: Brie, Jarlsberg, smoked hickory, or Swiss
    Salads: Caesar, romaine, tossed greens, or chicken
    Sandwiches: BLT, tuna, turkey, or Italian
    (a) If one is allowed to choose exactly one item from exactly two of the three different courses, how many selections would be possible?
    (b) If a couple is allowed to choose exactly two items from exactly two of the three different courses, how many selections would be possible?

12. A password protocol for a certain network requires that all passwords use digits or lower-case letters and consist of six to eight characters.
    (a) How many passwords are possible?
    (b) How many passwords are possible that include at least one digit and at least one letter?
    (c) How many passwords are possible that include at least two letters?
    (d) How many passwords are possible that contain no vowels?
    (e) How many passwords are possible that include at least one digit and at least one consonant?

13. A password protocol for a certain network requires that all passwords use digits or letters and consist of five to seven characters. The protocol is case sensitive.
    (a) How many passwords are possible?
    (b) How many passwords are possible that include at least one digit and at least one letter?
    (c) How many passwords are possible that include at least one digit, at least one lower-case letter, and at least one upper-case letter?
    (d) How many passwords are there that contain the string "CAT" (in any mixture of cases)?

14. In a certain state, of the 500 largest companies, 200 offer (free) health insurance (to all employees), 300 offer dental insurance, and 150 offer life insurance. Moreover, 150 offer both health and dental insurance, 100 offer health and life, 75 offer dental and life, and 50 offer all three types of coverage. How many companies offer none of these three coverages?

15. (a) Use the inclusion-exclusion principle to determine the number of positive integers less than 6000 that are divisible by at least one of the primes 3, 5, or 7.
    (b) Use the inclusion-exclusion principle to determine the number of positive integers less than 6000 that are divisible by none of the primes 3, 7, or 11.
    (c) Write and execute computer loops that will check your answers to (a) and (b).

16. (a) Use the inclusion-exclusion principle to determine the number of positive integers less than 4000 that are divisible by at least one of the numbers 4, 6, or 10.
    (b) Use the inclusion-exclusion principle to determine the number of positive integers less than 4000 that are divisible by none of the numbers 5, 6, or 15.
    (c) Write and execute computer loops that will check your answers to (a) and (b).

17. (a) Use the inclusion-exclusion principle to determine the number of positive integers less than 6000 that are divisible by at least one of the primes 3, 5, 7, or 11.
    (b) Write and execute a computer loop that will check your answers to (a).

18. Use (1) to write down an explicit formula for the inclusion-exclusion principle for five sets.

19. (a) Use the inclusion-exclusion principle to determine the number of positive integers less than 4000 that are divisible by at least one of the numbers 4, 6, 10, 15, or 20.
    (b) Use the inclusion-exclusion principle to determine the number of positive integers less than 4000 that are divisible by none of the numbers 2, 3, 5, 7, or 11.
    (c) Write and execute computer loops that will check your answers to (a) and (b).

20.   (a) Explain why among a group of 60 foreign exchange students from the United States, at least 2 came from the same state.
(b) What is the minimum number of cards that must be drawn from a shuffled standard deck of 52 cards to guarantee that there will be at least one pair? Provide an example to show that if one less than this number is drawn, a pair need not come up.

21.   (a) Explain why in a group of 21 men whose heights range from 5 feet to 6 feet 7 inches, there must be at least two whose height, rounded to the nearest inch, must be the same.
(b) What is the minimum number of cards that must be drawn from a shuffled standard deck of 52 cards to guarantee that there will be at least three cards of the same suit? Provide an example to show that if one less than this number is drawn, three same suit cards need not appear.

22.   Show that if 13 points are chosen on a circle of radius 1, then at least two of these points will be within a distance of 1/2 from one another.

23.   (a) Show that if 10 points are randomly selected in the interior of an equilateral triangle of side length 1, then there will be two of these points whose distance from one another is less than 1/3.
(b) Show the result of Part (a) is sharp by constructing an example of nine points within an equilateral triangle of side length 1 such that the distance between any pair is greater than or equal to 1/3.

24.   (a) Show that if five points are randomly selected within the interior of a square of side length 2, then there will be two of these points whose distance from one another is less than $\sqrt{2}$.
(b) Show that if nine points are randomly selected with the interior of a square of side length 2, then there will be three of these points such that the distance between any pair is less than $\sqrt{2}$.

25.   Given a positive integer $n > 1$, determine the minimum positive integer K_n,   such that if any K_n points are selected in the interior of an equilateral triangle of side length 1, then there must be at least two of these points that lie at a distance less than $1/n$ from each other.

26.   Prove that given any 11 positive integers, at least two of them will have their difference being divisible by 10.

27.   Prove that given any seven positive integers, at least two of them will have either their sum or their difference being divisible by 10.
Suggestion: Use the pigeonhole principle with the pigeonholes determined by the last digit of each integer in such a way that two integers in the same pigeonhole will have either their sum or their difference divisible by 10.

28.   Suppose that we have a list of 8 positive integers (with possible duplications) that add up to 20. Show that we can always draw a sublist (with possible duplications) from this list whose elements add up to four.
Suggestion: First show that the list must contain 1 or 2, then use the pigeonhole principle.

29.   At a medium-sized university there are 869 students taking an introductory statistics course this semester among 10 sections. What is the smallest possible enrollment in the largest section?

30.   How many truth tables are possible for logical statements containing n logical variables?

31.   Use mathematical induction to prove the inclusion-exclusion principle (1) for finite unions of finite sets:

$$|A_1 \cup A_2 \cup \cdots \cup A_n| = \sum_{i=1}^{n}|A_i| - \sum_{i_1 < i_2}|A_{i_1} \cap A_{i_2}| + \cdots$$
$$+ (-1)^{a+1}\sum_{i_1 < i_2 < \cdots < i_a}|A_{i_1} \cap A_{i_2} \cap \cdots \cap A_{i_a}| + \cdots$$
$$+ (-1)^{n+1}|A_1 \cap A_2 \cap \cdots \cap A_n|.$$

5.2: PERMUTATIONS, COMBINATIONS, AND THE BINOMIAL THEOREM

The Difference Between a Permutation and a Combination

Our next example will compare and contrast the concepts of a *permutation* and a *combination*. Although the example is small enough to count by brute-force, we will solve it with an approach that lends itself easily to generalization.

EXAMPLE 5.10: (*Motivating example for understanding the difference between a permutation and a combination*) Suppose that Mr. Vitali has interviewed four women: Alice, Betty, Christine, and Daisy, to fill three job openings at his Italian restaurant, and that all turned out to be equally qualified.
(a) In how many ways can Mr. Vitali hire a cashier, a cook, and a waitress from these four applicants?
(b) In how many ways can Mr. Vitali hire three of these four women to work as waitresses?

SOLUTION: Although the two questions are similar, there is one very important, yet perhaps subtle, difference. In the first question, the order in which Mr. Vitali hires/assigns the women is definitely important, since the jobs are all different. In the second question, however, order/assignment is not relevant, since the women are being hired for identical positions.

Part (a): (***Order matters: permutations***) We have already shown how to answer questions like this using the multiplication principle; the answer is given by:

$$\underset{\substack{\text{women to}\\\text{hire as}\\\text{cashier}}}{4} \cdot \underset{\substack{\text{women left}\\\text{to hire as}\\\text{cook}}}{3} \cdot \underset{\substack{\text{women left}\\\text{to hire as}\\\text{waitress}}}{2} = 24.$$

Each of these 24 outcomes can be viewed as an ordered triple, e.g., (B, C, D), and is called a *permutation (or rearrangement) of the four objects* A, B, C, D, *taken three at a time*. The obvious abbreviations are being used, e.g., the triple (B, C, D) would correspond to hiring B(etty) as the cashier, C(hristine) as the cook, and D(aisy) as the waitress.

Part (b): (***Order doesn't matter: combinations***) Consider a typical permutation of the 24 outcomes for Part (a), e.g., (B, C, D).   If we *rearrange* (or *permute*) the women in this list, we get a different outcome/permutation for Part (a), e.g., (C, D, B), would represent the outcome of hiring C as the cashier, D as the cook, and B as the waitress.   Thus different **permutations** correspond to different outcomes for Part (a).   However, for Part (b), all of these permutations of the list (B, C, D) would correspond to the same outcome for Part (b), since all of B, C, D would be hired as waitresses.   Thus, for Part (b), this outcome should really be represented as the set {B, C, D}, since order does not matter (in set notation). Such an object is called a *combination of the four objects* A, B, C, D, *taken three at a time*.

In order to arrive at the answer to Part (b), we first find out how many different permutations there are of (just) the ordered list (B, C, D).   The multiplication principle can easily give us the answer:

$$\underset{\substack{\text{choose one}\\\text{of B,C,D}\\\text{for 1st slot}}}{3} \cdot \underset{\substack{\text{choose one}\\\text{remaining two}\\\text{for 2nd slot}}}{2} \cdot \underset{\substack{\text{only one}\\\text{choice for}\\\text{last slot}}}{1} = 6.$$

This example is (intentionally) small enough so we can check this by listing all of these six permutations of (B, C, D):

(B, C, D), (B, D, C), (C, B, D), (C, D, B), (D, B, C), (D, C, B).

Each of these six outcomes of Part (a) morph into the single combination/set outcome {B, C, D} for Part (b).   In the same fashion, every other outcome of Part (b) will correspond to six different (permutations) of Part (a), and there is no overlapping since different outcomes of Part (b) will correspond to different sets. Therefore, if we (temporarily) let Y denote the answer to Part (b), and X denote the answer to Part (a), it follows that $6Y = X$, so that $Y = X/6 = 24/6 = 4$.   Again, because of the size of this example, it is easy to get this answer directly (each of the four outcomes corresponds to deciding which of the four women not to hire). These ideas are easily generalized, and this is what we do next.

DEFINITION 5.1:   A **permutation** of a set of distinct objects is any rearrangement of them (as an ordered list).   More generally, if $1 \le k \le n$, a **k-permutation** of a set of n distinct objects is any permutation of any k of these n objects.

In the preceding example, we saw a way to count the 24 3-permuations of the set of four women {A, B, C, D}. The next result generalizes this idea.

THEOREM 5.4:   The number of k-permutations taken from a set of n distinct objects ($1 \le k \le n$) is denoted by $P(n,k)$ and is given by the following formula:

$$P(n,k) = n(n-1)(n-2)\cdots(n-k+1). \tag{2}$$

(Note that the number of factors is k.)   This number is read as **the number of permutations of n objects taken k at a time**.

Proof: This is a simple application of the multiplication principle. We may view the task of forming a k-permutation as the k-step process of filling in the k ordered slots with objects taken from the n distinct objects (where each object can be used only once), working, say, from left to right. For the first slot, we have n objects to choose from. Once one has been selected, we move on to the second slot, where we may put any of the remaining $n - 1$ objects. For the third slot there will be $n - 2$ choices, and so on, until we get to the kth slot, when there will remain $n - k + 1$ choices of objects (convince yourself of this!).   □

The special case when $k = n$ is of such importance that it motivates the following definition.

DEFINITION 5.2: Let n be a positive integer. The number of permutations of n distinct objects, $P(n,n)$, by (1) equals $n(n-1)(n-2)\cdots 3\cdot 2\cdot 1$. This number is denoted $n!$, and is called **the factorial of n**, or just **n factorial**. By convention, we define $0! = 1$.

An easy manipulation of (2) allows us to rewrite $P(n,k)$ entirely in terms of factorials:

$$P(n,k) = n(n-1)(n-2)\cdots(n-k+1)$$
$$= \frac{n(n-1)(n-2)\cdots(n-k+1)[(n-k)(n-k-1)\cdots 2\cdot 1]}{[(n-k)(n-k-1)\cdots 2\cdot 1]} = \frac{n!}{(n-k)!}.$$

In summary:

$$P(n,k) = \frac{n!}{(n-k)!}. \tag{3}$$

Formula (3) is more useful for theoretical and analytic manipulations than for actually computing numbers of permutations. For example, to compute $P(1000,3)$ by (2) is easy: $1000\cdot 999\cdot 998$, whereas formula (2) would involve 2000 multiplications (if we were to compute the factorials directly).[5] Before giving some more examples of permutations, we first formally introduce combinations. After this is done we give some examples that will mix both concepts so as to help the reader to develop a sense to better distinguish between permutations and combinations.

---

[5] Of course, most mathematical software computing platforms have built-in functions for computing factorials, but factorials get so large so quickly that, as a general rule, it is best to use formula (2) rather than (3) when computing numbers of permutations.

DEFINITION 5.3:   If $0 \le k \le n$, a **k-combination** of a set of n distinct objects is any (unordered) subset that contains exactly k of these objects.

In the preceding example, we previewed a general method for counting combinations when we counted the six 3-combinations of the set {A, B, C, D} of four women. Here is the general result:

THEOREM 5.5:   The number of k-combinations taken from a set of n distinct objects $(0 \le k \le n)$ is denoted by $C(n,k)$ and is given by the following formula:

$$C(n,k) = \frac{P(n,k)}{k!} = \frac{n!}{k!(n-k)!}. \tag{4}$$

This number is read as **the number of combinations of n objects taken k at a time**, or simply as **n choose k**.

Proof: The only 0-combination is the empty set, so theorem is true for $k = 0$ (both sides equal 1). Next we assume that $k > 0$. Each k-combination $\{x_1, x_2, \cdots, x_k\}$ is just a subset of size k from the set of n objects under consideration, and gives rise to $k!$ permutations of its elements, by Theorem 2.3 (i.e., $P(k,k) = k!$).   Since k-permutations arising from different k-combinations must also be different (because their elements come from different sets), and since all k-permutations (of the n objects under consideration) must be obtainable in this way, we may conclude that $C(n,k) \cdot k! = P(n,k)$.   Dividing this equation by $k!$, and then using (3) produces (4). $\square$

Computing and Counting with Permutations and Combinations

As with formula (3), since the factorials get large so quickly, directly computing the factorials in (4) is not an efficient way to compute $C(n,k)$.   In particular, this can cause problems in floating point arithmetic computing systems (see Chapter 5 of [Sta-04]).   Computationally, it is best to cancel the largest factorial in the denominator of (4) with the same factors in the numerator.   For example, we would compute $C(200,2)$ as follows:

$$C(200,2) = \frac{200!}{2! \cdot 198!} = \frac{200 \cdot 199 \cdot \cancel{198!}}{2! \cdot \cancel{198!}} = \frac{200 \cdot 199}{1 \cdot 2} = 19{,}900.$$

Most computing platforms have built-in functions for computing factorials, permutations, and combinations.

As promised, we now give several examples to help provide the reader with some intuition on distinguishing between permutations and combinations. The key question to ask when trying to decide whether permutations or combinations are relevant is whether order matters—if it does, we are dealing with permutations, while if order does not matter, combinations are relevant. As in the last section, two (or more) counting principles might need to be combined to solve a given problem.

EXAMPLE 5.11: Answer each of the following counting questions.
(a) A recent Honolulu Marathon had 24,265 participants. How many top three finishes are (theoretically) possible?
(b) In how many ways can a committee of two democrats and two republicans be formed from a group of 60 republican senators and 40 democratic senators?
(c) Answer the question of Part (b) with the additional requirement that democratic senators K and L refuse to serve together.
(d) In how many ways can four math books, two computer science books, and five economics books be arranged on a shelf?
(e) In how many ways can the books in Part (d) be arranged on a shelf if books of the same subject need to be grouped together?

SOLUTION: Part (a): Order clearly matters here: $P(24265, 3) = 24265 \cdot 24264 \cdot 24263 \approx 1.485 \times 10^{13}$. (Of course, only a handful of these outcomes would be reasonably likely.)

Part (b): The order that the people are put on this committee is not important, so combinations are relevant here. We can form such a committee by first taking a 2-combination of the set of 60 republicans (this can be done in $C(60,2)$ ways), and then taking a 2-combination from the set of 40 democrats (this can be done in $C(40,2)$ ways). By the multiplication principle, there will be $C(60,2) \cdot C(40,2) = (60 \cdot 59 / 2!)(40 \cdot 39 / 2!) = 1,380,600$ such committees.

Part (c): We need to modify our solution of Part (b) to answer the present question. The part that needs modification is in computing the number of ways of choosing two democrats. There are two natural (and disjoint) cases: either a committee with neither K nor L, or a committee with (exactly) one of K or L. There will be $C(38,2) = 38 \cdot 37 / 2 = 703$ committees of the first type (since with neither K nor L, 38 democratic senators remain), and $C(2,1) \cdot C(38,1) = 2 \cdot 38 = 76$ committees of the latter type. Thus, we will have a grand total of $(60 \cdot 59 / 2!)(703 + 76) = 1,378,820$ such committees.

Part (d): We have to arrange $4 + 2 + 5 = 11$ different books in a row. Order clearly matters here: there are $11! = 39,916,800$ permutations of these 11 books.

Part (e): We first treat each type of book as a "block." Thus we have three blocks: the M block consisting of the four math books, the C block consisting of the two computer science books, and the E block consisting of the five economics books. We can view arranging the books on the shelf as first arranging the three blocks, for which there are 3! ways to do this, and then deciding how to permute the books in each block: for the M block, there are 4! ways to arrange the four math books, and similarly, there are 2! and 5! ways to arrange the books in the C and E blocks, respectively. Therefore, by the multiplication principle, the total number of ways of arranging the books in this fashion will be $3! \cdot 4! \cdot 2! \cdot 5! = 34,560.$

EXERCISE FOR THE READER 5.7: (*Circular permutations*) In Chinese restaurants, tables are often circular (so that everyone has an equally prominent seat), with a Lazy Susan in the center to facilitate access of the meal items; see Figure 5.7.
(a) If there are *n* seats around the table, in how many ways could *n* people be seated around the table?   Two seating arrangements are considered to be equivalent if the relative positions of all people are the same, i.e., if one arrangement can be obtained from the other by a rotation.
(b) How many arrangements are possible if *n* = 2*k*, where there are *k* men and *k* women and each man has a woman on either side?
(c) How many arrangements are possible as in Part (b) under the additional requirement that Jimmy and Sue should be seated next to one another?

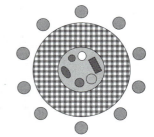

FIGURE   5.7:   A Chinese dinner table for Exercise for the Reader 5.7.

(d) How many arrangements are possible as in Part (b) under the additional requirement that Jimmy and Sue should **not** be seated next to one another?

Our next example concerns five-card poker hands, which are assumed to be randomly drawn cards from a (shuffled) standard deck of playing cards.[6]

EXAMPLE 5.12: A poker hand consists of a random drawing of five cards from the standard 52-card deck.
(a) How many different poker hands are possible?
(b) How many poker hands are possible that contain at least one pair?
(c) How many poker hands are possible that contain (only) a pair?
(d) How many poker hands are possible that contain no aces?
(e) How many poker hands are possible that contain at least one ace?

SOLUTION: Since order does not matter in poker hands, we are dealing with combinations in each part.

---

[6] The cards of a standard 52-card deck are evenly divided among the four *suits* (*clubs* and *spades*, which are black, and *diamonds* and *hearts*, which are red), and each suit has 13 denominations: (A)ce = 1, 2, …, 9, 10, (J)ack, (Q)ueen, (K)ing. The last three cards are *face cards*. A (five-card) *poker hand* is usually described in the most complimentary terms among the following possibilities, listed in order from least valuable (most common) to most valuable (most rare): high card, pair, two (separate) pairs, three of a kind, straight (five cards in sequence, ace can go before 2 or after king), flush (five cards of the same suit), full house (three of a kind and a pair), four of a kind, straight flush (straight plus flush), royal flush (10, J, Q, K, A, all of same suit). Thus, an example of a pair would be {8, 8, 2, 4, K}. Order does not matter in a poker hand.

Part (a): $C(52,5) = 52 \cdot 51 \cdot 50 \cdot 49 \cdot 48 / (5 \cdot 4 \cdot 3 \cdot 2 \cdot 1) = 2,598,960$.

Part (b): Although a poker hand is never made to order, a good way to count poker hands with certain properties is to realize their construction as the result of a sequence of choices. First, we choose the denomination for the pair, there are 13 ($= C(13,1)$) choices. Next, from the four cards of this chosen denomination, we need to choose two, and there are $C(4,2) = 4 \cdot 3/(1 \cdot 2) = 6$ ways to do this. Once this is done we need to choose three other cards from the 50 cards that are left. There are $C(50,3) = 50 \cdot 49 \cdot 48/(1 \cdot 2 \cdot 3) = 19,600$ ways to do this. The multiplication principle now gives us the total number of possibilities to be:
$$C(13,1) \cdot C(4,2) \cdot C(50,3) = 13 \cdot 6 \cdot 19,600 = 1,528,800.$$

Part (c): First, we choose the denomination for the pair; there are 13 choices. Next, from the four cards of this given denomination, we need to choose two, and there are $C(4,2) = 4 \cdot 3/(1 \cdot 2) = 6$ ways to do this. Once this is done we need to choose three different denominations for the remaining cards. There are $C(12,3)$ $12 \cdot 11 \cdot 10/1 \cdot 2 \cdot 3 = 220$ ways to do this. Finally for each of these chosen denominations, we have to choose one of the 4 cards.
$$C(13,1) \cdot C(4,2) \cdot C(12,3) \cdot 4^3 = 13 \cdot 6 \cdot 220 \cdot 4^3 = 1,098,240.$$

Part (d): Thinking of the poker hand as being dealt from a deck with the four aces removed, this gives the total number of such hands to be:
$$C(48,5) = 48 \cdot 47 \cdot 46 \cdot 45 \cdot 46 / (5 \cdot 4 \cdot 3 \cdot 2 \cdot 1) = 1,712,304.$$

Part (e): We give two methods for counting these poker hands:

Method 1: (*Direct counting*) We decompose into disjoint cases:
Number of hands with at least one ace = Number with exactly one ace + Number with exactly two aces + Number with exactly three aces + Number with all four aces.
To count the number of hands with exactly two aces, we use the multiplication principle as follows:

$$\underbrace{C(4,2)}_{\substack{\text{Number of ways} \\ \text{to choose 2 aces} \\ \text{from the the four}}} \cdot \underbrace{C(48,3)}_{\substack{\text{Number of ways} \\ \text{to choose 3 cards} \\ \text{from the non-aces}}}$$

The same idea works for other cases to yield the grand total to be:

$$\underbrace{C(4,1) \cdot C(48,4)}_{\text{Hands with exactly one ace}} + \underbrace{C(4,2) \cdot C(48,3)}_{\text{Hands with exactly two aces}} + \underbrace{C(4,3) \cdot C(48,2)}_{\text{Hands with exactly three aces}} + \underbrace{C(4,4) \cdot C(48,1)}_{\text{Hands with all four aces}}$$

$$= 778,320 + 103,776 + 4512 + 48 = 886,656.$$

Method 2: (*Using complements*) The complement of the set of poker hands containing at least one ace is simply the set of poker hands with no aces. So we can get the answer we want by subtracting the number of hands with no aces (which we figured out in Part (d)) from the total number of poker hands (that we computed in Part (a)):

$$\underbrace{2,598,960}_{\substack{\text{Total number of}\\\text{possible poker hands}}} - \underbrace{1,712,304}_{\substack{\text{Number of possible}\\\text{poker hands without}\\\text{aces}}} = \underbrace{886,656}_{\substack{\text{Number of possible}\\\text{poker hands with at}\\\text{least one ace}}}$$

This provides a nice check. In general, complements can often save time for counting problems involving phrases such as "at least" or "no more than."

EXERCISE FOR THE READER 5.8:   In a standard five-card poker hand, compute the total number of possible five-card poker hands that are:
(a) full houses.
(b) flushes.
(c) four of a kind hands.

At this point, it will be beneficial to observe a useful identity for combination numbers $C(n, k)$. Since $C(n, k)$ represents the total number of k-element subsets of a set with n elements, it follows that $\sum_{k=0}^{n} C(n,k)$ must be the total number of subsets of a set with n elements. But we know from Theorem 5.1 that this latter number is just 2^n. We have thus proved the following theorem:

THEOREM 5.6: The number of subsets of a set with n elements is

$$C(n,0) + C(n,1) + \cdots + C(n,n) = 2^n. \tag{5}$$

The Binomial Theorem

This is one of many combinatorial identities involving k-combination coefficients. Our next result is a theorem of algebra involving these coefficients. In such algebraic contexts, the k-combination coefficients are customarily referred to as *binomial coefficients*, and are denoted as follows:

NOTATION: For nonnegative integers n and k, with $k \le n$, the **binomial coefficient** $\binom{n}{k}$ is the number $C(n,k)$ of k objects taken n at a time, i.e., (from (4)):

$$\binom{n}{k} = \frac{n!}{k!(n-k)!}.$$

THEOREM 5.7: (*The Binomial Theorem*) If x and y are any numbers, and n is a nonnegative integer, then

$$(x+y)^n = \sum_{k=0}^{n} \binom{n}{k} x^k y^{n-k}. \tag{6}$$

Proof: There are several ways to prove this theorem; keeping in the spirit of this chapter, we will present a combinatorial proof. If we expand the left side of (5)

$$(x+y)^n = \underbrace{(x+y)(x+y)\cdots(x+y)}_{n \text{ factors}},$$

we can view the result as a sum of terms, each one arising by choosing either an x or a y from each of the n factors. Thus, each term that will arise must be of the form $x^k y^{n-k}$ for some nonnegative integer k between 0 and n (inclusive), and this will correspond to choosing x's from k of the factors, and y's from the remaining $n-k$ factors. Now, in how many such ways can the term $x^k y^{n-k}$ arise? This will simply be the number of ways that one can choose a set of k of the n factors to be designated as x-factors (and the remaining $n-k$ factors to be designated as y-factors). Since the order in which these factors are selected is unimportant (the product of the x's and y's will always work out to be $x^k y^{n-k}$), this number is simply $C(n,k) = \binom{n}{k}$. This completes the proof of (6). □

EXAMPLE 5.13: Use the binomial theorem to expand the following square and cubic polynomials: (a) $(x+y)^2$, and (b) $(a-2b)^3$.
(c) What is the coefficient of x^{10} in the expansion of $(x+2)^{16}$?

SOLUTION: Part (a): Using (5) directly with $n=2$, we obtain

$$(x+y)^2 = \binom{2}{0} x^0 y^{2-0} + \binom{2}{1} x^1 y^{2-1} + \binom{2}{2} x^2 y^{2-2} = y^2 + 2xy + x^2.$$

Part (b): Using (6) with $n=3$ (and with $x=a$ and $y=-2b$), we obtain

$$(a-2b)^3 = \binom{3}{0} a^0 (-2b)^{3-0} + \binom{3}{1} a^1 (-2b)^{3-1} + \binom{3}{2} a^2 (-2b)^{3-2} + \binom{3}{3} a^3 (-2b)^{3-3}$$

$$= -8b^3 + 3a \cdot 4b^2 + 3a^2(-2b) + a^3$$

$$= a^3 - 6a^2b + 12ab^2 - 8b^3$$

Part (c): The term in the right side of (6) corresponding to x^{10} would correspond to the index $k = 10$. The resulting coefficient (with $n = 16$ and $y = 2$ and removing x^{10}) is thus

$$\binom{16}{10} \cdot 2^{16-10} = \frac{16!}{10!6!} \cdot 2^6 = \frac{11 \cdot \cancel{12} \cdot 13 \cdot 14 \cdot \cancel{15} \cdot 16}{1 \cdot \cancel{2} \cdot \cancel{3} \cdot \cancel{4} \cdot \cancel{5} \cdot \cancel{6}} \cdot 2^{\cancel{6}4} = 512,512.$$

One very attractive way to obtain binomial coefficients is from the following so-called **Pascal's triangle** that is often first introduced in high school algebra courses:

$$
\begin{array}{ccccccccc}
& & & & 1 & & & & \\
& & & 1 & & 1 & & & \\
& & 1 & & 2 & & 1 & & \\
& 1 & & 3 & & 3 & & 1 & \\
1 & & 4 & & 6 & & 4 & & 1 \\
\end{array}
\qquad
\begin{array}{ccccccccc}
& & & & \binom{0}{0} & & & & \\
& & & \binom{1}{0} & & \binom{1}{1} & & & \\
& & \binom{2}{0} & & \binom{2}{1} & & \binom{2}{2} & & \\
& \binom{3}{0} & & \binom{3}{1} & & \binom{3}{2} & & \binom{3}{3} & \\
\binom{4}{0} & & \binom{4}{1} & & \binom{4}{2} & & \binom{4}{3} & & \binom{4}{4} \\
\end{array}
$$

The triangle of numbers goes on forever. Each of the outer diagonal entries is 1, corresponding to the identity $\binom{n}{0} = \binom{n}{n} = 1$. Each internal entry is obtained by adding the two entries that lie immediately to the above left and right of it. This follows from the following identity for binomial coefficients:

$$
\binom{n-1}{k-1} + \binom{n-1}{k} = \binom{n}{k}, \tag{7}
$$

valid whenever $0 < k < n$.

Proof of (7): We can give a nice combinatorial proof of (7) as follows: The right side of (7) is the number of different subsets of size k that one can choose from a set of n objects. We consider a set T with n elements and remove one of the elements, which we label as a, and we call the resulting set S (for smaller set). Thus, we can write $T = S \cup \{a\}$, where the union is disjoint (so S has $n-1$ elements). We know that $\binom{n}{k}$ corresponds to the number of subsets of T having k elements. Now, such a subset can either be a subset of S also with k elements (if it does not contain a), or consists of a together with a subset of S of size $k-1$ (if it does contain a). There are $\binom{n-1}{k}$ subsets of the first type (corresponding to subsets of S having k elements) and $\binom{n-1}{k-1}$ subsets of the second type (corresponding to subsets of S having $k-1$ elements). Since these two types of sets form a disjoint partition of the size k subsets of T, the identity (7) follows. □

EXERCISE FOR THE READER 5.9: Give a non-combinatorial proof of the identity (7) using factorial manipulations.

EXERCISE FOR THE READER 5.10: (a) Prove the following identity:

$$1 = \binom{m}{1} - \binom{m}{2} + \binom{m}{3} - \cdots \pm \binom{m}{m},$$

where m is any positive integer.

(b) Use the identity of Part (a) to prove the general case of the inclusion-exclusion principle (formula (1) from the last section):

$$| A_1 \cup A_2 \cup \cdots \cup A_n | = \sum_{i=1}^{n} | A_i | - \sum_{i_1 < i_2} | A_{i_1} \cap A_{i_2} | + \cdots$$

$$+ (-1)^{a+1} \sum_{i_1 < i_2 < \cdots < i_a} | A_{i_1} \cap A_{i_2} \cap \cdots \cap A_{i_a} | + \cdots$$

$$+ (-1)^{n+1} | A_1 \cap A_2 \cap \cdots \cap A_n |.$$

Suggestion: For Part (a), use the binomial theorem. For Part (b), consider a single element $x \in \cup A_i$, let m be the number of sets A_i to which x belongs. Use the identity of Part (a) to count the number of times the right side of (1) contributes to the count of the element x.

Multinomial Coefficients

The binomial coefficients are a special case of the so-called *multinomial coefficients*, which we motivate with the following example:

EXAMPLE 5.14: (*Motivating example for multinomial coefficients*) How many "words" can be created by rearranging the four letters in the word "look?" Here we take a word in the general sense to mean any sequence of four letters, regardless whether it has any meaning in any language.

SOLUTION: If the letters in "look" were all different, the answer would simply be the number of permutations of four objects, or $4! = 24$. Let us temporarily label the duplicate letter o's as o_1 and o_2, so they will be distinct. Then, any of the 24 possible permutations of the list "$l o_1 o_2 k$," say "$k o_1 o_2 l$," can have the two symbols o_1 and o_2 permuted ($k o_2 o_1 l$) and the result will be indistinguishable from the original permutation, once the artificial labels on the o's have been removed. On the other hand, any other permutation of "$k o_1 o_2 l$" would be distinguishable if the o-labels are detached. Thus, to get the number of distinguishable permutations of "look," we need to divide the total number of permutations $4!$ by this duplication number $2!$, to get $4!/2! = 12$.

We now state and prove the general result:

THEOREM 5.8: (*Permutations of Objects That Are Not All Distinguishable*) Suppose that n objects that are of k different types. Assume that there are

n_1 objects of Type 1, n_2 are of Type 2, ..., and n_k objects of Type k, where $n = n_1 + n_2 + \cdots + n_k$. The number of distinguishable permutations of these n objects is given by the **multinomial coefficient**

$$\binom{n}{n_1, n_2, \cdots, n_k} \equiv \frac{n!}{n_1! \, n_2! \cdots n_k!}.$$

This number also coincides with the number of ways to place n distinct objects into k distinguished groups with n_1 objects in the first group, n_2 in the second group, ..., and n_k in the last group.

We will give two different combinatorial proofs of this theorem; the first proceeds along the lines of the motivating example, while the second is a more direct proof.

Proof 1: We know that there are $n! = (n_1 + n_2 + \cdots + n_k)!$ permutations of these objects. For any such permutation, any of the n_1 objects of Type 1 can be permuted in any of the $n_1!$ possible ways and the resulting permutation will not be distinguishable from the original. The same holds true if we perform any of the $n_2!$ possible permutations of the Type 2 objects, any of the $n_3!$ possible permutations of the Type 3 objects, and so on. By the multiplication principle, it follows that each permutation thus correspond to a total of $n_1! \cdot n_2! \cdots n_k!$ permutations that are indistinguishable from one another. These sets of permutations partition the entire collection of permutations. Dividing the total number of permutations by this duplication number gives us the asserted number of distinguishable permutations. The latter statement can be justified in the same fashion. □

Proof 2: Imagine a row of n slots to fill with these n objects. Choose n_1 slots to be filled with objects of Type 1. There are $\binom{n}{n_1}$ ways to do this. From the remaining $n - n_1$ slots, we next choose n_2 of them to be filled with objects of Type 2. This can be done in $\binom{n - n_1}{n_2}$ ways. From the remaining $n - n_1 - n_2$ slots, we next choose n_3 of them to be filled with objects of Type 3. This can be done in $\binom{n - n_1 - n_2}{n_3}$ ways. Continuing in this fashion, the multiplication principle tells us that the number of distinguishable permutations of these n objects is given by:

$$\binom{n}{n_1} \cdot \binom{n-n_1}{n_2} \cdot \binom{n-n_1-n_2}{n_3} \cdots \cdot \binom{n-n_1-n_2-\cdots-n_{k-1}}{n_k}$$

$$= \frac{n!}{n_1!\,(n-n_1)!} \cdot \frac{(n-n_1)!}{n_2!\,(n-n_1-n_2)!} \cdot \frac{(n-n_1-n_2)!}{n_3!\,(n-n_1-n_2-n_3)!}$$

$$\cdots\cdots \frac{(n-n_1-n_2-\cdots-n_{k-1})!}{n_k!\,(n-n_1-n_2-\cdots-n_{k-1}-n_k)!}$$

$$= \frac{n!}{n_1!\,n_2!\,n_3!\cdots n_k!\,0!} = \frac{n!}{n_1!\,n_2!\cdots n_k!}.$$

The latter statement can be justified in the same fashion. □

EXERCISE FOR THE READER 5.11: How many distinguishable permutations are there of the word MISSISSIPPI?

In the basic Example 5.9, we would have $n = 4$ (four letters), $n_1 = n_3 = 1$ (corresponding to the unique letters "l" and "k") and $n_2 = 2$ (corresponding to the duplicated letter "o"), and the answer we obtained equals $\binom{4}{1,\,2,\,1} = \frac{4!}{1!\,2!\,1!} = \frac{24}{1\cdot 2\cdot 1} = 12$. The multinomial coefficients are so named because of the multinomial theorem that we will describe shortly. Note that when $k = 2$, the multinomial coefficient $\binom{n}{n_1,\,n_2}$ coincides with the binomial coefficients $\binom{n}{n_1} = \binom{n}{n_2}$.

The combinatorial proof that we gave for the binomial theorem generalizes naturally to the expansions of powers of multinomials. The general result is contained in the following theorem:

The Multinomial Theorem

THEOREM 5.9: (*The multinomial theorem*) If $x_1, x_2, \cdots, x_r$ are numbers, and n is a nonnegative integer, then

$$(x_1 + x_2 + \cdots + x_r)^n = \sum_{\substack{k_1+k_2+\cdots+k_r=n \\ k_i \text{ nonnegative integer}}} \binom{n}{k_1,\,k_2,\cdots,k_r} x_1^{k_1} x_2^{k_2} \cdots x_r^{k_r} \qquad (8)$$

The sum in the right-hand side of (8) can be viewed as being taken over all of the vectors $(k_1, k_2, \cdots, k_r)$ of nonnegative integers ($0 \le k_i \le n$) whose components add up to n.

EXERCISE FOR THE READER 5.12: Prove Theorem 5.9.

EXAMPLE 5.15: Use the multinomial theorem to expand $(x + 2y + 3z)^2$.

SOLUTION: We have $r = 3$ (trinomial) and $n = 2$ (with $x_1 = x$, $x_2 = 2y$ and $x_3 = 3z$).   The totality of vectors (k_1, k_2, k_3) corresponding to the terms in the right-hand side of (8) are as follows: $(k_1, k_2, k_3) = (2,0,0)$, $(0,2,0)$, $(0,0,2)$, $(1,1,0)$, $(1,0,1)$, and $(0,1,1)$. The first three of the corresponding multinomial coefficients all equal $2!/2! = 1$, while the last three equal $2!/1!1! = 2$. Computing with (8) (in the order that these vectors were listed), we now obtain

$$(x + 2y + 3z)^2 = x^2 + (2y)^2 + (3z)^2 + 2[x(2y) + x(3z) + (2y)(3z)]$$
$$= x^2 + 4y^2 + 9z^2 + 4xy + 6xz + 12yz.$$

EXERCISE FOR THE READER 5.13: Show how formula (8) can specialize to the binomial theorem (5).

EXERCISE FOR THE READER 5.14: What is the coefficient of $a^6 b^3 c^3 d^2$ in the expansion of $(2a - 3b + 4c - d)^{14}$?

We end this section with another useful counting argument.   At first glance, combinations might not seem very relevant, but with an ingenious artifice they may indeed be applied.

EXAMPLE 5.16: (*Motivating example for a partitioning argument*)   Joey has five identical chocolate bars that he plans to give to his three cousins, Abby, Billy, and Christy.   In how many different ways can he distribute these bars to his cousins?

SOLUTION: The different distributions of the five chocolate bars can be displayed graphically by laying out the five chocolate bars in a horizontal row, and inserting two partitions anywhere among the six slots between bars (or to the left/right of all of them).   This will partition the bars into three groups (some possibly empty): the group to the left of the two partitions, which we arbitrarily assign to be Abby's allotment, the group between the two partitions: Billy's allotment, and the group to the right of the two partitions: Christy's allotment. Figure 5.8 shows a particular allotment with this scheme.   Note that there is one less partition bar than the number of people to distribute to.

FIGURE 5.8: A possible distribution for the five (identical) chocolate bars to three people. The two barriers partition the distribution into three categories: (i) to the left of the first partition: no bars to Abby, (ii) between the two barriers: one bar to Billy, and (iii) to the right of the second partition: four bars to Christy.

Clearly, the different allotments of the five bars to the three individuals correspond to the number of different bar/partition diagrams. We can view this question as the count of the number of distinguishable arrangements of seven objects (five chocolate bars + two partition bars), where the five chocolate bars are identical as are the two partition bars. Thus, by Theorem 5.8, the number of such arrangements is $\binom{5+(3-1)}{5,\ 2} = \binom{7}{2} = 21$.

The reader should be able to prove the following general result, the task will be left as Exercise 36.

THEOREM 5.10: (*Distribution of Identical Objects to Different Places*) The number of ways to distribute n identical objects to d different (distinguishable) places is given by $\binom{n+(d-1)}{d-1}$.

EXERCISE FOR THE READER 5.15: (a) How many different nonnegative integer solutions are there (for x_1, x_2, x_3, x_4) in the equation
$$x_1 + x_2 + x_3 + x_4 = 12?$$
(b) How many different positive integer solutions are there for the equation in Part (a)?
Suggestion: For Part (a), consider the analogy of placing 12 identical balls into 4 different urns and view assigning each x_i a nonnegative integer as placing this number of balls into the ith urn. For Part (b), introduce new variables $y_i = x_i + 1$ which will be positive integers whenever the x_i's are nonnegative integers, and then use the method of Part (a).

EXERCISE FOR THE READER 5.16: How many terms are there in the sum (8) of the multinomial theorem?

EXERCISES 5.2:

1.   (a) Write down all of the permutations of the word CAT.
       (b) Write down all 2-permutations of the objects $\{A, B, C, D, E\}$.
       (c) Write down all 2-combinations of the objects $\{A, B, C, D, E\}$.

2.   (a) Write down all of the permutations of the list $(1, 2, 3)$.
       (b) Write down all 3-permuations of the objects $\{1, 2, 3, 4, 5\}$.
       (c) Write down all 3-combinations of the objects $\{1, 2, 3, 4, 5\}$.

3.   Compute each of the following quantities:

       (a) $P(3,3)$ (b) $C(5,5)$ (c) $P(52,3)$
       (d) $C(20,5)$ (e) $P(5,1)$ (f) $C(6,0)$

4.   Compute each of the following quantities:

       (a) $P(6,3)$ (b) $C(5,2)$ (c) $P(100,5)$
       (d) $C(200,5)$ (e) $P(8, 8)$ (f) $C(1000,0)$

5.   *Pizza Castle* offers 12 different toppings on their pizzas.
       (a) How many different three-topping pizzas can be ordered?
       (b) How many different pizzas can be made that include up to three toppings (no toppings is possible, this would be just a plain cheese pizza)?

6.   *Cold Cream* offers its ice cream sundae with a choice of one, two, or three scoops of ice cream, and a choice of exactly three different toppings from eight available toppings. (Regardless of the number of scoops ordered, three different toppings must be chosen.)
       (a) How many different sundaes can be ordered with vanilla ice cream?
       (b) How many different sundaes can be ordered if the ice cream can be chosen from 31 different flavors, but it must be the same flavor for each scoop?
       (c) How many different sundaes can be ordered if the ice cream can be chosen from 31 different flavors, and can be a different flavor for each scoop?

7.   (a) Suppose you have won five tickets to an upcoming LA Lakers basketball game. In how many ways can you invite 4 of your 12 best friends to come along?
       (b) Suppose you have seven different NBA basketball team T-shirts. In how many ways can you distribute these shirts to 7 of your 12 best friends?

8.   The computer science (CS) faculty at a certain university consists of 20 members, 13 of whom are senior faculty and 7 of whom are junior members. The total number of faculty at this university is 365. A hiring committee for a new computer science faculty member is to be composed of five faculty members. How many hiring committees are possible if:
       (a) They all come from the CS department?
       (b) The committee is composed of three senior and two junior CS faculty members?
       (c) The committee is composed of CS faculty with at least one junior member?
       (d) The committee contains exactly one outside (the computer science department) faculty member, at least one senior CS faculty member, and at least one junior CS faculty member?

9.   A math department is giving out awards to a particularly strong senior class. It has three different awards: Award I, worth $2000, Award II, worth $1000, and Award III, worth $500. It decides that it can distribute a total of five awards, to five students from the group of eight outstanding students A, B, C, D, E, F, G, and H. In how many ways can the awards be distributed if:
       (a) Five Award III's will be distributed.
       (b) Any combination of the three awards can be given.
       (c) Any combination of the three awards can be given, but students B and H should either both

get an award or both not get an award.

(d) Any combination of the three awards can be given, but students B and H should either both get the same award or both not get an award.

10. The French club at a certain university has 12 active members, 5 of whom are men. A yearbook photo will be taken with 8 of the 12 students lined up in a row. How many such photo arrangements are possible if:
    (a) There are no restrictions.
    (b) Half of the subjects must be women.
    (c) Half of the subjects must be women, the men all stand together, and the women all stand together.
    (d) Half of the subjects must be women, and the men should all stand together.
    (e) Half of the subjects must be women, and no two men stand together.
    (f) If there are more women than men, no two women stand together; otherwise no two men stand together.

11. At a five-year reunion of a college tennis team, 26 former teammates show up, 15 of whom are men.
    (a) If everyone shakes everyone else's hand, how many handshakes will there be?
    (b) If all men shake hands with one another and all the women hug one another, how many hugs and handshakes will there be?
    (c) If all women shake hands with one another and hug all of the men, how many hugs and handshakes will the women be involved with?

12. Count the number of possible three-card poker hands (consisting of three cards drawn from a shuffled standard deck of 52 cards) that contain:
    (a) At least two spades.
    (b) A flush (three cards of the same suit).
    (c) Three different suits.

13. Count the number of possible four-card poker hands (consisting of four cards drawn from a shuffled standard deck of 52 cards) that contain:
    (a) At most one heart.
    (b) Four different suits.
    (c) At least two cards of the same suit.

14. In the senatorial primary election of a certain year in Guam, there are 23 Democratic candidates and 16 Republican candidates. The rules for a voting ballot are that up to 15 candidates can be voted for, but only from one party. In how many ways can a ballot be (correctly) cast, assuming that voting for no candidates is an acceptable ballot (indeed, a political statement)?

15. To pass an exam, a law student must choose five of eight essay questions to answer. How many choices does he/she have? What if he/she is required to answer at least three of the first four questions?

16. How many permutations of the letters $\{A, B, C, D, E, F, G\}$ are there, such that:
    (a) A precedes B?
    (b) A precedes B, and C precedes D?
    (c) A precedes B, which in turn precedes C?
    (d) C, D, and E appear together in this order?
    (e) A and B are appear together in this order, as do C and D?

17. How many permutations of the letters $\{A, B, C, D, E, F, G\}$ are there, such that:
    (a) C and D are next to each other?
    (b) A, B, and C are next to each other?
    (c) C is between A and B?
    (d) F, A, and D are seated together in this order?
    (e) F, A, and D are seated together in this order, as are G and B?

18. (a) How many possible finishes are there in a three-car drag race if double and triple ties are possible?
     (b) How many possible finishes are there in a four-car drag race if double, triple, amd quadruple ties are possible?

19. (a) In how many different ways can 10 men be paired off to dance with 10 of 15 women?
     (b) Same question as (a) but with the additional requirement that Jack either dances with Cindy or sits out (in which case only nine couples would dance).
     (c) Same question as (a) but with the additional requirement that Jenny and Clair will either both dance or both sit out?

20. (a) Expand $(x-z)^5$.

     (b) Expand $(2x+3y)^6$.

     (c) What is the coefficient of x^{12} in the expansion of $(2x^3-5)^5$?

21. (a) Expand $(x+z)^7$.

     (b) Expand $(5x+y^3)^5$.

     (c) What is the coefficient of x^6y^6 in the expansion of $(3x^2-4y^3)^5$?

22. (a) Prove that for any positive integer n, we have $3^n = \sum_{k=0}^{n}\binom{n}{k}2^k$.

     (b) Obtain a similar expansion for x^n for any real number x.

     (c) From Part (b) obtain the expansion $\sum_{k=0}^{n}\binom{n}{k}(-1)^k = 0$.

23. Prove that for any positive integer n, we have $\sum_{k=0}^{n}\binom{n}{k}^2 = \binom{2n}{n}$.

     Suggestion: An elegant combinatorial proof can be achieved by considering a set T containing $2n$ elements, and splitting T into two n-element sets R and S. Now any k-combination of T must be expressible as a disjoint union of a j-combination of R and an $n-j$ combination of S, for some nonnegative integer j.

24. (a) Expand $(x+2y+3z)^4$.

     (b) Expand $(x-y^2-z^3+w)^3$.

     (c) What is the coefficient of $x^5y^3z^8$ in the expansion of $(5x-2y+3z^2)^{12}$?

25. (a) Expand $(2x-2y+5z)^4$.

     (b) Expand $(x+y+z+w)^4$.

     (c) What is the coefficient of $x^4y^6z^8w^{24}$ in the expansion of $(x+2y+3z^2+w^4)^{20}$?

26. How many distinguishable permutations are there of each of the following words?
     (a) CANADA   (b) SWEET
     (c) BANANA   (d) MATHEMATICS

27. How many distinguishable permutations are there of each of the following words?
     (a) YOYO   (b) LOLLIPOP
     (c) ELEMENTAL   (d) KAMEHAMEHA

28. Suppose that a boat runs colored flags up a vertical pole to make signals. The boat has three red flags, and six white flags.
(a) How many different signals can the boat's captain give using all nine of these flags?
(b) How many different signals can be made using exactly three flags? Assume that only the relative position of the flags matters, i.e., different gaps and/or positionings of the flags do not count as different signals.
(c) How many different signals can be made using from one to three flags? See Part (b) for the conventions.

29. Suppose that a boat runs colored flags up a vertical pole to make signals. The boat has three red flags, two green flags, and four yellow flags.
(a) How many different signals can the boat's captain give using all nine of these flags?
(b) How many different signals can be made using exactly three flags? Assume that only the relative position of the flags matters, i.e., different gaps and/or positionings of the flags do not count as different signals.
(c) How many different signals can be made using from one to three flags? See Part (b) for the conventions.

30. (a) In how many ways can 24 new (and identical) computers be distributed to the Math Laboratory, the Computer Laboratory, and the Physics Laboratory at a certain university?
(b) Same question as (a) but with the additional requirement that the Physics Laboratory must receive at least three computers, and the other two labs must receive at least one each.

31. Suppose that we have $15,000 to invest in (up to) three different mutual funds: A, B, and C, and that we can allocate investments in each fund in increments of $500.
(a) How many such investment allocations are possible?
(b) How many such allocations are possible if the minimum investments in funds A and C are $2500?

32. (a) How many solutions are there of the equation $x_1 + x_2 + x_3 + x_4 = 10$, where each $x_i \geq -1$ an integer?
(b) Repeat Part (a) if now only $x_1, x_3 \geq -1$ but $x_2, x_4 \geq -2$.

33. *Pumpkin's Donuts* sells eight different kinds of donuts. How many different dozens of donuts can be sold?

34. A wallet contains five traveler's checks that are taken from the following denominations: $1, $5, $50, $500.
(a) How many different combinations of travelers checks are possible?
(b) Do different combinations always result in different total dollar amounts?

35. (a) Give a combinatorial proof of the identity $\binom{n}{n-k} = \binom{n}{k}$, $0 \leq k \leq n$, and then give a noncombinatorial proof.
(b) For a positive integer n, what is the value of the integer k, $0 \leq k \leq n$, for which $\binom{n}{k}$ is at its maximum value? Show that the binomial coefficients increase as k increases to this value, and then decrease as k increases from this value to n.
Note: The identity in Part (a) corresponds to the left-right symmetry in Pascal's triangle.
Suggestion: For Part (b), deal separately with the cases in which n is even and n is odd.

36. Use mathematical induction to give another proof of the binomial theorem.

37. How many onto functions f are there with the following domains and codomains?
(a) $f : \{1,2,3,4,5\} \rightarrow \{1,2,3,4,5,6\}$
(b) $f : \{1,2,3,4,5,6\} \rightarrow \{1,2,3,4,5,6\}$
(c) $f : \{1,2,3,4,5,6\} \rightarrow \{1,2,3,4,5\}$
(d) $f : \{1,2,3,4,5,6,7\} \rightarrow \{1,2\}$

38.   (a)   If A is a finite set with n elements, show there are $2^n - 2$ different onto functions $f : A \to \{1, 2\}$.

(b) How many onto functions are there of the form $f : \{1, 2, 3, 4, 5, 6, 7\} \to \{1, 2, 3\}$?

Suggestion: For Part (a) notice that any nonconstant function is onto. For Part (b), use the result of Part (a) in counting separately the (disjoint) cases where $|f^{-1}(\{3\})| = 1, 2, 3, 4,$ or 5.

39.   Prove that for nonnegative integers n, m, and r, with $0 \le r \le \min(n, m)$, we have

$$\sum_{k=0}^{r} \binom{n}{k} \binom{n}{r-k} = \binom{n+m}{r}.$$

Suggestion: This result generalizes that of Exercise 23; the suggestion given there can be modified for the present needs.

40.   Prove Theorem 5.9.

41.   Prove the following identity: $\sum_{k=0}^{n} k \binom{n}{k} = n \cdot 2^{n-1}$.

Suggestion: An elegant combinatorial proof can be achieved by counting the number of ways to form a committee (from a group of n individuals), along with a distinguished chairperson. The left and right sides of the identity outline two different schemes for counting the total number of such committees.

42.   Prove the following identity: $\displaystyle\sum_{k=0}^{n} k \binom{n}{k}^2 = n \cdot \binom{2n-1}{n-1}$.

Suggestion: A combinatorial proof can be given using an idea similar to that given in the suggestion of the preceding exercise. This time, the committee is formed from two separate groups of individuals and the chairperson is taken from the first group.

43.   Give a combinatorial proof of the following identity, which is valid for $1 \le k \le n$:

$$k \binom{n}{k} = n \binom{n-1}{k-1}.$$

44.   Give a combinatorial proof of the following identity, which is valid for $1 \le k \le n$:

$$\binom{n}{k} = \sum_{j=k}^{n} \binom{j-1}{k-1}.$$

5.3: GENERATING FUNCTIONS

In this section we will develop a very effective tool for analyzing sequences relating to combinatorial problems. This tool is based on storing the terms of the sequence as coefficients of a formal power series.   The resulting formal power series is called the *generating function* for a given sequence. Generating functions can be manipulated by a set of natural rules, which are motivated by the ordinary arithmetic of polynomials, allowing one to operate on the whole sequence at once, and these concepts give them surprising power that can be used to solve seemingly intractable problems.   Many useful properties about the convergence of power series are proved in calculus books. This section will develop generating functions from a non-calculus perspective. Although some facts will be "borrowed" from calculus, our treatment will be entirely self-contained. We will thus bypass details concerning the convergence of series, and merely perform formal manipulations on them and show how to apply such manipulations to solve an assortment of combinatorial problems.

Generating Functions and Power Series

DEFINITION 5.4: For any sequence $a_0, a_1, a_2, \cdots$ of real numbers, the corresponding **(ordinary) generating function** of the sequence is the following formal infinite power series:

$$G(x) = a_0 + a_1 x + a_2 x^2 + \cdots = \sum_{n=0}^{\infty} a_n x^n.$$

We use the function notation on the left of the above equation, even though the infinite series may not define much of a function. Such an infinite series of increasing nonnegative powers of the variable x with real number coefficients is called a **power series**.

NOTE: Power series are studied in standard calculus courses (usually in the second semester). If we substitute $x = 0$ into any power series, the series becomes $a_0 + a_1 \cdot 0 + a_2 \cdot 0^2 + \cdots = a_0 + 0 + 0 + \cdots = a_0$. Although it is impossible to perform the infinite number of additions required in a general power series (when a nonzero number x is substituted), there are circumstances when the infinite sum makes sense (in which case we say it *converges*), and in such a case it can be evaluated to any degree of accuracy by adding up a sufficiently large finite number of its terms. For any power series there always exists a number R (called the *radius of convergence* of the power series) in the range $0 \le R \le \infty$, such that the power series converges for all x in the range $|x| < R$, but does not converge for any x in the range $|x| > R$. It is possible that $R = 0$ (in which case the series converges only when $x = 0$), but in all other cases the power series truly equals some function of x within the radius of convergence. In such cases, we identify the series with the function that it defines, i.e., both are considered to be the generating function of the sequence.

In this section the adjective "ordinary" (for generating functions) will be redundant since this will be the only sort of generating functions that will be considered. More extensive treatments consider other useful generating functions, such as exponential generating functions; see, for example, [Wil-90]. Throughout this section we shall treat generating functions essentially as formal objects, not concerning ourselves with the question of whether the infinite series converges for any nonzero numbers x.[7] We will present closed formulas for a few key

---

[7] For readers who have studied calculus, when the series converges for nonzero values of x, the resulting function will coincide with the generating function, and the series will be its so-called Taylor

generating functions (i.e., the function formula equals the power series within a positive radius of convergence), but the calculus details will be omitted.

EXAMPLE 5.16:   Determine the generating functions of the following sequences:

(a) $a_n = 1 \ (n = 0,1,2,\cdots)$

(b) $a_n = n! \ (n = 0,1,2,\cdots)$

(c) $a_0 = 1, \ a_1 = 1, \ a_2 = 1, \ a_3 = 1$

NOTE: For any *finite* sequence as in Part (c), in order to form the generating function, the remaining terms are assumed to be zero.   Thus the power series will just be a finite sum, and will define a polynomial.

SOLUTION:

Part (a):   The generating function is $G(x) = 1 + x + x^2 + \cdots = \sum_{n=0}^{\infty} x^n$.   Example 5.18 presents an explicit formula for the function defined by this power series.

Part (b):   The generating function is $G(x) = 1 + x + 2x^2 + \cdots + n!x^n + \cdots = \sum_{n=0}^{\infty} n!x^n$.

Part (c): The generating function is the polynomial $G(x) = 1 + x + x^2 + x^3$.

Algebraic identities sometimes allow us to express polynomials (finite power series) as single mathematical expressions. The next example will provide a few such simplifications.

EXAMPLE 5.17:   For each part, express the polynomial generating function defined by the sequence of coefficients as a single-term closed-form algebraic expression:

(a) $a_0 = 1, \ a_1 = 2, \ a_2 = 4, \ a_3 = 8, \ a_4 = 16$

(b) $a_0 = C(n,0), a_2 = C(n,1), \cdots, a_n = C(n,n),$ where n is a positive integer.

SOLUTION:

Part (a):   The generating function for the given sequence is $1 + 2x + 4x^2 + 8x^3 + 16x^4 = (2x)^0 + (2x)^1 + (2x)^2 + (2x)^3 + (2x)^4$, which is a finite geometric series (see Proposition 3.5 of Section 3.1), so by formula (3) of Chapter 3, it can be rewritten as: $((2x)^5 - 1)/(2x - 1) = (32x^5 - 1)/(2x - 1)$.

Part (b):   The generating function for the given coefficients is $\sum_{k=0}^{n} C(n,k) \, x^k$. Using the binomial theorem (formula (6) of Theorem 5.7):

---

series.   This gives another way to view the coefficients in terms of derivatives of the generating functions.

$$(x+y)^n = \sum_{k=0}^{n}\binom{n}{k} x^k y^{n-k} = \sum_{k=0}^{n} C(n,k)\, x^k y^{n-k},$$

and substituting $y = 1$, gives us $(1+x)^n = \sum_{k=0}^{n} C(n,k)\, x^k$. The left side is the factored form of the polynomial generating function for the sequence of binomial coefficients.

Such algebraic conversions for polynomials in the above example can be extended to a great variety of infinite power series by making use of some key (building block) closed-form expressions for power series that will be borrowed from calculus. Our next example provides an important generating function for a very simple sequence: the sequence of constant 1s:

EXAMPLE 5.18: Part (a): In calculus courses, it is proved that the generating function of the infinite sequence of 1s: $a_0 = 1, a_1 = 1, a_3 = 1\cdots$, is the function $1/(1-x)$, and furthermore that the power series equals the function when $|x| < 1$:

$$\frac{1}{1-x} = 1 + x + x^2 + \cdots = \sum_{n=0}^{\infty} x^n. \qquad (8)$$

Another useful generating function corresponds to the sequence of reciprocal factorials: $a_n = 1/n!$, and is the exponential function e^x. The following equality is valid for all real numbers x:

$$e^x = 1 + x + \frac{x^2}{2!} + \frac{x^3}{3!} + \cdots = \sum_{n=0}^{\infty} \frac{x^n}{n!}. \qquad (9)$$

Part (b): Use (8) to find a closed expression for the generating function of the sequence $a_n = \begin{cases} 1, & \text{if } n \text{ is even} \\ 0, & \text{if } n \text{ is odd} \end{cases}$.

SOLUTION: Part (b): The generating function of the sequence is $G(x) = 1 + 0 \cdot x + 1 \cdot x^2 + 0 \cdot x^3 + \cdots = 1 + x^2 + x^4 + x^6 \cdots$. If we make the substitution $x \mapsto x^2$ in (8) (which will be a valid equation if $|x| < 1$), we arrive at:

$$\frac{1}{1-x^2} = 1 + x^2 + x^4 + \cdots = \sum_{n=0}^{\infty} x^{2n},$$

which gives an explicit formula for the generating function at hand.

Arithmetic of Generating Functions

Polynomials, i.e., functions of the form

$$a_0 + a_1 x + a_2 x^2 + \cdots + a_d x^d = \sum_{n=0}^{d} a_n x^n,$$

are determined by a <u>finite</u> sequence of coefficients $a_0, a_1, a_2, \cdots, a_d$ (which we assume to be real numbers). They are the generating functions for their sequence of coefficients. They serve as good motivators for the arithmetic of more general formal power series (i.e., generating functions). When we write down a polynomial, unless all of its coefficients are zero (the zero polynomial) in the representation above, we may always assume that the *leading coefficient*, a_d, is nonzero. In this case the polynomial is said to have *degree d* (the highest power that appears with a nonzero coefficient).

Since polynomials (finite power series) define functions on the entire domain of real numbers, these generating functions can be identified with the functions that they represent.

Polynomials can be added/subtracted term-by-term, and two polynomials can be multiplied in the usual fashion that is taught in basic algebra using the rules $x^n x^m = x^{n+m}$ and $x^0 = 1$. For example, to multiply two general third degree polynomials,

$$(a_0 + a_1 x + a_2 x^2 + a_3 x^3) \cdot (b_0 + b_1 x + b_2 x^2 + b_3 x^3),$$

we could start off as follows:

$$(a_0 b_0) + (a_0 b_1 + a_1 b_0) x + (a_0 b_2 + a_1 b_1 + a_2 b_0) x^2 + (a_0 b_3 + a_1 b_2 + a_2 b_1 + a_3 b_0) x^3 + \cdots.$$

For higher powers (up to six), the coefficients are composed of fewer and fewer terms, and thus the computations are simplified. The degree four (next) term is:

$$(a_1 b_3 + a_2 b_2 + a_3 b_1) x^4.$$

If we adopt the convention indicated above that unlisted coefficients are assumed to be zero, we could rewrite this degree four term to fit the general pattern of the first three:

$$(a_0 b_4 + a_1 b_3 + a_2 b_2 + a_3 b_1 + a_4 b_0) x^4.$$

This latter approach, although more complicated for polynomials, leads to the definition of multiplication of formal power series that is included in Part (c) of the following definition:

DEFINITION 5.5: (*Arithmetic of Generating Functions/Formal Power Series*):
Suppose we have two generating functions $F(x) = \sum_{n=0}^{\infty} a_n x^n$, $G(x) = \sum_{n=0}^{\infty} b_n x^n$.
(a) The **sum** of these generating functions is

$$F(x) + G(x) = \sum_{n=0}^{\infty} (a_n + b_n)x^n.$$

(b) The **difference** of these generating functions is

$$F(x) - G(x) = \sum_{n=0}^{\infty} (a_n - b_n)x^n.$$

(c) The **product** of these generating functions is

$$F(x) \cdot G(x) = \sum_{n=0}^{\infty} (a_0 b_n + a_1 b_{n-1} + a_2 b_{n-2} + \cdots + a_n b_0)x^n = \sum_{n=0}^{\infty} \left(\sum_{k=0}^{n} a_k b_{n-k} \right) x^n.$$

Calculus books prove that if the series for $F(x)$ and $G(x)$ both have positive radii of convergence, then the formally defined series above really do correspond to the generating functions for the sum, difference, and product of $F(x)$ and $G(x)$, and the equations will be valid if $|x|$ is less than the minimum of the two radii of convergence.

EXAMPLE 5.19: Find a formula for the nth coefficient a_n of the sequence whose generating function is given by $e^x /(1-x)$.

SOLUTION: We will use Part (c) of Definition 5.5, along with equations (8) and (9):

$$\frac{e^x}{1-x} = e^x \cdot \frac{1}{1-x} = \left(1 + x + \frac{x^2}{2!} + \frac{x^3}{3!} + \cdots \right) \cdot \left(1 + x + x^2 + x^3 + \cdots \right)$$

$$= 1 \cdot 1 + (1 \cdot 1 + 1 \cdot 1)x + \left(1 \cdot 1 + 1 \cdot 1 + \frac{1}{2!} \cdot 1 \right) x^2 + \left(1 \cdot 1 + 1 \cdot 1 + \frac{1}{2!} \cdot 1 + \frac{1}{3!} \cdot 1 \right) x^3 + \cdots.$$

The pattern has quickly revealed itself, and we may write $e^x /(1-x) = \sum_{n=0}^{\infty} a_n x^n$,

where $a_n = \dfrac{1}{0!} + \dfrac{1}{1!} + \dfrac{1}{2!} + \cdots + \dfrac{1}{n!} = \sum_{k=0}^{n} \dfrac{1}{k!}.$

From Definition 5.5, it is clear that the generating function 0 (corresponding to the infinite sequence of zeros: $a_n = 0$) serves as the additive identity, and the generating function 1, corresponding to the sequence $a_1 = 1$, $a_n = 0$, if $n \neq 0$, serves as the multiplicative identity. Put differently, if $F(x)$ is any generating function, then $F(x) + 0 = F(x)$ and $F(x) \cdot 1 = F(x)$.

EXERCISE FOR THE READER 5.17: Obtain closed-form expressions for the following generating functions.

(a) $x^2 + x^3 + x^4 + \cdots = \sum_{n=2}^{\infty} x^n$
(b) $(1-x)\left(1 + 2x + \dfrac{4x^2}{2!} + \dfrac{8x^3}{3!} + \cdots \right).$

The Generalized Binomial Theorem

The binomial generating function that was obtained in Example 5.17:

$$(1+x)^n = \sum_{k=0}^{n} \binom{n}{k} x^k,$$

has a very useful extension to non-integer powers. In order to state the result, it is convenient to generalize the definition of the binomial coefficients $\binom{n}{k}$ for non-integer values of n. The following definition does this using a formula that is equivalent to the formula (4) for (ordinary) binomial coefficients in case n is a nonnegative integer.

DEFINITION 5.6: If a is a real number and k is a nonnegative integer, we define the generalized binomial coefficient $\binom{a}{k}$ by the formula:

$$\binom{a}{k} = \begin{cases} \dfrac{a(a-1)(a-2)\cdots(a-k+1)}{k!}, & \text{if } k > 0, \\ 1, & \text{if } k = 0. \end{cases}$$

EXAMPLE 5.20: Compute the generalized binomial coefficient $\binom{3/2}{4}$.

SOLUTION: Substituting $a = 3/2$ and $k = 4$ into the formula of the above definition gives:

$$\binom{3/2}{4} = \frac{(3/2)(1/2)(-1/2)(-3/2)}{4!} = \frac{3}{128}.$$

EXERCISE FOR THE READER 5.18:   If a is a negative integer: $a = -n$,   show that the generalized binomial coefficient $\binom{a}{k}$ can be expressed using ordinary binomial coefficients as follows: $\binom{a}{k} = (-1)^k C(n+k-1,k)$.

We are now ready to state the generalized binomial theorem that provides a generating function power series expansion for $(1+x)^a$, where a is any real number. Unless a is a nonnegative integer, the power series will be an infinite series.

THEOREM 5.11: (*The Generalized Binomial Theorem*) If a is any real number, and x is a real number with $|x| < 1$, then

$$(1+x)^a = 1 + \binom{a}{1}x + \binom{a}{2}x^2 + \binom{a}{3}x^3 + \cdots = \sum_{n=0}^{\infty}\binom{a}{n}x^n. \tag{10}$$

We point out that in case a is a positive integer, then (by Definition 5.6) $\binom{a}{k} = 0$ whenever $k > a$, so the expansion (10) is a finite series and (10) reduces to the ordinary binomial theorem (Theorem 5.7), but in all other cases, $\binom{a}{k} \neq 0$ for all positive integers k, so (10) is really an infinite series.

EXAMPLE 5.21: Apply the generalized binomial theorem to find power series expansion for the following function: $1/(1-x)^2$.

SOLUTION: If we substitute $a \mapsto -2, x \mapsto -x$ into (10), the following expansion that is valid for $|x| < 1$ results:

$$\frac{1}{(1-x)^2} = 1 + \binom{-2}{1}(-x) + \binom{-2}{2}(-x)^2 + \binom{-2}{3}(-x)^3 + \cdots = \sum_{n=0}^{\infty}\binom{-2}{n}(-x)^n.$$

Since $\binom{-2}{n} = \dfrac{-2(-3)(-4)\cdots(-2-n+1)}{1 \cdot 2 \cdot 3 \cdots n} = \dfrac{(-1)^n 2 \cdot 3 \cdot 4 \cdots (n+1)}{1 \cdot 2 \cdot 3 \cdots n} = (-1)^n(n+1),$

the above expansion becomes:

$$\frac{1}{(1-x)^2} = 1 + -2(-x) + 3(-x)^2 + -4(-x)^3 + \cdots = 1 + 2x + 3x^2 + 4x^3 + \cdots = \sum_{n=0}^{\infty}(n+1)x^n.$$

This expansion could have also been obtained by multiplying that of (8) by itself (using Definition 5.5(c)). Since it is often useful, we record it as numbered equation for future reference (the equality is valid when $|x| < 1$):

$$\frac{1}{(1-x)^2} = 1 + 2x + 3x^2 + 4x^3 + \cdots = \sum_{n=0}^{\infty}(n+1)x^n. \tag{11}$$

The following more general expansion will often be useful; its justification is similar to the argument in Example 5.21, and is left as the next exercise for the reader. If n is a positive integer and $|x| < 1$, then we have:

$$\frac{1}{(1-x)^a} = 1 + C(a,1)x + C(a+1,2)x^2 + C(a+2,3)x^3 + \cdots$$

$$= \sum_{n=0}^{\infty} C(n+a-1, a-1)x^n. \tag{12}$$

EXERCISE FOR THE READER 5.19: Establish the power series expansion (12).

EXERCISE FOR THE READER 5.20: Determine the power series expansion for the function $\sqrt{1+x/2}$.

Using Generating Functions to Solve Recursive Sequences

Now that we have presented a decent collection of generating functions, it is time to show their usefulness in combinatorics. We will begin by demonstrating some general schemes by which generating functions can be used to solve recursive relations. We begin with an easy example for which an explicit formula can quickly be deduced without much work.

EXAMPLE 5.22: Consider the recursively defined sequence:
$$\begin{cases} a_0 = 1 \\ a_n = 2a_{n-1} + 1 \ (n \geq 1). \end{cases}$$

An explicit formula for this sequence can be easily obtained, for example, by computing the first few terms: $a_1 = 3$, $a_2 = 7$, $a_3 = 15$, $a_4 = 31$, and discovering the pattern $a_n = 2^{n+1} - 1$, which can then be established by induction (see also Example 3.6). We will use this example to showcase how generating functions can be used to solve recursive relations. Instead of working with the terms of the sequence, we will consider the generating function for this sequence: $F(x) = \sum_{n=0}^{\infty} a_n x^n$, and use the given recurrence relation to obtain a closed from expression for this function. We multiply both sides of the recurrence relation by x^n, and then take the formal (infinite) sum of both sides in the range $n \geq 1$ where the recurrence is valid:

$$a_n = 2a_{n-1} + 1 \ (n \geq 1) \Rightarrow \sum_{n=1}^{\infty} a_n x^n = 2\sum_{n=1}^{\infty} a_{n-1} x^n + \sum_{n=1}^{\infty} 1 \cdot x^n.$$

Now let us look at each of these three formal sums and aim to find the corresponding generating functions. The first sum is just

$$\sum\nolimits_{n=1}^{\infty} a_n x^n = \sum\nolimits_{n=0}^{\infty} a_n x^n - a_0 = F(x) - 1.$$

Relating the second formal sum to $F(x)$ requires a slightly different manipulation:

$$2\sum\nolimits_{n=1}^{\infty} a_{n-1} x^n = 2x\sum\nolimits_{n=1}^{\infty} a_{n-1} x^{n-1} = 2x\sum\nolimits_{n=0}^{\infty} a_n x^n = 2xF(x).$$

The third formal sum is closely related to the expansion (8):

$$\sum_{n=1}^{\infty} x^n = \sum_{n=0}^{\infty} x^n - 1 = \frac{1}{1-x} - 1.$$

Thus, the formal series equation above corresponds to the following equation for the generating function $F(x)$:

$$F(x) - 1 = 2xF(x) + \frac{1}{1-x} - 1 \Rightarrow (1-2x)F(x) = \frac{1}{1-x} \Rightarrow F(x) = \frac{1}{(1-x)(1-2x)}.$$

This generating function determines the entire sequence (a_n). In order to obtain an explicit formula for the terms of the sequence, we use the *partial fractions method* [8] of algebra to expand the expression on the right side:

$$\frac{1}{(1-x)(1-2x)} = \frac{A}{(1-x)} + \frac{B}{(1-2x)}.$$

To determine the constants A and B on the right side, we first multiply both sides of the equation by the denominator on the left to obtain:

$$1 = A(1-2x) + B(1-x).$$

If we substitute $x = 1$ into this equation, we obtain $A = -1$, and substituting $x = 1/2$ produces $B = 2$. The original equation now gives:

$$F(x) = \frac{-1}{(1-x)} + \frac{2}{(1-2x)}.$$

Each of the terms on the right can easily be expanded using (8):

---

[8] The partial fractions expansion applies to any quotient of polynomials $P(x)/Q(x)$, where the degree of the numerator is less than that of the denominator, and the denominator $Q(x)$ is factored into powers of distinct linear factors $(x-a)^k$, where a is a complex number. Each factor of $Q(x)$ of form $(x-a)^k$ gives rise to a sum of terms of form: $\dfrac{A_1}{x-a} + \dfrac{A_2}{(x-a)^2} + \cdots + \dfrac{A_2}{(x-a)^k}$, where $A_1, A_2, \cdots, A_k$ are constants, in the partial fraction expansion. Each such term can be expanded into a power series using (12). If $P(x)$ were to have higher degree than $Q(x)$, a long division of polynomials could be used to rewrite $P(x)/Q(x)$ as a sum of a polynomial and $R(x)/Q(x)$, where the remainder $R(x)$ has smaller degree than $Q(x)$. Technical note: Although the examples and exercises given in this book will involve only real numbers, the methods that we develop still work in cases of complex numbers (because (12) remains valid if x is a complex number of modulus less than 1).

$$F(x) = \frac{1/2}{(1-x)} + \frac{2}{(1-2x)} = -1\sum_{n=0}^{\infty} x^n + 2\sum_{n=0}^{\infty} (2x)^n = \sum_{n=0}^{\infty} (2^{n+1} - 1)x^n.$$

Thus we have found that $a_n = 2^{n+1} - 1$.

Although the heavy machinery developed in this example was not really needed, the generating function method can be used in the same fashion to solve recurrences that are not so easily solved by other means. This is one of the beauties of the generating function method.

EXERCISE FOR THE READER 5.21: Use the generating function method of the previous example to solve the recurrence: $\begin{cases} a_0 = 1 \\ a_n = 3a_{n-1} - 1 \ (n \geq 1). \end{cases}$

Our next example involves a recurrence whose solution is not so amenable to discovery as in the preceding example.

EXAMPLE 5.23:   Use the method of generating functions (as developed in Example 5.22) to determine an explicit formula for the following recursively defined sequence:

$$\begin{cases} a_1 = 4 \\ a_n = 3a_{n-1} + 2n - 1 \ (n \geq 2). \end{cases}$$

SOLUTION: In order to facilitate the use of generating functions, we extend the definition of the sequence to define a_0 in a way that will make the recursion formula valid for $n = 1$ (and hence for $n \geq 1$). If we use the recursion formula with $n = 1$: $a_1 = 3a_0 + 2 \cdot 1 - 1$, substitute $a_1 = 4$, we obtain $a_0 = 1$.

Following the method of the preceding example, we let $F(x) = \sum_{n=0}^{\infty} a_n x^n$ be the generating function for the given sequence, multiply both sides of the recurrence relation by x^n, and then take the formal (infinite) sum of both sides in the range $n \geq 1$ (where the recurrence is valid):

$$a_n = 3a_{n-1} + 2n - 1 \ (n \geq 1) \Rightarrow \sum_{n=1}^{\infty} a_n x^n = 3\sum_{n=1}^{\infty} a_{n-1} x^n + 2\sum_{n=1}^{\infty} nx^n - \sum_{n=1}^{\infty} x^n.$$

Three out of the four sums on the right can be converted to closed-form expressions as in the preceding example. The second sum on the right can be converted using the expansion (11) as follows:

$$2\sum_{n=1}^{\infty} nx^n = 2x\sum_{n=1}^{\infty} nx^{n-1} = 2x\sum_{n=0}^{\infty} (n+1)x^n = \frac{2x}{(1-x)^2}.$$

Hence, the preceding equation involving four power series transforms into:

$$F(x) - 1 = 3xF(x) + \frac{2x}{(1-x)^2} - \frac{1}{1-x} + 1.$$

Solving this equation for $F(x)$ and converting to a single term gives us:

$$F(x) = \frac{2x^2 - x + 1}{(1-3x)(1-x)^2}.$$

The partial fractions expansion will have the form:

$$F(x) = \frac{2x^2 - x + 1}{(1-3x)(1-x)^2} = \frac{A}{1-3x} + \frac{B}{1-x} + \frac{C}{(1-x)^2}.$$

To determine the three constants A, B, C, we first clear out all denominators:

$$2x^2 - x + 1 = A(1-x)^2 + B(1-x)(1-3x) + C(1-3x).$$

Substituting $x = 1$ yields $C = -1$. Substituting $x = 1/3$ yields $A = 2$, and finally substituting $x = 0$ (or any third number) yields $B = 0$. Thus we have determined the partial fractions expansion of the generating function to be:

$$F(x) = \frac{2}{1-3x} - \frac{1}{(1-x)^2}.$$

Applying the expansions (8) and (11) allows us to determine the corresponding single power series expansion:

$$F(x) = \frac{2}{1-3x} - \frac{1}{(1-x)^2} = 2\sum_{n=0}^{\infty}(3x)^n - \sum_{n=0}^{\infty}(n+1)x^n = \sum_{n=0}^{\infty}(2 \cdot 3^n - n - 1)x^n.$$

We have thus arrived at the following closed formula for the given recursively defined sequence: $a_n = 2 \cdot 3^n - n - 1$.

EXERCISE FOR THE READER 5.22: Use the generating function method to solve the recurrence: $\begin{cases} a_0 = 1 \\ a_n = 2a_{n-1} + 3^n \ (n \geq 1). \end{cases}$

Whereas the methods of Section 3.2 (see Theorems 3.7, 3.8, and 3.9, and the examples given) could have also been applied to solve the recurrences of the preceding example, our next example is not amenable to the theory of Section 3.2, since the formula for the nth term involves all previous terms (rather than a fixed number of them).

EXAMPLE 5.24:   Obtain an explicit formula for the following recursively defined sequence:

$$\begin{cases} a_0 = 1 \\ \dbinom{n+4}{4} = \displaystyle\sum_{k=0}^{n} a_k a_{n-k} \ (n \geq 1). \end{cases}$$

SOLUTION: We first note that the recurrence equation $\dbinom{n+4}{4} = \displaystyle\sum_{k=0}^{n} a_k a_{n-k}$

remains valid when $n = 0$, and hence is valid for all nonnegative integers n. If, as usual, we let $F(x) = \sum_{n=0}^{\infty} a_n x^n$ be the generating function for the given sequence, we first observe from Definition 5.5(c) that sequence formed by the right side of the recursion formula has the generating function $F(x)^2$.   Also, from (12) we see

that the coefficients $\dbinom{n+4}{4}$ have the generating function $\dfrac{1}{(1-x)^5}$.   It follows

that if we multiply both sides of the recurrence relation by x^n, and then take the formal (infinite) sum of both sides in the range $n \geq 0$ (where the recurrence is

valid) the resulting equation $\displaystyle\sum_{n=0}^{\infty} \dbinom{n+4}{4} x^n = \sum_{n=0}^{\infty} \left(\sum_{k=0}^{n} a_k a_{n-k} \right) x^n$ corresponds to the

equation $(1-x)^{-5} = F(x)^2$,   from which it follows that $F(x) = (1-x)^{-5/2}$.   We may apply the generalized binomial Theorem 5.11 (formula (10)) to obtain the expansion of this generating function:

$$F(x) = 1 + \binom{-5/2}{1}(-x) + \binom{-5/2}{2}(-x)^2 + \binom{-5/2}{3}(-x)^3 + \cdots = \sum_{n=0}^{\infty} \binom{-5/2}{n}(-x)^n.$$

But since $\dbinom{-5/2}{n} = \dfrac{(-5/2) \cdot (-7/2) \cdots (-2n-3/2)}{n!} = (-1)^n \dfrac{5 \cdot 7 \cdots (2n+3)}{2^n n!}$,   we

obtain the desired explicit formula $a_n = \dfrac{5 \cdot 7 \cdots (2n+3)}{2^n n!}$.

EXERCISE FOR THE READER 5.23: Use the generating function method to derive the explicit formula given in equation (4) of Chapter 3 for the famous Fibonacci sequence (first introduced in Example 3.7) that is recursively defined

by: $\begin{cases} f_1 = 1, f_2 = 1, \\ f_n = f_{n-1} + f_{n-2} \ (n \geq 3). \end{cases}$

Using Generating Functions in Counting Problems

Generating functions can be used for a variety of counting problems. The hand computations that arise with such methods are typically quite laborious, so it is most convenient to make use of computers for such tasks. The computer implementation material at the end of this chapter contains some useful information in this regard.

In order to motivate the concepts, we begin by redoing the simple Example 5.16, which was used in Section 5.2 to motivate a general partitioning argument. This example will again serve as a motivating example for the application of generating functions to counting problems.

EXAMPLE 5.25: (*Motivating Example for Counting Techniques Using Generating Functions*) Joey has five identical chocolate bars that he plans to give to his three cousins, Abby, Billy, and Christy. Use generating functions to determine the number of different ways that Joey can distribute these bars among his cousins.

SOLUTION: Each cousin can receive from zero to five bars, and this gives rise to the generating polynomial $1(= x^0) + x + x^2 + x^3 + x^4 + x^5$. The exponent represents the number of candy bars the particular cousin receives; since there are no differences on how Joey can distribute the bars to his three cousins, the three generating functions are the same.

We claim that the number of solutions to the problem is the coefficient of x^5 in the product of the three generating polynomials: $(1 + x + x^2 + x^3 + x^4 + x^5)^3$, the latter being the generating function for the problem. The reason is that the x^5 term in the product is the sum of all terms of the form $x^A x^B x^C$, where x^A is taken from the first factor (and A represents the number of bars Alice receives), x^B is taken from the first factor (corresponding to Billy receiving B bars), x^C is taken from the third factor, and $A + B + C = 5$. There is thus a one-to-one correspondence between the ways that Joey can distribute the five bars among his three cousins, and the terms $x^A x^B x^C$ that arise in expanding the product of the three (in this case identical) generating polynomials of the three cousins. A computation (best done on a computer) shows the coefficient of x^5 in the expansion of $(1 + x + x^2 + x^3 + x^4 + x^5)^3$ to be 21, in agreement with the solution to Example 5.16.

The generating function approach to counting is much more versatile than some of the specialized techniques introduced in Section 5.2. The next example is a variation of the previous one that is not so clearly solvable using the techniques of

Section 5.2, but is easily solved by the same technique introduced in the solution of Example 5.25.

EXAMPLE 5.26:   Joey has five identical chocolate bars that he plans to give to his three cousins, Abby, Billy, and Christy. Use generating functions to determine the number of different ways that Joey can distribute these bars to his cousins under the following constraints: Abby must get at least one bar and Billy must get an even number of bars.

SOLUTION: There are no constraints on the number of bars that Christy can receive so her generating function is exactly as it was in Example 5.25: $F_C(x) = 1(= x^0) + x + x^2 + x^3 + x^4 + x^5$. Since Abby must receive at least one bar, her generating function is the same but without the $1(= x^0)$ term (since we need to omit the option of giving her zero bars): $F_A(x) = x + x^2 + x^3 + x^4 + x^5$. Finally, Billy's generating function is obtained from Christy's by removing all of the odd powers of x: $F_B(x) = 1(= x^0) + x^2 + x^4$. The generating function for the whole problem is the product of these three:

$$F_A(x) \cdot F_B(x) \cdot F_C(x) = (x + x^2 + x^3 + x^4 + x^5) \cdot (1 + x^2 + x^4) \cdot (1 + x + x^2 + x^3 + x^4 + x^5).$$

The number of ways in which Joey can distribute the five bars subject to the given constraints is the coefficient of x^5 in this product. This coefficient can be easily computed with the aid of a computer (or with a hand computation), and is 9.

NOTE: The generating functions for such counting problems actually contain much more information than is typically used. In Example 5.26, the expanded form of the generating function is:

$$F_A(x) \cdot F_B(x) \cdot F_C(x) = x + 2x^2 + 4x^3 + 6x^4 + 9x^5 + 11x^6 + 12x^7$$
$$+ 12x^8 + 11x^9 + 9x^{10} + 6x^{11} + 4x^{12} + 2x^{13} + x^{14}.$$

Each coefficient of a certain power x^n in this expansion gives the number of ways that Joey could give out n bars to his three cousins subject to the constraints represented by the generating functions (at most 5 bars can be given to any cousin since x^5 is the highest power appearing in each cousin's generating function). So for example, the coefficient of x is 1, corresponding to the fact that there is only one way for Joey to give out just one bar to his three cousins, due to the constraint that Abby must get at least one bar: Abby: 1, Billy: 0, Christy: 0. Similarly, the coefficient of x^2 being 2 corresponds to the fact that the following are the only ways that Joey could distribute two bars to his cousins under the specified constraints: (i) Abby: 1, Billy: 0, Christy: 1, or (ii) Abby: 2, Billy: 0, Christy: 0.

EXERCISE FOR THE READER 5.24:   A winner of a contest is allowed to (blindly) draw (and keep) exactly four bills from an urn that contains ten $1 bills, five $10 bills, and two $100 bills.   How many different combinations of bills could be chosen?   What if instead of four bills, six bills are chosen?

Note that in both of the above examples, we could have added higher powers of x: $x^6, x^7, \cdots$ to each of the three generating functions (Abby's, Billy's, and Christy's) because when we looked for the coefficient of x^5, such higher degree terms would not contribute to this lower degree term. Thus, we could have even used infinite series for these generating functions.   Since we do have explicit closed formulas for an assortment of infinite power series, it is sometimes convenient to use infinite power series for generating functions in counting problems.

EXAMPLE 5.27:   Use generating functions to give another proof of Theorem 5.10:   (*Distribution of identical objects to different places*)   The number of ways to distribute n identical objects to d different (distinguishable) places is given by $\binom{n+(d-1)}{d-1}$.

SOLUTION: Although the first idea for the generating function for the number of objects that are placed in the ith place $(1 \leq i \leq d)$ would be $1 + x + x^2 + \cdots + x^n$ (since there are only n objects in totality that can be placed), it will be more convenient to use the infinite power series: $1 + x + x^2 + \cdots$.   The resulting generating function for the whole counting problem is just the product of these d identical individual generating functions $(1 + x + x^2 + \cdots)^d = 1/(1-x)^d$.   The number of ways to distribute n identical objects to d different (distinguishable) places is then the coefficient of x^n in this expansion. By (12) (with $a = d$) this coefficient is $C(n + d - 1, d - 1)$, and the proof is complete.

EXERCISES 5.3:

NOTE: As pointed out in the section proper, calculating coefficients of generating functions is sometimes not feasible without the aid of a computer.   In cases where such situations arise in the exercises below, readers who do not have access to an appropriate computing system might choose to pass up hand computations of such coefficients. While symbolic and/or computer algebra systems are very well suited for the polynomial manipulations needed in such calculations, the computer implementation material at the end of this section will provide details on performing such computations on any standard computing platform.

1.   Write down the generating function for each of the following sequences:
    (a) $a_n = (-1)^n$ $(n = 0, 1, 2, \cdots)$ (b) $a_0 = 6, a_2 = 4, a_4 = 2, a_6 = 1$

2.   Write down the generating function for each of the following sequences:

(a) $a_n = 2^{n+3}$ $(n = 0,1,2,\cdots)$ (b) $a_1 = 2$, $a_2 = 4$, $a_3 = 2$, $a_8 = 4$

3.   For each of the following finite sequences, (i) write down the (polynomial) generating function, and (ii) if possible use either the binomial theorem (Theorem 5.7) or the formula for finite geometric series (Theorem 3.5) to express the function as a closed-form expression.

     (a) $a_n = (-1)^n$ $(n = 0,1,2,\cdots,10)$

     (b) $a_n = C(10,n)$ $(n = 0,1,2,\cdots,10)$

     (c) $a_0 = 20$, $a_1 = -40$, $a_2 = 80$, $a_3 = -160$, $a_4 = 320$

     (d) $a_n = C(5,n-2)$ $(n = 2,3,4,5,6,7)$

4.   For each of the following finite sequences, (i) write down the (polynomial) generating function, and (ii) if possible use either the binomial theorem (Theorem 5.7) or the formula for finite geometric series (Theorem 3.5) to express the function as a closed-form expression.

     (a) $a_0 = 6$, $a_1 = 3$, $a_2 = 3/2$, $a_3 = 3/4$, $a_4 = 3/8$

     (b) $a_n = C(10,10-n)$ $(n = 0,1,2,\cdots,10)$

     (c) $a_0 = 10$, $a_1 = 30$, $a_2 = 90$, $a_3 = 270$, $a_4 = 810$

     (d) $a_n = 2^n C(4,n)$ $(n = 0,1,2,3,4)$

5.   Obtain closed-form algebraic expressions for the generating functions defined by the following infinite sequences:

     (a) $a_n = (-1)^n$ $(n = 0,1,2,\cdots)$

     (b) $a_n = (-2)^n / n!$ $(n = 0,1,2,\cdots)$

     (c) $a_1 = 5$, $a_n = (-1)^n$ $(n = 0,2,3,\cdots)$

     (d) $a_n = 1/(n+2)!$ $(n = 0,1,2,3,\cdots)$

6.   Obtain closed-form algebraic expressions for the generating functions defined by the following infinite sequences:

     (a) $a_n = \begin{cases} 0, & \text{if } n = 0,2,4\cdots \\ -2, & \text{if } n = 1,3,5\cdots \end{cases}$

     (b) $a_n = \begin{cases} (-1)^{n/2}, & \text{if } n = 0,2,4\cdots \\ 0, & \text{if } n = 1,3,5\cdots \end{cases}$

     (c) $a_n = \begin{cases} 2, & \text{if } n = 0,1,2 \\ 0, & \text{if } n = 2,4,6\cdots \\ -2, & \text{if } n = 3,5,7\cdots \end{cases}$

     (d) $a_0 = a_1 = 1, a_n = 1/(n-2)!$ $(n = 2,3,4\cdots)$

     (e) $a_n = 3^{n-1}/(n+2)!$ $(n = 0,1,2,3,\cdots)$

7.   Determine the sequence corresponding to each of the following closed-form expressions for generating functions.

     (a) $x^3(x+5)^4$ (b) $(1-x)^3 - x^5$ (c) $1/(1+3x)$

     (d) $1/(1+x) - x/(1-2x)$ (e) $1/[(1+x)(1-2x)]$ (f) $e^{2x}(1-x^2)$

8.   Determine the sequence corresponding to each of the following closed-form expressions for generating functions.

     (a) $x^2(1-2x)^4$ (b) $(x+3)^4 + x^3$ (c) $x/(1+x^2)$

     (d) $1/(1+x^2) - x/(1-2x)$ (e) $x/[(1+x^2)(1-2x)]$ (f) $x^2(1+x)e^{-x}$

9. Determine the coefficient of x^8 in each of the following expansions.

   (a) $(1+x+x^2+x^3)(1+x^2+x^4+x^6)(1+x^3+x^6+x^9)$

   (b) $(1+x+x^2+x^3)^3$

   (c) $(1+2x+3x^2+4x^3+\cdots)(1-x^2+x^4-x^6+\cdots)$

   (d) $(1+x+x^2+x^3+\cdots)+(x+x^2+x^3+x^4+\cdots)^2+(x^2+x^3+x^4+\cdots)^3+\cdots$

10. Determine the coefficient of x^9 in each of the expansions of Exercise 9.

11. Suppose that the generating function of a certain sequence $\{a_n\}_{n=0}^{\infty}$ has a closed-form expression $F(x)$.

    (a) What is the sequence that has $x^3 F(x)$ as its generating function?

    (b) What is the sequence that has $(1-x)F(x)$ as its generating function?

    (c) What is the sequence that has $F(x)/(1-x)$ as its generating function?

12. Suppose that the generating function of a certain sequence $\{a_n\}_{n=0}^{\infty}$ has a closed-form expression $F(x)$.

    (a) What is the sequence that has $2x^2 F(x)$ as its generating function?

    (b) What is the sequence that has $(1+x)F(x)$ as its generating function?

    (c) What is the sequence that has $F(x)/(2+x^2)$ as its generating function?

13. Evaluate each of the following generalized binomial coefficients:

    (a) $\dbinom{-6}{3}$ (b) $\dbinom{3.5}{5}$

14. Evaluate each of the following generalized binomial coefficients:

    (a) $\dbinom{-1/2}{4}$ (b) $\dbinom{0.9}{5}$

15. (a) Determine the power series expansion for the following generating function: $1/\sqrt{x+1}$.

    (b) For each of the following closed-form generating functions, determine the first four terms of the corresponding power series expansion:

    (i) $(1+x)^{3.5}$ (ii) $e^x/\sqrt{1+x}$

16. (a) Determine the power series expansion for the following generating function: $\sqrt{1+x}$.

    (b) For each of the following two closed-form generating functions, determine the first four terms of the corresponding power series expansion:

    (i) $(3-5x)^{2.5}$ (ii) $e^{2x}\sqrt{1+x}$

17. Use the generating function method to find explicit formulas for each of the following recursively defined sequences:

    (a) $\begin{cases} a_0 = 3 \\ a_n = 2a_{n-1}+5\ (n \geq 1) \end{cases}$ (b) $\begin{cases} a_2 = 1 \\ a_n = 3a_{n-1}-1\ (n \geq 3) \end{cases}$ (c) $\begin{cases} a_0 = 1 \\ a_n = 2a_{n-1}+3n\ (n \geq 1) \end{cases}$

    (d) $\begin{cases} a_0 = 1,\ a_1 = 1, \\ a_n = 2a_{n-2}+5\ (n \geq 2) \end{cases}$ (e) $\begin{cases} a_0 = 1,\ a_1 = 2, \\ a_n = 2a_{n-2}+a_{n-1}\ (n \geq 2) \end{cases}$

18. Use the generating function method to find explicit formulas for each of the following recursively defined sequences:

    (a) $\begin{cases} a_0 = 1 \\ a_n = 4a_{n-1} - 1 \ (n \geq 1) \end{cases}$
    (b) $\begin{cases} a_{10} = 4 \\ a_n = 2a_{n-1} + 2 \ (n \geq 11) \end{cases}$
    (c) $\begin{cases} a_0 = 3 \\ a_n = 3a_{n-1} + 2n \ (n \geq 1) \end{cases}$

    (d) $\begin{cases} a_0 = 1, \ a_1 = 1, \\ a_n = 3a_{n-2} + 2 \ (n \geq 2) \end{cases}$
    (e) $\begin{cases} a_0 = 1, \ a_1 = 1, \\ a_n = 2a_{n-2} + 3a_{n-1} - 2 \ (n \geq 2) \end{cases}$

19. Use the generating function method to find explicit formulas for each of the following recursively defined sequences:

    (a) $\begin{cases} a_0 = 1 \\ 1 = a_n + 2a_{n-1} + 3a_{n-2} + \cdots + na_1 + (n+1)a_0 \ (n \geq 1) \end{cases}$

    (b) $\begin{cases} a_0 = 2 \\ n = a_n + 2a_{n-1} + 3a_{n-2} + \cdots + na_1 + (n+1)a_0 \ (n \geq 1) \end{cases}$

20. Use the generating function method to find explicit formulas for each of the following recursively defined sequences:

    (a) $\begin{cases} a_0 = 4 \\ 1 = a_n + a_{n-2} + a_{n-4} + \cdots \ (n \geq 1) \end{cases}$
    (b) $\begin{cases} a_0 = 4 \\ n+1 = a_n + a_{n-2} + a_{n-4} + \cdots \ (n \geq 1) \end{cases}$

For each counting problem in Exercises 21–26 below, do the following: (i) Write down a generating function for the problem, and indicate which coefficient will be the answer to the problem. (ii) Determine this coefficient, and hence the answer to the problem.
NOTE: As demonstrated in the section proper, generating functions for counting problems need not be unique.

21. (a) In how many ways can seven bills among $1, $5, or $10 bills be distributed to Jimmy?
    (b) How many combinations of rainy and sunny days can there be over 1 week (ignore the order of the days)?
    (c) In how many combinations can 10 drinks be ordered from the choices of beers, glasses of wine, or martinis?

22. (a) In how many ways can one choose six coins from a tin containing four pennies and six dimes?
    (b) How many combinations of eight stamps can be formed using 1¢, 3¢, or 5¢ stamps?
    (c) How many ways can one place 10 toppings on a pizza from among the following topping choices: pepperoni, artichoke, chicken, onions, cheese, and bell peppers (so multiple toppings must be selected, e.g., 4 artichoke, 4 chicken, and 2 bell pepper toppings; ignore the order of the toppings)?

23. (a) In how many ways can seven bills among $1, $5, and $10 bills be distributed to Jimmy if he must get at least one $10 bill, and an odd number of $5 bills?
    (b) How many combinations of rainy and sunny days can there be over 1 week (ignore the order of the days) if there are an odd number of rainy days and at most 5 sunny days?
    (c) In how many combinations can 10 drinks be ordered from the choices of beers, glasses of wine, or martinis, if there must be at least 2 martinis and there cannot be only one glass of wine?

24. (a) How many ways can one choose six coins from a tin containing four pennies and six dimes if an odd number of pennies were selected, and at least two dimes were selected?
    (b) How many combinations of eight stamps can be formed using 1¢, 3¢, or 5¢ stamps, if at least two 5¢ stamps are used, and at most five 1¢ stamps are used?
    (c) How many ways can one place 10 toppings on a pizza among the following topping choices: pepperoni, artichoke, chicken, onions, cheese, and bell peppers (so multiple toppings must be selected, e.g., 4 artichoke, 4 chicken, and 2 bell pepper toppings) if at least one topping must be artichoke, and at most five toppings are cheese?

25. In how many ways can 6 pieces of fruit be chosen from a basket that contains five (identical) apples, six oranges, two pineapples, and three bananas if at most one pineapple can be chosen, and if one banana is chosen then all must be chosen?

26. (a) In how many ways can a $100 bill be exchanged into smaller bills using any of the following denominations: $1, $5, $10, $20, $50?
    (b) How would the answer in Part (a) change if we could also use $2 bills?
    (c) In how many ways can a $1 bill be changed using coins of the following values: 50¢, 25¢, 10¢, 5¢, and/or 1¢?

NOTE: (*Partitions of Integers*)   A **partition** of a positive integer n is an (unordered) list of positive integers whose sum is n. The number of partitions of n is called the **partition function** and is denoted as $p(n)$. For example, the partitions of 4 are easily seen to be:

$$4, 3 + 1, \ 2 + 2, 2 + 1 + 1, \ \text{and } 1 + 1 + 1 + 1$$

so that $p(4) = 5$.

The **parts** of a partition are simply the integers that appear in it.
For each positive integer m, we define the related functions:

$\qquad p_m(n) \ = \ $ the number of partitions of n each of whose parts is at most m,

$\qquad q_m(n) \ = \ $ the number of partitions of n that consist of at most m parts.

For example, $p_2(4) = 3$ since three of the above listed partitions of four have all parts being at most two; also $q_2(4) = 3$ since exactly three of the above listed partitions of four have at most two Parts (4, 3 + 1, and 2 + 2). Since any part of a partition of n can be at most n and since there can be at most n parts, it follows that $p_m(n) = p(n) = q_m(n)$ whenever $m \ge n$. Although it is not immediately obvious, it turns out that $p_m(n) = q_m(n)$, as will be shown in Exercise 32.

Partitions are important in number theory and certain combinatorial problems. Although there is no efficient algorithm for computing the partition function, generating functions can be used in the study of partitions. This will be the subject of the Exercises 27–32.

27. (a) Show that a generating function for the sequence $\{p_m(n)\}_{n=1}^{\infty}$ (the number of partitions of n

    whose parts are each at most m) is given by $G_m(x) = \dfrac{1}{(1 - x)(1 - x^2)(1 - x^3)\cdots(1 - x^m)}$, and thus,

    when $m \ge n$ this also serves as a generating function of the partition function $p(n)$. Use this generating function to compute $p(n)$, for $n = 4, 5, 6$, and 8.

    (b)   Determine the value of m and the coefficient of $G_m(x)$ that will give the answer to the following counting problem, and then obtain the answer: *In how many ways can a postage of 15¢ be made using stamps of values 1¢, 2¢, 3¢, 4¢, and/or 5¢?*
    NOTE: According to the fact mentioned in the preceding note, the answer in Part (b) will also be the answer to the following problem: *In how many ways can a 15¢ postage be made using at most five stamps of values 1¢, 2¢, 3¢, 4¢, ...,14¢, or 15¢?*
    Suggestion: For both Parts (a) and (b), use the expansions $1/(1 - x^k) = 1 + x^k + x^{2k} + x^{3k} + \cdots$ (which follow from (8), for any positive integer k).

28. (a) Show that the function $G_m(x) = \dfrac{1}{(1 - x)(1 - x^2)(1 - x^3)\cdots}$ is a generating function of the

    partition function $p(n)$.

    (b)   Let $p_O(n)$ denote the number of partitions of n into odd integer Parts (i.e, as a sum of odd integers). Since $1 + 1 + 1 + 1$ and $3 + 1$ are the only such partitions of 4, we have $p_O(4) = 2$.

Find a generating function for $p_O(n)$, and then use it to compute $p_O(n)$ for $n = 4, 5, 6, 7$, and 10.

NOTE: Although the generating function defined in Part (a) involves an infinite product, each resulting term in the expansion (with the exception of 1, which is the product of an infinite number of 1's) involves only a finite product and there are only finitely many terms associated with each power of x.

29. (a) Let $p_D(n)$ denote the number of partitions of n into distinct Parts (i.e, no two parts of the partition are the same number). Since 4 and 3 + 1 are the only such partitions of 4, we have $p_D(4) = 2$. Find a generating function for $p_D(n)$.

    (b) Use your generating function of Part (a) to compute $p_D(n)$ for $n = 4, 5, 6, 7$, and 10.

30. (*Euler's Theorem*) Show that the function $p_O(n)$ of Exercise 28 is the same as the function $p_D(n)$ of Exercise 29 by showing they have the same generating function.

31. Use generating functions to show that every positive integer can be uniquely expressed as a sum of distinct powers of 2.

32. (*Star Diagram and a Proof that $p_m(n) = q_m(n)$*) Fill in the details of the following outline of a proof of the fact, mentioned above, that the number of partitions of a positive integer n into at most m parts is the same as the number of partitions of n into parts of size at most m, i.e., that $p_m(n) = q_m(n)$: Any partition of n can be represented by its *star diagram* that consists of n stars (or asterisks) grouped in rows corresponding to the parts of the partitions. For example, the partition $5 + 4 + 2$ of 11 has the following star diagram:

$$* \quad * \quad * \quad * \quad *$$
$$* \quad * \quad * \quad *$$
$$* \quad *$$

Each star diagram has a conjugate star diagram obtained viewing the columns as the parts rather than the rows (i.e., transposing rows and columns), and this gives rise to a *conjugate partition*. The conjugate partition of the preceding is $11 = 3 + 3 + 2 + 2 + 1$, which has the following star diagram:

$$* \quad * \quad *$$
$$* \quad * \quad *$$
$$* \quad *$$
$$* \quad *$$
$$*$$

Show that the conjugate operation bijectively maps all partitions of n into m parts into all partitions of n into parts of size at most m, thus proving that $p_m(n) = q_m(n)$.

NOTE: (*Sicherman Dice*) The next two exercises introduce an interesting problem relating to dice, along with a solution to this problem using generating functions. The problem asks whether it is possible to repaint the six faces of a pair of dice with positive integers (with duplications being allowed on the same die) in such a way that is different from the standard die (i.e., the six faces contain the numbers 1 through 6) and that when these modified dice are rolled together, the number of ways that the sum of the two numbers appearing as a given value (between 2 and 12) will be the same as for a pair of standard dice. Such a pair of dice were discovered by Colonel George Sicherman (a computer programmer) and first appeared in the literature in a 1978 *Scientific American* article by Martin Gardner. The next two exercises will use generating functions to find Sicherman's dice, and show that they are unique.

33. (*Rolling Generalized Dice*)
    (a) Explain why the function $D(x) = (x + x^2 + x^3 + x^4 + x^5 + x^6)(x + x^2 + x^3 + x^4 + x^5 + x^6)$ serves as a generating function for the problem of counting the number ways that a particular

outcome occurs when two dice are tossed and the outcome is viewed as the sum of the two numbers on the top faces (i.e., an integer between 2 and 12 inclusive). Compute the coefficient of x^4 of this generating function, and show that it is the same as the number of ways that a 4 (total) can be obtained by rolling two dice.

(b)   Next, suppose that the six faces of a pair of dice are repainted with numbers that are (not necessarily distinct) positive integers:

Die 1: $a_1 \le a_2 \le a_3 \le a_4 \le a_5 \le a_6$ $\qquad$ Die 2: $b_1 \le b_2 \le b_3 \le b_4 \le b_5 \le b_6$

Show that a generating function for the problem of counting the number of ways that these two dice can be tossed and the sum of the numbers appearing on the top faces adding up to a certain number (i.e., an integer between $a_1 + b_1$ and $a_6 + b_6$ inclusive) is given by:

$$(x^{a_1} + x^{a_2} + x^{a_3} + x^{a_4} + x^{a_5} + x^{a_6})(x^{b_1} + x^{b_2} + x^{b_3} + x^{b_4} + x^{b_5} + x^{b_6})$$

Check this result by expanding the function in the very simple case in which all numbers on Die 1 are 1s and all numbers on Die 2 are 2s.

34.   (*Sicherman Dice*) (a) Use the generating functions established in the preceding exercise to show that there is only one pair of Sicherman dice, namely those dice with the following face values: 1, 2, 2, 3, 3, 4, and 1, 3, 4, 5, 6, 8.

(b)   Verify that the 36 possible outcomes obtained by adding the numbers shown of the Sicherman dice given in Part (a) really do amount to the same number of outcomes for the numbers 2–12 as in the case for ordinary dice.

Suggestion: Use the fact that the generating function for a standard pair of dice (as given in Part (a) of the preceding exercise) factors as follows:

$$(x + x^2 + x^3 + x^4 + x^5 + x^6)^2 = x^2(1+x)^2(1+x+x^2)^2(1-x+x^2)^2.$$

We will also need the fact that polynomials with integer coefficients obey a unique factorization property similar to that of the integers.[9]   By equating this generating function to the generating function for a pair of Sicherman dice (from Part (b) of the previous exercise):

$$(x^{a_1} + x^{a_2} + x^{a_3} + x^{a_4} + x^{a_5} + x^{a_6})(x^{b_1} + x^{b_2} + x^{b_3} + x^{b_4} + x^{b_5} + x^{b_6})$$
$$= x^2(1+x)^2(1+x+x^2)^2(1-x+x^2)^2,$$

it follows (from unique factorization) that the individual generating functions for each of the two dice must satisfy:

$$x^{a_1} + x^{a_2} + x^{a_3} + x^{a_4} + x^{a_5} + x^{a_6} = x^{\alpha_1}(1+x)^{\alpha_2}(1+x+x^2)^{\alpha_3}(1-x+x^2)^{\alpha_4},$$

and

$$x^{b_1} + x^{b_2} + x^{b_3} + x^{b_4} + x^{b_5} + x^{b_6} = x^{\beta_1}(1+x)^{\beta_2}(1+x+x^2)^{\beta_3}(1-x+x^2)^{\beta_4},$$

where α_i, β_i are nonnegative integers that satisfy $\alpha_i + \beta_i = 2$, for $i = 1, 2, 3, 4$. Substitute $x = 0$ into both of the above equations to conclude that $\alpha_1 = 1 = \beta_1$. Substitute $x = 1$ into both of the above equations to conclude that $\alpha_2 = \alpha_3 = 1 = \beta_2 = \beta_3$. Finally, consider the generating function for the pair and show there are now three feasible solutions: (i) $\alpha_4 = 1 = \beta_4$, (ii) $\alpha_4 = 0, \beta_4 = 2$, or (iii) $\alpha_4 = 2, \beta_4 = 0$.

35.   (a) Establish the expansion

$$(x + x^2 + \cdots + x^9)(1 + x + x^2 + \cdots + x^9)^5 = x(1 - x^9)(1 - x^{10})^5(1 - x)^{-6}$$

and then show this function is a generating function for the problem of counting the number of positive six digit integers the sum of whose digits is n.

(b) How many positive six digit numbers have a digit sum of 22?

---

[9] The interested reader can find details on this topic in any good book on abstract algebra, such as [Hun-96]. The analog of the prime numbers in this polynomial factorization theory are so-called (integer coefficient) *irreducible* polynomials: they cannot be further factored into integer coefficient polynomials of smaller degree. Another relevant topic in abstract algebra is the existence of algorithms for performing such integer factorizations of polynomials. Such algorithms are built in to most symbolic/computer algebra systems.

(c) Find a generating function for the problem of counting the number of k-digit positive integers whose digits sum to n.

36.   (a) Find a generating function for the problem of counting the number of positive integers between 1 and $10^k - 1$ (inclusive) whose digits sum to n.

(b) How many positive integers in the range 1 to 999,999 have a digit sum of 22?
Suggestion: See Exercise 35(a).

APPENDIX TO SECTION 5.3: APPLICATION TO WEIGHTED DEMOCRACIES

We close this section with an application to weighted voting systems. It will provide an excellent illustration of how generating functions can sometimes be used to produce algorithms that are significantly more efficient than other methods for solving combinatorial problems.   The implementation of generating functions for the problem that we discuss is more sophisticated and less transparent than for the other counting problems we have discussed so far; the development is due to J. M. Bilbao, et. al [BFLL-00]. A different and more general method was subsequently discovered by V. Yakuba [Yak-08].

In many democratic voting systems, it is often equitable for different voters to have different powers of vote. For example, at a stockholder meeting, people who own greater numbers of shares are allowed proportionately stronger votes. Similarly, in the European Union, larger nations (like France or the UK) have five times as much voting power as some of the smaller nations (like Luxembourg). Let us first define the voting systems that we will be considering:

DEFINITION 5.7: A **weighted voting system** consists of the following:
A set of N **voters**: $V_1, V_2, \cdots, V_N$, a corresponding set of N **weights**: $w_1, w_2, \cdots, w_N$, which are assumed to be positive integers giving the number of votes controlled by the corresponding voters, and a **quota** q, which is a positive integer satisfying $\sum_{n=1}^{N} w_i \geq q > (1/2)\sum_{n=1}^{N} w_i$ equaling the minimum number of votes needed to pass a motion that is being voted on. A weighted voting system with these parameters will be denoted by $[q : w_1, w_2, \cdots, w_N]$.

EXAMPLE 5.28: Here are a few simple examples of weighted voting systems:
(a) [7: 5, 2, 1]. In this system, a motion wins if, and only if, V_1 and V_2 vote for it. V_3's vote is thus irrelevant; such a voter is called a **dummy voter**.
(b) [9: 5, 5, 4, 2, 1]. In this system, in order for a motion to pass, either V_1 or V_2 must vote for it.
(c) [10: 8, 3, 2, 2, 1]. In this system, voter V_1 has **veto power** since in order for any motion to pass it must have V_1's vote.

The above example makes it clear that some serious inequities might arise in weighted voting systems. The examples were small enough so that this was evident, but when the number of voters exceeds even a moderate size such as 30, the analysis of weighted voting systems can become extremely complex. Different notions of assigning a certain percentage of the "power" to the voters have been developed. One of the more widely accepted methods was developed by John Banzhaf III,[10] and is described in the following definition.

---

[10] John Banzhaf III (1940–) is a law professor at George Washington University. The "Banzhaf index" was actually invented by Lionel Penrose in a 1946 statistics paper, but went largely unnoticed by the scientific community.   Banzhaf rediscovered and popularized the index in a seminal 1965 paper entitled "Why weighted voting doesn't work," in which he demonstrated that a certain voting system in Nassau County, NY gave "power" to only three out of the six voting districts.

DEFINITION 5.8: (*Banzhaf Power Index*) Suppose that we are given a weighted voting system $[q : w_1, w_2, \cdots, w_N]$. A **coalition** is a nonempty set S of voters (who may vote together) and its **weight** $\text{wgt}(S)$ is simply the sum of the weights of its individual voters. A coalition is a **winning coalition** if its weights add up to at least q, otherwise it is a **losing coalition**. A voter in a winning coalition is **critical** (for that coalition) if without him/her the coalition would become a losing one. For each voter V_n, we define $c(V_n)$ to be the total number of winning coalitions in which V_n is critical, and we let $T = \sum_{n=1}^{N} c(V_n)$. The **Banzhaf index** of a voter V_n is defined to be $c(V_n)/T$.

EXAMPLE 5.29: Compute the Banzhaf indices of each voter in the weighted voting system [7: 5, 2, 1] of Example 5.28(a).

SOLUTION: We analyze all coalitions for the given weighted voting system:

Coalition	Weight	Winning?	Critical Voters
$\{V_1\}$	5	No	N/A
$\{V_2\}$	2	No	N/A
$\{V_3\}$	1	No	N/A
$\{V_1, V_2\}$	7	Yes	V_1, V_2
$\{V_1, V_3\}$	6	No	N/A
$\{V_2, V_3\}$	3	No	N/A
$\{V_1, V_2, V_3\}$	8	Yes	V_1, V_2

Thus we have $c(V_1) = c(V_2) = 1$, $c(V_3) = 0$, $T = 2$, so the Banzhaf indices of V_1, V_2 are both 1/2, and that of V_3 is zero. This can be interpreted by saying that for all practical purposes, V_1, V_2 share an equal amount of power in this weighted voting system, despite the fact that V_1 controls 150% more votes than V_2.

EXERCISE FOR THE READER 5.25: Compute the Banzhaf indices of each voter in the weighted voting system [9: 5, 5, 4, 2, 1] of Example 5.28(b).

The following exercise for the reader asks to compute the Banzhaf indices for the historically significant setting of the 1964 Nassau County voting system that motivated Banzhaf to develop his theory.

EXERCISE FOR THE READER 5.26: (a) Compute the Banzhaf indices of each voter in the Nassau County NY Board of Supervisors weighted voting system [59: 31, 31, 28, 21, 2, 2], then (b) identify any voters who are *dictators* (any voter whose weight is greater to or equal to the quota), dummies, or have veto power.

The brute-force approach used in Example 5.29 quickly becomes impractical since it generally requires checking through $2^N - 1$ coalitions. For example, to compute the Banzhaf indices for the 50 states and the District of Columbia with respect to the electoral votes accorded to each state (which are computed using the latest census figures), this approach would require looking at over 2 quadrillion coalitions, which, at the time of the writing of this book, would be intractable to perform in a reasonable amount of time even on a supercomputer.[11]

Generating functions can be used to render a much more efficient algorithm for computing Banzhaf indices. The method is a bit more involved than the previous generating function counting schemes that we have introduced; the details of its development will be the subject of the following example:

EXAMPLE 5.30: Develop a method involving generating functions to compute the Banzhaf indices of each voter in a weighted voting system $[q : w_1, w_2, \cdots, w_N]$.

SOLUTION: In case any of the weights w_n is at least equal to the quota q, then the corresponding voter is a dictator since his/her vote alone decides the overall decision. Since such weighed voting systems are not very interesting, we henceforth make the following:

Assumption: $w_n < q$ for each n $(1 \le n \le N)$.

We begin with the following generating function:

$$F(x) = (1 + x^{w_1})(1 + x^{w_2}) \cdots (1 + x^{w_N}).$$ (13)

We observe from (13) that since the term of highest degree in the expanded form of $F(x)$ is $x^{w_1 + w_2 + \cdots + w_N}$, it follows that the degree of $F(x)$ is $W \triangleq \sum_{n=1}^{N} w_i$. (The total weight of all votes.) We introduce notation for the coefficients of the expanded form of the generating function by means of the following equation:

$$F(x) = a_W x^W + a_{W-1} x^{W-1} + \cdots + a_2 x^2 + a_1 x + a_0.$$ (14)

We have already observed that $a_W = 1$; similarly (13) shows that $a_0 = 1$. The following observation is seen by comparing (13) and (14) and interprets the coefficients of the generating function in terms of the voting system:

Key Observation: a_i = the number of coalitions of weight i.

We first will develop a recursive formula for these important generating function coefficients. The idea will be to build up the generating function $F(x)$ through a sequence of N factor multiplications. More precisely, we define the following sequence of generating functions:

$$F_0(x) = 1$$
$$F_1(x) = (1 + x^{w_1})$$
$$F_2(x) = (1 + x^{w_1})(1 + x^{w_2})$$
$$F_3(x) = (1 + x^{w_1})(1 + x^{w_2})(1 + x^{w_3})$$
$$\vdots$$
$$F_N(x) = (1 + x^{w_1})(1 + x^{w_2}) \cdots (1 + x^{w_N}) = F(x).$$

---

[11] On his Web site, Banzhaf has a link to some data sets including computations of the 51 Banzhaf indices for this electoral college system. In contrast to the efficient exact method that we will introduce, Banzhaf employed an approximation method that used simulations to randomly generate 4.29 billion coalitions, and it took nearly 25 hours on his computer. Simulations will be explained in Section 6.2.

We note that

$$F_j(x) = F_{j-1}(x)(1 + x^{w_j}), \text{ for } 1 \le j \le N. \tag{15}$$

For each index j, $0 \le j \le N$, we define integers $a_i^{(j)}, (0 \le i \le W)$ to be the coefficients of the expansion of $F_j(x)$, and thus we may write:

$$F_j(x) = a_W^{(j)} x^W + a_{W-1}^{(j)} x^{W-1} + \cdots + a_2^{(j)} x^2 + a_1^{(j)} x + a_0^{(j)}. \tag{16}$$

If we define $a_i^{(j)} = 0$ if i is a negative integer, it follows by comparing coefficients of x^i in both sides of (15) that the following recursion formula is valid:

$$a_i^{(j)} = a_i^{(j-1)} + a_{i-w_j}^{(j-1)} \quad (0 \le i \le W, 1 \le j \le N). \tag{17}$$

We have thus developed an effective recursion scheme for computing the coefficients $a_i^{(j)}$ (and hence also the coefficients $a_i = a_i^{(N)}$), but how is this going to help us to compute the desired Banzhaf indices? We next show how these coefficients can help us to compute another set of coefficients from which we <u>will</u> be able to directly obtain the Banzhaf indices; this latter process will be the novelty of the method. To this end, we temporarily focus attention on a certain voter $V_n (1 \le n \le N)$. It will be convenient to introduce the following notation:

NOTATION: We let σ_n be the number of **swings** for voter V_n; in other words, this is the number of losing coalitions S such that $V_n \notin S$, that would become winning coalitions if V_n were to join.

Since $\text{wgt}(V_n) = w_n$, it follows that[12]

$$\sigma_n = |\{S \subseteq \{V_1, V_2, \cdots, V_N\} : V_n \notin S \text{ and } q - w_n \le \text{wgt}(S) < q\}|. \tag{18}$$

Notice that $\sigma_n = c(V_n)$. Each of these coefficients is the sum of the following coefficients:

$$b_i = |\{S \subseteq \{V_1, V_2, \cdots, V_N\} : V_n \notin S \text{ wgt}(S) = i\}|, \tag{19}$$

that is

$$\sigma_n = \sum_{i=q-w_n}^{q-1} b_i. \tag{20}$$

---

[12] Although, by definition, coalitions are required to be nonempty, the condition $S \ne \varnothing$ would be a redundancy in (18) since it is already required that $q - w_n < \text{wgt}(S)$ and we are assuming throughout the development that $w_n < q$ (thus the latter condition implies that $\text{wgt}(S) > 0$, so S cannot be empty).

From the σ_n 's the Banzhaf indices are easily computed:

$$\text{Banzhaf index of voter } V_n = \frac{\sigma_n}{\sigma_1 + \sigma_2 + \cdots + \sigma_N}. \tag{21}$$

We could compute the b_i 's in the same fashion that was shown for the a_i 's (using a corresponding generating function with one less weight), but there is a more efficient scheme.   If we define $W_n = W - w_n$ (i.e., the total voting weight less the weight of voter V_n), then reordering and taking V_n last, it follows that we may write:

$$
\begin{aligned}
F(x) &= a_W x^W + a_{W-1} x^{W-1} + \cdots + a_2 x^2 + a_1 x + a_0 \\
&= (b_{W_n} x^{W_n} + b_{W_n -1} x^{W_n -1} + \cdots + b_2 x^2 + b_1 x + b_0)(1 + x^{w_n})
\end{aligned} \tag{22}
$$

(This really just follows from the recursion formula (15), if we reorder the voters so that V_n is taken last.)   If, as was already done for the a_i 's, we define $b_i = 0$ whenever i is a negative integer, then by comparing coefficients of x^i in both sides of (22) we obtain the following recursion formula:

$$b_i = a_i - b_{i-w_n} \ (i = 0,1,2,...). \tag{23}$$

By appropriately combining the preceding recursion formulas, we arrive at the following algorithm for the computation of Banzhaf indices.

ALGORITHM   5.1:   (*Generating   Function   Based   Recursion   Algorithm   for Computing Banzhaf Indices in a Weighted Voting System*)
Input: A weighted voting system $[q : w_1, w_2, \cdots, w_N]$, where the weight w_n of voter V_n is less than the quota q (so there are no dictators).
Output: The corresponding Banzhaf indices of the N voters.

Step 1: (*Initialize known coefficients*)
Set $a_0 = a_W = 1$.   Set $a_0^{(0)} = 1$ and $a_i^{(0)} = 0$, for each $i \neq 0$.
Set $a_i^{(j)} = b_i = 0$, whenever index i is negative.

Step 2: (*Compute the $a_i^{(j)}$'s*)
FOR index $j = 1$ TO N
        FOR index $i = 0$ TO $i = W - 1$
                Set $a_i^{(j)} = a_i^{(j-1)} + a_{i-w_j}^{(j-1)}$ (using (17))
        END i FOR
END j FOR

Step 3: (*Record the a_i 's*)
FOR index $i = 1$ TO $i = W - 1$
        Set $a_i = a_i^{(N)}$
END i FOR

Step 4: (*Compute the* σ_n *'s*)

FOR index $n = 1$ TO N

        (*First find the needed* b_i *'s corresponding to voter* V_n)

        FOR index $i = 0$ TO $q - 1$

            Set $b_i = a_i - b_{i - w_n}$

        END i FOR

        (*Record* σ_n)

        $\sigma_n = \sum_{i = q - w_n}^{q-1} b_i$

END n FOR

Step 5: (*Compute the Banzhaf indices*)

Set $T = \sum_{n=1}^{N} \sigma_n$

FOR index $n = 1$ TO N

        Banzhaf index of $V_n = \sigma_n / T$

END n FOR

Of course, for small sized voting systems, the overhead of this algorithm would make it more cumbersome than the brute-force method. But notice that the computational steps 2, 4, and 5 respectively take at most NW, Nq, and $2N - 1$ mathematical operations (additions, subtractions, divisions), which total **less than** $3NW$ operations. Recall that N is the number of voters, and W is the total weight of the votes. Compare this with the amount of work needed in the brute-force approach that was used in Example 5.29. In general, all but the nonempty set of voters needs to be considered, and there are $2^N - 1$ of these. Also, for each coalition, a nontrivial amount of work needs to get done (check its weight and determine whether it is winning, and if it is, determine the critical voters). So this method requires **more than** 2^N mathematical operations. In the electoral college example that Banzhaf considers on his Web site, $N = 51$ (50 states and the District of Columbia), and $W = 538$ (total number of electoral votes). Thus, the brute-force approach would require (much) more than $2^{51} = 2.25... \times 10^{15}$ mathematical operations, whereas Algorithm 5.1 would require at most $3 \cdot 51 \cdot 538 = 82,314$ mathematical operations. Thus, although as Banzhaf had found, the brute-force approach is impossible to do in a reasonable amount of time (even with a computer), Algorithm 5.1 would be well suited for this example. The computer implementation material at the end of this chapter will consider some specific applications of Algorithm 5.1.

The following exercise for the reader asks to compute the Banzhaf indices both directly (by considering all coalitions) and by using the generating function method for the historically significant setting of the 1964 Nassau County voting system that motivated Banzhaf to develop his theory. The size of the example is small enough to do by hand, and for the speed of the generating function method to not yet be realized. Some larger examples that are feasible only with the latter method will be considered in the computer exercises and implementation material that follows this appendix.

EXERCISE FOR THE READER 5.27: Use Algorithm 5.1 to recompute the Banzhaf indices of each voter in the weighted voting system [7: 5, 2, 1] of Example 5.28(a).

In analyzing voting systems, particularly with generating function methods, it often occurs that the weights of the votes in the system can be reduced while preserving the essential features of the system. As a very simple example, everyone should immediately agree that the voting system [4: 2, 2, 2, 2] is equivalent to [2: 1, 1, 1, 1], in that all possible voting scenarios in either system would have the same result in either system. The following definition generalizes this concept.

DEFINITION 5.9: (*Equivalent Voting Systems*) Two weighted voting systems $[q : w_1, w_2, \cdots, w_N]$ and $[q' : w_1', w_2', \cdots, w_{N'}']$ are said to be equivalent if the number of voters is the same, and any coalition of voters in the first is winning if, and only if, it is winning in the second.

Since the running time of Algorithm 5.1 depends on the total weight W of all votes, when applying this algorithm, it is desirable to find a voting system equivalent to the one being analyzed and with as small a total weight as possible.

EXERCISE FOR THE READER 5.28: In 1994, the Nassau County NY Board of Supervisors modified their weighted voting system to the following: [65: 30, 28, 22, 17, 7, 6].
(a) Show that [15:7, 6, 5, 4, 2, 1] is an equivalent voting system.
(b) Apply the generating function method to compute the Banzhaf power indices using the original system, and then the equivalent system of Part (a).
(c) Show that there does not exist an equivalent voting system of smaller weight than that given in Part (a). Thus, the system given in Part (a) is called a *minimum equivalent voting system* to the given system.

Although it is desirable to have equivalent voting systems of reduced total weight, there does not seem to exist an efficient algorithm for their determination.

ADDITIONAL EXERCISES FOR THE APPENDIX TO SECTION 5.3:

1. For each voting system given, do the following. Compute all Banzhaf power indices by: (i) Analyzing each coalition for its critical voters, as in the solution of Example 5.29. (ii) Using generating functions. (iii) Then identify any voters who are dictators, dummies, or have veto power.
   (a) [4: 2, 1, 1, 1, 1]   (b) [12: 8, 5, 5, 3, 2]

2. For each voting system given, do the following. Compute all Banzhaf power indices by: (i) Analyzing each coalition for its critical voters, as in the solution of Example 5.29. (ii) Using generating functions. (iii) Then identify any voters who are dictators, dummies, or have veto power.
   (a) [6: 4, 2, 2, 1, 1]   (b) [22: 14, 12, 10, 8, 5]

3. For each statement below regarding voting systems, either explain why it is (always) true, or provide an example of a weighted voting system in which it is false.
   (a) If the Banzhaf power index of a voter is greater than 1/2, then the voter must be a dictator.
   (b) If there are N voters, then any voter with veto power must have Banzhaf power index at least $1/N$.

COMPUTER IMPLEMENTATIONS AND EXERCISES FOR SECTION 5.3

(*Polynomial Arithmetic on Computers*) Readers who are using computing platforms with so-called symbolic or computer algebra capabilities (such as MAPLE, Mathematica®, and MATLAB®) will be able to perform the polynomial multiplications and additions/subtractions needed to compute coefficients of generating functions. But it is not hard to use any computing system to perform this sort of arithmetic. We will show how to represent polynomials as vectors and some corresponding efficient schemes for performing arithmetic on them. In all of our applications, the polynomials will have integer coefficients, so this may be assumed in what follows.

It is often convenient to store a polynomial $f = a_n x^n + a_{n-1} x^{n-1} + \cdots a_1 x + a_0$ as the vector of its

coefficients: $f = \sum_{i=0}^{n} a_i x^i \sim [a_n, a_{n-1}, \cdots, a_1, a_0]$. This basic idea is so important that we repeat it with

emphasis:

Polynomial	Vector of Coefficients
$a_n x^n + a_{n-1} x^{n-1} + \cdots a_1 x + a_0$	$[a_n, a_{n-1}, \cdots, a_1, a_0]$

The addition and multiplication operations can be converted into corresponding operations on such vectors. Suppose that $g = \sum_{i=0}^{m} b_i x^i \sim [b_m, b_{m-1}, \cdots, b_1, b_0]$ is another polynomial. From the definition of addition of polynomials, we may write:

$$f + g \sim [c_N, c_{N-1}, \cdots, c_1, c_0], \quad \text{where} \quad N = \max(n, m) \text{ and } c_i = a_i + b_i, \tag{24}$$

for $1 \le i \le N$. (We adhere to the convention made in the section that unspecified coefficients are zero.)

To understand the vector version of polynomial multiplication, we first see how it will work if we multiply $f \ne 0$ by a **monomial,** which is a nonzero polynomial consisting of a single term: $b_k x^k$

$$f \cdot b_k x^k = (a_n x^n + a_{n-1} x^{n-1} + \cdots a_1 x + a_0) \cdot x^k = b_k [a_n x^{n+k} + a_{n-1} x^{n+k-1} + \cdots a_1 x^{1+k} + a_0 x^k]$$

In vector notation, this multiplication becomes:

$$[a_n, a_{n-1}, \cdots, a_1, a_0] \cdot [b_k, \underbrace{0, 0, \cdots, 0}_{k \text{ zeros}}] = b_k [a_n, a_{n-1}, \cdots, a_1, a_0, \underbrace{0, 0, \cdots, 0}_{k \text{ zeros}}]. \tag{25}$$

By (repeatedly) using the distributive law, a general polynomial multiplication can be broken down into a sum of multiplications of a polynomial by a monomial:

$$f \cdot g = f \cdot \sum_{i=0}^{m} b_i x^i = \sum_{i=0}^{m} f \cdot b_i x^i \Rightarrow f \cdot g \sim \sum_{i=0}^{m} b_i [a_n, a_{n-1}, \cdots, a_1, a_0, \underbrace{0, 0, \cdots, 0}_{i \text{ zeros}}]. \tag{26}$$

Thus, with this method of storing polynomials along with the associated algorithms (24) and (26) for their addition and multiplication, we have an efficient means for manipulating polynomials on computing platforms. This will serve as a basis for the computer implementation such polynomial arithmetic and thus in their use in computing coefficients of generating functions.

1. (*Program for Polynomial Addition*) (a) Write a program with syntax:
$$\texttt{Sum = PolyAdd(px,qx)}$$
that will add two polynomials. The two inputs, `px` and `qx` are vectors representing the polynomials to be added. The output, `Sum`, is a vector representing the sum of the inputted polynomials. If the sum is the zero polynomial, the output should be [0]; otherwise, the output should have a nonzero first component (so that the degree of the sum is one less than the length of the output vector).
(b) Run your program on the following polynomial additions:
(i) $(x^5 + x + 1) + (x^8 + x^6 + 4x^2 + 2)$
(ii) $(x^3 + 2x^2 + 1) + (x^8 - x^7 + x^6 - x^5 + x^4 - x^3 + x^2 - x + 1)$

2. (*Program for Polynomial Multiplication*) (a) Write a program with syntax:
$$\texttt{Prod =PolyMult(px,qx)}$$
that will multiply two polynomials. The two inputs, `px` and `qx` are vectors representing the

polynomials to be multiplied. The output, `Prod`, is a vector representing the product of the inputted polynomials. If the product is the zero polynomial, the output should be [0]; otherwise, the output should have a nonzero first component (so that the degree of the sum is one less than the length of the output vector).

(b) Run your program on the following polynomial multiplications:

(i) $(x^5 + x + 1) \cdot (x^8 + x^6 + 4x^2 + 2)$

(ii) $(x^3 + 2x^2 + 1) \cdot (x^8 - x^7 + x^6 - x^5 + x^4 - x^3 + x^2 - x + 1)$

3. Making use of the programs of either of the preceding two computer exercises, compute the coefficient of x^{10} in each of the following generating functions:

(i) $(x^8 - x^7 + x^6 - x^5 + x^4 - x^3 + x^2 - x + 1)^4$

(ii) $1/[(1-x)(1-x^2)(1-x^3)]$

4. (a) Determine the number of ways that \$1000 could be distributed using bills of any or all of the following denominations: \$1, \$5, \$10, \$20, \$50, \$100.

(b) Suppose that we roll 10 regular dice, and we add up all of the numbers that show (so the number will lie between 6 and 60, inclusive). How many different ways could a total of 15 show up? How about a total of 30?

5. (*Program for Partition Function Based on Generating Functions*) (a) Write a program with syntax:

$$pn = \texttt{Partition(n)}$$

that will input, n, a positive integer, and will output, pn, the number of partitions of *n*, i.e., $p(n)$. Use the generating function of Part (a) of Ordinary Exercise 27 with $m = n$, the expansion (8) (multiple times with powers of *x* replacing *x*), and repeatedly use the program of Computer Exercise 2.

(b) Use your function from Part (a) to compute the terms of the sequence $p(5), p(10), p(15), p(20), \cdots$ until it takes a term more than 1 minute to execute.

(c) Create a more efficient program than that of Part (a):

$$pn = \texttt{PartitionVer2(n)}$$

by modifying the program of Computer Exercise 2 so that coefficients of powers higher than x^n are ignored.

(d) Repeat Part (b) using instead the program of Part (c).

6. (*Program for Partition Function Not Based on Generating Functions*) (a) Write a program with syntax:

$$pn = \texttt{PartitionBrute(n)}$$

having the same input/output as that of Part (a) of the preceding computer exercise, but in this exercise do not base the program on the generating function approach. Use either a brute-force approach, or whatever other method you can think of.

(b) Use your function of Part (a) to compute the terms of the sequence $p(5), p(10), p(15), p(20), \cdots$ until a term takes more than 1 minute to execute. Compare with the performance of the generating function based program.

7. (*Program for Odd Partition Function Based on Generating Functions*) (a) Write a program with syntax:

$$pOddn = \texttt{OddPartition(n)}$$

that will input, n, a positive integer, and will output, pOddn, the number of odd partitions of *n*, i.e., the function $p_O(n)$ introduced in ordinary Exercise 28(b). Use the generating function that was found in Exercise 28(b).

(b) Use your function of Part (a) to compute the terms of the sequence $p_O(5), p_O(10), p_O(15), p_O(20), \cdots$ until a term takes more than 1 minute to execute.

(c) Create a more efficient program than that of Part (a):

$$pOddn = OddPartitionVer2(n)$$

by modifying the program of Computer Exercise 2 so that coefficients of powers higher than x^n are ignored.

(d) Repeat Part (b) using instead the program of Part (c).

NOTE: The remaining computer exercises deal with the material from the appendix to Section 5.3.

8. (*Program for Computation of Banzhaf Power Indices: Brute-force Version*) (a) Write a program with syntax:

$$BIndVec = BanzhafBrute(q, WVec)$$

whose two input variables correspond to the parameters of a weighted voting system: q, a positive integer representing the quota, and WVec, a vector of positive integers representing the voter weights. The output, BIndVec, will be the corresponding vector of Banzhaf power indices. The program should follow the brute-force method that was used in Example 5.29, i.e., by considering all coalitions, identifying critical voters for each, and keeping separate tallies for each voter.

(b) Use your function from Part (a) to compute the Banzhaf power indices for the weighted voting system of Example 5.29 and for the Nassau County Board of Supervisors system of Exercise for the Reader 5.26.

(c) For each positive even integer n, we define a weighted voting system $\mathscr{V}_n$ defined by:

$$\mathscr{V}_n = [n(n-1)/2; n, n-2, n-4, \cdots, 2]$$

Repeatedly apply your program of Part (a) to the systems $\mathscr{V}_n$ with $n = 10, 12, 14, \ldots$ until the program takes longer than one minute to execute.

9. (*Program for Computation of Banzhaf Power Indices: Generating Function Version*) (a) Write a program with syntax:

$$BIndVec = BanzhafGF(q, WVec)$$

whose input/output variables are the same as in the preceding computer exercise, but that implements Algorithm 5.1.

(b) Use your function from Part (a) to compute the Banzhaf power indices for the weighted voting system of Example 5.29 and for the Nassau County Board of Supervisors system of Exercise for the Reader 5.26.

(c) For each positive even integer n, we define a weighted voting system $\mathscr{V}_n$ defined by:

$$\mathscr{V}_n = [n(n-1)/2; n, n-2, n-4, \cdots, 2]$$

Repeatedly apply your program of Part (a) to the systems $\mathscr{V}_n$ with $n = 10, 12, 14, \ldots$ until the program takes longer than one minute to execute. Compare with the results of Part (c) of the preceding computer exercise.

10. (*Computation of Banzhaf Power Indices in Some Well-Known Voting Systems*) In this exercise, you are to apply your program from Computer Exercise 9 to compute the Banzhaf power indices of the well known weighted voting systems that are described below. You will need to obtain or download the relevant data from the internet.

(a) The European Community weighted voting system that was established by the Treaty of Nice that went into effect in 2004.

(b) The 51 voters (50 states and the District of Columbia) of the United States Electoral College.

Chapter 6: Discrete Probability and Simulation

Probabilistic thinking has its origins in gambling, where there have always been very strong incentives for knowing one's odds of winning a bet. In the seventeenth century, French nobleman and author Antoine de Méré did a lot of gambling in the (still famous) Monte Carlo casinos on the French Riviera. Although he was winning a lot of money, he tried to develop mathematical proportions for the chance that certain bets would win. After several failed attempts, he became acquainted with two of the most prominent mathematicians of the era, Blaise Pascal (Figure 6.1) and Pierre de Fermat (Figure 6.2). In 1654, he asked them: "which is more likely, (a) to get at least one 1 with four throws of a die, or (b) to

FIGURE 6.1: Blaise Pascal[1] (1623– 1662), French mathematician and physicist.

FIGURE 6.2: Pierre de Fermat[2] (1601–1665), French mathematician.

---

[1] Pascal's mother died when he was only three years of age. He received his primary education at home from his father. Pascal's father decided not to let Blaise study mathematics until the age of 15. But Pascal's curiosity prevailed, and he began clandestinely to study geometry when he was 12. He was not able to contain his excitement when he let his father know about one of his new discoveries. His father then gave in, let him study mathematics, and introduced him to some famous mathematicians of the day. Pascal's scientific abilities soon began to truly flourish. At age 16, he proved some important theorems in projective geometry that were well received by the mathematics community, and he continued to make important contributions to mathematics throughout his rather short life, despite recurring health problems. Pascal also did impressive work in other fields. In his early 20's, he invented a mechanical calculator (one of the first) for his father, who worked as a tax collector. He wrote many significant papers in physics. Pascal was a very religious man, and his *Pensées*, a collection of philosophical and religious thoughts, was widely read by the general public.

[2] After completing (the equivalent of) a bachelor's degree in mathematics, Fermat earned a law degree and followed an impressive career path as a government lawyer and council member in the city of Toulouse. He maintained his ardent interest in mathematics throughout his life and was involved in cutting-edge mathematics among the leaders in the field. His published works are relatively small in number, however, but this was due to his preference for solving new problems rather than taking the time to formally write up his work on problems that he finished, the fact that he held another very demanding full-time job notwithstanding. Fermat often made bold scientific claims, such as finding faults or simpler approaches to the famous development of optics by René Descartes (the founder of analytic geometry). This sometimes put him at odds with his subjects, and caused him difficulties in getting his work accepted, but in most cases, Fermat's assertions later proved to be correct.

get at least one pair of 1s (*snake eyes*) with 24 throws of two dice?" The answer, as we will soon be able to verify (Exercise 13), is (a), and the problem is known as *de Méré's paradox*. The ensuing exchange of letters between the two mathematicians gave birth to the modern theory of probability.

Probability has come a long way since its inception over 350 years ago, with its applications extending from gambling to areas such as insurance, medical technology, and weather prediction, to name a few. Methods have been developed to solve or at least better understand a great many questions that have arisen. Still, there are many important problems that are extremely difficult or impossible to solve analytically.

Simulation is an ideal tool to help one get an estimate of a numerical answer to a problem, without having to perform complicated analytic derivations. The idea is simple: We set up an experiment that will mimic the problem we are trying to solve, and we let the computer run through the experiment, repeatedly. Rather than a *deterministic algorithm*, where a given input will always produce the same output, a simulation relies on the important concept of *randomness*, and is the prototype for more general *randomized (or stochastic) algorithms*. The increasingly higher speeds of computers continue to make simulation play an important role in scientific research. Probability can be used to model either discrete phenomena or continuous phenomena (e.g., time) but we will primarily focus on discrete probability. Section 6.1 will introduce the fundamental concepts and theory of discrete probability. Section 6.2 will discuss randomness, random variables, and simulations.

6.1: INTRODUCTION TO DISCRETE PROBABILITY

Probability is the branch of mathematics having to do with rigorously assigning percentages of likelihoods for certain events that are subject to chance. The following questions are typical of those for which probability can be used to answer: What are the chances that a poker hand is a full house? What is the probability that a meteor weighing over 5000 pounds will strike the continental United States in the next 10 years? What is the probability that a certain 25-year-old male motorist in Los Angeles will have an accident over the next 12 months; how about for a female motorist of the same age in Cedar City (Utah)? What is the average time a customer has to wait in line to check out at a grocery store if five cashiers are working?[3] We begin by introducing the general concepts that will allow us to rigorously formulate such probabilistic questions.

---

[3] This question will require some more information, along the lines of: at what rate do customers arrive?, How fast are the customers served by the cashiers?, etc.

Experiments, Sample Spaces, and Events

DEFINITION 6.1: An **experiment** is any process that has an **outcome**. Each repetition of an experiment is called a **trial**. The **sample space**, S, of an experiment is the set of all possible outcomes.

We stipulate that exactly one outcome in the sample space occurs each time an experiment is performed. The following example describes some basic experiments.

EXAMPLE 6.1: We describe four experiments along with the associated sample spaces.

(a) Flip a penny and record if it lands on heads or tails: $S = \{H, T\}$.

(b) Throw a die and record the number of pips showing: $S = \{1, 2, 3, 4, 5, 6\}$.

(c) Flip a nickel and a penny, and record which side each coin shows: $S = \{HH, HT, TH, TT\}$. (The first letter corresponds to the nickel.)

(d) Randomly select 40 students and ask them their birthdays. Find out if there are at least two individuals with a common birthday. $S = \{Yes, No\}$.

DEFINITION 6.2: An **event** E is a subset of the sample space S, i.e., $E \subseteq S$. If the outcome of an experiment belongs to E, we say that the **event E has occurred**.

For any experiment, one goal of probability theory is to create a *probability function* P that assigns to each event E, the likelihood P(E) that E will occur. These numbers P(E)–the *probabilities*, should have certain properties that we will discuss momentarily. In case of a finite sample space for which each outcome is equally likely (such as each of the experiments in Parts (a) through (c) of the preceding example), the following definition of probability should seem quite reasonable.

Experiments with Equally Likely Outcomes

DEFINITION 6.3:[4] (*Probabilities in case of a finite sample space with each outcome being equally likely*) If E is an event in an experiment with finitely many, equally likely outcomes, then

$$P(E) = \frac{|E|}{|S|}, \tag{1}$$

where (recall) $|E|$ denotes the number of elements in the set E.

---

[4] This definition is due to Pierre Simon Laplace (1749–1827), an illustrious French mathematician, who introduced it in his expansive two-volume treatise: *Théorie Analytique des Probabilités*, the first edition of which appeared in 1812. This influential book was initially dedicated to Napoleon I (Bonaparte), but the dedication was deleted from future editions after Napoleon's defeat by the British in 1815.

With this definition, the computation of probabilities is reduced to counting the number of elements in events. We usually describe a coin (or a die) as **fair** if it is equally likely to turn out any outcome when flipped (or tossed). Thus, for example, when we flip a fair coin, $P(H) = 1/2 = P(T)$. Coins or dice that are not fair are usually called *biased* or *loaded*. Unless indicated otherwise, we will assume that all coins, dice, or other gambling devices are fair.

EXAMPLE 6.2: Compute the probability $P(E)$ for each event E described below:
(a) In the experiment of flipping a penny and a nickel, E is the event "at least one head."
(b) In the experiment of throwing a pair of dice, E is the event "the pips add up to 7 or 11" (i.e., the event of winning a basic game of *craps*).
(c) In the experiment of being dealt a five-card poker hand from a shuffled standard deck of 52 playing cards, E is the event of "getting a pair."[5]

SOLUTION: Part (a): Using the notation of Part (a) of Example 6.1, the event E can be expressed as {HT, TH, HH}, and thus $P(E) = |E|/|S| = 3/4$.

Part (b): The 36 elements of the sample space S are shown in Figure 6.3, and those that correspond to the sum of the pips being 7 or 11 (i.e., those in the event E) are shaded. Note that the elements of the sample space have been displayed in a fashion that make it quite simple to locate the outcomes in E.   Thus, since E consists of eight outcomes, we   may   conclude   that $P(E) = |E|/|S| = 8/36 = 2/9 \approx 22.2\%$.

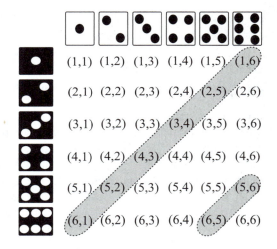

FIGURE 6.3: The sample space of the 36 outcomes in the experiment of rolling two dice. The shaded outcomes are those that will win a game of craps (seven or eleven).

Part (c): The sample space of this experiment is just the set of all possible five-card subsets of a deck of 52 cards (order does not matter in a poker hand), and so consists of $C(52,5) = 2,598,960$ elements.   The hand of "a pair" in poker has the form: $\{a, a, b, c, d\}$,

---

[5] The cards of a standard 52-card deck are evenly divided among the four *suits* (*clubs* and *spades*, which are black, and *diamonds* and *hearts* which are red), and each suit has 13 denominations: (A)ce = 1, 2, …, 9, 10, (J)ack, (K)ing, (Q)ueen. The last three cards are *face cards*. A (five-card) *poker hand* is usually described in the most complimentary terms among the following possibilities, listed in order from least valuable (most common) to most valuable (most rare): high card, pair, two (separate) pairs, three of a kind, straight (five cards in sequence, ace can go before 2 or after king), flush (five cards of the same suit), full house (three of a kind and a pair), four of a kind, straight flush (straight plus flush), royal flush (10, J, Q, K, A, all of same suit). Thus, an example of a pair would be {8, 8, 2, 4, K}.

where a, b, c, d represent cards of distinct denominations. Thus, the event E, consisting of all such "pair" hands, can be counted using the multiplication principle as follows: first select the denomination (a) of the pair, next select the pair, then select three different denominations (from the remaining 12), and finally, select a card from each. Thus,

$$|E| = 13 \cdot C(4,2) \cdot C(12,3) \cdot 4^3 = 1,098,240,$$

and we conclude that $P(E) = 1,098,240/2,598,960 \approx 0.4226$.

EXERCISE FOR THE READER 6.1: Compute the indicated probabilities:
(a) A group of three students is randomly selected from a class of 25 to go to a professional basketball game. What is the probability that Mike and Jimmy will both get to go?
(b) Compute the probability of being dealt a full house in a game of poker.
(c) If I flip a quarter six times, what is the probability that at least once the flip will turn up heads?

Kolmogorov's Axioms

The underlying theory of probability depends heavily on set theory, since, after all, events in probability are subsets of the sample space. During the early twentieth century, while modern set theory was developing in a form that would resolve many of the paradoxes that had previously arisen, there became a need to lay down a new foundation for probability. This was accomplished in 1933 by Russian mathematician Andrey Kolmogorov; see Figure 6.4.[6]

---

[6] Kolmogorov's parents were never married; his mother died during his birth, and his father, who was exiled, did not take any responsibility in his upbringing. He was raised by his mother's sister, and his family name is that of his maternal grandfather. Although he began his undergraduate studies at Moscow State University without any particular mathematical focus (he also studied Russian history and metallurgy), he was studying in Russia during a period when the Russian mathematics school was in its heyday. This is surprising given that Russia was in the Stalin era, an era marked by many years of oppression and rigid authoritarianism, particularly since the Russian revolution and lasting until Stalin's death in 1953. Kolmogorov's mathematical interests were so invigorated by the exciting lectures and research seminars of some of the most prominent mathematicians at the time (including Luzin, Egorov, Stepanov, and Aleksandrov), that he published eight extremely significant research papers while an undergraduate. We point out that it has always been exceptionally rare for undergraduates to publish (significant) research papers in pure mathematics. Furthermore, Kolmogorov's papers established results that were previously thought to be impossible by the international mathematical community. One of his eight undergraduate papers was his first one of what would be a seminal series in probability.
  Throughout his life, Kolmogorov continued to produce research that was deep, important, and interesting. In addition to probability, he did groundbreaking work in mathematical analysis and topology. He was a very popular and effective teacher, who advised 76 doctoral students, many of whom themselves later became prominent mathematicians. As a university professor, for many years he engaged in his tradition of Sunday walks, where all of his doctoral (and interested undergraduate) students, as well as doctoral students of other advisors were invited. These walks took place outside of the city, and were fondly remembered by all who were invited to participate. They always included discussions of interesting problems in mathematics in an informal setting, along with talk about the arts and general current events. Numerous honors and honorary degrees were bestowed upon Kolmogorov throughout his life.

Kolmogorov based his theory on three very intuitive axioms, from which all of probability theory could be developed. His elegant development has been likened to Euclid's axiomatic development of geometry in the latter's time-enduring textbook, *The Elements*. In order to state Kolmogorov axioms, we will need the following definition:

DEFINITION 6.4: Two events E and F are called **mutually exclusive** if they are disjoint as sets, i.e., if $E \cap F = \varnothing$. A (finite or infinite) sequence of events $E_1, E_2, \cdots$ is **pairwise mutually exclusive** if any pair of events taken from this sequence is mutually exclusive, i.e., if $i \neq j$ are indexes of two sets from this sequence, then E_i and E_j are mutually exclusive.

FIGURE 6.4: Andrey Nikolaevich Kolmogorov (1903–1987), Russian mathematician.

DEFINITION 6.5: (*Kolmogorov's Axioms of Probability*) Given any set S (a sample space), a **probability** on S is a set function P, that assigns to each event E (subset of S) its corresponding probability P(E) in such a way that the following axioms are satisfied:

(1) For any event E, $0 \leq P(E) \leq 1$. (*Any probability lies between 0% and 100%.*)

(2) $P(S) = 1$. (*It is a sure thing that any outcome will be in the sample space.*)

(3) If a (finite or countably infinite) sequence of events $E_1, E_2, \cdots$ is pairwise mutually exclusive, then

$$P(\bigcup_i E_i) = \sum_i P(E_i). \tag{2}$$

(*Probabilities of disjoint events can be added.*)

We point out that this definition does not even require the sample space to be a finite set. From these simple (and intuitive) axioms, many interesting consequences can be derived. Note that in the third axiom (equation (2)), if we take just a pair of mutually exclusive events: $E_1 = A$ and $E_2 = B$, the axiom states that $P(A \cup B) = P(A) + P(B)$.

EXERCISE FOR THE READER 6.2: Show that for a finite sample space for which all outcomes are equally likely, the formula (1) $P(E) = |E| / |S|$ (from Definition 6.3) satisfies Kolmogorov's axioms.

Probability Rules

The following theorem collects some useful rules of probability that can all be derived from Kolmogorov's axioms. Unless stated otherwise, we assume that we are in the context of a particular sample space S on which a probability function P is defined that satisfies Kolmogorov's axioms.

THEOREM 6.1: (*Probability Rules*)   Given events E and F, the following formulas are valid:

(a) (*Complementary Probability Rule*) $P(E) = 1 - P(\sim E)$.

(b) (*Monotonicity*) If $E \subseteq F$, then $P(E) \le P(F)$.

(c) (*Inclusion-exclusion Principle*) $P(E \cup F) = P(E) + P(F) - P(E \cap F)$.

Proof: We prove only Part (a); the proofs of the other two can be accomplished in a similar fashion and will be left as exercises. Note that the events E and its complement $\sim E$ are certainly mutually exclusive, and their union equals the sample space S. Using the axioms, we may thus obtain:

$$1 \underset{\text{Axiom 2}}{=} P(S) = P(E \cup \sim E) \underset{\text{Axiom 3}}{=} P(E) + P(\sim E) \implies 1 - P(\sim E) = P(E),$$

as desired.

EXERCISE FOR THE READER 6.3:   (a) Prove Part (c) of Theorem 6.1.
(b) Prove the following inclusion-exclusion principle for three events:

$$P(E \cup F \cup G) = P(E) + P(F) + P(G)$$
$$- P(E \cap F) - P(E \cap G) - P(F \cap G) + P(E \cap F \cap G).$$

Since events are sets, when dealing with a small number of them, Venn diagrams can be helpful. The following example will illustrate such a situation.

EXAMPLE 6.3: Suppose that the weather forecast estimates the probability of rain tomorrow as 70%, the probability of lightning tomorrow as 40%, and the probability of rain or lightning as 95%. Compute each of the following probabilities concerning tomorrow's weather:
(a) No lightning.
(b) Rain and lightning.
(c) Lightning but no rain.
(d) Neither rain nor lightning.

SOLUTION: Using obvious notations, we can write $P(R) = 0.7$, $P(L) = 0.4$, and $P(R \cup L) = 0.95$.

Part (a):   Using the complementary probability rule, we obtain: $P(\sim L) = 1 - P(\sim (\sim L)) = 1 - P(L) = 1 - 0.4 = 0.6$.

Part (b): Invoking Part (c) of Theorem 6.1 (and solving it for the probability of the intersection), we can write:

$$P(R \cap L) = P(R) + P(L) - P(R \cup L) = 0.7 + 0.4 - 0.95 = 0.15.$$

Part (c):   This could also be done with the probability rules, but let's instead show an alternative (more intuitive) approach that uses a Venn diagram. Using the probabilities that were given along with the one just found in Part (b), Axiom 3 lets us fill in the three probabilities within the circles shown in Figure 6.5. The 0.05 probability in the figure is that of the complement of the union. It can be obtained either by the complement rule (since we were given the probability of $R \cup L$). The required probability may now be

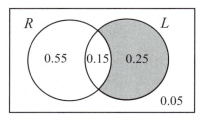

FIGURE 6.5: A Venn diagram with probabilities for the solution of Example 6.3.

simply read it off from the Venn diagram: $P(L \sim R) = 0.25$.

Part (d): The event "neither rain nor lightning" is just $\sim R \cap \sim L = \sim (L \cup R)$ (we have used De Morgan's law), so we can read the corresponding probability off the Venn diagram to be 0.05.

Our next example is a famous problem in probability; most people tend to greatly overestimate its answer when they first hear about it, and many are stupefied by the results.

EXAMPLE 6.4: (*The Birthday Problem*) We let $B(n)$ denote the probability that at least two people among a group of n people with randomly distributed birthdays will share a common birthday.[7] It is clear that $B(1) = 0$, $B(366) = 1$, and for all values of n in between, $B(n)$ must increase as n increases (if more people are in the room, there is a better chance of a common birthday). The problem is to find the first value of n for which $B(n) > 0.5$. In other words, how many people would be needed in a room for there to be a better than 50% chance of a common birthday?

SOLUTION: For each value of n, the sample space of this experiment can be viewed as the set of all vectors (ordered lists) $(b_1, b_2, \cdots, b_n)$ of possible birthdays of the n people in the room, and thus has 365^n elements. Let E denote the event that there is a common birthday in the list (i.e., $\exists i \neq j, b_i = b_j$). Describing (or counting) the elements of E is complicated, but the complementary event $\sim E$ is simply the event that there are no common birthdays in the list, and this is easy to count (using the multiplication principle). Thus using the complementary probability rule, we obtain:

---

[7] We ignore leap year birthdays (on February 29).

$$B(n) \equiv P(E) = 1 - P(\sim E) = 1 - \frac{|\sim E|}{|S|} = 1 - \frac{365 \cdot 364 \cdots (365 - n + 1)}{365 \cdot 365 \cdots 365}.$$

(The numerator on the right side was obtained using the multiplication principle: for the first person's birthday there are 365 possibilities, for the second since there are no common birthdays in any $\sim E$ outcome there are 364 possibilities, for the third there are 363, and so on.) It is now easily checked with a simple computer loop that $B(n)$ first exceeds 0.5 when $n = 23$. The plot of all of the values of $B(n)$ is shown in Figure 6.6.[8]

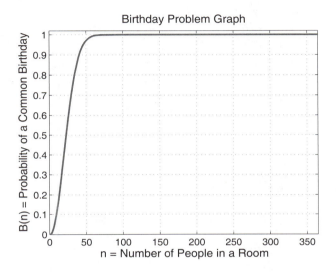

FIGURE 6.6: Plot of the probabilities for the birthday problem of Example 6.4.

NOTE: We point out that if we attempted to directly count the number of outcomes in E (rather than using the complementary probability rule), there would be an inordinate number of cases to consider; see Exercise 14 for an illustration of such a difficulty on a smaller scale.

EXERCISE FOR THE READER 6.4: A very active French Club at a certain university has been awarded three one-semester scholarships for some of its 10 members to attend *La Sorbonne*. The club votes to distribute the scholarships to its members by random selection. Compute the following probabilities:

(a) Alice will be chosen.   (b) Alice and Ben will be chosen
(c) Alice will not be chosen.   (d) Alice or Ben will not be chosen.

---

[8] While working at the University of Hawaii, the author was told of a professor who often taught probability to a class of about 40 students. On the first day of class he would routinely offer the following bet to anyone in the class who would care to take it: He would say to the class: "I don't know any personal information about you, but I will bet anyone $1 that there will be two people in this room who have the same birthday." Most of the class would jump at this seemingly "sucker bet." As Figure 6.6 shows, the probability of a common birthday among 40 people jumps up to about 90%. Thus, on average, nine times out of ten, the professor would collect a stack of one dollar bills from his wager. Some would argue this is taking advantage of the students, but he defended his ritual as teaching them a valuable and relatively inexpensive lesson about gambling.

We point out that probability functions can easily be constructed on any finite (or countably infinite) sample space S as follows: Let $p : S \to [0,1]$ be any function, i.e., for each $s_i \in S$, $p(s_i)$ is a real number between 0 and 1 (inclusive). If this function satisfies $\sum_i p(s_i) = 1$, then for any event E, the set function defined by $P(E) = \sum_{s_i \in E} p(s_i)$ is a probability function that satisfies Kolmogorov's axioms (see Exercise 29).   For example, if we apply this construction to the set $\{1, 2, 3, 4, 5\}$ using the function p specified by $p(1) = p(3) = p(5) = 1/9$, and $p(2) = p(4) = 1/3$, the construction would lead to a Kolmogorov probability function that assigns triple the probability to an even number that it does to an odd number, e.g., $P(\{1,2,3\}) = p(1) + p(2) + p(3) = 1/9 + 1/3 + 1/9 = 5/9$.   Of course, for such a construction to be useful, it must model some actual experiment.   We next move on to the important concepts of conditional probability and independence of events.   These will lead us to some interesting probability models.

Conditional Probability

Conditional probabilities are probabilities based on additional information being known. For example, if we randomly chose a student from a certain university and considered the event that the student can run a six minute mile (i.e., run a distance of one mile in a time of at most six minutes), this probability would be quite low, most likely under 5%. If however, we had additional information that the student selected was on the university's track team, this would greatly increase the probability, perhaps to the 90% range.   In medical science, conditional probabilities are used to assess certain risk factors based on family history, genetics, and lifestyle.   Insider stock trading is another example of conditional probabilities, where certain individuals who are privy to confidential information (for example, a not yet announced treatment breakthrough by a pharmaceutical company) use this information to profit by stock sales/purchases (at the expense of other stock investors).   Another group of notorious examples are the card-counting strategies designed to beat casinos in blackjack games.   Here is a simple but more specific example: Suppose that two balls are drawn (without replacement) from an urn containing three red and two black balls.   Let E be the event "the second ball drawn is black."   Then, a priori (not knowing anything about the first ball being drawn), $P(E) = 2/5$. However, suppose that we knew that the event F: "the first ball is black" has occurred. This decreases the chance of the second ball being black since at the second draw, the urn has only one black ball along with three red balls. Effectively, knowing that the event F has occurred has reduced the sample space for the computation of the event E. The *reduced sample space* is now F, and the number of outcomes in which E can occur (in the reduced sample space) is the number of outcomes in $E \cap F$. This allows us to conclude that the probability of E given that F has occurred is 1/4. This probability is called the conditional probability of E given F, and is written as $P(E \mid F)$. The reader should check that $P(E \mid F)$ will coincide with the ratio of the probabilities $P(E \cap F)/P(F)$. Exercise 26 will ask the reader to verify this formula in situations where all outcomes have equally likely probabilities (to which Laplace's definition applies). We thus take this as our general definition of conditional probability:

DEFINITION 6.6: Suppose that E and F are events with $P(F) > 0$. The **conditional probability of E given that F has occurred** is defined to be

$$P(E \mid F) = \frac{P(E \cap F)}{P(F)}. \tag{3}$$

We point out that in the setting of a finite sample space with equally likely outcomes, formula (1) allows us to rewrite formula (3) in the form

$$P(E \mid F) = \frac{|E \cap F|}{|F|}.$$

EXAMPLE 6.5: Compute the conditional probabilities for each of the following questions:
(a) Three dice are rolled. What is the conditional probability that (exactly) one of the dice shows a one, given that no two of the dice show the same number?
(b) The conditions at the (fictitious) *Galley Prison* are abysmal: 15% of the inmates have AIDS, 20% have TB, and 10% have both AIDS and TB. A prisoner is randomly chosen and is shown to have TB. What is the probability that he also has AIDS?

SOLUTION: Part (a): Let D represent the event that no two of the dice show the same number (D stands for "different"), and E be the event that exactly one die shows a one. Viewing the sample space as the set of all ordered triples (a, b, c) of possible outcomes of the three dice, from the multiplication principle (Section 5.1) we see that $|D| = 6 \cdot 5 \cdot 4$ and $|E \cap D| = 3 \cdot 5 \cdot 4$ (the factor 3 represents the choices of where the one can go in the ordered triple). Since all outcomes are equally likely, we conclude that

$$P(E \mid D) = \frac{|E \cap D|}{|D|} = \frac{3 \cdot \cancel{5} \cdot \cancel{4}}{6 \cdot \cancel{5} \cdot \cancel{4}} = \frac{1}{2}.$$

Part (b): With the obvious notations for the events, we can easily compute the desired conditional probability:

$$P(A \mid T) = \frac{P(A \cap T)}{P(T)} = \frac{0.10}{0.20} = \frac{1}{2}.$$

Thus, the additional information that the prisoner has TB greatly increases the chances that he has AIDS (from 15% to 50%).

Note that in Part (a) of the above example, the corresponding (unconditional) probability could be computed as $P(E) = \dfrac{|E|}{|S|} = \dfrac{3 \cdot 5 \cdot 5}{6 \cdot 6 \cdot 6} = \dfrac{25}{72} \approx 0.3742$. Thus, in both of the examples above, as well as in the motivating example, the conditional probability $P(E \mid F)$ turned out to be different from the corresponding (unconditional) probability $P(E)$. In other words, the additional knowledge that the event F has occurred changes the probability that event E will occur. Although we will formalize this concept a bit later, in cases where both events have positive (unconditional) probabilities, such pairs of events are called *dependent*, whereas the events E and F will be *independent* if $P(E \mid F) = P(E)$, i.e., the knowledge that event F has occurred does not change the probability that E has occurred.   Here is a simple example of two independent events. Suppose that we flip a fair coin 10 times, and let F be the event that the first nine flips turned up heads, and E be the event that the tenth flip turns up heads. Since knowing that we landed on heads the first nine times does not influence the 50% chance of getting heads on the tenth flip, we have $P(E \mid F) = P(E)$, and the events are independent.

Conditional probabilities are at the core of all of the (in)famous card-counting strategies that were developed to beat dealers in blackjack games in casinos.   The rules of blackjack are set up so that the dealer always has a higher probability of winning (the *house advantage*). But these probabilities are based on unconditional probabilities, assuming that each new card that comes up from the deck is equally likely. In a card counting strategy, players mentally keep a tally of the cards that have shown up in a game (until the decks get reshuffled or replaced), for example, a system might assign a "+1" to each high card with value greater than 8, and a " –1" to each card with value less than 5. If at a given point in the game the running total is, say –9, this means that nine more low cards have been used than high cards, and thus high cards will have a higher conditional probability of showing up next than a low card.   Such additional information has been successfully used to develop strategies that give players advantages over dealers, and this has resulted in casinos adopting certain countermeasures (reshuffling or replacing decks often, video surveillance with facial analysis software to spot and promptly "escort out" card counters, etc.), but it is still possible even at the time of the writing of this book to beat casinos at certain blackjack betting tables. We will return to this concept in the next section.[9]

---

[9] The first scientific study that both popularized card counting and proved that it worked was done by American mathematician Edward Oakley Thorp (1932 –) in his doctoral dissertation at UCLA. Thorp was an avid blackjack player, and was able to successfully test out his theory with practice in actual casinos (in "research trips" from Los Angeles to Las Vegas and Reno).   Thorp used computer simulations to prove that his strategies could change the house advantage of about 5% to a card counting player's advantage of about 1%. He subsequently transcribed his discoveries into a practical

EXERCISE FOR THE READER 6.5: The traffic management office of a city has collected data for New Year's Eve automobile accidents over the past several years, and has produced the following summary of accident probabilities based on the total number of cars driven in the city on previous New Year's Eves.

	Dry Roads	Snow/Ice
Accident	0.005	0.020
No Accident	0.725	0.250

Based on this data, compute the following (predicted) probabilities for this New Year's Eve:
(a) The probability that there will be snowy or icy road conditions.
(b) The probability of getting into an accident (if you will drive on New Year's Eve).
(c) The probability of getting into an accident if the roads are dry.
(d) The probability of getting into an accident if the roads are snowy or icy.

The Multiplication Rule

The definition of conditional probability (equation (3)) can be easily solved for $P(E \cap F)$ by simply multiplying both sides by $P(F)$:

$$P(E \mid F) = \frac{P(E \cap F)}{P(F)} \quad \Rightarrow$$

$$P(E \cap F) = P(E \mid F) \cdot P(F). \tag{4}$$

The resulting equation (4) is known as the **multiplication rule**, and it turns out to be particularly useful both in theory and in problem solving. The next example illustrates its simplicity for a problem that we by now know well how to solve directly.

EXAMPLE 6.6: Compute the probability of drawing two hearts from a well shuffled deck of playing cards (when only two cards are drawn)
(a) first directly, and then
(b) using the multiplication rule.

---

and less technical paperback book entitled "Beat the Dealer," [Tho-62], which became a New York Times bestseller and has sold over 700,000 copies. Successful Hollywood movies on the subject of card counting have also come out; for example, *Rain Man* (1988) and *21* (2008).
 After completing his Ph.D. in 1958, Thorp was awarded a postdoctoral position at M.I.T for three years and later moved on to accept a professorship at the University of California, Irvine, first in the mathematics department, and then in the finance department. Thorp later became interested in applying probability and statistics to the stock market. Thorp left his academic position in 1982 to start his own hedge fund company Edward O. Thorp & Associates.

SOLUTION: We let H_1 be the event of drawing a heart with the first card, and H_2 be the event of drawing a heart with the second card.

Part (a): $P(H_1 \cap H_2) = C(13,2)/C(52,2) = 78/1326 = 1/17$.

Part (b): The multiplication rule (4) tells us that $P(H_1 \cap H_2) = P(H_2 \mid H_1)P(H_1)$. Now since there are 13 hearts, $P(H_1) = 13/52$. After a heart has been removed there will be only 12 hearts and 51 cards total, so $P(H_2 \mid H_1) = 12/51$. Hence, $P(H_1 \cap H_2) = (13/52) \cdot (12/51) = 1/17$.

Some comments are in order. The solution to Part (b) only looks longer than that of Part (a) because we put in more detail (since it was our first example of the multiplication rule). Since set intersections is a commutative operation (i.e., $H_1 \cap H_2 = H_2 \cap H_1$), the multiplication rule (4) could also be written in the form: $P(H_1 \cap H_2) = P(H_1 \mid H_2)P(H_2)$. This form would be much more complicated to work with for this problem. The next exercise for the reader shows how the multiplication rule can (easily) be extended to intersections of greater numbers of sets. Exercise 32 will generalize this to any finite number of sets.

EXERCISE FOR THE READER 6.6: (a) Use the product rule (4) (for two sets) to establish the following analogue for three sets:

$$P(A \cap B \cap C) = P(A \mid (B \cap C)) \cdot P(B \mid C) \cdot P(C).$$

(b) Next obtain the analogue for four sets:

$$P(A \cap B \cap C \cap D) = P(A \mid B \cap C \cap D) \cdot P(B \mid C \cap D) \cdot P(C \mid D) \cdot P(D).$$

(The general pattern should now be clear.)

Conditioning and Bayes' Formula

We next move on to discuss two important and related consequences of the product rule: *conditioning* and *Bayes' formula*. Both of these methods allow one to use known conditional probabilities to compute unknown probabilities, both absolute and conditional.

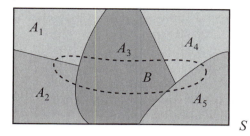

FIGURE 6.7: A mosaic diagram for a partition of a sample space S into disjoint pieces A_i along with an arbitrary event B.

Suppose that the sample space S can be **partitioned** into a pairwise mutually exclusive union of events $A_1, A_2, \cdots A_n$, i.e., S is the disjoint union of these sets (see Figure 6.7). Typically, in the setting of probability questions, one would have some good information relating to these events. Now, if B is an arbitrary event, then the events $A_1, A_2, \cdots A_n$ can also be used to decompose B into pairwise mutually exclusive events:

$$B = \bigcup_{i=1}^{n}(B \cap A_i).$$

If we use the fact that disjoint probabilities can be added (Kolmogorov's Axiom 3), and then use the product rule, we are led to:

$$P(B) = \sum_{i=1}^{n} P(B \cap A_i) = \sum_{i=1}^{n} P(B \mid A_i)P(A_i). \tag{5}$$

Use of formula (5) is known as **conditioning the event B on the events $A_1, A_2,$** $\cdots,\ A_n$. An important special case is when $n = 2$, so that $A_1 = F$, $A_2 = \sim F$, and (5) becomes:

$$P(B) = P(B \mid F)P(F) + P(B \mid \sim F)P(\sim F).$$

This special case is called "**conditioning on whether or not F has occurred.**" Conditioning is used to separate a difficult probability computation into cases that are more manageable. We will give examples shortly, but first we show how with one more small yet significant step we can obtain Bayes' formula that allows one to compute (difficult) conditional probabilities in terms of (simpler) reverse conditional probabilities. We continue to suppose that the sample space is partitioned by $A_1, A_2, \cdots, A_n$. By first using the definition of conditional probability and then using conditioning and the product rule, we arrive at **Bayes' formula**:

$$P(A_j \mid B) = \frac{P(A_j \cap B)}{P(B)} = \frac{P(B \mid A_j)P(A_j)}{\displaystyle\sum_{i=1}^{n} P(B \mid A_i)P(A_i)}. \tag{6}$$

This formula has such tremendous utility in probability and statistics that its creator, Thomas Bayes (Figure 6.8) is considered one of the founders of statistics. Although the above development looks quite straightforward, we emphasize that Bayes did not have the luxury of the clean axiomatic setting of probability in which to do his work. He had to work much harder for his results over an unexplored and Spartan landscape.

FIGURE 6.8:
Thomas Bayes[10]
(1702–1761),
English minister and
mathematician.

EXAMPLE 6.7: (*Quality control*) Three factories A, B, C all produce the same model of a computer. Their outputs consist of 20%, 35%, and 45%, respectively, of the total computers produced. Factory A has a 6% defect rate for the computers it produces, while the defect rates for factories B and C are 3% and 2%, respectively.

(a) If a computer is randomly selected, compute the probability that it will turn out to be defective.

(b) Suppose that the computer selected did turn out to be defective. Compute the probability that it came from factory A.

SOLUTION: We use the following natural notations for the relevant events: $D =$ the selected computer is defective, A, B, C = the selected computer came from factory A, B, C, respectively.

Part (a): Since A, B, C clearly partition the sample space S, we can condition on whether the computer came from factory A, B, or C:

$$P(D) = P(D \mid A)P(A) + P(D \mid B)P(B) + P(D \mid C)P(C)$$
$$= (.06)(.2) \quad + \quad (.03)(.35) \quad + \quad (.02)(.45)$$
$$= 0.0315$$

Part (b): We are asked to find $P(A \mid D)$. This at first might seem difficult, but since the problem gives us all of the relevant reverse conditional probabilities, we are set up perfectly to use Bayes' formula. Since we have already found $P(D)$ in Part (a), we do not need the full formula (6):

---

[10] Bayes, like his father, was a Nonconformist minister in England (his father was one of the first to be ordained).   Despite his very strong potential for university studies, Nonconformists were not allowed to study at Cambridge or Oxford, so he studied (logic and theology) at the University of Edinburgh. His career was dedicated entirely to the ministry, but his sharp mathematical skills and his avid interest in mathematics kept him active in mathematics as a hobby throughout his life.   He wrote two particularly insightful papers, courageously working on difficult topics for which he found the prevalent state of knowledge quite unsatisfactory.   One was on probability theory and the other on the theory of fluxions. Although he never published any papers during his lifetime (the two mentioned papers were published posthumously), his mathematical talents did not go unnoticed: he was elected as a Fellow of the Royal Society.   Bayes' probability theory was most certainly before its time. Pascal and Fermat had not even begun to put the subject on a solid foundation. Bayes' work was rediscovered by others, and its importance has been witnessed by generations of extensions and applications.

$$P(A \mid D) = \frac{P(A \cap D)}{P(D)} = \frac{P(D \mid A)P(A)}{P(D)} = \frac{(.06)(.2)}{.0315} = 0.3810.$$

Thus, knowing the computer is defective has nearly doubled the chances that it came from factory A.

EXAMPLE 6.8: (*Public Health: Should AIDS Tests Be Mandatory?*) Suppose that it is known that in a certain branch of the military, 1 in 1000 test positive for HIV (the virus that causes AIDS) and that there is a test that can be given to individuals that has the following results: $P(T+ \mid HIV) = 0.99$ and $P(T+ \mid \sim HIV) = 0.02$. Here the event HIV means that the individual being tested actually has the HIV virus, whereas the event T+ (positive test result) indicates that the test indicates the presence of the HIV virus in the individual. Thus the latter (conditional) probability is the chance that the test incorrectly diagnoses an individual with HIV even though that individual does not have it (such a test result is called a *false positive*).

(a) Find $P(T+)$.

(b) Find $P(HIV \mid T+)$.

(c) If 100,000 individuals are given the test, about how many will test positive, and of these, how many will actually have the HIV virus? Should mandatory AIDS testing be adopted?

SOLUTION: Part (a): We condition on whether or not an individual has HIV:

$$P(T+) = P(T+ \mid HIV)P(HIV) + P(T+ \mid \sim HIV)P(\sim HIV)$$
$$= (.99)(.001) \qquad + \qquad (.02)(.999)$$
$$= 0.02097$$

Part (b): We use Bayes' formula without the conditioning (that was done in Part (a)):

$$P(HIV \mid T+) = \frac{P(T+ \mid HIV)P(HIV)}{P(T+)} = \frac{(.99)(.001)}{.02097} = 0.04721.$$

Part (c): Of the $N = 100,000$ individuals, by Part (a), the number who will test positive will be approximately $N \cdot P(T+) \approx 2097$. By Part (b), the number of these 2097 individuals who test positive and who actually have HIV is approximately $2097 \cdot P(HIV \mid T+) \approx 99$. Thus, nearly 2000 of the 2097 who test positive for HIV really do not have it! This (false) shocking news could have a tremendous impact on the lives of these individuals, and one could argue that this is a heavy price to pay for the relatively small number of HIV cases that were correctly detected.[11]

---

[11] Most tests of serious diseases (e.g., tuberculosis) have similar problems with false positives, which always occur with a positive (conditional) probability. Another problem with medical detection tests

EXERCISE FOR THE READER 6.7:   A new car rental company opens for business with 20 cars, all of the same type. It runs a promotion that has put a $100 bill in the glove boxes of a randomly selected 5 of the 20 cars, and the first person to rent any of these special cars will get to keep the $100 bounty. On the first day, five randomly selected cars were rented and returned the next morning (and randomly mixed in with the rest of the fleet). The next day, four randomly selected cars (from the 20) were rented out. What is the probability that none of these latter four cars contains a $100 prize?

Suggestion: Condition on the number of cars that contain a $100 bill after the first day's rentals are returned.

Independent Events

We end this section with a discussion of the very important concept of independence of events. Intuitively, two events are independent if knowing that one has occurred does not have any influence on the knowledge of whether the other has occurred. For example, if we flip a (fair) coin nine times and get heads at each flip, the probability of getting heads on the tenth flip is still only 1/2—the tenth flip is independent of the first nine flips. In terms of conditional probabilities, events E and F are independent if $P(E\,|\,F) = P(E)$ and $P(F\,|\,E) = P(F)$.   Since $P(E\,|\,F) = P(E \cap F)/P(F)$, the first of the previous two equations (when multiplied by $P(F)$) produces the equation

$$P(E \cap F) = P(E) \cdot P(F). \tag{7}$$

Equation (7) also follows in the same fashion from $P(F\,|\,E) = P(F)$. In light of this symmetry, equation (7) is taken as the formal definition of independence of events E and F:

DEFINITION 6.7: Two events, E and F, are said to be **independent** if equation (7) holds. If E and F are not independent, we say they are **dependent**.

EXERCISE FOR THE READER 6.8:   We have already shown that two events E and F that satisfy either of the equations $P(E\,|\,F) = P(E)$ or $P(F\,|\,E) = P(F)$ will be independent (i.e., satisfy equation (7)). What about the converse? Compare and contrast the definition of E and F being independent with the validity of the equations $P(E\,|\,F) = P(E)$ (or $P(F\,|\,E) = P(F)$).

---

that we did not mention is that of false negative, where an individual has the disease but the disease is not detected by the test.

EXAMPLE 6.9: Suppose that a red die and a green die are rolled together. Determine whether the pairs of events below are independent.
(a) $E = \{$Red die shows 6$\}$, $F = \{$Green die shows 2$\}$.
(b) $E = \{$Red + Green shows 8$\}$, $F = \{$Green shows 2$\}$.

SOLUTION: Part (a): Intuitively, it is clear that the two events are independent. Let us verify this by showing that $P(E \cap F) = P(E) \cdot P(F)$. Certainly, $P(E) = 1/6 = P(F)$. Next, in the sample space of 36 outcomes (see Figure 6.3), only one of these is in $E \cap F$, so that $1/36 = P(E \cap F) = P(E) \cdot P(F)$.

Part (b): Our intuition is less clear here. Let's first look at the definition of independence: $P(E \cap F) = P(E) \cdot P(F)$. From Figure 6.3, we see that $P(E) = 5/36$, and clearly $P(F) = 1/6$. But since there is only one outcome in $E \cap F$ (Red = 6 and Green = 2), we get that $P(E \cap F) = 1/36 \neq 5/216 = P(E) \cdot P(F)$. Thus E and F are dependent.

PROPOSITION 6.2: If E and F are independent events, then so are each of the following pairs of events: (i) $\sim E$ and F, (ii) E and $\sim F$, (iii) $\sim E$ and $\sim F$.

Proof: Part (i): Since F is the disjoint union of the two sets $E \cap F$ and $(\sim E) \cap F$ (it is easy to see this, or just draw a Venn diagram), by Kolmogorov's Axiom 3, we can add the probabilities:

$$P(\sim E \cap F) + P(E \cap F) = P(F) \Rightarrow P(\sim E \cap F) = P(F) - P(E \cap F).$$

Using the fact that E and F are independent, we may conclude from the above that:

$$P(\sim E \cap F) = P(F) - P(E) \cdot P(F) = (1 - P(E)) \cdot P(F) \underset{\substack{\text{Complementary} \\ \text{Probability Rule}}}{=} P(\sim E) \cdot P(F).$$

The independence of $\sim E$ and F is thus established.

Part (ii): This is accomplished in the same fashion as (i), or (since the intersection operation is commutative) we can simply interchange the roles of E and F.

Part (iii): This part follows directly from Parts (i) and (ii), applied in succession. Alternatively, here is a direct proof:

$$
\begin{aligned}
P(\sim E \cap \sim F) &= P(\sim (E \cup F)) && \text{(De Morgan's law)} \\
&= 1 - P(E \cup F) && \text{(complementary probability rule)} \\
&= 1 - P(E) - P(F) + P(E \cap F) && \text{(inclusion-exclusion principle)} \\
&= 1 - P(E) - P(F) + P(E) \cdot P(F) && \text{(since } E \text{ and } F \text{ are independent)} \\
&= [1 - P(E)] \cdot [1 - P(F)] && \text{(algebra)} \\
&= P(\sim E) \cdot P(\sim F). && \text{(complementary probability rule)}
\end{aligned}
$$

$\square$

The above result should appear rather intuitive; for example, if the knowledge that the event E occurred has no bearing on the probability of F occurring, then knowing $\sim E$ occurred (in other words, knowing E did not occur) should also have no bearing on the probability of F occurring.   Our next exercise for the reader should serve as a caution to readers to not let their intuition about independence stray too far.

EXERCISE FOR THE READER 6.9:   (*Sex Distribution*) The sample space is the set of all families with three children; we consider each of the eight possible sex distributions: bbb, bbg, …, ggg to be equally likely (here b = boy, g = girl, and the oldest to youngest children are listed from left to right). Let

> E = the event that there is at most one girl
> F = the event that the first child is a girl, and
> G = the event that the family has children of both sexes.

Show that E and G are independent, that F and G are independent, but that $E \cap F$ and G are dependent.

NOTE: As a cautionary note, we point out that for families of either two or four children, the events E and G are no longer independent! (See Exercise 21.)

We next give the definition of independence for collections of more than two sets:

DEFINITION 6.8: Three events E, F, and G are said to be **independent** if each of the following equations holds:

$$P(E \cap F) = P(E) \cdot P(F), \ P(E \cap G) = P(E) \cdot P(G), \ P(F \cap G) = P(F) \cdot P(G), \ \text{and}$$
$$P(E \cap F \cap G) = P(E) \cdot P(F) \cdot P(G).$$

More generally, a finite collection of events $E_1, E_2, \cdots, E_n$ is said to be **independent** if for every subcollection $E_{i_1}, E_{i_2}, \cdots, E_{i_r}$ of these events (where $r \le n$ and $1 \le i_1 < i_2 < \cdots < i_r \le n$), we have

$$P(E_{i_1} \cap E_{i_2} \cap \cdots \cap E_{i_r}) = P(E_{i_1}) \cdot P(E_{i_2}) \cdots \cdot P(E_{i_r}).$$

Finally, an infinite collection of events is **independent** if every finite subcollection is independent.

Analogues of Proposition 6.2 hold for general collections of independent events. For example, one can show that any two events A and B resulting from combining different subcollections of events from an independent family of events with the standard set theoretic operations (unions, intersections, and complements) will be independent. From all of the conditions that need to hold for independence, it is clear that verification of independence can be an arduous task for even a moderately sized collection. Students are often tempted to assume independence since it makes it so much simpler to compute probabilities of intersections. One common situation where independence is present is with the repetition of an experiment under the same conditions (such as flipping a coin). Such repetitions are called **independent trials**, and any collection of events that depends on outcomes of disjoint sets of runs of the experiment will be independent. Independent trials will be the fundamental concept behind simulations, which will be developed in the next section. Another use of independence is in the construction of reliable systems. This is quite a general idea, but basically rests on the fact that any component of a system has a certain probability for failure. To increase reliability in vital systems, it is important to have back-up or contingency components. Having identical components perform independently is often a most effective strategy. The following example gives a basic illustration of this.

EXAMPLE 6.10: The three-way circuit shown in Figure 6.9 will allow information (or oil or electricity) to flow from point A to point B, provided that at least one of the three paths from A to B has all of its breakers closed. The number shown next to each breaker is the probability that the breaker will be open at any given time. The breakers function independently. Compute the probability that information can flow from A to B at any given time.

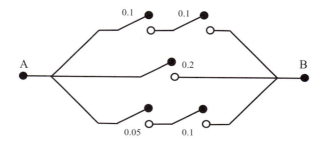

FIGURE 6.9: A three-way circuit for flow from A to B. Each of the five breakers shown functions independently, and their respective probabilities of being open at any given time are indicated. For the circuit to function, at least one of the three paths from A to B must have all of its breakers closed.

SOLUTION: We begin by introducing notations for some events needed to solve this problem. We let T be the event that the top linear path is functioning, i.e., that both of the two breakers on the top segment are closed. We let T_1 and T_2 denote the (sub)events that the first (left) and second (right) breakers on the top segment are closed (i.e., functioning properly). We let M denote the corresponding event for the middle linear path (which has only one breaker), and B, B_1, and B_2 denote the corresponding events for the bottom linear path. The probability that we wish to compute is $P(T \cup M \cup B)$. We begin by rewriting this expression into a more convenient form:

$$\begin{aligned}
P(T \cup M \cup B) &= 1 - P(\sim [T \cup M \cup B]) & \text{(complementary probability rule)} \\
&= 1 - P(\sim T \cap \sim M \cap \sim B]) & \text{(De Morgan's law)} \\
&= 1 - P(\sim T) \cdot P(\sim M) \cdot P(\sim B). & \text{(independence)}
\end{aligned}$$

Next, observe that each of the three probabilities in the last line is associated with the event that the corresponding linear path is not functioning. The complementary event probabilities are easier to compute since for a (single) linear path to be functioning, all breakers on it must be closed. For example,

$$P(\sim T) = 1 - P(T) = 1 - P(T_1) \cdot P(T_2) = 1 - (0.9) \cdot (0.9) = 0.1900.$$

The two probabilities (0.9) were the complementary probabilities for the top two breakers to be open (not functioning) that were given in Figure 6.9. In a similar fashion we compute:

$$P(\sim M) = 1 - (0.8) = 0.2000 \text{ and } P(\sim B) = 1 - (0.95) \cdot (0.9) = 0.1450.$$

Substituting these three probabilities into the earlier expression leads us to

$$P(T \cup M \cup B) = 1 - P(\sim T) \cdot P(\sim M) \cdot P(\sim B) = 1 - (0.1900)(0.2000)(0.1450)$$
$$= 0.9945.$$

Discrete Problems with Infinite Sample Spaces

Up to this point all of our examples have involved finite sample spaces. Discrete probability also includes situations where a discrete probability experiment may be repeated infinitely many or an arbitrarily large number of trials. As a simple example, consider the experiment of continuing to flip a coin until we get the first heads. In practice this will not take a lot of flips, but theoretically, the experiment could go on any number of trials. In order to compute associated probabilities for such problems, it is often necessary to compute a sum of infinitely many nonnegative numbers that turns out to be a finite number. We now cite two relevant formulas in this regard that will be sufficient for our purposes:

NOTE: Two infinite series formulas:[12]

(1)   Recall that an *infinite geometric series* is any infinite series of the form $\sum_{n=K}^{\infty} ar^n$, where K is an integer, and a and r (the *ratio*) are constant (real numbers) with $|r| < 1$. Such geometric series always converge and their sums are given by the following formula:

$$\sum_{n=K}^{\infty} ar^n = \frac{ar^K}{1-r} = \frac{\text{first term}}{1-\text{ratio}} \qquad (8)$$

(2) The *differentiated geometric series* summation below is valid for any numbers a and r (the *ratio*), where a and r are constant (real numbers) with $|r| < 1$.

$$\sum_{n=0}^{\infty} nar^{n-1} = \frac{a}{(1-r)^2}. \qquad (9)$$

---

[12] Infinite series results such as these are studied and proved in calculus books, so we will not take the time to explain their proofs, but only use them as tools to solve any probability problems in which they might arise.

For example, formula (8) tells us that $\sum_{n=0}^{\infty}(1/2)^n = \dfrac{1}{1-(1/2)} = 2,$ and the infinite

series $\sum_{n=0}^{\infty} n(1/2)^n$ can be summed by first factoring out a $(1/2)$ and then

invoking formula (9): $\sum_{n=0}^{\infty} n(1/2)^n = (1/2)\sum_{n=0}^{\infty} n(1/2)^{n-1} = (1/2)\dfrac{1}{(1-1/2)^2} = 2.$

Our next example will require summing an infinite number of probabilities, and will also preview the important notion of *expectation*, a topic that will be covered in the next section.

EXAMPLE 6.11: Suppose you are offered the following bet: we continue to roll a pair of dice repeatedly until either "snake eyes" (double 1s) appears, in which case you win and would get paid $10 or until <u>two</u> trials have turned up 7s (i.e., the pips added up to 7), in which case you lose and would have to pay $5. Should you accept this bet?

SOLUTION: Let W be the event that you win, L the event that you lose, and D (for draw) be the event that the game goes on forever. A game is determined by a sequence (possibly infinite) of integers between 2 and 12 (representing the sum of numbers showing on a trial of rolling a pair of dice). The sequences that correspond to the event W are of three types:

(i) $< 2 >$ (win on the first roll)

(ii) $< a_1, 2 >,\ < a_1, a_2, 2 >,\ < a_1, a_2, a_3, 2 >, \cdots$

$\qquad$ (where each $a_i \in \{3, 4, 5, ..., 11, 12\} \sim \{7\}$)

(iii) sequences as in (ii) but with exactly one $a_i = 7$

Since the trials are independent, the individual probabilities of each sequence can be decomposed into the products of the corresponding probabilities for each roll, and these latter probabilities can be easily seen from Figure 6.3. Thus, on any single roll, $P(2) = 1/36$, $P(7) = 6/36 = 1/6$, and $P(\{3,...,12\} \sim \{7\}) = 29/36$. Using independence, we can now compute the probabilities of each of the three (mutually exclusive) types of sequences:

(i) $P(< 2 >) = P(2) = 1/36$

(ii) $P(< a_1, 2 >) = (29/36)(1/36),\ \ P(< a_1, a_2, 2 >) = (29/36)^2(1/36),$

$\qquad P(< a_1, a_2, a_3, 2 >) = (29/36)^3(1/36), \cdots$

(iii) For any sequence $< a_1, a_2, a_3, \cdots, a_n, 2 >$ of this third type, since there are exactly n choices of where to put the (single) 7, the associated probability will be $P(< a_1, a_2, a_3, \cdots, a_n, 2 >) = n(29/36)^{n-1}(1/6)(1/36)$. Summing up the resulting infinite series, we obtain using (8) and (9):

$$P(W) = P(i) + P(ii) + P(iii)$$

$$= \sum_{n=0}^{\infty} (1/36)(29/36)^n + \sum_{n=1}^{\infty} n(1/216)(29/36)^{n-1}$$

$$= \frac{1/36}{1-29/36} + \frac{1/216}{(1-29/36)^2} = 13/49.$$

(We are working in exact arithmetic since we wish to account for all probabilities.)

In a similar fashion, the sequences corresponding to L (losing) all have the form $<a_1, a_2, a_3, \cdots, a_n, 7>$ where $n \geq 1$, exactly one $a_i = 7$, and the rest of them satisfy $a_i \in \{3, 4, 5, ..., 11, 12\} \sim \{7\}$. Summing up the associated probabilities gives us that

$$P(L) = \sum_{n=1}^{\infty} n(1/6)(29/36)^{n-1}(1/6) = \sum_{n=1}^{\infty} n(1/36)(29/36)^{n-1} = \frac{1/36}{(1-29/36)^2} = 36/49.$$

At this point, we notice that $P(W) + P(L) = 1$. Since D, W, and L are disjoint events (whose union makes up the entire sample space), it follows that $P(D) = 0$. Thus, although the event D seems to be rich in outcomes (i.e., it contains all outcomes of the form $<a_1, a_2, a_3, \cdots>$ where at most one a_i can equal 7, and the rest can be any numbers in $\{3, 4, 5, ..., 11, 12\} \sim \{7\}$), the probability of such an outcome (i.e., a game that goes on forever) actually occurring is zero.

Now let's decide if the bet that we are offered would be favorable to us. To analyze this, suppose that we play a large number n of these games. Then, the number of these games which we win should be approximated by $n \cdot P(W) = 13n/49$. Since we would win $10 for each of these games, our total winnings would be $130n/49$ (dollars). Similarly, the number of games that we will lose is approximated by $n \cdot P(L) = 39n/49$. For each of these lost games, we lose $5, so our total loss would be $195n/49$ (dollars). Subtracting these estimated losses from our estimated winnings gives $130n/49 - 195n/49 = -65n/49$. Thus, on average (divide this total loss by the number n of games played), we would lose 65/49, or about $1.33 per game. (In the vocabulary of the next section, our expected gain in placing this bet would be -1.33.) Thus, it would not be prudent for us to take this bet.

In real life problems, it must be decided whether or not certain events are independent. Independence hypotheses can have significant effects on the analysis of many problems and hence also on the resulting impacts on the solutions to real-life problems. The following exercise for the reader will give a real-life problem where the reader will need to decide whether it is reasonable to assume independence.[13]

---

[13] Here is a true story that should help to emphasize the possibilities of dangerous impacts of unwarranted independence assumptions. Up until the mid-1980s, the aviation industry assumed that

EXERCISE FOR THE READER 6.10: Professor Billingsley has booked a trip from Chicago to Bangkok. He will fly from Chicago to San Francisco, from San Francisco to Tokyo, and from Tokyo to Bangkok. At each stop he (and his luggage) change planes, and the probabilities that the luggage gets lost at each of the airports are p_C for Chicago, p_S for San Francisco, and p_T for Tokyo. When he arrives in Bangkok, Professor Billingsley finds that his luggage did not make it. Compute the probabilities that the luggage was lost in (i) Chicago, (ii) in San Francisco, and (iii) in Tokyo.

EXERCISES 6.1:

1. (a) What is the probability that a roll of a single die shows an odd number or a 6?
    (b) What is the probability that two heads result when a penny and a nickel are flipped?
    (c) What is the probability that at least two heads will result if a fair coin is tossed four times? What if it is instead tossed seven times?
    (d) Suppose that traffic engineers have collected the following data for the past 200 weekdays on the number of cars that enter a particularly busy Los Angeles intersection during the afternoon rush hour period from 5 pm to 6 pm:

Number of Cars Entering	Frequency
Under 10,000	10
10,001–15,000	59
15,001–20,000	96
20,001–25,000	32
Over 25,000	3

Based on this data, estimate the probability that on the next workday the number of cars entering this intersection between 5 pm and 6 pm will be over 25,000 (a bad traffic day), and also the probability it will be at most 15,000.
    (e) What is the probability that a roll of two dice results in two numbers that add up to an even number?
    (f) What is the probability that a family with five children has exactly two girls? Assume that at each birth, a boy and a girl are equally likely.
    (g) What is the probability that a card drawn from a well shuffled deck is a jack or a red card?

2. (a) What is the probability that a single roll of a die results in a three or a four?
    (b) What is the probability that a single card drawn from a shuffled deck will be a face card (J,

---

airplane jet engines functioned independently in their safety calculations. In 1983, not long after an *Eastern Airlines* flight from Miami to Nassau took off with 172 people on board, one of its three engines shut down due to low oil pressure. As the plane immediately turned back for an emergency landing, the two other engines' low pressure warning lights began to flash, and soon after both remaining engines shut down and the plane rapidly began losing altitude at 13,000 feet. Thanks to the valiant efforts of the crew, they managed to restart one of the engines at about 4000 feet, and the plane was able to land without any casualties. Immediately after the landing, an NTSB investigation found that the same pair of mechanics had serviced all three engines and did not replace the O-ring seals. Ordinarily (at the time), the chance of malfunction of a jet engine during a flight was roughly 1/10,000. If the engines are assumed to function independently, the chance of all three engines malfunctioning simultaneously would thus be $(1/10,000)^3$, or one in one trillion. After this incident, the FAA ruled that engines on commercial airplane engines must be separately serviced by different teams of mechanics.

Q, or K)?

(c) What is the probability that when two cards are randomly drawn from a shuffled deck, they are both aces?

(d) Suppose that a loaded die is labeled with the numbers 1, 1, 2, 3, 5, 6. If two such dice are rolled, what is the probability that the sum of the numbers showing is odd?

(e) What is the probability that heads result exactly twice when a penny is flipped three times? What if it is flipped five times?

(f) What is the probability that a card drawn from a well shuffled deck is a 6 or an ace?

3.   Suppose that a three-card hand is randomly drawn from a shuffled deck of cards.

(a) What the probability that the hand contains a pair? (That is, a set of cards of the form {a, a, b}, where a and b are different denominations.)

(b) What is the probability that the hand contains a (at least one) king or a queen?

(c) What is the probability that the hand is a flush?

(d) What is the probability that the hand consists of three face cards (jacks, queens, or kings)?

(e) What is the probability that the hand consists of three-of-a-kind (i.e, three cards of the same denomination)?

(f) What is the probability that the hand consists of a straight (i.e., three consecutive cards in the following denomination sequence: A,2,3,4,5,6,7,8,9,10,J,Q,K,A)?

4.   An experiment consists of randomly drawing three marbles from an urn containing 10 marbles of which five are green, four are red, and one is black.

(a) What is the probability that the draw contains the black marble?

(b) What is the probability that all marbles drawn are green?

(c) What is the probability that all three marbles drawn are of the same color?

(d) What is the probability of drawing two red and one green marble?

(e) What is the probability of drawing a red or a black marble?

(f) What is the probability of drawing marbles of the three different colors?

5.   If two cards are randomly drawn from a well shuffled deck of playing cards, find the probability that

(a) The first card is a jack.   (b) The second card is a spade.

(c) A pair is drawn.   (d) The hand contains at least one jack.

Note: Students often find questions like the one in Part (b) rather confusing. It is perhaps helpful to observe that without any further information being given (say, about the first card), the chance that the second card is a spade should equal (by randomness) the chance that the first (or any other card in the deck) be a spade.

6.   (a) What is the probability that a five-card poker hand "has" three of a kind?

(b) What is the probability that a five-card poker hand is a flush?

(c) What is the probability that a five-card poker hand is a straight?

7.   (a) What is the probability that a five-card poker hand contains four of a kind?

(b) What is the probability that a five-card poker hand "has" two pairs?

(c) What is the probability that a five-card poker hand is a straight flush?

8.   (*Quality Control*) (a) A lot of 80 computers, nine of which are defective, is inspected by a quality control worker as follows: five computers are randomly selected and tested. The lot is rejected if one or more is found defective. Compute the probability that the lot will be rejected.

(b) How would the answer to Part (a) change if instead 10 are inspected?

9.   (*Public Health*) A public health official is screening small villages in an Asian country for the presence of a virulent disease that initially does not show any symptoms. One village of 550 people has 12 that are infected with this disease. The official randomly selects 20 people and tests for the disease. If no one tests positive, the village is declared disease-free.

(a) What is the probability that the village erroneously gets declared disease-free?

(b) How would the answer to Part (a) change if instead of 20 people, 50 were tested?

10. (*de Méré's Paradox*) Verify de Méré's paradox: namely, that it is more likely to roll at least one with four throws of a single die than it is to roll at least one pair of 1s with 24 rolls of a pair of dice.

11. Compute each of the indicated conditional probabilities:

    (a) Suppose that A and B are events with $P(A) = 0.4$, $P(B) = 0.6$, and $P(A \cap B) = 0.2$. Compute $P(A \mid B)$ and $P(A \mid \sim B)$.

    (b) A fair coin is tossed four times. Compute the probability that at least three tosses will land as heads given that the first toss landed as heads. How would this probability change if it were instead given that the first toss landed as tails?

    (c) In a certain state, 60% of the residents are Democrats, 55% favor a mortgage stimulus package, and 40% are Democrats who favor the stimulus package. What is the probability that a given (randomly selected) Democrat favors the stimulus package? Suppose a randomly selected resident turns out to favor the stimulus package. What is the probability that he/she is a Democrat?

    (d) Suppose that a pair of dice is rolled. What is the probability that at least one of the die lands on five, given that the sum of the dice is s, when $s = 2, 3, \cdots, 12$?

12. Compute each of the indicated conditional probabilities:

    (a) Suppose that A and B are events with $P(A) = 0.3$, $P(B) = 0.8$, and $P(A \cap B) = 0.1$. Compute $P(A \mid B)$ and $P(\sim A \mid B)$.

    (b) An urn contains four red balls and seven black balls. Two balls are removed and not replaced. What is the probability that the second ball is red, given that the first ball is black? What is the probability that the second ball is red, given that both balls are the same color?

    (c) In a certain college, 60% of the students are females, 25% of the students are business majors, and 15% are female business majors. What is the probability that a given (randomly selected) female student is a business major? What is the probability that a randomly selected business major is female?

    (d) Suppose that the cards in a shuffled deck are turned over one-by-one. Suppose a king turns up as the very first card. What is the probability that the next card turned over will also be king? Suppose a king first turns up as the 15th card. What is the probability that the 16th card will also be a king?

13. (*Conditional versus Unconditional Probabilities*) In each part, two events, A and B of an experiment are indicated. You should first try to guess whether the conditional probability $P(A \mid B)$ is greater than, less than, or equal to the corresponding unconditional probability $P(A)$ (so in the last case A and B will be independent). Then explicitly compute the two probabilities to check your intuition.

    (a) A and B are events with $P(A) = 0.4$, $P(B) = 0.4$, and $P(A \cap B) = 0.2$.

    (b) In a small town, 20% of the adult residents work at the nuclear power plant, 0.10 % of the residents have cancer, and 0.05% of the residents have cancer and are working at the nuclear plant. A is the event "has cancer," and B is the event "works at the nuclear power plant."

    (c) A coin is flipped four times. A is the event "a run of three consecutive flips with the same outcome HHH or TTT has occurred," and B is the event "at least one flip was tails."

    (d) A die is thrown three times. A is the event "at least two numbers are the same," and B is the event "at least one six was rolled."

    Suggestion for Part (d): After using the definition of conditional probability, work with complements. Both De Morgan's law and the inclusion-exclusion principle will be useful with this approach.

14. (*Conditional versus Unconditional Probabilities*) Repeat the directions of Exercise 13 for each of the following pairs of events.

    (a) A and B are events with $P(A) = 0.8$, $P(B) = 0.95$, and $P(A \cap B) = 0.7$.

    (b) On a small island, 30% of the inhabitants are left-handed, 10% have blue eyes, and 0.1% are blue-eyed left-handers. A is the event "is left-handed," and B is the event "has blue eyes."

    (c) In a game of poker between two people, A is the event "Player #1 has a flush," and B is the event "Player #2 has four of a kind."

(d) A coin is flipped four times. *A* is the event "all four flips landed as heads," and *B* is the event "a run of two consecutive flips with the same outcome HH or TT has occurred."

15. (*Sampling without Replacement*) Urn A contains three red balls and two black balls, while urn B contains eight red balls and three black balls. Suppose that a coin is flipped. If it lands heads, then two balls are drawn randomly without replacement from urn A. If the coin lands tails then two balls will be drawn from urn B in the same fashion. Compute the following probabilities:
(a) The probability that the second ball is red.
(b) The probability that both balls are red.
(c) The probability that the balls were drawn from urn A, given that both balls drawn are red.
(d) The probability that at least one ball drawn is red.
(e) The probability that the balls were drawn from urn B, given that at least one ball drawn is red.

16. (*Sampling with replacement*) Urn A contains three red balls and two black balls, while urn B contains eight red balls and three black balls. Suppose that a coin is flipped. If it lands on heads, then a ball is randomly drawn from urn A, its color noted, it then gets replaced in urn A, the balls get reshuffled, and a second ball is drawn. If the coin lands tails then two balls will be drawn from urn B in the same fashion. Compute the following probabilities:
(a) The probability that the second ball is red.
(b) The probability that both balls are red.
(c) The probability that the balls were drawn from urn A, given that both balls drawn are red.
(d) The probability that at least one ball drawn is red.
(e) The probability that the balls were drawn from urn B, given that at least one ball drawn is red.

17. A gambler has four nickels in his pocket. Two of them are fair coins, one of them has heads on both sides, and the last one is unfair with an 80% chance of landing on heads.
(a) If he randomly draws a coin from his pocket and flips it, what is the probability that it lands on heads?
(b) Given that the coin selected as in Part (a) lands as tails, what is the probability it is one of the two fair coins?
(c) Given that the coin selected as in Part (a) lands as heads, what is the probability that it is the two-headed nickel?
(d) Given that the coin selected as in Part (a) lands as tails, what is the probability that it is the two-headed nickel?

18. Urn A contains three $20 bills and five $5 bills, urn B contains five $50 bills and two $20 bills, and urn C contains eight $100 bills. Without knowing this information, a contestant picks an urn at random and from it randomly removes (without replacement) three bills.
(a) What is the probability that the contestant drew at least $100 (total)?
(b) Find the conditional probability that the contestant drew from urn A, given that the three bills the contestant drew added up to less than $100.

19. Suppose we have three cards of the same size and texture that are colored either white or black on each side, with the color combinations as shown in Figure 6.10, and that you are offered the following straight-up bet: the cards are shuffled and one is randomly drawn and laid out on the table to expose a white face. You will win $5 if the other side is also white, but if it's black you lose $5. Is this a fair bet or would it be either advantageous or foolish for you to accept it?

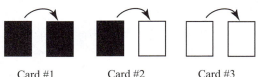

Card #1   Card #2   Card #3

FIGURE 6.10: The faces of the three cards for Exercise 19. Both sides of card #1 are black, both sides of Card #3 are white, while card #2 has one black side and one white side.

20. A 50,000 pound meteor is heading towards a city and four missiles are shot at it. If any of these missiles hits the meteor, the latter will shatter and will not cause any serious damage. The missiles has independent probabilities of 40%, 60%, 70%, and 90%, respectively, of hitting the meteor.
(a) What is the probability that the meteor gets hit by at least one missile?
(b) Given that the meteor was hit by at least one missile, what is the probability that the missile with 90% accuracy made a hit? Find also the corresponding conditional probability for the missile with 40% accuracy.

21. (*Detecting independent events can often be nonintuitive*) As in Exercise for the Reader 6.9, we consider the sample space of possible strings of sex distributions for families, but now with either two children or four children. Thus the sample space of two-children families will be {bb, bg, gb, gg}, and that for four-child families will be the 16 four-character sequences obtainable with the alphabet {b, g}. Show that the following events are dependent:

E = the event that there is at most one girl
G = the event that the family has children of both sexes.

Recall in Exercise for the Reader 6.9 we saw that these events are independent for three-child families.

22. Consider the circuit shown in Figure 6.11 for moving data from point A to point B. The five breakers operate independently and the probability that each is open at any given time is as indicated in the figure. At any given time, data will flow from A to B if there is a path through closed breakers.
(a) If the 0.05 (vertical) breakers are open (i.e., removed), compute the probability that data will flow (from A to B) in this parallel circuit.
(b) Compute the conditional probability that if data flows in the simplified circuit of Part (a) then the top breaker is closed.
(c) Compute the probability that data will flow from A to B in the complete circuit shown in Figure 6.11.
(d) Compute the conditional probability that if data flows in the complete circuit of Figure 6.11, then the top breaker is closed.
(e) Compute the conditional probability that if data flows in the complete circuit of Figure 6.11, then at least one of the two vertical (0.05) breakers is closed.

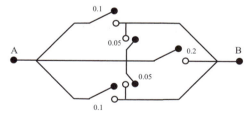

FIGURE 6.11: A three-way circuit with a crossover for the data flow problems of Exercise 22.

23. Suppose that a penny is flipped repeatedly until either it lands as tails for the first time or it has landed as heads four times.
(a) What is the probability that this experiment ends with the last flip being heads?
(b) What is the probability that this experiment ends with the last flip being tails?
(c) Find the conditional probability that this experiment ends with the last flip being heads, given that the first flip lands as heads.
(d) Find the conditional probability that this experiment ends with the last flip being tails given that the first flip lands as heads.

24. Suppose that a penny is flipped repeatedly until either two consecutive heads (HH) appear or three consecutive tails (TTT) appear.

(a) What is the probability that this experiment ends with the last flip being heads?

(b) What is the probability that this experiment ends with the last flip being tails?

(c) Find the conditional probability that this experiment ends with the last flip being heads, given that the first flip lands as heads.

(d) Find the conditional probability that this experiment ends with the last flip being tails, given that the first flip lands as heads.

25. An experiment involves flipping a penny until it lands on heads.

(a) Describe the sample space.

(b) Describe the event E_n that the experiment takes n flips, and compute $P(E_n)$.

(c) Compute $P(\bigcup_n E_n)$. Is there an outcome in the sample space that lies outside $\bigcup_n E_n$? If there is, find one; otherwise explain why one cannot exist.

26. Recall from the definitions of independent events that for two events only one condition needs to be checked, whereas for three events four conditions needs to be checked.

(a) How many conditions need to be checked to verify that a collection of four events is independent? Answer the same question for five events.

(b) Find a formula for the number of conditions that needs to be checked to verify that a collection of n events is independent, where $n > 1$ is an integer.

27. A salesman got 16 parking tickets while making his routes in the same city. All of the tickets were given on Wednesdays. Find the probability of this event. Should he feed the parking meters only on Wednesdays? Assume that he works the same number of hours on each of the five workdays.

28. If a group of 30 women stand in a row, compute the probability that Alice and Gail will have exactly one woman in between them. More generally, for each $r < 29$, compute the probability that Alice and Gail have exactly r women in between them. For this latter question, which value of r gives the greatest probability, i.e., what is the most likely number of people that Alice and Gail will have standing in between them?

29. A drawer contains n different pairs of socks that were randomly thrown in after being laundered. If $2s$ socks are randomly removed from the drawer (with $s \leq n$), compute the following probabilities:

(a) no matched pairs of socks.   (b) exactly one matched pair.   (c) exactly two matched pairs.

30. (*A comparison example for the usage of the complementary probability rule*) This exercise will demonstrate how much more complicated a direct approach (Parts (b) and (c)) can be compared with an application of the complementary probability rule (Part (a)) to a birthday problem.

(a) Use the complementary probability rule to compute the probability that among five randomly chosen persons at least two were born on the same weekday (of the 7 days of the week).

(b) Show that any outcome in the event E of a common birth weekday for the problem in Part (a) must occur in one of the following mutually exclusive cases: (here a, b, c, d denote different days of the week) $\{a, a, b, c, d\}$, $\{a, a, a, b, c\}$, $\{a, a, a, a, b\}$, $\{a, a, a, a, a\}$, $\{a, a, b, b, c\}$, or $\{a, a, b, b, b\}$.

(c) For each of the five (sub)events of Part (b), count the number of outcomes contained in it, and then add up all of these numbers. Use this to compute the probability of E, and check with the answer you obtained in Part (a).

31. (*Variations on the birthday problem*) For each of the following events, which are variants of that of the birthday problem (Example 6.4), do the following: (i) Find the smallest number n of people that are needed in order that the probability of the event occurring first exceeds 50%. (ii) Plot a graph (cf. Figure 6.6) of the probability of the birthday-related event occurring versus n, the number of people in the room. As in the original birthday problem, assume that the peoples' birthdays are randomly distributed over a 365-day year.

(a) The event that someone has a birthday on May 19.

(b) The event that (at least) two people share May 19 as their common birthday.

(c) (*Triple birthday problem*) The event that at least three people share the same birthday (any day of the year).

32. Prove Part (b) of Theorem 6.1.

33. Suppose that S is a finite or countably infinite set (a sample space) and that $p : S \to [0,1]$ is a function satisfying $\sum_i p(s_i) = 1$. Show that the set function defined for any event $E \subseteq S$ by $P(E) = \sum_{s_i \in E} p(s_i)$ is a probability function that satisfies Kolmogorov's axioms.

34. Verify that the definition of conditional probability $P(E \mid F)$ given in Definition 6.6 coincides with the ratio of the number of outcomes in E that are also in F over the number of outcomes in the reduced sample space F, in the case that all outcomes of the sample space are equally likely.

35. (*Conditional probability satisfies Kolmogorov's axioms*) Let F be an event with $P(F) > 0$ in any probability system that satisfies Kolmogorov's axioms (Definition 6.5). Show that the conditional probability function $s(E) \equiv P(E \mid F)$ also satisfies Kolmogorov's axioms.

36. (*The general product rule*) Prove the following general version of the multiplication rule for events $E_1, E_2, \cdots, E_n$:
$$P(E_1 \cap E_2 \cap \cdots \cap E_n) = P(E_1) \cdot P(E_2 \mid E_1) \cdot P(E_3 \mid E_1 \cap E_2) \cdot \cdots \cdot P(E_n \mid E_1 \cap E_2 \cap \cdots \cap E_{n-1}).$$
Suggestion: Use induction.

37. (*Bonferroni's inequality*) (a) Assume $P(E) = 0.8$ and $P(F) = 0.7$, and prove that $P(E \cap F) \geq 0.5$.

(b) Prove *Bonferroni's inequality* for two sets: $P(E \cap F) \geq P(E) + P(F) - 1$.

(c) Use mathematical induction to prove *Bonferroni's inequality* for n sets:
$$P(E_1 \cap E_2 \cap \cdots \cap E_n) \geq P(E_1) + P(E_2) + \cdots + P(E_n) - 1.$$

38. Suppose that $E_1, E_2, \cdots, E_n$ is a collection of independent events. Prove that
$$P(E_1 \cup E_2 \cup \cdots \cup E_n) = 1 - [1 - P(E_1)] \cdot [1 - P(E_2)] \cdot \cdots \cdot [1 - P(E_n)].$$

39. (*Inclusion-exclusion principle for a collection of n sets*) (a) Show that both the two-set inclusion-exclusion principle (of Theorem 6.1(c)) and the three-set version (of Exercise for the Reader 6.3) are special cases of the following *n-set inclusion-exclusion principle*:
$$P(E_1 \cup E_2 \cup \cdots \cup E_n) = \sum_{i=1}^{n} P(E_i) - \sum_{i_1 < i_2} P(E_{i_1} \cap E_{i_2}) + \cdots$$
$$+ (-1)^{a+1} \sum_{i_1 < i_2 < \cdots < i_a} P(E_{i_1} \cap E_{i_2} \cap \cdots \cap E_{i_a}) + \cdots + (-1)^{n+1} \sum_{i_1 < i_2 < \cdots < i_n} P(E_{i_1} \cap E_{i_2} \cap \cdots \cap E_{i_n}).$$

In this notation, the typical summand $\sum_{i_1 < i_2 < \cdots < i_a} P(E_{i_1} \cap E_{i_2} \cap \cdots \cap E_{i_a})$ means that we should take the sum over all of the $C(n,a)$ possible subcollections of size a of the events $\{E_1, E_2, \cdots, E_n\}$. In other words, to compute $P(E_1 \cup E_2 \cup \cdots \cup E_n)$, we first add up the $C(n,1)$ probabilities of single sets E_i, then we subtract off the $C(n,2)$ probabilities of all double intersections $E_i \cap E_j$ $(i < j)$, then add on the $C(n,3)$ probabilities of all triple intersections $E_i \cap E_j \cap E_k$ $(i < j < k)$, and so on.

(b) Use mathematical induction to prove the *n*-set inclusion-exclusion principle.

(c) Give a noninductive (combinatorial) proof of the *n*-set inclusion-exclusion principle.

(d) Use the *n*-set inclusion-exclusion principle to solve the following famous *matching problem*:[14] A frazzled secretary has just stuffed *n* letters into envelopes and sealed them, after which she realizes that the corresponding address labels that were waiting for her at the printer were in a different (mixed up) order than the sealed envelopes. Being late for an important engagement, she randomly sticks the *n* labels onto the *n* envelopes and drops them in the outgoing mail slot on her way out, smiling goodbye to the manager. Find the probability that at least one letter will go to its correct address.

Suggestions and Comments: For (c), consider a single outcome x of the sample space. If $x \in \bigcup_{i=1}^{n} E_i$, the trick will be to show that (with cancellations) x will be counted exactly once in the right-side expression of the inclusion-exclusion principle. Let $N = N(x)$ be the largest positive integer such that x is in N of the sets $\{E_1, E_2, \cdots, E_n\}$. Then, in the right side of the inclusion-exclusion formula, x will be counted $C(N,1) = N$ times with $\sum_{i=1}^{n} P(E_i)$; next, since x will lie in exactly $C(N, 2)$ sets of the form $E_i \cap E_j$ $(i < j)$, it will be counted $-C(N, 2)$ times with $-\sum_{i_1 < i_2} P(E_{i_1} \cap E_{i_2})$, and in general x will be counted $(-1)^a C(N,a)$ times with $(-1)^{a+1} \sum_{i_1 < i_2 < \cdots < i_a} P(E_{i_1} \cap E_{i_2} \cap \cdots \cap E_{i_a})$, as long as $a \leq N$. When $a > N$, such terms will not count x since x cannot lie in an intersection of more than N E_i's. Use the binomial theorem to show that these resulting counts add up to one.

The answer to Part (d) is $1 - 1/2! + 1/3! - 1/4! + \cdots + (-1)^{n+1}/n!$. This expansion looks similar to the Taylor series expansion of e^x from calculus: $e^x = 1 + x/1! + x^2/2! + x^3/3! + \cdots$ that converges for all real numbers x. If we substitute $x = -1$, and compare with the probability, we see that the probability expansion equals the first n terms of the series $1 - e^{-1} = 0.6321...$, and thus for large values of n, the probabilities of the matching problem are all approximately equal to this number. Many people find it surprising that the probabilities of the matching problem stabilize like this for large values of n.

6.2: RANDOM NUMBERS, RANDOM VARIABLES, AND BASIC SIMULATIONS

Many problems in mathematics and other sciences are either very difficult or impossible to solve analytically. This is a reality that any scientist is well aware of. Even for problems that do admit exact solutions, it is often difficult and time-consuming to obtain them. **Simulation** is the technique of running trials of artificial experiments that replicate the actual phenomena under study, in order to

---

[14] This famous problem dates back to the early history of probability. It can be presented in several different settings; ours is rather contemporary. The answer depends on n, and it would be instructive for the reader to guess at what happens to this probability as n gets larger before going on to solve it explicitly.

obtain approximations of answers to the questions that need to be answered. Simulations are run on computers and the outcomes can be supported with probability and statistical theories. With the ever-increasing power and convenience of computers, simulations are often much more economical (and safe) than running actual tests; examples include NASA's testing of a new space shuttle, or performances of automobiles in crash tests. Simulations can be run to test designs of objects before they are actually built; such tests can save much time and energy of going through complete production processes of designs that are destined to fail.

Probabilities as Relative Frequencies

The calculation of probabilities provides a very nice class of problems to which computer simulations naturally can be applied. This follows from a very intuitive theorem from probability that may be roughly stated as follows: If we (independently) repeat any experiment a very large number n of trials and tally the total number of these trials $c_E(n)$ in which an event E occurred, then P(E) will be very close to the ratio $c_E(n)/n$. We next give a more rigorous statement of this theorem, which is a special case of the so-called *Law of Large Numbers*.

THEOREM 6.3: (*Probabilities as Relative Frequencies*) Suppose that an experiment is repeated, independently and continually, and that at each trial (say the nth, $n = 1, 2, 3, 4, \ldots$) a count $c_E(n)$ is kept of the total number of trials from the first through the nth for which the event E has occurred. It then follows that with probability 1, we have

$$\lim_{n \to \infty} \frac{c_E(n)}{n} = P(E). \tag{10}$$

Some comments about this theorem are in order. We need to explain the meaning of the clause "with probability 1" appearing in the last sentence. We do so using an example. Suppose, for simplicity, the experiment under consideration consists of a single flip of a fair coin, and the event E is "heads." Here, we are to continue to repeat this experiment, and $c_E(n)$ will simply be the count of the number of flips that were heads among the first n flips. Equation (10) involves the limit of $c_E(n)/n$ as $n \to \infty$. We can think of a (much) bigger mega-experiment whose outcomes consist of the infinite sequences of head/tail outcomes each resulting from a single (but infinite) series of repetitions of the coin flip experiment. If we denote the sample space of this mega-experiment by $\mathscr{S}$, then a typical element $\omega \in \mathscr{S}$ consists of the resulting infinite head/tail sequence from a single infinite series of coin flips, e.g.,

$$\omega = < H, H, T, T, T, H, T, T, \cdots > .$$

For this element of the sample space, we have $c_E(1) = 1$, $c_E(2) = 2$, $c_E(3) = c_E(4) = c_E(5) = 2$, etc. Thus, a more proper notation for $c_E(n)$ would be $c_{E,\omega}(n)$, since this function of n depends on the particular element $\omega \in \mathscr{S}$ that represents a mega-coin flip experiment. From the probabilities associated with any individual trial (coin flip) (P(H) = P(T) = 1/2), the associated mega-sample space $\mathscr{S}$ naturally inherits a probability function. For example, the event $\mathscr{E}$ consisting of all elements in $\mathscr{S}$ that begin with a sequence of four heads:

$$\mathscr{E} = < \text{H, H, H, H, *, *, *, *, } \cdots >,$$

where the "*" denotes either "H" or "T", would have probability $P(\mathscr{E}) = (1/2)^4$ corresponding to the requirements only on the first four flips.[15]   Theorem 6.3 states that the set of infinite series of coin flip sequences that do not satisfy (10) must have probability zero. This is not to say that it is impossible to violate (10). For example, the sample point $< H,H,H,H,H,H,H,H, \cdots > \in \mathscr{S}$ corresponding to getting heads all the time, while theoretically possible, is <u>extremely</u> unlikely. This would even be more incredible than a human being who always got heads whenever he/she flipped a coin. (Such an individual could become very rich as a gambler.)   For a proof of Theorem 6.3, we refer the interested reader to any good book on probability theory, for example, [Bil-85], Section 6.6 of [Chu-79], or Chapter 8 or [Ros-02].

Theorem 6.3 is a cornerstone for the use of simulations to approximate probabilities. It tells us that we can estimate any probability by repeating an experiment a large number of times and recording the relative frequency of the event under consideration. The only drawback is that the estimate (10) is only a limit; it does not tell us how large we need to take n in order to achieve the desired accuracy.   Later we will develop some techniques that will be helpful in determining when to stop a simulation. After all, computers (as well as humans) can only do a finite number of experiments in a finite amount of time.

Random Numbers and Random Variables

The main ingredient to running any simulation is the ability to (repeatedly) generate *random outcomes* of an experiment so that the outcomes arise according to the true probabilities of the experiment. Before the time of computers, random outcomes and numbers were generated by purely mechanical means: coin flips, spinning wheels, drawing (numbered) balls from an urn, etc. In order for a

---

[15] The space $\mathscr{S}$ is very large (and uncountable) and it is actually impossible to define a probability function (that satisfies the three Kolmogorov axioms) on all subsets of $\mathscr{S}$. The probability can be defined for a very large class of (so-called *measurable*) subsets of $\mathscr{S}$. These measurable sets include all reasonable sets that would ever come up in applications. Further discussion of these topics would require more advanced prerequisites than this book assumes. A fine (albeit rather advanced) reference for probability in this more advanced setting is the book by Billingsley [Bil-85].

computer to produce random numbers, it must be programmed to do so. An oxymoron thus presents itself: a sequence of truly random numbers should be "unpredictable," whereas any output of a computer program is entirely predictable. So technically, a computer can generate only what we call **pseudorandom numbers**, which although obtained in a deterministic fashion, satisfy all of the important statistical tests for true randomness. Most scientific computing platforms contain such "random number generators" that will randomly generate a real number between 0 and 1 on demand. We treat such a generator as our black box.[16]

CONVENTION: Throughout the rest of the text we assume that we have access to a random number generator (on our computing platform), which we refer to as `rand`. Each time we call on it, `rand` will generate a (new) pseudorandom real number in the interval [0,1].

The random numbers generated by `rand` are equally likely to land anywhere in [0,1]. They are thus said to be **uniformly distributed** on [0,1]. They can be used to generate random outcomes for virtually any discrete (or continuous) experiment. In order to better understand how this can be done, we now digress to discuss the important related concept of a random variable, and its associated distribution. When performing an experiment, we are often interested in some numerical quantity (or function) of the outcome. Since such a numerical quantity depends on a random outcome (for each trial of an experiment), we call such a function a *random variable*:

DEFINITION 6.9: A **random variable**, usually denoted as X, Y, Z, etc., is a real-valued function with domain being the sample space, i.e., $X : S \to \mathbb{R}$.[17] A random variable X is called **discrete** if it has either a finite or countably infinite set of outcomes (i.e., its range is at most countable).[18] The **probability distribution function (pdf)** (also known as the **probability mass function**) of a discrete random variable X is the function $p(= p_X)$ defined as follows:

$$p : \text{range}(X) \to [0,1] :: p(x) = P(X = x).$$

---

[16] One such method for pseudorandom number generation was the linear congruential method that was discussed in Section 4.2.

[17] **Technical aside:** It is also required that for any real number α, the set $X^{-1}([\alpha,\infty)) \equiv \{s \in S \mid X(s) \geq \alpha\}$ be an event. For finite or countably infinite sample spaces this condition is redundant; even in general probability settings (satisfying Kolmogorov's axioms), all but the most obscure functions $S \to \mathbb{R}$ will satisfy this condition. In what follows we tacitly forgo further comments on such matters. Interested readers can find more details in [Bil-85].

[18] A set S is *countable* if its elements can be listed as a finite or infinite sequence: $S = \{s_1, s_2, \cdots, s_n\}$ or $S = \{s_1, s_2, \cdots\}$. This condition can be rephrased in terms of functions by stating that there exists a one-to-one mapping $f : S \to \mathbb{Z}_+$, or that there exists an onto mapping $g : \mathbb{Z}_+ \to S$. A set is uncountable if it is not countable. Examples of countable sets include the integers $\mathbb{Z}$, the rational numbers $\mathbb{Q}$, and any subsets of these (or any countable) sets. The real numbers $\mathbb{R}$, or any interval of real numbers like [0,1] is an uncountable set. Countably infinite sets are the "smallest" infinite sets.

As a very simple example, consider the experiment of rolling a single die, and let X be the random variable equaling the number of pips that show.   This discrete random variable has range range(X) = {1,2,3,4,5,6}. The probability distribution function of X is the function $p = p_X : \{1,2,3,4,5,6\} \to [0,1]$ defined by $p(i) = P(X = i) = 1/6$, since each of the six possible outcomes is equally likely. A graph of this pdf is shown a bit later in Figure 6.14(a).

EXERCISE FOR THE READER 6.11: Consider the experiment of flipping a coin until either it lands as heads, or four flips have landed as tails. Let X be the random variable equaling the number of flips needed for this experiment to terminate. Find the range and the probability distribution function of X.

For a discrete random variable X, we can enumerate its range: $x_1, x_2, x_3, \cdots$ and since these values cover all possible outcomes of the experiment, their associated probabilities must add up to 1: $\sum_i p(x_i) = 1.$

We point out that it is sometimes convenient for us to allow random variables to take on ∞ as a value (see Exercise 25 for an example), but the event that such a random variable will be infinite will always be zero, so it will be practically impossible for this event to occur.

Binomial Random Variables

EXAMPLE 6.12: (*Binomial random variables*)   We consider an experiment with only two outcomes, which we refer to as a *success* or a *failure*. Thus, we may write: P(*success*) = p and P(*failure*) = q = 1 – p (since $p + q$ = 1). For example, if the experiment is flipping a coin, and we label "heads" as a *success*, then $p = q$ = 1/2. If the experiment is rolling a pair of dice and a *success* is considered winning a game of craps (the dice adding up to either 7 or 11), then p = 2/9 and q = 7/9. We repeat our experiment for a total of n trials, and we let

X = total number of *successes* in the n independent trials.

X is called a **binomial random variable** with parameters n (number of trials) and p (probability of success); it is denoted as $X \sim \mathcal{B}(n, p)$. The range of X is {0, 1, 2, ..., n}.   By independence, the probability of any particular outcome of the n trials with k successes (and thus $n - k$ failures) will be $p^k q^{n-k}$.   Since there are $C(n, k)$ such sequences, we get that

$$P(X = k) = C(n,k) p^k q^{n-k} = C(n,k) p^k (1-p)^{n-k}. \tag{11}$$

By the binomial Theorem 5.7 (of Section 5.2), $\sum_{k=0}^{n} P(X = k) = \sum_{k=0}^{n} C(n,k) p^k (1-p)^{n-k} = (p+1-p)^n = 1,$ as expected.   For example, in the setting where the trials represent playing games of craps (as above), the probability of winning five of $n = 6$ games is thus $P(X = 5) = C(6,5)(2/9)^5 (7/9)^1$ $= 0.0025...$ (about 1/4%), whereas the probability of losing five of six games would be $P(X = 1) = C(6,1)(2/9)^1 (7/9)^5 = 0.3795...$ (or about 38%).

Most discrete simulations basically require the generation of values of a relevant discrete random variable. This will be accomplished using the uniformly distributed random variable generated by `rand`. Such random variables are not discrete (since their ranges are uncountable), but do fall under the umbrella of the following definition:

Continuous Random Variables

While we will be primarily concerned with discrete random variables, it will be helpful to have a basic understanding of continuous random variables. In particular, our random number generator rand is an example of a continuous random variable.

DEFINITION 6.10: A **continuous random variable** is a random variable for which there exists a nonnegative (**probability**) **density function** (**pdf**) $f (= f_X)$ defined on the set of real numbers that has the following property for any set A of real numbers:

$$P(X \in A) = \begin{cases} \text{The area under the graph of } f, \text{ that} & {}_{19} \\ \text{lies above the } x\text{-axis and over the set } A. \end{cases} \tag{12}$$

Probability density functions always have total area equal to 1 (= 100%) under their graphs and over the x-axis since this would correspond to the event $A = (-\infty, \infty)$, which must cover the whole sample space. Geometrically, equation (12) says that the probability that X lands in a set A is just the proportion of the total area under the density curve that lies over the set A.

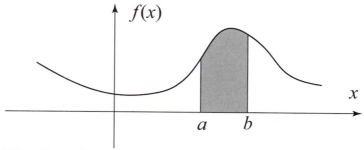

FIGURE 6.12: Illustration of a typical probability density function for a continuous random variable X. The shaded region represents the probability $P(a \le X \le b)$.

---

[19] In calculus, this area is called the integral of the function f over the set A and is denoted as $\int_A f(x)dx$.

We point out some properties of a density function f for a continuous random variable. Generally speaking, by equation (12) (and Figure 6.12) X is more likely to land in regions where $f(x)$ is large. We summarize some basic properties of continuous random variables and their density functions; while these should be quite intuitive, formal justifications may be given.[20]

1. $f(x) \geq 0$ (*Density functions are nonnegative.*)
2. $P(-\infty < X < \infty) = 1$ (*The total area under a density curve is always 1.*)
3. $P(X = a) = 0$ (*The probability of any exact value is zero.*)

DEFINITION 6.11: The **cumulative distribution function (cdf)** of a random variable X is the real-valued function $F(= F_X)$ defined by

$$F : \mathbb{R} \to [0,1] :: F(x) = P(X \leq x).$$

In case X is a discrete random variable with range $\{x_1, x_2, \cdots\}$, in terms of the corresponding probability distribution function p, we may write

$$F(x) = P(X \leq x) = \sum_{x_i \leq x} p(x_i),$$

while for a continuous random variable, equation (12) gives us that

$$F(x) = P(X \in (-\infty, x]) = \begin{cases} \text{The area under the graph of } f \text{ that} \\ \text{lies above the horizontal and to the left of } x. \end{cases}$$

NOTE: (*For readers who have studied calculus; others may ignore*) If we differentiate this equation, written as $F(x) = \int_{-\infty}^{x} f(t)dt$, (using the fundamental theorem of calculus), we obtain

$$F'(x) = f(x),$$

i.e., the derivative of the cdf is the pdf.[21]

For any cumulative distribution function (for either discrete or continuous random variables) it is not difficult to verify that $F : \mathbb{R} \to [0,1]$ is a nondecreasing function, with $F(x)$ decreasing to 0 as x moves out to $-\infty$, and $F(x)$ increasing to 1 as x moves out to ∞ (see Exercise 28). The cdf of a discrete random variable will be a step function with jumps only at the (at most countable) values of its range: $x_1, x_2, x_3, \cdots$. Also, at each x_i, the amount of the jump of F will equal

---

[20] Sometimes we will deal with density functions that are piecewise continuous, meaning that they are continuous, except at a finite number of points (at which they have jumps). The most general functions that can serve as density functions are so-called *measurable functions*. Details on this more sophisticated general approach can be found in [Bil-85].
[21] This relation holds at all values of x at which f is continuous. For piecewise continuous functions, it will fail only at those values of x for which f has a jump discontinuity.

$p(x_i)$.   For example, the cumulative distribution function for the discrete random variable X equaling the number showing on a single roll of a die has jumped by 1/6 at each of the numbers {1,2,3,4,5,6} in its range. The graph is shown in Figure 6.13.

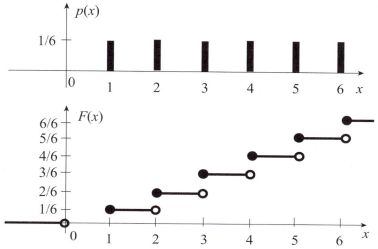

FIGURE 6.13: Graphs of (a) (top) the pdf $f(x)$, and (b) (bottom) the cdf $F(x)$ for the random variable X = the number showing when a single die is rolled.

EXERCISE FOR THE READER 6.12:   Sketch plots for both the probability mass function and the cdf for the binomial random variable $\mathcal{B}(6, 2/9)$ (corresponding to the six trials of craps of Example 6.12).

Uniform Random Variables

EXAMPLE 6.13: (*Uniform random variables*) For any interval $[a, b]$ (with $a < b$), the   continuous   random variable with pdf

$$f(x) = \begin{cases} 1/(b-a), & \text{if } a \leq x \leq b \\ 0, & \text{otherwise} \end{cases},$$

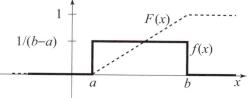

is said to be **uniformly distributed** over the interval $[a, b]$, we denote this as $X \sim \mathcal{U}(a, b)$. The pdf and cdf for such a uniform random variable are shown in Figure 6.14. Note that the area under $f(x)$ is a

FIGURE 6.14: Graphs of the pdf $f(x)$ and the cdf $F(x)$ for the uniform distribution $\mathcal{U}$ (a, b) on an interval $[a, b]$.

rectangle with width $b - a$ and height $1/(b - a)$ (and so has area equal to one). Since for $a < x < b$, the area under the pdf to the left of a is also a rectangle with height width $x - a$ and height $1/(b - a)$, the following formula for the cdf thus follows:

$$F(x) = \begin{cases} 0, & \text{if } x < a \\ (x-a)/(b-a), & \text{if } a \leq x \leq b \\ 1, & \text{if } x > b. \end{cases}$$

Our generic random number generator `rand` has distribution $\mathcal{U}(0, 1)$.

Setting up a Simulation

We will soon introduce some more important concepts relating to random variables, but at this point, since we have sufficient background, without further ado, we show how to set up a simulation for a simple probability experiment.

NOTE: To actually run a simulation, the reader will need a computer. Those readers who are not using computers can still read through and learn the material on how to set up a simulation and be able to understand the results of actual computer simulations that we display.   In our development of simulations that follows, we will occasionally give some computing advice; such passages may be ignored by noncomputer users.

EXAMPLE 6.14: (*The simplest simulation: coin flips*) Set up a simulation and (if you are using a computer) run it on your computing platform that will simulate 25,000 flips of a coin and record the proportion of heads (the simulated approximation that a coin flip results in heads). Also, keep a running record of the ratio of heads over the total number of flips and plot this ratio versus the number of flips.

SOLUTION: To use our $\mathcal{U}(0, 1)$-random number generator `rand`, if a generated random number is less than $1/2$, we count the trial as a heads, while if it is greater than or equal to $1/2$, we count the trial as a tails.[22]   The scheme is illustrated in Figure 6.15. The following pseudocode describes how to implement such a simulation. Since we will need to record all of the cumulative ratios of the number of heads over the number of flips, we will use a vector (that we call `frequencies`) for this purpose.

---

[22] We arbitrarily assigned the middle value $1/2$ to correspond to tails. At first glance, this may seem that we are giving a slight preference to tails, but recall that the probability that any continuous random variable takes on an exact value is zero.

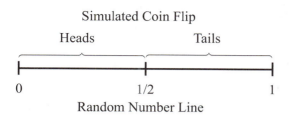

Simulated Coin Flip

Heads   Tails

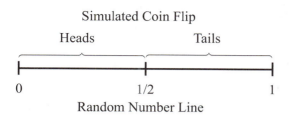

0   1/2   1

Random Number Line

FIGURE 6.15: Scheme for using random numbers in [0, 1] to simulate coin flips.

```
Initialize: headcount = 0,
frequencies (a vector)
FOR n = 1 TO n = 25000
  flip = rand
  IF flip < 1/2
    UPDATE
headcount = headcount + 1
  END IF
  SET
frequencies(n) = headcount/n
END FOR
OUTPUT frequencies(25000)
(overall relative frequency)
```

The final approximating ratio (`headcount/25000`) for a run of this simulation turned out to be 0.4967.   A plot of the record of all cumulative frequencies obtained is shown in Figure 6.16.   Although quite simple, this example nicely displays some general features of running a simulation. Due to the randomness of the experiment, the relative frequencies do not tend smoothly to the limiting value, but do so in a jagged fashion, often crossing the exact value. Repetitions of the simulations would result in different data and graphs (you should do some experiments) corresponding to different elements in the mega-sample space.

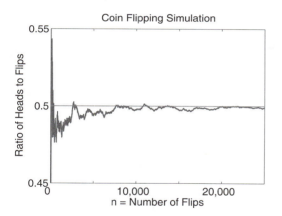

FIGURE 6.16: Complete graphical record of all relative frequencies of the heads outcome during a simulation of 25,000 coin flips.

For a required accuracy of the estimate, the question of how far must we go is always a critical one. The reader should run some experiments to see what transpires in this example if we use larger numbers of trials. Approximately how large a value of N can be used in each part before your computing platform begins to take more than 5 minutes? Run times can be cut down if we forgo storing the vector of cumulative frequencies.   Such experiments will motivate the need to develop some theory to help us to decide when to end a simulation. Initially, it will be helpful to perform several trials of a simulation with the same number of trials each, and then to display the results in a histogram. This allows one to get a rough idea of the consistency of the results. Since trials are independent, these histograms will tend to take the general bell-shape-curve (normal distribution) that is studied in statistics.

EXAMPLE 6.15:   (*Simulating Craps Games*) This example will involve simulating repeated trials of the game of craps discussed earlier: throw two dice; you win if the pips add up to either seven or eleven. We have shown that the probability of winning is $2/9 = 0.2222\ldots$ (an elementary computation).

(a) Simulate $K = 10$ sets of $N = 1000$ craps games each, and record the proportion of winning games for each of the 10 sets of simulated data. Plot the results in a histogram and record the final approximation resulting from formula (10).

(b) Repeat Part (a) but this time simulate $K = 1000$ sets of $N = 5000$ craps games each.

SOLUTION: Simulating a single throw of two dice can be achieved by using a scheme similar to the one employed in the last example, but this would involve breaking up the unit interval into six equally spaced subintervals and using a five-case if branch (`rand <= 1/6`, or `1/6 < rand <=2/6`, etc.).   It will be a bit simpler to program if we first multiply the `rand` by six, since the resulting value will now be a randomly generated number uniformly distributed in the interval [0, 6], and then round upward to the next greatest integer. For example, if `rand` were equal to $0.2301,\ldots$, then `6*rand = 1.3806` which rounds up to 2, and this would represent rolling a 2 in the simulated roll of two dice. Your computing platform might have built-in functions that round up (or down) to the nearest integer (these are the *ceiling* and *floor* functions that were introduced in Section 2.1); if not, you might want to build them.

The pseudocode below is given for Part (a), but is easily modified for Part (b) by changing the parameters for K and N.

```
Initialize:   K = 10  (# of sets of trials)
              N = 1000  (# of trials per set)
              P  (a vector to record data for K groups of trials)

FOR k = 1 TO k = K
  set WinCounter = 0  (initialize win counter for kth set of N trials)

  FOR n = 1 TO n = N
    set dice1 = ceiling(6*rand), dice2 = ceiling(6*rand)
    set sum = dice1 + dice2
   IF sum = 7 OR  sum = 11  (winning craps trial)
      update WinCounter = WinCounter + 1;  (augment win counter)
   END IF
  END n FOR
  set P(k) = WinCounter/N;  (Simulated probability from kth set of)
                              (N trials)
END k FOR
OUTPUT mean(P)  (mean of the vector P is overall relative frequency of wins)
```

The outputted final approximations are as follows, and the corresponding two histograms are shown in Figure 6.17.[23]

For Part (a): `mean(P)` →ans = 0.2231
For Part (b): `mean(P)` →ans = 0.22225764000000

These results compare reasonably well with the exact answer $2/9 = 0.\overline{2}$.

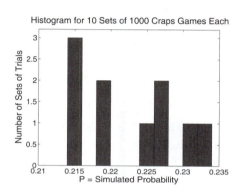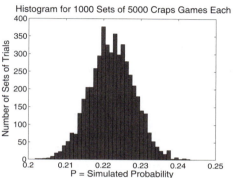

FIGURE 6.17: (a) (left) Histogram from the set of 10 simulated probabilities for a winning game of craps using 1000 games each. (b) (right) Corresponding histogram with 1000 simulated probabilities where each is computed using 5000 games.

Caution: Your results and histograms will vary each time you run this simulation, depending on the state of your random variable generator. In particular, if you repeat this same experiment, and look at the resulting histograms and means of P, you will most likely see different results on each run.[24]

Some comments are in order. Note that in Part (b), five million simulations were performed.   Unlike with Part (a), the computation in Part (b) took a more noticeable amount of time (perhaps at least a few seconds, depending on the machine you are using and the processes that you are running at the time). We thus begin early to notice the limits of our computer power. Generally speaking, (at the time of the writing of this book) most personal computers can handle on the order of about 10 million computations before they start to take several seconds in real time.   The histogram in Figure 6.17(b) starts to exhibit the familiar bell-

---

[23] Many computing platforms have tools for creating histograms (if the scientific computing platform you are using does not have such a tool, the readily available Excel software by Microsoft could be used). If it is not feasible for you to create histograms on your system, you may skip those portions of exercises asking for them. Even with the same data sets, histograms vary, depending on the number of "data bins" that are being used.

[24] Once again we remind you that your computer's random number generator is based on an internal program. As such, it is commonly the case that the sequence of random numbers generated upon restarting your computing platform may always be the same. There are usually easy ways to safeguard against this, for example, by setting the state of the random number generator to depend on the computer's internal clock. See the help menu on your random number generator for details.

shaped curve.   As K (the number of sets of data) goes to infinity while N (the number of trials per set) is held fixed, the histogram will approach a certain symmetric bell-shaped curve that is centered around the mean, whereas if K is held fixed and N goes to infinity, the shape of the bell becomes narrower (and taller). These facts can be proved by using the very important *central limit theorem*, but the main ramification for us is that histograms formed by repeated trial runs will provide very good indicators on how many trials to use in a simulation (through some trial and error runs) and a good idea of the accuracy of our resulting approximation.[25]

The idea used in the last example for generating random dice throws can be generalized in several directions. The following three computer exercises (which noncomputer readers should read, but may skip doing) include the creation of some very useful simulation tools.

(COMPUTER) EXERCISE FOR THE READER 6.13:   (a) Write a program having the following syntax:
```
v = randvecsimulator(n, min, max)
```
where the inputs are: n, a positive integer, min and max, two integers satisfying min $\leq$ max, and whose output is a $1 \times n$ vector v, whose components v(i) are integers randomly chosen from the range min $\leq$ v(i) $\leq$ max.
(b) Use your program to generate data with the inputs: min = −5, max = 5, and with n = 5, and then with n = 20.
(c) Use the program with min = −5, max = 5, and several large values of n to create some histograms corresponding to the generated vectors .   Do your histograms seem to be tending to a general shape? Explain why or why not.

We next discuss how to simulate experiments in which <u>not</u> all outcomes have equal probabilities.   If there are finitely many outcomes $\omega_1, \omega_2, \cdots, \omega_n$, with corresponding probabilities $p_1, p_2, \cdots, p_n$, then, since $\sum_{i=1}^{n} p_i = 1$, we can break up the unit interval into mutually exclusive subintervals $I_1, I_2, \cdots, I_n$ having these probabilities   as   their   lengths.   For   example,   we   could   take $I_1 = [0, p_1), I_2 = [p_1, p_1 + p_2), \cdots, I_n = [p_1 + p_2 + \cdots + p_{n-1}, 1]$,   then,   a   random number r generated by rand will lie in exactly one of these subintervals, say $r \in I_i$. This will correspond to randomly generating the outcome ω_i. The next two exercises for the reader are designed to gradually enable the reader to program such a scheme.

EXERCISE FOR THE READER 6.14: Consider again the game of simple craps. Each game (experiment) is played by throwing a pair of dice, and observing the

---

[25] In fact, the central limit theorem can be used to develop procedures for deciding when a simulation can be halted, given a desired accuracy goal. Such developments are beyond the scope of this brief introduction but can be found in any good book on simulation, such as the one by Sheldon Ross [Ros-02].

sum of the numbers showing on the two dice. This will be an integer s in the range [2, 12]. The game is won if $s = 7$ or 11.

(a) Compute the (exact) probabilities $p_i \equiv P(s = i)$, for $i = 2, 3, \ldots, 12$.

(b) (COMPUTER) Write a program with syntax `[v, wins, Pwin] = craps(N)`, having a single input variable `N`, a positive integer corresponding to the number of craps games to be simulated, and three output variables `v`, `wins`, and `Pwin`. The variable `v` will be a length `N` vector where each component is an integer in the range [2, 12] corresponding to a simulated outcome of `N` games of craps. The variable `wins` will denote the total number of the `N` simulated games that were winning games of craps, and the last variable `Pwin` is the resulting simulated probability of winning a game of craps. The program should be written in such a way that only `N` random numbers are generated and the scheme in the above paragraph is used to generate `N` corresponding random craps game outcomes.

NOTE: For storage space considerations, the reader may wish to create, instead of the length `N` vector `v`, an 11-component vector whose components are the frequencies of the 11 possible values (2,3,...,12) that occur.

(c) Run your program to recreate simulated data and histograms for the parameters given in Part (b) of Example 6.15, and compare the results.

(COMPUTER) EXERCISE FOR THE READER 6.15:   (a) Write a general simulation program with the following syntax:

$$v = \texttt{GeneralFiniteSimulator(N, SampleSpc, ProbVec)}$$

This program will be able to simulate any number `N` of random outcomes of any experiment having a finite sample space. The second input variable `SampleSpc` is a vector of distinct integers `[a1  a2  ... an]` corresponding to the (n) possible outcomes of the experiment (these integers correspond to the outcomes $\omega_1, \omega_2, \cdots, \omega_n$ of the experiment), and `ProbVec` is the corresponding vector of probabilities `[p1  p2  ... pn]`. Thus, for our previous coin flipping simulation, we would have `SampleSpc` = [0 1] and `ProbVec` = [1/2 1/2]. For the craps game experiment of the preceding exercise for the reader, we would have `SampleSpc` = [2 3 ... 12]. The output `Vec` is a length `N` vector of simulated outcomes of `N` independent trials of the experiment.

(b) Use this program to re-do Part (b) of Example 6.14.

(c) Use this program to re-do Part (b) of Example 6.15.

Generating Random Permutations and Random Subsets

Since the number of permutations grows exorbitantly fast relative to the number of objects, it is often useful to be able to generate random permutations of a given list of objects. There are other practical applications of random permutations. For example, in a medical experiment, the group of participants must be randomly split up into various control groups of certain sizes $n_1, n_2, \cdots$. This can be accomplished once the names are randomly permuted, for then we could simply designate the first n_1 individuals of the permuted list to be in the first group, the next n_2 individuals to be in the second group, etc. We next explain an algorithm for generating a random permutation of the set $\{1, 2, ..., n\}$.

ALGORITHM 6.1: Generating a Random Permutation of $\{1, 2, ..., n\}$:
Step 1: Set $v(i) = i$, for $i = 1, 2, ..., n$, so the vector v is the original (unpermuted) increasing order list.[26] We will permute a series of pairs of elements of the vector v and when we are done the result will be a random permutation. Initialize $k = n$.
Step 2: Choose a random element from the set $\{v(1), v(2), ..., v(k)\}$, and permute this element with $v(k)$.
Step 3: Decrease k by 1: $k = k - 1$. If $k > 1$ go back to step 2, otherwise proceed to step 4.
Step 4: The resulting vector v is a random permutation of the original.

Some remarks are in order. For the permutation generated by the above algorithm to be random means that each of the $n!$ possible permutations is equally likely to be produced. That this is indeed the case is easily established in several different ways; see, for example, Exercise 29.   This program can also be used to generate random k-combinations (subsets of size k) of a set of n elements.   In this case the algorithm can be halted as soon as the (last) k positions of the permutation have been determined, or, if $k > n/2$, we can just fill the last $n - k$ positions and take the random k-combination to be the complement of the resulting $(n - k)$-combination.

(COMPUTER) EXERCISE FOR THE READER 6.16: (a) Write a program that implements Algorithm 6.1 to generate random permutations of the set $\{1, 2, ..., n\}$, and having the following syntax:

$$v = \texttt{RandPerm(n)}$$

The output v is a length n vector of the resulting random permutation. Run your program twice each with $n = 1, 6, 12$, and 50.
(b) (*Operation Counts*) How many arithmetical/logical operations are   required with an input value of n?

Although this algorithm may not seem like the most natural one to generate random permutations, it runs very inexpensively. In Exercise 30 we will compare it to some more natural permutation-generating algorithms.

EXERCISE FOR THE READER 6.17: In a game that you play, an unfair coin is repeatedly flipped, until either two consecutive heads are flipped (in which case you win the game) or a total of 3 tails are flipped (in which case you lose the game). The flips of the coin are independent, and for each one we have P(Heads) = 2/3 and P(Tails) = 1/3.

---

[26] Actually, the algorithm will work regardless of which permutation of $\{1, 2, ..., n\}$ is initially stored as the vector v.

(a) (COMPUTER) Run a simulation with 1 million trials to approximate the probability of winning this game.

(b) Compute this probability exactly using probability rules, and compare the exact answer with the estimate you obtained in Part (a).

(COMPUTER) EXERCISE FOR THE READER 6.18: In this exercise we consider both a simulated approximation to, and the exact computation of the probability that solves the following problem:

> (*Boarding Pass Problem*) Suppose that 100 people are boarding an airplane, each one having a boarding pass for one of the 100 seats. The first passenger to board has lost his boarding pass, and randomly chooses one of the 100 vacant seats in which to sit. The remaining 99 passengers (none of whom have lost their boarding passes) continue to board in order and to take their assigned seats. If a certain passenger finds his or her seat is occupied, then he/she randomly selects one of the remaining vacant seats. This continues until all passengers have boarded. What is the probability that the last passenger will get his/her assigned seat?

(a) Run 10,000 sets of 10,000 trials to simulate this problem, create a histogram of the results, and compute the overall simulated probability for the 100 million trials.

(b) Exactly how many outcomes are there in the sample space of (a single trial of) this experiment? Do each of these outcomes have the same probability? Would a brute-force approach be feasible to compute the answer to the boarding pass problem? Either perform such a calculation or explain why it is not feasible.

(c) Compute exactly the answer to the boarding pass problem (using the rules and theorems of probability).

Suggestion: In writing the simulation programs of Part (a), although it is tempting to make use of the program that you wrote in Exercise For The Reader 6.15, and this will certainly work, you will find that it will cut down on the speed of the program if it calls on other programs. Thus your simulation programs will be more robust if they are self-contained. The reader is encouraged to perform some comparisons so as to convince him/herself of this fact.

Expectation and Variance of a Random Variable

We next discuss two important concepts regarding random variables, the *expectation* and the *variance*. Both are numerical quantities; the first one is a measure of central tendency of the random variable, and the second is a measure of the dispersion of the values. We will restrict our development to the setting of discrete random variables.

DEFINITION 6.12: Suppose that X is a discrete random variable with probability mass function $p(x)$, and the x_i's are the values of X for which $p(x_i) > 0$. Thus, $\sum p(x_i) = 1$. The **expectation** (or the **expected value**) of the random variable X is given by

$$E[X] = \sum x_i p(x_i), \tag{13}$$

provided that this sum converges absolutely. The expectation $E[X]$ is also denoted as μ_X, or simply as μ. The **variance** of X is then defined to be

$$\mathrm{Var}(X) = E[(X - \mu)^2], \tag{14}$$

provided that this quantity is finite. The variance is also denoted as σ^2, and its square root σ is called the **standard deviation** of X.

The expectation is thus a weighted average of the different values that a random variable can assume, with the weights equaling the corresponding probabilities that the values will occur. Although the outcomes of a random variable are "random," the expectation gives an estimate for the long-term average outcome of the random variable. For example, in the setting of X being the payoff in a gambling bet, the expected value gives the very important long-term expected payoff of the bet. This follows from Theorem 6.3; see Exercise 31 and Theorem 6.4 below. As a specific example, suppose we have an urn that contains 30 $1 bills and 2 $100 bills, and we randomly draw a bill from the urn. If we were offered the opportunity to pay $5 to draw a bill like this and keep it, would it be prudent to take this opportunity? To answer this, we let X = the value of the bill drawn, and we compute (using (13))

$$E[X] = \$1 \cdot P(X = 1) + \$100 \cdot P(X = 100) = \$1 \cdot 30 / 32 + \$100 \cdot 2 / 32 \approx \$7.19.$$

This means that if we play this game a large number of times, the long-term average amount that we draw would approach $7.19. Less the $5 it cost us to play this game, this would amount to a net win of $2.19, so we should play this game.

The following theorem is the analogue of Theorem 6.3 for viewing expectations as long-term averages.

THEOREM 6.4: (*Expectations as Long-Term Averages*) Suppose that an experiment is repeated, independently and continually, and that at each trial (say the nth, n = 1, 2, 3, 4, ...) X_n is the value of a certain random variable X that is associated with the experiment. It then follows that with probability 1, we have

$$\lim_{n \to \infty} \frac{X_1 + X_2 + \cdots + X_n}{n} = E[X]. \tag{15}$$

This theorem has the practical application of showing how simulations can be used to estimate expectations: simply replicate the experiment on the computer a large number n times, and let $s_n = X_1 + X_2 + \cdots + X_n$ be the cumulative sum of the values of the random variable X for all of the n trials, then we estimate E[X] by s_n / n. Exercise for the Reader 6.19 (and its solution) will provide an illustration of this method.

In the same way, the variance is a weighted average of the squares of the deviations of the random variable's values from its mean. Since the outcomes of most probabilistic experiments can be measured in terms of an appropriate random variable, simulations can be employed to estimate any sort of numerical outcome of an experiment. Expectations are often very useful, but they can be difficult to compute analytically. The next example should help the reader get a better feel for expectations.

EXAMPLE 6.16: For each of the following random variables, compute the indicated expectations and variances:

(a) For the experiment of flipping a coin, let X be the random variable that equals 1 if heads turns up and 0 if tails turns up. Compute $E[X]$ and $Var(X)$.

(b) For any experiment and event A, let 1_A denote the **indicator function** of A, i.e., $1_A = 1$ if A occurs and otherwise $1_A = 0$. (Thus the random variable in Part (a) is the indicator function for the outcome of heads in a coin toss.) Compute $E[1_A]$ and $Var(1_A)$.

(c) For the experiment of rolling two dice, let X be the sum of the two numbers appearing. Compute $E[X]$.

SOLUTION: Part (a): There are only two values of X: $X = 0$ and $X = 1$, so by definition: $E[X] = 0 \cdot P(X = 0) + 1 \cdot P(X = 1) = 0 \cdot (1/2) + 1 \cdot (1/2) = 1/2 \equiv \mu.$ Knowing the expectation $\mu = 1/2$, we may now compute the variance $Var(X) =$
$$(0 - \mu)^2 \cdot P(X = 0) + (1 - \mu)^2 \cdot P(X = 1) = (-1/2)^2 \cdot (1/2) + (1/2)^2 \cdot (1/2) = 1/4.$$

Part (b): As in Part (a), there are only two values for this random variable, so $E[1_A] = 0 \cdot P(1_A = 0) + 1 \cdot P(1_A = 1) = 0 + 1 \cdot P(A) = P(A),$ and
$$\begin{aligned}
Var(1_A) &= (0 - \mu)^2 \cdot P(1_A = 0) + (1 - \mu)^2 \cdot P(1_A = 1) \\
&= (-P(A))^2 \cdot (1 - P(A)) + (1 - P(A))^2 \cdot (P(A)) \\
&= P(A) \cdot (1 - P(A))[P(A) + (1 - P(A))] \\
&= P(A) \cdot (1 - P(A)).
\end{aligned}$$

Part (c): As was (easily) obtained in our solution to Exercise for the Reader 6.14, the probability mass function for this experiment is shown in the following table:

k	2	3	4	5	6	7	8	9	10	11	12
$p(k)$	1/36	2/36	3/36	4/36	5/36	6/36	5/36	4/36	3/36	2/36	1/36

From these values, we use (13) to get the expectation:
$$\begin{aligned}
E[X] = \sum_{k=2}^{12} k \cdot p(k) \\
= \{2 \cdot 1 + 3 \cdot 2 + 4 \cdot 3 + 5 \cdot 4 + 6 \cdot 5 + 7 \cdot 6 + 8 \cdot 5 + 9 \cdot 4 + 10 \cdot 3 + 11 \cdot 2 + 12 \cdot 1\} / 36 \\
= 252 / 36 = 7.
\end{aligned}$$

Using (13), (14), and the result just obtained, we may write:

$$\text{Var}[X] = \sum\nolimits_{k=2}^{12} (k-7)^2 \cdot p(k) = \{(-5)^2 \cdot 1 + (-4)^2 \cdot 2 + (-3)^2 \cdot 3 + (-2)^2 \cdot 4$$
$$+ (-1)^2 \cdot 5 + 0^2 \cdot 6 + 1^2 \cdot 5 + 2^2 \cdot 4 + 3^3 \cdot 3 + 4^2 \cdot 2 + 5^2 \cdot 1\} / 36$$
$$= 35/6.$$

EXERCISE FOR THE READER 6.19: A carnival offers the following betting game that costs $1 to play: An urn contains three red balls, two blue balls, and seven green balls. Three balls are randomly drawn from the urn.   If they are all red, you win $10 (less the one dollar you paid to play), if they are all different colors, you win $2, and in all other cases you lose (the one dollar you paid to play).
(a) (Computer) Run a simulation with 1 million trials to estimate your expected winnings (losses) for playing this game, by adding up the total winnings of all these games and dividing by the number (1 million) of games played.
(b) Compute the expectation exactly, and compare with your simulated answer in Part (a).
Suggestion: For Part (a) model the 12 balls by the set of integers $\{1, 2, \ldots, 12\}$, where we let balls #1 – #3 be "red," balls #4 – #5 be blue, and the remaining balls #6 – #12 be green. A random draw of three balls is thus a random 3-combination from the set $\{1, 2, \ldots, 12\}$.

Independence of Random Variables

The expectation and variance operators possess many important properties.   To state some of these, we first generalize the concept of independence to random variables:

DEFINITION 6.13: (*Independence of Random Variables*) (i) (*Two random variables*)   If X and Y are two random variables, we say that X and Y are **independent** if the following relationship holds for any two sets of real numbers A and B:

$$P(X \in A, Y \in B) = P(X \in A) \cdot P(Y \in B). \tag{16}$$

Equation (16) simply states that the two events $X \in A$ and $Y \in B$ must be independent, and this must be true for any sets of real numbers A and B.   This amounts to a lot of things to check. It can be shown using Kolmogorov's axioms that the following weaker version of (16) is actually equivalent to it. [27]

$$P(X \le a, Y \le b) = P(X \le a) \cdot P(Y \le b). \tag{17}$$

---

[27] Many texts simply use (17) as the definition of independent random variables. Technical aside: When (16) is used, it really only needs to be checked (and has meaning) when A and B are so-called *measurable* sets of real numbers; see [Bil-86] for more details and the corresponding proof of the equivalence of (16) and (17), as well as the analogue for more variables (i.e., the proof that (18) and (19) are equivalent). In practice (and everywhere in this book) all sets that arise are measurable, so we will not concern ourselves with this technical restriction.

The difference is that (17) requires only that the independence relation of (16) be checked when A and B are semi infinite intervals of the form $\{x \le a\}$. We will assume the equivalence of (16) and (17) whenever we need it; see the footnote below for a reference to for a proof.

(ii) (*Finite numbers of random variables*) If $X_1, X_2, \cdots, X_n$ are random variables, we say that they are **independent** provided that the following relationship holds for any n sets $A_1, A_2, \cdots, A_n$ of real numbers:

$$P(X_1 \in A_1, X_2 \in A_2, \cdots, X_n \in A_n) = P(X_1 \in A_1) \cdot P(X_2 \in A_2) \cdots \cdot P(X_n \in A_n). \quad (18)$$

Equivalently ,

$$P(X_1 \le a_1, X_2 \le a_2, \cdots, X_n \le a_n) = P(X_1 \le a_1) \cdot P(X_2 \le a_2) \cdots \cdot P(X_n \le a_n), \quad (19)$$

must hold for any real numbers $a_1, a_2, \cdots, a_n$.

(iii) (*Infinite collections of random variables*) An infinite collection of random variables is **independent** if every finite subcollection of it is independent.

In practice, (16) and (18) are more powerful formulations to use, whereas in verifying independence, (17) and (19) are more manageable. In the case of discrete random variables, the following result further simplifies the criterion for independence.

PROPOSITION 6.5: If X and Y are discrete random variables, then they are independent if and only if the following equation is valid for any real numbers a and b:

$$P(X = a, Y = b) = P(X = a) \cdot P(Y = b). \quad (20)$$

Proof: If X and Y are independent, then (20) easily comes from (16) using the singleton sets $A = \{a\}$ and $B = \{b\}$. Conversely, suppose that (20) holds; we will prove (17). Fix real numbers a, and b. We let x_i and y_j denote those countably many values which X and Y (respectively) can attain with positive probability, and we let p_X and p_Y denote the corresponding probability mass functions. We may thus write:

$$P(X \le a) \cdot P(Y \le b) = \left(\sum_{x_i \le a} p_X(x_i) \right) \cdot \left(\sum_{y_j \le b} p_Y(y_j) \right)$$

$$= \sum_{x_i \le a} \sum_{y_j \le b} p_X(x_i) p_Y(y_j)$$

(the order of summation may be changed-absolute convergence)

$$= \sum_{x_i \le a} \sum_{y_j \le b} P(X = x_i) P(Y = y_j) \quad \text{(definition of probability mass function)}$$

$$= \sum_{x_i \le a} \sum_{y_j \le b} P(X = x_i, Y = y_j) \quad \text{(by (20))}$$

$$= \sum_{x_i \leq a} P(X = x_i, \, Y \in \bigcup_{y_j \leq b} \{y_j\}) \quad \text{(by (2): disjoint probabilities can be added)}$$

$$= P(X \in \bigcup_{x_i \leq a} \{x_i\}, \, Y \in \bigcup_{y_j \leq b} \{y_j\}) \quad \text{(by (2): disjoint probabilities can be added)}$$

$$= P(X \leq a, \, Y \leq b).$$

The proof is complete. $\square$

Just as events that depend on disjoint independent trials will be independent, so will associated random variables. This will be the main use of independent random variables in the context of simulations. The following theorem contains an extremely useful property of expectation involving sums of (constant multiples of) random variables. We stress that the result <u>does not</u> require the random variables to be independent.

Linearity of Expectation

THEOREM 6.6: (*Linearity of Expectation*)[28] If $X_1, X_2, \cdots, X_n$ are (discrete) random variables, and $a_1, a_2, \cdots, a_n$ are real numbers then

$$E[a_1 X_1 + a_2 X_2 + \cdots + a_n X_n] = a_1 E[X_1] + a_2 E[X_2] + \cdots + a_n E[X_n]. \tag{21}$$

NOTE: (*Regarding the following proof*) In cases where any of the random variables in this result have countably infinite ranges, the proof below (and some others that follow) will involve some manipulations of infinite sums (also called infinite series) of positive numbers. Readers who are not experienced with such sums may assume that all random variables have finite ranges so that all of the sums in the proof below are finite sums.

Proof: We will prove that for any two random variables X and Y, and any real number a, we have:

$$(i) \;\; E[X + Y] = E[X] + E[Y], \;\; \text{and} \;\; (ii) \;\; E[aX] = aE[X]. \tag{22}$$

These two identities combine at once to give $E[aX + bY] = aE[X] + bE[Y]$ (if b is also a real number), which in turn can easily be used to obtain (21) using mathematical induction (Exercise 32). We will prove (ii) first. If $a = 0$, both sides of (22)(*ii*) are clearly 0, so we assume that $a \neq 0$. In this case, we have $p_{aX}(ax_i) \equiv P(aX = ax_i) = P(X = x_i) \equiv p_X(x_i)$, so that using the definition (13) of expectation, we have:

---

[28] As mentioned earlier, we are restricting our treatment of expectation and variance to discrete random variables. These definitions can be extended to the setting of continuous random variables, and all of the results of this theorem (and most other theorems regarding expectation and/or variances) continue to be valid for continuous random variables.

$$E[aX] = \sum ax_i p_{aX}(ax_i) = a\sum x_i p_{aX}(ax_i) = a\sum x_i p_X(x_i) = aE[X].$$

The proof of (22)(*i*) will be facilitated by the following formula that gives a representation of any discrete random variable as a weighted sum of indicator functions:

$$X = \sum_i x_i 1_{A_i}, \tag{23}$$

where $A_i = X^{-1}(x_i) = \{s \in S : X(s) = x_i\}$. The validity of this formula follows from the definitions of the sets A_i: These sets partition the sample space, so that for any $s \in S$, the sum of functions on the right side of (23), when evaluated at s, has only one nonzero term determined by which set A_i's belongs to. To prove (22)(*i*), we let $Y = \sum_j y_j 1_{B_j}$ be the corresponding representation of Y. Noting that $1_{A_i} \cdot 1_{B_j} = 1_{A_i \cap B_j}$, and $1 = \sum_i 1_{A_i} = \sum_j 1_{B_j}$, we may write:

$$X + Y = \sum_i x_i 1_{A_i} + \sum_j y_j 1_{B_j} = \sum_i \sum_j x_i 1_{A_i \cap B_j} + \sum_j \sum_i y_j 1_{A_i \cap B_j} = \sum_i \sum_j (x_i + y_j) 1_{A_i \cap B_j}.$$

It now follows that[29]

$$\begin{aligned}
E[X+Y] &= \sum_i \sum_j (x_i + y_j) P(A_i \cap B_j) \\
&= \sum_i \sum_j x_i P(A_i \cap B_j) + \sum_i \sum_j y_j P(A_i \cap B_j) \\
&= \sum_i x_i \sum_j P(A_i \cap B_j) + \sum_j y_j \sum_i P(A_i \cap B_j) \\
&= \sum_i x_i P(A_i) + \sum_j y_j P(B_j) \quad \text{(by Kolmogorov's axiom (2))} \\
&= E[X] + E[Y]. \quad \square
\end{aligned}$$

Properties of Variances

Using the linearity of expectation, we can easily derive a useful alternative formula for computing variances (after we have computed the expectation $\mu = E[X]$):

$$\begin{aligned}
\text{Var}(X) &= E[(X-\mu)^2] = E[X^2 - 2\mu X + \mu^2] \\
&= E[X]^2 - 2\mu E[X] + \mu^2 E[1] = E[X]^2 - 2\mu^2 + \mu^2 = E[X^2] - (E[X])^2.
\end{aligned}$$

In summary, we have proved the following identity:

---

[29] By absolute convergence, the summations can be performed in any order.

$$\mathrm{Var}(X) = E[X^2] - (E[X])^2. \tag{24}$$

Unlike the expectation, the variance need not be additive, but it turns out to be if the variables are independent. In order to obtain the general result, it is helpful to introduce the following concept relating to two random variables:

DEFINITION 6.14: Suppose that X and Y are (discrete) random variables with respective means μ_X and μ_Y. The **covariance** of X and Y is defined by

$$\mathrm{Cov}(X,Y) = E[(X - \mu_X)(Y - \mu_Y)].$$

(Note that $\mathrm{Var}(X) = \mathrm{Cov}(X,X)$.)

Using linearity of expectation, we may obtain an alternative formula for the covariance:

$$\mathrm{Cov}(X,Y) = E[(X - \mu_X)(Y - \mu_Y)] = E[XY - \mu_Y X - \mu_X Y + \mu_X \mu_Y]$$
$$= E[XY] - \mu_Y E[X] - \mu_X E[Y] + \mu_X \mu_Y = E[XY] - \mu_X \mu_Y.$$

In summary, we have established the following identity:

$$\mathrm{Cov}(X,Y) = E[XY] - E[X]E[Y]. \tag{25}$$

Our next result collects some useful facts about variances.

PROPOSITION 6.7: If X and Y are (discrete) random variables, and a is a real number then
(a) (*Variance of Sum*) $\mathrm{Var}(X + Y) = \mathrm{Var}(X) + \mathrm{Var}(Y) + 2\mathrm{Cov}(X,Y)$.
(b) (*Variance is Additive for Independent Random Variables*) If X and Y are independent, then $\mathrm{Var}(X + Y) = \mathrm{Var}(X) + \mathrm{Var}(Y)$.
(c) $\mathrm{Var}(aX) = a^2 \mathrm{Var}(X)$.

Proof: Part (a): Starting with the definition of variance, we obtain:

$$\mathrm{Var}(X + Y) = E[((X + Y) - E(X + Y))^2]$$
$$= E[(X + Y - \mu_X - \mu_Y)^2] \qquad \text{(linearity of expectation)}$$
$$= E[(X - \mu_X)^2 + (Y - \mu_Y)^2 + 2(X - \mu_X)(Y - \mu_Y)] \text{ (expanding)}$$
$$= E[(X - \mu_X)^2] + E[(Y - \mu_Y)^2] + 2E[(X - \mu_X)(Y - \mu_Y)]$$
$$\qquad\qquad\qquad \text{(linearity of expectation)}$$
$$= \mathrm{Var}(X) + \mathrm{Var}(Y) + 2\mathrm{Cov}(X,Y). \qquad \text{(definition of covariance)}$$

Part (b): In light of the formula in Part (a) and formula (25) for the covariance, it suffices to show that if X and Y are independent, then $E[XY] = E[X]E[Y]$.

Starting with the definition of expectation (with some terms left possibly uncombined), we obtain

$$E[XY] = \sum_i \sum_j x_i y_j P(X = x_i, Y = y_j)$$
$$= \sum_i \sum_j x_i y_j P(X = x_i) P(Y = y_j) \qquad \text{(using independence)}$$
$$= \sum_i x_i P(X = x_i) \sum_j y_j P(Y = y_j) \qquad \text{(regrouping)}$$
$$= E[X] \cdot E[Y]. \qquad \text{(definition of expectation)}$$

The proof of Part (c) is left to the next exercise for the reader. □

EXERCISE FOR THE READER 6.20: (a) Prove Part (c) of Proposition 6.5. (b) Show that the covariance is *linear* (in each variable):

$$\text{Cov}(aX_1 + bX_2, Y) = a\text{Cov}(X_1, Y) + b\text{Cov}(X_2, Y).$$

(c) Extend the formula of Part (a) of Proposition 6.5 to the following formula for the variance of a sum of n random variables:

$$\text{Var}(X_1 + X_2 + \cdots + X_n) = \sum_{i=1}^n \text{Var}(X_i) + 2\sum_{i>j} \text{Cov}(X_i, X_j). \qquad (26)$$

EXAMPLE 6.17: Compute the mean and the variance of the binomial random variable $X \sim \mathcal{B}(n, p)$.

SOLUTION: Proceeding directly from the definition would involve technical computations (see the next exercise for the reader and Exercise 33). A much more natural way to perform these calculations is to use linearity. We let $X_i \ (1 \le i \le n)$ denote the outcome on the ith trial, i.e., $X_i = 1$, if the ith trial is a success, and otherwise $X_i = 0$. Then clearly $X = \sum_{i=1}^n X_i$ and the X_i's are independent, since the trials are independent. Furthermore, the X_i's are identically distributed; their means and variances are easily computed:

$$E[X_i] = 0 \cdot P[X_i = 0] + 1 \cdot P[X_i = 1] = 0 \cdot q + 1 \cdot p = p.$$

Similarly, $E[X_i^2] = 0^2 \cdot P[X_i = 0] + 1^2 \cdot P[X_i = 1] = p.$ Thus, by Part (b) of Theorem 6.6, we may obtain

$$\text{Var}(X_i) = E[X_i^2] - (E[X_i])^2 = p - p^2 = p(1 - p) = pq.$$

Using linearity (Part (a) of Theorem 6.6), we conclude that

$$E[X] = E\left[\sum_{i=1}^n X_i\right] = \sum_{i=1}^n E[X_i] = \sum_{i=1}^n p = np.$$

Also, since the X_i's are independent, their variances may be added to obtain the variance of their sum (Part (c) of Theorem 6.6):

$$\mathrm{Var}[X] = \mathrm{Var}\left[\sum_{i=1}^{n} X_i\right] = \sum_{i=1}^{n} \mathrm{Var}[X_i] = \sum_{i=1}^{n} pq = npq.$$

EXERCISE FOR THE READER 6.21: Let X be a binomial random variable, $X \sim \mathcal{B}(n, p)$. Compute E[X] directly from the definition of the expectation.

Any discrete random variable with finite range can be simulated using the program `GeneralFiniteSimulator` of Exercise for the Reader 6.15. Actually, for all practical purposes, <u>any</u> discrete random variable X can be simulated with a similar program. Our next example provides such a simulation scheme for a so-called *Poisson* random variable:

Poisson Random Variables

DEFINITION 6.15: A random variable X is said to be a **Poisson random variable with parameter** λ, if for some $\lambda > 0$, it has probability mass function given by

$$P(X = i) = e^{-\lambda} \frac{\lambda^i}{i!} \quad (i = 0, 1, 2, \cdots). \tag{27}$$

NOTE: It follows from the infinite series expansion for e^x (from calculus) $e^x = \sum_{i=0}^{\infty} x^i / i!$ that (27) indeed defines a probability mass function. Readers who have studied calculus will know why this is true; others should take it as an identity that can be manipulated just like with any finite sum.

Poisson random variables are used to model the number of arrivals over a fixed period of time where the expected number is known and arrivals are independent of one another. For example, the number of arrivals of customers to a business, the number of arrivals of auto insurance claims to a company, or even the number of earthquakes to occur in a certain geographical region. The parameter λ is just the expected number of arrivals per time period, as we will show in the following example.

EXAMPLE 6.18: (a) Write a program to simulate a Poisson random variable with parameter $\lambda > 0$.
(b) Run your program of Part (a) to simulate 1 million trials, find the resulting (simulated) mean, and use the data to plot the resulting estimated probability density function.
(c) Show analytically that a Poisson random variable with parameter $\lambda > 0$, has mean and variance both equal to λ.

SOLUTION: Part (a): The following pseudocoded program is a minor modification of the `GeneralFiniteSimulator` of the solution to Exercise for the Reader 6.15. To save on (computer) computation time, we will use the following recursive formula for the Poisson probabilities that directly follows from (27).

$$P(X = 0) = e^{-\lambda}, \quad P(X = i+1) = P(X = i) \cdot \lambda / (i+1) \quad (i = 0, 1, 2, \cdots). \tag{28}$$

```
Program:  PoissonRVSimulator
INPUT:    lambda <positive parameter>
          N <number of trials for simulation>
set r = rand
set Prob = exp(-lambda) <initialize Poisson probability>
set CumProb = Prob      <initialize cummulative probability>
set X = 0               <initialize value of Poisson random variable>
WHILE r > CumProb
        update X = X + 1
        update Prob = Prob*lambda/X, CumProb = CumProb + Prob
END
OUTPUT:   X
```

Some comments are in order. Although it is plausible that the while loop in the above program might need to go through any given number of iterations in a single run of this program, the probabilities in (27) decay so rapidly that this number of iterations will almost surely not get too large. For example, when $\lambda = 5$, it is *practically impossible* that X will equal 100.[30] A related issue arises if we are using a floating point arithmetic system: In an execution of the above program, if the value of `CumProb` were to get within the machine epsilon of 1 (that is about 10^{-16} in an IEEE double precision standard system; see Section 4.5), these numbers become indistinguishable from 1 and the system will end the while loop, so this places an upper bound on how large the outputted value of X can be.

Part (b): Rather than store the complete length 1 million vector of the record of all trials, we instead store a frequency vector `Tracker` whose ith component will be the number of trails for which the simulated Poisson trial took on the value $i-1$. From the comments mentioned in Part (a), it will suffice to initialize this `Tracker` vector to have 100 components of zeros:

```
set Tracker = 100 length vector of 0's
FOR n = 1 TO 1000000
     set X = PoissonRVSimulator(5)
     update Tracker(X) = Tracker(X) + 1
END
OUTPUT:  SimPMF = Tracker/1000000
```
(*Next from the data in Tracker, we compute the mean.*)
`CumSum = 0` (*initialize cumulative sum*)

---

[30] If X is a Poisson random variable with parameter $\lambda = 5$, then (27) can be used to show that $P(X \geq 100) \approx 2 \times 10^{-32}$. So this means that we would need to go through $5 \times 10^{31-6} = 5 \times 10^{25}$ groups of 1 million simulations of such Poisson random variables to see one in which at least one outcome of the random variables was at least 100.

```
FOR i = 1 TO 100
     CumSum = CumSum + (i-1)*Tracker(i)
END
OUTPUT:   CumSum/1000000   (estimate for mean)
```

For a typical run of this program, a plot of the simulated probability mass function (from the outputted vector SimPMF) is shown, along with the actual probability mass function plot, in Figure 6.18. The simulated mean value of the values of X over the 1 million trials is 5.0027, and this compares well with the theoretical mean of 5 that we will derive next. We point out that among the 1 million simulated Poisson random variables, the largest value attained was 20.

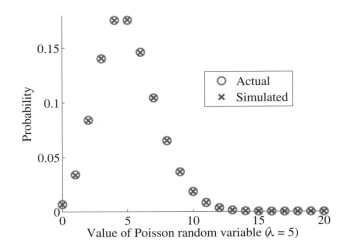

FIGURE 6.18: Comparison of the analytical probability mass function (o's) with the simulated pdf from the generation of 1 million simulated Poisson random variables with parameter $\lambda = 5$.

Part (c): This part will be accomplished using the infinite series expansion (from calculus) for e^x that was mentioned earlier: $e^x = \sum_{i=0}^{\infty} x^i / i!$ (*).   From (27), and the definition of expectation we obtain:

$$E(X) = \sum_{i=0}^{\infty} i\,P(X = i) = \sum_{i=0}^{\infty} i\,e^{-\lambda} \frac{\lambda^i}{i!} = e^{-\lambda} \sum_{i=1}^{\infty} \frac{\lambda^i}{(i-1)!} = \lambda e^{-\lambda} \sum_{i=1}^{\infty} \frac{\lambda^{i-1}}{(i-1)!}$$

$$\underset{\substack{\text{Subst.}\\ j=i-1}}{=} \lambda e^{-\lambda} \sum_{j=0}^{\infty} \frac{\lambda^j}{j!} \underset{\text{By (*)}}{=} \lambda e^{-\lambda} e^{\lambda} = \lambda.$$

To compute the variance, we will use formula (24) $\mathrm{Var}(X) = E[X^2] - (E[X])^2$. We proceed as above to compute $E[X^2]$:

$$E(X^2) = \sum_{i=0}^{\infty} i^2 \, P(X=i) = \sum_{i=0}^{\infty} i^2 \, e^{-\lambda} \frac{\lambda^i}{i!} = \sum_{i=1}^{\infty} \frac{i e^{-\lambda} \lambda^i}{(i-1)!} = \underset{\underset{j=i-1}{\text{Subst.}}}{=} \lambda \sum_{j=0}^{\infty} \frac{(j+1) e^{-\lambda} \lambda^j}{j!}$$

$$= \lambda \sum_{j=0}^{\infty} \frac{j e^{-\lambda} \lambda^j}{j!} + \lambda \sum_{j=0}^{\infty} \frac{e^{-\lambda} \lambda^j}{j!} \underset{\text{By (*)}}{=} \lambda^2 \sum_{j=1}^{\infty} \frac{e^{-\lambda} \lambda^{j-1}}{(j-1)!} + \lambda e^{-\lambda} e^{\lambda} \underset{\text{By (*)}}{=} \lambda^2 + \lambda.$$

It now follows from (24) that $\mathrm{Var}(X) = E[X^2] - (E[X])^2 = \lambda^2 + \lambda - (\lambda)^2 = \lambda$.

EXERCISE FOR THE READER 6.22: (*Typhoon Predictions*) According to *FEMA* (*The Federal Emergency Management Agency*) records, there were 10 major typhoons that struck the Western Pacific island of Guam within a recent 41-year period. Such typhoons caused major disaster on the island such as (six foot diameter concrete) power line poles being knocked down like toothpicks. In the aftermath, most island residents had no water for several days and no electricity for several weeks.
(a) Use a Poisson model with this data to predict the probability that there will be at least one typhoon to strike Guam in the next year.
(b) Use the same model to predict the probability that exactly two major typhoons will strike Guam in the next year.
(c) Use a similar Poisson model to predict the probability that a major typhoon will strike Guam in the next three months.

EXERCISES 6.2:

1. For each of the following random variables X find (i) the range range(X), (ii) the probability distribution function (pdf) $p = p_X$, (iii) the cumulative distribution function (cdf) $F = F_X$, (iv) the expectation $\mu = \mu_X = E[X]$, and (v) the variance $\sigma^2 = \sigma_X^2 = \mathrm{Var}(X)$.
   (a) $X =$ the total number of heads in the experiment of flipping a coin four times.
   (b) $X =$ the number of green balls drawn in the experiment of drawing two balls <u>with replacement</u> from an urn containing 4 red balls, 2 blue balls, and 1 green ball. This means that we draw one ball, note its color, and then throw it back in and reshuffle the balls before drawing the second ball
   (c) $X =$ the number of green balls drawn in the experiment of drawing two balls <u>without replacement</u> from an urn containing 4 red balls, 2 blue balls, and 1 green ball. This means that we draw one ball from the urn, and then draw a second ball.
   (d) Urn 1 contains 6 $1 bills and 10 $20 bills and urn 2 contains 2 $1 bills and 3 $50 bills. The experiment consists of flipping a coin. If it lands heads, we draw 1 bill from urn 1, and if it lands tails we draw 1 bill from urn 2. $X =$ the value of the bill drawn.

2. For each of the following random variables X find (i) the range range(X), (ii) the probability distribution function (pdf) $p = p_X$, (iii) the cumulative distribution function (cdf) $F = F_X$, (iv) the expectation $\mu = \mu_X = E[X]$, and (v) the variance $\sigma^2 = \sigma_X^2 = \mathrm{Var}(X)$.
   (a) $X =$ the total number of "snake eyes" (pairs of 1s) rolled when a pair of dice is rolled.
   (b) $X =$ the number of green balls drawn in the experiment of drawing two balls <u>with replacement</u> from an urn containing 2 red balls, 3 blue balls, and 4 green balls. This means that we draw one ball, note its color, and then throw it back in and reshuffle the balls before drawing

the second ball.

(c) X = the number of green balls drawn in the experiment of drawing two balls <u>without replacement</u> from an urn containing 2 red balls, 3 blue balls, and 4 green balls. This means that we draw one ball from the urn, and then draw a second ball.

(d) Urn 1 contains 6 $1 bills and 10 $20 bills, urn 2 contains 2 $1 bills and 3 $50 bills, and urn 3 contains 1 $5 bill and 10 $100 bills. The experiment consists of throwing a fair die. If it lands as 1, 2, or 3, we draw 1 bill from urn 1, if it lands as 4 or 5, we draw 1 bill from urn 2, and if it lands on 6, we draw <u>2 bills</u> from urn 3. X = the total value of the bills drawn.

3. Suppose that a fair die is tossed 10 times.
    (a) Compute the probability that the number 6 shows up exactly 2 times.
    (b) Compute the probability that the number 6 shows up at least 2 times.
    (c) Compute the probability that the number 6 never shows up in these 10 throws.

4. Suppose that a coin is flipped 100 times.
    (a) Compute the probability that the coin will land as heads at least 10 times.
    (b) Compute the probability that the coin will land as heads exactly 50 times.

5. You decide to throw a huge party on March 4 and invite 200 people. To impress your guests, you announce that you will give a bottle of champagne and a box of exotic chocolates to any guest who has a birthday on that day.
    (a) What is the probability that you will have to give out exactly k champagne bottles and boxes of chocolates?
    (b) What is the probability that you will have to give out no more than 5 champagne bottles and boxes of chocolates?

6. Suppose that an important "yes" or "no" message needs to be transmitted using only binary codes (1 for yes, or 0 for no), but that the transmission channel is noisy and there is a 25% chance that any given binary digit will get wrongly transmitted. To improve the chance that the message gets correctly read, an odd length string of 1's will get sent for "yes" and an odd length string of 0's get sent for "no." If the length of the string is N (an odd integer), the message gets interpreted as a "yes" if most of the N digits of the transmitted string are 1's and "no" otherwise.
    (a) Compute the probabilities that messages will get correctly interpreted if $N = 3$ length strings are used, and if $N = 5$ length strings are used.
    (b) What is the smallest value of N so that if length N strings are used, there will be at least a 95% chance that messages will be correctly interpreted? How about if we want the success probability to be 99.5%?

7. In an experiment two dice are thrown, and the minimum of the two numbers showing is recorded as a random variable X. Compute the probability mass function (pdf) for X, and then plot the graph of the corresponding cumulative distribution function (cdf).

8. In an experiment two dice are thrown, and the maximum of the two numbers showing is recorded as a random variable X. Compute the probability mass function (pdf) for X, and then plot the graph of the corresponding cumulative distribution function (cdf).

9. Re-do Exercise 7 under the assumption that the dice are biased to land on 1 with probability 0.25, and the remaining numbers are equally likely to turn up.

10. Re-do Exercise 8 under the assumption that the dice are biased to land on 3 with probability 0.25, and the remaining numbers are equally likely to turn up.

11. In an experiment two dice are thrown, a red one and a white one. Let R denote the number that shows on the red die, W denote the number that shows on the white die, and X denote the minimum of the two numbers that show, i.e., $X = \min(R, W)$. Compute the following expectations and variances:

    (a) E(X).   (b) Var(X).   (c) E($R + X$).   (d) Var (R) + Var(X).   (e) Var ($R + X$).

12. Let R and W be as in Exercise 11, and let Y denote the maximum of the two numbers that show, i.e., $Y = \max(R, W)$. Compute the following expectations and variances:

    (a) $E(Y)$.   (b) $\text{Var}(Y)$.   (c) $E(W + Y)$.   (d) $\text{Var}(W) + \text{Var}(Y)$.   (e) $\text{Var}(W + Y)$.

13. Suppose that in a lot containing 500 computers, 32 of them are defective. If a random sample of 10 is taken from the 500 and tested, what is the expected number of defective computers that will be found? What is the corresponding standard deviation?

14. Three balls are randomly removed (without replacement) from an urn that contains three red balls and seven black balls. What is the expected number of red balls that will be drawn? What is the corresponding standard deviation?

15. An urn contains three red balls and seven black balls. A ball is randomly drawn, its color noted, and then it is replaced in the urn, and the balls are mixed up. This is repeated twice more (so a total of three balls are drawn, with replacement). What is the expected number of red balls that will be drawn? What is the corresponding standard deviation?

16. A certain country has 40 Olympic athletes, and of these 6 have performance enhancing drugs in their systems that are detectable with official tests. On five consecutive days during a competition period, an athlete is randomly selected from these 40 and tested for performance enhancing drugs. (Thus the same athlete could get selected more than once.) What is the expected number of days on which the selected athlete tests positive for performance enhancing drugs?

17. In an experiment two dice are thrown, a red one and a white one. Let R denote the number that shows on the red die, W denote the number that shows on the white die, X denote the minimum of the two numbers that show, i.e., $X = \min(R, W)$ and $Y = \max(R, W)$. Determine whether the following pairs of random variables are independent. Explain your answers.

    (a) R and W.   (b) R and X.   (c) R and $R + W$.   (d) X and Y.

18. Let R, W, X, and Y be the random variables of Exercise 17. Determine whether the following pairs of random variables are independent. Explain your answers.

    (a) R and $W - X$.   (b) W and Y.   (c) W and $R + W$.   (d) $X + Y$ and XY.

19. Consider the experiment of flipping a coin until it lands on heads. Let X be the random variable equaling the number of flips required. Compute $E[X]$.

20. Suppose that we have a shuffled deck of cards and that we continue to draw cards until we draw the ace of spades. Let X be the random variable equaling the number of cards that are drawn. Compute $E[X]$.

21. Suppose that a cereal company has an ongoing promotion where each box of their cereal contains one of five special tokens. If you collect all five you win a free box of cereal.
    (a) Suppose that each of the five tokens is equally likely to appear. If someone has bought 10 boxes of cereal, what is the expected number of different tokens he/she will have collected?
    (b) Re-do Part (a) in the general setting where the person has bought N boxes of cereal, where N is any positive integer.
    (c) Re-do Part (a) if one of the tokens is 1000 times more rare than each of the other four (so the other four are equally likely to be found in a box of cereal).
    (d) Re-do Part (b) with the assumptions changed as in Part (c).
    Suggestion: Let X be the number of distinct tokens that one has collected after buying the indicated number of boxes of cereal, and for each i, $1 \le i \le 5$, and let X_i equal 1 if the ith type of token has been collected; otherwise $X_i = 0$. Note that $X = \sum_{i=1}^{5} X_i$.

22. (*Web site Traffic*) A certain international business's Web site has an average of 10 hits per minute. The website can handle at most 50 hits in a single minute; otherwise the server will

crash and need to be reconfigured (resulting in some lost business). The number of hits to this website in any given time period can be modeled with a Poisson random variable.
(a) Compute the probability that the website will have no visits over the next minute.
(b) Compute the probability that the website will have at most 20 visits over the next minute.
(c) Compute the probability that the website will crash over the next minute (due to more than 50 visits).

23. (*Geiger Counters*) High levels of radiation can go undetected for long periods of time and cause long-term health problems. A Geiger counter is a device that measures radiation, often in units known as microSieverts/hour (μSv/h), by detecting and counting radioactive particles that enter it (and usually they emit a clicking noise for each particle detected). Radiation occurs naturally around the world at levels ranging from about 0.05 μSv/h to 0.12 μSv/h. Generally radiation levels above 1 μSv/h are cause for concern (e.g., possible emissions from nuclear waste or a dirty bomb). At the high extreme, the 134 workers at the Chernobyl nuclear power plant during its meltdown received 800,000 to 16 million μSv/h radiation, and of these 28 died within three months of the exposure. Emission of radioactive particles in a given time period tends to follow a Poisson distribution. Working in units of 0.01 μSv and taking the mean of the number of nuclear particles detected by a Geiger counter in a one hour period to be 0.08 μSv, compute the following probabilities
(a) The probability that the Geiger counter will give a reading of at most 0.02 μSv/h.
(b) The probability that the Geiger counter will give a reading at most 0.05 μSv/h.
(c) The probability that the Geiger counter will give a reading above 0.1 μSv/h.

24. Prove that if X_1 is a Poisson random variable with parameter $\lambda_1 > 0$, and if X_2 is a Poisson random variable with parameter $\lambda_2 > 0$, and these random variables are independent, then $X_1 + X_2$ is a Poisson random variable with parameter $\lambda_1 + \lambda_2$.

Suggestion: $P(X_1 + X_2 = i) = \sum_{j=1}^{i} P(X_1 = j, X_2 = i - j) = \sum_{j=1}^{i} P(X_1 = j)P(X_2 = i - j)$.

25. (*Geometric Random Variables*) The distributions of this exercise arise in the same setting as did the binomial random variables (Example 6.12). Consider repeated independent trials of an experiment that results in either a *success*, with probability p, or *failure*, with probability $q = 1 - p$. Here we assume that $0 < p < 1$. The random variable X that equals the number of trials needed until the first success is called a **geometric random variable** with parameter p.

(a) Show that the probability mass function for X is given by:
$$P(X = n) = (1 - p)^{n-1} p, \quad n = 1, 2, 3, \cdots, \quad P(X = \infty) = 0, \,^{31}$$
and verify that $\sum_{n=1}^{\infty} P(X = n) = 1$.

(b) Show that the cumulative distribution function (cdf) of X satisfies $F(n) = 1 - p^n$, for $n = 1, 2, 3, \cdots$.

(c) Show that $E[X] = 1/p$ and $\text{Var}(X) = p/(1-p)^2$.

Suggestion: You will need some infinite series expansions (from calculus). For Part (a), use the infinite (geometric) series expansion $\sum_{n=0}^{\infty} x^n = 1/(1-x)$ (valid for $-1 < x < 1$). For Part (b), use the finite geometric series expansion (see Proposition 3.5). In Part (c), to get the expectation, you will need the differentiated geometric series: $(d/dx)\sum_{n=0}^{\infty} x^n = [1/(1-x)]'$, or

---

[31] Thus, although it is theoretically possible that an infinite sequence of failures might occur (i.e., $X = \infty$), the probability that this will ever happen is zero.

$\sum_{n=0}^{\infty} n x^{n-1} = 1/(1-x)^2$. To compute $E[X^2]$ (for the variance), you will also need the twice

differentiated geometric series $(d/dx)\sum_{n=0}^{\infty} n x^{n-1} = [1/(1-x)^2]'$, or

$\sum_{n=0}^{\infty} n(n-1)x^{n-1} = 3/(1-x)^3$.

26. (*Negative Binomial Random Variables*) The distributions of this exercise arise in the same setting as did the binomial random variables (Example 6.12). Consider repeated independent trials of an experiment that results in either a *success*, with probability p, or *failure*, with probability $q = 1 - p$. Here we assume that $0 < p < 1$, and that r is a positive integer. The random variable X that equals the number of trials needed until the cumulative total number of successes reaches r is called a **negative binomial random variable** with parameters p and r.

(a) Show that the probability mass function for X is given by:

$$P(X = n) = \binom{n-1}{r-1} p^r (1-p)^{n-r}, \quad n = 1, 2, 3, \cdots, \quad P(X = \infty) = 0.$$

(b) Show that $E[X] = r/p$ and $\text{Var}(X) = r(1-p)/p^2$.

Suggestion: For Part (a), count the number of n-trial experiments where there are exactly r successes among the first $n-1$ trials to be $\binom{n-1}{r-1}$. Although Part (b) can be done directly, a more elegant and efficient approach is to realize that $X = X_1 + X_2 + \cdots + X_r$, where X_i is the number of trials after the $(i-1)$th success until the ith success, for $i > 1$ (X_1 is the number of trials until the first success). By independence, each X_i is a geometric random variable (Exercise 17) with parameter p. Use Theorem 6.6 in conjunction with the results of Exercise 17(c).

27. (*For Readers Who Have Studied Calculus: Properties of Density Functions for Continuous Random Variables*) Suppose that $f : \mathbb{R} \to \mathbb{R}$ is a piecewise continuous (probability) density function for a continuous random variable X, i.e., the relationship is governed by equation (12) $P(X \in A) = \int_A f(x)dx$. Establish the validity of the following three properties that were mentioned in the section:

1. $f(x) \geq 0$ (*Density functions are nonnegative.*)

2. $\int_{-\infty}^{\infty} f(x)dx = 1$ (*The total area under a density curve is always one.*)

3. $P(X = a) = \int_a^a f(x)dx = 0$ (*The probability of an exact amount is zero.*)

Suggestions: For Part (a), suppose that $f(x_0) < 0$, for some number x_0. Since f is piecewise continuous, it is either left-continuous or right-continuous at x_0. If f is right continuous at x_0, then there is a number $\delta > 0$, such that $f(x) < f(x_0)/2$ (a negative number) for any x in the interval $x_0 \leq x \leq x_0 + \delta$. But then it follows (from (12) that $P(\{x : x_0 \leq x \leq x_0 + \delta\}) = \int_{x_0}^{x_0+\delta} f(x)dx < 0$ —a contradiction (since probabilities cannot be negative). The proof when f is left continuous at x_0 is similar. For Part (b), since X is real-valued, we have $1 = P(S) = P(X \in (-\infty, \infty)) = \int_{-\infty}^{\infty} f(x)dx$ (using (12)).

28. (*For Readers Who Have Studied Calculus: Properties of Cumulative Distribution Functions*) Suppose that $F : \mathbb{R} \to [0, 1]$ is the cdf of a (continuous or discrete) random variable X (see Definition 6.11). Prove the following:

(a) F is nondecreasing, i.e., $x_1 < x_2 \Rightarrow F(x_1) \leq F(x_2)$.

(b) $\lim_{x \to -\infty} F(x) = 0$ and $\lim_{x \to \infty} F(x) = 1$.

Suggestion: Here is an outline of a proof of the second limit of Part (b). The other limit can be done in a similar fashion. Let $E_0 = (-\infty, 0]$, and for each positive integer n, define $E_n = (n-1, n]$. Since these sets are pairwise mutually exclusive and their union equals $\mathbb{R}$, it follows from Kolmogorov's axiom (2) that $1 = P(S) = P(\bigcup_{n=0}^{\infty} E_i) = \sum_{n=0}^{\infty} P(E_i)$. But, using (12) and the definition of an infinite series, the right side of this equals $\lim_{N \to \infty} \sum_{n=0}^{N} P(E_i) = \lim_{N \to \infty} \int_{-\infty}^{N} f(x) dx = \lim_{N \to \infty} F(N)$. The result now follows from Part (a).

29. (a) Prove that the permutations generated by Algorithm 6.1 given in this section are indeed random.
(b) Suppose that we modify Algorithm 6.1 by allowing the index ℓ to decrease rather than increase, i.e., by replacing Step 2 with the following:

Iterative Step 2′: FOR $\ell = k$ TO 2, repeat the following step:
Generate a random integer j in the range $1 \le j \le \ell$, and interchange the jth and the ℓth element of VEC.

Show that the resulting modified algorithm also produces random permutations.

Suggestion: For Part (a), let $\{\mu(i)\}_{i=1}^{k}$ denote a permutation generated by the algorithm, i.e., $\mu(i) = \text{VEC}(i)$, where VEC is the final value of VEC in the algorithm. Let σ be an arbitrary permuation of $\{1, 2, \ldots, k\}$. It must be shown that $P(\mu = \sigma) = 1/k!$. Proceed using induction on k. For Part (b), set $n = 3$, and show that $P(\mu(3) = 3) = 2/9 \ne 1/3$, where $\{\mu(i)\}_{i=1}^{3}$ is a permutation generated by the (modified) algorithm.

30. (*Another Scheme for Generating Random Permutations*) Consider the following algorithm:
Step 1: Generate n random numbers using rand n times: $U_1, U_2, \cdots, U_n$.

Step 2: Sort these numbers to be in increasing order: $U_{\mu(1)}, U_{\mu(2)}, \cdots, U_{\mu(n)}$.

Step 3: The resulting vector of indices $\{\mu(i)\}_{i=1}^{n}$ is a random permutation of $\{1, 2, \cdots, n\}$.

For example, when $n = 3$, and we generated the following random numbers: $U_1 = 0.9501$, $U_2 = 0.2311$, $U_3 = 0.6068$, then since $U_2 < U_3 < U_1$, the resulting permutation of $\{1, 2, 3\}$ would be $(2, 3, 1)$.
Prove that this algorithm works.[32]
Suggestion: First prove that if X and Y are independent uniformly distributed random variables, then $P(X = Y) = 0$. One way to prove the this is as follows: For any positive integer n, partition the interval $[0, 1]$ into n disjoint intervals of equal length $I_1, I_2, \cdots, I_n$. Using independence of X and Y, we may write: $P(X = Y) = P\left(\bigcup_i \{X \in I_i\} \cap \{Y \in I_i\}\right) \le$
$\sum_i P(\{X \in I_i\} \cap \{Y \in I_i\}) = \sum_i P(X \in I_i)P(Y \in I_i) = n(1/n)^2 = 1/n$.
Since n was arbitrary, it follows that $P(X = Y) = 0$.

31. (*Expectation for a Gambling Game = Long-Term Average Payoff of Game*) Suppose that a certain random game of chance has K possible outcomes s_i $(1 \le i \le K)$, each with a

---

[32] Although the random numbers $U_1, U_2, \cdots, U_n$ will have no duplications (with probability one), in case there are duplications, we simply agree to make a random choice of which index to put first. This settles any possible contingencies.

corresponding probability p_i and payoff x_i. Thus the "payoff" random variable for the game is X, where $X(s_i) = x_i$ $(1 \le i \le K)$. Suppose that the game is played repeatedly, and the cumulative payoff after n games is $A(n)$.

(a) Prove that with probability 1 (in the sense of Theorem 6.3) we have $\lim_{n \to \infty} A(n)/n = E[X]$. Thus, $E[X]$ is the long term average gain/loss of the game. For this reason $E[X]$ is often referred to as the *value of the game*; the game is *fair* if $E[X] = 0$. A positive value of $E[X]$ indicates a longterm average gain, so the game is favorable (to the player), while a negative value of $E[X]$ indicates an unfavorable bet.

(b) A certain state's *Powerball Lottery* game works as follows: Forty ping-pong balls numbered 1 through 40 are mixed up and five of them are randomly drawn (using a special transparent glass popping machine that is shown on television) without replacement. Each play card costs $1 and on which the player puts down a set of five numbers taken from the numbers 1 through 40. If all five of your numbers get drawn, your ticket wins $200,000, if four out of five of your numbers get drawn, you win $1000, and if three out of five of your numbers get drawn you win $20. What is the value of this game?

Note: Although the "possible" payoffs for many such state lotteries (and many other casino bets) are quite attractive, the value of such games invariably turns out to be negative. Thus, in effect, anyone who plays is (on average) making a donation to the state coffers (or casino owner pockets and additional taxes to state coffers). Many poor people are lured into the false promises of such bets in areas where gambling is legal, often betting away their food and rent money. This is why the issue of whether to allow lotteries or casino gambling in states is often a very controversial one.

Suggestion: For Part (a), let $c_i(n)$ be the number of times in n games that outcome s_i has occurred $(1 \le i \le K)$. Note that $A(n) = \sum_{i=1}^{k} x_i c_i(n)$. Apply Theorem 6.3 to get that $c_i(n)/n \to p_i$ (for each index i, with probability 1). For Part (b), note that the payoff amount should always have $1 (the price of the ticket) subtracted. Thus, if the ticket does not win, the payoff is $-\$1$.

32. Complete the proof of Theorem 6.6(a) by using mathematical induction and what was proved in the text: $E[aX + bY] = aE[X] + bE[Y]$, to prove (21):

$$E[a_1 X_1 + a_2 X_2 + \cdots + a_n X_n] = a_1 E[X_1] + a_2 E[X_2] + \cdots + a_n E[X_n].$$

33. (a) Using only the definition of expectation, compute $E[X^2]$, where X is a binomial random variable with parameters n and p, i.e., $X \sim \mathscr{B}(n, p)$.
(b) Use the result of Part (a), the fact that $E[X] = np$ (which was established in two ways in the text) and (22): $Var(X) = E[X^2] - (E[X])^2$, to compute the variance of X ($X \sim \mathscr{B}(n, p)$).

34. (*Developing a Winning Strategy*) Consider the following game:
The host has written two different amounts of money and sealed them in two envelopes. The player is allowed to choose one of the envelopes, open it and see the amount of money. At this point the player can either opt to win the amount of money that he/she has just seen, or instead choose to switch to the other envelope and and win whatever amount that is written inside the second envelope. The player will, of course, not be allowed to switch back to the first envelope's amount if he/she has opted to change envelopes.
Develop a strategy for the player that will give him/her a greater than 50% chance of selecting the envelope with the larger amount of money, and prove your result.
Note: No information about the distribution of the two amounts of money written inside the envelopes is needed to develop such a strategy.

COMPUTER EXERCISES 6.2:

NOTE: Although the use of histograms in conjunction with simulations has been rather informal in this section, they are nonetheless useful in illustrating the overall accuracies of simulated estimates. In the next section we will formalize much of this intuition into an effective scheme for putting precise levels of confidence on the results of our simulations. The experience of histograms in this section and the following computer exercise set should help the reader to develop a useful intuition that will help to make the relevant developments of the next section much more understandable.

1. An urn contains 3 red balls, 4 white balls, and 10 black balls.
   (a) Simulate $N = 1000$ drawings of 2 balls with replacement to approximate the probability that both balls drawn are white. (After the first ball is randomly drawn and its color recorded, it is replaced and the balls are mixed up before the second ball is randomly drawn.)
   (b) Repeat Part (a) with $N = 100,000$ drawings.
   (c) Repeat Part (a) with $N = 10,000,000$ drawings.
   (d) Create a histogram for $K = 1000$ sets of $N = 10$ simulated drawings of the experiment in Part (a), and then another with $K = 1000$ sets of $N = 1000$ drawings.
   (e) Create a histogram for $K = 100$ sets of $N = 10,000$ simulated drawings of the experiment in Part (a), and then another with $K = 100$ sets of $N = 100,000$ drawings. Compare these histograms with those of Part (d).
   (f) Compute the exact probability and compare.

2. An urn contains 3 red balls, 4 white balls, and 10 black balls.
   (a) Simulate $N = 1000$ drawings of 2 balls without replacement to approximate the probability that both balls drawn are white.
   (b) Repeat Part (a) with $N = 100,000$ drawings.
   (c) Repeat Part (a) with $N = 10,000,000$ drawings.
   (d) Create a histogram for $K = 1000$ sets of $N = 10$ simulated drawings of the experiment in Part (a), and then another with $K = 1000$ sets of $N = 1000$ drawings.
   (e) Create a histogram for $K = 100$ sets of $N = 10,000$ simulated drawings of the experiment in Part (a), and then another with $K = 100$ sets of $N = 100,000$ drawings. Compare these histograms with those of Part (d).
   (f) Compute the exact probability and compare.

3. Repeat all parts of Computer Exercise 2 if the urn contains 3 red balls, 10 white balls, and 8 black balls.

4. Repeat all parts of Computer Exercise 1 if the urn contains 3 red balls, 10 white balls, and 8 black balls.

5. In Parts (a) through (c) below, run 100,000 simulated trials to estimate the probabilities in question.
   (a) If 10 dice are rolled, find the probability that the maximum number rolled is a 6.
   (b) If 10 dice are rolled, find the probability that the maximum number rolled is a 5.
   (c) If 10 dice are rolled, find the probability that the sum of the numbers rolled is at least 50.
   (d) For each of the probabilities of Parts (a) through (c), create a histogram for $K = 100$ sets of $N = 100$ simulated trials of the experiment, and then another with $K = 100$ sets of $N = 100,000$ drawings.

6. In Parts (a) through (c) below, run 100,000 simulated trials to estimate the probabilities in question.
   (a) If 8 dice are rolled, find the probability that the maximum number rolled is a 6.
   (b) If 6 dice are rolled, find the probability that the maximum number rolled is a 5.
   (c) If 5 dice are rolled, find the probability that the sum of the numbers rolled is at least 25.
   (d) For each of the probabilities of Parts (a) through (c), create a histogram for $K = 100$ sets of $N = 100$ simulated trials of the experiment, and then another with $K = 100$ sets of $N = 100,000$

drawings.

7. An urn contains three red balls, and three green balls. An experiment proceeds as follows:
Step 1: Randomly draw a ball from the urn, and proceed to Step 2.
Step 2: *Case 1:* If the drawn ball is green, replace it along with a new green ball into the urn, and mix up the balls. Then return to Step 1.
Case 2: If the drawn ball is red, proceed to Step 3.
Step 3: Record the total number of balls that were drawn and end the experiment.
Let the random variable X be the total number of balls that are drawn in this experiment.
(a) Compute analytically the probability mass function of X.
(b) Simulate $N = 100,000$ trials of this experiment, estimate the probability $P(X = 2)$, and check with the exact answer of Part (a).
(c) Simulate $N = 100,000$ trials of this experiment, estimate the probability $P(X > 4)$, and check with the exact answer that is obtainable from the answer to Part (a).
(d) Create a histogram for $K = 1000$ sets of $N = 10$ simulated trials of the experiment to estimate $P(X = 2)$, and then another with $K = 1000$ sets of $N = 1000$ trials.
(e) Create a histogram for $K = 100$ sets of $N = 10,000$ simulated trials of the experiment to estimate $P(X > 4)$, and then another with $K = 100$ sets of $N = 100,000$ drawings.
(f) Use simulations to estimate $E[X]$, and use some histograms to support the accuracy of your estimate.

8. Repeat all parts of Computer Exercise 7 with the same experiment, but with the urn now initially containing one red ball and three green balls, and declaring the experiment to have ended if 1000 green balls have been drawn.
Note: It can be shown that without this limiting restriction, $E[X]$ would be infinite, whereas in Computer Exercise 7, $E[X] < \infty$ (without the restriction).

9. (a) Create a histogram for $K = 1000$ sets of $N = 10$ simulated trials of the experiment in (ordinary) Exercise 11 to estimate the expected value, and then another with $K = 1000$ sets of $N = 1000$ trials. What are the overall means of the simulated expected values for each of the two sets of data?
(b) Repeat the instructions of Part (a), but change the histogram parameters to: $K = 100$ sets of $N = 10,000$ simulated trials of the experiment, and then $K = 100$ sets of $N = 100,000$ trials.

10. Repeat both parts of Computer Exercise 9, but this time to estimate the expected value in (ordinary) Exercise 12.

11. Repeat both parts of Computer Exercise 9, but this time to estimate the expected value in (ordinary) Exercise 13.

12. Repeat both parts of Computer Exercise 9, but this time to estimate the expected value of (ordinary) Exercise 13.

13. This computer exercise will look at some questions related to the cereal company of (ordinary) Exercise 17 that has an ongoing promotion where each box of their cereal contains one of five special tokens. If you collect all five you win a free box of cereal. Suppose that each of the five tokens is equally likely to appear.
(a) If someone has bought 10 boxes of cereal, use simulations to estimate the expected number of different tokens he/she will have collected. Aim for an accuracy of 0.2 (that is, the true answer should be within 0.2 of your estimate), and use histograms to support the quality of your estimate.
(b) Continue to suppose that each of the five tokens is equally likely to appear. Use simulations to estimate the expected number of boxes of cereal one would need to buy in order to collect all five tokens. Aim for an accuracy of 0.2, and use histograms to support the quality of your estimate.

14. Change the assumptions of the token collecting problem in the preceding exercise to be that one of the tokens is 100 times more rare than each of the other four, but the other four are equally likely to be found in a box of cereal.

(a) Suppose someone has bought 10 boxes of cereal. Use simulations to estimate the expected number of different tokens he/she will have collected. Aim for an accuracy of 0.2 (that is, the true answer should be within 0.2 of your estimate), and use histograms to support the quality of your estimate.

(b) Use simulations to estimate the expected number of boxes of cereal one would need to buy in order to collect all five tokens. Aim for an accuracy of 1, and use histograms to support the quality of your estimate.

15.   (a) Use simulations to estimate the expected number of times that one would need to flip a fair coin before accumulating a total of 12 heads. Achieve accuracy to the nearest flip, and use histograms to support the quality of your estimate. Compare your estimate with the analytical answer from (ordinary) Exercise 20.

(b) Repeat Part (a) but this time assume the coin is biased, with only 0.3 probability of landing as heads.

16.   (a) Write a program that generates random permutations and is based on the algorithm of (ordinary) Exercise 24. Let the syntax be `permvec = PermProg(vec)`, where the input `vec` is any (ordered) vector and the output `permvec` is another vector that is a random permutation of `vec`. The algorithm requires a vector sort, so you can either use your system's built-in program for this (if available) or an appropriate program from Chapter 5.

(b) Use a simulation to test the effectiveness of this program as follows: For the vector [1, 2, 3, $\cdots$, 10] perform $N = 10,000$ random permutations, and plot an associated histogram for the random variable $X =$ the place (location) of the number 1 in the randomly permuted vector. Thus, for example, if the permuted vector is [6, 4, 3, 10, 2, 7, 5, 1, 9, 8], then X would equal 8 for this vector (since 1 is located in the 8th slot). What distribution does this random variable have? Does your histogram support this?

(c) Compare the runtimes of the program that you created in Part (a) alongside a corresponding program that is based on Algorithm 6.1. Use test vectors of the form [1, 2, $\cdots$, K], with $K = 10^3, 10^4, 10^5, \cdots$ until one of the runtimes starts to take longer than 30 seconds on your machine.

17.   (*A Fair Way to Flip Even an Unfair Coin*) Suppose that we have a coin, which when flipped will land heads with probability p, $0 < p < 1$. The following scheme involving flipping this coin will always result in a "heads" outcome with probability exactly 0.5:

Step 1: Flip the coin twice.

Step 2: If the flips both land as heads or both land as tails, return to Step 1, otherwise proceed to Step 3.

Step 3: Take the result of the last flip as the result of the experiment.

(a) Run some sets of simulations and generate some histograms that will help to support this claim when $p = 0.55$.

(b) Repeat Part (a) when $p = 0.75$.

(c) Repeat Part (a) when $p = 0.05$.

Note: Since it is never really known if any real-life coin is truly fair, this procedure should always be used when one wants to guarantee a completely fair coin flip. To prove that this procedure is indeed fair, note that since the procedure terminates whenever a new pair of flips are different, we have (letting $q = 1 - p$): P("heads") = P(TH | TH or HT) = $qp/(qp + pq)$ = 1/2.

18.   (*A Scheme to Reverse the Odds of an Unfair Coin*) Suppose that we have a coin, which when flipped will land heads with probability p, $0 < p < 1$. Consider the following scheme for obtaining a "heads" or "tails" outcome from flipping the coin that will change the probability of obtaining heads to $1 - p$:

Continue flipping the coin until both heads and tails have appeared. Take the result of the last flip.

(a) Run some sets of simulations and generate some histograms that will help to support this claim when $p = 0.55$.

(b) Repeat Part (a) when $p = 0.75$.

(c) Repeat Part (a) when $p = 0.05$.

(d) Can you analytically prove that this algorithm reverses the probability of obtaining heads?

19. (a) Use simulations with supporting histograms to approximate each of the following probabilities with an accuracy goal of 0.05.
    (i) If we throw a die three times, the probability the numbers get larger with each throw.
    (ii) The corresponding probability that the numbers get smaller with each throw.
    (iii) The corresponding probability that the numbers do not decrease.
    (b) Compute each of the above probabilities exactly, either analytically or by using a brute-force computer program, and compare with the simulated approximations of Part (a).
    Note: An acceptable outcome for the event in (i) would be: 2, 3, 5; the outcome 2, 2, 6 would be acceptable for the event in (iii) but not for the event in (i).

20. (a) Use simulations with supporting histograms to approximate each of the following probabilities with an accuracy goal of 0.05.
    (i) The probability of getting a run of at least 6 straight heads in the course of 10 consecutive flips of a fair coin.
    (ii) Compute the probability of the same event if there are 15 flips.
    (b) Compute each of the above probabilities exactly, either analytically, or by using a brute-force computer program, and compare with the simulated approximations of Part (a).

21. Use simulations to approximate the expected value of the *Powerball Lottery* game that was described in (ordinary) Exercise 25(b). Aim for an estimate that is accurate with a margin of error allowance of 2 cents, and use histograms to support the accuracy of your estimates. Compare with the analytical answer (that was obtained in Exercise 25(b)).

22. (*Simulation for the Birthday Problem*) (a) Write a program with syntax

    ```
    P = birthdaysimulator(N,n,d)
    ```

    that computes the simulated probability P that a common birthday will occur among n people in a room whose birthdays are randomly chosen among d possibilities. The first input variable N is the number of trials, and the second input variable n is the number of people in the room. The third input variable, d, is optional, and denotes the number of possible birthdays. The default value for d is 365.
    (b) Run this program using n = 23, n = 40, and n = 180 (and d = 365) to obtain 1000 simulated values for each of the three corresponding probabilities where each of these simulated value was obtained of 1000 trials (N = 1000). Create histograms and compute each of the means for your three data sets, then find the errors of each of these simulated probabilities when they are compared with the exact answers that were computed in Example 6.4.
    Suggestion: For Part (a), a random trial of the experiment can be modeled by generating a length n vector vec whose entries are random integers in the range {1, 2, ..., d} –representing birthdays of n people. One way to check for duplicated entries in this vector (i.e., a common birthday) is: let vecs = the corresponding sorted vector, and check if the vectors [vecs 0] and [0 vecs] have any equal entries.

23. (*Variation of the Birthday Problem*)   Devise a strategy using simulations to compute the minimum number of people, N3, that would need to be a room so that the probability of at least three people having a common birthday would exceed 50%. Perform such a program and give the resulting estimate. Compare your resulting estimate with the exact answer that was found in (ordinary) Exercise 27(c).
    Suggestion: A slight modification of the suggestion to the preceding exercise will work here; simply change the last step of that suggestion to check if the vectors [0 0 vecs] and [vecs 0 0] have any equal entries.

24. (*Variation of the Birthday Problem*)   Devise a strategy using simulations to compute the minimum number of people, N4, that would need to be a room so that the probability of at least four people having a common birthday would exceed 50%. Perform such a computation. See the suggestions of the preceding two exercises for ideas.

25. Ten cards numbered 1 through 10 are shuffled. Set up some simulations to approximate the answers to the following questions.

(a) What is the probability that someone who guesses the order of the 10 cards will get at least 6 correct?

(b) How would the answer to Part (a) change if the guessing is done card-by-card where the $(n+1)$st card is guessed after having seen the result of the first n cards?

(c) How would the answers to Parts (a) and (b) change if, instead of 10 different cards, there were 20, and the problem was to compute the probabilities of guessing at least 12 correctly (same percentage of correct guesses).

26. (*Gambler's Ruin: Fair Game*)   Suppose that a gambler starts off with $10, and continues to play a fair game with a $10 bet until he either reaches his goal of $100 or goes broke.   Use simulations to approximate the probabilities of the following events.

(a) The gambler reaches his $100 goal (without going broke).

(b) The gambler goes broke (before reaching his $100 goal).

(c) The gambler's fortune reaches at least $80 before eventually going broke (without reaching his $100 goal).

27. (*Shipping Port Logistics*)   Suppose that cargo ships arrive at a port that has a single processing station.   Three types of cargo ships can arrive at the port for unloading: small, mid-sized, and jumbo -ized, and the time (in days) required to unload each type of ship is 1 day, 3 days, and 5 days, respectively.   Cargo ships always arrive on different days, with gaps between arrival days ranging from 0 days to 5 days. The probabilities for these gaps are: 15% for a 0-day gap, 25% for a 1-day gap, 30% for a 2-day gap, and 10% each for a 3-, 4-, or 5-day gap.   The ships arrive independently of one another with each arrival having a 30% chance of being a small ship, a 40% chance of being a mid-size ship, and 30% chance of being a jumbo ship. A newly arriving ship will proceed to the station to get served if service is available; otherwise, the ship will wait in the port area to get served on a first-come, first-served basis.

(a) Set up a simulation program that will run for a large number of days, and track the average time that an incoming ship must wait to get served.   Plot this average versus the number of days that the simulation runs to see if it is stabilizing.

(b) Modify your model to also track the percentage of idle time for the processing station.

28. (*Shipping Port Logistics*)   Suppose that cargo ships arrive at a port which has two crane-equipped processing stations: older Station I, and the newer and larger Station II. Three types of cargo ships can arrive into the port for unloading: small, midsized, and jumbo sized, and the time (in days) required to unload each type of ship is 2 days, 3 days, and 5 days (respectively) in Station I, and 1 day, 2 days, and 3 days (respectively) in Station II. Cargo ships always arrive on different days, with gaps between arrival days ranging from 0 days to 5 days.   The probabilities for these gaps are: 35% for a 0-day gap, and 20% for a 1-day gap, 15% for a 2-day gap, and 10% each for a 3-, 4-, or 5-day gap.   The ships arrive independently of one another with each arrival having a 30% chance of being a small ship, a 40% chance of being a mid-size ship, and 30% chance of being a jumbo ship. A newly arriving ship will proceed to Station II, if it is available; if not, it will proceed to Station I, if it is free; otherwise, the ship will wait in the port area until one of the stations becomes available, and will get served on a first-come, first-served basis.

(a) Set up a simulation program that will run for a large number of days, and track the average time that an incoming ship must wait to get served.   Plot this average versus the number of days that the simulation runs to see if it is stabilizing.

(b) Modify your model to also track the percentage of idle time for each of Stations I and II, that is, the percentage of time from the start of the simulation until through the last day of the simulation that each of the stations is not engaged in an unloading operation.

29. Suppose a game is played where three dice are rolled repeatedly until either three 6's show on a single roll (win), or two separate rolls have resulted in at least a pair of 5's and a pair of 4's (lose).

(a) Simulate 1 million trials of this game and approximate the probability of winning it.

(b) Modify your code in Part (a) so that it will also compute the average number of throws per game, as well as the maximum number of throws that occurred in the 1 million simulated games.

Chapter 7: Complexity of Algorithms

Two main aspects of technology have much to do with enhancing the quality of our lives in many ways: Advances in computing hardware and advances in algorithms and the corresponding software programs that run on various sorts of hardware. The first sort of advance has been rather well predicted by *Moore's law*, which states that computing speeds on the latest computers double approximately every 18 months.[1] Algorithm advances, on the other hand, depend on less predictable theoretical breakthroughs, and tend to have much greater impacts. Such advances boil down to the discovery of a new and more efficient way to solve an important problem. Up to this point the reader has had many encounters with algorithms and instances where two different algorithms designed for the same problem can have considerably different performances and capacities. An **algorithm** can be defined to be any procedure that is designed to perform a certain task or obtain the solution to a certain problem. When problems have inputs, an algorithm for its solution will produce the solution as the output corresponding to the given inputs. Algorithms

FIGURE 7.1: Abu Muhammad al-Khwarizmi (c. 780–c. 850), Persian mathematician.

were named in honor of the Persian mathematician al-Khwarizmi[2] (Figure 7.1) who wrote a book on basic arithmetic algorithms in the decimal system. Although

---

[1] This was first observed in the 1960s by Gordon Moore (1929–), one of the founders of the *Intel Corporation*.

[2] Not much is known about the life of Al-Khwarizmi, but he worked most of his life with a group of scholars for the *House of Wisdom* in what is now Baghdad, Iraq. The group took great efforts translating well-known scholarly works; one such book was a Hindu book on decimal (base 10) numbers that was presented to the group by an Indian diplomatic officer. al-Khwarizmi was so impressed with this decimal number system (the one used today) that he went further and wrote another book about algorithms to perform many mathematical operations in decimal notation. (Think about trying to do things like multiplication using Roman numerals.) al-Khwarizmi subsequently wrote another very important book on algebra, where he showed how to solve many sorts of first and second degree equations. In this book he gave many real-world applications of algebra. Interestingly, at that time, all algebraic expressions were expressed in words (the present-day notation having not yet been invented). Both books were translated into Latin, and had such a tremendous influence on the

human beings can use algorithms to solve problems, we are generally slow and prone to making mistakes, and so it is more efficient to employ computers for implementing algorithms.   In our work so far, we have already encountered several important algorithms in connections with other topics.   Apart from working correctly (i.e., producing the correct answer to a given instance of a problem), an algorithm should also be efficient.   In the 1970s a new method for measuring efficiency of algorithms, known as *computational complexity*, was developed by Stanford computer scientist Donald Knuth. This chapter will consist of two sections.   Section 7.1 will introduce some algorithms for searching and sorting.   These algorithms will form a nice basis to lead into the development of complexity theory of algorithms, which will be the topic Section 7.2. Complexity theory provides a platform-independent framework for measuring and comparing the performance of algorithms.

7.1: SOME ALGORITHMS FOR SEARCHING AND SORTING

Problems involving sorting and searching are very fundamental and occur frequently. At first glance, these tasks might seem quite elementary, but there is a great deal of theory and algorithms available for sorting and searching.   In particular, the technologies behind many of the state-of-the-art internet search engines use intricate algorithms for searching and sorting according to page rank. Evidence of this fact is illustrated by the existence of an entire 800-page volume (volume 3) of the famous (presently) three-volume magnum opus *The Art of Computer Programming* [Knu-97] by the illustrious computer scientist Donald Knuth.[3]   Knuth even states, "I believe that virtually every important aspect of programming arises somewhere in the context of sorting and searching." We will present two search algorithms; one of them (the linear search) is slower but more versatile, and the other (the binary search) assumes that the elements of the list are sorted, but performs much more rapidly.   Sorting is typically a more difficult task than searching, so we save it for last.   We will present four different sorting algorithms in the text, and the exercises will present another.

---

international scientific communities that Al-Khwarizmi was honored by the introduction of the words algorithm and algebra, in recognition of the impact of his two books.

[3] A brief biography of Donald Knuth appears later in this chapter. Although Knuth has made numerous contributions, he considers these books to be his greatest work and it is his lifelong mission to continue with its writing to a complete five-volume set. Microsoft chairman Bill Gates said of this work, "You should definitely send me a résumé if you can read the whole thing."

The Linear Search Algorithm

We begin with a simple algorithm for the solution of the general search problem. Given a list of integers, $a_1, a_2, \cdots, a_n$, suppose we wish to determine whether an integer t appears in this list. The simplest approach is to simply start from the leftmost integer in the list and compare it with t, and move on to successive integers in the list until a match is found (if ever).[4]   Here is the pseudocoded version of this algorithm:

ALGORITHM 7.1: (*The Linear Search Algorithm*)

Input: An unsorted list (vector) of integers: $a_1, a_2, \cdots, a_n$, and a target integer t.

Output: An index integer $k \in \{0, 1, 2, \cdots, n\}$, such that if $k > 0$, then $a_k = t$, while if $k = 0$, then t does not appear in the list $a_1, a_2, \cdots, a_n$.

FOR $i = 1$ TO n
    IF $a_i = t$
        OUTPUT $k = i$, EXIT Algorithm
    END IF
END FOR
OUTPUT $k = 0$ (if algorithm has not already terminated), and EXIT Algorithm

In the worst case (if t does not appear in the list or appears at the end) this algorithm could require n iterations in the for loop (i.e., as many comparisons as there are elements of the list); in the best case we need only one iteration—Why? If we are searching a list of distinct integers in which the target appears, then on average the   algorithm will require $(n+1)/2$ iterations, since it is equally probable that the target appears in any of the n slots of the list—see Exercise 21 for a formal justification of this fact using concepts from Chapter 6.   In case the integers in the list are sorted (say from smallest to largest), there are much more efficient algorithms.

EXERCISE FOR THE READER 7.1: Develop an algorithm that will determine the *maximum value* in a list of integers $a_1, a_2, \cdots, a_n$, along with an index i corresponding to the first location in this list where the maximum occurs, i.e., the outputs will be $\max(a_1, a_2, \cdots, a_n)$, and the smallest index i such that $a_i = \max(a_1, a_2, \cdots, a_n)$. What simple change would be required in your algorithm for it to determine the *minimum value* rather than the maximum value?

---

[4] For generality, we allow duplications in the search lists, which means there may be more than one place where the target can be found. Our search algorithms will be designed to find just one place in the list where the target occurs, not all of them.

The Binary Search Algorithm

For sorted lists, the binary search algorithm is a good example of what is often called a "divide and conquer" algorithm, where the search list gets iteratively divided (say into two pieces of half the size of the original search list) until the search is complete. Before formally stating the algorithm in pseudocode, we motivate and explain it with a small example.

EXAMPLE 7.1: Suppose that we wish to "search" for the target number 88 in the sorted list of integers:

$$[2, 6, 16, 22, 23, 32, 35, 44, 52, 59, 66, 69, 84, 88, 91, 94]$$

The binary search would proceed as follows: The list has 16 elements. We split the list into a left half and a right half, each having 8 elements.

$$[2, 6, 16, 22, 23, 32, 35, 44] \qquad [52, 59, 66, 69, 84, 88, 91, 94]$$

We compare the last/largest element (44) of the left list with the target (88) and since $88 > 44$, it must be in the right list if it appears at all. In the next iteration, we dissect the right list into two smaller lists:

$$[52, 59, 66, 69] \qquad [84, 88, 91, 94]$$

We compare the last/largest element (69) of the left list with the target (88) and since $88 > 69$, it must be in the right list if it appears at all. We next split the right list into two: $[84, 88]$ $[91, 94]$. Since $88 \not> 88$, the search is now restricted to the left interval, which in turn gets split into two single number lists: $[84]$ $[88]$. Since $88 > 84$, we arrive at the one single term list $[88]$, which can no longer be split. We check to see if the target equals the number (it does), so the search is complete, and we can exit the algorithm.

In general, suppose that we are given a sorted list of integers, $a_1 \le a_2 \le \cdots \le a_n$, and that we wish to determine whether a target integer t appears in this list. We first compare t with the term a_m, where the index m (think "middle") is taken to be $m = \lfloor (n+1)/2 \rfloor = \text{floor}((n+1)/2)$. Note that if n is odd, m is the middle index, while if n is even, m is the index of the last/largest element in the left half list (as in Example 7.1). If $t > a_m$, we restrict the search to the portion of the sequence to the right of a_m: $a_{m+1}, a_{m+2}, \cdots, a_n$; otherwise we restrict to the left portion of the sequence: $a_1, a_2, \cdots, a_m$. This procedure gets iterated until the restricted list becomes a single element, at which point a comparison with the target is made. Note that at the first iteration, the size of the left/right list portion is at most $\lceil n/2 \rceil = \text{ceil}(n/2)$. We are now ready to formally state this algorithm in pseudocode.

ALGORITHM 7.2: (*The Binary Search Algorithm*)

Input: A sorted list (vector) of integers: $a_1 \leq a_2 \leq \cdots \leq a_n$, and a target integer t.

Output: An index integer $k \in \{0, 1, 2, \cdots, n\}$, such that if $k > 0$, then $a_k = t$, while if $k = 0$, then t does not appear in the list $a_1, a_2, \cdots, a_n$.

Step 1: Initialize search interval endpoint indices:

INITIALIZE: LeftIndex = 1, RightIndex = n

Step 2: Iteratively reduce search interval to a single number:

WHILE LeftIndex < RightIndex
      SET m = floor([LeftIndex + RightIndex]/2)

      IF $t > a_m$
          UPDATE LeftIndex = $m + 1$
      ELSE
          UPDATE RightIndex = m
      END IF
END WHILE

Step 3: Check if target equals the number

IF $t = a_m$
      OUTPUT $k = m$
ELSE
      OUTPUT $k = 0$
END IF

EXERCISE FOR THE READER 7.2: Apply the binary search algorithm to search for the target number 46 in the sorted list of integers:

$$[5, 9, 11, 18, 27, 29, 39, 44, 46, 56, 67, 72, 85, 86, 92, 98]$$

The Selection Sort Algorithm

Given a large list of names or integers, it is often convenient to put them in order (alphabetical or numerical). Sorting algorithms are designed to do this, and there are a wide variety of such algorithms. The first sorting algorithm that we present is the most basic one, one whose correctness is obvious. Given a sequence of unsorted integers: $a_1, a_2, \cdots, a_m$, the *selection sort* first compares a_1 and a_2, then compares the smaller of these two with a_3, then compares the smaller of these two with a_4, and so on through the rest of the sequence until we have found the smallest element. We remove this element from the sequence and put it as the first element of the sorted list, and then find the smallest element of the residual sequence (which will in turn be the second smallest element of the original sequence). We continue in this fashion until the residual sequence is empty, and the sorted sequence will have been produced. Here is the pseudocoded version of this algorithm.

ALGORITHM 7.3: (*The Selection Sort Algorithm*)

Input: An unordered list (vector) of integers: $a_1, a_2, \cdots, a_n$.

Output: An increasing reordering of these integers, i.e., an increasing sequence $b_1 \le b_2 \le \cdots \le b_n$ of integers[5] such that for each index i, $b_i = a_j$, for some index j.

INITIALIZE: UnsortedList = $[a_1, a_2, \cdots, a_n]$, numberUnsorted = n
        SortedList = [ ] (empty list)
FOR i = 1 TO numberUnsorted − 1
        Initialize b = UnsortedList(1)
        FOR j = 2 TO numberUnsorted
           Update b = min(b, UnsortedList(j))
        END j FOR
        Remove b from UnsortedList, Append b to SortedList
        UPDATE: numberUnsorted = numberUnsorted −1
END i FOR
(*Note: After this loop is executed, UnsortedList will contain just one remaining element—the largest element.*)

Append the last remaining element of UnsortedList to SortedList

OUTPUT: SortedList, and EXIT Algorithm

Here is a brief recap of how the selection sort works: use $n − 1$ comparisons to determine the smallest element of the list, move it to the head of the sorted list, next use $n − 2$ comparisons to find the smallest element of the remaining list (= the second smallest of the original list), and move this to the end of the sorted list. Then use $n − 3$ comparisons to find the third smallest element, and so on. This makes it clear that the number of comparisons required in Algorithm 7.3 is always

$$(n-1)+(n-2)+\cdots+2+1 = \frac{n(n-1)}{2},$$

where we have used formula (1) of Chapter 3.

---

[5] Many authors will call such a sequence *nondecreasing*, and reserve the term *increasing* for sequence of distinct integers with $b_1 < b_2 < \cdots < b_n$. But nondecreasing seems to simply mean a sequence that is not decreasing, with no other nice property, so we prefer the more suggestive terminology even for nondistinct integers (duplicated terms).

The Bubble Sort Algorithm

The *bubble sort* algorithm is modeled by the physical behavior of bubbles in a flask: larger bubbles tend to sink below the smaller ones. Roughly speaking, the bubble sort algorithm operates as follows: In the first pass through the list, larger elements are permuted with smaller ones. Here, "larger" means "greater than" in the underlying ordering. The net result of the first pass is that the largest element of the list will be at the end. (But some large elements that were encountered before the largest may have been moved down the list.) This process is repeated until we arrive at a pass in which no permutations are required. At this point the list will be ordered.

EXAMPLE 7.2: Figure 7.2 illustrates the bubble sort algorithm applied to the following list of integers [5, 3, 1, 2, 4]. In the first pass, the heaviest item (5) sinks to the bottom. Restarting on the top of the list, the second pass first permutes 1 and 3, and then 2 and 3. After this, the items are sorted, and so in the third pass no permutations are needed, so the algorithm stops.

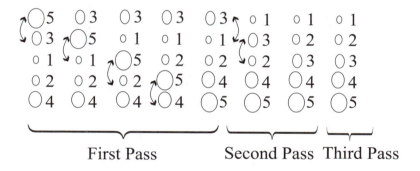

FIGURE 7.2: Illustration of the bubble sort algorithm of Example 7.2. The larger numbers are represented by larger bubbles, which sink to the bottom of the list.

EXERCISE FOR THE READER 7.3: Apply the bubble sort algorithm to sort (in alphabetical order) the following list of names:

Joey, Tara, Amanda, Teresa, Patrick, Dennis.

EXERCISE FOR THE READER 7.4: We have shown that the number of comparisons needed when the selection sort algorithm is applied to a list of n distinct integers is always the same, i.e., $n(n-1)/2$. If the bubble sort algorithm is applied to such a list, is the number of comparisons always the same? Explain your answer.

The following is a pseudocode formulation of the bubble sort algorithm:

ALGORITHM 7.4: (*The Bubble Sort Algorithm*)

Input: An unordered list (vector) of integers: $a_1, a_2, \cdots, a_n$.

Output: An increasing reordering of these integers, i.e., an increasing sequence $b_1 \le b_2 \le \cdots \le b_n$ of integers such that for each index i, $b_i = a_j$, for some index j.

FOR $i = 1$ TO $n - 1$
        FOR $j = 1$ TO $n - i$
            IF $a_j > a_{j+1}$
                Interchange a_j, a_{j+1}
            END IF
        END j FOR
        IF no interchanges were needed,
            OUTPUT current list, and EXIT
        END IF
END i FOR

The Quick Sort Algorithm

When we show how to compare efficiencies of algorithms in the next section, we will find that the selection sort and bubble sort algorithms are of about the same efficiency.   The *quick sort* algorithm that we will now introduce can be significantly more efficient on average. We will make and justify a more precise statement in the next section. Here we only introduce the algorithm. Like the binary search algorithm, the quick sort is an example of a divide and conquer algorithm. The following small example will show how the quick sort works; it will be followed by a pseudocoded algorithm.

EXAMPLE 7.3: Suppose that we need to sort the following unordered list of integers: 22 84 16 65 32 19   7   51 78   15.   The quick sort algorithm works as follows: Begin by selecting any element of the list, say 51. We then compare all other elements of the list with 51. Those that are less than 51 are placed in brackets on the left of 51, and those that are greater than 51 are placed in brackets on the right of 51.   After this is done we are left with:

[22 16 32 19 7 15] 51 [84 65 78].

We now iterate this procedure on any bracketed lists that contain more than one number, until all brackets contain a single integer. Starting with the left bracketed list, if we choose 16, we next arrive at:

[7 15] 16 [22 32 19] 51 [84 65 78].

If we use 7 in the first brackets, 22 in the second, and 84 in the third, simultaneously performing this operation on each of the three bracketed expressions yields:

7 [15] 16 [19] 22 [32] 51 [65 78] 84.

One final iteration is required; if we use 65 the final result is

$$7 \; [15] \; 16 \; [19] \; 22 \; [32] \; 51 \; 65 \; [78] \; 84,$$

which is (ignoring the brackets) the sorted list.

Note that the quick sort algorithm has some flexibility depending on which numbers in the bracketed lists are used. One could always take the first element in the list (for definiteness) or use a random selection.

EXERCISE FOR THE READER 7.5: If the quick sort is applied to a list of (unsorted) integers $a_1, a_2, \cdots, a_n$, what is the maximum number of comparisons that would be needed (worst case)?

EXERCISE FOR THE READER 7.6: (a) Apply the quick sort algorithm to sort (in alphabetical order) the following list of names:

    Joey, Louise, Dan, Tara, Steve, Amanda, Teresa, Sandra, Patrick, Dennis.

When selecting an element in the brackets, always use the leftmost element.
(b) Repeat the search in Part (a) but this time always selecting the rightmost element in the bracketed vectors.

Below we present a pseudocode formulation for the recursive quick sort algorithm. In Step 2 of this algorithm, we have not specified how elements get selected from the list. Technically this does not completely specify the algorithm, but rather gives a *meta-algorithm*, which gives a variety of algorithms depending on which selection scheme is used. Since differences in the selection scheme do not in general make much of a difference in the efficiency of the quick sort algorithm, we will leave it to the reader to use whatever selection scheme is most convenient.

ALGORITHM 7.5: (*The Quick Sort Algorithm*)

Input: An unordered list (vector) of integers: $[a_1, a_2, \cdots, a_n]$.

Output: An increasing reordering of these integers, i.e., an increasing sequence $b_1 \le b_2 \le \cdots \le b_n$ of integers such that for each index i, $b_i = a_j$, for some index j.

Step 1: If the number of elements n in List $= [a_1, a_2, \cdots, a_n]$ is 1, output List, and exit the algorithm.

Step 2: Select an element x from the List, and remove x from List.

Step 3: Form vectors LessList and GreaterList as follows:
Run through the elements y of List (remember x was deleted from List in Step 2):
If $y \le x$, append y to LessList, otherwise append y to GreaterList.

Step 4: (*Recursive Step*) Apply Quick Sort to LessList and GreaterList, and let their outputs be: LessListOut, GreaterListOut.

Step 5: OUTPUT: the concatenation [LessListOut, x, GreaterListOut] (and remove all internal brackets) and EXIT algorithm.

The Merge Sort Algorithm

Of the sorting algorithms thus far presented, when we analyze them in the next section we will see that on average the bubble sort and quick sort outperform the selection sort, but in the worst case the former two will not do better than the latter. Our next sorting algorithm is a recursive divide and conquer algorithm that unconditionally performs better than these three. In fact, the efficiency of this so-called merge sort algorithm can be shown to be the best possible for any general sorting algorithm (that now exists or will ever be created in the future). Thus, the merge sort is what we may call a sorting algorithm with optimal performance. Again, all of these comments will be substantiated in the next section. Here we will simply present the algorithms. The basic idea of the merge sort is as follows:

Given a list of (unsorted) integers $[a_1, a_2, \cdots, a_n]$, we split the list into two sublists of about the same size (give or take one element in case n is odd). We then recursively apply the algorithm to the two smaller sublists, until we are down to sublists with only one element, and then merge the sublists back into a <u>sorted</u> list. The merge sort is the most complicated of the search algorithms we present, but it is also the one with the best performance. We give first a simple example, and then provide a more detailed explanation of the algorithm.

EXAMPLE 7.4: We will use the merge sort algorithm to sort the elements of the vector [22 84 16 65 32 19 7 51 78 15].

We begin by splitting the list into two sublists:

    (split)   [22 84 16 65 32] [19 7 51 78 15].

These in turn are each split into two smaller lists:

    (split)   [22 84] [16 65 32] [19 7] [51 78 15].

Splitting all lists once more now gives:

    (split)   [22] [84] [16] [65 32] [19] [7] [51] [78 15].

Since we are down to sublists of size at most two, we skip the last splitting operation and proceed to sorting the two element lists and then putting things back together into a single sorted list as follows:

    →   (sort)   [22] [84] 16 [32 65] [19] [7] [51] [15 78]

    →   (merge)   [22 84] [16 32 65] [7 19] [15 51 78]

    →   (merge)   [16 22 32 65 84] [7 15 19 51 78]

    →   (merge)   [7 15 16 19 22 32 51 65 78 84]

We point out that the merging operation merged lists back into sorted versions of the same larger lists from which they were created; this is due to the recursive nature of the algorithm: pieces are reassembled in order (merged) in the same way they were split.   Figure 7.3 shows a helpful schematic diagram of the complete process:

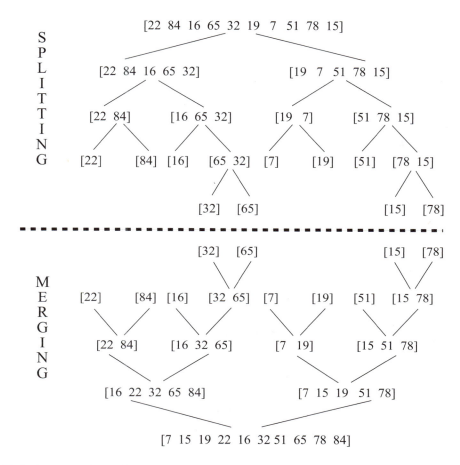

FIGURE 7.3: Illustration of the merge sort algorithm applied to an unsorted list (top).

It will be convenient to separately state the merging algorithm that performs the second half of the work (the harder part) of the merge sort. When we merged lists in the preceding example, we did not specify how it was done; the following algorithm gives an efficient recursive algorithm for merging two sorted lists into a single sorted list.   The basic idea behind the algorithm is this: If we have two sorted lists $[a_1, a_2, \cdots, a_n]$, $[b_1, b_2, \cdots, b_m]$ that we wish to combine (merge) into a single sorted list, the first element of the merged list will be the smaller of a_1 or b_1; we then sequester this element into the newly created list, and recursively apply this procedure again to the resulting smaller lists to eventually build the merged list to completion.

ALGORITHM 7.6: (*The Merging Algorithm*)

Input: Two sorted lists (vectors) of integers: $[a_1, a_2, \cdots, a_n]$, $[b_1, b_2, \cdots, b_m]$.
(We allow the possibility that either of these lists is empty, i.e., $n = 0$ or $m = 0$.)
Output: A single sorted list of these $n + m$ integers, i.e., an increasing sequence

$$\text{Merge}([a_1, a_2, \cdots, a_n], [b_1, b_2, \cdots, b_m]) = [c_1, c_2, \cdots, c_{n+m}],$$

where $c_1 \leq c_2 \leq \cdots \leq c_{n+m}$ are the same integers as those appearing in the inputted lists counting repetitions.

IF $n = 0$
        OUTPUT $[b_1, b_2, \cdots, b_m]$
ELSE IF $m = 0$
        OUTPUT $[a_1, a_2, \cdots, a_n]$
ELSE IF $a_1 \leq b_1$
        OUTPUT $[a_1 \ \text{Merge}([a_2, a_3, \cdots, a_n], [b_1, b_2, \cdots, b_m])]$
ELSE
        OUTPUT $[b_1 \ \text{Merge}([a_1, a_2, \cdots, a_n], [b_2, b_3, \cdots, b_m])]$
END IF

We point out that each call of the merging algorithm requires at most $n + m - 1$ comparisons. This upper bound would occur in the worst case where one of the original lists runs down to a single element before the other list becomes empty. Having created this merge algorithm, it is a simple matter to write a corresponding recursive algorithm for the merge sort:

ALGORITHM 7.7: (*The Merge Sort Algorithm*)

Input: An unsorted list (vector) of integers: $[a_1, a_2, \cdots, a_n]$.

Output: $\text{MergeSort}([a_1, a_2, \cdots, a_n]) = [b_1, b_2, \cdots, b_m]$, an increasing reordering of these integers, i.e., an increasing sequence $b_1 \leq b_2 \leq \cdots \leq b_n$ of integers such that for each index i, $b_i = a_j$, for some index j.

IF $n = 1$
        OUTPUT $[a_1, a_2, \cdots, a_n]$
ELSE
OUTPUT $\text{Merge}(\text{MergeSort}([a_1, a_2, \cdots, a_{\text{floor}(n/2)}], [a_{\text{floor}(n/2)+1}, a_{\text{floor}(n/2)+2}, \cdots, a_n]))$
END IF

EXERCISE FOR THE READER 7.7: Create a diagram, similar to Figure 7.3, that illustrates the merge sort algorithm to sort (in alphabetical order) the following list of names:

Joey, Louise, Dan, Tara, Steve, Amanda, Teresa, Sandra, Patrick, Dennis.

A Randomized Algorithm for Computing Medians

We close this section by developing an efficient algorithm for finding the *median* of a list of numbers. The medians is an important statistical quantity that provides a useful measure of the center of a list of data. If we have a sorted list of data, and if there is an odd number of data values (for example the list: 1,3,5,7,9) then the median is the middle data value (in the example the median would be 5), and if there is an even number of data values—so there is no middle data value (for example, 1,3,5,7), the median is defined to be the average of the two middle data values (so in the example the median would be $(3 + 5)/2 = 4$). One way to think of the median is that (if the data values are all different) then 50% of the rest of the data will lie below the median, and 50% will lie above it. The following is a more formal definition:

DEFINITION 7.1: Given a list of integers (or, more generally real numbers) $a_1, a_2, \cdots, a_n$, the **median** M of the list defined as follows: Let $b_1 \leq b_2 \leq \cdots \leq b_n$ be the corresponding sorted list.

Case 1: If $n = 2k + 1$ is odd, the median is defined to be $M = b_{k+1} = b_{\text{ceil}(n/2)}$.

Case 2: If $n = 2k$ is even, the median is defined to be $M = \dfrac{b_k + b_{k+1}}{2} = \dfrac{b_{n/2} + b_{n/2+1}}{2}$.

EXERCISE FOR THE READER 7.8: (a) Find the medians of each of the following lists: (i) 1 2 3 (ii) 1 3 2 (iii) 1 2 3 4
(b) Give an example of a list where the mean is over 10 times greater than the median.

Many tasks involving data are often made much easier if the data is sorted (imagine having to look up a word in an *unordered dictionary* where the words are listed in random order, rather than in the usual alphabetical order). Although the merge sort algorithm is the most efficient of all sorting algorithms, it still remains true that sometimes performing a sort before another algorithm may take significantly more work than using an alternative approach for the problem at hand. Indeed, if we sort the data in order to obtain a median, we are obtaining a lot more information than we really need. We proceed now to develop a much more efficient divide and conquer algorithm that will directly compute the median of a set of unsorted data. This algorithm will be an example of a **randomized algorithm** because some of its steps will involve making random choices. The average performance of randomized algorithms can sometimes be shown to be superior to that of deterministic algorithms where the random selections are replaced by certain deterministic schemes. The analysis of randomized algorithms will involve probability, and we will elaborate on this in the next section. Chapter

9 will delve much more deeply into some important examples of randomized algorithms.

The randomized algorithm for computing medians will be a simple adaptation of the following general randomized algorithm for computing the kth smallest element of an unsorted list, which the reader should notice has some commonality with the quick sort Algorithm 7.5. Before formally stating the algorithm, we will motivate it with an example.

EXAMPLE 7.5: Suppose that we would like to find the seventh smallest element of the list [32, 84, 16, 78, 65, 32, 22, 7, 51, 78, 15, 2, 19]. Note that since the list has 13 elements, the seventh smallest element is the median. We initialize CurrentList to be the inputted list. If we pick any number a from CurrentList, say $a = 22$, we can form three corresponding lists from it:

LessList = elements of CurrentList less than a = [16, 7, 15, 2, 19]
EqualList = elements of CurrentList equal to a = [22]
GreaterList = elements of CurrentList greater than a = [32, 84, 78, 65, 32, 51, 78]

This step involved 13 comparisons, but it immediately narrows down the search to a smaller list. Indeed, LessList contains five elements, which must be the five smallest elements of the original list, EqualList contains the next smallest element, i.e., the sixth smallest, and GreaterList contains the seventh smallest through the largest element of the list. Thus we may now restrict our search to finding the smallest element of GreaterList and recursively apply this procedure with CurrentList updated to be this GreaterList. If we pick the element $a = 65$, we arrive at the following breakdown of CurrentList (= the previous GreaterList):

LessList = elements of the list less than a = [32, 32, 51]
EqualList = elements of the list equal to a = [65]
GreaterList = elements of the list greater than a = [84, 78, 65, 78]

We have now narrowed the search to finding the smallest element of LessList, so we update CurrentList to be this LessList. If we "pick" the element $a = 32$, we arrive at the following breakdown of CurrentList (= the previous LessList):

LessList = elements of the list less than a = [ ]
EqualList = elements of the list equal to a = [32, 32]
GreaterList = elements of the list greater than a = [84, 78, 65, 78]

Since the smallest element has to be the smallest element of the new EqualList, and all elements of EqualList are (always) the same, the search terminates at 32 (which is the median of the original list).

The example illustrates the basic idea of our recursive algorithm. The only remaining detail is to explain how the element a is selected in the CurrentList. Ideally we would like to choose a in such a way that LessList and GreaterList are approximately of the same size (so that the lists will shrink as quickly as possible and the algorithm will converge rapidly). In the best case, if we ever select a to be the element that we are searching for, the algorithm would terminate at that point, but this coincidence usually does not occur early on in a search. Similarly, the worst case would occur by, say, always choosing a to be the largest element. Without any guiding knowledge a priori, random selections often turn out to be the best strategy, and this is exactly what will be done in the following algorithm.

ALGORITHM 7.8: *(Randomized Algorithm for Finding the kth Smallest Element in an Unordered List)*

Input: An unsorted list (vector) of integers: $[a_1, a_2, \cdots, a_n]$, and a positive integer $k \leq n$.

Output: Smallest($[a_1, a_2, \cdots, a_n], k$) = the kth smallest element of the unordered list.

IF $n = 1$
>   OUTPUT a_1, and EXIT algorithm

ELSE

Randomly select an index in the range $\{1, 2, \ldots, n\}$, and set $a = a_i$.

Form three new lists:

>   LessList = the list of elements from $[a_1, a_2, \cdots, a_n]$ that are less than a

>   EqualList = the list of elements from $[a_1, a_2, \cdots, a_n]$ that are equal to a

>   GreaterList = the list of elements from $[a_1, a_2, \cdots, a_n]$ that are greater than a

Let the number of elements in these three lists be ℓ, e, g, respectively.

IF $k \leq \ell$
>   SET CurrentList = LessList, set Ind = k

ELSE IF $\ell < k \leq \ell + g$
>   OUTPUT EqualList(1), and EXIT algorithm.

ELSE
>   SET CurrentList = GreaterList, set Ind = $k - \ell - e$

END IF

OUTPUT Smallest(CurrentList, Ind)

Although randomized algorithms are amenable to theoretical analysis (as will be done in the next section), the computer exercises at the end of this section will guide the reader through some experiments that will nicely demonstrate the efficiency of using Algorithm 7.8 for finding medians compared with finding them by first sorting the data.

EXERCISE FOR THE READER 7.9: Devise a simple algorithm for computing medians of lists of data that calls on Algorithm 7.8.

EXERCISES 7.1:

1. Apply the binary search algorithm (Algorithm 7.2) to perform each of the following searches. Show all steps with the sequence of smaller lists.

    (a) The target $t = 21$ in the list:　[21, 23, 30, 33, 39, 42, 68, 68, 70, 72, 73, 80, 83, 86, 91, 99].
    (b) The target $t = 69$ in the list:　[3, 4, 8, 9, 10, 13, 26, 36, 48, 69, 84, 87, 87, 94].
    (c) The target $t = 69$ in the list:
    [3, 4, 5, 10, 15, 16, 19, 20, 25, 31, 32, 40, 46, 53, 56, 61, 62, 69, 80, 82, 86, 92, 96].

2. Apply the binary search algorithm (Algorithm 7.2) to perform each of the following searches. Show all steps with the sequence of smaller lists.

    (a) The target $t = 95$ in the list:　[7, 10, 18, 28, 30, 33, 37, 40, 60, 67, 72, 82, 84, 85, 93, 95].
    (b) The target $t = 25$ in the list:　[1, 14, 18, 25, 26, 31, 35, 36, 43, 49, 70, 81, 84, 87, 90, 97, 100].
    (c) The target $t = 81$ in the list:
    [2, 3, 4, 12, 15, 25, 27, 40, 45, 46, 57, 58, 59, 60, 63, 65, 68, 72, 76, 78, 81, 82, 85, 87, 92, 93].

3. Apply the selection sort algorithm (Algorithm 7.3) to sort each of the following lists:

    (a) [86, 3, 33, 23, 77, 46, 38, 86, 36].
    (b) Jimmy, Cathy, Tommy, Karina, Linda, Teresa, Loise, Sandra.
    (c) [46, 97, 7, 60, 28, 15, 5, 33, 42, 40, 89, 85, 43, 98, 97].

4. Apply the selection sort algorithm (Algorithm 7.3) to sort each of the following lists:

    (a) [86, 3, 33, 23, 77, 46, 38, 3, 36].
    (b) Alice, Mallory, Eve, Bob, Peggy, Helen, Janice.
    (c) [62, 76, 1, 7, 22, 0, 45, 32, 62, 93, 13, 79, 65, 51, 46, 29, 25, 99, 73, 24].

5. Apply the bubble sort algorithm (Algorithm 7.4) to sort each of the lists of Exercise 3. For each part, show each of the passes.

6. Apply the bubble sort algorithm (Algorithm 7.4) to sort each of the lists of Exercise 4. For each part, show each of the passes.

7. Apply the quick sort algorithm (Algorithm 7.5) to sort each of the lists of Exercise 3, first (i) by always selecting in Step 2 the leftmost element in the lists, and then (ii) by always selecting in Step 2 the rightmost element in the lists.

8. Apply the quick sort algorithm (Algorithm 7.5) to sort each of the lists of Exercise 3, first (i) by always selecting in Step 2 the leftmost element in the lists, and then (ii) by always selecting in Step 2 the rightmost element in the lists.

9. Apply the merge sort algorithm (Algorithm 7.7) to sort each of the lists of Exercise 3, first (i) by writing down the progression of lists, as in Example 7.4, and then (ii) by creating a diagram as in Figure 7.3.

10. Apply the merge sort algorithm (Algorithm 7.7) to sort each of the lists of Exercise 3, first (i) by writing down the progression of lists, as in Example 7.4, and then (ii) by creating a diagram as in Figure 7.3.

11. Suppose that $[a_1, a_2, \cdots, a_n]$ is a list of n identical numbers.

    (a) Determine the minimum number of comparisons and the maximum number of comparisons that the binary search (Algorithm 7.2) would require to find this number.
    (b) Determine the minimum number of comparisons and the maximum number of comparisons that the bubble sort (Algorithm 7.4) would require to sort this list.

12. Suppose that $[a_1, a_2, \cdots, a_n]$ is a list of n identical numbers.

    (a) Determine the minimum number of comparisons and the maximum number of comparisons that the selection sort (Algorithm 7.3) would require to sort this list.
    (b) Determine the minimum number of comparisons and the maximum number of comparisons that the quick sort (Algorithm 7.5) would require to sort this list.
    (c) Determine the minimum number of comparisons and the maximum number of comparisons that the merge sort (Algorithm 7.7) would require to sort this list.

13. (*Quartiles of a List of Data*) Informally, the *first quartile Q1* and the *third quartile Q3* of a list of data $a_1, a_2, \cdots, a_n$ are just the medians of the left half of the sorted data (below the location of the median), and of the right half of the sorted data, respectively. Here is a more formal definition: Let $b_1 \le b_2 \le \cdots \le b_n$ be the corresponding sorted list.

    Case 1: If $n = 2k + 1$ is odd, the first quartile $Q1$ is defined to be the median of the left half of the data: $b_1, b_2, \cdots, b_k$, and the third quartile $Q3$ is defined to be the median of the right half of the data: $b_{k+2}, b_{k+3}, \cdots, b_n$.

    Case 2: If $n = 2k$ is even, the first quartile $Q1$ is defined to be the median of the left half of the data: $b_1, b_2, \cdots, b_k$, and the third quartile $Q3$ is defined to be the median of the right half of the data: $b_{k+1}, b_{k+2}, \cdots, b_n$.

    (a) Compute the first and third quartiles for the data lists in Exercises 1(a), 3(a), and 3(c).
    (b) Explain how one would use Algorithm 7.8 to compute the quartiles $Q1$, $Q3$, from an unsorted list of integers $a_1, a_2, \cdots, a_n$. Mainly: What indices does one need to input in Algorithm 7.8 in order to compute these quartiles?

14. (*List Insertion*) (a) Develop an algorithm that is based on the linear search Algorithm 7.1 and inputs a sorted list of integers $a_1, a_2, \cdots, a_n$, along with another integer x, and whose ouput will be a sorted list $b_1, b_2, \cdots, b_{n+1}$, whose elements consist of those of the original list, along with the integer x, which is inserted into the correct location of the list.
    (b) Develop an algorithm with same inputs and output as the one in Part (a), except that it is based on the binary search Algorithm 7.2.

15. (*Insertion Sort Algorithm*) This exercise develops yet another algorithm, called the **insertion sort algorithm** for sorting an (unordered) list of data $a_1, a_2, \cdots, a_n$. The insertion sort algorithm works by building the sorted list by removing elements, one at a time, from the original list, and inserting them into the correct location of the sorted list. In general, the way the elements get selected from the original list can be by any scheme, but we will simply always select the leftmost element. The pseudocode for the algorithm is as follows:

```
IF n = 1
    OUTPUT a₁, EXIT algorithm
ELSE IF a₁ ≤ a₂
    Initialize UnsortedList = [a₃, a₄, ···, aₙ],  SortedList = [a₁, a₂],  s = 2
    (s will always be the number of elements in SortedList)
ELSE
    Initialize UnsortedList = [a₃, a₄, ···, aₙ],  SortedList = [a₂, a₁],  s = 2
END IF
FOR i = 3 TO n
    Ind = 1
    WHILE  SortedList(Ind) < aᵢ  AND  Ind ≤ s
        Update:  Ind = Ind + 1
    END
```

Update UnsortedList by removing a_i

Update SortedList by inserting a_i at location specified by Ind

END

(a) Apply the insertion sort algorithm to sort each of the lists of Exercise 3. For each sort, show all steps of the algorithm.

(b) What is the minimum number of comparisons needed by the insertion sort algorithm to sort a list of n elements? Describe all lists for which such a minimum number of comparisons would be needed.

16.   (*Insertion Sort Algorithm*)   (a) Apply the insertion sort algorithm (of Exercise 15) to sort each of the lists of Exercise 4. For each sort, show all steps of the algorithm.

(b) What is the maximum number of comparisons needed by the insertion sort algorithm to sort a list of n elements? Describe all lists for which such a minimum number of comparisons would be needed.

17.   Suppose that $[a_1, a_2, \cdots, a_n]$ is a list of n numbers, and that with the exception of (at most) 5% of the numbers, the list is sorted, i.e., if a certain set of $0.05n$ numbers is removed from the list, the resulting sublist will be sorted. Determine the maximum number of comparisons that would be required if the following sorting algorithms were to be applied to this list:

(a) The selection sort (Algorithm 7.3).

(b) The bubble sort (Algorithm 7.4).

(c) The quick sort (Algorithm 7.5).

18.   Suppose that $[a_1, a_2, \cdots, a_n]$ is a list of n numbers, and that with the exception of (at most) 5% of the numbers, the list is sorted, i.e., if a certain set of $0.05n$ numbers is removed from the list, the resulting sublist will be sorted. Determine the maximum number of comparisons that would be required if the following sorting algorithms were to be applied to this list:

(a) The merge sort (Algorithm 7.7).

(b) The insertion sort (Exercise 15).

19.   Provide an algorithm for sorting a list of n integers $[a_1, a_2, \cdots, a_n]$ that satisfy $m \le a_i \le M$ that does not use any comparisons. What is the maximum number of mathematical operations that will be required by your algorithm to sort such a list?

20.   Explain why if the bubble sort Algorithm 7.4 is applied to sort a list with n elements, and if k passes are made, then the total number of comparisons is given by

$$(n-1) + (n-2) + \cdots + (n-k) = nk - C(k+1, 2).$$

21.   Assume that we have a randomly generated unsorted list of n distinct integers: $[a_1, a_2, \cdots, a_n]$, that contains a target integer t. Let X be the number of iterations that the linear search Algorithm 7.1 uses to find the target in the list. Show that the expected value of X is given by $E[X] = (n+1)/2$.

Suggestion: Use the definition of expectation and the fact that the probability that t appears in any of the slots is $1/n$. Use formula (1) from Chapter 3.

22.   Suppose that we apply the merge sort Algorithm 7.7 to a list of integers $[a_1, a_2, \cdots, a_n]$, where $n = 2^k$. Let $C(k)$ denote the number of comparisons needed in the worst case.

(a) Compute $C(k)$ for $k = 0, 1, 2,$ and 3.

(b) Obtain a recurrence relation for $C(k)$.

(c) Solve the recurrence relation that you obtained in Part (b).

(d) What sort of estimates for the worst case number of comparisons can you obtain when the merge sort algorithm is applied to lists with an arbitrary number of elements, i.e., not simply a power of 2?

COMPUTER EXERCISES 7.1:

NOTE: The theme of this exercise set will be two-fold. First, the reader will gain some experience in writing programs for search and sort algorithms that involve the manipulation of vectors and lists. After these programs are written, later computer exercises will ask the reader to perform some experiments on randomly generated lists that will provide empirical comparisons of some of these search and sort algorithms. These experiments will nicely motivate the theoretical approach for such comparisons that will be the theme of the next section.

1. (*Program for the Linear Search Algorithm*) (a) Write a program with the following syntax:

    $$\texttt{Ind}\ \ =\ \texttt{LinearSearch(List,t)}$$

    The input `List` is a vector of integers, the input `t` is an integer, and the output `Ind` is a nonnegative integer that will be zero if `t` does not appear in `List`, and otherwise will be the smallest index i for which $\texttt{List}(i) = \texttt{t}$. Your program should follow Algorithm 7.1.
    (b) Apply your program to the each of the searches of (ordinary) Exercise 1.

2. (*Program for the Binary Search Algorithm*) (a) Write a program with the following syntax:

    $$\texttt{Ind}\ \ =\ \texttt{BinarySearch(SortList,t)}$$

    The input `SortList` is a vector of integers that are sorted from smallest to largest, the input `t` is an integer, and the output `Ind` is a nonnegative integer that will be zero if `t` does not appear in `SortList`, and otherwise will be the smallest index i for which $\texttt{List}(i) = \texttt{t}$.
    (b) Apply your program to the each of the searches of Ordinary Exercise 1.

3. (*Program for Random Generation of Sorted Lists*) (a) Write a program with the following syntax:

    $$\texttt{SortList}\ \ =\ \texttt{SortListGenerator(n,maxGap)}$$

    The inputs `n` and `maxGap` are positive integers, the output `SortList` is a length n vector (list) of positive integers such that $\texttt{SortList}(1)$ is a randomly generated positive integer in the range $1 \le \texttt{SortList}(1) \le \texttt{maxGap}$, and for an index $i > 1$, $\texttt{SortList}(i)$ is formed by randomly generating a positive integer in the range $[1, \texttt{maxGap}]$ and adding it to $\texttt{SortList}(i-1)$.
    (b) Run your program of Part (a) with the following input settings: $\texttt{n} = 6, \texttt{maxGap} = 10, \texttt{n} = 10, \texttt{maxGap} = 6$.
    Suggestion: For Part (a), see the computer exercises of Section 6.2 for information on generating random integers in a specified range.

4. (*Empirical Performance Comparisons Between the Linear and Binary Search Algorithms*)
    In this exercise you will compare the run times of the linear search program of Computer Exercise 1 with those of the binary search program of Computer Exercise 2 on some randomly generated <u>sorted</u> lists and targets.
    (a) Use the program of Computer Exercise 3 to generate sorted lists starting with size $n = 2$ and `maxgap` = 5, and doubling n to 4, 8, 16, …., until the linear search takes more than one minute to execute. Compare the runtimes on each of these inputs of the linear search and the binary search.
    (b) Repeat Part (a) but with `maxgap` changed from 5 to 20. Are there any significant differences in the runtimes for each program compared with those of Part (a)?

5. (*Program for the Selection Sort Algorithm*) (a) Write a program with the following syntax:

    $$\texttt{SortList}\ \ =\ \texttt{SelectionSort(List)}$$

The input `List` is a vector of integers, and the output `SortList` is a reordering of the integers in the input into an increasing sequence. The program should operate according to the selection sort Algorithm 7.3.

(b) Apply your program to each of the sorting problems of Ordinary Exercise 3(a)(c).

6.   (*Program for the Bubble Sort Algorithm*) (a) Write a program with the following syntax:

$$\text{SortList} \ = \ \text{BubbleSort(List)}$$

The input `List` is a vector of integers, and the output `SortList` is a reordering of the integers in the input into an increasing sequence. The program should operate according to the bubble sort Algorithm 7.4.

(b) Write a program with the following syntax:

$$\text{SortListAll} \ = \ \text{BubbleSortShow(List)}$$

that operates in exactly the same way as the program in Part (a), except that the output `SortListAll` is a matrix whose ith row gives the resulting vector after the ith pass through the bubble sort algorithm. Thus the last row of the output matrix should be the same as the output of the program in Part (a), i.e., the sorted list.

(c) Apply your programs of Parts (a) and (b) to the sorting problem of Example 7.2, and check that the results agree with those of the example.

(d) Apply both programs to each of the sorting problems of Ordinary Exercise 3(a)(c).

7.   (*Program for the Quick Sort Algorithm*) (a) Write a program with the following syntax:

$$\text{SortList} \ = \ \text{QuickSort(List)}$$

The input `List` is a vector of integers, and the output `SortList` is a reordering of the integers of the input into an increasing sequence. The program should operate according to the quick sort Algorithm 7.5, where the leftmost element in the list is always selected.

(b) Write a program with the following syntax:

$$\text{SortList, IterCount} \ = \ \text{QuickSortShow(List)}$$

that operates in exactly the same way as the program in Part (a) except it has a second output `IterCount` that gives the number of iterations that were used.

(c) Apply both programs to each of the sorting problems of Ordinary Exercise 3(a)(c).

8.   (*Program for the Merge Sort Algorithm*) (a) Write a program with the following syntax:

$$\text{SortList} \ = \ \text{Merge(SortListA, SortListB)}$$

The inputs `SortListA`, `SortListB` are two sorted lists and the output `SortList` is a reordering of the elements of both lists into a single increasing sequence. The program should operate according to the merging Algorithm 7.6.

(b) Write a program with the following syntax:

$$\text{SortList} = \text{MergeSort(List)}$$

The input `List` is a vector of integers, and the output `SortList` is a reordering of the integers in the input into an increasing sequence. The program should operate according to the merge sort Algorithm 7.7.

(c) Apply both programs to each of the sorting problems of Ordinary Exercise 3(a)(c).

9.   (*Program for Random Generation of Unsorted Lists*) (a) Write a program with the following syntax:

$$\text{List} \ = \ \text{ListGenerator(n,min,max)}$$

The inputs $n > 0$, `min`, and `max` are integers, with $\text{min} \leq \text{max}$. The output `List` is a length n vector (list) of positive integers that have been randomly generated from the range [`min`, `max`]. Run your program with the following sets of input values: $(\text{n,min,max}) = (6, 1, 10)$, $(10, 0, 1)$, and $(12,1,100)$.

(b) Write a program with the following syntax:

```
List  = ListGeneratorDistinct(n,min,max)
```

The inputs, output, and functionality are exactly as in Part (b), except that the elements of the output list should be distinct integers (no duplications). One necessary requirement is that `max` $-\min \geq n-1$. Run your program with the following sets of input values: `(n,min,max)` = (6, 1, 10), (10, 1, 100), and (12,1,100).

Suggestion: See the computer exercises of Section 6.2 for information on generating random integers in a specified range, with or without duplications.

10. (*Empirical Performance Comparisons Between Sorting Algorithms*)
In this exercise you will compare the runtimes of the various sorting algorithms that were introduced in the section proper, on unsorted lists of increasing size that will be randomly generated using Computer Exercise 9.
(a) Use the program of Computer Exercise 9(a) to generate unsorted lists with the parameter settings $(n,\texttt{min},\texttt{max})=(k,1,k^2)$, where k runs through the values 2, 4, 8, 16, etc. (doubling at each iteration). For each list run each of the following sorting programs:
(i) `SelectionSort` (Computer Exercise 5)
(ii) `BubbleSort` (Computer Exercise 6(a))
(iii) `QuickSort` (Computer Exercise 7)
(iv) `MergeSort` (Computer Exercise 8(b))
Record the runtimes of each, and continue this process until one of the algorithms takes longer than two minutes to execute. Display the results in graphical form (or with a table).
(b) Weed out the slowest program(s), after doing Part (a), and continue the list generation and run time comparisons with the rest until the next slowest program takes longer than two minutes to execute. Display the results in graphical (or tabular) form.
(c) (*Final Round*) If two programs remain, continue the competition with larger lists.

11. (*Program for Randomized Algorithm that Computes the kth Smallest Element of an Unordered List*) (a) Write a program with the following syntax:

```
element  = Smallest(List,k)
```

The input `List` is a vector of integers, the input `k` is a positive integer at most equal to the length of `List`, and the output `element` is the *k*th smallest element of `List`. The program should operate according to the randomized Algorithm 7.8.
(b) Apply your program of Part (a) to find the smallest, second smallest and third smallest elements of the lists of Ordinary Exercise 3(a)(c).

12. (*Empirical Performance Comparisons Between Finding Medians Using the Merge Sort Algorithm Versus Finding them Using the Randomized Algorithm 7.8*)
In this exercise, you will perform experiments that compare the run times of the following two algorithms for computing the median of a list of numbers:
(i) Apply the merge sort program (Computer Exercise 8(b)), which is generally the most efficient of the sorting algorithms, to sort the list and then compute the median using the Definition 7.1.
(ii) Apply the randomized Algorithm 7.8 in conjunction with the strategy from the solution of Exercise for the Reader 7.9.
Use the program of Computer Exercise 9(a) to generate unsorted lists starting with the parameter settings $(\texttt{n},\texttt{min},\texttt{max})=(k,1,k^2)$, where k runs through the values 2, 4, 8, 16, etc. (doubling at each iteration). For each list run the above two algorithms to compute the medians of the lists. Record the run times of each, and continue this process until one of the algorithms takes longer than two minutes to execute. Display the results in graphical form (or with a table).

7.2: GROWTH RATES OF FUNCTIONS AND THE COMPLEXITY OF ALGORITHMS

In this section we will introduce a uniform yardstick by which the speed and efficiency of algorithms may be measured. This will provide a fair way to compare the efficiencies of different algorithms that are designed to solve the same problem. The method will be platform-independent. One naïve way to compare the performance of two algorithms is to write programs for them, run them both on the same computer on sets of the same inputs corresponding to problems of increasing size, and to keep track of and compare the times it takes for the programs to execute. Indeed, we have employed this direct approach in some of the previous computer exercise sets. There are several drawbacks to this approach. First, execution times depend on the particular machine and computing platform that we are using and are not so easily translated to other platforms and machines. The architecture of the particular computing platform comes into play: some systems may be relatively faster, say, in executing a for loop with comparisons than a for loop involving additions or other mathematical operations. The RAM or memory caching is also an important factor. Thus it would be a horrendously inordinate task to take all of these factors into consideration so as to put together an overall comparison of any two given algorithms. In addition to this, since computing speed doubles approximately every 18 months (according to Moore's law), the time estimates obtained by such a method quickly become dated. In the next paragraph, we will briefly and informally summarize the approach that will be used to measure the speed of algorithms. In the remainder of this section we will elaborate and formalize these concepts.

A Brief and Informal Preview

Suppose that we have two algorithms: ALG1 and ALG2 that are designed to solve a given problem whose input size is quantified by some positive integer n. As a simple example, the reader may wish to view the problem as that (considered in the previous section) of sorting an inputted list of integers, where n is the number of elements of the list. The approach will consist of two steps:

Step 1: Count all computer operations that are used to execute each of the two algorithms on such an input. In the count, all computer operations are considered equal: comparisons, additions/subtractions, multiplications/divisions, etc. These counts sometimes vary; for example, we may count according to the *worst-case* (the most common) or *average-case*. The counting techniques that we have learned in Chapter 5 and in Sections 3.1 and 3.2 often play a crucial role in such analyses. Once this is done, we obtain corresponding operation count functions for the two algorithms. This Step 1 is typically the most difficult of the two steps.

Step 2: After Step 1 has been completed, we will have obtained two operation count functions of the input size variable n for each of the two algorithms. We will then compare the growth of these functions as n gets large. This will be accomplished by simplifying the specific count functions by what are called *big-O estimates*.[6] Part of the task in this section will be to rank the different growth rates among some commonly used big-O estimates. For example, say that, for the two algorithms we are comparing, we have obtained the following two operation count functions:

$$f_1(n) = 3n^2 + 44n + 25, \quad f_2(n) = n^3 + 5,$$

which give the number of computer operations required when ALG1, ALG2, repectively, are used on a problem of input size n. So the next question is: Which algorithm is more efficient? The technical answer is that it will depend on the value of n: for small values of n, the value of $f_1(n) = 3n^2 + 44n + 25$ is greater than that of $f_2(n) = n^3 + 5$, because the lower order terms $44n$, 25 of the first function will dominate all of the terms of the latter function, and so ALG2 will be more efficient than ALG1. But as n gets larger, the highest order term n^3 of the latter function will quickly dominate all other terms of both functions, and ALG1 is the more efficient algorithm. Figure 7.4 shows two graphs that help to illustrate these comparisons; in particular, $f_1(n) \le f_2(n)$, for $n \ge 9$.

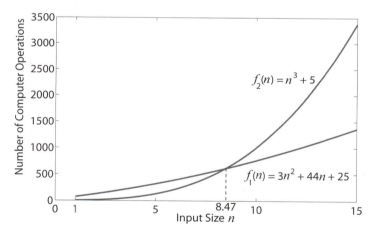

FIGURE 7.4: Comparison of graphs of two functions for the running time of two algorithms, depending on the input size n.

When we quantify the running times of algorithms, we will always be interested in the growth of the corresponding computation counting function as n gets large. When we do this, we may ignore all but the fastest growing term of the function. Thus, for $f_2(n) = n^3 + 5$, the constant term 5 becomes insignificant compared to the third order term n^3, as n gets large. Similarly, the function $f_1(n) = 3n^2 + 44n + 25$ can be estimated to grow like its highest order term n^2. Notice in this latter estimate, we even went a step further and removed the

---

[6] Such big-O estimates are often employed in Step 1 to simplify some of the counting arguments.

multiplicative constant 3 from the highest order term. This can be justified by Moore's law, since the difference of a multiplicative constant can be made up for by an appropriate number of years of improvements in computer speeds.

Big-O Notation

The estimation scheme for measuring growth of functions that was done in Step 2 of the above informal example will now be made precise using so-called big-O notation. The basic idea is that the big-O notation allows us to estimate any function of n by only retaining its highest order (= fastest growing term), and discarding lower order terms as well as any multiplicative constants. We give first the formal definition; it will be followed by several examples.

DEFINITION 7.2: Suppose that we have two nonnegative functions f, g, whose domains are the positive integers $\mathbb{Z}_+$. We say that $f(n)$ is **big-O** of $g(n)$, and write this as $f(n) = O(g(n))$, if there exists a positive constant C and a positive integer k such that the following inequality holds for each $n \geq k$:

$$f(n) \leq C \cdot g(n).$$

We say that the constants C and k are **witnesses** to the relationship $f(n) = O(g(n))$.

NOTE: We need only one pair of witnesses C, k to establish a big-O relationship. Notice that if C, k serve as witnesses, then so will any other pair C', k' that satisfy $C' \geq C, k' \geq k$, so the collection of witnesses will be infinite.

The helpful way to visualize big-O relationships is as follows: $f(n) = O(g(n))$ if, and only if the graph of $f(n)$ <u>eventually</u> lies below the graph of $C \cdot g(n)$ (a constant multiple of $g(n)$'s graph). The word "eventually" here simply means for n sufficiently large (i.e., beyond some point on the x-axis): $n \geq k$, for some positive integer k. In terms of growth, the big-O relationship $f(n) = O(g(n))$ expresses that fact that the function $f(n)$ grows at most as fast as (or is dominated by) the function $g(n)$. Other apt expressions for the big-O relationship $f(n) = O(g(n))$ are:

"$g(n)$ grows at least as fast as $f(n)$," or

"$g(n)$ dominates $f(n)$," or

" $f(n)$ is dominated by $g(n)$."

In order to establish a big-O relationship, the basic scheme is to identify the highest order terms, and then to choose the witnesses appropriately large, so that the highest order terms dominate the remaining terms. This approach is illustrated in the following example.

EXAMPLE 7.6: For each of the following big-O statements, either find witnesses that will prove it is valid, or explain why it is invalid (false).

(a) TRUE or FALSE? $n^2 + 1000n = O(0.001n^2 + 1)$.

(b) TRUE or FALSE? $1000n^3 = O(n^{3.01})$.

(c) TRUE or FALSE? $2^n = O(10n^{10})$.

SOLUTION: Part (a): Answer: True. Justification: The highest order (= fastest growing) terms of the functions on both sides are the terms involving n^2. Since these terms are the same (the constant multiples do not matter) this indicates that the stated big-O relationship should be true. Before we prove it, let us first prove the simpler big-O relationship where the lower order terms are ignored: $n^2 = O(0.001n^2)$. Indeed, with the witnesses $k = 1$ and $C = 1000$, the required inequality from Definition 7.2 becomes: $n^2 \le 1000 \cdot 0.001n^2 = n^2$, which is true for all n. That was easy. Now how do we deal with the lower order terms? Although the constant 1000 on the term $1000n$ is large, if $n \ge 1000$, then $1000n \le n^2$, so it follows that if $n \ge 1000$, then $n^2 + 1000n \le n^2 + n^2 = 2n^2 = 2000 \cdot 0.001n^2 \le 2000(0.001n^2 + 1)$. This proves that $k = 1000$ and $C = 2000$ are witnesses that prove the big-O relationship $n^2 + 1000n = O(0.001n^2 + 1)$ to be valid.

Part (b): Answer: True. Justification: The (only) term on the left is of order n^3 which is lower order than the $n^{3.01}$ term on the right (since the one on the right has a higher exponent), so we should be able to find witnesses to prove this big-O relationship. To find witnesses, we rewrite $n^{3.01} = n^{0.01+3} = n^{0.01}n^3$. Since $n^{0.01} \ge 1$, we may prove the relationship by using the witnesses $k = 1$ and $C = 1000$.

Part (c): Answer: False. Justification: The term on the left of $2^n = O(10n^{10})$ is an *exponential term* (i.e., a term of the form b^n with $b > 1$ a constant; here $b = 2$). The term on the right is a *power term* (i.e., a term of the form n^p, with $p > 0$ a constant; here $p = 10$).[7] As a general rule, any exponential term dominates any power term, but not the other way around, i.e., exponential terms grow strictly faster than polynomial terms. This general rule, and some other useful ones will be proved in Proposition 7.1 below. From this fact, it follows that the big-O relationship $2^n = O(10n^{10})$ is false.

---

[7] Note that in our definition of a power term, the constant exponent p is allowed to be any positive number. In particular, p need not be an integer. In case p is an integer, the power term is a polynomial that is defined by $n^p = \underbrace{n \cdot n \cdots \cdots n}_{p \text{ times}}$. In the general case (when p is not necessarily an integer), the definition is $n^p = \exp(p \ln n)$.

CAUTION: **Despite their notation, big-O relationships are <u>not</u> equations!** There is a definite abuse of the equal sign notation in big-O notation. In particular **the big-O relation is not symmetric**. For example, the big-O relationship $n^2 = O(n^3)$ is obviously true (with witnesses $C = k = 1$), but the symmetric relationship $n^3 = O(n^2)$ is false because n^3 grows (strictly) faster than n^2. (In terms of witnesses, note that corresponding equation of Definiton 7.1: $n^3 \le C \cdot n^2$, cannot hold, no matter what constant C we have; just take n larger than C to see this.)   Nonetheless, the big-O notation is often convenient and such conveniences have kept it mathematical and comuter science writing.

EXERCISE FOR THE READER 7.10:   (TRUE or FALSE?) For each of the following big-O statements, either find witnesses that will prove it is valid, or explain why it is invalid (false):

(a) $10^n = O(n!)$.

(b) $200(1 + 1/n) = O(1)$.

The following proposition summarizes some useful big-O relations.

CAUTION: Parts of the following proof are quite technical, but the statements of the proposition are not; so readers with a less technical background are advised to at least remember the results of the proposition but may wish to skip over some of the details of the proof.[8]

PROPOSITION 7.1: (*Some Basic Growth Rates*) Each of the following big-O growth relations is strict, i.e., the symmetrically reversed big-O relations for each part are all false.   Moreover, in any of the big-O results $f(n) = O(g(n))$ below, we will show that for <u>any</u> positive constant C, the inequality $g(n) \ge Cf(n)$ holds for sufficiently large n. In words, the dominations below are all *strict*.

(a) (*Logarithms are strictly dominated by power terms*) If p is any positive number, then $\log(n) = O(n^p)$.

(b) (*Power terms are strictly dominated by power terms with higher powers*) If $0 < p < p'$ are positive numbers then $n^p = O(n^{p'})$.

---

[8] Although this big-O concept does not formally require calculus, the ideas are often seen in second semester calculus courses and can be more easily understood using the concept of a limit. Since we do not assume calculus, our development will be independent of calculus. In some of the proofs, we will need to use some properties of logarithm and exponential functions that are usually seen in precalculus courses. For example, $y = e^x$ is a strictly increasing positive function that gets arbitrarily large as $x \to \infty$.   Any power can be expressed in terms of the exponential function using the identity $b^p = e^{p \ln b}$. Logarithmic identities such as $\log(x^p) = p \log(x)$ (valid for any logarithm function) will also be used. For more details on such identities, readers may consult any book on precalculus or on calculus. Needless to say, readers who have had second semester calculus will have an easier time digesting the big-O concepts.

(c) (*Power terms are strictly dominated by exponential terms*) If $p > 0$ and $b > 1$ are positive numbers then $n^p = O(b^n)$.

NOTE: Although Part (a) is formulated for the common logarithm $\log(n) = \log_{10}(n)$, it follows from the change of base formula $\log_b(n) = \log(n)/\log(b)$, that any other logarithm function is a constant multiple of the common logarithm, so Part (a) is really valid for any logarithm $\log_b(n)$ ($b > 1$).

Proof: Part (a): We start by recalling the basic inequality $x < e^x$, which is valid for any real number x; see Figure 7.5.[9] The desired big-O estimate

$$\log(n) = O(n^p)$$

is equivalent to the validity of the inequality

$$\log(n) \leq Cn^p,$$

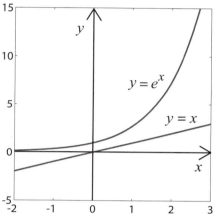

for $n \geq k$, where k is some positive integer, and $C > 0$ is some positive constant. If we rewrite the power on the right in terms of the exponential function, and replace the logarithm on the left by the natural logarithm (which is just a constant multiple of the common logarithm, so the constant may be absorbed in the constant C), the inequality becomes:

FIGURE 7.5: Graphical illustration of the fundamental inequality $x < e^x$. (Axis scales are different.)

$$\ln(n) \leq Ce^{p\ln n}.$$

If we substitute $x = p\ln(n)$ into the inequality $x < e^x$, we obtain $p\ln(n) < e^{p\ln(n)}$, which when divided by p gives $\ln(n) \leq (1/p)e^{p\ln n}$. Since this is valid for any $p > 0$ and positive integer n, it follows that $\ln(n) \leq Ce^{p\ln n}$, holds for any positive integer n, if we take $C = 1/p$.

In order to show that the symmetric big-O relation is false, i.e., that $n^p \neq O(\log(n))$, we need to show that for any fixed constant $C > 0$, the inequality $e^{p\ln n} \leq C\ln(n)$ cannot hold for all positive integers $n \geq k$. To do this, we begin

---

[9] Actually, it is true that $x + 1 \leq e^x$ for any real number x (with equality only when $x = 0$), but we will not need this stronger inequality.

with the inequality $\ln(n) \le (1/p)e^{p\ln n}$ that was proved above to be valid for all positive integers n. We also use the logarithmic identity $\ln(n^\ell) = \ell\ln(n)$. If we replace n with n^ℓ in the first inequality, and use this latter identity we obtain that $\ell\ln(n) \le (1/p)e^{p\ell\ln n} = (1/p)(e^{p\ln n})^\ell$. If n is chosen large enough so that $e^{p\ln n} > 2$ and held fixed, then the right side more than doubles each time ℓ is increased by one, while in order for the left side to double, ℓ would need to be doubled. From this it follows that if ℓ is taken large enough, then $C\ell\ln(n) < (1/p)e^{p\ell\ln n} = (e^{p\ln n})^\ell$, i.e., $(n^\ell)^p > C\ln(n^\ell)$, as we wished to show. Notice that since n did not need to be an integer in this proof; the inequality is true for all sufficiently large n.

Part (b): If $0 < p < p'$, then for any positive integer n, we have $p\ln(n) \le p'\ln(n)$, and hence (since $y = e^x$ is an increasing function) $n^p = e^{p\ln(n)} \le e^{p'\ln(n)} = n^{p'}$. This proves that $k = 1$ and $C = 1$ are witnesses to the big-O relationship $n^p = O(n^{p'})$.

In order to show that the symmetric big-O relation is false, i.e., that $n^{p'} \ne O(\log(n^p))$, we will show that for any fixed constant $C > 0$, the inequality $n^{p'} > Cn^p$ holds for arbitrarily large positive integers n. (Note: The constant C here is playing a very different role than it does in Defintion 7.2.) Indeed, since $n^{p'} = n^{(p'-p)+p} = n^{(p'-p)}n^p$ the former inequality follows from the fact that since $p' - p > 0$, the power function $n^{(p'-p)} = e^{(p'-p)\ln n} > C$ for all sufficiently large positive integers n.

Part (c):   From the strict domination result of Part (a), it follows that for sufficiently large n, we have $n \ge (p/\ln b)\ln n$.   Multiplying by $\ln b$, and then taking exponentials gives $b^n \ge n^p$, for sufficiently large n. This is the asserted big-O relation $n^p = O(b^n)$. The proof that the symmetric big-O relationship is false is similar and will be left to the reader. □

NOTATION: Although it is not standard in undergraduate texts, the following very intuitive notation from mathematical analysis will be helpful in differentiating between various growth rates:
If $f(n), g(n)$ are two functions on the positive integers (or more generally, the set of positive integers greater than or equal to k, for some positive integer N). We write

$f(n) \precsim g(n)$ to mean $f(n) = O(g(n))$, i.e., $f(n)$ is **dominated** by $g(n)$,

and we write

$f(n) \prec g(n)$ if, additionally, for any positive constant C, the inequality $g(n) \ge Cf(n)$ is true for sufficiently large n. i.e., $f(n)$ is **strictly dominated** by $g(n)$.

Thus, the results of Proposition 7.1 can be elegantly (and intuitively) rephrased as follows:

$$\log(n) \prec n^p \prec n^{p'} \prec b^n,$$

where $p < p'$ are any positive real numbers, and $b > 1$ is any real number.

The big-O estimates of Proposition 7.1 are among the most commonly used, but it is also common to have a logarithm (or a power of a logarithm) multiplied by a power of n. How these growth rates fit in with those of the proposition is described in the following example.

EXAMPLE 7.7: Show that for any positive numbers $p < p'$ and s, we have:

$$n^p \prec (\log n)^s n^p \prec n^{p'}.$$

SOLUTION: The first strict domination follows from the fact that as n increases to infinity, $\log n$ and hence also $(\log n)^s$ also increases to infinity. For the second domination, if we write $p' - p = 2r$, then Proposition 7.1(a) tells us that $\log n \prec n^{r/s}$, from which it follows that $(\log n)^s \prec n^r$. The latter domination implies that $(\log n)^s \leq Cn^r$ if $n \geq k$, where C and k are positive constants (witnesses). Muliplying both sides of this inequality by n^p gives us $(\log n)^s n^p \leq Cn^{p+r}$ if $n \geq k$, which may be restated as $(\log n)^s n^p \precsim n^{p+r}$. But by Proposition 7.1(b), since $p' > p + r$, we have $n^{p+r} \prec n^{p'}$. Combining this strict domination with the previous domination gives us $(\log n)^s n^p \prec n^{p'}$, which is the desired strict domination.

EXERCISE FOR THE READER 7.11: Show that for positive numbers $1 < b < b'$, we have $b^n \prec (b')^n$.

Combinations of Big-O Estimates

When an operation count is made for an algorithm, it is typical that several different quantities need to be added to obtain the value of the operation count function. For this reason it is important to learn how to compare the various terms and to recognize those of the fastest growth (which dominate lower order terms). The following result is a rather direct consequence of the definitions, but is often useful when one needs to consolidate individual big-O estimates.

PROPOSITION 7.2: Suppose that $f(n), g(n), h(n), k(n)$ are nonnegative functions on the positive integers.

(1) If $f(n) \precsim h(n)$, $g(n) \precsim h(n)$ then $f(n) + g(n) \precsim h(n)$. If both of the first two dominations are strict, then so is the latter domination.

(2) If $f(n) \precsim h(n)$, $g(n) \precsim k(n)$ then $f(n) \cdot g(n) \precsim h(n) \cdot k(n)$. If either of the first two dominations is strict, then so is the latter domination.

NOTE: In terms of big-O notation, the first statement of (1) can be written as $f(n) = O(h(n))$ and $g(n) = O(h(n))$, then $f(n) + g(n) = O(h(n))$. Likewise, the first statement of (2) may be expressed as follows: If $f(n) = O(h(n))$ and $g(n) = O(k(n))$, then $f(n) \cdot g(n) = O(h(n) \cdot k(n))$.

We will prove only (1); the proof of (2) is left as Exercise for the Reader 7.12.

Proof of (1): The dominations $f(n) \precsim h(n)$, $g(n) \precsim h(n)$ may be expressed as:

$$f(n) \leq C_1 h(n), \text{ for all } n \geq k_1, \text{ and}$$
$$g(n) \leq C_2 h(n), \text{ for all } n \geq k_2,$$

where C_1, k_1 and C_2, k_2 are corresponding sets of witnesses. If we put $C = C_1 + C_2$ and $k = \max(k_1, k_2)$, then for $n \geq k$, since both of the previous two inequalities hold, we may write:

$$f(n) + g(n) \leq C_1 h(n) + C_2 h(n) = (C_1 + C_2)h(n) = Ch(n).$$

We have thus shown that the pair C, k are witnesses for the domination (big-O relation) $f(n) + g(n) \precsim h(n)$.

To show the strict domination statement, given any positive integer M, the strict dominations $f(n) \prec h(n)$, $g(n) \prec h(n)$ tell us that there exists an integer k such that if $n \geq k$, then $h(n) \geq (2M)f(n)$, $h(n) \geq (2M)g(n)$, from which it follows that $h(n) \geq M(f(n) + g(n))$. Since M could be chosen arbitrarily large, we have proved the strict domination. □

EXERCISE FOR THE READER 7.12: Prove Part (2) of Proposition 7.2.

EXERCISE FOR THE READER 7.13: Suppose that $f(n)$ is a polynomial in n of degree D, i.e., $f(n) = a_D n^D + a_{D-1} n^{D-1} + \cdots + a_1 n + a_0$, where $a_D > 0$. Show that $f(n) = O(n^D)$.

Big-Omega and Big-Theta Notation

Big-O notation allows one to express a statement that one function (on the left side) is dominated by another function (on the left side). While such statements are the most common in analyzing algorithms, it is sometimes required to make more precise statements about the growth of a function, or to give a function which our function dominates. Such statements can be made using the related notation that is described in the following two definitions.

DEFINITION 7.3: Suppose that we have two nonnegative functions f, g, whose domains are the positive integers $\mathbb{Z}_+$. We say that $f(n)$ is **big-Omega** of $g(n)$, and write this as $f(n) = \Omega(g(n))$, if there exists a positive constant C and a positive integer k such that the following inequality holds for each $n \geq k$:

$$f(n) \geq C \cdot g(n).$$

We say that the constants C and k are **witnesses** to the relationship $f(n) = \Omega(g(n))$.

Observe that $f(n) = \Omega(g(n)) \Leftrightarrow g(n) = O(f(n))$. Put in the "domination" terminology, $f(n) = \Omega(g(n))$ means that $g(n)$ is dominated by $f(n)$, or $f(n)$ grows at least as fast as $g(n)$. From these observations, it follows that the two relations $f(n) = O(g(n))$, $f(n) = \Omega(g(n))$ together imply that $f(n), g(n)$ dominate one another, or grow at the same rate. The big-Theta notation of the next definition is reserved for this situation.

DEFINITION 7.4: Suppose that we have two nonnegative functions f, g, whose domains are the positive integers $\mathbb{Z}_+$ and that satisfy $f(n) = O(g(n))$ and $f(n) = \Omega(g(n))$. We say that $f(n)$ is **big-Theta** of $g(n)$, and write this as $f(n) = \Theta(g(n))$.

As was pointed out for the big-O notation, we caution the reader that **big-Omega and big-Theta relations are not equalities!** In the next example, we sharpen the result of Exercise for the Reader 7.13 from a big-O estimate to a big-Theta estimate.

EXAMPLE 7.8: Suppose that $f(n)$ is a polynomial in n of degree D, i.e., $f(n) = a_D n^D + a_{D-1} n^{D-1} + \cdots + a_1 n + a_0$, where $a_D > 0$. Show that $f(n) = \Theta(n^D)$.

SOLUTION: Since the big-O half $f(n) = O(n^D)$ of the big-Theta estimate $f(n) = \Theta(n^D)$ has been established in the solution of Exercise for the Reader 7.13, we have left to show the big-Omega estimate $f(n) = \Omega(n^D)$.   By Proposition 7.1, we have $|a_j| n^j \prec n^D$, for each $j < D$. It follows that for $j < D$, and for n sufficiently large, we have $|a_j| n^j \leq \dfrac{a_D}{2D} n^D$, whenever $n \geq k$, where k is a positive integer that does not depend on j. It follows that whenever $n \geq k$, we have

$$f(n) = a_D n^D + a_{D-1} n^{D-1} + \cdots + a_1 n + a_0$$
$$\geq a_D n^D - |a_{D-1}| n^{D-1} - \cdots - |a_1| n - |a_0|$$
$$\geq a_D n^D - D \cdot \frac{a_D}{2D} n^D$$
$$= \frac{a_D}{2} n^D.$$

This is the big-Theta estimate that was needed.

The following exercise for the reader will provide an opportunity for the reader to assess his/her understanding of several of growth rate concepts and notations that have been introduced.

EXERCISE FOR THE READER 7.14: Classify each statement as either true or false and then justify your answer.
(a) $n = \Omega(n^2)$.
(b) If $f(n) = \Omega(g(n))$ and $g(n) = \Omega(h(n))$, then $f(n) = \Omega(h(n))$.
(c) $1 + 2 + 3 + \cdots + n = \Theta(n^2)$.

The next exercise for the reader gives some simple and useful rules for combining big-Theta estimates.

EXERCISE FOR THE READER 7.15: Prove the following general rules in which all functions are assumed to be nonnegative functions on the integers.
(a) If $f(n) \precsim g(n)$ then $g(n) = \Omega(g(n) + f(n))$.
(b) If $f(n) = \Theta(g(n))$ then $f(n) \cdot h(n) = \Theta(g(n) \cdot h(n))$.

Complexity of Algorithms

With the technical development of growth rates of integer functions behind us, we are now ready to illustrate some examples of the complete program of complexity analysis of algorithms. As mentioned earlier, the first and most difficult step is to count the number of computer operations, usually in the worst case, but sometimes also in the best case or in the average case, as a function of the input size n of the algorithm. After this is done, the resulting count function should be estimated with the simplest and slowest growing big-O estimate possible, usually taken from one of the growth functions of Propositions 7.1 or 7.2.   Of course, if $f(n) = O(g(n))$, and if $h(n)$ grows faster than $g(n)$, (i.e., $g(n) \precsim h(n)$), then $f(n) = O(h(n))$, so a most informative big-O complexity estimate would be one of slowest possible growth.   Ideally, a big-Theta estimate would be the most accurate, but it is not always feasible to obtain one. The following examples will illustrate some of the techniques in performing such complexity analyses.

EXAMPLE 7.9: (*Complexity Analysis of the Linear Search Algorithm 7.1*) Determine the (a) worst-case, (b) best-case, and (c) average-case complexity of the linear search Algorithm 7.1 for finding a target element t that appears in an unsorted list of n distinct elements.

SOLUTION: Part (a): As pointed out after the statement of Algorithm 7.1, the maximum number of comparisons needed to search for an element is n, so the best big-O estimate is thus $n = O(n)$. It is also a big-Theta estimate $n = \Theta(n)$.

Part (b): In the best case, the target appears as the first element of the list so only one comparison is needed, so the best big-O estimate is thus $O(1) = \Theta(1)$.

Part (c): As pointed out after the statement of Algorithm 7.1, the average number of comparisons for the linear search algorithm to find an element that appears in an unsorted list of n elements is $(n+1)/2 = O(n), = \Theta(n)$. (We have tacitly used the result of Example 7.8.) Thus, the worst-case number of comparisons is roughly double the average-case complexities, but the big-O/big-Theta summaries find the operation counts to be indistinguishable. The best-case complexity is better than both of these, although the best case is not very likely to occur.

EXAMPLE 7.10: (*Complexity Analysis of the Selection Sort Algorithm 7.3*) Determine the (a) worst-case, (b) best-case, and (c) average-case complexity of the selection sort Algorithm 7.3 for sorting a list of n elements.

SOLUTION: As was shown after the statement of the algorithm, the number of comparisons required for the selection sort algorithm to sort a list of n elements is always

$$(n-1) + (n-2) + \cdots + 2 + 1 = \frac{n(n-1)}{2}$$

(even if the list is already sorted). This means that the worst, best, and average cases all have the same operation counts and hence the same complexity $O(n(n-1)/2) = O(n^2), = \Theta(n^2)$. (We have tacitly used the result of Example 7.8.)

EXAMPLE 7.11: (*Complexity Analysis of the Bubble Sort Algorithm 7.4*) Determine the (a) worst-case, (b) best-case, and (c) average-case complexity of the bubble sort Algorithm 7.4 for sorting a list of n elements.

SOLUTION: At the ith pass of the bubble sort algorithm, since the bottom i elements (i.e., the heaviest i elements) will be sorted, a total of $n-i$ comparisons will be needed. The algorithm proceeds until either $n-1$ passes have been completed or no permutations are needed. Thus the maximum number of comparisons will be $(n-1) + (n-2) + \cdots + 2 + 1 = n(n-1)/2$ (we used formula (1) of Chapter 3). Thus the worst-case complexity is $n(n-1)/2 = O(n^2), = \Theta(n^2)$.

In the best case, only one pass is needed (and no permutations will occur since the list will already have been sorted), so the best-case complexity is $n - 1 = O(n), = \Theta(n)$.

The average-case complexity analysis is more complicated. In an average-case analysis, we let X be the number of computer operations required by the algorithm on an input of size n. Since X is a random variable, we seek to compute its expected value E[X]. For the bubble sort algorithm, the only computer operations will be comparisons. It will be very difficult to compute E[X], but we will estimate it with upper and lower bounds that will be sufficient to obtain a common big-Theta complexity estimate. Since we have already pointed out the worst-case upper bound $X \leq n(n-1)/2$, it follows that $E(X) \leq n(n-1)/2$, and we have left to obtain a lower bound. This can be accomplished by considering what are called *inversions* in a permutation. Given a list (vector) $a_1, a_2, \cdots, a_n$ (thought of as an input for the bubble sort algorithm) of the integers $1, 2, \cdots, n$, an *inversion* is any pair a_i, a_j with $i < j$ that is out of order, i.e., $a_i > a_j$. Throughout the course of the bubble sort algorithm any inversion in the initial list will give rise to an interchanging of numbers and hence at least one resulting comparison. It follows that E[X] is bounded below by the expected number of inversions in a random permutation of $1, 2, \cdots, n$. But for any pair a_i, a_j with $i < j$ in a random permutation $a_1, a_2, \cdots, a_n$ it is equally likely that $a_i > a_j$ or $a_i < a_j$. Hence, the expected number of inversions in a random permutation of $1, 2, \cdots, n$ is exactly half of the number of pairs a_i, a_j with $i < j$, which yields the lower bound $E(X) \geq C(n, 2)/2 = n(n-1)/4$. In summary, we have proved that

$$\frac{n(n-1)}{4} \leq E(X) \leq \frac{n(n-1)}{2}.$$

Since both sides of this double inequality have the same big-O/big-Theta estimate, we arrive at the average case complexity estimate $E(X) = O(n^2), = \Theta(n^2)$.

Whereas the big-O/big-Theta worst-case and average-case complexity estimates were identical for each of the algorithms in the above examples, our next example will show that the quick sort algorithm is significantly faster on average than in the worst-case. Indeed, as we will soon see, the average-case complexity of the quick sort algorithm is the best possible average complexity of any sort algorithm. The proof in the example will require as an additional tool the following big-Theta growth rate estimate, a proof of which can be found in any calculus book.[10]

---

[10] For readers who wish to check into this, calculus books rarely use big-Theta notation, so this formula will probably be written in another form such as $1 + \dfrac{1}{2} + \dfrac{1}{3} + \cdots + \dfrac{1}{n-1} + \dfrac{1}{n} \sim \log n$. Actually, the approximation is stronger than a standard big-Theta approximation in the sense that the ratio

$$1 + \frac{1}{2} + \frac{1}{3} + \cdots + \frac{1}{n-1} + \frac{1}{n} = \Theta(\log n) \tag{1}$$

EXAMPLE 7.12: (*Complexity Analysis of the Quick Sort Algorithm 7.5*) Determine the (a) worst-case and (b) the average-case complexity of the quick sort Algorithm 7.5 for sorting a list of n elements.

SOLUTION: The worst case for the quick sort algorithm occurs when each selected value is always either the largest or smallest in the list from which it is selected. In this case the number of comparisons will be $(n-1) + (n-2) + \cdots + 2 + 1 = n(n-1)/2$. Thus the worst-case complexity is $n(n-1)/2 = O(n^2)$, $= \Theta(n^2)$.

To perform the average-case complexity analysis, we assume that $a_1, a_2, \cdots, a_n$ is a random permutation of $1, 2, \cdots, n$. We let X be the number of computer operations required by the quick sort algorithm on an input list of n elements. Since X is a random variable, we seek to compute its expected value E[X]. For each pair of integers $1 \le i < j \le n$, we let $X_{i,j}$ denote the random variable defined as follows:

$$X_{i,j} = \begin{cases} 1, & \text{if } i \text{ and } j \text{ are ever compared in the course of the quick sort} \\ 0, & \text{otherwise} \end{cases}$$

Since two any two elements of the list are compared at most once, it follows that $X = \sum_{i<j} X_{i,j}$, and so by the linearity of expectation $E(X) = \sum_{i<j} E(X_{i,j})$. The problem has thus been reduced to computing the simpler expected values $E(X_{i,j})$. To compute these expectations, it will help to answer the following question: At what point in the quick sort algorithm can the value of $X_{i,j}$ be determined, i.e., at what point in the algorithm can it be determined whether i and j get compared? Initially, List contains all of the integers $1, 2, \cdots, n$, and hence also i and j. For all subsequent iterations where the selected integer (for comparison) is either less than i or greater than j, i and j will continue to remain together in List. For example, if a number less than i is selected, then both i and j will be moved to GreaterList. It is the iteration when the selected integer (for comparisons) first appears in the interval $\{i, i+1, i+2, \cdots, j-1, j\}$ where the value of $X_{i,j}$ is determined. There are two cases:

$\left[1 + \dfrac{1}{2} + \dfrac{1}{3} + \cdots + \dfrac{1}{n-1} + \dfrac{1}{n} \right] \Big/ \log n$ gets as close as we wish to 1 as n gets large. In the notation of

calculus: $\displaystyle \lim_{n \to \infty} \left[1 + \dfrac{1}{2} + \dfrac{1}{3} + \cdots + \dfrac{1}{n-1} + \dfrac{1}{n} \right] \Big/ \log n = 1.$

Case 1: The selected integer lies strictly between i and j. In this case i will be placed in LessList, j in GreaterList, so i and j will never be compared and $X_{i,j} = 0$.

Case 2: The selected integer is either i or j. In this case i and j will be compared (at exactly this iteration) and so $X_{i,j} = 1$.

Since Case 2 has two chances out of $|\{i, i+1, i+2, \cdots, j-1, j\}| = j - i + 1$ of occurring, it follows that $P(X_{i,j} = 1) = \dfrac{2}{j-i+1}$. We may now compute

$$E(X) = \sum_{i<j} E(X_{i,j}) = \sum_{i<j} \frac{2}{j-i+1}$$

$$= 2\sum_{i=1}^{n-1} \sum_{j=i+1}^{n} \frac{1}{j-i+1}$$

$$= 2\sum_{i=1}^{n-1} \left[\frac{1}{2} + \frac{1}{3} + \cdots + \frac{1}{n-i+1} \right]$$

$$= 2 \begin{bmatrix} \frac{1}{2} + \frac{1}{3} + \cdots + \frac{1}{n-1} + \frac{1}{n} \\ + \frac{1}{2} + \frac{1}{3} + \cdots + \frac{1}{n-1} \\ \ddots \\ + \frac{1}{2} + \frac{1}{3} \\ + \frac{1}{2} \end{bmatrix}$$

$$= 2\left\{ (n-1)\left[1 + \frac{1}{2} + \frac{1}{3} + \cdots + \frac{1}{n-1} + \frac{1}{n} \right] - (n-1) - \frac{n-2}{n} - \frac{n-3}{n-1} - \cdots - \frac{1}{3} \right\}.$$

From this computation, it follows that

$$2(n-1)\left[1 + \frac{1}{2} + \frac{1}{3} + \cdots + \frac{1}{n-1} + \frac{1}{n} \right] - 4(n-1) \le E(X)$$

$$\le 2(n-1)\left[1 + \frac{1}{2} + \frac{1}{3} + \cdots + \frac{1}{n-1} + \frac{1}{n} \right].$$

Using the big-Theta estimate (1) in conjunction with the result of Exercise for the Reader 7.15, we see that both sides of the above double inequality are $\Theta(n\log(n))$, and hence $E(X) = \Theta(n\log(n))$. This average-case complexity estimate for the quick sort is far superior to the $\Theta(n^2)$ worst-case complexity that was established in Part (a).

Optimality of the Merge Sort Algorithm

In the complexity analysis of many recursive algorithms, such as divide and conquer algorithms, it is often straightforward to obtain a recurrence relation for the operation counting function. Big-O and/or big-Theta estimates can sometimes be gleaned directly from such recurrence relations. We illustrate this general technique when we obtain a worst-case complexity estimate of $O(n \log n)$ for the merge-sort Algorithm 7.7. Before we do this we will prove a preliminary result that shows a worst-case complexity estimate of at least $O(n \log n)$ for <u>any</u> sorting algorithm. More precisely: if a sorting algorithm has wort-case complexity estimate of $O(f(n))$, then $n \log n = O(f(n))$, i.e., $f(n) = \Omega(n \log(n))$. With these two results the merge sort will have been established as an optimal sorting algorithm.

PROPOSITION 7.3: Any algorithm that sorts an unordered list of n distinct elements has worst-case complexity at least $O(n \log n)$.

Proof: The number of permutations of a list of n elements is $n!$. A certain sequence of comparisons will be needed by the algorithm to sort any of these permutations. The result of such a sequence of comparisons can be identified with a finite string: $00111100\ldots$, where the ith bit of the string is 1, say if the left element of the ith comparison is greater than the right element; otherwise it is 0. Different permutations must correspond to different bit strings. Thus the number of bit strings that result in this way must be at least $n!$. The worst-case complexity of the sorting algorithm would correspond to one of these binary strings of maximum length. Let M be this maximum length, where M is a function of the input size n: $M = M(n)$. The number of binary strings of length at most M is:

$$2^0 \,(\text{length } 0) + 2^1 \,(\text{length } 1) + 2^2 \,(\text{length } 2) + \cdots + 2^M \,(\text{length } M) = \frac{2^{M+1} - 1}{2 - 1},$$

where we have used the the formula for a sum of a finite geometric series (Proposition 3.5). This quantity is $\Omega(2^M)$. Since the number of identified binary strings must be at least $n!$, we must have $n! \precsim 2^M$. Taking logarithms gives $\log(n!) \precsim \log(2^M) = M \log(2) \precsim M.$ [11]   Since we are aiming to show that $M \succsim n \log n$, it will suffice to show that $\log(n!) \succsim n \log n$. Using the rules for the

---

[11] More formally, $n! \precsim 2^M$ may be written as $n! \le C 2^M$ for all sufficiently large n (where C is some positive constant). Taking logs on both sides gives that (since logs with base > 1 are increasing functions) $\log(n!) \le \log(C 2^M) = \log C + M \log 2$, which implies that $M \succsim \log(n!)$.

logs of products and quotients and the fact that logs (with base $b > 1$) are increasing, we obtain:

$$\log(n!) \geq \log n + \log(n-1) + \log(n-2) + \cdots + \log\left(\lfloor n/2 \rfloor\right)$$
$$\geq (n/2)\log(n/2)$$
$$= (n/2)\log n - (n/2)\log 2$$
$$= \Theta(n\log n).$$

(In the equality, we used the result of Exercise for the Reader 7.15.)   The proof of Proposition 7.3 is now complete. □

PROPOSITION 7.4: The merge sort Algorithm 7.7 applied to an unordered list of n distinct elements has complexity $O(n\log n)$.

Proof: We let $f(n)$ denote the number of comparisons that the merge-sort algorithm uses to sort a list of n elements. Among other things, we will show that this number depends only on the size n of the list, not on the list itself. Since we have already pointed out the requisite merge Algorithm 7.6 requires at most $n+m$ comparisons to merge two sorted lists of lengths n and m, it follows from the recursive step of Algorithm 7.7 that $f(n) = f\left(\lfloor n/2 \rfloor\right) + f\left(\lceil n/2 \rceil\right) + n$.   This recursion formula is a bit awkward to solve, but since we seek only an upper bound (rather than an explicit formula) for $f(n)$, we will instead use it to prove the following upper bound, which gives the desired $O(n\log n)$ complexity estimate.

Claim: $f(n) \leq 12n\log n + 1$.

Proof of Claim: We use induction on n.
Basis Step: $n = 1$: Since $f(1) = 0$, the inequality is trivial.
Induction Step: We assume $n > 1$, and that the inequality is true for all positive integers less than n. We must show that the inequality is true for n. Using the recursion formula for $f(n)$, in conjunction with the induction hypothesis, and the simple inequality $\lceil n/2 \rceil \leq 2n/3$, we obtain:

$$f(n) = f\left(\lfloor n/2 \rfloor\right) + f\left(\lceil n/2 \rceil\right) + n$$

$$\leq 12\lfloor n/2 \rfloor\log\left(\lfloor n/2 \rfloor\right) + 12\lceil n/2 \rceil\log\left(\lceil n/2 \rceil\right) + n + 2$$

$$\leq 12\lfloor n/2 \rfloor\log(2n/3) + 12\lceil n/2 \rceil\log(2n/3) + n + 2$$

$$\leq 12n[\log(2/3) + \log n] + n + 2$$

$$\leq 12n\log n - 2.11n + n + 2$$

$$\leq 12n\log n + 1.$$

This completes the proof of the claim, and hence also of Proposition 7.4. □

EXERCISE FOR THE READER 7.16: Determine the (a) worst-case and (b) average-case complexity of the randomized Algorithm 7.8 for finding the kth smallest element of an unsorted list of n elements.
Suggestion: Consider what happens when selection elements land in the middle 50% of elements in the list. What is the probability that such an element gets selected?

The Classes P and NP

It has become customary among scientists who study algorithms to consider problems for which algorithms exist with polynomial worst-case complexity $O(n^p)$, i.e., algorithms that work in **polynomial time**, to be **tractable problems**, and other problems to be **intractable**. Common special cases of polynomial complexity include **linear complexity** $O(n)$ and **quadratic complexity** $O(n^2)$. Thus problems for which there are algorithms with any lower order worst-case complexity such as logarithmic $O(\log n)$ or combinations such as $O(n \log n)$ would also be considered tractable. Complexities that are higher order than polynomial complexities are often collectively called **exponential** complexities, even though they are not necessarily of the form $O(b^n)$. For example, an algorithm with the factorial complexity $O(n!)$ would thus be said to have exponential complexity, even though $b^n \prec n!$.

Practically speaking, the above definition of a tractable problem is not always realistic, since a large exponent and multiplicative constant could make for an extremely slow algorithm. For example, if we had a polynomial time algorithm that for an input of size n required $2,000,000n^{5000}$ computer operations, and an exponential algorithm that took $(1.0001)^n$ computer operations to process an input of size n, then in order for the polynomial time algorithm to work faster than the exponential algorithm, we would need to have an input of size $n > 1$ billion! Despite this sort of possibility, for most important tractable problems, efficient algorithms have been found.

In our work thus far, we have had numerous experiences with tractable and intractable problems. For example, the searching and sorting problems are tractable since polynomial time algorithms have been found for these, as shown in this section. Intractability is more a temporary definition. Just because a polynomial-time algorithm is unknown today for a certain problem (making it intractable), one may get discovered tomorrow, and this would change the problem's classification to tractable. Famous examples of intractable problems are that of finding a prime factor of a positive integer n (where the input size is the number of digits of n, i.e., $\log n$), and the discrete logarithm problem. Indeed, all public key cryptography systems are based on intractable problems.

Among intractable problems, some have the special property that given a possible solution to the problem, it is possible to check whether is actually is a solution using a polynomial time algorithm. Both the prime factorization problem and the discrete logarithm intractable problems have this property. For example, given a purported prime factor of an integer n, it can simply be checked by dividing n by it and seeing if there is no remainder. (In the exercises, we will show that the division algorithm works in polynomial time.) A special notation has been adopted that distinguishes this sort of intractable problem from a tractable problem when the problems are restricted to so-called **decision problems**, which are problems that have a yes or no answer. One example of a decision problem is the primality problem: Given a positive integer n, is n prime? The primality problem was intractable for centuries, but, as pointed out in Chapter 3, a polynomial algorithm for it was discovered in 2004. Many non-decision problems give rise to similarly difficult decision problems. For example, the prime factorization problem, which remains intractable, and is believed to be permanently intractable, gives rise to the following simpler (but still intractable) decision problem: Given positive integers $n > k$, does n have a prime factor less than k? Decision problems are extremely common in discrete structures, usually phrased in the form: Does there exist a certain discrete structure (among an inordinately large set of discrete structures) that has a certain property?

DEFINITION 7.5: (*The Classes P and NP*) A tractable decision problem is said to belong to the **class P**. The **class NP** consists of all decision problems (tractable or not) for which there is a polynomial-time algorithm that can check whether a purported solution to the problem is actually a solution.

NOTE: *P* stands for **polynomial**, but *NP* does not stand for non-polynomial, but rather **non-deterministic polynomial**.

There is an interesting special subclass of *NP* problems that are known as *NP* **complete** problems. These *NP* problems have the property that if a polynomial-time algorithm can be found for any of them, then it could be translated into a polynomial-time algorithm to solve any other *NP* complete problem. The first *NP* complete problem was discovered in the 1970s, and since then a diverse collection of famous interesting and difficult intractable problems has been shown to be *NP* complete.   The discrete logarithm problem and knapsack problems of Sectin 4.5 are both *NP* complete.   Interested readers may consult the book [GaJo-79] by Michael Garey and David Johnson for a full account of *NP* completeness including examples and proofs of the *NP* completeness of a large assortment of problems spanning a number of different areas. Since such famously difficult *NP* complete problems have withstood centuries of efforts by many of the best mathematicians to find a polynomial time algorithm, it is very strongly believed that no such algorithm can ever exist, or in different notation, the $P \neq NP$.[12]   A

---

[12] The question of whether $P = NP$ is one of the seven original millennium problems of the Clay Mathematics Institute. The institute offers a (US) $1 million prize to any scientist who is able to solve one of the millennium problems. In 2010, the institute awarded $1 million to Russian mathematician

non-decision problem that can be shown to be at least as hard to solve as a certain NP complete decision problem is called an **NP hard** problem.

EXERCISES 7.2:

1. (TRUE/FALSE?) For each of the following big-O statements, determine whether it is true or false.
    (i) Justify your true/false answers directly using Definition 7.2, by either establishing witnesses (for a true answer), or explaining why witnessness cannot exist (for a false answer).
    (ii) Check to see whether any of the theory developed in the section can directly justify your true/false answer, and explain the details of such justifications.

    (a) $150 = O(n)$.   (b) $2n + 10 = O(n)$.

    (c) $2n^3 + 10n = O(n^2)$.   (d) $(\log n)\sqrt{n} = O(n^2)$.

    (e) $n^2(1/2)^n = O(1)$.   (f) $(n^4 + 10n)/(2n^2 + 4) = O(n^2)$.

2. (TRUE/FALSE?) For each of the following big-O statements, determine whether it is true or false.
    (i) Justify your true/false answers directly using Definition 7.2, by either establishing witnesses (for a true answer), or explaining why witnessness cannot exist (for a false answer).
    (ii) Check to see whether any of the theory developed in the section can directly justify your true/false answer, and explain the details of such justifications.

    (a) $5\sqrt{n} = O(n)$.   (b) $n^{10} = O(2^n)$.

    (c) $n \log n(\log(\log n)) = O(n^{1.5})$.   (d) $2n^2 + 10n = O(n(\log n)^{10})$.

    (e) $n^n = O(n!)$.   (f) $(n^2 \log n)/(5n + 4) = O(n)$.

3. (TRUE/FALSE?) Classify each of the dominance/strict dominance relations as either true or false. Justify your answers.

    (a) $150 \preceq \log n$.   (b) $2n + 10 \preceq n$.

    (c) $2n^3 + 10n + 5 \prec n^3$.   (d) $\log(\log n) \prec \log n$.

    (e) $n^2 2^n \prec 3^n$.   (f) $(n^2 + 4n)/(2n + 3) \preceq n$.

4. (TRUE/FALSE?) Classify each of the dominance/strict dominance relations as either true or false. Justify your answers.

    (a) $150n \preceq n \log n$.   (b) $(2n)^{2n} \preceq 3^n$.

    (c) $(2n)^{2n} \preceq 3^{3n}$.   (d) $n^5 + n^4 \log n + n^2(1/3)^{2n} \preceq n^5$.

    (e) $\log n^{2n} \preceq n^2$.   (f) $(400n + 4)/(3n + 2) \preceq 1$.

5. (i) For each expression below, determine, if one exists, the best possible big-O estimate of the form $O(n^a b^n (\log n)^c)$, where the nonnegative numbers a, b, c are restricted to the range $0 \le a, c;\ 1 \le b$.
    (ii) Is the best possible big-O estimate in (i) also a big-Theta estimate? Justify your answer.

Grigoriy Perelman for proving the Poincaré conjecture, and now only six millennium problems remain unsolved. Interestingly, Perelman turned down this $1 million prize, stating that his contribution was no greater than that of American mathematician Richard Hamilton, who had suggested the outline for Perelman's program.

(iii) For each expression below, determine, if one exists, the best possible big-O estimate of the form $O(n^a b^n)$, where the nonnegative numbers a, b are restricted to the range $0 \le a; 1 \le b$.

(iv) Is the best possible big-O estimate in (iii) also a big-Theta estimate? Justify your answer.

(a) $2n + 5n^4 + 7$.

(b) $2n + 7\log n$.

(c) $2n^2 \log n + 7(\log n)^2 n$.

(d) $(2^2 n^3 + 3^3 n^2)^2$.

(e) $\ln 6^n + \sqrt{n} \log n$.

(f) $(n + 4n^3)/(n + 3)$.

6. (i) For each expression below, determine, if one exists, the best possible big-O estimate of the form $O(n^a b^n (\log n)^c)$, where the nonnegative numbers a, b, c are restricted to the range $0 \le a, c; 1 \le b$.

(ii) Is the best possible big-O estimate in (i) also a big-Theta estimate? Justify your answer.

(iii) For each expression below, determine, if one exists, the best possible big-O estimate of the form $O(n^a b^n)$, where the nonnegative numbers a, b are restricted to the range $0 \le a; 1 \le b$.

(iv) Is the best possible big-O estimate in (iii) also a big-Theta estimate? Justify your answer.

(a) $2n^{1.25} + n^{1.50} + 3n^{1.75}$.

(b) $2n^2 (\log(n))^3 + 7n(\log n)^4$.

(c) $\ln n + \log_2(n^4)$.

(d) $n\log(n)\log(\log(n))$.

(e) $2^n 3^{2n} 4^{3n}$.

(f) 2^{n^2}.

7. (a) Suppose that we have a sorting program with complexity $\Theta(n^2)$ that spends exactly 0.1 seconds to sort any list with 1000 items. Assuming that that the time $T(n)$ required by this program to sort a list with n items is proportional to n^2, how much time would it take this progam to sort a list with 100,000 items? With 1 million items?

(b) Suppose that we have a sorting program with complexity $\Theta(n\log n)$ that spends exactly 0.1 seconds to sort any list with 1000 items. Assume that the time $T(n)$ required by this program to sort a list with n items is proportional to $n\log n$, how much time would it take this progam to sort a list with 100,000 items? With 1 million items?

8. (a) Suppose that we have a primality checking program with complexity $\Theta(1.5^n)$ that spends exactly 0.0001 seconds to determine whether any $n = 9$ digit number is prime. Assuming that the time $T(n)$ required by this program to check the primality of any n digit integer is proportional to 1.5^n, how much time would it take this progam to check the primality of a 50 digit number? What about a 1000 digit number?

(b) Suppose that we have a primality checking program with complexity $\Theta(n^3)$ that spends exactly 0.0001 seconds to determine whether any $n = 9$ digit number is prime. Assume that the time $T(n)$ required by this program to check the primality of any n digit integer is proportional to n^3, how much time would it take this progam to check the primality of a 50 digit number? What about a 1000 digit number?

9. (TRUE/FALSE?) Determine whether each of the following relations is true or false. Justify your answers.

(a) $150 = \Omega(n)$.

(b) $2n + 10 = \Omega(n)$.

(c) $(2n + 5)^3 = \Omega(n^2)$.

(d) $(\log n)\sqrt{n} = \Omega(n^2)$.

(e) $\log n^3 = \Theta(\log n)$.

(f) $\ln(n^4 + 10n) = \Omega(\log n)$.

10. (TRUE/FALSE?) Determine whether each of the following relations is true or false. Justify your answers.

(a) $(n+1)^2 = \Omega(n)$.

(b) $3^{n+12} = \Omega(3^n)$.

(c) $n! = \Omega(10^n)$.

(d) $5^{\log n} = \Omega(n)$.

(e) $n^3 2^n = \Omega(4^n)$.

(f) $(\log n)^{\log n} = \Omega(n / \log n)$.

11. Let r be a positive real number and let $f(n)$ be the sum of the corresponding geometric series with ratio r up to the nth power: $f(n) = 1 + r + r^2 + \cdots + r^n$. Show that

$$f(n) = \begin{cases} \Theta(1), & \text{if } r < 1 \\ \Theta(n), & \text{if } r = 1 \\ \Theta(r^n), & \text{if } r > 1. \end{cases}$$

In words, as far as big-Theta is concerned, the sum of a finite geometric series is the first term if the ratio is less than 1, the number of terms if the ratio is 1, and the last term if the ratio is greater than 1.

12. (a) Let k be a positive integer. Show that $1^k + 2^k + \cdots + n^k = O(n^{k+1})$.

(b) Does the result of Part (a) remain true if the big-O is replaced by a big-Theta?

13. (TRUE/FALSE?) Classify each statement below as either true of false. For those that you assert as true, give a proof; for those that you assert as false, give a counterexample. All functions appearing in these statements are assumed to be positive functions on the set of positive integers.

(a) If $f(n) = O(g(n))$, and p is a positive integer, then $f(n)^p = O(g(n)^p)$.

(b) If $f(n) = O(g(n))$, then $\log(f(n)) = O(\log(g(n)))$. Assume here that the functions both grow to infinity as n gets larger and larger.

(c) $\log n! = O(n \log n)$.

14. (TRUE/FALSE?) Classify each statement below as either true of false. For those that you assert as true, give a proof; for those that you assert as false, give a counterexample. All functions appearing in these statements are assumed to be positive functions on the set of nonnegative integers.

(a) The relation $f(n) = \Theta(g(n))$ is an equivalence relation on the set of all positive real-valued functions on the nonnegative integers.

(b) If $f(n) = O(g(n))$, then $\log(2^{f(n)}) = O(2^{g(n)})$.

(c) $\log n! = \Omega(n \log n)$.

15. (*Complexity Analysis of the Insertion Sort Algorithm*) Determine the (a) worst-case, (b) best-case, and (c) average-case complexity of the insertion sort algorithm of Exercise 15 of Section 7.1 for sorting a list of n elements.

16. Determine the (a) worst-case, (b) best-case, and (c) average-case complexity of the algorithm described in the solution to Exercise for the Reader 7.1 for determining the maximum of a list of n integers.

17. (a) Determine a good big-O estimate for the worst-case complexity of the iterative algorithm of Example 3.8(a) (in Section 3.2) for computation of the nth Fibonnaci number f_n. To measure the efficiency of the algorithm, take the size of the input n to be $\lfloor \log_2(n) \rfloor = \Theta(\log n)$, which, from Section 4.1 is the size of the bit string in the binary representation of a positive integer n.

(b) Determine a good big-O estimate for the worst-case complexity of the recursive algorithm of

Example 3.8(b) for computation of the nth Fibonnaci number f_n. To measure the efficiency of the algorithm, take the size of the input n to be $\lfloor \log_2(n) \rfloor = \Theta(\log n)$, which, from Section 4.1 is the size of the bit string in the binary representation of a positive integer n.

18.   Determine a good big-O estimate for the worst-case complexity of the recursive algorithm of Example 3.9 (in Section 3.2) for calculating the determinant of a square $n \times n$ matrix (based on cofactor expansion). To measure the efficiency of the algorithm, take the size of the input to be n^2, the number of entries in the matrix. Is your big-O estimate actually a big-Theta estimate?

19.   Determine a good big-O estimate for the worst-case complexity of Algorithm 4.2 (in Section 4.1) for the addition of two base b integers. To measure the efficiency of the algorithm, take the size of the input to be K, the length of the inputted strings. Is your big-O estimate actually a big-Theta estimate?

20.   Determine a good big-O estimate for the worst-case complexity of Algorithm 4.3 (in Section 4.1) for the subtraction of two base b integers. To measure the efficiency of the algorithm, take the size of the input to be K, the length of the inputted strings. Is your big-O estimate actually a big-Theta estimate?

21.   Determine a good big-O estimate for the worst-case complexity of Algorithm 4.4 (in Section 4.1) for the mutiplication of two base b integers. To measure the efficiency of the algorithm, take the size of the input to be the length of the larger of the inputted strings. Is your big-O estimate actually a big-Theta estimate?

22.   (a) Determine a good big-O estimate for the worst-case complexity of Algorithm 4.5 (in Section 4.2) for fast modular exponentiation. To measure the efficiency of the algorithm, take the size of the input n to be $\lfloor \log_2(x) \rfloor = \Theta(\log x)$, which, from Section 4.1 is the size of the bit string in the binary representation of the positive integer exponent x. Is your big-O estimate actually a big-Theta estimate?
           (b) Repeat Part (a) for best-case complexity.
           (c) Repeat Part (a) for average-case complexity.

NOTE: (*Complexity of the Euclidean Algorithm 3.1*) The next three exercises will develop big-O estimates for the Euclidean Algorithm 3.1. It will follow that when the Euclidean algorithm is applied to two positive integers $a \geq b > 0$, the worst-case complexity is $O((\log a)^2 \log b)$. This is quite fast, as our experience thus far with the Euclidean algorithm has shown. Exercise 23 will show that if the Euclidean Algorithm is applied to two positive integers $a \geq b > 0$, then the number of iterations is $O(\log b)$. The lion's share of the work rests with the integer divisions, which are discussed in Exercises 24 and 25.

23.   (*Lamé's Theorem*) Prove the following:
           Lamé's Theorem: If the Euclidean Algorithm is applied to an integer division $a \div b$, where $a \geq b > 0$ are positive integers, then the number of iterations is $O(\log b)$.

           Suggestion: Using the notation of Algorithm 3.1, if we set $r_0 = b$ and $r_{-1} = a$, then the Euclidean algorithm consists of $n + 1$ applications of the division algorithm, and these can all be expressed as $r_{i-1} = q_i r_i + r_{i+1}$ $(i = 0, 2, \cdots, n)$. Solve each of these for the last remainder: $r_{i+1} = q_i r_i - r_{i-1}$ $(i = 0, 2, \cdots, n)$. The second-to-last of these equations (the last remainder is 0), certainly gives $r_n \geq f_2$, where f_k denotes the kth term of the Fibonacci sequence (Example 3.7). Iteratively apply the remaining equations to successively obtain that $r_{n-1} \geq f_3$, $r_{n-2} \geq f_4$, $\cdots$, $b = r_1 \geq f_{n+1}$. Then apply equation (4) from Chapter 3.

24. (*A Simple Brute-Force Algorithm for Integer Division*) The following is a very simple algorithm for performing the division algorithm (Proposition 3.14) to compute the quotient q and remainder r in an integer division $a \div d$, where a, d are integers, with $d > 0$. It basically continues to subtract d from $|a|$, once at a time, keeping track of the number of times we subtract (the final result will be the quotient) and the results of the subtraction (the final result will be the remainder) until we arrive at a remainder smaller than d. If $a < 0$, we need one final subtraction, and to adjust the remainder.

    Step 1: Initialize Quot = 0, Rem = $|a|$.

    Step 2:
    WHILE REM $\geq d$
          UPDATE: Rem = Rem − d
          UPDATE: Quot = Quot + 1
    END WHILE

    Step 3: (*Make needed adjustments if dividend is negative*)
    IF $a < 0$
          UPDATE: Rem = d − REM
          UPDATE: Quot = − (Quot + 1)
    END IF

    Step 4: OUTPUT: quotient = Quot and remainder = Rem

    (a) Explain why the above algorithm correctly performs the division algorithm.

    (b) Show that the above algorithm has worst-case complexity $O(d \log(|a|))$.

25. (*An Improved Algorithm for Integer Division*) The following is a recursive algorithm for performing the division algorithm (Proposition 3.14) to compute the quotient q and remainder r in an integer division $a \div d$, where a, d are integers, with $d > 0$. This exercise will show that it has worst-case complexity $O((\log(|a|))^2)$. This is typically much better than the algorithm of Exercise 24, especially in cases where the d and a both have approximately the same number n of binary digits, in which case the respective complexity estimates are $O(n^2)$, $O(n \cdot 2^n)$.

    Since recursive algorithms call on themselves, we specify a name and syntax for this program:

    (Quot, Rem) = FastDivision(*a*,*d*)
    IF $a = 0$
        OUTPUT Quot = 0, Rem = 0, EXIT Algorithm
    ELSE
        (Quot, Rem) = FastDivision(floor(*a*/2), *d*)
        UPDATE Quot = 2 · Quot, Rem = 2 · Rem
        IF a is odd
            UPDATE Rem = Rem + 1
        END IF
        IF Rem $\geq d$
            UPDATE Quot = Quot + 1, Rem = Rem − d
        END IF
    END IF
    OUTPUT Quot, Rem

    (a) Explain why the above algorithm correctly performs the division algorithm.

    (b) Show that the above algorithm has worst-case complexity $O((\log(|a|))^2)$.

26. (*Banzhaf Indices by Brute Force; This Exercise Depends on the Appendix to Section 5.3*)
    (a) Determine a good big-O estimate for the worst-case complexity of the brute-force algorithm (illustrated in Example 5.29) for computing the Banzhaf power indices for an inputted weighted voting system $[q : w_1, w_2, \cdots, w_N]$. Use $N + \log_2(W)$ to measure the size of the input, where $N =$

the number of voters and $W = \sum_{n=1}^{N} w_i$ is the sum of the voter weights. Is your big-O estimate actually a big-Theta estimate?

(b) Repeat Part (a) but for best-case complexity.

27.   (*Banzhaf Indices by Generating Functions; This Exercise Depends on the Appendix to Section 5.3*) (a) Determine a good big-O estimate for the worst-case complexity of the generating function-based Algorithm 5.12 for computing the Banzhaf power indices for an inputted weighted voting system $[q : w_1, w_2, \cdots, w_N]$.   Use $N + \log_2(W)$ to measure the size of the input, where $N =$ the number of voters and $W = \sum_{n=1}^{N} w_i$ is the sum of the voter weights. Is your big-O estimate actually a big-Theta estimate?

28.   Show that the function $f(n)$ is defined on the set of all positive integers by the recurrence

relation: $f(n) = \begin{cases} 1, & \text{if } n = 1 \\ f(\lfloor n/2 \rfloor) + n^2, & \text{if } n > 1 \end{cases}$, satisfies $f(n) = O(n^2)$.

(b) Does the function in Part (a) satisfy $f(n) = \Omega(n^2)$?

29.   Show that the function $f(n)$ is defined on the set of all positive integers by the recurrence

relation: $f(n) = \begin{cases} 2, & \text{if } n = 1, 2 \\ 2f(\lfloor n/3 \rfloor) + n, & \text{if } n > 2 \end{cases}$, satisfies $f(n) = O(n)$.

(b) Does the function in Part (a) satisfy $f(n) = \Omega(n)$?

30.   Use a recursion formula as in the proof of Propostion 7.4 to analyze the complexity of the binary search Algorithm 7.2 to search for a target number that appears in a sorted list of length n.

Chapter 8: Graphs, Trees, and Associated Algorithms

A very useful discrete structure is that of a graph. A *graph* consists of a finite set of *vertices* along with a set of *edges*, each of which connects a pair of vertices. Although the formal definition will be given in terms of sets, if the number of vertices is not too great, a graph can be effectively represented geometrically. Figure 8.1 displays some examples of graphs.

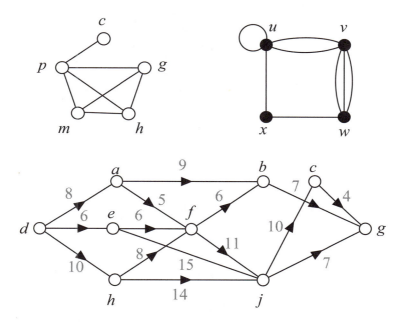

FIGURE 8.1: Some examples of graphs: (a) (upper left) a simple graph, (b) (upper right) a general graph, (c) a directed graph with edge weights.

Graphs are ideal tools for modeling an assortment of discrete problems and objects. The most basic (and most important) variety of a graph is what is called a *simple graph* in which each edge connects a distinct pair of vertices, and there can be at most one edge between any pair of vertices. An example of a simple graph is shown in Figure 8.1(a). The five vertices are labeled, and there are seven edges indicated by line segments between various pairs of vertices. We point out that the intersection of the edges {p, h} and {m, g} is not a vertex of the graph, just an incidental intersection. This graph serves as a model for the following problem:

495

Problem: A certain university math department has six professors: Petersen (chair), Cayley, Euler, Kuratowski, Menger, and Tutte. The five departmental committees are:

Curriculum: Cayley, Kuratowski
Grade Inflation: Euler, Menger
Hiring: Petersen, Euler, Menger
Math Projects: Petersen, Menger, Tutte
Prizes: Cayley, Euler, Kuratowski, Tutte

How many different monthly (nonoverlapping) meeting times are needed to assure there will be no time conflicts?

In the graph of Figure 8.1(a), the five vertices represent the five committees (and are labeled in the obvious fashion), and two vertices are connected by an edge if, and only if the corresponding committees have a member in common. In the setting of graphs, this problem is an instance of the *graph coloring problem*: Determine the smallest number N of colors needed to color the vertices such that each vertex can be painted using one of the N colors in such a way that if two vertices are joined by an edge, then they must be painted different colors. Since any two of the bottom four vertices of the graph (p, g, m, and h) are joined by an edge, they must all be painted four different colors, so $N \geq 4$. But the fifth vertex c can be painted the same color as either of the bottom two (m or h), or g, so it follows that $N = 4$.

In certain network applications there may be multiple links between any given pair of vertices, and simple graphs are not adequate models. General graphs are allowed to have multiple edges between a pair of vertices, and also to have *self-loops*, which are edges connecting a vertex to itself. The graph shown in Figure 8.1(b) is a general graph; there is one self-loop at vertex u, two multi-edges between vertices u and v, and three multi-edges between vertices v and w. Sometimes it is useful for the edges of a graph to have directions; such graphs are called *directed graphs*, and Figure 8.1(c) gives an example of a directed graph with the additional feature that the edges have weights. Such a directed graph with edge weights can be used to represent a network flow problem, where each edge represents a pipe, the directions represent the flow directions, and the weights represent the flow capacities. The important *network flow problem* for this network is to determine the maximal amount of flow that can be routed from vertex d (the *source*) to vertex g (the *sink*), using the pipes in this network.

Various concepts and definitions relating to graphs will be given in Section 8.1. The idea of a path in a graph is very intuitive but extremely useful. Section 8.2 will be devoted to paths in graphs along with related concepts and some applications. A *tree* is a special type of simple graph that appears in numerous applications and theoretical developments; Section 8.3 will be dedicated to trees. Chapter 9 will be devoted to further applications of graphs, and to the development of associated alogorithms.

8.1: GRAPH CONCEPTS AND PROPERTIES

Simple Graphs

We begin by using the language of sets to give the formal definition of a simple graph.

DEFINITION 8.1: A **simple graph** is an ordered pair of sets $G = (V, E)$, where V is a nonempty set of **vertices** (or **nodes**) of G, and E, the set of **edges** of G, is a set of two-element pairs (2-combinations) of vertices. Thus, each edge of G can be expressed as $\{u, v\}$, where u, and v are distinct vertices, i.e., $u, v \in V$, $u \neq v$. The vertices u and v determining an edge $\{u, v\}$ are called the **endpoints** of the edge. The edge $\{u, v\}$ is said to **join** u and v, and the edge is said to be **incident** to either of its endpoints. Any two vertices in G that are joined by an edge are said to be **adjacent**, and are called **neighbors**. A vertex with no neighbors is called **isolated**.

NOTE: The vertex and edge sets of a graph G will sometimes be denoted as $V(G)$ and $E(G)$, respectively, when we wish to emphasize the dependence on G—for example, when more than one graph is under consideration.

It is often helpful to think of a simple graph as a social network in the following way: the vertices are a set of people, and two people are joined by an edge if, and only if they are *acquaintances*. So if two vertices are not adjacent, the meaning is that the people they represent are *strangers* to each other. What makes graphs to be particularly rich in theory and applications is that they provide effective means to visualize abstract set-theoretic relationships. Such visualizations make graph theory much more intuitive than pure set theory, which gives rise to numerous applications. We emphasize that, for a given graph (with more than one vertex), although the vertex and edge sets are uniquely determined, it can be drawn in infinitely many different ways.

EXAMPLE 8.1: Figure 8.2 shows three different drawings of the same graph on four vertices. The graph represents a social network with four vertices {Tom, Jan, Eve, Bob}, and four edges {Tom, Jan}, {Jan, Bob}, {Bob, Eve}, and {Eve, Tom}.

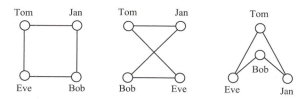

FIGURE 8.2: Three different drawings of the same four vertex graph.

Determining whether two given graph drawings represent the same graph is a special case of the famously difficult *graph isomorphism problem* that we will come back to later on in this section.

General Graphs

In order to formally define general graphs, it is convenient to view the edge set by means of a function with co-domain $\mathscr{P}(V)$, the power set of the set V (of vertices). Although the power set of V is the set of all subsets of V, the only ones that will represent edges will be one-, or two-element subsets. The following definition provides the details of this scheme.

DEFINITION 8.2: A **general graph** is a triple $G = (V, E, \Phi)$, where V is a nonempty set of **vertices** (or **nodes**) of G, E is the set of **edges** of G, and Φ, called the **edgemap**, is a function $\Phi : E \to \mathscr{P}(V)$, where $|\Phi(e)| = 1$ or 2, for each $e \in E$. The vertices in $\Phi(e)$ are called the **endpoints** of the edge e. An edge e having only one endpoint (i.e., $|\Phi(e)| = 1$) is called a **self-loop**. Two edges, e_1, e_2 that have the same endpoints (i.e., $\Phi(e_1) = \Phi(e_2)$) are called **parallel edges** or **multiedges**.[1]

When a general graph is specified by a drawing in which the edges are labeled, all three objects of the triple $G = (V, E, \Phi)$ can be explicitly deduced, and so, in practice, these objects are rarely all written out. The following example will illustrate these points.

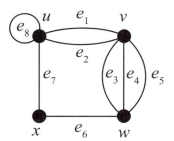

FIGURE 8.3: A completely labeled multigraph.

EXAMPLE 8.2: The multigraph shown in Figure 8.3 has vertex set $V = \{u, v, w, x\}$, edge set $E = \{e_1, e_2, e_3, e_4, e_5, e_6, e_7, e_8\}$, and edgemap Φ defined by:

$$\Phi(e_1) = \Phi(e_2) = \{u, v\},$$
$$\Phi(e_3) = \Phi(e_4) = \Phi(e_5) = \{v, w\},$$
$$\Phi(e_6) = \{w, x\},$$
$$\Phi(e_7) = \{u, x\},$$
$$\Phi(e_8) = \{u\}.$$

The edge e_8 is the only self-loop of this multigraph.

---

[1] We caution the reader that the vocabulary in graph theory is quite vast, and unfortunately not always consistent. For example, what we call a *general graph* other authors will sometimes call a *pseudograph*, or a *multi-graph*, or even just a *graph*. We will aim to use vocabulary that is both widely used and intuitive. Whenever reading a new book or paper on graph theory, one must make sure of the vocabulary that is being used.

Of course, a simple graph is a special case of a general graph. Since a simple graph has no parallel edges (or self-loops), its edgemap is redundant (i.e., the edges are uniquely described by their endpoints). Thus, any comments or theorems about general graphs will apply also to simple graphs. For this reason, we will adopt the following convention:

CONVENTION: Unless we specifically indicate otherwise, a "graph" will always be taken to mean a general graph. Also, the edgemap of a graph will often not be explicitly mentioned.

Degrees, Regular Graphs, and the Handshaking Theorem

DEFINITION 8.3: If v is a vertex of a graph G, then the **degree** of v, denoted $\deg(v)$ (or $\deg_G(v)$, if we wish to emphasize the dependence on G), is the number of edges incident to v, with any self-loops counted twice. A simple graph in which all vertices have the same degree k is called a **regular graph** or, more precisely, a **k-regular** graph.

Thus, for a simple graph, the degree of a vertex is simply the number of neighbors that is has. For a general graph that has been drawn, the degree of a vertex is simply the number of rays emanating from it, if we zoom in sufficiently close (so that no self-loops can be seen).

EXAMPLE 8.3: (a) Find the degree of each vertex of the graph in Example 8.1. (b) Do the same for the graph of Example 8.2.

SOLUTION: Part (a): Each vertex has degree 2.
Part (b): The self-loop e_8 contributes 2 to $\deg(u)$, while the other three edges e_1, e_2, e_7 that are incident to it each contribute 1, so we obtain $\deg(u) = 5$. In a similar fashion, we compute $\deg(v) = 5$, $\deg(w) = 4$, and $\deg(x) = 2$.

Notice that when we add up the degrees of the vertices for each of the two graphs in the preceding example, we obtain 8 and 16. These numbers are precisely twice the number of edges in the respective graphs. This is no coincidence, as the following simple, but important theorem shows.

THEOREM 8.1: (*The Handshaking Theorem*) For any graph G, we have

$$\sum_{v \in V(G)} \deg_G(v) = 2|E(G)|. \tag{1}$$

The **Handshaking theorem** terminology stems from the application to social networks, represented by simple graphs: Suppose the vertices of a simple graph represent the set of people at a party, and an edge between a pair of vertices represents the fact that the two endpoints shook hands. Thus, every edge represents a single handshake (between the two people that are joined by the edge), and the degree of a person is the number of people that he/she shook hands with. This is essentially the proof of the theorem for simple graphs; not much more work is needed to prove the result for general graphs.

Proof of Theorem 8.1: If an edge e has two endpoints u and v, then it will contribute one to each of $\deg(u)$ and $\deg(v)$. If e is a self-loop incident to a vertex u, then it will contribute two to $\deg(u)$. Thus, in any case, an edge of the graph will always contribute two to the sum on the left of (1), and the identity follows.

An important consequence of the handshaking theorem is the following:

COROLLARY 8.2: In any graph G, the number of vertices of odd degree must be even.

Proof: Let us partition the vertex set $V(G)$ into $V_{odd} = $ {vertices of odd degree} and $V_{even} = $ {vertices of even degree}. The sum on the left side of (1) can be decomposed by summing separately over each of these two sets of vertices:

$$\sum_{v \in V(G)} \deg_G(v) = \sum_{v \in V_{odd}} \deg_G(v) + \sum_{v \in V_{even}} \deg_G(v) \equiv S_{odd} + S_{even}. \qquad (2)$$

Now, each of the terms being added in the second sum S_{even} on the right side of (2) is an even integer, and so (since a sum of even integers is even) S_{even} is even. Since the handshaking theorem tells us that the whole sum in (2) must be even, this forces the first sum S_{odd} on the right side of (2) to be even. But each of the terms being added in is an odd number, so this sum will be even if, and only if the number of terms being added, namely $|S_{odd}|$, is even. This is what we were aiming to prove. □

EXERCISE FOR THE READER 8.1: (a) Use the handshaking theorem to answer the following question: If there are 100 people at a party, and everyone shakes hands with everyone else, how many handshakes take place?
(b) In a party with 99 people, is it possible that everyone shakes hands with exactly three other people?

Before we introduce further graph theoretic concepts, it will be helpful to become familiar with a few important families of graphs. When confronted with new definitions and/or questions about graphs, it is helpful to have such examples on hand, both to help better understand the concepts, and also to run "experiments" in testing any new conjectures.

Some Important Families of Simple Graphs

DEFINITION 8.4: (*Some Important Families of Simple Graphs*)   (a)   For each positive integer n, the **complete graph on n vertices**, denoted as K_n, is the simple graph having n vertices, and (exactly one) edge between any pair of distinct vertices. Drawings of some complete graphs are shown in Figure 8.4.

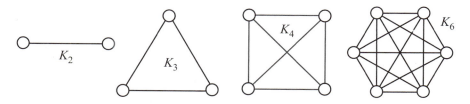

FIGURE 8.4: Drawings of a few complete graphs.

The complete graph K_n can be thought of as an n-person party in which everyone shakes hands with everyone else (the edges represent the handshakes).

(b)   For each positive integer $n \geq 2$, the **n-path graph**, denoted by P_n, is the simple graph defined by the following vertex and edge sets:

$$V(P_n) = \{v_1, v_2, \cdots, v_n\},$$
$$E(P_n) = \{\{v_1, v_2\}, \{v_2, v_3\}, \{v_3, v_4\}, \cdots, \{v_{n-1}, v_n\}\}.$$

The n-path graph is drawn in Figure 8.5.

FIGURE 8.5: A drawing of the n-path graph P_n.

(c)   For each positive integer $n \geq 3$, the **n-cycle graph**, denoted by C_n, is the simple graph defined by the following vertex and edge sets:

$$V(C_n) = \{v_1, v_2, \cdots, v_n\},$$
$$E(C_n) = \{\{v_1, v_2\}, \{v_2, v_3\}, \{v_3, v_4\}, \cdots, \{v_{n-1}, v_n\}, \{v_n, v_1\}\}.$$

Figure 8.6 shows drawings of some n-cycle graphs.

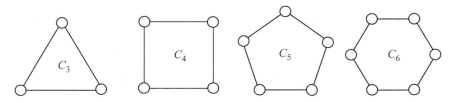

FIGURE 8.6: Drawings of a few n-cycle graphs.

We point out that $P_2 = K_2$ and $C_3 = K_3$.

(d)   For each positive integer $n \geq 1$, the **n-hypercube**, denoted by Q_n, is the simple graph having vertex set $V(Q_n) = \{\text{length } n \text{ bit strings}\}$, and having an edge between two of its vertices if, and only if, the corresponding bit strings differ by exactly one bit.

For a specific example, when $n = 2$, there are four length 2 bit strings: $V(Q_2) = \{00, 01, 10, 11\}$.   There are four pairs of these vertices that differ by exactly one bit: $E(Q_2) = \{\{00,01\}, \{01,11\}, \{11,10\}, \{10,00\}\}$.   Drawings of the first three hypercubes are shown in Figure 8.7.   We next show how the hypercubes can be constructed inductively.

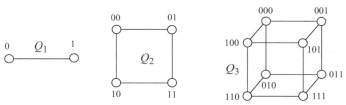

FIGURE 8.7: Drawings of the first three hypercubes.

Let n be a positive integer.   To construct Q_{n+1} from Q_n, we may proceed as follows: Construct two copies of the graph Q_n, with the vertices labeled as length n binary strings.   Next, on one of these two copies, add the prefix "0" to each vertex bit string, and on the other, add the prefix "1" to each vertex label.   Finally, add edges that join all corresponding pairs of vertices (that differ only in the first new bit).   The reader is encouraged to visually verify this construction when $n = 1$ and $n = 2$, by examining Figure 8.7.

EXERCISE FOR THE READER 8.2:   (a) Draw a picture of the 4-hypercube Q_4.

(b) Determine the number of vertices and edges in Q_n.

Bipartite Graphs

An important class of graphs are those for which all of vertices can be painted either "black" or "white" in such a way that adjacent vertices will never have same color. For example, if we consider the graph whose vertices are the boys and girls at a school dance, and whose edges link each boy to the girls whom he danced with during the evening, then this graph will satisfy this property if we paint the boy vertices "black" and the girl vertices "white." We enunciate first a formal definition for this class, and then give some more examples.

DEFINITION 8.5: A graph G is **bipartite** if the vertex set V can be partitioned into two subsets: $V = U \cup W$, such that each edge of G has one endpoint in U and one endpoint in W. The pair U, W is called a **(vertex) bipartition** of G.

So U could be the vertices that get painted "black" and W those that get painted "white," or vice-versa. In particular, if a graph is bipartite, any given vertex could be painted either "black" or "white," and the remaining vertices could be painted in an appropriate fashion.

EXAMPLE 8.4: (a) For which values of n is the n-cycle C_n bipartite?

(b) For which values of n is the complete graph K_n bipartite?

SOLUTION: Part (a): As we progress around any cycle, say clockwise, the bipartite condition forces us to alternate the colors as we paint the vertices encountered. This painting will all work fine, until perhaps when we get to painting the last vertex encountered. The color it gets painted (which is forced) must be different from the color that the first vertex was painted. This will happen exactly when n is even. This idea is illustrated in Figure 8.8, for the first few n-cycles.

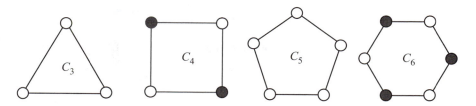

FIGURE 8.8: The n-cycle C_n is bipartite if, and only if n is even.

Part (b): Clearly, K_2 is bipartite. If n is at least 3, then K_n contains a copy of C_3, which we know from Part (a) is not bipartite.[2] It follows that K_n is also not

---

[2] This argument previews the concept of a *subgraph*. After we formally introduce subgraphs, this sort of argument can be made with fewer words.

bipartite. To make this more precise, let u, v, w be any three vertices of K_n. Since $\{u,v\}, \{v,w\}, \{w,u\}$ are edges of K_n, if K_n were bipartite, and if we color u "black," then v would have to be "white" (since $\{u,v\}$ is an edge), and then w would have to be "black" (since $\{v,w\}$ is an edge), and so it would follow that u would have to be "white" (since $\{w,u\}$ is an edge). This is a contradiction.

EXERCISE FOR THE READER 8.3: (a) For which values of n is the n-path P_n bipartite?
(b) For which values of n is the hypercube Q_n bipartite?

Since each edge of a graph poses a constraint for a bipartition (the joined vertices need to be painted different colors), it is generally true that more edges tend to make it more difficult for a graph to be bipartite. There are, however, simple families of so-called complete bipartite graphs that have, in one sense, the maximal set of edges. These are introduced in the following definition:

DEFINITION 8.6: For positive integers n and m, the **complete bipartite graph** $K_{n,m}$ is the simple graph defined by the following vertex and edge sets:

$$V(K_{n,m}) = \{u_1, u_2, \cdots, u_n\} \cup \{v_1, v_2, \cdots, v_m\}$$
$$E(K_{n,m}) = \{\{u_i, v_j\} : 1 \le i \le n, \, 1 \le j \le m\}.$$

Thus all u-vertices are adjacent to each of the v-vertices, but no two u-vertices are adjacent, nor are any two v-vertices. Figure 8.9 shows drawings of some of these graphs.

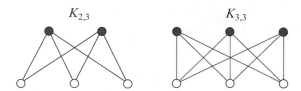

$K_{2,3}$ $K_{3,3}$

FIGURE 8.9: Drawings of the complete bipartite graphs (a) (left) $K_{2,3}$ and (b) (right) $K_{3,3}$ with the corresponding vertex bipartition being indicated by a "black" and "white" vertex coloring.

Degree Sequences

Given any graph, it is a simple matter to obtain the *degree sequence* $(d_1, d_2, \cdots, d_n)$ of its vertices $v_1, v_2, \cdots, v_n$ (i.e., $d_i = \deg(v_i)$). But what about the reverse problem: Given a sequence of nonnegative integers $(d_1, d_2, \cdots, d_n)$, is there always a graph that realizes this sequence as its degree sequence? By the handshaking theorem, this is impossible unless the sum $\sum_{i=1}^{n} d_i$ is even. It is not difficult to show that if $\sum_{i=1}^{n} d_i$ is even, then there always exists a general graph having $(d_1, d_2, \cdots, d_n)$ as its degree sequence; see Exercise 11. A more interesting problem asks to find a simple graph with a given degree sequence, and this motivates the following definition:

DEFINITION 8.7: A sequence of nonnegative integers $(d_1, d_2, \cdots, d_n)$ is called a **graphic** sequence if (some permutation of) it is the degree sequence of some simple graph G. Such a simple graph is said to **realize** the sequence.

For example, $(2, 2, 2)$ is graphic, since it is the degree sequence of K_3. Obviously, $(2, 0, 0)$ is not graphic. (Why?) We are going to develop a simple recursive algorithm that will determine whether a given sequence $(d_1, d_2, \cdots, d_n)$ is graphic. At the heart of the algorithm is the fact that if we have a graphic sequence, then we may assume that the vertex with the highest degree has as its neighbors all the vertices of the next highest degrees. This will allow us to recursively delete the vertex of highest degree of the given sequence, to obtain a new sequence of smaller total degree, where the latter sequence is graphic if, and only if the former is. Once the algorithm terminates, it will be clear whether the final sequence is graphic, and the original sequence will have the same property. Let us begin with the key theorem about graphic sequences. The theorem and the associated algorithm were discovered independently by Havel [Hav-55], and Hakimi [Hak-62]. Since the degree of any vertex in any n-vertex simple graph can be at most $n - 1$, in the statement of the theorem we specifically rule out this trivial sort of non-graphic sequence by assuming that $d_i < n$ for each index i.

THEOREM 8.3: (*Degree Sequences of Simple Graphs*) (a) Suppose that we have a graphic sequence $(d_1, d_2, \cdots, d_n)$, with $n > d_1 \geq d_2 \geq \cdots \geq d_n$. Then there is simple graph G with $V(G) = \{v_1, v_2, \cdots, v_n\}$, with $\deg(v_i) = d_i$ $(1 \leq i \leq n)$, and such that the neighbors of v_1 are $v_2, v_3, \cdots, v_{d_1+1}$.

(b) A length n sequence $(d_1, d_2, \cdots, d_n)$ of nonnegative integers with $n > d_1 \geq d_2 \geq \cdots \geq d_n$ is graphic if, and only if, the length $n - 1$ sequence $(d_2 - 1, d_3 - 1, \cdots, d_{d_1+1} - 1, d_{d_1+2}, \cdots, d_n)$ is graphic.

Proof: Part (a): Since the given sequence is graphic, there exist simple graphs that realize it. Among all such realizations, choose one, G, with the property that the number of neighbors N that v_1 has among the set $\{v_2, v_3, \cdots, v_{d_1+1}\}$ is as large as possible. Our aim is to show that $N = d_1$. We will assume that $N < d_1$, and proceed by the method of contradiction. Since $N < d_1$, there exists an index k, with $2 \le k \le d_1 + 1$, such that the vertex v_k is not adjacent to v_1, and (since $\deg(v_1) = d_1$) there exists an index $\ell > d_1 + 1$, such that the vertex v_ℓ is adjacent to v_1. Now, since $k < \ell$, v_k has at least as many neighbors as does v_ℓ, and since one of the latter's neighbors (v_1) is not a neighbor of the former, it follows that v_k has a neighbor v_j that is not among the neighbors of v_ℓ. We create a new graph G^* by performing some minor edge exchanges with the edges of G. More precisely, the graph G^* is defined by the following vertex and edge sets:

$$V(G^*) = V(G),$$
$$E(G^*) = E(G) \cup \{\{v_1, v_k\}, \{v_\ell, v_j\}\} \sim \{\{v_k, v_j\}, \{v_1, v_\ell\}\}.$$

The construction is illustrated in Figure 8.10.

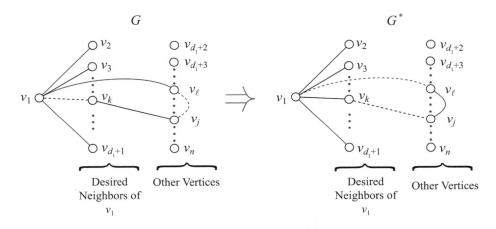

FIGURE 8.10: Illustration of the edge exchange in the proof of Theorem 8.3.

An important point of this construction is that the degree sequence of G^* is the same as that for G. Thus, the graph G^* is a realization of the degree sequence $(d_1, d_2, \cdots, d_n)$, with the property that v_1 has $N + 1$ neighbors among the set $\{v_2, v_3, \cdots, v_{d_1+1}\}$. This contradicts the maximality of N, and thus completes the proof of the first assertion.

Part (b): If the monotone decreasing sequence $(d_1, d_2, \cdots, d_n)$ is graphic, then by Part (a), there exists simple graph G, with $V(G) = \{v_1, v_2, \cdots, v_n\}$, satisfying $\deg(v_i) = d_i$ $(1 \le i \le n)$, and such that the neighbors of v_1 are $v_2, v_3, \cdots, v_{d_1+1}$. Consider the simple graph $\hat{G}$ obtained from G by deleting vertex v_1, along with all of the d_1 edges that are incident to v_1. The graph $\hat{G}$ will have degree sequence $(d_2 - 1, d_3 - 1, \cdots, d_{d_1+1} - 1, d_{d_1+2}, \cdots, d_n)$, as desired. Conversely, if we start off with a graphic sequence $(d_2 - 1, d_3 - 1, \cdots, d_{d_1+1} - 1, d_{d_1+2}, \cdots, d_n)$, we let $\hat{G}$

be a corresponding realizing graph, with its vertices labeled as $v_2, v_3, \cdots, v_n$. We create a new graph G from $\hat{G}$ by adding one new vertex v_1, along with all of the following d_1 edges $\{v_1, v_2\}, \{v_1, v_3\}, \cdots, \{v_1, v_{d_1+1}\}$. This graph is a realization of the degree sequence $(d_1, d_2, \cdots, d_n)$. □

Part (b) of the above theorem yields the following recursive algorithm:

ALGORITHM 8.1: (*Recursive Graphic Sequence Determination Algorithm*)
Input: A sequence $(d_1, d_2, \cdots, d_n)$ of nonnegative integers with $n > d_1 \geq d_2 \geq \cdots \geq d_n$.
Output: An answer "TRUE" or "FALSE," depending, respectively, on whether the given sequence is graphic or not.

WHILE $d_1 > 0$

    Let $(d_1', d_2', \cdots, d_{n-1}')$ be a monotone decreasing permutation of
    $(d_2 - 1, d_3 - 1, \cdots, d_{d_1+1} - 1, d_{d_1+2}, \cdots, d_n)$

    UPDATE $n = n - 1$, $(d_1, d_2, \cdots, d_n) = (d_1', d_2', \cdots, d_n')$
    IF any $d_i < 0$
        THEN Output "FALSE," EXIT
    ELSE IF $d_1 = 0$
        THEN Output "TRUE," EXIT
    END IF
END WHILE

In the course of the algorithm, if the sequences $(d_1, d_2, \cdots, d_n)$ are all recorded, they can be used (in the reverse order) to construct a realizing simple graph (in case the inputted sequence is graphic). This technique will be illustrated in the following example.

EXAMPLE 8.5: Apply Algorithm 8.1 to the sequence (4, 4, 3, 3, 2, 2). In case the sequence turns out to be graphic, use the sequences produced in the course of the algorithm to create a drawing of a realizing graph.

SOLUTION: For each iteration of the algorithm, we indicate the two resulting sequences: First (i) the sequence resulting from the operation $(d_1, d_2, \cdots, d_n) \rightarrow (d_2 - 1, d_3 - 1, \cdots, d_{d_1+1} - 1, d_{d_1+2}, \cdots, d_n)$, and then (ii) the monotone decreasing permutation (even when it is not needed).[3]

---

[3] Although we follow Algorithm 8.1 until it terminates, it follows from Theorem 8.3 that if, at any iteration, we can easily see that the resulting sequence is or is not graphic, the algorithm may be terminated at this step with the corresponding conclusion being made.

Initial Sequence: (4, 4, 3, 3, 2, 2)
Iteration #1: (i) (3, 2, 2, 1, 2), (ii) (3, 2, 2, 2, 1)
Iteration #2: (i) (1, 1, 1, 1), (ii) (1, 1, 1, 1)
Iteration #3: (i) (0, 1, 1), (ii) (1, 1, 0)
Iteration #4: (i) (0, 0), (ii) (0, 0) → "TRUE"

Thus the Algorithm 8.1 has determined the original sequence to be graphic. We now show how we can work backwards with the permuted sequences (ii) to draw a representing simple graph for this degree sequence. We start with the last permuted sequence of Iteration #4: (0,0). A realizing graph for this sequence is the graph consisting of two isolated vertices; see the leftmost graph of Figure 8.11. Next, we move on to the permuted sequence of Iteration #3: (1,1,0). To realize this sequence, we add one new vertex to the first graph, and one edge from it to either of the first two vertices; see the middle graph of Figure 8.11.[4] The third graph in Figure 8.11 is a realization of the sequence of Iteration #2: (1,1,1,1). The newly added vertex (d) needs to be connected to vertex b of the previous graph.

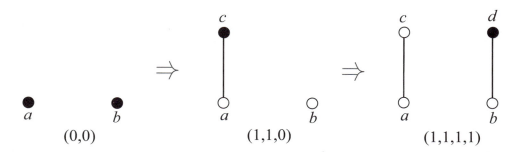

FIGURE 8.11: The first three steps in "growing" a realizing graph for the graphic sequence of Example 8.5. We start of with two isolated vertices (left), corresponding to the graphic sequence (0,0). At each step, one new vertex is adjoined to the existing graph so that the next sequence is realized. When new vertices are added, they are colored black, while carryover vertices from the previous graph are colored white.

The last two steps of this graph "growing" procedure are shown in Figure 8.12, with the final graph on the right being a realization of the original sequence.

---

[4] Students sometimes have difficulty getting used to this process, since there are many ways to place the new vertices as they arise, and often different choices for vertices to select as neighbors for the newly added vertices. As long as the sequences are correctly represented, any (correct) drawing will do just fine. A reasonable policy is to proceed in a way that makes the drawings as clean and simple as possible. One safe scheme is to place the n vertices according to the vertices of a regular n-gon.

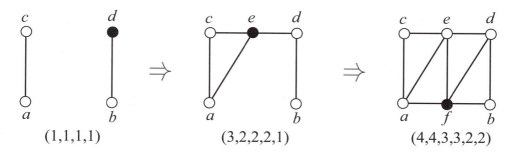

FIGURE 8.12: The last two steps of the "growing" of a realizing graph for the sequence of Example 8.5, continued from Figure 8.11.

EXERCISE FOR THE READER 8.4: Repeat the directions of Example 8.5 for the sequence (5,5,5,3,3,3,3,3).

EXERCISE FOR THE READER 8.5: In Algorithm 8.1, would it make any difference if we switched the order of the two conditions (and their consequences) inside the IF branch? Carefully justify your answer.

Subgraphs

Analogous to the concept of a subset in set theory is the concept of a subgraph in graph theory:

DEFINITION 8.8: Suppose that $G = (V, E)$ is a graph. Another graph $H = (V', E')$ is said to be a **subgraph** of G, if $V' \subseteq V$ and $E' \subseteq E$. More generally, we use this terminology, or simply say **G contains a copy of H** if the vertices of H can be relabeled with some of G's vertex labels in such a way that all of the edges of H are edges of G.

EXAMPLE 8.6: (a) Each of the first two graphs in either Figure 8.11 or Figure 8.12 is a subgraph of any graph to the right of it. Any of the three graphs in Figure 8.11 is a subgraph of any of the three graphs of Figure 8.12.
(b) The complete graph K_6 (shown in Figure 8.4) contains copies of K_1, K_2, K_3, K_4, K_5, and K_6 (a graph is always a subgraph of itself), and also has copies of C_1, C_2, C_3, C_4, C_5, and C_6. Any labeling of vertices of these subgraphs with the labeling of the vertices of K_6 will suffice to demonstrate these subgraph relations.

The next exercise for the reader will give the reader an opportunity to gain some familiarity with one of the single most important graphs in graph theory. The **Petersen graph**, shown in Figure 8.14, is named after the Danish mathematician

FIGURE 8.13: Julius Peter Christian Petersen (1839–1910), Danish mathematician.

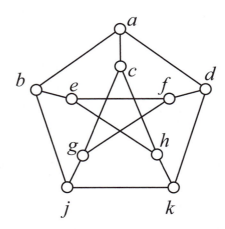

FIGURE 8.14: The Petersen graph.

Julius Petersen (Figure 8.13).[5] Despite its simplicity, the Petersen graph has an assortment of interesting features; it is often used in testing graph theoretic conjectures, and as a source of counterexamples. An entire book of over 350 pages has been written about the Petersen graph, see [HoSh-93].

EXERCISE FOR THE READER 8.6: Which of the following are subgraphs of the Petersen graph: $K_3, C_4, C_5, C_6, C_8, P_5, P_6$?

---

[5] Julius Petersen was brought up on very modest means, but managed to carve out a reputation as the founding father of Danish mathematics; his obituary garnered front page headlines in Copenhagen in when he was dubbed the "Hans Christian Andersen of science." As a schoolboy, he was notably talented in mathematics, and spent a great deal of time solving difficult problems (when his teachers ran out of new problems, he created his own). In 1854 at the age of 15, he had to drop out of school because of financial hardship, and took a job at his uncle's grocery. This uncle passed away about a year later and left Petersen with a modest inheritance that allowed him to pick up his studies again. He passed the Copenhagen College of Technology entrance examination and began his university studies there in 1856. Two years later he wrote his first book (on logarithms), which would turn out to be the first of an impressive lineup of scientific books, some of which were translated (from Danish) into several other languages and remained in print over 100 years after their release.

As his finances started to run out, Petersen took on work teaching at a prestigious private school from 1859 through 1871, where he taught six days a week for seven hours each day. He was recognized as a caring and effective teacher, although he had discipline problems with the school children. In 1862, he began his graduate studies in mathematics at the University of Copenhagen, and got married in the same year. Despite his arduous work schedule and his familial responsibilities, he completed his master's degree in 1866 and his doctorate in 1871. His doctoral thesis won him the university's Gold Medal award. He was immediately awarded a position at the College of Technology, and later became a professor at the University of Copenhagen. With the added time that his new and well-deserved positions afforded him, Petersen's productivity flourished. He put tremendous energy into his lectures, many of which evolved into significant textbooks on various subjects. His research work had a far-reaching span, including areas in algebra, mechanics, economics, cryptography, and, of course, graph theory. His first paper on graph theory was published in 1891, and gave the subject a tremendous impetus. He tended to treat problems that he found interesting with a fresh approach, avoiding reading other works on the subject.

Isomorphism of Simple Graphs

In the more relaxed version of the definition of a subgraph given in Definition 8.8, we allowed H to be a subgraph of G if the vertices of H could be relabeled with labels of some of the vertices of G in such a way that H could be "imbedded" in G. In case two graphs (with the same number of vertices) can be made to be the same by relabeling vertices of one of them, then the graphs are called *isomorphic*. In general, it is a very difficult (NP complete) problem to decide whether two graphs are isomorphic, or more generally, whether one graph is a subgraph of another graph. We next provide some more formal definitions leading to the concept of graph isomorphism in the setting of simple graphs.

DEFINTION 8.9: Suppose that $G = (V, E)$ and $G' = (V', E')$ are simple graphs, and that $f : V \to V'$ is a one-to-one (vertex) function.

(a) The function f is said to **preserve adjacency** if for any pair of vertices $u, v \in V$, we have $\{u, v\} \in E \implies \{f(u), f(v)\} \in E'$. In other words, if u and v are neighbors in G, then their images $f(u)$ and $f(v)$ must be neighbors in G'.

(b) The function f is said to **preserve non-adjacency** if for any pair of vertices $u, v \in V$, we have $\{u, v\} \notin E \implies \{f(u), f(v)\} \notin E'$. In other words, if u and v are not neighbors in G, then their images $f(u)$ and $f(v)$ must not be neighbors in G'.

(c) The function f is said to be a **graph isomorphism** from G to G' if it is bijective, and preserves both adjacency and non-adjacency. In this case we say that the graphs G and G' are **isomorphic**, and write this as $G \cong G'$. If no such isomorphism exists, we say that G and G' are not isomorphic and write $G \not\cong G'$.

NOTE: Although it is technically incorrect, we sometimes adopt the common practice of denoting a vertex bijection in any of the above definitions as a function between the graphs, and write it as $f : G \to G'$. The concept of isomorphism can be extended to general graphs; see Exercises 41–44 and the note and definition immediately preceding them; see also [GrYe-06].

We first give a very simple example:

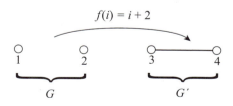

$f(i) = i + 2$

1 2 3 4

G G'

FIGURE 8.15: The vertex function of Example 8.7.

EXAMPLE 8.7: Consider the function $f : G \to G'$ from the graph G consisting of two isolated vertices, and the graph $G' = K_2$, which is defined in Figure 8.15. Indicate whether this function preserves adjacency and whether it preserves non-adjacency. Is it a graph isomorphism?

SOLUTION: Since G has no edges (and hence no adjacencies to preserve) the adjacency preserving condition is vacuously true, so f preserves adjacency. But f does not preserve non-adjacency: vertices 1 and 2 are not adjacent in G, but their images $f(1) = 3$, $f(2) = 4$ are adjacent in G'. Therefore f is not a graph isomorphism.

EXERCISE FOR THE READER 8.7: Suppose that G and G' are simple graphs. Show that if $f : G \rightarrow G'$ is a graph isomorphism, then:

(a) $\deg_G(v) = \deg_{G'}(f(v))$, for each $v \in V(G)$, and

(b) G and G' must have the same degree sequences.

As alluded to near the beginning of the chapter, isomorphic graphs are really the same objects as far as graph theory is concerned. The fact that two graphs are isomorphic may be masked by different drawings and/or labelings of vertices.

EXAMPLE 8.8: For each part, determine whether the given pair of graphs is isomorphic.
(a)   The two graphs shown in Figure 8.16.
(b)   The two graphs shown in Figure 8.17.

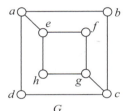

FIGURE 8.16: Graphs for isomorphism question of Example 8.8(a).

SOLUTION: Part (a): Although the two graphs have the same degree sequences, no vertices of degree 2 are adjacent in G, whereas H has two pairs of degree 2 vertices that are adjacent (t and s, and x and w). It follows that G and H cannot be isomorphic.

Part (b): One way to show that two graphs are isomorphic is to "distort" one or both of the graphs so that their equivalence becomes clear. Figure 8.18 shows a two-part distortion of the graph H that will make it clear how to define an isomorphism $f : G \rightarrow H$.

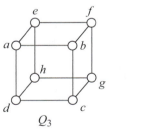
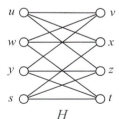

FIGURE 8.17: Graphs for isomorphism question of Example 8.8(b).

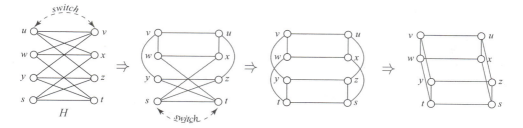

FIGURE 8.18: A distortion of the graph of Figure 8.17 into a graph that is clearly the hypercube Q_3.

Comparing the final distorted graph of Figure 8.18, with the labeled graph of Q_3 of Figure 8.17, an isomorphism easily presents itself: Define $F : G \to H$ by mapping the vertices e, f, a, b, h, g, d, c of Q_3 to the vertices v, u, y, z, w, z, t, s of H, respectively. Although there are many other isomorphisms between these two graphs, they make up a rather small percentage of the vertex bijections, as the reader will be asked to demonstrate in the next exercise for the reader.

EXERCISE FOR THE READER 8.8: Show that roughly 0.1% of the vertex bijections between the graphs Q_3 and H of Example 8.8(b) are graph isomorphisms.
Suggestion: Use the multiplication principle to count the number of isomorphisms.[6]

EXERCISE FOR THE READER 8.9: Show that the isomorphism relation $G \cong H$ is an equivalence relation on the set of all simple graphs.
Remark: This fact allows us to view the set of all simple graphs as broken up into nonisomorphic types (the different equivalence classes), and gives further justification to the fact mentioned earlier that any drawing of a graph is by no means a unique way of representing it.

EXERCISE FOR THE READER 8.10: (a) Up to isomorphism, how many different three vertex simple graphs are there? Draw representatives for each of them.
(b) Repeat Part (a) for four-vertex simple graphs.

---

[6] Since Q_3 and H are, as far as graph theory is concerned, the same object, the number of isomorphisms between them is the same as the number of graph isomorphisms from Q_3 to itself. An isomorphism from a graph to itself is called an *automorphism*. Since Q_3 has a great deal of symmetry, it has a lot more automorphisms than is typical. In particular, the 0.1% figure (for the percentage of vertex bijections that correspond to automorphisms) above is rather high for an eight-vertex graph.

In order for two graphs G and G' to be isomorphic, it is necessary, first of all, that they have the same number of vertices, and also have the same degree sequences. (Thus, regular graphs tend to be the most difficult ones for isomorphism questions.)   More generally, since a graph isomorphism preserves adjacency, if one of the graphs has a certain graph theoretic property that the other one does not have, then the graphs are not isomorphic. Apart from checking degree sequences, another common method to show that graphs are not isomorphic is to show that one contains a certain number of copies of a particular subgraph and the other does not.   Some other graph theoretic invariants that can be used to detect nonisomorphism will be introduced in the next section. There are, unfortunately, no known necessary and sufficient conditions to determine whether two graphs are isomorphic. The only way to show that two graphs are isomorphic is to provide a graph isomorphism between them. Since for graphs with n vertices, there are $n!$ different vertex bijections between them, and it takes some time to determine whether a given vertex bijection is an isomorphism, a brute-force search for a graph isomorphism is hopelessly impractical for all but the smallest values of n. Before moving on, we will point out one more method that is sometimes useful to decide whether two simple graphs are isomorphic.

The Complement of a Simple Graph

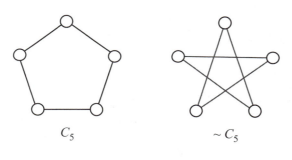

DEFINITION 8.10: If $G = (V, E)$ is a simple graph, then the **complement** of G, denoted by $\sim G$, is the simple graph defined by the following vertex and edge sets:

$$V(\sim G) = V, \ E(\sim G) = \sim E,$$

where $\sim E$ denotes all of the 2-combinations of vertices of V that are not in E.

FIGURE 8.19: The 5-cycle graph C_5 and its complement $\sim C_5$.

Figure 8.19 shows an example of a graph and its complement.

EXERCISE FOR THE READER 8.11:   Suppose that G and H are simple graphs. Show that a vertex bijection $f : G \to H$ is an isomorphism if, and only if the same vertex mapping is an isomorphism between the complement graphs $f : \sim G \to \sim H$.   In particular, $G \cong H$ if, and only if $\sim G \cong \sim H$.

In a graph isomorphism problem in which two graphs have a lot of edges (compared to their complements), it is usually easier to look at the complements.

Representing Graphs on Computers

As we have indicated, graphs have numerous applications, and we will be developing several algorithms to solve some important applied problems about graphs. In order to be able to use computers to assist in executing such algorithms, we need some ways to represent graphs that can be easily inputted into computers. (Although graph drawings are useful for human eyes, these are inadequate for efficient computer implementations.) We next indicate some effective ways to represent graphs using matrices.

The most common way to represent graphs is using so-called *adjacency matrices*. These matrices are always square matrices, and so are easily manipulated on computers.

DEFINITION 8.11: Given an n-vertex graph G, with some specified ordering of the vertices, the corresponding **adjacency matrix** is the $n \times n$ matrix A (or $A(G)$ to indicate the dependence on G) whose (i,j) entry a_{ij} is the number of edges in $E(G)$ that join the ith vertex and the jth vertex.

Thus, the entries in any adjacency matrix are nonnegative integers. If G is a simple graph, the entries can be only zeros and ones, and the diagonal entries must all be zeros. (Why?)

EXAMPLE 8.9: (a) Write down the adjacency matrix for the general graph shown in Figure 8.3 using the following ordering of the vertices: u, v, w, x.
(b) Create a drawing for the simple graph that is represented by the following adjacency matrix:

$$A = \begin{bmatrix} 0 & 1 & 1 & 0 \\ 1 & 0 & 0 & 1 \\ 1 & 0 & 0 & 1 \\ 0 & 1 & 1 & 0 \end{bmatrix}.$$

SOLUTION: Part (a): The general graph and the corresponding adjacency matrix are shown in Figure 8.20. We point out that the edge labels (given in Figure 8.3) are irrelevant for the adjacency matrix.

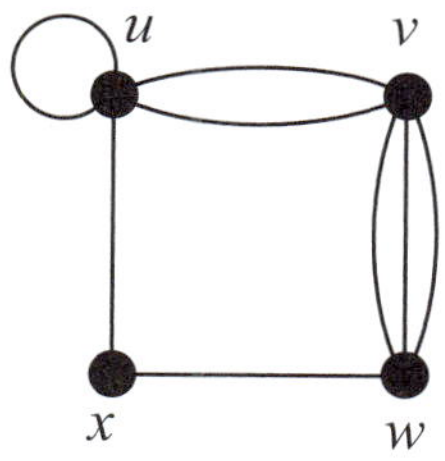

$$A = \begin{array}{c} \\ u \\ v \\ w \\ x \end{array} \begin{array}{cccc} u & v & w & x \\ \begin{bmatrix} 1 & 2 & 0 & 1 \\ 2 & 0 & 3 & 0 \\ 0 & 3 & 0 & 1 \\ 1 & 0 & 1 & 0 \end{bmatrix} \end{array}$$

FIGURE 8.20: A general graph and the adjacency matrix corresponding to the vertex ordering u, v, w, x.

Part (b): Labeling the vertices as 1, 2, 3, 4, a drawing of the graph, which is just C_4, is shown in Figure 8.21 along with its adjacency matrix.

$$A = \begin{bmatrix} 0 & 1 & 1 & 0 \\ 1 & 0 & 0 & 1 \\ 1 & 0 & 0 & 1 \\ 0 & 1 & 1 & 0 \end{bmatrix}$$

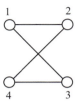

FIGURE 8.21: The adjacency matrix of Example 8.9(b), along with a drawing of the corresponding graph.

EXERCISE FOR THE READER 8.12: Create a drawing for a general graph that has the following adjacency matrix:

$$A = \begin{bmatrix} 0 & 1 & 2 & 3 \\ 1 & 1 & 2 & 3 \\ 2 & 2 & 2 & 4 \\ 3 & 3 & 4 & 3 \end{bmatrix}.$$

We emphasize that adjacency matrices depend on the ordering of the vertices, and so are not unique. Since adjacency matrices contain all of the information about the structure of a graph, it follows that two graphs are isomorphic, if, and only if, the adjacency matrix of one can be made into the adjacency matrix of the other by applying some permutation to (both) the rows and the columns. Although such an isomorphism testing algorithm can be easily coded into a computer program (see Computer Exercises 16–17), it amounts to nothing more than a brute-force algorithm and is by no means a practical algorithm for even modest-sized graphs.

The next definition provides another matrix representation for graphs, where the rows still represent the vertices, but the columns now correspond to the edges of the graph.

DEFINITION 8.12: Given a graph G, with some specified ordering of its n vertices, and of its m edges, the corresponding **incidence matrix** is the $n \times m$ matrix B (or $B(G)$ to indicate the dependence on G) whose (i,j) entry b_{ij} is 1 if the ith vertex is incident to the jth edge (which is not a self-loop), $b_{ij} = 2$ if the jth edge is a self-loop incident on the ith vertex, and otherwise $b_{ij} = 0$.

NOTE: In the case of a simple graph with a specified ordering of the vertices, it is sometimes convenient to use the **lexicographic order** on the edges. In the lexicographic order (introduced in Section 2.2), the edges are written so that the vertex with the smaller index is always listed first. The edges are ordered by first comparing the first vertex, and in cases of a tie, then looking at the second vertices. For example, the lexicographic ordering of the six edges of a complete graph on the four vertices: a, b, c, d (ordered alphabetically) would be:

$$\{a,b\}, \{a,c\}, \{a,d\}, \{b,c\}, \{b,d\}, \{c,d\}.$$

EXAMPLE 8.10: (a) Write down the incidence matrix for the general graph shown in Figure 8.3 using the following ordering of the vertices: u, v, w, x.
(b) Write down the incidence matrix for the edge-labeled graph shown in Figure 8.22 using the alphabetical ordering of the vertices.
(c) Write down the incidence matrix for the graph of Figure 8.22 using the alphabetical ordering on the vertices and the induced lexicographic orderering on the edges.

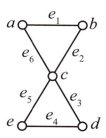

FIGURE 8.22: The edge-labeled graph for Example 8.10(b).

SOLUTION: Part (a): The multigraph and the corresponding incidence matrix are shown in Figure 8.23.

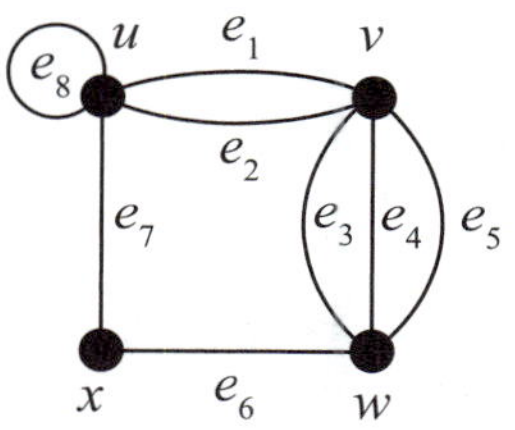

$$B = \begin{array}{c} \\ u \\ v \\ w \\ x \end{array} \begin{array}{cccccccc} e_1 & e_2 & e_3 & e_4 & e_5 & e_6 & e_7 & e_8 \\ \left[\begin{array}{cccccccc} 1 & 1 & 0 & 0 & 0 & 0 & 1 & 2 \\ 1 & 1 & 1 & 1 & 1 & 0 & 0 & 0 \\ 0 & 0 & 1 & 1 & 1 & 1 & 0 & 0 \\ 0 & 0 & 0 & 0 & 0 & 1 & 1 & 0 \end{array}\right] \end{array}$$

FIGURE 8.23: The graph of Example 8.10(a) and its incidence matrix corresponding to the vertex ordering u, v, w, x.

Part (b): The incidence matrix is:

$$B = \begin{array}{c} \\ a \\ b \\ c \\ d \\ e \end{array} \begin{array}{cccccc} e_1 & e_2 & e_3 & e_4 & e_5 & e_6 \\ \left[\begin{array}{cccccc} 1 & 0 & 0 & 0 & 0 & 1 \\ 1 & 1 & 0 & 0 & 0 & 0 \\ 0 & 1 & 1 & 0 & 1 & 1 \\ 0 & 0 & 1 & 1 & 0 & 0 \\ 0 & 0 & 0 & 1 & 1 & 0 \end{array}\right] \end{array}$$

Part (c): The edges are now ordered as: $\{a,b\} = e_1$, $\{a,c\} = e_6$, $\{b,c\} = e_2, \{c,d\} = e_3, \{c,e\} = e_5, \{d,e\} = e_4$. The resulting incidence matrix, which can be obtained by appropriately permuting the columns of the matrix in the solution of Part (b) shown below:

$$
\begin{array}{c}
\quad\; e_1 \; e_6 \; e_2 \; e_3 \; e_5 \; e_4 \\
\begin{array}{c} a \\ b \\ c \\ d \\ e \end{array}
\left[
\begin{array}{cccccc}
1 & 1 & 0 & 0 & 0 & 0 \\
1 & 0 & 1 & 0 & 0 & 0 \\
0 & 1 & 1 & 1 & 1 & 0 \\
0 & 0 & 0 & 1 & 0 & 1 \\
0 & 0 & 0 & 0 & 1 & 1
\end{array}
\right]
\end{array}
$$

We point out some properties of incidence matrices for any general graph:

(1) All entries are zeros, ones, or twos.
(2) Each column will have either two 1s or a single 2, in which case the corresponding edge is a self-loop, and the rest of the colum entries are 0s.
(3) The degree of a vertex of the graph equals the sum of the entries of the corresponding row of the incidence matrix.
(4) Any matrix with properties (1) and (2) is the incidence matrix of some graph.

For simple graphs with a large number of vertices and a moderate number of edges,[7] the incidence and adjacency matrices can get quite large and require a lot of space to store primarily zero entries. It such cases, it may be more efficient to code the graph using an *adjacency list* or an *edge list*, as described in the following definition.

DEFINITION 8.13: If G is a simple graph, the **adjacency list** of G, with respect to some ordering of its n vertices, is an n-row list whose rows are indexed by the vertices (in order), and in which each row lists (in order) all other vertices that have the row index vertex as a neighbor. The **edge list** of G, with respect to some ordering of its n vertices, is an $m \times 2$ matrix of all of the edges of $E(G)$ (where $m = |E(G)|$), listed in lexicographic order, where each edge is listed only once (with the smaller indexed vertex listed first).

EXAMPLE 8.11:   Write down the adjacency list and the edge list for the simple graph shown in Figure 8.22 using the alphabetical ordering of the vertices. (The edge labels can be ignored here.)

SOLUTION: Figure 8.24 shows the graph along with the adjacency and edge lists. Of course, for such a small example, the potential economy of storage is not realized.

---

[7] An n-vertex simple graph can have at most $C(n,2) = n(n-1)/2 = \Theta(n^2)$ edges. So if such a graph had $O(n)$ edges (which we consider to be moderate), the incidence matrix would only have $O(n)$ nonzero entries.

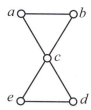

Edge List:	
a	b
a	c
b	c
c	d
c	e
d	e

Adjacency List:	
Vertex:	Neighbors:
a	b, c
b	a, c
c	a, b, d, e
d	c, e
e	c, d

FIGURE 8.24: The adjacency and edge lists representing the simple graph shown, according to the alphabetical ordering on the vertices.

We point out that although the vertex set is explicit in adjacency lists, it may be ambiguous in edge lists in case there are isolated vertices. Thus, when specifying an edge list for any graph with isolated vertices, it is necessary to state the vertex list as well. In the computer exercises, vertex sets will always be assumed to be of the form $\{1, 2, \ldots, n\}$, so it will suffice to simply specify the number of vertices n along with an edge list.

Directed Graphs (Digraphs)

Having gained some familiarity with general graphs, we next move on to describe directed graphs, in which edges are allowed to have associated directions. After giving the definition, we will briefly summarize how some of the concepts and results for graphs can be adapted for directed graphs. The definition will be analogous to Definition 8.2 for general graphs (in order to allow for parallel edges); but here the edgemap will have codomain $V \times V$, since each directed edge will be characterized by an <u>ordered</u> pair of vertices (unlike for a general graph where an edge was characterized by an unordered set of one or two vertices).

DEFINITION 8.14: A **directed graph** (or **digraph**) is a triple $\vec{G} = (V, E, \Delta)$, where V is a nonempty set of **vertices** (or **nodes**) of $\vec{G}$, E is the set of (**directed**) **edges** of $\vec{G}$, and Δ, called the **edgemap**, is a function $\Delta : E \to V \times V$. If $\Delta(e) = (u, v)$, then vertex u is called the **tail** and vertex v is called the **head** of edge e. We also say that the edge e is **directed from u to v**. If $u = v$, then the edge e is called a **directed self-loop**. Two edges, e_1, e_2, with the same heads and tails (i.e., $\Delta(e_1) = \Delta(e_2)$) are called **parallel edges**.

A directed graph can be drawn in a similar fashion to how graphs are drawn, except that an arrow is added to each directed edge from the tail vertex to the head vertex.[8] We point out that although directions on self-loops can be drawn with

---

[8] Although it may seem more general to allow some edges to be undirected (for example, think of a network of city streets in which some are one-way streets and others are two-way), such undirected edges can be accounted for by adding two oppositely directed edges (with the same vertices) in place of each undirected edge.

either a clockwise or with a counterclockwise sense, self-loops can have only one direction (since they have the same head and tail vertex). Two directed graphs are shown in Figure 8.25.

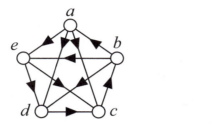 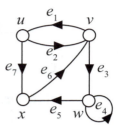

FIGURE 8.25: Two directed graphs. (a) (left) A tournament, (b) (right) A directed graph with two parallel edges and one self-loop, with edge labels.

DEFINITION 8.15:   The **underlying graph** of a digraph $\vec{G} = (V, E, \Delta)$ is the graph G obtained from $\vec{G}$ by removing all directions from the edges. More precisely, $G = (V, E, \Phi)$, where the (undirected) edgemap Φ is defined from the (directed) edgemap Δ by: $\Delta(e) = (u, v) \Rightarrow \Phi(e) = \{u, v\}$.   A **simple digraph** is a digraph whose underlying graph is a simple graph.

When digraphs are presented by drawings, edgemaps are rarely used and, as with simple graphs, they are redundant for simple digraphs (as long as the directions on the edges are specified in some way).

EXERCISE FOR THE READER 8.13: Write out the edgemap Δ for the digraph specified in Figure 8.25(b).

As there were several important families of graphs, digraphs also have some special varieties that have been studied extensively. The next definition describes one such type of digraph.

DEFINITION 8.16:   A **tournament** is a simple directed graph whose underlying graph is a complete graph.

The digraph of Figure 8.25(a) is a tournament on five vertices, since the underlying graph is K_5. A tournament on n vertices can be thought of as the results of some match (say a tennis match) where the n vertices represent n players who entered in the tournament, and each pair of players played exactly one match. An edge directed from one vertex to another means the tail vertex beat the head vertex. Thus, for the tournament in Figure 8.25(a), player a lost to player b, but beat each of the other players.

Degrees of vertices in directed graphs are separated into those leaving a vertex and those entering a vertex.

DEFINITION 8.17:   Given a directed graph $\vec{G} = (V, E, \Delta)$, and a vertex $v \in V$,

1.   The **indegree** of v, denoted $\deg^-(v)$, is the number of directed edges having v as the head (i.e., that "enter into" the vertex v).
2.   The **outdegree** of v, denoted $\deg^+(v)$, is the number of directed edges having v as the tail (i.e., that "go out from" the vertex v).

Note that directed self-loop at vertex v contributes one to $\deg^-(v)$ and to $\deg^+(v)$. The dependence on $\vec{G}$ can be emphasized by denoting the indegree and outdegree as $\deg_{\vec{G}}^-(v)$, and $\deg_{\vec{G}}^+(v)$, respectively.

Since each (directed) edge of a digraph contributes exactly one to both $\deg^-(v)$ and $\deg^+(v)$, the following digraph version of the handshaking theorem follows:

THEOREM 8.4 (*The Handshaking Theorem for Digraphs*):   For any digraph $\vec{G}$, we have

$$\sum_{v \in V(\vec{G})} \deg_{\vec{G}}^-(v) = \sum_{v \in V(\vec{G})} \deg_{\vec{G}}^+(v) = \left| E(\vec{G}) \right| \qquad (3)$$

EXERCISE FOR THE READER 8.14:   Compute the indegrees and outdegees of each of the vertices of the digraph of Figure 8.25(b), and verify that identity (3) is valid.

It is easy to extend the concept of isomorphism of graphs to directed graphs. Since we have restricted ourselves to simple graph isomorphisms in the text, we will restrict our focus in the following definition to simple digraphs.[9]

DEFINITION 8.18:   Two simple digraphs $\vec{G}$ and $\vec{H}$ are **isomorphic**, if there exists an isomorphism between the underlying graphs $f : G \to H$ that also preserves edge directions. That is, if $e = (v, u)$ is a (directed) edge in $E(\vec{G})$, then $(f(v), f(u))$ is the corresponding directed edge in $E(\vec{H})$.

Subdigraphs are defined analogously to subgraphs (Definition 8.8), with the additional assumption that edge directions must be preserved. As was the situation

---

[9] As we mentioned earlier, general graph isomorphisms will be defined in the exercises, and with this definition, the following definition for simple digraph isomorphisms will work equally well for general digraph isomorphisms.

for graphs, the *isomorphism problem* for digraphs (determine whether two digraphs are isomorphic) is equally difficult, and there is no known practical algorithm to solve it. A common way to proceed is to compare both digraphs to look for some graph theoretic property that is not shared by both, and this will rule out an isomorphism. There is no magic list of such properties that will provide a guarantee that the graphs are not isomorphic, but one should only attempt to find an isomorphism after thoroughly checking through all of the easily computed invariants (in/out degree sequences, subdigraph containments, etc.).

EXAMPLE 8.12: Determine whether the digraphs shown in Figure 8.26 are isomorphic.

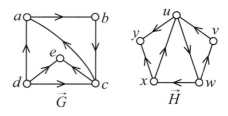

SOLUTION: These digraphs are not isomorphic. One way to see this is note that $\vec{G}$ has a vertex d with $\deg_{\vec{G}}^-(d) = 0$,

whereas $\vec{H}$ has no such vertex.

FIGURE 8.26: The digraph isomorphism question of Example 8.12.

EXERCISE FOR THE READER 8.15: Determine whether the underlying graphs of the two digraphs of Figure 8.26 are isomorphic.

EXERCISE FOR THE READER 8.16: Is it possible to assign directions on the two remaining edges on the two graphs in Figure 8.27, so that the two resulting digraphs will be isomorphic (as digraphs). If yes, place appropriate arrows on the two diagonal edges in each graph and explain why the two

FIGURE 8.27: Two partially defined digraphs for Exercise for the Reader 8.16.

digraphs are isomorphic. Otherwise, explain why this is not possible.

Adjacency and incidence matrices, as well as adjacency and edge lists can be easily adapted to represent digraphs. Here is a summary of the modifications that are needed in the corresponding definitions for graphs:

For the representation of general digraphs:
- Adjacency matrices: As in Definition 8.11, but the (i,j) entry a_{ij} of the adjacency matrix is now the number of <u>directed</u> edges in $E(G)$ that are directed from the ith vertex and the jth vertex.
- Incidence matrices: As in Definition 8.12, but the (i,j) entry b_{ij} of the incidence matrix is -1 if the ith vertex is the tail of the jth edge, 1 if it is a head, and (as before) 2 if the jth edge is a self-loop incident on the ith vertex, and otherwise $b_{ij} = 0$.

For the representation of simple digraphs:

- Adjacency list: As in Definition 8.13, but each row lists all other vertices that are a head of an edge having the row vertex as its tail.
- Edge list: As in Definition 8.12, but edges are listed as the ordered pairs that represent them (listed in lexicographic order).

EXERCISE FOR THE READER 8.17: (a) Find the adjacency matrix of the tournament of Figure 8.25(a) (using lexicographic order on the vertices).
(b) Find the incidence matrix of the digraph of Figure 8.25(b) (using lexicographic order on the vertices).
(c) Find the edge list of the simple digraph $\vec{G}$ of Figure 8.26(a).

Some Graph Models for Optimization Problems

As we indicated at the outset of this chapter, because of their versatility, graphs have been successfully applied to model numerous problems in mathematics, computer science, and an assortment of other scientific fields. Much of this modeling has led to the discovery of novel graph algorithms that have become the state-of-the-art methodology in numerous problem settings. We will close this section by developing a small sampling of some graph models for optimization problems. *Optimization problems* in discrete structures form one of the most important areas of applications. Basically, an optimization problem asks to find a certain discrete structure among a (usually extremely large) collection that has the maximum or minimum value of some quantity under consideration.

EXAMPLE 8.13: (*Some Graph Models*)
(a) (*Shortest Paths in a Network*)
Consider a network of cities that are linked by certain available air routes (e.g., that for a shipping company or commercial airline flights). A very small example of such a network is shown in Figure 8.28 by a digraph, where a bidirectional edge between a pair of cities represents two oppositely directed edges. In such an application, the edges have weights (not shown in the figure) that represent costs, distances, times, or some other commodity that we are interested in minimizing. Given an ordered pair of cities in the network

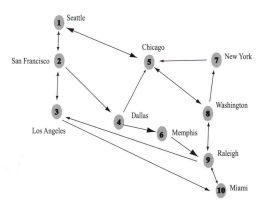

FIGURE 8.28: A digraph model for the available flights in a small network of cities.

(say San Francisco, CA and Raleigh, NC in Figure 8.28), the problem is to find a path (or route) going from the first city to the second city that has the smallest possible total cost (obtained by adding the costs of the segments in the route). In

Section 8.3, we will develop an efficient algorithm to solve such problems. Such programs are used, for example, in airfare search engines when you request the lowest priced airfare for a certain itinerary. Such websites deal with very large networks involving several competing airlines, and the price matrix can vary by day.

(b) (*Maximum Matchings in Graphs*) A *matching* in a graph is any set of edges, such that no two of which share a common vertex. The *maximum matching problem* asks to find a matching of maximum possible size in a given graph. The problem has numerous applications, and we will develop an efficient algorithm to solve it in this chapter. One such application is to the *job assignment problem*: Given a set of jobs: $J_1, J_2, \cdots, J_M$, a set of people (or machines) $P_1, P_2, \cdots, P_N$, that can perform some of these jobs, along with the list of jobs that each P_i is capable of performing, the problem is to assign certain people to certain jobs for which they are qualified in such a way that each job gets assigned to at most one person, and the maximum possible number of jobs has been assigned. This problem is easily modeled as a maximum matching problem on a bipartite graph G, whose vertices are $J_1, J_2, \cdots, J_M$ and $P_1, P_2, \cdots, P_N$, and G will have an edge between P_i and J_j if, and only if person P_i is qualified to do job J_j. See Figure 8.29.

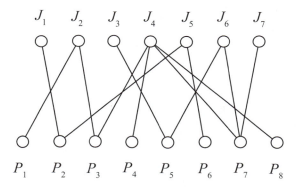

FIGURE 8.29: A job assignment graph for Example 8.13(b).

(c) (*The Traveling Salesman Problem*) There are many pairs of graph theoretic problems that appear somewhat similar in difficulty, and for which efficient algorithms are known for one, but not for the other. The *traveling salesman problem* involves a weighted complete graph. Traditionally, the vertices represent a collection of cities that the salesman must visit, and the weights on the edges represent distances. The problem is to find a permutation of the cities, such that if the salesman visits these cities in this order, and then returns to his home city, the total distance traveled will be a minimum.

The traveling salesman problem has numerous applications, such as deciding on an optimal drill sequence in the production of microchips (where millions of precision holes need to be drilled in a silicon wafer, and it is desired to minimize the time that the drill needs to travel between the holes that it drills). Although this problem may not seem to be much more difficult than the shortest path problem in Part (a), it is one of the most famously difficult problems in combinatorial optimization. Its difficulty and applications have motivated much research and spurred the discovery of new methods that have widespread applications. The traveling saleman problem is known to be NP complete, so the general consensus is that an efficient algorithm for it does not exist.

EXERCISES 8.1:

1. For each of the four simple graphs shown in Figure 8.30, do the following:
   (a) List the vertices.
   (b) List the edges.
   (c) Find the degrees of all vertices.
   (d) Verify the validity of the handshaking theorem.

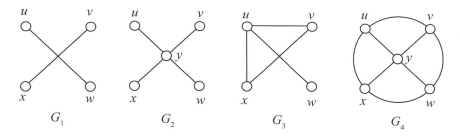

FIGURE 8.30: Four simple graphs for Exercise 1.

2. Repeat each part of Exercise 1 for the following simple graphs:
   (i) The bipartite graph $K_{2,3}$ (see Figure 8.9 (a)).

   (ii) The Petersen graph (see Figure 8.14).
   (iii) The bowtie graph of Figure 8.22.
   (iv) The job assignment graph of Figure 8.29.

3. For each of the four general graphs shown in Figure 8.31, do the following:
   (a) List the vertices.
   (b) List the edges.
   (c) Find the degrees of all vertices.
   (d) Verify the validity of the handshaking theorem.
   (e) Label the edges, and then write down the edgemap function Φ.

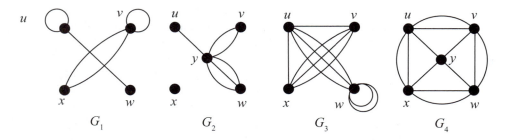

FIGURE 8.31: Four simple graphs for Exercise 3.

4.   Repeat each part of Exercise 3 for each of the two eight-vertex general graphs that are shown in Figure 8.32.

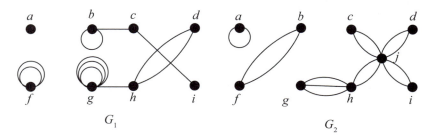

FIGURE 8.32: Two eight-vertex general graphs for Exercise 4.

5.   Draw a simple graph with the following degree sequence, and indicate whether the graph belongs to one of the special families introduced in this section.   Try to do each of these without using Algorithm 8.1.

    (a) (2, 2, 2, 2, 2)   (b) (2, 2, 2, 1, 1)
    (c) (2, 2, 1, 1, 1,1)   (d) (3, 3, 3, 3)

6.   Draw a simple graph with the following degree sequence, and indicate whether the graph belongs to one of the special families introduced in this section.   Try to do each of these without using Algorithm 8.1.

    (a) (2, 2, 2, 0, 0)   (b) (3, 3, 2, 2, 2)
    (c) (1, 1, 1, 1, 1,1)   (d) (4, 3, 2, 1, 1)

7.   For each sequence below, determine whether it is graphic. For those that are graphic, draw a simple graph that realizes it; for those that are not graphic, explain why (either using Algorithm 8.1 or some other justification).

    (a) (4, 4, 3, 3, 2, 1)   (b) (6, 5, 4, 3, 2,1)
    (c) (5, 5, 3, 3, 3, 3)   (d) (4, 4, 4, 4, 1, 1)

8.   For each sequence below, determine whether it is graphic. For those that are graphic, draw a simple graph that realizes it; for those that are not graphic, explain why (either using Algorithm 8.1 or some other justification).

    (a) (4, 4, 4, 3, 2, 1)   (b) (5, 5, 5, 5, 5, 1, 1)
    (c) (4, 4, 4, 4, 4, 4)   (d) (5, 5, 4, 4, 1, 1)

9.   (a) Is it possible that in a party of seven people, everyone shakes hands with exactly three other people?

(b) Prove that if there are an even number $n > 2$ of people at a party, then it is always possible for everyone to shake hands with exactly three other people.

Suggestion for Part (b): To construct a 3-regular graph on n-vertices, line the vertices around a circle, and start with the n-cycle graph. Add some appropriate edges to this graph to obtain the desired graph.

10. (a) Is it possible that in a party of seven people, everyone shakes hands with exactly four other people?

(b) Prove that if there are $n > 4$ people at a party, then it is always possible for everyone to shake hands with exactly four other people.

Suggestion for Part (b): Follow the suggestion given in Exercise 9.

11. (*Realization of Sequences with General Graphs*) As indicated in the text, as long as a sequence of nonnegative integers satisfies the necessary consequence of the handshaking lemma that the number of vertices of odd degree is even (Corollary 8.2), there is always a general graph that will realize it. The following simple construction for drawing such a general graph will demonstate this assertion, assuming we have such an admissible degree sequence $(d_1, d_2, \cdots, d_n)$.

Step 1: Draw the the correct number of vertices anywhere on your paper $v_1, v_2, \cdots, v_n$.

Step 2: On each vertex v_i draw in $\lfloor d_i / 2 \rfloor$ self-loops, i.e., this is the maximum number of self-loops that v_i can have without exceeding its desired degree d_i.

Step 3: Now the remaining degrees available on each vertex is either 1 (if the vertex originally had odd degree) or 0 (if it originally had even degree). The number of vertices with remaining degree 1 must therefore be even, and so can be paired up. Pair these vertices up and join each pair of vertices with a single edge. The resulting multigraph will now have the desired degree sequence.

For each sequence below, if possible, use the above algorithm to create a drawing of a multigraph that realizes it.

(a) $(6, 6, 3, 3, 3, 3, 2)$ (b) $(3, 3, 3, 3, 3, 3, 3)$

(c) $(5, 5, 4, 4)$ (d) $(10, 10, 2, 2)$

12. Repeat the directions of Exercise 11 for each of the following degree sequences.

(a) $(5, 5, 3, 3, 1)$ (b) $(6, 6, 3, 3, 2)$

(c) $(12, 8, 4, 0)$ (d) $(6, 6, 5, 5, 4, 4, 3, 3)$

13. Classify the following statement as one of: **Always True, Sometimes True,** or **Never True**, and justify your answer: A subgraph of a bipartite graph is bipartite.

14. Classify the following statement one of: **Always True, Sometimes True,** or **Never True**, and justify your answer: A bipartite graph cannot have a self-loop.

15. (a) Prove the following:

Proposition: If G is a simple graph with at least two vertices, then at least two vertices of G must have the same degree.

(b) Is the statement in Part (a) true for general graphs?

(c) Prove the following social proposition: Assume that we have a party with at least two people, and that each pair of guests can be classified as acquaintances or strangers. Then there will be at least two people at the party who will have the same number of acquaintances, among other guests at the party.

Suggestion for Part (a): In any simple graph with n vertices, observe that there cannot be both a vertex of degree 0 and a vertex of degree $n - 1$.

16. (a) Prove that for any positive integer $n > 1$, the complete bipartite graph $K_{n,n}$ contains copies of the cycles C_k, for $k = 4, 6, \cdots, 2n$.

(b) Prove that a bipartite graph can never contain a copy of a cycle of odd length, i.e., if G is bipartite, and $k > 1$ is an odd integer, then C_k is not a subgraph of G.

17.   Determine any subgraph relations between the following three graphs: $P_n, C_n, K_{n,n}$, where $n \geq 3$.

In each of Exercises 18–21, a pair of graphs is given. Determine whether these graphs are isomorphic. If they are isomorphic, provide an isomorphism between them; if they are not isomorphic, provide a rigorous argument to justify your assertion.

18.

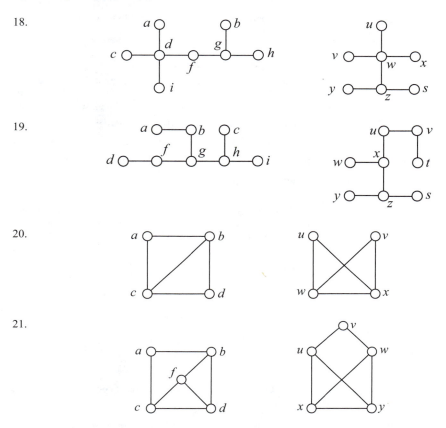

19.

20.

21.

In each of Exercises 22–25, a series of graphs is given. Find any isomorphism relations between graphs in the series. Justify all answers with isomorphisms or proofs of nonisomorphisms. (You may wish to use the fact that graph isomorphism is an equivalence relation.)

22.

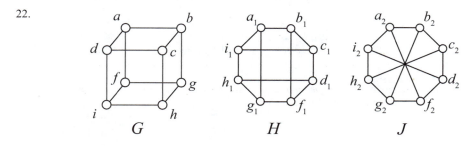

G H J

23.

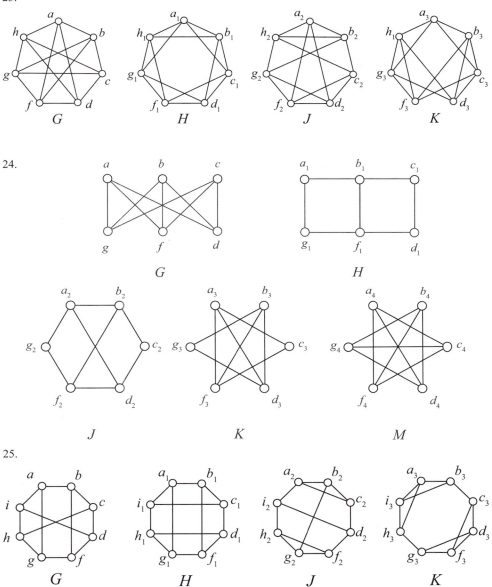

24.

25.

26. (a) Suppose that G and H are simple graphs, and that $f : G \to H$ is a vertex bijection mapping. Show that if $f : G \to H$ preserves adjacency, then when viewed as a mapping of the complement graphs, $f : {\sim}G \to {\sim}H$, this mapping preserves non-adjacency on the complement graphs.

(b) Prove that if $f : G \to H$ preserves non-adjacency, then when viewed as a mapping of the complement graphs, $f : {\sim}G \to {\sim}H$, this mapping preserves adjacency on the complement graphs.

27. If a simple graph G has degree sequence $(d_1, d_2, \cdots, d_n)$, what is the degree sequence of the complement graph ${\sim}G$?

28. For each of the four simple graphs of Exercise 2, do the following tasks. Use the alphabetical ordering of the vertices. In cases where the ordering on the vertices is not specified, you will need to specify some ordering. Use the lexicographic ordering of the edges.
    (a) Write down the adjacency matrix.
    (b) Write down the incidence matrix.
    (c) Write down the adjacency list.
    (d) Write down the edge list.

29. Repeat the directions of Exercise 28 for each of the four simple graphs of Exercise 1 (see Figure 8.30).

30. Repeat the directions of Parts (a) and (b) of Exercise 28 for each of the two general graphs of Exercise 4 (see Figure 8.32).

31. Repeat the directions of Parts (a) and (b) Exercise 28 for each of the two general graphs of Exercise 3 (see Figure 8.31).

32. (a) Show that the adjacency matrix A of a general graph G is always symmetric.
    (b) Find necessary and sufficient conditions for the adjacency matrix of a digraph to be symmetric.
    Note: Recall that a square matrix A is said to be symmetric if the matrix equals its transpose, i.e., $A = A'$.

33. For each of the four directed graphs shown in Figure 8.33, do the following:
    (a) List the vertices.
    (b) List the edges.
    (c) Find the indegrees and outdegrees of all vertices.
    (d) Verify the validity of the handshaking theorem.
    (e) Label the edges, and then write down the edgemap function Δ.

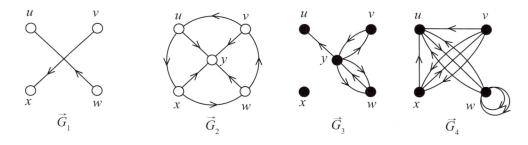

FIGURE 8.33: Four digraphs for Exercise 33.

34. Repeat the directions of Exercise 33 for each of the four digraphs of Figure 8.34.

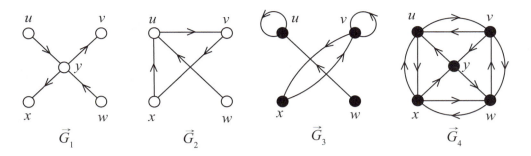

FIGURE 8.34: Four digraphs for Exercise 34.

35. For each of the four directed graphs of Exercise 33 (see Figure 8.33), do the following. The vertices should be ordered alphabetically, and the edges lexicographically.
    (a) Write down the adjacency matrix.
    (b) Write down the incidence matrix.
    (c) Write down the adjacency list for each simple digraph.
    (d) Write down the edge list for each simple digraph.

36. Repeat the directions of Exercise 35 for each of the four directed graphs of Exercise 34 (see Figure 8.34).

In Exercises 37–40, a series of directed graphs is given. (a) Find any (digraph) isomorphism relations between graphs in the series. Justify all answers with isomorphisms or proofs of nonisomorphisms. (You may wish to use the fact that digraph isomorphism is an equivalence relation.) (b) Determine any isomorphism relations that exist among the corresponding underlying graphs.

37.

38.

39.

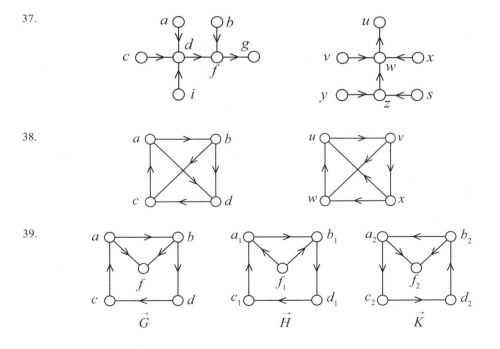

40.

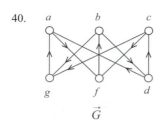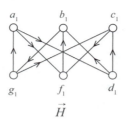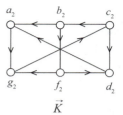

NOTE: In fulfilling a promise that we made in the section proper, we now give the definition of graph isomorphism for general graphs. (Recall that Definition 8.9 defined isomorphisms between simple graphs only.)

DEFINITION 8.19: We say that a pair of general graphs, $G = (V, E, \Phi)$ and $G' = (V', E', \Phi')$, are **isomorphic**, if there exists a pair of *bijective* functions: a *vertex function* $f : V \to V'$, and an *edge function* $g : E \to E'$, such that for each edge $e = \{u, v\} \in E$, we have that $g(e) = \{f(u), f(v)\} \in E'$.

We allow the case $u = v$, which corresponds to the edge being a self-loop. In less formal language, this simply states that the edges of the two graphs can be put into a one-to-one correspondence (by the function g) in such a way that the endpoints in a pair of corresponding edges are matched by the vertex function f.[10]

With Definition 8.19, our definition of digraph isomorphisms (Definition 8.18) now extends verbatim to the setting of general digraphs.

41.   Prove that general graph isomorphism (Definition 8.19) is an equivalence relation.

42.   Prove that isomorphism of general directed graphs (Definition 8.19 and Definition 8.18) is an equivalence relation.

In Exercises 43-44, a series of directed graphs is given. (a) Find any (digraph) isomorphism relations between graphs in the series. Justify all answers with isomorphisms or proofs of nonisomorphisms. (You may wish to use the fact that digraph isomorphism is an equivalence relation.) (b) Determine any isomorphism relations that exist among the corresponding underlying (general) graphs.

43.

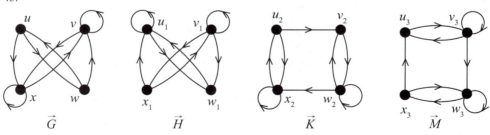

---

[10] The last defining condition of Definition 8.9 can be formalized using the edgemaps as follows: For each edge $e \in E$, if $\Phi(e) = \{u, v\}$, then $\Phi'(g(e)) = \{f(u), f(v)\}$, but this additional abstraction will not be needed.

44.

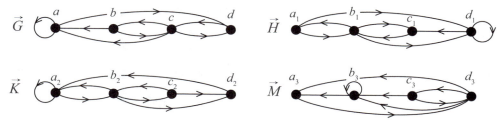

45. Classify the following statement as one of: **Always True, Sometimes True, or Never True**, and justify your answer: Suppose that $\vec{G}$ and $\vec{H}$ are simple directed graphs, and that G and H are their corresponding underlying graphs. If a vertex mapping $f : \vec{G} \rightarrow \vec{H}$ is a digraph isomorphism, then the same function, when viewed as a vertex mapping $f : G \rightarrow H$, is a graph isomorphism.

46. Suppose that G and H are simple graphs, and that $f : G \rightarrow H$ is a one-to-one vertex mapping. Show that:
    (a) If f preserves adjacency, then $\deg(f(v)) \geq \deg(v)$, for each vertex $v \in V(G)$.
    (b) If f preserves non-adjacency, then $\deg(f(v)) \leq \deg(v)$, for each vertex $v \in V(G)$.
    (c) Conclude that if f is an isomorphism, then $\deg(f(v)) = \deg(v)$, for each vertex $v \in V(G)$.

47. Suppose that G_1 and G_2 are simple graphs, and that $f : G_1 \rightarrow G_2$ is a one-to-one vertex mapping. Suppose also that H is a graph. Show that: If f preserves adjacency, and G_1 contains a copy of H, then so does G_2.

48. (a) Find all simple graphs on three vertices that are isomorphic to their complement graph.
    (b) Find all simple graphs on four vertices that are isomorphic to their complement graph.
    (c) Find all simple graphs on five vertices that are isomorphic to their complement graph.

49. Show that if a simple graph G and its complement graph $\sim G$ are isomorphic, then the number of vertices of G is congruent to zero, or one (mod 4), i.e., $|V(G)| = 4k$ or $4k + 1$, for some integer k.

50. From the proposition of Exercise 15, any simple graph with at least two vertices will always have two vertices of the same degree.
    (a) Up to isomorphism find all simple graphs on $n = 2$ and $n = 3$ vertices that have exactly two vertices of the same degree.
    (b) Repeat Part (a) for $n = 4$ and $n = 5$.
    (c) Use a recursive construction to find, up to isomorphism, all simple graphs with $n \geq 2$ vertices such that exactly two vertices have the same degree.
    Suggestion: Take note of the fact that such graphs cannot have 0 or $n - 1$ as their double degree.

51. Show that any simple graph having n vertices and m edges will always have a vertex of degree at least $2m/n$.

52. (*Second Degrees in Simple Graphs*) Recall that once an ordering on the vertices $v_1, v_2, \cdots, v_n$ of a graph is specified, the corresponding degree sequence $(d_1, d_2, \cdots, d_n)$ is well defined, and that if two graphs are isomorphic, their degree sequences must be permutations of one another. For a simple graph G, we recall that the degree d_i of a vertex v_i is the number of neighbors of v_i. We define the *second degree* of a vertex of a simple graph to be the total number of vertices that are neighbors of at least one neighbor of the vertex, but different from the vertex (so a vertex is

not counted among its own second neighbors).

(a) Compute the sequence of second degrees for the n-path graph and P_n.

(b) Compute the sequence of second degrees for n-cycle graph and C_n.

(c) Find a pair of simple graphs with the same degree sequence, but whose second degree sequences are not permutations of one another.

(d) Find a pair of simple graphs with the same second degree sequence, but whose degree sequences are not permutations of one another.

(e) Explain why if the second degree sequences of two simple graphs are not permutations of one another, then the graphs are not isomorphic.

53. (a) Prove that if G is a simple graph with six vertices, then either G or its complement graph $\sim G$ must contain a copy of K_3.

(b) Prove the following social proposition: *Assume that we have a party with six people, and that each pair of guests can be classified as acquaintances or strangers. Then the party will contain a group of three people who are either mutual acquaintances, or mutual strangers.*

(c) Does the conclusion of Part (b) continue to hold for parties of five people?

54. Up to isomorphism, how many simple graphs are there that have five vertices and three edges?

55. (a) Up to isomorphism, how many simple digraphs are there that have three vertices and three edges, but no self-loops?

(b) Up to isomorphism, how many general digraphs are there that have three vertices and three edges, but no self-loops?

56. Find a formula for the number of labeled simple digraphs on n vertices.

Note: The intention of this exercise is as follows: If $n = 6$, say, place six vertices, with labels 1,2,3,4,5,6 around a circle. Count the number of ways that directed edges can be added to produce different labled simple graphs. For example, the three graphs with the single edges (i) (1,2), (ii) (2,1), and (iii) (2,3) would all be counted separately, even though they are isomorphic.

57. An *automorphism* on a simple graph $G = (V, E)$ is an isomorphism from G to itself. More precisely, an automorphism of G is a permutation of the vertices $f : V \to V$ that preserves both adjacency and non-adjacency. The identity map $1_V : V \to V :: 1_V(v) = v$ is trivially always an automophism of G.

(a) Show that the composition of two automorphisms of G is an automorphism of G.

(b) Show that the inverse (permutation) of an automorphism of G is an automorphism of G.

Note: With the results of this exercise, it follows that the set of all automorphisms of a graph G forms what is called in abstract algebra a *group*. Under the vertex permutation of any automorphism, the permuted vertices are, for all graph theoretic purposes, equivalent to their preimage vertices. In general, the more symmetry that a graph possesses, the richer its automorphism group will be.

58. Describe the set (group) of all automorphisms of the following graphs: (See the previous exercise.)

(a) The path graphs P_3, P_4, and P_5.

(b) The cycle graphs C_3, C_4, and C_5.

(c) The complete graphs K_3, K_4, and K_5.

(d) The n-path graph P_n.

(e) The n-cycle graph C_n.

(f) The complete graph K_n.

NOTE: We end this set with a few exercises that deal with two concepts in graph theory that have importance in both theory and applications. First we give the relevant definitions.

DEFINITION 8.20: Suppose that $G = (V, E)$ is a simple graph.

(1) A set of vertices $S \subseteq V$ is called **independent** if no two vertices belonging to S are adjacent. The **independence number** $\alpha(G)$ is the number of vertices in an independent set of largest possible size.

(2) A **clique** is a maximal set of vertices $T \subseteq V$ with the property that any two vertices belonging to T are adjacent. The **clique number** $\omega(G)$ is the number of vertices in a clique of largest possible size.

The adjective "maximal" in (2) means that the set T has the stated property, and that T is not a proper subset of a (larger) set of vertices that also the (clique) property. The following example illustrates these two concepts.

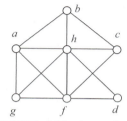

EXAMPLE 8.14: Let G be the graph shown in Figure 8.35. The maximal independent sets of this graph are the following: $\{a, c, d\}, \{b, g, d\}, \{b, f\}, \{c, g, d\}$, and $\{h\}$. Since the largest such set has three vertices (there are three such sets), we have $\alpha(G) = 3$.

The cliques of G are the following:

FIGURE 8.35: A graph for Example 8.14.

$\{a, b, h\}, \{a, f, g, h\}, \{b, c, h\}, \{c, f, h\}$, and $\{d, f, h\}$. Since the largest of these has four elements, we have $\omega(G) = 4$.

The problems of computing the independence and clique numbers of a graph are both NP hard (in the above example, we computed them by exhaustively finding all of the relevant maximal sets, proceeding in alphabetical order on the vertices).

59. For each of the following simple graphs do the following: (i) Find all maximal independent sets, (ii) find the independence number, (iii) find all cliques, (iv) find the clique number.
    (a) The two graphs of Exercise 20.
    (b) The graph G of Exercise 22.
    (c) The graph H of Exercise 22.
    (d) The graph J of Exercise 22.

60. Repeat the directions of Exercise 59 for the following graphs:
    (a) The two graphs of Exercise 21.
    (b) The graph G of Exercise 24.
    (c) The graph H of Exercise 24.
    (d) The graph J of Exercise 24.
    (e) The graph K of Exercise 24.

61. Compute the independence number for each of the following graphs:
    (a) The n-path graph P_n.   (b) The n-cycle graph C_n.
    (c) The complete graph K_n.   (d) The Petersen graph.
    (e) The hypercube Q_n.

62. Compute the clique number for each of the graphs given in Exercise 59.

63. Let G be a simple graph and $\sim G$ be its complement. Prove the following identities.
    (a) $\alpha(G) = \omega(\sim G)$.   (b) $\omega(G) = \alpha(\sim G)$.

64. Let G be a simple graph on n vertices, and let $\Delta = \max\{\deg(v) : v \in V(G)\}$ (i.e., Δ is the largest degree of any of the vertices). Prove that $\alpha(G) \geq n/(\Delta + 1)$.

COMPUTER EXERCISES 8.1:

1. (*Program for Graphic Sequence Determination Algorithm with Brief Output*) (a) Write a program `Out = GraphSequence(Seq)` that inputs a sequence `Seq` of nonnegative integers, and outputs a truth value `Out` that is either "1" or "0" according as to whether the inputted sequence can (1) or cannot (0) be realized by a simple graph. The program should use Algorithm 8.1, and, just like the algorithm, should use a while loop structure.
    (b) Run your program on the sequences of Example 8.5 and Exercise for the Reader 8.4, and give the outputs.
    (c) Run your program on the sequences of ordinary Exercises 7 and 8, and give the outputs.

2. (*Program for Graphic Sequence Determination Algorithm with Full Output*) (a) Write a program `Out = GraphSequenceShow(Seq)` that inputs a sequence `Seq` of nonnegative integers, and outputs a truth value `Out` that is either "Yes" or "No" according as to whether the inputted sequence can, or cannot be realized by a simple graph. In addition, in either case, the Yes/No output should be followed by a printout of the permuted degree sequences that were produced at the various iterations of the algorithm. The program should use Algorithm 8.1.
    (b) Run your program on the sequences of Example 8.5 and Exercise for the Reader 8.4, and give the outputs.
    (c) Run your program on the sequences of ordinary Exercises 7 and 8, and give the outputs.

3. (*Recursive Program for Graphic Sequence Determination Algorithm with Brief Output*) Although Algorithm 8.1 is recursive in that it repeatedly reduces the given problem to a smaller instance, the algorithm is written with a while loop. (a) Write a program `Out = GraphSequenceRecursive(Seq)` having the same syntax, input, and outputs as the program of Computer Exercise 1, but that is written in a way so the program calls on itself (rather than using a while loop).
    (b) Run your program on the sequences of Example 8.5 and Exercise for the Reader 8.4, and give the outputs.
    (c) Run your program on the sequences of Exercises 7 and 8, and give the outputs.

4. (*Recursive Program for Graphic Sequence Determination Algorithm with Full Output*) (a) Write a program `Out = GraphSequenceShowRecursive(Seq)` having the same syntax, input, and outputs as the program of Computer Exercise 2, but that is written in a way so the program calls on itself (rather than using a while loop).
    (b) Run your program on the sequences of Example 8.5 and Exercise for the Reader 8.4, and give the outputs.
    (c) Run your program on the sequences of ordinary Exercises 7 and 8, and give the outputs.

5. (a) Run your program of Computer Exercise 1 on the sequence of threes $(3,3,\cdots,3)$, with length n ranging from 100 to 200. For which values of these n is the sequence graphic? (Compare with the result of ordinary Exercise 9.)
    (b) Run your program on the sequence of fours $(4,4,\cdots,4)$, with length n ranging from 100 to 200. For which values of these n is the sequence graphic? (Compare with the result of ordinary Exercise 10.)
    (c) Run your program on the sequence of fives $(5,5,\cdots,5)$, with length n ranging from 100 to 200. For which values of these n is the sequence graphic? Make a conjecture as to which values of n the sequence of fives will be graphic (i.e., for which there exists a 5-regular simple graph).

6. Compare the performance of the programs `GraphSequence` of Computer Exercise 1, and `GraphSequenceRecursive` of Computer Exercise 3 by testing both on long sequences of fives $(5,5,\cdots,5)$. Continue doubling the length of the sequence (starting out with length $n = 6$), until the execution time exceeds five seconds. Indicate how large the sequence becomes (n) when this first occurs for each of the two algorithms.

7. (*Program to Compute Degree Sequences of Graphs*) (a) Write a program with syntax `DegSeq = DegreeSequence(AdjMat)`, with input `AdjMat` being the adjacency matrix of a graph, with respect to some ordering of the vertices, and whose output `DegSeq` is the corresponding sequence of degrees of the vertices.
(b) Run your program on the adjacency matrix of Example 8.9(a) and Exercise for the Reader 8.4, and give the outputs.
(c) Create adjacency matrices for the simple graphs in Figure 8.30 (ordering the vertices alphabetically), and then run your program of Part (a) on each.
(d) Create adjacency matrices for the two general graphs in Figure 8.32 (ordering the vertices alphabetically), and then run your program of Part (a) on each.

8. (*Program to Compute Degree Sequences of Graphs*) (a) Write a program with syntax `DegSeq = DegreeSequenceI(IncMat)`, with input `IncMat` being the incidence matrix of a graph, with respect to some ordering of the vertices and edges, and whose output `DegSeq` is the corresponding sequence of degrees of the vertices.
(b) Run your program on the incidence matrices of Example 8.10(a)(b), and give the outputs.
(c) Create incidence matrices for the simple graphs in Figure 8.30 (ordering the vertices alphabetically, and the edges lexicographically), and then run your program of Part (a) on each.
(d) Create incidence matrices for the two general graphs in Figure 8.32 (ordering the vertices alphabetically, and the edges any way you wish), and then run your program of Part (a) on each.

9. (*Program to Convert Edge Lists to Adjacency Matrices for Simple Graphs*) (a) Write a program with syntax `AdjMat = EdgeList2AdjMatrix(EdgeList, n)`, with input `EdgeList` being the edge list of a simple graph, where the vertices are labeled with positive integers, ordered with the usual order on numbers. The second output n is the number of vertices. The output `AdjMat` is the corresponding adjacency matrix of the graph.
(b) Run your program on the edge list of Figure 8.24 (where the vertices a, b, c, d, e, are relabeled as 1, 2, 3, 4, 5).
(c) Write a small program to create an edge list for the 10-path graph P_{10}, and then run the program of Part (a). Give both outputs.
(d) Write a small program to create an edge list for the 10-cycle graph C_{10}, and then run the program of Part (a). Give both outputs.
(e) Will your program or Part (a) work for digraphs? If it does, explain why; otherwise write a related program for digraphs.
Note: In case it is not convenient for your computing platform to implement lists (or if you simply prefer to avoid them), you may simply store edge lists as two-column matrices.

10. (*Program to Convert Edge Lists to Incidence Matrices for Simple Graphs*) (a) Write a program with syntax `IncMat = EdgeList2IncMatrix(EdgeList, n)`, with input `EdgeList` being the edge list of a simple graph, where the vertices are labeled with positive integers, ordered with the usual order on numbers. The second output n is the number of vertices. The output `IncMat` is the corresponding incidence matrix of the graph, with the lexicographic ordering on the edges.
(b) Run your program on the edge list of Figure 8.24 (where the vertices a, b, c, d, e, are relabeled as 1, 2, 3, 4, 5).
(c) Write a small program to create an edge list for the 10-path graph P_{10}, and then run the program of Part (a). Give both outputs.
(d) Write a small program to create an edge list for the 10-cycle graph C_{10}, and then run the program of Part (a). Give both outputs.
Note: In case it is not convenient for your computing platform to implement lists (or if you simply prefer to avoid them), you may simply store edge lists as two-column matrices. Notice also that the lexicographic ordering of the edges specified for the incidence matrix coincides with the way the edges are ordered in the edge list.

11. (*Program to Convert Adjacency Lists to Adjacency Matrices for Simple Graphs*) (a) Write a program with syntax `AdjMat = AdjList2AdjMatrix(AdjList)`, with input `AdjList`

being the adjacency list of a simple graph, where the vertices are labeled with positive integers, ordered with the usual order on numbers. The output `AdjMat` is the corresponding adjacency matrix of the graph.

(b) Run your program on the adjacency list of Figure 8.24 (where the vertices a, b, c, d, e are relabeled as 1, 2, 3, 4, 5).

(c) Write a small program to create an adjacency list for the 10-path graph P_{10}, and then run the program of Part (a). Give both outputs.

(d) Write a small program to create an adjacency list for the 10-cycle graph C_{10}, and then run the program of Part (a). Give both outputs.

Note: In case it is not convenient for your computing platform to implement variable length lists (or if you simply prefer to avoid them), you may store adjacency lists as matrices, where the number of columns is the maximum degree of the vertices, and rows corresponding to smaller numbers of vertex neighbors are filled with zeros.

12. (*Program to Convert Adjacency Lists to Incidence Matrices for Simple Graphs*) (a) Write a program with syntax `IncMat = AdjList2IncMatrix(AdjList)`, with input `AdjList` being the adjacency list of a simple graph, where the vertices are labeled with positive integers, ordered with the usual order on numbers. The output `IncMat` is the corresponding incidence matrix of the graph.

(b) Run your program on the adjacency list of Figure 8.24 (where the vertices a, b, c, d, e are relabeled as 1, 2, 3, 4, 5).

(c) Write a small program to create an adjacency list for the 10-path graph P_{10}, and then run the program of Part (a). Give both outputs.

(d) Write a small program to create an adjacency list for the 10-cycle graph C_{10}, and then run the program of Part (a). Give both outputs.

Note: In case it is not convenient for your computing platform to implement variable length lists (or if you simply prefer to avoid them), you may store adjacency lists as matrices, where the number of columns is the maximum degree of the vertices and rows corresponding to smaller number of vertex neighbors are filled with zeros.

13. (*Program to Convert Adjacency Matrices to Incidence Matrices*) (a) Write a program with syntax `IncMat = AdjMatrix2IncMatrixSimple(AdjMat)`, with input `AdjMat` being the adjacency matrix of a simple graph, where the vertices are labeled with positive integers, ordered with the usual order on numbers. The output `IncMat` is the corresponding incidence matrix of the graph, where the edges are labeled lexicographically.

(b) Run your program on the adjacency matrix of Example 8.9(b).

(c) Write a small program to create an adjacency matrix for the 10-path graph P_{10}, and then run the program of Part (a). Give both outputs.

(d) Write a small program to create an adjacency matrix for the 10-cycle graph C_{10}, and then run the program of Part (a). Give both outputs.

(e) Modify your program of Part (a), into a new program with syntax `IncMat = AdjMatrix2IncMatrix(AdjMat)` that is able to handle adjacency matrices of general graphs.

(f) Run your program of Part (e) on the incidence matrix of the solution of Part (a) of Example 8.9. Give the output.

(g) Run your program of Part (e) on the incidence matrices for the two general graphs of Figure 8.32 (where the vertices are ordered alphabetically). Give the output.

Note: For Part (e), the edges are ordered lexicographically so multiple edges will be adjacent and their columns in the incidence matrix will be identical. Thus, the way in which multiple edges are ordered amongst themselves is not relevant.

14. (*Program to Convert Incidence Matrices to Adjacency Matrices*) (a) Write a program with syntax `AdjMat = IncMatrix2AdjMatrix(IncMat)`, with input `IncMat` being the incidence matrix of a graph, where the vertices are labeled with positive integers, ordered with the usual order on numbers, and the edges lexicographically. The output `AdjMat` is the

corresponding adjacency matrix of the graph.

(b) Run your program on the incidence matrix of Example 8.10(c).

(c) Run your program on the incidence matrices of ordinary Exercise 8.28(b).

15. (*Program to Compute Second Degree Sequences of Simple Graphs*) (a) Write a program with syntax `SecDegSeq = SecondDegreeSequence(AdjMat)`, with input `AdjMat` being the adjacency matrix of a simple graph, with respect to some ordering of the vertices, and whose output `DegSeq`, is the corresponding sequence of second degrees of the vertices. See ordinary Exercise 52 for the relevant definition.

(b) Run your program on adjacency matrices for the three eight-vertex simple graphs of ordinary Exercise 22. Do the outputs tell you anything about whether the graphs are isomorphic?

(b) Run your program on adjacency matrices for the four seven-vertex simple graphs of ordinary Exercise 23. Do the outputs tell you anything about whether the graphs are isomorphic?

(c) Write a small program to create an adjacency matrix for the n-path graph P_n. Use this program, along with the program of Part (a) to compute the second degree sequences of the n-path graphs for $n = 10, 11$, and 12. Do you observe any pattern?

(d) Write a small program to create an adjacency matrix for the n-cycle graph C_n. Use this program, along with the program of Part (a) to compute the second degree sequences of the n-cycle graphs for $n = 10, 11$, and 12. Do you observe any pattern?

16. (*Brute-Force Isomorphism Checker for Simple Graphs*) (a) Write a program
        `Out, Perm = IsomorphismChecker(AdjMat1, AdjMat2)`
that inputs a pair of adjacency matrices `AdjMat1, AdjMat2` that correspond to two simple graphs having the same number of vertices (with some specified ordering of their vertices), and outputs a truth value `Out` that is either "1" or "0" according as to whether the inputted graphs are isomorphic (1 = True, 0 = False). The second output variable `Perm` will be a vector representing a permutation corresponding an isomorphism, in case the graphs are isomorphic; otherwise it will be the single number 0. The program should go through all vertex permutations on the first matrix to check to see if it can be transformed into the second matrix. In case an isomorphism is found, the program exits and outputs also the (first) vertex permutation that rendered the matrices equal (and hence produces a graph isomorphism).

(b) Apply your program to the isomorphism problem of ordinary Exercise 20.

(c) Apply your program to the isomorphism problem of ordinary Exercise 24.

(d) Apply your program to the isomorphism problem of ordinary Exercise 25.

17. (*Brute-Force Isomorphism Checker for Simple Digraphs*) (a) Will your program or Part (a) of the previous computer exercise work for digraphs? If it does, explain why; otherwise write a related program for digraphs.

(b) Apply your program to the isomorphism problem of ordinary Exercise 38.

(c) Apply your program to the isomorphism problem of ordinary Exercise 39.

(d) Apply your program to the isomorphism problem of ordinary Exercise 40.

8.2: PATHS, CONNECTEDNESS, AND DISTANCES IN GRAPHS

This section concerns accessibility of the vertices of a graph through paths along edges. As in the last section, the definitions will be very intuitive. Informally (given a drawing of a graph), a vertex w of a graph is *reachable* from another vertex v, if it is possible to put our finger down at vertex v, and trace it (without lifting it up) along certain edges of the graph so as to end at the vertex w. The resulting sequence of edges used is called a *path* from v to w. For graphs this relation is symmetric, but for digraphs the edge directions need to be respected, so reachability need not be symmetric. Once we have a reachability relationship, it is natural to look for a path that is as efficient as possible, and this leads to the important concept of *distance* between vertices. For a graph to be *connected*, it will mean that any two vertices are reachable from one another. Some graphs are more connected than others and may possess other sorts of traversability properties that will turn out to be useful in applications. We now turn to a more formal development of these concepts.

Paths, Circuits, and Reachability in Graphs

DEFINITION 8.21: Suppose that $G = (V, E)$ is a graph, and $v, w \in V$ are a pair of vertices. A **path** in G from v to w is an alternating sequence of vertices and edges:

$$P = < v_0, e_1, v_1, e_2, v_2, \cdots, v_{k-1}, e_k, v_k >,$$

such that the endpoints of edge e_i are the vertices $\{v_{i-1}, v_i\}$, for $1 \leq i \leq k$, $v_0 = v$, and $v_k = w$. We say the path P *passes through* the vertices $v_0, v_1, v_2, \cdots, v_{k-1}, v_k$, and *traverses* the edges $e_1, e_2, \cdots, e_k$, and that the path has **length** k, since it traverses k edges. If such a path exists, we say that the vertex w is **reachable** from the vertex v in G. A path having positive length ($k > 0$) from any vertex to itself is called a **circuit**. A path (or circuit) is called **simple** if it never traverses the same edge twice. A simple circuit that does not pass through the same vertex twice (except for the initial and final vertex) is called a **cycle**.[11]

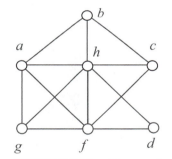

FIGURE 8.35: The graph of Example 8.15.

Observe that in the case of a simple graph, the above path can be specified by the sequence of vertices $< v_0, v_1, v_2, \cdots, v_{k-1}, v_k >$, since the corresponding edges are uniquely determined. Also note that any vertex is reachable from itself by the path $< v_0 >$ of length zero.

EXAMPLE 8.15: In the graph G of Figure 8.35 (reproduced here), the following are two paths from vertex a to vertex d:

$$P = < a, b, c, h, g, f, d >, \quad Q = < a, h, d >.$$

---

[11] Thus, in a simple graph, a cycle is really just a copy of one of the *n*-cycle graphs C_n, for some $n \geq 3$. Note that the assumption that a cycle does not pass twice through the same edge follows from the fact that it does not visit a vertex twice (except for the endpoints). Cycles in general graphs may pass through the same vertices multiple times, due to self-loops and parallel edges.

Both of these paths are simple, P has length six and Q has length two. Since a and d are not adjacent, there are no shorter paths between them than Q.

The following are two cycles in G:

$$C = <a,b,c,h,d,f,g,a>, \quad D = <a,b,h,a>.$$

The first one has the additional property that it passes through all of the vertices of G. In particular, it is clear that any two vertices of G are reachable from one another. Here is an example of a circuit in G that is not a simple circuit: $<a,b,a>$. Can the reader find a simple circuit in G that is not a cycle? Here is one such example: $<a,b,h,f,g,h,a>$. (The reader should carefully verify with the definitions that this is indeed such an example.)

Paths, Circuits, and Reachability in Digraphs

DEFINITION 8.22: Suppose that $\vec{G} = (V,E)$ is a directed graph and $v,w \in V$ is a pair of vertices. A **path** in $\vec{G}$ from v to w is an alternating sequence of vertices and edges:

$$P = <v_0, e_1, v_1, e_2, v_2, \cdots, v_{n-1}, e_n, v_n>,$$

such that each e_i has tail vertex v_{i-1} and head vertex v_i, for $1 \le i \le n$, $v_0 = v$ and $v_n = w$. The path is also said to **join** v to w. The path has **length** n, since it traverses n edges. If such a path exists, we say that the vertex w is **reachable** from the vertex v in $\vec{G}$. A path having positive length ($n > 0$) from any vertex to itself is called a **circuit**. A path (or circuit) is called **simple** if it never traverses the same edge twice.[12]

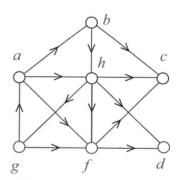

FIGURE 8.36: The digraph of Example 8.16.

EXAMPLE 8.16: In the digraph $\vec{G}$ of Figure 8.36, the following are two paths from vertex a to vertex d:

$$P = <a,b,h,a,f,d>, \quad Q = <a,h,d>.$$

---

[12] Once again we caution the reader about the unfortunate inconsistencies in graph theory terminology in the literature. For example, in some books, what we have called a path will be called a "walk," what we have called a circuit will be called a "closed walk"; the word "path" sometimes is reserved for what we call a path but with the additional condition that there is no repetition of vertices. The reader should always pay special attention to the terminology and definitions when reading any new book or paper on graph theory.

Both of these paths are simple, P has length five and Q has length two. Since a and d are not adjacent, there are no shorter paths between them than Q. Notice that no vertex (other than c itself) is reachable from the vertex c, since all edges incident to c are directed towards c.

NOTE: The notation for paths/circuits is the same for digraphs as it was for graphs. This is because paths and circuits have associated directions. The only difference is that in digraphs these directions must agree with the edge directions of the digraph.

Connectedness and Connected Components

DEFINITION 8.23: A graph is **connected** if for every pair of vertices v and w, there is a path from v to w. A digraph is **(strongly) connected** if for every pair of vertices v and w, there is a directed path from v to w. A digraph is **weakly connected** if its underlying graph is connected.

If we view a graph as a network of two-way roads (the edges) between some towns (the vertices), connectivity of the graph simply corresponds to the property that it is always possible to drive (on the roads) between any pair of cities. For a digraph, the same interpretation holds if the edges are viewed as one-way roads. As pointed out in Example 8.15, the graph of that example is connected. Since this graph is also the underlying graph of the digraph of Figure 8.36, it follows that this digraph is weakly connected. It was pointed out in Example 8.16 that the vertex c is not reachable from any other vertex, so this digraph is not strongly connected. We point out that as far as reachability and connectedness questions are concerned, parallel edges and self-loops are irrelevant and can be ignored.

In terms of graph drawings, a helpful way to think of connectedness is that the only way it could fail for a certain graph G is if it is possible to create a drawing of G that is literally split up into two or more disjoint "pieces." This concept is formalized in the following definition.

DEFINITION 8.24: A **(connected) component** of a graph $G = (V, E)$ is a maximal connected subgraph. For every vertex $v \in V$, the **component of v**, denoted by $C(v)$, is the (unique) component of G that contains v.

The components of a graph partition the graph into disjoint subgraphs (see Exercise for the Reader 8.18). Thus, a graph is connected if it consists of a single connected component.

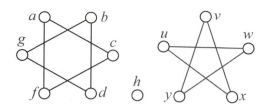

FIGURE 8.37: The non connected graph of Example 8.17.

EXAMPLE 8.17: Find the components of the graph of Figure 8.37.

SOLUTION: There are four components: the two copies of K_3 on the leftmost portion, with vertices $\{a, c, f\}$ and $\{b, d, g\}$, the isolated vertex $\{h\}$, and the copy of the 5-cycle on the right with vertices $\{v, x, u, w, y\}$.

EXERCISE FOR THE READER 8.18: Given a simple graph $G = (V, E)$, show that the relation on the vertex set V specified by:

$$v \sim w \quad \Leftrightarrow \quad w \text{ is reachable (by a path in } G) \text{ from } v,$$

is an equivalence relation on V, and the equivalence classes are the vertex sets of the various components of G. Conclude that $v \sim w \quad \Leftrightarrow \quad C(v) = C(w)$.

We are now ready to discuss the important concept of distances in graphs. The definitions that will follow all involve the concept of optimal paths between pairs of vertices. By default, we will declare the length of each unweighted edge of a graph to be one, so the length of a path will simply be the number of edges it traverses. In applications, edges can have weights representing actual distances, prices (of airline tickets, for example), or times, and the definitions can be modified accordingly. Such extensions will be examined in Section 9.2.

Distances and Diameters in Graphs

DEFINTION 8.25: Suppose that $G = (V, E)$ is a graph, and $v, w \in V$ is a pair of vertices. The **distance** from v to w, denoted $\text{dist}(v, w)$ (or by $\text{dist}_G(v, w)$ if we wish to emphasize the dependence on G), is the length of a shortest path from v to w, provided that these vertices are reachable from one another. In case the vertices are not reachable, we define $\text{dist}(v, w) = \infty$. The **diameter** of the graph G, $\text{diam}(G)$, is the maximum possible distance between pairs of vertices of G, i.e.,

$$\text{diam}(G) = \max_{v, w \in V(G)} \{\text{dist}_G(v, w)\}.$$

EXAMPLE 8.18: (a) For the graph of Figure 8.35 (Example 8.15), find $\text{dist}(b, d)$.

(b) Determine the diameter of the graph of Figure 8.35.

(c) Determine the diameter of the n-cycle graphs, $\text{diam}(C_n)$, when $n \geq 3$.

SOLUTION: Part (a): Since b and d are not adjacent, the distance between them must be at least two. But since the path $<b, h, d>$ has length two, it follows that $\text{dist}(b,d) = 2$.

Part (b): Since the vertex h is adjacent to all other vertices, it follows that any pair of vertices can be joined by a length-two path going through h as in Part (a). This implies $\text{diam}(G) \leq 2$, but since G has pairs of nonadjacent vertices (as in Part (a)), it follows that $\text{diam}(G) \geq 2$, so we may conclude that $\text{diam}(G) = 2$.

Part (c): For any pair of vertices v, w in C_n, the distance $\text{dist}(v, w)$ will be the length of the shorter (simple) path from v to w (proceeding clockwise or counter clockwise if the cycle is drawn as in Figure 8.8).   Since the sum of the lengths of these two paths is n (the total number of edges), it follows that the distance will be a maximum when the vertices are "oppositely" positioned on the cycle so that each of the two paths has length as close as possible to $n/2$.   In case n is even, both paths can have length $n/2$, and this will be the resulting diameter, and when n is odd, the shorter path will have $(n-1)/2$ edges, and the longer path will have $(n+1)/2$ edges, so the diameter will be $(n-1)/2$. In summary, we have proved that

$$\text{diam}(C_n) = \begin{cases} n/2, & \text{if } n \text{ is even} \\ (n-1)/2, & \text{if } n \text{ is odd} \end{cases} = \left\lfloor \frac{n}{2} \right\rfloor.$$

EXERCISE FOR THE READER 8.19: Determine the diameters of the following simple graphs:

(a)   The n-path P_n, for $n \geq 2$.

(b)   The n-hypercube graph Q_n, for $n \geq 1$.

Concepts like the diameter of a graph are often difficult for students since they tacitly involve a mixture of both maximums and minimums.   Indeed, the definition of the diameter can be written as:

$$\text{diam}(G) = \max_{v, w \in V(G)} \left\{ \min \{ \text{length}(P) \mid P \text{ is a path in } G \text{ joining } v \text{ to } w \} \right\}.$$

This sort of phenomenon will reappear at several junctures in the remainder of this chapter.   In the digraph setting, distances and related concepts are defined in exactly the same way, except that edge directions need to be respected.   The following definition describes some concepts about graphs that are related to distances and diameters.

Eccentricity, Radius, and Central Vertices

DEFINTION 8.26: Suppose that G is a graph and $v \in V(G)$. The **eccentricity** of the vertex v, denoted $ecc(v)$, is the distance from v to a farthest vertex from v, i.e.,

$$ecc(v) = \max_{w \in V(G)} \{dist(v, w)\}.$$

The **radius** of the graph G, denoted $rad(G)$, is the minimum of all vertex eccentricities, i.e.,

$$rad(G) = \min_{v \in V(G)} \{ecc(v)\}.$$

A **central vertex** z of the graph G is a vertex of minimum eccentricity, i.e., $ecc(z) = rad(G)$.

Observe that the diameter of a graph can also be expressed in terms of eccentricities: $diam(G) = \max_{v \in V(G)} \{ecc(v)\}$. In network applications, central vertices often correspond to optimal locations for central hubs or supply centers, since the worst-case distance to any other node (vertex) would be minimized.

EXAMPLE 8.19: (a) Find the radius and all central vertices of C_4.

(b) Find all central vertices of the graph G shown in Figure 8.38.

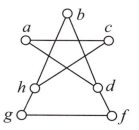

FIGURE 8.38: The graph G of Example 8.19.

SOLUTION: Part (a): In C_4, since each vertex has eccentricity 2, $rad(C_4) = 2$, and every vertex is a central vertex.

Part (b): The graph is left-right symmetric, so $ecc(a) = ecc(c)$, $ecc(h) = ecc(d)$, and $ecc(g) = ecc(f)$. Among the vertices b, a, h, and g, the largest distance between them is $dist(a, g) = 3$, and all other distances are at most 2. Thus, $ecc(a) = ecc(c) = ecc(g) = ecc(f) = 3$, $ecc(b) = ecc(h) = ecc(d) = 2$, and the central vertices of G are b, h, and d.

EXERCISE FOR THE READER
8.20: For the "witch" graph of
Figure 8.39, find the following:

(a) dist(c, j)
(b) The diameter
(c) ecc(c), ecc(j)
(d) The radius
(e) All central vertices

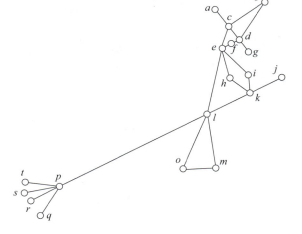

FIGURE 8.39: The "witch" graph
for Exercise for the Reader 8.20.

When computing distances between vertices, the paths on which distances are
realized must be simple and have no repeated vertices (why?). The following
simple result will be useful; informally it states that any path can always be pruned
down into a simple path with no repeated vertices.

PROPOSITION 8.5: If $P = <v_0, e_1, v_1, e_2, v_2, \cdots, v_{n-1}, e_n, v_n>$ is a path from a
vertex $v = v_0$ to another vertex $w = v_n$ in a graph or a digraph, then P contains a
simple subpath $P' = <v_0', e_1', v_1', e_2', v_2', \cdots, v_{m-1}', e_m', v_m'>$ with no repeated
vertices, that also joins $v = v_0'$ to $w = v_n'$. This means that P' is a simple path
and with $m \leq n$, and with its edge and vertex sequences: $<e_1', e_2', \cdots, e_{m-1}', e_m'>$,
$<v_0', v_1', v_2', \cdots, v_{m-1}', v_m'>$ being subsequences of the corresponding sequences of
P.

Proof: If the given path P had a repeated vertex, say $v_i = v_j = u$, where
$0 \leq i < j \leq n$, then the portion of P from v_i to v_j: $<v_i, e_{i+1}, v_{i+1}, e_{i+2}, v_{i+3}, \cdots,$
$e_j, v_j>$ would be a circuit that could be deleted from P to render a shorter path
from v to w. If we iteratively apply this process we will eventually be left with a
path from v to w that contains no repeated vertices, and hence also no repeated
edges, as desired. ◻

The next example gives an application of paths to social network and collaboration
graphs.

EXAMPLE 8.20: (*Minimal Length Paths in Social Network/Collaboration Graphs*)

(a) (*Actor/Movie Graph*) A popular (and very difficult) game for movie aficionados is known as the *Six Degrees of Kevin Bacon*, and is based on the constantly evolving Internet Movie Database (http://us.imdb.com/), which contains detailed information including over 1 million actors in major and independent motion pictures from around the world. The game is based on the large and constantly evolving actor (or actress) collaboration graph, where the set of vertices are the set of all actors appearing in the database, and two actors are joined by an edge if they both appeared in the same movie. Kevin Bacon (born in 1958) is a famous Hollywood actor who appeared in numerous movies beginning with *Animal House* in 1978. Others he has appeared in include *JFK*, *A Few Good Men,* and *Frost/Nixon.* The original *Six Degrees of Kevin Bacon* game works as follows: given any actor, the goal is to determine the distance to Kevin Bacon in this graph or, failing that, to find a shorter path in the collaboration graph than any of the other people playing can find (without resorting to the online database, of course). The distance of an actor to Kevin Bacon is called the **Bacon number** of the actor.[13] For example, at the time of the writing of this section, the Bacon number of Denzel Washington is two. They are linked by the path: (Denzel Washington, Josh Brolin)—both in the movie *American Gangster*, and (Josh Brolin, Kevin Bacon) —both were in the movie *Hollow Man.* There is a webpage (http://oracleofbacon.org/) that computes Bacon numbers (and displays details of optimal paths) and, more generally, distances between any two actors in terms of the latest data on the Internet Movie Database. The site also computes other interesting data concerning the movie collaboration graph including eccentricities and more.

(b) (*Mathematical Collaboration Graph*) The concept of a Bacon numbers was preceded by the analogous concept of **Erdös numbers**, based on the mathematical journal paper collaboration graph of the widely published mathematician Paul Erdös (see Section 5.1 for a brief biography). In the mathematical journal paper collaboration graph, the vertices are all mathematicians (who have published at least one journal paper), and two mathematicians are joined by an edge if, and only if, they have coauthored at least one paper. Paul Erdös is the only mathematician with Erdös number zero. Any mathematician who coauthored a paper with Erdös has Erdös number 1; anyone who has coauthored a paper with someone with Erdös number 1, but not with Erdös, has Erdös number 2, and so on. A mathematician has Erdös number ∞ if he/she does not have a finite Erdös

---

[13] As new movies are released, new vertices and edges get added to this movie collaboration graph, so Bacon numbers can decrease with time, and are well-defined only with respect to a certain time frame. The Bacon number of Kevin Bacon is zero. Any actor who made a movie with Kevin Bacon has Bacon number 1. For example, Kiefer Sutherland and Tom Cruise have Bacon number 1 because they both appeared with Kevin Bacon in *A Few Good Men.* These Bacon numbers can never change with time (Why?). An actor like Denzel Washington who has not (yet) made a movie with Keven Bacon, but made one with someone whose Bacon number is 1 (i.e., someone who made a movie with Kevin Bacon) has Bacon number 2, but this could go down to 1 in the future if the two ever make a movie together. The highest Bacon number of any actor in the Internet Movie Database is 8 (at the time of this writing).

number. Some websites have been set up to compute Erdös numbers based on extensive databases of mathematical papers, and, more generally, the distance between any two mathematicians in the collaboration graph. For example, at the time of this writing, the author's Erdös number is three, since he coauthored a paper with Wayne S. Smith, who, in turn, coauthored a paper with Andrew Odlyzko, an Erdös coauthor.

(c) (*The Human Web Graph*) Another constantly evolving graph has vertex set consisting of all living human beings, and an edge links two people if they are aquainted. Distances in this graph measure the social distance between two individuals, and thus have important ramifications in fields such as in epidemiology, and in the spread of information. Although this graph is too unwieldy for a complete analysis, it has been conjectured that its diameter is very small, perhaps six or seven. This would mean that, say a certain villager in Timbuktu is just six or seven introductions away from a certain car mechanic in Tuscaloosa, Alabama. The first large-scale partial justification of such conjectures was conducted by Microsoft researchers in 2006 where they analyzed email correspondences sent in a single month in 2006 between 180 million people in their network, and found that the average distance (using email correspondences as links) between any two people in their network was 6.6, and that 78% of all pairs of people are connected with seven or fewer edges.

Adjacency Matrices and Distance Computations in Graphs and Directed Graphs

By their very definition, adjacency matrices (for both graphs and digraphs) contain information about all pairs of vertices that are separated by distance one (i.e., that are connected by an edge). The following interesting result shows that powers of adjacency matrices contain useful information concerning larger distances between pairs of vertices.

THEOREM 8.6: (*Powers of Adjacency Matrices Encode Distances*) Suppose that A is the adjacency matrix of a certain graph or digraph with respect to a certain ordering of the vertices: $v_1, v_2, \cdots, v_n$. If k is a nonnegative integer and $B = (b_{ij})$ is the kth power of A, i.e., $B = A^k = A \cdot A \cdots A$ (k times), then b_{ij} is the total number of paths in the graph (or directed paths in the digraph) that start at v_i, end at v_j, and have length k.

Before proving this result, we first give an application to shipping logistics that will demonstrate its utility. In our example, the underlying network is intentionally made small, with only five cities, in order to facilitate hand computation yet clearly showcase all of the concepts. The same ideas could be applied to much larger networks; for example a *FedEx* logistics network of 500 cities would result in a 500 by 500 matrix, for which computing platforms could easily compute powers in split second speed.

EXAMPLE 8.21: (*Application to Shipping Logistics*) A Pacific Rim air shipping company has connecting flights between five cities as indicated by the directed graph of Figure 8.40.

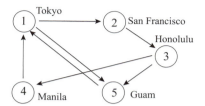

(a) Find the adjacency matrix A, compute $B = A^2$, and interpret the entries b_{31}, b_{12} in terms of the shipping network.

FIGURE 8.40: A small air shipping network.

(b) Compute $C = A^3$, and interpret the entries c_{14}, c_{12} in terms of the shipping network.

(c) What is the significance of the diameter of the directed graph in terms of the shipping network?

(d) Explain how the adjacency matrix could be used to compute the diameter of this (or any) directed (or undirected) graph, and perform the computation for this shipping network.

SOLUTION: Part (a): We use the given ordering of the vertices to form the adjacency matrix A:

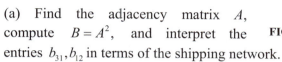

$$
A = \begin{array}{c} Rows = \\ Origins \end{array}
\begin{array}{c}
\#1 \Rightarrow \\ \#2 \Rightarrow \\ \#3 \Rightarrow \\ \#4 \Rightarrow \\ \#5 \Rightarrow
\end{array}
\begin{bmatrix}
0 & 1 & 0 & 0 & 1 \\
0 & 0 & 1 & 0 & 0 \\
0 & 0 & 0 & 1 & 1 \\
1 & 0 & 0 & 0 & 0 \\
1 & 0 & 0 & 0 & 0
\end{bmatrix}.
$$

Next we form the square of the adjacency matrix:

$$
B = A^2 = A \cdot A =
\begin{bmatrix}
0 & 1 & 0 & 0 & 1 \\
0 & 0 & 1 & 0 & 0 \\
0 & 0 & 0 & 1 & 1 \\
1 & 0 & 0 & 0 & 0 \\
1 & 0 & 0 & 0 & 0
\end{bmatrix}
\begin{bmatrix}
0 & 1 & 0 & 0 & 1 \\
0 & 0 & 1 & 0 & 0 \\
0 & 0 & 0 & 1 & 1 \\
1 & 0 & 0 & 0 & 0 \\
1 & 0 & 0 & 0 & 0
\end{bmatrix}
=
\begin{bmatrix}
1 & 0 & 1 & 0 & 0 \\
0 & 0 & 0 & 1 & 1 \\
2 & 0 & 0 & 0 & 0 \\
0 & 1 & 0 & 0 & 1 \\
0 & 1 & 0 & 0 & 1
\end{bmatrix}
$$

Theorem 8.6 tells us that because row 3, column 1 entry b_{31} equals 2, there are exactly two routes from Honolulu (#3) to Tokyo (#1) that use exactly two flights. This fact is easily confirmed by glancing at the network: the two itineraries are: Honolulu → Guam → Tokyo, and Honolulu → Manila → Tokyo.

This can be restated in the language of graph theory as follows: there are two paths of length 2 in the directed graph from vertex #3 to vertex #1: $3\to5\to1$, and $3\to4\to1$.

The entry $b_{12} = 0$ similarly tells us that although there is a direct flight from Tokyo (#1) to San Francisco (#2), the journey cannot be made using two flights.

Part (b):

$$C = A^3 = A \cdot A^2 = \begin{bmatrix} 0 & 1 & 0 & 0 & 1 \\ 0 & 0 & 1 & 0 & 0 \\ 0 & 0 & 0 & 1 & 1 \\ 1 & 0 & 0 & 0 & 0 \\ 1 & 0 & 0 & 0 & 0 \end{bmatrix} \cdot \begin{bmatrix} 1 & 0 & 1 & 0 & 0 \\ 0 & 0 & 0 & 1 & 1 \\ 2 & 0 & 0 & 0 & 0 \\ 0 & 1 & 0 & 0 & 1 \\ 0 & 1 & 0 & 0 & 1 \end{bmatrix} = \begin{bmatrix} 0 & 1 & 0 & 1 & 2 \\ 2 & 0 & 0 & 0 & 0 \\ 0 & 2 & 0 & 0 & 2 \\ 1 & 0 & 1 & 0 & 0 \\ 1 & 0 & 1 & 0 & 0 \end{bmatrix}$$

In light of Theorem 8.6, the entry $c_{14} = 1$ tells us that there is precisely one route that goes from City #1 (Tokyo) to City #4 (Manila) and that uses exactly three flights. The reader can verify using Figure 8.40 that this unique route is:

$$\text{#1(Tokyo)}\to\text{#2(San Francisco)}\to\text{#3(Honolulu)}\to\text{#4(Manila)}.$$

The entry $c_{12} = 1$ indicates that there is precisely one route that goes from City #1 (Tokyo) to City #2 (San Francisco) and that uses exactly three flights. The reader can verify using Figure 8.40 that this unique route is:

$$\text{#1(Tokyo)}\to\text{#5(Guam)}\to\text{#1(Tokyo)}\to\text{#2(San Francisco)},$$

demonstrating that these matrices count literally <u>all</u> possibilities. Of course, programs can be easily designed to weed out such inefficient routes.

Part (c): The diameter of a directed graph is the maximum of the distances between pairs of vertices. This translates to the maximum number of flights that would be needed to ship from any city in the network to any other.

Part (d): We make one preliminary observation that might help readers to discover the relevant algorithm. From the facts:

(i) The entries of A tell us (by definition of an adjacency matrix) the number of (directed) paths of length one from the row vertex to the column vertex.
(ii) The entries of A^2 tell us (by Theorem 8.6) the number of (directed) paths of length two from the row vertex to the column vertex.

It follows that the entries of the matrix $A + A^2$ tell us the number of (directed) paths of length one or two from the row vertex to the column vertex.

By the same token, it follows that the entries of the matrix sum $A + A^2 + \cdots + A^k$ tell us the number of (directed) paths of length 1, 2, $\cdots$, or k, from the row vertex to the column vertex. Thus, the diameter will be the smallest nonnegative exponent k such that all non-diagonal entries of the sum $A + A^2 + \cdots + A^k$ are nonzero.[14] Thus, we need only continue adding successive powers of the adjacency matrix until the result has no zero entries, except possibly on the main diagonal. Here are the computations for the example on hand:

$$A + A^2 = \begin{bmatrix} 1 & 1 & 1 & 0 & 1 \\ 0 & 0 & 1 & 1 & 1 \\ 2 & 0 & 0 & 1 & 1 \\ 1 & 1 & 0 & 0 & 1 \\ 1 & 1 & 0 & 0 & 1 \end{bmatrix}, \quad A + A^2 + A^3 = \begin{bmatrix} 1 & 2 & 1 & 1 & 3 \\ 2 & 0 & 1 & 1 & 1 \\ 2 & 2 & 0 & 1 & 3 \\ 2 & 1 & 1 & 0 & 1 \\ 2 & 1 & 1 & 0 & 1 \end{bmatrix},$$

$$A + A^2 + A^3 + A^4 = \begin{bmatrix} 4 & 2 & 2 & 1 & 3 \\ 2 & 2 & 1 & 1 & 3 \\ 4 & 2 & 2 & 1 & 3 \\ 2 & 2 & 1 & 1 & 3 \\ 2 & 2 & 1 & 1 & 3 \end{bmatrix}.$$

It follows that the diameter of the network is four. Note that there was just one zero nondiagonal entry of $A + A^2 + A^3$ that necessitated adding another power of the adjacency matrix: This was the row 5, column 4 entry; it corresponds to the fact that it is not possible to use the network of Figure 8.40 to ship from Guam (City #5) to Manila (City #4) using at most three flights.

We next formalize the useful algorithm that was developed in the preceding example. We make one small cosmetic enhancement concerning the addition of powers of the adjacency matrix: If instead of looking at the sums: $A, A + A^2, A + A^2 + A^3, \cdots$ and terminating as soon as no entries, except possibly on the main diagonal, are zero, we may equivalently look at the sums $I, I + A, I + A + A^2, I + A + A^2 + A^3, \cdots$ and terminate as soon as all of the matrix entries are nonzero. Since the identity matrix $I = A^0$ counts the number of zero length paths going from the row vertex to the column vertex, this modification naturally corresponds to simply accounting for the zero length paths. Note that if the process does not terminate when we get to adding A^n (n = the number of vertices), then the graph/digraph is not connected.[15]

ALGORITHM 8.2: Computation of Diameter of a Directed Graph or a Graph Using the Adjacency Matrix:

Input: The n by n adjacency matrix A of a graph or a directed graph with $n > 2$ vertices, with respect to some ordering of the vertices.[16]
Output: The diameter diam of the (directed) graph.

---

[14] The main diagonal entries correspond to reaching a vertex from itself. Since the distance from a vertex to itself is zero, these distances need not be considered in computing diameters. Likewise, since the shipping network is not a tour company, shipping from a city to itself is not a relevant issue.
[15] It follows from Proposition 8.5 that the diameter of a connected (di)graph with n vertices is at most $n - 1$.
[16] In the trivial case of a graph (or digraph) with just one vertex, the diameter is plainly zero.

Step 1: If all nondiagonal entries are nonzero (this means every vertex is reachable from every other vertex by a path of length 1) exit the algorithm and output the diameter to be 1.

Step 2: (*Iterative Step*) Continue adding increasing powers of A:

$I + A + A^2$, $I + A + A^2 + A^3$,... until either the power being added equals n, or we obtain a matrix that has no nonzero entries. Let M denote the exponent of the last power that was added.

Case 1: If $M < n$, output diam = M and exit.

Case 2: If $M = n$, output diam = ∞ (graph or digraph is not connected) and exit.

EXERCISE FOR THE READER 8.21:
(a) Apply Algorithm 8.2 to compute the diameter of the graph representing the shipping network shown in Figure 8.41.
(b) By inspecting Figure 8.41 and using Theorem 8.6, find the row 2 column 4 entry of the matrix $A + A^2 + A^3$. Then check this answer with the matrices of Part (a).

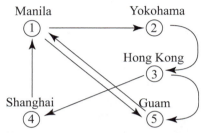

FIGURE 8.41: The small shipping network of Exercise for the Reader 8.21.

EXERCISE FOR THE READER 8.22: (a) Develop an algorithm based on adjacency matrices that is similar to Algorithm 8.2 and that will compute the eccentricity of an inputted vertex in a graph or a digraph.
(b) Apply your algorithm to compute the eccentricity of the vertex #1 (Tokyo) in the digraph of Example 8.21.

NOTE: In applications it is common to use the term **network** to refer to the graph or digraph that models a given applied problem. In this context the vertices are usually called **nodes** and the edges (directed or undirected) are called **links**.

While the shipping network of Example 8.21 was easily analyzed by direct inspection, when the number of vertices increases to a couple of dozen or so, visual inspection is no longer a feasible approach for the computation of diameters and related quantities. The reader may wish to try to compute (by visual inspection) the diameter of the 23-city shipping network shown in Figure 8.42. Using Algorithm 8.1 with the aid of a computer, however, this problem is quickly solved, and the diameter of the graph is found to be 8.

Let us do a preliminary worst-case complexity estimate for Algorithm 8.2 for an inputted adjacency network of n vertices. Each matrix multiplication requires the computation of n^2 entries, each of which requiring n multiplications and $n-1$ additions, and thus $O(n^3)$ mathematical operations. Each matrix addition requires only $O(n^2)$ mathematical operations. Since there will be at most n matrix multiplications and additions, this gives an upper bound of $n(O(n^3) + O(n^2)) = O(n^4)$ for the worst-case complexity of Algorithm 8.2. If our computing system can perform 10^7 mathematical operations in less than 10 seconds, this means that we could expect to use Algorithm 8.2 to compute diameters of networks involving up to $\sqrt[4]{10^7} \approx 56$ nodes (assuming the big-O proportionality constant to be 1).

AIR FREIGHT DAILY SCHEDULE

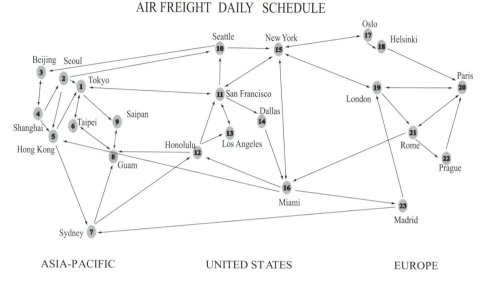

FIGURE 8.42: A directed graph representing a rather small international shipping network.

Of course, this worst-case analysis grossly overestimates the work typically required in using Algorithm 8.2. For example, the diameter is often much smaller than n, and many of the matrix multiplications involve zero entries. Also, there are faster algorithms for matrix multiplication than the standard $O(n^3)$ algorithm—see Section 4.4. In Section 8.5 we will present a more efficient algorithm for computing distances and diameters in networks.

We have focused our application of Theorem 8.6 on prototypical transportation networks, but networks can be used to model an assortment of other real-life phenomena including: computer networks, telephone networks, pipeline networks, and road networks. In some network applications, apart from accessibility and efficiency, there are other relevant considerations that include capacity (e.g., in a computer or pipeline network, we are interested not only in the time to transport a unit of data or liquid from one node to another, but also the amount of this quantity that can be transported). Such capacity problems will be considered in Section 8.5. At present, we move on to a brief discussion concerning network reliability. But we first fulfill our promise to prove theorem 8.6:

Proof of Theorem 8.6: We proceed by induction on k.

Basis Step: $k = 1$: This case simply follows from the definition of the adjacency matrix $A^1 = A$, whose row i column j entry gives the number of edges (length one paths) from vertex i to vertex j.

Induction Step: Let k be a positive integer. We assume that the theorem is true for A^k and must show it is also true for $A^{k+1} = AA^k$. We let $C = A^{k+1}$ and $B = A^k$. We need to show that the row i column j entry of C, c_{ij}, is equal to the following quantity: $\text{Paths}_{k+1}(i,j) \triangleq$ the number of paths in the network from vertex i to vertex j that have length $k+1$. By the definition of matrix multiplication, we have: $c_{ij} = a_{i1}b_{1j} + a_{i2}b_{2j} + \cdots + a_{in}b_{nj}$. Each path from vertex i to vertex j that has length $k+1$ can be broken down into a length 1 path from vertex i to some vertex ℓ, followed by a length k path from vertex ℓ to vertex j. By the multiplication principle, the number of such paths that visit vertex ℓ on the first node is $\text{Paths}_1(i,\ell) \cdot \text{Paths}_k(\ell,j)$ and by the inductive hypothesis, this latter product is the same as $a_{i\ell}b_{\ell j}$. From these facts, it now follows that:

$$\text{Paths}_{k+1}(i,j) \triangleq \sum_{\ell=1}^{n} \text{Paths}_1(i,\ell) \cdot \text{Paths}_k(\ell,j) = \sum_{\ell=1}^{n} a_{i\ell}b_{\ell j} = c_{ij}. \quad \square$$

Edge and Vertex Cuts in Connected Graphs/Digraphs

Whereas small diameters increase the efficiency of connected networks, another important aspect about networks concerns their reliability and vulnerability in case some nodes (vertices) or links (edges) fail. In practice networks are subject to occasional breakdown or maintenance of certain parts. Properly designed networks should anticipate such events and be able to function efficiently with alternate paths/routes. In order to properly discuss such breakdowns, we formally define the vertex and edge deletion operations for a digraph or a graph. Roughly speaking, when a vertex gets deleted from a (di)graph, all edges that are incident to the vertex need also to get deleted (if a node is down, we cannot realistically use it in our routes), but when an edge gets deleted, all other vertices and edges remain intact.

DEFINITION 8.27: Suppose that $G = (V, E)$ is a graph with a vertex $v \in V$ and a set of vertices $U \subseteq V$. We define the **vertex deletion graph** and **vertex set deletion graph** to be the following subgraphs of G:

$G - v$: vertices: $V \sim \{v\}$, edges: $\{e \in E : v \text{ is not an endpoint of } e\}$.

$G - U$: vertices: $V \sim U$, edges: $\{e \in E : \{\text{endpoints of } e\} \cap U = \varnothing\}$.

If $f \in E$ is an edge and $F \subseteq E$ is a set of edges, we define the **edge deletion graph** and **edge set deletion graph** to be the following subgraphs of G:

$G - f$: vertices: V, edges: $E \sim \{f\}$.
$G - F$: vertices: V, edges: $E \sim F$.

The corresponding definitions for digraphs are identical; simply change each occurrence of G to $\vec{G}$.

The following example illustrates these concepts.

EXAMPLE 8.22: Let G be the graph shown in Figure 8.43, with the following edge labels: $f_1 = \{x, y\}$, $f_2 = \{x, z\}$, $f_3 = \{y, z\}$. Draw each of the following deletion graphs:

(a) $G - \{s, t, u\}$

(b) $G - v$

(c) $G - f_3$

(d) $G - \{f_1, f_2\}$

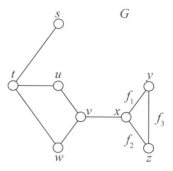

FIGURE 8.43: The simple graph of Example 8.22.

SOLUTION: The four deletion graphs are drawn in Figure 8.44.

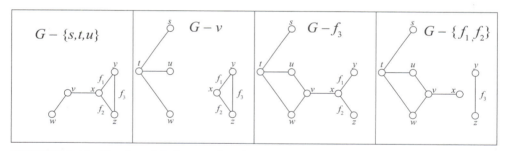

FIGURE 8.44: The four deletion graphs of Example 8.22.

Observe that the deletion graphs of Parts (b) and (d) literally broke the graph into pieces (the resulting graphs were no longer connected like the original graph), whereas those of Parts (a) and (c) remained connected. The former phenomenon is important; we single it out in the following definition:

DEFINITION 8.28: A set U of vertices of a graph G is called a **vertex cut** if $G - U$ has more components than G. In case $U = \{v\}$ is a single vertex, the vertex v is called a **cut vertex**.

A set F of edges of a graph G is called an **edge cut** if $G - F$ has more components than G. In case $F = \{f\}$ is a single edge, the edge f is called a **cut edge** or a **bridge**.

The definitions for directed graphs are the same, but keep in mind that the (strong) connectivity definition for digraphs is used.

Thus, in the preceding example, v is a cut vertex and $\{f_1, f_2\}$ is an edge cut.

EXERCISE FOR THE READER 8.23:　Find all cut vertices and bridges for the graph of Example 8.22.

A **minimum edge cut** of a connected network (having at least one edge) is an edge cut of smallest possible cardinality.　Reliable networks have large-sized minimum edge cuts (i.e., the network will remain connected even if a large number of links are down).　The existence of a bridge results in a minimum edge cut of size 1, an unstable network.　The next exercise for the reader considers the opposite extreme.

EXERCISE FOR THE READER 8.24:　Find the size of a minimum edge cut for the complete graph K_n, if $n \geq 2$.

EXERCISE FOR THE READER 8.25:　Prove that if f is an edge of a connected simple graph G, then either f is a bridge or f belongs to a cycle in G.

Characterization of Bipartite Graphs Using Cycles

We end this section by proving the following theorem, which gives a useful alternative way of viewing bipartite graphs:

THEOREM 8.7: (*Characterization of Bipartite Graphs*) A graph $G = (V, E)$ is bipartite if, and only if G has no cycles of odd length.

Proof: The "only if" part is quite straightforward:　In a bipartite graph G with an associated coloring of the vertices, if we traverse any cycle (starting at any vertex), the vertices must toggle between black and white colors as we pass through each edge.　So if a cycle had odd length, this would mean that when we get to the last vertex before returning to the starting vertex, we would have an even number of toggles, so this last vertex would have the same color as the starting vertex.　But the last vertex is adjacent to the starting vertex, so this would be impossible.

For the "if part," we assume that G has no cycles of odd length, and we need to show that G is bipartite.　We may assume that G has at least two vertices and that G is connected. (Why?)　We fix a vertex $u \in V$, and we define a partition of sets $X \cup Y = V$ as follows:

$$X = \{x \in V : \mathrm{dist}(u, x) \text{ is even}\}, \quad Y = \{y \in V : \mathrm{dist}(u, y) \text{ is even}\}.$$

It suffices to prove the following:

Claim: (X, Y) is a bipartition of V.

Proof of Claim: We must show that if two vertices w, z are in the same set (i.e., $w, z \in X$ or $w, z \in Y$) then $\{w, z\} \notin E$. We assume that $w, z \in X$; the corresponding proof for the case $w, z \in Y$ is similar. We will proceed by the method of contradiction: so we assume that $\{w, z\} \in E$.

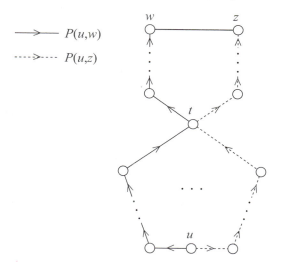

Let $P(u, w)$ and $P(u, z)$ be paths in G that realize $\text{dist}(u, x)$ and $\text{dist}(u, z)$, respectively. By definition of X, both of these paths have even lengths. Let t be the last common meeting vertex of $P(u, w)$ and $P(u, z)$; see Figure 8.45.

FIGURE 8.45: Minimal length paths in the proof of Theorem 8.7.

Note that $t = u$ is a possibility. By minimality of $P(u, w)$ and $P(u, z)$, it follows that the number of edges of each of $P(u, w)$ and $P(u, z)$ between u and t must be the same (so either they are both even or both odd), and so the parities of the remaining portions $P(t, w)$ and $P(t, z)$ must also be the same (i.e., both even or both odd). It follows that the path obtained by taking $P(t, w)$, followed by the path $< w, z >$, and then finally followed by the reverse path $P(z, t)$ of $P(t, z)$ produces a cycle of odd length in G. With this contradiction, the proof of the claim as well as the theorem is now complete. $\square$

EXERCISES 8.2:

1. For each of the following vertex sequences of the simple graph G of Figure 8.46 answer each of the following questions: (i) Does the sequence represent a path in G? For those that are paths, (ii) What is the length of the path? (iii) Is the path simple? (iv) Is the path a circuit?

    (a) $< t, u, y, x >$ (b) $< t, w, u, y >$

    (c) $< v, y, x, t, u, y >$ (d) $< t, u, y, u, t >$

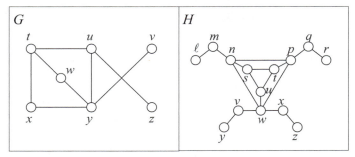

FIGURE 8.46: Two simple graphs for Exercises 1 and 2.

2.   For each of the following vertex sequences of the simple graph H of Figure 8.46 answer each of the following questions: (i) Does the sequence represent a path in H? For those that are paths, (ii) What is the length of the path? (iii) Is the path simple? (iv) Is the path a circuit?

   (a) $< n, p, q, r, q >$ (b) $< u, w, t, p >$

   (c) $< w, u, t, s, n, w >$ (d) $< \ell, m, n, p, t, s, n >$

3.   For each of the following vertex sequences of the simple directed graph $\vec{G}$ of Figure 8.47 answer each of the following questions: (i) Does the sequence represent a path in $\vec{G}$? For those that are paths, (ii) What is the length of the path? (iii) Is the path simple? (iv) Is the path a circuit?

   (a) $< v, y, u >$ (b) $< u, y, v, u, y, v >$

   (c) $< w, v, u, y, v, x, z, w >$ (d) $< v, z, y >$

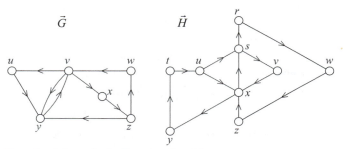

FIGURE 8.47: Two directed graphs for Exercises 3 and 4.

4.   For each of the following vertex sequences of the simple directed graph $\vec{H}$ of Figure 8.48 answer each of the following questions: (i) Does the sequence represent a path in $\vec{H}$? For those that are paths, (ii) What is the length of the path? (iii) Is the path simple? (iv) Is the path a circuit?

   (a) $< t, u, s, r, w, z, x >$ (b) $< u, s, v, z, u >$

   (c) $< z, x, y, t, u, x >$ (d) $< v, z, s, r, w, z, x, s, v >$

5.   Determine whether the simple graph G of Figure 8.48 is connected and find the components.

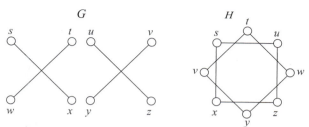

FIGURE 8.48: Two graphs for Exercises 5 and 6.

6.   Determine whether the simple graph H of Figure 8.48 is connected and find the components.

7.   Determine whether each of the directed graphs of Figure 8.49 is connected and whether each is weakly connected.

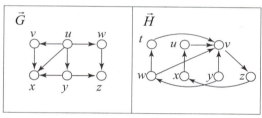

FIGURE 8.49: Two directed graphs for Exercise 7.

8.   Determine whether each of the directed graphs of Figure 8.50 is connected and whether each is weakly connected.

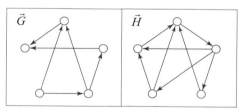

FIGURE 8.50: Two directed graphs for Exercise 8.

9.   Find the following for the simple graph G of Figure 8.46:
     (a) dist(t,u), dist(v,z), dist(x,w).
     (b) The diameter diam(G) and the radius rad(G).
     (c) ecc(t), ecc(v).
     (d) All central vertices.

10. Find the following for the simple graph H of Figure 8.46:
     (a) dist(v,p), dist(u,n), dist(z,y).
     (b) The diameter diam(H) and the radius rad(H).
     (c) ecc(z), ecc(u).
     (d) All central vertices.

11. (a) Explain the meaning of the components in the actor/movie graph of Example 8.20(a).
    (b) Is it possible for a component to consist of a single actor? Explain.

12. (a) Explain the meaning of the components in the mathematical collaboration graph of Example 8.20(b).
    (b) Is it possible for a component to consist of a single mathematician? Explain.
    (c) What can be said about the Erdös number of a mathematician who lies in the same component as Erdös in this graph?

13. Determine the diameter, the radius, and all the central vertices of the following simple graphs:
    (a) The n-path graph P_n, $n \geq 2$ (Definition 8.4).
    (b) The Petersen graph (Figure 8.14).
    (c) The complete graph K_n, $n \geq 1$ (Definition 8.4).

14. Determine the diameter, the radius, and all the central vertices of the following simple graphs:
    (a) The n-cycle graph C_n, $n \geq 3$ (Definition 8.4).
    (b) The complete bipartite graph $K_{n,m}$, $n,m \geq 1$ (Definition 8.6).
    (c) The graph J of Exercise 25 in Section 8.1.

15. (a) Write down the adjacency matrix A for the three-city shipping network of Figure 8.51.
    (b) For the matrix A of Part (a), use Theorem 8.6 and the figure to interpret and compute the row 1 column 2 entry of the matrix $A + A^2 + A^3$. (Do not perform any matrix multiplications.)
    (c) Use Algorithm 8.2 to compute the diameter of the three-city shipping network of Figure 8.51.
    (d) Write down the adjacency matrix A for the five-city shipping network of Figure 8.51.
    (e) For the matrix A of Part (d), use Theorem 8.6 and the figure to interpret and compute the row 2 column 3 entry of the matrix $A + A^2 + A^3$. (Do not perform any matrix multiplications.)
    (f) Use Algorithm 8.2 to compute the diameter of the five-city shipping network of Figure 8.51.

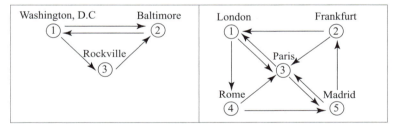

FIGURE 8.51: Two shipping networks for Exercise 15.

16. (a) Write down the adjacency matrix A for the four-city shipping network of Figure 8.52.
    (b) For the matrix A of Part (a), use Theorem 8.6 and the figure to interpret and compute the row 2 column 3 entry of the matrix $A + A^2 + A^3$. (Do not perform any matrix multiplications.)
    (c) Use Algorithm 8.2 to compute the diameter of the four-city shipping network of Figure 8.52.
    (d) Write down the adjacency matrix A for the six-city shipping network of Figure 8.52.
    (e) For the matrix A of Part (d), use Theorem 8.6 and the figure to interpret and compute the row 4 column 3 entry of the matrix $A + A^2 + A^3$. (Do not perform any matrix multiplications.)
    (f) Use Algorithm 8.2 to compute the diameter of the six-city shipping network of Figure 8.52

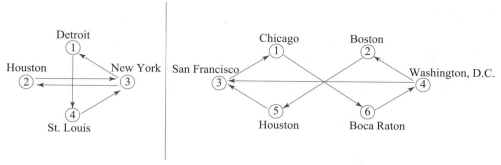

FIGURE 8.52: Two shipping networks for Exercise 16.

17. For the graph G of Figure 8.46, do the following:
    (a) Draw each of vertex deletion graphs: $G-v$, $G-w$, $G-\{w,y\}$.

    (b) With the edge labels: $f_1 = \{t,u\}$, $f_2 = \{w,y\}$, $f_3 = \{y,v\}$, draw each of the following edge deletion graphs: $G-f_1$, $G-\{f_2,f_3\}$, $G-\{f_1,f_2,f_3\}$.
    (c) Find all cut-vertices.
    (d) Find all bridges.
    (e) What is the size of a minimum edge cut?

18. For the graph H of Figure 8.46, do the following:
    (a) Draw each of vertex deletion graphs: $H-m$, $H-u$, $H-\{\ell,r,w\}$.

    (b) With the edge labels: $f_1 = \{t,u\}$, $f_2 = \{v,y\}$, $f_3 = \{w,x\}$, draw each of the following edge deletion graphs: $H-f_1$, $H-\{f_2,f_3\}$, $H-\{f_1,f_2,f_3\}$.
    (c) Find all cut-vertices.
    (d) Find all bridges.
    (e) What is the size of a minimum edge cut?

19. Draw an example of a simple graph that has paths of arbitrarily long length.

20. Draw an example of a simple directed graph that has (directed) paths of arbitrarily long length.

21. (a) What is the minimum number of edges that a simple graph with four vertices must have in order that it be connected? Give an example of such a graph.
    (b) How many isomorphism classes of connected simple graphs of four vertices and the minimum number of edges (see Part (a)) are there?
    (c) What is the minimum number of edges that a simple graph with n vertices must have in order that it be connected? Give an example of such a graph.

22. (a) Prove that every simple graph with at least one vertex has at least one noncut vertex.
    (b) Prove that every simple graph with at least two vertices has at least two noncut vertices.
    (c) Give an example of a simple graph with at least three vertices that does not have three noncut vertices.

23. (a) Use the relevant definitions to prove the following inequalities for any connected graph G:
    $$\text{rad}(G) \le \text{diam}(G) \le 2 \cdot \text{rad}(G).$$
    (b) Do the inequalities of Part (a) remain valid without the connectedness assumption? Provide either counterexample(s) or a proof.

24. (a) What is the maximum number of edges that a simple graph with five vertices could have in order that it have (at least) two components? Give an example of such a graph.
    (b) How many isomorphism classes of simple graphs of five vertices, two components, and the

maximum number of edges (as in Part (a)) are there?

(c) What is the maximum number of edges that a simple graph with seven vertices could have in order that it have (at least) two components? Give an example of such a graph.

(d) How many isomorphism classes of simple graphs of seven vertices, two components, and the maximum number of edges (as in Part (a)) are there?

25. For each of the following graphs, determine the number of cut vertices, the number of bridges, and the size of a minimum edge cut:

(a) The n-path graph P_n, $n \geq 2$ (Definition 8.4).

(b) The complete bipartite graph $K_{n,m}$, $n, m \geq 1$ (Definition 8.6).

(c) The complete graph K_n, $n \geq 1$ (Definition 8.4).

26. For each of the following graphs, determine the number of cut vertices, the number of bridges, and the size of a minimum edge cut:

(a) The n-cycle graph C_n, $n \geq 3$ (Definition 8.4).

(b) The graph of Figure 8.35.

(c) The graph J of Exercise 25 in Section 8.1.

27. For each graph below, find the largest value of k for which the graph contains a copy of the k-cycle C_k, $k \geq 3$, or indicate that the graph contains no cycles.

(a) The complete graph K_n, $n \geq 1$ (Definition 8.4).

(b) The complete bipartite graph $K_{n,m}$, $n, m \geq 1$ (Definition 8.6).

(c) The Petersen graph (Figure 8.14).

28. The *girth* of a simple graph that contains a copy of at least one cycle C_n, $n \geq 3$, is defined to be the length n of the shortest cycle it contains. The girth of a simple graph that does not contain any cycles is undefined. Determine the girths of each of the following graphs.

(a) The complete graph K_n, $n \geq 3$ (Definition 8.4).

(b) The complete bipartite graph $K_{n,m}$, $n, m \geq 3$ (Definition 8.6).

(c) The Petersen graph (Figure 8.14).

NOTE: (***Paths/Cycles can sometimes help to resolve isomorphism questions***) As indicated in Section 8.1, checking to see whether two graphs contain copies of other graphs as subgraphs can often be helpful in showing that two graphs are not isomorphic. Looking at (simple) path and cycle subgraphs is a special case. In case two graphs contain simple paths or cycles that are long, this fact can often be useful in constructing an isomorphism between the graphs, if they are indeed isomorphic. This is done by matching vertices in the corresponding cycles, so that their degrees and perhaps other properties are preserved. Exercises 29–32 will expound on these ideas.

Each of Exercises 29–32 presents a series of graphs. By considering paths and/or cycles, determine whether any pairs of the following three graphs are isomorphic. For each pair, either provide a specific isomorphism (or use the fact that isomorphism is an equivalence relation) or explain why the graphs are not isomorphic.

29.   G   H   K

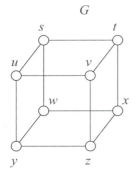   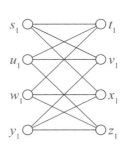   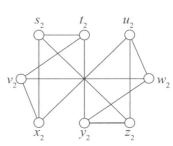

30.

G   H   K

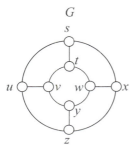   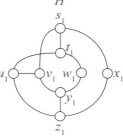   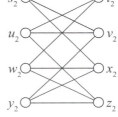

NOTE: (***Structure of the Petersen graph***) The next two exercises involve the Petersen graph. There is an interesting structure of the Petersen graph that is often helpful, for example, in establishing isomorphisms. If the vertices are viewed as the $10 = C(5,2)$ two-element subsets of $\{1,2,3,4,5\}$, then the Petersen graph results by joining two vertices if, and only if, the two-element sets they represent are disjoint. Figure 8.53 shows one such vertex labeling for the Petersen graph.

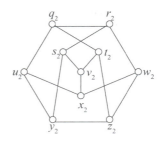

FIGURE 8.53: The set-theoretic structure of the Petersen graph.

31.   P   G   H

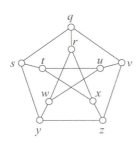   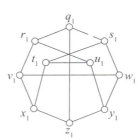   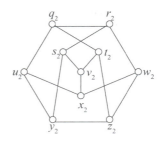

32.

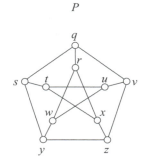
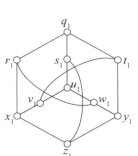
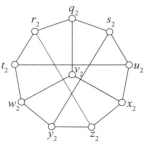

33. How many isomorphism classes are there of five-vertex simple graphs with two components? Draw representatives for each of the classes.

34. How many isomorphism classes are there of six-vertex simple graphs with two components? Draw representatives for each of the classes.

35. Prove that the number black/white bipartition vertex colorings of a bipartite graph G is $2^{C(G)}$, where $C(G)$ is the number of components of G.

36. Prove that any simple graph having at least as many edges as it has vertices will have at least one cycle, i.e., a copy of C_n, for some $n \geq 3$.

37. Prove the following proposition: *A simple graph G is bipartite if, and only if every subgraph H of G has an independent set consisting of at least half of its vertices.*
    Note: See Definition 8.20 in the Exercises of Section 8.1 for the definition of an independent set of vertices.
    Suggestion: For the "if" part, use the method of contradiction. Make use of the result of Exercise 61(d) of Section 8.1.

38. Prove the following proposition: *If a graph G has exactly two vertices u,v of odd degree, then there is a path in G joining u to v.*
    Suggestion: Proceed by the method of contradiction. Corollary 8.2 will be helpful.

39. Prove that if G is a simple graph, then either G or its complement $\sim G$ (or both) will be connected.

40. (*Line Graphs*) The **line graph** $L(G)$ of a simple graph G has the edges of G, $E(G)$, as its vertices, i.e., $V(L(G)) = E(G)$. Two vertices of $L(G)$ are adjacent if, and only if the corresponding edges of G are incident to a common vertex of G. Figure 8.54 shows an example of a simple graph G along with a drawing of its line graph $L(G)$.

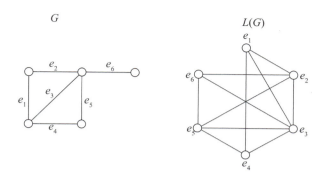

FIGURE 8.54: An example of a simple graph G and a drawing of its corresponding line graph $L(G)$.

(a) Create drawings of the following line graphs: $L(P_2), L(P_3), L(C_3), L(K_4)$.

(b) Show that $L(C_n) \cong C_n$.

(c) Show that $\sim L(K_5) \cong$ the Petersen graph.

(d) Show that for any simple graph G, the number of edges of $L(G)$ is $\sum_{v \in V(G)} C(\deg(v), 2)$.

(e) Prove that a simple graph G is isomorphic to its line graph $L(G)$ if, and only if G is 2-regular.

(f) Show that $L(K_3) \cong K_3$, and $L(K_{1,3}) \cong K_3$.

(g) Show that if G and H are simple graphs, neither of which is isomorphic to either K_3 or $K_{1,3}$, and if their line graphs are isomorphic: $L(G) \cong L(H)$, then $G \cong H$.

41. (*Adjacency Matrix-Based Algorithm for Determination of the Components of a Graph*)
(a) Develop an algorithm based on adjacency matrices that will compute the components of a simple graph.
(b) Apply your algorithm of Part (a) to the six vertex graph consisting of two disjoint copies of K_3, using the following ordering of the vertices: the first three vertices are those of one copy of K_3 and the last three are those of the second copy.
(c) Repeat Part (b) but with the vertices being ordered by alternately taking a vertex from each copy of K_3.

42. (*Adjacency Matrix-Based Algorithm for Determination of the Components of a Digraph*) The *components* of a digraph are defined similarly to those of a graph, except that the underlying equivalence relation is that two vertices should be reachable from one another.
(a) Develop an algorithm based on adjacency matrices that will compute the components of a digraph.
(b) Apply your algorithm of Part (a) to the six vertex digraph $\vec{G}$ of Figure 8.47 (with the vertices ordered alphabetically).
(c) Apply your algorithm of Part (a) to the seven vertex digraph $\vec{H}$ of Figure 8.49 (with the vertices ordered alphabetically).

43. (*Adjacency Matrices Can Count 3-Cycle Subgraphs*) Let A be the adjacency matrix for a simple graph with some vertex ordering, and let $B = A^2$ and $C = A^3$.
(a) Show that b_{ii} is the degree of vertex # i.
(b) Show that c_{ii} is twice the number of three cycles that pass through vertex # i.

COMPUTER EXERCISES 8.2:

1. (*Program for Algorithm 8.2*) (a) Write a program for computing the diameter of a network (i.e., either a graph or a digraph) based on Algorithm 8.2. The syntax should be as follows:

    $$\text{diam = NetworkDiameterAlg8\_2(A)}$$

    where the input A is the adjacency matrix for the network, and the output diam is the diameter of the network.

    (b)   Check your program with the results obtained for the networks of Example 8.21 and Exercise for the Reader 8.21.

2. (*Program for Computing Vertex Eccentricities in a Graph*) (a) Write a program for computing all of the vertex eccentricities in a network (i.e., either a graph or a digraph) that is based on computing powers of the adjacency matrix. The syntax should be as follows:

    $$\text{EccVec = NetworkEccentricities(A)}$$

    where the input A is the adjacency matrix for the network, and the output EccVec is a vector of the corresponding vertex eccentricities.

    (b) Check your program on the networks of Example 8.22 and Exercise for the Reader 8.21.
    Suggestion: The main idea for the program is contained in the solution of Exercise for the Reader 8.21. But the program should be structured so that internally it computes the sequence $I + A + A^2$, $I + A + A^2 + A^3$,... only once, reading off any new eccentricities of vertices discovered as each new power of A is added.

3. (*Some Small Network Questions for the Programs of Computer Exercises 1 and 2*) Run your programs of Computer Exercises 1 and 2 to compute the diameters and vertex eccentricities for each of the following networks:
    (a) The seven-city network shown below:

    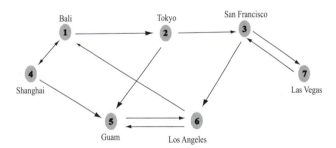

    (b) The 10-city network shown below:

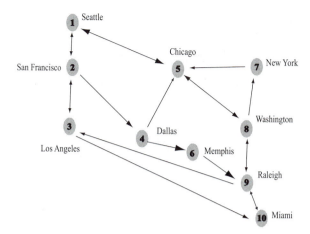

(c) The 10-node network shown below:

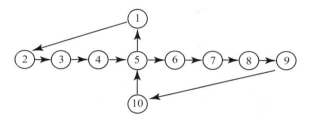

(d) The 23-city network of Figure 8.42.

NOTE: (***Diameter Reduction in Networks***) We again use the familiar setting of shipping networks to motivate the topic of diameter reduction in a network. Suppose that the manager of a certain air shipping network feels that the diameter of the network is too large (for example, think of the 23-city network of Figure 8.42, which has diameter 8), and wants to consider the feasibility of adding some new flights in order to reduce the diameter. Suppose first that the manager would like to know how much of a reduction is possible with the addition of a single new route (i.e., a single new arrow in the network diagram), and furthermore, wishes to know all of the new routes that would yield this optimal reduction.[17] Since adding one new edge to a digraph network corresponds to changing the corresponding off diagonal entry in the adjacency matrix from 0 to 1 (for a graph network two entries in the matrix need to be changed corresponding to the two directions between the cities), the brute-force approach would be to simply use Algorithm 8.2/Computer Exercise 1 (or whatever other program we have for computing network diameters) to compute the diameters of all of the different modified adjacency matrices that arise in this way and keep track of the routes that give rise to the lowest diameters. The following two computer exercises will put these ideas into practice.

4. (*Diameter Reduction in a Digraph Network—One New Directed Edge*)
   (a) Write a program with the following syntax:

   ```
   OldDiam,NewDiam,Edges = DiamReduction1NewArrow(A)
   ```
   where the input A is the adjacency matrix for a digraph network, and the outputs are as follows: OldDiam is the diameter of the original network, NewDiam is the lowest diameter of a new digraph obtained from the original by adding one new (directed) edge, and Edges is a two-column matrix whose rows are all of the directed edges that will realize this diameter reduction.

---

[17] Although from a mathematical perspective, the existence of one such new route is usually sufficient knowledge, from a business perspective it is often necessary to know all options. Indeed, establishing a new route between a pair of cities requires negotiations and cost, and some options may be more feasible than others.

In the case that `NewDiam = OldDiam` (i.e., no diameter reduction is possible), `Edges` should be the empty matrix.

(b) Run your program on the networks of Example 8.22 and Exercise for the Reader 8.21.

(c) Run your program on each of the networks of Computer Exercise 3.

5. (*Diameter Reduction in a Digraph Network—Two New Directed Edges*)
   (a) Write a program with the following syntax:

   $$\texttt{OldDiam, NewDiam, Edges = DiamReduction2NewArrows(A)}$$

   where the input A is the adjacency matrix for a digraph network, and the outputs are as follows: `OldDiam` is the diameter of the original network, `NewDiam` is the lowest diameter of a new digraph obtained from the original by adding two new (directed) edges, and `Edges` is a four-column matrix whose rows are all of pairs of directed edges that will realize this diameter reduction. In that case that `NewDiam = OldDiam` (i.e., no diameter reduction is possible), `Edges` should be the empty matrix.

   (b) Run your program on the networks of Example 8.22 and Exercise for the Reader 8.21.

   (c) Run your program on each of the networks of Computer Exercise 3.

   (d) Does each pair of edges that realizes a maximum reduction of diameter with two new directed edges necessarily contain an edge corresponding to a maximum reduction of diameter using one new directed edge? Explain your answer.

NOTE: (***Random Graphs***) Randomly generated graphs (or digraphs) are often useful for an assortment of applications and hypothesis testing. The concept of how to randomly generate a graph on n vertices is quite simple: There are $C(n,2)$ possible edges, corresponding to pairs of distinct vertices. For each of these possible edges, we flip a fair coin; if it turns out heads, we include the edge; otherwise we do not. This process is easily programmed on a computer using adjacency matrices and random number generator. Because of its importance, we highlight the procedure as the following algorithms, which have a more general feature of being able to specify the target edge density (corresponding to changing the fair coin to a biased one in the above explanation).

ALGORITHM 8.3A: Random Generation of a Simple Digraph with Specified Edge Density:

Input: A positive integer n representing the number of vertices of the random digraph to be generated, and a positive number p, $0 < p < 1$, representing the edge density.

Output: The n by n adjacency matrix A of a simple digraph that is randomly generated, where the directed edges between pairs of vertices are independently selected with probability p.

Step 1: Initialize the matrix A as the n by n matrix of zeros.
Step 2: (*Random Edge Selection*) Proceed through each of the non-diagonal entries of A, and with probability p change the entry from 0 to 1. This is accomplished by the following more precise nested for loop:

FOR $i = 1$ TO n (i is the row index)
        FOR $j = 1$ TO n (j is the column index)
             Set U = rand (i.e., generate a random real number in (0,1) from the uniform distribution)
                IF $U < p$ (the probability of this event is p) AND $i \neq j$ (don't change diagonal entries)
                    Set $A(i,j) = 1$ (i.e., we add a directed edge from vertex #i to vertex #j)
                END IF
        END j FOR
END i FOR

The corresponding algorithm for graphs is similar, we just need to maintain $A(i,j) = A(j,i)$.

ALGORITHM 8.3B: Random Generation of a Simple Graph with Specified Edge Density:

Input: A positive integer n representing the number of vertices of the random graph to be generated, and a positive number p, $0 < p < 1$, representing the edge density.

Output: The n by n adjacency matrix A of a simple graph that is randomly generated, where the edges between pairs of vertices are independently selected with probability p.

Step 1: Initialize the matrix A as the n by n matrix of zeros.
Step 2: (*Random Edge Selection*) Proceed through each of the above-diagonal entries of A, and with probability p change the entry (and the corresponding below-diagonal entry) from 0 to 1. This is accomplished by the following more precise nested for loop:

FOR $i = 1$ TO n (i is the row index)
      FOR $j = i + 1$ TO n (j is the column index)
         Set U = rand (i.e., generate a random real number in (0,1) from the uniform distribution)
         IF $U < p$ (the probability of this event is p)
            Set $A(i,j) = 1$ and $A(j,i) = 1$ (i.e., we add an edge between vertex #i to vertex #j)
         END IF
      END j FOR
END i FOR

6.   (*Program for Random Generation of Simple Graphs with Specified Edge Density*)
    (a) Write a program with the following syntax:

                  `A = RandomSimpleGraph(n,p)`

    where the input n is a positive integer corresponding to the number of vertices of the random simple graph that will be generated, the input p is the edge selection probability (i.e., the edge density), and output A is an adjacency matrix produced by an application Algorithm 8.3B.
    (b) Run your program with parameter settings n = 100, and p = 0.15, and count the number of edges in the resulting graph. Repeat 20 times.

7.   (*Program for Random Generation of Simple Digraphs with Specified Edge Density*)
    (a) Write a program with the following syntax:

                  `A = RandomSimpleDigraph(n,p)`

    where the input n is a positive integer corresponding to the number of vertices of the random simple digraph that will be generated, the input p is the edge selection probability (i.e., the edge density), and output A is an adjacency matrix produced by an application Algorithm 8.3A.
    (b) Run your program with parameter settings n = 100 and p = 0.15, and count the number of directed edges in the resulting graph. Repeat 20 times.

8.   (*Estimating the Probability that a Randomly Generated Simple Graph is Connected*)
    (a) Use your program of Computer Exercise 6 to randomly generate 100 10-vertex simple graphs with edge density $p = 0.5$, and for each of them use your program of Computer Exercise 1 to check whether the graph is connected (i.e., the diameter is finite). Use these results to estimate the probability that a randomly generated 10-vertex simple graph (with edge density $p = 0.5$) is connected.
    (b) Repeat Part (a), but with the number of vertices changed to 20.
    (c) Repeat Part (a), but with the edge density changed to $p = 0.10$.
    (d) Repeat Part (a), but with the edge density changed to $p = 0.10$, and the number of vertices changed to 20.
    (e) It is clear that for a fixed number n of vertices, the probability that a randomly generated graph on n vertices with edge density p is connected increases with p. In the extreme cases when $p = 0$ (no edges) this probability is zero, and $p = 1$ (complete graph) this probability is one. Do some more computer experiments to estimate the value of p that will make this probability equal to 1/2 in the case of a 10-vertex graph.

9.   (*Estimating the Probability that a Randomly Generated Simple Digraph is Connected*)
    (a) Use your program of Computer Exercise 7 to randomly generate 100 10-vertex simple digraphs with edge density $p = 0.5$, and for each of them use your program of Computer Exercise 1 to check whether the digraph is connected (i.e., the diameter is finite). Use these results to estimate the probability that a randomly generated 10-vertex simple digraph (with edge density $p = 0.5$) is (strongly) connected.

(b) Repeat Part (a), but with the number of vertices changed to 20.

(c) Repeat Part (a), but with the edge density changed to $p = 0.10$.

(d) Repeat Part (a), but with the edge density changed to $p = 0.10$, and the number of vertices changed to 20.

(e) It is clear that for a fixed number n of vertices, the probability that a randomly generated digraph on n vertices with edge density p is connected increases with p. In the extreme cases when $p = 0$ (no edges) this probability is zero, and $p = 1$ (complete digraph) this probability is one. Do some more computer experiments to estimate the value of p that will make this probability equal to 1/2 in the case of a 10-vertex digraph.

10. (*Adjacency Matrix-Based Program for Determination of the Components of a Graph*)
(a) Write a program for computing all of the components of a simple graph. Base the program on computing powers of the adjacency matrix. The syntax should be as follows:

$$\texttt{CompVec = Components(A)}$$

where the input A is the adjacency matrix for the network, and the output `CompVec` is a vector of positive integers that are associated with the vertices (in the order corresponding to the adjacency matrix) and defined as follows: vertex #1 is assigned the (component) number 1, as are all other vertices in the same component as vertex #1. The next vertex in the graph not in the same component as vertex #1 (i.e., in case the graph is not connected) will be assigned (component) number 2, as will be all other vertices that are also in this component. The scheme continues in this fashion.

(b) Check your program against the results of Parts (b) and (c) of ordinary Exercise 41.

(c) Use the program of Part (a) in conjunction with the `RandomSimpleGraph` program of Computer Exercise 6 to estimate the expected number of components of a random 20-vertex simple graph with edge density $p = 0.03$. Generate at least 100 graphs.

Note: For Part (a), the solution to ordinary Exercise 41 describes the relevant algorithm.

8. 3: TREES

A connected simple graph with no cycles is called a *tree*. *Forests* are simple graphs with no cycles, and so each component of a forest is a tree. Trees are very important graphs, and they arise in numerous applications. The most familiar example of a tree is a family tree, such as the one pictured in Figure 8.55.

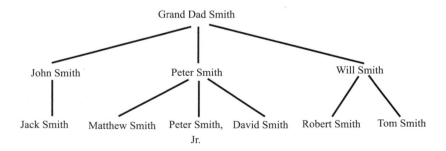

FIGURE 8.55: An example of a family tree. This paternal tree for three generations of the Smith family has 10 vertices (people) and (like any tree) one less edge.

The chain of command in a military organization (or a corporation) is modeled by a tree. The folder structure in any computer operating system or network has the structure of a tree. Trees are the basis for some of the most efficient search algorithms and sorting algorithms. For these reasons and others that we will soon see, trees are the most important graphs in computer science. Despite their simplicity, trees remain a very active area of research in graph theory.

Basic Concepts About Trees

DEFINITION 8.29: A **forest** is a simple graph that contains no cycles, i.e., contains no n-cycle $C_n (n \geq 3)$ as a subgraph. A **tree** is a connected forest.

Figure 8.56 shows three simple graphs. Despite its appearance, G is not a tree since it is itself a cycle, T has no cycles and is connected, so it is a tree. F has no cycles, so it is a forest but not a tree, since it is not connected.

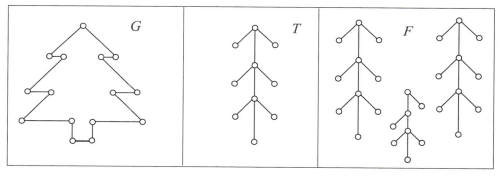

FIGURE 8.56: G is not a tree, T is a tree, and F is a forest consisting of three trees.

EXAMPLE 8.23: Which of the graphs shown in Figure 8.57 are trees?

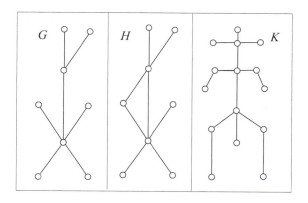

FIGURE 8.57: Three simple graphs for Example 8.23.

SOLUTION: All three graphs are connected, but since H contains a 3-cycle (i.e., a triangle) it is not a tree. Both G and K are cycle-free, and so are trees.

NOTE: (*General graphs with multiedges and/or self-loops are never trees*) Trees (and forests) must be simple graphs. For a simple graph, the shortest possible cycle is C_3. For a general graph, any pair multiedges that connect a certain pair of vertices gives rise to a two-cycle (start at one vertex, take one edge to the other, then take a different edge back to the first vertex), and similarly a self-loop gives rise to a one-cycle. Thus, general graphs that are not simple can never be trees/forests.

EXERCISE FOR THE READER 8.26: Up to isomorphism, how many different five-vertex trees are there? Draw representatives of each isomorphism class.

The following definition separates the vertices of a tree into two types:

DEFINITION 8.30: A vertex of a tree or forest is called a **leaf** if it has degree 1, otherwise it is called an **internal vertex**.

For example, the tree G of Figure 8.57 has six leaves and two internal vertices.

One simple yet very important fact about trees is that if a tree has at least two vertices, then it will always have at least two leaves. This will be formally stated and proved as our next theorem. This fact often turns out to be extremely useful when one needs to prove results about trees. To see this, we first observe that whenever a leaf (and the sole edge that is adjacent to it) is deleted from a tree, the resulting graph will still be connected and cycle free (since it is a subgraph of the original tree), thus it is also a tree. Proofs by induction can be implemented by deleting a leaf from a tree of $n + 1$ vertices to obtain a tree of n vertices (to which the inductive hypothesis can be applied).

THEOREM 8.8: A tree with at least two vertices must have at least two leaves.

Proof: Let T be a tree with at least two vertices. Consider a simple path P in T with no repeated vertices and that is of maximum possible length: $P = <u = v_0, v_1, \cdots, v_k = v>$. We claim that u, v must be leaves. Indeed, if the initial vertex u were not a leaf, then it would be incident to some edge $\{u, w\}$ different from the first edge $\{u, v_1\}$ of P. This edge cannot appear in P, since if it did, the segment of P from the initial vertex u to the second appearance of u would be a cycle in the tree T. Also, the vertex w cannot appear in P since if it did a cycle could be formed. Thus, we can attach this edge to the left end of P to obtain a longer simple path with no repeated vertices in T, contradicting the property that P has maximum possible length. Thus u is a leaf; a similar argument shows that v is a leaf as well. □

We illustrate the utility of this theorem by proving the following proposition that tells us the number of edges in trees and forests.

PROPOSITION 8.9: (a) A tree on n vertices has $n-1$ edges.
(b) A forest on n vertices that has k components has $n-k$ edges.

Proof: Part (a): We use induction on the number n of vertices of a tree.

Basis Step: A tree with $n=1$ vertex must have 0 $(=n-1)$ edges (as does any simple graph on 1 vertex).

Inductive Step: Assume the result is true for trees on n vertices, where $n \geq 1$, and let T be a tree with $n+1$ vertices. Let u be a leaf of T; by Theorem 8.8, T has at least two leaves. The vertex deletion graph $T-u$ is a tree on n vertices, so by the inductive hypothesis, it has $n-1$ edges. But T has one additional edge (namely the unique edge joining the leaf u to the rest of T), thus T has $n-1+1=(n+1)-1$ edges, as was needed to show.
Part (b): Let F be a forest with k components: $T_1, T_2, \cdots, T_k$. Since each T_i is a tree, we know it has one less edge than it has vertices, by the just proved Part (a) of the proposition. It follows that:
$$| E(F)| = | E(T_1)| + | E(T_2)| + \cdots + | E(T_k)|$$
$$=(| V(T_1)| -1) + (| V(T_2)| -1) + \cdots + (| V(T_k)| -1)$$
$$= | V(T_1)| + | V(T_2)| + \cdots + | V(T_k)| -k = | V(F)| -k,$$

as we wished to show. □

EXERCISE FOR THE READER 8.27: Use Proposition 8.9 to prove that a simple graph with n vertices that has k components must have at least $n-k$ edges.

The following theorem gives several equivalent definitions of trees; it contains several useful properties of trees that provide insights into understanding their structure.

THEOREM 8.10: (*Equivalent Definitions of Trees*) If T is a simple graph on n vertices, then the following statements are logically equivalent:
1. T is a tree.
2. T is connected and has $n-1$ edges.
3. T has no cycles and has $n-1$ edges.
4. Any two vertices of T are joined by a unique simple path in T.
5. T is connected, and every edge is a bridge (i.e., deleting an edge renders the graph disconnected).
6. T has no cycles and if we add any new edge to T (between two of its vertices) the resulting simple graph will have a unique cycle.

The reader would do well to attempt to write direct proofs for any or all parts of these equivalences before (or instead of) reading the following proof.

Proof: $1 \Rightarrow 2$: This implication follows directly from the definition of a tree and Proposition 8.9(a).

$2 \Rightarrow 3$: We will accomplish the proof by method of contradiction: We assume that statement 2 is true, and that T has a cycle $T_1 = <v_1, v_2, \cdots, v_k, v_1>$. Since T_1 has k vertices and k edges, by Proposition 8.9(a), it cannot be the whole tree T. Let U denote the set of remaining vertices of T and let $\ell = |U|$. Thus $n = k + \ell$. Since T is connected, there exists a vertex $u_1 \in U$ that is adjacent to some vertex in T_1. We define the subgraph T_2 of T to be T_1, along with the new vertex u_1 and with a new edge that joins u_1 to T_1. Thus T_2 has as many edges as it has vertices, so once again it cannot be the whole tree T. By repeating this construction, we obtain a sequence of subgraphs of T: $T_1, T_2, \cdots, T_\ell$ with the property that $|E(T_i)| = |V(T_i)| = k + i$, for each i, $1 \le i \le \ell$. Thus, T_ℓ, and hence also T, must have at least $n = k + \ell$ edges—a contradiction.

$3 \Rightarrow 4$: Since T has no cycles, the same must be true of each of its (connected) components; thus each component of T is a tree, and so T is a forest. But since T has $n-1$ edges, it follows from Proposition 8.9(b) that T has only one component, so T is a tree. Next, suppose that two vertices v, w of T were joined by two different paths: $P = <v = v_0, v_1, \cdots, v_k = w>, Q = <v = v_0', v_1', \cdots, v_\ell' = w>$, which by Theorem 8.5, may be assumed to have no repeated edges or vertices. Since P and Q are different, there must be a first vertex x after which they split, and then a first vertex y where they come back together. The subpath of P: $P(x,y)$ from x to y, followed by the subpath of Q: $Q(y,x)$ (in the reverse direction) from y to x then forms a cycle in T—a contradiction!

$4 \Rightarrow 5$: We assume that any two vertices of T are joined by a unique path, and we must show that every edge is a bridge. Indeed: Any edge is a path joining its two endpoints, so deleting it would disconnect the two vertices in T.

$5 \Rightarrow 6$: Assuming statement 5, since every edge of a cycle is not a bridge, T cannot have any cycle. We have left to show the second part of statement 6: If we add a new edge $e = \{u,v\}$ to T, the resulting graph $T + e$ will have a unique cycle. Since by item 5, T is connected, there is already a path P in T joining u to v, and by Proposition 8.5, we may assume that P is simple with no repeated vertices. Thus, P followed by e forms a cycle in $T + e$. If there were another cycle in $T + e$, then since T contains no cycles, this cycle must also be expressible in the form of a simple path Q in T with no repeated vertices from u to v, followed by e. Just as in the proof of $3 \Rightarrow 4$, we can now use P and Q to find a cycle in T—a contradiction!

$6 \Rightarrow 1$: We have only to prove that T is connected, assuming statement 6. But this is easy: Given two different nonadjacent vertices u, v, let e be the new edge $\{u,v\}$. By item 6, if we add this edge to T, there will be a unique cycle. If we remove e from this cycle, what will be left is a path (in T) from u to v. $\square$

After reading through the proofs of the above six implications or, better, having come up with some alternative proofs, readers who still feel a bit uneasy with these ideas should try to write direct proofs of some of the 24 other implications. The following exercise for the reader requests proofs of two of these.

EXERCISE FOR THE READER 8.28: Using only definitions and material before the statement of Theorem 8.10, write proofs of the following implications involving items of Theorem 8.10: (a) $3 \Rightarrow 2$ (b) $5 \Rightarrow 3$.

Rooted Trees and Binary Trees

For an assortment of applications it is often convenient to single out a particular vertex of a tree as the "root" of the tree. When this is done the tree is called a rooted tree. Since there is a unique simple path joining any two vertices of a tree, the distance from a vertex to the root is simply the number of edges in this unique path. Rooted trees are usually thought to have all edges directed "away" from the root, and in drawing rooted trees, the root is usually placed on top. The family tree of Figure 8.55 is an example of a rooted tree, with Grand Dad Smith being the root. Although rooted trees are technically directed graphs, since direction arrows are rarely drawn on them, we consider rooted trees as trees with special attributes rather than as digraphs. The following definition includes some additional terminology pertaining to rooted trees. This terminology stems from genealogy, since family trees were the first examples of rooted trees.

DEFINITION 8.31: A **rooted tree** is a tree with a designated vertex selected to be the **root**. All edges are thought to be directed away from the root. If a vertex u immediately precedes a vertex v in the path from the root, then u is called the **parent** of v and v is a **child** of u. If two vertices have the same parent, they are called **siblings**. A vertex w is an **ancestor** of a vertex v if w lies on the unique path from the root to v, and in this case we also say that v is a **descendant** of w. The **depth** of a vertex is its distance from the root (in particular, the depth of the root is zero). The **height** of a rooted tree is maximum depth of the rooted tree.

EXAMPLE 8.24: An example of a rooted tree is shown in Figure 8.58. The root is r, which has three children: a, b, c. These depth one vertices in turn have the following children: vertex a is the parent of only one child d; vertex b is the parent of three children, the siblings e, f, and g, and vertex c is the parent of four children, the siblings h, i, j, k. The height of the rooted tree is three, and is realized by two depth three vertices

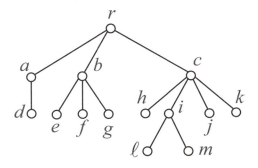

FIGURE 8.58: The rooted tree of Example 8.24.

(siblings), which are the children ℓ, m of vertex i. The root is always the (unique) ancestor of all other vertices, i.e., all other vertices are descendants of the root. Vertex a has only one descendant: its child d, while vertex c has six descendants. The number of ancestors of any vertex in a rooted tree is the same as the depth of the vertex. For example, the depth two vertex e has two ancestors: its parent b and the root. The depth three vertex m has these three ancestors: i, c, and r. This rooted tree has nine leaves (in rooted trees, leaves are the vertices with no children), and the remaining five vertices all have children, and are internal vertices. This way of drawing a rooted tree is typical: the root goes on top, and vertices of the same depth are lined up horizontally.

EXERCISE FOR THE READER 8.29: Suppose that T is a rooted tree with root r and height h. Without knowing any more specific information about T, use the vocabulary of Definition 8.31 to describe each of the following quantities as accurately as possible.
(a) The eccentricity of the root: $\text{ecc}(r)$.
(b) The shortest path between two vertices (which is unique in any tree).

The following definition provides a fine-tuned classification of rooted trees. Such rooted trees turn out to be cornerstones of numerous searching and sorting algorithms, and we will soon elaborate on this concept.

DEFINITION 8.32: A rooted tree in which every internal vertex has m or fewer children is called an ***m*-ary tree**. A rooted tree in which every internal vertex has exactly m children is called **full**. A full rooted tree in which all leaves have the same depth is called a **complete *m*-ary tree**. In the (very important) special case where $m = 2$, the terms **binary tree**, **complete binary tree**, and **full binary tree** are used.

EXAMPLE 8.25: The rooted tree of Figure 8.58 is a 4-ary tree that is not complete. It is not a 3-ary tree since vertex c has four children.

EXAMPLE 8.26: Figure 8.59 shows two binary trees, the one on the left is full since every internal vertex has two children, the one on the right is complete since it has the additional property that all leaves have the same depth.

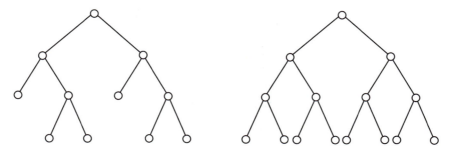

FIGURE 8.59: Two binary trees of height 3; the binary tree on the left is full but not complete; the one on the right is complete.

Models with Rooted Trees

We next describe a sampling of examples of rooted trees that are adapted for specific applications.

EXAMPLE 8.27: (*Decision Trees*) Figure 8.60 shows part of a 26-ary decision tree that serves as the basis for an automated document spell checker program. For each typed word, such programs scan down the decision tree, using each letter to decide which edge to take down. Vertices corresponding to words in the dictionary of all words in the program are given a special designation (in the figure they are colored black). If an encountered word does not correspond to a word in the dictionary, an indication is provided that the typed word is possibly misspelled, and usually some nearby correctly spelled words are provided as suggested replacements, along with the option of adding the typed word to the spell checker's dictionary so that future warnings will not appear when the same word is entered.

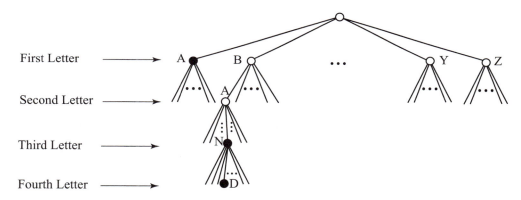

FIGURE 8.60: Part of a decision tree serving as the basis for a spell checker program. The black vertices correspond to words in the spell checker's dictionary. The figure shows the following dictionary words: a, ban, band.

Typically, the way the a given word gets searched for in the tree (by the computer algorithm) is to go through the first few letters to narrow down the portion of the tree in which the word could lie, and then perform a linear search on all descendant words. This is very similar to how a human being uses a real dictionary (in book form): use the first few letters to navigate to the page/column of the book which must contain the word (this corresponds to navigating down the rooted tree), then scan through the entries to find the word (this corresponds to a linear search). Such trees are the basis of many much more complicated search engines.

EXAMPLE 8.28: (*Binary Tree Representation of Algebraic Expressions*) In the design of computer algebra systems, algebraic expressions are often stored as binary trees. The leaves of the trees are the numbers and/or variables appearing in the expression, and the internal vertices with two children denote binary operations that have two inputs, e.g., algebraic operators such as $+$, $-$, $/$, $*$, $\wedge$ (addition, subtraction, division, multiplication, exponentiation), etc., while internal vertices with single children denote single variable (*unary*) operators such as $\sqrt{}$, log( ), etc. The operations at lower levels (i.e., greater depths in the tree) get done first; the final operation thus corresponds to the root of the binary tree. Figure 8.61 shows examples of binary tree representations of the three algebraic expressions $1 + xy^3$, $\dfrac{1 - xy}{y + x^2}$, and $\sqrt{x + \sqrt{x+1}}$.

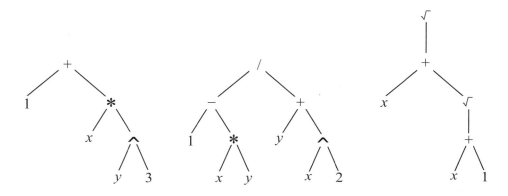

FIGURE 8.61: Three binary trees representing the algebraic expressions $1 + xy^3$, $\dfrac{1 - xy}{y + x^2}$, and $\sqrt{x + \sqrt{x+1}}$, respectively.

EXERCISE FOR THE READER 8.30: Draw binary trees that represent each of the following algebraic expressions:
$$(x + 2y)(x - 2y), \quad x^2 + y^2 + z^2 \text{ and } \sin(\cos(x+2) - 3y).$$

EXAMPLE 8.29: (*Multistage Tournament Structures*) Rooted trees are often used to structure single elimination tournaments where teams or individuals are paired to play in a sequence of stages (e.g, the World Cup, the NBA basketball playoffs, or any tennis tournament), with the winners of one stage progressing to the next. In the final match, there will be two players and the winner will be placed at the root of the tree (thus the root of the tree is the last vertex to be labeled). Figure 8.62 shows such a tree for a tournament involving 14 players: $P_1, P_2, \cdots, P_{14}$. It was necessary to give two of the players a *bye* in the first round; this is usually done in a fair way, such as by awarding any available byes to the players with the strongest records entering into the tournament.

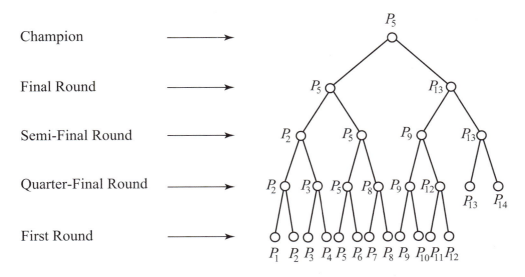

Champion $\longrightarrow$

Final Round $\longrightarrow$

Semi-Final Round $\longrightarrow$

Quarter-Final Round $\longrightarrow$

First Round $\longrightarrow$

FIGURE 8.62: A multi-stage tournament tree involving 14 players. Player P_5 is the overall winner (the champion); players P_{13}, P_{14} were given byes on the first round.

Such single-elimination tournament trees can always be organized so that each leaf represents a player/team, each internal vertex represents a match between two players/teams, and the byes are of depth one less than the maximum depth.[18] General m-ary trees that have the property that all leaves are at the maximum possible depth or one less are called **balanced**.

Properties of Rooted Trees

The following result is sometimes useful in estimating sizes in problems and algorithms involving rooted trees. The first statement pertains to general m-ary trees, and then a stronger result is given for full m-ary trees.

THEOREM 8.11: If T is an m-ary tree with height h and having n vertices, then the following inequality holds:

$$h+1 \leq n \leq \frac{m^{h+1}-1}{m-1}. \tag{3}$$

If T is complete, then the right upper bound is attained: $n = \dfrac{m^{h+1}-1}{m-1}$, and T has exactly m^k vertices of depth k, $0 \leq k \leq h$.

---

[18] To see this, note that if we start with a complete binary tree of a given depth, then for each pair of children leaves (at maximum depth) that we delete, the parent of both will be a new leaf, so the number of leaves goes down by one.

Proof: We first deal with the case in which T is a complete m-ary tree. An easy induction argument proves that T has exactly m^k vertices of depth k, $0 \le k \le h$:

Basis Step: $k = 0$. There is only $m^0 = 1$ vertex of depth zero: the root vertex.

Induction Step: We assume that $0 \le k < h$, and that we know T has m^k vertices of depth k, and we must deduce that T has m^{k+1} vertices of depth $k + 1$. The depth $k + 1$ vertices are precisely the children of the m^k vertices of depth k, and since T is complete, each of these vertices has m children, so there are a total of $m \cdot m^k = m^{k+1}$ vertices of depth $k + 1$.

Using the formula for the sum of a finite geometric series (Proposition 3.5/formula (3) of Chapter 3), combined with what was just proved, the total number of vertices of T is: $n = m^0 + m^1 + \cdots + m^h = \dfrac{m^{h+1} - 1}{m - 1}$.

Next we assume that T is a general m-ary tree. Letting n_k denote the number of vertices of depth k, it is clear that $n_k \ge 1$ for each k, $0 \le k < h$, since there must be at least one path from the root to a depth k vertex. This implies $n = n_0 + n_1 + \cdots + n_h \ge 1 + 1 + \cdots + 1 \,(h+1 \text{ times}) = h + 1$, which is the first inequality of (3). The second inequality follows from what has already been proved about the number of vertices in a complete m-ary tree, and the fact that the number of vertices in any m-ary tree cannot exceed the number in the complete m-ary tree of the same height. $\square$

The following consequence of Theorem 8.11 for general m-ary trees becomes more precise in case the tree is balanced (and thus applies to single-elimination tournament trees).

COROLLARY 8.12: Suppose that T is an m-ary tree with height h that has ℓ leaves. Then $h \ge \lceil \log_m \ell \rceil$. In case T is balanced then equality holds: $h = \lceil \log_m \ell \rceil$. (Recall that $\lceil x \rceil = \text{ceil}(x)$ —the ceiling function, is the smallest positive integer that is greater than or equal to x.)

Proof: Since Theorem 8.11 tells us that complete m-ary trees have m^h vertices (which are the leaves) of (maximum) depth h, it certainly follows that there are at most m^h leaves of depth h. But for any leaf of depth less than h, since leaves have no children, this leaf will reduce the number of possible leaves, compared to the complete m-ary tree of the same height, by at least $m - 1$, since internal vertices in a complete m-ary tree have m children. It follows that $\ell \le m^h$, so applying $\log_m$ to both sides gives $\log_m \ell \le \log_m m^h = h$. Since h is an integer, it follows that

$h \geq \lceil \log_m \ell \rceil$. If the tree is balanced, all leaves will have depth h or $h-1$, so it follows that $m^{h-1} < \ell \leq m^h$, and thus $h = \lceil \log_m \ell \rceil$.

One application of this corollary shows that in putting together a single-elimination tournament involving N players (or teams), the number of rounds required will be $\lceil \log_2 N \rceil$. Thus, with 46 players, we would need $\lceil \log_2 46 \rceil = \lceil 5.52... \rceil = 6$ rounds.

EXERCISE FOR THE READER 8.31: Suppose that we need to schedule a single-elimination grand slam tennis tournament involving $N = 73$ players.
(a) How many rounds will be needed?
(b) How many byes will there be in the first round?
(c) How many tennis matches will there be in the tournament?

In the case of any full m-ary tree, there is a simple relationship between the number of internal vertices and the number of leaves. Our next result provides this relationship; it will be followed with some consequences and applications.

THEOREM 8.13: Suppose that T is a full m-ary tree having n vertices and i internal vertices (and thus $\ell = n - i$ leaves). Then

$$n = mi + 1. \tag{4}$$

Proof: With the exception of the root, every vertex in T is the child of a unique parent. Since the parents are exactly the internal vertices and each internal vertex has m children, it follows that the number of children is mi. Adding 1 for the root, we see that (4) gives the total number of vertices of T. □

CONSEQUENCE OF THEOREM 8.13: In any full m-ary tree T, if we know any one of the three numbers: $n =$ the number of vertices, $i =$ the number of internal vertices, and $\ell =$ the number of leaves, then we can use (4) and the equation $n = \ell + i$ to determine the other two numbers. This idea will be demonstrated in the following example.

EXAMPLE 8.30: If there are 100 players who enter in a single elimination grand slam tennis tournament, how many matches will take place?

SOLUTION: Single elimination tournaments are modeled by full binary trees ($m = 2$), as shown in Example 8.29. The number of players is just the number of leaves, so $\ell = 100$. The number of matches is the number i of internal vertices. In equation (4) $n = mi + 1$, we know $m = 2$, but we do not know the other two variables. But since we know $\ell = 100$, the equation $n = \ell + i$ becomes $n = 100 + i$, which, when substituted into (4) now gives $100 + i = 2i + 1 \Rightarrow i = 99$.

Alternative Solution: Here is a slick way to see that the number of matches is one less than the number of players in any single elimination tournament: Note that the tournament starts with ℓ players, and each of the i matches eliminates exactly one player. Since there is only one player who does not get eliminated, it follows that $i = \ell - 1$.

Another alternative and simple approach was used in the solution to Exercise for the Reader 8.31, but the solution using (4) generalizes to *m*-ary trees as follows:

OBSERVATION: In any full binary tree, the number of leaves is one more than the number of internal vertices, i.e., $\ell = i + 1$. (*Proof:* Formula (4) of Theorem 8.13 becomes $n = 2i + 1$, but since $n = \ell + i$, this becomes $\ell + i = 2i + 1$ $\Rightarrow \ell = i + 1$. $\square$)

EXERCISE FOR THE READER 8.32: An email campaign is initiated to spread the word about a major political event. The organizer sends an email to five people with the information, along with instructions to send a similar email the next day to five new people that do not yet appear on the database list (maintained by the organizer). The process stops once emails have been sent to (at least) 5000 people.
(a) How many days would it take for 5000 people to receive emails?
(b) How many people sent emails throughout the course of the campaign?

Ordered Tree Traversal Algorithms

When trees are used to store data, in order to efficiently store such trees on computers and efficiently access data, it is important to have standard methods for ordering the vertices. We will present three very prevalent algorithms that will accomplish this task. They are called *tree traversal algorithms* since each has a geometric realization. The algorithms will first be presented in recursive form since such a format is more suitable for computer implementation. The basic assumption is that children are labeled from left to right.[19] Recursive algorithms on ordered trees are based on viewing such trees recursively, where the children of the root: $r_1, r_2, \cdots, r_s$ (labeled left to right) each determine corresponding *child subtrees* $T_1, T_2, \cdots, T_s$ of the whole tree determined by it and all of its descendants—see Figure 8.63.

---

[19] When they are drawn this is a natural ordering and will be helpful in the geometric realizations of the tree traversal algorithms.

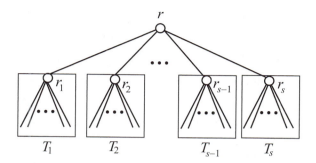

FIGURE 8.63:   In the recursive view of a rooted tree T, the tree is made up of root r, edges to its children $r_1, r_2, \cdots, r_s$, and the child subtrees $T_1, T_2, \cdots, T_s$ determined by descendants of the children.

The three recursive tree traversal algorithms are called *preorder*, *inorder*, and *postorder*, and the names indicate the order in which the root is visited in relation to the subtrees $T_1, T_2, \cdots, T_s$. Each recursive algorithm will be illustrated on the rooted tree of Figure 8.58, reproduced here for convenience, before moving on to the next.

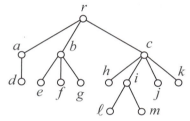

FIGURE 8.58 <Reproduced>.

NOTATION: In each algorithm, an ordered list (or vector) LIST will be constructed by adding one vertex at a time. We use the phrase "Append v to LIST" to mean we stack the vertex v at the right end of the vector LIST. For example, if LIST = [b, g, a], then appending c to LIST would produce LIST = [b, g, a, c].

The *preorder traversal* begins by visiting the root vertex r, and then recursively applying the algorithm to visit the subtrees $T_1, T_2, \cdots, T_s$ from left to right.

ALGORITHM 8.4: (*Preorder Traversal of a Rooted Tree*)
Input: A rooted tree T.
Output: An ordered list, PreOrder(T), of all vertices of T (in preorder).

IF T consists of its single root vertex (r)
    OUTPUT PreOrder(T) = [r]
OTHERWISE
    SET $T_1, T_2, \cdots, T_s$ be the child trees (from left to right)
    OUTPUT PreOrder(T) = [r, PreOrder(T_1), PreOrder(T_2), $\cdots$, PreOrder(T_s)]
END

NOTE: This, as with other recursive algorithms, takes some time to get comfortable with. This is due to the fact that when the recursive Step 3 gets applied to one of the original root's children trees T_i, we first reapply Step 2 to this child tree and its root r_i will be different from the root r of the whole tree T (even though r is what is written in the description of Step 2). Also, T_i will have its own children and child trees: $T_{i,1}, T_{i,2}, \cdots, T_{i,t}$. These will generally be different in number and form from the original root's children, $T_1, T_2, \cdots, T_s$, despite the fact that the recursive algorithm uses static notation.[20]

Thus, each tree T_i has all of its vertices appended to LIST before the processing of the next child tree.

EXAMPLE 8.31: Apply Algorithm 8.4 to the rooted tree of Figure 8.58.

SOLUTION: Before appending the subtree traversals, we start with LIST = [r], and the three root children subtrees T_1, T_2, T_3 that are shown in Figure 8.64.

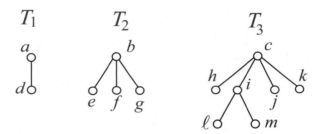

FIGURE 8.64:   The three children subtrees T_1, T_2, T_3 of the rooted tree of Figure 8.58.

In the recursive step:

Processing of the first child subtree T_1: Since the root of T_1 is a, this is the next vertex to get appended to the list: LIST = [r, a]. Also, a has only one child subtree consisting of its own root vertex d. We then recursively the algorithm to this one-vertex tree to yield LIST = [r, a, d]. T_1 is now completely traversed, and we move on to T_2.

Processing of the second child subtree T_2: Since the root of T_2 is b, this is the next vertex to get appended to the list: LIST = [r, a, d, b]. The vertex b has three child subtrees each consisting only of its own root vertex: e, f, g. The recursive step is then applied to each of these 1-vertex trees (in order) resulting in the update: LIST = [r, a, d, b, e, f, g]. T_2 is now completely traversed, and we move on to T_3.

---

[20] The good news is that this turns out not to be a problem when coding such algorithms into computer programs. The computer exercises at the end of this section will provide the reader ample opportunity to understand why this is true.

Processing of the third child subtree T_3: Since the root of T_3 is c, this is the next vertex to get appended to the list: LIST = $[r, a, d, b, e, f, g, c]$. The vertex c has four child subtrees; only the second one has more than one vertex; the other three are processed as above. The second tree has root i (which gets appended to LIST first), followed by the processing of its two child trees, which are single vertices ℓ, m (in order). Thus, T_3's vertices are appended to LIST in the following order: h, i, ℓ, m, j, k.

The algorithm outputs LIST as PreOrder(T) = $[r, a, d, b, e, f, g, c, h, i, \ell, m, j, k]$.

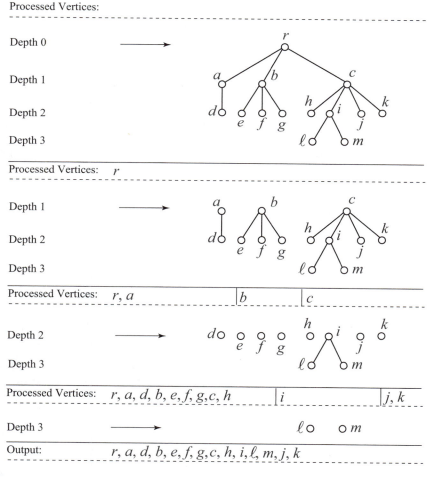

FIGURE 8.65: An illustration of the parallelized implementation of the preorder algorithm; the child trees are processed simultaneously by increasing depth.

NOTE: In this and the other two tree traversal algorithms, the child subtree vertices can be processed *in parallel* (i.e., simultaneously), by moving down one depth level at a time. Figure 8.65 illustrates such an implementation on the rooted tree of Figure 8.58; the reader should compare with the solution of Example 8.31.

The *inorder traversal* begins by processing the leftmost child subtree T_1 of the root vertex r, next visits the root r (of the whole tree), then recursively applies the algorithm to visit the remaining child subtrees $T_2, \cdots, T_s$ from left to right.

ALGORITHM 8.5: (*Inorder Traversal of a Rooted Tree*)
Input: A rooted tree T, with root r, and root child trees $T_1, T_2, \cdots, T_s$.
Output: An ordered list, InOrder(T), of all vertices of T (in inorder).

IF T consists of its single root vertex (r)
    OUTPUT InOrder(T) = [r]
OTHERWISE
    SET $T_1, T_2, \cdots, T_s$ be the child trees (from left to right)
    OUTPUT InOrder(T) = [InOrder(T_1), r, InOrder(T_2), $\cdots$, InOrder(T_s)]
END

EXAMPLE 8.32: Apply Algorithm 8.5 to the rooted tree of Figure 8.58.

SOLUTION: The three root child subtrees are shown in Figure 8.64.

Step 1: Processing of the first child subtree T_1: Since T_1 has only one vertex (d) below its root, this former vertex gets visited before its root resulting in LIST = [d, a].

Step 2: The root of T gets appended → LIST = [d, a, r].

Step 3: Processing of the second child subtree T_2: The three child subtrees of T_2 are single vertices: e, f, g. We first append vertex e to LIST, next the root (b) of T_2, and finally the remaining vertices f, g → LIST = [d, a, r, e, b, f, g].

Step 4: Processing of the third child subtree T_3: First, the leftmost child tree (of T_3), which is a single vertex (h) gets appended, and we next append the root c of T_3. The second child tree of T_3 consists of a root (i) and two children: ℓ, m (in order). These children are single vertex trees, so the three vertices get appended inorder: ℓ, i, m. The remaining two children trees of T_3 are single vertices j, k, which get appended in this same order. In summary, the vertices of T_3 are visited in this order: h, c, ℓ, i, m, j, k.

The algorithm outputs InOrder(T) = [d, a, r, e, b, f, g, h, c, ℓ, i, m, j, k].

EXERCISE FOR THE READER 8.33: Create a figure similar to Figure 8.65 showing a parallelized implementation of the Algorithm 8.5 to redo Example 8.32. (Just as described in the note preceding Figure 8.35, the child trees should be recursively processed simultaneously, with increasing depth.)

The *postorder traversal* processes the root r last, after all its children subtrees $T_1, T_2, \cdots, T_s$ have been recursively processed from left to right.

ALGORITHM 8.6: (*Postorder Traversal of a Rooted Tree*)
Input: A rooted tree T, with root r, and root child trees $T_1, T_2, \cdots, T_s$
Output: An ordered list, PostOrder(T), of all vertices of T (in postorder).

IF T consists of its single root vertex (r)
    OUTPUT InOrder(T) = [r]
OTHERWISE
    SET $T_1, T_2, \cdots, T_s$ be the child trees (from left to right)
    OUTPUT PostOrder(T) = [PostOrder(T_1), PostOrder(T_2), $\cdots$, PostOrder(T_s), r]
END

EXAMPLE 8.33: Apply Algorithm 8.6 to the rooted tree of Figure 8.58.

SOLUTION: A parallelized approach is summarized in Figure 8.66.

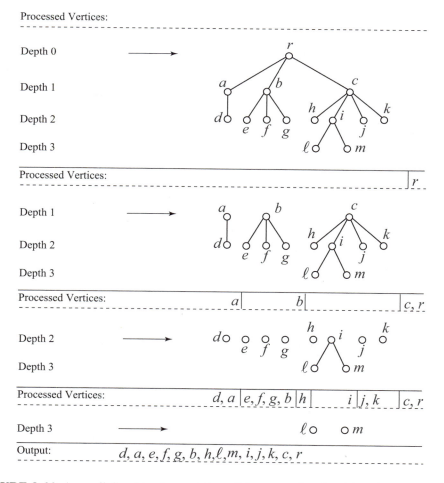

FIGURE 8.66: A parallelized implementation of the postorder algorithm for Example 8.33 where the child trees are processed simultaneously by increasing depth.

Compare this with the corresponding Figure 8.65 for the preorder traversal; the only difference is that as we proceed down increasing depth levels, the roots of children trees always get placed to the right in the postorder, whereas in the preorder they get placed on the left.

NOTE: (*Easy Scheme to Remember the Three Traversals for Binary Trees*) In the case of a binary tree (where each vertex has at most two children), the three traversal algorithms can be summarized simply as follows:

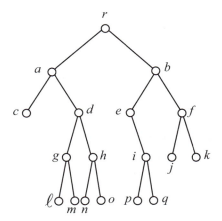

>   **Preorder: root, left, right**
>   **Inorder: left, root, right**
>   **Postorder: left, right, root**

EXERCISE FOR THE READER 8.34: Determine (a) the preorder, (b) the inorder, and (c) the postorder vectors of vertices for the binary tree of Figure 8.67:

FIGURE 8.67: The binary tree of Exercise for the Reader 8.67.

Now that the reader has had some experience working with the three recursive algorithms of Algorithms 8.4–8.6, we will next illustrate an alternative geometric method for computing these three vertex traversals. While the recursive algorithms are most suitable for computer implementations, the following geometric method is quite convenient for hand computing traversals for moderately sized rooted trees.

ALGORITHM 8.7: (*Geometric Algorithm for Finding the Preorder, Inorder, and/or Postorder Traversals of a Rooted Tree*)
NOTE: This algorithm is formulated for hand calculations. The computer exercises of this section will develop implementations for the recursive algorithms of Algorithms 8.4–8.6.

Input: A drawing of a rooted tree T.
Output: Ordered lists of the vertices of T in the preorder, inorder, and/or postorder traversals.

Step 1: Starting at the root, draw a curve counterclockwise traversing alongside all edges of the tree.

Step 2: As the curve is traversed, list the vertices in the following orders:
For Preorder: List each vertex the <u>first</u> time it is encountered.
For Inorder: List each leaf the first time it is encountered, and each internal vertex the second time it is encountered.
For Postorder: List each vertex the <u>last</u> time it is encountered.

NOTE: (*Alternative Scheme for Postorder*) By symmetry, it is perhaps more convenient to use the following variation for the postorder: If we proceed around the loop clockwise (the opposite direction), the vertices can be listed the first time they are encountered, <u>but</u>, with this variation they should be listed from right to left. This corresponds to the simple fact that the postorder of a rooted tree is the preorder of the left-right reflected tree.

EXAMPLE 8.34: Apply Algorithm 8.7 to find the pre-, in-, and postorder traversals of the rooted tree of Figure 8.58.

SOLUTION: *Step 1:* A drawing of the curve around the tree is shown in Figure 8.68.

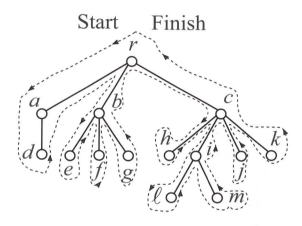

FIGURE 8.68: A curve drawn counterclockwise around the rooted tree of Figure 8.58.

Step 2: As the curve is traversed, we list the vertices in the orders indicated by the algorithm:

For Preorder: List each vertex the <u>first</u> time it is encountered:

$$[r,a,d,b,e,f,g,c,h,i,\ell,m,j,k]$$

For Inorder: List each leaf the first time it is encountered, and each internal vertex the second time it is encountered:

$$[d,a,r,e,f,g,b,h,c,\ell,m,i,j,k]$$

For Postorder: We use the alternative scheme by traversing the path clockwise and stacking the vertices from right to left (the first time they are encountered):

$$[d,a,e,f,g,b,h,\ell,m,i,j,k,c,r]$$

These results all agree (or course) with those of Examples 8.31–8.33.

Binary Search Trees

We will next show a very powerful method for searching large databases using the structure of a binary tree. Such binary search trees are the method of choice for the computer storage of large sets of information due to the speed of being able to search and update the structure. All that is needed is that the records being stored in the vertices of the binary tree each have associated labels that are ordered (e.g., numbers or names).[21]   The vertices of a binary search tree cannot be placed arbitrarily, but must meet the following condition:

DEFINITION 8.33:   A **binary search tree** is a binary tree where the vertex labels are ordered, and the label on any vertex must be greater than the labels of the vertices on the left child subtree (i.e., the left descendants), and less than the labels of all vertices on the right child subtree (i.e., the right descendants).

EXAMPLE 8.35:   Figure 8.69 illustrates a binary search tree where the vertex labels are integers. If the label 36 were to be changed to 28, the tree would no longer be a binary search tree since the root (32) would have a right descendant with a smaller vertex label. If we needed to add one new vertex with label 30 to the binary search tree of Figure 8.69, the conditions of Definition 8.33 would force its location to be the right child of the vertex 27.

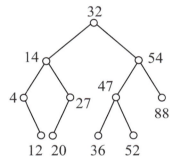

FIGURE 8.69: A binary search tree.

We are going to present a very efficient search algorithm for binary trees, which at the same time will tell us the location of the new vertex needed to store a new label (that is not located in such a search) with respect to the existing tree. We motivate the idea with an artificially small example. Suppose that a company has an existing set of nine customers: Manning, Sanchez, Smith, Beckham, Hernandez, Gretzky, Pierce, James, and Woods.

The company would like to start a computerized database for this set of customers, along with their records. With any such system, the customer's records would be linked to their names (or whatever other attribute is used to index their records). In this example, we will simply use alphabetical order of the last name. Although searching for records can be done quickly with an alphabetized list, the process of adding a new record is not so efficient (on a computer platform) since all records after the new record would need to be reindexed. Another scheme would be to simply store records in the order they are received. This method makes it easy to add new records—just stack them on the end. But searches are time-consuming

---

[21] More detailed records can be associated with these labels and any record can be "pulled up" once its label is located.

(imagine a company with 1 million customer records). If we instead store the record names as vertices of a binary tree, it turns out to be very easy both to search for existing records, and to add new ones. Figure 8.70 shows the nine steps used to build such a binary tree using the list of nine names given above.

The first name Manning is designated the root (at this point, the tree has no other vertices). The next name Sanchez comes after Manning so is placed as the right child of the root. The next name Smith is greater than both Manning and Sanchez, so must be placed as the right child of Sanchez. Beckham comes before Manning, so is placed as the root's left child. Hernandez comes before Manning (so must be in the root's left child tree), but after Beckham, so is placed as the right child of Beckham. Like Hernandez, Gretzky comes before Manning and after Beckham. It also comes before Hernandez, so it will be placed as the left child of Hernandez. Pierce comes after Manning, and before Sanchez, and so is placed as the left child of Sanchez. In the same fashion, James get placed as the right child of Hernandez, and Woods gets placed as the right child of Smith.

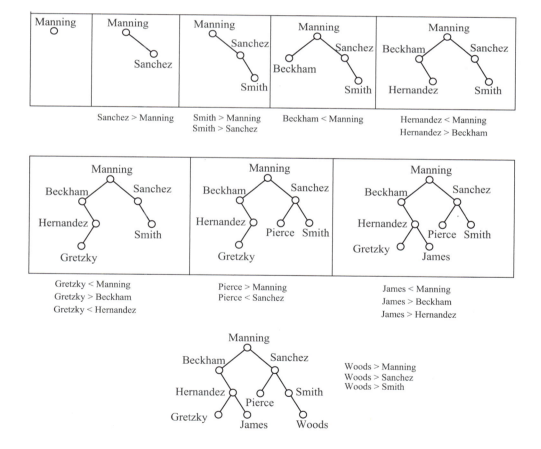

FIGURE 8.70: The steps used to construct a small binary search tree.

The following algorithm formalizes this binary search tree building procedure. It serves as both a search algorithm and as a new vertex insertion algorithm, in case the vertex label being searched for does not yet appear in the binary search tree. It is just a formally worded account of the following natural scheme for searching for a given record in a binary tree, and inserting it as a new vertex v in case it does not yet appear in the tree: We distinguish a vertex u of a binary search tree from its label label(u), the former thought to encode the location of the vertex, and the latter used when searching for it. If the existing tree is null (empty), we insert v as the root. Otherwise, we begin the search at the root r: if label(v) = label(r), the search is complete, if label(v) < label(r), (by the definition of a binary search tree, if v appears in the tree it must be in the left child subtree), apply this algorithm recursively to the left child subtree; in the remaining case, label(v) > label(r), (by the definition of a binary search tree, if v appears in the tree it must be in the right child subtree), apply this algorithm recursively to the right child subtree. The algorithm will either locate v in the tree or insert v as a new vertex so that the expanded tree remains a binary search tree.

ALGORITHM 8.8: (*Recursive Search and/or Insertion in a Binary Search Tree*)

Input: A binary search tree T, in which the vertices have some ordered labels and a LABEL.

Output: Either a vertex v in the tree with label(v) = LABEL, or the creation of a new vertex v (as new left or right child of a vertex in T), which when labeled with LABEL, the expanded tree remains a binary search tree.

Step 1: Compare LABEL with label(root), where "root" denotes the root of the tree.
If LABEL = label(root), output root as the location of vertex in the tree with the searched for LABEL, and exit the algorithm. Otherwise proceed to Step 2.

Step 2:
Case 1: If LABEL < label(root), proceed to the left child subtree.
Case 2: If LABEL > label(root), proceed to the right child subtree.
If this subtree is null, then LABEL does not appear in the tree, so create a new vertex v at this location and assign its label to be LABEL. Exit the algorithm.
If this subtree is not null, go back to Step 1 (using this smaller subtree in place of the previous tree).

EXAMPLE 8.36: Apply Algorithm 8.8 to the binary search tree of Figure 8.70 (the final nine-vertex tree) and with LABEL = Robbins.

SOLUTION: In the first application of Step 1, we find Robbins > Manning, and then Step 2 tells us to move on the right child subtree rooted at Sanchez. In the next application of Step 1, we find Robbins < Sanchez, and then Step 2 tells us to proceed to the left child subtree of Sanchez, which consists of the single (root) vertex Pierce. In the final application of Steps 1 and 2, since Robbins > Pierce, and Pierce has no (right) child, we insert a new vertex located as the right child of Pierce and with label Robbins. The resulting augmented tree is shown in Figure 8.71.

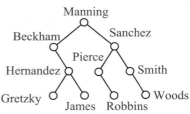

FIGURE 8.71: The tree resulting by applying Algorithm 8.8 to the binary search tree of Figure 8.70 with LABEL = Robbins.

EXERCISE FOR THE READER 8.35:   Use Algorithm 8.8 to create a binary tree to store the following sequence of integers: 39, 55, 22, 12, 35, 17, 44, 42, 25.

COMPLEXITY ANALYSIS OF ALGORITHM 8.8:   The following analysis will quantify our earlier comments regarding the efficiency of Algorithm 8.8.   Assume that we have an existing binary search tree T with n labeled vertices and height h. To facilitate the analysis, we introduce an associated full binary tree $\hat{T}$ by adding children (without labels) to any vertex of T that has fewer than two children, as well as to all (labeled) leaves of T. The process of forming $\hat{T}$ from T is illustrated in Figure 8.72.

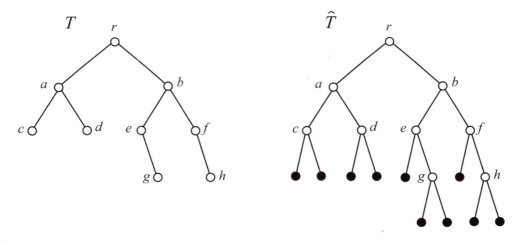

FIGURE 8.72: Modification of a labeled binary tree T into a full binary tree $\hat{T}$. The added vertices are shown in black; after the completion, each labeled vertex has two children.

We will estimate the number of comparisons needed to locate a certain LABEL, or determine that LABEL does not appear in T. The worst case is when we need to go all the way down through the height of T to either locate LABEL or determine that it does not appear, and this would require $h + 1$ comparisons (with vertices down a path of the tree, starting with the root). The same path in the completed tree $\hat{T}$ would either land us at the vertex whose label is LABEL, or place us at the unlabeled vertex that should be added to T.   For general binary search trees, Corollary 8.12 tells only that $h \geq \lceil \log_2 \ell \rceil$, where ℓ is the number of leaves.

Since the leaves in $\hat{T}$ are the unlabeled vertices, and the internal vertices are the n labeled vertices (of T), it follows from the observation made after Example 8.30 that $\hat{T}$ has $n + 1$ leaves, and thus $h \geq \lceil \log_2 \ell \rceil = \lceil \log_2 (n+1) \rceil$. In case the tree T is also balanced, Corollary 8.12 tells us this is an equality. Since balanced binary search trees thus have this controlled complexity estimate, it would be ideal to modify Algorithm 8.8 so that the binary search trees are always balanced. Although we will not delve deeper into this topic, such enhancements are indeed possible.   For example, if we had such a balanced binary search tree that contained 1 trillion labeled vertices, and we apply Algorithm 8.8 to search for or insert a certain label, the maximum number of comparisons required would thus be only $1 + \lceil \log_2 (n+1) \rceil = 1 + \lceil \log_2 (10^{12} + 1) \rceil = 41$.

Representing Rooted Trees on Computers

Since trees are simple graphs, each of the four representation schemes developed in Section 8.1 is applicable, of course, if we are interested in describing only the graphical structure. For rooted trees, the schemes of Section 8.1 for representing directed graphs would be more appropriate, but there are more effective and efficient ways to capture the important properties of rooted trees. For example, none of these previous representation schemes is designed to single out a root vertex. Also, in many applications, the order of the children is an important characteristic of a rooted tree (for example, in search trees). Apart from these problems, trees have so few edges (one less than the number of vertices), that adjacency matrices would waste a lot of memory; for example an adjacency matrix used to store a 1000 vertex tree would require 1 million entries, but less than 2000 would be nonzero! Of the representations of Section 8.1, adjacency lists and edge lists generally tend to be the most suitable for storing rooted trees, and these can be adapted (by ordering the vertices appropriately, e.g., placing the root first) to contain all of the needed information.   We will introduce some new data structures that are particularly designed to take advantage (and to nicely display characteristics) of the special structure of rooted trees.   Names of the following representation schemes are not standard, so we do not enunciate them as formal definitions but rather through an example.

EXAMPLE 8.37: We describe some data structure representation schemes for rooted trees, and show how each would work for the binary tree T of Figure 8.72. Throughout, we use the notation that n is the number of vertices of some rooted tree T.

(a) (*Parent List*) For applications of rooted trees in which the order of the children is not important, a *parent list* of the vertices is often sufficient. Such a parent list can be formed as a two-column matrix P, where the first column gives the vertex labels in some convenient order (preferably sorted in some way to facilitate searches), and the second column lists the corresponding label of the parent of each vertex. The root vertex is usually listed first; since it is the only vertex with no parent, and the second column entry for the root is either a nonvertex character (e.g., 0) or "NULL." The parent list for the binary tree of Figure 8.72 is shown in Table 8.1.

Vertex	Parent
r	NULL
a	r
b	r
c	a
d	a
e	b
f	b
g	e
h	f

TABLE 8.1: Parent list for the rooted tree of Figure 8.72.

(b) (*Children List*) Another convenient way to store a rooted tree is to create a list of the children of each of the vertices. If the rooted tree is a complete m-ary tree, this list can be stored as an n-row, $(m+1)$-column matrix. For general m-ary trees, we can use an n-row, $(m+1)$-column matrix, but rows for the vertices with fewer than m children will need to have additional column entries filled either with "NULL" or some obvious non-vertex characters. If the children are listed from left to right, then all of the information about the rooted tree will be contained in the children list. As in Part (a), the vertices should be listed in some convenient sorted order and the root should be listed first. Such a representation for the binary tree of Figure 8.72 is shown in Table 8.2.

Vertex	Left Child	Right Child
r	a	b
a	c	d
b	e	f
c	NULL	NULL
d	NULL	NULL
e	NULL	g
f	NULL	h
g	NULL	NULL
h	NULL	NULL

TABLE 8.2: Children list for the rooted tree of Figure 8.72.

Each of these representations has its own advantages and weaknesses. For example, the children list makes it easy to distinguish leaves as the vertices with no children, whereas this is less evident from the parent list. The parent list can be created from the children list (see the computer exercises of this section) but not vice versa if the order of the children is needed. Of course, in some applications it may be helpful to work with both representations.

EXERCISE FOR THE READER 8.36: Write down the parent list and the children list representations for the rooted tree of Figure 8.58.

EXERCISES 8.3:

1.   For each of the following graphs, indicate if it is a forest and/or a tree.

(a)   (b)   (c)   (d)

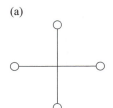   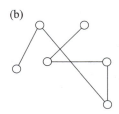   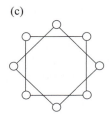   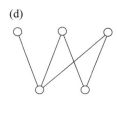

2.   For each of the following graphs, indicate if it is a forest and/or a tree.

(a)   (b)   (c)   (d)

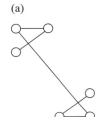   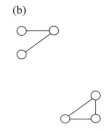   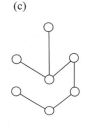   

3.   (a) How many edges does a tree with 100 vertices have?
      (b) How many vertices does a tree with 100 edges have?
      (c) How many edges does a forest with 100 vertices and 4 components have?
      (d) How many vertices does a forest with 100 edges and 10 components have?
      (e) Can a simple graph with 100 edges and 4 components have 110 vertices?

4.   (a) How many edges does a tree with 1000 vertices have?
      (b) How many vertices does a tree with 1000 edges have?
      (c) How many edges does a forest with 1000 vertices and 40 components have?
      (d) How many vertices does a forest with 1000 edges and 100 components have?
      (e) Can a simple graph with 1000 edges and 40 components have 1100 vertices?

5.   For which values of the positive integer parameter n are the following graphs trees?
      (a) The n-path graph P_n $(n \geq 2)$.   (b) The complete graph K_n $(n \geq 1)$.

6.   For which pairs of positive integers n, m is the complete bipartite graph $K_{n,m}$ a tree?

7.   (a) Up to isomorphism, how many different six-vertex trees are there? Draw representatives of
      each isomorphism class.
      (a) Up to isomorphism, how many different four-vertex forests are there? Draw representatives
      of each isomorphism class.

8.   (a) There are a total of 11 nonisomorphic trees with 7 vertices. Draw representatives of each
      isomorphism class.
      (b) Up to isomorphism, how many different five-vertex forests are there that have exactly two
      components? Draw representatives of each isomorphism class.

9. Answer the following questions pertaining to the rooted tree of Figure 8.73(a) with designated root vertex r:
    (a) What are the vertices of depth two?
    (b) Which vertex is the parent of i?
    (c) Which vertices are the children of a?
    (d) Which vertices are ancestors of p?
    (e) Which vertices are descendants of b?
    (f) What is the height of the rooted tree?
    (g) What is the diameter of the rooted tree?
    (h) Which are the internal vertices?

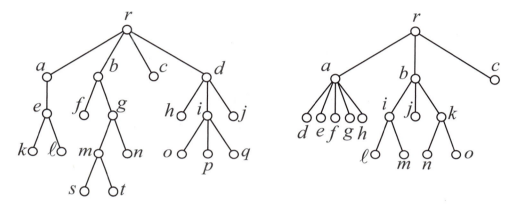

FIGURE 8.73: (a) (left), (b) (right), two rooted trees.

10. Answer the following questions pertaining to the rooted tree of Figure 8.73(b) with designated root vertex r:
    (a) What are the vertices of depth three?
    (b) Which vertex is the parent of i?
    (c) Which vertices are the children of a?
    (d) Which vertices are ancestors of m?
    (e) Which vertices are descendants of b?
    (f) What is the height of the rooted tree?
    (g) What is the diameter of the rooted tree?
    (h) Which are the internal vertices?

11. Draw binary trees that represent each of the following algebraic expressions:

    (a) $(1+x)(1-xy)$ (b) $\ln\left(\dfrac{1+x}{1-x}\right)$ (c) $(xy+yz)^2(x+z)^3$

12. Draw binary trees that represent each of the following algebraic expressions:

    (a) $x+5(1+y)^4$ (b) $\dfrac{1-x^2}{(1+x)^2}$ (c) $(x+y)(x^2+y^2)^3$

13. Suppose that 25 teams enter a single elimination basketball tournament.
    (a) How many rounds will the tournament have?
    (b) How many byes will there be?
    (c) How many games will be played?

14. Suppose that 50 boxers enter a single elimination tournament.
    (a) How many rounds will the tournament have? A round here is a tournament round as opposed to one of the 12 rounds within a single boxing match.

      (b) How many byes will there be?

      (c) How many matches will be boxed?

15.   In a small Pacific island news spreads by word of mouth/gossip. The village has 163 inhabitants. One villager gets some news by ham radio about a former resident who has won a big lottery. This villager tells 5 people about the news, and the next day, each of these people will tell 5 more people, and so on.

      (a) What is the smallest number of days needed for the news to spread to the whole village?

      (b) If we know that after the first day, each person spreading the news will tell it to at least two new people (until everyone knows), what would be the greatest number of days needed for the news to spread to the entire village?

16.   Suppose that T is a 4-ary tree with height 10.

      (a) What is the largest number of vertices that T can have?   For such a tree, how many of the vertices are internal?

      (b) What is the smallest number of vertices that T can have?   For such a tree, how many of the vertices are internal?

17.   Perform each of the following tasks on the following rooted trees:

      (a) The binary tree of Figure 8.69.   (b) The rooted tree of Figure 8.73(a).

      (i)   Apply Algorithm 8.4 to find the preorder ordering of the vertices of T.

      (ii)   Repeat (i) using Algorithm 8.7.

      (iii)   Apply Algorithm 8.5 to find the inorder ordering of the vertices of T.

      (iv)   Repeat (iii) using Algorithm 8.7.

      (v)   Apply Algorithm 8.6 to find the postorder ordering of the vertices of T.

      (vi)   Repeat (v) using Algorithm 8.7.

18.   Repeat the instructions of Exercise 17 on each of the following rooted trees:

      (a) The binary tree of Figure 8.71.   (b) The rooted tree of Figure 8.73(b).

19.   Apply Algorithm 8.8 to construct a binary search tree for each of the following lists:

      (a) Kansas, Missouri, Texas, Oklahoma, Iowa, Michigan, Ohio, Illinois

      (b) 77, 88, 66, 55, 99, 44, 60, 48, 62, 85

20.   Apply Algorithm 8.8 to construct a binary search tree for each of the following lists:

      (a) Nevada, Colorado, California, Washington, Oregon, Wyoming, Arizona, New Mexico

      (b) 123, 88, 64, 139, 154, 100, 22, 108, 62, 90

21.   Construct a parent list table and a children list table for the rooted tree of Figure 8.73(a).

22.   Construct a parent list table and a children list table for the rooted tree of Figure 8.73(b).

23.   In a rooted tree that represents the structure of folders and files in an electronic storage device (with the root being the main storage device folder), interpret the following concepts in terms of folder/file language. Note: A vertex can be either a file or a folder.

      (a) The parent of a vertex

      (b) Two vertices that are siblings

      (c) A child of a vertex

      (d) The ancestors of a vertex

      (e) The descendants of a vertex

      (f) The depth of a vertex

      (g) The height of the tree

      (h) The internal vertices

24.   Suppose that a binary search tree is to be built up using Algorithm 8.8 on a list of 100 names that is presented in some order.

      (a) What is the maximum height of a binary search tree that could be produced in this way?

      (b) Under what circumstances (in terms of how the names have been ordered) will the binary tree be of the maximum possible height?

(c) What is the minimum height of a binary search tree that could be produced in this way?

(d) Under what circumstances (in terms of how the names have been ordered) will the binary tree be of the minimum possible height?

NOTE: Exercises 25–28 develop decision trees corresponding to strategies for solving a class of problems involving a group of objects with identical appearances, one of which has a slightly different weight. The problem is to outline a strategy to identify this object using only a balance scale in such a way that the number of weighings that might be needed is kept to a minimum. We will use the context of a group of coins among which there is one counterfeit coin that is either heavier or lighter than the rest.

EXAMPLE 8.38: Suppose that we have five coins and a balance scale. Given that one of the coins is counterfeit and has a different weight than the others, develop a strategy using a decision tree for determining the counterfeit coin, and whether it is heavier or lighter than the rest.

SOLUTION: Since each weighing of the scale can have three possible outcomes: left heavy, balance, or right heavy, any strategy may be described by a 3-ary tree. Since there are 10 possible outcomes (i.e., for each of the five coins, it can be either heavier or lighter than the others), and these 10 outcomes must appear as (terminal) leaves of any decision tree, it follows that any strategy would require at least three weighings (a depth two 3-ary tree has at most 9 leaves). Figure 8.74 shows a decision tree for the problem that uses (at most) three weighings, and thus represents an optimal strategy.

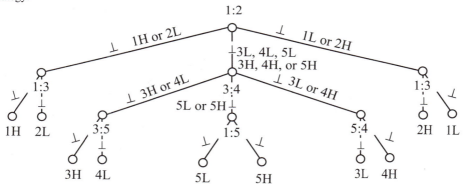

FIGURE 8.74: A decision tree for Example 8.38 for the use of a weight balance to detect the counterfeit coin among five coins and determine whether it is lighter or heavier than the others. Each internal vertex represents a weighing; left branches (and the left tilted symbol ⊥) indicate the left side is heavier, vertical branches (and the untilted symbol ⊥) indicate equal weights, and right branches (and the right tilted symbol ⊥) indicate the right side is heavier. Information gained on the counterfeit coin is written along corresponding edges.

Note that in Figure 8.74 we list only information gained on which are candidate coins that could be counterfeit, but there is additional information on some coins that are not counterfeit. For example on the first weighing, if the scale tilts (i.e., we go to either the left or right child of the root vertex), then it is determined that either coin 1 or coin 2 must be counterfeit, so it follows that coins 3–5 must be authentic coins.

25. Suppose that in the problem of Example 8.38 it is initially known that the counterfeit coin is heavier than the other four coins.

    (a) What is the total number of outcomes of the new problem of simply determining the counterfeit coin? What is the minimum depth that a 3-ary decision tree would need to have so that it could have at least as many leaves as there are final outcomes of this counterfeit coin problem?

    (b) Determine a strategy for the problem of Part (a) and write down the corresponding decision tree that requires the least possible number of weighings.

26.   Suppose we have six coins including one counterfeit coin that either weighs less or more than the other five.   We seek a strategy to detect the counterfeit coin, and to determine whether it is lighter or heavier than the rest.
        (a) What is the total number of outcomes? What is the minimum depth that a 3-ary decision tree would need to have so that it could have at least as many leaves as there are final outcomes of this counterfeit coin problem?
        (b) Determine a strategy for the problem of Part (a) and write down the corresponding decision tree that requires the least possible number of weighings.

27.   Suppose we have eight coins including one counterfeit coin that is heavier than the others.   We seek a strategy to detect the counterfeit coin, and to determine whether it is lighter or heavier than the rest.
        (a) What is the total number of outcomes? What is the minimum depth that a 3-ary decision tree would need to have so that it could have at least as many leaves as there are final outcomes of this counterfeit coin problem?
        (b) Determine a strategy for the problem of Part (a) and write down the corresponding decision tree that requires the least possible number of weighings.

28.   Suppose we have 13 coins including one counterfeit coin that is heavier than the others.   We seek a strategy to detect the counterfeit coin, and to determine whether it is lighter or heavier than the rest.
        (a) What is the total number of outcomes? What is the minimum depth that a 3-ary decision tree would need to have so that it could have at least as many leaves as there are final outcomes of this counterfeit coin problem?
        (b) Determine a strategy for the problem of Part (a) and write down the corresponding decision tree that requires the least possible number of weighings.

29.   (*Polish/Prefix Notation for Algebraic Expressions*)   The Polish mathematician Jan Łukasiewicz (1878–1956) developed the following scheme for representing algebraic expressions that avoids the need for parentheses.   This notation has come to be called *Polish notation*, or more suggestively *prefix* notation, and works as follows: operators (which are either binary or unary according to whether they take two input operands or a single input operand) are written immediately <u>before</u> operands.   An expression in prefix notation may be processed from left to right, where each time an operator is read, it is applied recursively to the operand(s) immediately following it on the right. Here are a few examples:

Prefix Notation	Corresponding Algebraic Expression
$+\ x\ y$	$x + y$
$\wedge + x\ y\ 2$	$(x+y)^2$
$\sqrt{\ } + 1 \sqrt{\ } * x\ y$	$\sqrt{1 + \sqrt{xy}}$

        In the second prefix string, for example, the operator "$\wedge$" is binary, so is applied to the next two operands. The next symbol in the prefix string is "+" which itself is another binary operator, so it gets applied to the two operands that immediately follow (this is part of the recursive evaluation of the "$\wedge$" operation) to produce "$x + y$," which will be the first operand of "$\wedge$." The second is the next symbol "2" in the string, and the final evaluation produces $(x+y)^2$.

        (a) Evaluate each of the following prefix notation expressions:
        (i) $-\ 4\ 6$ (ii) $+ * \ 2\ 4\ 6$ (iii) $\sqrt{\ }\ /\ + 13\ 12 \wedge 6\ 2$

        (b) Find the algebraic expressions represented by the following prefix notation expressions:
        (i) $\wedge\ 3\ y$ (ii) $+ - \ x\ y\ z$ (iii) $\ln\ /\ * x\ y + 1 \wedge x\ 2$
        (c) Find prefix notation expressions for each of the three algebraic expressions of Exercise 11.
        (d) For the binary tree representations found in Exercise 11 for the three algebraic expressions, determine (and write down) the preorder of the vertices.
        (e) Prove that if the preorder algorithm (Algorithm 8.4) is applied to a rooted tree representing an

algebraic expression, the resulting expression is the same as the prefix notation of the expression.
Suggestion for Part (e): Use induction.

30. (*Reverse Polish/Postfix Notation for Algebraic Expressions*)   The *reverse Polish* or *postfix* notation for an algebraic expression works very much like the Polish/prefix notation scheme of the previous exercise, except that the operators are written immediately <u>after</u> their operand(s). An expression in postfix notation may be processed from left to right, where each time an operator is read, it is applied recursively on the operand(s) immediately preceding it on the left. Here are a few examples:[22]

Postfix Notation	Corresponding Algebraic Expression
$x \ y + 2 \ ^\wedge$	$(x+y)^2$
$x \ y + 2 \ x + \ ^\wedge$	$(x+y)^{2+x}$
$x \ y * \ \log \ x \ y \ 2 \ ^\wedge + \ \sqrt{\ } \ /$	$\dfrac{\log(xy)}{\sqrt{x+y^2}}$

(a) Evaluate each of the following postfix notation expressions:

(i) $4 \ 3 \ 5 + *$ (ii) $1 \ 3 + 2 \ / \ 3 \ ^\wedge$ (iii) $12 \ 3 - \sqrt{\ } \ 7 * 3 + 3 \ /$

(b) Find the algebraic expressions represented by the following prefix notation expressions:

(i) $2 \ y \ 5 \ ^\wedge \ *$ (ii) $a \ b * c \ - d \ /$ (iii) $x \ 2 \ ^\wedge \ y \ 3 \ ^\wedge + \log \sqrt{\ }$

(c) Find postfix notation expressions for each of the three algebraic expressions of Exercise 12.

(d) For the binary tree representations found in Exercise 12 for the three algebraic expressions, determine (and write down) the postorder of the vertices.

(e) Prove that if the postorder algorithm (Algorithm 8.4) is applied to a rooted tree representing an algebraic expression, the resulting expression is the same as the postfix notation of the expression.
Suggestion for Part (e): Use induction.

31. Prove that every tree is bipartite.

32. Prove that a tree with an even number of edges must have at least one vertex of even degree.

33. Write proofs of the following implications involving items of Theorem 8.10. You may use ideas introduced in the proof of Theorem 8.10, but not any of the results of the theorem.

(a) $1 \Rightarrow 4$ (b) $4 \Rightarrow 1$ (c) $2 \Rightarrow 5$ (d) $5 \Rightarrow 2$

34. Write proofs of the following implications involving items of Theorem 8.10. You may use ideas introduced in the proof of Theorem 8.10, but not any of the results of the theorem.

(a) $2 \Rightarrow 4$ (b) $4 \Rightarrow 2$ (c) $3 \Rightarrow 6$ (d) $6 \Rightarrow 3$

35. (a) Prove that a connected n-vertex simple graph with n edges contains exactly one cycle.

(b) Does a connected n-vertex simple graph with $n+1$ edges contain exactly two cycles? Either provide a proof or give a counterexample.

---

[22] Reverse Polish notation (RPN) was introduced in the 1950s by Australian philosopher Charles L. Hamblin (1922–1985). It was heralded as a most efficient system for computer implementation since (i) it reduces the keystrokes by avoiding the use of parentheses along with the chore of having to balance them, and (ii) it logically corresponds to the human thought process when one does arithmetic in one's mind. For example to compute the quotient $(8+7)/(2+3)$ in our minds, we would first compute $8+7 = 15$, and then $2 + 3 = 5$, store these results (in our memory) and then compute their quotient $15/5 = 3$. This is exactly how RPN is organized: $8 \ 7 + 2 \ 3 + /$, with the quotient being performed (and read) last. Postfix notation is frequently used in compiling computer arithmetic systems. The HP Corporation was the first to use RPN in their electronic calculators in 1968, and has continued to offer it in some of their models ever since. Most winners of time-based calculator competitions have used such calculators.

36.   Show that a tree with a vertex of degree d will have at least d leaves.
      Suggestion: Use induction.

37.   Up to isomorphism, determine all trees T whose complement graph $\sim T$ is also a tree.

38.   Prove that in any tree every vertex of degree greater than one is a cut vertex.

39.   Suppose that T is an n-vertex tree in which the maximum vertex degree is three. Prove that the number of degree three vertices is no more than $\lfloor (n-2)/2 \rfloor$.

      Suggestion: Letting n_i denote the number of vertices of degree i, (justify and then) use the equations $n_1 + n_2 + n_3 = n$, $n_1 + 2n_2 + 3n_3 = 2(n-1)$.

40.   Suppose that T is an n-vertex tree and let n_i denote the number of vertices of T that have degree i. Show that the number of leaves of T is given by $2 + n_3 + 2n_4 + 3n_5 + 4n_6 + \cdots$.

      Suggestion: Justify   and   then   use   the   equations $n = n_1 + n_2 + n_3 + \cdots$, $2(n-1) = n_1 + 2n_2 + 3n_3 + \cdots$.

41.   (*Degree Sequences of Trees*)   There is no simple criterion to check whether a given sequence of nonnegative integers is a degree sequence for a simple graph. (Recall that Algorithm 8.1 was used for this determination.)   For trees, however, there is the following very simple characterization, which this exercise asks the reader to prove:
      Proposition: *A sequence of positive integers* $(d_1, d_2, \cdots, d_n)$, *with* $n \geq 1$, *is the degree sequence of a tree if, and only if* $\sum_{i=1}^{n} d_i = 2n - 2$.
      Suggestion: Use induction.

42.   (*Central Vertices of Trees*)   This exercise will provide an outline of a proof by induction of the following result:
      Proposition: *A tree T has either one or two central vertices, and if there are two, then they are adjacent.*
      (a) (*Basis Step*) Observe that the result is true for the tree consisting of a single vertex.
      (b) (*Induction Step*) Assume that the proposition is true for all trees having at most n vertices, where n is a positive integer, and that T is a tree on $n+1$ vertices.   Let $\hat{T}$ the graph obtained from T by deleting all of its leaves (and the edges adjacent to them).   Show that the central vertices of $\hat{T}$ are the same as those of T, and apply the inductive hypothesis to complete the proof of the proposition.

COMPUTER EXERCISES 8.3:

NOTE: We will assume throughout that vertex labels of trees are positive integers. This removes any technical issues in comparing two vertices since most computing systems have built-in functionality to compare two integers (i.e., given two different integers a, b, determine whether $a < b$ or $a > b$).   Any examples of the section referred to in this set whose vertex labels are not integers should be given integer labels using a natural scheme, such as the following: $a = 1$, $b = 2$, ..., $z = 26$.[23]   This set of computer exercises will develop some programs for rooted trees. The basic data structures will be the parent list and the children list, which are described in Example 8.37. In particular, the parent list will

---

[23] Readers who have more experience with programming may wish to extend the functionality of some of the programs to include the comparisons of strings of letters using lexicographic order.

be a two-column matrix where the first column lists the vertices (in order) and the second column their corresponding parents. The children list will always be an $m + 1$ column matrix where m is the maximum number of children (i.e., the smallest value of m such that the rooted tree is an m-ary tree). The first column lists all the vertices in order (but putting the root first), and the remaining columns list the corresponding children (in order) with some convenient label (such as "0" or "null") used to fill empty slots in cases where there are less than m children; we will use 0s to fill such slots.

1. (*Children List to Parent List*) (a) Write a program that will input a children list matrix for a rooted tree and will output the corresponding parent list matrix. The syntax should be as follows:

    ```
    ParentList = ChildrenList2ParentList(ChildrenList)
    ```

    The input `ChildrenList` should be an $m + 1$ column matrix (of integers), and the output `ParentList` a two-column matrix as described in the preceding note.
    (b) Apply your program to the binary tree of Figure 8.69.
    (c) Apply your program to the binary tree of Figure 8.72.
    (d) Apply your program to the 4-ary tree of Figure 8.68.

2. (*Parent List to Depth List*) (a) Write a program that will input a parent list matrix for a rooted tree and will output the corresponding *depth list* two-column matrix, whose first column lists the vertices (in the same order as the inputted parent list matrix) and whose corresponding second column entries give the nonnegative integer depths of the vertices. The syntax should be as follows:

    ```
    DepthList = ParentList2DepthList(ParentList)
    ```

    (b) Apply your program to the binary tree of Figure 8.69.
    (c) Apply your program to the binary tree of Figure 8.72.
    (d) Apply your program to the 4-ary tree of Figure 8.68.
    Suggestion: Rather than trace the path back to the root for each vertex, it is much more efficient to progress down through the successive generations of vertices, assigning the common depth to each generation.

3. (*Program to Compute All Descendants of an Inputted Vertex in a Rooted Tree*) (a) Write a program that determines all descendants of an inputted vertex in an inputted rooted tree. The syntax should be as follows:

    ```
    DescendantList = DescendantFinder(ChildrenList, vertex)
    ```

    The input `ChildrenList` should be as described in the note at the beginning of this computer exercise set, and the second input `vertex` should be an (actual) vertex number. The output will be a vector (possibly empty) of all vertex labels that are descendants of the inputted vertex.
    (b) Apply your program to the binary tree of Figure 8.69 with the root vertex, and with a couple of vertices of different depths.
    (c) Apply your program to the binary tree of Figure 8.72 with the root vertex, and with a couple of vertices of different depths.
    (d) Apply your program to the 4-ary tree of Figure 8.68 with the root vertex, and with a couple of vertices of different depths.

4. (*Program to Compute All Ancestors of an Inputted Vertex in a Rooted Tree*) (a) Write a program that determines all descendants of an inputted vertex in an inputted rooted tree. The syntax should be as follows:

    ```
    AncestorList = AncestorFinder(ChildrenList, vertex)
    ```

    The input `ChildrenList` should be as described in the note at the beginning of this computer exercise set, and the second input `vertex` should be a vertex number. The output will be a vector (possibly empty) of all vertex numbers that are ancestors of the inputted vertex.
    (b) Apply your program to the binary tree of Figure 8.69 with the root vertex, and with a couple of vertices of different depths.
    (c) Apply your program to the binary tree of Figure 8.72 with the root vertex, and with a couple

of vertices of different depths.

(d) Apply your program to the 4-ary tree of Figure 8.68 with the root vertex, and with a couple of vertices of different depths.

5.   (*Rooted Tree to Children Subtrees*) (a) Write a program that will input a children list matrix for a rooted tree and will output children list matrices for each of the children subtrees of the root. The syntax should be as follows:

```
ChildChildrenLists = ChildTrees(ChildrenList)
```

The input `ChildrenList` corresponds to the inputted tree, the output `ChildChildren Lists` is an ordered list of *s* matrices each being a children list for one of the child subtrees of the root of the inputted tree. Here *s* denotes the number of children of the root of the inputted tree; see Figure 8.63.

(b) Apply your program to the binary tree of Figure 8.69.

(c) Apply your program to the binary tree of Figure 8.72.

(d) Apply your program to the 4-ary tree of Figure 8.68.

Suggestion: The most straightforward data structure for the output variable in this program is a three-dimensional array, where the third dimension is the index corresponding to the child subtree represented by each two-dimensional matrix. An ordered list of matrices is really a three-dimensional array. Here the first two dimensions of the three-dimensional array are the common dimensions of the children list matrices, and the third dimension determines the number of matrices in the list. In order to create a well defined array, use the same number of rows for each of the outputted children tree matrices. The number of rows will be the depth of the original tree, which can be computed by calling on the programs of Computer Exercises 1 and 2. Unallocated rows in outputted child tree matrices can be filled with zeros.

6.   (*Preorder Traversal of a Rooted Tree-Recursive Program*) (a) Write a program that will input a children list matrix for a rooted tree and will output a vector containing the preorder list of the vertices. The program should be recursive and based on Algorithm 8.4. The syntax should be as follows:

```
PreOrder = PreOrderAlg(ChildrenList)
```

(b) Apply your program to the binary tree of Figure 8.69.

(c) Apply your program to the binary tree of Figure 8.72.

(d) Apply your program to the 4-ary tree of Figure 8.68.

Suggestion: One possible strategy to create a program that captures Algorithm 8.4 is to have the program `PreOrderAlg` consist of a single line, which calls on a recursive program that requires two inputs:

```
PreOrder = PreOrderRecursiveAlg(ChildrenList,List)
```

where the second input variable represents a partial list that is being built up in going from one iteration to the next. The former program should call on the latter with the second input variable being an *empty list*, since the first time we enter into the recursive algorithm the list is empty. Thus, the single line of code for the first program will look something like: `PreOrder = PreOrderRecursiveAlg(ChildrenList, [])`. A possible pseudocode for the latter recursive program is as follows:

```
IF ChildrenList has nonzero entries (i.e., the tree is nonempty)
   Append the first entry (the root) to List
END (IF)
IF ChildrenList has more than one nonzero entry (i.e., the root
                                            has children)
           Set s = the number of children
           Set ChildChildrenLists = ChildTrees(ChildrenList)
           (i.e., apply the program of Computer Exercise 5)
           FOR i = 1 TO s
               Set CMi = ith Child Tree Children List
               (i.e., the ith matrix of ChildChildrenLists)
               Set PreOrder = PreOrderRecursiveAlg(CMi, PreOrder)
```

```
                    END (FOR)
          END (IF)
```

7. (*Inorder Traversal of a Rooted Tree-Recursive Program*) (a) Write a program that will input a child list matrix for a rooted tree and will output a vector containing the inorder list of the vertices. The program should be recursive and based on Algorithm 8.5. The syntax should be as follows:

```
     InOrder = InOrderRecursiveAlg(ChildrenList)
```

(b) Apply your program to the binary tree of Figure 8.69.
(c) Apply your program to the binary tree of Figure 8.72.
(d) Apply your program to the 4-ary tree of Figure 8.68.
Suggestion: Use the ideas of the suggestion for Computer Exercise 6.

8. (*Postorder Traversal of a Rooted Tree-Recursive Program*) (a) Write a program that will input a child list matrix for a rooted tree and will output a vector containing the postorder list of the vertices. The program should be recursive and based on Algorithm 8.6. The syntax should be as follows:

```
     PostOrder = PostOrderRecursiveAlg(ChildrenList)
```

(b) Apply your program to the binary tree of Figure 8.69.
(c) Apply your program to the binary tree of Figure 8.72.
(d) Apply your program to the 4-ary tree of Figure 8.68.
Suggestion: Use the ideas of the suggestion for Computer Exercise 6.

9. (*Recursive Search and/or Appending in a Binary Search Tree*) (a) Write a program that will input a child list matrix `ChildrenList` for a rooted tree and a label, `LABEL`, and will output either the row index of the child list matrix in which the label appears as the first column entry (i.e., the label was located in the tree), or will ouput a new `ChildrenList` matrix, with one additional row and corresponding to the proper insertion of a new vertex in the tree that carries the inputted label. The program should follow Algorithm 8.8, and the syntax should be as follows

```
   Output = BinarySearchTreeFinderInserter(ChildrenList, LABEL)
```

(b) Apply your program to the binary search tree creation problem of Exercise for the Reader 8.35.
(c) Apply your program to the binary search tree creation problem of ordinary Exercise 19(b).
Suggestion: Use the ideas of the suggestion for Computer Exercise 6.

APPENDIX TO SECTION 8.3: APPLICATION OF ROOTED TREES TO DATA COMPRESSION AND CODING; HUFFMAN CODES

Computers and other digital devices store data as binary sequences, i.e., as sequences of 0s and 1s. The most fundamental way to accomplish this is to establish a correspondence between the characters that appear in the data and strings of an appropriately large fixed length N. The popular ASCII system, for example uses $N = 8$ length bit strings to store the $2^N = 256$ symbols in its alphabet. These include upper- and lowercase letters, digits, and punctuation marks, as well some additional symbols and keyboard control symbols. Such a fixed length representation system makes it straightforward and unambiguous to pass back and forth between actual data (at the human interface) and the corresponding binary representation of it (for the computer interface) since the binary data always is processed in blocks of size N. To establish some relevant vocabulary, we begin with the following definition.

DEFINITION 8.34: A **binary code** for an alphabet of symbols $\mathscr{A}$ is simply a one-to-one function $\mathscr{A} \to \{$finite length bit strings$\}$. Functional notation is not usually employed in this setting. The bitstring associated with any symbol in $\mathscr{A}$ is called its **codeword**, and the juxtaposition of codewords produced from a string of characters in $\mathscr{A}$ is simply called the **code** of the string.

It is often the case that when certain types of data (or files) need to be stored, some characters appear more frequently than others. For example, in ordinary written English, the letter "e" occurs much more often than the letter "z." In order to store such data efficiently, it makes sense to use shorter binary codes for less frequently used symbols. The following simple example shows one problem that must be addressed with any variable length binary code.

EXAMPLE 8.39: Consider the binary code defined on the alphabet $\mathscr{A} = \{a,b,c,d,e\}$ by:

$$a \mapsto 1, b \mapsto 10, c \mapsto 11, d \mapsto 01, e \mapsto 101.$$

As with <u>any</u> binary code, any string of symbols in $\mathscr{A}$ can be easily coded by substituting individual characters with their corresponding codes. For example, the code of "abba" would be "110101." With the reverse (decoding) operation, however, there can be problems. For example, the code "101" could have either come from the string "ad," "e," or from "ba."

The problem that was illustrated in the Example 8.39 stemmed from the fact that some codewords are prefixes of others, i.e., the codeword for "a" is "1," which is also the first bit of the codewords for "b," "c," and "e." Binary codes that avoid this problem are called *prefix codes* (but perhaps a better name would be *no-prefix codes*):

DEFINITION 8.35: A binary code in which no codeword occurs as a prefix of another is called a **prefix code**.

EXAMPLE 8.40: Consider the binary code defined on the alphabet $\mathscr{A} = \{a,b,c,d,e\}$ by:

$$a \mapsto 1, b \mapsto 000, c \mapsto 001, d \mapsto 010, e \mapsto 011.$$

Since no codeword of any of the five alphabet letters occurs as a prefix of another, this is a prefix code.

It is easy to see how any binary string produced by a prefix code can be unambiguously decoded: simply process the binary digits from left to right until a codeword is detected, record the corresponding symbol in $\mathscr{A}$, and start over with the next bit. Continue until the entire code is decoded. For example, using the prefix code of Example 8.40, the code 10000010101 would be decoded with the following steps:

$$10000010101 \to \text{a}\textbf{0000010101} \to \text{ab}\textbf{0010101} \to \text{abc}\textbf{0101} \to \text{abcd}\textbf{1} \to \text{abcda}$$

EXAMPLE 8.41: (*Using Binary Trees to Construct Prefix Codes/Prefix Code Trees*) Constructing prefix codes can be easily done by constructing full binary *prefix code trees* as follows:
Step 1: Given an alphabet $\mathscr{A}$ of $\ell > 1$ symbols, we draw (any) full binary tree with ℓ leaves.
Step 2: Label the leaves in a one-to-one fashion with the symbols of $\mathscr{A}$.
Step 3: Take the codeword for each symbol in $\mathscr{A}$ to be the sequence of 0s and 1s resulting from the unique path going from the root to the corresponding leaf in the tree, using a 0 for each left child path and a 1 for each right child path.

As a specific example, consider the problem of creating a prefix code for the alphabet $\mathscr{A} = \{a,b,c,d,e,f,\%,\&,\$,?\}$. There are many 10-leaf complete binary trees, and for each there are in turn 10! ways to label the leaves with the alphabet symbols. Figure 8.75 shows one such tree, along with the resulting codeword/character correspondence.

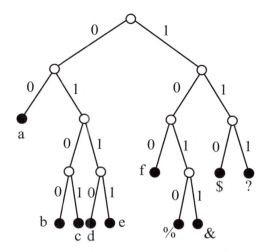

character	codeword
a	00
b	0100
c	0101
d	0110
e	0111
f	100
%	1010
&	1011
$	110
?	111

FIGURE 8.75: A full binary tree used to create a prefix code, which is shown on the right.

The following simple example illustrates how prefix codes can compress data.

EXAMPLE 8.42: Suppose that we have a certain data file consisting of a string of length 1 million of characters from the alphabet $\mathscr{A} = \{a,b,c,d,e\}$. Suppose also that these characters occur in the string with the following frequencies:

character	a	b	c	d	e
frequency	0.6	0.1	0.15	0.12	0.03

(a) Estimate the length of the code used to represent the data string if a fixed length binary representation code is used.

(b) Estimate the length of the code used to represent the data string if the prefix code of Example 8.40 is used.

SOLUTION: Part (a): Since there are five characters in the alphabet, a fixed length representation code would require a codeword length N of at least 3. Thus the million character string would have a code of at least 3 million bits.

Part (b): Since the codeword for "a" has one bit, and the codewords of the other letters have three, using the table of frequencies of the letters we may estimate the number of bits of the code resulting from this prefix code to be: $1{,}000{,}000(0.6 \times 1 + 0.1 \times 3 + 0.15 \times 3 + 0.12 \times 3 + 0.03 \times 3) = 1.8$ million.

Thus, the data will have been compressed to 60% of its original size.

Example 8.42 motivates the following definition that will help us to compare the compression rates of different full binary tree constructions of prefix codes:

DEFINITION 8.36: Suppose that T is a full binary tree with ℓ leaves $u_1, u_2, \cdots, u_\ell$ which have been assigned positive weights $\mathrm{wgt}(u_1), \mathrm{wgt}(u_2), \cdots, \mathrm{wgt}(u_\ell)$. The **average weighted depth** of T is defined by:

$$\mathrm{wgt}(T) = \sum_{i=1}^{\ell} \mathrm{wgt}(u_1) \cdot \mathrm{depth}(u_i). \tag{5}$$

When T represents a prefix code for a coding of a file or string with a known distribution of characters, and we take the weights of the leaves as the corresponding frequencies, then the average weighted depth $\mathrm{wgt}(T)$ is just the average number of bits per character in the code, so the length of the binary code is just $N \cdot \text{wgt}(T)$, where N is the number of characters in the original string. This is exactly what was done in the estimates of Example 8.42.

The above examples have clearly demonstrated two things:

1. Constructing prefix codes for a given alphabet is easy (as shown in Example 8.41).
2. Prefix codes can be designed to compress data files if we have information on the distribution of the characters in the alphabet.

But one major problem remains: Since there are many full binary trees with a certain number $\ell > 1$ of leaves, and in turn many ways to label them with characters in the alphabet that we wish to code, we would like to know how to do this in a way that would lead to the most efficient data compression. Since there are only finitely many full binary trees, given the frequencies, each of these along with the $\ell!$ ways of labeling the leaves with symbols could, in principle, be checked. But this number grows so quickly with ℓ that such a brute-force approach is completely impractical. We will next describe an efficient algorithm that accomplishes the task. The resulting prefix codes that it produces are known as *Huffman codes*, named after David A. Huffman, who discovered the algorithm. Huffman made the discovery in 1951, while taking an electrical engineering course as a graduate student at MIT. The professor had offered students the option of completing a term paper project in lieu of the final exam.[24]

FIGURE 8.76: David A. Huffman (1925–1999), American scientist.

Huffman codes remain a state-of-the-art coding procedure that continues to be used in digital storage and communication devices. One example is HD television. Another is the *JPEG* (Joint Photographic Experts Group) system, which is used to compress electronic photograph files. JPEG compression works in two stages; the first stage significantly reduces the color information contained in individual pixels of an electronic photograph file in a way that is not noticeable by the human eye (this stage is based on the fact that human eyes have a much greater sensitivity to changes in light than to subtle changes in color). The second stage involves a Huffman encoding procedure on the suitably reduced pixel alphabet resulting from the first stage. When the file is viewed by a recipient, the Huffman code of the second stage is

---

[24] In his early years of education, David Huffman fell behind his classmates; he later attributed this to some problems in his family that eventually led to his parents' divorce. His mother tried to help and even took a job as a mathematics teacher in a school for troubled students so he could study there. After beginning his studies in this new school, his teachers quickly began to realize that he was quite a child prodigy. He began to flourish, soon moving ahead of his classmates to graduate early and then earning his B.S. degree in electrical engineering from the Ohio State University at the age of 18. He subsequently served in the Navy, and then returned to complete an M.S. degree, also at Ohio State, after which he moved on to MIT to earn a Ph.D. in electrical engineering. He then served on the faculty at MIT for 14 years and became a full professor there before moving to the University of California–Santa Cruz to accept an offer and challenge to serve as the founding chair of the new computer science department.

Huffman made significant contributions to information theory, digital communications, and electric circuits, but his best known discovery is Huffman codes. Some of his motivation to begin work on this interesting coding problem came from learning that his professor and the illustrious Claude Shannon (see Chapter 2 for a biography) had struggled with it. He had been working hard on the project for a couple of months. A week before the end of the term he thought that he would not be able to complete it, so decided to abandon his work and to start preparing for the final exam. But just as he threw his research notes into the wastebasket, a spark of inspiration came upon him that let him to a key discovery to complete the project!

Huffman's hobbies included outdoor activities like backpacking, body surfing, and scuba diving, which he continued to enjoy throughout his later years.

reversed. Whereas the first stage cannot be reversed, the human eye would not easily be able to tell the difference. A similar use of Huffman codes occurs in the popular *MP3* system of audio compression. As with JPEG, MP3 consists of two stages: In the first stage a compression based on human hearing capabilities is used to significantly reduce the alphabet of sound symbols. After this, a "lossless" Huffman code is employed for a final compression.

Huffman's algorithm is a recursive one; the inputs are the ℓ symbols in the alphabet $\mathscr{A}$ that we wish to code, along with the associated frequencies of the symbols. The algorithm starts with a forest of ℓ isolated vertices with vertex weights assigned as the frequencies of the symbols. At each iteration, a new tree is created by merging two trees of the forest whose weights have a minimum sum. The procedure is quite simple, as the following algorithm shows. What was difficult was showing that the algorithm really produces a tree of minimum possible average weighted depth (and thus produces the prefix code with best possible data compression). We will prove this result after stating the algorithm and giving an example.

ALGORITHM 8.9: (*Huffman Coding*)

Input: A set of $\ell > 1$ symbols $u_1, u_2, \cdots, u_\ell$ along with a corresponding set of positive number weights: $\mathrm{wgt}(u_1), \mathrm{wgt}(u_2), \cdots, \mathrm{wgt}(u_\ell)$. (In applications, the weights are the frequencies.)

Output: A prefix code $u_i \mapsto c_i$ ($1 \le i \le \ell$) (called the **Huffman code**) corresponding to a full binary tree (called the **Huffman tree**) with ℓ leaves labeled with the weights such that the tree has minimum possible average weighted depth.

Step 1: Initialize F to be the forest of ℓ isolated vertices assigned the weights $\mathrm{wgt}(u_1), \mathrm{wgt}(u_2)$, $\cdots, \mathrm{wgt}(u_\ell)$.

Step 2: (*Recursive Step*)
WHILE F is not a tree
        Find two rooted trees T, T' in F whose weights are the smallest. Merge these two trees into a new rooted tree by introducing a new root vertex, and placing T as its left child subtree, and T' as the right child subtree. Assign this merged tree's weight to be the sum of the weights of T, T'.
END (WHILE)

Step 3: Read the resulting Huffman code from the rooted tree produced in Step 2 (as shown in Example 8.41).

NOTE: In the recursive step of the above algorithm, in case there are ties for the two lightest trees, any choice can be made, whatever is most convenient in the computation/program. We also point out that the weights of the trees in the algorithm are not the average weighted depths of formula (5); the tree weights of this algorithm are an artifice that make the algorithm work correctly.

EXAMPLE 8.43: Apply the Huffman coding algorithm to the following alphabet of symbols with corresponding frequencies:

character	a	b	c	d	e	f	g
frequency	0.3	0.1	0.15	0.12	0.03	0.2	0.1

SOLUTION: The progression of steps of the algorithm is shown in Figure 8.77 (next page), where the (total) weight of each tree is listed over its root vertex, and the two root vertices of the trees that will be merged in the next iteration are colored black. From the resulting Huffman tree produced by the final iteration, we may read off the following Huffman code:

character	a	b	c	d	e	f	g
Huffman codeword	00	0100	011	100	0101	11	101

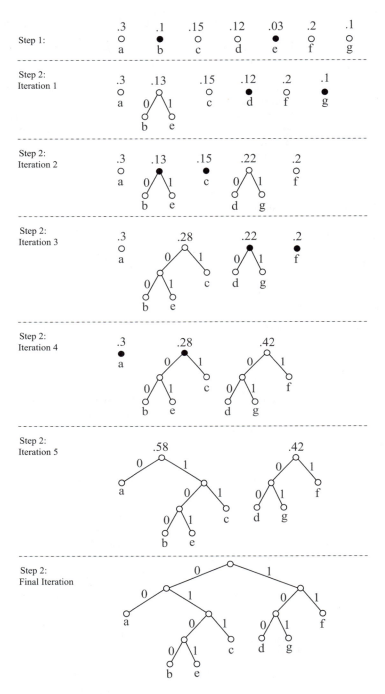

FIGURE 8.77:   The application of the Huffman coding algorithm (Algorithm 8.9) to the character/frequency date of Example 8.43. In each iteration the root vertices of two of the lightest trees that will be merged in the next iteration are colored in black.

The following theorem legitimizes Algorithm 8.9 by showing that the Huffman tree has the smallest possible average weighted depth among all full binary trees whose leaves are the given symbols, with their assigned frequencies.

THEOREM 8.14: Suppose that we are given a set of $\ell > 1$ symbols $u_1, u_2, \cdots, u_\ell$ along with a corresponding set of positive number weights: $\mathrm{wgt}(u_1), \mathrm{wgt}(u_2), \cdots, \mathrm{wgt}(u_\ell)$. A **Huffman tree** produced by Algorithm 8.9 produces a tree of minimum average weighted depth among all full binary trees with ℓ leaves labeled with the weights $\mathrm{wgt}(u_1), \mathrm{wgt}(u_2), \cdots, \mathrm{wgt}(u_\ell)$.

Proof: We proceed by induction on $\ell > 1$.
Basis Step: $\ell = 2$ In this case the Huffman tree H consists of a root with two children (the leaves), with the latter two vertices being assigned the weights $\mathrm{wgt}(u_1), \mathrm{wgt}(u_2)$. Since there are no other full binary trees with two leaves, all will have the same average weighted depth as given by (5):

$$\mathrm{wgt}(T) = \sum_{i=1}^{\ell} \mathrm{wgt}(u_1) \cdot \mathrm{depth}(u_i) = \mathrm{wgt}(u_1) + \mathrm{wgt}(u_2).$$

Inductive Step: We assume that the theorem holds for any symbol/weight sets of size ℓ, where $\ell \geq 2$ is a fixed integer. The task is to use this assumption to show that the theorem is true for all symbol/weight sets of size $\ell + 1$. So let us be given such a set of parameters: $u_1, u_2, \cdots, u_{\ell+1}$; $\mathrm{wgt}(u_1), \mathrm{wgt}(u_2), \cdots, \mathrm{wgt}(u_{\ell+1})$. Reindexing the symbols if necessary, we may assume that the weight sequence is monotone decreasing: $\mathrm{wgt}(u_1) \geq \mathrm{wgt}(u_2) \geq \cdots \geq \mathrm{wgt}(u_\ell) \geq \mathrm{wgt}(u_{\ell+1})$. We introduce a new data set with ℓ symbols and weights: $u_1', u_2', \cdots, u_\ell'$; $\mathrm{wgt}(u_1'), \mathrm{wgt}(u_2'), \cdots, \mathrm{wgt}(u_\ell')$ defined by:

$$u_i' = \begin{cases} u_i, & \text{if } i < \ell \\ u^*, & \text{if } i = \ell \end{cases}, \quad \mathrm{wgt}(u_i') = \begin{cases} \mathrm{wgt}(u_i), & \text{if } i < \ell \\ \mathrm{wgt}(u_\ell) + \mathrm{wgt}(u_{\ell+1}), & \text{if } i = \ell \end{cases},$$

where u^* denotes any new symbol that has not already appeared in this proof. From Algorithm 8.9, it follows that there is a very simple relationship between a Huffman tree H for the original data set $u_1, u_2, \cdots, u_{\ell+1}$; $\mathrm{wgt}(u_1), \mathrm{wgt}(u_2), \cdots, \mathrm{wgt}(u_{\ell+1})$ and a Huffman tree H' for the modified data set $u_1', u_2', \cdots, u_\ell'$; $\mathrm{wgt}(u_1'), \mathrm{wgt}(u_2'), \cdots, \mathrm{wgt}(u_\ell')$: We can obtain H from H' by changing the leaf of the vertex u_ℓ' to an internal vertex with left and right children $u_\ell, u_{\ell+1}$ and assigning these new leaf vertices their corresponding weights. This relationship is shown in Figure 8.78.

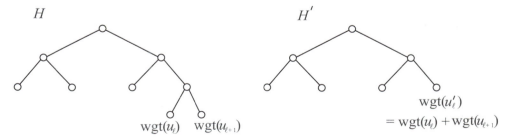

FIGURE 8.78: Illustration of the difference between the two Huffman trees H and H' in the proof of Theorem 8.14.

Next we compare the average weighted depths $\mathrm{wgt}(H), \mathrm{wgt}(H')$. In the formula (5) used to compute these two quantities, all of the $\ell - 1$ common leaf vertices of the two trees have the same weights and so make the same contribution to their respective sums in (5). Thus, we need only compare the remaining

terms $\mathrm{wgt}(u_\ell)\cdot\mathrm{depth}(u_\ell)+\mathrm{wgt}(u_{\ell+1})\cdot\mathrm{depth}(u_{\ell+1})$ of $\mathrm{wgt}(H)$ with $\mathrm{wgt}(u_\ell')\cdot\mathrm{depth}(u_\ell')$ of $\mathrm{wgt}(H')$. But since $\mathrm{wgt}(u_\ell)+\mathrm{wgt}(u_{\ell+1})=\mathrm{wgt}(u_\ell')$, and $\mathrm{depth}(u_\ell)=\mathrm{depth}(u_{\ell+1})=\mathrm{depth}(u_\ell')+1$, it follows that $\mathrm{wgt}(u_\ell)\cdot\mathrm{depth}(u_\ell)+\mathrm{wgt}(u_{\ell+1})\cdot\mathrm{depth}(u_{\ell+1})=\mathrm{wgt}(u_\ell')\cdot\mathrm{depth}(u_\ell')+\mathrm{wgt}(u_\ell)+\mathrm{wgt}(u_{\ell+1})$.
We have thus obtained the following equality:

$$\mathrm{wgt}(H)=\mathrm{wgt}(H')+\mathrm{wgt}(u_\ell)+\mathrm{wgt}(u_{\ell+1}). \tag{6}$$

Next, we let T_{OPT} be a tree of minimum average weighted depth among all full binary trees with $\ell+1$ leaves labeled with the weights $\mathrm{wgt}(u_1),\mathrm{wgt}(u_2),\cdots,\mathrm{wgt}(u_{\ell+1})$. Consider an internal vertex v of T_{OPT} that is of maximum depth. We may assume that the two children of v are labeled with the two lightest weights $\mathrm{wgt}(u_\ell),\mathrm{wgt}(u_{\ell+1})$. Indeed, if they were not, we could switch the labels with these lightest children and the resulting change could not make average weighted depth any larger (since weights with larger depths are multiplied by larger numbers by (5)). Now we form an ℓ leaf tree T_{OPT}' (with leaf weights) from T_{OPT} in exactly the same way that H' is related to H (see Figure 8.78). The same argument used in the preceding paragraph to obtain (6) thus shows that

$$\mathrm{wgt}(T_{\mathrm{OPT}})=\mathrm{wgt}(T_{\mathrm{OPT}}')+\mathrm{wgt}(u_\ell)+\mathrm{wgt}(u_{\ell+1}). \tag{7}$$

We are now nicely set up to complete the proof: Since the Huffman tree H' is optimal for the data $u_1',u_2',\cdots,u_\ell';\ \mathrm{wgt}(u_1'),\mathrm{wgt}(u_2'),\cdots,\mathrm{wgt}(u_\ell')$, and T_{OPT}' is a full binary tree whose leaves have the same weight sets, it follows that

$$\mathrm{wgt}(T_{\mathrm{OPT}}')\geq\mathrm{wgt}(H'). \tag{8}$$

If we use this inequality in conjunction with (5) and (7), we may obtain:

$$\begin{aligned}\mathrm{wgt}(T_{\mathrm{OPT}})&=\mathrm{wgt}(T_{\mathrm{OPT}}')+\mathrm{wgt}(u_\ell)+\mathrm{wgt}(u_{\ell+1}) &&\text{(by (7))}\\&\geq\mathrm{wgt}(H')+\mathrm{wgt}(u_\ell)+\mathrm{wgt}(u_{\ell+1}) &&\text{(by (8))}\\&=\mathrm{wgt}(H) &&\text{(by (6))}\end{aligned}$$

The Huffman tree H has thus been shown to be optimal, and the proof is complete. $\square$

EXERCISES FOR THE APPENDIX TO SECTION 8.3:

1. Use the prefix code shown in Figure 8.75 to decode each of the following binary sequences:
    (a) 0101000100111
    (b) 0001000001101010
    (c) 10001110111011010111000001100111

2. Use the Huffman code that was created in Example 8.43 to decode each of the following binary sequences:
    (a) 011001010101
    (b) 1010011110101
    (c) 110101010110001011011010101000101010111

3. Draw a prefix code tree of depth two for the alphabet $\mathscr{A} = \{a,b,c,d\}$, and then write down a corresponding prefix code.

4. Draw a prefix code tree of depth three for the alphabet $\mathscr{A} = \{a,b,c,d\}$, and then write down a corresponding prefix code.

5. Apply the Huffman coding algorithm to the following alphabet of symbols with corresponding frequencies:

character	a	b	c	d
frequency	0.25	0.25	0.25	0.25

6. Apply the Huffman coding algorithm to the following alphabet of symbols with corresponding frequencies:

character	a	b	c	d
frequency	0.7	0.15	0.08	0.07

7. Apply the Huffman coding algorithm to the following alphabet of symbols with corresponding frequencies:

character	a	b	c	d	1	2	3
frequency	0.1	0.35	0.05	0.12	0.08	0.18	0.12

8. Apply the Huffman coding algorithm to the following alphabet of symbols with corresponding frequencies:

character	a	b	c	d	e	f	g
frequency	0.15	0.2	0.15	0.12	0.05	0.18	0.15

9. Apply the Huffman coding algorithm to create a prefix code for the following phrase that will result in a binary code having the smallest possible number of bits: *cats bats gnats and rats* (Treat the space between words as a symbol.)

10. Apply the Huffman coding algorithm to create a prefix code for the following phrase that will result in a binary code having the smallest possible number of bits: *grand stand band is dandy* (Treat the space between words as a symbol.)

11. (a) What is the largest possible depth that a Huffman tree produced by Algorithm 8.9 could have on an inputted set of 10 symbols?
    (b) Provide an example of frequencies for 10 symbols with which such a maximum depth tree would be realized.

12. (a) Let $\ell > 1$ be a positive integer. What is the largest possible depth that a Huffman tree produced by Algorithm 8.9 could have on an inputted set of ℓ symbols?
    (b) Provide an example of frequencies for ℓ symbols with which such a maximum depth tree would be realized.

13. Explain how every prefix code is realizable as a Huffman code (i.e., as an output of Algorithm 8.9) using as input the sequence of its symbols and some corresponding sequence of frequencies.

14. As was pointed out in the section proper, when ties arise between equal weighted trees in Algorithm 8.9, choices must be made and thus Huffman trees and codes need not be unique. But are the lengths of the binary strings representing the symbols of any Huffman code uniquely determined? Explain your answer.

COMPUTER EXERCISES FOR THE APPENDIX TO SECTION 8.3:

NOTE: We will briefly outline a scheme to implement the Huffman coding algorithm (Algorithm 8.9) on a computer. We will assume that the reader has gone through the relevant computer exercises of this section. The main new technique that is needed to program the Huffman coding algorithm is an implementation on the merging of two trees. This is quite easy to do if we work with the children representation matrices, especially since all trees involved in the Huffman algorithm will be full binary trees. As usual, we will label all vertices of trees with positive integers.

1.   (*Program for the Huffman Merging of Two Full Binary Trees*) (a) Write a program that will input two children list matrices for two full binary trees T_1, T_2 whose vertex labels have nothing in common, and will output the children list for the merged tree T obtained by introducing a new root vertex for T, and making T_1, T_2 the left and right child subtrees, respectively, of this root. The syntax should be as follows:

    ```
    MergedTreeChildrenList = HuffmanTreeMerge(T1, T2, RootLabel)
    ```

    The inputs T1,T2 are the three-column children list matrices for the two inputted full binary trees, and the output MergedTreeChildrenList is the children list matrix for the Huffman merged tree. The third input RootLabel is the label that will be assigned to the root of the merged tree.
    (b) Apply your program to replicate the merging of the two trees in the final iteration of Step 2 in Figure 8.77. You will need to assign the vertices integer labels to create their children lists.
    Suggestion: First create a new label for the root of the *T*, and create the corresponding first row of the output matrix (it will be something like: [rootlabelT rootLabelT1 rootLabelT1]). The remaining rows of the output matrix will simply be all of the rows of the two inputted matrices, taken in any order.

2.   (*Program for the Huffman Codes*) (a) Write a program whose input is a vector of at least two vertex weights VertexWgts, which are any positive numbers, and whose output Codes is a matrix whose rows list the corresponding Huffman codes as produced by Algorithm 8.9.

    ```
    Codes = HuffmanCodes(VertexWgts)
    ```

    (b) Apply your program to redo Example 8.43.
    (c) Apply your program to redo ordinary Exercises 5 and 7.
    (d) Many tables have been published on the distribution of the letters of the English alphabet in standard written English, for example, Table 1.1 shows the frequencies that were computed by Beker and Piper, [BePi-82].[25] Apply your program to compute a corresponding Huffman code.

    Suggestion: This program should call on that of the preceding computer exercise. The number of columns of the output matrix should be the height of the Huffman tree. Each row entry of the output matrix will consist of either a 0, a 1, or a -1, the latter being used to fill any unused slots of the row after the binary code has been completely listed.

---

[25] Of course, there will be variations in frequencies depending on the text corpus being examined. For example, the distributions in emails, brief text messages, and computer codes would each have distinguishing characteristics. But for most written English that is not completely informal, the distribution given in Table 1.1 works remarkably well.

TABLE 1.1 Frequencies of the letters of the English alphabet.

Letter	Probability	Letter	Probability
a	.082	n	.067
b	.015	o	.075
c	.028	p	.019
d	.043	q	.001
e	.127	r	.060
f	.022	s	.063
g	.020	t	.091
h	.061	u	.028
i	.070	v	.010
j	.002	w	.023
k	.008	x	.001
l	.040	y	.020
m	.024	z	.001

Chapter 9: Graph Traversal and Optimization Problems

In this chapter we will discuss a wide range of practical problems that are best formulated in terms of simple graphs. The problems have numerous applications, making it very important to have schemes for solving instances of them. We will describe efficient algorithms when they exist, and explain the state of affairs for problems for which no efficient algorithms are believed to exist. The chapter will consist of three sections, each of which considering its own theme of problems. In Section 9.1 we will describe two graph traversal problems, each of which has historical significance. The first one asks: Given a simple graph G, does there exist a simple path, or a simple circuit in G that traverses every edge of G exactly once? This problem was considered by the Swiss mathematician Leonhard Euler, and his work on the problem is considered the origin of the subject of graph theory. The second related problem asks: Given a simple graph G, does there exist a path or a circuit in G that visits each vertex exactly once (except for the endpoints, in the case of a circuit)? This problem was considered by the Irish mathematician Sir William Rowan Hamilton. Although these problems might seem quite similar at first, one of them can be solved with an efficient (and simple) algorithm, while the other has been shown to be NP complete, so no efficient algorithm exists for it and none is believed to exist. Can you guess which is the harder problem?

Section 9.2 will consider some general optimization problems that are modeled with simple graphs, where in some cases it will be assumed that the edges have associated weights. We will present an efficient algorithm for the determination of shortest edge weighted paths. The unweighted case of this problem was discussed in Section 8.2, where an algorithm was presented that was based on adjacency matrices. The algorithm that we present here will be much more efficient. We will also present an efficient algorithm for the maximum matching problem that was described in Section 8.1.

Section 9.3 will develop some important theory and an efficient algorithm for network flow problems, which involve weighted directed graphs where the edge weights represent flow capacities (for liquid, data, etc.), and the problem is to find the maximum quantity of flow that can be routed in the network from a designated source vertex to a designated sink vertex.

9.1: GRAPH TRAVERSAL PROBLEMS

<u>Euler Paths and Tours and the Origin of Graph Theory</u>

Having already gone through an extensive chapter covering many graph theoretic concepts, we have more than adequate preparation to fully describe the problem that led to the origin of the graph theory, along with its general solution, which constituted the first theorem of the subject.

Graph theory has its roots in a 1736 visit to the old Prussian city of Königsberg[1] by the illustrious mathematician Leonhard Euler (see Section 3.3 for a portrait and brief biography). The river Pregel runs through the city center as shown in Figure 9.1. There are seven bridges connecting various land portions. The Prussian Emperor Frederick the Great told Euler of an innocently simple sounding puzzle that the local people were not able to resolve: Was it possible to plan a walk where each bridge would be crossed exactly once? The problem aroused Euler's interest, and he was able to solve it by proving a much more general theorem.

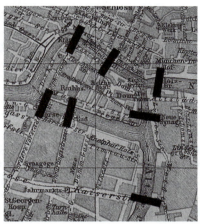

FIGURE 9.1: A (1910) map of the seven bridges over the Pregel River in the old city of Königsberg; the bridges are emphasized.

Euler modeled the bridges in Königsberg by the graph shown in Figure 9.2. The four vertices represent the land masses carved out by the Pregel, and the edges are the bridges.

In the (modern) vocabulary of Chapter 8, the Königsberg problem thus asks whether there exists a simple path in the (general) graph of Figure 9.2 that traverses all of the edges of the graph. Before we state Euler's general theorem, we give a general definition for the relevant concept and a related one:

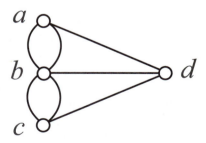

FIGURE 9.2: A graph representing the seven bridges of Königsberg; the edges represent the bridges and the vertices represent the land masses carved out the by river.

---

[1] Königsberg was renamed Kaliningrad, after it was taken over by the USSR after WWII.

DEFINITION 9.1: Suppose that G is a connected graph. An **Euler path** in G is a simple path in G that traverses every edge. An **Euler tour** in G is a simple circuit in G that traverses every edge. We say that G is **Eulerian** if it admits an Euler tour.

In the language of Definiton 9.1, the Königsberg problem asks whether the graph of Figure 9.2 admits an Euler path. Note that since a circuit is a path, any Euler tour is also an Euler path, so the Königsberg problem asks for a weaker requirement than the graph be Eulerian. We point out that the assumption that G is connected is an essential prerequisite for Euler paths/tours, since neither could exist in a disconnected graph. The following theorem provides a complete classification of graphs that admit either an Euler path or tour. After proving the theorem, we will present a simple and efficient algorithm that can be used to find these objects when they exist.

THEOREM 9.1: (*Classification of Graphs that Admit Euler Paths/Tours*) Suppose that G is a connected graph with at least two vertices.

(1) G is Eulerian if, and only if every vertex of G has even degree.
(2) G admits an Euler path if, and only if G has at most two vertices of odd degree. Futhermore, if there are exactly two vertices of odd degree, any Euler path must start at one of these vertices and end at the other.

We point out that from Corollary 8.2, it follows that having at most two vertices of odd degree is equivalent to having zero or two vertices of odd degree. Before proving this theorem, we apply it to the Königsberg bridge problem and to a larger problem.

EXAMPLE 9.1: (a) Show that it is not possible to plan a walk in ancient Königsberg in such a way that each of the seven bridges is crossed exactly once.
(b) For the river/bridge configuration of Figure 9.3, is it possible to plan a walk in such a way that each of the seven bridges is crossed exactly once? If such a walk is possible, are there any restrictions on where it must start and end?

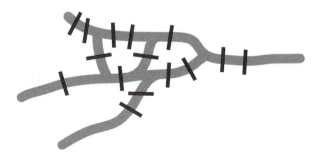

FIGURE 9.3: The river/bridge configuration of Example 9.1(b).

SOLUTION: Part (a): Since the degrees of the four vertices of the Königsberg graph of Figure 9.2 are 3, 3, 3, 5, it follows from Theorem 9.1 that this graph does not have an Euler path, proving a negative answer to the Königsberg bridge problem.

Part (b): If we label the five land masses carved out by the river, for example, as shown in Figure 9.4(a), we can easily form the corresponding graph by taking the land masses as the vertices and the bridges as the edges; see Figure 9.4(b).

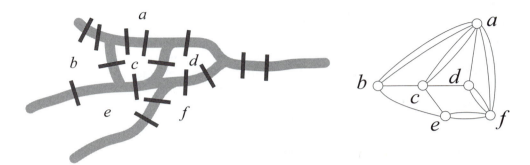

FIGURE 9.4: (a) (left) Labels are assigned to the land masses determined by the rivers of Figure 9.3. (b) (right) Taking the vertices to be the labeled land masses of (a), and edges to the bridges, we obtain the representing graph.

Vertices a and c have odd degrees (7 and 5, respectively) and all other vertices have even degree, so it follows from Theorem 9.1 that the graph has an Euler path (which must have a and c as its terminal vertices) but it does not have an Euler tour. Thus, it is possible to plan a walk over the bridge network of Figure 9.3 that traverses each bridge exactly once, but such a walk must start at one of the land masses a or c, and end at the other one.

EXERCISE FOR THE READER 9.1: If one new bridge could be constructed across one of the Pregel River segments of Figure 9.1, could it be done in such a way that the problem of Part (a) would have an affirmative answer (using all eight bridges)? If yes, could it furthermore be done so as to guarantee that the walk could be organized to end at the same land mass where it started?

Proof of Theorem 9.1:
Step 1: We will first show how (1) and (2) are equivalent.
We first assume that (1) is true, and use it to prove that (2) is true. In one direction, suppose that we have a graph G that contains an Euler path P. If this P had different terminal vertices $u \neq v$, then we consider the graph $\hat{G}$ defined by $V(\hat{G}) = V(G)$, and $E(\hat{G}) = E(G)$ along with a new edge with endpoints u and v.[2]

It follows that circuit $\hat{P}$ formed by joining this new edge of $\hat{G}$ to the path P produces an Euler tour in $\hat{G}$. By (1) it follows that every vertex of $\hat{G}$ has even degree, and so u, v have odd degree in G (since one edge incident to each is

---

[2] In case G already has an edge with endpoints u,v, this new edge in $E(\hat{G})$ is a different parallel edge.

removed) and all other vertices even degree in G. For the converse, we assume that G is a graph with exactly two vertices u, v of odd degree. The graph $\hat{G}$ defined above will have only even degree vertices, so by (1) will have an Euler tour $\hat{P}$. If we delete the non-G edge from $\hat{P}$, we are left with an Euler path in G. Thus we have shown that $(1) \Rightarrow (2)$. A very similar argument shows that $(2) \Rightarrow (1)$.

Step 2: We prove (1).

First we assume that G is Eulerian, and must show that every vertex has even degree. Let v be any vertex of G, and let C be an Euler tour of G. Every time C traverses v, it must traverse two different edges (one for entering and one for exiting), and thus each such "pass" contributes 2 to $\deg(v)$. But since C accounts for all edges of G, these even contributions must add up to the total (even) degree of v.

We have left to prove the converse: Assuming that every vertex of a connected graph G has even degree, we must show that G has an Euler tour. We will proceed by (strong) induction on the number of edges of G, $|E(G)|$.

Basis Step: $|E(G)| = 2$. (This is the smallest number of edges that a graph satisfying the hypotheses of Theorem 9.1 may have.) In this case G must consist of two vertices joined by a pair of parallel edges between them. The two edges (traversed in opposite directions) form an Euler tour.

Induction Step: Assume validity for $|E(G)| \le k$, and let G be a graph with $k + 1$ edges.

First, it follows from Theorem 8.8 that G must have cycles. Let C be a simple circuit of G with maximum possible length. We will show that C is an Euler tour by assuming it is not and reaching a contradiction. Consider the edge deletion graph $H = G - E(C)$. Although H need not be connected, each of its vertices (which are the vertices of G) must have even degree (in H). This follows from the fact that a circuit was removed; by the argument of the preceding paragraph, the degree reductions for each affected vertex are all even. Let K be a nontrivial component of H. By the induction hypothesis, it follows that K has an Euler tour D. Since G is connected, K (or any component of H) must share common vertices with C; let u be a vertex that is common to both circuits; see Figure 9.5.

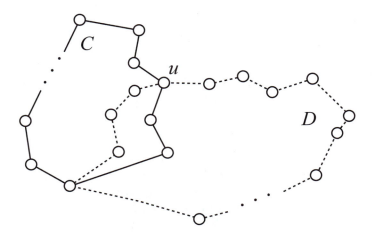

FIGURE 9.5: Two intersecting circuits in the proof of Theorem 9.1.

We can splice the two circuits C and D together to form a larger one quite simply as follows: Start a vertex u (common to both), traverse the complete circuit C (so we are back again at u), then traverse the complete circuit D. Since C and D are simple circuits with no edges in common, this spliced circuit is a simple circuit with more edges than C, a contradiction to the fact that C was chosen to have the maximum possible number of edges. □

Now that we have Theorem 9.1 to tell us the complete story on whether a given graph admits an Euler tour or path, the next natural question is: How does one produce such an object once it is known to exist? The proof of Theorem 9.1 leads to an efficient algorithm for the construction of Euler paths/tours, which essentially works as follows: Suppose that a connected graph G with at least two vertices is Eulerian. Initially we label all edges of G as "unused," and initiate an empty list of "TourEdges." Pick any vertex u of G. Start scanning through unused edges to find one e_1 incident to u. Append e_1 to TourEdges and remove e_1 from the "unused" edges. Next, we scan through the remaining unused edges to find one e_2 incident to the other endpoint of e_1. Append e_2 to TourEdges and remove e_2 from the "unused" edges. We continue this process. Since all degrees are even, it must eventually lead us back to u. If unused edges incident to u remain, we restart the process, and continue to do so until there are no longer any unused edges incident to u. We then move on to the next vertex u_{next} visited by TourEdges and repeat the process, but now inserting the edges we obtain (which will constitute a cycle starting and ending at u_{next}) into TourEdges, directly between the two edges incident to u_{next}. We then move on to the next vertex and proceed in the same fashion. Since G is connected, this process will eventually use all unused edges, and hence produce an Euler tour. The following algorithm formalizes this process, and its correctness is justified by the first part of the proof of Theorem 9.1. As an alternative, there is another algorithm (*Fleury's algorithm*) that is described as Algorithm 9.3 in the notes of the exercise set at the end of this section. The latter algorithm is simpler to describe and understand, but generally is less efficient than Algorithm 9.1A.

ALGORITHM 9.1A: Construction of Euler Tours in Eulerian Graphs

Input: An (ordered) edge list EdgeList for a connected Eulerian graph with at least two vertices, and an initial vertex u.

Ouput: An (ordered) list TourEdges of the edges, corresponding to an Euler tour.

Step 1: Initialize UnusedEdges = Edges, and
TourEdges = [ ] (empty list)
CurrentVertex = u
EndVertex = u
  CurrentInsertionPoint = End
DoneVertices = [ ]

Step 2: (*Iterative Step*)
WHILE UnusedEdges is nonempty
   Scan UnusedEdges for one that is incident to EndVertex
  IF such an edge is found, call it CurrentEdge
    Insert CurrentEdge to TourEdges after CurrentInsertionPoint
    UPDATE CurrentInsertionPoint to be after CurrentEdge in TourEdges
    UPDATE EndVertex = other endpoint of CurrentEdge
    Remove CurrentEdge from UnusedEdges
  ELSE IF no such edge exists
     Append CurrentVertex to DoneVertices
     Update CurrentVertex as next vertex in TourEdges that
                       is not in DoneVertices
           (*such a vertex exists while UnusedEdges is nonempty*)
     UPDATE EndVertex = CurrentVertex
     UPDATE CurrentInsertionPoint to be after first edge in TourEdges
       incident to CurrentVertex
  END IF
END WHILE

Although the algorithm is written in a formal pseudocode fashion (that is easy to translate into a computer program), its idea is quite simple:

Continue iterating the process of building a tour by scanning through the unused edges until we get back to the starting vertex. We restart the process (moving on to a new vertex in which to splice a new simple circuit) whenever unused edges remain.

This algorithm can be easily adapted to construct Euler paths for connected graphs that have exactly two vertices of odd degree.

ALGORITHM 9.1B: Construction of Euler Paths in Connected Graphs that have Exactly Two Vertices of Odd Degree

Input: An edge list EdgeList for a connected graph with exactly two vertices u, v of odd degree.

Ouput: An ordered list of the edges, EulerPathEdges, corresponding to an Euler path from u to v.

Step 1: Append a new edge $\{v, u\}$ to the beginning EdgeList.

Step 2: Apply Algorithm 9.1A with inputs EdgeList and initial vertex v to produce an Euler tour TourEdges. Note that $\{v, u\}$ (the edge added in Step 1) is the first edge in TourEdges.

Step 3: Remove the first edge $\{v, u\}$ from TourEdges to produce the output EulerPathEdges.

NOTE: In cases of parallel edges for general graphs, since we are not concerned with distinguishing any edges with the same endpoints, the output of the algorithm can be listed, without any ambiguity, as a sequence of vertices. Also, when this algorithm is applied, it is helpful to distinguish the orders of the endpoints of an edge, since these will determine the corresponding path of vertices determined by the EulerPathEdges.

EXAMPLE 9.2: Apply Algorithm 9.1B to construct an Euler path for the graph of Figure 9.4(b), with initial vertex a.

SOLUTION: We order the edges lexicographically (for scanning purposes). We preappend a new edge $\{c, a\}$ (between the only two vertices of odd degree) to EdgeList. Thus, when (in Step 2 of Algorithm 9.1B) we apply Algorthm 9.1A, the intial (ordered) list of UnusedEdges is as follows:[3]

$$\{c,a\},\{a,b\},\{a,b\},\{a,c\},\{a,c\},\{a,d\},\{a,f\},\{a,f\},\{b,c\},\{b,e\},\{c,d\},\{c,e\},$$
$$\{d,f\},\{d,f\},\{e,f\},\{e,f\}$$

As we proceed throughout the iterative Step 2 of Algorithm 9.1A we show the edge sequences inserted into TourEdges each time the CurrentVertex and CurrentInsertionPoint change:

CurrentVertex = c, CurrentInsertionPoint = End:
TourEdges: $\{c,a\},\{a,b\},\{b,a\},\{a,c\},\{c,a\},\{a,d\},\{d,c\},\{c,b\},\{b,e\},\{e,c\}$

CurrentVertex = a, CurrentInsertionPoint = after first edge $\{c,a\}$
TourEdges:
$\{c,a\}, \ \{a,b\},\{b,a\},\{a,c\},\{c,a\},\{a,d\},\{d,c\},\{c,b\},\{b,e\},\{e,c\}$

$\overbrace{\{a,f\},\{f,a\}}$

---

[3] In using either Algorithm 9.1A or B, it is not nessessary to distinguish between different parallel edges as long as they are listed the appropriate number of times in EdgeList. Thus, the "edgemap" concept of Section 8.1 is not needed here.

CurrentVertex = f, CurrentInsertionPoint = after first edge $\{a,d\}$
TourEdges:
  $\{c,a\},\{a,f\},\{f,a\}\ \{a,b\},\{b,a\},\{a,c\},\{c,a\},\{a,d\},\{d,c\},\{c,b\},\{b,e\},\{e,c\}$

  $\overbrace{\{f,d\},\{d,f\},\{f,e\},\{e,f\}}$

Figure 9.6 shows the order in which the edges have been chosen.

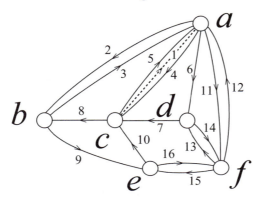

FIGURE 9.6: Illustration of Algorithm 9.1B applied to the graph of Figure 9.4(b).

The final Step 3 of Algorithm 9.1B tells us to remove the first edge in the preceding edge sequence to produce the following Euler path in the orginal graph: $\langle a, f, d, f, e, f, a, b, a, c, a, d, c, b, e, c\rangle$. For this small example, it would have been quite easy to complete the tour with a single spliced cycle after the first iteration (with current vertex = c) was completed by looking at the remaining edges on the graph, but we followed the order indicated by the algorithm.

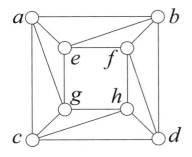

FIGURE 9.7: The Eulerian graph for Exercise for the Reader 9.2.

EXERCISE FOR THE READER 9.2: Apply Algorithm 9.1A to construct an Euler tour for the Eulerian graph shown in Figure 9.7. Use a as the initial vertex and the lexicographic ordering on the edges.

EXERCISE FOR THE READER 9.3: Show that the complexity of Algorithm 9.1A is $O(|E|)$, where E is the set of edges of the Eulerian graph. Recall this means that the number of logical operations to execute Alogrithm 9.1A is bounded by $k\,|\,E\,|$, for some constant $k > 0$ that is independent of the graph.

The result of Exercise for the Reader 9.3 shows that Algorithm 9.1 is very efficient.   In the computer exercises the reader will have the opportunity to compare its performance with that of Fleury's algorithm, which is another popular algorithm for constructing Euler tours.

Euler Paths and Tours in Digraphs

The definitions of Eulerian graphs, paths, and tours, all extend almost verbatim to the setting of directed graphs, the only change being the obvious one that paths need to respect the edge directions.   Theorem 9.1 and the corresponding Algorithms 9.1AB also have natural analogues to the digraph setting. The digraph version of Theorem 9.1 is given in the following result.

THEOREM 9.2:   (*Classification of Digraphs That Admit Euler Paths/Tours*) Suppose that $\vec{G}$ is a connected digraph with at least two vertices.

(1) $\vec{G}$ is Eulerian if, and only if every vertex $v \in V(\vec{G})$ has its indegree equal to its outdegree, i.e., $\deg^-(v) = \deg^+(v)$.

(2) $\vec{G}$ admits an Euler path from a vertex u to a different vertex v if, and only if $\deg^-(u) + 1 = \deg^+(u), \deg^-(v) = \deg^+(v) + 1$, and $\deg^-(w) = \deg^+(w)$, for all other vertices $w \in V(\vec{G})$.

A proof of Theorem 9.2 can be achieved by a routine modification of our proof of Theorem 9.1, and this task will be left as Exercise 30.   Algorithms 9.1A and B need no changes in wording to work for digraphs; we simply need to respect the orders of the edges.   The idea will be illustrated in the following example.

EXAMPLE 9.3: For each digraph below, (a) determine whether it is Eulerian, and find an Euler tour if one exists.   (b) In case the digraph is not Eulerian, determine whether there is an Euler path, and find one if one exists.

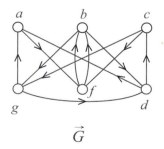
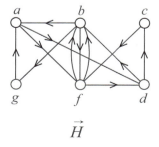

FIGURE 9.8: The two digraphs of Example 9.3.

SOLUTION: In the digraph $\vec{G}$, all vertices except a and c have indegree and outdegree equal to two. Since the vertices a, c, have indegrees 2, 1, and outdegrees 1, 2, respectively, it follows from Theorem 9.2 that $\vec{G}$ is not Eulerian, but does admit an Euler path from c to a. To obtain such an Euler path, we first adjoin the directed edge (a,c) to $\vec{G}$. In the resulting digraph $\vec{G} + (a,c)$, all vertices now have indegree and outdegree equal to two, so we know there is an Euler tour. We proceed by using the modification of Algorithm 9.1A and B, obtained by using ordered edges rather than unordered edges. We initialize the UnusedEdges using the lexicographic order:

$(a,c),(a,f),(b,d),(b,g),(c,f),(c,g),(d,a),(d,c),(f,b),(f,b),(g,a),(g,d)$

CurrentVertex $= a$, CurrentInsertionPoint $=$ End:
TourEdges: $(a,c),(c,f),(f,b),(b,d),(d,a),(a,f),(f,b),(b,g),(g,a)$

UnusedEdges: $(c,g),(d,c),(g,d)$ CurrentVertex: c

CurrentInsertionPoint: After Edge (a,c)
Cycle to be Inserted: $(c,g),(g,d),(d,c)$
Final Euler Path (after deleting first edge (a,c) from TourEdges:
$< a,c,g,d,c,f,b,d,a,f,b,g,a >$

Since all vertices in digraph $\bar{H}$ have their indegrees equal to their outdegrees, it follows from Theorem 9.2(1) that $\bar{H}$ is Eulerian. A corresponding Euler tour is easily obtained from the "directed" modification of Algorithm 9.1A: we begin by ordering the edges lexicographically:

$(a,d),(a,f),(b,a),(b,f),(b,g),(c,f),(d,b),(d,c),(f,b),(f,b),(f,d),(g,a)$

After two iterations of Step 2 of the algorithm, an Euler tour is produced:[4]
CurrentVertex $= a$, CurrentInsertionPoint $=$ End:
TourEdges: $(a,d),(d,b),(b,a),(a,f),(f,b),(b,f),(f,b),(b,g),(g,a)$

UnusedEdges: $(c,f),(d,c),(f,d)$ CurrentVertex: d

CurrentInsertionPoint: After Edge (a,d)
Cycle to be Inserted: $(d,c),(c,f),(f,d)$
Final Euler Path (after deleting first edge (a,c) from TourEdges:
$< a,d,c,f,d,b,a,f,b,f,b,g,a >$

---

[4] When doing the algorithm by hand, it helps to list the original ordering of the edges once, and cross them off the list as they are added to the TourEdges list.

Application of Eulerian Digraphs: De Bruijn Sequences

We will next describe an application of Eulerian digraphs to coding and communications. We begin with the relevant definition.

DEFINITION 9.2: For a positive integer n, a **(2,n)-de Bruijn sequence**[5] is a length 2^n binary string such that every possible length n bitstring occurs exactly once as a substring, where wrapping around is allowed.   More generally, if $b > 1$ is a positive integer, a **(b,n)-de Bruijn sequence** is a length b^n string in the alphabet $\{0,1,\cdots,b-1\}$ such that every possible length n string in this alphabet occurs exactly once as a substring, where wrapping around is allowed.

Observe that no string of length less than b^n could have the de Bruijn property (Why?). We begin with some simple examples of de Bruijn sequences.

EXAMPLE 9.4:   (a) The reader can easily check that the following four sequences are (2,n)-de Bruijn sequences for $n = 1, 2, 3, 4$:

       01   0110   01110100   0000100110101000

(b)   0123 is clearly a (4,1)-de Bruijn sequence, while 011220021 can readily be checked to be a (3,2)-de Bruijn sequence.

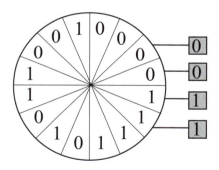

FIGURE 9.9: The four sensors on the right read the position 0011 of the drum whose 16 sectors are labeled with a (2,4)-de Bruijn sequence.

De Bruijn sequences have some very important applications in coding theory and communications.   They can be used to efficiently crack codes/passwords in cases where an enter key is not required. They can also be used to identify the position of a rotating drum. Figure 9.9 shows how such a scheme could be implemented on a drum using 16 sectors labeled with the bits of a (2,4)-de Bruijn sequence. Greater precision can be achieved by using larger de Bruijn sequences.

From the small scale Example 9.4, it is not at all clear that de Bruin sequences always exist, much less how to find them when they do

---

[5] De Bruijn sequences are named after Nicolaas Govert de Bruijn (1918–), a Dutch mathematician. During his career de Bruijn worked both in academia and in industry, and his research contributions were deep and widespread, ranging from combinatorics and discrete mathematics (his Erdös number is 1) to optimal control and mathematical analysis. He received numerous accolades and honors for his work.

exist. Both issues were positively resolved in 1946 when de Bruijn proved their existence using a constructive algorithm on certain associated Eulerian digraphs. The whole idea stems from the following insightful concept of a digraph associated with a given set (p,n) of de Bruijn parameters:

DEFINITION 9.3: For a given pair of positive integers $b > 1, n \geq 1$, the **de Bruijn digraph** $D_{b,n}$ has vertices

$$V(D_{b,n}) = \{\text{length } n-1 \text{ strings in the alphabet } \{0,1,2,\cdots,b-1\}\}.$$

For each vertex $d_1 d_2 \cdots d_{n-1}$, there will be b outgoing edges; one to each vertex of the form $d_2 d_3 \cdots d_{n-1} d_n$, where $d_n \in \{0,1,2,\cdots,b-1\}$.

EXAMPLE 9.5: Create drawings of the following de Bruijn digraphs:
(a) $D_{2,3}$ (b) $D_{3,2}$

SOLUTION: Drawings of these two de Bruijn digraphs are shown in Figure 9.10.

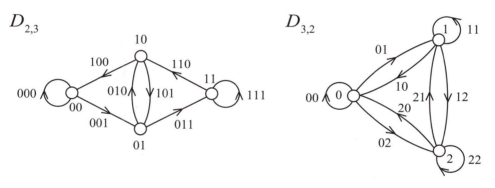

FIGURE 9.10: Drawings of the two de Bruijn digraphs $D_{2,3}$ and $D_{3,2}$.

Our next result will record the fact that any de Bruijn digraph $D_{b,n}$ is Eulerian, and from this fact will follow a simple algorithm for constructing corresponding de Bruijn sequences.

PROPOSITION 9.3: (*De Bruijn Digraphs Are Eulerian*) For a given pair of positive integers $b > 1, n \geq 1$, the de Bruijn digraph $D_{b,n}$ is Eulerian.

Proof: By Definition 9.3, it directly follows that $\deg^+(v) = \deg^-(v) = b$, for any vertex $v \in V(D_{b,n})$, so by Theorem 9.2, it suffices to show that $D_{b,n}$ is (strongly) connected. Consider two vertices: $v = d_1 d_2 \cdots d_{n-1}, v' = d_1' d_2' \cdots d_{n-1}' \in V(D_{b,n})$. The sequence of vertices

$$< v = d_1 d_2 \cdots d_{n-1}, \ d_2 \cdots d_{n-1} d_1', \ d_3 \cdots d_{n-1} d_1' d_2', \cdots$$
$$\cdots, \ d_{n-1} d_1' \cdots d_{n-3}' d_{n-2}', \ d_1' d_2' \cdots d_{n-1}' = v' >$$

determines a path in $D_{b,n}$ that joins v to v'. □

We next state the corresponding algorithm that allows one to easily construct a (b,n)-de Bruijn sequence from a corresponding Eulerian digraph $D_{b,n}$.

ALGORITHM 9.2: Construction of de Bruijn Sequences

Input: A pair of positive integers $b > 1, n \geq 1$.
Ouput: A (b,n)-de Bruijn sequence.

Step 1: Construct the corresponding de Bruijn digraph $D_{b,n}$.

Step 2:   Construct   a   (directed)   Euler   tour   starting   with   vertex $v_1 = 00 \cdots 0$: $T = \langle v_1, v_2, \cdots, v_{b^n+1} = v_1 \rangle$.[6]   The directed modification of Algorithm 9.1A may be used to complete this step.

Step 3: From the vertices of the tour constructed in Step 2 (each of which is a string of length $n-1$ in the alphabet $\{1,2,\cdots,b-1\}$) form a corresponding sequence $S = \langle s_1, s_2, \cdots, s_{b^n} \rangle$, where s_i is the first character of the corresponding string v_i.

Before giving the simple proof of the correctness of the algorithm (and hence the existence of de Bruijn sequences for any parameters), we give an example of its use.

EXAMPLE 9.6: Use Algorithm 9.2 to construct
(a) a $(2,3)$-de Bruijn sequence, and
(b) a $(3,2)$-de Bruijn sequence.

SOLUTION: *Step 1:* The relevant de Bruijn digraphs have been contructed in Example 9.8. These digraphs are sufficiently small to easily construct Euler tours (by consulting Figure 9.10) without resorting to the formal algorithm:

Part (a): *Step 2:* Euler tour in $D_{2,3}$: $T = \langle 00,00,01,10,01,11,11,10,00 \rangle$.

Step 3: Reading off the first digits of all but the last element of the tour T, we obtain the $(2,3)$-de Bruijn sequence: $S = \langle 0,0,0,1,0,1,1,1 \rangle$. The reader may easily check that this is indeed a $(2,3)$-de Bruijn sequence. Notice that it is different from $(2,3)$-de Bruijn sequence given in Example 9.4 (even if we try wrapping it

---

[6] We could have started the tour at any vertex; we only chose this one for definiteness.

around). Indeed, although the de Bruin digraphs are unique, there are as many de
Bruijn sequences as there are Euler tours in a de Bruijn digraph.

Part (b): Step 2: Euler tour in $D_{3,2}$: $T = \langle 0,0,1,1,2,2,0,2,1,0 \rangle$.

Step 3: Since the first digits of the string element of this sequence are the whole
elements, we obtain the corresponding (3,2)-de Bruijn sequence by simply
removing the last element of T: $S = \langle 0,0,1,1,2,2,0,2,1 \rangle$. The reader may easily
check that this is indeed a (3,2)-de Bruijn sequence.

EXERCISE FOR THE READER 9.4: (a) Draw the de Bruin digraph $D_{2,4}$.
(b) Use Algorithm 9.2 to construct a (2,4)-de Bruijn sequence.

COROLLARY 9.4: For a given pair of positive integers $b > 1, n \geq 1$, (b,n)-de
Bruijn sequences exist and can be constructed using Algorithm 9.2.

Proof: Let T be an Euler tour constructed by Algorithm 9.2. Given a length n
string $w = d_1 d_2 \cdots d_n$ in the alphabet $\{0,1,2,\cdots,b-1\}$, T must traverse the directed
edge w, and in so doing pass from the vertex $d_1 d_2 \cdots d_{n-1}$ to $d_2 \cdots d_n$. From the
latter vertex, it must move on to some vertex of the form $d_3 \cdots d_{n-1} f_1$. In turn,
from this vertex, it must next visit a vertex of the form $d_4 \cdots d_{n-1} f_1 f_2$. Since T will
continue traversing vertices whose first characters successively run through the
sequence $d_1, d_2, \cdots, d_n$, it follows that this sequence is a subsequence of the
sequence of first characters constructed in Step 3 of Algorithm 9.2. □

Hamilton Paths and Tours

If instead of searching for an (Euler) tour in a graph that traverses each edge
exactly once, we wish to find a tour that visits every vertex exactly once, there is
unfortunately no nice result like Theorem 9.1 that tells us when such a tour exists,
nor is there a nice algorithm like Algorithm 9.1 that shows how to construct one,
once it is known to exist.

DEFINITION 9.4: Suppose that G is a connected graph. A **Hamilton path** in G
is a simple path in G that passes through every vertex exactly once. A **Hamilton
tour** in G is a cycle in G that passes through every vertex exactly once. We say
that G is **Hamiltonian** if it admits a Hamilton tour.

In other words, an n-vertex connected graph G is Hamiltonian if it contains a copy
of the n-cycle C_n as a subgraph, and any such subgraph will constitute a Hamilton
tour.

FIGURE 9.11: Sir William Rowan Hamilton (1805–1865), Irish mathematician.

Note that if G has a Hamilton tour $<u = v_0, v_1, \cdots, v_n, v_{n+1} = u>$, then it also has a Hamilton path: simply delete the last edge of the tour: $<u = v_0, v_1, \cdots, v_n>$. The problem of determining whether a graph is Hamiltonian has been shown to be NP complete, and is thus among the most difficult problems in discrete mathematics. Hamiltonian graphs are named in honor of the Irish mathematician Sir William Rowan Hamilton[7] (Figure 9.11), who introduced the problem of finding a Hamilton tour through a game that he invented. This game along with a graph theoretic interpretation is explained in the following example.

EXAMPLE 9.7: (*The Icosian Game*) The Icosian Game, invented by Hamilton, consists of trying to find a "trip around the world" that visits each of the vertices of the three dimensional dodecahedron (a 12-faced regular polyhedron, each face being a regular pentagon) shown in Figure 9.12 such that all traveling is done along the edges, and every vertex is visited exactly once. In other words, the problem is to find a Hamilton tour of this three-dimensional graph. Does this problem have a solution?

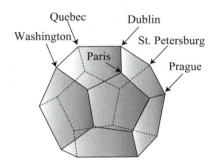

FIGURE 9.12 The Icosian game of Example 9.7 asks to find a cycle of the graph shown that visits each vertex (city) exactly once and which travels only along the edges.

Two versions of the game were available: The portable version: a wooden dodecahedron, with pegs for the vertices, which were marked with the capitals of major world cities, and a long piece of string for players to attempt to form a

---

[7] Hamilton was the greatest Irish scientist to have lived so far. He was astoundingly intelligent and is known to have mastered arithmetic at the age of three and to have been able to read at the age of four, not only English, but also Latin, Greek, and Hebrew! By age 10 he was fluent in the major European languages and began to study Eastern languages. In his teens he switched his focus to science and at age 21 he was appointed as an astronomer at Trinity College in Dublin. His contributions ranged in an impressive diversity from optics (important in astronomy) to mathematics, physics, and even critically acclaimed poetry. In 1857, his work in abstract algebra led him to invent a mathematical game, called "The Icosian Game" which we describe in Example 9.7. In 1859, Hamilton sold the rights to this game to a London game company for £50. The game was not quite as successful as *Rubik's Cube*, but it did help to popularize mathematics. Hamilton married in 1833, but his wife, being handicapped, was unable to tend to homemaking and the marriage did not go well. In the last two decades of his life, he became reclusive and an alcoholic, and he died at age 60 of gout. Among his personal items were found a great number of significant and unpublished works, often amid remains of half-eaten dinners.

round-the-world tour (Hamilton tour). The parlor version was a flat wooden board with a two-dimensional graph of Figure 9.13 engraved on the surface along with pegs for the vertices.

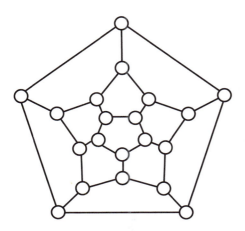

FIGURE 9.13 The two-dimensional version of the Icosian game.

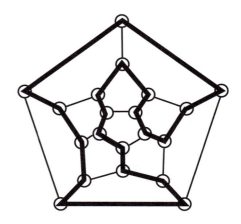

FIGURE 9.14 A Hamilton tour for the graph of Figure 9.13.

The graph of Figure 9.13 is isomorphic to the edge-vertex frame graph of the dodecahedron of Figure 9.12. The isomorphism can be seen by imagining the edge-frame of the dodecahedron to be stretchable, and by stretching the bottom face of the dodecahedron into a larger pentagon, and then flattening the rest of the (3D) edges into this large pentagon. A solution to this puzzle is provided by the Hamilton tour of Figure 9.14.

EXERCISE FOR THE READER 9.5: Is the Petersen graph (Figure 8.14) Hamiltonian? Does it admit a Hamilton path?

EXERCISE FOR THE READER 9.6: Explain why for graphs with at least three vertices, self-loops, and parallel edges have no effect on whether the graph is Hamiltonian. Thus, the question about whether a given graph is Hamiltonian is really a question about simple graphs.

Application of Hamiltonian Graphs: Gray Codes

The next example gives an application of Hamiltonian graphs to the construction of certain binary codes designed to minimize transmission errors when data is transmitted over noisy channels.

EXAMPLE 9.8: (*Application of Hamiltonian Graphs: Gray Codes*) For each positive integer N, a **Gray code of order N** is an ordering of the set of all 2^N bit strings of length N, such that any two adjacent strings in the sequence, as well as the first and last strings in the sequence differ in exactly one bit. For example, the sequence <00, 10, 11, 01> is a Gray code of order 2. Gray codes were invented in the 1940's by Frank Gray, a scientist at AT&T Bell Labs, as a means for transmitting digital signals to reduce errors. Gray codes are usually implemented so that nearby bit strings represent nearby actual objects (e.g., pixel colors in a digital image, or sound frequencies in an audio image). Thus, for example, if digital images from a satellite of a far away planet are being sent back to Earth using Gray codes of order 24 (8 bits for each color in an RGB pixel representation), any length 24 string that had one of its binary digits corrupted by cosmic radiation would be read as a very nearby color, and so such pixels along with the "big pictures" would not be noticeably affected.

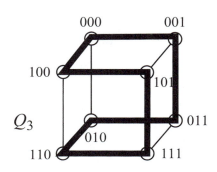

FIGURE 9.15   A Hamilton tour for the hypercube Q_3.

Since adjacent vertices of the N-dimensional hypercube Q_N differ in exactly one bit, it follows that if it can be shown that Q_N is Hamiltonian, then Gray codes of order N will exist (just use the vertex ordering determined by a Hamilton tour). For example, Figure 9.15 shows a Hamilton tour for Q_3. The resulting Gray code (if we start the tour at the vertex 000) is: <000, 001, 011, 010, 110, 111, 101, 100>. Using the recursive construction of Q_N, it easily follows that any Q_N, with $N \geq 2$, is Hamiltonian. Indeed, if we have a Hamilton tour for Q_N, we delete any single edge of the tour in both copies of Q_N that are used to recursively construct Q_{N+1}, and then join corresponding endpoints of the two resulting subgraph Hamilton paths of Q_N to form a Hamilton tour of Q_{N+1}. It thus follows that Gray codes of order N exist for any positive integer N.

EXERCISE FOR THE READER 9.7: Use the construction described in Example 9.8 to construct a Gray code of order 4.

Sufficient Conditions for a Graph to be Hamiltonian

Although there are no efficient methods to determine whether a given graph is Hamiltonian, there are some useful sufficient conditions that can be used to show a given graph is Hamiltonian, and some necessary conditions that can be used to show that a graph is not Hamiltonian. We will present some of these results here. There exist Hamiltonian graphs that violate all of these sufficient conditions, and there exist non-Hamiltonian graphs that satisfy all of these necessary conditions. Since there are no useful results that provide both necessary and sufficient conditions to show that a graph is Hamiltonian, the task of determining whether a graph is Hamiltonian is usually approached by attempting to construct a Hamilton tour, and either succeeding or arriving at a contradiction in the process. This strategy will be illustrated at the end of this section.

The sufficient conditions that we present are based on the following general principle: If additional edges are added to a Hamiltonian graph, the resulting graph will still be Hamiltonian (i.e., the same tour will work for the modified graph). More edges increase the likelihood of being able to find Hamilton tours. The following two results quantify this principle; the first is due to Gabriel Dirac (1952) and the second is a stronger result that was proved by Oystein Ore in 1960.

THEOREM 9.5: (*Dirac's Theorem*) If G is a simple graph with $n \geq 3$ vertices each of which has degree at least $n/2$, then G is Hamiltonian.

THEOREM 9.6: (*Ore's Theorem*) If G is a simple graph with $n \geq 3$ vertices such that $\deg(u) + \deg(v) \geq n$ for each pair of nonadjacent vertices u and v, then G is Hamiltonian.

It is clear that Dirac's theorem is an easy corollary of Ore's theorem. Exercise 33 will guide the reader through a proof of Ore's theorem. Any cycle graph C_n with $n > 4$ clearly violates the conditions of Ore's (and of Dirac's) theorems, but is clearly Hamiltonian. This simple example shows that the conditions of both theorems are not necessary conditions.

Necessary Conditions for a Graph to be Hamiltonian

We will provide two useful results that are often helpful in ruling out the possibility that a given graph is Hamiltonian. It is often more difficult to prove that a certain discrete structure does not exist than to prove it does, since the latter task is complete as soon as one can find just one example (of perhaps several). We remind the reader that necessary conditions (that are not sufficient) can be used to show that a graph is non-Hamiltonian but not to show that a graph is Hamiltonian. When applying the second proposition below, however, it can sometimes be helpful to construct a Hamilton tour, if there is one. The results are quite simple to prove, but a drawback is that it is not always so straightforward to apply them.

PROPOSITION 9.7: If G is a simple Hamiltonian graph and S is any nonempty set of vertices of G, then the vertex deletion graph $G - S$ can have at most $|S|$ components.

Proof: Let C be a Hamilton tour in G. Then, as for any cycle C_n, $C - S$ has exactly $|S|$ components. But since G has the same vertex set as C, and all of the latter's edges, any component of $C - S$ is a subset of a component of $G - S$. It follows that $G - S$ has no more components than $C - S$ has, i.e., at most most $|S|$ components. $\square$

EXAMPLE 9.9: Determine whether the graph G of Figure 9.16 is Hamiltonian.

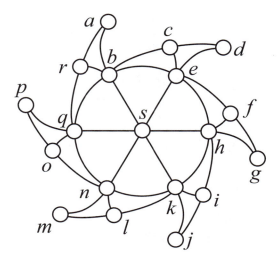

FIGURE 9.16: The simple graph G of Example 9.9.

SOLUTION: Consider the following set of six vertices: $S = \{b, e, h, k, n, q\}$. The vertex deletion graph $G - S$ consists of six disjoint copies of P_2 and one isolated vertex (s), and thus has more components than $|S|$. It follows from Proposition 9.7 that G cannot be Hamiltonian.

When we need to determine whether a given graph is Hamiltonian, and it does not satisfy any known sufficient conditions, one strategy is to attempt to construct a Hamilton tour. Our next proposition lists three simple consequences that must hold in the construction of a Hamilton tour of a simple graph. The reader should convince him/herself why each of these follow from the definitions.

PROPOSITION 9.8: Suppose that we wish to construct a Hamilton tour C of a simple graph G. The following conditions must hold.
    (1) C must include both edges that are incident to any vertex of degree 2.
    (2)   If two edges incident to a certain vertex have been shown to necessarily be part of C, then all other edges that are incident to this vertex may be deleted.

(3) If at any point during the contruction of a C, a cycle is formed by edges that necessarily belong to C and this cycle does not visit every vertex, then G cannot be Hamiltonian.

EXAMPLE 9.10: Determine whether each of the following graphs is Hamiltonian. In case of a Hamiltonian graph, provide a Hamilton tour.
(a) The graph G of Figure 9.17.
(b) The graph H of Figure 9.17.

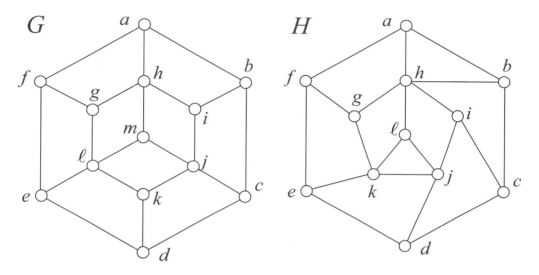

FIGURE 9.17: The graphs G and H of Example 9.10.

SOLUTION: For ease of notation, we will denote edges simply by the string of their vertices. Thus for example, we use ab to denote the edge $\{a, b\}$.

Part (a): We will attempt to construct a Hamilton tour C and will use Proposition 9.8. There are no degree two vertices for which we can apply Part (1) of the proposition. We can avoid having to initially consider separate cases by looking first at vertex m. C must contain two of the three edges incident to m, but since the graph is symmetric with respect to these three edges, we may assume that C contains the edges hm and ℓm, so by Part (2) of the proposition we may delete the edge mj. Next, since the vertex g will have two of its three incident edges included in C, by Part (3), exactly one of gh or $g\ell$ must be included. By symmetry we may assume C includes gh, along with the third edge fg, and hence delete $g\ell$. Since h now has two incident edges included in C, we may delete the remaining two edges ah and hi. In doing this on paper, it is helpful to highlight the edges that are shown to be included in C and to place an "X" mark over those that get deleted. Figure 9.18 illustrates the status of the attempted construction up to this

point. From Figure 9.18, it is now
clear that C must contain the edges:
fa, ab, bi, ij, jc, cd, and we can delete
edges bc and jk. In light of the last
deleted edge, C must contain the other
two edges incident to k: dk and $k\ell$.
But the edges found to necessarily
belong to C now form a cycle:

$< \ell, m, h, g, f, a, b, i, j, c, d, k, \ell >$

Since this cycle does not include the
vertex e, we may conclude from Part
(3) of the proposition that G is not
Hamiltonian.

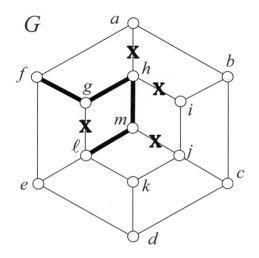

FIGURE 9.18: Attempted construction of a
Hamilton path using Proposition 9.5.

Part (b): Unlike the graph G, H does
not possess a vertex about which all
of the edges are symmetrically related
to one another, so we are forced to
begin the construction with a separation into cases. The vertex ℓ does have two
of its three edges, $\ell j, \ell k$ that are symmetrically related (although the drawing of H
in Figure 9.18 is not geometrically symmetric with respect to these two edges,
what is relevant here are the edge/vertex relationships). This reduces the number
of initial cases from three to two:

Case 1: A Hamilton tour C contains the edges $h\ell, \ell k$.

Case 2: A Hamilton tour C contains the edges $\ell j, \ell k$.

Let us begin with Case 1: (Once again, the reader is advised to create a drawing
like the one in Figure 9.18 to track this attempted construction.) By Part (3) of the
proposition, C must contain exactly one of the two edges gh, gk. Since these two
cases are not symmetric, we split into two subcases:

Subcase 1a: C contains gk, but not gh: So C must also contain gf, and by Part (2)
of the proposition, we can also delete the edges ke, kj. From the first of these
deletions, it now follows that C contains ef and ed. From here, in turn, we may
delete af, and conclude that C contains ah and ab. From these two included
edges, we may delete the edges hb, hi, and then conclude that C contains the edges
ij, ic, bc, jd, and we now have constructed a Hamilton tour:

$$< \ell, k, g, f, e, d, j, i, c, b, a, h, \ell >.$$

Since H has been shown to be Hamiltonian, we need not proceed through the
remaining cases.

EXERCISE FOR THE READER 9.8:
(a) Does the graph G of Figure 9.17 admit a Hamilton path?
(b) In the solution of Part (b) of Example 9.10, what would have happened if we
began with Subcase 1b?

EXERCISE FOR THE READER 9.9: Determine whether each of the following graphs is Hamiltonian. In case of a Hamiltonian graph, provide a Hamilton tour.
(a) The graph G of Figure 9.19.
(b) The graph H of Figure 9.19.

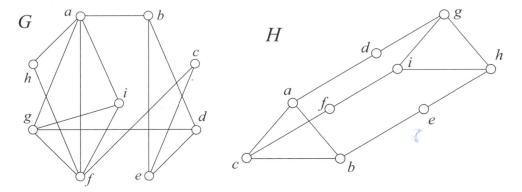

FIGURE 9.19: The graphs G and H of Exercise for the Reader 9.9.

EXERCISES 9.1:

In Exercises 1–4, two graphs are given. For each, (a) determine whether the graph is Eulerian, and (b) determine whether the graph admits an Euler path.

1.

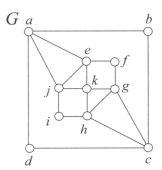
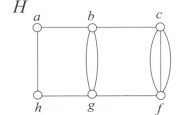

2.

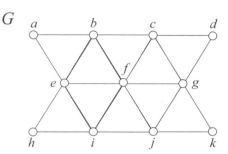 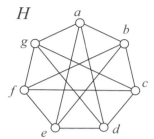

3.

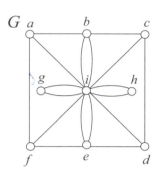 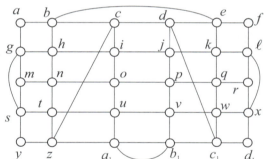

4.

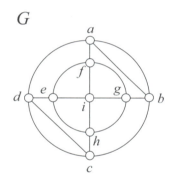

5.   For each family of graphs below, indicate the values of the parameters for which the graph is (i) Eulerian, (ii) admits an Euler path.

(a) The *n*-path graph P_n, $n \geq 2$ (Definition 8.4).

(b) The complete graph K_n, $n \geq 2$ (Definition 8.4).

6.   For each family of graphs below, indicate the values of the parameters for which the graph is (i) Eulerian, (ii) admits an Euler path.

(a) The *n*-cycle graph C_n, $n \geq 3$ (Definition 8.4).

(b) The complete bipartite graph $K_{n,m}$, $n, m \geq 1$ (Definition 8.6).

7.   Among the graphs of Exercises 1 and 3, for those that were determined to be Eulerian, use Algorithm 9.1A to construct an Euler tour. For those that are not Eulerian but admit an Euler path, apply Algorithm 9.1B to construct one. Use the lexicographic ordering of the edges.

8.   Among the graphs of Exercises 2 and 4, for those that were determined to be Eulerian, use Algorithm 9.1A to construct an Euler tour. For those that are not Eulerian but admit an Euler path, apply Algorithm 9.1B to construct one. Use the lexicographic ordering of the edges.

In Exercises 9–10, two digraphs are given. For each, (a) determine whether the digraph is Eulerian, and (b) determine whether the graph admits an Euler path. If the digraph is Eulerian, use the digraph analogue of Algorithm 9.1A (as in Example 8.3) to construct an Euler tour. If the digraph is not Eulerian but admits an Euler path, use the digraph analogue of Algorithm 9.1B to construct an Euler path.

9.

10.

11. Draw the de Bruijn digraph $D_{4,2}$ and then use Algorithm 9.2 to construct a (4,2)-de Bruijn sequence.

12. Draw the de Bruijn digraph $D_{3,3}$ and then use Algorithm 9.2 to construct a (3,3)-de Bruijn sequence.

In Exercises 13–16, two graphs are given. For each, either construct a Hamilton tour, or prove that the graph is not Hamiltonian.

13.

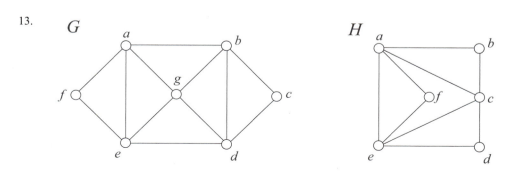

14.

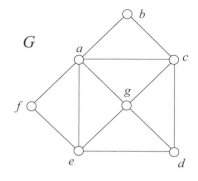

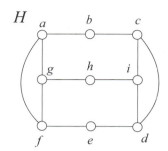

15.

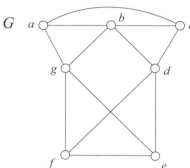

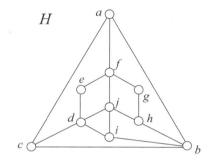

16.

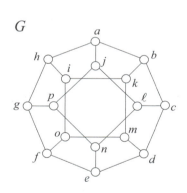

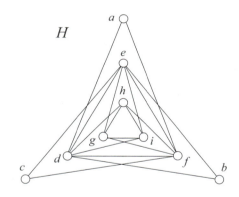

17.   We define the rectangular "city block" grid graph $R_{m,n}$, where m and n are positive integers, to be the graph with vertex set $V = \{(i, j) : 1 \le i \le m+1, 1 \le j \le n+1\}$, and edge set E defined by $\{(i, j),(i', j')\} \in E$ if, and only if $|i - i'| + |j - j'| = 1$.   Figure 9.20 depicts $R_{3,1}$, $R_{4,3}$, and $R_{m,n}$.

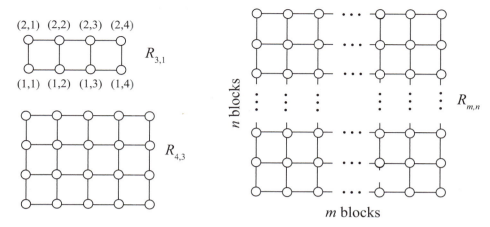

FIGURE 9.20: The grid graphs of Exercise 17.

(a) Show that $R_{3,4}$ is Hamiltonian.

(b) Show that $R_{3,3}$ is Hamiltonian.

(c) Show that $R_{4,4}$ is not Hamiltonian.

(d) Find necessary and sufficient conditions on n and m for $R_{m,n}$ to be Hamiltonian (and prove they are correct).

(e) Suppose that 36 students are seated in a 6 by 6 square grid and the teacher would like to rearrange them so that every student moves to an adjacent seat, either directly in front, behind, to the left or to the right. Is such a rearrangement possible? How about if there were 49 students arranged in a 7 by 7 square grid?

18. For each of the two floor plans below, determine whether or not it is possible to plan a walk through the rooms so that each door is used exactly once. In case it is possible indicate such a walk by providing the sequence of rooms in the order they are visited.

19. For each of the graphs of Exercises 13 and 15, either find a Hamilton path, or prove that one does not exist.

20. For each of the graphs of Exercises 14 and 16, either find a Hamilton path, or prove that one does not exist.

21. Show that if any vertex is deleted from the Petersen graph, the resulting graph will be Hamiltonian, i.e., letting P denote the Petersen graph, if $v \in V(P)$, then (the vertex deletion graph) $P - v$ is Hamiltonian.

22. For which values of $n \geq 3$ is the complement of the n-cycle graph $\sim C_n$ Hamiltonian?

23. In each part, provide an example of a simple connected graph G having the indicated properties.
    (a) G is Eulerian, but not Hamiltonian.
    (b) G is Hamiltonian, but not Eulerian.
    (c) G is both Eulerian and Hamiltonian.
    (d) G is neither Eulerian nor Hamiltonian.

24. Prove that if every vertex of a simple graph G has degree two, then G is Hamiltonian.

25. Prove that if a graph contains a cut vertex then it cannot be Hamiltonian.

26. Suppose that G is a four-regular graph with eight vertices.
    (a) Show that G must be Hamiltonian.
    (b) If $v \in V(G)$, is it necessarily the case that the vertex deletion graph $G - v$ is Hamiltonian? Justify your answer.
    (c) If $u, v \in V(G)$, is it necessarily the case that the vertex deletion graph $G - \{u, v\}$ is Hamiltonian? Justify your answer.

NOTE: (*Fleury's Algorithm for Euler Tours*) An alternative algorithm to Algorithm 9.1A for determining an Euler tour of an Eulerian graph is Fleury's algorithm. We will state this algorithm in a way that will make it applicable to either finding an Euler tour of an Eulerian graph, or to finding an Euler path in case one exists.

ALGORITHM 9.3: (*Fleury's Algorithm for Constructing Euler Paths and Tours*)

Input: A simple connected graph G with two vertices $s, f \in V(G)$ (s = starting vertex, f = finishing vertex). If $s = f$ it is assumed that every vertex in G has even degree (so by Theorem 9.1 G is Eulerian), while if $s \neq f$ it is assumed that s, f have odd degree and all other vertices have even degree (so by Theorem 9.1, G admits an Euler path from s to f).
Ouput: An (ordered) list of vertices of G, Path, corresponding to either an Euler tour (if $s = f$) or an Euler path (if $s \neq f$) from s to f.
NOTE: We formulate the algorithm in case G is Eulerian. The modification for non-Eulerian graphs having an Euler tour is done in the same fashion as the way Algorithm 9.1B was constructed from Algorithm 9.1A.

Step 1: Initialize UnusedEdges = Edges (the edges could be listed in any convenient order), EndVertex = s, and Path = [s].

Step 2: (*Iterative Step*) While UnusedEdges is nonempty, scan UnusedEdges for an edge incident to EndVertex, first try to find an edge that is not a bridge for the edge set deletion graph $G - \{TourEdges\}$ (i.e., edges whose removal would leave the unused edges portion of the graph disconnected); if none is available then use a bridge. Append this edge to TourEdges, delete it from UnusedEdges, and update EndVertex to be the other endpoint of the new added edge, and append this new EndVertex to Path.

A proof of the fact that Fleury's algorithm works indeed will be outlined in Exercise 29. Although at first glance this algorithm might seem simpler than Algorithm 9.1, since it requires the determination of bridges of graphs, which is a nontrivial task, this algorithm is not very often used in practice (for large graphs). Nonetheless for small graphs it is an easy algorithm to learn and use by hand.

27. Apply Fleury's Algorithm 9.3 to the graph of Figure 9.4(b) using a for the starting vertex and c for the finishing vertex. Use the lexicographic ordering on the edges for scanning purposes.

28. Apply Fleury's Algorithm 9.3 to construct an Euler tour for the Eulerian graph of Figure 9.7 using a for the starting and finishing vertex. Use the lexicographic ordering on the edges for scanning purposes.

29. (*Proof that Fleury's Algorithm Works*) This exercise contains an outline of the proof of correctness of Fleury's Algorithm 9.3. We assume that we have a simple connected graph G with two vertices $s, f \in V(G)$ (s = starting vertex, f = finishing vertex). If $s = f$ it is assumed that every vertex in G has even degree (so by Theorem 9.1 G is Eulerian), while if $s \neq f$ it is assumed that s, f have odd degree and all other vertices have even degree (so by Theorem 9.1, G admits an Euler path from s to f). Thus, to take care of both cases, we need to show that Fleury's algorithm produces a simple path that traverses all of the edges in G.

(a) Show that either $\deg(s) = 1$, or there is an edge incident to s that is not a bridge of G.

(b) Using Part (a), use mathematical induction on the number of edges of G to prove that Fleury's algorithm produces a simple path that traverses all of the edges of G.

Suggestion: For Part (a) in case $\deg(s) > 1$, assume that only bridges are incident to s, and use Corollary 8.2 to arrive at a contradiction. For the inductive step in Part (b), separate into the two cases of Part (a). In case s has a non-bridge edge e that is incident to it, apply the inductive hypotheses to the edge deletion subgraph $G - e$ with s changed to the other endpoint of e.

30. Prove Theorem 9.2.

31. Explain why if a bipartite graph G is Hamiltonian, then the number of vertices must be even.

32. Let G be the Petersen graph, and let v be any vertex of G. Show that the vertex deletion graph $G - v$ is Hamiltonian.
Suggestion: Use symmetry to significantly reduce the number of cases to consider.

33. (*Proof of Ore's Theorem*) Proceed through the following outline of a proof of Ore's Theorem 9.6:

(a) Suppose that G is a simple graph on $n \geq 3$ vertices. Suppose that two vertices u, v of G for which $\{u, v\} \notin E(G)$ satify $\deg(u) + \deg(v) \geq n$. Show that G is Hamiltonian if, and only if, the edge added graph $\hat{G} = G + \{u, v\}$ is Hamiltonian, where $\hat{G}$ is defined by $V(\hat{G}) = V(G)$ and $E(\hat{G}) = E(G) \cup \{\{u, v\}\}$.

(b) Use the result of Part (a) to prove Ore's theorem.

Suggestions for Part (a): For the nontrivial implication, start with a Hamilton tour C of $\hat{G}$, which may be assumed to include the edge $\{u, v\}$ (otherwise it would be a Hamilton tour of G). Write $C = <v_0 = u, v_1 = v, v_2, \cdots, v_{n-1}, v_n = u>$, and define U to be the G-neighbors v_i of u for which $1 < i < n - 1$, and V to be the G-neighbors v_i of v for which $1 < i < n - 1$. Show that $|U| = \deg_G(u) - 1$, $|V| = \deg_G(v) - 1$, and then use the hypothesis to show that $|U \cap V| \geq 1$. Use a vertex in $U \cap V$ to modify C into a Hamilton tour for G.

34. (*Sufficient Condition for a Simple Graph to have a Hamilton Path*) Prove the following: Suppose that G is a simple graph on $n \geq 3$ vertices, and that $\deg(u) + \deg(v) \geq n - 1$, for any pair of nonadjacent vertices u, v. Prove that G admits a Hamilton path.
Suggestion: Introduce a new graph $\hat{G}$ with one new vertex $\hat{u}$, defined by $V(\hat{G}) = V(G) \cup \{u\}$, and $E(\hat{G}) = E(G) \cup \{\{\hat{u}, v\} : v \in V(G)\}$. Use Ore's theorem to show that $\hat{G}$ is Hamiltonian and show how to create a Hamilton path in G from a Hamilton tour in $\hat{G}$.

9.2: TREE GROWING AND GRAPH OPTIMIZATION ALGORITHMS

Discrete optimization problems involve finding an object among a large set of similar discrete structures that has a certain extremal (i.e., either maximum or minimum) property. For some important discrete optimization problems efficient algorithms have been developed for their solution, but other such important problems have been shown to be NP hard. For such hard discrete optimization problems that arise in applications without good exact solution algorithms, the next best thing is to develop a so-called *heuristic* (*algorithm*). A heuristic algorithm is one that does not necessarily produce an optimal solution, but comes with a performance guarantee that assures that the solution it produces will be optimal up to some fixed factor. For example, in a minimization problem seeking to find a discrete structure that minimizes a certain positive function f, defined on a specified set of admissible graphs, a heuristic with a performance guarantee factor of 2, would produce an admissible graph G for which $f(G) \geq 2 \min f$ $= 2f(G_{\mathrm{OPT}})$, where G_{OPT} is an admissible graph that minimizes the function f. In words, the graph produced from the heuristic is good (or not too bad) in the sense that it may not have a minimum value of f, but has no more than twice this minimum value. The section will begin by presenting some simple tree growing algorithms that will produce certain tree subgraphs in an inputted simple graph. These algorithms are all based on the same idea, but differ in the way that vertices are selected in the tree growing process. We will then show how these tree growing algorithms can be adapted to solve an assortment of important graph optimization problems. The problems in this section are: the minimum distance problem in edge-weighted graphs, the minimum spanning tree problem in edge-weighted graphs, and two very important vertex search programs: the *depth-first* and the *breadth-first search*.   In the latter part of this section we will discuss the traveling salesman problem, which is one of the most important and extensively studied problems in discrete optimization.   We will show why the problem is NP hard, and discuss some heuristics that have been developed, along with their corresponding performance guarantees.

Minimum Spanning Tree for an Edge-Weighted Graph

Many graph optimization problems involve the determination of a tree that is a subgraph of a given connected graph and that has a certain extremal property. Recall from Section 8.3 that a tree is a simple connected graph that contains no cycles. For reference, we collect some of the relevant definitions.

DEFINITION 9.5: A subgraph H of a connected simple graph G is called a **spanning subgraph** of G if H is connected, and the vertices of H coincide with the vertices of G, i.e., $V(H) = V(G)$.

An **edge-weighted graph** is a graph in which each edge has a positive real number weight associated with it.

A **minimum spanning tree** for an edge-weighted connected graph is a spanning subgraph that is a tree with the property that the sum of all of its edge weights is as small as possible where the minimum is taken over all spanning subgraphs.

We point out that a spanning subgraph of G whose edge weights have a minimum sum must be a tree, since if it were not, it would have a cycle, and any edge of such a cycle could be deleted and would produce a subgraph that was still spanning but had smaller total edge weight. This is not possible if we already started with a spanning subgraph of minimal weight.

Suppose that the vertices in a connected edge-weighted graph represent different cities that need to be connected into a network (say with a fiber-optic cable or a pipeline). As a simple example, consider the edge-weighted graph of Figure 9.21.

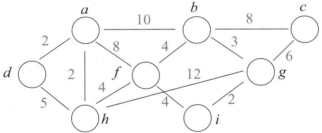

FIGURE 9.21: An edge-weighted simple graph.

Since cable or pipeline can be very expensive and difficult to lay, the weights on the edges may represent the amount of money in tens of thousands of dollars that it would cost to lay the material on the main roadways connecting different pairs of cities.   Depending on the roadways or nonroadways connecting the cities, the prices need not depend only on the distance been a pair of cities. Various types of roadways may require different types of laying procedures (e.g., a road versus a bridge). The county needs to decide the cheapest way to lay the material so that all of the cities will be connected.   Thus, in the above terminology, we seek a minimum spanning tree of the given edge-weighted graph.   The brute-force approach of checking the weights of all spanning trees would be horribly inefficient, even for this small example.   Indeed, in general it can be shown that an n-vertex connected graph may have as many as n^{n-2} spanning trees.

Research on the minimal spanning tree problem dates back to the work of Czech mathematician Otakar Borůvka (1899–1995) in his work on the laying of electric lines in Southern Moravia in what is now the Czech Republic.[8]   We will soon present a very efficient algorithm for construction minimum spanning trees that falls under the framework of the tree growing meta-algorithm that we present next. Another popular algorithm for constructing minimum spanning trees but which is not a tree growing algorithm, is Kruskal's algorithm; this algorithm will be presented in the exercises.

Tree Growing Meta-Algorithm

We will present a general tree growing algorithm that can be easily adapted to solve the problem of determining a minimum spanning tree of a connected edge-weighted graph.   It is quite a versatile algorithm that can be easily adapted to solve other important graph optimization problems.   The algorithm starts with a single vertex of a connected graph G, and continues adding edges, one by one, thus growing the tree T one edge at a time.   In order to describe our general tree growing algorithm, it will be convenient to introduce the following terminology. This definition will involve a tree T that is a subgraph of a (not necessarily edge-weighted) connected graph G. T should be thought of as the tree that is being grown in the algorithm, one edge at a time.

DEFINITION 9.6: Suppose that T is a tree, which is a subgraph of a connected graph G. Any edge of T is called a **tree edge**, while an edge of $E(G) \sim E(T)$ is called a **non-tree edge**. Similarly, any vertex of T is called a **tree vertex**, while any vertex in $V(G) \sim V(T)$ is called a **non-tree vertex**. A **boundary edge** for T is any non-tree edge for which one endpoint is a tree vertex and the other is a non-tree vertex.

EXAMPLE 9.11: Figure 9.22 shows a tree subgraph T (whose edges are printed with thick lines) of a connected graph G.

---

[8] Borůvka published two papers on the subject in 1926, but they were both written in the Czech language. Recently, they have been translated into English by Czech scholars. These papers are available for open access free downloads. Simply run an internet search on your favorite search engine for "Boruvka" (do not worry about the accent) to obtain them. Borůvka gave the first algorithm for constructing minimal spanning trees; it will be presented and examined in the exercises of this section.

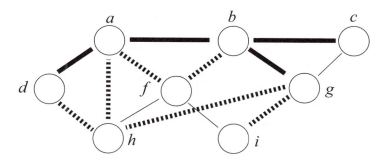

FIGURE 9.22: A tree subgraph T (edges drawn with thick lines) of the connected graph of Example 9.11. The boundary (non-tree) edges are drawn with thick dashed lines.

These thick drawn edges are the tree edges, and all remaining edges are the non-tree edges. The tree vertices are those connected by the tree edges: a,b,c,d,g. The remaining vertices are the non-tree vertices. The boundary edges are the following non-tree edges: $\{a, f\}, \{a, h\}, \{b, f\}, \{d, h\}, \{g, h\}, \{g, i\}$. Note that when any of these boundary edges is added to the tree, the resulting subgraph will again be a tree. Once this is done, the boundary edges of the new tree will change.

EXAMPLE 9.12: If the boundary edge $e = \{g, h\}$ of the tree T of Example 9.11 is added, the resulting tree $T + e$ is shown (with thick lines) along with its new set of boundary edges (thick dashed lines) in Figure 9.23.

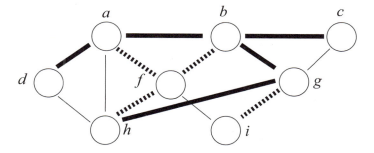

FIGURE 9.23: A tree subgraph $T + e$ (edges drawn with thick solid lines) obtained from the tree T of Example 9.11 by adding the boundary edge $e = \{g, h\}$. The boundary (non-tree) edges are drawn with thick dashed lines.

Notice that in updating the boundary edges, some previous boundary edges for T (notably $\{a, h\}$, and $\{d, h\}$) are no longer boundary edges for $T + e$, and that one new boundary edge $\{f, h\}$ was introduced. In general, when a new boundary edge is added, the only modifications in the boundary edges (that need to be removed or added) will have an endpoint at the new non-tree vertex that was added (in the above example, this was vertex h), and only those edges with endpoints at the added vertex need to be considered when updating the set of boundary edges.

All of the above facts remain true in the general setting of Definition 9.6. In particular,   if e is a boundary edge, then $T + e$ will again be a tree.   (*Proof:* Since the only new vertex added, the non-tree endpoint of e, is reachable from the tree endpoint of e, and this tree endpoint is reachable from all other vertices of T, it follows that $T + e$ is connected. The addition of the boundary edge e to T cannot create a cycle, since only one of its endpoints belongs to T. □)   This is the essential (yet simple) idea behind our basic tree growing algorithm: Starting with a single vertex tree, boundary edges are added one by one until a spanning tree is obtained. Different adaptations of this scheme will depend on how boundary edges are chosen to be added. We may now state our general tree growing algorithm. We will intentionally leave the iterative step of the algorithm where the boundary edge gets selected for adding to the existing tree as not completely specified. Of course, for this to be a true algorithm, each step needs to be completely specified, so this would be more appropriately termed a *meta-algorithm*. An assortment of specific algorithms can be obtained from it by specifying the scheme by which boundary edges are selected. One simple scheme would be to randomly select a boundary edge at each iteration.

ALGORITHM 9.4: Tree Growing/Vertex Discovery Meta-Algorithm for Connected Graphs

Input: A connected graph G and a starting (root) vertex $v \in V(G)$.

Output: An ordered list of the vertices DiscoveryList and a spanning tree T for G.

Step 1: INITIALIZE: DiscoveryList = $[v]$, $T = [v]$, and BdyEdges = {all edges of $E(G)$ incident to v}.

Step 2: (*Iterative Step*)
WHILE BdyEdges $\neq \varnothing$
> **Next Edge Selection:** Select an edge e from BdyEdges
> SET u = the non-tree endpoint of e.
> Append u to DiscoveryList
> UPDATE BdyEdges by removing any existing ones that have u as an endpoint (and thus form a cycle in $T + e$), and by adding new edges of $E(G)$ that have u as one endpoint and another endpoint not in DiscoveryList.
> UPDATE $T = T + e$

END WHILE

Step 3: OUTPUT: DiscoveryList, T.

By modifying the "Next Edge Selection" task of Step 2 in the above algorithm, we can obtain an assortment of specialized tree construction/vertex discovery algorithms that can be tailored to solve some very important problems.

Prim's Algorithm for Minimum Spanning Trees

The first specialization of the meta-tree growing algorithm that we present is Prim's algorithm.[9] It is an efficient algorithm for constructing minimum spanning trees of an inputted connected graph. It is a very simple "greedy" adaptation of the tree growing meta-algorithm (Algorithm 9.4), where a boundary edge of lightest weight is selected at each iteration.

ALGORITHM 9.5: Prim's Algorithm for Constructing a Minimum Spanning Tree of an Edge-Weighted Connected Graph

Input: An edge-weighted connected graph G and a starting (root) vertex $v \in V(G)$.

Output: A minimum spanning tree T for G.

Apply the tree growing meta-algorithm (Algorithm 9.4) with the following specification of the **Next Edge Selection** *part of Step 2:*
A boundary edge from BdyEdges of minimum weight is selected. In cases of ties, the selection is made according to some default priority.[10]

Output: The spanning tree T produced by the meta-algorithm.

In terms of discovery vertices, Prim's algorithm simply selects the next discovery vertex to be a non-tree vertex that is reachable from a current tree vertex by an edge of smallest possible weight (i.e., a cheapest new vertex to add). Although Prim's algorithm is quite simple to state and use, as with any algorithm, the correctness needs to be ascertained. The following proposition, a proof of which is outlined in Exercise 27, easily implies that Prim's algorithm does indeed produce a minimum spanning tree.

PROPOSITION 9.9: Suppose that G is a connected edge-weighted graph. If T_ℓ denotes the tree constructed after ℓ iterations of the tree growing step in Prim's algorithm (corresponding to the iterative Step 2 of Algorithm 9.4), where $1 \le \ell \le |V(G)| - 1$, then T_ℓ is a subgraph of a minimum spanning tree of G.

Since T_ℓ is a tree with ℓ edges (and hence $\ell + 1$ vetices), the correctness of Prim's algorithm follows from the proposition with $\ell = |V(G)| - 1$. Another algorithm

---

[9] Prim's algorithm is named after the American mathematician Robert C. Prim (1921–). Prim studied electrical engineering in his undergraduate years but earned his PhD in mathematics from Princeton in 1949. Most of his career was spent in industrial mathematical positions, including posts with the General Electric Company, the US Naval Ordnance Lab, and Bell Telephone Laboratories. He published his minimum spanning tree algorithm in 1957 [Pri-57].

[10] It will not matter which default priority scheme is used in cases of ties (e.g., resolving ties by comparing non-tree vertices according to some predetermined ordering of the vertices, or simply by random selection); the trees produced by Prim's algorithm will always be minimum spanning trees.

for constructing minimum spanning trees that is not a tree growing algorithm will be described in the exercises.

EXAMPLE 9.13: Apply Prim's algorithm (Algorithm 9.5) to construct a minimum spanning tree for the edge-weighted graph of Figure 9.21. Use vertex a as the starting vertex and resolve any ties using the alphabetical ordering of the non-tree vertices, and then (if necessary) tree vertices.

SOLUTION: The minimum spanning tree produced by the algorithm is shown in Figure 9.24 (with thick lines). The order in which edges were added in the tree growing process is indicated by the discovery numbers of the vertices, which are shown in bold-faced numbers inside the circles representing the vertices in the figure. The starting vertex a has discovery number 0. Initially, the boundary edges are $\{a, b\}$, $\{a, d\}$, $\{a, f\}$, $\{a, h\}$, with corresponding weights 10, 2, 10, 2, respectively. The tie is resolved by choosing the non-tree vertex d as the next discovered vertex and hence the edge $\{a, d\}$ as the first added edge. This process continues to produce the indicated minimum spanning tree, which has weight 23.

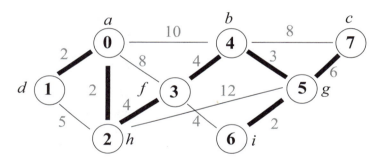

FIGURE 9.24: The minimum spanning tree subgraph T (edges drawn with thick lines) obtained by Prim's algorithm in Example 9.13. The bold-faced numbers inside the vertices denote their discovery numbers, i.e., the order in which they were grown onto the spanning tree, with the initial vertex a having a discovery number zero.

Although it was not needed to describe the minimum spanning tree, the vertex discovery numbers shown in Figure 9.24 make it easy to see how Prim's algorithm progressed. They simply correspond to the DiscoveryList output of Algorithm 9.4: DiscoveryList = $[a, d, h, f, b, g, i, c]$.

EXERCISE FOR THE READER 9.10: Apply Prim's algorithm (Algorithm 9.5) to construct a minimum spanning tree for the edge-weighted graph of Figure 9.25. Use vertex a as the starting vertex and resolve any ties using the alphabetical ordering of the non-tree vertices, and then (if necessary) tree vertices.

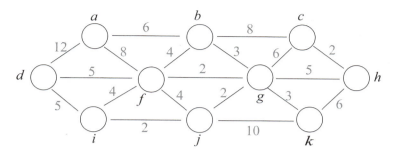

FIGURE 9.25: The edge-weighted graph for Exercise for the Reader 9.10.

COMPLEXITY/PERFORMANCE OF PRIM'S ALGORITHM: If Prim's algorithm is efficiently implemented, its performance time on a connected graph with E edges and V vertices can be made $O(E \log(V))$. Interested readers may refer to Chapter 23 of [CLRS-01] for a detailed justification of this estimate. The computer implementation at the end of this chapter will provide details on some reasonably efficient implementations of Prim's algorithm.

Dijkstra's Algorithm for Shortest Distances in an Edge-Weighted Connected Graph

Edge-weighted graphs can naturally be used to model networks of cities, where the vertices represent the cities and the edge weights represent distances (or transportation costs or travel times) between corresponding pairs of cities. It is an important problem to find a shortest path (or least expensive route) between any pair of cities. The unweighted case has been considered in Section 8.2, where a matrix-based algorithm for computing (non edge-weighted) network diameters was introduced. This algorithm can be easily adapted to compute distances between pair of vertices, but is not very well suited for finding shortest paths that realize these distances. With a slight variation on the "Next Edge Selection" scheme that was used to obtain Prim's algorithm, the tree growing meta-algorithm gives rise to a very efficient algorithm, known as Dijkstra's algorithm,[11] for

---

[11] Edsger W. Dijkstra (1930–2002) was a Dutch computer scientist who studied physics at the University of Amsterdam where he obtained all of his university education and his PhD in 1959. He became very interested in computer programming during his studies well before this was a recognizable line of work. In fact, in 1957, when he listed computer programming as his profession on his marriage license, the authorities would not accept it. (He was still able to get married, but officially changed his profession to physicist.) He became a professor of mathematics at the Eindhoven University of Technology (in the Netherlands) in 1962. In 1973 he became a research fellow for the Burroghs Corp. In 1984, he moved to the US to become a professor of computer science at the University of Texas at Austin. In the late 1950s he was one of the original contributors to the ALGOL computing language. This high level language served as a model for future computing languages because of its ease of use and precision. He came up with his minimal path algorithm in 1959 and had subsequently made many other substantial contributions to computer science, including work in

solving the shortest path problem.   While Prim's algorithm selects the next discovery vertex that was closest to an existing tree vertex (as measured by the edge weight from a tree vertex to the non-tree vertex), Dijkstra's algorithm makes the selection by looking at the <u>total</u> distance (or weight) from the starting vertex (by adding all the weights of the edges on the unique path in the existing tree along with the new boundary edge). After vertices are "discovered" the distances to the starting root vertex will not change and will be the shortest distances to the the starting vertex.   Since these shortest distances correspond to the sum of the edge weights along the unique paths in the outputted tree, they can be easily computed after tree growing algorithm is complete, or more simply, they can be recorded as the algorithm progresses.

Compared with Prim's algorithm, Dijkstra's algorithm requires a bit more bookkeeping during the iterative steps. Apart from the vector DiscoveryList and a corresponding vector DistanceList that will store the distances of vertices to the root vertex as the vertices are discovered, we will also make use of two other vectors that we will briefly describe before stating the algorithm.

1. CurrDist will be a vector that will store current upper estimates for the distances of vertices to the root. This vector is initialized to have CurrDist(v) = 0 and CurrDist$(w) = \infty$, for all other vertices w. Each time a new vertex is discovered, we need to check whether distance estimates (stored in this vector) of any undiscovered vertices that are adjacent to the newly discovered vertex can be decreased by going through the new vertex. If any are, their estimates in CurrDist need to be updated.

2. Pred will be a vector that stores the predecessor vertices in the minimum distance tree as the vertices are discovered, with the convention that Pred(v) = Null (since the root vertex does not have a predecessor). All components of this vector are initially set to "Null." Anytime the CurrDist vector is updated for a given vertex, the corresponding component of Pred is updated to the predecessor that gives rise to the new shorter distance. Just as with CurrDist, once a vertex is "discovered," its component in Pred will no longer change. Note that the vector Pred completely specifies the rooted tree (it is analogous to the "parent list" computer form of representing a rooted tree that was introduced in Example 8.37). Dijkstra's algorithm applies equally well for graph and digraph networks; in the digraph setting we only need that every vertex be reachable from the starting vertex (this is a weaker assumption than the digraph being strongly connected).

ALGORITHM 9.6: Dijkstra's Algorithm for Computing Shortest Distances from a Specified Root Vertex in an Edge-Weighted Connected Graph or Digraph

NOTE: We will formulate the algorithm in the setting of graphs; the only change needed for digraphs is that edge directions need to be respected when determining

---

"deadlock avoidance" in networks. He has also written several expository texts. In recognition of his achievements, he received the Turing Award in 1972, a very prestigious honor for computer scientists.

boundary edges at each iteration of the tree growing. In the digraph setting, it will be assumed that all vertices are reachable from the starting vertex.

Input: An edge-weighted connected graph G and a starting (root) vertex $v \in V(G)$.

Output: A shortest distance tree T for G from v, along with a vector DistanceList, whose components give the distances of the vertices to the root vertex v.

Initialize: DistanceList $= [0]$ (this list will store the shortest distances of vertices to the root vertex as the vertices are discovered)
Initialize the following vectors of length $|V(G)|$:

CurrDist$(v) = 0$ and CurrDist$(w) = \infty$, for all other vertices w.

Pred$(w) =$ Null, for all vertices w.

Apply the tree growing meta-algorithm (Algorithm 9.4) with the following specification of the **Next Edge Selection** *part of Step 2:*
A boundary edge from BdyEdges is selected that has the property that its weight added to the weights of the tree edges in the unique path from the starting vertex v to the non-tree endpoint is as small as possible. In cases of ties, the selection is made according to some default priority.[12]

As each vertex is discovered, update its predecessor (in the tree) in the Pred Vector, update its distance to v in DistanceList, and check all vertices adjacent to the newly discovered vertex to see if the distance stored in CurrDist can be decreased by going through the newly discovered vertex. In cases where a decrease occurs, update the corresponding components of the CurrDist and Pred vectors accordingly. Record its corresponding shortest distance to the root vertex in a separate vector DistanceList corresponding to DiscoveryList.

Output: The spanning tree T (which can be outputted as the vector Pred), the DiscoveryList vector produced by the meta-algorithm, along with the vector DistanceList of corresponding shortest distances.

Note that Dijkstra's algorithm does more than compute the shortest distance between a pair of vertices: it computes the distances from the inputted starting vertex to all other vertices, along with a corresponding minimum distance tree. Geometrically, the algorithm can be viewed as follows: Think of the edges of the graph as being hollow clear tubes, and think of a green fluid flowing out from the origin vertex, progressing along all unused edges at unit speed, where the edge weights represent the tube distances to be covered by the flow. Vertices will be discovered in the order that the green fluid reaches them, and the distances to the root will be the time that it took the green fluid to reach them. We soon will prove

---

[12] As with Prim's algorithm, it will not matter which default priority scheme is used in cases of ties (e.g., resolving ties by comparing non-tree vertices according to some predetermined ordering of the vertices, or simply by random selection); the trees produced by Dijkstra's algorithm will always be minimum distance trees.

a proposition to demonstrate the correctness of Dijkstra's algorithm, but we first give an example of its implementation. Although a computer program would best work directly with the vectors described in the above algorithm, we will explain the iterative steps graphically in the following example.

EXAMPLE 9.14:   Apply Dijkstra's algorithm (Algorithm 9.6) to construct a minimum distance tree for the edge-weighted graph of Figure 9.21 using d as the root vertex.   List also the distances of all vertices to the root.   In the tree growing algorithm, resolve any ties using alphabetical order on the non-tree vertices, and then (if necessary) tree vertices.

SOLUTION: We give a graphical display of each iterative step, and each time a vertex is discovered, a directed (thick) edge will be added from the predecessor vertex to the newly discovered vertex, and the resulting shortest distance will be indicated inside square brackets next to the vertex. As with Prim's algorithm, the discovery numbers will be indicated in bold inside the circles corresponding to each vertex. Also, the values of CurrDist will be indicated adjacent to each vertex and listed in square brackets. The current predecessors of undiscovered vertices that have finite CurrDist estimates will be indicated by thin directed edges. Thus, directed edges to vertices will be made thick only once the vertex is discovered.

We now describe the iterative Step 2 of the tree growing algorithm:

First Iteration: In this case, the newly discovered vertex is the root vertex d, the adjacent vertices are a and h and the CurrDist updates would be

CurrDist (a) = CurrDist (d) + wgt($\{a,d\}$) = 0 + 2 = 2, and
CurrDist (h) = CurrDist (d) + wgt($\{d,h\}$) = 0 + 5 = 5.

This is summarized geometrically in Figure 9.26.   The predecessor of newly discovered vertex a is the root, and thus the root is the permanent value of the predecessor of a, so this is indicated by a thick directed edge.

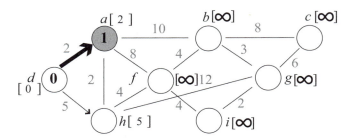

FIGURE 9.26: A geometric illustration of the first iteration of Dijkstra's algorithm applied in Example 9.14. Entering into the first iteration, only the initial vertex d is "discovered," and the first iteration discovers the vertex a (shaded); the two discovery numbers are indicated in the vertex circle. The current distance estimates for all vertices are shown in square brackets. The nonfinal predecessor of vertex h (based on the current distance estimate) is indicated by a thin directed edge.

Second Iteration: All undiscovered vertices adjacent to the latest discovered vertex (a) need to have their current distances to the root checked with the length of the new path going through vertex a. The newly accessible vertices b and f previously had their current distance estimates to be ∞:

CurrDist (b) = CurrDist (a) + wgt($\{a,b\}$) = 2 + 10 = 12, and
CurrDist (f) = CurrDist (a) + wgt($\{a,f\}$) = 2 + 8 = 10.

The previously accessible vertex h needs to have it previous current distance estimate (5) compared with the length of the new path going through vertex a, which is CurrDist (a) + wgt($\{a,h\}$) = 2 + 2 = 4. Since this is smaller than the current distance estimate to the root, and is the smallest among all undiscovered vertices, vertex h will be the newly discovered vertex in this iteration. Its predecessor needs to be updated from the vertex d to the vertex a (corresponding to a shortest path to the root). This second iteration is summarized geometrically in Figure 9.27.

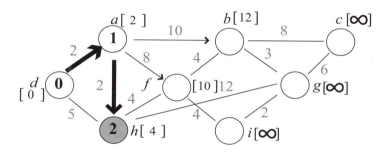

FIGURE 9.27: A geometric illustration of the first iteration of Dijkstra's algorithm applied in Example 9.14. The conventions of Figure 9.26 are used here as well.

The remaining five iterations are shown in Figure 9.28. The DistanceList vector can be read off from the final iteration of Figure 9.28: DistanceList = [2, 12, 20, 0,8,14, 4, 12] (for the vertices in alphabetical order $[a,b,c,d,f,g,h,i]$).

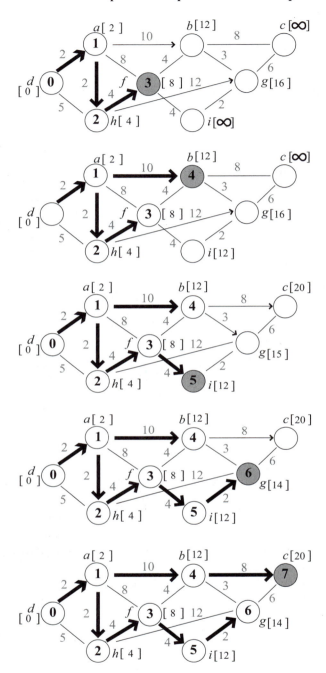

FIGURE 9.28: A geometric illustration of iterations three through seven of Dijkstra's algorithm applied to Example 9.14. The shortest path distances to each vertex from the root vertex d are indicated (in the final iteration figure) in square brackets, while corresponding shortest paths are obtained by tracing back the tree (the thick arrows) to the root vertex.

A few comments are in order: Note that in the fourth iteration, there was a tie between vertex b and vertex i for having the shortest distance to the root. Alphabetical order had us choose vertex b first. Also, the existing predecessor to b gave an equally short path as the newly dectected path through vertex f, and since the new path did not have smaller length, the predecessor of b was not changed.

EXERCISE FOR THE READER 9.11: Apply Dijkstra's algorithm (Algorithm 9.6) to construct a minimum distance tree for the edge-weighted graph of Figure 9.29 using a as the root vertex. List the distances of all vertices to the root. In the tree growing algorithm, resolve any ties using alphabetical order on the non-tree vertices, and then (if necessary) tree vertices.

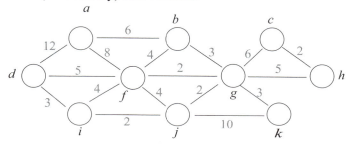

FIGURE 9.29: The edge-weighted graph for Exercise for the Reader 9.11.

EXERCISE FOR THE READER 9.12: Apply Dijkstra's algorithm (Algorithm 9.6) to construct a minimum distance tree for the edge-weighted digraph of Figure 9.30 using a as the root vertex. List the distances of all vertices to the root. In the tree growing algorithm, resolve any ties using alphabetical order on the non-tree vertices.

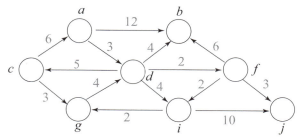

FIGURE 9.30: The edge-weighted digraph for Exercise for the Reader 9.12.

We next fulfill an earlier promise and prove the correctness of Dijkstra's algorithm. The following result proves more; it shows that at any stage in the construction the paths and distances determined by the growing tree give shortest paths and distances to all discovered vertices. This stronger fact was tacitly assumed in our explanation and implementation of Dijkstra's algorithm. We state it only for non-directed graphs; the result and proof are easily modified for the case of directed graphs.

PROPOSITION 9.10: Suppose that G is a connected edge-weighted graph, with a specified root vertex v. If T_ℓ denotes the tree constructed after ℓ iterations of the tree growing step in Dijkstra's algorithm (corresponding to the iterative Step 2 of Algorithm 9.4), where $1 \leq \ell \leq |V(G)| - 1$, then for any vertex $w \in T_\ell$, the unique path from v to w in T_ℓ is a shortest path between these two vertices in G. Moreover, the vertices of $V(G) \sim V(T_\ell)$ are no closer to v than any of the vertices of T_ℓ.

Proof: We will proceed by induction on ℓ.

Basis Step: When $\ell = 1$, T_ℓ consists of v and a closest vertex to it (in G), along with the corresponding edge that connects them.

Inductive Step: We assume that $1 \leq \ell < |V(G)| - 1$, and that T_ℓ has the indicated property. We need to show the same is true for $T_{\ell+1}$. Since $V(T_{\ell+1}) = V(T_\ell) \cup \{z\}$, where z is the $(\ell+1)$st discovered vertex, it suffices to show that the unique path from v to z in $T_{\ell+1}$ is a shortest path between these two vertices in G. We let w denote the predecessor of z in $T_{\ell+1}$. By the inductive hypothesis, $\mathrm{dist}_{T_{\ell+1}}(v, w) = \mathrm{dist}_G(v, w)$. Next, we let

$$P = \langle v_0 = v, v_1, \cdots, v_k = z \rangle$$

be a shortest path in G from v to z. Let j be the smallest index such that $v_{j+1} \notin V(T_\ell)$. Notice that $0 < j < k$. By the way Dijkstra's algorithm selects boundary edges, z is a non-tree vertex of T_ℓ that is closest to v, and thus must be at least as close to v as is the non-tree vertex v_{j+1}. This tells us that

$$\mathrm{dist}_{T_{\ell+1}}(v, z) \leq \mathrm{dist}_G(v, v_{j+1}) \leq \mathrm{dist}_G(v, z),$$

and thus equality must hold throughout, and the proof is complete. □

COMPLEXITY/PERFORMANCE OF DIJKSTRA'S ALGORITHM: We will provide a liberal upper estimate of the amount of work needed to perform Dijkstra's algorithm on an edge-weighted graph or digraph with V vertices. The tree growing algorithm runs through $V - 1$ iterations. At the ith iteration, there will be at most $V - i$ boundary edges, and for each the edge weight needs to get added to an existing distance (to the tree endpoint) and this sum gets compared with the value of CurrDist for the undiscovered vertex. The latter value is updated if necessary. So we have a constant amount $O(1)$ of work for each vertex. This gives an upper bound complexity estimate of

$$\sum_{i=1}^{V-1}(V-i)O(1) = O(1)[1 + 2 + 3 + \cdots + (V-1)] = O(1)(V-1)V/2 = O(V^2),$$

where we have used formula (1) of Example 3.1.

This $O(V^2)$ complexity estimate is the best estimate for general graphs, but for sparse graphs (i.e., where the number of edges is closer to V than to $V(V-1)/2$) it will perform even faster. Depending on the density of edges, there are implementations of Dijkstra's algorithm to minimize the resulting complexity. Interested readers may refer to Chapter 24 of [CLRS-01] and Chapter 3 of [DPV-08] for more details. Dijkstra's algorithm remains one of the best general algorithms for computing shortest paths and distances.

Dijkstra's algorithm can be adapted to answer any questions relating to diameters and distances in edge-weighted digraphs and graphs, and hence also for non edge-weighted digraphs and graphs. The following exercise for the reader provides some examples.

EXERCISE FOR THE READER 9.13: Explain how Dijkstra's algorithm could be adapted to perform the following tasks for a simple graph G.
(a) Determine whether G is connected.
(b) Compute the diameter $\text{diam}(G)$.
(c) Determine the number of components of G.

Depth-First Searches and Breadth-First Searches

We introduce yet another pair of adaptations of the meta-algorithm for tree growing (Algorithm 9.4) that turn out to be extremely useful for an assortment of applications in computer science and engineering. Although they are not exclusively used for or motivated by optimization problems, they are very natural instances of the tree growing meta-algorithm. The names of the algorithms, *depth-first search* and *breadth-first search*, are intuitive but rather traditional. They are used for many other purposes than simple vertex searches. As with any tree growing algorithm, they may be viewed as vertex discovery algorithms; the order of discovery is determined by the way in which the boundary edge selection step is done. The basic difference in how these two algorithms operate is as follows: A depth-first search aims to discover as its next vertex one that is adjacent to the most recently discovered vertex, backtracking whenever necessary. Breadth-first searches aim to discover all vertices adjacent to the root vertex first (the level-one vertices), and then proceed to find all vertices that are two edges away from the root, then three edges away, and so on. Thus a depth-first search will proceed as deeply as possible along each edge emanating from the root, and backtrack as necessary, while the breadth-first search "fans out" to search first for the closest vertices to the root. Both are very simple to describe and run very efficiently. Since both algorithms care only about vertex accessibility, multi-edges and self-loops are irrelevant, and so we assume that we are dealing with a simple graph or a digraph without parallel edges or self-loops. We now formally state both algorithms.

ALGORITHM 9.7: Depth-First Search on a Simple Graph or Digraph *G*

Input: A simple graph or digraph *G* and a starting (root) vertex $v \in V(G)$.

Output: A rooted spanning tree *T* for the component of *G* with root vertex *v*, along with DiscoveryList, a vector ordering the vertices of *G* by their discovery numbers.

NOTE: In case *G* is connected, *T* will be a spanning tree (for all of *G*).[13]

Apply the tree growing meta-algorithm (Algorithm 9.4) with the following specification of the **Next Edge Selection** *part of Step 2:*
A boundary (directed) edge from BdyEdges is selected that has the property that its tree endpoint has the <u>largest</u> possible discovery number, i.e., was the most recently discovered vertex, among all tree endpoints of boundary edges. In cases of ties, the selection is made according to some default priority.

OUTPUT the spanning tree *T*, and the DiscoveryList vector produced by the meta-algorithm.

ALGORITHM 9.8: Breadth-First Search on a Simple Graph or Digraph *G*

Input: A simple graph or digraph *G* and a starting (root) vertex $v \in V(G)$.

Output: A rooted spanning tree *T* for the component of *G* with root vertex *v*, along with DiscoveryList, a vector ordering the vertices of *G* by their discovery numbers.

NOTE: In case *G* is connected, *T* will be a spanning tree (for all of *G*).

Apply the tree growing meta-algorithm (Algorithm 9.4) with the following specification of the **Next Edge Selection** *part of Step 2:*
A boundary (directed) edge from BdyEdges is selected that has the property that its tree endpoint has the <u>smallest</u> possible discovery number, i.e., was the earliest discovered vertex, among all tree endpoints of boundary edges. In cases of ties, the selection is made according to some default priority.

OUTPUT the spanning tree *T*, and the DiscoveryList vector produced by the meta-algorithm.

As usual, we will illustrate these tree growing algorithms graphically. The computer implementation material at the end of this section will provide readers with tools for writing corresponding programs.

EXAMPLE 9.15: Apply depth-first and breadth-first searches (Algorithms 9.7 and 9.8) to the simple graph shown in Figure 9.21, where the edge weights are

---

[13] Although the meta-Algorithm 9.4 was stated for connected graphs, it will still function without this assumption, but the tree it produces will no longer be a spanning tree (for the whole graph). Also, in the case of a digraph that is not strongly connected, the output will be a tree that contains all vertices that are reachable from the starting vertex.

ignored.   Start both searches at the vertex a.   In the tree growing algorithm, resolve any ties using alphabetical order on the non-tree vertices, and then (if necessary) tree vertices.

.

SOLUTION: The results of the two searches are shown in Figure 9.31.

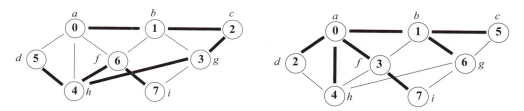

FIGURE 9.31: Comparison of the depth-first search tree (left) and the breadth-first search tree (right) for Example 9.15, with root (starting) vertex a.   Discovery numbers are indicated in bold, and tree edges are drawn with thick lines.

EXERCISE FOR THE READER 9.14:   Apply depth-first and breadth-first searches (Algorithms 9.7 and 9.8) to the simple graph shown in Figure 9.29, where the edge weights are ignored.   Start both searches at the vertex a.   In the tree growing algorithm, resolve any ties using alphabetical order on the non-tree vertices, and then (if necessary) tree vertices.

EXERCISE FOR THE READER 9.15:   Apply depth-first and breadth-first searches (Algorithms 9.7 and 9.8) to the simple digraph shown in Figure 9.30, where the edge weights are ignored.   Start both searches at the vertex a.   In the tree growing algorithm, resolve any ties using alphabetical order on the non-tree vertices, and then (if necessary) tree vertices.

COMPLEXITY/PERFORMANCE OF DEPTH-FIRST AND BREADTH-FIRST SEARCH ALGORITHMS: Suppose that G is a simple graph with V vertices and E edges.   Each of the algorithms must explore each edge (in the connected component of the root) twice, once for the tree endpoint and once for the non-tree endpoint. For each of the corresponding tasks, a constant amount $O(1)$ of work is required. This gives a linear time (in the number E of edges) complexity estimate of $O(E)$. The estimate for digraphs is the same and the analysis is similar.

NOTE: Breadth-first searches are very efficient algorithms for distance-related problems in non edge-weighted graphs.
As we have pointed out, Dijkstra's algorithm is the best general algorithm for solving distance-related problems in edge-weighted graphs and digraphs.   For example, to detect connectedness of a given graph or digraph on V vertices, the $O(V^2)$ running time is far better than the $O(V^4)$ running time of Algorithm 8.2, which was only applicable to non edge-weighted graphs.   For non edge-weighted connected graphs and digraphs, the breadth-first tree produced by Algorithm 9.8 is

a shortest distance tree from the root vertex, as is the one produced by Dijkstra's algorithm. The corresponding distances to the root are easily computed from this tree in linear time in the number of vertices $O(V)$. Since connected graphs have at least $V-1$ edges, it follows that the breadth-first search can do the job of Dijkstra's algorithm with running time $O(E)$, where E denotes the number of edges. This is superior to the $O(V^2)$ whenever graph is sparse.

The Traveling Salesman Problem

The *traveling salesman problem* (*TSP*) is one of the most famous problems in discrete mathematics. It is computationally very difficult and has numerous applications. Consequently, it has been extensively studied, has spurred the development of many new research methods and techniques, and continues to be a proving ground for new computational methods. Here we will introduce the reader to this problem along with a few aspects and milestones about it.

In its classical formulation, the traveling salesman problem assumes that we have a saleman who needs to plan a trip from his home city through a set of other cities so that the total distance traveled will be a minimum. This can easily be formulated in the language of edge-weighted graphs, where the vertices would represent the cities that need to be visited (including the home city) and the edge weights would represent the distances. Using the language of Section 9.1, a salesman's tour would correspond to a (directed) Hamilton tour in the edge-weighted graph.

DEFINITION 9.7: Given an edge-weighted complete graph G on n vertices, the **traveling salesman problem** (**TSP**) asks to find a Hamilton tour of minimum possible weight. More generally, the TSP may be formulated in any edge-weighted connected graph G by introducing very large (or infinite) weights on edges that do not belong to G.

EXERCISE FOR THE READER 9.16: Show that any algorithm that can solve the traveling salesman problem can be adapted to determine whether a connected graph G is Hamiltonian. Since the latter problem is NP complete, it follows that the traveling salesman problem is at least as difficult to solve as the Hamiltonian tour problem. Since the Hamilton tour problem is NP complete, it follows that the TSP is NP hard.
Suggestion: Consider the corresponding complete graph K on the vertices of G. Assign edge weights on K by setting the weight of each edge in $E(G)$ to be 1, and all other edge weights to be 2.

The traveling salesman problem originated in the early part of the twentieth century; in its original formulation, the weights represented distances, but in its numerous applications, the weights may take on any quantity that needs to be minimized. When we go out to run errands on any given Saturday, we informally need to solve a small-scale traveling salesman problem. How to best plan a trip which starts and ends at home and will include a stop at the grocery store, the hardware store, the post office, and perhaps also the bicycle shop is a form of the traveling salesman problem. When FedEx sends its delivery trucks out to deliver each truckload of parcels across town, both traffic and distance must be taken into consideration in weighing the different routes. But in the end, what needs to be found is a reasonably good solution to a traveling salesman problem. In the manufacturing of microchips and circuit boards, thousands of tiny holes need to be drilled (with a laser drill) at specified locations on a silicon wafer. Planning the drill sequence so as to minimize total distance needed to be traveled by the drill head is yet another form of the traveling salesman problem. Very large scale integrated circuits ("VLSI circuits") actually need millions of holes to be drilled in the circuit board, showing that traveling salesman problems with huge numbers of cities really do need to be solved. Numerous other applications span over areas such as x-ray crystallography, circuit wiring, robotics, order picking (in a warehouse), DNA mapping and archaeological dating (see [AMOR-95]), and the list goes on. Because of its importance, the traveling salesman problem has been and still is being extensively studied. It is surprising that such a simple sounding problem as the traveling salesman problem still does not have an efficient algorithm for its solution. Indeed, this important problem has provided motivation for much of the important work that has been done in the development of combinatorial algorithms and complexity analysis. A great deal has been written about the traveling salesman problem, and readers wishing to learn more about it might wish to consult either of the following two excellent books that have been written on the TSP: [ABCC-06], [LLRS-85]. The first of these is of course more up to date, and nicely traces many major developments relating to the TSP. The latter carefully develops many of the tools that have been successfully applied to the TSP (many of which were created under the motivation of this problem). Both books give extensive explanations of the relevant algorithms.

The following example is a very small instance of the TSP:

EXAMPLE 9.16: Suppose that we have a network of five cities as shown in the graph of Figure 9.32. The dollar amounts associated to each edge are the airfare prices for a single individual who purchases a one-way ticket in either direction on the path. Marv, a Pacific regional sales manager is based in Honolulu, and wishes to plan a trip to visit each of the five cities in the network (each one exactly once) and then return to Honolulu. What is the least expensive way for him to make this journey?

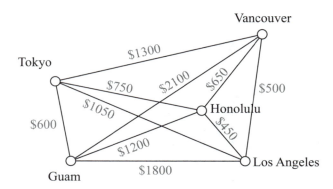

FIGURE 9.32: Air ticket prices for the traveling salesman problem of Example 9.16.

SOLUTION: Since this problem is so small, it can be solved by the brute-force approach of going through all possible (Hamilton) tours, and checking the prices of each. For quick reference, it will be convenient to number the cities in the network (= vertices in the graph). We adopt the following numbering scheme:

   1 = Honolulu, 2 = Tokyo, 3 = Vancouver,   4 = Guam, and 5 = Los Angeles

We gave Marv's home base (Honolulu) the number 1; the other number assignments were arbitrary. Since flights are available between any pairs of these cities, basically the choices Marv has will correspond exactly to listings of these five numbers, starting with 1, but the remaining four numbers can be permuted (i.e., rearranged) in any way. Thus, for example, the route represented by the sequence: [1 5 4 2 3] would correspond to the journey:

   Honolulu→Los Angeles→Guam→Tokyo→Vancouver(→Honolulu).

We omitted the final "1" in the numerical vector since it is understood that Marv will return to Honolulu at the end of the journey. Thus, the number of routes to consider is the number of ways to permute four objects (in this case the cities 2, 3, 4, and 5) and this number is 4! = 24. We could then use the pricing information to compute the price of each of these 24 possible routes for Marv and in this way find the most economical option. Actually, these 24 tours pair up in a natural way: each tour has the same price as its oppositely directed tour (just take the flights in the reverse order), so we really need to check only 12 different tours.
Going through this checking process produces the following cheapest tours:
   1   3   5   2   4   1 ⟹
Honolulu→ Vancouver → Los Angeles→ Tokyo→ Guam(→Honolulu)
and the corresponding reverse order tour
   1   4   2   5   3   1 ⟹
Honolulu→ Guam→Tokyo→ Los Angeles→Vancouver(→Honolulu).

Each of these tours costs $4000.

The brute-force method is perfectly reliable, and was feasible for this example, but for larger numbers of cities it quickly becomes impractical. The number of possible routes to check for the n-city traveling salesman problem, namely $(n-1)!/2$, becomes inordinately large, extremely quickly. Thus the brute-force method is not a viable option except on small-sized problems, even if we use a supercomputer. Further evidence is provided in Table 9.1.

TABLE 9.1: Comparisons of the number of traveling salesman routes and corresponding computing times on standard computers and supercomputers for networks containing 5, 10, 15, 20, and 25 cities.

Number of Cities	Number of Traveling Salesman Tours $(n-1)!/2$	Time to do the brute-force method on Intel 10 GHz box with 6GB RAM	Time to do the brute-force method on supercomputer[14]
5	12	Under one second	Under one second
10	20160	Under one minute	Under one second
15	3,113,510,400	Several months	Under one minute
20	3,201,186,852,863,997	About 18 years	Under one week
25	12,926,008,369,442,623,324,160	About 93 million years	About three years

Due to the importance of the TSP and the limitations of the brute-force method, scientists have been searching for more efficient exact algorithms, as well as heuristics for the traveling salesman problem. A seminal paper in this direction was written by George Dantzig, Ray Fulkerson, and Selmer Johnson in 1954 [DaFuJo-54], in which they solved a 42-city (within the contiguous United States) traveling salesman problem of finding the shortest possible road tour. Their method used linear programming, a powerful tool in discrete mathematics that was actually pioneered by George Dantzig some years earlier. The optimal TSP tour for their problem is illustrated in Figure 9.33.[15]

---

[14] Comparisons with supercomputers are a bit difficult. The biggest advantage of the current "super-computers" is not in their clock speeds, but in their ability to access memory quickly. In ballpark figures, it takes an Intel box 60 cycles to access a random byte of memory, whereas a Cray T3E can do it in two cycles. This also changes the way you program. Another advantage of supercomputers is that they run on relatively lean operating systems. Windows XP/Vista/7 and MacOS all eat up a lot of processor speed because of their graphical user interface. Finally, most supercomputers today have many processors (1024 give or take a power of two) in them. So assuming that the code can parallelize, it will go 1024 times faster. All of these things require that the programmer knows how to take advantage of them. So, it is possible to come up with a test in which an Intel box machine beats a Cray T3E. We use a very rough comparison factor of 30,000 for the speed of a supercomputer versus that of an Intel box.

[15] The author would like to kindly acknowledge Bill Cook, a professor at Georgia Tech and an author of the book [ABCC-06] for kindly providing data sets for the optimal TSP solutions that were used to create Figures 9.33 and 9.44. Professor Cook maintains a very impressive and extensive webpage on the TSP: http://www.tsp.gatech.edu/

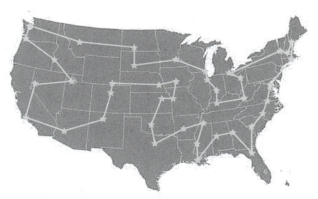

FIGURE 9.33: Illustration of the optimal tour for the 42-city traveling salesman problem found in 1954 by Dantzig, Fulkerson, and Johnson.

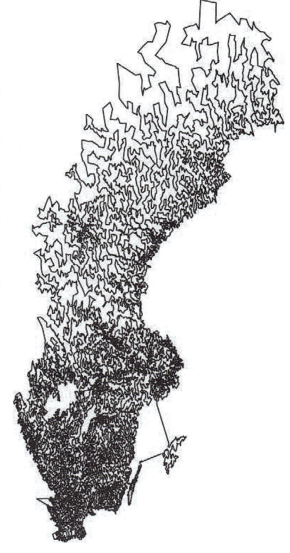

Since the time of the Dantzig, Fulkerson, and Johnson paper, scientists in industry and academia alike have been competing in a traveling salesman problem olympics. In 1980, the world record for (optimally) solving a traveling salesman problem was a 318-city problem, in 1985 it jumped to 532, and in 1988 it jumped to a whopping 2,392. At the turn of the second millenium, the record broke 13,000 cities. Figure 9.34 shows the record breaking optimal tour that was found in 2004 (the 50th anniversary since the "granddaddy" paper) of <u>all</u> of the 24,978 cities and towns in Sweden. As of 2010, the largest TSP that was optimally solved contained roughly 100,000 cities.

FIGURE 9.34:   Illustration of the 2004 record breaking traveling salesman optimal tour of 24,978 cities and towns in Sweden found by a team with W. Cook and four others.

While increasing computer power helps increase the sizes of problems that can be solved, it is algorithmic improvements that have had the most significant impact on progress with the TSP. Indeed, according to Moore's law, computing capabilities double every 18 months, so theoretically, the same technology used in the 1954 discovery, if applied using 2004 computing power (50 years later) would be able to solve a TSP that had 33 1/3 times the complexity. Because of the factorial complexity of the TSP, this would not even allow it to solve a 43-city problem rather than a 42-city problem (in the same amount of time), a far cry from the 2004 record. Many of the improved algorithms are based on extensions of the linear programming ideas of Dantzig, Fulkerson, and Johnson.

Insertion Heuristics for the Traveling Salesman Problem

Despite the current state of affairs of being able to optimally solve TSPs with over 100,000 cities, such calculations are extremely laborious and costly, and moreover many applied problems (e.g., in circuit design) require faster solutions of much larger problems. It is therefore important to have more efficient heuristic algorithms for the TSP. We will close this section by describing a family of heuristics for the traveling salesman problem known as *insertion heuristics*. These heuristics build up a tour from a sequence of cycles of increasing length; at each iteration a new vertex is inserted in the best place of the existing tour. The ways in which the new vertices get selected give rise to different variations of the insertion heuristics. Before presenting the algorithm, let us explain how to best insert a new vertex, once selected, into an existing tour.

Suppose that we have an existing (incomplete) Tour $= \ <v_1, v_2, \cdots, v_k, v_1>$, and we must insert a new vertex v_{new} into this tour (how this new vertex has been

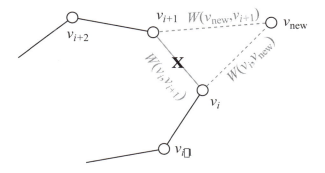

FIGURE 9.35: Illustration of the iterative step of the insertion heuristic; the crossed out edge of the current tour is replaced by the two new edges (dashed) going to the new vertex.

selected is presently not a concern). Since the current tour has k vertices, there are k places in which the new vertex can be inserted to create the new tour. Each such possible insertion involves deleting one of the k existing edges, $[v_i \ v_{i+1}]$ and replacing it with the two new edges $[v_i \ v_{new}]$, $[v_{new} \ v_{i+1}]$ (see Figure 9.35).[16]

From Figure 9.35, it is clear that the **weight differential** in making the insertion is the weight of the two new added edges, less that of the deleted edge:

$$\underbrace{W(v_i, v_{new}) \ + \ W(v_{new}, v_{i+1})}_{\text{sum of two new edge weights}} \ - \ \underbrace{W(v_i, v_{i+1})}_{\text{deleted edge weight}}. \tag{1}$$

(Here, $W(v_i, v_j)$ denotes the weight of the edge $\{v_i, v_j\}$.) At each iteration of the insertion method, the place where the new vertex is inserted is the one that minimizes the weight differential in (1). As usual, in case of ties, any choice is fine. Next, we make a formal statement of this general meta-heuristic. All specific insertion heuristics may be obtained from this meta-heuristic by specifying the "Next Vertex Selection" scheme.

ALGORITHM 9.9: Insertion Meta-Heuristic for the Traveling Salesman Problem on an Edge-Weighted Complete Graph G

Input: A complete edge-weighted graph G along with a matrix W of edge weights and a starting vertex v_1.
Output: A cycle TOUR that visits all vertices exactly once and its WEIGHT.

Step 1: Initialize: TOUR = $< v_1 >$, REMAIN = {all other vertices}, and WEIGHT = 0.

Step 2: **Next Vertex Selection:** Select a vertex v_{new} from REMAIN
Set TOUR = $< v_1 \ v_{new} \ v_1 >$. Delete v_{new} from REMAIN, and update WEIGHT to be $2W(v_1, v_{new})$.

Step 3: (*Iterative Step*).
WHILE REMAIN $\neq \varnothing$

> **Next Vertex Selection:** Select a vertex v_{new} from REMAIN
> Find an edge $\{v_i, v_{i+1}\}$ in TOUR = $< v_1, v_2, \cdots, v_k, v_1 >$ for which the insertion of v_{new} would minimize (1), insert this v_{new} between v_i and v_{i+1} in TOUR and reindex. Delete v_{new} from REMAIN, and add the corresponding weight differential (1) to WEIGHT.

END WHILE

Step 3: OUTPUT TOUR, WEIGHT

---

[16] For convenience, we let $v_{k+1} = v_1$ so that all edges of the current tour may be expressed in the form $[v_i \ v_{i+1}]$.

Three common specializations of this insertion meta-algorithm are listed in the next algorithm. In stating some of these algorithms, it will be helpful to extend the notion of the weight (or distance) between two vertices v and w, in an edge-weighted graph, $W(v, w)$, to the (lowest) weight (or distance) between a vertex v, and a set S of vertices. We denote this latter weight/distance by $W(v, S)$ and define it to be the smallest possible weight/distance between v and a vertex in S. Put differently,

$$W(v, S) \equiv \min\{W(v, u) \mid u \text{ is in } S\}. \tag{2}$$

This is analogous to the definition of the (shortest) distance from a point to a line (or other set) in Euclidean geometry.

ALGORITHM 9.10: Three Specific Insertion Heuristics for the Traveling Salesman Problem on an Edge-Weighted Complete Graph G

Input: A complete edge-weighted graph G, along with a matrix W of edge weights, and a starting vertex v_1.

Output: A cycle TOUR that visits all vertices exactly once and its WEIGHT.

Apply the insertion meta-heuristic Algorithm 9.9, with the following specifications for the "**Next Vertex Selection**" in Steps 2 and 3.

A. Nearest Insertion: Choose v_{new} to be as close as possible to a vertex in the current TOUR, i.e., v_{new} is chosen from REMAIN so that $W(v_{new}, \text{TOUR})$ is as small as possible.

B. Furthest Insertion: Choose v_{new} to be as far as possible from all the vertices in the current TOUR, i.e., v_{new} is chosen from REMAIN so that $W(v_{new}, \text{TOUR})$ is as large as possible.

C. Cheapest Insertion: Choose v_{new} so that the weight differential caused by adding v_{new} to the current tour (in the best possible place so as to minimize (1)) is as small as possible, i.e., v_{new} is chosen from REMAIN so that $\min\{W(v_i, v_{new}) + W(v_{new}, v_{i+1}) - W(v_i, v_{i+1}) \mid <v_i, v_{i+1}> \text{ is an edge in current TOUR}\}$ is as small as possible.

NOTE: As usual, in cases of ties, each of the above next vertex selections is made according to some default priority. But here, unlike with the previous non-heuristic optimization algorithms, different choices may lead to final tours of different weights.

EXAMPLE 9.17: Apply the nearest insertion heuristic to the traveling salesman problem of Example 9.16, using Honolulu as the starting vertex.

SOLUTION: We adopt the vertex numbering scheme that was used in Example 9.16:

1 = Honolulu, 2 = Tokyo, 3 = Vancouver, 4 = Guam, and 5 = Los Angeles

In Step 1 of the meta-heuristic, we initialize TOUR = <1>, WEIGHT = 0, and REMAIN = {2, 3, 4, 5}.　　In Step 2, we find that the least expensive place to go from Honolulu is Los Angeles, update TOUR be the cycle <1, 5, 1>, WEIGHT = $450 + $450 = $900, and REMAIN = {2,3,4}.

Step 3: First Iteration: We look for the cheapest place to go to from either of the two cities that we have in our existing tour. This would be city #3 (Vancouver). The price differential in placing this city either between 1 and 5 or between 5 and 1 is the same, namely $500 + $650 – $450 = $700. We randomly choose to put it between 1 and 5 to get the next cycle TOUR = <1, 3, 5, 1>, whose ticket price would be $900 (previous cycle price) + $700 (price differential) = $1600 = WEIGHT. We now have two remaining cities: REMAIN = {2, 4}.

Step 3: Second Iteration: We next notice that the cheapest (remaining) city to go from any of the three cities #1, #3, or #5 in our current cycle <1, 3, 5, 1> would be Tokyo (#2) at $750 (from #1). We look at the various possible price differentials when Tokyo (#2) is inserted at different locations in our cycle:
Between Honolulu (#1) and Vancouver (#3): Price differential = $750 + $1300 – $650 = $1400.
Between Vancouver (#3) and Los Angeles (#5): Price differential = $1300 + $1050 – $500 = $1850.
Between Los Angeles (#5) and Honolulu (#1): Price differential = $1050 +$750 – $450 = $1350.
Thus the last option is the cheapest so we insert Tokyo (#2) within the segment Los Angeles (#5) → Honolulu (#1) to produce our next cycle: TOUR = <1, 3, 5, 2, 1>, whose ticket price is $1600 (previous cycle price) + $1350 (price differential) = $2950 = WEIGHT. Only one city remains: REMAIN = {4}.

Step 3: Third Iteration:　　With only one more city, Guam (#4), to add in our cycle, we are now at the final iteration. We only have to find the best place to insert Guam into our existing cycle so as to minimize the price differential. For such a small network, this can be done by eye (and by ignoring possibilities that give too large a price differential), but we include here all the possibilities:
Between Honolulu (#1) and Vancouver (#3):　　Price differential = $1200 + $2100 – $650 = $2650.
Between Vancouver (#3) and Los Angeles (#5):　　Price differential = $2100 + $1800 – $500 = $3400.
Between Los Angeles (#5) and Tokyo (#2):　　Price differential = $1800 + $600 – $1050 = $1350.
Between Tokyo (#2) and Honolulu (#1): Price differential = $600 + $1200 – $750 = $1050.

Since the last option is the most economical, this gives the final cycle: TOUR = <1, 3, 5, 2, 4, 1>, whose ticket price is $2950 (previous cycle price) + $1050 (price differential) = $4000 = WEIGHT.

This is the same as the optimal price that was found using the brute-force method. In our implementation of this algorithm, there was one random choice made (when there were two choices of where in the existing cycle [1 5 1] to insert Vancouver (#3)). If we instead made the other choice, we would have, in the end, constructed the mirror image final cycle: [1 2 5 3 1].

EXERCISE FOR THE READER 9.17: Apply the furthest insertion heuristic to the traveling salesman problem of Example 9.16, using Honolulu as the starting vertex.

RUNNING TIMES FOR INSERTION HEURISTICS: By means of efficient programming, when applied to an n-city TSP, the nearest and furthest insertion heuristics will run in $O(n^2)$-time, whereas the cheapest insertion heuristic will run in $O(n^2 \log(n))$-time.

Performance Guarantees for Insertion Heuristics for the Traveling Salesman Problem

Despite the success of the nearest insertion heuristic on the (very small) TSP in Example 9.17, since we are working with heuristic algorithms, they may not (and typically do not) produce optimal TSP tours. We will give a theorem that provides performance guarantees for each of the insertion heuristics of Algorithm 9.10. The theorem does require one basic assumption that the network must satisfy the **triangle inequality**. This means that the cost of going from one city to another is always cheapest if one goes directly (on the edge joining the two cities); see Figure 9.36:

$$W(A,B) \leq W(A,C) + W(C,B) \text{ for all vertices } A, B, C.$$

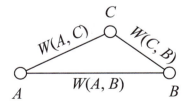

FIGURE 9.36: The triangle inequality stipulates that the cheapest/shortest way to go from one vertex (A) to another vertex (B) on an edge-weighted graph is by using the edge connecting them.

The triangle inequality is a very natural assumption. It is automatic for networks where weights between edges represent actual distances, and can be made to work in most all practical applications. For example, if a certain air carrier charged $1400 to make a shipment from Guam to Honolulu, and another carrier charged $400 to ship from Guam to Taipei, and $800 to ship from Taipei to Honolulu, we could put this lower price of $1200 in place of the direct shipment price on the edge from Guam to Honolulu. We caution the reader that in edge-weighted graphs that violate the triangle inequality, performance guarantees can fail badly, as is illustrated by the following example:

EXAMPLE 9.18: Figure 9.37 shows a four-city complete edge-weighted graph. If we apply the nearest neighbor insertion heuristic starting at vertex A, the resulting tour $<A, B, C, D, A>$ is nearly 1000 times more expensive than the optimal tour $<A, C, D, B, A>$! For four-city weighted networks satisfying the triangle inequality, however, the following theorem would guarantee the performance ratio to be at most 1.5.

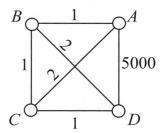

FIGURE 9.37: A network violating the triangle inequality.

THEOREM 9.11: (*Performance Guarantees for Traveling Salesman Problem Insertion Heuristics*) Suppose that we have an instance of the traveling salesman problem that has n vertices and satisfies the triangle inequality. We let OPT denote the weight of an optimal tour (solution) of the TSP.

1. If any implementation of insertion meta-heuristic (Algorithm 9.9) is used and produces a tour of weight INSERT, then we have the upper bound

$$\frac{\text{INSERT}}{\text{OPT}} \leq \text{ceil}(\log_2(n)) + 1. \tag{3}$$

2. If the nearest, furthest, or cheapest insertion heuristics (Algorithm 9.10) is used and produces a tour of weight SPINSERT, then we have the upper bound

$$\frac{\text{SPINSERT}}{\text{OPT}} \leq 2 \cdot \left(1 - \frac{1}{n}\right) < 2. \tag{4}$$

Furthermore, for each $n \geq 6$ and any of the three insertion heuristics, there exist TSPs with n vertices for which the inequality in (4) is an equality.

The theorem shows that although the cheapest insertion heuristic is more complicated and requires more work than the other two, the performance guarantees are the same. An estimate such as (4) that can actually be attained is called a **sharp** estimate. The estimate in Part (3) is the best known estimate for the general insertion method, but there does not seem to be any known example of a TSP where the ratio INSERT/OPT can get larger than 4. Performances may vary depending on the particular instance of the TSP. Numerous experiments, however, have been carried out by researchers by randomly generating TSPs and these have consistently found that the furthest insertion method outperforms the nearest and cheapest insertion heuristics; see, for example, the paper [JüReRi-95]. The complete proof of Theorem 9.11 is rather lengthy; a good reference for the proof along with other related results is the paper [RoStLe-77]. The exercises will develop some of the main ideas in the setting of an even simpler heuristic.

In general, more complicated (and time-consuming) heuristics are available to solve the traveling salesman problem with any desired degree of accuracy. At the time of this writing, with a moderately sized network of desktop computers, traveling salesmen problems with over one million cities can be solved in under an hour with a performance guarantee of 3.5%. This may be an option which is feasible to VLSI circuit designers who want to be efficient and practical at the same time. An extra 3.5% over the optimal time is an acceptable figure to deal with, since the savings gained by increased accuracy (say to 2%) would need to be contrasted with the additional computer time and more complicated programs that would be needed.

EXERCISES 9.2:

1. Answer the following questions about the edge-weighted bipartite graph $K_{3,3}$ shown in Figure 9.38(a):

   (a) Let T be the tree determined by the edge $\{b, f\}$. Find all boundary edges.

   (b) For the tree in Part (a), which boundary edge e should be added in a tree growing algorithm if we use alphabetical order on non-tree endpoints to determine the selection? Find the new set of boundary edges for the tree $T + e$.

   (c) Let T be the tree determined by the edges $\{a, f\}$, $\{f, c\}$. Find all boundary edges.

   (d) For the tree in Part (c), which boundary edge e should be added in a tree growing algorithm if we choose one of smallest weight (and resolve ties using alphabetical order of the non-tree endpoints)? Find the new set of boundary edges for the tree $T + e$.

2. Answer the following questions about the edge-weighted bipartite graph Q_3 shown in Figure 9.38(b):

   (a) Let T be the tree determined by the single vertex $\{f\}$. Find all boundary edges.

   (b) For the tree in Part (a), which boundary edge e should be added in a tree growing algorithm if we use alphabetical order on non-tree endpoints to determine the selection? Find the new set of boundary edges for the tree $T + e$.

   (c) Let T be the tree determined by the edges $\{b, d\}$, $\{d, i\}$. Find all boundary edges.

   (d) For the tree in Part (c), which boundary edge e should be added in a tree growing algorithm if we choose one of smallest weight (and resolve ties using alphabetical order of the non-tree endpoints)? Find the new set of boundary edges for the tree $T + e$.

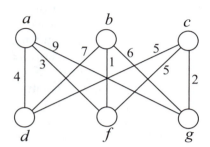
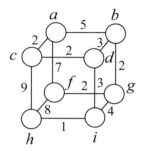

FIGURE 9.38: Two edge-weighted graphs: (a) (left) the complete bipartite graph $K_{3,3}$; (b) (right) the three-dimensional hypercube Q_3.

3.   (a) Apply Prim's algorithm (Algorithm 9.5) to construct a minimum spanning tree for the edge-weighted graph $K_{3,3}$ of Figure 9.38(a). Use vertex a as the starting vertex and resolve any ties using the alphabetical ordering of the non-tree vertices, and then (if necessary) tree vertices.
     (b) Repeat Part (a) using b as the starting vertex.
     (c) Repeat Part (a) using c as the starting vertex.

4.   (a) Apply Prim's algorithm (Algorithm 9.5) to construct a minimum spanning tree for the edge-weighted graph Q_3 of Figure 9.38(b). Use vertex a as the starting vertex and resolve any ties using the alphabetical ordering of the non-tree vertices, and then (if necessary) tree vertices.
     (b) Repeat Part (a) using g as the starting vertex.
     (c) Repeat Part (a) using f as the starting vertex.

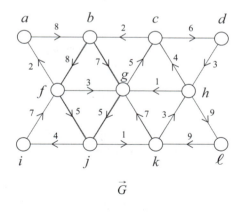
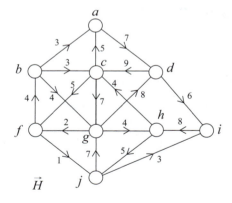

FIGURE 9.39: Two edge-weighted directed graphs: (a) (left) $\vec{G}$; (b) (right) $\vec{H}$.

5.   (a) Apply Prim's algorithm (Algorithm 9.5) to construct a minimum spanning tree for the underlying graph of edge-weighted digraph $\vec{G}$ of Figure 9.39(a) (i.e., ignore the edge directions). Use vertex a as the starting vertex and resolve any ties using the alphabetical ordering of the non-tree vertices, and then (if necessary) tree vertices.
     (b) Repeat Part (a) using b as the starting vertex.
     (c) Repeat Part (a) using d as the starting vertex.

6.   (a) Apply Prim's algorithm (Algorithm 9.5) to construct a minimum spanning tree for the underlying graph of edge-weighted digraph $\vec{H}$ of Figure 9.39(b) (i.e., ignore the edge directions). Use vertex a as the starting vertex and resolve any ties using the alphabetical ordering of the non-tree vertices, and then (if necessary) tree vertices.
    (b) Repeat Part (a) using f as the starting vertex.
    (c) Repeat Part (a) using h as the starting vertex.

7.   (a) Apply Dijkstra's algorithm (Algorithm 9.6) to construct a minimum distance tree for the edge-weighted graph $K_{3,3}$ of Figure 9.38(a) using a as the root vertex. List the distances of all vertices to the root. Resolve any ties in the tree growing algorithm by using the alphabetical ordering of the non-tree vertices, and then (if necessary) tree vertices.
    (b) Repeat Part (a) using b as the root vertex.
    (c) Repeat Part (a) using c as the root vertex.

8.   (a) Apply Dijkstra's algorithm (Algorithm 9.6) to construct a minimum distance tree for the edge-weighted graph Q_3 of Figure 9.38(b) using a as the root vertex. List the distances of all vertices to the root. Resolve any ties in the tree growing algorithm by using the alphabetical ordering of the non-tree vertices, and then (if necessary) tree vertices.
    (b) Repeat Part (a) using g as the root vertex.
    (c) Repeat Part (a) using f as the root vertex.

9.   (a) Apply Dijkstra's algorithm (Algorithm 9.6) to construct a minimum distance tree for the underlying graph of edge-weighted digraph $\vec{G}$ of Figure 9.39(a) (i.e., ignore the edge directions) using a as the root vertex. List the distances of all vertices to the root. Resolve any ties in the tree growing algorithm by using the alphabetical ordering of the non-tree vertices, and then (if necessary) tree vertices.
    (b) Repeat Part (a) using h as the root vertex.
    (c) Repeat Part (a) using j as the root vertex.
    (d) Apply Dijkstra's algorithm (Algorithm 9.6) to construct a minimum distance tree for the digraph $\vec{G}$ of Figure 9.39(a) (i.e., this time do not ignore the edge directions) using a as the root vertex. List the distances of all vertices to the root. Resolve any ties in the tree growing algorithm by using the alphabetical ordering of the non-tree vertices, and then (if necessary) tree vertices.
    (e) Repeat Part (d) using h as the root vertex.
    (f) Repeat Part (d) using j as the root vertex.

10.   (a) Apply Dijkstra's algorithm (Algorithm 9.6) to construct a minimum distance tree for the underlying graph of edge-weighted digraph $\vec{H}$ of Figure 9.39(b) (i.e., ignore the edge directions) using a as the root vertex. List the distances of all vertices to the root. Resolve any ties in the tree growing algorithm by using the alphabetical ordering of the non-tree vertices, and then (if necessary) tree vertices.
    (b) Repeat Part (a) using b as the root vertex.
    (c) Repeat Part (a) using i as the root vertex.
    (d) Apply Dijkstra's algorithm (Algorithm 9.6) to construct a minimum distance tree for the digraph $\vec{H}$ of Figure 9.39(b) (i.e., this time do not ignore the edge directions) using a as the root vertex. List the distances of all vertices to the root. Resolve any ties in the tree growing algorithm by using the alphabetical ordering of the non-tree vertices, and then (if necessary) tree vertices.
    (e) Repeat Part (d) using b as the root vertex.
    (f) Repeat Part (d) using i as the root vertex.

11.   Apply the depth-first and breadth-first searches (Algorithms 9.7 and 9.8) to the following graphs and digraphs, where the edge weights are ignored. Start both searches at the vertex a. In the tree growing algorithm, resolve any ties using alphabetical order on the non-tree vertices, and then (if necessary) tree vertices.

(a) The labeled graph $K_{3,3}$ of Figure 9.38(a).

(b) The <u>underlying graph</u> of edge-weighted digraph $\vec{G}$ of Figure 9.39(a) (i.e., ignore the edge directions).

(c) The <u>digraph</u> $\vec{G}$ of Figure 9.39(a) (i.e., this time <u>do not</u> ignore the edge directions).

12. Apply the depth-first and breadth-first searches (Algorithms 9.7 and 9.8) to the following graphs and digraphs, where the edge weights are ignored. Start both searches at the vertex a. In the tree growing algorithm, resolve any ties using alphabetical order on the non-tree vertices, and then (if necessary) tree vertices.

(a) The labeled graph Q_3 of Figure 9.38(b).

(b) The <u>underlying graph</u> of edge-weighted digraph $\vec{H}$ of Figure 9.39(b) (i.e., ignore the edge directions).

(c) The <u>digraph</u> $\vec{H}$ of Figure 9.39(b) (i.e., this time <u>do not</u> ignore the edge directions).

13. (*Scheduling Errands*) In Figure 9.40(a), a map of main streets in a small city is given with the locations of some places indicated. Each city block has a length of one mile.

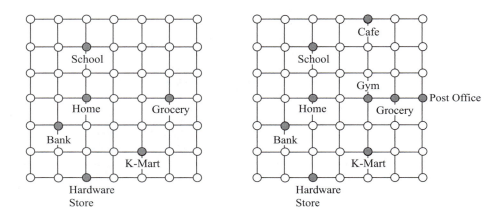

FIGURE 9.40: Maps of some locations at the intersections of major streets in a small city. Each block is one mile long: (a) (left) with six stops to make, and (b) (right) with nine stops.

(a) Draw a complete weighted graph having six vertices that represent the landmarks of Figure 9.40(a) and with the weight on each edge between a pair of vertices being the shortest driving distance between them (driving on the city streets).

(b) (*This part is a Computer Exercise*)[17] Suppose that you start at "Home" and need to run some errands that involve visiting each of the other five locations indicated on the map, and then return home. Apply the brute-force algorithm to find a tour with minimum driving distance. How many actual driving routes (following the streets in the city grid of Figure 9.40(a)) are there that will meet the optimal cycle's minimum total mileage?

Partial Answers: shortest tour distance: 18 miles; 4 different tours realize this minimum.

(c) Label the vertices of the graph as follows: 1= Home, 2 = Bank, 3 = School, 4 = Hardware Store, 5 = K-Mart, and 6 = Grocery. Apply (by hand) the nearest insertion heuristic to obtain a "good" route. What is the excess percentage of the total weight of the resulting cycle compared to that of the optimal tour obtained in Part (b)?

---

[17] Computer users may consult the material at the end of the section for implementation details. Since the answers are given, this part may be skipped by non-computer users.

(d) Repeat Part (c) with the cheapest insertion method.

14. (*Scheduling Errands*) In Figure 9.40(b), a map of main streets in a small city is given with the locations of some places indicated. Each city block has a length of one mile.

(a) Draw a complete weighted graph having nine vertices that represent the landmarks of Figure 9.40(b) and with the weight on each edge between a pair of vertices being the shortest driving distance between them (driving on the city streets).

(b) (*This Part is a Computer Exercise*)[18] Suppose that you start at "Home" and need to run some errands that involve visiting each of the other five locations indicated on the map, and then return home. Apply the brute-force algorithm to find a tour with minimum driving distance. How many actual driving routes (following the streets in the city grid of Figure 9.40(b)) are there which will meet the optimal cycle's minimum total mileage?

Partial Answers: shortest tour distance: 22 miles; 8 different tours realize this minimum.

(c) Label the vertices of the graph as follows: 1 = Home, 2 = Bank, 3 = School, 4 = Hardware Store, 5 = K-Mart, 6 = Grocery, 7 = Gym, 8 = Cafe, and 9 = Post Office. Apply (by hand) the nearest insertion heuristic to obtain a "good" route. What is the excess percentage of the total weight of the resulting cycle compared to that of the optimal tour obtained in Part (b)?

(d) Repeat Part (c) with the cheapest insertion method.

NOTE: (*Upgrades versus Maintenance Costs*) The costs relating to maintaining any piece of machinery versus the price of upgrading it over time can be modeled with a directed graph. Determining the optimal time(s) for an upgrade (so as to minimize long-term costs) can then be viewed as a shortest path problem. For a semi-specific example, suppose that we have just bought a new car (time $t = 0$), and we know how much to expect for maintenance costs over the next five-year period. Suppose that we also know the trade-in values of the car for a new model during each of the next five years. (Such information might be supplied to us by a dealer.) We assume that the car gets traded in at the end of the five year period. We can view all of our different options over the next five years as paths in a directed graph as follows: There will be six vertices labeled as 0, 1, 2, 3, 4, 5 for the time in years elapsed since we bought our new car. We put directed edges from any vertex to any higher numbered vertex. The directed network is shown in Figure 9.41. The significance of an edge going from vertex #i to vertex #j (with $i < j$) is that we would have purchased a new car at time i, paid maintenance costs on it for $j - i$ years, and then traded it in at time j. The weight of such an edge would be the cost of purchasing the new car less the trade-in value, plus the maintenance costs over the years in between i and j. A most economical upgrade strategy over the next five years would correspond to a path of minimal weight originating at vertex 0 and terminating at vertex 5.

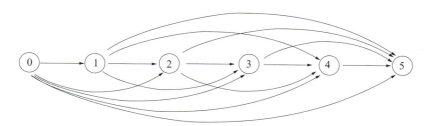

FIGURE 9.41: A digraph representing the problem of minimizing maintenance/upgrade costs over a five-year period.

15. (*Automobile Upgrades versus Maintenance Costs*) Otis, a carpenter, has just purchased a pick-up truck for $20,000, which he plans to use for hauling. Based on his intended usage, the dealer supplies him with the following estimated annual maintenance costs and trade-in values over the next five years.

---

[18] Computer users may consult the material at the end of the section for implementation details. Since the answers are given, this part may be skipped by non-computer users.

Age of Truck (in years)	Annual Maintenance Cost	Trade-In Value
0	$600	—
1	$1,500	$16,000
2	$2,000	$13,000
3	$3,000	$11,000
4	$4,500	$8,000
5	$6,000	$6,000

(a) Based on this data, determine the best course of action as far as when Otis should trade in his truck (possibly several times) so as to minimize the total expenditures over this five year period. Assume that the price of similar trucks will remain at $20,000 over this five year period, and that he trades in his truck at the end of the 5 year period.

(b)   If we extend the digraph of Figure 9.41 to be over n years (so it will have $n + 1$ vertices), how many (directed) edges will this digraph have?

16.   (*Photocopy Machine Upgrades versus Maintenance Costs*) A university mathematics department is planning its budget over the next six years. It has just purchased a heavy-duty full service photocopy machine for its faculty to make copies of exams, handouts, and research papers. The machine costs $12,000, and based on its usage, the photocopy company gives the following estimates on annual maintenance costs and trade-in values. The price of a new machine goes up 6% every year.

Age of Machine (in years)	Annual Maintenance Cost	Trade-In Value
0	$500	—
1	$1,000	$10,000
2	$1,200	$9,000
3	$1,200	$7,000
4	$2,200	$6,000
5	$3,400	$5,000
6	$4,800	$4,000

(a) Determine the most economical photocopier purchasing/maintenance plan over the next six years assuming that the copier gets traded in at the end of the six year period.

(b) What effect would increasing all of the trade-in values by $1000 have on the strategy found in Part (a)?

(c) What effect would decreasing all of the trade-in values by $1000 have on the strategy found in Part (a)?

17.   (*Network Reliability*)   The edge-weighted graph below represents a certain communication network where a weight $W(i,j)$ on an edge $\{i, j\}$ represents the probability that the communication link from vertex i to vertex j does not fail.

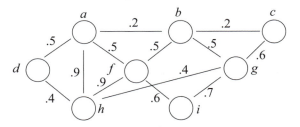

Assuming the probabilities that the various links fail are independent of one another, determine a most reliable path on which to send a communication from vertex d to vertex c. What is the probability that this communication path does not fail?

Suggestion: Consider the edge-weighted graph with the same vertices and edges, but with the weights modified to be $\tilde{W}(i, j) = -\log_{10} W(i, j)$.

18. (*Network Reliability*) The edge-weighted digraph below represents a certain communication network where a weight $W(i,j)$ on a directed edge (i, j) represents the probability that the directed communication link from vertex i to vertex j does not fail, and it is assumed that link failures are independent.

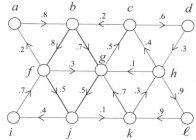

(a) Determine a most reliable communication path in this network on which to route a communication from vertex a to vertex ℓ. What is the probability that this communication path does not fail?

(b) Determine a most reliable communication path in this network on which to route a communication from vertex ℓ to vertex a. What is the probability that this communication path does not fail?

Suggestion: Consider the edge-weighted digraph with the same vertices and edges, but with the weights modified to be $\bar{W}(i, j) = -\log_{10} W(i, j)$.

19. (*Maximum Spanning Trees*) A **maximum spanning tree** of an edge-weighted connected graph is a spanning tree (the sum of whose edge weights is) of maximum possible weight.

(a) Explain how to modify Prim's algorithm (Algorithm 9.5) so that its output will be a maximum spanning tree rather than a minimum spanning tree.

(b) Use the algorithm of Part (a) to construct maximum spanning trees for the following edge-weighted graphs, with ties being resolved by using alphabetical order on the non-tree vertices, and then (if necessary) tree vertices.

(i) The edge-weighted graph of Figure 9.21, with starting vertex a.

(ii) The edge-weighted graph of Figure 9.25, with starting vertex a.

20. Prove that if an edge-weighted graph has an edge whose weight is smaller than all other edges, then this edge must be part of a minimum spanning tree.

21. Give an example to show that Prim's algorithm does not always work to produce a minimum spanning (directed) tree if applied to a connected edge-weighted digraph with an arbitrarily selected starting vertex. We assume, of course, that the edge directions are respected in the tree growing process.

22. Suppose that we apply Prim's algorithm to a connected edge-weighted digraph with n vertices n times, by varying the starting vertex through all possible vertices of the digraph (and respecting edge directions in the tree growing process). Of the n (directed) spanning trees created, is it necessarily true that one with the smallest weight will be a minimum spanning (directed) tree? Either prove that it is or construct a counterexample to show that it may not be.

NOTE: (***Kruskal's Algorithm for Minimum Spanning Trees***) An alternative algorithm to Prim's Algorithm 9.5 constructing a minimum spanning tree in an edge-weighted connected graph is Kruskal's algorithm. It was discovered in 1956 by American mathematician Joseph B. Kruskal (1928–).[19] The

---

[19] Kruskal's story should be an encouraging one to students. He discovered his algorithm as a second-year graduate student at Princeton. He did not know if his short paper was worthy of publication, but his professors encouraged him to submit it to a professional journal. After earning his PhD in 1954, he

idea is very simple. We build the minimal spanning tree edge by edge. We begin with an edge of minimal weight. The iterative step will then add new edges which are chosen to be of minimal weight (of the edges that remain) and such that when a new edge is added no cycle is formed. We keep this up until $n - 1$ edges are added (where n is the number of vertices). It may seem quite remarkable that such a simple approach actually leads to a minimal spanning tree, but it always will, and the algorithm is an efficient one; having a similar complexity upper bounds to Prim's algorithm.

ALGORITHM 9.11: (*Kruskal's Algorithm for Minimum Spanning Trees in Edge-Weighted Connected Graphs and Digraphs*)
NOTE: We will formulate the algorithm in the setting of graphs; the only change needed for digraphs is that edge directions need to be respected when determining boundary edges at each iteration.

Input: An edge-weighted connected graph G with $n > 1$ vertices.
Output: A minimum spanning tree T for G.

Step 1: We begin by sorting the edges of the graph: $e_1, e_2, \cdots, e_m$ in such a way that the corresponding weights of these edges $w_1, w_2, \cdots, w_m$ satisfy: $w_1 \le w_2 \le \cdots \le w_m$.[20] Initialize a collection of edges $\mathscr{E} = \{e_1\}$. If $m = 1$, output $T = \mathscr{E}$ and exit the algorithm, otherwise: Initialize Index = 2, Set $G_{\mathscr{E}}$ to be the subgraph of G determined by the edges of $\mathscr{E}$ and their endpoints.

Step 2: (*Iterative Step*)
WHILE $|\mathscr{E}| < n$

   IF e_{Index} does not form a cycle in $G_{\mathscr{E}}$

      UPDATE $\mathscr{E} = \mathscr{E} \cup \{e_{\text{Index}}\}$ (i.e., add the edge e_{Index} to the collection $\mathscr{E}$)

      UPDATE $G_{\mathscr{E}} = G_{\mathscr{E}} + e_{\text{Index}}$
   END IF
   UPDATE Index = Index + 1
END WHILE

OUTPUT: The spanning tree $T = G_{\mathscr{E}}$

Note that Kruskal's algorithm is not a tree growing algorithm (so it does not fall under the framework of the meta-tree growing Algorithm 9.4), since the subgraphs $G_{\mathscr{E}}$ need not be trees (although they are forests—Why?). If we are using Kruskal's algorithm by hand, it is quite easy to implement by simply drawing the graph G, and highlighting the edges that get picked up by $\mathscr{E}$ (say in green). This makes it straightforward to visually check whether a new edge will create a cycle in the existing forest. The computer implementation material at the end of this section will provide a very efficient scheme to implement this cycle-checking task in the creation of a computer program. With such efficient computer implementation, it can be shown that the performance time of Kruskal's algorithm on an edge-weighted connected graph (or digraph) with V vertices and E edges is $O(E \log(V))$, the same as that for Prim's algorithm. The interested reader may consult Chapter 23 of [CLRS-01] for a justification of this fact.

23. Apply Kruskal's algorithm to find a minimal spanning tree for each of the following edge-weighted graphs:
    (a) The graph of Figure 9.21.
    (b) The edge-weighted graph $K_{3,3}$ of Figure 9.38(a).

    (c) The underlying graph of the edge-weighted digraph $\vec{G}$ of Figure 9.39(a).

---

worked in academia for five years before taking an industrial job at Bell Laboratories where he spent the rest of his career.
[20] As usual, in cases of edges of equal weights, any default order can be used to sort them.

24. Apply Kruskal's algorithm to find a minimal spanning tree for each of the following edge-weighted graphs:
    (a) The graph of Figure 9.29.
    (b) The edge-weighted graph Q_3 of Figure 9.38(a).

    (c) The underlying graph of the edge-weighted digraph $\vec{H}$ of Figure 9.39(b).

25. Provide an algorithm that will input an edge-weighted graph and one of its edges, and will output a spanning tree that will contain the inputted edge and have minimum weight over all such spanning trees.

26. Prove the correctness of Kruskal's algorithm, i.e., prove that the tree produced by it is always a minimum spanning tree of an inputted connected edge-weighted graph.
    Sketch of Proof: Show first that the output of the algorithm is a tree T, and let M be any minimum spanning tree. If $T = M$, there is nothing to prove, so assume that $T \neq M$. Let e be the first edge in the construction of T that is not in M. Since $M + e$ must contain a cycle, some edge f of this cycle cannot belong to T (because both T and M are trees). Consider the graph M_1 obtained from M by exchanging and replacing edge f with edge e. Show that M_1 is again a minimum spanning tree. Continue this edge exchange procedure to obtain a (finite) sequence of minimum spanning trees $M_1, M_2, \cdots, M_k$ with $M_k = T$.

27. Prove Proposition 9.9, and thus establish the correctness of Prim's algorithm.
    Suggestion: Proceed by induction on ℓ. In the inductive step, let T be a minimum spanning tree that contains T_ℓ, and let e denote the boundary edge that is grown onto T_ℓ to produce $T_{\ell+1}$. In (the nontrivial) case $e \notin E(T)$, since T is a tree, the graph T' obtained from T by adjoining edge e will contain a cycle. Let f be the first edge encountered in this cycle in going from the tree endpoint that is not a boundary edge (in the tree growing construction of $T_{\ell+1}$ from T_ℓ). Look at the graph obtained from T by exchanging edge f with edge e.

28. Suppose that T is the rooted tree created by performing a depth-first search on a graph G. Prove that the pre-order traversal of T coincides with the discovery order of the vertices in the depth-first search.

29. Describe an efficient algorithm based on the depth-first search algorithm that will input a graph G with a certain ordering of its n vertices, and whose output will be a length n vector COMP of positive integers in the set $\{1, 2, \ldots, C\}$, where C is the number of components of G, such that COMP(i) = COMP(j), if, and only if, the ith vertex and the jth vertex lie in the same component of G.

30. Describe all simple graphs G for which the depth-first search starting with at least one of its vertices will produce the same rooted tree as the breadth-first search algorithm.

31. Prove that a vertex v of a connected graph G is a cut vertex if, and only if there exist two distinct vertices u and w (both different from v) such that every path from u and w in G must pass through v.

32. (*An Efficient Algorithm for Finding Cut Vertices*) Prove the following proposition, which immediately yields an efficient algorithm for determining cut vertices of a graph.
    Proposition: Suppose that the depth-first search (Algorithm 9.7) is applied to a connected graph G with starting vertex v to produce a rooted tree T. Then v is a cut vertex of G if, and only if v has more than one child in T.
    Suggestion: Use the result of Exercise 31.

NOTE: (*Nearest Neighbor Heuristic for the Traveling Salesman Problem*) The simplest of all heuristics for the traveling salesman problem is a greedy algorithm known as the *nearest neighbor* *heuristic*. We initialize the tour as the inputted starting vertex, and at each iteration, we find a vertex which is nearest to the most recently added vertex (i.e., we proceed along a lightest edge to the next unvisited vertex). Once all vertices have been visited, we return to the starting vertex to complete the tour. The formal algorithm is as follows:

ALGORITHM 9.12: Nearest Neighbor Heuristic for the Traveling Salesman Problem on an Edge-Weighted Complete Graph *G*

Input: A simple graph *G* along with a matrix *W* of edge weights and a starting vertex v_1.

Output: A cycle TOUR that visits all vertices exactly once and its WEIGHT.

Step 1: Initialize: TOUR = $< v_1 >$, CurrentVertex = v_1, REMAIN = {all other vertices}, and WEIGHT = 0

Step 2: (*Iterative Step*).
WHILE REMAIN $\neq \varnothing$

      FIND a vertex v_{new} in REMAIN such that $W(\text{CurrentVertex}, v_{new})$ is as small as possible.

      Add this weight to WEIGHT,

      APPEND v_{new} to TOUR (after CurrentVertex),

      UPDATE CurrentVertex = v_{new},

      Delete v_{new} from REMAIN
END WHILE

APPEND v_1 to TOUR (after CurrentVertex, to complete the tour), and add $W(\text{CurrentVertex}, v_1)$ to WEIGHT

Step 3: OUTPUT: TOUR, WEIGHT

The nearest neighbor heuristic has a performance guarantee that is a bit better than that of Theorem 9.11 for general insertion heuristics; see Exercise 37 below.

33.   Apply the nearest neighbor heuristic to the instance of the traveling salesman problem of Example 9.16, using Honolulu as the starting vertex.

34.   Upon his graduation from business school, Jack plans a vacation in France. He flies into Paris, rents a car, and plans to visit friends in each of the five cities shown in Figure 9.42 below.

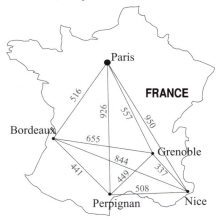

FIGURE 9.42: A map showing driving distances between some cities in France. Distances are given in km.

(a) Using the brute-force method, find both the shortest and the longest cycles that Jack could drive in order to visit each city exactly once. (Jack may be interested in the longest cycle if, for example, he is looking for an excuse to extend his vacation.) How do the resulting distances compare?

(b) Label the vertices of the graph as follows: 1 = Paris, 2 = Grenoble, 3 = Bordeaux, 4 = Perpignan, and 5 = Nice, and apply (by hand) the nearest neighbor algorithm to obtain a "good" route. What is the excess percentage of the total distance of the resulting cycle compared to the optimal tour obtained in Part (a)?

(c) Repeat Part (b) with the nearest insertion method.

(d) Repeat Part (b) with the furthest insertion method.

(e) Repeat Part (b) with the cheapest insertion method.

35.   (a) Fix a positive integer n. Construct a TSP for which the nearest neighbor heuristic produces a tour of weight greater than 10^n times the weight of an optimal tour.

(b) Repeat Part (a) for the nearest insertion method.

(c) Repeat Part (a) for the furthest insertion method.

(d) Repeat Part (a) for the cheapest insertion method.

NOTE: Optimality results in (edge-weighted) graph theory are often proved by the artifice of introducing nonnegative functions on the vertices of the graph, which bear certain relationships with the edge weights. The following lemma is typical of results that will help prove optimization results; Exercise 37 will use it to prove a performance guarantee for the nearest neighbor heuristic. The proof of the lemma is outlined in Exercise 36.

LEMMA 9.12: Suppose that we have a function f that assigns to each vertex v of an n-vertex edge-weighted graph a nonnegative number $f(v)$ having the following properties:

(i) $W(v,w) \geq \min(f(v), f(w))$ for all vertices v and w (as usual, W is the edge weight matrix).

(ii) $f(v) \leq (1/2)\text{OPT}$ for each vertex v (as in Theorem 9.11, OPT is the weight of an optimal TSP tour).

Then $\sum f(v) \leq (1/2)(\text{ceil}(\log_2 n)) + 1)\text{OPT}$, where the sum is taken over all vertices.

We point out that the triangle inequality need not be assumed for the validity of this lemma.

36.   (*Proof of Lemma 9.12*)   Complete the following outline to prove Lemma 9.12. We fix an optimal cycle C for the graph that visits all vertices and has minimal weight.

(a)   Write the vertices of the graph as $v_1, v_2, \cdots, v_n$ where the indexing is done so that $i \leq j \Rightarrow f(v_i) \geq f(v_j)$. For each index k between 1 and n (inclusive), consider the complete subgraph G_k determined by the vertices $v_1, v_2, \cdots, v_{\min(2k,n)}$. In G_k we form the cycle C_k that visits all of the vertices of G_k in the same order that C visits them. Use the triangle inequality to show that the weight of C_k, $\text{WGT}(C_k)$ is less than OPT, the weight of the C.

(b) Show that $\text{WGT}(C_k) \geq \sum\limits_{i=1}^{\min(2k,2)} c_i f(v_i)$, where c_i is the number of edges of C_k that join vertex v_i to another vertex v_j (of G_k) with $i > j$ (so that $f(v_i) = \min(f(v_i), f(v_j))$).   Next, noting that each c_i can equal 0, 1, or 2, and the sum of all of the c_i 's is $\min(2k,n)$ (the number of edges of C_k), deduce the lower bound $\sum\limits_{i=1}^{\min(2k,2)} c_i f(v_i) \geq 2 \sum\limits_{i=k+1}^{\min(2k,2)} c_i f(v_i)$.

(c) Combine the results of Parts (a) and (b) to produce the inequality $\text{OPT} \geq 2 \sum\limits_{i=k+1}^{\min(2k,2)} c_i f(v_i)$, for any k, $1 \leq k \leq n$.

(d) Apply the inequality of Part (c) successively using $k = 2^0$, $k = 2^1$, $k = 2^2$, $\cdots$ and combine the results to obtain $\text{ceil}(\log_2(n)) \cdot \text{OPT} \geq 2 \sum_{i=2}^{n} f(v_i)$. Next, incorporate the assumption (b) of

the lemma into the inequality just obtained, and complete the proof of the lemma.

37.   (*A Performance Guarantee for the Nearest Neighbor Heuristic—Algorithm 9.12*)   Apply
      Lemma 9.12 to prove the first part of the following:

      Proposition: If the nearest neighbor heuristic (Algorithm 9.12) is used on an *n*-vertex instance
      of the traveling salesman problem that satisfies the triangle inequality, and produces a tour of
      weight NEARNBR, then we have the upper bound

$$\frac{\text{NEARNBR}}{\text{OPT}} \leq \frac{\text{ceil}(\log_2(n)) + 1}{2}. \tag{5}$$

      On the other hand, there exist TSPs with arbitrarily large numbers *n* of vertices such that the
      ratio on the left side of (5) can be greater than $(1/3)\log_2(n+1) + 4/9$.

      Suggestion: Base the construction of the function *f* on a given instance of the nearest neighbor
      heuristic as follows: Set $f(v)$ equal to the minimal distance of *v* to its nearest neighbor that
      was selected in the heuristic.

      NOTE: A family of examples that exhibit the performance ratio $(1/3)\log_2(n+1) + 4/9$ can be
      found in [RoStLe-77]; since the construction is rather involved (3 1/2 pages), we omit the
      details here.

38.   The traveling salesman problem asks to find a Hamilton tour of minimum weight in a complete
      edge-weighted graph.   Give an example to show that if we change the requirement of finding a
      Hamilton tour to finding a (closed) circuit that visits every vertex, then a solution circuit need
      not necessarily be Hamilton tour.   What happens if we assume that the triangle inequality
      holds?

COMPUTER EXERCISES 9.2:

NOTE: In this section, more than in most others thus far, the data structures needed are rather
complicated. Particular data structure implementations of the algorithms can have significant impacts
on the efficiency of the algorithms. We will not elaborate on the optimal data structures for each of the
algorithms, but encourage the reader to examine and compare some different possibilities in the course
of writing programs for the following computer exercises. For more details on the use of optimal data
structures on an assortment of algorithms, the reader may consult [CLRS-01], or any good book on data
structures, such as [Bra-08]. The following specifies our general syntax conventions regarding graphs,
digraphs, and trees throughout this set of computer exercises:

*We will assume that the inputted graphs are presented as edge lists (see Section 8.1), where the edges
are listed in lexicographic order, and the vertices are {1,2,..,n}. Trees that are constructed will be
presented using parent lists (see Example 8.37 of Section 8.3).*

Computer Exercises 1–3 will create some programs that will execute integral parts of the tree growing
meta-algorithm (Algorithm 9.4).   These programs lie at the core of the iterative step of the tree
growing meta-algorithm.   We first deal with the setting of graphs; the corresponding digraph programs
will be very similar with some minor modifications.

1.   (*Program for Determining the Set of "Unused" Edges within an Iteration of the Tree Growing
      Meta-Algorithm 9.4 for Graphs*) (a) Write a program with the following syntax:

```
UnusedEdges  = UnusedEdgeFinder(EdgeListG, EdgeListT, BdyEdges)
```

The input `EdgeListG`, is the edge list (a two-column matrix) of a graph G, the input `EdgeListT` is the list of the edges of a subtree T, the input `BdyEdges` is a list of the edges of the graph that are boundary edges of T. The output `UnusedEdges` is a list of edges of the graph that remain potential boundary edges in future iterations of the meta-tree growing algorithm, i.e., that do not appear as an edge in T or as a boundary edge, and do not have both endpoints in the tree.

(b) Apply your program to the graph and subtree of Figure 9.22, with the vertices being relabeled as integers in alphabetical order ($a = 1$, $b = 2$, $c = 3$, $d = 4$, $f = 5$, ...). Your program should identify exactly two unused edges: {5, 7} and {5, 8}. (Why?)

(c) Apply your program to the graph and subtree of Figure 9.23, with the vertices being relabeled as integers in alphabetical order ($a = 1$, $b = 2$, $c = 3$, $d = 4$, $f = 5$, ...). Your program should identify exactly one unused edge: {5, 8}. (Why?)

In principle, when we initiate the tree growing algorithm with a tree being a single starting vertex, the unused edges are simply all edges of G that are not boundary edges. The updates can be easily done in going from one iteration to the next, so in principle, the more costly program of Computer Exercise 1 need not be used. The program of the following computer exercise is organized for this strategy.

2. (*Program for Updating the Boundary Edges Set (BdyEdges) in Algorithm 9.4 for Graphs*) (a) Write a program with the following syntax:

```
NewBdyEdges, UnusedEdges
    = BdyEdgesUpdate(EdgeListT, BdyEdges, e, UnusedEdges)
```

The first input `EdgeListT` is the list (a two-column matrix) of the edges of a subtree T of an underlying graph G, the input `BdyEdges` is a list of the edges of the graph that are boundary edges of T, the input e is a boundary edge T (i.e., a single row vector of `BdyEdges`), and the final input is a list of the edges of G that have not yet been used (i.e., appear as a tree edge, boundary edge, or have both endpoints in the tree). The first output `NewBdyEdges` is a list of edges of the graph (or digraph) that are (new) boundary edges of $T + e$. The second output `UnusedEdges` is the updated list of unused edges.

(b) Apply your program to the graph and subtree of Figure 9.22, with the vertices being relabeled as integers in alphabetical order ($a = 1$, $b = 2$, $c = 3$, $d = 4$, $f = 5$, ...), the boundary edges as indicated by thick dashed lines, and particular boundary edge e as indicated in Example 9.12. Check the output of your program with the result of Example 9.12.

For most uses of the meta-Algorithm 9.4, the edges need to be compared or analyzed in some way. For this reason it is useful to have all edges of the underlying graph G indexed. This can be accomplished by storing the edge lists as three-column matrix where the first two columns represent (as usual) the endpoints of the edges, while the third column entries represent the edge indices. For example, one three-column representation of the edge list of the graph G, the subtree T, and the boundary edges of Figure 9.22 is shown on the left.

EdgeListG =			EdgeListT =			BdyEdges =		
1	2	1	1	2	1	1	5	3
1	4	2	1	4	2	1	7	4
1	5	3	2	3	5	2	5	6
1	7	4	2	6	7	4	7	9
2	3	5				6	7	12
2	5	6				6	8	13
2	6	7						
3	6	8						
4	7	9						
5	7	10						
5	8	11						
6	7	12						
6	8	13						

For simplicity, the edges of G are indexed by their row numbers; notice that the indices of edges of EdgeListT and BdyEdges correspond to those of G. The next computer exercise asks the reader to rewrite the program of the Computer Exercise 2 so that such an edge index scheme is used.

3. (*Program for Updating the Boundary Edges Set (BdyEdges) in Algorithm 9.4 for Graphs with Edge Indices*) (a) Write a program with the following syntax:

```
NewBdyEdges, UnusedEdges
    = BdyEdgesUpdateInd(EdgeListT, BdyEdges, e, UnusedEdges)
```

The inputs/outputs and functionality of the program are exactly as in Computer Exercise 2, except that all edge list matrices now have three columns, where the third column entries represent edge indices.

(b) Apply your program to the graph and subtree of Figure 9.22, with the vertices being relabeled as integers in alphabetical order ($a = 1$, $b = 2$, $c = 3$, $d = 4$, $f = 5$, ...), the boundary edges as indicated by thick dashed lines, and particular boundary edge e as indicated in Example 9.12. Check the output of your program with the result of Example 9.12.

Suggestion: For Part (a), start with the program that you wrote for Computer Exercise 2. Each time an edge set is updated, say BdyEdges, define a corresponding two-column matrix BdyEdgesEndpts consisting just of the endpoints of the edges (i.e., indices are removed), and apply the search portions of the Computer Exercise 2 program to these truncated matrices.

4.   (*Program for Prim's Algorithm 9.5 for Minimum Spanning Trees*) (a) Write a program with the following syntax:

```
[ParentList DiscoveryList] = Prim(EdgeListG, Weights, startV, n)
```

that will perform Prim's Algorithm 9.5 on an edge-weighted graph G. The inputs are an indexed list of the edges of G EdgeListG, a vector of the corresponding edge weights Weights, the number of the starting vertex startV, and the number of vertices n. It is assumed that the vertices are numbered 1,2, ..., n. There will be two outputs: an n row two-column matrix ParentList giving the parent list of the tree, and the vector DiscoveryList giving the discovery numbers of the vertices (in order). Do not be concerned about how ties are resolved in this program; just as long as a boundary edge of lightest weight is always selected, Prim's algorithm is guaranteed to produce a minimum spanning tree.

(b) Apply your program to construct a minimum spanning tree for the edge-weighted graph of Figure 9.21, with the vertices relabeled as integers in alphabetical order ($a = 1$, $b = 2$, $c = 3$, $d = 4$, $f = 5$, ...), and using 1 as the starting vertex. Check that the weight of the tree produced by your program agrees with that of the one constructed in Example 9.13 (although the trees may be different). Repeat for the edge-weighted graph of Figure 9.25 with starting vertex $a = 1$, and compare the weight of the tree produced with that of the solution of Exercise for the Reader 9.10.

(c) Apply your program to the problems of Ordinary Exercise 5, and compare with the solutions of that exercise.

(d) Write a program Prim_OrderTieBreaks that has exactly the same inputs/outputs and functionality as the program of Part (a), with the only change being that any ties (in choosing a lightest boundary edge) are resolved by choosing the one with the lowest non-tree vertex label.

(e) Run your program of Part (d) on each of the minimum spanning tree problems of Parts (b) and (c) and check that the trees produced agree exactly with those obtained in the solutions in the text proper.

5.   (*Program for Dijkstra's Algorithm 9.6 for Computing Minimum Distances in Edge-Weighted Graphs*) (a) Write a program with the following syntax:

```
[ParentList DiscoveryList DistanceList]
        = Dijkstra(EdgeListG, Weights, startV, n)
```

that will perform Dijkstra's Algorithm 9.6 on an edge-weighted graph G. The inputs are an indexed list of the edges of G, EdgeListG, a vector of the corresponding edge weights Weights, the number of the starting vertex startV, and the number of vertices n. It is assumed that the vertices are numbered 1,2, ..., n. There will be three outputs: an n row two-column matrix ParentList giving the parent list of the tree, the vector DiscoveryList giving the discovery numbers of the vertices (in order), and a vector DistanceList giving the shortest distances of each vertex to the starting vertex. Do not be concerned about how ties are resolved in this program.

(b) Apply your program to construct a minimum distance tree for the edge-weighted graph of Figure 9.21, with the vertices relabeled as integers in alphabetical order ($a = 1$, $b = 2$, $c = 3$, $d =$

$4, f = 5, \ldots$), and using 1 as the starting vertex. Check that the DistanceList vector produced by your program agrees with the one produced in Example 9.14. Repeat for the edge-weighted graph of Figure 9.29 with starting vertex $a = 1$, and compare the output with that of the solution of Exercise for the Reader 9.11.

(c) Apply your program to the problems of Ordinary Exercise 9(a),(b),(c), and compare with the solutions of that exercise.

(d) Write a program `Dijkstra_TieBreaks` that has exactly the same inputs/outputs and functionality as the program of Part (a), with the only change being that any ties (in choosing a lightest boundary edge) are resolved by choosing the one with the lowest non-tree vertex label, and (if necessary) the lowest tree label.

(e) Run your program of Part (d) on each of the minimum spanning tree problems of Parts (b) and (c) and check that the trees produced agree exactly with those obtained in the solutions.

6. (*Program for Dijkstra's Algorithm 9.6 for Computing Minimum Distances in Edge-Weighted Digraphs*) (a) Write a program with the following syntax:

```
[ParentList DiscoveryList DistanceList]
       = DijkstraDigraph(EdgeListG, Weights, startV, n)
```

that will perform Dijkstra's Algorithm 9.6 on an edge-weighted digraph G. The inputs are an indexed list of the edges of G, `EdgeListG`, a vector of the corresponding edge weights `Weights`, the number of the starting vertex `startV`, and the number of vertices n. It is assumed that the vertices are numbered $1, 2, \ldots, n$. There will be three outputs: an n row two-column matrix `ParentList` giving the parent list of the tree, the vector `DiscoveryList` giving the discovery numbers of the vertices (in order), and a vector `DistanceList` giving the shortest distances of each vertex to the starting vertex. Do not be concerned about how ties are resolved in this program.

(b) Apply your program to construct a minimum distance spanning tree for the edge-weighted digraph of Figure 9.30, with the vertices relabeled as integers in alphabetical order ($a = 1$, $b = 2$, $c = 3$, $d = 4$, $f = 5$, $\ldots$) and using 1 as the starting vertex. Check that the DistanceList vector produced by your program agrees with the one obtained in the solution of Exercise for the Reader 9.12.

(c) Apply your program to the problems of Ordinary Exercise 9(d)(e)(f), and compare with the solutions of that exercise.

7. (*Program for the Depth-First Search Algorithm 9.7 for Graphs*) (a) Write a program with the following syntax:

```
[ParentList DiscoveryList]
         = DepthFirstSearch(EdgeListG, startV, n)
```

that will perform the depth-first search Algorithm 9.7 on a graph G. The inputs are an indexed list of the edges of G, `EdgeListG`, the number of the starting vertex `startV`, and the number of vertices n. It is assumed that the vertices are numbered $1, 2, \ldots, n$. There will be two outputs: an n row two column matrix `ParentList` giving the parent list of the tree, and the vector `DiscoveryList` giving the discovery numbers of the vertices (in order). Do not be concerned about how ties are resolved in this program.

(b) Apply your program to the graph of Figure 9.21, where the edge weights are ignored and with the vertices being relabeled as integers in alphabetical order ($a = 1$, $b = 2$, $c = 3$, $d = 4$, $f = 5, \ldots$), and using 1 as the starting vertex. Compare the DiscoveryList vector produced by your program with the one produced in Example 9.15.

(c) Apply your program to the graph of Figure 9.29 (ignore the edge weights) with starting vertex $a = 1$. Compare the DiscoveryList vector produced by your program with the one produced in the solution of Exercise for the Reader 9.14.

(d) Write a program `DepthFirstSearch_TieBreaks` that has exactly the same inputs/outputs and functionality as the program of Part (a), with the only change being that any ties (in choosing a lightest boundary edge) are resolved by choosing the one with the lowest non-tree vertex label.

(e) Run your program of Part (d) on each of the graphs of Parts (b) and (c) and check that the trees produced agree exactly with those produced in Example 9.15 and the solution of Exercise for the Reader 9.14.

8. (*Program for the Breadth-First Search Algorithm 9.8 for Graphs*) (a) Write a program with the following syntax:

```
[ParentList DiscoveryList]
        = BreadthFirstSearch(EdgeListG, startV, n)
```

that will perform the depth-first search Algorithm 9.8 on a graph *G*. The inputs are an indexed list of the edges of *G*, EdgeListG, the number of the starting vertex startV, and the number of vertices n. It is assumed that the vertices are numbered 1, 2, ..., *n*. There will be two outputs: an *n* row two-column matrix ParentList giving the parent list of the tree, and the vector DiscoveryList giving the discovery numbers of the vertices (in order). Do not be concerned about how ties are resolved in this program.
(b) Apply your program to the graph of Figure 9.21, where the edge weights are ignored and with the vertices relabeled as integers in alphabetical order ($a = 1$, $b = 2$, $c = 3$, $d = 4$, $f = 5$, ...) and using 1 as the starting vertex. Compare the DiscoveryList vector produced by your program with the one produced in Example 9.15.
(c) Apply your program to the graph of Figure 9.29 (ignore the edge weights) with starting vertex $a = 1$. Compare the DiscoveryList vector produced by your program with the one produced in the solution of Exercise for the Reader 9.14.
(d) Write a program BreadthFirstSearch_TieBreaks that has exactly the same inputs/outputs and functionality as the program of Part (a), with the only change being that any ties (in choosing a lightest boundary edge) are resolved by choosing the one with the lowest non-tree vertex label.
(e) Run your program of Part (d) on each of the graphs of Parts (b) and (c) and check that the trees produced agree exactly with those produced in Example 9.15 and the solution of Exercise for the Reader 9.14.

9. (*Modifications of Above Depth-First and Breadth-First Programs for Digraphs*)
Write programs DepthFirstSearchDig, BreadthFirstSearchDig with the same syntax and functionality as the programs DepthFirstSearch, BreadthFirst Search of Computer Exercises 7, 8 for graphs, except that they operate in the digraph setting. Apply your program to the problems of Ordinary Exercise 11(c) and compare the results with the solutions obtained for these problems.

10. (a) (*Random TSP Generator*) Write a program that will randomly generate Euclidean distance TSPs and that has the following syntax: [W, Points] = TSPRand(n,d). Here, the first input n is an integer greater than 1, and the second optional input d is a positive number (default value 10). The program will generate n points randomly distributed in the square $0 \le x, y \le d$. These points will be the vertices of the complete graph for the TSP. The first output variable is the $n \times n$ edge-weight matrix W whose entries give the Euclidean distance between corresponding pairs of points. The second output variable, Points is an $n \times 2$ matrix whose rows give the *x*- and *y*-coordinates of the vertices. Figure 9.43 shows two examples of randomly generated TSP graphs that were generated by such a program.
(b) Do some trial runs with small values of n (3, 4, and 5) and check that the price matrix P indeed gives the correct distances between the points in Points.
Note: Recall the (Euclidean) distance between two points (x_1, y_1) and (x_2, y_2) in the plane is given by $\text{dist} = \sqrt{(x_2 - x_1)^2 + (y_2 - y_1)^2}$. The Euclidean distance satisfies the triangle inequality.

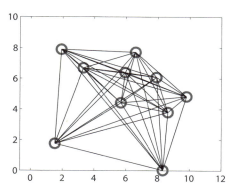 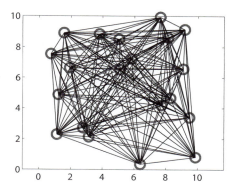

FIGURE 9.43: Two graphs for randomly generated TSPs constructed with the program of Computer Exercise 10; the one on the left has $n = 10$ vertices, and the one on the right has $n = 20$ vertices (both graphs were generated using the default value $d = 10$).

11. (*Programs for the Traveling Salesman Problem*) (a) Write a program with the following syntax:

$$[\texttt{lowprice, Rts}] = \texttt{TSP\_Brute(W)}$$

that will solve the traveling salesman problem on an edge-weighted complete n vertex graph G, using the brute-force method. The only input will be a (symmetric) $n \times n$ matrix W, specifying the edge weights. It is assumed that the vertices are numbered 1, 2, ..., n. There will be two outputs: a number `lowprice` giving the lowest price (or weight) of a Hamilton tour, and an n column matrix `Rts`, each row of which specifies an optimal Hamilton tour (as a permutation of $\{1, 2, ..., n\}$ starting with vertex 1).

(b) Write a program with the following syntax:

$$[\texttt{Tour, weight}] = \texttt{TSPNearNbr(W,start)}$$

that will apply the nearest neighbor heuristic (Algorithm 9.12) to the traveling salesman problem on an edge-weighted complete n vertex graph G. The inputs will be a (symmetric) $n \times n$ matrix W, specifying the edge weights, and a starting vertex number `start`. It is assumed that the vertices are numbered 1, 2, ..., n. There will be two outputs: a length n vector specifying the TSP tour produced by the heuristic, and a number `weight` giving the corresponding weight.

(c) Write a program with the following syntax:

$$[\texttt{Tour, weight}] = \texttt{TSPNearInsertion(W,start)}$$

that will apply the nearest insertion heuristic (Algorithm 9.10A) to the traveling salesman problem on an edge-weighted complete n vertex graph G. The inputs will be a (symmetric) $n \times n$ matrix W, specifying the edge weights, and a starting vertex number `start`. It is assumed that the vertices are numbered 1, 2, ..., n. There will be two outputs: a length n vector specifying the TSP tour produced by the heuristic, and a number `weight` giving the corresponding weight.

(d) Write a program with the following syntax:

$$[\texttt{Tour, weight}] = \texttt{TSPCheapestInsertion(W,start)}$$

that will apply the nearest insertion heuristic (Algorithm 9.10C) to the traveling salesman problem on an edge-weighted complete n vertex graph G. The inputs will be a (symmetric) $n \times n$ matrix W, specifying the edge weights, and a starting vertex number `start`. It is assumed that the vertices are numbered 1, 2, ..., n. There will be two outputs: a length n vector specifying the TSP tour produced by the heuristic, and a number `weight` giving the corresponding weight.

12. (*Experimental Comparisons of TSP Heuristics*)   In this exercise you will be using the program of Computer Exercise 10 to generate TSPs and run some experiments that will compare relative performances of TSP heuristics using the programs of Computer Exercise 11.

(a) Randomly generate 100 five-vertex TSPs (using the default value d = 10 is fine), and apply to each the nearest neighbor heuristic, the nearest insertion heuristic, the cheapest insertion heuristic, and the brute-force method. Create corresponding vectors of the weights of the tours produced as follows:

vec1: Nearest Neighbor Insertion Weight/Furthest Insertion Weight
vec2: Nearest Insertion Weight/Furthest Insertion Weight
vec3: Cheapest Insertion Weight/Furthest Insertion Weight
vec4: Brute-Force Method (optimal weight)

Compare the ratios of the components of each of the first three vectors with the corresponding entries of vec4. Comment on the results of these experiments and their relation to the performance guarantee theory presented in the text and ordinary exercises.

(b) Next, randomly generate 100 five-vertex TSPs (using the default value d = 10 is fine), and apply the above three heuristics (but not the brute-force method) to create corresponding vectors of 100 weights each: vec1, vec2, vec3. Form three new vectors corresponding to the ratios vec1:vec2, vec1:vec3, vec2:vec3 and comment on the relative performance of the three heuristics.

(c) Repeat Part (b) with additional experiments with larger TSPs (e.g., some 100-vertex TSPs). How do the results compare with those of Parts (a) and (b)?

9.3: NETWORK FLOWS

In this section we will be examining problems involving optimizing the flow of a substance or objects through an edge-weighted directed network. Applications are numerous.   In case vertices represent telecommunication or internet routing junctions and the weights on the edges represent capacities (or bandwidths) for data flow, an important problem is to maximize the flow of data from one point to another by sending different parts of it over different paths in the network. If the vertices represent road intersections of major routes in a metropolitan area, the edges represent the routes, and the edge weights represent the vehicle capacities per hour, a natural problem is to find the maximum flow from/to a certain suburb to/from the city during rush hours.   If the network represented is a pipeline network (for, say, water, oil, or sewage) and the edge weights represent maximum fluid flows in the various pipes, the maximal flow problem would be vital to analyze supply or removal issues for the fluid. Anyone who drives through traffic knows that traffic flow is largely dependent on the worst bottlenecks of the routes. We will prove a famous result, known as the *Ford–Fulkerson maximum flow, minimum cut theorem* that formalizes this notion and equates maximum flows in a network with minimal cuts in the network.   Given an origin vertex and a destination vertex in a directed network, a *cut* is simply a set of edges whose removal would cause the network to be disconnected.   Ford and Fulkerson's proof of their groundbreaking theorem is a constructive one that yields an efficient algorithm for determining maximum flows and minimum cuts.   We will present several applications, including the maximum matching problem for bipartite graphs.   In particular, we will use the Ford–Fulkerson theorem to give a proof of a famous result about maximum bipartite matchings known as *Hall's marriage theorem*.

Flow Networks

We proceed now to establish some notation and concepts for network flow problems.

DEFINITION 9.8: A **flow network** consists of an edge-weighted directed graph with a designated **source vertex** s (start) and a designated **sink vertex** t (finish). All other vertices will be called **transit vertices**. We assume that there is at least one path in the network from s to t, i.e., that t is reachable from s. The weights on the edges: $w_{ij} = W(i, j) > 0$ are positive rational numbers called the **edge-flow capacities,** and correspond to the maximum flow (of objects or substance per unit time) along the corresponding directed edge (i, j). An **admissible flow** in this network is a sequence of nonnegative rational $f_{ij} = F(i, j) \geq 0$, the **flow numbers**, indexed with the directed edges of the network that satisfy the following properties:

(i) $f_{ij} \leq w_{ij}$ for all directed edges (i, j) (*respect of edge capacities*), and

(ii) $\sum_k f_{ki} = \sum_\ell f_{i\ell}$ for all $i \neq s, t$ (*conservation of flow, i.e., the flow into vertex i must equal the flow out of vertex i*)

In (ii), the first sum is taken over all k such that (k, i) is a directed edge in the network and represents the **inflow** into vertex i; analogously, in the second sum ℓ runs through all vertices such that (i, ℓ) is a directed edge of the network and represents the **outflow** from vertex i. The **value** of an admissible flow F, denoted by val(F), is the total flow out of s. By the conservation of flow, this flow must also equal the total flow into t (there is nowhere else for the flow to escape to). In summary, for an admissible flow F, we may write:

$$\text{val}(F) = \sum_\ell f_{s\ell} \ (\equiv \text{net flow out of } s) = \sum_k f_{kt} \ (\equiv \text{net flow into } t). \tag{6}$$

The **maximum flow** in a flow network equals the largest possible value of val(F), over all admissible flows F. The main problem for flow networks is the following:

MAXIMUM FLOW PROBLEM: Given a flow network, determine the maximum flow from the source vertex to the sink vertex and find an admissible flow that achieves this maxium flow.

EXAMPLE 9.19: Figure 9.44 shows a simple flow network with six vertices. Figure 9.45 shows an admissible flow F on this network, with the flow number f_{ij} for a directed edge (i, j) listed in parentheses next to the corresponding flow capacity.

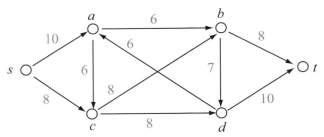

FIGURE 9.44: A flow network with six vertices; the numbers listed next to each directed edge are the edge-flow capacities.

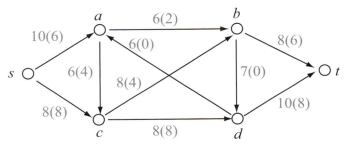

FIGURE 9.45: An admissible flow for the flow network of Figure 9.44; the numbers listed next to each directed edge are the edge-flow capacities; the corresponding flow numbers are shown in parentheses.

The value of the flow F, val(F), is 14, equaling (by (6)) the net flow out of s, or the net flow into t.

Cuts in Flow Networks

In the flow network of Example 9.19, it is clear that the value of any flow cannot exceed the sum of the edge capacities for all edges emanating from s (10 + 8 = 18), and likewise cannot exceed the sum of the edge capacities for all edges going into t. These are just two "bottlenecks" in the network. The following definition will formalize the concept of a "bottleneck," and the main result of this section will show that the maximum flow in any flow network will always equal the minimum over all such "bottlenecks."

DEFINITION 9.9: A **cut** of a flow network is a partition of the vertices $\mathcal{V}$ of the network into two (disjoint) sets: $(\mathcal{V}_1, \mathcal{V}_2)$ such that the source vertex s lies in $\mathcal{V}_1$ and the sink vertex t lies in $\mathcal{V}_2$. The set $\mathcal{V}_1$ is called the **s-side** of the cut, and $\mathcal{V}_2$

is called the *t*-side of the cut. The **capacity** $W(\mathcal{V}_1, \mathcal{V}_2)$ of the cut $(\mathcal{V}_1, \mathcal{V}_2)$ of a flow network is the sum of flow capacities of all the edges that have the tail endpoint in $\mathcal{V}_1$ and the head endpoint in $\mathcal{V}_2$. In symbols:

$$W(\mathcal{V}_1, \mathcal{V}_2) = \sum \{W(i, j) : i \in \mathcal{V}_1, j \in \mathcal{V}_2\}.$$

For any admissible flow $f_{ij} = F(i, j) \geq 0$, or, more generally, for any numerical valued function on the edges, we use the same notation to represent the **total flow** from the vertex set in $\mathcal{V}_1$ to the vertex set $\mathcal{V}_2$:

$$F(\mathcal{V}_1, \mathcal{V}_2) = \sum \{F(i, j) : i \in \mathcal{V}_1, j \in \mathcal{V}_2\}.$$

As with admissible flows, the direction of flow is important for the definition of a cut. We note that, in particular, an edge directed from $\mathcal{V}_2$ to $\mathcal{V}_1$ does not contribute to the capacity $W(\mathcal{V}_1, \mathcal{V}_2)$.[21] Examples of two cuts for the flow network in Figure 9.44 may be constructed by setting:

For the first cut: $(\mathcal{V}_1, \mathcal{V}_2)$: set $\mathcal{V}_1 = \{s, a, c\}$,
for the second cut: $(\mathcal{U}_1, \mathcal{U}_2)$: set $\mathcal{U}_1 = \{s, c\}$.

These two cuts are illustrated in Figure 9.46. Their capacities are computed as follows:

$$W(\mathcal{V}_1, \mathcal{V}_2) = W(a,b) + W(c,b) + W(c,d) = 6 + 8 + 8 = 22,$$
$$W(\mathcal{U}_1, \mathcal{U}_2) = W(s,a) + W(c,b) + W(c,d) = 10 + 8 + 8 = 26.$$

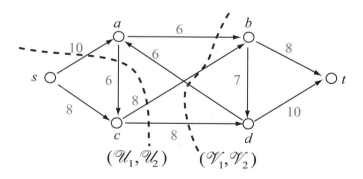

FIGURE 9.46: Illustration of cuts corresponding to $\mathcal{U}_1 = \{s, c\}$ and to $\mathcal{V}_1 = \{s, a, c\}$ for the flow network of Figure 9.44; the numbers listed next to each directed edge are the edge-flow capacities.

It is sometimes helpful to imagine a cut in a flow network to be a river that runs across and cuts off flow through the relevant edges.

---

[21] Note that $\mathcal{V}_2$ is simply the complement $\sim \mathcal{V}_1$ of $\mathcal{V}_1$ with respect to the set of all vertices $\mathcal{V}$.

Notice also that for the flow of Figure 9.45 (flow values in parentheses) if we compute the flows from the s-sides of the cuts to the t-sides:

$$F(\mathcal{V}_1, \mathcal{V}_2) = F(a,b) + F(c,b) + F(c,d) = 2 + 4 + 8 = 14,$$
$$F(\mathcal{U}_1, \mathcal{U}_2) = F(s,a) + F(c,b) + F(c,d) = 6 + 4 + 8 = 18,$$

the flows from the t-sides of the cuts to the s-sides:

$$F(\mathcal{V}_2, \mathcal{V}_1) = F(d,a) = 0,$$
$$F(\mathcal{U}_2, \mathcal{U}_1) = F(a,c) = 4,$$

and then subtract the corresponding quantities, we obtain

$$F(\mathcal{V}_1, \mathcal{V}_2) - F(\mathcal{V}_2, \mathcal{V}_1) = 14 - 0 = 14 = \text{val}(F),$$
$$F(\mathcal{U}_1, \mathcal{U}_2) - F(\mathcal{U}_2, \mathcal{U}_1) = 18 - 4 = 14 = \text{val}(F).$$

These latter equations turn out to be true in general and are a consequence of the conservation of flow. Indeed, it follows that for any admissible flow $f_{ij} = F(i, j)$, the total flow $F(\mathcal{V}_1, \mathcal{V}_2)$ from the s-side of a cut $(\mathcal{V}_1, \mathcal{V}_2)$ to the t-side, less the amount $F(\mathcal{V}_2, \mathcal{V}_1)$ that flows from the t-side back to the s-side, must equal the value $\text{val}(F)$ of the flow. We record this useful fact in the following lemma:

LEMMA 9.13: For any cut $(\mathcal{V}_1, \mathcal{V}_2)$ and flow $f_{ij} = F(i, j)$ in a flow network, we have $\text{val}(F) = F(\mathcal{V}_1, \mathcal{V}_2) - F(\mathcal{V}_2, \mathcal{V}_1)$.

From this lemma, we may immediately deduce the following half of the maximum flow/minimum cut theorem:

COROLLARY 9.14: For any cut $(\mathcal{V}_1, \mathcal{V}_2)$ and flow $f_{ij} = F(i, j)$ in a flow network, we have $\text{val}(F) \leq W(\mathcal{V}_1, \mathcal{V}_2)$. It follows that the maximum flow is at most equal to the minimum capacity of any cut in a flow network. Moreover, this fact gives the following:

Certificate of Optimality: If we find a flow and a cut such that the value of the flow is the capacity of the cut, then the flow is a maximum flow and the cut is a minimum cut.

Proof:

$$
\begin{aligned}
\text{val}(F) = F(\mathcal{V}_1, \mathcal{V}_2) - F(\mathcal{V}_2, \mathcal{V}_1) \qquad & \text{(By Lemma 9.13)} \\
\leq F(\mathcal{V}_1, \mathcal{V}_2) \qquad & \text{(Since } F(\mathcal{V}_2, \mathcal{V}_1) \geq 0) \\
\leq W(\mathcal{V}_1, \mathcal{V}_2). \qquad & \text{(Since } F(i, j) \leq W(i, j) \text{ for any edge } (i, j))
\end{aligned}
$$

The last two sentences follow from the first inequality since the former is valid for all flows and cuts, including any maximum flow and minimum cut. ☐

The Maximum Flow/Minimum Cut Theorem

It turns out that the inequality of the last sentence of the preceding corollary is actually an equality, i.e., the maximum flow in any flow network equals the minimum capacity of the cuts. This remarkable fact was discovered by the American mathematicians Lester R. Ford (1886–1975) and David R. Fulkerson (1924–1976).[22] We state this result as the following theorem:

THEOREM 9.15: (*The Ford–Fulkerson Maximum Flow/Mimimal Cut Theorem*) In any flow network, the value of a maximum flow coincides with the capacity of a minimum cut.

This theorem has a tremendous variety of applications and theoretical usefulness. We will examine some of these shortly, but we first present the original proof, which is a constructive one that yields a very efficient algorithm for finding maximal flows in any flow network.

(Constructive) Proof: Suppose that we are given a flow network with designated start and end vertices s, t, and edge capacities $w_{ij} = W(i, j)$. We first assume that the edge capacities are positive integers. After developing the constructive algorithm for this case, we will then show how it can be easily adapted to deal with the general case.

We will be working with *integral flows*, i.e., admissible flows $f_{ij} = F(i, j)$ whose values are nonnegative integers. The construction will start with any admissible flow and proceed through a sequence of iterations, each of which produces an admissible flow with a larger value than the preceding flow, until this is no longer possible. The flow of the final interation will thus be a maximum flow.

Initial Step: We initialize the construction with any admissible integral flow $f_{ij} = F(i, j)$; for example, we could always start with the *zero flow* $f_{ij} = 0$. The current flow is initially set to be the initial flow.

Iterative Step: We next show how the iterative construction will work to bring us from the current flow either to a next flow with larger value, or to a certification that the current flow is a maximum flow. The key idea will be to use the current flow to introduce a corresponding **augmented digraph** that is defined as follows: The vertex set of the augmented digraph is just the vertex set $\mathcal{V}$ of the original flow network. The directed edges of the augmented digraph are of the following two types:

---

[22] Ford and Fulkerson were industrial mathematicians working for the RAND Corp. They published their work in a seminal paper [FoFu-56], and later wrote a textbook on the subject of network flows [FoFu-62].

(i)   (*Forward Edges*) If (i, j) is an edge of the current flow network for which $w_{ij} > f_{ij}$, we include it as an edge of the augmented digraph and assign its weight to be the unused (slack) edge capacity $A(i, j) = w_{ij} - f_{ij} > 0$, i.e., the available excess capacity for flow on this edge.

(ii)   (*Backward Edges*) If (i, j) is an edge of the current flow network for which $f_{ij} > 0$, we include the oppositely directed edge (j, i) as an edge of the augmented digraph and assign its weight to be $A(j, i) = f_{ij} > 0$, i.e., this is the amount that the current flow can be reduced on the edge (i, j).

We further stipulate that the augmented digraph will contain no backward edges directed into the source vertex s, and no backward edges directed from the sink vertex t. In summary, if f sends a positive amount of flow along an edge (i, j), then the augmented digraph contains the edge (i, j); if f sends any flow (even zero flow) that is less than the edge capacity $W(i, j)$, then the augmented digraph contains the backward edge (j, i). Figure 9.47 shows the augmented digraph for the network flow of Figure 9.45.

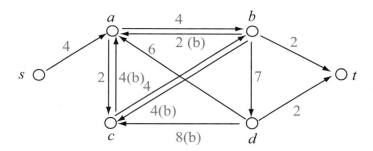

FIGURE 9.47: The augmented digraph for the network flow of Figure 9.45 with edge weights $A(i, j) = a_{ij}$. The weights of the backward edges have been indiced by "(b)."

What makes an augmented digraph useful is that the corresponding flow on the original network can be augmented (increased) if, and only if there is a path in the augmented digraph from s to t, and such a path can be easily used to increase the flow in the corresponding original network flow. Before describing the general principle, let us point out a simple example using the augmented digraph of Figure 9.47. There are several paths from s to t, and each one will give rise to a different way to augment the corresponding network flow of Figure 9.45. For example, the path $< s, a, b, t >$ on the augmented digraph of Figure 9.47 contains only forward edges whose weights indicate the amount by which the flows on the corresponding edges of Figure 9.45 can be increased. Since the minimum of the four amounts is $\min(4, 4, 2) = 2$, if we increase the flow on each of these three edges of Figure 9.45 by 2: $f_{sa} = 6 \to 8, f_{ab} = 2 \to 4, f_{bt} = 6 \to 8$, and all other edge flows are left the same, then it is easy to check the this new flow is admissible and has a value 2 units larger than that of Figure 9.45. We next consider the following path in Figure 9.47 $< s, a, b, d, c, b, t >$, which contains exactly one backward edge $< d, c >$. Since the minimum of the weights of the edges of this path (in the augmented digraph of Figure 9.47) is 2, the corresponding flow of Figure 9.45 can be augmented by 2 by adding 2 units to the weights of all edges corresponding to forward edges in this path: $f_{sa} = 6 \rightarrow 8, f_{ab} = 2 \rightarrow 4, \quad f_{bd} = 0 \rightarrow 2, f_{cb} = 4 \rightarrow 6, \quad f_{bt} = 6 \rightarrow 8$, by decreasing by 2 the flow on the reverse edge corresponding to the backward edge: $f_{cd} = 8 \rightarrow 6$, and leaving the flow on all other edges the same. The reader can easily check that this modified flow is admissible and has a value of two greater than that of Figure 9.45. We next enunciate the underlying important general principle and prove its correctness in the following Claim:

Claim: If there exists a path in the augmented digraph from s to t, we will be able to increase the value of the current flow; otherwise the current flow is a maximum flow.

Proof of Claim: If $P = < v_0 = s, v_1, v_2, \cdots, v_k = t >$ is a path in the augmented digraph, which we may assume to be simple with no repeated vertices (see Proposition 8.5), we will use it to define a flow $\tilde{F}(i, j) = \tilde{f}_{ij}$ of greater value than the current flow $F(i, j) = f_{ij}$ as follows:

$$\tilde{f}_{ij} = \begin{cases} f_{ij}, & \text{if neither } (i, j) \text{ nor } (j, i) \text{ is an edge of the augmented digraph path } P \\ f_{ij} + 1, & \text{if } (i, j) \text{ is a forward edge of the augmented digraph path } P \\ f_{ij} - 1, & \text{if } (j, i) \text{ is a backward edge of the augmented digraph path } P \end{cases}$$

Since the current flow values and the capacities are assumed to be integers, and since the current flow value on each forward edge of P is less than the capacity, the new flow value $f_{ij} + 1$ will still be less than or equal the capacity. Also, on each backward edge (j, i) of P, since (i, j) is an edge of the flow network on which the current flow is positive, the new flow value $f_{ij} - 1$ will still be nonnegative. In order to show that the new flow is admissible, it remains to show that the conservation constraints are satisfied. The only vertices that need checking are the internal vertices of the path P (the rest have exactly the same inflows and outflows as the current flow). So we consider an internal vertex v_i $(0 < i < k)$ of P. We consider separate cases depending on whether the augmented digraph edges incident to this vertex are forward or backward edges:

Case 1: $(v_{i-1}, v_i), (v_i, v_{i+1})$ *are both forward edges:* In this case the two edges are also edges of the flow network on which the current flow is less than capacity, and on which the new flow has increased the flow by one unit. Since one edge goes into v_i and the other goes out, and all other edges incident to v_i have the same flow values as the current flow, the conservation of flow has been maintained.

Case 2: $(v_{i-1}, v_i), (v_i, v_{i+1})$ *are both backward/non-forward edges:* In this case the oppositely directed edges $(v_i, v_{i-1}), (v_{i+1}, v_i)$ are edges of the flow network on which the current flow is positive, and on which the new flow has decreased the flow by one unit. Since one edge goes into v_i and the other goes out, and all other edges incident to v_i have the same flow values as the current flow, the conservation of flow has been maintained.

Case 3: *One of* $(v_{i-1}, v_i), (v_i, v_{i+1})$ *is a forward edge and the other is a backward/non-forward edge:* These two edges correspond to two edges in the flow network that are either both entering v_i or both directed away from v_i. Since the new flow is one greater on one edge and one less on the other, and since the new flow values are the same on all other edges incident to v_i, it follows that the conservation of flow has been maintained.

Thus we have shown that the new flow $\tilde{F}(i, j) = \tilde{f}_{ij}$ is admissible. Also since the first edge of P must be a forward edge of the augmented digraph, it follows that $val(\tilde{F}) = val(F) + 1$.

Next, suppose that there is no path in the augmented digraph from s to t. We will show that the current flow F must be a maximum flow. This will be accomplished by finding a cut $(\mathcal{V}_1, \mathcal{V}_2)$ of the flow network that has capacity val(F). It would then follow from Corollary 9.14 that F is a maximum flow. We define $\mathcal{V}_1$ to consist of s and all vertices that are reachable from s in the augmented digraph, and $\mathcal{V}_2 = \mathcal{V}_1 \sim \mathcal{V}_1$. Since there is no path in the augmented digraph from s to t, we know that $t \in \mathcal{V}_2$. We consider any pair of vertices $i \in \mathcal{V}_1, j \in \mathcal{V}_2$. Since there is a path $P(s,i)$ in the augmented digraph from s to i, it follows that the edge (i, j) is not an edge of the augmented digraph since if it were it could be juxtaposed with $P(s, i)$ to create a path from s to j, and this would contradict the fact that $j \in \mathcal{V}_2$. It follows that if (i, j) is an edge of the flow network, then $f_{ij} = w_{ij}$, and if the opposite edge (j, i) is an edge of the flow network, then $f_{ji} = 0$. It now follows that

$$F(\mathcal{V}_1, \mathcal{V}_2) = W(\mathcal{V}_1, \mathcal{V}_2), \ F(\mathcal{V}_2, \mathcal{V}_1) = 0.$$

Hence from Lemma 9.13, we deduce that the value of the flow val(F) $= F(\mathcal{V}_1, \mathcal{V}_2) - F(\mathcal{V}_2, \mathcal{V}_1)$ is the capacity of the cut $W(\mathcal{V}_1, \mathcal{V}_2)$. The claim is thus proved.

Iterative Step (cont.): At each iteration, we form the augmented digraph and check to see whether there is a path from s to t. If there is, the construction in the claim shows how to obtain a larger flow $\tilde{F}$ from it and the current flow F. In this case we update the current flow to be $\tilde{F}$, and repeat the iterative step. If there is no path from s to t in the augmented digraph, the claim tells us that the current flow is maximum, so we may output the current flow as the maximum flow and we may stop the iterative steps. This completes our constructive proof of the maximum flow-minimum cut theorem in the case of positive integer capacities. In case the capacities are positive rational numbers, we simply need to factor out a least common denominator, and then solve the related problem with the numerators (integers). At the end, we divide the flow numbers by the least common denominator to obtain a solution to the original problem. □

We will provide a more practical formulation of the constructive algorithm in the above proof, but we first make one simple observation. Although the algorithm will always determine a maximal flow, if we increase the flow by only one unit at each iteration, it may take a lot of time for the algorithm to execute (think about a situation where the maximal flow is over a million!). Instead of always sending one additional unit of flow at each iteration, we would do better to send the largest amount of flow possible on the path. The following example, which was informally previewed earlier, illustrates an implementation of this idea.

EXAMPLE 9.20: Figure 9.48(a) shows a path (in thick edges) from s to t in the augmented digraph of Figure 9.47. Recall that the edge weights on the augmented digraph represent excess capacities corresponding to the admissible network flow of Figure 9.45 from which the augmented digraph was constructed. Since the smallest excess capacity of the path edges is 2, that means that the flow values on the corresponding network edges of Figure 9.45 can be modified by 2 units. Since the affected edges are all forward edges, the flows are increased by two on these edges (if there were any backward edges, the flows would have been decreased by 2 on them). The resulting flow is shown in Figure 9.48(b); it has a value 2 greater than the flow of Figure 9.45.

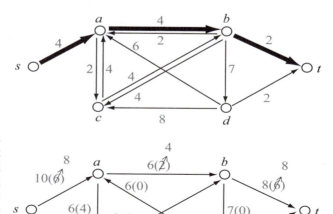

FIGURE 9.48: (a) (top) A path from s to t (thick edges) in the augmented digraph for the network flow of Figure 9.45. (b) (bottom) A resulting improved flow from the flow of Figure 9.45, using excess capacities from the path of (a).

EXERCISE FOR THE READER 9.18: Taking as the "current flow" the flow of Figure 9.48(b), construct the corresponding augmented digraph, and then apply the method of Example 9.20 to create an improved flow.   Show that this improved flow is a maximum flow for the flow network.

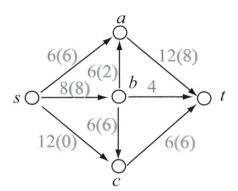

EXERCISE FOR THE READER 9.19: For the five-vertex flow network and flow illustrated in Figure 9.49, construct the corresponding augmented digraph, and then apply the method of Example 9.20 to create an improved flow.

FIGURE 9.49: A flow on a five vertex-flow network for Exercise for the Reader 9.19.

The Ford–Fulkerson Maximum Flow/Minimum Algorithm

Although it is indeed constructive, the proof of Theorem 9.15 requires some additional details in order to render a precise algorithm.   Ford and Fulkerson developed such an algorithm, but the efficient formulation that we present below is due to Jack Edmonds and Richard Karp [EdKa-72]. The virtue of this approach is that it will quickly and effectively determine whether the current flow may be augmented, and if it can, it will find a flow augmenting path in the augmenting digraph, along with the maximum amount of flow that can be sent through it. The approach will not consider the entire augmenting digraph.[23]   Starting with the source vertex, vertices are first labeled and then scanned for neighbors that are reachable in the augmented digraph. Except for the source, each labeled vertex will have a unique parent vertex in this scheme (resulting from the scan). Each label is a list of three items. The source s has a special label: (source, $+$, ∞), any other labeled vertex v will have a label of the form $(u, \pm, f)$, where u is the parent vertex that was scanned and led to v, the sign $\pm$ is determined by whether the augmented digraph edge that led from u to v in the scanning process was a forward edge (meaning that the current flow may be increased on this edge, in which case a "$+$" is used) or if the edge that leads from u to v was a backward edge (meaning that the current flow may be decreased on the opposite edge in the flow network,

---

[23] In the language of trees, the implemtation will construct a rooted tree (with root vertex being the source of the flow network), the edges of which are taken from the augmented digraph of the current flow. One might at first think that it would be prudent to search for a flow augmenting path that would give the greatest additional amount of flow, if the current flow is not maximum. But this would entail a great deal of additional work at each iteration, and the resulting algorithm would be much less efficient than the one we will present.

in which case a "–" is used). The third component f of the label denotes the minimum of all flow increase/decrease amounts on the unique sequence of scanned augmented digraph edges leading back from u to the source. This process gets iterated until either the sink vertex is labeled, or no new vertices can be labeled. In the former case, a flow augmenting path can be determined, while in the latter the current flow will be maximum. We will assume that the flow capacities are integers since most applications fall in this setting; the proof of Theorem 9.15 shows the simple modification needed to deal with the setting in which flow capacities are (non-integer) rational numbers.

ALGORITHM 9.13: The Ford–Fulkerson Algorithm for the Maximal Flow Problem (Implementation by Edmonds–Karp)

INPUT: A flow network with positive integer flow capacities $w_{ij} = W(i, j)$, designated source vertex s, and designated sink vertex t.
OUTPUT: A maximum admissible flow $f_{ij} = F(i, j)$, from the source s to the sink t.

Step 1: Initialize the current flow to be any admissible (integer-valued) flow $F(i, j) = f_{ij}$ from the source to the sink. The zero flow may be used.

Step 2: (*Initialize Label and Scan Vectors*)
Label the source vertex s with the label (source, +, ∞).
Initialize a vector UnlabeledVertices to be all other vertices than s.
Initialize another vector UnscannedVertices = [s]. (These will be the vertices that have been labeled, but not yet scanned.) When we "**pop**" a vertex from UnscannedVertices, we always take and remove the leftmost one; when we "**append**" a vertex to UnscannedVertices, we always append it on the right. Thus, the earlier a vertex gets labeled, the earlier it will be scanned.

Step 3: (*Iterative Step: Scan and Label*)
WHILE UnscannedVertices $\neq \varnothing$ AND sink vertex $t \in$ UnlabeledVertices
    Pop a vertex v from UnscannedVertices, and Read Label(v) = $(u, \pm, f)$.
    For each vertex w in UnlabeledVertices
    IF (v,w) is an edge for which $f_{vw} < w_{vw}$
        Set Label(w) = $(v, +, m)$, where $m = \min(w_{vw} - f_{vw}, f)$
        Remove w from UnlabeledVertices, Append w to UnscannedVertices
    ELSE IF (w,v) is an edge for which $f_{wv} > 0$
        Set Label(w) = $(v, -, m)$, where $m = \min(f_{wv}, f)$
        Remove w from UnlabeledVertices, Append w to UnscannedVertices
    ELSE
        Do not label w
    END IF
END WHILE

Step 4: (*Iterative Step: Augmenting the Flow*) If the sink t is unlabeled, then the current flow is a maximum flow, exit the algorithm.

If the sink t is labeled, initialize vertex $= t$, and proceed as follows to increment the flow value by f, where f is the third component of Label(t), by backtracking on edges to the source.

WHILE vertex $\neq s$
    IF Label(vertex) $= (u, +, g)$
        Increment $F(u,$ vertex$) = F(u,$ vertex$) + f$, Update vertex $= u$
    ELSE IF Label(vertex) $= (u, -, g)$
        Decrement $F($vertex$, u) = F($vertex$, u) - f$, Update vertex $= u$
    END IF
END WHILE

Step 5: Remove all labels except for Label(s), reset the vectors UnlabeledVertices to be all vertices other than s, and UnscannedVertices $= [s]$.   Return to Step 3 (Step 4 will have exited the algorithm if a maximum flow was found).

NOTE: The order in which the vertices are scanned in this algorithm is not important; any convenient default ordering scheme may be used. Also, the way the program is set up, the scanning process first checks for a forward edge in the augmented digraph and uses it if one exists.   A slightly more effective implementation would check the existence of both forward and backward edges, and, if there are edges of both types, choose the one with the greatest amount of excess capacity.

The following example illustrates a by-hand implementation of Algorithm 9.13; the algorithm is presented in a fashion that makes it readily codable into a computer program; the computer exercises section will provide further details.

EXAMPLE 9.21:   Consider the flow network of Figure 9.50. Taking the initial flow to be the zero flow, apply the Ford–Fulkerson Algorithm 9.13 to construct a maximal flow.   In the scanning Step 3, process the unlabeled vertices in alphabetical order.

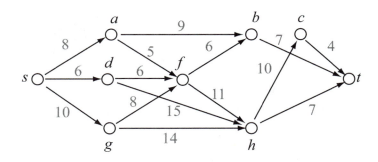

FIGURE 9.50: The flow network of Example 9.21.

SOLUTION: As we proceed through the steps of the algorithm, we include graphical summaries of the current flow statuses. As was done earlier, we denote current positive flows along edges by numbers in parentheses next to the flow capacities of Figure 9.50 (which do not change). *Step 1: (Initializing the Flow)* was done in the statement of the example.

Step 2: (Initialize Label and Scan Vectors)
Label(s) = (source, +, ∞).
UnlabeledVertices = $[a, b, c, d, f, g, h, t]$.
UnscannedVertices = $[s]$.

Step 3: Scan and Label: Iteration #1:
In this first iteration, we will give a detailed account of the scanning/labeling process; in subsequent iterations, we will only list the labels.

We pop vertex s from UnscannedVertices.
By examination of Figure 9.51, we introduce the following labels:
Label(a) = (s, +, 8), Label(d) = (s, +, 6), Label(g) = (s, +, 10).

The corresponding updated vectors are:
UnlabeledVertices = $[b, c, f, h, t]$, UnscannedVertices = $[a, d, g]$.

Next, we pop vertex a from UnscannedVertices.
By examination of Figure 9.51, we introduce the following labels:
Label(b) = (a, +, 8), Label(f) = (a, +, 5).

The corresponding updated vectors are:
UnlabeledVertices = $[c, h, t]$, UnscannedVertices = $[d, g, b, f]$.

Next, we pop vertex d from UnscannedVertices.
By examination of Figure 9.51, we introduce the following label:
Label(h) = (d, +, 6).

The corresponding updated vectors are:
UnlabeledVertices = $[c, t]$, UnscannedVertices = $[g, b, f, h]$.

Next, we pop vertex g from UnscannedVertices.
By examination of Figure 9.51, this vertex will lead to no new labels, so we only update UnscannedVertices = $[b, f, h]$.

Next, we pop vertex b from UnscannedVertices.
By examination of Figure 9.51, we introduce the following label:
Label(t) = (b, +, 7).

Since the sink t has been labeled, we move on to Step 4.

Step 4: Augmenting the Flow: Iteration #1:

Backtracking using the labels of Step 3 from the sink to the source, and since the edges are all flow increasing edges with final flow value = 7, we increment the current flow by seven on the following edges: $F(b,t) = 7$, $F(a,b) = 7$, $F(s,a) = 7$.

The new current flow after this first iteration is shown in Figure 9.51; this will be the current flow as we enter into iteration #2.

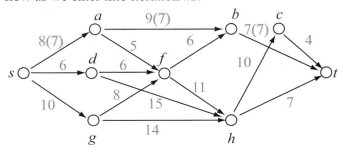

FIGURE 9.51: The flow in the Ford–Fulkerson algorithm of Example 9.21, as we enter into iteration #2; the value of the flow is 13.

Step 5: We remove all labels except Label(s), and reset the vectors:
UnlabeledVertices $= [a, b, c, d, f, g, h, t]$.
UnscannedVertices $= [s]$.

Step 3: Scan and Label: Iteration #2:
Here are the new labels that are created (see Figure 9.51):

Label(a) = (s, +, 1), Label(d) = (s, +, 6), Label(g) = (s, +, 10);
Label(b) = (a, +, 1), Label(f) = (a, +, 1);
Label(h) = (d, +, 6);
Label(c) = (h, +, 6), Label(t) = (h, +, 6).

Step 4: Augmenting the Flow: Iteration #2:
Backtracking using the labels of Step 3 from the sink to the source, and since the edges are all flow increasing edges with final flow value = 6, we increment the current flow by six on the following edges: $F(h,t) = 6$, $F(d,h) = 6$, $F(s,d) = 6$.

The new current flow after this second iteration is shown in Figure 9.52; this will be the current flow as we enter into iteration #3; Step 5 is repeated as before.

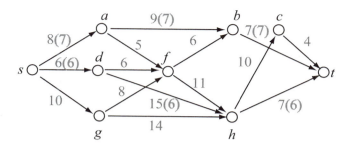

FIGURE 9.52: The flow in the Ford–Fulkerson algorithm of Example 9.21, as we enter into iteration #3; the value of the flow is 13.

Step 3: Scan and Label: Iteration #3:
Here are the new labels that are created (see Figure 9.52):

Label(a) = (s, +, 1), Label(g) = (s, +, 10);
Label(b) = (a, +, 1), Label(f) = (a, +, 1);
Label(h) = (g, +, 10);
Label(c) = (h, +, 10); Label(d) = (h, −, 6)
Label(t) = (h, +, 1).

Step 4: Augmenting the Flow: Iteration #3: $F(h,t) = 7$, $F(g,h) = 1$, $F(s,g) = 1$.

The remaining two iterations of Step 3 are shown next to the current flows entering into the iteration. Thus new current flow after this third iteration is shown in Figure 9.53; and to the right of it the vertex labels of the fourth iteration are indicated.

Iteration #4:

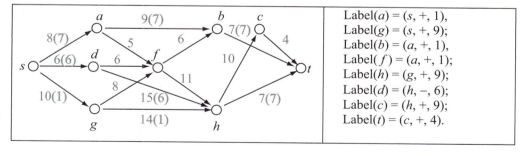

Label(a) = (s, +, 1),
Label(g) = (s, +, 9);
Label(b) = (a, +, 1),
Label(f) = (a, +, 1);
Label(h) = (g, +, 9);
Label(d) = (h, −, 6);
Label(c) = (h, +, 9);
Label(t) = (c, +, 4).

FIGURE 9.53: The flow (on the left) in the Ford–Fulkerson algorithm of Example 9.21, as we enter into iteration #4; the value of the flow is 14. The vertex labels created in iteration #4 are shown on the right.

Step 4: Augmenting the Flow: Iteration #4:
$F(c,t) = 4$, $F(h,c) = 4$, $F(g,h) = 5, F(s,g) = 5$.

~~~~~~

Iteration #5: Although it is quite clear from Figure 9.54 that the current flow is maximum (since the flow on all edges leading to *t* are at capacity), for completeness we provide the details of the final iteration.

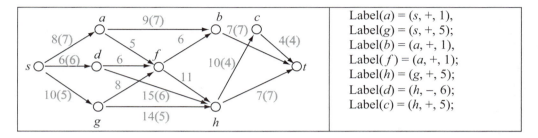

Label(a) = (s, +, 1),
Label(g) = (s, +, 5);
Label(b) = (a, +, 1),
Label(f) = (a, +, 1);
Label(h) = (g, +, 5);
Label(d) = (h, −, 6);
Label(c) = (h, +, 5);

FIGURE 9.54: The flow (on the left) in the Ford–Fulkerson algorithm of Example 9.21, as we enter into iteration #5; the value of the flow is 18. The vertex labels created in iteration #5 are shown on the right.

Since *t* did not get labeled in this iteration, it follows that the flow of Figure 9.54 is a maximum flow, and the algorithm terminates.

NOTE: (*Minimum Cuts from Algorithm 9.13*) After Algorithm 9.13 is applied to construct a maximal flow of a flow network, a corresponding minimum cut (which is guaranteed to exist by Theorem 9.15) can easily be obtained as follows: In the final iteration of the vertex labeling step (Step 3), set $\mathcal{V}_1$ = {all labeled vertices}, and $\mathcal{V}_2 = \sim\mathcal{V}_1$ = {all unlabeled vertices}. Then it follows from the proof of Theorem 9.15, that $(\mathcal{V}_1, \mathcal{V}_2)$ is a minimum cut. In Example 9.21, the only unlabeled vertex is the sink *t*, and $\mathcal{V}_2 = \{t\}$ corresponds to the minimum cut that was already mentioned.

NOTE: (*Performance/Complexity of Algorithm 9.13*) In their paper [EdKa-72], Edmonds and Karp showed that Algorithm 9.13 will converge in at most $(V^3 - V)/4$ iterations for any network graph with *V* vertices. Observe that this estimate does not depend on the flow capacities. It follows that when applied to a flow network with *V* vertices and *E* edges, the worst-case complexity runtime of Algorithm 9.13 is $O(EV^3)$. Compare this with the brute-force method of searching for a minimum cut: The number of cuts in a network with *V* vertices is simply the number of subsets of the $\mathcal{V} \sim \{s\}$, which is 2^{V-1}. Also, for each cut there is a lot of computing to do, on average amounting to locating and adding up capacities of a large number of edges. We may thus roughly estimate the brute-force minimum cut algorithm to have exponential complexity $O(2^V \cdot E)$.

EXERCISE FOR THE READER 9.20: Apply the Ford–Fulkerson algorithm to construct a maximal flow of the flow network of Figure 9.55. Use the initial flow indicated in the figure, and in the scanning step 3, process the unlabeled vertices in alphabetical order.

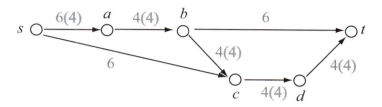

FIGURE 9.55: The flow network with initial flow for Exercise for the Reader 9.20.

EXERCISE FOR THE READER 9.21: Apply the Ford–Fulkerson algorithm to construct a maximal flow of the flow network of Figure 9.56. Use the zero flow initially, and in the scanning step 3, process the unlabeled vertices in alphabetical order. Determine also a minimum cut.

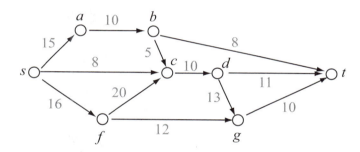

FIGURE 9.56: The flow network for Exercise for the Reader 9.21.

Applications of Maximum Flows

Applications of finding maximal flows in networks are surprisingly diverse. The maximum flow/minimal cut theorem (Theorem 9.15) gives rise to a particularly rich and diverse collection of applications, and the efficient Algorithm 9.13 provides effective and practical solution tools for specific instances. We will illustrate some applications. In particular, we will show how Algorithm 9.13 can be applied to the problem finding maximum matchings in a bipartite graph. This problem was first introduced in Section 8.1, in the context application of matching individuals to jobs (Example 8.13). In the context of this problem, the Ford–Fulkerson Theorem 9.15 can be used to prove an interesting result concerning bipartite matchings known as *Hall's marriage theorem*.

EXAMPLE 9.22: (*Trucking Logistics*) Suppose a company has a fleet of 10 delivery trucks T1, T2, ..., T10, that need to transport some crates to various destinations. There are 12 different types of crates C1, C2, ..., C12 with a quantity of 5 crates of each type. The company wants each truck to carry no two crates of the same type. The crates are all of the same size and the capacity of the first 4 trucks is 10 crates while the capacity of the remaining 6 smaller trucks is 4 crates. Represent this problem as a maximal network flow problem and determine the maximum number of crates that can be shipped in a single shipment with this fleet.

SOLUTION: With larger networks, complete graphical representations tend to become too clustered to be of much use. Figure 9.57 shows only the pattern of the edges in the network.

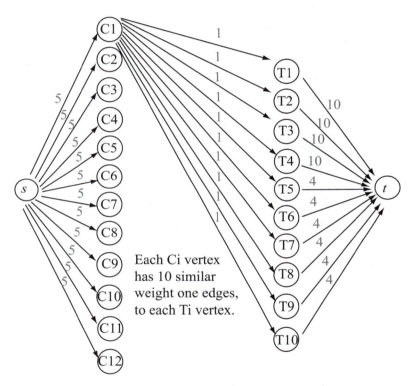

FIGURE 9.57: The flow network for the trucking problem of Example 9.22. There are too many edges to make a complete geometric illustration of the network practical.

There are $2+12+10 = 24$ vertices and $12+12 \cdot 10+10 = 142$ edges in this network. The size of the problem is too large for a "by hand" solution using Algorithm 9.13, but not for a computer assisted solution. The natural pattern of the edges makes it quite a simple matter to store the network in computer format. The computer exercises will guide the reader through such a solution that will show it is possible to ship all 60 containers and provide a specific shipping arrangement (via the maximum flow numbers).

Our next example showcases a well-known problem from operations research. Specific examples tend to be too large to solve without a computer, so we will simply state the general problem and show how the maximum flow/minimum cut theorem sheds some very useful information.

EXAMPLE 9.23: (*The Supply and Demand Problem*) Suppose that we have a network of suppliers $S = \{s_1, s_2, \cdots, s_k\}$ and consumers $C = \{c_1, c_2, \cdots, c_\ell\}$. For example, the suppliers could be companies or warehouses, and the consumers could be households or stores, respectively. Each period (e.g., day, week, month, etc.) supplier s_i would like to send out $\text{supp}(s_i)$ units of a certain commodity (e.g, any good, service, or utility), and consumer c_j desires to consume $\text{dem}(c_j)$ units of this commodity. Moreover, for each pair of supplier s_i and consumer c_j there is a capacity $\text{cap}(s_i, c_j)$ of how much of the commodity s_i can send to c_j each period. The **supply and demand problem** asks whether it is possible to satisfy the desires of all of the suppliers and consumers.

In order for the supply and demand problem to have a solution, it is obviously necessary that the total supply equals the total demand:

$$\sum_{i=1}^{k} \text{supp}(s_i) = \sum_{i=1}^{\ell} \text{dem}(c_j). \tag{7}$$

Assuming that (7) holds, we will use Theorem 9.15 to do much better and obtain a necessary and sufficient condition for the supply and demand problem to have a solution. We introduce an appropriate flow network: The vertex set will be $S \cup C \cup \{s, t\}$. The (directed) edges will be of three types:

(i) For each $s_i \in S$, there is an edge (s, s_i) with capacity $W(s, s_i) = \text{supp}(s_i)$.

(ii) For each $c_j \in C$, there is an edge (c_j, t) with capacity $W(c_j, t) = \text{dem}(c_j)$.

(iii) For each pair $s_i \in S$, $c_j \in C$, there is an edge (s_i, c_j) with capacity $W(s_i, c_j) = \text{cap}(s_i, c_j)$.

The diagram would look similar to that of Figure 9.57. It is now clear that the supply and demand problem has a solution if, and only if, the maximum flow of the above network has value given by the number in (7). (Note that (7) is obviously an upper bound for the maximum flow since it is the value of the cut $(\{s\}, \sim\{s\})$, and of the cut $(\sim\{t\}, \{t\})$.) The following result provides a necessary and sufficient condition for the supply and demand problem to have a solution:

PROPOSITION 9.16: (*Solvability of the Supply and Demand Problem*) Suppose that we have a supply and demand problem whose data satisfies (7). Then the problem has a solution in which all supplies will be used and demands will be met if, and only if, the following condition holds in the associated flow network:

For every subset of vertices $T \subset S \cup C$, we have

$$\sum_{s_i \in S \cap T} \text{supp}(s_i) + W(\sim T, T) \ge \sum_{c_j \in C \cap T} \text{dem}(c_j). \tag{8}$$

In the vocabulary of the problem, condition (8) states that for any set of vertices T, the total supply inside T, plus the total capacity for sending flow from outside T into T must be at least as large as the demand within T.

Proof: For a subset of vertices $T \subset S \cup C$, we consider the cut $(\mathcal{V}_1, \mathcal{V}_2)$ specified by $\mathcal{V}_1 = \{s\} \cup \sim T$, $\mathcal{V}_2 = \{t\} \cup T$. Note that all cuts are expressible in this fashion. The edges directed from $\mathcal{V}_1$ to $\mathcal{V}_2$ are of three types: edges of type (i) (s, s_i) where $s_i \in T$, edges of type (ii) (c_j, t) where $c_j \in \sim T$, and edges of type (iii) (s_i, c_j) where $s_i \in \sim T, c_j \in T$. It follows that the capacity of this cut is given by:

$$W(\mathcal{V}_1, \mathcal{V}_2) = \sum_{s_i \in S \cap T} \text{supp}(s_i) + \sum_{c_j \in C \cap \sim T} \text{dem}(c_j) + W(\sim T, T).$$

Since (using (7)), in order for the problem to have a solution, it is necessary that these cut capacities have a minimum equal to

$$\sum_{i=1}^{k} \text{supp}(s_i) = \sum_{i=1}^{\ell} \text{dem}(c_j) = \sum_{c_j \in C \cap \sim T} \text{dem}(c_j) + \sum_{c_j \in C \cap T} \text{dem}(c_j),$$

it follows that (8) must be necessary. Conversely, if we take $T = S \cup C$, then inequality (8) becomes equation (7), which means that we will have a minimal cut whose capacity shows the supply and demand problem has a solution. $\square$

The next exercise for the reader provides a problem that is easily generalized to a method for counting the number of edge-disjoint paths from one specified vertex to another in any graph or digraph.

EXERCISE FOR THE READER 9.22:   (*Counting Edge Disjoint Paths in Graphs/Digraphs: Networks with Undirected Edges*) Figure 9.58 shows a simple graph that represents a network of military supply routes. The military commander is tasked with determining the maximum possible number of edge disjoint routes within the network that start at vertex s (the military base) and end at vertex t (the enemy target location). The other vertices in the network denote secure transit points (for fuel, supplies, rest, etc.), and the edges denote possible routes between the vertices. These routes are subject to being compromised by the enemy, which is why the commander wishes to send a maximum number of convoys on different paths that share no common edges. Model this problem as a maximum flow problem.

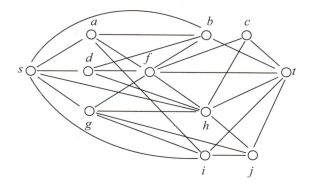

FIGURE 9.58: The military supply route graph for Exercise for the Reader 9.22.

Maximum Matchings in Bipartite Graphs

Matchings in graphs were first informally introduced in Example 8.13(b), where they were motivated by the job assignment problem. Since we will be discussing them in greater detail here, we begin with a formal definition:

DEFINITION 9.10: A **matching** in a simple graph G is any set of edges $\mathcal{M}$ $\subseteq E(G)$ such that no two edges in $\mathcal{M}$ are incident to a common vertex. A **maximum matching** is a matching of maximum possible cardinality.

EXAMPLE 9.24: (*Maximal Matchings Versus Maximum Matchings*) From earlier definitions, the reader may recall that the adjective "maximal" indicates that an object has a specified property and that no strictly larger object that contains the first object can still have the property. Figure 9.59(a) shows an example of the matching $\mathcal{M} = \{\{a,d\},\{b,f\},\{c,g\}\}$, which is obviously a maximum matching, since all vertices are covered by the endpoints of its edges. Figure 9.59(b) shows an example of another matching $\mathcal{N} = \{\{a,b\},\{\{f,g\}\}$, which has only two edges so is not maximum matching. Notice, however that if any new edge is added to $\mathcal{N}$, the result will no longer be a matching. This means that $\mathcal{N}$ is a maximal matching.

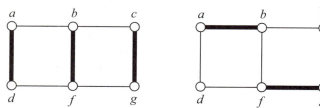

FIGURE 9.59: (a) (left) A maximum matching (indicated by the three heavy edges) in a simple graph. (b) (right) A maximal matching that is not a maximum matching.

EXERCISE FOR THE READER 9.23:   For the bipartite graph of Figure 9.60, do
each of the following:
(a) Find all maximum matchings.
(b) Is there a maximal matching that is not a maximum matching?   Either provide
an example or explain why one does not exist.

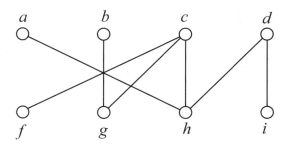

FIGURE 9.60: The bipartite graph of Exercise for the Reader 9.23.

Finding maximum matchings is a problem that arises in numerous applications,
and fortunately there are efficient algorithms for this task. We will show that the
maximum flow algorithm can be used to find maximum matchings in the very
important case of a bipartite graph. We will demonstrate this idea in the following
example, and then we will use the maximum flow/minimum cut theorem to obtain
an interesting and not at all obvious theoretical result about such maximum
matchings.   Both the example and the theorem will be motivated by the following
problem:

DEFINITION 9.11: (*The Marriage Problem*): Suppose that we have a set of n
single women $w_1, w_n, \cdots, w_n$ and ℓ single men $m_1, m_2, \cdots, m_\ell$.   Each woman w_i
has a certain set of men $M_i \subseteq \{m_1, m_2, \cdots, m_\ell\}$ such that for each $m_j \in M_i$, the
woman w_i and the man m_j are completely compatible and suitable for marriage.
The marriage problem asks to marry off as many of the women as possible (each
man or woman being only allowed to marry at most one person—no polygamy).
This problem is easily modeled as a maximum matching problem by the bipartite
graph whose vertices are $\{w_1, w_n, \cdots, w_n\} \cup \{m_1, m_2, \cdots, m_\ell\}$, and that has an edge
$\{w_i, m_j\}$ for each woman–man pair for which $m_j \in M_i$.

EXAMPLE 9.25:   Figure 9.61 is a bipartite graph representing a marriage
problem involving 10 women (left) and 10 men (right). Even with this small size,
the solution, i.e., the maximum number of women who can be married off to a
compatible husband, is not so trivially done "by hand."

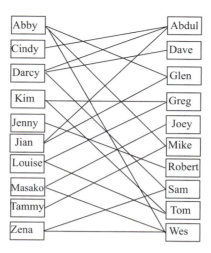

FIGURE 9.61: The marriage problem of Example 9.25.

Any marriage problem, or more generally any matching problem on a bipartite graph, can be easily solved using the maximum flow algorithm. To form the corresponding flow network, we make all of the edges in the bipartite graph to be directed edges in the flow network (in marriage problems, they can be directed from the women to the men vertices). We also introduce a source vertex s, and an edge from s to each woman vertex, and a sink vertex t, along with one edge from each man vertex to t. The capacities of all edges are set to 1. The flow network associated to the marriage graph of Figure 9.61 is shown in Figure 9.62.

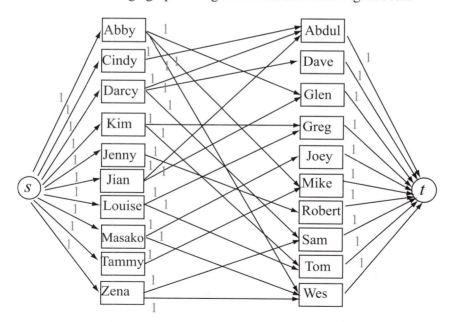

FIGURE 9.62: The flow network associated with the marriage graph of Figure 9.61.

It is now quite clear that a maximum flow in the flow network for a marriage problem (or, more generally for any bipartite matching problem) is equivalent to a maximum matching: we simply take the matching edges to be those edges between women and men on which a flow of 1 is being sent. Thus we may apply the Ford–Fulkerson Algorithm 9.13 to solve any bipartite matching problem. Notice that each flow augmentation could increase the flow by only one unit (since no edges have greater capacity). Thus, a bipartite matching problem involving m vertices may take up to floor($m/2$) iterations of the algorithm to solve.   The computer exercises at the end of this section will guide the reader through solving such problems of moderate sizes (including the marriage problem of the above example).   We give here an exercise for the reader involving a smaller marriage problem that is suitable for hand computation.

EXERCISE FOR THE READER 9.24: Use the Ford–Fulkerson Algorithm 9.13 to solve the marriage problem of Figure 9.63.   Use the zero flow initially.

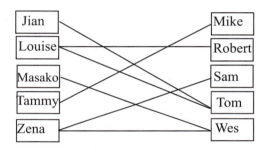

FIGURE 9.63: The marriage problem of Exercise for the Reader 9.24.

Hall's Marriage Theorem

We have demonstrated how any bipartite maximum matching problem can be recast as a maximum flow problem and thus efficiently solved using the Ford–Fulkerson maximal flow algorithm. We will next show how the maximum flow-minimum cut theorem can shed additional insight on such problems.   We will use Theorem 9.15 to prove a theorem about bipartite matching originally proved in 1935 by Philip Hall using a very different approach.   Hall's theorem was motivated by and is formulated in terms of the marriage problem. In any marriage problem, one obvious necessary condition in order that it be possible for all of the women to be married off is that for any set of k women $W \subseteq \{w_1, w_n, \cdots, w_n\}$, the set of all men that these k women collectively consider to be suitable husbands must have at least k men in it, i.e., $\left| \bigcup_{w_i \in W} M_i \right| \geq k$. It is not at all obvious, and thus quite surprising that this condition is also sufficient. This is the essence of Hall's marriage theorem.   In order to state it in its most general form, it will be convenient to introduce the following terminology:

DEFINITION 9.12: Suppose we have a bipartite graph G with vertex bipartion $V(G) = X \cup Y$. A matching $\mathcal{M}$ on G is called **X-saturating** if every vertex in X is an endpoint of an edge in $\mathcal{M}$, i.e., $|\mathcal{M}| = |X|$.

Note that an X-saturated matching on a bipartite graph is clearly a maximum matching.

THEOREM 9.17: (*Hall's Marriage Theorem*) Suppose that G is a bipartite graph with vertex bipartion $V(G) = X \cup Y$. There exists an X-saturating matching on G if, and only if for every subset $S \subseteq X$, the collective set of all neighbors of vertices in X must be at least as large as X. If we denote the set of all neighbors of vertices in X by $N(S)$, then this condition may be written as $|N(S)| \geq |S|$.

EXERCISE FOR THE READER 9.25: Rephrase Hall's marriage theorem in the language of the job assignment problem of Example 8.13(b).

Proof of Theorem 9.17: To make the proof easier to read, we will view G as a marriage problem graph, where X is the set of women, and Y is the set of men. We consider the associated flow network. We will make a slight modification to this network that will help to weed out cuts that are not minimal: we will reset the capacity of all edges that join a woman to a man to have capacity equal to $|X| + 1$. Notice that any admissible flow in this modified network would also be admissible in the original, and thus it is equivalent to work on this modified flow network. This observation follows from the simple fact that each woman vertex has a maximum inflow of one, so the amount of flow on any edge joining a woman to a man can take only two possible values: 0 or 1. It suffices to show that the capacity of any cut in this network is at least $|X|$. Indeed, by Theorem 9.15, from this it would follow that the maximum flow of this network is at least $|X|$. We consider any cut $(\mathcal{V}_1, \mathcal{V}_2)$ of the flow network where, as usual, $\mathcal{V}_1$ denotes the s-side of the cut. Let $W = \mathcal{V}_1 \cap X$, i.e, the women in $\mathcal{V}_1$. As in the statement of the theorem, the set $M = N(W)$ is then the set of all men collectively accepted as suitable by the women in W.

Now, if any man in M was in $\mathcal{V}_2$, then the capacity of the cut would be at least $|X| + 1$, so we may assume that all men in M lie in the s-side of the cut. Next, suppose that y is a man not in M, i.e., $y \in Y \sim M$. Since none of the women in W find y to be a suitable candidate for a husband, it follows that there are no edges from any vertex in $\mathcal{V}_1$ to y, so the capacity of $(\mathcal{V}_1, \mathcal{V}_2)$ will be smallest if all such men lie in $\mathcal{V}_2$ (if y belongs to $\mathcal{V}_1$, then the cut $(\mathcal{V}_1, \mathcal{V}_2)$ would contain the edge from y to t). In other words, we have shown that if $(\mathcal{V}_1, \mathcal{V}_2)$ is a minimum cut, then we may assume that for some set of women $W \subseteq X$, we may write:

$$\mathcal{V}_1 = \{s, W, N(W)\}, \text{ and } \mathcal{V}_2 = \{t, X \sim W, Y \sim N(W)\}.$$

Since this cut consists of edges from s to women in $X \sim W$ and from men in $N(W)$ to t, it follows that the capacity of the cut is given by:

$$W(\mathscr{V}_1, \mathscr{V}_2) = |X \sim W| + |N(W)| = |X| - |W| + |N(W)|.$$

But by the assumption of the theorem, we know that $|N(W)| \geq |W|$, and it thus follows from the above equation that $W(\mathscr{V}_1, \mathscr{V}_2) \geq |X|$, as was needed to show. $\square$

EXERCISES 9.3:

1. Figure 9.64(a)(b) shows two flow networks, where the flow capacities are the numbers not in parentheses above each edge. Indicate whether the numbers in parentheses determine an admissible flow for the network. If the flow is admissible, give the value of the flow; if it is not, explain which constraint(s) are violated.

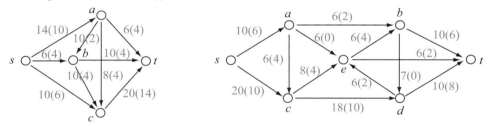

FIGURE 9.64: Two flow networks (a) (left) (b) (right) with flow capacities being indicated by the first numbers above each edge. The second set of numbers (in parentheses) are flow values.

2. Figure 9.65(a)(b) shows two flow networks, where the flow capacities are the numbers not in parentheses above each edge. Indicate whether the numbers in parentheses determine an admissible flow for the network. If the flow is admissible, give the value of the flow; if it is not, explain which constraint(s) are violated.

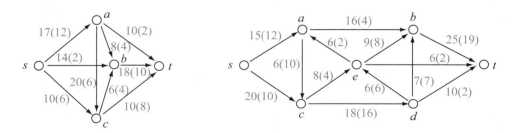

FIGURE 9.65: Two flow networks (a) (left) (b) (right) with flow capacities being indicated by the first numbers above each edge. The second set of numbers (in parentheses) are flow values.

3. (a) Determine the capacities of each of the following cuts of the flow network of Figure 9.64(a). (Ignore the numbers in parentheses above each edge.): $(\{s\}, \{a,b,c,t\})$, $(\{s,a\}, \{b,c,t\})$, $(\{s,a,b,c\}, \{t\})$.

(b) Determine the capacities of each of the following cuts of the flow network of Figure 9.64(b). (Ignore the numbers in parentheses above each edge.): $(\{s\}, \sim \{s\})$, $(\{s,a,e,d\}, \{b,c,d,t\})$, $(\{s,e\}, \sim \{s,e\})$.

4. (a) Determine the capacities of each of the following cuts of the flow network of Figure 9.65(a). (Ignore the numbers in parentheses above each edge.): $(\{s\}, \{a,b,c,t\})$, $(\{s,b\}, \{a,c,t\})$, $(\{s,a,b,c\}, \{t\})$.

(b) Determine the capacities of each of the following cuts of the flow network of Figure 9.65(b). (Ignore the numbers in parentheses above each edge.): $(\sim \{s\}, \{t\})$, $(\{s,b,e,d\}, \{a,c,d,t\})$, $(\sim \{c,t\}, \sim \{c,t\})$.

5. (a) Use trial and error to determine a maximum flow of the flow network of Figure 9.64(a). Find a corresponding minimum cut, and use the certificate of optimality (from Corollary 9.14) to verify.

(b) Repeat Part (a) for the flow network of Figure 9.64(b).

6. (a) Use trial and error to determine a maximum flow of the flow network of Figure 9.65(a). Find a corresponding minimum cut, and use the certificate of optimality (from Corollary 9.14) to verify.

(b) Repeat Part (a) for the flow network of Figure 9.65(b).

7. Use the Ford–Fulkerson Algorithm 9.13 to construct maximum flows for the following networks using the indicated initial flows. Also, using the method described in the text, use the result of the final iteration of the algorithm to provide a corresponding minimum cut, and verify that the capacity of your cut equals the value of your flow. In the scanning step of the algorithm, process the unlabeled vertices in alphabetical order. Show the labels and flow augmentations in each iteration. In Figure 9.64, ignore the numbers in parentheses above each edge.

(a) The flow network of Figure 9.64(a); using the initial flow specified by $f_{sc} = f_{ct} = 10$.

(b) The flow network of Figure 9.64(a); using the initial flow specified by $f_{sb} = f_{bt} = 4$.

(c) The flow network of Figure 9.64(a); using the zero flow for the initial flow.

(d) The flow network of Figure 9.64(b); using the initial flow specified by $f_{sa} = 10$, $f_{ae} = f_{et} = f_{ab} = f_{bt} = 5$.

(e) The flow network of Figure 9.64(b); using the initial flow specified by $f_{sc} = f_{ce} = f_{et} = 5$.

(f) The flow network of Figure 9.64(b); using the zero flow for the initial flow.

8. Use the Ford–Fulkerson Algorithm 9.13 to construct maximum flows for the following networks using the indicated initial flows. Also, using the method described in the text, use the result of the final iteration of the algorithm to provide a corresponding minimum cut, and verify that the capacity of your cut equals the value of your flow. In the scanning step of the algorithm, process the unlabeled vertices in alphabetical order. Show the labels and flow augmentations in each iteration. In Figure 9.65, ignore the numbers in parentheses above each edge.

(a) The flow network of Figure 9.65(a); using the initial flow specified by $f_{sb} = f_{bt} = 2$.

(b) The flow network of Figure 9.65(a); using the initial flow specified by $f_{sa} = f_{at} = 5$.

(c) The flow network of Figure 9.65(a); using the zero flow for the initial flow.

(d) The flow network of Figure 9.65(b); using the initial flow specified by $f_{sa} = f_{ab} = f_{bt} = 15$.

(e) The flow network of Figure 9.65(b); using the initial flow specified by $f_{sc} = f_{ce} = f_{et} = 6$.

(f) The flow network of Figure 9.65(b); using the zero flow for the initial flow.

9. Apply the Ford–Fulkerson Algorithm 9.13 to write down the vertex labels for just the first three iterations along with the corresponding flow values for following networks using the indicated initial flows.

(a) The flow network of Figure 9.66(a); using the initial flow specified by $f_{sa} = f_{at} = 10$.

(b) The flow network of Figure 9.66(a); using the initial flow specified by $f_{sg} = f_{gd} = f_{dt} = 5$.

(c) The flow network of Figure 9.66(a); using the zero flow for the initial flow.

(d) The flow network of Figure 9.66(b); using the initial flow specified by $f_{sa} = f_{at} = 4$.

(e) The flow network of Figure 9.66(b); using the initial flow specified by $f_{sc} = f_{cb} = f_{bt} = 3$.

(f) The flow network of Figure 9.66(b); using the zero flow for the initial flow.

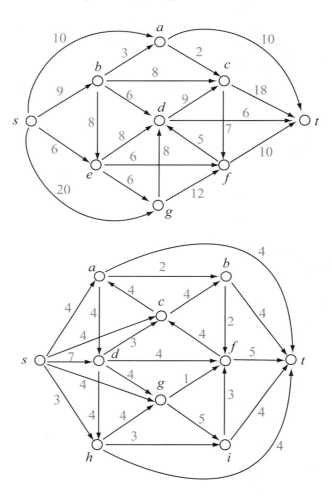

FIGURE 9.66: Two flow networks (a) (top) (b) (bottom) for Exercise 9 with flow capacities being indicated by the numbers above each edge.

10. Apply the Ford–Fulkerson Algorithm 9.13 to write down the vertex labels for <u>just the first three iterations</u> along with the corresponding flow values for following networks using the indicated initial flows.

(a) The flow network of Figure 9.67(a); using the initial flow specified by $f_{sa} = f_{at} = 9$.

(b) The flow network of Figure 9.67(a); using the initial flow specified by $f_{sd} = f_{df} = f_{ft} = 6$.

(c) The flow network of Figure 9.67(a); using the zero flow for the initial flow.

(d) The flow network of Figure 9.67(b); using the initial flow specified by $f_{sc} = f_{cb} = f_{bt} = 4$.

(e) The flow network of Figure 9.67(b); using the initial flow specified by $f_{sj} = f_{jk} = f_{kt} = 3$.

(f) The flow network of Figure 9.67(b); using the zero flow for the initial flow.

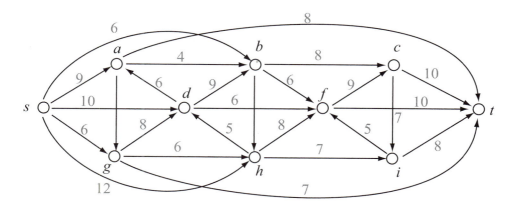

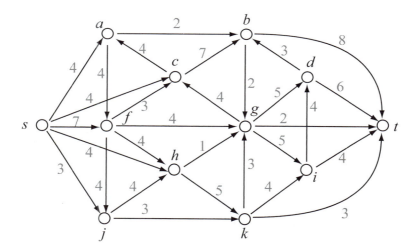

FIGURE 9.67: Two flow networks (a) (top) (b) (bottom) for Exercise 9 with flow capacities being indicated by the numbers above each edge.

11. In all of the flow networks described in this section, there were never any multi-edges. Would the theory and Algorithm 9.13 still work for networks with the presence of multi-edges? Would the use of multi-edges increase the scope of applications? Explain your answers.

12. Suppose that we needed to solve a maximum flow problem involving multiple sources and sinks. Explain how the Ford–Fulkerson maximum flow Algorithm 9.13 could be adapted to apply to such networks.
    Suggestion: Introduce a super-source vertex with edges directed to each of the individual source vertices whose capacities equal the sum of the capacities of all outgoing edges from the individual sources. A super-sink vertex should be similarly introduced.

13. (a) Suppose that we needed to solve a maximum flow problem in which some or all of the transit vertices had capacity constraints, i.e., each restricted transit vertex has a limit on how

much flow can pass through it. Explain how such a network may be modeled by an ordinary flow network to which the Ford–Fulkerson maximum flow Algorithm 9.13 may be applied.

(b) Apply the method of Part (a) to solve the maximum flow problem of Example 9.21 where each transit vertex is restricted to at most nine units of flow going through it; use the zero flow for the initial flow.

(c) Apply the method of Part (a) to solve the maximum flow problem of Exercise 7(e) where each transit vertex is restricted to at most 12 units of flow going through it.

Exercises 14–16 are related to an application to machine shop scheduling, which is described in the following note and example and is due to Ahuja, Magnanti, and Orlin [AhMaOr-93].

NOTE: (*Machine Shop Scheduling*) We will describe an application of maximum flows to the following problem:

MACHINE SHOP SCHEDULING PROBLEM: Suppose that we have a set of J jobs that need to get completed in a certain machine/job shop. These J jobs are represented by three sequences:

The *release times*: $\{r_1, r_2, \cdots, r_J\}$ giving the times that each job is ready to be released for processing.

The *due times*: $\{d_1, d_2, \cdots, d_J\}$ giving the times that each job is due for completion.

The *processing times*: $\{p_1, p_2, \cdots, p_J\}$ giving the times that each job requires for completion.

The obvious necessary condition $d_j \geq r_j + p_j$ is assumed to hold for each job j. The machine shop is assumed to have M equivalent machines, which at any given time can work on only one job. Job j can be sent to a machine m only after its release time r_j and before its due time d_j. It is allowed to switch the processing of a job j from one machine to another at any time (and we assume such transitions take a negligible amount of time), but at any time, job j can be processed by at most one machine.   A solution to the machine shop scheduling problem is any time-dependent assignment of the jobs to the machines so that all of the above constraints are satisfied. Note that the problem is not to complete all jobs as quickly as possible, but rather to see if it is possible to complete all of them by their required due times. The following small example will illustrate how machine shop scheduling problems can be modeled by maximum flow problems, and thus solved by Algorithm 9.13.

EXAMPLE 9.26:   Suppose that we wish to solve the machine shop scheduling problem with $M = 2$ machines, and whose jobs are described by the following data:

j = job number	1	2	3	4	5
r_j = release time	0	1	4	0	2
d_j = due time	4	5	8	6	8
p_j = processing time	3	2	3	2	5

Model this problem as a maximum flow problem.

SOLUTION: One key ingredient in the modeling process is to separate the timeline into discrete intervals during which no release times or due times occur. This can be achieved by sorting the times in the set $\{r_j\} \cup \{d_j\}$ = $\{0, 1, 2, 4, 5, 6, 8\}$. In this case, we have the following six disjoint time intervals: $T(0,1)$, $T(1,2)$, $T(2,4)$, $T(4,5)$, $T(5,6)$, $T(6,8)$. We introduce an appropriate flow network as follows: The vertices will be of the following types:

(i) A source vertex s, and a sink vertex t.
(ii) For each job j, a job vertex Job(j).
(iii) A vertex for each of the disjoint time intervals $T(a,b)$ constructed above.
The corresponding network edges are of the following types:

(i) For each job j, an edge from the source s to Job(j) with capacity p_j.

(ii) For each of the disjoint time interval vertices $T(a,b)$, an edge from it to the sink t with capacity $M(b-a)$ corresponding to the amount of machine time available during the time interval.

(iii) For each job j, and each time interval vertex $T(a,b)$, with $a \le r_j, d_j \le b$, an edge from Job(j) to $T(a,b)$ with capacity $b-a$, equaling the amount of time that can be allocated to job j by (all of) the machines during the time interval.

Figure 9.68 illustrates this flow network for Example 9.26.

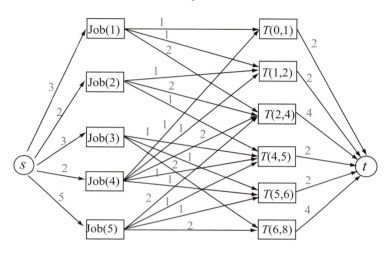

FIGURE 9.68: The flow networks for the machine shop scheduling problem of Example 9.26.

14. (*Machine Shop Scheduling*) (a) Show that a machine shop scheduling problem has a solution if, and only if the maximal flow of the associated network as described in the above note and example has value $\sum_{j=1}^{J} p_j$.

(b) Explain why, in order for a machine shop scheduling problem to have a solution, it is necessary that $\sum_{j=1}^{J} p_j \le M \cdot \max_{1 \le j \le J} d_j$. Then give an example in which this condition holds but for which there is no solution.

15. (*Machine Shop Scheduling*) Apply Algorithm 9.13 to find a maximum flow for the network flow problem of the machine shop scheduling problem of Example 9.25 (Figure 9.68). Does the problem have a solution? If not, add an additional machine, modify the following flow network accordingly, and apply Algorithm 9.13 once again to check for a solution.
NOTE: If doing this exercise by hand, just do the first three iterations, if using a computer (as explained in the computer exercises material), complete the algorithm.

16. (*Machine Shop Scheduling*) Suppose that we wish to solve the machine shop scheduling problem whose jobs are described by the following data:

j = job number	1	2	3	4	5
r_j = release time	1	8	3	12	5
d_j = due time	10	20	12	20	20
p_j = processing time	8	9	7	6	13

(a) According to the necessary condition of Part (b) of Exercise 14, what is the minimum number M of machines that would be required for this problem to have a solution? Draw the associated flow network.

(b) Apply Algorithm 9.13 to find the maximum flow for the flow network that you constructed in Part (a). Does the corresponding machine shop scheduling problem have a solution? If not, add an additional machine, modify the following flow network accordingly, and apply Algorithm 9.13 once again to check for a solution.

NOTE: If doing this exercise by hand, just do the first three iterations, if using a computer (as explained in the computer exercises material), complete the algorithm.

17. (*Counting Disjoint Paths*) Apply the Ford–Fulkerson Algorithm 9.13 to count the maximum number of disjoint paths from vertex s to vertex t in Figure 9.58; see Exercise for the Reader 9.22. Start with the zero flow. Give a maximum set of such disjoint paths.

18. For each of the two graphs G and H below, do the following:
(a) Find a maximum matching.
(b) Find a maximal matching that is not a maximum matching, if one exists.

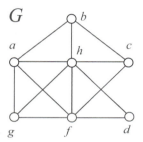 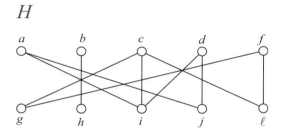

19. For each of the two graphs G and H below, do the following:
(a) Find a maximum matching.
(b) Find a maximal matching that is not a maximum matching, if one exists.

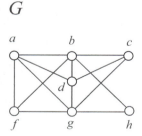 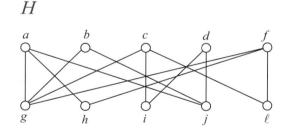

20. Apply Algorithm 9.13 to solve the marriage problem represented in Figure 9.69(a). Start with the zero flow.

21. Apply Algorithm 9.13 to solve the marriage problem represented in Figure 9.69(b). Start with the zero flow.

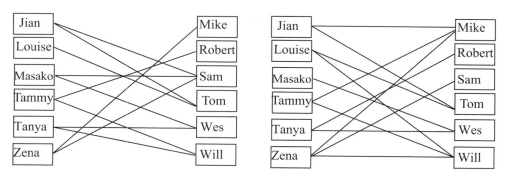

FIGURE 9.69: Two small marriage problems (a) (left), (b) (right).

22. Apply Algorithm 9.13 to solve the job assignment problem of Example 8.13(b) (which is illustrated in Figure 8.29). Start with the zero flow.

23. A certain solitaire card game asks the player to lay all 52 cards face up in a 4 row, 13 column array and then select one card from each column in such a way that the 13 cards selected will include all 13 denominations (A, 2, 3, ..., 10, J, Q, K).

Show that it is always possible to win this game.
Suggestion: Use Hall's marriage theorem.

24. Suppose that we have an instance of the marriage problem with an equal number of men and women. Under which of the following additional assumptions will it be possible to marry off everyone to a compatible mate? For each, either give a justification why it is always possible, or provide a counterexample.
(a) Each woman is compatible with exactly one man, and each man is compatible with exactly one woman.
(b) Each woman is compatible with exactly two men, and each man is compatible with exactly two women.
(c) Each woman is compatible with exactly a men, and each man is compatible with exactly a women, where a is a positive integer (at most n).
(d) Each woman is compatible with at least a men, and each man is compatible with at least a women, where a is a positive integer (at most n).
Suggestion: It is recommended that the reader think through and attempt each part, in order, before reading on. Here is a suggestion for a general approach for Part (c): In order to apply Hall's marriage theorem, consider a set W of women. By assumption there are $a|W|$ edges going from these women to the men. Since each man has degree exactly a, these edges cannot be incident to less than a men.

25. For each of the following requirements, either provide an example of a flow network that satisfies it, or explain why one cannot exist.
(a) A flow network with a unique maximum flow.
(b) A flow network with exactly two different maximum flows.
(c) A flow network with exactly three different maximum flows.

COMPUTER EXERCISES 9.3

NOTE: Throughout the following computer exercises, we shall adhere to the following scheme for representing flow networks:
It is assumed that the vertices are numbered from 1 to $n + 2$, where vertex 1 is the source (s), vertex $n + 2$ is the sink (t), and the n vertex numbers in-between represent the transit vertices. Flow networks will be unambiguously represented by their edge list and edge capacities. The edge list, as described in Section 8.1, is a two-column matrix whose rows represent the directed edges. We denote this matrix as EdgeList. In general, different graphs may have the same edge lists (since isolatated vertices do not appear as endpoints of edges), but in a flow network all vertices appear as edge endpoints, so there is no ambiguity. The edge capacities will be represented by a column vector EdgeWgts. Any flow will be represented by a column vector Flow of the same size as EdgeWgts that specifies the values of a flow in the network. We assume any such flow vector satisfies the edge constraints (i.e., Flow $\leq$ EdgeWgts), but not necessarily the conservation of flow constraints. Since the latter takes more work to check, the first computer exercise below with ask the reader to write a program to check whether an inputted flow vector (that satisfies the edge capacity constraints) satisfies the conservation of flow constraints (and thus is an admissible flow). Whenever any of these variables appears in any of the following computer exercises, their meaning will not be repeated.

1. (*Program for the Linear Search Algorithm*) (a) Write a program with the following syntax:

   Flag, VertexFlows = FlowChecker(EdgeList,EdgeWgts,Flow)

   whose purpose is to check whether an inputted flow is admissible for an inputted flow network. The first output Flag will equal 1 if the inputted flow Flow is admissible; otherwise it will be zero. The second output VertexFlows will be a 2-column matrix whose ith row entries will be the total inflow into vertex i, and the total outflow from vertex i, respectively, resulting from the inputted flow Flow. Thus, Flag = 1, if, and only if, the first and second column entries of VertexFlows are the same.
   (b) Apply your program to verify the admissible flow of Example 9.19.
   (c) Apply your program to the flows of Ordinary Exercise 1.
   (d) If an admissible flow is inputted, which row(s) of the output VertexFlows will give the value of the flow?

2. (*Program for the Cut Capacity Calculator*) (a) Write a program with the following syntax:

   cutCapacity = CutCapacity (EdgeList,EdgeWgts,sSideCut)

   whose purpose is to compute the capacity of an inputted flow network. The input sSideCut is a vector of the vertex numbers on the s-side of a cut. In the notation of the text, this would correspond to the set $\mathcal{V}_1$ of a cut ($\mathcal{V}_1$, $\mathcal{V}_2$). The output cutCapacity is the corresponding capacity of the cut; in the notation of the text this is $W(\mathcal{V}_1, \mathcal{V}_2)$.
   (b) Apply your program of Part (a) to compute the capacities for each of the cuts of Ordinary Exercise 3.

3. (*Brute-Force Program for Finding All Minimum Cuts in a Flow Network*) (a) Write a program with the following syntax:

   MinCuts, minCap = MinCutFinderBrute(EdgeList,EdgeWgts)

   whose purpose is to determine all cuts of minimum capacity of an inputted flow network. The output MinCuts is a vector each row of which gives the numbers of the vertices that constitute the s-side of a minimum cut. Since the number of such vertices in different minimum cuts may vary, the number of columns of this matrix is determined by a minimum cut with the greatest number of s-side vertices. The second output is simply the capacity of (all of) these minimum cuts.

(b) Apply your program of Part (a) to the flow network of Example 9.21. Check that the minimum cut obtained in the solution to that example is one of those found by the program, and that the second output `minCap` agrees with the maximum flow value that was found in that example, as guaranteed by the maximum flow-minimum cut Theorem 9.15.

(c) Apply your program of Part (a) to the flow networks of Figures 9.64(a) and (b). Check that the second output `minCap` agrees with the maximum flow value that was found in the solution of Ordinary Exercise 7.

(c) Apply your program of Part (a) to the flow networks of Figures 9.65(a) and (b). Check that the second output `minCap` agrees with the maximum flow value that was found in the solution of Ordinary Exercise 9.

4. (*Program for the Generating Augmented Digraphs Corresponding to Flows in Flow Networks*)
    (a) Write a program with the following syntax:

    ```
    AEdgeList, AEdgeCaps = AugDigraph(EdgeList,EdgeWgts, FlowInit)
    ```

    whose purpose is to determine the augmented digraph corresponding to an inputted flow network with initial flow specified by the third input variable `FlowInit`. The two output variables specify the edge list and corresponding edge weights of the augmented digraph, which was defined in the proof of Theorem 9.15. The directions of the edges (forward or backward) will be determined by storing weights (in the output variable) to be the actual weights for forward edges, and the negatives of the actual weights in the case of backward edges.

    (b) Apply your program of Part (a) to the flow network and initial flow of Figure 9.45 (with the vertex numbering being done in alphabetical order). Check that the outputted augmented digraph agrees with the one shown in Figure 9.47.

5. (*Program for the Edmonds–Karp Adaptation of the Ford–Fulkerson Maximum Flow Algorithm 9.13*) (a) Write a program with the following syntax:

    ```
    maxFlow, Flow = FordFulkerson(EdgeList,EdgeWgts, FlowInit)
    ```

    whose purpose is to determine a maximum flow of an inputted flow network. The third input, `FlowInit`, is an optional initial flow that will be used; its default is the zero flow. The first output `maxFlow` is a positive integer giving the value of a maximum flow, and the second output `Flow` is a column vector specifying the values of a corresponding maximum flow on the edges (corresponding to the input matrix `EdgeList`). The structure of the program should follow Algorithm 9.13. For definiteness, set the program up so that in Step 3 of Algorithm 9.13, unlabeled vertices are scanned in numerical order.

    (b) Apply your program of Part (a) to the flow network of Example 9.21 (with the vertex numbering being done in alphabetical order). Check that the outputted flow (and its capacity) agree exactly with the one obtained in the solution to the example.

    (c) Apply your program of Part (a) each part of Ordinary Exercise 7, and check the outputs with the answers obtained in that exercise.

    (d) Apply your program of Part (a) each part of Ordinary Exercise 9, and check the outputs with the answers obtained in that exercise.

6. (*Program for the Edmonds–Karp Adaptation of the Ford–Fulkerson Maximum Flow Algorithm 9.13—Minimum Cut Finder*) (a) Write a program with the following syntax:

    ```
    maxFlow, MinCut = FordFulkersonCut(EdgeList,EdgeWgts, FlowInit)
    ```

    whose purpose is to determine a minimum cut, along with the value of a maximum flow of an inputted flow network. The third input, `FlowInit`, is an optional initial flow that will be used; its default is the zero flow. The first output `maxFlow` is a positive integer giving the value of a maximum flow, and the second output `MinCut` is a vector specifying the numbers of the vertices that constitute the *s*-side of a minimum cut. The structure of the program should follow Algorithm 9.13 (see the note following the statement of the algorithm for the minimum cut finding adaptation). For definiteness, set the program up so that in Step 3 of Algorithm 9.13, unlabeled vertices are scanned in numerical order.

    (b) Apply your program of Part (a) to the flow network of Example 9.21. Check that the first output `maxFlow` agrees with the capacity of the cut found by the program and with the

maximum flow determined in the solution of Example 9.21.
(c) Apply your program of Part (a) each part of Ordinary Exercise 7, and check the `maxFlow` outputs with the answers obtained in that exercise.
(d) Apply your program of Part (a) each part of Ordinary Exercise 9, and check the `maxFlow` outputs with the answers obtained in that exercise.

7. (*Program for the Supply and Demand Problem*) (a) Write a program with the following syntax:

    ```
    maxFlow, FlowMatrix = SupplyAndDemandSolver(Supp, Dem, Cap)
    ```

whose purpose is to solve, as best as possible, the supply and demand problem of Example 9.23. The first input, `Supp`, is a vector of the supply amounts, the second input, `Dem`, is a vector of the demand amounts (both assumed to be positive integer vectors). The third input is a corresponding matrix of nonnegative integers whose (i, j) entry is the maximum amount that supplier #i may send to demander/consumer #j. The first output `maxFlow` is a positive integer giving the total amount that the suppliers can send to the consumers subject to the inputted constraints, and the second output, `FlowMatrix`, is a matrix whose (i, j) entry is the amount that supplier #i sends to demander/consumer #j in a way that will achieve the maximum flow.
(b) Apply your program of Part (a) to solve the supply and demand problem with the following data: Supp = [20, 18, 16, …, 4, 2]; Dem = [16, 16, 16, 16, 16, 6, 6, 6, 6, 6], and no capacity constraints (or just set Cap(i, j) = 20 for each pair of indices i, j). Can all supplies be allocated and all demands met (i.e., does this supply and demand problem have a solution)?
(c) Redo Part (b) but with the change that Cap(i, j) = 10 for each pair of indices i, j.
(d) Redo Part (b) but with the change that Cap(i, j) = 5 for each pair of indices i, j.
(e) Redo Part (b) but with the change that Cap(i, j) = 2 for each pair of indices i, j.

Chapter 10: Randomized Search and Optimization Algorithms

Randomness is a natural phenomenon on which many natural processes are based. Although the evolution of living species depends on many factors, randomness plays a vital role. For example, human beings are created by the mating of their parents, and in this process when the genes of the offspring are formed, some genes will be inherited from the mother, some from the father, but others will be different from both, being the result of a random (non-predictable) mutation. The power of genetics gives rise to the "survival of the fittest" whereby the most effective traits tend to be preserved and adapt to environmental changes. Ant species are particularly well known for their organization and effectiveness in foraging for food sources. Of all creatures on Earth, ants have the greatest total biomass. Experiments have shown that their effectiveness in searching for food is essentially a combination of two main concepts: randomness and the use of pheromones, which are chemicals that ants leave on trails and can be detected by other ants. Successful random trails tend to get reinforced with larger amounts of pheromones that allow the whole ant colony to benefit from the discoveries of individuals.

For centuries, applied mathematics has been using mathematics to model natural phenomena, and to make useful predictions. With access to increasingly powerful computers, scientists began to explore the ideas of modeling computation and algorithms on some successful natural phenomena. The results have been very encouraging, having led, for example, to the creation of the successful field of *artificial intelligence*. *Randomized algorithms* have the common thread in that integral parts of their functionality are based on random selections. We have already witnessed the power of randomization in some of the specialized algorithms that we have seen in earlier chapters. Examples include the randomized primality tests (Appendix to Section 3.3) and the randomized algorithm for computing medians (Section 7.2). We have also seen the power of randomizations in simulations (Section 6.2). In this final chapter, we will give a brief introduction to *randomized search and optimization algorithms*. These are general multi-purpose algorithms that use randomness as a key ingredient and that are designed for searches and optimization. All the algorithms we consider will be heuristic algorithms that aim to find an optimal solution, but will usually find one that is of good quality but not necessarily optimal. They are typically applied to intractable problems where finding an exact solution (to a reasonably sized problem) is not a feasible option. The basic problem for which such an algorithm would be applied is to find (search for) a discrete structure that has a certain distinguishing property among an inordinately large set of discrete structures. This problem can be metaphorically thought of as the problem of finding a very small needle in a very large haystack. Oftentimes, the distinguishing property that we are searching for is that some function of this discrete structure be either a minimum or a maximum, and in this setting we have an *optimization problem*. A nice example of such an intractable optimization problem is the traveling salesman problem that was discussed in Section 9.2.

Section 10.1 will discuss some general concepts about randomized search and optimization algorithms, along with brief descriptions and examples of some well-known families of such algorithms. Section 10.2 will delve more deeply into one particular family of such algorithms known as *genetic algorithms*. Genetic algorithms are optimization algorithms that are modeled on natural evolution, with genetics and the associated random phenomena playing an integral role. Since their invention in the 1970s, they have been successfully used in a wide variety of applications.

Before we go on, we point out that this chapter is quite different from those that came before it in three main respects:

- First, since all algorithms developed are randomized, answers to most exercises will vary (even if done by the same person on the same computer at different times since the random numbers generated will vary).
- Second, while the use of computers has been recommended but optional in all previous chapters, and much could be experienced and learned without them, readers who wish to fully experience this chapter would be at a serious disadvantage if they were to proceed without a computer. Indeed, although many interesting things could be learned from simply reading this chapter, the difference in going through this chapter with or without a computer would be similar to the difference in touring a tropical rainforest by hiking through it versus by a helicopter tour.
- Third, there will be quite a bit less theory in this chapter than in previous ones. Theoretical results about randomized algorithms are always based on probability. Although we will give proofs of a few basic results, theorems about more complicated randomized algorithms are not always feasible (or possible) to obtain. In such cases, we will use concepts of probability to motivate certain portions of the algorithm, rather than give comprehensive theorems about the algorithms. Since general theorems tend to be difficult to obtain, the results concerning a given randomized algorithm tend to be conservative guarantees, which are not very impressive. Indeed, it is the record of performance of such algorithms that has led to their success.

To reflect these large differences from previous chapters, the end-of-section exercise sets will not be separated into ordinary and computer exercises; there will simply be a single set. Additionally, since answers to such exercises will often vary, there will be no answers/brief solutions to the end-of-section exercises in Appendix C, but Appendix B will include complete solutions for most all exercises for the reader.

10.1: RANDOMIZED SEARCH AND OPTIMIZATION: AN OVERVIEW

Random Search Algorithms

The most basic example of a randomized algorithm is the *random search*. The following is a general definition. We categorize our description of a random search as a definition rather than as an algorithm, since the setting is quite general and implementation details will vary.

DEFINITION 10.1: A **random search algorithm** iteratively generates random discrete structures $D_1, D_2, \cdots$ within a set $\mathscr{D}$ of interest until a desired one is found (if ever). In case there is some optimality function $F : \mathscr{D} \to \mathbb{R}$ that we are trying to either maximize or minimize, a record is kept of a best candidate for an optimal discrete structure that is found during the random search. After the first random discrete structure D_1 is generated, we initialize RECORD = D_1 and VALUE = $f(D_1)$. In all future generations, after generating $D_i \in \mathscr{D}$, we check to see whether $f(D_i)$ is better than VALUE (i.e., $f(D_i) <$ VALUE for a minimization problem, $f(D_i) >$ VALUE for a maximization problem). If it is, we then update RECORD = D_i and VALUE = $f(D_i)$.

The mechanics of how the discrete structures are generated will depend on the set $\mathscr{D}$ under consideration, but it must be assured that any element of $\mathscr{D}$ has an equal chance of being generated. Random searches are extremely versatile and usually easy to program. They tend to be used as a first approach to a problem and as a baseline by which to compare the performance of other more specialized algorithms. Random searches are blind searches in that they use no information about the structures under consideration or (in case of an optimization problem) the function that is being optimized. More sophisticated randomized search algorithms will make use of the additional structure of the problem at hand.

We will hold off giving an example of a random search algorithm until we describe another randomized optimization algorithm. Then we will compare the performance of the two algorithms on the same problem.

Hill Climbing Algorithms

Roughly speaking, a randomized hill climbing algorithm works to solve a discrete optimization problem as follows: A random element is generated. Its functional value (using the function that we are trying to optimize) is compared with those of "nearby" elements of the discrete structure. The best element is recorded. If an improvement is realized by one of the nearby elements, the records are updated, and the process gets repeated, this time looking at the nearby elements of the newly found best element. Once the process fails to find a better element among the nearby elements (i.e., a local optimal structure has been found), the record is retained, and the process gets repeated with a new randomly generated discrete structure. At each repetition, the overall best structure is updated. Thus, the hill climbing algorithm combines randomness with a local search.   Here is a metaphorical way to view randomized hill climbing: Think of a complicated mountain range that is under a thick cloud cover. In order to find the highest peak, a plane drops off an individual who parachutes down to some random location in the range. The individual then tries to climb to the highest local peak, measures it with an altimeter, and sends the data back. This process gets repeated.   We next give a more formal definition.   As we did with random searches, we list our description of hill climbing algorithms as a definition rather than as an algorithm. The implementation details provided in the definition, especially for the local search, are particularly general.

DEFINITION 10.2: A **hill climbing algorithm** for finding a discrete structure in some admissible set $\mathscr{D}$ that maximizes (or minimizes) a function $F : \mathscr{D} \to \mathbb{R}$ works as follows: For each $D \in \mathscr{D}$, there is defined a neighborhood $N(D)$ that is a relatively small subset of $\mathscr{D}$ (thought of as other discrete structures in $\mathscr{D}$ that are near D). The construction of these neighborhoods is specific to the problem at hand, but the remainder of the algorithm works the same for all problems.

NOTE: We will formulate the procedure for a minimization problem; the formulation for a maximization problem is similar (just reverse the signs of the inequalities).

We let T = the number of main iterations to be executed.

Step 0: INITIALIZE: IterCount = 1
INITIALIZE: OverallRecord = $\varnothing$, OverallRecordValue = ∞, $i = 1$

Step 1: Randomly generate a discrete structure $D \in \mathscr{D}$.
INITIALIZE: Record = D,   Value = $f(D)$

Step 2: Write $N(D) = \{D^1, D^2, \cdots, D^{n(D)}\}$.
SET: OldValue = Value
FOR $j = 1$ TO $n(D)$

   IF $f(D^j) <$ Value

      UPDATE Record $= D^j$, Value $= f(D^j)$

   END IF
END FOR

Step 3:
IF Value < OldValue
  UPDATE D = Record
  GO BACK to Step 2
OTHERWISE
  GO TO Step 4
END IF

Step 4:
IF Value < OverallRecordValue
 UPDATE: OverallRecord = Record, OverallRecordValue = Value
END IF
IF IterCount < T
  UPDATE IterCount = IterCount + 1 and GO BACK to Step 1
OTHERWISE
  OUTPUT: OverallRecord, OverallRecordValue, and EXIT algorithm
END IF

Hill climbing algorithms may also be referred to as randomized local search algorithms. Unlike a random search algorithm, which blindly searches through the set of discrete structures, hill climbing uses a relatively small amount of work to make sure that we will not be missing any better elements that are nearby the ones that get randomly generated. Our next example will compare the two methods on a small instance of the traveling salesman problem, which was described in detail in Section 9.2. Readers who are adept with computers are encouraged to follow along and run the algorithms on their own.

CAUTION: As with any randomized algorithms, the results will vary, since they basically will depend on different sets of random numbers that were generated to execute the algorithms.

EXAMPLE 10.1: *(Random Search versus Hill Climbing on a Traveling Salesman Problem)* We will present a fair comparison of a random search and a hill climbing search for an optimal solution to a relatively small **traveling** salesman problem with 16 cities. The data for the problem is available for downloading on the book's Web page.[1] In order to make the comparison fair, we will run each of the algorithms so that the total number of tours generated is about the same. We will do the hill climbing search first, and then run a random search using the same number of iterations.

The key to implementing a hill climbing search is to find an effective way to assign neighbors to elements in the set of discrete structures under consideration. This is a very important step since, as we will soon see, it has a significant effect on the performance of the algorithm. In our local search, which we refer to as *2-vertex permuation*, we will look at all tours obtained from a randomly generated tour by permuting two vertices (other than vertex #1, which plays a special role). This local search is closely related to the so-called 2-*opt algorithm* for determining neighbors of a given TSP tour. The neighbors of a given tour from the 2-opt algorithm are obtained as follows: We first delete two edges of the tour, and then we reconnect the resulting two components of the tour with two new edges so that a new graph is formed. See Figure 10.1 for an illustration of this 2-opt operation, where the tour <1,2,3,4,5,6,1> gets transformed to the tour <1,2,5,4,3,6,1>. It is clear that for each two edges deleted, there is only one such way to create a new TSP tour, if the deleted edges do not share a common endpoint. (If we delete two edges sharing a common endpoint, there is no other way to create a different tour by adding two edges.) Clearly each such 2-opt modification corresponds to permuting two vertices in the tour vertex sequence, but the converse is not quite true (consider two edges that share an endpoint). Since an n-vertex TSP tour has $C(n-1,2) = (n-1)n/2$ pairs of distinct vertices (excluding vertex #1), it follows that the number of neighbors of a given tour in our local search is $\Theta(n^2)$.

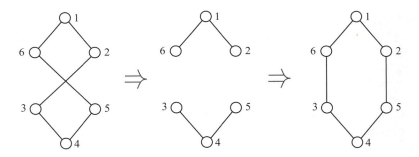

FIGURE 10.1: Illustration of how the 2-opt procedure starts with a cycle tour (left), deletes two edges (middle), and reconnects the two pieces of the tour with new edges to form a new tour (right).

---

[1] The data for this problem originates from the TSPLIB Web site (http://comopt.ifi.uni-heidelberg.de/software/TSPLIB95/) which maintains an extensive set of downloadable benchmark TSP data sets of varying sizes. We used the data set named "ulysses16." A somewhat elaborate procedure (involving spherical trigonometry and nonstandard rounding procedures) is needed to convert the raw data coordinates for these problems into an integer valued distance matrix (which the TSPLIB group assumes is used compute optimal tours). The book's Web page includes downloadable files of the distance matrix, an optimal tour, and the optimal distance.

We will not provide the pseudocodes here, but they can be found, along with the pseudocode for a 2-opt hill climbing algorithm, in the solution to Exercise for the Reader 10.1. Figure 10.2 shows the graph of the "Values" found as a function of the number of tours that were considered. The spikes correspond to generating a new random tour after the previous local search algorithm has stagnated. Also shown in the same graph is the result of the corresponding random search.

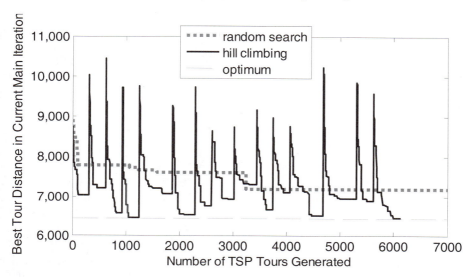

FIGURE 10.2: Performance comparison graphs for the random search (thick dashed line) and the 2-vertex permutation hill climbing search (solid line) for the TSP of Example 10.1. The optimal shortest distance is shown by the thin dashed line. The random search graph indicates the cumulative best tour's total distance, while that for the hill climbing algorithm resets the best tour distance each time the the local search is complete.

It is instructive to compare the two graphs of Figure 10.2. Although results will always vary because the algorithms depend on random numbers, the relative performances of Figure 10.2 are rather typical (as repeated runs will demonstrate). Notice that all but one of the 15 main iterations of the hill climbing algorithm, the best tour found was better than the best overall tour found by the random search. The overall best tour found by the hill climbing algorithm had a total distance of 6875, that for the random search was 7389. By contrast, the optimal tour distance is 6859 (from the TSPLIB website mentioned earlier; this optimal distance has been theoretically verified), and so the best hill climbing tour is only 0.2% longer than the optimal tour, while that found by the random search is 7.7% larger! Figure 10.3 shows graphs of the best tours found in each of these two runs by the two algorithms.

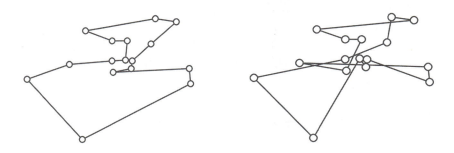

FIGURE 10.3: Graphs of the best TSP tours that were found in the computer runs of the hill climbing algorithm (left), and the random search algorithm (right) for the 16-city TSP of Example 10.1.

Apart from the local search component, the rest of the hill climbing algorithm is quite generic. Devising an appropriate local search/neighbor scheme that captures the essence of the problem requires ingenuity, and sometimes trial and error. The local search method used in a hill climbing algorithm can make or break its success. If we change the local search procedure from the 2-vertex permutation that was used in Example 10.1 to the 2-opt algorithm, the hill climbing algorithm becomes even significantly more effective. In order to demonstrate this, we will need to compare on performances on a more difficult (larger) TSP; this will be the subject of the following exercise for the reader.

EXERCISE FOR THE READER 10.1: Perform a fair comparison of

   (a) a random search and a hill climbing search,
   (b) the hill climbing algorithm with the 2-vertex permutation local search component (of Example 10.1), and
   (c) the hill climbing algorithm with the 2-opt local search component.

to search for an optimal solution of the 22-city benchmark TSP "Ulysses22" (whose data sets may be downloaded from the book's Web page). Perform 100 runs of each algorithm (each with a similar number of tours) and store the best distances of each run for the two algorithms as (length 100) vectors. Compare the means, minimums, maximums, and the *standard deviations*[2] of the two vectors.

---

[2] The standard deviation of a vector $x = [x_1, x_2, \cdots, x_n]$ is defined to be $\sqrt{\sum_{i=1}^{n}(x_i - \bar{x})^2 / (n-1)}$. It serves as a standard statistical measure of the spread of the data values that are stored in the vector.

The *k*-Opt Local Search Algorithm for Traveling Salesman Problems

The most effective algorithms that have been developed (randomized or not) for solving large TSPs (either optimally or with a feasible solution of very good quality) have relied heavily on effective local searches. The 2-opt local search idea that was introduced in Example 10.1 is generalized in the following definition.

ALGORITHM 10.1: The *k*-Opt Algorithm for the Traveling Salesman Problem on an Edge-Weighted Complete Graph *G*

Input: A complete edge-weighted graph *G* along with a matrix *W* of edge weights, and a starting vertex v_1, an initial tour *T*, and a positive integer *k*, with $1 < k \leq n$.

Output: Another tour TOUR.

Step 0: INITIALIZE: TOUR = *T*
TOUR = *T*.

Step 1: SET E_k = {*k*-element subsets of the tour edge set *E*(TOUR)}.

Step 2: For each set $\{e_1, e_2, \cdots, e_k\} \in E_k$, proceed one-by-one through all tours *T* that contain the edges $E(\text{TOUR}) \sim \{e_1, e_2, \cdots, e_k\}$.

IF $wgt(T) < wgt(\text{TOUR})$

   UPDATE TOUR = *T* and GO BACK to Step 1.
END

Step 3: OUTPUT: TOUR and EXIT algorithm.

Note that when *k* = 2, Algorithm 10.1 works slightly differently than the local search strategy that we used in Example 10.1. Apart from the fact that the neighboring tours in the latter were a bit more general, the main difference is that in the local search of Example 10.1, all two edge subsets are considered (even as the original tour gets updated), while in the *k*-opt algorithm, the set of neighbors changes as soon as an improving tour is found. The next exercise for the reader will ask to perform a computer experiment to compare the performance of the two algorithms. As an implementation note, we point out that in the comparison step $wgt(T) < wgt(\text{TOUR})$, rather than completely evaluate $wgt(T)$, we really need only to check whether the sum of the weights of the new edges of *T* is less than the sum of the weights of those edges $\{e_1, e_2, \cdots, e_k\}$ of TOUR that they replaced. This can save considerable amounts of computing time, particularly when *N* is much larger than *k*.

EXERCISE FOR THE READER 10.2:   Consider the 3-opt algorithm on an instance of the TSP. Determine the number of possible tours that can be obtained by deleting a set of three edges $\{e_1, e_2, e_3\}$ of a given tour T in each of the following three cases:

Case 1: No two of the three edges in $\{e_1, e_2, e_3\}$ share a common endpoint.

Case 2: Two of the three edges in $\{e_1, e_2, e_3\}$ share one common endpoint.

Case 3: The three edges constitute a 3-path in T.

We say that a tour T is **k-opt** if it cannot be improved by the k-opt algorithm. By the construction, any outputted tour of Algorithm 10.1 will be a k-opt tour. A **successful k-opt move** is any tour T in Algorithm 10.1 that satisfies the IF condition of Step 2 (and thus leads to an improvement in tour length). Figure 10.4 shows an example of a successful 3-opt move.

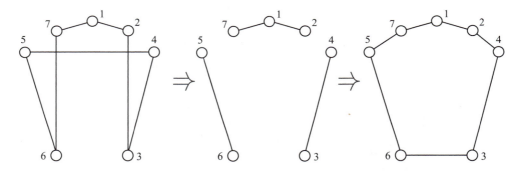

FIGURE 10.4: An example of a successful 3-opt move; the resulting tour (on the right) is easily seen to be a 3-opt tour (in fact it is easy to see that it is the (unique) optimal TSP tour), if ordinary (Euclidean) distances are used.

Of course, the larger k is, the more effective the k-opt algorithm will be. Indeed, if $k' > k$, then any k'-opt move is also a k-opt move (simply leave the additional edges fixed).   The amount of work needed to execute the k-opt algorithm increases exponentially with k.   Indeed, the N-opt algorithm (applied to an N-city TSP tour) will optimally solve the TSP since it is just the brute-force algorithm that checks through all possible tours (Why?).

To give some idea of the gap between theory and practice in such algorithms, we give some experimental results regarding the k-opt local search algorithm.   The results are quite encouraging. We follow these by a sampling of some theoretical results concerning worst-case performance of the k-opt algorithm.

SOME EMPIRICAL RESULTS ABOUT THE *k*-OPT ALGORITHM 10.1:
1. The 3-opt algorithm performs significantly better than the 2-opt. For example, a single application of the 3-opt algorithm applied to a randomly generated tour of a certain 48-city TSP that satisfies the triangle inequality (the problem hk48 from the earlier mentioned TSPLIB Web page) has about a 5% chance of finding the

optimal tour.   By independence, it follows that with T master iterations of a hill climbing algorithm applied to this TSP, the chance that an optimal tour will be detected is (using the complementary probability rule) $1 - (.95)^T$.   It follows that with $T = 10$ iterations, the chance of detecting an optimal tour goes up to 40%, with $T = 50$, it rises to over 90%.

2. The significant amount of additional complexity usually does warrant using the 4-opt algorithm over the 3-opt algorithm.

3.   The k-opt algorithm appears as a component of some of the most efficient algorithms for the TSP. One of the first such TSP heuristics was devised by Shen Lin and Brian W. Kernighan [LiKe-73] in which the k-opt algorithm is iteratively applied with varying values of k.

SOME THEORETICAL RESULTS ABOUT THE k-OPT ALGORITHM 10.1 (reference: [ChKaTo-99]):

1.   For each k, there exist arbitrarily large integers N and corresponding N-city TSP problems and k-opt tours, whose lengths exceed the length of an optimal tour by a factor of at least $\sqrt[4k]{n}/4$.   For example, when $N = 50$ and $k = 4$, this is over 40%.

2. For each k, the worst-case running time of the k-opt algorithm is exponential.

Randomized Hill Climbing Algorithms

The hill climbing Algorithm 10.1 can be further randomized in the local search by choosing successive neighbors to examine in random order, rather than using a specified ordering. This scheme will introduce one more parameter, since we will need to specify how many times a randomly chosen neighbor will be tested.

DEFINITION 10.3:   A **randomized hill climbing algorithm** for finding a discrete structure in some admissible set $\mathscr{D}$ that maximizes (or minimizes) a function $F : \mathscr{D} \to \mathbb{R}$ works in the same fashion as the hill climbing algorithm of Definition 10.2 but has one additional input parameter M, and has the following change to Step 2 of definition (for minimization problems):

New Step 2: Write $N(D) = \{D^1, D^2, \cdots, D^{n(D)}\}$.

SET: OldValue = Value

FOR $k = 1$ TO M

    Let j be a randomly generated integer in the range $1 \le j \le n(D)$.

   IF $f(D^j) <$ Value

      UPDATE Record $= D^j$,   Value $= f(D^j)$

   END IF

END FOR

A Brief Discussion on Some Other Randomized Heuristic Algorithms

We end this section with a brief discussion of an assortment of other randomized heuristic algorithms that have achieved particularly impressive success since their invention. Each of these algorithms has, at several junctures of its history, held the record among all algorithms on some extremely large benchmark problems (such as in the traveling salesman problem "Olympics"). For size considerations, we will avoid giving specific examples and details for these algorithms. Indeed, each of these algorithm families has an assortment of additional features and extensions that allow for greater adaptability and effectiveness on assortments of discrete optimization problems. Opportunities for implementation will appear in the exercises. The next section will provide a much more elaborate treatment of *genetic algorithms*, which are another sort of randomized algorithm.

Tabu Search

The word "Tabu" is a Polynesian word meaning spiritually prohibited.[3] Like the hill climbing algorithm, a key component of a tabu search is the assignment of a neighborhood $N(D)$ for each $D \in \mathscr{D}$, the set of discrete structures under consideration. The construction of these neighborhoods is specific to the problem at hand; the remainder of the basic tabu search algorithm works the same for all problems. The main difference between a tabu search and a hill climbing algorithm is that the tabu search maintains a **tabu list** of elements of $\mathscr{D}$ to which the search is not permitted to be moved. The tabu list changes with the iterations, but it is basically designed to keep the algorithm from stagnating at local optimum value by giving it the ability to walk away from such local extrema to experiment with other possible search areas. This important difference is metaphorically illustrated in Figure 10.5.

One very basic scheme is to maintain a tabu list consisting of the M most recently visited structures (where M is a positive integer parameter). Tabu search was invented in the 1970s by University of Colorado Professor Fred Glover. Glover coauthored an authoritative book [GlLa-97] on Tabu search, which describes many implementation details and successes of the method.

---

[3] The corresponding English word "taboo" was adopted by English explorer Captain James Cook during a 1777 visit to the island nation of Tonga.

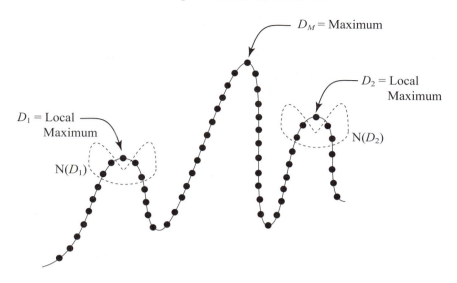

FIGURE 10.5: Illustration of the main pitfall of the hill climbing algorithm, if the hill climbing algorithm reaches a local maximum (D_1 or D_2) it becomes trapped in a cycle, since it restricts its search to the corresponding local neighborhood ($N(D_1)$ or $N(D_2)$). The tabu search is designed to experiment by walking away from this neighborhood and perhaps discovering a new maximum (ultimately the sought-after overall optimal structure D_M).

Ant Colony Optimization

Ant colony optimization (ACO) algorithms are randomized optimization algorithms that experiment through the search space using procedures that replicate the very effective food foraging behavior of ants. It was invented in the early 1990s by Marco Dorigo in his Ph.D thesis where he applied it to the traveling salesman problem, and it has since been elaborated and adapted to provide quality solutions for a number of NP complete problems, including the job-shop scheduling problem, the vehicle routing problem, and the set covering problems. A nice general reference is the book coauthored by Dorigo [BoDoTh-99].

Ant colony optimization is based on the following two main aspects of ant foraging behavior:

1. In their search for food, individual ants explore paths with a degree of randomness.
2. As an ant wanders on a certain path, it lays down a trace of a certain chemical known as a **pheromone**. The resulting pheromone trails are detected by other ants and increase the chances that they will use previously used paths.

Pheromones evaporate over time, so the longer a path from the nest to the food source, the longer it will take for the ant to return to the food source and hence the weaker the pheromone trail will be. This reinforces better trails and helps ants to find the shortest paths between food sources, see Figure 10.6.

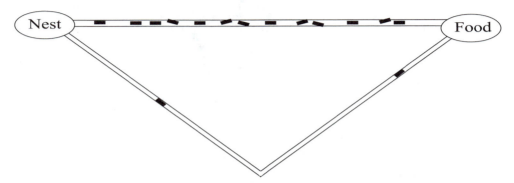

FIGURE 10.6: Ant colonies use a combination of randomness and pheromone trails to discover food sources and to establish the shortest paths to them from the colony nests.

The following example outlines the structure of the ACO algorithm applied to the traveling salesman problem.

EXAMPLE 10.2: (*Description of an implementation of the ACO algorithm applied to a traveling salesman problem*) Suppose that we have an instance of the traveling salesman problem presented by an N-vertex complete graph $G = K_N$, and a corresponding edge weight function W. We explain a basic implementation of the ant colony optimization algorithm for this problem. The development follows the original approach by Dorigo. The performance was promising but not quite state-of-the-art in comparison to other competing algorithms. Subsequently, refinements were made on the ACO implementation for the TSP as well as other important problems that rendered an algorithm that was able to compete with and sometimes outperform some of the other leading heuristic algorithms.

A group of m artificial ants $\alpha_1, \alpha_2, \cdots, \alpha_m$ is introduced, m being a parameter. Another parameter that needs to be specified is the number of (time) iterations T. For each iteration $t = 1, 2, \cdots, T$, the m ants are initially randomly distributed among the N vertices, and each will build its own TSP tour. In the process of building its tour, when an ant α_k decides which city to visit next, it will employ a specific probabilistic transition rule. Before giving the formula for this rule, we first point out that it will depend on the following three factors:

1. An ant will not visit a city it has already visited (unless it is completing the tour). This can be implemented by building a separate tabu list for each ant consisting of the cities the ant visits, and which get purged in passing from one generation to the next.

2. The reciprocals of the distances $1/W(i,j)$, which can be thought of as the *visibilities*. Note that if the decision on which city to visit next depended only on this item and item 1, we would arrive at a greedy local heuristic similar to those that were introduced in Section 9.2.

3. The concentration of pheromone on the edge from i to j: $\tau_{ij}(t)$. These quantities get updated at each iteration, and they depend on the cumulative uses by all of the ants, as well as any evaporation that may have occurred. We will explain later how the $\tau_{ij}(t)$'s get updated from one generation to the next.

The following formula describes the so-called transition rule that gives the probability $p_{\alpha_k}(i,j)$ that ant α_k will decide to go from vertex i to vertex j in any iteration tour it constructs. In this formula, we let $T_{\alpha_k}(i)$ denote the (tabu list of the) cities ant i has already visited when it reaches city i.

$$p_{\alpha_k}(i,j) = \begin{cases} \dfrac{\left(\tau_{ij}(t)\right)^a \left(1/W(i,j)\right)^b}{\displaystyle\sum_{j \in T_{\alpha_k}(i)} \left(\tau_{ij}(t)\right)^a \left(1/W(i,j)\right)^b}, & \text{if } j \notin T_{\alpha_k}(i) \\ 0, & \text{if } j \in T_{\alpha_k}(i) \end{cases} \tag{1}$$

In this formula the parameters a, b are nonnegative real numbers that control the relative importance of the pheromone intensity versus the visibilities in the determination of these probabilities. The extreme case in which $a = 0$ corresponds to a greedy local search algorithm (that does not use any information about the pheromone trails). The other extreme case in which $b = 0$ corresponds to the local information being ignored and basing the decisions entirely on the global pheromone information.

At the completion of an iteration, each ant lays a pheromone trail on each of the edges of its tour T_{α_k} of quantity

$$\Delta\tau_{\alpha_k} = Q/\text{length}(T_{\alpha_k}), \tag{2}$$

where $Q > 0$ is another parameter. Thus the shorter trails will be given higher levels of pheromone.

Finally, in order to prevent stagnation, the pheromone must be allowed to decay, and this is done with the geometric decay indicated by the following formula:

$$\tau_{ij}(t+1) = (1-\rho)\tau_{ij}(t) + \sum_{(i,j) \in T_{\alpha_k}} \Delta\tau_{\alpha_k}, \tag{3}$$

where $0 < \rho < 1$ is a parameter. Initially, we assume that there is a trace amount of pheromone on all of the edges: $\tau_{ij}(0) = \tau_0$, where $\tau_0 > 0$ is a parameter. The exercises at the end of this section will provide the reader with the opportunity of testing this algorithm and to fine tune the parameter levels for specific instances of the TSP. Dorigo suggests the following default parameter settings:

m (number of ants) = N (number of cities), $a = 1$, $b = 5$, $\rho = 0.5$, $Q = 100$, $\tau_0 = 10^{-6}$.

Simulated Annealing

The word "annealing" comes from physics and has to do with the cooling of liquids or solids. At high temperatures, the molecules in a substance exhibit a high degree of randomized motion, but as the temperature cools, the molecules tend to converge to configurations of low energy. Indeed, if a metal is heated to a sufficiently high temperature, and the cooling process is controlled to be sufficiently slow, then the atoms will arrange themselves into crystalline structures of minimal energy. It is this molecular physical phenomenon on which simulated annealing (SA) algorithms are based. The SA algorithm can be thought of as tracing a path in the set of all admissible discrete structures with the usual aim of heading towards and ultimately obtaining an optimum discrete structure. As with the tabu search (and unlike the hill climbing algorithm), simulated annealing allows the path to sometimes progress to structures of lower value, and this helps to keep it from getting stuck at a local optimum discrete structure that is not globally optimum. Simulated annealing dates back to the 1980s, but one of its key components is the so-called *Metropolis algorithm*, which dates back to the 1950s.

DEFINITION 10.4: A **simulated annealing (SA) algorithm** for finding a discrete structure in some admissible set $\mathscr{D}$ that maximizes (or minimizes) a function $F : \mathscr{D} \to \mathbb{R}$ works as follows:

Requirements:

1. (*Local neighborhoods for each admissible discrete structure*) As with the hill climbing algorithm, we assume that for each $D \in \mathscr{D}$, there is defined a neighborhood $N(D)$ that is a relatively small subset of $\mathscr{D}$ (thought of as other discrete structures in $\mathscr{D}$ that are near D).

2. (*Cooling schedule*) This will depend on the particular problem at hand and consists of the following components:
(i) Starting temperature: THigh. Should be set high enough to allow a great deal of movement in $\mathscr{D}$.
(ii) Temperature decrement: One common scheme is to have the temperature decay geometrically: $T(k+1) = (1-\rho) \cdot T(k)$, where $0 < \rho < 1$ is a parameter.

With some problems, practitioners have found a more rapid decay such as $T(k+1) = (1/\log(2k)) \cdot T(k)$, or $T(k+1) = (1/k) \cdot T(k)$, to be more effective.

(iii) Final temperature: TLow. Should be set to a small positive number, that is small enough so that not much movement is allowed in the near final iterations. TLow should not be set to zero since $T(k)$ may never reach zero with geometric and other decrement schemes.

3. (*Maximum number of iterations*) *M* This parameter is set in advance and determines the running time of the algorithm. The cooling schedule must take into consideration the value of *M* (otherwise TLow may not be reached or may be reached too soon).

Here are the steps of SA for a minimization problem, to which the physical interpretation corresponds:

Step 1: Set the initial temperature $T = $ THigh, and randomly generate a discrete structure $D \in \mathscr{D}$.
INITIALIZE: Record $= D$, Value $= f(D)$, $i = 1$.

Step 2:
FOR $k = 1$ TO M (*M* is a parameter)
    Write $N(\text{Record}) = \{D^1, D^2, \cdots, D^{n(D)}\}$.
    Let j be a randomly generated integer in the range $1 \le j \le n(D)$.
    SET: OldValue $=$ Value, $\Delta V = f(D^j) - $ Value
    Let j be a randomly generated integer in the range $1 \le j \le n(D)$.
    IF $\Delta V < 0$
       UPDATE Record $= D^j$, Value $= f(D^j)$
    ELSE
       Let U be a randomly generated real number in $(0,1)$ (*see Section 6.2*)
       IF $U \le \exp(-\Delta V / T)$
       UPDATE Record $= D^j$, Value $= f(D^j)$
    END IF
    IF $T \le$ TLow
       OUTPUT: OverallRecord, OverallRecordValue, and EXIT algorithm
    OTHERWISE
       Lower T according to the cooling schedule.
    END IF
END FOR

EXERCISES AND COMPUTER EXERCISES 10.1:

NOTE: As pointed out earlier, in both sections of this chapter, the exercises and computer exercises will be combined into a single set. Answers will vary due to the random nature of the problems, so answers will not be given in Appendix C for this chapter (but Appendix B still contains complete solutions for most of the exercises for the reader).

1.   Apply the randomized hill climbing algorithm of Definition 10.3 to the TSP of Example 10.1. Use the same number of master iterations (15) and $M = 15$ (so the number of evaluations of both will be the same). Compare the results of 30 repetitions of each algorithm (randomized hill climbing versus hill climbing), and discuss any observed empirical differences in performance.

2.   Apply the randomized hill climbing algorithm of Definition 10.3 to re-do Exercise for the Reader 10.1 with the "Ulyssess 22" TSP.

3.   Download the data for the 52-city TSP "Berlin52" from the book's Web page, and repeat what was done in Example 10.1 for this problem. In particular create a plot as in Figure 10.2, and compare with the optimal tour length of 7542 (from the TSPLIB Web page).

4.   Repeat the instructions of Exercise for the Reader 10.1 for the 52-city TSP "Berlin52" that was discussed in the preceding Exercise, and compare with the optimal tour length of 7542 (from the TSPLIB Web page).

5.   Repeat Exercise 4, but this time use a randomized hill climbing algorithm (Definition 10.3).

6.   (a) Write a program for the 3-opt algorithm.
      (b) Perform a fair comparison of the 3-opt and the 2-opt hill climbing algorithms using the 52-city TSP "Berlin52" from the book's Web page, which has optimal tour length 7542.

7.   (*Random Searches for a Tournament Problem*) Recall that a tournament is a simple directed graph whose underlying graph is a complete graph (Definition 8.16). For each positive integer k, a tournament with at least k vertices is said to have *property* S_k if for every set of k players, there is a player that beats all other players in the set. It can be shown that for any positive integer k, if n is sufficiently large, then there exist tournaments on n vertices with property S_k. In fact, it has been proved that if

$$C(n,k) \cdot (1 - 2^{-k})^{n-k} < 1, \tag{4}$$

then there exists a tournament on n vertices that has property S_k.[4]

(a) Show that with $k = 2$, formula (4) shows that there exist tournaments on n players (vertices) that have property S_2, whenever $n > 20$.

(b) Apply a random search to find a tournament on 21 players that has property S_2.

(c) Apply a random search to find a tournament on 22 players that has property S_2.

(d) The theorem above is a necessary condition. For example, when $k = 2$ and $n = 20$, the

---

[4] This result can be proved using the so-called *probabilistic method*; see Theorem 1.2.1 in the book of the same title by Noga Alon and Joel Spencer [AlSp-00]. The probabilistic method is an elegant technique that was developed by Paul Erdös for proving that certain discrete structures exist by assigning a natural probability on the set of discrete structures under consideration, and showing that the set of structures of interest have positive probability, and so the set must be nonempty (thus proving existence). Readers who have done well with the probability material in Chapter 6 are encouraged to check out this beautiful book. It is a difficult read, but a rewarding one.

condition (4) is not satisfied so that the theorem is not applicable. Apply a random search to find a tournament on 20 players that has property S_2.

(e) What is the smallest integer n for which you can use a random search (that will run in less than 4 minutes) to find a tournament on n vertices that has property S_2?

(f) Using the growth rate theory of Section 7.2, show that $C(n,k) \prec (1 - 2^{-k})^{k-n}$, and deduce that condition (4) will indeed be satisfied for any given positive integer k, provided that n is sufficiently large.

(g) Find the smallest positive integer n such that (4) is true for $k = 3$. Would a random search be feasible to search for an n vertex tournament on this number of vertices that has property S_3? If yes, perform such a search; if not, explain.

Note: A random tournament can be formed by (using the computer to simulate) flipping a coin for each edge of the complete graph on the number of players to determine who wins (i.e., the direction of the edge). A player that beats all others would correspond to a vertex of zero indegree.

8. (*Random Searches for Sum-Free Subsets of a Set of Integers*) A set $S = \{a_1, a_2, \cdots, a_k\}$ of nonzero integers is said to be *sum-free* if the sum of any two integers in S is not an element of S, i.e., $a_i, a_j \in S \Rightarrow a_i + a_j \notin S$. For example, the set $\{2,5,6\}$ is sum free but the set $\{2,4,7\}$ is not, since $2 + 2 = 4$ is in the set. Erdös proved that any set $A = \{a_1, a_2, \cdots, a_n\}$ of n nonzero integers will always have a sum-free subset that has size greater than $n/3$. (The theorem and proof appear as Theorem 1.4.1 in [AlSp-00].)

(a) Use a random search to find a sum-free subset of $\{1,2,3,4,5,6,12,24,48\}$ with at least four elements.

(b) Use an exhaustive search to determine how many sum-free subsets of the set in Part (a) there are that contain at least four elements, and what is the maximum number of elements that such a sum-free subset can contain?

(c) Use a random search to find a sum-free subset of $\{-5, -4, -3, -2, -1, 1, 2, 3, 4\}$ with at least four elements.

(d) Use an exhaustive search to determine how many sum-free subsets of the set in Part (c) there are that contain at least four elements, and what is the maximum number of elements that such a sum-free subset can contain?

(e) Use a random search to find a sum-free subset of $\{-9, -9, \cdots, -1, 1, 2, \cdots, 9\}$ with at least seven elements.

(f) Use a random search to find a sum-free subset of $\{-20, -19, \cdots, -1, 1, 2, \cdots, 19\}$ with at least fourteen elements.

Suggestion: See Algorithm 6.1 and the paragraph that follows it for an efficient algorithm to generate finite subsets of a given set of a specified size.

9. (*Random Search for Maximum Independent Sets*) We recall from Definition 8.20 (in the Exercises of Section 8.1) that a set of vertices S of a simple graph G is called *independent* if no two vertices of S are adjacent (in G). The problem of finding a *maximum independent set* of vertices a given graph G (i.e., an independent set of maximum cardinality) is a very difficult (NP hard) problem.

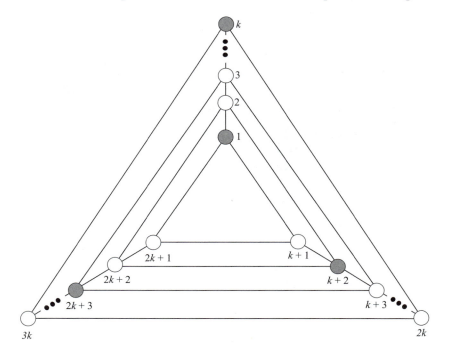

For a positive integer k, we define T_k to be the graph with $3k$ vertices illustrated above. It consists of k copies of the complete triangle graph K_3 that are joined together in a series with corresponding vertices of adjacent copies being joined by an edge. We use the labeling scheme shown in the figure. It is clear that a maximum independent set of vertices of T_k consists of k vertices, one from each triangle graph, provided they are chosen in such a way that the vertices chosen of adjacent triangles are in different positions.

(a) How many maximum independent sets does T_k have?

(b) What is the probability that a randomly chosen subset of vertices of T_k will be a maximum independent set?

(c) How would the answer to Part (b) change if we chose a random set of vertices with the correct number (k) of vertices?

(d) Apply a random search to find an independent set of vertices of T_{30} with the goal of finding one of largest possible cardinality. At each generation, randomly generate an arbitrary set of vertices (of any size). Run your program with a number of iterations so that the runtime will be at least 20 seconds.

(e) Redo Part (d) but this time restrict the subsets being generated to be of size at most 30.

(f) Apply a random search to find an independent set of vertices of T_{50} with the goal of finding one of largest possible cardinality. At each generation, randomly generate an arbitrary set of vertices (of any size). Run your program with a number of iterations so that the runtime will be at least 20 seconds.

(g) Redo Part (f) but this time restrict the subsets being generated to be of size at most 30.

(h) Redo Part (f) but this time restrict the subsets being generated to be of size at most 50.

Suggestion: See the suggestion for Exercise 2 for generating random subsets of a given size. To generate a random subset in a range, say 2 to 20, first generate a random integer in this range and then generate a random subset of this size (any set with one vertex is always independent, so we need not waste time generating and checking single vertex sets). The vertex labeling scheme shown in the figure makes the edge labeling very easy.

10. (*Hill Climbing for Maximum Independent Sets*) Refer to Exercise 9 for the definitions needed in this exercise, including the graph T_k. For an independent set of vertices S of a graph G, we define the neighborhood $N(S)$ to consist of all vertex subsets of G that contain S and have exactly one additional vertex, i.e., vertex sets of the form $S \cup \{v\}$, where $v \in V(G) \sim S$.

(a) Apply the hill climbing algorithm to find an independent set of vertices of T_{30} with the goal of finding one of largest possible cardinality. At each generation, randomly generate an arbitrary set of vertices (of any size). Set the program up with a number of iterations so that the runtime will be at least 20 seconds. Compare the results with the corresponding random search of Exercise 9(d).

Note: Technically speaking, this is not exactly a hill climbing algorithm (according to Definition 10.2) since the sets generated are not always independent sets and so will not be admissible and $N(S)$ will not be defined. So hill climbing really only takes place in cases where the randomly generated set S is independent.

(b) Redo Part (a) but this time restrict the subsets being generated to be of size at most 30. Compare the results with the corresponding random search of Exercise 9(e).

(c) Apply the hill climbing algorithm to find an independent set of vertices of T_{50} with the goal of finding one of largest possible cardinality. At each generation, randomly generate an arbitrary set of vertices (of any size). Set the program up with a number of iterations so that the runtime will be at least 20 seconds. Compare the results with the corresponding random search of Exercise 9(f).

(d) Redo Part (c) but this time restrict the subsets being generated to be of size at most 50. Compare the results with the corresponding random search of Exercise 9(g).

(e) Redo Part (c) but this time restrict the subsets being generated to be of size at most 30. Compare the results with the corresponding random search of Exercise 9(h).

Suggestion: See the suggestion for Exercise 9. Also, some time can be saved in the for loop of Step 2 of the hill climbing algorithm by exiting the loop as soon as a single neighbor is an independent set (why?).

11. (*Discretization of Continuous Functions*) For each even positive integer n, we define a function $F_n : [0,1] \to \mathbb{Z}$ by setting $F_n(x) = k$, if k is even and $1/k \le x \le 1/(k-1)$, and otherwise $F_n(x) = 0$. The graph of F_8 is shown in the figure below. (The vertical line portions are not part of the graph; they were added to emphasize the drops and local maxima.) For large values of n, the functions F_n become good examples for testing randomized optimization algorithms since they have many local maxima, and those with larger values occur on smaller sets. Although optimizing this function falls under the category of continuous optimization, rather than discrete optimization, in this exercise we will show a general method to convert such continuous optimization problems into discrete optimization problems.

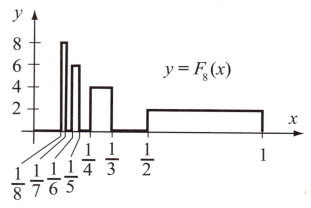

(a) How many local maxima does the function F_n have?

(b) Given any rational-valued function $F : (0,1) \to \mathbb{Q}$, [5] and given a positive integer D, we define the *discretization* of F to D decimals to be the restriction of the function F to all real numbers that are expressible in the form $x = 0.d_1 d_2 \cdots d_D$, where each d_i is a digit in the set $\{0, 1, \cdots, 9\}$, i.e., whose decimal expansions require at most D digits after the decimal point. We denote these points as $(0,1)_D$ and the corresponding restricted function as $F_D : (0,1)_D \to \mathbb{Q}$. Give a scheme for randomly generating points in $(0,1)_D$ so that each point has the same probability of being selected.

(c) Explain why the maximum (minimum) values of any function $F : (0,1) \to \mathbb{Q}$, if they exist, are always at least as large (at least as small) as the maximum (minimum) values of F_D, where D is any positive integer.

(d) Show that for any positive integer n, the maximum value of the function F_n (described above) will equal the maximum value of the disretization $(F_n)_D$, provided that D is sufficiently large. How large must D be to ensure that these maxima will equal when $n = 8$? How about if $n = 100$? How about if $n = 1000$?

(e) Give an example of a function $F : (0,1) \to \mathbb{Z}_+$, such that the maximum value of F_D is always less than the maximum value of F, no matter how large D is.

Suggestion for Part (a): Use the methods of Section 6.2 to randomly generate a nonnegative integer ℓ in the range $0 < \ell < 10^D$, and take $x = \ell / 10^D$.

12. (*Random Search Applied to Maximizing a Difficult Family of Functions*) Read Exercise 5 before proceeding with this exercise. We let $F_n : [0,1] \to \mathbb{Z}$ be the function described in Exercise 5.

(a) For a given positive integer n, compute the probability that a single randomly generated real number x in the interval $(0,1)$ (produced by the `rand` generator of Section 6.2) will maximize the function F_n. Find this value explicitly when $n = 10$, $n = 100$, and $n = 1000$.

(b) For a given positive integer n, compute the probability that at least one of M randomly generated real numbers x in the interval $(0,1)$ (produced by the `rand` generator of Section 6.2) will maximize the function F_n. Find this value explicitly when $n = 50$, with each of $M = 100$, $M = 10,000$, and $M = 1,000,000$. Find this value explicitly when $n = 1000$, with each of $M = 100$, $M = 10,000$, and $M = 1,000,000$.

(c) Perform a random search for the maximum value of the discretized function $(F_{50})_D$, where D is a positive integer at least two larger than the minimum required for the maximum of $(F_{50})_D$ to be 100 (i.e., that of F_{50}). Use $M = 10,000$ random numbers. Repeat this experiment 200 times. How does the number of trials for which the true maximum (100) was found, compare with the corresponding probabilistic estimate of Part (b)?

(d) Perform a random search for the maximum value of the discretized function $(F_{1000})_D$, where D is a positive integer at least two larger than the minimum required for the maximum of $(F_{1000})_D$ to be 2000 (i.e., that of F_{1000}). Use $M = 10,000$ random numbers; what is the maximum value that was found? Repeat with $M = 1,000,000$ random numbers.

(e) Explain why even though D is sufficiently large so that the maximum value of the function F_n (described above) will equal the maximum value of the discretization $(F_n)_D$, a random search on the latter may have a significantly smaller probability of success (to find the maximum) for $(F_n)_D$, than would a randomly generated real number to locate the maximum of F_n. Next explain why if D is at least two larger than the minimum value of the last sentence

---

[5] The assumption that the co-domain consists of rational numbers is simply to assure that the numbers can be represented on a computer. The same ideas will hold for real-valued functions, if the computer programs work with floating point representations (see Section 4.4).

then the probability of success (to find the maximum) for $(F_n)_D$ would be roughly the same as that for which a randomly generated real number would locate the maximum of F_n.

Suggestion for Part (a): Use the methods of Section 6.2 to randomly generate a nonnegative integer ℓ in the range $0 < \ell < 10^D$, and take $x = \ell / 10^D$.

13. (*Tabu Search Applied to Maximizing a Difficult Family of Functions*) Read Exercises 5 and 6 before proceeding with this exercise. We let $F_n : [0,1] \to \mathbb{Z}$ be the function described in Exercise 5. This exercise will set up a simple tabu search scheme for finding the maximum of the corresponding discretization $(F_n)_D$, where D of an appropriate size to ensure the maximum value of the discretization $(F_n)_D$ is the same as that of F_n. We first define the required neighborhoods of a number $x = 0.d_1 d_2 \cdots d_D \in (0,1)_D$. Actually, we define a series of such neighborhoods $N_\ell(x)$ that depend on a positive integer parameter $\ell \le D$. It will be easier to understand if we show some special cases before presenting the general definition.

$N_1(0.d_1 d_2 \cdots d_{D-1} d_D) = \{0.d_1 d_2 \cdots d_{D-1} e_D : e_D \in \{0,1,\cdots,9\}\}$.

$N_2(0.d_1 d_2 \cdots d_{D-1} d_D) = \{0.d_1 d_2 \cdots d_{D-2} e_{D-1} e_D : e_{D-1}, e_D \in \{0,1,\cdots,9\}\}$.

In general, $N_\ell(x)$ consists of all numbers in $(0,1)_D$ obtainable by changing any or all of the last ℓ digits in the expansion $x = 0.d_1 d_2 \cdots d_D \in (0,1)_D$.

(a) Explain why $N_D(x) = (0,1)_D$, and for each increase in ℓ by 1, the number of elements in $N_\ell(x)$ increases tenfold.

(b) Perform a tabu search to find the maximum of the discretized function $(F_{50})_D$, where D is the positive integer exactly one larger than the minimum required for the maximum of $(F_{50})_D$ to be 100 (i.e., that of F_{50}). Use the neighborhoods $N_2(0.d_1 d_2 \cdots d_{D-1} d_D)$, and first let the digit d_{D-1} vary through all 10 possible values before varying the digit d_D. Let the tabu list be (up to) the 10 most recent numbers generated, and restart the process anytime the maximum is increased. Use a varying amount of master generations (of initially randomly generated seed numbers in $(0,1)_D$). Compare the results with those found with the random search of Exercise 6(c) with a comparable number of generations.

(c) Repeat Part (b) with the following modification: Let the final two digits $d_{D-1}d_D$ vary through the 100 different values in order: 00, 01, 02, ..., 99. Compare the effectiveness of both schemes, and explain any significant differences in performance.

(d) Perform a tabu search for the maximum value of the discretized function $(F_{1000})_D$, where D is the positive integer at exactly larger than the minimum required for the maximum of $(F_{1000})_D$ to be 2000 (i.e., that of F_{1000}). First use the scheme of Part (b) with $N_2(0.d_1 d_2 \cdots d_{D-1} d_D)$, and then a corresponding scheme with $N_3(0.d_1 d_2 \cdots d_{D-1} d_D)$. Compare the results with those found with the random search of Exercise 6(d) with a comparable number of generations.

(e) Explain why making D several units larger than that required so that the maximum value of the function F_n (described above) will equal the maximum value of the discretization F_D, can seriously hinder the success of the tabu search schemes of Parts (b) and (d).

10.2: GENETIC ALGORITHMS

Genetic algorithms were created and extensively developed during the 1960s and 1970s by University of Michigan Professor John Holland; see Figure 10.7. Holland holds a joint professorship in engineering, computer science, and psychology, and his multidisciplinary background lends testament to the vast applicability of his invention. The field of genetic algorithms is a branch of evolutionary computation that has flourished into its own discipline with regular conferences, and with applications spanning across the spectrum of sciences. Genetic algorithms are designed to search for solutions to problems where the set of feasible solutions is inordinately large so as to rule out brute-force searches. They are heuristic in nature, and

FIGURE 10.7: John Holland (1929–), American computer scientist.

are designed to find good solutions, but not always optimal ones. They have so many parameters and modifications that their theory has much catching up to do with many new effective adaptations that have proved to be successful with applications. The basic ideas of a genetic algorithm are motivated by Charles Darwin's theory of natural selection, "survival of the fittest," which he introduced in his monumental 1859 treatise *On the Origin of Species by Means of Natural Selection*. The components of a genetic algorithm are quite simple and can be summarized as follows. We begin with a scheme for encoding all elements of the feasible (inordinately large) solution space, often as bit strings (similar to DNA strings in genetics). The problem to be solved will usually be to search for some element of the solution space that is the best in terms of a desirable certain property. We then introduce a nonnegative *fitness function* that measures the quality of the particular element of the solution space. (We should have a way to compute this function in terms of the encoded representations.) The process begins with a randomly selected population of *chromosomes* (of a moderate size) from the solution space. The chromosomes are represented by strings of *genes* that can take on several values (in biology/evolution the possible values or forms of a gene are called *alleles*). These elements will then mate to produce the next generation. The selection for mating is determined by use of the fitness function: more fit individuals are more likely to procreate, while the less fit individuals will procreate less or not at all. The reproduction is usually done with one or both of the following two steps. First, the bit strings of a given mating pair have their genes *crossed over* and/or swapped in some fashion. Subsequently, each gene of the offspring has a probability for a *mutation*. The idea of the mating is that better solutions tend to combine to give even more fit future generations, just as in nature, where features that hinder fitness tend to die off whereas features that help organisms to survive or flourish in an environment tend to persist in future generations. The mutation operation is important as it prevents the scheme from prematurely converging to a locally optimal solution; in nature, mutations create diversity that in the same way can help species acquire new useful traits that help survivability.

Motivating Example

We begin with an example of a very simple optimization problem that will allow us to introduce some of the main facets of a genetic algorithm.

EXAMPLE 10.3: (*Motivating example for a genetic algorithm*) The problem is to use a genetic algorithm to determine the maximum value of the function

$$f(x) = \frac{100}{[(x-178)/50]^2 + 1} + \frac{160}{[(x-384)/100]^2 + 1},$$

over the set of <u>integers</u> $\{0,1,2,\cdots,511\}$. This problem has a very small solution space and so can easily be solved by brute-force, so we will be able accurately assess the performance of the genetic algorithm(s) that we implement.

SOLUTION: We outline the general steps that could be done in the construction of any genetic algorithm, and then work through the example at hand.

Step 1: Encode the elements of the solution space. Elements of solution space here are quite simple to represent with binary bit strings, we simply represent the integers in this set $\{0,1,2,\cdots,511 = 2^9 - 1\}$ as nine-digit bit strings (of zeros and ones) $[b_1 \ b_2 \ \cdots b_9]$ $(b_i \in \{0,1\})$, where an integer's associated **chromosome** is simply the bit string of its base-2 (binary) expansion. We use a tilde $\sim$ to denote this correspondence, so, for example, $85 \sim [0\ 0\ 1\ 0\ 1\ 0\ 1\ 0\ 1]$ (i.e., $85 = 0\cdot 2^8 + 0\cdot 2^7 + 1\cdot 2^6 + 0\cdot 2^5 + 1\cdot 2^4 + 0\cdot 2^3 + 1\cdot 2^2 + 0\cdot 2^1 + 1\cdot 2^0$). Thus, each chromosome is made up of nine (ordered) **genes**, and each gene has two possible **alleles** (0 or 1). In general, the encoding step can be a creative and intricate process. This will become more apparent when we begin to examine some of the wide range of problems to which genetic algorithms can be applied.

Step 2: Introduce the fitness function and show how it can be computed on encoded elements of the solution space. A **fitness function** for a genetic algorithm is simply a nonnegative function on elements of the solution space whose values are larger for better feasible solutions. In our case, since the problem at hand is to maximize a nonnegative function, a natural choice for the fitness function is simply the function $f(x)$ that was given, and this will be what we use. We stress that this is not the only choice. For example, the functions $2f(x)$, $f(x)+500$, $f(x)^2$, and $\sqrt{f(x)}$, could equally well serve as fitness functions, and each would correspond to a different genetic algorithm.[6] In the sequel we will encounter some situations in which such modifications can be helpful. This begins to show the diversity of modifications that one has in constructing genetic algorithms. In our case, the value of the function $f(x)$ at a bit string $[b_1 \ b_2 \ \cdots \ b_9]$ would be obtained by substituting $\sum_{i=1}^{9} 2^{9-i} b_i$ into the formula for x.

Step 3: Randomly generate an initial population from the solution space. We have experience going back to Chapter 6 with this sort of random generation task. We must decide, however, the size N of the initial population. For simplicity of the mating process, we stipulate only that N be even. (The population size is yet another parameter for the genetic algorithm.) There is even the possibility that the population size can vary from generation to generation. As one would expect, larger populations tend to have better chances of locating optimal solutions, but the amount of work per generation also increases. For now, to keep things simple, we will work with an initial population size of $N = 10$, and all future populations will be of this same size.   The ten rows of the matrix A below represent the chromosomes of the randomly generated initial population; the entries of A were generated by using the computer's random number generator (rand) multiplied by two, and then taking the floor of the results.

$\rightarrow$ A =

0	0	1	1	0	0	0	0	0	← chromosome #1
0	1	1	0	0	0	1	1	1	← chromosome #2
1	1	1	1	1	0	1	0	0	← chromosome #3
0	1	1	0	0	0	1	0	1	← chromosome #4
0	1	0	0	0	0	1	0	0	← chromosome #5
1	1	0	0	1	1	0	0	1	← chromosome #6
0	0	1	1	1	0	1	1	1	← chromosome #7
1	1	0	1	1	1	1	1	1	← chromosome #8
1	0	1	0	1	0	1	0	1	← chromosome #9
1	0	0	1	1	1	0	1	0	← chromosome #10

Step 4:   Formulate the mating selection process.   Since our mating selection process will be identical in passing from one generation to the next (although this need not be the case in general), we need only explain how the mating selection will be done for the initial population of the $N = 10$ chromosomes constructed above (*generation zero*). Since selection depends on fitness, we first compute the fitness of each of the chromosomes of the initial population. For example, the fitness of chromosome #1 is given by $f(0 \cdot 2^8 + 0 \cdot 2^7 + 1 \cdot 2^6 + 1 \cdot 2^5) = f(96) =$ 44.3179. The fitness levels of the remaining chromosomes are computed in the same fashion (computationally, one should implement a program that converts binary expansions to their integer equivalents first; we refer to Section 4.1); the

---

[6] More generally, each of these examples falls under the category of a composition $\varphi \circ f(x)$, where $\varphi : [0, \infty) \rightarrow [0, \infty)$ is a (nonnegative) strictly increasing function. The examples mentioned in the text come from using $\varphi(t) = 2t$, $t + 500$, t^2, and $\sqrt{t}$, respectively.

results are shown in Table 10.1. Table 10.1 also shows that number of times each chromosome is selected to mate. These mating numbers will now be explained.

#	Chromosome	Fitness f_i	Proportion of Total Fitness $f_i / \sum_j f_j$	Mating Frequency
1	0 0 1 1 0 0 0 0 0 ~ 96	44.3179 = f_1	0.04292	0
2	0 1 1 0 0 0 1 1 1 ~ 199	121.1837 = f_2	0.11736	2
3	1 1 1 1 1 0 1 0 0 ~ 500	70.5672 = f_3	0.06834	2
4	0 1 1 0 0 0 1 0 1 ~ 197	122.9621 = f_4	0.11908	1
5	0 1 0 0 0 0 1 0 0 ~ 132	75.9270 = f_5	0.07353	0
6	1 1 0 0 1 1 0 0 1 ~ 409	155.0636 = f_6	0.15017	1
7	0 0 1 1 1 0 1 1 1 ~ 119	61.7429 = f_7	0.05980	1
8	1 1 0 1 1 1 1 1 1 ~ 447	117.8789 = f_8	0.11416	0
9	1 0 1 0 1 0 1 0 1 ~ 341	143.6327 = f_9	0.13910	2
10	1 0 0 1 1 1 0 1 0 ~ 314	119.2896 = f_{10}	0.11553	1

TABLE 10.1: Fitness levels for generation zero of the genetic algorithm of Example 10.3. The mating frequencies we determined by randomly selecting 10 chromosomes (see the spinning wheel in Figure 10.8) based on total fitness proportions.

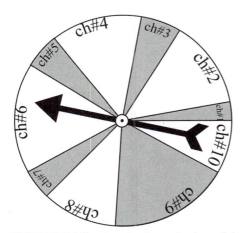

FIGURE 10.8: A spinning wheel model for the fitness-proportionate mate selection scheme.

Of course, in natural evolution, many factors are present to determine the survivability and strength of a given organism. In genetic algorithms, the fitness level should be thought of as an overall summary of such factors. This is why the choice the fitness function is such an important aspect of creating a genetic algorithm. Even with the fitness levels known, we still need to ascertain how they will be used to determine a selection scheme for mating. One natural selection scheme that is particularly simple to implement is to make $N = 10$ random drawings (with replacement) from the chromosome pool where the probability that a particular chromosome #i gets selected equals its fitness level divided by the sum of all of the fitness levels $f_i / \sum_j f_j$. Intuitively this can be visualized by using a spinning wheel, where the probabilities $f_i / \sum_j f_j$ of selection make up the percentages of landing on a corresponding sector, which would mean the corresponding chromosome gets selected; see Figure 10.8.[7]　Such a scheme is easy to program directly, or can be accomplished using the program GeneralFiniteSimulator of Exercise for the Reader 6.14.　Table 10.1 summarizes these probabilities as well as the resulting mating pool selected from a typical run of such a selection process.　The actual sequence of randomly selected chromosomes that were generated by such a scheme is as follows:[8]

$$6 \quad 1 \quad 10 \quad 6 \quad 8 \quad 2 \quad 2 \quad 9 \quad 9 \quad 5$$

This (unsorted) sequence will be important for the next step.

We assume that we have access to a program with syntax:

$$n = \texttt{bin2int(vec, k)},$$

that converts binary string vectors (of zeros and ones) into their integer equivalents.　The first input is vec, a vector of zeros and ones representing a binary string, the optional second input variable k is a nonnegative integer that will specify the length of the output vector, and the output n is a nonnegative integer represented by the binary string vec.　Without k being specified, the output will be a vector of minimal length.　(When $n = 0$ the output should be v = [0]), while if k is specified, the length of the output vector should be at least equal to this minimal needed length, and　the output vector will be prepadded with additional zeros, as necessary.　For example the output of int2bin(3) should be [1 1], while that of int2bin(3,5) should be [0 0 0 1 1].　The reader may refer to Section 4.1 for the details needed to write such a program.

EXERCISE FOR THE READER 10.3: Write a program with syntax

$$M = \texttt{MatingPool(P,fitness)}$$

that has two input variables: a population matrix P whose rows represent binary strings of the chromosomes making up a population in a genetic algorithm, and fitness, a vector of fitness values whose ith entry equals the fitness of the chromosome represented by the ith row of P (= P(i,:)).　The output M is a

---

[7] Since our general definition of a fitness function assumes only that it is nonnegative, it is possible that some chromosomes in the initial population have a zero fitness levels $(f_i = 0)$, or even that the total fitness of all elements in the initial population is zero $\sum_j f_j = 0$.　In the first case, this simply means that such individuals would never be selected to mate.　We will assume that the latter case does not occur (otherwise our mating process would not make sense); if it is does, either choose another initial population or modify the fitness function to be a bit more generous.

[8] The results were randomly obtained and will of course vary even on the same computing platform. Note that chromosome #3, which had only about 7% of the total fitness, got selected twice whereas chromosome #8 with about 11.5% of the total fitness did not get selected at all for mating.　This is just the effect of chance, and part of the reason biological phenomena cannot be completely predictable. Nonetheless, if one adds up the total fitness percentages of the mating pool, it will be nearly 110%, exceeding that of the original population.　Such generational trends are typical for genetic algorithms, and their combinations　are what make the methods succeed.　Just as in evolution, occasionally inferior choices are made and organisms with defects are created, but the long-run averages tend to improve through generations.

matrix of the same size as P, whose rows represent the randomly selected chromosomes making up the mating pool. More precisely, a spinning wheel procedure (see Figure 10.8, or just use the `GeneralFiniteSimulator` program of Exercise for the Reader 6.15) should be used to select elements of the current population (rows of P) to make up (in the order selected) the rows of M. Of course, the chromosomes of P may appear more than once (or not at all) in the mating pool matrix M. Run your program on the initial population matrix appearing in the text, and run it once more (using the same P). Compare the resulting mating pools.

Step 5: Mating Part I: Crossover. The ten chromosomes (with duplications) selected for mating in the preceding step will be randomly paired off, and each pair will produce two offspring for the next generation. The offspring will be produced by crossing over (swapping) portions of the chromosome strings of the parents. This can be done in a number of ways (again adding to the already great variety in setting up a genetic algorithm), but we begin with a simple **single-point crossover** that is illustrated in Figure 10.9.

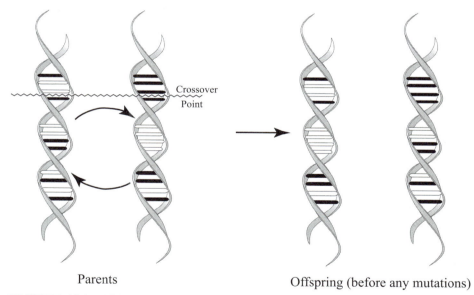

Parents   Offspring (before any mutations)

FIGURE 10.9: Illustration of a single-point crossover. A crossover point is randomly selected for each mating pair of parents. The offspring result from swapping the portions of the chromosomes appearing after the crossover point.

As for the random pairing off, we may simply read off subsequent pairs from the unordered list that was (randomly) generated in the preceding step. Thus, the five pairs of parents that each produce two offspring for the next generation are as follows:[9]

Parents: {6   1} {10   6} {8   2} {2   9} {9   5}

The crossover point for two chromosomes of length ℓ will be randomly chosen (for each mating pair) from the set $\{1, 2, \cdots, \ell - 1\}$. Thus, for the above five pairs of parents, we will need to (randomly) choose five crossover locations from the set $\{1, 2, \cdots, 8\}$. Simply filling in a length five vector with the entries `ceiling(9*rand)` will do this for us: here is the result of such an operation: (again, these results are random, and will vary):

Crossover Points: 5   6   2   4   8

Let us explain how the crossover will work for the first pair of parents, Chromosome #6 and Chromosome #1 (with crossover point = 5, meaning after the fifth genes of the parents).

parents:

$$\begin{cases} 1 & 1 & 0 & 0 & 1 & \boxed{1 \quad 0 \quad 0 \quad 1} \\ 0 & 0 & 1 & 1 & 0 & 0 \quad 0 \quad 0 \quad 0 \end{cases} \Rightarrow$$

offspring:

$$\begin{cases} 1 & 1 & 0 & 0 & 1 & 0 & 0 & 0 & 0 \\ 0 & 0 & 1 & 1 & 0 & \boxed{1 \quad 0 \quad 0 \quad 1} \end{cases}$$

These two offspring replace the parents in the next generation, but they are subject to possible mutations in the next step.

We give now the pseudocode for the single point crossover operator described above: `C = Crossover(M)`

```
Initialize: M = matrix whose rows represent chromosomes
                  in mating pool  (in random order)
            N = number of rows of M  (size of population, even)
            len = number of columns of M  (chromosome length)
            C = M  (initialize output matrix to equal M)
```

```
FOR i = 1 TO N/2
    c = ceiling((len-1)*rand))  (crossover point location)
    (need to swap the tails of the (2i-1)th and (2i)th chromosomes)
    (we attach the tail of the (2i)th chromosome onto the (2i-1)th)
    C(2*i-1,(c+1):len) = M(2*i, (c+1):len)
    (now attach the  tail of the (2i-1)th onto the (2i)th)
    C(2*i,(c+1):len) = M(2*i-1, (c+1):len)
END   i FOR
```

We will hold off running this program until we explain the second part of the mating process:

---

[9] Although it has not occurred here, it is certainly possible to have two identical parents in one of the mating pairs. In such a case, any crossover will produce a pair of offspring identical to the parents. A subsequent mutation, however, could eventually change these offspring.

Step 6: Mating Part II: Mutation. After a set of offspring have been produced by mating and crossover, the **mutation operator** simply runs through each of the genes of the offspring chromosomes and randomly mutates each allele value (from 0 to 1 or vice versa) with (gene-by-gene) **mutation probability** p_m (yet another parameter). As mentioned earlier, as in real life, the mutation operator safeguards against the genetic algorithm missing some fruitful combinations of genes and can prevent the algorithm from converging too rapidly to a locally optimal solution. The mutation probability p_m is often rather small (on the order of 1/1000), but optimal values of it will vary depending on the nature of the problem being solved, the fitness function, etc. Sometimes, the value of this parameter is allowed to evolve as the genetic algorithm progresses, in much the same way as the population becomes more fit from generation to generation. In this introductory example, we simply use the fixed value $p_m = 1/1000$, meaning that (since a generation consists of 90 genes) roughly one gene in every eleven generations will be mutated. Coding such a mutation operator is rather straightforward and will be left as the following exercise for the reader.

EXERCISE FOR THE READER 10.4: Write a program with syntax:

```
P = Mutation(C,pm)
```

that takes two input variables, C = a matrix (assumed to have just been outputted by the crossover operator), and pm = the gene-by-gene probability of a random mutation, and an output matrix P, having the same size as C, where the rows correspond to chromosome strings of the population of C, after being acted on by the mutation operator.

We now apply the foregoing mating scheme to the initial population to obtain the following generation 1 population matrix:

$\rightarrow$P =

1	1	0	0	1	0	0	0	0	$\leftarrow$ chromosome #1
0	0	1	1	0	1	0	0	1	$\leftarrow$ chromosome #2
1	0	0	1	1	1	0	0	1	$\leftarrow$ chromosome #3
1	1	0	0	1	1	0	1	0	$\leftarrow$ chromosome #4
1	1	1	0	0	0	1	1	1	$\leftarrow$ chromosome #5
0	1	0	1	1	1	1	1	1	$\leftarrow$ chromosome #6
0	1	1	0	1	0	1	0	1	$\leftarrow$ chromosome #7
1	0	1	0	0	0	1	1	1	$\leftarrow$ chromosome #8
1	0	1	0	1	0	1	0	0	$\leftarrow$ chromosome #9
0	1	0	0	0	0	1	0	1	$\leftarrow$ chromosome #10

As it turned out, the mutation operator did not change any of the 90 entries of the matrix. The reader can verify that this mean fitness has increased from 103.3 to 117.9, and the maximum fitness has increased from 155.1 to 160.8, in passing from generation 0 to generation 1.

Step 7: Continue to let the genetic algorithm run until some convenient stopping criterion has been met. There are many sorts of stopping criteria, a simple one being to just let the algorithm run for a fixed number of generations. Another would be to stop when, say, the maximum fitness has not made more than a specified amount of improvement during the preceding 10 (or so) iterations. The result of the algorithm will simply be the most fit chromosome that was encountered in during the run.

The Basic Genetic Algorithm

In terms of the operators introduced above, we now provide the outline for genetic algorithm that runs through G generations.

Algorithm 10.2: A basic genetic algorithm that runs for G generations with mutation and crossover.
It is assumed that we have a fitness function f and an encoding scheme, as well as a mating selection scheme has been established.

Step 1: Generate the initial population matrix A whose rows are binary strings representing the initial population chromosomes, and set g = 1 (g = generation counter). Form the vector `fitness`, whose *i*th entry is the fitness of the chromosome corresponding to the *i*th row of A. Record the row of A corresponding to the chromosome of maximum fitness, CMax, and its corresponding fitness level MaxFit. (In case of a tie, choose any such chromosome.)

Step 2: If g = G, go to Step 4. If g = 1, set P = A.   Next, let M = MatingPool(P,fitness).

Step 3: (*Mating*) Let C = crossover(M), and then set P = mutation(C,pm). Update the `fitness` vector. Record the maximum fitness level of the new population matrix. If it exceeds MaxFit, update CMax (to be an appropriate row of P) and MaxFit (to be this new highest fitness level).

Step 4: Output CMax and MaxFit.

Before proceeding with the example at hand, we stress that the results will vary since all of the programs depend on random numbers generated within the program.[10] It will thus not be feasible for the reader to attempt to replicate this or any future generations of any genetic algorithm. In terms of the module programs thus far introduced, the above algorithm, applied to the example at hand, can be put into pseudocode as follows:

```
pm = 1/1000, N = 10, s = 9, G = 225  (initialize parameters)
P = floor(2*rand(N,s))    (generate initial population)
FOR i = 1 TO N
       fitness(i) = f(bin2int(P(i,:)))   (create fitness vector)
```

---

[10] Thus, it is not quite proper to refer to the generation one, two, etc. population of a genetic algorithm, since these populations can vary depending on the computer being used or even the particular run on the same computer. A more proper wording would be to refer to "a generation one population," etc.

```
END (i FOR)
MaxFit = max(fitness), imax = index (where fitness(imax) = maximum)
CMax = P(imax, :)    (chromosome of maximum fitness)
FOR g = 1 TO G
    M = MatingPool(P,fitness)
    C = Crossover(M)
    P = Mutation(C,pm)  (next generation population matrix)
    FOR i = 1 TO N
        fitness(i) = f(bin2int(P(i,:)))   (create fitness vector)
    END i FOR
    IF   max(Fitness) > MaxFit
       MaxFit = max(fitness)  (update maximum fitness)
       imax = index  (where fitness(imax) = maximum)
       CMax = P(imax, :)    (update chromosome of maximum fitness)
    END IF
END g FOR
OUTPUT:  MaxFit, CMax
```

The initial population is graphically compared with the generation 1 and generation 10 populations in Figure 10.10a. Figure 10.10b gives two plots of the maximum fitness of each generation using two different mutation parameters: pm = 1/1000 and pm = 1/500.

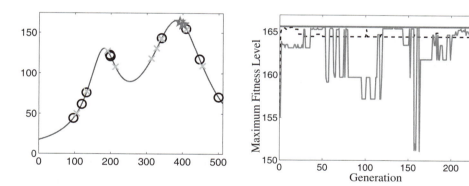

FIGURE 10.10: (a) (left) Graph of the (fitness) function $f(x)$ that is to be maximized in Example 10.3, along with the initial population (o's), the generation 1 population (x's), and the generation 10 population (stars). (b) (right) Plots of the maximum fitness for each generation for 225 generations of the genetic algorithm of Example 10.3 (dashed graph), along with that for another 225 generations of the same algorithm but with the mutation probability changed from 1/1000 to 1/100 (solid graph). The thick horizontal line represents the maximum fitness.

A few comments are in order. In Figure 10.10(a), it may appear that the generation 10 population (indicated with stars) is smaller than the other two populations shown. All of the populations are indeed of the same size 10, it is just that the generation 10 chromosomes have converged closer to the maximum and so not all are distinguishable. In Figure 10.10(b), notice how much more volatile the maximum fitness levels are when the mutation rate is changed from 1/1000 to 1/100. This indicates a greater degree of experimentation. As a result, the latter genetic algorithm did actually discover the maximum of $f(x)$, whereas the former did not (although it came quite close). In fact, observe that the 1/1000-mutation rate genetic algorithm seemed to have its maximum fitness converge to a suboptimal level. Basically this happened because the mutation rate was too low for significant experimentation, and because the fitness function does not change much for near optimal levels. These two features caused the domination of the population by well-fit, but not quite optimal chromosomes.

Cloning and Inversions

Many other variations of genetic algorithms are possible; for example, in addition to using modified fitness functions, a fitness function can even be allowed to change during the execution of the algorithm (depending, say, on the mean fitness values or the range of fitness values of the current population).   Another modification would be to **clone** the highest (or two, or three highest) fitness level chromosomes in a generation so that they will automatically be reproduced (cloned) in the next generation. This would make the graphs in Figure 10.10(b) nondecreasing (Why?).   Such variations and others will be addressed in the exercises, and in the following two exercises for the reader.

EXERCISE FOR THE READER 10.5:   (*Inversion*)   Another feature that is commonly implemented in genetic algorithms is the so-called **inversion operator**. The inversion operator can be applied to the new population (after crossover), either directly before or after the mutation operator, and works as follows: Each chromosome in the new population will be acted on by the inversion operator with probability $p_{inv} > 0$ (a parameter).   The inversion operator acts on a chromosome of length ℓ as follows: first a (left) index, a, is randomly chosen from the set $\{1, 2, \cdots, \ell - 1\}$,   next, a right index, b, is randomly chosen from the set $\{a+1, a+2, \cdots, \ell\}$. The portion of the chromosome from the ath gene through the bth gene is then replaced by the same portion in the reverse order. For example, if the chromosome being inverted is [0 1 1 1 0 0 0 1 0] and the left and right indices randomly selected are $a = 2$ and $b = 7$, then the inversion operator would act as follows:

$$[0\ 1\ 1\ 1\ 0\ 0\ 0\ 1\ 0] \quad \Rightarrow \quad [0\ 0\ 0\ 0\ 1\ 1\ 1\ 1\ 0]$$

(a) Write a program with syntax:

$$P = \text{Inversion}(C, pinv)$$

that takes two input variables, C = a matrix (assumed to have been just outputted by the crossover operator) and `pinv` = the chromosome-by-chromsome probability of a random inversion, and an output matrix P, having the same size as

C, where the rows correspond to chromosome strings (rows of C), each after being acted on by the inversion operator with probability pinv.
(b) Rewrite your genetic algorithm code of Example 10.3 to include the inversion operator immediately following the crossover, and run it using some different values of pinv along with pm = 1/1000 or 1/100. Can you find values of pinv, where the results appear to improve those without using inversion? For comparison, create some plots corresponding to those of Figure 10.10b.

EXERCISE FOR THE READER 10.6: (*Cloning*) (a) Modify the genetic algorithm code of Example 10.3 so that each time we move from one generation to the next, a chromosome (CMax) of maximal fitness in the current generation is selected to be cloned and appear in the next generation. For simplicity, make CMax appear as the first row of the next population matrix (thus replacing the first new offspring produced). Run your modified code using each of the parameters pm = 1/1000 and 1/100. Create some plots corresponding to those of Figure 10.10b, and compare with the performance of Example 10.3.

Application to Ramsey Numbers

Example 10.3 was carefully designed to facilitate the binary encoding of the solution space and to have a natural choice of a fitness function. Genetic algorithms have vast applicability, as long as one is able to develop an encoding scheme for the solution space where the crossover operation will make sense, and establish a fitness function that captures well the essence of the problem. We will next give an example where both of these tasks will be a bit less natural. The application area has to do with estimating so-called *Ramsey numbers*. These numbers were introduced by the short-lived English mathematician Frank P. Ramsey (1903–1930). Ramsey numbers have been the focus of much attention and activity in combinatorial mathematics, and many questions about them remain unsolved.

DEFINITION 10.5: For any pair of integers $a, s > 1$, we say that a positive integer N has the **(a, s) Ramsey property** if, whenever there is a party with N people present, in this party there will always exist either a group of a mutual acquaintances, or a group of s mutual strangers.[11] The **Ramsey number $R(a, s)$** is defined to be the minimum number N that has the (a, s) Ramsey property (if such a numbers exist).

For a fixed pair (a, s), it is clear from the definition that if N has the (a, s) Ramsey property, and $K > N$, then K also has the (a, s) Ramsey property (at a K-person party, simply focus on any N of the people). Similarly, if N does not have the (a,

---

[11] We assume that any pair of people at the party will either be acquaintances or strangers (to one another).

s) Ramsey property, and $K < N$, then K also cannot have the (a, s) Ramsey property (Why?). Ramsey proved that $R(a,s) < \infty$, for all $a, s > 1$, but very few Ramsey numbers are known exactly. For example, from the definition it is clear that $R(2, 2) = 2$, and also that $R(2, n) = n$, for any integer $n > 1$. It is not difficult to show (and we will do this below) that $R(3, 3) = 9$. Although $R(4, 4)$ is known to be 18, the value of $R(5, 5)$ is still unknown. At the time of this writing, the best information known about this latter Ramsey number is that $43 \leq R(5,5) \leq 49$. In general, an upper bound for a Ramsey number, e.g., $R(5,5) \leq 49$, is proved by showing any party of 49 people must satisfy the Ramsey condition (i.e., there must be either five mutual acquaintances or five mutual strangers at the party). A lower bound, e.g., $43 \leq R(5,5)$, is proved by constructing a counterexample, i.e., constructing a 42-person party in which there are no groups of five mutual acquaintances and no groups of five mutual strangers. Genetic algorithms can be applied to the problem of finding such counterexamples, and thus to the task of obtaining lower bounds for Ramsey numbers. As usual, the heart of the matter will be to decide on a suitable encoding scheme, as well as an appropriate fitness function.

Conceptually, Ramsey numbers can be recast in the setting of graphs. Indeed, consider any six-vertex graph, such as the one shown in Figure 10.11, where two vertices (persons) are connected by an edge if they are acquainted; otherwise they are strangers. The fact that $R(3, 3) = 6$ (or really just the fact that $R(3,3) \geq 6$) tells us that such a graph (thought to represent a six-person party) must have either three people who are mutual friends, or three who are mutual strangers. For the graph of Figure 10.11, it is easily seen that (only) the latter case occurs; for example, the vertices 1, 2, and 5, are mutual strangers.[12] To

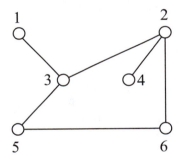

FIGURE 10.11: A six-person party graph; edges represent acquaintance relationships.

give the flavor (on a very small scale) of how the task of exactly computing a Ramsey number proceeds, we will compute $R(3, 3)$ to prove the following theorem:

THEOREM 10.1: $R(3, 3) = 9$.

---

[12] Graph theory has some relevant concepts in which we can elegantly restate the condition that Ramsey numbers need to satisfy: In any graph, a set of vertices $v_1, v_2, \cdots, v_k$ is called a *clique* if any pair of them is joined by an edge of the graph (i.e., are acquaintances). A set of vertices is called an *independent set* if no pair of them is joined by an edge (i.e., all are mutual strangers). Thus, $R(a,s)$ is the smallest number such that any graph with at least $R(a,s)$ vertices will contain either a clique of size a or an independent set (of vertices) of size s.

Proof:[13]   (a) (*Upper bound*)   We show that $R(3, 3) \leq 6$ by proving that any six-person party will have either three mutual acquaintances or three mutual strangers. Consider such a party, and let's pick one of the guests—call him/her A. Now the other five guests can be separated into two groups: those who know A and those who are strangers to A. One of these groups must contain at least three people (this is rather clear, but also follows from the generalized pigeonhole principle). *Case 1:* (At least) three people are acquainted with A; call them B, C, and D. If these latter three people are mutual strangers, then they satisfy one of the required Ramsey conditions. Otherwise, at least two of them, say B and C, will be acquaintances. But then together with A, B and C will form a set of three mutual acquaintances, and the Ramsey condition is satisfied. *Case 2:* Three people are all strangers to A. This case is dealt with in an analogous fashion.

(b)   (*Lower bound—via a genetic algorithm*)   We show that $R(3, 3) > 5$ by constructing a counterexample of a five-vertex graph that has neither a three-vertex clique nor a three-vertex independent set. Although this can be done by direct trial and error or brute-force, we will employ a genetic algorithm to obtain such an example. The solution space is the set of all five-vertex graphs. Such graphs can have at most $C(5,2) = 10$ edges, and so can be represented by length 10 binary strings. Such a scheme is more precisely explained with the following example:

Assign reference numbers to (possible) edges of a five vertex graph as follows: (shaded blocks represent edges of a particular example).

Reference #	1	2	3	4	5	6	7	8	9	10
Edge	(1,2)	(1,3)	(1,4)	(1,5)	(2,3)	(2,4)	(2,5)	(3,4)	(3,5)	(4,5)

Corresponding binary string: $e = [1\ 0\ 1\ 1\ 1\ 0\ 0\ 0\ 1\ 0]$. The entries corresponding to edges included in the graph equal 1 and are shaded, i.e., this vector will represent the five vertex graph (with vertices labeled from 1 to 5) whose edges are precisely those from the above table that correspond to indices i where $e(i) = 1$. The corresponding graph is shown in Figure 10.12. This graph has three mutual strangers (e.g., vertices 2, 4, and 5), and so is not a counterexample to show $R(3, 3) > 5$.

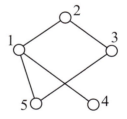

FIGURE 10.12: A 5-vertex graph represented by the (edge) vector $e = [1\ 0\ 1\ 1\ 1\ 0\ 0\ 0\ 1\ 0]$.

We formulate our nonnegative fitness function as follows: Each five-vertex graph (represented by a length 10 binary sequence as shown above) has a total of $C(5,3) = 10$ three-vertex subsets. In order for such a five-vertex graph to serve as a counterexample (to show $R(3, 3) > 5$), it would need to be the case that the Ramsey condition fails for each of the 10 three-vertex subsets. We will let the

---

[13] The proof of this theorem actually appeared as one of the problems on the famous *Putnam Exam.* The (William Lowell) Putnam exam is an extremely challenging mathematics exam that is annually given to a select group of undergraduates.

fitness function of the graph simply be the count of the number of three-vertex subsets for which the Ramsey condition fails. Thus, a five-vertex graph will be a counterexample to show that $R(3, 3) > 5$ precisely if it has a fitness value of 10. We need a convenient way to evaluate the fitness level of a five-vertex graph. It will be useful to work with the following *lookup vector*: VL = [1 5 8 10], which gives the reference numbers where a new first vertex appears in the edge ordering in the table above. Indeed, two vertices $i < j$ of the graph are acquainted if and only if there is an edge joining them, and this in turn is equivalent to the corresponding entry of the representing vector e equaling 1. But the reference number corresponding to the edge (i, j) is precisely $VL(i) + j - i - 1$ (see the table above). This fact is the basis for the following pseudocoded program for computing the fitness of a five-vertex graph that is represented by a length 10 binary string e.

Pseudocode for program `fitness = Ramsey3_3_fit(e)`, that inputs a length 10 binary vector `e` (representing a 5-vertex graph) and outputs its `fitness` (an integer in {0, 1, ..., 10} as described above):

```
Initialize:  e  (length 10 binary vector)
             VL = [1   5   8   10]   (look up vector)
             fitness = 0   (initial fitness for e)
          T = matrix  of  the  10  3-combinations  of  the  5-
vertex set
             {1, 2, 3, 4, 5}  (each row of T is an increasing list
                                of three integers from the set {1, 2, 3, 4, 5})
FOR i = 1 TO 10
          comb = T(i,:)   (we will run through all 10 3-combinations)
          (we need to check whether comb fails the Ramsey
                Condition, if yes, we bump up fitness by 1)
          a12 = e(VL(comb(1)) + comb(2) - comb(1) - 1)
          a13 = e(VL(comb(1)) + comb(3) - comb(1) - 1)
          a23 = e(VL(comb(2)) + comb(3) - comb(2) - 1)
   (a12 = 1 if and only if comb(1) and comb(2) are acquainted, similarly for a13, and a23)
          IF   a12 + a13 + a23 > 0 AND a12 + a13 + a23 < 3
   (the first condition says at least two of comb(1), comb(2), comb(3) are acquainte;, the
                        second says at least two are mutual strangers)
                fitness = fitness + 1
          END IF
END i FOR
OUTPUT:  fitness
```

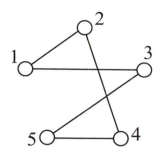

FIGURE 10.13: A counterexample to show that 5 does not have the (3,3) Ramsey property.

With the above encoding scheme and fitness program, it is a simple matter to code a basic genetic algorithm (Algorithm 10.2) to seek a desired - example. Indeed the sample space is so small (there are only $2^{10} = 1024$ graphs) one can even use brute-force to find out how many counterexamples there are. Using a mutation probability of $p_m = 0.05$, an initial population of size $N = 10$, and $G = 250$ generations resulted in a counterexample represented by the sequence $e = [1\ 1\ 0\ 0\ 0\ 1\ 0\ 0\ 1\ 1]$. The reader should run such a program.[14] The corresponding graph, shown in Figure 10, is easily verified to be a counterexample (although such verification is not necessary since the program checked it already). For example, each person (vertex) has only two acquaintances, but in all cases those two acquaintances are mutual strangers. Thus there cannot be a group of three mutual acquaintances. Similarly, there can be no group of three mutual strangers. □

EXERCISE FOR THE READER 10.7: Of the 1024 possible five-vertex graphs, how many of them could serve as counterexamples to show that $R(3, 3) > 5$?

Whereas a counterexample needed to show $R(3, 3) > 5$ (in fact, all such counterexamples) could be found by brute-force (on any computer), the sample space in our next example will be far too large for this.

EXAMPLE 10.4: (a) Use a basic genetic algorithm to search for a counterexample to show that $R(3, 5) > 13$.[15]

(b) Using the fitness function implemented in Part (a), perform a random search for a counterexample (using about the same amount of computer time as the genetic algorithm of Part (a)), and record the highest fitness level graph encountered. Compare the result with that of Part (a).

SOLUTION: Here the solution space consists of all 13-vertex graphs, and there are $2^{C(13,2)} = 2^{78} = 3.022 \times 10^{23}$ of these, dashing any hopes that a brute-force method might be successful.

Part (a): The encoding scheme and the fitness function will be analogous to the one introduced in the proof of Theorem 10.1. A 13-vertex graph has $C(13,2) = 78$ possible edges, and so will be represented by a length 78 binary vector. We

---

[14] It will most probably work, but if it does not, try again, perhaps with some modifications (like changing the mutation parameter, using a larger number of generations, cloning, etc.). The counterexample you get may be different from the one we obtained.
[15] It is known that $R(3, 5) = 14$. This was first proved in an important 1955 paper by R. E. Greenwood and A. M Gleason ([GrGl-55]).

take $\{1, 2, 3, \ldots, 13\}$ to be the vertex set. The 78 entries of a representing vector correspond to the natural ordering of 2-combinations of the vertex set:

$$\{1,2\},\ \{1,3\},\ \{1,4\},\ \ldots,\ \{1,13\},\ \{2,3\},\ \{2,4\},\ \ldots,\ \{2,13\},\ \{3,4\},\ \ldots,\ \{12,13\}.$$

The fitness function will be computed as follows: For a given 13-vertex graph, we first search through all $C(13,3) = 286$ 3-combinations of vertices and count how many of these are <u>not</u> mutual acquaintance networks (cliques). Denote this total as T_1. Next we search through all $C(13,5) = 1287$ 5-combinations of vertices and count how many of these are <u>not</u> mutual stranger networks (independent sets). Denote this latter total as T_2. In order not to (arbitrarily) give more weight to the T_2 total, we propose using $(T_1 + 2T_2/9)/2$. (Note that $(2/9)\cdot 1287 = 286$.) The corresponding reference look-up vector will be:

$$VL = [1 \quad 13 \quad 24 \quad 34 \quad 43 \quad 51 \quad 58 \quad 64 \quad 69 \quad 73 \quad 76 \quad 78].$$

(Note: This, and other such look-up vectors, can easily be generated by a simple computer loop; see Exercise 7.)   The resulting fitness function is easily programmed like the one in the proof of Theorem 10.1; here separate loops should be used to compute T_1 and T_2. We applied (just the basic) genetic algorithm (Algorithm 10.2) using the parameters: $N = 200$ (population size), $G = 5000$ (generations), and $p_m = 1/200$ (mutation probability), with the programming just like the pseudocode given for Algorithm 10.2. A single run of this program did not find a counterexample, but came quite close. The maximum fitness of an example found in this (single run of) the genetic algorithm was 285.56 out of a possible 286 (which would correspond to a counterexample).

Part (b): In order to run a random search that would take about the same amount of work as the genetic algorithm used in Part (a), we need to give a rough estimate of the amount of work used to execute this genetic algorithm. For each iteration of the genetic algorithm, the lion's share of the work (by far) is the evaluation of the fitness values of each string in the population. So we will compare the genetic algorithm to a random search algorithm in which we perform 5000 iterations, in each of which we randomly generate 200 graphs (represented by their binary strings). Thus a total of $200\cdot 5000 = 100,000$ random graphs will be generated. At each iteration, we compute the fitness values of each of the 200 graphs, and update the cumulative record maximum, if necessary. Here is a pseudocode for this random search algorithm:

Pseudocode for program [MaxFit CMax] = RandomSearch3_5, that runs through 5000 trials of generating 200 random 13-vertex graphs, and updating the maximum fitness value encountered (MaxFit), and a corresponding graph (CMax), as needed with each generation. The two output variables will be the maximum fitness value encountered (MaxFit) and a binary sequence (CMax) representing a corresponding graph with this fitness value. We will use the Fitness program explained earlier.

```
Initialize Paramters:  G = 5000, N = 200
P = floor(2*rand(N,s))    (generate initial population)
FOR i = 1 TO N
     fit(i) = Fitness(P(i,:)))   (create fitness vector)
END i FOR
MaxFit = max(fit), imax = index  (where fit(imax) = maximum)
CMax = P(imax, :)   (graph of maximum fitness)
FOR g = 1 TO G
     P = floor(2*rand(N,s))  (next N graphs are randomly generated)
     FOR i = 1 TO N
          fit(i) = Fitness(P(i,:)))   (create fitness vector)
     END (i FOR)
     IF  max(fit) > MaxFit
        MaxFit = max(fit)  (update maximum fitness)
        imax = index   (where Fitness(imax) = maximum)
        CMax = P(imax, :)   (update chromosome of maximum fitness)
     END IF
END g FOR
OUTPUT:  MaxFit, CMax
```

Figure 10.14 compares the cumulative performances of the basic genetic algorithm with the random search algorithm, along with a modified genetic algorithm that uses cloning, and inversion, as well as a modified mutation operator (see Exercise 23(d)). None of these algorithms found a counterexample, although the latter genetic algorithm came quite close: it reached (rather quickly) a fitness level of 285 7/9, just 2/9 off from the counterexample fitness level of 286. When a device called fitness scaling (see Exercise 24 and the note preceding it) was added to the latter algorithm (as in Exercise 24(d)), a counterexample was found (in under 5000 generations), and again for 15 times out of 20 subsequent runs of the same program. The first such counterexample found is shown in Figure 10.14b. From the fitness function, we deduce that the corresponding 13-vertex graph would have no three vertex cliques, and of all of the 1287 possible 5-vertex subsets, only two of them would be independent sets. The difficulty of the modified genetic algorithm in reaching a counterexample suggests that the *fitness landscape* is quite complicated, and that the slightly less than optimal solutions in the later generations have significant differences from a counterexample graph (however many there may be), i.e., differences that are not readily obtained from a few mutations and/or an inversion.

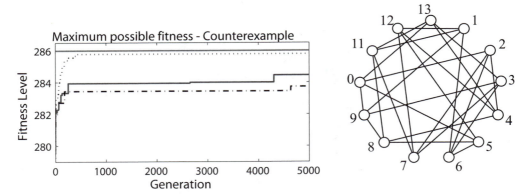

FIGURE 10.14: (a) (left) Comparsion of the three methods of Example 10.4 in search of a counterexample for $R(3, 5) > 13$. Dash-dot (lower curve) is the random search method. The solid (next lower curve) is the basic genetic algorithm, and the dotted curve is a more elaborate genetic algorithm. The solid horizontal line on top is the required fitness level for a counterexample. (b) (right) A counterexample found when the latter genetic algorithm was enhanced by fitness scaling.

The exercises will provide some additional illustrations of variety of problems to which genetic algorithms can be effectively applied.

Theoretical Underpinnings

With all of the parameters and other adjustment controls that one may impose on a genetic algorithm, it is no wonder that a coherent theory describing the performance of genetic algorithms is unlikely to ever catch up with all of the new and interesting variations. The most striking theoretical results that have been achieved were developed in the setting of (very) basic genetic algorithms, with only crossover and (sometimes) mutations. Even in such basic settings, rather sophisticated mathematical tools were needed. For example, Michael D. Vose and Gunar E. Liepins [VoLi-91] used a dynamical systems approach to analyze equilibria and fixed points of the crossover and mutation operators. A very different approach was taken by physicists Adam Prügel-Bennett and Jonathan Shapiro (see [PBSh-94] and [PBSh-97]) who took more of a macroscopic approach using statistical mechanics to obtain exact formulas for the so-called *cumulants* (statistical quantities associated with each generation) of the population for (just) the crossover operator. A nice synopsis of these two approaches can be found Melanie Mitchell's book [Mit-98]. Here we give a brief discussion of Holland's original theoretical development. Although this theory is rather limited in precise results, it does provide insights into how Holland viewed his invention to work.

Holland's point of view was that genetic algorithms discover and combine good building blocks for chromosomes. Holland called these building blocks schema, and they are defined as follows:

DEFINITION 10.6: For a genetic algorithm operating on length ℓ binary chromosomes, a schema $\mathbb{S}$ is a subset of chromosomes that can be represented by a length ℓ string whose elements are taken from the set $\{0, 1, *\}$. The asterisk symbol represents a *wild card* or a "don't care" slot.

Thus for example the length 5 schema $\mathbb{S} = 0**11$ represents the set of four length 5 chromosomes whose first gene is 0, and whose last two genes are 1s, i.e., the set $\mathbb{S}$ $=\{00011, 00111, 01011, 01111\}$. Any chromosome in a schema $\mathbb{S}$ is said to be an instance of $\mathbb{S}$. This schema is said to have **order** 3 since it has three **defined genes** (nonasterisks). The **defining length** of a schema is the distance between its outermost defined genes. Thus the defining length of the schema $0**11$ is 4.

Not all subsets of chromosomes are schema, since there are only 3^{ℓ} length ℓ schema, compared to $2^{2^{\ell}}$ subsets of length ℓ chromosomes (Exercise 26). Every chromosome belongs to 2^{ℓ} different schema (Exercise 26), and Holland viewed the genetic algorithm as processing information on all schema associated with a given generation's population. Indeed, each such schema represented by (at least one chromosome in a) certain generation's population has an average fitness estimate gotten by simply taking the average over all chromosomes in a population that belong to it. Although these individual schema averages are not explicitly computed, we will soon see that a genetic algorithm performs as if it were actually keeping track of all of these averages. We now proceed to make this notion explicit as we develop Holland's **schema theorem**:

We let $m(\mathbb{S}, g)$ denote the number of instances of the schema $\mathbb{S}$ at generation g (in the running of a genetic algorithm). Our goal is to estimate $m(\mathbb{S}, g + 1)$ in terms of $m(\mathbb{S}, g)$. We will do this by examining how each aspect of the basic genetic algorithm (mating, crossover, mutation) influences this transition, one at a time. We let $f_{avg}(\mathbb{S}, g)$ denote the average observed fitness of $\mathbb{S}$ at generation g, i.e.,

$$f_{avg}(\mathbb{S}, g) = \sum_{x \in \mathbb{S}} f(x)/m(\mathbb{S}, g)$$ (the sum is taken over all generation g strings that

are instances of $\mathbb{S}$). Under the assumption of fitness proportionate mating selection (as described in Example 10.3), the expected number of times that a chromosome x will appear in the mating pool is $f(x)/f_{avg} \left(= N \cdot f(x) / \sum f(x_i)\right)$.

First we just consider mating, and temporarily let $m(\mathbb{S}, g + 1)$ denote the number of instances of $\mathbb{S}$ in the mating pool for generation $g + 1$. The expectation $E[m(\mathbb{S}, g + 1)]$ of this quantity is simply the expected number of appearances that generation g instances of $\mathbb{S}$ make in the mating pool. Summing up the expected individual appearances that were explained in the preceding paragraph, we arrive at

$$E[m\,(\mathfrak{S}, g+1)] = \sum_{x \in \mathfrak{S}} f(x)/f_{avg}$$

$$= [f_{avg}(\mathfrak{S},g)\cdot m(\mathfrak{S},g)]/f_{avg} = [f_{avg}(\mathfrak{S},g)/f_{avg}]\cdot m(\mathfrak{S},g).$$

This important formula (the equality of the first and last quantities) shows the important fact that schema that are above/below average fitness are expected to have an increase/decrease of instances in the mating pool (and the proportion of expected increase/decrease is made precise). It is noteworthy that this fact holds true despite the fact that schema average fitness levels are not explicitly computed in a genetic algorithm.

EXERCISES AND COMPUTER EXERCISES 10.2:

1. Consider the problem of maximizing the function $f(x) = 15\cos(x/60) + (x/150)^2$ over the set of integers $\{0, 1, 2, \ldots, 511\}$.
    (a) Apply the genetic algorithm of Example 10.3 with 225 generations, with $p_m = 1/1000$, and with all other parameters as in the example. Compare the maximum fitness obtained with the true answer, and create a plot of the maximum fitness at each generation (compare with Figure 10.10(b)).
    (b) Repeat Part (a) using $p_m = 1/100$.
    (c)   Repeat Parts (a) and (b) invoking also the inversion operator with $p_{inv} = 1/10$. (See Exercise for the Reader 10.5.)
    (d)   Repeat Parts (a) and (b) invoking also the inversion operator with $p_{inv} = 1/4$. (See Exercise for the Reader 10.5.)
    (e) Repeat Parts (a) and (b) invoking the cloning procedure described in Exercise for the Reader 10.6.
    (f)   Which of the above variations (or any others that you may have examined) of the genetic algorithm seem to work best for this problem? Explain your answer.

2. Repeat all parts of Exercise 1 with the function being changed to $f(x) = ([x - 200]/100)^2$.

3. Repeat all parts of Exercise 1 with the function being changed to $f(x) = 3 + \dfrac{100}{10 + (x - 325)^2}$.

4. Run suitable modifications of the genetic algorithm of Example 10.3 to obtain plots similar to those shown in Figure 10.10, using the two mutation parameters: $p_m = 1/1000$ and $p_m = 1/100$.   In particular, you will need to create a vector of the maximum fitness level at each generation so that this vector can be plotted.

5. Consider the problem of maximizing the function $f(x) = (5000x - x^2)\sin^2(x/440)/1500^2$ over the set of integers $\{0, 1, 2, \ldots, 4095\}$.
    (a) Apply the genetic algorithm of Example 10.3 with the following parameters: $N = 50$ (population size), $G = 500$ (generations), $p_m = 1/1000$, and with all other parameters as in the

example. Compare the maximum fitness obtained with the true answer, and create a simultaneous plot showing (i) the maximum fitness at each generation, (ii) the mean (average) fitness at each generation, and (iii) the minimum fitness at each generation.

(b)   Repeat Part (a) using $p_m = 1/100$.

(c)   Repeat Parts (a) and (b) invoking also the inversion operator with $p_{inv} = 1/10$. (See Exercise for the Reader 10.5.)

(d)   Repeat Parts (a) and (b) invoking also the inversion operator with $p_{inv} = 1/4$. (See Exercise for the Reader 10.5.)

(e)   Repeat Parts (a) and (b) invoking the cloning procedure described in Exercise for the Reader 10.6.

(f)   Repeat Parts (a) and (b) invoking a cloning procedure similar to that described in Exercise for the Reader 10.6, except that now two chromosomes will be cloned in going from one generation to the next: one of the highest possible fitness levels, and another of the next highest possible fitness level. (Thus the two chromosomes cloned will always have different fitness levels.)

(g)   Which of the above variations (or any others that you may have examined) of the genetic algorithm seem to work best for this problem? Explain your answer.

6.   Repeat all parts of Exercise 5 with the function being changed to $f(x) = ([x - 2200]/1000)^2$.

7.   Repeat all parts of Exercise 5 with the function being changed to

$$f(x) = \frac{(x - 2300)^2}{1000} \cos^2\left(\frac{x - 1500}{1000}\right).$$

8.   Write a program with syntax `VL = lookupvector(N)` that takes as input an integer $N > 1$, and whose output `VL` is a vector of length $N - 1$ whose ith entry is the index for which the 2-combination $\{i, i + 1\}$ appears in the canonical ordering of all 2-combinations of $\{1, 2, \ldots, N\}$:

$\{1, 2\}, \{1, 3\}, \ldots, \{1, N\}, \{2, 3\}, \{2, 4\}, \ldots, \{2, N\}, \{3, 4\}, \ldots, \{N - 1, N\}$.

Run your program with the inputs: $N = 5, 10, 13$, and 17.

9.   (a)   Use a basic genetic algorithm (with a fitness function analogous to the one implemented in Example 10.4) to search for a counterexample to show the Ramsey number bound $R(3, 4) > 8$. Make the parameters G (number of generations) and N (population size) small enough so that the program will run on your platform in less than 15–20 minutes.

(b)   Using the fitness function implemented in Part (a), perform a random search for a counterexample (using about the same amount of computer time as the genetic algorithm of Part (a)), and record the highest fitness level graph encountered. Compare the result with that of Part (a).

(c)   Experiment with the algorithm in Part (a) using three different values for the mutation parameter p_m to try to improve performance.

(d)   Repeat Parts (a) and (c) invoking also the inversion operator with $p_{inv} = 1/10$. (See Exercise for the Reader 10.5.)

(e)   Repeat Parts (a) and (c) invoking also the inversion operator with $p_{inv} = 1/4$. (See Exercise for the Reader 10.5.)

(f)   Repeat Parts (a) and (b) invoking a cloning procedure similar to that described in Exercise for the Reader 10.6, except that now two chromosomes will be cloned in going from one generation to the next: one of the highest possible fitness levels, and another of the next highest possible fitness level. (Thus the two chromosomes cloned will always have different fitness levels.)

(g)   Which of the above variations (or any others that you may have examined) of the genetic algorithm seem to work best for this problem? Explain your answer.

Note: It is known that $R(3, 4) = 9$ (original reference: [GrGl-55]). In any of the above parts, if a counterexample is actually found by your genetic algorithm, record it and also the number of generations it took to find it (you can modify your program to stop if and when a counterexample is found). In subsequent parts, an improved algorithm would be one that finds a counterexample in fewer generations.

10.   Repeat all parts of Exercise 9 with the goal being changed to search for a counterexample to show the Ramsey number bound $R(3, 5) > 13$.

11.   Repeat all parts of Exercise 8 with the goal being changed to search for a counterexample to show the Ramsey number bound $R(3, 6) > 17$.
     Note: It is known that $R(3, 6) = 18$. This was first discovered by G. Kéry in 1964 [Ker-64].

12.   Repeat all parts of Exercise 10 with the goal being changed to search for a counterexample to show the Ramsey number bound $R(3, 6) > 17$. (See the note to Exercise 11).

13.   Given the length 15 vector $v = [1,\ 2,\ -3,\ 4,\ -5,\ 6,\ -7,\ 8,\ -9,\ 10,\ -20,\ 50,\ 70,\ -80,\ 90]$,
     set up a genetic algorithm that will search to maximize the dot product $v \bullet w \equiv \sum_{i=1}^{15} v(i)w(i)$ over
     all length 15 vectors w each of whose entries is either -1 or 1. Experiment with several of the parameters of the genetic algorithm with the aim of maximizing such dot products. Compute the exact maximum value of these dot products by hand, and compare this exact result with what your genetic algorithms gave you. Is the exact maximum attained using a unique vector w, or are there several ways that it can be attained?

14.   Repeat Exercise 13 using instead the length 15 vector $v = [-7,\ -6,\ -5,\ \cdots, 5,\ 6,\ 7]$.

NOTE: (*Eight queens problem*) In the game of chess, a queen (symbolized by one of the eight icons in Figure 10.15) is allowed to move any number of spaces vertically, horizontal, or diagonally, on an 8×8 chessboard. The queen is the most powerful piece in chess. The problem is to place one queen in each column in such a way that no two queens can attack each other (by being able to reach one another in a single move).[16]

FIGURE 10.15:   A placement of eight queens on a chessboard.

15.   (*Eight queens problem*) In the game of chess, a queen (symbolized by one of the eight icons in Figure 10.15) is allowed to move any number of spaces vertically, horizontal, or diagonally, on an 8×8 chessboard. The queen is the most powerful piece in chess.   The problem is to place one queen in each column in such a way that no two queens can attack each other (by being able to reach one another in a single move).[17]
     (a) Encode a given deployment of eight queens by a length eight octal string sequence, where the ith entry of string is simply the row number (counting from the top) in which the ith queen is placed (except that row 8 corresponds to base 8 entry $= 0$). For example, the deployment of Figure 10.16 would be encoded by the following string: [7 3 4 0 1 6 3 1].[18]   Let the fitness function of a given deployment be the number of pairs of queens (out of the $C(8,2) = 28$ possible pairs) that cannot attack one another. Use a basic genetic algorithm with a size $N = 50$

---

[16] The colors of the queens are not relevant here.
[17] The colors of the queens are not relevant here.
[18] The corresponding binary encoding scheme would use length 24 binary strings. The main difference with binary encoding versus octal encoding for this problem is in the crossover operation: For binary encoding there is a 2/3 chance that the crossover point will correspond to being inside a row number for a queen, and thus will mutate that queen's position, whereas for octal encoding this will never occur. See Exercises 17–18 for a comparison.

population and run it for $G = 250$ generations to search for a solution of the eight queens problem. Experiment with a few different of mutation probabilities.

(b) Repeat Part (a) but with cloning of an individual of top fitness in passing from one generation to the next (Exercise for the Reader 10.6).

16. (*The n rooks problem*) The n rooks problem is similar to the n queens problem discussed in the section, except that n rooks rather than n queens are placed in each column of an $n \times n$ chessboard. A rook can attack another piece only by moving vertically or horizontally. This problem is easier than the n queens problem, for it is easy to see that a solution simply amounts to making sure there is only one rook in each row (and in each column).

(a) Use a genetic algorithm with octal (base 8) encoding to search for a solution to the 8 rooks problem. Experiment with various modifications such as cloning and inversion, but use a population size of $N = 20$.

(b) Redo Part (a), but this time using binary encoding. Which encoding scheme seems to work better (with all other parts of the genetic algorithm being the same)?

(c) What is the probability that a random (blind) placement of rooks (one in each column) will result in a solution of the 8 rooks problem?

17. (*The n queens problem*) This is a generalization of the 8 queens problem of Exercise 15. Let n be a positive integer and consider an $n \times n$ chessboard. The n queens problem asks to one place n queens on the board, one in each column, in such a way that no pair of queens can attack one another.

(a) Write a program to compute the number of solutions of the n queens problem for $n = 1, 2, \ldots, 7$. Interestingly, (you should find that) this number does not always increase with n. Compute also the corresponding probabilities that a blind (random) move would yield a solution.

(b) Use a genetic algorithm with cloning and inversion to seek a solution to the 9 queens problem. Unlike for the 8 queens problem, binary coding would not be suitable here.

18. (*The n queens problem*) Refer to the preceding exercise for definitions.

(a) Use a genetic algorithm with hexadecimal (base 16) encoding to search for a solution to the 16 queens problem. Experiment with various modifications such as cloning and inversion.

(b) Redo Part (a), but this time using binary encoding. Which encoding scheme seems to work better (with all other parts of the genetic algorithm being the same)?

19. (*An optimization problem for a function of two variables*) Finding global minima or maxima of functions of more than one variable can be a difficult optimization problem. Several testing functions have been developed in order to test (and/or compare) optimization methods. One such function is the so-called *Goldstein–Price function* ([GoPr-71], see Figure 10.16):

$$F(x,y) = \left(1 + (x + y + 1)^2 (19 - 14x + 3x^2 - 14y + 6xy + 3y^2)\right) \cdot$$
$$\left(30 + (2x - 3y)^2 (18 - 32x + 12x^2 + 48y - 36xy + 27y^2)\right)$$

The problem is to find the minimum value of this function in the region $-2 \le x, y \le 2$. This function is just a polynomial, but computing its global minimum value, which is known to be 3, is a difficult task. Although this is a continuous optimization problem, we will be solving a discretized version obtained by *pixilating* the region into $2^{12} = 4096$ (equally spaced) x- and y-coordinates:

$$\begin{cases} x(i) = -2 + 4(i - 1)/(2^{12} - 1), & 1 \le i \le 2^{12} \\ y(j) = -2 + 4(i - 1)/(2^{12} - 1), & 1 \le j \le 2^{12}. \end{cases}$$

The Goldstein–Price function

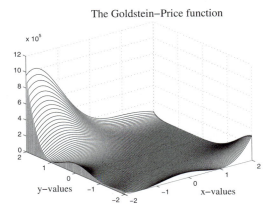

FIGURE 10.16: The Goldstein–Price function.

Encode these points $(x(i), y(j))$ using binary $e = e(i,j)$ strings of length 24, where the first 12 coordinates correspond to the binary representation of $i - 1$, and the remaining 12 coordinates correspond to the binary representation of $j - 1$. For the fitness function (keep in mind this is a minimization problem), for each generation: $e_1, e_2, \cdots, e_N$ we use

$$\text{fit}(e_i) = \max_{1 \leq j \leq N} F(e_j) - F(e_i).\ [19]$$

(a) Use a basic genetic algorithm with a size $N = 50$ population and run it for $G = 250$ generations to approximate the minimum value of $F(x,y)$. Experiment with a few differen mutation probabilities.

(b) Repeat Part (a) but now with the following encoding scheme change: the first coordinate of the string representing $(x(i), y(j))$ is the first component of the length 12 binary string representing $i - 1$, the second coordinate is the first component of the length 12 binary string representing $j - 1$, the third and fourth coordinates correspond to the second coordinates of the $i - 1$ and $j - 1$ binary strings, and so on.

(c) Use computer loops to determine the actual minimum value of the Goldstein-Price function on the given pixilated grid.

20. (*Optimization of a function of two variables*) (a) Modify the genetic algorithm of Part (a) Exercise 19 used to minimize the Goldstein-Price function, but now pixilate the region into $2^{20} = 1,048,576$ (equally spaced) x- and y-coordinates. In this setting, there are over a trillion points in the (discretized) solution space, making a brute-force check of the true minimum an impractical task.

(b) Repeat Part (a), but this time using the encoding scheme of Part (b) of Exercise 19.

(c) Count the total number of chromosomes tested in Part (a), and perform a random search (for the minimum) through the solution space of Part (a) using this same number of functional evaluations. How does the result here compare with what you got in Parts (a) and (b)?

21. (*Optimization of a function of two variables*) (a) Modify the genetic algorithm of Part (a) Exercise for the Reader 10.12 to seek the minimum value of the so-called *six hump camelback*

---

[19] Thus, since $\max_{1 \leq j \leq N} F(e_j)$ can change with each generation, we see that the fitness function is not a single formula, but one that can change with each generation. We are abusing notation a bit: when we write an expression like $F(e_j)$, we really mean $F(x(e_j), y(e_j))$, where $x(e_j), y(e_j)$ are the x- and y-coordinates corresponding to the string e_j.

function (see [DiSz-78]), $f(x,y) = (4 - 2.1x^2 + x^4/3)x^2 + xy + 4(y^2 - 1)y^2$ over the region $-3 \le x \le 3,$ $-2 \le y \le 2.$ Pixilate using 2^{20} equally spaced points for each coordinate. (b) Repeat Part (a), but this time using the encoding scheme of Part (b) of Exercise 19.

(b) Repeat Part (a), but this time using the encoding scheme of Part (b) of Exercise 19.

(b) Count the total number of chromosomes tested in Part (a), and perform a random search (for the minimum) through the solution space of Part (a) using this same number of functional evaluations. How does the result here compare with what you got in Parts (a) and (b)?

Note: To five decimals, the global minimum value of the six hump camelback function is $-1.0316,$ and this occurs at $(x,y) = \pm(0.0898, -0.7126).$

22. (*Optimization of a function of two variables*) Consider the function
$$f(x,y) = \frac{3}{1 + x^2 + y^2} - \frac{10}{4 + (x+1.5)^2 + y^2},$$
over the region $-4 \le x, y \le 4.$

(a) Explain why over this region, we have $-2.5 \le f(x,y) \le 3.$

(b) Pixilate the region using 2^{12} equally spaced x- and y-coordinates.

(c) Use a genetic algorithm as in Part (a) of Exercise 19 to search for the global minimum value of $f(x,y)$ over the indicated region.

(d) Repeat Part (c), but change the fitness function to $\text{fit}(e) = 3 - f(x(e), y(e))$ (nonnegative by Part (a)).

(e) Repeat Part (c) using both inversion (Exercise for the Reader 10.5) and cloning (Exercise for the Reader 10.6).

(f) Repeat Part (d) using both inversion (Exercise for the Reader 10.5) and cloning (Exercise for the Reader 10.6).

23. (*Optimization of a function of three variables*) (a) Use a basic genetic algorithm to search for the minimum value of the function $f(x,y,z) = x^2 + y^2 + z^2$ over the region $-2 \le x,y,z \le 2.$ Pixelate each of the coordinates into 2^{12} equally spaced coordinates. Use a population of $N = 100,$ and let it run for $G = 500$ generations. Experiment with a few different mutation probabilities.

(b) Repeat Part (a) using also inversion (see Exercise for the Reader 10.5). Experiment with a few different inversion probabilities.

(c) Repeat Part (a) invoking also the cloning procedure described in Exercise for the Reader 10.6.

(d) Perform a random search through $NG = 50,000$ of the $(2^{12})^3$ points of the solution space of Part (a) to search for a global minimum.

Note: The minimum value of the function is clearly 0, occurring (only) at $(x,y,z) = (0,0,0).$

24. (*Optimization of a function of three variables*) Repeat all parts of Exercise 20 with the function changed to $f(x,y,z) = x^2 - y^3 \cos(z^2)$ over the region $-2 \le x,y,z \le 2.$

25. (*An optimization problem for zero-one matrices*) A **zero-one matrix** is simply a matrix whose entries are only 0s and/or 1s. For a zero-one matrix $A,$ we define its *norm*, denoted $\|A\|,$ to be simply the sum of all (nonzero) elements of $A,$ in other words, $\|A\|$ is simply the number of entries of A that equal 1. If A and B are zero-one matrices of sizes $n \times m$ and $m \times r,$ respectively, then we can define the **Boolean product** of A and $B,$ denoted $A \odot B,$ to be the $n \times r$ zero-one matrix C whose i-j entry $C(i,j),$ is taken to be 1 if the corresponding entry of the ordinary matrix product AB is nonzero, and 0 otherwise. For a positive integer $n,$ we let T_n denote the lower-triangular zero-one $n \times n$ matrix having all 1s on and below the main diagonal, and all 0s above the main diagonal:

$$T_n = \begin{bmatrix} 1 & 0 & \cdots & 0 & 0 \\ 1 & 1 & \ddots & & \vdots \\ \vdots & \vdots & \ddots & 0 & 0 \\ 1 & 1 & \cdots & 1 & 0 \\ 1 & 1 & \cdots & 1 & 1 \end{bmatrix}.$$

For a given n, we consider the following optimization problem: For two zero-one matrices A, and B of size $n \times n$ such that $A \odot B = T_n$, we seek to minimize the sum $\|A\| + \|B\|$.

(a) Write two programs with syntaxes: `C = BooleanProd(A,B)` and `x = BooleanNorm(A)` for computing the Boolean product and the norm, respectively. The first one inputs two zero-one matrices A and B of compatible sizes (so that $A \odot B$ is defined), and the latter inputs any zero-one matrix A. Use your programs to compute the quantities $A \odot B$ and

$$\|A \odot B\| \text{ when } A = \begin{bmatrix} 1 & 1 & 0 \\ 0 & 1 & 1 \\ 1 & 1 & 0 \end{bmatrix} \text{ and } B = \begin{bmatrix} 1 & 0 & 1 & 0 \\ 1 & 1 & 1 & 0 \\ 1 & 1 & 0 & 1 \end{bmatrix}.$$

(b) Use a brute-force method (on the computer) to solve the above-mentioned optimization problem in the case $n = 2$. Give an example of a resulting optimal pair of zero-one matrices A and B, as well as the corresponding minimal norm sum $\|A\| + \|B\|$. Is the optimal pair unique?

(c) Develop a genetic algorithm that will aim to find a solution to the above-mentioned optimization problem in the case $n = 2$. Run your genetic algorithm and report the results. Your algorithm should obtain a pair of $n \times n$ zero-one matrices A and B, whose Boolean product is T_n. Report the pair that was found by your algorithm, along with the norm sum. Compare this norm sum with the true optimal value found in Part (b).

(d) Repeat Part (c) (except for the last question) in the case $n = 4$.

(e) Repeat Part (c) (except for the last question) in the case $n = 8$.

Suggestion and Note: To get a feel for the problem, it would be instructive to "try" to solve the optimization problem in case $n = 2$ by hand. For the genetic algorithms, I suggest using single binary strings to encode pairs of zero-one matrices A and B whose Boolean product is not necessarily equal to T_n. For your fitness function, you might use something like

$[2n^2 - \|A\| - \|B\|] + \alpha L$, where α is a positive constant, and L is counts the number of entries of $A \odot B$ that equal the corresponding entries of T_n. Thus, we allow pairs A and B in the solution space even if their Boolean product does not equal T_n, but a penalty is assessed (in the form of a lower fitness value) for each entry that is amiss. You will need to experiment with some different values of the parameter α; different values may work better for different values of n. The minimum norm sums are as follows: for $n = 2$, it is 5, for $n = 4$, it is 12, and for $n = 8$, it is 29; see [Ton-04].

26. (*Another mutation scheme*) The mutation operator introduced in the text spreads mutations uniformly over all genes. An alternate scheme would be to randomly select chromosomes for mutation, according to some probability $p_{select} > 0$, and for each selected chromosome, a number of genes are selected for mutation.

(a) Write a program `P = MutationV2(C, pselect, NumM)`, where C and P are as in the original mutation operator (Exercise for the Reader 10.4), `pselect` is the chromosome selection probability $p_{select} > 0$, described above, and `NumM` is a nonnegative integer corresponding to the number of random mutations that will take place on each selected chromosome (row of C). For each selected chromosome, the program should run through `NumM` iterations of randomly selecting a gene (from all possible genes) and mutating its allele.

(b) Write a program `P = MutationV2(C, pselect, vec)`; the first two inputs and the output variable are as in Part (a), and the third input `vec` is a vector of nonnegative integers. For each selected chromosome, the program should run through `NumM` iterations of randomly selecting a gene (from all possible genes) and mutating its allele, where `NumM` is randomly selected from the entries of vec (each entry being equally likely to be selected).

(c) Modify the basic genetic algorithm of Example 10.4 to replace the mutation operator with that of Part (a) with `pselect` = 0.02 and `NumM` = 14, and also using the cloning scheme of Exercise for the Reader 10.6. Run your program a few times and compare the results with those shown in Figure 10.14a.

(d) Modify the basic genetic algorithm of Example 10.4 to replace the mutation operator with that of Part (b) with `pselect` = 0.02 and `vec` = [7 14 28], and also using the cloning scheme of Exercise for the Reader 10.6. Run your program a few times and compare the results with those shown in Figure 10.14a.

(e) Repeat Part (d), but this time use a cloning scheme that clones two individuals to automatically survive into the next generation, one of maximal fitness and another of next highest fitness.

NOTE: (*Fitness scaling*) We have mentioned that the choice of a fitness function in the formulation of a genetic algorithm is an important control that can significantly affect the performance of the algorithm. Under the assumption of fitness-proportionate mating (the mating scheme that we introduced in the text), it is clear late in a genetic algorithm, when the fitness levels tend to converge towards an (often) near but sometimes sub-optimal value, the selection often can become more or less random among these near suboptimal chromosomes, and mating and crossover operations can stagnate. One useful idea is to linearly scale the fitness values, i.e., the vector F of raw fitness values becomes $aF + b$, where a and b are constants that depend on the generation. It is desirable to do this in such a way that the average fitness of a given generation remains the same after scaling, but the maximum fitness works out to be a fixed multiple C of the average fitness (the range $1.2 \leq C \leq 2$ has apparently been successful in applications, see Chapter 3 of [Gol-89]). This will help maintain a healthy competition for mating, even late in a genetic algorithm. The only problem with this approach is that the scaled fitness values determined may sometimes be negative. There are some ways to correct this defect, and the following algorithm, developed by David Goldberg ([Gol-89]) gives one such implementation.

Algorithm 10.3: Linear fitness scaling in a genetic algorithm.

Input: A vector of N (nonnegative) raw fitness values: F and a scalar $C > 1$.

Output: A vector of N corresponding scaled (nonnegative) fitness values: $\hat{F}$ with the property that $\hat{F}$ and F have the same average value $a(F)$ (i.e., $\Sigma F(i)/N = \Sigma \hat{F}(i)/N$), and, if possible, $\max(\hat{F}) = Ca(F)$.

Step 1: Compute the maximum, average, and minimum values of F: `Fmax` = `max(F)`, `Fmin` = `min(F)`, `Favg` = `avg(F)`.

Step 2: If `Fmin` > `(C*Favg - Fmax)/(C - 1)` (linear scaling will work)
Set `dx` = `Fmax - Favg`, `a` = `(C - 1)*Favg/dx`, `b` = `Favg*(Fmax - C*Favg)/dx` and go to Step 4. Otherwise go to Step 3.

Step 3: (scale as much as possible)
  Set `dx` = `Favg - Fmin`, `a` = `Favg - Fmin`, `b` = `-Fmin*Favg/dx`
Step 4: Set $\hat{F} = aF + b$

See Exercise 25 for more details on why linear scaling may not always be possible and why Step 3 is used in cases where it is not possible.

27. (*Fitness Scaling*) (a) Write a program with syntax `Fhat = fitscale(F, C)` that inputs a vector F of nonnegative fitness values and a number $C > 1$, and outputs a vector `Fhat` of scaled fitness values obtained using (the above) Algorithm 10.3.

(b) Modify the basic genetic algorithm of Example 10.4 to incorporate the `fitscale` function being implemented with `C` = 2 on each generation (with the result being fed into the `MatingPool` function) and also using the cloning scheme of Exercise for the Reader 10.6. Also, use $T_1 + T_2$ for the total (raw) fitness. Run your program a few times and compare the results with those shown in Figure 10.14a.

(c) Repeat Part (b) changing C to be 1.2.

(d) Repeat Part (c), but this time use a cloning scheme that clones two individuals to automatically survive into the next generation, one of maximal fitness and another of next highest fitness.

28. (*Fitness Scaling*)   (a)   Show that the criterion in Step 2 of Algorithm 10.3 is satisfied precisely when the minimum (raw) fitness value Fmin (of the current population) is at least equal to the number α shown in Figure 10.17. Thus, the scaled fitness values determined by the line of Figure 10.17 will be nonnegative and will satisfy the desired conditions for the scaled fitness.

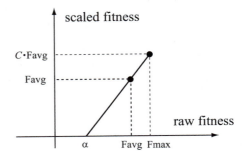

(b) In case (see Figure 10.17) $Fmin \leq \alpha$, show that Step 3 of Algorithm 10.3 uses instead the line passing through the points (0, Fmin) and (Favg, Favg) (placed in Figure 10.17) to scale the raw fitness values into the scaled fitness values.

FIGURE 10.17:   (Linear) fitness scaling attempts use the line shown to transform fitness values.   For this to work we need $Fmin \geq \alpha$.

Show further that in this case, we will have the maximum scaled fitness being (strictly) less than CFmax, but the former will be as large as possible under the constraint that (Favg, Favg) is on the scaling line (and since fitness values need to be nonnegative).

29. (a)   Show that there are only 3^{ℓ} length ℓ schema, compared to $2^{2^{\ell}}$ subsets of length ℓ chromosomes.

(b) Show that every chromosome belongs to 2^{ℓ} different schema.

Appendix A: Pseudocode

This appendix provides a brief summary of the *pseudocode* that is sometimes used in the text to present certain algorithms.   Pseudocode is not an official programming language; it is intended for human to read rather than for computers. Pseudocode is not a universal language; we choose ours with the first priority being to facilitate human understanding first, and to a lesser extent to make it easy to translate into computer programs.   As with computer programs, our pseudocode will implement some fundamental logical constructs.   Since computers operate on purely logical principles, it is advised that readers read at least as far as Section 1.1 in the text proper before looking over this short appendix.

For simplicity, this appendix will use (real) number data in the illustrations. The text will introduce the reader to a wide variety of data structures to which these pseudocode constructs may be applied. More examples involving pseudocode can be found in the computer exercise sets that appear at the ends of several sections.

Basic Commands and Their Meanings

Algorithms will be discussed thoroughly throughout the text, but for the purposes of this appendix, an algorithm consists of a list of commands, which, when applied to an given input, will produce a corresponding output. The list of commands are executed in order, until all have been executed, or the command "EXIT" or "STOP" has been encountered. Algorithms often involve variables whose values change throughout the course of the algorithm. When a variable is introduced with an initial value, the command "SET" or "INITIALIZE" will be used. Once an algorithm has completed its tasks, we use the pseudocode command "OUTPUT" to indicate what information the algorithm should produce.

When the value of a variable needs to be changed in the course of an algorithm, the command "UPDATE" is used. For example, if the variable x has been stored as 3, the command:

```
UPDATE x = x^2
```

would cause the value of x to be updated to $3^2 = 9$. In mathematical notation, the above command might seem troubling: if $x = 3$, the mathematical equation $x = x^2$ becomes $3 = 9$, which is nonsense.   The equals sign in this context represents an assignment operator, rather than an equality check.   An alternative notation that is sometimes used in place of the above is the following:

```
UPDATE x -> x^2
```

If Branches

It often occurs in the course of an algorithm that contingencies occur, and the algorithm must "branch off" with different lists of commands, depending on which case occurs.   Such contingencies are dealt with using an "If branch."

The basic form of an if branch is as follows:

```
IF <condition>
   <commands list 1>
ELSE   (or "OTHERWISE")
   <commands list 2>
END IF
```

The "ELSE" line and the list of commands "<commands list 2>" that follow it are optional.   When an if branch as above is encountered, the truth value of the "condition" is checked.   If it tests true, the commands in "<commands list 1>" are executed in order, and the then the remainder of the if branch is bypassed. If there is an "ELSE" portion, and the "condition" tested false, then only the commands in "<commands list 1>" are executed. If branches must always terminate with an "END" command, to indicate the limit on the commands to which the branch applies.

EXAMPLE A.1:   Below is the pseudocode for a function that inputs a real number x and output its absolute value $|x| = \begin{cases} x, & \text{if } x \geq 0 \\ -x, & \text{if } x \geq 0 \end{cases}$.

```
IF x >= 0
   OUTPUT x
ELSE
   OUTPUT -x
END IF
```

If branches can be designed to accommodate any number of different cases. The general form of an if branch is as follows:

```
IF <condition 1>
   <commands list 1>
ELSE IF <condition 2>
   <commands list 2>
```

```
ELSE IF <condition 3>
   <commands list 3>
...
ELSE IF <condition N>
   <commands list N>
ELSE
   <commands list last>
END IF
```

For Loops and While Loops

For loops and while loops allow an algorithm to run through a series of commands, repeatedly. In the case of a for loop, the commands are executed a fixed number of times, whereas in a while loop the commands are executed until some specified condition is met. In both, the variables can change values from one iteration (= cycle through the commands of the loop) to the next. Here is the basic structure of each type of loop:

For Loop:	While Loop:

`FOR n = a TO b`
   `<commands list>`
`END FOR`
	`WHILE <condition>`
	   `<commands list>`
	`END WHILE`

In the for loop, "n" is the **counter**, and it will run through the specified initial value "a," and proceed in increments of 1 until it reaches (or exceeds) the final specified value "a." For each value of n, the list of commands "`<commands list>`," constituting the **body** of the loop, will get executed (in order), and then n automatically advanced to its next value (i.e., `n -> n + 1`). The counter must be a variable and it does not have to be called n (we could instead call it `counter`, `count`, `month`, or whatever is convenient). The counter may, or may not appear in the body of the loop. Notice that, as with if branches, both for and while loops require an "`END`" after the body of the loops (to complete the loop). Here is a simple example of a for loop.

EXAMPLE A.2: Write a for loop to compute the sum of the squares of all integers from 2 to 20:
$$2^2 + 3^2 + 4^2 + \cdots + 20^2.$$

SOLUTION: We will first initialize a variable $\mathtt{Sum}$ for the cumulative sum throughout the loop (its initial value is zero):

Line	Pseudocode
1	SET Sum = 0; (*initialize sum*)
2	FOR n = 2 TO 20
3	UPDATE Sum = Sum + n^2
4	END
5	OUTPUT Sum (*display the final sum*)

This for loop operates as follows: The counter n starts off at 2 (n = 2), the body then resets the cumulative sum from its initial value (Sum = 0) to this value + n^2, or $0 + 2^2 = 4$. Next, n gets incremented by 1 to its next value in the vector (n = 3), the body now updates the cumulative sum to be its previously stored value (4) plus $n^2 = 3^3 = 9$. So Sum now gets reset to be 13. Next n = 4, and this all continues until n has reached its terminal value (20) and the body is executed for the last time.[1] When this loop is entered into the command window, the only output one sees is: →Sum = 2869.

We point out a couple of additional features introduced in the pseudocode of Example A.2. Comments to the code are included in parentheses in italic font (in lines 1 and 5 of the program). Such comments are ignored by computer programs and are included only to help (human) readers understand the program. Also note that in line 4, the "END FOR" was simply written as "END," since there was only one loop. In an actual program, the "FOR" part of "END FOR" is considered a comment; computer programs only need a single "END" to close the body of any if branch, or for/while loop. Such "END" specifier comments will be more useful in cases where a program invokes multiple for/while loops and if branches, with some being nested within others.

Writing loops and more general programs to solve a given problem sometimes takes several attempts and some debugging, even for experienced programmers, but especially for beginners. Here are a few good general suggestions for writing a program to solve a problem. First, you must totally understand the problem and should do some "small" cases (or smaller versions of it) by hand. Then write your program/loop for such a small case in a way that will make it easy to generalize. For example, suppose that we needed to write a for loop to compute the sum $1^2 + 3^2 + 5^2 + \cdots + 501^2$. Since sum is too large to do by hand, it might be a good idea to first write a program that computes the smaller corresponding sum $1^2 + 3^2 + 5^2 + 7^2$, and then check whether it gives the correct answer (that you <u>can</u> get by hand). If it does, your loop should be set up so you can make a minimal change (say changing 7 to 501) so that it can compute the original "big" sum. If it fails to get the correct answer (for the "smaller" sum), there is no way it will work correctly for the "larger" sum, so you have to debug your program. Sometimes

---

[1] Any time you are having difficulty understanding how a loop operates, a good idea is to (temporarily) allow additional outputs in the body (and maybe even put in some extra commands to display additional data). The additional output can often serve as a check and help one to detect any bugs.

the error will be a basic syntax error where the loop does not even run or give any output. Most computing platforms will often give you useful feedback to correct such problems. If the loop executes, but gives a wrong answer, start by displaying some additional outputs within the body of the loop so you can run the loop again and view more intermediate calculations to help you find which step the problem is occurring. Carefully examine this intermediate output data, and follow each step with hand calculations to try to find the error. If you need more feedback, you can (temporarily, for diagnostic purposes) add extra variables in the loop so that their values can be displayed. Perseverance is very important here. The experience that one gains by going through this trial, error, and debugging process (sometimes several times for a given program/loop) is very valuable in learning how to write programs. Once you have your program finally working correctly for the smaller problem, remember to go back and surpass the intermediate outputs so that when you run the program on the bigger problem, you won't be flooded with a lot of unnecessary data.

In a while loop, a `<condition>` is entered next to the `WHILE` operator. The loop continues to get iterated as long as the condition tests true. Such a `<condition>` usually depends on the variables in the body and so can change with the iterations (as it should). The `<condition>` can be made up using some of the equality or inequality operators:=, >, <, >=, >=, ~=(not equal), as well as any set of logical operators that are discussed in Section 1.1 (for OR, AND, etc.). In this introduction, we illustrate only one of the first type of operators. All of the variables appearing in `<condition>` must be initialized before the loop is entered (otherwise your computing platform will not recognize them and will give an error message). For loops are simpler than while loops; the latter are useful when one does not know initially how many iterations of the body of the loop are needed, as in the next example.

EXAMPLE A.3: Suppose, starting at his 25th birthday, Michael deposits $5000 at the beginning of every year into a retirement annuity that pays 9% interest per year, compounded annually He wants to retire when his annuity first reaches or exceeds $1 million. In how many years will he be able to retire with this plan?

SOLUTION:

Line	Pseudocode
1	SET Balance = 5000 (*initialize Balance*)
2	SET year = 0 (*initialize year counter*)
3	WHILE Balance < 1000000
4	UPDATE Balance = (1.09)*Balance + 5000
5	UPDATE year = year + 1; (*update year counter*)
6	END
7	OUTPUT year, Balance (*This will display the year and the*
8	*corresponding balance that first broke $1 million.*)

When this code is entered, the following output results: year $=$ 34.00, Balance $=$ 1078553.77.   Thus, Michael will be able to retire at age 59, with a nest egg of $1,078,553.77.

Unlike for loops that run through a specified number of iterations and then stop, one must be careful when writing a while loop to avoid **infinite loops**. An infinite loop occurs when a while loop is written so that the <condition> will never be satisfied no matter how many iterations run.   If you enter such a loop into a computer program, the computer will dutifully proceed to execute it in vain, and will never stop unless you realize what has happened and force it to exit. Infinite loops are usually easy to detect since the corresponding computer run will eventually seem to be taking too long to execute.

Appendix B: Solutions to All Exercises for the Reader

CHAPTER 1: MATHEMATICAL FOUNDATIONS

EFR 1.1: (a) The only situation where the contrapositive $\sim Q \to \sim P$ will be false is when the hypothesis ($\sim Q$) is true and the conclusion ($\sim P$) is false. This is equivalent to P being true and Q being false, which is precisely the only situation when the implication $P \to Q$ is false. Thus the two are logically equivalent. (Truth tables would also do the job, but would shed less light.)
(b) $P \vee \sim P$
(c) $P \wedge (Q \wedge \sim Q)$

EFR 1.2: (a) $\sim (\sim P \wedge \sim Q)$ (directly verify by comparing cases where statements are false).
(b) From Example 1.2, $P \to Q \equiv \sim P \vee Q$. Using Part (a), the latter is equivalent to $\sim (\sim (\sim P) \wedge \sim Q) \equiv \sim (P \wedge \sim Q)$.

EFR 1.3: (a) A truth table is constructed below with the relevant compound statements being highlighted. Since the truth values are identical, the equivalence is established.

P	Q	R	$P \vee Q$	$(P \vee Q) \vee \sim R$	$P \vee \sim Q$	$(P \vee \sim Q) \wedge R$	$[(P \vee \sim Q) \wedge R] \to P$
T	T	T	T	T	T	T	T
T	T	F	T	T	T	F	T
T	F	T	T	T	T	T	T
T	F	F	T	T	T	F	T
F	T	T	T	T	F	F	T
F	T	F	T	T	F	F	T
F	F	T	F	F	T	T	F
F	F	F	F	T	T	F	T

(b) We can prove the equivalence of Part (a) directly using equivalences from Theorem 1.1. Below is such a proof. It certainly takes more creativity to come up with such a proof than was needed with the rote proof done in Part (a), but such an effort is rewarded with a deeper understanding of the whole theory.

$$[(P \vee \sim Q) \wedge R] \to P \equiv \sim [(P \vee \sim Q) \wedge R] \vee P \text{ (implication as disjunction: Theorem 1.1 II(a))}$$
$$\equiv [(\sim P \wedge Q) \vee \sim R] \vee P \text{ (both of De Morgan's laws: Theorem 1.1 I(d))}$$
$$\equiv [(\sim P \wedge Q) \vee P] \vee \sim R \text{ (associativity and commutativity: Theorem 1.1 I(a)(b))}$$
$$\equiv [(\sim P \vee P) \wedge (Q \vee P)] \vee \sim R \text{ (distributivity: Theorem 1.1 I(c))}$$
$$\equiv [\mathbf{T} \wedge (Q \vee P)] \vee \sim R \text{ (tautology: Theorem 1.1 III(a))}$$
$$\equiv (Q \vee P) \vee \sim R \text{ (identity law: Theorem 1.1 I(g))} \quad \square$$

(c) $(P \vee Q) \wedge (\sim P \vee R) \equiv Q \vee R$ is readily seen to be invalid; for example if P and R are false, but Q is true, the left and right sides have opposite truth values. The implication $(P \vee Q) \wedge (\sim P \vee R) \Rightarrow Q \vee R$ is true, as the reader can easily verify.

EFR 1.4: Using logical notation, the argument takes on the following form:

(i) $W \wedge N \to P$, (ii) $\sim I \to W$, (iii) $(I \vee \sim N) \to H$, (iv) $\sim I$, (v) $\sim H$ $\therefore$ P

From (ii) and (iv) modus ponens (Theorem 1.2 (c)) gives us (vi) W. From (iii) and (v), modus tollens gives us (vii) $\sim (I \vee \sim N)$, which by De Morgan's law (Theorem 1.1 I(d)) and double negation (Theorem 1.1 I(e)) is equivalent to (viii) $\sim I \wedge N$. By subtraction (Theorem 1.2 (b)) (viii) produces (ix) N, and when this is combined with (vi) and modus ponens is used with (i) we obtain the desired conclusion P.

EFR 1.5:   To use the method of proof by contradiction, in addition to the hypotheses (i) $P \rightarrow (Q \vee R)$, (ii) $\sim R \rightarrow P$, and (iii) $\sim Q$, we are also allowed to assume the negation of the conclusion: (iv) $\sim R$. Modus ponens using (ii) and (iv) gives us (v) P. Using modus ponens once again with (i) and (v) produces (vi) $Q \vee R$. Next using commutativity and disjunctive syllogism with (iii) and (vi), we obtain (vii) R, which contradicts (iv). The proof is now complete.

EFR 1.6:   No. In formal wording, the statement tells us that the barber shaves a man if and only if that man does not shave himself. *Case 1: The barber does not shave himself.* By the statement it would follow that the barber does indeed shave himself—a contradiction. *Case 2: The barber shaves himself.* The statement would again contradict this. In either case we are led to a contradiction. Therefore, such a situation is not possible.

EFR 1.7:   We make use of the symbolic notation introduced in the solution of Example 1.6.
(b) Using equivalence (3) we can write: $\sim \forall x\,(W(x) \wedge H(x)) \equiv \exists x \sim (W(x) \wedge H(x))$ De Morgan's law allows the last statement to be expressed alternatively as: $\exists x\, \sim W(x) \vee \sim H(x)$.   In words, this negation could be translated as: "At least one of the marathoners is either not warmed up or not hydrated."
(c) Using formula (3) and the double negation law, we easly obtain the negation: $\sim \forall x\,(\sim D(x)) \equiv$ $\exists x \sim (\sim D(x)) \equiv \exists x\, D(x)$. Translated into English, the negation simply states that at least one of the players under six feet tall can dunk a basketball.
(d) Again equivalence (3) tells us that $\sim \forall x\,(P(x) \rightarrow [T(x) \wedge D(x)]) \equiv \exists x \sim (P(x) \rightarrow [T(x) \wedge D(x)])$.
To facilitate a smooth translation into English, we negate the implication and then use De Morgan's law (see Theorem 1.1 II (b) and I (d), resp.), to obtain the equivalent form: $\exists x\, P(x) \wedge [\sim T(x) \vee \sim D(x)]$.
In English, this becomes: "At least one person got into tonight's performance who either did not have a ticket or was not dressed properly."

EFR 1.8:   We introduce the predicate $K(a,b)$ to mean "a knows b," where the universe of discourse is all people (in this or some fictitious college). In symbols, the statement can be written as follows:

$$\forall a[\,K(a,\text{Jimmy}) \vee \exists b(K(a,b) \wedge K(b,\text{Jimmy}))].$$

To negate this, we will use both (2) and (3) as well as two applications of De Morgan's law; we do it in several steps (the reader should explain the validity of each of these steps).

$$\sim \forall a[\,K(a,\text{Jimmy}) \vee \exists b(K(a,b) \wedge K(b,\text{Jimmy}))] \equiv \exists a \sim [\,K(a,\text{Jimmy}) \vee \exists b(K(a,b) \wedge K(b,\text{Jimmy}))]$$
$$\equiv \exists a[\sim K(a,\text{Jimmy}) \wedge \sim \exists b(K(a,b) \wedge K(b,\text{Jimmy}))]$$
$$\equiv \exists a[\sim K(a,\text{Jimmy}) \wedge \forall b \sim (K(a,b) \wedge K(b,\text{Jimmy}))]$$
$$\equiv \exists a[\sim K(a,\text{Jimmy}) \wedge \forall b(\sim K(a,b) \vee \sim K(b,\text{Jimmy}))]$$

In terms of the original language, this negation can be expressed by saying: *There is someone who does not know Jimmy and does not know anyone who knows Jimmy.*

EFR 1.9:   The following Venn diagrams show both sides of the commutative and associative laws for the symmetric difference, repectively.

(a) $A \Delta B = B \Delta A$

(b) $A \Delta (B \Delta C) = (A \Delta B) \Delta C$

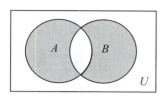

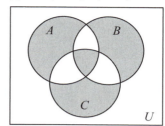

A logical proof could also be accomplished (for Part (b), the reader may wish to split it up into proving the two subset inclusions). We point out that for both parts, the symmetric difference consisted of all regions that overlapped an odd number of the given sets (here exactly one set or exactly three sets).

EFR 1.10: Thanks to the elegant quantifier enhanced definition (4) of unions of a collection of sets, the proof can be achieved quite expediently as follows:

$x \in A \cap (\bigcup_k B_k)$

$\quad \Leftrightarrow x \in A \ \wedge \ \exists k \ x \in B_k$ (definition (4))

$\quad \Leftrightarrow \exists k [x \in A \ \wedge \ x \in B_k]$ (by definition of universal quantification)

$\quad \Leftrightarrow x \in \bigcup_k (A \cap B_k)$ (definition (4))

EFR 1.11: (a) Employing the strategy used in Example 1.12, the subsets of {Red, White, Blue} of of two types: either they do not contain the (new) element "Blue," or they do contain "Blue." Subsets of the first type are exactly the four members of $\mathscr{P}(\{Red, White\})$ that were found in Example 1.12, i.e., $\varnothing$, {Red}, {White}, {Red, White}.   The subsets that contain the new element "Blue" must be obtainable as one of these four sets with the element "Blue" adjoined, namely the sets {Blue}, {Red, Blue}, {White, Blue}, {Red, White, Blue}.   These two selections of four different subsets of {Red, White, Blue} cover all possible cases, so the power set of {Red, White, Blue}, i.e., $\mathscr{P}(\{Red, White, Blue\}) = \{\varnothing, \{Red\}, \{White\}, \{Red, White\}, \{Blue\}, \{Red, Blue\}, \{White, Blue\},$ {Red, White, Blue}}.

(b) Let's review what we have discovered so far: From Example 1.12, we say that power set of the empty set (which contains $n = 0$ elements) had $1 = 2^0$ element, the power set of a one element set ($n = 1$) contains $2 = 2^1$ elements, the power set of a two element set ($n = 2$) contains $4 = 2^2$ elements, and from Part (a) above, the power set of a three element set contains $8 = 2^3$ elements.[1] More importantly, the strategy introduced in Example 1.12 and used in Part (a) above shows a general way to get subsets of a new set with one additional element from those of another set for which the power set is already known, and that one new element causes the number of subsets to double. These facts together lead us to the conjecture that a set with n elements will (always) have 2^n subsets. This conjecture is indeed true, but to make this proof legitmate we will need to use the principle of mathematical induction, which will be introduced a bit later in this chapter. This fact is the reason for the alternative notation 2^A for the power set $\mathscr{P}(A)$.

EFR 1.12: Although the sets $(A \times B) \times C$ and $A \times B \times C$ look very much alike and have the same number of elements (which is easily verified), they are different. Elements of $(A \times B) \times C$ are ordered

---

[1] It should be clear that the number of subsets of a power set with a given number (say two) of elements does not depend at all on the nature of the particular elements, thus the number of subsets of each of the sets {Red, White}, {7, 9}, and {Paris, Helsinki} is the same (4) even though the subsets will be different objects.

pairs where the first element is an ordered pair, while elements of $A \times B \times C$ are ordered triples. A simple example will illustrate these points. If $A = B = C = \{0,1\}$, the element $(1,1,1)$ of $A \times B \times C$ would correspond to the (different) element $((1,1),1)$ of $(A \times B) \times C$.

CHAPTER 2: RELATIONS AND FUNCTIONS, BOOLEAN ALGEBRA AND CIRCUIT DESIGN

EFR 2.1: There are numerous ways to organize charts and tables for the given relation. We present two such possibilities. In the table (left), the entries in the leftmost column denote the first element of the relation R and those of the topmost row denote the second element of the relation. For each such pair (a, b), the table entry is either 1 (if aRb) or (if $a\!\!\not{R}b$). In the

R	1	2	3	4	5	6
1	0	0	0	1	1	1
2	0	0	0	0	0	0
3	1	0	0	0	0	0
4	1	0	0	0	0	0
5	1	1	0	0	0	0
6	1	1	0	0	0	0

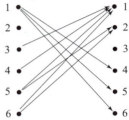

diagram (right), aRb if and only if an arrow connects the number a on the left to the number b on the right. We may now easily conclude that $2R5$ is false, while $6\!\!\not{R}6$ is true.

EFR 2.2: (a) Using the definitions of pre-images and intersections, we may proceed as follows:
$x \in f^{-1}(B_1 \cap B_2) \Leftrightarrow f(x) \in B_1 \cap B_2 \Leftrightarrow f(x) \in B_1$ and $f(x) \in B_2 \Leftrightarrow x \in f^{-1}(B_1)$ and $x \in f^{-1}(B_2) \Leftrightarrow$
$x \in f^{-1}(B_1) \cap x \in f^{-1}(B_2)$.

(b) Here is a simple example to demonstrate that the corresponding equality can fail for images. Consider the function $f : \mathbb{R} \to \mathbb{R} :: f(x) = x^2$, and the intervals $A_1 = [-2,-1]$, and $A_2 = [1,2]$. Since these intervals are disjoint, we have $f(A_1 \cap A_2) = f(\varnothing) = \varnothing$. On the other hand, $f(A_1) = [1,4]$
$= f(A_2)$, so that $f(A_1) \cap f(A_2) = [1,4] \neq \varnothing = f(A_1 \cap A_2)$.

(c) The example we just gave for Part (b) shows that the inclusion $f(A_1) \cap f(A_2) \subseteq f(A_1 \cap A_2)$ can fail, so the only hope is that we might be able to salvage the reverse inclusion $f(A_1 \cap A_2) \subseteq f(A_1) \cap f(A_2)$. Indeed, this latter relation is true and can easily be proved:
$y \in f(A_1 \cap A_2) \Rightarrow y = f(a), \exists a \in A_1 \cap A_2 \Rightarrow y = f(a), \exists a \in A_1$ and $y = f(a), \exists a \in A_2 \Rightarrow$
$y \in f(A_1)$ and $y \in f(A_2) \Rightarrow y \in f(A_1) \cap f(A_2)$. Unlike for the proof in Part (a), the arrows here are not all reversible. The reader should figure out exactly which one(s) are not.

EFR 2.3: (a) The function $g(k) = 101 - k$ defines a bijection from $\mathbb{Z}$ onto itself. *Proof:* (i) g is one-to-one: $g(k_1) = g(k_2) \Rightarrow 101 - k_1 = 101 - k_2 \Rightarrow k_1 = k_2$. (ii) g is onto: Given $n \in \mathbb{Z}$, if we take $k = 101 - n$, then $g(k) = g(101 - n) = 101 - (101 - n) = n$.

(b) The function defines an injection from $\mathbb{Z}$ to $\mathbb{Z}$ but is not onto. *Proof:* (i) h is one-to-one: $h(k_1) = h(k_2) \Rightarrow k_1^3 = k_2^3 \Rightarrow k_1 = k_2$ (since we can take cube roots of both sides). (ii) h is not onto: The equation $(h(k) =)k^3 = 2$ has no integer solutions, so, 2 is not in the range of h, and thus h is not onto.

Note: If we view the formula $h(x) = x^3$ as defining a function from $\mathbb{R}$ to $\mathbb{R}$, we get a function that is both one-to-one and onto. Thus the domain and codomain can be just as important as the formula for these properties.

EFR 2.4: *Proof:* Part (a): Necessity ($\Rightarrow$): We assume that $f : \{1, 2, \cdots, n\} \to \{1, 2, \cdots, n\}$ is a one-to-one function. It follows that each of $f(1), f(2), \cdots, f(n)$ must be different elements of the codomain $\{1, 2, \cdots, n\}$. Thus the images are n different elements in a set with a total of n elements, hence the codomain is completely covered by these images, i.e., f is onto.
Part (b): Sufficiency ($\Leftarrow$): We prove the contrapositive. Assume that f is not one-to-one. We will deduce that f is not onto. Since f is not one-to-one, this means that there are two different elements i and j in the domain $\{1, 2, \cdots, n\}$ which f takes to the same image, i.e., $f(i) = f(j)$. Therefore, the list of n images under f $f(1), f(2), \cdots, f(n)$ contains at least one duplication, and so can consist of at most $n - 1$ elements. Thus, f cannot be onto since the range does not cover all of the codomain.

EFR 2.5: (a) Using the two tabular form representations, we compute the different images under the compostion:
$$(\sigma \circ \tau)(1) = \sigma(\tau(1)) = \sigma(3) = 2, \ (\sigma \circ \tau)(2) = \sigma(\tau(2)) = \sigma(4) = 1, \ (\sigma \circ \tau)(3) = \sigma(\tau(3)) = \sigma(5) = 5,$$
$$(\sigma \circ \tau)(4) = \sigma(\tau(4)) = \sigma(6) = 3, \ (\sigma \circ \tau)(5) = \sigma(\tau(5)) = \sigma(1) = 4, \ (\sigma \circ \tau)(6) = \sigma(\tau(6)) = \sigma(2) = 6.$$
Summarizing in tabular form gives: $\sigma \circ \tau : \begin{pmatrix} 1 & 2 & 3 & 4 & 5 & 6 \\ 2 & 1 & 5 & 3 & 4 & 6 \end{pmatrix}$.

Computing in the same fashion the reverse-order composition:
$$(\tau \circ \sigma)(1) = \tau(\sigma(1)) = \tau(4) = 6, \ (\tau \circ \sigma)(2) = \tau(\sigma(2)) = \tau(6) = 2, \ (\tau \circ \sigma)(3) = \tau(\sigma(3)) = \tau(2) = 4,$$
$$(\tau \circ \sigma)(4) = \tau(\sigma(4)) = \tau(1) = 3, \ (\tau \circ \sigma)(5) = \tau(\sigma(5)) = \tau(5) = 1, \ (\tau \circ \sigma)(6) = \tau(\sigma(6)) = \tau(3) = 5.$$
produces the tabular form: $\tau \circ \sigma : \begin{pmatrix} 1 & 2 & 3 & 4 & 5 & 6 \\ 6 & 2 & 4 & 3 & 1 & 5 \end{pmatrix}$.

(b) $(g \circ F)(n, m) = g(F(n, m)) = g(3n - 4m^3) = 2 \cdot (3n - 4m^3) - 5 = 6n - 8m^3 - 5.$ This function is not one-to-one, since, for example, $(g \circ F)(36, 3) = -5 = (g \circ F)(0, 0)$.

EFR 2.6: (a) Yes. *Proof:* (i) Reflexivity: If $a \in S$, then $(a, a) \in R_1$ and $(a, a) \in R_2$ so $(a, a) \in R_1 \cap R_2$. (ii) Symmetry: If $(a, b) \in R_1 \cap R_2$, then $(b, a) \in R_1 \cap R_2$, since both R_1 and R_2 are symmetric. (iii) Transitivity: $(a, b), (b, c) \in R_1 \cap R_2$, then $(a, c) \in R_1 \cap R_2$, since both R_1 and R_2 are transitive.
(b) No.   Transitivity can fail. Here is a simple counterexample. Take $S = \{1, 2, 3\}$. Take $R_1 = \{(1,1), (2,2), (3,3), (1,2), (2,1)\}$ and $R_2 = \{(1,1), (2,2), (3,3), (2,3), (3,2)\}$. It is readily verified that both R_1 and R_2 satifies the axioms of an equivalence relation. The relation $R_1 \cup R_2$ contains $(1,2)$ and $(2,3)$, but not $(1,3)$, so it is not transitive. We point out that $R_1 \cup R_2$ will always be both reflexive and symmetric, as the reader can easily verify.

EFR 2.7: To prove that equivalence modulo m is an equivalence relation on $\mathbb{Z}$, simply replace every occurrence of 12 with m in the proof in the solution to Example 2.7(c). The proof that the equivalence classes are precisely $[0], [1], [2], \cdots, [m-1]$, can be accomplished by replacing all occurrence of 11 and 12, by $m - 1$ and m, respectively in the solution to Example 2.8.

EFR 2.8: Let s, t, and u be three length 5 bit strings. Certainly s has the same number of 1s as itself, i.e., $s \sim s$, and we have reflexivity. If $s \sim t$, then the numbers of 1s in s and t are the same, so $t \sim s$ as well, and we have symmetry. If $s \sim t$ and $t \sim u$, then s and t have the same number of 1s, and t and u

have the same number of 1's, thus so do s and u, so $s \sim u$, and transitivity is established. Thus we have an equivalence relation. The elements that are equivalent to [01101] are those length 5 bit strings that have exactly three 1's (and two 0's). These are: [00111], [01011], [01101] (itself), [01110], [10011], [10101], [10110], [11001], [11010], [11100].

EFR 2.9: (a) *Proof of Equivalence Relation*:   Let x, y, and $z \in A$. (i) Reflexivity: Since $f(x) = f(x)$, we have $x \sim x$. (ii) Symmetry: If $x \sim y$, then $f(x) = f(y)$, so $f(y) = f(x)$, and thus $y \sim x$. (iii) Transitivity: If $x \sim y$ and $y \sim z$, then $f(x) = f(y)$ and $f(y) = f(z)$, so $f(x) = f(z)$, and thus $x \sim z$.
(b)   *Proof that the equivalence classes are the level sets*:   By definition, $[x] = \{y \in A \mid y \sim x\}$ $= \{y \in A \mid f(y) = f(x)\} = f^{-1}(f(x))$.

EFR 2.10: Let (a,b), (a',b'), and $(\hat{a},\hat{b})$ be arbitrary elements of $A \times B$. Since $\preceq_B$ is reflexive, we have $a = a$ and $b \preceq_B b$, and thus $(a,b) \preceq (a,b)$, which means that $\preceq$ is reflexive. Next, assume that $(a,b) \preceq (a',b')$ and $(a',b') \preceq (a,b)$. If $a \neq a'$, then the fact that $(a,b) \preceq (a',b')$ forces $a \preceq_A a'$ and the fact that $(a',b') \preceq (a,b)$ forces $a' \preceq_A a$. But then antisymmetry of $\preceq_A$ would lead to the contradiction that $a = a'$. Thus, it must indeed be the case that $a = a'$. From this, the two order relations lead us to $b \preceq_B b'$ and $b' \preceq_B b$, and so since $\preceq_B$ is antisymmetric, we conclude that $b = b'$. We may now conclude that $(a',b') = (a,b)$, so that $\preceq$ is antisymmetric.   Finally, assume that $(a,b) \preceq (a',b')$ and $(a',b') \preceq (\hat{a},\hat{b})$. Let's split into two cases: *Case 1*: $a \neq a'$. Therefore, since the first ordering $(a,b) \preceq (a',b')$ forces $a \preceq_A a'$, and since the second ordering $(a',b') \preceq (\hat{a},\hat{b})$ forces $a' \preceq_A \hat{a}$, we are led to conclude (using transitivity of $\preceq_A$) $a \preceq_A \hat{a}$. Notice that a could not equal $\hat{a}$, because then the relations $a \preceq_A a'$ and $a' \preceq_A \hat{a}$ would force (by antisymmetry of $\preceq_A$) $a = a'$ — contradicting the Case 1 assumption.   The desired relation $(a,b) \preceq (\hat{a},\hat{b})$ now follows. *Case 2*: $a = a'$. The ordering $(a,b) \preceq (a',b')$ now forces $b \preceq_B b'$. If also $a' = \hat{a}$, then the second ordering $(a',b') \preceq (\hat{a},\hat{b})$ would force $b' \preceq_B \hat{b}$, so transitivity of $\preceq_B$ would yield $(a' = \hat{a}, \text{ and}) \ b \preceq_B \hat{b}$, so that $(a,b) \preceq (\hat{a},\hat{b})$, as desired. The only remaining situation is where $a' \neq \hat{a}$. In this case, since the latter relation $(a',b') \preceq (\hat{a},\hat{b})$ gives $a' \preceq_A \hat{a}$, and we know (Case 2 assumption) that $a = a'$, the desired relation $(a,b) \preceq (\hat{a},\hat{b})$ once again follows. The proof that the the lexicographic order is a partial order is now complete.

EFR 2.11: Let $(P, \preceq_P)$, $(Q, \preceq_Q)$, and $(R, \preceq_R)$ be arbitrary posets. (i) Reflexivity: The identity map $i_P : P \to P :: i_P(x) = x$, is clearly an order-preseving bijection (which equals its own inverse), so, it's a poset isomorphism. Thus we have $(P, \preceq_P) \sim (P, \preceq_P)$.   (ii) Symmetry: Assume that $(P, \preceq_P) \sim (Q, \preceq_Q)$. This means there is a poset isomorphism $f : (P, \preceq_P) \to (Q, \preceq_Q)$. The inverse map $f^{-1} : (Q, \preceq_Q) \to (P, \preceq_P)$ is also a bijection, and is order preserving and so is its inverse satisfies $(f^{-1})^{-1} = f$.   Thus we have $(Q, \preceq_Q) \sim (P, \preceq_P)$.   (iii) Transitivity: Assume that both $(P, \preceq_P) \sim (Q, \preceq_Q)$ and $(Q, \preceq_Q) \sim (R, \preceq_R)$.   This is tantamount to the existence of two poset isomorphisms: $f:(P,\preceq_P)\to(Q,\preceq_Q)$ and $g:(Q,\preceq_Q)\to(R,\preceq_R)$. It suffices to show that the composition $g\circ f:(P,\preceq_P)\to(R,\preceq_R)$ is a poset isomorphism. Compositions of bijections are bijections, so we just need to show this composition preserves order. So assume that $a,b\in P$ with $a\preceq_P b$. Then $f(a)\preceq_Q f(b)$, since f is order preserving. But now, since g is order preserving, we can conclude that $g(f(a))\preceq_Q g(f(b))$, i.e., that $(g\circ f)(a)\preceq_Q (g\circ f)(b)$, which is precisely what we needed to show.

EFR 2.12: (a) *Proof that divisibility is a partial order:* Suppose that a, b, and $c\in A$. (i) Reflexivity: Since $a=1\cdot a$, we have $a\,|\,a$. (ii) Antisymmetry: If $a\,|\,b$ and $b\,|\,a$, then we can write $b=ka$ and $a=nb$, for some $k,n\in\mathbb{Z}$. Substitution gives $b=knb$, and dividing this by b gives $kn=1$. Therefore, $k=n=1$, so $a=b$. (iii) Transitivity: If $a\,|\,b$ and $b\,|\,c$, then we can write $b=ka$ and $c=nb$, for some $k,n\in\mathbb{Z}$. Substitution gives $c=(nk)a$, so that $a\,|\,c$.
(b) The Hasse diagram for this poset when $A=\{1,2,3,4,5,6,10,12,20,30,60\}$ is shown on the right.

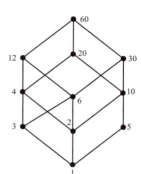

EFR 2.13: (a) The following table establishes the consensus law:

x	y	z	xy	$\bar{x}z$	yz	$xy+\bar{x}z+yz$	$xy+\bar{x}z$
1	1	1	1	0	1	1	1
1	1	0	1	0	0	1	1
1	0	1	0	0	0	0	0
1	0	0	0	0	0	0	0
0	1	1	0	1	1	1	1
0	1	0	0	0	0	0	0
0	0	1	0	1	0	1	1
0	0	0	0	0	0	0	0

(b) We now prove the consensus law by setting each side equal to zero and solving for the variables. First, $xy+\bar{x}z+yz=0\Leftrightarrow$ either $y=z=0$ (and x is either 0 or 1), or $x=0$, $y=1$, and $z=0$, or $x=1$, $y=0$, and $z=1$. Next, $xy+\bar{x}z=0\Leftrightarrow$ either $y=z=0$ (and x is either 0 or 1), or $y=1$, $x=0$, and $z=0$, or $z=1$, $x=1$, and $y=0$. These two solution sets are identical, so the two Boolean expressions are equivalent.

(c) Below is a proof of the consensus law using Theorem 2.4. We freely use the commutative and associative laws of Boolean addition and multiplication (Theorem 2.4(a,b)).

$$\begin{aligned}
xy+\bar{x}z &= x(y+yz)+\bar{x}(z+yz) &&\text{(absorption—Theorem 2.4(f))}\\
&= xy+xyz+\bar{x}z+\bar{x}yz &&\text{(distributivity—Theorem 2.4(c))}\\
&= xy+\bar{x}z+(x+\bar{x})yz &&\text{(distributivity—Theorem 2.4(c))}\\
&= xy+\bar{x}z+1\cdot yz &&\text{(unit identity—Theorem 2.4(i))}\\
&= xy+\bar{x}z+yz &&\text{(identity law—Theorem 2.4(g))}\quad\square
\end{aligned}$$

EFR 2.14: We could either work directly with the formula using Boolean arithmetic (Theorem 2.4), or construct a table of values. For this problem both approaches take a comparable amount of work. We indicate how the Boolean algebra manipulation would work, and then go on to (completely) construct a table to solve the problem.

$$
\begin{aligned}
\overline{x\overline{y}z + \overline{z}w} &= \overline{x\overline{y}z} \cdot \overline{\overline{z}w} && \text{(De Morgan's law—Theorem 2.4(d))}\\
&= (\overline{x} + \overline{\overline{y}} + \overline{z}) \cdot (\overline{\overline{z}} + \overline{w}) && \text{(De Morgan's law—Theorem 2.4(d))}\\
&= (\overline{x} + y + \overline{z}) \cdot (z + \overline{w}) && \text{(double complementation—Theorem 2.4(e))}\\
&= xz + yz + \overline{z}z + \overline{w}x + \overline{w}y + \overline{w}\overline{z} && \text{(distributivity—Theorem 2.4(c))}\\
&= (w + \overline{w})x(y + \overline{y})z + (w + \overline{w})(x + \overline{x})yz + 0 + \overline{wx}(y + \overline{y})(z + \overline{z}) + \overline{w}(w + \overline{w})y(z + \overline{z})\\
&\quad + \overline{w}(x + \overline{x})(y + \overline{y})\overline{z} && \text{(unit/zero identity, identity laws—Theorem 2.4(g,i))}
\end{aligned}
$$

The distributive law would then need to be used and would result in 20 minterms that would then need to be combined. The interested reader can continue with this method and check to see that it would give the same result as we obtain below by constructing a table of values:

w	x	y	z	$\overline{y}$	$x\overline{y}z$	$\overline{z}$	$\overline{z}w$	$x\overline{y}z + \overline{z}w$	$\overline{x\overline{y}z + \overline{z}w}$
1	1	1	1	0	0	0	0	0	1
1	1	1	0	0	0	1	1	1	0
1	1	0	1	1	1	0	0	1	0
1	1	0	0	1	0	1	1	1	0
1	0	1	1	0	0	0	0	0	1
1	0	1	0	0	0	1	1	1	0
1	0	0	1	1	0	0	0	0	1
1	0	0	0	1	0	1	1	1	0
0	1	1	1	0	0	0	0	0	1
0	1	1	0	0	0	1	0	0	1
0	1	0	1	1	1	0	0	1	0
0	1	0	0	1	0	1	0	0	1
0	0	1	1	0	0	0	0	0	1
0	0	1	0	0	0	1	0	0	1
0	0	0	1	1	0	0	0	0	1
0	0	0	0	0	0	1	0	0	1

From the table, we may now glean the following sum-of-products decomposition:

$$\overline{x\overline{y}z + \overline{z}w} = wxyz + w\overline{x}yz + w\overline{x}\overline{y}z + wx\overline{y}\overline{z} + \overline{w}xyz + \overline{w}x\overline{y}z + \overline{w}\overline{x}yz + \overline{w}\overline{x}y\overline{z} + \overline{w}\overline{x}\overline{y}z + \overline{w}\overline{x}\overline{y}\overline{z}.$$

EFR 2.15: (a) $[(x+1)(yz)]^d = x \cdot 0 + (y+z) \ (= y + z)$, $[x + \overline{y} + z + 0]^d = x\overline{y}z \cdot 1 \ (= x\overline{y}z)$.

(b) $f^d(w, x, y, z) = \left[\overline{x\overline{y}z + \overline{z}w}\right]^d = (x + \overline{y} + z)(\overline{z} + w).$

EFR 2.16: Taking duals of the consensus law $xy + \overline{x}z + yz = xy + \overline{x}z$ produces the following identity: $(x + y)(\overline{x} + z)(y + z) = (x + y)(\overline{x} + z).$

EFR 2.17:

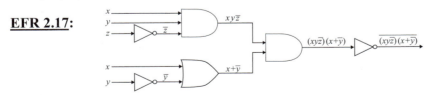

EFR 2.18: (a) The equivalence is just De Morgan's law (Theorem 2.4 (d)).
(b) The completed Karnaugh map for the function of Table 2.3 is as shown on the right. Only 1-cell blocks can be used to cover the 1-cells, and the two such required blocks give the expression $xy + \overline{x}\overline{y}$.

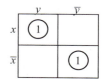

EFR 2.19: The completed Karnaugh map shown on the right shows that four 1-cell blocks are required to cover the 1-cells. This means that the corresponding simplified Boolean expression produced by the Karnaugh method coincides with the original expression, and the number of logical gates in the circuits would be the same.

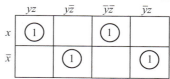

EFR 2.20: (a) The reader can create a table of values for the given Boolean function, and then be able to read off the following sum-of-products representation:
$f(w,x,y,z) = \overline{w}xyz + \overline{w}xy\overline{z} + \overline{w}x\overline{y}\overline{z} + \overline{w}\,\overline{x}y\overline{z} + \overline{w}xy\overline{z}$. The Karnaugh diagram shown (below on the left) was obtained by placing 1s in each cell corresponding to a minterm of this representation, and placing Xs in the cells of the don't care minterms that were given.

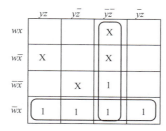
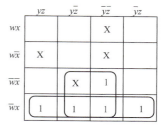

There are two permissible ways to cover the 1's with (maximal) 4-cell blocks. The one in the middle gives the representing expression $\overline{w}x + \overline{y}\overline{z}$, while the one on the right gives the expression $\overline{w}x + \overline{w}\overline{z}$.

(b) The two expressions found in Part (a) are not equivalent, but they differ only on inputs that we don't care about, so both do indeed represent the given function for all practical purposes.

CHAPTER 3: THE INTEGERS, INDUCTION, AND RECURSION

EFR 3.1: Let $S(n)$ denote the statement contained in equation (2), for $n \in \mathbb{Z}_+$.

1. *Basis Step*: $S(1)$ is obtained by substituting $n = 1$ into (2): $1^2 = 1(1+1)(2 \cdot 1 + 1)/6$, or $1 = 1 \cdot 2 \cdot 3 / 6$, which is true.

2. *Inductive Step*: Let $k \in \mathbb{Z}_+$. Our task to is assume that $S(k)$ is true, and use this (inductive hypothesis) do deduce that $S(k+1)$ is true, i.e., that $\sum_{i=1}^{k+1} i^2 = \dfrac{(k+1)(k+1+1)(2[k+1]+1)}{6}$, or, in

simpler terms: $\sum_{i=1}^{k+1} i^2 = \dfrac{(k+1)(k+2)(2k+3)}{6}$ (*). Breaking off the last term from the summation

on the left, and invoking the inductive hypothesis on the remaining terms, we obtain: $\sum_{i=1}^{k+1} i^2 =$

$\sum_{i=1}^{k} i^2 + (k+1)^2 = \dfrac{k(k+1)(2k+1)}{6} + (k+1)^2$. The proof will be complete if we can show that the

right side of the latter equation is the same as the right side of (*). A bit of algebraic manipulation is all that is needed:

$$\frac{k(k+1)(2k+1)}{6} + (k+1)^2 = \frac{k(k+1)(2k+1) + 6(k+1)^2}{6} = \frac{(k+1)[k(2k+1) + 6(k+1)]}{6}$$

$$= \frac{(k+1)[2k^2 + 7k + 6]}{6} = \frac{(k+1)(k+2)(2k+3)}{6}. \quad \Box$$

NOTE: Although the algebra was a bit messy, it was guided by the fact that we knew where we wanted to go with it. Such algebraic manipulations are common in proofs by induction that verify formulas.

EFR 3.2: Letting T_n denote the sum of the first n odd integers, i.e., $T_n = 1 + 3 + 5 + \cdots + (2n-1)$ $= \sum_{i=1}^{n}(2i-1)$, let's compute the first few values of this sequence until we notice some sort of pattern:

$$T_1 = 1, \quad T_2 = 1 + 3 = 4, \quad T_3 = 1 + 3 + 5 = 9, \quad T_4 = 1 + 3 + 5 + 7 = 16, \quad T_5 = 1 + 3 + 5 + 7 + 9 = 25, \ldots$$

It is appearing that these sums turn out to be the squares of consecutive positive integers, i.e.,

$$T_1 = 1^2, \quad T_2 = 2^2, \quad T_3 = 3^2, \quad T_4 = 4^2, \quad T_5 = 5^2, \ldots$$

We thus conjecture that $T_n = n^2$, and proceed to prove this by induction.

1. *Basis Step*: $T_1 = 1^2$, we have already observed this fact.

2. *Inductive Step*: We assume that $T_k = k^2$, and must then prove that $T_{k+1} = (k+1)^2$, where $k \in \mathbb{Z}_+$. We do this by breaking T_{k+1} up into two pieces, one of which is T_k, on which we can apply the inductive hypothesis $T_k = k^2$:

$$T_{k+1} = \sum_{i=1}^{k+1}(2i-1) = \underbrace{\sum_{i=1}^{k}(2i-1)}_{=T_k} + (2[k+1]-1) \underset{\substack{by\ inductive \\ hypothesis}}{=} \underbrace{k^2}_{=T_k} + 2k + 1.$$

The last term is a perfect square $(k+1)^2$, which is exactly what we needed it to be. $\quad \Box$

EFR 3.3: The flaw occurs in the inductive step, in the case $k = 1$. When the horse X is exchanged for the remaining horses (there is only one other horse Y, which is brown), it is still all by itself so we cannot conclude that it is also brown. We point out that the basis step and the inductive step argument are all valid except in the single case when $k = 1$. One such failure can destroy an inductive proof, as is quite obvious from this example (since the result is so obviously false).

EFR 3.4: Part (a): The induction proof that we give for unions is identical to that for intersections (only the symbols need to be switched), so we give only the proof for unions. Let $S(n)$ denote the statement that when a union of n sets $A_1, A_2, \cdots, A_n$ is taken with the sets written in the same order but parentheses are used to indicate the order in which unions of pairs of sets are taken, the result will always equal the union obtained by working from left to right, i.e., to the set $(\cdots(((A_1 \cup A_2) \cup A_3) \cup A_4) \cdots \cup A_n)$. The truth of $S(n)$ thus implies that no matter how parentheses are grouped in a union of n sets, the resulting union will be the same. We need to show that $S(n)$ is true for all $n \in \mathbb{Z}_+, n \geq 3$.

1. *Basis Step*: $S(3)$ is simply Theorem 1.3(b): $(A \cup B) \cup C = A \cup (B \cup C)$; in this case, there are only two ways to group the parentheses (in a three-set union).

2. *Inductive Step*: Assuming that $S(k)$ is true, where $k \geq 3$ is a positive integer, we need to show that $S(k+1)$ is also true. So we consider a union (with parentheses inserted) of $k+1$ sets $A_1, A_2, \cdots, A_{k+1}$ (where the left to right appearance of these sets is as they are written). In this union, some pair of adjacent sets, say A_j and A_{j+1} will have their union taken first (these will be enclosed in the innermost pair of parentheses). Let $B = A_j \cup A_{j+1}$. If we replace this parenthesized union by B in the original union expression, we have a union of k sets: $A_1, A_2, \cdots, A_{j-1}, B, A_{j+2}, \cdots, A_{k+1}$, so the

inductive hypothesis allows us to write this union as $(\cdots(\cdots(A_1 \cup A_2)\cdots A_{j-1}) \cup B) \cup \cdots A_{k+1})$. Now, in this latter expression we replace B with $A_j \cup A_{j+1}$ to obtain:

$$(\cdots \boxed{(\cdots(A_1 \cup A_2)\cdots A_{j-1}) \cup (A_j \cup A_{j+1}))} \cup \cdots A_{k+1}).$$

Now, within the boxed portion, we can use $S(3)$ to change the parentheses to look like:

$$(\cdots \boxed{((\cdots(A_1 \cup A_2)\cdots A_{j-1}) \cup A_j) \cup A_{j+1})} \cup \cdots A_{k+1}),$$

and this is the desired form. □

Part (b):   We let $S(n)$ denote the statement that the given identity $\sim (A_1 \cup A_2 \cup \cdots \cup A_n) = \sim A_1 \cap \sim A_2 \cap \cdots \cap \sim A_n$ is valid for any n sets. We use induction to show that $S(n)$ is true for all $n \in \mathbb{Z}_+$. In what follows below, we freely use the associative rules for unions and intersections for a finite number of sets that were established in Part (a).

1. *Basis Step*: $S(1)$ simply states that $\sim (A_1) = \sim A_1$, for any set A_1. This is trivially true (since the parentheses are redundant).

2. *Inductive Step*: Assuming that $S(k)$ is true, where k is a positive integer, we need to show that $S(k+1)$ is also true. So let $A_1, A_2, \cdots, A_k, A_{k+1}$ be sets, we let A denote the set $A_1 \cup A_2 \cup \cdots \cup A_k$. By De Morgan's law (Theorem 1.3(d)), we can write $\sim (A \cup A_{k+1}) = \sim A \cap \sim A_{k+1}$.   The inductive hypothesis tells us that $\sim A[= \sim (A_1 \cup A_2 \cup \cdots \cup A_k)] = \sim A_1 \cap \sim A_2 \cap \cdots \cap \sim A_k$. Combining the last two equations gives us that:

$$\sim (A_1 \cup A_2 \cup \cdots \cup A_k \cup A_{k+1}) = \sim (A \cup A_{k+1}) = (\sim A_1 \cap \sim A_2 \cap \cdots \cap \sim A_k) \cap \sim A_{k+1}$$
$$= \sim A_1 \cap \sim A_2 \cap \cdots \cap \sim A_k \cap \sim A_{k+1}. \quad □$$

EFR 3.5: As in the solution to Example 3.4, we take $S(n)$ to denote the statement that a postage value of n cents can be achieved using only three cent or five cent stamps. Our goal is to show that $\forall n \geq 8[S(n)]$, but now we use ordinary induction. The basis step (proof that $S(8)$ is true) is the same as in the example. For the inductive step, we let k be an integer greater than seven, we assume the postage value k cents can be achieved using five cent or three cent stamps, and we use this fact that the postage value $k+1$ cents can be so obtained. We separate into cases:

Case 1: The combination of stamps for the k cent postage includes a five cent stamp. Here we simply replace one five cent stamp with two three cent stamps. The resulting combination will now add up to $k+1$ cents, as desired.

Case 2: The combination of stamps for the k cent postage includes at least 3 three cent stamps. Here we simply replace 3 three cent stamps with 2 five cent stamps. The resulting combination will now add up to $k+1$ cents, as desired.

The above two cases cover all possibilities since if we are outside of both of them, the combination for the k cent postage could only include a maximum of six cents (two three cent stamps), but this is not possible since we are assuming that k > 7.

EFR 3.6: To establish this theorem about breaking chocolate bars, we use strong mathematical induction. Let $S(n)$ be the statement that a rectangular chocolate grid containing n squares will always take $n-1$ (horizontal or vertical) breaks (along the grid lines) to break into n squares. We need to show that $\forall n \geq 1 [S(n)]$.

1. *Basis Step*: S(1) says that if the chocolate bar had only $n=1$ piece, it would take $n-1=0$ breaks to break it into 1 piece. This is obvious since no breaks are needed (or possible).

2. *Inductive Step*: Let $k \in \mathbb{Z}_+$, we assume that $S(j)$ is true for each $j \leq k$. Using this assumption, our task is to show that $S(k+1)$ is true. So, suppose that we have a chocolate bar containing

$k+1$ squares. We start off with one break that will break the bar into two smaller bars with, say a, and b squares, respectively. Thus, a and b are positive integers with $a+b=k+1$. By the strong inductive hypotheses, breaking each of these smaller bars into a and b pieces, will take $a-1$ and $a-1$ breaks, respectively. Putting this all together, we have proved that no matter how it gets done, breaking a $k+1$ square chocolate bar into its $k+1$ pieces will always take $1+(a-1)+(b-1)=a+b-1=(k+1)-1$ breaks. This is what we needed to prove. $\square$

EFR 3.7: Part (a): If we let $n=0$ correspond to the initial year that Rose gets hired, then we have $r_0 = 32,500$. Now, by what was stated, to get from a given year's salary r_n to the next year's salary r_{n+1}, we first take 103.5% of the given year's salary $1.035 r_n$, and then add 1000 to this amount, thus we have the recursion formula: $r_{n+1} = 1.035 r_n + 1000$.

Part (b): To (attempt) to obtain an explicit formula for r_n, we use the method (introduced in Example 3.6) of repeatedly substituting the recursion formula into itself.

$$\begin{aligned} r_n &= 1.035 r_{n-1} + 1000 & &= 1.035(1.035 r_{n-2} + 1000) + 1000 \\ &= 1.035^2 r_{n-2} + 1000(1+1.035) & &= 1.035^2(1.035 r_{n-3} + 1000) + 1000(1+1.035) \\ &= 1.035^3 r_{n-3} + 1000(1+1.035+1.035^2) & &= 1.035^3(1.035 r_{n-4} + 1000) + 1000(1+1.035+1.035^2) \\ &\;\;\vdots & &\;\;\vdots \\ &= 1.035^n r_{n-n} + 1000(1+1.035+1.035^2 + \cdots + 1.035^{n-1}) \\ &= 1.035^n r_0 + 1000(1+1.035+1.035^2 + \cdots + 1.035^{n-1}). \end{aligned}$$

The parenthesized sum is a geometric series, so formula (3) of Proposition 3.5 with the variable assignments $(a \to 1, r \to 1.035, n \to n-1)$ allows us to express this sum as $(1.035^n - 1)/0.035$. Substituting this and the intitial value $r_0 = 32,500$ into the expression we derived for r_n produces the following explicit formula: $r_n = 1.035^n r_0 + 1000 \cdot (1.035^n - 1)/0.035$.

EFR 3.8: The only mathematical operations used in the computation of elements of the Fibonacci sequence are additions.

Part (a): As was explained in the text (see the top part of Figure 3.6), the iterative algorithm requires to one addition to compute each f_n, when $n \geq 3$ (no additions are required to compute f_1, f_2 since these sequence elements are given). So to compute f_n, $n \geq 3$ using the iterative algorithm, two additions will be required to compute each of $f_3, f_4, \cdots, f_n$, so this works out to $2 \cdot (n-3+1) = 2n-4$ arithmetic operations.

Part (b): In discovering the general pattern it is helpful to draw information flow diagrams (as in the lower one of Figure 3.6) for some small values of n to go along with the calculations we now perform (and the reader is encouraged to do this). We let c_n denote the number of arithmetic operations (= the number of additions) that the recursive program `fibonnacciRec` of Example 3.8 uses to compute f_n, when $n \geq 3$. Computing f_3 calls on the two known values f_1, f_2 and adds them up, so $c_3 = 1$. To get c_4, we note that to compute $f_4 = f_3 + f_2$, the recursive program calls on itself to produce f_3, which required $c_3 = 1$ addition and the (known) value f_2 (which required no additions) and added them up (one more addition), so $c_4 = 1+1 = 2$. Assuming now that $n \geq 5$, and that we have figured out all values c_k with $k < 5$, to compute $f_n = f_{n-1} + f_{n-2}$, the recursive algorithm would call on itself to first evaluate f_{n-1}, which will take c_{n-1} additions, and f_{n-2}, which will take c_{n-2} additions, and then add these up, which will take one more addition, so $c_n = c_{n-1} + c_{n-2} + 1$. This sequence starts off the same way as the Fibonacci sequence but one unit of time later (i.e., $f_1 = f_2 = 1, f_3 = 2$, whereas

$c_3 = 1, c_4 = 2$), but it grows faster (and so will eventually catch up) because of the extra term of 1 being added in the recursion formula compared to that for the Fibonnaci. The methods we give at the end of this section will show how we can obtain an explicit formula for c_n.

Part (c): To compute f_{30}, the recursive algorithm would require c_{30} mathematical operations. To obtain this number a simple computer for loop is more feasible than a hand calculation. (Another method would be to read ahead at the end of this section and derive an explicit formula for c_n, but this would also take much more work than running a simple for loop.) Here is the pseudocode for such a for loop:

```
cn_oneb4 = 2; cn_twob4 = 1  <initialize first two values of cn>
FOR  n = 5 TO 30
   cn_new = cn_oneb4 + cn_twob4 <obtain new value of cn>
   cn_twob4 = cn_oneb4, cn_oneb4 = cn_new <update two most recent values of cn>
END <n FOR>
OUTPUT cn_new
```

This program gives us $c_{30} = 832039$. To see how far the iterative program would go with the same amount of computation, we set the number of additions, $2n - 4$, that we determined in Part (a) that the iterative program needs to compute f_n, set it equal to 832039, and solve for n to get 416021.5, so with the same amount of work, the iterative program could compute f_n, for any value of n running up through 416,021!

EFR 3.9: Using the recursion formula, $1! = 1 \cdot 0! = 1 \cdot 1 = 1, , 2! = 2 \cdot 1! = 2 \cdot 1 = 2, , 3! = 3 \cdot 2! = 3 \cdot 2$ $(= 3 \cdot 2 \cdot 1) = 6, 4! = 4 \cdot 3! = 4 \cdot 6 (= 4 \cdot 3 \cdot 2 \cdot 1) = 24$, and $5! = 5 \cdot 4! = 5 \cdot 24 (= 5 \cdot 4 \cdot 3 \cdot 2 \cdot 1) = 120$. From the patterns that we recognized and put in parentheses in these computations, it seems that in general we should have $n! = n \cdot (n-1) \cdot (n-2) \cdots 2 \cdot 1$, for $n \ge 1$. We let $S(n)$ denote this statement, and proceed to use mathematical induction to prove that $S(n)$ is true for all $n \ge 1$.
1. *Basis Step*: $S(1)$ is just the equation $1! = 1$, which was already observed in the above calculations.
2. *Inductive Step*: We assume that $S(k)$ is true, i.e., that $k! = k \cdot (k-1) \cdot (k-2) \cdots 2 \cdot 1$. From this, and the recursive definition (with $n = k+1$), $(k+1)! = (k+1) \cdot k!$, we conclude that $(k+1)! = (k+1) \cdot (k) \cdot (k-1) \cdots 2 \cdot 1$, which is $S(k+1)$. $\square$

EFR 3.10: Part (a): Plugging the basis step values into the recursion formula produces $a_4 = a_3 + a_2 a_1 = 2 + 3 \cdot 1 = 5$. By the same token be get $a_5 = a_4 + a_3 a_2 = 5 + 2 \cdot 3 = 11$ and $a_6 = a_5 + a_4 a_3 = 11 + 5 \cdot 2 = 21$.
Part (b): The program to evaluate the nth term of this sequence is a very simple one to write:

```
INPUT   n   <index of term to evaluate, a positive integer>
a_rec = 2, a_old = 3, a_oldest = 1
IF n = 1
   OUTPUT an = a_oldest
ELSE IF n = 2
   OUTPUT an = a_old
ELSE IF n = 3
   OUTPUT an = a_rec
ELSE
   FOR k = 4 TO n
      set anew = a_rec + a_old*a_oldest
      set a_oldest = a_old, a_old = a_rec, a_rec = a_new
   END <k FOR>
   OUTPUT an = anew
END <IF>
```

Despite the simplicity of the sequence (and the corresponding program), the terms of the sequence grow <u>explosively fast</u>.   On a standard floating point arithmetic computing platform, if the above program is called to evaluate a_{15}, the answer would overflow to infinity. This can be circumvented using symbolic computation, and one could find that $a_{15} = 9.612... \times 10^{1519}$. Even the exact value of a_9, which is 561,993,796,032,558,961,827,630, would not be obtainable in a floating point arithmetic system (working in IEEE double precision standard arithmetic).   A floating point system would yield an answer such as 5.6199×10^{23} for a_9. For more information of floating point arithmetic, see Chapter 5 of [Sta-05].

EFR 3.11: Part (a):

```
INPUT n <positive integer>
IF n =1, OUTPUT y = 2
ELSE OUTPUT y =  EvenRec(n-1) + 2
```

Part (b):

```
INPUT n <positive integer>
set y = 2
FOR k = 2 TO n
    y = y + 2
END <FOR>
OUTPUT y
```

Part (c):   In this case, both the recursive and iterative programs require exactly $n-1$ arithmetic operations (all additions).

EFR 3.12:   The characteristic equation of the recurrence relation $a_n = 4a_{n-1} - 4a_{n-2}$ is $r^2 - 4r + 4 = 0$, which factors as $(r-2)^2 = 0$. Thus, there is one characteristic double root $\lambda = 2$. By Theorem 3.6, the general solution of the recurrence is given by $a_n = x_1 2^n + x_2 n2^n$, where x_1, x_2 are constants.   Substituting the initial conditions $a_0 = 1, a_1 = 0$ produces: $1 = a_0 = x_1$, and so $0 = a_1 = 1 \cdot 2 + x_2 \cdot 2$, so $x_2 = -1$, and the explicit solution of the given recursively defined sequence is thus $a_n = 2^n - n2^n$. Both formulas will produce $a_7 = -768$.

EFR 3.13:   The characteristic equation of the recurrence relation $a_n = a_{n-1} + a_{n-2} - a_{n-3}$ is $r^3 - r^2 - r + 1 = 0$.   The left side easily factors: $r^3 - r^2 - r + 1 = r^2(r-1) - r(r-1) = (r^2-1)(r-1)$ $= (r+1)(r-1)^2$.   The characteristic roots of the recurrence are $\lambda = -1, 1, 1$, so by Theorem 3.7, the general solution of the recuurence is given by $a_n = x(-1)^n + (y+nz)(1)^n = x(-1)^n + y + nz$,   where x, y, z are constants. Substituting the initial conditions $a_0 = 1, a_1 = 0, a_2 = 1$, produces

$$\begin{cases} 1(= a_0) = & x + y \\ 0(= a_1) = -x + y + & z \\ 1(= a_2) = & x + y + 2z. \end{cases}$$

Isolating y in the first equation gives $y = 1 - x$ (*); isolating z in the second equation and using (*) gives $z = x - y = x - (1-x) = 2x - 1$ (**). Substituting (*) and (**) into the third equation gives $1 = x + (1-x) + 2(2x-1)$, or $1 = 4x - 1$, so $x = 1/2$. By (*) and (**), we now also obtain $y = 1/2$ and $z = 0$, so the explicit solution of the given recurrence is thus $a_n = [(-1)^n + 1]/2$.

(b) With any of these formulas, one readily obtains that $a_{10} = 1$.

(c) The characteristic polynomial of the homogeneous portion of the recurrence relation $a_n = 4a_{n-1} - 4a_{n-2}$ is $r^2 - 4r - 4 = (r-2)^2$. Since the inhomogeniety function $f(n) = 2^n n$ is a first degree polynomial (in n) times an exponential whose base is a characteristic root, Case 2 of Theorem 3.9 tells us that we can look for a particular solution of the inhomogeneous recurrence relation to have the form $a_n^p = n^2(dn+e)2^n$, where d and e are constants. We need to determine values of d and e that will make this solve the recurrence equation (we will deal with the initial conditions later). Direct substitution of this expression into the recurrence equation gives:

$$n^2(dn+e)2^n = 4(n-1)^2(d[n-1]+e)2^{n-1} - 4(n-2)^2(d[n-2]+e)2^{n-2} + 2^n n.$$

Dividing both sides by 2^n, expanding the right side and collecting like terms leads us to:

$$
\begin{aligned}
dn^3 + en^2 &= 2d(n-1)^3 + 2e(n-1)^2 - d(n-2)^3 - e(n-2)^2 + n \\
&= 2d[n^3 - 3n^2 + 3n - 1] + 2e[n^2 - 2n + 1] - d[n^3 - 6n^2 + 12n - 8] - e[n^2 - 4n + 4] + n \\
&= dn^3 + en^2 + [6d - 4e - 12d + 4e + 1]n + [-2d + 2e + 8d - 4e] \\
&= dn^3 + en^2 + [-6d + 1]n + [6d - 2e].
\end{aligned}
$$

To make this an identity (i.e., to make the generic particular solution form solve the recurrence), we need to choose $d = 1/6$ and $e = 1/2$. We have thus obtained the following particular solution for the recurrence relation: $a_n^p = n^2(n/6 + 1/2)2^n = n^2(n+3)2^{n-1}/3$. Since (by Theorem 3.6), the general solution of the associated homogeneous recurrence equation is $a_n^h = x2^n + yn2^n$, where x and y are constants, it follows (from Theorem 3.8) that the general solution of the given inhomogeneous recurrence equation is $a_n^p + a_n^h = n^2(n+3)2^{n-1}/3 + x2^n + yn2^n$. To determine x and y, we apply the initial conditions to this expression:

$$1 = a_0 = a_0^p + a_0^h = 0 + x \cdot 1 + 0 \Rightarrow x = 1, \text{ and so}$$
$$2 = a_1 = a_1^p + a_1^h = 1 \cdot 4 \cdot 1/3 + 1 \cdot 2 + y \cdot 1 \cdot 2 = 4/3 + 2 + 2y \Rightarrow y = -2/3.$$

We have now completely derived an explicit solution for the given recursively defined sequence: $a_n = n^2(n+3)2^{n-1}/3 + 2^n - n2^{n+1}/3$.

(d) Both this and the original recursion formula will yield the value $a_6 = 3264$.

EFR 3.14: (a) $16{,}000 = 2^7 \cdot 5^3$, $42{,}757 = 11 \cdot 13^2 \cdot 23$
(b) $\gcd(100, 76) = 4$, $\gcd(16000, 960) = 320$

EFR 3.15: (a) $\text{lcm}(12, 28) = 84$, $\text{lcm}(100, 76) = 1900$
(b) Using the indicated notation, we let $g = p_1^{\sigma_1} \cdot p_2^{\sigma_2} \cdots p_n^{\sigma_n}$. Since for each index i, we have $\alpha_i \geq \sigma_i$, it follows that $c \equiv p_1^{\alpha_1 - \sigma_1} p_2^{\alpha_2 - \sigma_2} \cdots p_n^{\alpha_n - \sigma_n}$ is a nonnegative integer that satisfies:

$$g \cdot c = p_1^{\sigma_1} p_2^{\sigma_2} \cdots p_n^{\sigma_n} \cdot p_1^{\alpha_1 - \sigma_1} p_2^{\alpha_2 - \sigma_2} \cdots p_n^{\alpha_n - \sigma_n} = p_1^{\alpha_1} p_2^{\alpha_2} \cdots p_n^{\alpha_n} = a,$$

and thus $g \mid a$. A similar argument shows that $g \mid b$, and thus $g \mid \gcd(a,b)$. Conversely, if k is any common divisor of a and b, then the prime factorization of k can only contain primes from the list $p_1, p_2, \cdots, p_k$. For any index i, the exponent of p_i in the prime factorization of k cannot exceed α_i (since $k \mid a$) and it cannot exceed β_i (since $k \mid b$), so it follows that this exponent cannot exceed $\min(\alpha_i, \beta_i) = \sigma_i$. From these facts it follows (just as above) that $k \mid g$, and so $g = \gcd(a,b)$.

Next we let $\ell = p_1^{\mu_1} \cdot p_2^{\mu_2} \cdots p_n^{\mu_n}$. Since for each index i, we have $\alpha_i \leq \mu_i$, it follows that $c \equiv p_1^{\mu_1 - \alpha_1} p_2^{\mu_2 - \alpha_2} \cdots p_n^{\mu_n - \alpha_n}$ is a nonnegative integer that satisfies:

$$a \cdot c = p_1^{\alpha_1} p_2^{\alpha_2} \cdots p_n^{\alpha_n} \cdot p_1^{\mu_1 - \alpha_1} p_2^{\mu_2 - \alpha_2} \cdots p_n^{\mu_n - \alpha_n} = p_1^{\mu_1} p_2^{\mu_2} \cdots p_n^{\mu_n} = \ell,$$

and thus $a \mid \ell$. A similar argument shows that $b \mid \ell$, and thus $\text{lcm}(a,b) \mid \ell$. Conversely, if k is any common multiple of a and b, then the prime factorization of k must contain all primes from the list

$p_1, p_2, \cdots, p_k$. For any index i, the exponent of p_i in the prime factorization of k must be at least α_i (since $a \mid k$) and it must be at least β_i (since $b \mid k$), so it follows that this exponent must be at least $\max(\alpha_i, \beta_i) = \mu_i$. From these facts it follows (just as above) that $\ell \mid k$, and so $\ell = \text{lcm}(a,b)$. □

(c) The result follows from multiplying the expressions g and ℓ for $\gcd(a, b)$ and $\text{lcm}(a, b)$ of Part (b), and using the fact that $\max(\alpha_i, \beta_i) + \min(\alpha_i, \beta_i) = \alpha_i + \beta_i$.

EFR 3.16: (a) Letting $q = \text{floor}(a/d)$ and $r = a - qd$, we certainly have that q and r are integers that satisfy $dq + r = a$. Since $q = \text{floor}(a/d) \le a/d$, if follows that $qd \le a$, and so $r = a - qd \ge 0$. Also, since $q = \text{floor}(a/d) > a/d - 1$, it follows that $qd > (a/d - 1)d = a - d$, so that $r = a - qd < d$.

(b) (i) quotient: $q = \text{floor}(123/5) = 24$, remainder: $r = a - qd = 123 - 24 \cdot 5 = 3$.

(ii) quotient: $q = \text{floor}(-874/15) = -59$, remainder: $r = a - qd = -874 - (-59) \cdot 15 = 11$.

EFR 3.17: $\begin{cases} 91 = 1 \cdot 65 + 26 \\ 65 = 2 \cdot 26 + 13 \\ 26 = 2 \cdot 13 + 0 \end{cases}$ so $\gcd(65,91) = 13$. $\begin{cases} 1665 = 1 \cdot 910 + 755 \\ 910 = 1 \cdot 755 + 155 \\ 755 = 4 \cdot 155 + 135 \\ 155 = 1 \cdot 135 + 20 \\ 135 = 6 \cdot 20 + 15 \\ 20 = 1 \cdot 15 + 5 \\ 15 = 3 \cdot 5 + 0 \end{cases}$ so $\gcd(1665,755) = 5$.

EFR 3.18: (a) Solving the second-to-last equation of the first set in the preceding solution for the remainder 13 gives $13 = 65 - 2 \cdot 26$. Next, we substitute into this the expression for 26 obtained by solving the first equation: $26 = 91 - 1 \cdot 65$, to obtain $13 = 65 - 2 \cdot 26 = 65 - 2 \cdot (91 - 1 \cdot 65) = 65 - 2 \cdot 91 + 2 \cdot 65 = 3 \cdot 65 - 2 \cdot 91$. In summary: $13 = \gcd(65,91) = 3 \cdot 65 - 2 \cdot 91$.

Similarly, we start with the second-to-last equation of the second set of the solution to EFR 3.17, solving for the remainder: $5 = 20 - 1 \cdot 5$. We then successively work our way up the list of equations, substituting each remainder in turn, and expressing in terms of the two most recent dividends:

$$5 = 20 - 1 \cdot 15 = 20 - 1 \cdot (135 - 6 \cdot 20) = 7 \cdot 20 - 1 \cdot 135$$
$$= 7 \cdot (155 - 1 \cdot 135) - 1 \cdot 135 = 7 \cdot 155 - 8 \cdot 135$$
$$= 7 \cdot 155 - 8 \cdot (755 - 4 \cdot 155) = -8 \cdot 755 + 39 \cdot 155$$
$$= -8 \cdot 755 + 39 \cdot (910 - 1 \cdot 755) = 39 \cdot 910 - 47 \cdot 755$$
$$= 39 \cdot 910 - 47 \cdot (1665 - 1 \cdot 910) = -47 \cdot 1665 + 86 \cdot 910.$$

In summary: $5 = \gcd(1665,910) = -47 \cdot 1665 + 86 \cdot 910$.

EFR 3.19: *Proof:* We will prove this by the method of contradiction. By the fundamental theorem of arithmetic and since n is assumed not to be a perfect square, we may write $n = p^j Q$, where p is a prime, j is a positive odd exponent, and Q is a positive integer that is relatively prime to p. (Q is just the product of the other prime factors in the unique factorization of n.) If $\sqrt{n}$ were rational, it could be written as a fraction of integers a/b, which we may assume is in lowest terms. This means that $b\sqrt{n} = a$. Squaring both sides gives that $p^j Q b^2 = a^2$. Since $p^j \mid p^j Q b^2$, we get that $p^j \mid a^2$, and since $j = 2k + 1$ is odd, it follows from the unique factorization of a that p^{k+1} is a factor of a. Since p is relatively prime to both Q and b, the equation $p^j Q b^2 = a^2$ together $p^{k+1} \mid a$ implies that $p^{j+1} = p^{2k+2} = (p^{k+1})^2 \mid p^j$, a contradiction. The result is therefore established. □

EFR 3.20: *Proof of Theorem 3.18:* We begin by recasting the assumptions $a \equiv a' \pmod m$ and $b \equiv b' \pmod m$ as $a - a' = km$ and $b - b' = \ell m$, for some integers k, ℓ.

(a) Since $a + b - (a' + b') = (a - a') + (b - b') = km + \ell m = (k + \ell)m$, we obtain $a + b \equiv a' + b' \pmod m$.

(b) Since $-a - (-a') = -a + a' = -(a - a') = (-k)m$, it follows that $-a \equiv -a' \pmod m$.

(c) Since $a \cdot b - a' \cdot b' = a \cdot b - a \cdot b' + a \cdot b' - a' \cdot b' = a(b - b') + (a - a')b = a(\ell m) + (km)b = [a\ell + kb]m$, it follows that $a \cdot b \equiv a' \cdot b' \pmod m$.

(d): This part follows from Parts (a) and (b), since $a - b = a + (-b)$.

(e): This part follows from Part (c), since exponentiation is a sequence of multiplications. $\square$

EFR 3.21: *Method 1:* (*Fast Modular Exponentiation*) We begin with $18^2 \equiv 5 \pmod{29}$, and continue to square both sides until the exponents exceed at least half of the desired exponent:

$$
\begin{aligned}
18^4 &\equiv 5^2 \equiv 25, \\
18^8 &\equiv 25^2 \equiv 625 \equiv 16, \\
18^{16} &\equiv 16^2 \equiv 256 \equiv 24, \\
18^{32} &\equiv 25, \\
18^{64} &\equiv 16, \\
18^{128} &\equiv 24, \\
18^{256} &\equiv 25, \\
18^{512} &\equiv 16.
\end{aligned}
$$

We will now be able to use the above powers to compute the desired power of 18 (mod 29). This is because $8052 = 512 + 256 + 32 + 2$, as the reader can easily check. It follows that we may compute

$$18^{802} = 18^{512} \cdot 18^{256} \cdot 18^{32} \cdot 18^2 \equiv 16 \cdot 25 \cdot 25 \cdot 5 \equiv 4 \pmod{29}.$$

Method 2: By Fermat's little theorem, $18^{28} \equiv 1 \pmod{29}$. If we apply the division algorithm to the integer division of 802 by 28, we obtain $802 = 28 \cdot 28 + 18$. It follows that $18^{802} \equiv (18^{28})^{28} \cdot 18^{18} \equiv 1^{80} \cdot 4 \equiv 4 \pmod{29}$.

EFR 3.22: By factoring each input into primes and then using (1) of Proposition 3.20, we obtain:

$$
\begin{aligned}
\phi(15) &= \phi(3 \cdot 5) = (3 - 1) \cdot 3^0 \cdot (5 - 1) \cdot 5^0 = 8 \\
\phi(20) &= \phi(2^2 \cdot 5) = (2 - 1) \cdot 2^1 \cdot (5 - 1) \cdot 5^0 = 8 \\
\phi(208) &= \phi(2^4 \cdot 13) = (2 - 1) \cdot 2^3 \cdot (13 - 1) \cdot 13^0 = 96 \\
\phi(2208) &= \phi(2^5 \cdot 3 \cdot 23) = (2 - 1) \cdot 2^4 \cdot (3 - 1) \cdot 3^0 \cdot (23 - 1) \cdot 23^0 = 704 \\
\phi(6624) &= \phi(2^5 \cdot 3^2 \cdot 23) = (2 - 1) \cdot 2^4 \cdot (3 - 1) \cdot 3^1 \cdot (23 - 1) \cdot 23^0 = 2112
\end{aligned}
$$

EFR 3.23: *Proof of Euler's Theorem:* As in the proof of Fermat's little theorem, we will denote the set A of integers mod m using square brackets: $[k]$ represents the set of all integers that are congruent to $k \pmod m$. We will construct a function with domain and codomain both being the set A of integers mod m that are relatively prime to m: $f : A \to A$, defined by $f([x]) = [ax]$. By Theorem 3.18, this definition will give the same output no matter which representative we use of $[x]$, so it is a well-defined function on elements of A. But we still need to check that the images, which *a priori* could be any integers mod m, are actually in the set A (i.e., so the co-domain of the function can be taken to be A). Indeed, if $[ax]$ did not belong to A, this would mean that ax is not relatively prime to m. But since both a and x are relatively prime to m, their product ax also must be, so indeed $[ax] \in A$. Next we will show that f is one-to-one. Suppose that $f([x]) = f([y])$. This means that $[ax] = [ay]$ or $ax \equiv ay \pmod m$. By definition, this means that $m \mid (ax - ay)$ or $m \mid a(x - y)$. But since $\gcd(m, a) = 1$, it follows that each of the prime powers in the factorization of m, must divide into the latter factor $x - y$, and thus, so must their product m, so $m \mid x - y$, which means that $[x] = [y]$, so f is one-to-one.

Since f is a one-to-one function of the set $A = \{[a_1], [a_2], \cdots, [a_K]\}$ (note that $K = \phi(m)$) to itself, it follows that the images of f: $f([a_1])$, $f([a_2])$, $\cdots$, $f([a_K])$ are simply a relisting of the elements of A, in perhaps a different order. It follows that if we multiply representatives from these to listings of the set A, we will get the same result (mod n):

$$a_1 \cdot a_2 \cdots a_K \equiv (a \cdot a_1) \cdot (a \cdot a_2) \cdots (a \cdot a_K) \equiv a^{K-1}(a_1 \cdot a_2 \cdots a_K) \pmod{m}.$$

This equation implies that $p \mid [a^{K-1}(a_1 \cdot a_2 \cdots a_K) - a_1 \cdot a_2 \cdots a_K]$ or $m \mid [(a^{\phi(m)} - 1)(a_1 \cdot a_2 \cdots a_K)]$, and the same argument used above shows that since m has no common factors with $a_1 \cdot a_2 \cdots a_K$, we must have $m \mid (a^{\phi(m)} - 1)$, so that $a^{\phi(m)} \equiv 1 \pmod{m}$, as we wished to prove. □

EFR 3.24:   Part (a): Since gcd(7,58) = 1, and $\phi(58) = \phi(2 \cdot 29) = (2-1) \cdot 2^0 \cdot (29-1) \cdot 29^0 = 28$, Euler's theorem tells us that $7^{28} \equiv 1 \pmod{58}$. Using the division algorithm for the integer division of 8486 by 28 gives $8486 = 303 \cdot 28 + 2$, and consequently $7^{8486} \equiv (7^{28})^{303} \cdot 7^2 \equiv 1 \cdot 49 \equiv 49 \pmod{58}$.

Part (b): Finding the last three digits of any number is the same as the answer we would get by converting it to an integer modulo 1000. Thus, we wish to find $13^{2017} \pmod{1000}$. Since gcd(13,1000) = 1 and $\phi(1000) = \phi(2^3 \cdot 5^3) = (2-1) \cdot 2^2 \cdot (5-1) \cdot 5^2 = 400$, Euler's theorem tells us that $13^{400} \equiv 1 \pmod{1000}$. Applying the division algorithm to the given exponent divided by 400 gives $2017 = 5 \cdot 400 + 17$, hence $13^{2017} \equiv (13^{400})^5 \cdot 13^{17} \equiv 1 \cdot 13^{17} \equiv 933 \pmod{1000}$. (In the last computation we used the method of Example 3.17.)

EFR 3.25:   (a) The prime factorization of 334 is $2 \cdot 167$, so by Theorem 3.23(1), 334 has $\phi(\phi(334)) = \phi(166) = 82$ primitive roots.

(b) By Proposition 3.22, the order of any integer relatively prime to 334 (mod 334) must divide $\phi(334) = 166 = 2 \cdot 83$, so the only possible orders are 1, 2, 83, and 166 (in which case we have a primitive root). To find the smallest primitive root, we go through positive integers a relatively prime to 334, starting with 3, computing (as in Method 1 of the solution to Example 3.17) the modular powers $a^2, a^{83} \pmod{334}$ until we find an integer whose order is 166: $3^2 \equiv 9, 3^{83} \equiv 1 \pmod{334}$ so $\text{ord}_{334}(3) = 83$, $5^2 \equiv 25, 5^{83} \equiv 333 \pmod{334}$ so $\text{ord}_{334}(5) = 166$, and 5 is the smallest primitive root (mod 334).

CHAPTER 4: NUMBER SYSTEMS

EFR 4.1: (a) [1101001111] (base 2) $\sim 2^9 + 2^8 + 2^6 + 2^3 + 2^2 + 2^1 + 2^0 = 847$.   [777] (base 8) $\sim 7 \cdot 8^2 + 7 \cdot 8^1 + 7 \cdot 8^0 = 511$.   [123ABC] (hex) $\sim 1 \cdot 16^5 + 2 \cdot 16^4 + 3 \cdot 16^3 + 10 \cdot 16^2 + 11 \cdot 16^1 + 12 \cdot 16^0 = 1,194,684$.

(b) (i) *Step 1*: $R = 122$. Step 2: Largest power of 2 not exceeding $R = 122$ is $2^6 = 64$, so ($K = 6$) and $c_6 = 1$. Update $R \rightarrow 122 - 64 = 58$, Step 2 is repeated: Now, $k = 5$ is the largest exponent such that $2^k \le 58 (= R)$, so we set $c_5 = 1$, and update $R \rightarrow 58 - 32 = 26$.

Step 2 is repeated: $k = 4$, $c_4 = 1$, $R \rightarrow 26 - 16 = 10$.

Step 2 is repeated: $k = 3$, $c_3 = 1$, $R \rightarrow 10 - 8 = 2$.

Step 2 is repeated: $k = 1$, $c_1 = 1$, $R \rightarrow 2 - 2 = 0$.

So 122 $\sim$ [1111010] (base 2).

(ii) *Step 1*: $R = 9675$. Step 2: Largest power of 32 not exceeding $R = 9675$ is $32^2 = 1024$, so ($K = 2$) and $c_2 = \text{floor}(9675/1024) = 9$. Update $R \rightarrow 9675 - 9 \cdot 32^2 = 459$. Step 2 is repeated: Now, $k = 1$ is the largest exponent such that $32^k \leq 459(= R)$, so we set $c_1 = \text{floor}(459/32) = 14$ and update $R \rightarrow 459 - 14 \cdot 32 = 11$. Step 2 is repeated: $k = 0$, $c_0 = 11$, $R \rightarrow 11 - 11 = 0$. So $9675 \sim [9 \ 14 \ 11]$ (base 32).

(iii) *Step 1*: $R = 52,396$. Step 2: Largest power of 16 not exceeding $R = 52,396$ is $16^3 = 4096$, so ($K = 3$) and $c_3 = \text{floor}(52396/4096) = 12$. Update $R \rightarrow 52,396 - 12 \cdot 16^3 = 3244$. Step 2 is repeated: Now, $k = 2$ is the largest exponent such that $16^k \leq 3244(= R)$, so we set $c_2 = \text{floor}(3244/16^2) = 12$ and update $R \rightarrow 3244 - 12 \cdot 16 = 172$.

Step 2 is repeated: $k = 1$, $c_1 = 10$, $R \rightarrow 172 - 10 \cdot 16 = 12$.

Step 2 is repeated: $k = 0$, $c_0 = 12$, $R \rightarrow 12 - 12 = 0$.

So $52,396 \sim [12 \ 12 \ 10 \ 12] = [CCAC]$ (hex).

EFR 4.2: (a) The (binary) addition algorithm is summarized in the diagram on the right. The resulting addition $[101111] + [001111] = [111110]$ (base 2) can be translated term-by-term into an ordinary integer addition (with the same method used in Part (a) of the solution to EFR 6.1): $47 + 15 = 62$.

	1	*1*	*1*	*1*	
1	0	1	1	1	1
0	0	1	1	1	1
1	1	1	1	1	0

(b) The (hexadecimal) addition algorithm is summarized in the diagram below. In each step, the letters are converted to their integer equivalents to perform the required addition. Here is a summary: E + 2 = 14 + 2 = 16, so the right-most entry in the table is 0, and the carry 1 is put over 4 (in italics) in the next column to the left. Next, we add 1(the carry) + 4 + A = 1 + 4 + 10 = 15 = F. F gets entered into the second from right entry, and there is no carry. The third from right slot addition is now D + A = 13 + 10 = 23, so we have a carry of 1, and enter 23 − 16 = 7. The final addition is now 1(the carry) + 7 + 1 = 9. The resulting addition $[7D4E] + [1AA2] = [97F0]$ (base 16) can be translated term by term into an ordinary integer addition (with the same method used in Part (a) of the solution to EFR 4.1): $32,078 + 6818 = 38,896$.

	1		*1*	
7	D	4	E
1	A	A	2
9	7	F	0

EFR 4.3: (a) The (binary) subtraction algorithm is summarized in the diagram on the right. Note the first borrow was needed on the second-from-right digit subtraction 0 − 1. We borrowed one from the next preceding digit which gave us 2 + 0 − 1 = 1. The resulting subtraction $[101101] + [001111] = [011110]$ (base 2) translates into the following integer subtraction: $45 − 15 = 30$.

(b) The (hexadecimal) subtraction algorithm is summarized in the diagram below. In each step, the letters are converted to their integer equivalents to perform the required subtraction. Here is a summary: E − 2 = 14 − 2 = 12 = C, so the right-most entry in the table is C, and there is no borrow to put over 4 in next column to the left. Next, we need to subtract 4 − A = 4 − 10, so we need to borrow 1 from the previous digit, and this adds 16 so the subtraction is now feasible: 16 + 4 − 10 = 10 = A. So A gets entered into the second from right entry, and there is no borrow − 1 is put over the preceding column. The third from right slot subtraction is now − 1(the borrow) + D − A = − 1 + 13 − 10 = 2, so we enter 2, and there is no borrow. The final subtraction is now 7 − 1 = 6. The resulting subtraction $[7D4E] − [1AA2] = [62AC]$ (hex) translates into the following integer subtraction: $32,078 − 6818 = 25,260$.

-1	*-1*	*-1*	*-1*		
1	0	1	1	0	1
0	0	1	1	1	1
0	1	1	1	1	0

	-1		
7	D	4	E
1	A	A	2
6	2	A	C

EFR 4.4:   (a)   The (binary) multiplication algorithm is summarized in the diagram on the right. Note that as with all base 2 multiplications there are no carries. In the additions of the digits, there were carries, which we omitted. The resulting multiplication $[1111] \times [1111] = [11100001]$ (base 2) translates into the following integer multiplication: $15 \times 15 = 225$.

```
        1  1  1  1
    ×   1  1  1  1
    ───────────────
        1  1  1  1
     1  1  1  1
  1  1  1  1
1  1  1  1
───────────────────
1  1  1  0  0  0  0  1
```

(b)   The base 7 multiplication algorithm is summarized in the diagram below. Here is how the computations went: We start by multiplying the last digit ($i = 0$) of the second number: 0, by the first number, and we get all zeros, so this is the first (unshifted) row in the multiplication table results. Next, we multiply the second to middle digit ($i = 1$) of the second number: 2, by the first number [262]. Starting with the last digit ($j = 0$), we have $2 \cdot 2 = 4$ (base 7) and there is no carry, so we simply put 4 in at the right of the second row in the multiplication table results, which is padded with one 0 at the right (one unit shift). Next ($j = 1$), we multiply $2 \cdot 6 = 12 = 1 \cdot 7 + 5$ (base 7) so we put 5 to the left of 4 in the second row of the multiplication table results, and put the carry above 2 in the next column (in the figure we have 4/1, the 4 pertains to a carry that will occur in the next digit multiplication). We now add our carry to the last multiplication for this digit ($j = 2$), to get $1 + 2 \cdot 2 = 5$ (base 7), so

```
        4/1  1/
         2   6   2
    ×    5   2   0
    ─────────────────
         0   0   0
     5   5   4
  2  0   3   3
  ─────────────────
  2  1  2   1   4   0
```

we place a 5 to the left of the last 5 that was entered. We now move on to the left digit ($i = 2$) of the second number: 5. First we multiply ($j = 0$) $5 \cdot 2 = 10 = 1 \cdot 7 + 3$ (base 7), so we enter 3 at the right of the third row in the multiplication table results, which is padded with two 0's at the right (two unit shift), and put the carry (1) on top of the column for the next digit (carries for this multiplication are written before the "/").   We add the carry (1) to the next multiplication: ($j = 1$) $1 + 5 \cdot 6 = 31 = 4 \cdot 7 + 3$ (base 7), so we put 3 to the left of the just entered 3 in the third row of the multiplication table results, and put the carry above 4 in the next column. In the final ($j = 2$) multiplication, we add the carry (4) $4 + 5 \cdot 2 = 14 = 2 \cdot 7 + 0$, so we enter 0 to the left of the 3 that was just entered, and since this was the last digit, we enter the final carry 2 to the left of this 3. Finally, we add up (in base 7) these three computed (and shifted rows), which can be done in two steps using Algorithm 4.2. The resulting multiplication $[262] \times [520] = [212140]$ (base 7) translates into the following integer multiplication: $142 \times 259 = 36{,}778$.

EFR 4.5:   Apply the division algorithm to write $a = 2q_a + r_a$ and $b = 2q_b + r_b$,   where $0 \le r_a, r_b < 2$. Thus a is even (odd) if $r_a = 0$ (1), and similarly for b. Since $a - b = 2(q_a - q_b) + r_a - r_b$, we see that $a \equiv b \pmod 2 \Leftrightarrow 2 \mid (r_a - r_b)$, but since r_a, r_b can only take on the values 0 or 1, it follows that $2 \mid (r_a - r_b)$ if, and only if r_a, r_b are both 0 or both 1, i.e., $a \equiv b \pmod 2$ if, and only if a and b have the same parity. □

There are two congruence classes (mod 2): the even integes and the odd integers.

EFR 4.6:   Working mod 12, we have: $11 + 8 \equiv 19 \equiv 7$, $5 \cdot 8 = 40 \equiv 4$, and $11^2 \equiv (-1)^2 \equiv 1$, and these computations give the following equalities in $\mathbb{Z}_{12}$: $11 + 8 = 7$, $5 \cdot 8 = 4$, and $11^2 = 1$. We will check to see whether an element $b \in \mathbb{Z}_{12}$ exists such that $5b = 1$ by multiplying 5 by all elements until (if ever) we find one that works: Working in $\mathbb{Z}_{12}$, we have: $5 \cdot 1 = 5, 5 \cdot 2 = 10, 5 \cdot 3 (\equiv 15) = 3$, $5 \cdot 4 (\equiv 20) = 8, 5 \cdot 5 (\equiv 25) = 1$. So $b = 5$ has the indicated property.

EFR 4.7: Using Algorithm 4.1, we obtain the binary expansion for the exponent: $225 \sim [11100001]$ (base 2). We initialize cumulative product: $a = 1$, and the cumulative square $s = 289$. We now proceed through the iterations of Step 2:

With $k = 0$, the corresponding binary digit of the exponent is 1, so we multiply the cumulative product by s to get 289 (mod 311): $a = 289$. We update cumulative square (mod 311) for the next iteration : $s \rightarrow s^2 = 289^2 \equiv 173$.

With $k = 1$, the corresponding binary digit of the exponent is 0, so we do not multiply the cumulative product by s. We update cumulative square (mod 311) for the next iteration : $s \rightarrow s^2 = 173^2 \equiv 73$.

With $k = 2$, the corresponding binary digit of the exponent is 0, so we do not multiply the cumulative product by s. We update cumulative square (mod 311) for the next iteration : $s \rightarrow s^2 = 73^2 \equiv 42$. 73

With $k = 3$, the corresponding binary digit of the exponent is 0, so we do not multiply the cumulative product by s. We update cumulative square (mod 311) for the next iteration : $s \rightarrow s^2 = 42^2 \equiv 209$.

With $k = 4$, the corresponding binary digit of the exponent is 0, so we do not multiply the cumulative product by s. We update cumulative square (mod 311) for the next iteration : $s \rightarrow s^2 = 209^2 \equiv 141$.

With $k = 5$, the corresponding binary digit of the exponent is 1, so we multiply the cumulative product by s to get $a = 40749 \equiv 8$ (mod 311). We update cumulative square (mod 311) for the next iteration: $s \rightarrow s^2 = 141^2 \equiv 288$.

With $k = 6$, the corresponding binary digit of the exponent is 1, so we multiply the cumulative product by s to get $a = 2304 \equiv 127$ (mod 311). We update cumulative square (mod 311) for the next iteration: $s \rightarrow s^2 = 288^2 \equiv 218$.

With $k = 7$ (last iteration) the corresponding binary digit of the exponent is 1, so we multiply the cumulative product by s to get $a = 27,686 \equiv 7$ (mod 311).

So we conclude that the final answer is: $289^{225} \equiv 7 \pmod{311}$.

EFR 4.8: (a) Initial vectors: $U = [1155, 1, 0]$, $V = [862, 0, 1]$.
Since $V(1) = 862$ is positive, we update the vectors:
$W = U - \text{floor}(U(1)/V(1))V = [293, 1, -1]$, $U = V = [862, 0, 1]$, $V = W = [293, 1, -1]$.
Since $V(1) = 293$ is positive, we update the vectors:
$W = U - \text{floor}(U(1)/V(1))V = [276, -2, 3]$, $U = V = [293, 1, -1]$, $V = W = [276, -2, 3]$.
Since $V(1) = 276$ is positive, we update the vectors:
$W = U - \text{floor}(U(1)/V(1))V = [17, 3, -4]$, $U = V = [276, -2, 3]$, $V = W = [17, 3, -4]$.
Since $V(1) = 17$ is positive, we update the vectors:
$W = U - \text{floor}(U(1)/V(1))V = [4, -50, 67]$, $U = V = [17, 3, -4]$, $V = W = [4, -50, 67]$.
Since $V(1) = 4$ is positive, we update the vectors:
$W = U - \text{floor}(U(1)/V(1))V = [1, 203, -272]$, $U = V = [4, -50, 67]$, $V = W = [1, 203, -272]$.
Since $V(1) = 1$ is positive, we update the vectors:
$W = U - \text{floor}(U(1)/V(1))V = [0, -862, 1155]$, $U = V = [1, 203, -272]$, $V = W = [0, -862, 1155]$.
Since $V(1) = 0$, the algorithm terminates.

We can read off the answer from the last updated vector U: $d = 1$, $x = 203$, and $y = -272$; it can be readily checked that these numbers satisfy: $d = 1155x + 862y$.
(b) In light of Proposition 4.2, since $\gcd(862, 1155) = 1$, and since $y = -272 \equiv 883 \pmod{1155}$, we have $862^{-1} = 883$ in $\mathbb{Z}_{1155}$.

EFR 4.9: (a) Since $d = \gcd(123, 456) = 3 | 12$, we can follow Algorithm 4.7 to obtain the $d = 3$ solutions of the congruence.
Step 1: We solve the modified congruence: $(123/3)y \equiv (12/3) \pmod{456/3}$, i.e., $41y \equiv 4 \pmod{152}$. Using the extended Euclidean algorithm (Algorithm 4.6) as in the solution to EFR 4.8, we compute $41^{-1} = 89$. We multiply both sides of the modified congruence by this inverse to solve it:

$$y \equiv 41^{-1} \cdot 41y \equiv 89 \cdot 4 \equiv 52 (\text{mod} 152).$$

Step 2: We may now list the $d = 3$ solutions of the original congruence: 52, 52 + 456/3, 52 + $2 \cdot 456/3 = \{52, 204, 356\}$. (The reader may wish to check each of these.)

(b) Since $15x + 4 \equiv 20 \pmod{25}$ is equivalent to $15x \equiv 16 \pmod{25}$, and since $d = \gcd(15, 25) = 5 \nmid 16$, we know that the congruence has no solution.

EFR 4.10: (a) Since $6x + 2 \equiv 5 \pmod{9}$ is equivalent to $6x \equiv 3 \pmod{9}$, and since $d = \gcd(6, 9) = 3 \mid 3$, we know that the congruence has $d = 3$ solutions. We may obtain them using Algorithm 4.7:

Step 1: We solve the modified congruence: $(6/3)y \equiv (3/3) \pmod{9/3}$, i.e., $2y \equiv 1 \pmod{3}$. Since $2 \cdot 2 = 4 \equiv 1 (\text{mod} 4)$, we know $2^{-1} = 2 (\text{mod} 3)$, and we may solve this modified congruence by multiplying both sides by this inverse to obtain: $y \equiv 2^{-1} \cdot 1 \equiv 2 \cdot 1 \equiv 2 (\text{mod} 9)$.

Step 2: The three solutions of the original congruence are $\{2, 2 + 9/3, 2 + 2 \cdot 9/3\} = \{2, 5, 8\}$.

(b) Since $6x + 2 \equiv 3 \pmod{9}$ is equivalent to $6x \equiv 1 (\text{mod} 9)$, and since $d = \gcd(6, 9) = 3 \nmid 1$, we know that the congruence has no solution.

(c) Since $\gcd(5, 9) = 1$, we know the congruence has a unique solution. Since $5 \cdot 2 \equiv 10 \equiv 1 (\text{mod} 9)$, we know $5^{-1} = 2$ in $\mathbb{Z}_9$, and this inverse can be used to solve the congruence:

$$5x + 2 \equiv 3 \pmod{9} \Rightarrow 5x \equiv 1 \pmod{9} \Rightarrow 5^{-1} \cdot 5x \equiv 2 \cdot 1 \pmod{9} \Rightarrow x \equiv 2 \pmod{9}.$$

EFR 4.11: In order to apply the Chinese remainder theorem, we first need to put the third congruence of the system, $3x \equiv 4 (\text{mod} 7)$, into proper form. Since $3 \cdot 5 \equiv 15 \equiv 1 (\text{mod} 7)$, $3^{-1} = 5 \pmod{7}$, and we can multiply both sides of this third congruence by this inverse to convert it into $x \equiv 6 (\text{mod} 7)$. We may now apply the solution scheme of the Chinese remainder theorem to the

equivalent system of congruences: $\begin{cases} x \equiv 0 (\text{mod} 2) \\ x \equiv 2 (\text{mod} 5) \\ x \equiv 6 (\text{mod} 7). \end{cases}$

Since the moduli are pairwise relatively prime, (10) (in the proof of the Chinese remainder theorem) provides us with a scheme for obtaining a simultaneous solution. We first set $N = 2 \cdot 5 \cdot 7 = 70$. With $b_1, b_2, b_3 = 0, 2, 6$ and $n_1, n_2, n_3 = 2, 5, 7$, in order to use (10), we must first determine e_1, e_2, e_3 from their defining equations: $e_i(N / n_i) \equiv 1 \pmod{n_i}$.

For e_1: $e_1 \cdot 35 \equiv 1 \pmod{2} \Leftrightarrow e_1 \cdot 1 \equiv 1 \pmod{2} \Leftrightarrow e_1 \equiv 1 \pmod{2}$.

For e_2: $e_2 \cdot 14 \equiv 1 \pmod{5} \Leftrightarrow e_2 \cdot 4 \equiv 1 \pmod{5} \Leftrightarrow e_2 \equiv 4 \pmod{5}$. (Since $2^{-1} = 3 \pmod{5}$.)

For e_3: $e_3 \cdot 10 \equiv 1 \pmod{7} \Leftrightarrow e_3 \cdot 3 \equiv 1 \pmod{7} \Leftrightarrow e_3 \equiv 5 \pmod{7}$.

Now we have all that we need to apply (10) to obtain the desired solution:

$$x = \sum_{i=1}^{3} b_i e_i (M / n_i) = 0 \cdot 1 \cdot (35) + 2 \cdot 4 \cdot (14) + 6 \cdot 5 \cdot (10) = 412 \equiv 62 \pmod{70}.$$

The general solution to the original system of congruences is the set of all integers that are congruent to 62 (mod 70).

EFR 4.12: (a) Since for each index j, $m_j | \text{lcm}(m_1, m_2, \cdots, m_k)$, the implication $\text{lcm}(m_1, m_2, \cdots, m_k) | b \Rightarrow m_j | b$ follows from Proposition 4.3. We write $\text{lcm}(m_1, m_2, \cdots, m_k) = p_1^{\alpha_1} p_2^{\alpha_2} \cdots p_n^{\alpha_n}$, where $p_1 < p_2 < \cdots < p_n$ are the distinct primes appearing in the unique factorization. For each index i, we must have $p_i^{\alpha_i}$ appearing as a factor of at least one m_j (since otherwise a lower exponent could be used in the lcm). Thus, from the assumptions

$m_2 \mid b, \cdots, m_k \mid b$ we may infer that $p_i^{\alpha_i} \mid b$, for each index i, $1 \leq i \leq n$. Thus each prime p_i appears in the prime factorization of b, with exponent at least α_i. This means that $\text{lcm}(m_1, m_2, \cdots, m_k) = p_1^{\alpha_1} p_2^{\alpha_2} \cdots p_n^{\alpha_n}$ must divide b. $\square$

(b) This is a direct consequence of the result of Part (a) with $b = ax - c$ (using the definition of a congruence). $\square$

EFR 4.13: We had reduced the Hindu puzzle to finding the smallest positive solution of the system

(6): $\begin{cases} x \equiv 1 \ (\text{mod } 4) \\ x \equiv 1 \ (\text{mod } 5) \\ x \equiv 1 \ (\text{mod } 6) \\ x \equiv 0 \ (\text{mod } 7) \end{cases}$ We may apply Proposition 4.5 to the first three of these congruences to convert

them into the single equivalent congruence: $x \equiv 1 \ (\text{mod } \text{lcm}(4,5,6))$. Since $\text{lcm}(4,5,6) = \text{lcm}(2^2, 5, 2 \cdot 3) = 2^2 \cdot 5 \cdot 3 = 60$, the original system is equivalent to: $\begin{cases} x \equiv 1 \ (\text{mod } 60) \\ x \equiv 0 \ (\text{mod } 7) \end{cases}$, which is now

amenable to the algorithm of the Chinese remainder theorem (since the moduli are relatively prime): We first set $N = 60 \cdot 7 = 420$. With $b_1, b_2 = 1, 0$ and $n_1, n_2 = 60, 7$, in order to use (10), we must first determine e_1, e_2 by their defining equations: $e_i(N / n_i) \equiv 1 \ (\text{mod } n_i)$.

For e_1: $e_1 \cdot 7 \equiv 1 \ (\text{mod } 60) \Leftrightarrow e_1 \equiv 43 \ (\text{mod } 60)$ (using the extended Euclidean algorithm).

For e_2: $e_2 \cdot 60 \equiv 1 \ (\text{mod } 7) \Leftrightarrow e_2 \cdot 4 \equiv 1 \ (\text{mod } 7) \Leftrightarrow e_2 \equiv 2 \ (\text{mod } 7)$. (Since $4^{-1} = 2 \ (\text{mod } 7)$.)

Now we have all that we need to apply (10) to get the desired solution:

$$x = \sum_{i=1}^{2} b_i e_i (M / n_i) = 1 \cdot 43 \cdot (7) + 0 \cdot 2 \cdot (60) = 301 \ (\text{mod } 420).$$

Putting this solution in the context of the original problem (of Example 4.12), we conclude that the smallest number of eggs that the woman could have had is 301.

EFR 4.14: $(AB)C = \left(\begin{bmatrix} 2 & -4 \\ 1 & 6 \end{bmatrix} \begin{bmatrix} 8 & 0 \\ -4 & 1 \end{bmatrix} \right) \begin{bmatrix} 3 & 7 \\ 5 & 5 \end{bmatrix} = \begin{bmatrix} 32 & -4 \\ -16 & 6 \end{bmatrix} \begin{bmatrix} 3 & 7 \\ 5 & 5 \end{bmatrix} = \begin{bmatrix} 76 & 204 \\ -18 & -82 \end{bmatrix}$

$A(BC) = \begin{bmatrix} 2 & -4 \\ 1 & 6 \end{bmatrix} \left(\begin{bmatrix} 8 & 0 \\ -4 & 1 \end{bmatrix} \begin{bmatrix} 3 & 7 \\ 5 & 5 \end{bmatrix} \right) = \begin{bmatrix} 2 & -4 \\ 1 & 6 \end{bmatrix} \begin{bmatrix} 24 & 56 \\ -7 & -23 \end{bmatrix} = \begin{bmatrix} 76 & 204 \\ -18 & -82 \end{bmatrix} = (AB)C.$

EFR 4.15: In the course of proving both identities, we assume that the sizes of the matrices involved are compatible so that both sides of the identities are defined. (i) The (i,j) entry of $A(B+C)$ is the dot product of the ith row of A: $[a_{i1} \ a_{i2} \cdots a_{im}]$ and the jth column of

$B + C$: $\begin{bmatrix} b_{1j} + c_{1j} \\ b_{2j} + c_{2j} \\ \vdots \\ b_{mj} + c_{mj} \end{bmatrix}$, and so equals $\sum_{k=1}^{m} a_{ik}(b_{kj} + c_{kj}) = \sum_{k=1}^{m} a_{ik}b_{kj} + \sum_{k=1}^{m} a_{ik}c_{kj}$. Since the last two sums

give the (i,j) entries of AB and AC, respectively, the identity $A(B+C) = AB + AC$ is thus established.

$\square$

(ii) The (i,j) entry of $\alpha(A+B)$ is $\alpha(a_{ij} + b_{ij}) = \alpha a_{ij} + \alpha b_{ij}$. Since the first term on the right is the (i,j) entry of αA, and the second term is the (i,j) entry of αB, the identity $\alpha(A+B) = \alpha A + \alpha B$ is proved.

$\square$

EFR 4.16: We assume that $A = [a_{ij}]$ is an $n \times m$ matrix.

(i) The (i,j) entries of AI_m is the dot product of the ith row of A: $[a_{i1} \ a_{i2} \cdots a_{im}]$ with the jth column

of I_m: $\begin{bmatrix} 0 \\ \vdots \\ 0 \\ 1 \\ 0 \\ \vdots \\ 0 \end{bmatrix} \leftarrow$ row j, and so equals a_{ij}, which is the (i,j) entry of A. This proves that $AI_m = A$.

(ii) The (i,j) entries of $I_n A$ is the dot product of the ith row of I_n: $[0 \ \cdots \ 0 \ \underset{\underset{\text{column } i}{\uparrow}}{1} \ 0 \ \cdots \ 0]$ with

the jth column of A: $\begin{bmatrix} a_{1j} \\ a_{2j} \\ \vdots \\ a_{nj} \end{bmatrix}$, and so equals a_{ij}, which is the (i,j) entry of A. This proves that

$I_n A = A$. $\square$

EFR 4.17: (a) If the ith row of A is the zero vector (all entries are zero), and B is any matrix with the same number of columns as A has rows, then, since any entry of the ith row of the matrix product AB will be obtained as a dot product using the ith row of A (which is the zero vector), the result will be zero. This proves that the ith row of AB is also the zero vector. In particular, if A were invertible, and we took $B = A^{-1}$, this would imply that the ith row of AA^{-1} is the zero vector. But this would contradict the property that $AA^{-1} = I$ (since the identity matrix does not have any rows of zeros). This contradiction shows that no matrix with a row of zeros can be invertible. $\square$

(b) If the jth column of the matrix A were all zeros, and B is any matrix with the same number of columns as A has rows, then, since any entry of the jth column of the matrix product BA will be obtained as a dot product using the jth column of A (which is the zero vector), the result will be zero. This proves that the jth column of BA is also the zero vector. In particular, if A were invertible, and we took $B = A^{-1}$, this would imply that the jth column of $A^{-1}A$ is the zero vector. But this would contradict the property that $A^{-1}A = I$ (since the identity matrix does not have any rows of zeros). This contradiction shows that no matrix with a column of zeros can be invertible. $\square$

EFR 4.18:

$$\det\begin{bmatrix} 5 & -6 & 9 \\ -12 & 2 & 7 \\ 2 & 3 & -7 \end{bmatrix} = 5\det\begin{bmatrix} 2 & 7 \\ 3 & -7 \end{bmatrix} - (-6)\det\begin{bmatrix} -12 & 7 \\ 2 & -7 \end{bmatrix} + 9\det\begin{bmatrix} -12 & 2 \\ 2 & 3 \end{bmatrix}$$
$$= 5(2 \cdot (-7) - 7 \cdot 3) + 6((-12) \cdot (-7) - 7 \cdot 2) + 9((-12) \cdot 3 - 2 \cdot 2)$$
$$= -175 + 420 - 360$$
$$= -115.$$

EFR 4.19: Since $\det(M) = \det\left(\begin{bmatrix} 2 & 6 \\ 3 & -9 \end{bmatrix}\right) = 2 \cdot (-9) - 6 \cdot 3 = -36 \neq 0$, the matrix M is invertible,

and its inverse is (by Theorem 4.9): $M^{-1} = \dfrac{1}{\det(M)}\begin{bmatrix} -9 & -6 \\ -3 & 2 \end{bmatrix} = -\dfrac{1}{36}\begin{bmatrix} -9 & -6 \\ -3 & 2 \end{bmatrix} = \dfrac{1}{36}\begin{bmatrix} 9 & 6 \\ 3 & -2 \end{bmatrix}$

or $\begin{bmatrix} 1/4 & 1/6 \\ 1/12 & -1/18 \end{bmatrix}$. Since $\det(N) = \det\left(\begin{bmatrix} 2 & 6 \\ 3 & 9 \end{bmatrix}\right) = 2 \cdot 9 - 6 \cdot 3 = 0$, the matrix N is not invertible.

EFR 4.20: (a) Working mod 3, we have $A \equiv \begin{bmatrix} 2 & 7 \\ 4 & 1 \end{bmatrix} \equiv \begin{bmatrix} 2 & 1 \\ 1 & 1 \end{bmatrix}$, $B \equiv \begin{bmatrix} 1 & 2 \\ 9 & 8 \end{bmatrix} \equiv \begin{bmatrix} 1 & 2 \\ 0 & 2 \end{bmatrix}$, and we compute:

$$A + B \equiv \begin{bmatrix} 2 & 1 \\ 1 & 1 \end{bmatrix} + \begin{bmatrix} 1 & 2 \\ 0 & 2 \end{bmatrix} \equiv \begin{bmatrix} 3 & 3 \\ 1 & 3 \end{bmatrix} \equiv \begin{bmatrix} 0 & 0 \\ 1 & 0 \end{bmatrix}, \quad AB \equiv \begin{bmatrix} 2 & 1 \\ 1 & 1 \end{bmatrix} \cdot \begin{bmatrix} 1 & 2 \\ 0 & 2 \end{bmatrix} \equiv \begin{bmatrix} 2 & 6 \\ 1 & 4 \end{bmatrix} \equiv \begin{bmatrix} 2 & 0 \\ 1 & 1 \end{bmatrix} \pmod 3.$$

Since $\det(B) \equiv 1 \cdot 2 - 2 \cdot 0 \equiv 2$ is invertible mod 3 (with $2^{-1} = 2$), it follows from Theorem 4.10, that

$B^{-1} \equiv \det(B)^{-1} \cdot \begin{bmatrix} 2 & -2 \\ -0 & 1 \end{bmatrix} \equiv 2 \cdot \begin{bmatrix} 2 & 1 \\ 0 & 1 \end{bmatrix} \equiv \begin{bmatrix} 4 & 2 \\ 0 & 2 \end{bmatrix} \equiv \begin{bmatrix} 1 & 2 \\ 0 & 2 \end{bmatrix} \pmod 3.$

(b) Working mod 10, we compute:

$$A + B \equiv \begin{bmatrix} 2 & 7 \\ 4 & 1 \end{bmatrix} + \begin{bmatrix} 1 & 2 \\ 9 & 8 \end{bmatrix} \equiv \begin{bmatrix} 3 & 9 \\ 13 & 1 \end{bmatrix} \equiv \begin{bmatrix} 3 & 9 \\ 3 & 1 \end{bmatrix}, \quad AB \equiv \begin{bmatrix} 2 & 7 \\ 4 & 1 \end{bmatrix} \cdot \begin{bmatrix} 1 & 2 \\ 9 & 8 \end{bmatrix} \equiv \begin{bmatrix} 65 & 60 \\ 13 & 16 \end{bmatrix} \equiv \begin{bmatrix} 5 & 0 \\ 3 & 6 \end{bmatrix} \pmod{10}.$$

Since $\det(B) \equiv 1 \cdot 8 - 2 \cdot 9 \equiv 0 \pmod{10}$, it follows from Theorem 4.10 that B^{-1} does not exist (mod 10).

EFR 4.21: Using cofactor expansion (Algorithm 4.8), we compute $\det(A) = -39$.

(a) Since $-39 \equiv 6 \pmod{15}$ and $\gcd(6, 15) = 3$, it follows from Theorem 4.10(1) that A is not invertible (mod 15).

(b) Since $-39 \equiv 9 \pmod{16}$ and $\gcd(9, 16) = 1$, it follows from Theorem 4.10(1) that A is invertible (mod 16). The classical adjoint matrix is

$$\text{adj}(A) = \begin{bmatrix} \det\begin{pmatrix} 4 & 5 \\ 0 & 1 \end{pmatrix} & -\det\begin{pmatrix} 1 & 3 \\ 0 & 1 \end{pmatrix} & \det\begin{pmatrix} 1 & 3 \\ 4 & 5 \end{pmatrix} \\ -\det\begin{pmatrix} 8 & 5 \\ 5 & 1 \end{pmatrix} & \det\begin{pmatrix} 1 & 3 \\ 5 & 1 \end{pmatrix} & -\det\begin{pmatrix} 1 & 3 \\ 8 & 5 \end{pmatrix} \\ \det\begin{pmatrix} 8 & 4 \\ 5 & 0 \end{pmatrix} & -\det\begin{pmatrix} 1 & 1 \\ 5 & 0 \end{pmatrix} & \det\begin{pmatrix} 1 & 1 \\ 8 & 4 \end{pmatrix} \end{bmatrix} = \begin{bmatrix} 4 & -1 & -7 \\ 17 & -14 & 19 \\ -20 & 5 & -4 \end{bmatrix} \equiv \begin{bmatrix} 4 & 15 & 9 \\ 1 & 2 & 3 \\ 12 & 5 & 12 \end{bmatrix}.$$

Since $\det(A)^{-1} \equiv 9^{-1} \equiv 9 \pmod{16}$, it follows that

$$A^{-1} = \det(A)^{-1} \cdot \text{adj}(A) = 9 \cdot \begin{bmatrix} 4 & 15 & 9 \\ 1 & 2 & 3 \\ 12 & 5 & 12 \end{bmatrix} \equiv \begin{bmatrix} 4 & 7 & 1 \\ 9 & 2 & 11 \\ 12 & 13 & 12 \end{bmatrix} \pmod{16}.$$

EFR 4.22: For a 2×2 matrix $A = \begin{bmatrix} a & b \\ c & d \end{bmatrix}$, each of the submatrices A_{ij} is the 1×1 submatrix of A obtained by deleting row i and column j, and since the determinant of a 1×1 matrix is simply the number (inside the brackets), we obtain:

$$\text{adj}(A) = \begin{bmatrix} \det(A_{11}) & -\det(A_{21}) \\ -\det(A_{12}) & \det(A_{22}) \end{bmatrix} = \begin{bmatrix} d & -b \\ -c & a \end{bmatrix}.$$

The special cases now readily follow from Proposition 4.11.

EFR 4.23: The $\mathbb{Z}_{26}$ vector corresponding to the plaintext is (see Table 4.2):

[19 0 17 8 3 22 23 6 23 22 13 20 0 13 5 7 7 20].

Regrouping it into a three row ciphertext matrix gives: $C = \begin{bmatrix} 19 & 8 & 23 & 22 & 0 & 7 \\ 0 & 3 & 6 & 13 & 13 & 7 \\ 17 & 22 & 23 & 20 & 5 & 20 \end{bmatrix}$. The

corresponding plaintext (uncoded) matrix U is given by $U = A^{-1}C$. We use Proposition 4.5 to compute

$A^{-1} \equiv \begin{bmatrix} 1 & 1 & 25 \\ 0 & 25 & 1 \\ 25 & 0 & 1 \end{bmatrix} (\mod 26)$, so that

$$U = A^{-1}C \equiv \begin{bmatrix} 1 & 1 & 25 \\ 0 & 25 & 1 \\ 25 & 0 & 1 \end{bmatrix} \cdot \begin{bmatrix} 19 & 8 & 23 & 22 & 0 & 7 \\ 0 & 3 & 6 & 13 & 13 & 7 \\ 17 & 22 & 23 & 20 & 5 & 20 \end{bmatrix} \equiv \begin{bmatrix} 2 & 15 & 6 & 15 & 8 & 20 \\ 17 & 19 & 17 & 7 & 18 & 13 \\ 24 & 14 & 0 & 24 & 5 & 13 \end{bmatrix} (\mod 26).$$

Using Table 3.1 to process the entries of U (in the prescribed order) produces the plaintext message: "Cryptography is fun." (Spaces were inserted and the additional n was deleted.)

EFR 4.24:   (a) $fl(1.22 \times 8.64) = fl(1.2 \times 8.6) = fl(10.32) = 10$.

(b) Adding from left to right: the first addition gives $fl(9+1) = 10$, and so all of the subsequent additions would get chopped (since we are working with only two significant digits).
Adding from right to left: $fl(0.06 + 0.05) = .11$, $fl(0.72 + 0.11) = .83$, $fl(0.87 + 0.83) = 1.7$,
$fl(0.4 + 1.7) = 2.1$, $fl(0.7 + 2.1) = 2.8$, $fl(1 + 2.8) = 3.8$, $fl(9 + 3.8) = fl(12.8) = 12$.

EFR 4.25:   (a) Multiplication (as well as addition) is commutative in floating point arithmetic because the multiplying (or adding) two numbers in opposite order will produce the same real number in ordinary arithmetic, and both will chop (or round) to the same floating point number (whatever value of s is being used).
(b) Multiplication is generally not associative in floating point arithmetic. Counterexamples that involve over/underflows only on one side are easily constructed. Counterexamples that do not rely under/overflows can be quickly found with some trial and error. Here is such an example in $s = 2$ significant digit (chopped) floating point arithmetic: $fl(fl(21 \cdot 22) \cdot 24) = fl(460 \cdot 24) = 11,000$, but $fl(21 \cdot fl(22 \cdot 24)) = fl(21 \cdot 520) = 10,000$.

EFR 4.26:   (a)   By Proposition 3.22(2), for any integer $a \in \mathbb{Z}_{79}^{\times}$, $\text{ord}_{79}(a) \mid \phi(79) = 78 = 2 \cdot 3 \cdot 13$,
so the only possibilities for $\text{ord}_{79}(a)$ are (1,) 2, 3, 6, 13, 26, 39, and 78, and we need only look at these modular powers of a to check if a is a primitive root.   Starting with $a = 2$, we have: $2^2 \equiv 4, 2^3 \equiv 8, 2^6 \equiv 64, 2^{13} \equiv 55, 2^{39} \equiv 1(\mod 79)$,   so $\text{ord}_{79}(2) = 39$.   Next with $a = 3$, we have: $3^2 \equiv 9, 3^3 \equiv 27, 3^6 \equiv 18, 3^{13} \equiv 24, 3^{39} \equiv 78(\mod 79)$, so $\text{ord}_{79}(3) = 78$, and $g = 3$ is the smallest primitive root.
(b) Bob (computes and) sends Alice the number $B = g^b \equiv 3^{33} \equiv 57(\mod 79)$, and Alice sends Bob the number $A = g^a \equiv 3^{51} \equiv 71(\mod 79)$.   On her end, Alice computes the common (secret) key as $K = B^a \equiv 57^{51} \equiv 61(\mod 79)$, and Bob computes it as $K = A^b \equiv 71^{33} \equiv 61(\mod 79)$.

EFR 4.27:   (a) Since $\phi(n) = (p-1)(q-1) = 66 \cdot 36 = 2376$, e will be the smallest integer greater than 1000 that is relatively prime to $\phi(n) = 2376$. Since e must be odd, we begin with 1001 and compute $\gcd(1001, \phi(n)) = 11$, and then move on to check $\gcd(1003, \phi(n)) = 1$, so $e = 1003$ is a legitimate encryption exponent. To encrypt the plaintext $P = 2012$, we raise P to the power of e (mod n) (using Algorithm 4.5) to obtain $C \equiv P^e \equiv 2012^{1003} \equiv 2095(\mod n)$.

(b) We use the extended Euclidean Algorithm 3.1 to compute $d \equiv e^{-1} \equiv 1843 \pmod{\phi(n)}$, which is the decryption exponent. To decrypt the ciphertext, we need to raise it to the power of decryption exponent $d = 1843 \pmod{n = 2479}$: $2095^{1843} \equiv 2012 \pmod{2479}$—the original plaintext message!

EFR 4.28: (a) By Proposition 3.22(2), for any integer $k \in \mathbb{Z}_{1231}^{\times}$, $\mathrm{ord}_{1231}(k) \mid \phi(1231) = 1230 = 2 \cdot 3 \cdot 5 \cdot 41$, so the only possibilities for $\mathrm{ord}_{1231}(k)$ are (1,) 2, 3, 5, 6, 10, 15, 41, 82, 123, 205, 246, 410, and 1230, and we need only look at these modular powers of k to check if k is a primitive root. Starting with $k = 701$, a series of simple exponentiations shows that all but the last of these modular powers is different from 1, so $\mathrm{ord}_{1231}(k) = 1230$, and so $g = 701$ is the desired primitive root.

(b) Bob's public key is $B \equiv g^b \equiv 701^{954} \equiv 143 \pmod{1231}$, and Alice's is $A \equiv g^a \equiv 701^{212} \equiv 990 \pmod{1231}$. Alice computes $C \equiv B^a P \equiv 144 \cdot 44 \equiv 181 \pmod{1231}$. Thus, the entire ciphertext would be $(A, C) = (990, 181)$.

(c) The decryption exponent is $p - 1 - b = 276$, and $A^{276} \equiv 143^{276} \equiv 966 \pmod{1231}$, and so $d_K((A, C)) \equiv A^{p-1-b} C \equiv 966 \cdot 181 \equiv 44 \pmod{1231}$, as expected.

EFR 4.29: (a) $a_2 = 5 > 3 = a_1$, $a_3 = 9 > 8 = a_1 + a_2$, $a_4 = 18 > 17 = a_1 + a_2 + a_3$, $a_5 = 36 > 35 = a_1 + a_2 + a_3 + a_4$, $a_6 = 100 > 71 = a_1 + a_2 + a_3 + a_4 + a_5$. This shows that $[a_1, a_2, a_3, a_5, a_5, a_6]$ is superincreasing.
(b) We use Algorithm 4.12:
Step 1: $S = 27$, Index = 6.
Step 2: Since $S < a_6 = 100$, we set $x_6 = 0$, and update Index $\rightarrow 6 - 1 = 5$.
Step 2: (Second iteration) Since $S < a_5 = 36$, we set $x_5 = 0$, and update Index $\rightarrow 4 - 1 = 4$.
Step 2: (Third iteration) Since $S \geq a_4 = 18$, we set $x_4 = 1$, update $S \rightarrow 27 - 18 = 9$ and update Index $\rightarrow 4 - 1 = 3$.
Step 2: (Fourth iteration) Since $S \geq a_3 = 9$, we set $x_3 = 1$, update $S \rightarrow 9 - 9 = 0$ and update Index $\rightarrow 4 - 1 = 2$.
Since S is now 0, the remaining two steps clearly set $x_2 = x_1 = 0$, and thus the unique solution to the given superincreasing knapsack problem is $[x_1, x_2, x_3, x_5, x_5, x_6] = [0, 0, 1, 1, 0, 0]$.

EFR 4.30: (a) Since the binary expansion of $\sum_{i=1}^{k} 2^i$ is a vector of k 1s, it follows (from the binary expansion development of Section 4.1) that this integer is one less than the next binary integer, which is a single 1 followed by a k 0s. Since this latter binary vector represents the integer 2^{k+1}, it follows that the given sequence is superincreasing (but just barely, since each term is exactly one more than the sum of the previous terms).
(b) This can be accomplished using (strong) mathematical induction. *Basis Step:* $i = 1$: Since a_1 is a positive integer, it must be at least $1 = 2^{i-1}$. *Inductive Step:* We assume $a_i \geq 2^{i-1}$ for all indices i up to and including some fixed index k. It suffices to show that the relation holds for the index $k + 1$: $a_{k+1} \geq 2^{(k+1)-1}$. But the superincreasing assumption tells us that $a_{k+1} > a_k + \cdots + a_2 + a_1$. By our inductive hypothesis, $a_k + \cdots + a_2 + a_1 \geq \sum_{i=1}^{k} 2^{i-1}$, and by the solution of Part (a), this latter sum equals $2^k - 1$. It therefore follows that $a_{k+1} \geq 2^k = 2^{(k+1)-1}$, as we needed to show. $\square$

EFR 4.31: (a) Working mod $m = 201$, we have $w \cdot [a_1 \ a_2 \ a_3 \ a_4 \ a_5 \ a_6] \equiv$ $77 \cdot [3, 5, 9, 18, 36, 100] \equiv [30, 184, 90, 180, 159, 62]$. Thus, this latter vector is Bob's public key $[b_1 \ b_2 \ b_3 \ \cdots b_n]$.

(b) The ciphertext is $f_b([1,1,1,0,0,0]) = x_1 b_1 + x_2 b_2 + x_3 b_3 + x_4 b_4 + x_5 b_5 + x_6 b_6 = 1 \cdot 30 + 1 \cdot 184 + 1 \cdot 90 + 0 \cdot 180 + 0 \cdot 159 + 0 \cdot 62 = 304 = s$.

(c) Using the extended Euclidean algorithm (Algorithm 4.6), we compute $w^{-1} \equiv 47$. (This need only be computed once, and could be supplied with the rest of Bob's private key.) Since $w^{-1} \cdot s \equiv 47 \cdot 304 \equiv 17 \pmod{m}$, the plaintext will be the solution of the superincreasing knapsack problem with weight vector $[a_1 \ a_2 \ a_3 \ a_4 \ a_5 \ a_6]$, and knapsack weight $s' = 17$. Algorithm 4.12 quickly produces the original plaintext.

CHAPTER 5: COUNTING TECHNIQUES, COMBINATORICS, AND GENERATING FUNCTIONS

EFR 5.1: Part (a): $12 \cdot 11 \cdot 10 \cdot 9 \cdot 8 = 95{,}040$ Part (b): With neither K nor S, there are (as in Part (a) but with the remaining 10 players to choose from) $10 \cdot 9 \cdot 8 \cdot 7 \cdot 6 = 30{,}240$ possible lineups. Now for the (disjoint) case in which exactly one of K or S appears, we use the multiplication principle as follows:

$$\underset{\substack{\text{choices for}\\\text{K or S}}}{2} \cdot \underset{\substack{\text{choices for}\\\text{where to put}\\\text{K or S}}}{5} \cdot \underset{\substack{\text{ways to fill the}\\\text{remaining four slots}\\\text{with remaining 10 players}}}{10 \cdot 9 \cdot 8 \cdot 7} = 50{,}400.$$

This gives a total of 80,640 lineups.

EFR 5.2: Invoking the notation of the solution of Example 5.6, we are looking for $|D_2 \cup D_3 \cup D_5 \cup D_{11}|$. Using the inclusion-exclusion principle along with the facts pointed out in the solution of Example 5.6 that $D_n \cap D_m = D_{\text{lcm}(n,m)}$ (and its easy extension to larger intersections by induction), and $|D_n| = \lfloor 3600/n \rfloor$, we obtain:

$$
\begin{aligned}
|D_2 \cup D_3 \cup D_5 \cup D_{11}| &= |D_2| + |D_3| + |D_5| + |D_{11}| - \left[|D_6| + |D_{10}| + |D_{22}| + |D_{15}| + |D_{33}| + |D_{55}|\right] \\
&= + \left[|D_{30}| + |D_{66}| + |D_{110}| + |D_{165}|\right] - |D_{330}| \\
&= \lfloor 3600/2 \rfloor + \lfloor 3600/3 \rfloor + \lfloor 3600/5 \rfloor + \lfloor 3600/11 \rfloor - \left[\lfloor 3600/6 \rfloor + \lfloor 3600/10 \rfloor \right. \\
&= + \lfloor 3600/22 \rfloor + \lfloor 3600/15 \rfloor + \lfloor 3600/33 \rfloor + \lfloor 3600/55 \rfloor\right] + \left[\lfloor 3600/30 \rfloor + \right. \\
&= + \lfloor 3600/66 \rfloor + \lfloor 3600/110 \rfloor + \lfloor 3600/165 \rfloor\right] - \lfloor 3600/330 \rfloor \\
&= 1800 + 1200 + 720 + 327 - [600 + 360 + 163 + 240 + 109 + 65] \\
&= + [120 + 54 + 32 + 21] - 10 \\
&= 2727.
\end{aligned}
$$

EFR 5.3: Each of the three strings of letters can be made into $10^3 = 1000$ license plates, and these are all different. Thus by the complement principle, we need only subtract the total number (3000) of these plates from the total number of Hawaii plates that was found in Example 5.2 $(17{,}576{,}000)$ to get the answer to this question: $17{,}573{,}000$.

EFR 5.4: Part (a): We subdivide the equilateral triangle into four smaller ones with side length 1/2 (see figure). Since there are five points in the larger triangle, at least two must lie in a single smaller triangle. Since the diameter of any triangle is its longest side length, it follows that two such points will lie at a distance of at most 1/2 from each other.

Part (b): The only way for two points on a triangle to have distance between them equaling the length of the (longest) side length is for the two points to be endpoints of such a side. Thus, for the two points of Part (a) to have separation distance equal to 1/2, they would have to lie at the endpoints of a side of one of the four smaller triangles, and this would put them on an edge of the original (larger) triangle.

EFR 5.5: Let $a_1, a_2, \cdots, a_{51}$ be the 51 positive integers. For each a_i, we write $a_i = 2^{p_i} b_i$, where b_i is an odd integer. Since there are exactly 50 odd integers between 1 and 100, it follows from the pigeonhole principle that at least two of the a_i's must share the same odd integer: i.e., $b_i = b_j$, for two indices $1 \le i \ne j \le 51$. It follows that either $a_i \mid a_j$ (if $p_i \le p_j$) or $a_j \mid a_i$. $\square$

EFR 5.6: By Proposition 2.2, any integer must fall into one of n equivalence classes in the equivalence relation of congruence modulo n. Therefore by the pigeonhole principle, in a set of $n + 1$ integers, there must be two, a and b, that lie in the same equivalence class (mod n). Thus $a \equiv b$ (mod n), and this means (by definition) that n divides $a - b$.

EFR 5.7: Part (a): Since arrangements that are obtainable from one another by rotations are considered equivalent, in counting all arrangements, we can place one particular person, call him/her X in some particular seat (since in any arrangement, X could always be brought to this seat with a rotation), and proceed, say clockwise, to fill the remaining $n-1$ seats with the remaining $n-1$ people. There are $(n-1)!$ such permutations of the remaining people, and all give nonequivalent seating arrangements of the whole group.

Part (b): We use the idea of Part (a). Take X as a man, then the multiplication principle tells us that to fill the remaining $n-1 = 2k-1$ seats with alternating men and women, there are $k \cdot (k-1) \cdot (k-1) \cdot (k-2) \cdot (k-2) \cdots 2 \cdot 2 \cdot 1 \cdot 1 = k! (k-1)!$ ways to do this.

Part (c): Put Jimmy in a particular seat. Now there are two choices where to put Sue, either counterclockwise or clockwise next to Jimmy. After this choice has been made, of the $n-2$ remaining seats, the multiplication principle tells us there will be $(k-1) \cdot (k-1) \cdot (k-2) \cdot (k-2) \cdots 2 \cdot 2 \cdot 1 \cdot 1$ $= [(k-1)!]^2$ ways to fill these. Thus (again by the multiplication principle), there are $2 \cdot [(k-1)!]^2$ different seating arrangements where Jimmy and Sue are sitting next to each other.

Part (d): Put Jimmy in a particular seat. Next, fill the counterclockwise and clockwise seats next to Jimmy; by the multiplication principle, there are $(k-1) \cdot (k-2)$ ways to do this with women other than Sue. Now, the remaining $n-3$ seats (clockwise) can be filled in any way (alternating the $k-1$ men and $k-2$ women), so there are a total of $(k-1) \cdot (k-2) \cdot (k-2) \cdot (k-3) \cdots 2 \cdot 1 \cdot 1 = (k-1)! (k-2)!$ ways to do this. Combining these counts (again with the multiplication principle) tells us that there are a total of $(k-2) \cdot [(k-1)!]^2$ different arrangements.

NOTE: The reader should observe that the answers in Parts (c) and (d) should add up to the answer in Part (b), and verify that this is the case.

EFR 5.8: Part (a): The multiplication principle allows us to count the number of full houses as follows:

$$\underbrace{13}_{\substack{\text{Number of ways} \\ \text{to choose denomination} \\ \text{of the three of a kind}}} \cdot \underbrace{12}_{\substack{\text{Number of ways} \\ \text{to choose denomination} \\ \text{of the two of a kind}}} \cdot \underbrace{C(4,3)}_{\substack{\text{Number of ways} \\ \text{to choose 3 cards} \\ \text{from first denom.}}} \cdot \underbrace{C(4,2)}_{\substack{\text{Number of ways} \\ \text{to choose 2 cards} \\ \text{from second denom.}}} = 13 \cdot 12 \cdot 4 \cdot 6 = 3744.$$

Part (b): As in Part (a), the multiplication principle yields the number of flushes to be

$$\underbrace{4}_{\substack{\text{Number of ways}\\ \text{to choose the suit}\\ \text{of the flush}}} \cdot \underbrace{C(13,5)}_{\substack{\text{Number of ways}\\ \text{to choose 5 cards}\\ \text{from this suit}}} \cdot = 4 \cdot 1287 = 5148.$$

Part (c): Once again, the multiplication principle can be used to tally the number of four of a kind poker hands

$$\underbrace{13}_{\substack{\text{Number of ways}\\ \text{to choose denomination}\\ \text{of the four of a kind}}} \cdot \underbrace{C(4,4)}_{\substack{\text{The number of ways}\\ \text{to choose (all) four cards}\\ \text{from this denom.} = 1.}} \cdot \underbrace{48}_{\substack{\text{Number of cards}\\ \text{to choose}\\ \text{to complete the hand}}} = 624.$$

EFR 5.9: We use the identity (4) to translate binomial coefficient into factorial expressions, manipulate these and then translate back:

$$\binom{n-1}{k-1} + \binom{n-1}{k} = \frac{(n-1)!}{(k-1)!(n-1-[k-1])!} + \frac{(n-1)!}{(k)!(n-1-k)!} \quad = \frac{(n-1)!}{(k-1)!(n-k)!} + \frac{(n-1)!}{(k)!(n-1-k)!}$$

$$= \frac{k \cdot (n-1)!}{k \cdot (k-1)!(n-k)!} + \frac{(n-1)!(n-k)}{(k)!(n-1-k)!(n-k)} \quad = \frac{[k+n-k] \cdot (n-1)!}{k!(n-k)!} = \frac{n \cdot (n-1)!}{k!(n-k)!}$$

$$= \frac{n!}{k!(n-k)!} = \binom{n}{k}.$$

EFR 5.10: Part (a): If we apply the binomial theorem to expand $0 = (-1+1)^m$, i.e., use $x = -1$ and $y = 1$ in (6), we obtain

$$0 = (-1+1)^m = \sum_{k=0}^{m} \binom{m}{k}(-1)^k (1)^{m-k} = \sum_{k=0}^{m} \binom{m}{k}(-1)^k = \binom{m}{0} - \binom{m}{1} + \binom{m}{2} - \cdots \pm \binom{m}{m}.$$

Since $\binom{m}{0} = 1$, subtracting the other terms from both sides of the above equation produces the desired identity.

Part (b): Consider $x \in \bigcup_{i=1}^{n} A_i$, and suppose that x lies in exactly m of the A_i 's. Then the counts for x from all of the terms of the right-hand side of (1) will only go up to $a = m$ (corresponding to intersections of m A_i 's). The contribution of x from the summation in the ath term (with $1 \le a \le m$)

$$(-1)^{a+1} \sum_{i_1 < i_2 < \cdots < i_a} |A_{i_1} \cap A_{i_2} \cap \cdots \cap A_{i_a}|$$

will equal to the number of ways of choosing the a indices $i_1 < i_2 < \cdots < i_a$ from $\{1,2,\cdots,n\}$ so that they all correspond to sets A_i which contain x. Since there are m such A_i 's, it follows that the contribution of x from this term must be $(-1)^{a+1}\binom{m}{a}$. Adding up all of these contributions (from $a = 1$ to $a = m$) gives exactly the sum $\binom{m}{1} - \binom{m}{2} + \cdots \pm \binom{m}{m}$, which, by Part (a) simply equals 1. Since this is true for any x in $\bigcup_{i=1}^{n} A_i$, while for any x outside $\bigcup_{i=1}^{n} A_i$, the contribution of x on the right (and the left) is clearly zero, the identity is established.

EFR 5.11: Of the 11 letters in the word MISSISSIPPI, only four are different: there is one M, four I's, four Ss, and two Ps. Thus, by Theorem 2.8, the number of distinguishable permutations of these 11 letters is $\binom{11}{1,4,4,2} = \frac{11!}{1!4!4!2!} = \frac{\cancel{4!}5\cdot\cancel{6}\cdot7\cdot\cancel{8}\cdot9\cdot10\cdot11}{\cancel{4!}\,\cancel{2}\cdot\cancel{3}\cdot\cancel{4}\cdot\cancel{2}} = 34{,}650.$

EFR 5.12: We give a combinatorial proof similar to what was done in our proof of the binomial theorem (Theorem 5.7). If we expand the left-hand side of (8):

$$(x_1 + x_2 + \cdots + x_r)^n = \underbrace{(x_1 + x_2 + \cdots + x_r)\cdot(x_1 + x_2 + \cdots + x_r)\cdots\cdots(x_1 + x_2 + \cdots + x_r)}_{n \text{ factors}},$$

the resulting expansion will consist of all terms that are products of the form $z_1 z_2 \cdots z_n$, where each z_i is a single term selected from the ith factor above (so each z_i must be one of $x_1, x_2, \cdots, x_r$). Thus, each term in the expansion will be of the form $x_1^{k_1} x_2^{k_2} \cdots x_r^{k_r}$, where the exponents are nonnegative integers that add up to n. By Theorem 5.8, the number of occurrences of this term in the expansion will equal $\binom{n}{k_1,\, k_2, \cdots, k_r}$. Putting these facts together shows that the expansion equals the right-hand side of (8), as desired. □

EFR 5.13: In the multinomial identity (8):

$$(x_1 + x_2 + \cdots + x_r)^n = \sum_{\substack{k_1+k_2+\cdots+k_r=n \\ k_i \text{ nonnegative integer}}} \binom{n}{k_1,\, k_2, \cdots, k_r} x_1^{k_1} x_2^{k_2} \cdots x_r^{k_r},$$

if we specialize to the case that $r = 2$, and put $x_1 = x$ and $x_2 = y$, and use the fact that $\binom{n}{k_1, k_2} = \binom{n}{k_1}$, and $k_2 = n - k_1$, it becomes: $(x+y)^n = \sum_{\substack{k_1+k_2=n \\ k_i \text{ nonnegative integer}}} \binom{n}{k_1} x^{k_1} y^{n-k_1}$. The proof is completed by noticing that this last summation is equivalent to $\sum_{k_1=0}^{n} \binom{n}{k_1} x^{k_1} y^{n-k_1} = \sum_{k=0}^{n} \binom{n}{k} x^k y^{n-k}$.

EFR 5.14: By the multinomial theorem, the full term is $\binom{14}{6,3,3,2}(2a)^6(-3b)^3(4c)^3(-d)^2$. Working out the coefficient, we get $-123{,}986{,}903{,}040$.

EFR 5.15: Part (a): Following the suggestion, we view the problem as the equivalent problem of counting how many ways we can distribute 12 identical balls into 4 different urns: Urn 1, Urn 2, Urn 3, and Urn 4. Thus, x_i represents the number of balls that are placed in Urn i. By Theorem 5.10, it follows that the number of ways this can be done (= the number of nonnegative integer solutions to the given equation) is $\binom{12+(4-1)}{4-1} = \binom{15}{3} = 455$.

Part (b): We introduce new variables $y_i = x_i + 1$, which will represent positive integers, and let the x_i's still represent nonnegative integers. Thus, the number of solutions of the equation $y_1 + y_2 + y_3 + y_4 = 12$ (in positive integers) is the same as the number of solutions of the equation $(x_1 + 1) + (x_2 + 1) + (x_3 + 1) + (x_4 + 1) = 12$ or $x_1 + x_2 + x_3 + x_4 = 8$ (in nonnegative integers). By the method developed in the solution of Part (a), this latter equation has $\binom{8+3}{3} = \binom{11}{3} = 165$ solutions, and so this is the number of positive integer solutions of the given equation.

EFR 5.16: The number of terms in the sum on the right-hand side of (8) (the multinomial theorem) is just the number of nonnegative integer solutions of the equation $k_1 + k_2 + \cdots + k_r = n$. By the method of the solution to Part (a) of the preceding exercise for the reader, this number is $\binom{n+r-1}{r-1}$.

EFR 5.17: Part (a): Using (8), we obtain $x^2 + x^3 + x^4 + \cdots = x^2(1 + x + x^2 + \cdots) = x^2/(1-x)$.

Part (b): Substituting $x \mapsto 2x$ in (9), it becomes $e^{2x} = 1 + 2x + \dfrac{4x^2}{2!} + \dfrac{8x^3}{3!} + \cdots$, so the given generating

function is $(1-x)e^{2x} = 1 + 2x + \dfrac{4x^2}{2!} + \dfrac{8x^3}{3!} + \cdots$.

EFR 5.18: Using Definition 5.6, for $k > 0$, we may write:

$$\binom{a}{k} = \binom{-n}{k} = \frac{(-n)(-n-1)(-n-2)\cdots(-n-k+1)}{k!}$$
$$= (-1)^k \frac{(n)(n+1)(n+2)\cdots(n+k-1)}{k!}$$
$$= (-1)^k \frac{(n+k-1)!}{(n-1)!k!} = (-1)^k C(n+k-1,k).$$

EFR 5.19: This follows directly from the generalized binomial theorem with $a = -N$, along with the result of Exercise for the Reader 5.18 (with k changed to n, and n changed to N).

EFR 5.20: With $a = 1/2$, (10) becomes $(1+x)^{1/2} = \sum_{n=0}^{\infty} \binom{1/2}{n} x^n$. Using Definition 5.6, for $k > 1$ we may write:

$$\binom{1/2}{k} = \frac{(1/2)(1/2-1)(1/2-2)\cdots(1/2-k+1)}{k!} = \frac{(1/2)(-1/2)(-3/2)\cdots([3-2k]/2)}{k!}$$
$$= \frac{(-1)^{k-1}}{2^k} \frac{(2k-3)(2k-5)\cdots 3\cdot 1}{k!} = \frac{(-1)^k}{2^k} \frac{(2k-3)(2k-5)\cdots 3\cdot 1\cdot(-1)}{k!}.$$

The last manipulation was done because the final formula is also valid when $k = 1$. Putting this all together, we may write:

$$(1+x)^{1/2} = 1 + x/2 - x^2/8 + x^3/16 - 5x^4/128 + \cdots = 1 + \sum_{n=1}^{\infty} \frac{(-1)^n}{2^n} \frac{(2n-3)(2n-5)\cdots 3\cdot 1\cdot(-1)}{n!} x^n.$$

Substituting $x \mapsto x/2$ in this expansion leads us to the desired expansion:

$$\sqrt{1+x/2} = 1 + x/4 - x^2/16 + x^3/32 - 5x^4/256 + \cdots = 1 + \sum_{n=1}^{\infty} \frac{(-1)^n}{4^n} \frac{(2n-3)(2n-5)\cdots 3\cdot 1\cdot(-1)}{n!} x^n.$$

EFR 5.21: We introduce the generating function for the sequence: $F(x) = \sum_{n=0}^{\infty} a_n x^n$. We multiply both sides of the recurrence relation by x^n, and then take the formal (infinite) sum of both sides in the range $n \geq 1$ where the recurrence is valid:

$$a_n = 3a_{n-1} - 1 \; (n \geq 1) \Rightarrow \sum_{n=1}^{\infty} a_n x^n = 3 \sum_{n=1}^{\infty} a_{n-1} x^n - \sum_{n=1}^{\infty} 1 \cdot x^n.$$

In the same fashion as in the solution of Example 5.22, this equation translates into the following algebraic equation for the generating function:

$$F(x) - 1 = 3xF(x) - \frac{1}{1-x} + 1 \Rightarrow (1-3x)F(x) = 2 - \frac{1}{1-x} = \frac{1-2x}{1-x} \Rightarrow F(x) = \frac{1-2x}{(1-x)(1-3x)}.$$

The partial fractions expansion of the right side is:

$$\frac{1-2x}{(1-x)(1-3x)} = \frac{A}{1-x} + \frac{B}{1-3x}.$$

To determine the constants A and B on the right side, we first multiply both sides of the equation by the denominator on the right to obtain:

$$1 - 2x = A(1-3x) + B(1-x).$$

If we substitute $x = 1$ into this equation, we obtain $A = 1/2$, and substituting $x = 1/3$ produces $B = 1/2$. The original equation now gives:

$$F(x) = \frac{1}{2}\left\{\frac{1}{1-x} + \frac{1}{1-3x}\right\}.$$

Each of the terms on the right can easily be expanded using (8) (a special case of (12)):

$$F(x) = F(x) = \frac{1}{2}\left\{\frac{1}{1-x} + \frac{1}{1-3x}\right\} = (1/2)\sum_{n=0}^{\infty} x^n + (1/2)\sum_{n=0}^{\infty} (3x)^n = (1/2)\sum_{n=0}^{\infty} (3^n + 1)x^n.$$

Thus we have found that $a_n = (3^n + 1)/2$.

EFR 5.22: Following the method of the Examples 5.22 and 5.23, we let $F(x) = \sum_{n=0}^{\infty} a_n x^n$ be the generating function for the given sequence, multiply both sides of the recurrence relation by x^n, and then take the formal (infinite) sum of both sides in the range $n \geq 1$ (where the recurrence is valid):

$$a_n = 2a_{n-1} + 3^n \ (n \geq 1) \Rightarrow \sum_{n=1}^{\infty} a_n x^n = 2\sum_{n=1}^{\infty} a_{n-1} x^n + \sum_{n=1}^{\infty} 3^n x^n.$$

The first two sums we have already seen, while the third sum on the right can be converted using the expansion (8) with the substitution $x \mapsto x/2$:

$$\sum_{n=1}^{\infty} 3^n x^n = \sum_{n=0}^{\infty} (3x)^n - 1 = \frac{1}{1-3x} - 1.$$

Hence, the preceding power series equation transforms into the following algebraic equation for the generating function:

$$F(x) - 1 = 2xF(x) + \frac{1}{1-3x} - 1 \Rightarrow F(x) = \frac{1}{(1-2x)(1-3x)}.$$

The partial fractions expansion will have the form:

$$F(x) = \frac{1}{(1-2x)(1-3x)} = \frac{A}{1-2x} + \frac{B}{1-3x}.$$

To determine the constants A, B, we first clear out all denominators:

$$1 = A(1-3x) + B(1-2x).$$

Substituting $x = 1/2$ yields $A = -2$, and substituting $x = 1/3$ yields $B = 3$. Thus we have determined the partial fractions expansion of the generating function to be:

$$F(x) = \frac{-2}{1-2x} + \frac{3}{1-3x}.$$

Applying the expansion (8) twice, we obtain

$$F(x) = \frac{-2}{1-2x} + \frac{3}{1-3x} = -2\sum_{n=0}^{\infty}(2x)^n + 3\sum_{n=0}^{\infty}(3x)^n = \sum_{n=0}^{\infty}(3^{n+1} - 2^{n+1})x^n.$$

We have thus arrived at the following closed formula for the given recursively defined sequence:

$$a_n = 3^{n+1} - 2^{n+1}.$$

EFR 5.23: Although the Fibonacci recurrence relation involves two prior terms, the generating function approach will be the same. We do make one small cosmetic adjustment because it is convenient to have our sequence indices begin at $n = 0$. We introduce the shifted sequence $e_n = f_{n+1}$, which is defined for all $n \geq 0$, and inherits the following recursive definition from that of f_n:

$$\begin{cases} e_0 = 1, e_1 = 1, \\ e_n = e_{n-1} + e_{n-2} \, (n \geq 2). \end{cases}$$

It suffices to obtain an explict formula for e_n, since the corresponding formula for the Fibonacci sequence and then be read off from the relation: $f_n = e_{n-1}$.

We let $F(x) = \sum_{n=0}^{\infty} e_n x^n$ be the generating function for sequence e_n, multiply both sides of the recurrence relation by x^n, and then take the formal (infinite) sum of both sides in the range $n \geq 2$ (where the recurrence is valid):

$$e_n = e_{n-1} + e_{n-2} \; (n \geq 2) \Rightarrow \sum_{n=2}^{\infty} e_n x^n = \sum_{n=2}^{\infty} e_{n-1} x^n + \sum_{n=2}^{\infty} e_{n-2} x^n.$$

Using the first two coefficients: $e_0 = 1, e_1 = 1$, we may transform each of these three sums in terms of the generating function:

$$\sum_{n=2}^{\infty} e_n x^n = F(x) - x - 1, \quad \sum_{n=2}^{\infty} e_{n-1} x^n = x\sum_{n=2}^{\infty} e_{n-1} x^{n-1} = x(F(x)-1), \quad \sum_{n=2}^{\infty} e_{n-2} x^n = x^2\sum_{n=2}^{\infty} e_{n-2} x^{n-2} = x^2 F(x).$$

The preceding power series equation thus transforms into the following algebraic equation for the generating function:

$$F(x) - x - 1 = x(F(x)-1) + x^2 F(x) \Rightarrow F(x) = \frac{1}{1-x-x^2}.$$

The denominator on the right has two real roots: $x = (-1 \pm \sqrt{5})/2$, which we temporarily will denote by r_+ and r_-. The partial fractions decomposition of $F(x)$ will take the form:

$$F(x) = \frac{1}{1-x-x^2} = \frac{A}{1-x/r_+} + \frac{B}{1-x/r_-}.$$

To determine the constants A, B, we first clear out all denominators:

$$1 = A(1 - x/r_-) + B(1 - x/r_+).$$

Substituting $x = r_+$ yields $A = \dfrac{1}{1 - r_+/r_-} = \dfrac{r_-}{r_- - r_+} = \dfrac{r_-}{-\sqrt{5}}$, and in a similar fashion, substituting $x = r_-$ yields $B = \dfrac{r_+}{\sqrt{5}}$. Thus we have determined the partial fractions expansion of the generating function to be:

$$F(x) = \frac{1}{\sqrt{5}} \left\{ \frac{-r_-}{1 - x/r_+} + \frac{r_+}{1 - x/r_-} \right\}.$$

Applying the expansion (8) twice, we obtain

$$F(x) = \frac{1}{\sqrt{5}} \left\{ \frac{-r_-}{1 - x/r_+} + \frac{r_+}{1 - x/r_-} \right\} = \frac{1}{\sqrt{5}} \left\{ -r_- \sum_{n=0}^{\infty} (x/r_+)^n + r_+ \sum_{n=0}^{\infty} (x/r_-)^n \right\} = \frac{1}{\sqrt{5}} \sum_{n=0}^{\infty} (-r_-(1/r_+)^n + r_+(1/r_-)^n)x^n.$$

We have thus arrived at a closed formula for the given recursively defined sequence, which we can simplify by noting that $r_+ r_- = -1$:

$$e_n = \frac{1}{\sqrt{5}} \left(-\frac{r_-}{r_+^{\,n}} + \frac{r_+}{r_-^{\,n}} \right) = \frac{1}{\sqrt{5}} \left(\frac{-r_-^{\,n+1} + r_+^{\,n+1}}{r_+^{\,n} r_-^{\,n}} \right) = \frac{1}{\sqrt{5}} \left(\frac{-[(-\sqrt{5}-1)/2]^{n+1} + [(\sqrt{5}-1)/2]^{n+1}}{(-1)^n} \right)$$

$$= \frac{1}{\sqrt{5}} \left(\frac{\sqrt{5}+1}{2} \right)^{n+1} - \frac{1}{\sqrt{5}} \left(\frac{1-\sqrt{5}}{2} \right)^{n+1}.$$

Finally, since $f_n = e_{n-1}$, this translates to the following explicit formula for the Fibonacci sequence:

$$f_n = \frac{1}{\sqrt{5}} \left(\frac{\sqrt{5}+1}{2} \right)^n - \frac{1}{\sqrt{5}} \left(\frac{1-\sqrt{5}}{2} \right)^n,$$

which was first introduced in Chapter 3.

EFR 5.24: For both questions being asked, we may take the generating function for the number of $1 bills drawn to be $F_O(x) = 1(=x^0) + x + x^2 + x^3 + x^4 + x^5 + x^6$. Since there are only five $10 bills, the corresponding generating function for the number of $10 bills drawn is $F_T(x) = 1 + x + x^2 + x^3 + x^4 + x^5$, and since there are only two $100 bills, the generating function for the number of these bills drawn is $F_H(x) = 1 + x + x^2$. The generating function for the both questions is the product of these three:

$$F_O(x) \cdot F_T(x) \cdot F_H(x) = (1 + x + x^2 + x^3 + x^4 + x^5 + x^6) \cdot (1 + x + x^2 + x^3 + x^4 + x^5) \cdot (1 + x + x^2).$$

The number of ways that four bills can be drawn is the coefficient of x^4 in this product, which can be computed to be 12, and the number of ways that six bills can be drawn is the coefficient of x^6 in this product, which can be computed to be 17.

EFR 5.25: As was pointed out in Example 5.28, any winning coalition must contain either V_1 or V_2; below we consider only the winning coalitions:

1. Four winning coalitions of the form $\{V_1, V_2\} \cup S$, where S is any subset of $\{V_4, V_5\}$:
Critical Voters: V_1, V_2

2. Four winning coalitions of the form $\{V_1, V_3\} \cup S$, where S is any subset of $\{V_4, V_5\}$:

Critical Voters: V_1, V_3

3. Four winning coalitions of the form $\{V_2, V_3\} \cup S$, where S is any subset of $\{V_4, V_5\}$:

Critical Voters: V_2, V_3

4. Four winning coalitions of the form $\{V_1, V_2, V_3\} \cup S$, where S is any subset of $\{V_4, V_5\}$:

Critical Voters: None

It follows that $c(V_1) = c(V_2) = c(V_3) = 4$, $c(V_4) = c(V_5) = 0$, $T = 12$, so the Banzhaf indices of V_1, V_2, V_3 are each 1/3, and those of V_4, V_5 are zero.

EFR 5.26: Part (a): Rather than going through all $2^6 - 1 = 65$ coalitions, this example is small enough to take advantage of its special structure. Clearly any winning coalition must include at least two of V_1, V_2, V_3; and conversely any coalition containing two of these three voters will be winning. Here is a summary count of these winning coalitions along with critical voters:

1. Eight winning coalitions of the form $\{V_1, V_2\} \cup S$, where S is any subset of $\{V_4, V_5, V_6\}$:

Critical Voters: V_1, V_2

2. Eight winning coalitions of the form $\{V_1, V_3\} \cup S$, where S is any subset of $\{V_4, V_5, V_6\}$:

Critical Voters: V_1, V_3

3. Eight winning coalitions of the form $\{V_2, V_3\} \cup S$, where S is any subset of $\{V_4, V_5, V_6\}$:

Critical Voters: V_2, V_3

4. Eight winning coalitions of the form $\{V_1, V_2, V_3\} \cup S$, where S is any subset of $\{V_4, V_5, V_6\}$:

Critical Voters: None

It follows that $c(V_1) = c(V_2) = c(V_3) = 16$, $c(V_4) = c(V_5) = c(V_6) = 0$, $T = 48$, so the Banzhaf indices of V_1, V_2, V_3 are each 1/3, and those of V_4, V_5, V_6 are zero.

Part (b): None of the voters are dictators or have veto power; but V_4, V_5, V_6 are dummies.

EFR 5.27: *Step 1:* The total weight is $W = 5 + 2 + 1 = 8$. The number of voters is $N = 3$.
Set $a_0 = a_W = 1$. Set $a_0^{(0)} = 1$, and $a_1^{(0)} = a_2^{(0)} = \cdots = a_8^{(0)} = 0$.
Note: $a_i^{(0)}$ are the coefficients of $F_0(x) = 1$.

Step 2: (Compute the $a_i^{(j)}$'s)

FOR index $j = 1$, FOR index $i = 0$ TO $i = W - 1$, Set $a_i^{(1)} = a_i^{(0)} + a_{i-5}^{(0)}$: (since $w_1 = 5$)

$a_0^{(1)} = a_0^{(0)} + a_{0-5}^{(0)} = 1 + 0 = 1, a_1^{(1)} = a_1^{(0)} + a_{1-5}^{(0)} = 0 + 0 = 0, \cdots, a_5^{(1)} = a_5^{(0)} + a_{5-5}^{(0)} = 0 + 1 = 1,$
$a_6^{(1)} = a_6^{(0)} + a_{6-5}^{(0)} = 0 + 0 = 0, a_7^{(1)} = a_7^{(0)} + a_{7-5}^{(0)} = 0 + 0 = 0.$

Note: $a_i^{(1)}$ are the coefficients of $F_1(x) = 1 + x^{w_1} = 1 + x^5$.

FOR index $j = 2$, FOR index $i = 0$ TO $i = W - 1$, Set $a_i^{(2)} = a_i^{(1)} + a_{i-2}^{(1)}$: (since $w_2 = 2$)

$a_0^{(2)} = a_0^{(1)} + a_{0-2}^{(1)} = 1 + 0 = 1, a_1^{(2)} = a_1^{(1)} + a_{1-2}^{(1)} = 0 + 0 = 0, a_2^{(2)} = a_2^{(1)} + a_{2-2}^{(1)} = 0 + 1 = 1,$
$a_3^{(2)} = 0 + 0 = 0, a_4^{(2)} = 0 + 0 = 0, a_5^{(2)} = 1 + 0 = 1, a_6^{(2)} = 0 + 0 = 0, a_7^{(2)} = 0 + 1 = 1.$

Note: $a_i^{(2)}$ are the coefficients of $F_2(x) = (1 + x^{w_1})(1 + x^{w_2}) = (1 + x^5)(1 + x^2) = 1 + x^2 + x^5 + x^7$.

FOR index $j = 3$, FOR index $i = 0$ TO $i = W - 1$, Set $a_i^{(3)} = a_i^{(2)} + a_{i-1}^{(2)}$: (since $w_3 = 1$)

$a_0^{(3)} = 1 + 0 = 1, a_1^{(3)} = 0 + 1 = 1, a_2^{(3)} = 1 + 0 = 1, a_3^{(3)} = 0 + 1 = 1, a_4^{(3)} = 0 + 0 = 0,$
$a_5^{(3)} = 1 + 0 = 1, a_6^{(3)} = 0 + 1 = 1, a_7^{(3)} = 1 + 0 = 1.$

Note: $a_i^{(3)}$ are the coefficients of $F_3(x) = (1+x^5)(1+x^2)(1+x) = 1 + x + x^2 + x^3 + x^5 + x^6 + x^7$.

Step 3: (*Record the* a_i *'s*) Set $a_i = a_i^{(N)}$ ($i = 1$ TO $i = W-1$)

$$a_1 = a_1^{(3)} = 1, a_2 = a_2^{(3)} = 1, a_3 = 1, a_4 = 0, a_5 = 1, a_6 = 1, a_7 = 1.$$

Step 4: (*Compute the* σ_n *'s*)

FOR index $n = 1$, $b_i = a_i - b_{i-5}$ ($i = 0$ TO $i = q - 1 = 6$)

$$b_0 = a_0 - b_{0-5} = 1, b_1 = a_1 - b_{1-5} = 1, b_2 = 1, b_3 = 1, b_4 = 0, b_5 = 0, b_6 = 0$$

$$\sigma_1 = \sum_{i=q-w_1}^{q-1} b_i = b_2 + b_3 + b_4 + b_5 + b_6 = 2$$

FOR index $n = 2$, $b_i = a_i - b_{i-2}$ ($i = 0$ TO $i = q - 1 = 6$)

$$b_0 = a_0 - b_{0-2} = 1, b_1 = a_1 - b_{1-2} = 1, b_2 = 0, b_3 = 0, b_4 = 0, b_5 = 1, b_6 = 1$$

$$\sigma_2 = \sum_{i=q-w_2}^{q-1} b_i = b_4 + b_5 + b_6 = 2$$

FOR index $n = 3$, $b_i = a_i - b_{i-1}$ ($i = 0$ TO $i = q - 1 = 6$)

$$b_0 = a_0 - b_{0-1} = 1, b_1 = a_1 - b_{1-1} = 0, b_2 = 1, b_3 = 0, b_4 = 0, b_5 = 1, b_6 = 0$$

$$\sigma_3 = \sum_{i=q-w_3}^{q-1} b_i = b_6 = 0$$

Step 5: (*Compute the Banzhaf indices*)

Set $T = \sum_{n=1}^{N} \sigma_n = 2 + 2 + 0 = 4$

Banzhaf index of $V_1 = \sigma_1 / T = 2/4 = 0.5$, Banzhaf index of $V_2 = \sigma_2 / T = 2/4 = 0.5$,

Banzhaf index of $V_3 = \sigma_3 / T = 0/4 = 0$.

CHAPTER 6: DISCRETE PROBABILITY AND SIMULATION

EFR 6.1: (a) The event E consisting of all groups of three students containing both Mike and Jimmy contains 23 outcomes (i.e., we need only select one more member of the group since two have already been selected from the total of 25 students). On the other hand, the total number of three-element sets that can be selected from a group of 25 students is $C(25,3) = 2300$. Hence $P(E) = 23/2300 = 1/100$.

(b) The hand of a full house in poker has the form: $\{a, a, a, b, b\}$, where a and b represent cards of different denominations. Using the multiplication principle, we can count the number of full house poker hands to be $13 \cdot C(4,3) \cdot 12 \cdot C(4,2) = 3744$. We may thus conclude that the probability of being dealt a full house is $3744 / C(52,5) \approx 0.00144$.

(c) By the multiplication principle, the sample space, consisting of all six-letter words made up using only the letters "H" and/or "T" (e.g., the word HHTTHT corresponds to getting heads on the first, second, and fifth with the remaining flips being tails) consists of $2^6 = 64$ elements. The event E that at least one flip results in heads contains most all of these elements. The only outcome not in E is the one that has all tails: TTTTTT. Thus E contains 63 outcomes (E is the complement of the singleton event {TTTTTT}). Hence, $P(E) = 63/64$.

EFR 6.2: Since any event E is a subset of the sample space S, we have $0 \le |E| \le |S|$, so that $P(E) = |E| / |S|$ must be a number between 0 and 1 (inclusive), and this gives axiom 1. Also, $P(S) = |S| / |S| = 1$, so axiom 2 is also valid. For axiom 3, if $E_1, E_2, \cdots$ is a sequence of pairwise mutual exclusive sets, then the number of elements in the union $\bigcup_i E_i$ of the sequence must equal to the sum of the numbers of elements in each set (there is no overlapping because of the pairwise mutually exclusiveness assumption). This gives us that

$$P(\bigcup_i E_i) = |\bigcup_i E_i| / |S| = (\sum_i |E_i|) / |S| = \sum_i (|E_i| / |S|) = \sum_i P(E_i),$$

thus establishing axiom 3.

EFR 6.3: (The reader is encouraged to draw a Venn diagram to better understand this proof.) We can decompose the event $E \cup F$ into a pairwise disjoint union $A \cup B \cup C$, where $A = E \sim F$, $B = E \cap F$, and $C = F \sim E$. Note also that we may express each of E and F as the disjoint unions $E = A \cup B$ and $F = B \cup C$. We can now use axiom 3 (twice) to obtain the desired result:

$$\begin{aligned}
P(E \cup F) = P(A \cup B \cup C) &\underset{\text{Axiom 3}}{=} P(A) + P(B) + P(C) \\
&= [P(A) + P(B)] + [P(B) + P(C)] - P(B) \\
&\underset{\text{Axiom 3}}{=} P(E) + P(F) - P(E \cap F).
\end{aligned}$$

(b) As in Part (a), a Venn diagram can be helpful here. Rather than proving this three-set version directly (using the axioms), we begin by using the result for two sets (Part (a)):

$$P(E \cup F \cup G) = P([E \cup F] \cup G) \underset{\text{By part (a)}}{=} P([E \cup F]) + P(G) - P([E \cup F] \cap G).$$

For the first term on the right, we again use Part (a): $P(E \cup F) = P(E) + P(F) - P(E \cap F)$. For the last term on the right, we first use the set (distributive law) equality $[E \cup F] \cap G = [E \cap G] \cup [F \cap G]$ and then apply Part (a) to the result:

$$P([E \cap G] \cup [F \cap G]) = P(E \cap G) + P(F \cap G) - P([E \cap G] \cap [F \cap G]).$$

Since the set inside the last probability is simply $E \cap F \cap G$, we can now obtain the desired identity by substituting the resulting two expressions into the the initial equation that we wrote down.
Note: This idea can be combined with mathematical induction to generalize the inclusion-exclusion principle to any (finite) number of sets. The general formula, which is a bit complicated, will be examined in the exercises.

EFR 6.4: The sample space S for this problem ("experiment") consists of all three-member subsets of the French Club, and so contains $C(10,3) = 120$ elements. For each part, we use the multiplication principle to count the elements in the corresponding event.
(a) $P(A) = C(9,2)/C(10,3) = 36/120 = 3/10$. (Only two more members need to be chosen from the remaining nine.)
(b) $P(A \cap B) = C(8,1)/C(10,3) = 8/120 = 1/15$. (Only one more member needs to be chosen from the remaining eight.)
(c) $P(\sim A) = 1 - P(A) = 1 - 3/10 = 7/10$. (Using Part (a)'s result.)
(d) $P(\sim A \cup \sim B) = P(\sim (A \cap B)) = 1 - P(A \cap B) = 1 - 1/15 = 14/15$. (We have used De Morgan's law in the first equality, and also the result of Part (b).)

EFR 6.5: There are really two essentially different events in this problem: D = snow or icy road conditions, and A = having an accident. Snow or icy road conditions is the event $\sim D$. Each of the ratios given in the table is a probability of an intersection of D or $\sim D$ with A or $\sim A$; for example, the upper right entry of the table states that $P(A \cap \sim D) = 0.020$. Now we proceed to compute the desired probabilities:
(a) $P(\sim D) = P(\sim D \cap A) + P(\sim D \cap \sim A) = 0.020 + 0.250 = 0.270$. (We have used the axiom that disjoint event probabilities can be added.)
(b) $P(A) = P(D \cap A) + P(\sim D \cap A) = 0.005 + 0.020 = 0.025$.

(c) $P(A \mid D) = \dfrac{P(A \cap D)}{P(D)} = \dfrac{0.005}{1 - P(\sim D)} = \dfrac{0.005}{1 - 0.270} = 0.007$.   (In the denominator we used the complementary probability rule and our answer to Part (a).)

(d) $\quad P(A|\sim D) = \dfrac{P(A \cap \sim D)}{P(\sim D)} = \dfrac{0.020}{0.270} = 0.074.$

Notice that snow or icy road conditions increase the probability of getting into an accident more than 10 times!

EFR 6.6: (a) We first invoke (4) $P(E \cap F) = P(E \mid F) \cdot P(F)$ taking $E = A$ and $F = B \cap C$ to obtain $P(A \cap B \cap C) [= P(A \cap (B \cap C))] = P(A \mid B \cap C) \cdot P(B \cap C)$. We use (4) again with $E = B$ and $F = C$ to (directly) obtain $P(B \cap C) = P(B \mid C) \cdot P(C)$. Combining this with the first equation produces the asserted multiplication rule.
(b) We can use the three-set result of Part (a) to conclude that:

$$P(A \cap B \cap C \cap D)[= P(A \cap B \cap (C \cap D))] = P(A \mid B \cap (C \cap D)) \cdot P(B \mid C \cap D) \cdot P(C \cap D).$$

We can finish by applying (4) to write $P(C \cap D) = P(C \mid D) \cdot P(D)$, and then substitute this into the previous equation.

EFR 6.7: At the start of the second day, there can be either 0, 1, 2, 3, 4, or 5 cars that still contain a $100 bill; call these events: $A_0, A_1, A_2, A_3, A_4,$ and $A_5,$ respectively. These events partition the sample space of possible outcomes of this experiment. We denote by N the event that none of the cars rented out on the second day contain a $100 bill. Conditioning on the A_i 's allows produces the equation $P(N) = \sum_{i=1}^{5} P(N \mid A_i)P(A_i)$. Since the first day's five cars can be chosen in $C(20,5)$ ways, it follows that the number of these outcomes that are in A_i must be $C(5,i) \cdot C(15, 5-i)$ (i cars are chosen from the 5 cars with $100 bills, and the remaining $5 - i$ cars are chosen from the 15 cars that do not have $100 bills). Now, if event A_i has taken place, then on the second day there will be $5 - i$ cars with $100 bills, while the remaining $15 + i$ cars will not contain $100 bills. Consequently, $P(N \mid A_i) = \dfrac{C(5-i, 0) \cdot C(15+i, 4)}{C(20, 4)}$, and we may now obtain:

$$P(N) = \sum_{i=0}^{5} P(N \mid A_i)P(A_i) = \sum_{i=0}^{5} \frac{C(5-i, 0) \cdot C(15+i, 4)}{C(20, 4)} \cdot \frac{C(5, i) \cdot C(15, 5-i)}{C(20, 5)} \approx 0.4108.$$

EFR 6.8: If E and F are independent events and $P(E) \neq 0$, then the independence equation $P(E \cap F) = P(E) \cdot P(F)$ implies (by dividing both sides by $P(E)$) that $P(F \mid E) = P(F)$. Similarly, as long as $P(F) \neq 0$, then independence of E and F implies $P(E \mid F) = P(E)$. If $P(E) = 0$, then $P(F \mid E)$ is not even defined so that the equation $P(F \mid E) = P(F)$ certainly cannot hold. A similar comment holds if $P(F) = 0$. We point out also that $P(E \cap F) \leq P(E), P(F)$ (by monotonicity), so if either $P(E) = 0$ or $P(F) = 0$, then $P(E \cap F) = 0$. Thus, if either E or F has zero probabililty, then the pair will be independent, but at least one of the two equations $P(F \mid E) = P(F)$ or $P(E \mid F) = P(E)$ cannot hold. In all other cases, these two equations are equivalent to independence. In particular, the independence equation is slightly more general (and versatile) than either of the two equations $P(F \mid E) = P(F)$ or $P(E \mid F) = P(E)$.

EFR 6.9: Clearly $P(E) = 4/8 = 1/2$ and $P(G) = 6/8 = 3/4$. Since $E \cap G = \{bbg, bgb, gbb\}$, we have $P(E \cap G) = 3/3 = (1/2) \cdot (3/4) = P(E) \cdot P(G)$, and so E and G are independent. Also, since $P(F) = 1/2$, and $F \cap G = \{gbb, gbg, ggb\}$, we have $P(F \cap G) = 3/8 = (1/2) \cdot (3/4) = P(F) \cdot P(G)$, and so E and G are independent. For the last part, note that $E \cap F \cap G = E \cap F$, so that $P([E \cap F] \cap G) = P(E \cap F) = 1/2 \neq (1/2) \cdot (3/8) = P(E \cap F) \cdot P(G)$, and this shows the dependence of $E \cap F$ and G.

EFR 6.10: Each of these three probabilities that are asked for is a conditional probability. Let C be the event that the luggage was lost in Chicago, S the event that it was lost in San Francisco, T the event that it was lost in Tokyo, and finally L be the event that Professor Billingsley's luggage did not make it to Bangkok. We assume that the events C, S, and T are independent (since each deals with a different luggage handling system and crew at different locations). We first compute $P(L)$:

$$
\begin{aligned}
P(L) &= 1 - P(\sim L) & \text{(complementary probability rule)}\\
&= 1 - P(\sim[C \cup S \cup T]) & \text{(since } C, S, \text{ and } T \text{ partition } L)\\
&= 1 - P(\sim C \cap \sim S \cap \sim T) & \text{(De Morgan's law)}\\
&= 1 - P(\sim C) \cdot P(\sim S) \cdot P(\sim T) & \text{(independence)}\\
&= 1 - (1 - p_C) \cdot (1 - p_S) \cdot (1 - p_T). & \text{(complementary probability rule)}
\end{aligned}
$$

Using this, we can now compute the three desired conditional probabilities:

$$
P(C \mid L) = \frac{P(C \cap L)}{P(L)} = \frac{P(C)}{P(L)} = \frac{p_C}{1 - (1 - p_C) \cdot (1 - p_S) \cdot (1 - p_T)}.
$$

$$
P(S \mid L) = \frac{P(S \cap L)}{P(L)} = \frac{P(\sim C \cap S)}{P(L)} = \frac{P(\sim C) \cdot P(S)}{P(L)} = \frac{(1 - p_C) p_S}{1 - (1 - p_C) \cdot (1 - p_S) \cdot (1 - p_T)}.
$$

$$
P(T \mid L) = \frac{P(T \cap L)}{P(L)} = \frac{P(\sim C \cap \sim S \cap T)}{P(L)} = \frac{P(\sim C) \cdot P(\sim S) \cdot P(T)}{P(L)} = \frac{(1 - p_C)(1 - p_S)(1 - p_T)}{1 - (1 - p_C) \cdot (1 - p_S) \cdot (1 - p_T)}.
$$

EFR 6.11: The sample space is $S = \{H, TH, TTH, TTTH, TTTT\}$, so range$(X) = \{1,2,3,4\}$. The pdf $p = p_X$ has the following corresponding values:

$$
p(1) = P(H) = 1/2, \ p(2) = P(TH) = 1/4, p(3) = P(TTH) = 1/8, p(4) = P(\{TTT^*\}) = 1/8.
$$

EFR 6.12: We first compute the values of the probability mass function. Formula (9) becomes $p(k) \equiv P(X = k) = C(6,k)(2/9)^k(7/9)^{n-k}$, and yields the following numerical values for the probability mass function:

k	0	1	2	3	4	5	6
$p(k)$	0.2214	0.3795	0.2711	0.1033	0.0221	0.0025	0.0001

From this probability mass function data, we easily obtain its graph (top graph on right), and then the corresponding cdf $F(x)$ (bottom graph on right). The values of $p(5)$ and $p(6)$ are so much smaller than $p(1)$ that it makes graphical depiction of the two functions difficult. In our graph of $F(x)$, the jumps were so (relatively) miniscule that they may have appeared non-existent (were it not for the solid dots that we added). Nevertheless, a cdf of any discrete probability mass function always has jumps where the latter function has positive values.

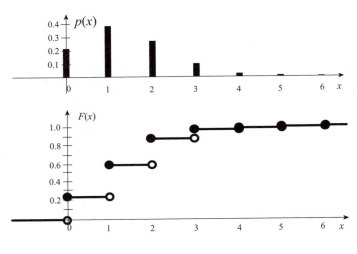

EFR 6.13: (a) We briefly outline the main idea, and then present the pseudocode for a program. The number of integers from min to max (inclusive) is N = max − min + 1. To produce such random integers, we can multiply rand by N (this will produce a random real number in the interval [0, N], then take the ceiling (this will result in a random integer in the set {1, 2, ... , N}), and finally add min − 1 to this integer (this will result in a random integer in the desired range). Here now is the pseudocode for the whole program:

```
Accept inputs:   n, min, max  (positive integer, integer, integer)
Initialize:   v  (length n output vector)
N = max - min + 1
FOR i = 1 TO n
     v(i) = ceiling(N*rand) + min - 1
END i FOR
OUTPUT v
```

(b) and (c): The histograms should be tending towards the uniform distribution since the random integers are equally likely to land anywhere within the indicated range.

EFR 6.14: (a) We first collect the values of the probability mass function for the sum of the pips when two dice are rolled. The values below are obtained as in Example 6.2 (see Figure 6.3):

k	2	3	4	5	6	7	8	9	10	11	12
$p(k)$	1/36	2/36	3/36	4/36	5/36	6/36	5/36	4/36	3/36	2/36	1/36

(We intentionally left all of the probabilities as nonreduced fractions; this will facilitate writing the program.)

(b) Although it would simplify matters to work with random integers in the range {2, 3, ..., 12} (which could easily be achieved using the program of the preceding exercise for the reader), in general probabilities need not have a nice common denominator (or even be rational numbers), so we will implement a more versatile approach of working directly with random real numbers in [0, 1]. This approach will preview the program that will be developed in the following exercise for the reader. The above probability mass function will be stored as two (length 11) vectors:

$$S \quad = [2 \ 3 \ 4 \ 5 \ 6 \ 7 \ 8 \ 9 \ 10 \ 11 \ 12] \quad \text{(for Sample space)}$$
$$Prob = [1 \ 2 \ 3 \ 4 \ 5 \ 6 \ 5 \ 4 \ 3 \ 2 \ 1]/36 \quad \text{(for Probabilities)}$$

Each simulation trial will proceed along the lines of the general procedure described in the text: we generate a random number r = rand. We then implement a while loop as follows: we first check if $r \leq P = \text{Prob}(1) = 1/36$; if it is we set the outcome of the trial to be S(1) = 2 (and note that this game of craps was lost); otherwise, we update P = P + Prob(2), and check whether $r \leq P$. If it is, we set the outcome of the trial to be S(2) = 3 (and record a lost game of craps). Otherwise, we update P = P + Prob(3), and continue. Eventually this process will end with some outcome in S. The pseudocode for the entire program is as follows:

```
Accept input:   N (positive integer)
Initialize:   v  (length N output vector),
              wins = 0  (integer), Pwin  (real number)
              S   = [2  3  4  5  6  7  8  9  10  11  12]
              Prob = [1  2  3  4  5  6  5  4  3   2   1] /36


FOR k = 1 TO N    (play N games of craps)
   r = rand
   P = Prob(1)    (initialize cumulative probability)
   ind = 1  (initialize vector index)
   WHILE r > P
      ind = ind + 1
      P = P + Prob(ind)
```

```
      END   (while)
      v(k) = S(ind)   (record outcome of kth game)
      IF  v(k) = 7 OR  v(k) = 11  (winning game)
         wins = wins + 1
      END IF
END k FOR
Pwin = wins/N
OUTPUT v  wins  Pwin
```

(c) The results and histograms should be similar to those obtained in Example 6.15.

EFR 6.15: (a) The program that was written in the preceding exercise for the reader will extend with minor modifications to the more general program of this exercise. The only changes are: (i) the vectors S and Prob are changed to `SampleSpc` and `ProbVec` and are now accepted as inputs, and there is now only one output (the vector v).

EFR 6.16: (a) The program can be coded quite simply with a for loop. The pseudocode is below.

```
Accept input:   n (positive integer)
Initialize:  v = [1   2   ... n]  (length n output vector)

FOR k = n TO 1 in decrements of -1
   RandInd=ceiling(rand*k);  (a random integer in the set {1,2,...,k})
   Temp = v(RandInd)
   v(RandInd) = v(k)
   v(k) = Temp
END k FOR
```

(b) The analysis is quite simple for this algorithm. At each of the n iterations of the loop, a random number is generated (an arithmetical operation), and then two values of the vector v are reset. So a total of $3n$ mathematical operations are used.

EFR 6.17: (a) Below is a pseudocode for a corresponding simulation program (using 1 million trials):

```
INITIALIZE:  WinCount=0   (counter for number of games won)
FOR T = 1 TO 1000000
      Flag = 0 (initialize flag, it will be set = 1 when game is won, = 2 if game is lost)
      TCount = 0  (initialize tail counter, game will be lost if this reaches 3)
      ConHeadCount = 0  (initialize consecutive head counter, game will
                          be won if this reaches 2)
      WHILE flag ?= 0  (continue flipping coin)
            r = rand
            IF r < 1/3
                Flip = 0  (tails)
            ELSE
                Flip = 1  (heads)
            END IF
            IF Flip ?= 0
                TCount = TCount + 1, ConHeadCount = 0
            ELSE
                IF ConHeadCount ?= 0
                    ConHeadCount = 1
                ELSE
                    ConHeadCount = 2
                END  IF
```

```
              END   IF
              IF TCount ?= 3    (game has just been lost)
                   Flag = 2
              ELSE IF ConHeadCount ?= 2  (game has just been won)
                   Flag = 1, WinCount = WinCount + 1
                 END IF
          END WHILE
  END FOR
  OUTPUT:  WinProb = WinCount/1000000
```

Three runs of this program produced the results: WinProb = 0.8284, 0.8289, 0.8282 (results will vary); this is in good agreement with the exact result of Part (b).

(b) The event W of winning such a game consists of all strings in the H(ead)/T(ail) alphabet that end in HH and contain at most two Ts. These outcomes are easily listed:

$$W = \{\text{HH, HTHH, HTHTHH, HTTHH, THH, THTHH, TTHH}\}.$$

Using independence and the fact that $P(H) = 2/3$ (and $P(T) = 1/3$), we can compute $P(W)$ by adding up the disjoint probabilities listed:

$$P(W) = (2/3)^2[1 + (2/3)(1/3) + (2/3)^2(1/3)^2 + (2/3)(1/3)^2 + 1/3 + (2/3)(1/3)^2 + (1/3)^2]$$
$$= 604/729 \approx 0.8285.$$

EFR 6.18: (a) Below is a pseudocode for a program with syntax:
$$P = \text{BoardingPassFunction}(N)$$
The input N is a positive integer equaling the number of trials to run in a simulation of the boarding pass problem. The output P is the resulting simulated probability. The way the program is organized is to let the original seat reservation of the ith entering passenger by i $(1 \le i \le 100)$.

```
Initialize:  Counter = 0
FOR TRIAL = 1:N
    set randseat = ceiling(100*rand)   (seat for passenger #1)
    IF randseat = 1 (passenger #1 takes correct seat)
        set Passenger = 101  (will mean all 100 get their seat)
        update Counter = Counter + 1
    ELSE
        Passenger = randseat  (next displaced passenger)
    END IF
    WHILE Passenger < 100
        randseat = randseat + ceiling(rand*(101-randseat))
        (this is a random seat number for the next passenger whose)
        (seat was taken, it will lie in the range from his/her seat #)
        (+ 1 to 101; 101 corresponds to his/her new seat = seat #1)
        (so then all remaining passengers get their orig. seats)
        IF randseat = 101
            Counter = Counter + 1, EXIT WHILE loop
        END IF
        update Passenger = randseat  (next displaced passenger)
    END WHILE
OUTPUT P = Counter/N
```

(b) Let c_n $(1 \le n \le 100)$ denote the number of different seating arrangements if Passenger #1 takes seat #n. Certainly $c_1 = c_{100} = 1$ since in the first situation everyone gets their assigned seat, and in the second, all but Passengers #1 and #100 do. Next, we notice that $c_{99} = 1 + c_{100}$, since in this case either Passenger #99 takes Seat #1 (so everyone but #1 and #99 get their seats) or Passenger #99 takes Seat #100; there is only $1 = c_{100}$ such arrangement. In general, by separating into the cases of where the first displaced passenger sits, we may establish the recursive formula $c_n = 1 + c_{n+1} + c_{n+2} + \cdots + c_{100}$, valid for $2 \le n \le 99$. Summing up this recursively defined sequence with a computer loop gives the total number of seating arrangements = $\sum_{n=1}^{100} c_n \approx 3.1692 \times 10^{29}$, far too large for a brute-force counting program.

(c) We use the notation $p \to s$ to mean "passenger #p takes seat #s." Let $E = "100 \to 100"$; we will prove that $P(E) = 1/2$. Conditioning on which seat Passenger #1 takes, we may write: $P(E) = \sum_{n=1}^{100} P(E \mid 1 \to n) P(1 \to n) = (1/100) \sum_{n=1}^{100} P(E \mid 1 \to n)$. We let P_n denote the conditional probability $P(E \mid 1 \to n)$. Since it is clear that $P_1 = 1$ and $P_{100} = 0$, it suffices to show that $P_n = 1/2$, for each remaining index n, $1 < n < 100$. We accomplish this by induction on n, starting at $n = 99$ (basis step) and working our way down to $n = 2$. If $1 \to 99$, then each of the passengers #2 through #98 will get their assigned seats, while passenger #99 will randomly choose either of the two remaining seats, #1, or #100. In the first case #100 gets his/her assigned seat, in the latter he/she does not. Thus, $P_{99} = 1/2$. Now assume that $k > 2$ is a positive integer, $k < 99$, and that $P_{k+1} = P_{k+2} = \cdots = P_{99} = 1/2$. (This is the (strong) inductive hypothesis.) If $1 \to k$, then passengers #2 through #($k-1$) will all get their assigned seats so that passenger k will be the first displaced passenger and will randomly select one of the $100 - (k-1) = 101 - k$ available seats. To compute P_k, we condition on which of these seats gets selected: $P_k = P(E \mid k \to 1) P(k \to 1) + \sum_{n=k+1}^{100} P(E \mid k \to n) P(k \to n)$ (we are assuming here that first $k - 1$ passengers have already been seated, so we are working in a reduced sample space). Now certainly $P(E \mid k \to 1) = 1$, $P(E \mid k \to 100)$, and $P(k \to n) = 1/(101 - k)$ (random selection). So it suffices to check that each of the remaining conditional probabilities equal 1/2. Indeed, if $k \to n$ (with $k+1 \le n \le 99$) then in addition to what we already know about the seated passengers, passengers #($k + 1$) through #($n - 1$) will also get their assigned seats, and passenger n will be the next displaced passenger. When passenger #n arrives, the situation will be exactly the same (as far as seat availability) as it was in the condition for P_n: only seats #1 and #($n + 1$) through #100 will be available, so it follows from the inductive hypothesis that $P(E \mid k \to n) = 1/2$, as we needed to show. □

EFR 6.19: (a) Below is a pseudocode for a simulation that follows along the lines of the suggestion. We will invoke the program `RandPerm` of Exercise for the Reader 6.15. We could also write a more efficient program that uses Algorithm 6.1 to generate 3-combinations directly (actually, a shorter version of it, as indicated in the paragraph following Algorithm 6.1); we leave such a task (as well as the performance comparisons) to the interested reader.

```
Initialize:  CumWin  = 0
(variable to store the cumulative winnings for all games played)
FOR Game = 1 TO 1000000
        set Shuffle = RandPerm(12)
            (corresponds to shuffling the 12 balls)
        set Draw = Shuffle(1:3)
            (first three components of the Shuffle Vector)
        set SDraw = sort(Draw)
            (the numbers of the balls in the draw are now sorted)
            IF SDraw(1) < 4 AND  SDraw(2) < 4 AND SDraw(3) < 4
            (3 red balls)
            update CumWin = CumWin + 10
      ELSE  IF  SDraw(1)<4  AND  SDraw(2)>3  AND  SDraw(2)<6  AND
SDraw(3)>5
(all balls are different colors)
(it is this part where sorting the draw has helped)
```

```
             update CumWin = CumWin + 2
          END IF
          update CumWin = CumWin - 1
```
(pay the $1 it cost to play the game)
```
END FOR
OUTPUT:  Expected Win = CumWin/1000000
```

(b) Let X be the random variable of the net winnings in one round of this carnival game. We need to compute $E[X]$. Taking into account the cost of the game, we have Range$(X) = \{-1, 1, 9\}$, and we have

$$E[X] = 1 \cdot P(X = 1) + 9 \cdot P(X = -1) - 1 \cdot P(X = -1)$$
$$= 1 \cdot C(3,1) \cdot C(2,1) \cdot C(7,1)/C(12,3) + 9 \cdot C(3,3)/C(12,3) - 1[1 - P(X = 1,9)]$$
$$\approx -0.5727.$$

Thus, in the long term, playing this carnival game repeatedly will result in a net average loss of about 57¢ per game.

EFR 6.20: (a) Using (24) and the linearity of expectation, we obtain:

$$\mathrm{Var}(aX) = E[(aX)^2] - (E[aX])^2 = E[a^2 X^2] - (aE[X])^2 = a^2 (E[X^2] - (E[X])^2) = a^2 \mathrm{Var}(X).$$

(b) Using (25), along with linearity of expectation, we obtain:

$$\mathrm{Cov}(aX_1 + bX_2, Y) = E[(aX_1 + bX_2)Y] - E[aX_1 + bX_2]E[Y]$$
$$= E[aX_1 Y + bX_2 Y] - (aE[X_1] + bE[X_2])E[Y]$$
$$= aE[X_1 Y] - aE[X_1]E[Y] + bE[X_2 Y] - bE[X_2]E[Y]$$
$$= a\mathrm{Cov}(X_1, Y) + b\mathrm{Cov}(X_2, Y).$$

(c) We prove (25) by induction on n. 1. *Basis Step:* When n = 1, the identity becomes $\mathrm{Var}(X_1) = \mathrm{Var}(X_1)$, which is certainly true. 2. *Inductive Step:* Assume that (25) is true for $n = k$, a positive integer, i.e., $\mathrm{Var}(X_1 + X_2 + \cdots + X_k) = \sum_{i=1}^{k} \mathrm{Var}(X_i) + 2\sum_{i>j} \mathrm{Cov}(X_i, X_j)$. We must show that the corresponding identity when $n = k + 1$ is also valid. We have:

$$\mathrm{Var}(X_1 + X_2 + \cdots + X_k + X_{k+1}) = \mathrm{Var}([X_1 + X_2 + \cdots + X_k] + X_{k+1}) \qquad \text{(Regrouping)}$$
$$= \mathrm{Var}(X_1 + X_2 + \cdots + X_k) + \mathrm{Var}(X_{k+1}) + 2\mathrm{Cov}([X_1 + X_2 + \cdots + X_k], X_{k+1}) \quad \text{(Proposition 7.7(a))}$$
$$= \sum_{i=1}^{k} \mathrm{Var}(X_i) + 2\sum_{k \geq i > j} \mathrm{Cov}(X_i, X_j) + \mathrm{Var}(X_{k+1}) + 2\mathrm{Cov}([X_1 + X_2 + \cdots + X_k], X_{k+1})$$

$$\text{(Inductive Hypothesis)}$$

$$= \sum_{i=1}^{k+1} \mathrm{Var}(X_i) + 2\sum_{k \geq i > j} \mathrm{Cov}(X_i, X_j) + 2\sum_{i=1}^{k} \mathrm{Cov}(X_i, X_{k+1}) \quad \text{(Linearity of Covariance-part(b))}$$

$$= \sum_{i=1}^{k+1} \mathrm{Var}(X_i) + 2\sum_{i>j} \mathrm{Cov}(X_i, X_j) \quad \text{(Regrouping and using } \mathrm{Cov}(X,Y) = \mathrm{Cov}(Y,X)). \ \square$$

EFR 6.21: Assuming $X \sim \mathscr{B}(n, p)$ and letting $q = 1 - p$, we have:

$$E[X] = \sum_{k=0}^{n} kP(X = k) = \sum_{k=0}^{n} k \binom{n}{k} p^k q^{n-k} = \sum_{k=1}^{n} k \frac{n!}{k!(n-k)!} p^k q^{n-k}$$

$$= \sum_{k=1}^{n} \not{k} \frac{n \cdot (n-1)!}{\not{k} \cdot (k-1)!(n-k)!} p^k q^{n-k} = np\sum_{k=1}^{n} \frac{(n-1)!}{(k-1)!(n-1-[k-1])!} p^{k-1} q^{n-1-(k-1)}$$

$$\underset{k \to k+1}{=} np\sum_{k=0}^{n-1} \frac{(n-1)!}{k!(n-k)!} p^k q^{n-1-k} = np \underbrace{\sum_{k=0}^{n-1} \binom{n-1}{k} p^k q^{n-1-k}}_{=(p+q)^{n-1}=1 \ (\text{Binomial Thm.})} = np. \ \square$$

EFR 6.22: According to the given data, typhoons occur at a rate of 10/41 typhoons/year. Modeling the (random variable of the) number of typhoons in the next year, X with a Poisson random variable, we have $X \sim \mathcal{P}(10/41)$, i.e., the mean number of major typhoons per year in Guam is 10/41.

(a) $P(X \geq 1) = 1 - P(X = 0) = 1 - e^{-10/41}(10/41)^0 / 0! \approx 0.2164 (= 21.64\%)$.

(b) $P(X = 2) = e^{-10/41}(10/41)^2 / 2! \approx 0.0233 (= 2.33\%)$.

(c) Since the annual mean rate of typhoons is 10/41, the corresponding rate for a three month period is one fourth of this, so the random variable Y modeling the number of occurrences over a three-month period satisfies $Y \sim \mathcal{P}(5/82)$. $P(Y \geq 1) = 1 - P(Y = 0) = 1 - e^{-5/82}(5/82)^0 / 0! \approx 0.0592 (= 5.92\%)$.

CHAPTER 7: COMPLEXITY OF ALGORITHMS

EFR 7.1:
Initialize max $= a_1$

FOR $i = 2$ TO n

  IF $a_i > $ max

     UPDATE max $= a_i$

  END IF
END FOR
OUTPUT max

The corresponding algorithm for finding the minimum can be created from the above by reversing the sign of the inequality in the IF condition. The variable name "max" should also be changed to "min," although this is not necessary for the program to function correctly.

EFR 7.2: We provide details for each step of Algorithm 7.2.
Step 1: Initialize search interval endpoint indices: INITIALIZE: LeftIndex = 1, RightIndex = 16
Note that $t = 46$ and $n = 16$ (part of the inputs).

Step 2: Iteratively reduce search interval to a single number:
First Iteration: $m = $ floor([1+16]/2]) = 8 (Note that $a_8 = 44$)

UPDATE LeftIndex = 9 (Since $t > a_8$)

Second Iteration: $m = $ floor([9+16]/2]) = 12 (Note that $a_{12} = 72$)

UPDATE RightIndex = 12 (Since t is not greater than a_{12}.)

Third Iteration: $m = $ floor([9+12]/2]) = 10 (Note that $a_{10} = 56$)

UPDATE RightIndex = 10 (Since t is not greater than a_{10}.)

Third Iteration: $m = $ floor([9+10]/2]) = 9 (Note that $a_9 = 46$)

UPDATE RightIndex = 9 (Since t is not greater than a_9.)

OUTPUT $k = 9$ (Since $t = a_9$)

EFR 7.3: The first three passes of the bubble sort algorithm are shown below. In the fourth pass (not shown), since the items are sorted, the algorithm will output the sorted list and quit.

Joey	Joey	Joey	Joey	Amanda	Amanda	Amanda	Amanda	Amanda
Tara	Amanda	Amanda	Amanda	Joey	Joey	Joey	Joey	Dennis
Amanda	Tara	Tara	Tara	Tara	Patrick	Patrick	Dennis	Joey
Teresa	Teresa	Patrick	Patrick	Patrick	Tara	Dennis	Patrick	Patrick
Patrick	Patrick	Teresa	Dennis	Dennis	Dennis	Tara	Tara	Tara
Dennis	Dennis	Dennis	Teresa	Teresa	Teresa	Teresa	Teresa	Teresa

                    First Pass   Second Pass   Third Pass

EFR 7.4: The number of comparisons needed by the bubble sort algorithm can vary, depending on the ordering of the inputted list. In the best case, only one pass is needed (which will require $n-1$ comparisons); in the worst case a full $n(n-1)/2$ comparisons will be needed.

EFR 7.5: The worst case for the quicksort occurs when at each iteration either the greatest or the smallest element of the list is selected. This means that we will have a total number of comparisons equal to: $n-1$ (1st iteration) $+ n-2$ (2nd iteration) $+ \cdots + 2 + 1 = n(n-1)/2$.

EFR 7.6: Rather than working on one bracket list at a time, at each iteration we will break down any bracketed lists (with more than one element) that remain:
Initially List = Joey, Louise, Dan, Tara, Steve, Amanda, Teresa, Sandra, Patrick, Dennis

First Iteration: Select x = Joey, and compare to form
[Dan, Amanda, Dennis] Joey [Louise, Tara, Steve, Teresa, Sandra, Patrick]

Second Iteration: Select x = Dan for the first bracket, x = Louise for the second bracket, and compare to form
[Amanda] Dan [Dennis] Joey Louise [Tara, Steve, Teresa, Sandra, Patrick]

Third Iteration: Select x = Tara, and compare to form
Amanda, Dan, Dennis, Joey, Louise [Sandra, Patrick] Tara, [Steve, Teresa]

Fourth Iteration: Select x = Sandra for the first bracket, x = Steve for the second bracket, and compare to form
Amanda, Dan, Dennis, Joey, Louise [Patrick] Sandra, Tara, Steve [Teresa]
Ignoring brackets, we now have the sorted list.

EFR 7.7:

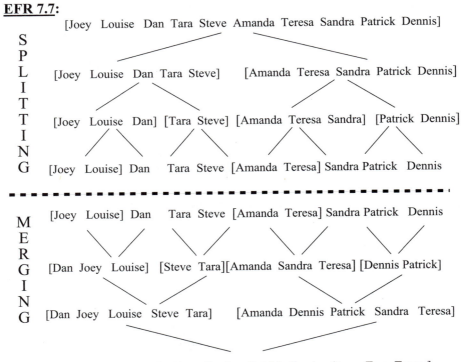

EFR 7.8: Part (a):
(i) The list is sorted, so median is the middle element 2.
(ii) We first sort the list and get the same lists as in (i), so median is the middle element 2.
(iii) The list is sorted, but it has an even number of elements, so the median is the average of the middle two elements $(2+3)/2 = 2.5$.

Part (b): The list 0, 1, 100 has median 1, and mean $= (0+1+100)/3 = 33.66\cdots$ that is over ten times as great as the median.

EFR 7.9: We simply need to combine Algorithm 7.8 with Definition 7.1:
INPUT: A list of integers $a_1, a_2, \cdots, a_n$.
OUTPUT: The median M of the list.

Step 1: Determine an index needed to locate the median in sorted list:
If n is odd, SET MedInd $=$ ceil($n/2$), otherwise set MedInd $= n/2$.

Step 2: Apply Algorithm 7.8 to compute median:
If n is odd, OUTPUT $M = $ Smallest($[a_1, a_2, \cdots, a_n]$, MedInd), and EXIT algorithm,
Otherwise OUTPUT
$M = $ Smallest $[$Smallest($[a_1, a_2, \cdots, a_n]$, MedInd) + Smallest($[a_1, a_2, \cdots, a_n]$, MedInd $+1)]/2$, and EXIT algorithm.

EFR 7.10: Part (a): Answer: True. Justification: If $n \geq 10$, then to find witnesses, we rewrite $10^n = 10^9 \cdot \underbrace{10 \cdot 10 \cdots \cdot 10}_{n-9 \text{ times}} \leq 10^9 \cdot 10 \cdot 11 \cdots \cdot n < 10^9 \cdot n!$. This proves that $k = 9$ and $C = 10^9$ are witnesses to the big-O relationship $10^n = O(n!)$.

Part (b): Answer: True. Justification: The inequality $200(1 + 1/n) \leq 400 = 400 \cdot 1$, which is valid for any positive integer n, proves that $k = 1$ and $C = 400$ are witnesses to the big-O relationship $200(1 + 1/n) = O(1)$.

EFR 7.11: The strict domination $b^n \prec (b')^n$ directly follows from the equation $(b')^n = b^n \cdot (b'/b)^n$ because $b'/b > 1$, and hence the factor $(b'/b)^n$ is itself an exponential growth term that strictly increases without bound (to infinity).

EFR 7.12: The dominations $f(n) \precsim h(n)$, $g(n) \precsim h(n)$ may be expressed as:
$$f(n) \leq C_1 h(n), \text{ for all } n \geq k_1, \text{ and}$$
$$g(n) \leq C_2 h(n), \text{ for all } n \geq k_2,$$

where C_1, k_1 and C_2, k_2 are corresponding sets of witnesses. If we put $C = C_1 \cdot C_2$ and $k = \max(k_1, k_2)$, then for $n \geq k$, since both of the previous two inequalities hold, we may write:
$$f(n) \cdot g(n) \leq C_1 h(n) \cdot C_2 k(n) = (C_1 \cdot C_2) h(n) \cdot k(n) = Ch(n) \cdot k(n).$$

We have thus shown that the pair C, k are witnesses for the domination (big-O relation) $f(n) \cdot g(n) \precsim h(n) \cdot k(n)$.

To show the strict domination statement, given any positive integer M, if, say the first domination was strict: $f(n) \prec h(n)$, then there exists an integer ℓ such that if $n \geq \ell$, then $h(n) \geq MC_2 f(n)$. If we set $k' = \max(k, \ell)$, then for $n \geq k'$, we have $h(n) \cdot k(n) \geq MC_2 f(n) \cdot \dfrac{1}{C_2} g(n) = Mf(n) \cdot g(n)$. Since M could be chosen arbitrarily large, we have proved the strict domination. $\square$

EFR 7.13: By repeatedly applying Proposition 7.2(1), we obtain the following strict dominations:
$$|a_1|n + |a_0| \prec |a_2|n^2,$$
$$|a_2|n^2 + |a_1|n + |a_0| \prec |a_3|n^3,$$
$$\vdots$$
$$|a_{D-1}|n^{D-1} + \cdots + |a_1|n + |a_0| \prec a_D n^D.$$
From the last of these, one more application of Proposition 7.2(1) gives us the nonstrict domination $f(n) = a_D n^D + a_{D-1} n^{D-1} + \cdots + a_1 n + a_0 \precsim a_D n^D$, which is equivalent to the big-O estimate $f(n) = O(n^D)$.

EFR 7.14: Part (a): Answer: False. Justification: Proposition 7.1(b) tells us that n^2 strictly dominates n.

Part (b): Answer: True. Justification: The estimates $f(n) = \Omega(g(n))$, $g(n) = \Omega(h(n))$ may be expressed as:
$$f(n) \geq C_1 g(n), \text{ for all } n \geq k_1 \text{ and}$$
$$g(n) \geq C_2 h(n), \text{ for all } n \geq k_2,$$

where C_1, k_1 and C_2, k_2 are corresponding sets of witnesses.   If we put $C = C_1 \cdot C_2$ and $k = \max(k_1, k_2)$, then for $n \geq k$, since both of the previous two inequalities hold, we may write:

$$f(n) \geq C_1 g(n) \geq C_1[C_2 h(n)] = Ch(n).$$

We have thus shown that the pair C, k are witnesses for the big-Omega estimate $f(n) = \Omega(h(n))$.

Part (c):   Answer: True.   Justification: From   formula (1) of Section 3.1, we know that $1 + 2 + 3 + \cdots + n = n(n+1)/2 = (1/2)n^2 + n/2$.   The big-Theta estimate is now a direct consequence of the result of Example 7.8.

EFR 7.15:   Part (a):   From the trivial dominations $g(n) \preceq g(n)$, $f(n) \preceq g(n)$, Proposition 7.2(1) tells us that $g(n) + f(n) \preceq g(n)$, i.e., $g(n) + f(n) = O(g(n))$, which is equivalent to the big-Omega estimate $g(n) = \Omega(g(n) + f(n))$.

Part (b):   The big-Theta estimate $f(n) = \Theta(g(n))$ may be expressed as follows: there a positive integer k and two positive constants C, D such that for each $n \geq k$, we have $Cg(n) \leq f(n) \leq Dg(n)$. Multiplying both sides of this inequality by the nonnegative number $h(n)$ gives us $Cg(n)h(n) \leq f(n)h(n) \leq Dg(n)h(n)$ for each $n \geq k$,   which is equivalent to $f(n) \cdot h(n) = \Theta(g(n) \cdot h(n))$.

EFR 7.16:   We will deal with the case in which $k \leq n/2$.   The case in which $k > n/2$ is dealt with in a similar fashion.

Part (a):   The worst case in Algorithm 7.8 occurs when the element selected for comparisons at each iteration is the largest element of the list. In this case the target list goes down in size by only one element at each iteration, so the number of comparisons will be: $n - 1 + (n-2) + \cdots + k = O(n^2)$.

Part (b):   In the average case, the list size will typically go down by a fixed proportion on each iteration. To see this, we note that if the selected element is chosen from the middle 50% of the items (i.e., from the 25% to the 75% largest elements), then the largest the size of LessList and GreaterList could be is 75% of the original list size.   So the next question is how many random selections are necessary in order to land a choice that is in the middle 50% of the numbers in a given list.   This is the same as the expected number of times N_H we would need to flip a fair coin in order to land a heads.

Certainly N_H is at least one (to land a head, we need at least one flip).   Since coin flips are independent, if we landed a tails on the first flip (which happens with probability 1/2), then the expected number of additional flips to land a head will also be $E(N_H)$.   It thus follows that $E(N_H) = 1 + (1/2)E(N_H)$, from which it immediately follows that $E(N_H) = 2$.   Thus, after two iterations of the algorithm, we would expect to choose an element that lies in the middle 50%, and the amount of work needed to split a list of n elements into LessList and GreaterList is $O(n)$, it follows that the operations count function $f(n)$ for Algorithm 7.8 will satisfy $E[f(n)] = E[f([3/4]n)] + O(n)$, which translates into the inequality $E[f(n)] \leq E[f([3/4]n)] + Cn$, for some positive constant C.

Iterating this inequality j times until $(3/4)^j n < 2$, we obtain:

$$\begin{aligned}
E[f(n)] &\leq E[f([3/4]n)] + Cn \\
&\leq E[f([3/4]^2 n)] + Cn + C[3/4]n \\
&\leq E[f([3/4]^3 n)] + Cn + C[3/4]n + C[3/4]^2 n \\
&\vdots \\
&\leq E[f(2)] + Cn + C[3/4]n + C[3/4]^2 n + \cdots + C[3/4]^{j-1} n \\
&= O(n).
\end{aligned}$$

(In the last big-O estimate we used the formula for the sum of a geometric series.) Since the average number of comparisons is at least $\Omega(n)$ (i.e., consider the optimal case when the list gets cut in half at each iteration), it follows that $E[f(n)] = \Theta(n)$.

CHAPTER 8: GRAPHS, TREES, AND ASSOCIATED ALGORITHMS

EFR 8.1: Part (a): This handshaking problem is modeled by the simple graph with 100 vertices (the people) and having an edge between each pair of vertices (the handshakes). Thus for each vertex v we have $\deg(v) = 99$, and by the handshaking theorem, we may compute the total number of handshakes to be $|E| = (1/2)\sum_{v \in V} \deg(v) = (1/2) \cdot 100 \cdot 99 = 4950$.

Part (b): No. Reason: With the graph model of Part (a), the situation would lead to a (simple) graph with all 99 vertices (an odd number) having odd degree 3, but this is impossible by Corollary 8.2.

EFR 8.2: Part (a): The drawing of Q_4 shown at the right was created using the recursive scheme indicated in the paragraph preceding this exercise for the reader. We began with two copies of Q_3, and we prefixed the vertex labels in one with "0," and with "1" in the other. Then new edges are drawn between corresponding vertices in the two copies.

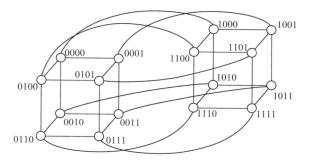

Part (b): The number of vertices $|V(Q_n)|$ of Q_n is the number of binary strings of length n, which by the multiplication principle is 2^n. Each vertex has degree n, since there are exactly n length n bit strings that differ from a given length n bit string in exactly one bit. It follows from the handshaking theorem that $|E(Q_n)| = (1/2)\sum_{v \in V} \deg(v) = (1/2) \cdot 2^n \cdot n = n2^{n-1}$.

EFR 8.3: Part (a): The n-path P_n is bipartite for each $n \geq 2$ (i.e., for each n for which the n-path was defined). A bipartition can be obtained quite simply: Start with the left vertex, color it black, color the next one white, and continue alternating color assignments as we progress through the remaining vertices. In the notation of Definition 8.4, with $V(P_n) = \{v_1, v_2, \cdots, v_n\}$, this corresponds to coloring the vertex v_i black when i is odd and white when i is even.

Part (b): The n-hypercube Q_n is bipartite for each positive integer n. A bipartition can be obtained as follows: color each vertex v black if its corresponding bit string is *odd* (i.e., has an odd number of 1's), while if the bit string is *even* (i.e., has an even number of 1's), color it white. Since (the bit strings of) any two adjacent vertices Q_n must differ in exactly one bit, it follows that adjacent vertices must have different parity, and so must have different colors. This shows we have a bipartition.

EFR 8.4: We proceed in the same fashion of the solution to Example 8.5:

> Initial Sequence: (5,5,5,3,3,3,3,3)
> Iteration #1: (i) (4,4,2,2,2,3,3), (ii) (4,4,3,3,2,2,2)
> Iteration #2: (i) (3,2,2,1,2,2), (ii) (3,2,2,2,2,1)
> Iteration #3: (i) (1,1,1,2,1), (ii) (2,1,1,1,1)
> Iteration #4: (i) (0,0,1,1), (ii) (1,1,0,0)

Iteration #5: (i) (0, 0, 0), (ii) (0, 0, 0) → "TRUE"

Algorithm 8.1 thus shows that the original sequence is graphic. Below is a sequence of graphs that realize the sequences (ii) in each of the iterations above, working our way from the last to the first iteration, so that the last graph realizes the given sequence; with the new vertices being added at each iteration being colored in black, while carryover vertices are colored white. We remind the reader that many different drawings are possible. Sometimes it is helpful to make several attempts at the drawings (by readjusting the placement of the vertices) if initial attempts turn out to give messy pictures.

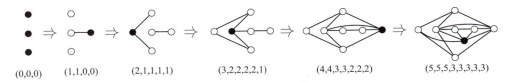

(0,0,0)   (1,1,0,0)   (2,1,1,1,1)   (3,2,2,2,2,1)   (4,4,3,3,2,2,2)   (5,5,5,3,3,3,3,3)

EFR 8.5: Yes, switching the order of the two conditions in the IF branch would produce an incorrect algorithm. The reason is that a nongraphic sequence may produce a zero first entry d_1 at some iteration of the algorithm. For example, the monotone sequence (2,2,1) is clearly not graphic (one way to see this is that by Corollary 8.2 a simple graph cannot have a single vertex of odd degree). But the first iteration of the Algorithm 8.1 produces the sequence $(0,-1)$. With the conditions of the IF branched switched, the modified algorithm would output "TRUE," and EXIT at this point, incorrectly indicating that the given sequence is graphic.

EFR 8.6: Throughout we will use the vertex labels of Figure 8.14. The Petersen graph clearly contains no triangles, hence no copy of K_3. The following are copies of cycles of length 5, 6, and 8.
$C_5 = <a,d,k,h,c,a>$, $C_6 = <a,d,k,j,g,c,a>$, $C_8 = <a,d,k,j,b,e,h,c,a>$.   Since C_8 contains copies of P_i, for any $i < 8$, it follows that the Petersen graph contains P_5, P_6 (and P_7) as subgraphs. The fact that C_4 is not a subgraph of the Petersen graph can be justified as follows: We separate the 10 vertices of the Petersen graph into the outside vertices: a, d, k, j, b and the inside vertices: c, h, e, f, g. Since the only edges connecting vertices of the same type (of these two types) form a single cycle subgraph C_5, it follows that if the Petersen graph did contain a copy of C_4, then the vertices such a 4-cycle must contain both inside and outside vertices. Assume that the Petersen graph contained a copy of C_4. By symmetry, and since it must contain both inside and outside vertices, we may assume it contains the edge $\{a,c\}$. But then it would also contain either edge $\{c, g\}$ or edge $\{c, h\}$ (but not both; see Figure 8.14). If it contained $\{c, g\}$, then it would have to contain two more edges of the form $\{g, *\}$, $\{*, a\}$, where * represents the fourth vertex. This vertex * would have to be different from c, and have both vertex g and vertex a as neighbors. But (see Figure 8.14), no such vertex exists, so we have a contradiction. A similar contradiction is obtained in case the subgraph were to contain the edge $\{c, h\}$.
NOTE: Later in this section we will introduce another useful way to view the Petersen graph that will make it easier to see some of its properties.

EFR 8.7: Part (a): Let $v, w \in V(G)$. Since f preserves adjacency and non-adjacency, it follows that w is a neighbor of v in G if, and only if, $f(w)$ is a neighbor of $f(v)$ in G'. Since G, G' are simple graphs, and f is a bijection between their vertex sets, we may conclude that
$$\deg_G(v) = |\{w \in V(G) : w \text{ is a neighbor of } v \text{ in } G\}|$$
$$= |\{w' \in V(G') : w \text{ is a neighbor of } f(v) \text{ in } G'\}| = \deg_{G'}(f(v)). \ \square$$

Part (b): This follows directly from Part (a) since $f : V(G) \to V(G')$ is a bijection.

EFR 8.8: The number of bijections between the set of 8 vertices $V(Q_3)$ and itself is $8! = 40,320$. Referring to the vertex labeling of Q_3 given in Figure 8.17, we see that if $F:V(Q_3) \to V(Q_3)$ is an isomorphism, then there are at most 8 choices for $F(a)$. (By symmetry of the drawing of the graph Q_3 given in Figure 8.17, we see that all vertices share the same adjacency properties, so there are exactly 8 choices.) Once a choice is made for $F(a)$, there are $3!$ ways to assign the three neighbors of a: e, b, d to the three neighbors of $F(a)$. At this point, since c is the unique common neighbor of b and d, it follows that $F(c)$ must be the unique common neighbor of $F(b)$ and $F(d)$. Similarly $F(h)$ is uniquely determined. Now consider the final two vertices: g and f each have two neighbors that have already been assigned, and their images $F(g)$ and $F(f)$ are uniquely determined as the unique remaining neighbors of the corresponding pairs. This shows that the number of isomorphisms $F:V(Q_3) \to V(Q_3)$ is $8 \cdot 6 = 48$, a fraction of $48/8! = 0.012$ of the total number of vertex bijections.

EFR 8.9: *Proof that isomorphism is an equivalence relation on the set of simple graphs:*
We assume that $G = (V, E)$, $G' = (V', E')$, $G'' = (V'', E'')$ simple graphs.

(i) *Reflexivity*: The identity function $i_V : V \to V :: i_V(v) = v$ clearly preserves adjacency and non-adjacency in G, so is an isomorphism, thus $G \cong G$.

(ii) *Symmetry*: Assume that $G \cong G'$. This means there exists an isomorphism $f : V \to V'$. Since f preserving adjacency is equivalent to $f^{-1} : V' \to V$ preserving non-adjacency (logical contrapositive), and f preserving non-adjacency is equivalent to f^{-1} preserving adjacency, it follows that $f^{-1} : V' \to V$ is also an isomorphism. This means $G' \cong G$.

(iii) *Transitivity*: Assume that $G \cong G'$ and that $G' \cong G''$. This means there exists isomorphisms $f : V \to V'$ and $g : V' \to V''$. Since compositions of adjacency preserving functions (and non-adjacency preserving functions) also preserve adjacency (non-adjacency), it follows that the composition $g \circ f : V \to V''$ is also an isomorphism, which proves that $G \cong G''$. $\square$

EFR 8.10: Part (a): The four isomorphism classes of three vertex graphs are shown below:

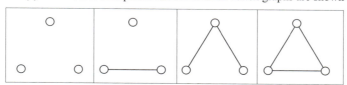

Part (b): The eleven isomorphism classes of four vertex graphs are shown below:

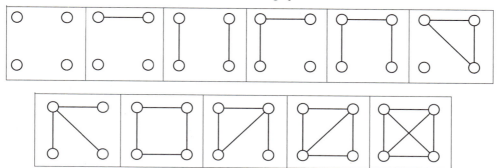

EFR 8.11: Since in order for two graphs to be isomorphic, it is necessary that there exist a bijection between their vertex sets, the result readily follows from the following lemma:

Lemma: Suppose that G and H are simple graphs, and that $f : V(G) \rightarrow V(H)$ is a bijection.

(a) When viewed as a mapping $G \rightarrow H, f$ preserves adjacency if, and only if f^{-1} preserves adjacency when viewed as a mapping $\sim H \rightarrow \sim G$.

(b) When viewed as a mapping $G \rightarrow H$, f preserves non-adjacency if, and only if f^{-1} preserves non-adjacency when viewed as a mapping $\sim H \rightarrow \sim G$.

Proof of Lemma: Part (a): Let $v, w \in V(G)$. If v, w are adjacent in G, and if f preserves adjacency, then $f(v), f(w)$ will be adjacent in H. The contrapositive of that is that if $f(v), f(w)$ are non-adjacent in H, then $f^{-1}[f(v)], f^{-1}[f(w)] = v, w$ are non-adjacent in G. Since non-adjacency in a graph is equivalent to adjacency in its complement graph, the latter statement is in turn equivalent to: If $f(v), f(w)$ are adjacent in $\sim H$, then $f^{-1}[f(v)], f^{-1}[f(w)] = v, w$ are adjacent in $\sim G$. This condition means that f^{-1} preserves adjacency when viewed as a mapping $\sim H \rightarrow \sim G$.

Part (b): Simply toggle all occurrences of the words adjacent/non-adjacent and adjacency/non-adjacency in the above proof of Part (a), and the result will be a proof of Part (b). □

EFR 8.12:

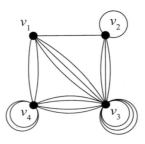

EFR 8.13:   The multigraph shown in Figure 8.25(b) has vertex set $V = \{u, v, w, x\}$, edgeset $E = \{e_1, e_2, e_3, e_4, e_5, e_6, e_7\}$, and edgemap $\Delta : E \rightarrow V \times V$ defined by:

$\Delta(e_1) = (v, u), \ \Delta(e_2) = (u, v), \ \Delta(e_3) = (v, w), \ \Delta(e_4) = (w, w), \ \Delta(e_5) = (w, x), \ \Delta(e_6) = (x, v), \ \Delta(e_7) = (u, x).$

EFR 8.14: The indegrees are: $\deg^-(u) = 1, \deg^-(v) = 2, \deg^-(w) = 2, \deg^-(x) = 2$ and they add up to 7, which is the number of edges. The outdegrees are: $\deg^+(u) = 2, \deg^+(v) = 2, \deg^+(w) = 2, \deg^+(x) = 1$ and since they also add up to 7, the handshaking theorem is thus confirmed.

EFR 8.15: The underlying graphs are isomorphic; an example of an isomorphism is given by the following vertex map: $a \mapsto x, b \mapsto y, c \mapsto u, d \mapsto w, e \mapsto v$. It is easily checked that this bijection preserves both adjacency and non-adjacency, and hence is a (simple) graph isomorphism.

EFR 8.16:   This is possible. To explain, it will be helpful to label the vertices of the two graphs. The first step is to observe that all of the indegrees/outdegrees of the left partial digraph are at least one (already), and this is why we have assigned one the direction of one of the edges on the right partial digraph (otherwise x would have indegree 0 and v would have outdegree 0). Next, since the left digraph already has a directed 4-cycle (a,b,c,d,a), we need to make sure we add a direction to the remaining edge on the right graph to form a directed 4-cycle; the figure shows (in bold) the resulting directed 4-cycle after an edge direction was added. We are not yet

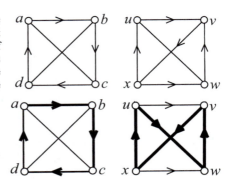

done; we still need to assign directions to the left partial digraph's remaining edges and exhibit an isomorphism. The directed 4-cycles that we have already can help us to take care of both of these tasks. If we follow both 4-cycles (indicated in bold in the preceding figure), starting say at vertex a on the left digraph and vertex u on the right digraph, we arrive at the following vertex correspondence: $a \mapsto u, b \mapsto w, c \mapsto v, d \mapsto x$. By its construction, this vertex bijection already preserves directions of four of the six directed edges on both graphs. We use this mapping to determine the directions on the remaining two edges on the left digraph so their directions will also be preserved (and hence we will have an isomorphism): The directed edge (u,v) would correspond to (a,c). Similarly the directed edge (x,w) should correspond to (d,b).

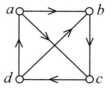

EFR 8.17: Part (a): Although not necessary, we label the rows/columns to make the adjacency matrices easier to read:

$$A = \begin{array}{c} \\ a \\ b \\ c \\ d \\ e \end{array} \begin{array}{c} \begin{array}{ccccc} a & b & c & d & e \end{array} \\ \left[\begin{array}{ccccc} 0 & 0 & 1 & 1 & 1 \\ 1 & 0 & 0 & 1 & 1 \\ 0 & 1 & 0 & 0 & 0 \\ 0 & 0 & 1 & 0 & 0 \\ 0 & 0 & 1 & 1 & 0 \end{array} \right] \end{array}.$$

Part (b): Incidence matrix:

$$B = \begin{array}{c} \\ u \\ v \\ w \\ x \end{array} \begin{array}{c} \begin{array}{ccccccc} e_1 & e_2 & e_3 & e_4 & e_5 & e_6 & e_7 \end{array} \\ \left[\begin{array}{ccccccc} 1 & -1 & 0 & 0 & 0 & 0 & -1 \\ -1 & 1 & -1 & 0 & 0 & 1 & 0 \\ 0 & 0 & 1 & 2 & -1 & 0 & 0 \\ 0 & 0 & 0 & 0 & 1 & -1 & 1 \end{array} \right] \end{array}$$

Part (c): Edge list:

Tail	Head
a	b
b	c
c	a
c	e
d	a
d	c
d	e

EFR 8.18: *Proof that vertex reachability is an equivalence relation on the vertex set of a simple graph:*

We assume that $G = (V, E)$ is a simple graph, and that $u, v, w \in V$.

(i) *Reflexivity*: The zero length path $P = \langle u \rangle$ shows that vertex u is reachable from itself.

(ii) *Symmetry*: Assume that vertex v is reachable from vertex u. This means there exists a path $P = \langle v_0, e_1, v_1, e_2, v_2, \cdots, v_{n-1}, e_n, v_n \rangle$ in G from $u = v_0$ to $v = v_n$. The reverse path $\langle v_n, e_n, v_{n-1}, e_{n-1}, v_{n-2}, \cdots, v_1, e_1, v_0 \rangle$ will then be a path in G from v to u, showing that u is reachable from v.

(iii) *Transitivity*: Assume that vertex v is reachable from vertex u, and that vertex w is reachable from from vertex v. This means there exist a path $P = \langle v_0, e_1, v_1, e_2, v_2, \cdots, v_{n-1}, e_n, v_n \rangle$ in G from $u = v_0$ to $v = v_n$, and a path $Q = \langle w_0, f_1, w_1, f_2, w_2, \cdots, w_{m-1}, f_m, w_m \rangle$ in G from $v = w_0$ to $w = w_m$. The juxtaposed path $PQ = \langle v_0, e_1, v_1, e_2, v_2, \cdots, v_{n-1}, e_n, v_n = w_0, f_1, w_1, f_2, w_2, \cdots, w_{m-1}, f_m, w_m \rangle$ will then be a path in G from u to w, showing that w is reachable from u. □

Since the equivalence classes determined by any equivalence relation partition the underlying set, it follows that for this reachability equivalence relation, the equivalence class of a vertex v, which is the set of all vertices that are reachable from v, could not contain any other vertices it is not connected. On the other hand, if u, w are also in this equivalence class, then u and w are both reachable from v, so by symmetry and transitivity, it follows that w is reachable from u. This shows that the equivalence class is connected, and thus indeed is the component of v.

EFR 8.19: Part (a): $\mathrm{diam}(P_n) = n - 1$ *Proof:* Recall that P_n consists of n vertices: $v_1, v_2, \cdots, v_n$ with edges between successive vertices.   It follows that the only paths in P_n are sequences of successive vertices, so that $\mathrm{diam}(P_n) = \mathrm{dist}(v_1, v_n) = n - 1$. □

Part (b): $\mathrm{diam}(Q_n) = n$ *Proof:* Viewing the vertices of Q_n as length n bit strings, we recall that two vertices are adjacent if, and only if their bit strings differ in exactly one bit. It thus follows that $\mathrm{dist}(000\cdots0, 111\cdots1) = n$. Indeed, since the bit strings differ in all of their bits, any path joining these two vertices would need to have length at least n, on the other hand the following path:
$$< 000\cdots0, 100\cdots0,\ 110\cdots0,\ 1110\cdots0,\ 111\cdots10,\ 111\cdots1 >$$
has length n. A similar construction shows that the distance between any pair of vertices is the number of bit entries in which they differ. This proves (more than) $\mathrm{diam}(Q_n) = n$. □

EFR 8.20: Part (a): There are three shortest paths in the graph joining c to j: $< c, e, \ell, k, j >$, $< c, e, h, k, j >$, $< c, e, i, k, j >$ and thus $\mathrm{dist}(c, j) = 4$.

Part (b): The longest path needed to join two vertices is 6, e.g., $\mathrm{dist}(g, j) = 6$, and this is thus the diameter.

Part (c): $\mathrm{ecc}(c) = 4$, $\mathrm{ecc}(j) = 6$

Part (d): $\mathrm{ecc}(e) = 3$, and all other vertices have larger eccentricities, thus the radius is 3.

Part (e): By Part (d) e is the unique central vertex.

EFR 8.21: Part (a): The adjacency matrix is $A = \begin{bmatrix} 0 & 1 & 0 & 0 & 1 \\ 0 & 0 & 1 & 0 & 0 \\ 0 & 0 & 0 & 1 & 1 \\ 1 & 0 & 0 & 0 & 0 \\ 1 & 0 & 0 & 0 & 0 \end{bmatrix}$.

$$I + A + A^2 = \begin{bmatrix} 2 & 1 & 1 & 0 & 1 \\ 0 & 1 & 1 & 1 & 1 \\ 2 & 0 & 1 & 1 & 1 \\ 1 & 1 & 0 & 1 & 1 \\ 1 & 1 & 0 & 0 & 2 \end{bmatrix}, \quad I + A + A^2 + A^3 = \begin{bmatrix} 2 & 2 & 1 & 1 & 3 \\ 2 & 1 & 1 & 1 & 1 \\ 2 & 2 & 1 & 1 & 3 \\ 2 & 1 & 1 & 1 & 1 \\ 2 & 1 & 1 & 0 & 2 \end{bmatrix},$$

$$I + A + A^2 + A^3 + A^4 = \begin{bmatrix} 5 & 2 & 2 & 1 & 3 \\ 2 & 3 & 1 & 1 & 3 \\ 4 & 2 & 3 & 1 & 3 \\ 2 & 2 & 1 & 2 & 3 \\ 2 & 2 & 1 & 1 & 4 \end{bmatrix}.$$

It follows from Algorithm 9.2 that the diameter of the shipping network is 4. Note that the final (fourth power of the incidence matrix needed to be added because there was a single nonzero entry (the row 5, column 4 entry) in $I + A + A^2 + A^3$. The meaning is that we really need four flights to get from Guam to Shanghai, but any other journey can be done in at most three flights.

Part (b): Theorem 8.6 tells us that the row 2 column 4 entry of the matrix $A + A^2 + A^3$ is the number of ways that we can ship from city 2 (Yokohama) to city 4 (Shanghai) using at most three flights. Examining the figure, we easily see that there is no direct flight, one route to ship using two flights (2 → 3 → 4), and no way to ship using exactly three flights, so it follows that this entry is $0 + 1 + 0 = 1$, and this agrees with the corresponding entry of the matrix $I + A + A^2 + A^3$ of Part (a). (The (2,4) entry of the identity matrix is 0, so does not change the entry of $A + A^2 + A^3$.)

EFR 8.22:   Part (a): The following <u>modifications</u> to Algorithm 8.2 will yield an algorithm that computes the eccentricity of an inputted vertex:

Input: Add a vertex number i as an additional input.

Output: Change the output to $\mathrm{ecc}(i)$.

Step 1: If the ith row of A has all nonzero entries (except the ith entry), output ecc$(i) = 1$ and terminate the algorithm.

Step 2: Continue adding increasing powers of A: $I + A + A^2 + \cdots + A^k$, until either $k = n$, or the ith row of the matrix sum has no zero entries except possibly the ith entry.

Part (b): All of the needed sums of increasing powers of the adjacency matrix A were already computed in the solution of Example 8.21 ($I = A^0$ was not added, but this makes no difference since we care only about nondiagonal entries of the sums).   From these calculations we see that Step 2 would terminate when $k = 3$. Thus ecc$(Tokyo) = 3$, as is easily confirmed by inspecting Figure 8.40.

EFR 8.23:   Cut vertices: t, v, x; Bridges: $\{s,t\}, \{v,x\}$

EFR 8.24:   Answer: $n-1$ Reason: If we delete all $n-1$ edges incident to a specific vertex, there will be two components: the isolated vertex and a copy of K_{n-1}. Thus we have given an edge cut of size $n-1$. It remains to show that if we delete fewer than $n-1$ edges from K_n, the resulting graph, which we denote by G, will be connected. Consider any ℓ vertices of G, with $1 \le \ell \le n/2$. If these vertices were not connected to the other vertices of G, this would mean that at least $\ell \cdot (n-\ell)$ edges of K_n were deleted. Since the function $f: \mathbb{R} \to \mathbb{R}$ defined by $f(x) = x(n-x)$ is a downward opening parabola with x-intercepts at $x = 0$ and $x = n$, by symmetry it follows that $f(x)$ increases as x moves from 0 to $n/2$ (and then decreases as x moves from $n/2$ onward to the right).   In particular, $\ell \cdot (n-\ell) = f(\ell) \ge f(1) = n-1$. This is a contraction to the assumption that fewer than $n-1$ were removed. $\square$

EFR 8.25:   *Proof:*   Let u, v be the endpoints of f, and assume f is not a bridge. Then $G - f$ is connected, so there is a path P in $G - f$ that joins u and v. By Proposition 8.5, we may assume that P is simple and contains no repeated vertices. Adjoining the edge f to (either) end of the path P thus forms a cycle in G. $\square$

EFR 8.26:   There are three nonisomorphic five vertex trees; representatives of these three isomorphism classes are shown below:

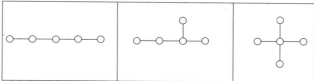

EFR 8.27:   *Proof:*   Consider a simple graph G with n vertices that has k components. If G has any cycle, delete any edge e_1 that is part of such a cycle, and consider the resulting edge deletion graph $G_1 \triangleq G - e_1$.   If G_1 has a cycle let e_2 be any edge of G_1 that is part of such a cycle, and let $G_2 \triangleq G_1 - e_2$. We continue this process of deleting edges that are parts of cycles until we arrive at a graph G_ℓ resulting from G from deleting ℓ edges and that has no cycles. Since at each step only an edge belonging to a cycle was deleted, it follows that no component of G was broken. Thus, G_ℓ is a forest with k components, so by Propositon 8.9, it will have $n - k$ edges. But G has at least as many edges as G_ℓ. $\square$

EFR 8.28:   Part (a): *Proof of* $3 \Rightarrow 2$: Since T has no cycles it is a forest on n vertices. Since it has $n-1$ edges, it must have one component by Propositon 8.9(a), i.e., it is connected. $\square$

Part (b): *Proof of* $5 \Rightarrow 2$: Since every edge is a bridge, T can have no cycles, so it's a tree and by Proposition 8.9(a) it has $n-1$ edges. □

EFR 8.29: Part (a): The eccentricity of the root of a rooted tree is the height of the tree.
Part (b): The shortest path between two vertices in a rooted tree is obtained finding their first common ancestor, and then juxtaposing the path from one vertex up the this ancestor with the path from the ancestor down to the other vertex.

EFR 8.30: Binary tree representations of the algebraic expressions:

$$(x+2y)(x-2y), \qquad x^2+y^2+z^2, \text{ and} \qquad \sin(\cos(x+2)-3y).$$

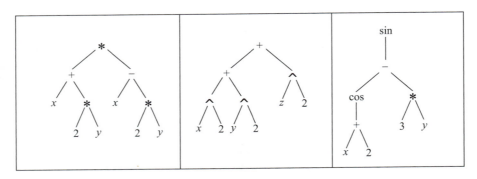

EFR 8.31: Part (a): $\lceil \log_2 73 \rceil = \lceil 6.18... \rceil = 7$.
Part (b): Here is the general rule for computing the number of byes: Letting N be the number of players in a (single-elimination) tournament and h be the height of a corresponding binary tournament tree, we may write $N = 2^{h-1} + j$, where $1 \le j \le 2^{h-1}$. The number of byes will be $2^{h-1} - j$. (*Proof:* Each of the depth $h-1$ vertices of the tree must have at least one child, so the first 2^{h-1} vertices of depth $h-1$ are assigned, say as left children. The remaining j vertices are usually assigned from left to right as corresponding left children. This leaves $2^{h-1} - j$ byes. □)
For $N = 73 = 64 + 9$, we will have $64 - 9 = 55$ byes.
Part (c): In any single elimination tournament there is exactly one loser for each match. Since each player/team, with the exception of the overall champion, loses exactly once, it follows that the number of matches is one less than the number of players/teams. For the example at hand, there are $73 - 1 = 72$ matches.

EFR 8.32: Part (a): The process is modeled by a complete 5-ary tree, where the vertices are the individual people, and the edges represent the emails that were sent. By Theorem 8.11, after h days, the number of vertices will be $(5^{h+1} - 1)/4$, which will be one more than the number of emails that were sent (since the root is the only vertex that did not receive an email). Thus, the number of days needed in order for 5000 people to receive emails will be the smallest positive integer h such that $(5^{h+1} - 1)/4 \ge 5001$. Either by solving this equation directly, or by simply substituting $h = 1, 2, 3, ...$ until the quantity exceeds 5000, we see that 6 days are required (after which 19,530 emails will have been sent).
Part (b): The number of people who sent emails will be the number of internal vertices i, so by formula (4) of Theorem 8.18, this number satisfies $n = 5i + 1$, where n is the total number of vertices, which by Part (a) is 19,531. It follows that $i = 3906$.

EFR 8.33:

Processed Vertices:

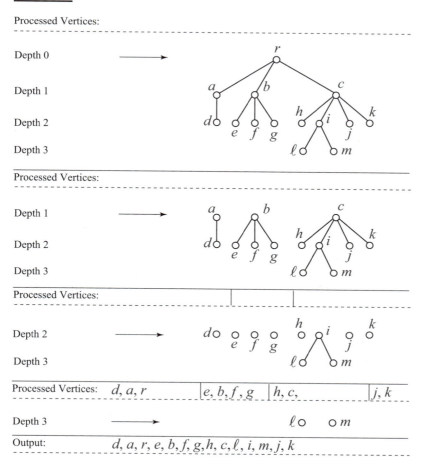

| Processed Vertices: | d, a, r | e, b, f, g | $h, c,$ | j, k |

Depth 3   ⟶ ℓ ○   ○ m

Output: $d, a, r, e, b, f, g, h, c, \ell, i, m, j, k$

EFR 8.34: In each part we will make use of the abbreviated schemes of the preceding note, and we will use the following notation: For a vertex x of any rooted tree, in the course of applying any one of these iterative algorithms, $T(x)$ denotes the unprocessed subtree that is rooted at x and includes all descendants of x (and the edges connecting these to x). The implementations below process all child subtrees in parallel, according to increasing depths of the vertices.

Part (a): Preorder:
$[r, T(a), T(b)] \rightarrow [r, a, c, T(d), b, T(e), T(f)] \rightarrow [r, a, c, d, T(g), T(h), b, e, T(i), f, j, k]$
$\rightarrow [r, a, c, d, g, \ell, m, h, n, o, b, e, i, p, q, f, j, k]$

Part (b): Inorder:
$[T(a), r, T(b)] \rightarrow [c, a, T(d), r, T(e), b, T(f)] \rightarrow [c, a, T(g), d, T(h), r, T(i), e, b, j, f, k]$
$\rightarrow [c, a, \ell, g, m, d, n, h, o, r, p, i, q, e, b, j, f, k]$

Part (c): Preorder:
$[T(a), T(b), r] \rightarrow [c, T(d), a, T(e), T(f), b, r] \rightarrow [c, T(g), T(h), d, a, T(i), e, j, k, f, b, r]$
$\rightarrow [c, \ell, m, g, n, o, h, d, a, p, q, i, e, j, k, f, b, r]$

EFR 8.35:

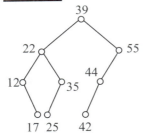

EFR 8.36:

Vertex	Parent	Vertex	Child 1	Child 2	Child 3	Child 4
r	NULL	r	a	b	c	NULL
a	r	a	d	NULL	NULL	NULL
b	r	b	e	f	g	NULL
c	r	c	h	i	j	k
d	a	d	NULL	NULL	NULL	NULL
e	b	e	NULL	NULL	NULL	NULL
f	b	f	NULL	NULL	NULL	NULL
g	b	g	NULL	NULL	NULL	NULL
h	c	h	NULL	NULL	NULL	NULL
i	c	i	ℓ	m	NULL	NULL
j	c	j	NULL	NULL	NULL	NULL
k	c	k	NULL	NULL	NULL	NULL
ℓ	i	ℓ	NULL	NULL	NULL	NULL
m	i	m	NULL	NULL	NULL	NULL

CHAPTER 9: GRAPH TRAVERSAL AND OPTIMIZATION PROBLEMS

EFR 9.1: By Theorem 9.1, placing one new bridge anywhere (so long as it connects two of the four land masses) in the map of Figure 9.2 would result in making two of the four vertices in the graph of Figure 9.3 increase their degrees by one, and hence have even degrees. Thus, by Theorem 9.1, the resulting graph has an Euler path. To make it Eulerian (and thus contain an Euler tour), we would need to install a second new bridge (or tear down an appropriate existing bridge) to make all of the degees of the resulting graph of Figure 9.3 to be even.

EFR 9.2: We proceed through the iterative Step 2 of Algorithm 9.1A and show the edge sequences inserted into TourEdges each time the CurrentVertex and CurrentInsertionPoint change:

CurrentVertex = a, CurrentInsertionPoint = End:
TourEdges: $\{a,b\},\{b,c\},\{c,d\},\{d,a\},\{a,e\},\{e,b\},\{b,f\},\{f,d\},\{d,h\},\{h,c\},\{c,g\},\{g,a\}$

CurrentVertex = e, CurrentInsertionPoint = after first edge $\{a,e\}$
TourEdges: $\{a,b\},\{b,c\},\{c,d\},\{d,a\},\{a,e\},\{e,b\},\{b,f\},\{f,d\},\{d,h\},\{h,c\},\{c,g\},\{g,a\}$

$$\overbrace{\{e,f\},\{f,h\},\{h,g\},\{g,e\}}$$

The following Euler tour has thus been produced: $<a,b,d,c,a,e,f,h,g,e,b,f,d,h,c,g,a>$.

EFR 9.3: Since each edge gets removed from UnusedEdges as soon as it is appended to TourEdges, and since the processing of each edge occurs in constant $O(1)$ time, it follows that the algorithm will work in linear time on the number of edges $|E|$, i.e., with complexity $O(|E|)$. We have assumed that the edges have been inputted in such a way (e.g., lexicographically) so that our computing can locate the first edge in UnusedEdges that is incident to a specific vertex (EndVertex) in a negligible amount of time.

EFR 9.4: Part (a):

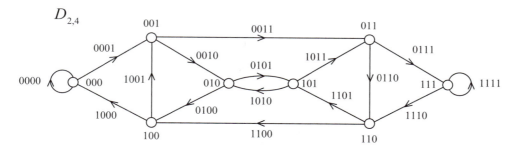

Part (b): We follow Algorithm 9.2: An Euler tour in $D_{2,4}$ starting at 000 is easily constructed (by inspection, or using the directed version of Algorithm 9.1A):

$$<000,000,001,011,111,111,110,100,001,010,101,011,110,101,010,100,000>$$

Reading off the first characters of all but the last element of the above sequence of de Bruijn edges produces a desired de Bruijn sequence: $<0,0,0,0,1,1,1,1,0,0,1,0,1,1,0,1>$.

EFR 9.5: The Petersen graph is not Hamiltonian. We will prove this by assuming that a Hamilton tour exists, starting to construct one, and eventually reaching a contradiction. This method previews some ad hoc strategies that will be elaborated on at the end of this section. To facilitate our arguments, we call the "outer" edges of the Petersen graph to be the five edges of the outer pentagon, the "inner

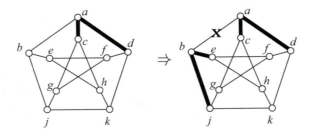

edges" to be the five edges that join two of the five vertices in the interior, and the "cross edges" to be those five edges that connected a perimeter vertex to an interior vertex. If the Petersen graph did admit a Hamilton tour, then such a tour would need to include one outer edge and an adjacent cross edge. By symmetry, we may assume the tour includes the edges *ac* and *ad*; see the first figure below, where these edges have been thickend to indicate their inclusion in the Hamilton tour. Next, since a Hamilton tour (or any cycle subgraph) passes through exactly two edges of at each of its vertices, and since it already passes through two edges incident to vertex *a*, it follows that the edge *ba* will not be part of the tour (we indicate this by putting an "X" over this edge in the second figure). Since there are only two remaining edges incident to vertex *b*, the Hamilton tour must pass through both of them, and these edges are thickened in the second figure.

In order to continue this strategy, we need to separate into some cases.

Case 1: The tour passes through *cg*:
We then know it cannot pass through *ch* (since there are already two tour edges incident to vertex *c*). It next follows that the tour must pass through the remaining two edges *eh*, *hk*, incident to *h*; see the first figure below. Next, notice that the tour cannot be part of Hamilton cycle, since it would already form a smaller cycle from the edges we have so far (*jkhebj*). So we can delete this edge, forcing the Hamilton tour to include the edges *dk* and *gj*, which we have darkened in the second figure below. The edges found so far to be part of a Hamilton tour, now form a cycle on a smaller set of vertices—a contradiction!

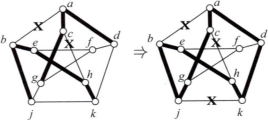

Case 2: The tour passes through *ch*:
A similar argument to that in Case 1 is summarized by the figure and likewise leads to a contradiction.

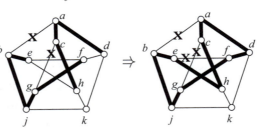

EFR 9.6: A self-loop could never be part of a Hamilton tour for any Hamiltonian graph with at least two vertices, since its inclusion would mean that the tour visited the endpoint more than once. Similarly, for any Hamiltonian graph with at least three vertices, a pair of parallel edges could never both be part of a Hamilton tour (in either direction) since this would mean that the tour would visit both endpoints at least twice.

EFR 9.7: We first construct a Hamilton tour for Q_4 using the indicated inductive scheme. We begin by creating two copies of Q_3 and in each we draw in Hamilton paths obtained from the Hamilton tour of Figure 9.15 by deleting the vertical edge that joins 101 to 111. In the corresponding merged drawing of Q_4, we then join the endpoints of the two Hamilton paths to produce a Hamilton tour of Q_4, as shown in the figure below.

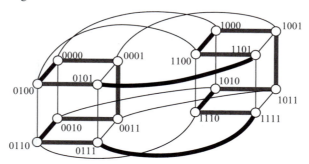

Following this tour (and starting at the vertex 0000) produces the following Gray code of order 4:

$$< 0000, 0001, 0011, 0010, 0110, 0111, 1111, 1110, 1010, 1011, 1001, 1000, 1100, 1101, 0101, 0100 >$$

EFR 9.8: Part (a): Here is a Hamilton path: $< m, \ell, g, f, a, h, i, b, c, j, k, d, e >$.

Part (b): *Subcase 1b:* C contains gh, but not gk: So C must also contain gf, and by Part (2) of the proposition, we can also delete the edges ha, hb, hi. From these deletions, it now follows that C contains fa, ab, bc, ci, and (since ℓj can be deleted) jd, de. From here, in turn, we may delete kj, and conclude that C contains ke, which completes the following Hamilton tour:

$$< \ell, k, e, d, j, i, c, b, a, f, g, h, \ell >.$$

EFR 9.9: Part (a): Suppose that C is a Hamilton tour for G. By Proposition 9.8(1), C must include the edges cf, ce, hf, ha. By Propostion 9.8(2), we may delete gf, gi, and since now gi, ai are the only two edges remaining that are incident to vertex i, they must be also included. Now that vertex a has two of its incident edges included, we may delete the other two: ag, ab. From the latter deleted edge, we infer that C must include be and bd. Since vertex g now only has one included edge, the only remaining edge gd must be included. We now have formed the following Hamilton tour of G (thus showing G is Hamiltonian): $< a, i, g, d, b, e, c, f, h, a >$.

Part (b): Suppose that C is a Hamilton tour for H. By Proposition 9.8(1), C must include the edges ad, dg, cf, fe, be, eh. Now, exactly one more of the remaining unused edges incident to vertex a must be included. By symmetry, we may assume that ac is included. This forces ab and bc to be deleted (since now vertex c already has two of its incident edges included). Now vertex b has only one of its incident edges included, and the other two have been deleted. It follows that H is not Hamiltonian.

EFR 9.10: The result of Prim's algorithm is illustrated in the figure below. When vertex d was discovered, the edge from f was chosen over the same weight edge from i because of the order scheme. The weight of this minimum spanning tree (indicated by thick edges) is 30.

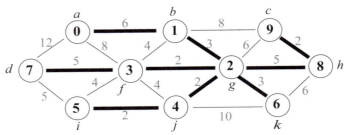

EFR 9.11: The figure below shows the result of Dijkstra's algorithm with the usual notation conventions: discovery numbers are indicated in bold in the vertex circles, distances to the starting vertex are indicated in square brackets, and the minimum distance tree's edges are drawn with thick edges with directions going from parent to child.

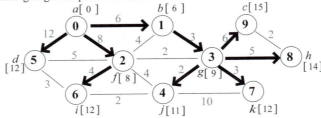

EFR 9.12: The figure below shows the result of Dijkstra's algorithm with the usual notation conventions: discovery numbers are indicated in bold in the vertex circles, distances to the starting vertex are indicated in square brackets, and the minimum distance tree's edges are drawn with thick edges with directions going from parent to child.

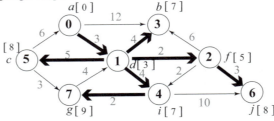

EFR 9.13: Part (a): Run Dijkstra's algorithm on the graph G with all edge weights taken to be one, and with any starting vertex. The graph will be connected if, and only if all distances are finite.

Part (b): Run Dijkstra's algorithm on the graph G with all edge weights taken to be one, and with each of the vertices taken as the starting vertex. This will compute the distances between all pairs of vertices in G, so the maximum of these distances will be the diameter of G.

Part (c): Run Dijkstra's algorithm on the graph G with all edge weights taken to be one, and with any starting vertex. All vertices whose distances are finite will constitute a component of G. If other vertices remain, run Dijkstra's algorithm again, but with a starting vertex taken to be any vertex not in the component that was just determined. The vertices with finite distances will constitute another component of G. Continue iterating this process until all componenets have been determined.

EFR 9.14: The following figure shows the resulting depth-first search tree (left) and the breadth-first search tree (right), with root (starting) vertex a. Discovery numbers are indicated in bold, and tree edges are drawn with thick lines.

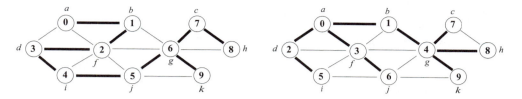

EFR 9.15: The figure below shows the resulting depth-first search directed tree (left) and the breadth-first search directed tree (right), with root (starting) vertex *a*. Discovery numbers are indicated in bold, and tree edges are drawn with thick lines.

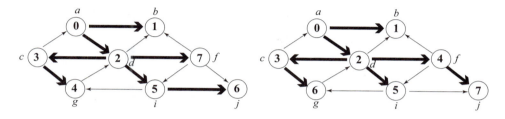

EFR 9.16: If we apply a TSP algorithm to the edge-weighted graph as described in the suggestion, a minimum weight tour will avoid the non-edges of *G* exactly if the original graph has a Hamilton tour. Put differently, if the original graph were not Hamiltonian, then a solution of the TSP would need to use some of those weight two edges that are not edges of *G*. Thus, the graph will be Hamiltonian if, and only if the output of any TSP algorithm is a tour with all edge weights being equal to one.

EFR 9.17: We use the vertex numbering scheme of Example 9.17. The mechanics of the furthest insertion heuristic will be the same as those of the nearest insertion heuristic in Example 9.17, except at each iteration, a furthest away city is chosen. Summarize the results of the steps:

Step 1: Same initialization as in Example 9.17: TOUR = <1>, WEIGHT = 0, and REMAIN = {2, 3, 4, 5}.

Step 2: The furthest away (pricewise) city from Honolulu is Guam (#4) at $1200, so we update as follows: TOUR = <1, 4, 1>, WEIGHT = 2400, and REMAIN = {2, 3, 5}.

Step 3: First Iteration: Furthest city from existing tour: Vancouver (#3), at a price of $650. The price differential in placing this city either between 1 and 4 or between 4 and 1 is the same, namely $650 + $2100 – $1200 = $1550. We randomly choose to put it between 1 and 4 to get the next cycle TOUR = <1, 3, 4, 1>, whose ticket price would be $2400 (previous cycle price) + $1550 (price differential) = $3950 = WEIGHT. We now have two remaining cities: REMAIN = {2, 5}.

Step 3: Second Iteration: Furthest city from existing tour: Tokyo (#2), at a price of $600. The price differentials of the possible insertion locations are as follows:
Between 1 and 3: $750 + $1300 – $650 = $1400.
Between 3 and 4: $1300 + $600 – $2100 = –$200.
Between 4 and 1: $600 + $750 – $1200 = $150.
Thus by inserting between 3 and 4, we actually will bring the price down: TOUR = <1, 3, 2, 4, 1>, WEIGHT = $3750, and REMAIN = {5}.

Step 3: Final Iteration: The price differentials of the possible insertion locations (for city #5) are as follows:
Between 1 and 3: $450 + $500 – $650 = $300.
Between 3 and 2: $500 + $1050 – $1300 = $250. (Minimum)
Between 2 and 4: $1050 + $1800 – $600 = $2250.
Between 4 and 1: $1800 + $450 – $1200 = $1050.
TOUR = <1, 3, 5, 2, 4, 1>, WEIGHT = $4000.

EFR 9.18: In the left figure below, a path is shown from s to t (thick edges) in the augmented digraph for the given network flow. The right figure below shows a resulting improved flow from the flow of Figure 9.48(b), using excess capacities from the path in the left figure.

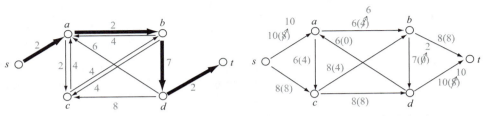

EFR 9.19: In the left figure below, a path is shown from s to t (thick edges) in the augmented digraph for the given network flow. The right figure below shows a resulting improved flow from the flow of Figure 9.49, using excess capacities from the path in the left figure.

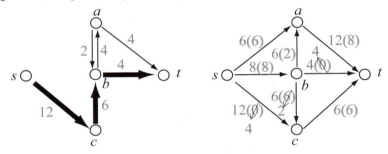

EFR 9.20: In the first iteration of the scan and labeling Step 3, the following labels are created: Label(a) = (s, +, 1), Label(g) = (s, +, 10); Label(b) = (a, +, 1), Label(f) = (a, +, 1); Label(h) = (g, +, 10); Label(c) = (h, +, 10); Label(d) = (h, –, 6), Label(t) = (h, +, 1). The resulting augmented flow is shown in the figure below. The labels for the second iteration of Step 3 are: Label(a) = (s, +, 2), Label(c) = (s, +, 2). Since t did not get labeled, it follows that the flow in the figure is maximum; its value is 8. Note that the labeled vertices in the final iteration along with s will determine a corresponding minimum cut: ({s,a,c},{b,d,t}).

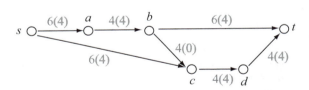

EFR 9.21: In the first iteration of the scan and labeling Step 3, the following labels are created: Label(a) = (s, +, 15), Label(c) = (s, +, 8), Label(f) = (s, +, 16); Label(b) = (a, +, 10); Label(d) = (c, +, 8); Label(g) = (f, +, 12); Label(t) = (b, +, 8). The resulting augmented flow is shown in the first figure below; it has value 8.
Second iteration labels: Label(a) = (s, +, 7), Label(c) = (s, +, 8), Label(f) = (s, +, 16); Label(b) = (a, +, 2); Label(d) = (c, +, 8); Label(g) = (f, +, 12); Label(t) = (d, +, 8). The resulting augmented flow is shown in the second figure; it has value 8.

Third iteration labels: Label(a) = (s, +, 7), Label(f) = (s, +, 16); Label(b) = (a, +, 2); Label(c) = (f, +, 16), Label(g) = (f, +, 12); Label(d) = (c +, 2); Label(t) = (g, +, 10). The

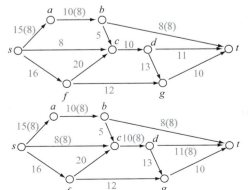

resulting augmented flow is shown in the third figure; it has value 26.

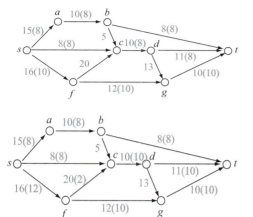

Fourth iteration labels: Label(a) = (s, +, 7), Label(f) = (s, +, 6); Label(b) = (a, +, 2), Label(c) = (f, +, 6), Label(g) = (f, +, 2); Label(d) = (c +, 2); Label(t) = (d, +, 2). The resulting augmented flow is shown in the fourth figure; it has value 28.

Fifth iteration labels: Label(a) = (s, +, 7), Label(f) = (s, +, 4); Label(b) = (a, +, 2), Label(c) = (f, +, 4), Label(g) = (f, +, 2). The resulting augmented flow is shown in the fourth figure; it has value 28. Since t did not get labeled, it follows that the flow in the figure is maximum; its value is 28.

EFR 9.22: All edges emanating from the source will be directed away from s, and assigned a capacity of 1, all edges terminating at the sink will be assigned a capacity of 1, and will be directed towards t. For all other remaining *internal* edges, they will be replaced by two oppositely directed edges with capacity 1. If the Ford-Fulkerson algorithm is applied to this flow network (which is different from those considered thus far since the underlying graph is not simple), for any pair of oppositely directed edges, only one can be assigned a flow because vertices can be labeled only once in the algorithm. Since the proof of Theorem 9.13 is easily seen to extend to be valid in this setting, it follows that a maximum flow found by the Ford-Fulkerson algorithm will correspond to a maximum number of disjoint paths from s to t.

EFR 9.23: Part (a): There is only one maximum matching: $\{\{a,h\},\{c,f\},\{b,g\},\{d,i\}\}$.
Part (b): Here is a maximal matching that is not a maximum matching: $\{\{b,g\},\{c,h\},\{d,i\}\}$.

EFR 9.24: The flow network is illustrated in the figure below.

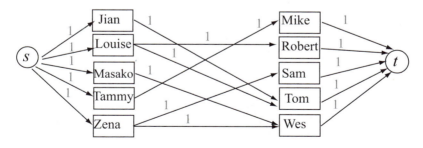

To simplify notation in the implementation of the algorithm, we use the following vertex numbering scheme: the five women vertices have numbers 2–6 (assigned in alphabetical order), the five men vertices have numbers 7–11. In the first iteration of the scan and labeling Step 3, the following labels are created:
Label(2) = (s, +, 1), Label(3) = (s, +, 1), Label(4) = (s, +, 1), Label(5) = (s, +, 1), Label(6) = (s, +, 1); Label(10) = (2, +, 1); Label(8) = (3, +, 1); Label(11) = (4, +, 1); Label(7) = (5, +, 1); Label(9) = (6, +, 1); Label(t) = (10, +, 1).

The resulting flow is completely described by the internal edge {2,10}, i.e., matching Jian and Tom. In the second iteration, the flow is augmented by 1 by the internal edge {3,8}, i.e., additionally matching Louise and Robert. In the third iteration Masako and Wes are matched. The fourth iteration matches Tammy and Mike, and in the fifth iteration Zena and Sam. This marriage problem thus has a great solution—everyone gets married.

EFR 9.25: Given a set of jobs J and a set of people P, all jobs in J can be assigned to qualified people in P (one job to a one person) if, and only if for each subset $K \subseteq J$, the corresponding subset of people $S \subseteq P$, that are qualified to do at least one job in K is at least as large as K, i.e., $|S| \geq |K|$.

CHAPTER 10: RANDOMIZED SEARCH AND OPTIMIZATION ALGORITHMS

EFR 10.1: We will provide pseudocode for running the algorithms in this exercise.

Step 1: We first download the distance matrix for this problem and store it as "AUlysses22"

Step 2: Tours will be stored as vectors of the form Tour = $[1, t_2, t_3, \cdots, t_N, 1]$, i.e., we may assume that the first and last city of the tour is 1, and $t_2, t_3, \cdots, t_N$ is a permutation of the remaining cities $2, 3, \cdots, N$. The following program computes the distance of an inputted tour:

```
dist = TourDistance(Tour)
INITIALIZE dist = 0
FOR i = 2 TO N
    (record distance of two cities and add to dist)
    disti = AUlysses22(Tour(i-1), Tour(i))
    dist = dist + disti
END FOR
OUTPUT: dist
```

Step 3: Program and run the Hill Climbing/2-vertex permuation Algorithm (of this EFR). We will use 100 iterations.

```
SET T = 100   (number of iterations)
INITIALIZE count = 0   (will count number of tour distances computed)
INITIALIZE OverallRecordTour = 1, overallRecordDist = Infinity
FOR j = 1 TO T
    SET flag = 0   (will be used to exit 2-vertex perm procedure when it stagnates)
    (generate random perm of 2,3, ..., N—see Chapter 6)
    SET PrePerm = a random permutation of 2,3, ...,N
    (form tour)
    SET Tour = [1 PrePerm 1]
    (compute distance of tour using program of Step 2)
    dist = TourDistance(Tour)
    UPDATE count = count + 1
    (initialize record distance and tour)
    SET recordDist = dist
    SET RecordTour = Tour
    SET recordDistOld = recordDist
    (Next, we start the 2-vertex iterative improvement. We go through all pairs of vertices
    different from vertex #1, of which there are O(n^2)), and look at the tour obtained by
    switching them.)
    WHILE flag = 0
        FOR a = 2 TO N-2,   FOR c = a+1 TO N
```

```
            SET Tour2Opt = Tour
            SET Tour2Opt([a c]) = Tour([c a])
            (compute distance of this 2Opt tour)
            dist = TourDistance(Tour2Opt)
            UPDATE count = count + 1
            (compare with best tour so far and update if improvment)
            IF dist < recordDist
                UPDATE recordDist = dist
                UPDATE RecordTour = Tour2Opt
                UPDATE Tour = Tour2Opt
            END IF
          END c FOR, END a FOR
          IF recordDist < recordDistOld  (2-opt did not stagnate)
            UPDATE recordDistOld = recordDist;
          ELSE (2-opt stagnated)
            UPDATE flag = 1
          END IF
       END WHILE
    IF recordDist < overallRecordDist
       UPDATE overallRecordDist = recordDist
       UPDATE OverallRecordTour = RecordTour
    END IF
    HillClimbing2VertexPerm(j) = recordDist
END j FOR
```

A run of this program shows that 439,375 tour evaluations were performed. The minimum, maximum, mean and standard deviation of the vector `HillClimbing2VertexPerm` were computed and will be included in a table that follows.

Step 4: Program and run the Hill Climbing/2-opt permuation Algorithm (of this EFR). We will use 100 iterations.

The structure for this program is similar to that of Step 3, the only change in that we replace the 2-vertex permuation local search in the while loop with a 2-opt search. Here is a basic outline on how this 2-opt search can be coded:

We go through all pairs of edges that do not share a common endpoint (there are $O(n^2)$), and look at the tour obtained by switching these edges in the current tour `Tour` as follows:

```
SET e1 = [Tour(a), Tour(a+1)]
SET e2 = [Tour(b), Tour(b+1)],
 (where we assume that 1 <=a <=N-2, and a+1 < b <= N)
```
The 2-opt tour resulting from removing these edges from Tour is the tour Tour2Opt, defined as follows:
```
Tour2Opt[1,2, ..., a] = Tour[1,2, ..., a]
Tour2Opt[a+1] = Tour[b]
Tour2Opt[a+2, ..., b] = Tour[b-1, b-2, ..., a+1]
Tour2Opt[b+1, ...,N+1] = Tour[b+1, ...,N+1]
```

A run of the resulting program shows that 345,667 tour evaluations were performed. The minimum, maximum, mean and standard deviation of the vector of record distances were computed and will be included in a table that follows.

Step 4: Program and run the random search algorithm. We will use 100 iterations with 5000 random tour generations each. This will result in a total of 5 million tour distance evaluations, which is comparable, but greater than the corresponding numbers for the hill climbing algorithms.

```
    SET T = 100  (number of iterations)
    FOR t = 1 TO T
    INITIALIZE count = 0 (will count number of tour distances computed)
    INITIALIZE RecordTour = 1, recordDist = Infinity
```

```
   FOR j = 1 TO 5000
      (generate random perm of 2,3, ..., N—see Chapter 6)
      SET PrePerm = a random permutation of 2,3, ...,N
      (form tour)
      SET Tour = [1 PrePerm 1]
      (compute distance of tour using program of Step 2)
      dist = TourDistance(Tour)
      UPDATE count = count + 1
      IF dist < recordDist
      UPDATE recordDist = dist
      UPDATE RecordTour = Tour
   END IF
   END j FOR
   RandomTour(t) = recordDist
END t FOR
```

The statistical results of runs of the above three programs are shown in the table below. Notice that although the 2-vertex permutation hill climbing significantly outperformed the random search, its best value was still about 37% above the optimal value. Notice also that the 2-opt hill climbing algorithm actually found an optimal tour!

Algorithm:	Minimum	Maximum	Mean	Standard Deviation
Hill Climbing/2-Vertex Perm	9632	16,074	12,723	1282
Hill Climbing/2-Opt	7035*	7463	7198.6	96.4
Random Search	20,867	24,886	23,369	845

*Optimal Value

EFR 10.2: (a) In case no two of the removed edges share a common endpoint, it follows the the removal of the three edges will leave disconnected paths, each of length at least 1 (so each having at least two vertices). We schematically illustrate the original edges relating to these disconnected paths in the figure below. (The three paths are $\vec{P_4}\vec{P_1}$, $\vec{P_2}$, $\vec{P_3}$. The path $\vec{P_4}\vec{P_1}$ was split up since the figure is illustrated on a line segment.) Below, the five possible ways to reattach these three edges to the left/right ends of these cycles are indicated.

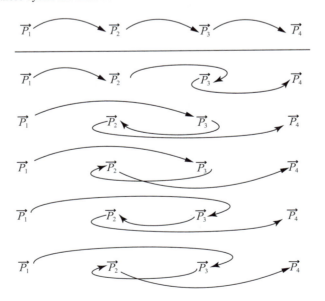

(b) The case in which two of the removed edges share a common vertex is different than the first case since one of the three components that remain will consist of a single vertex (labeled as v in the figure below), so there is no difference in reattaching a removed edge to the left or right side. The three possible new reattachements are shown in the figure below.

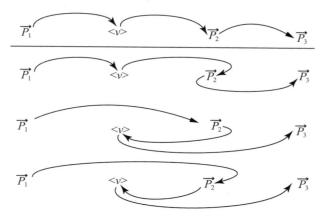

(c) The third case is shown in the figure below, there is just one new way to reattach the edges.

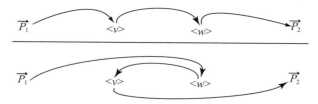

EFR 10.3: Here is the pseudocode for the program: `M = MatingPool(P,f)`

```
Initialize:   P  (initial population matrix)
              f  (fitness function)
              M = P  (initially set mating pool matrix = P)
              N = number of rows of P  (size of population)
(compute fitness levels of initial population)
FOR i = 1 TO N
     f(i) = f(bin2int(P(i,:)))
END (i FOR)
(from the just computed fitness vector, we form the vector of corresponding fitness proportions)
Prop = f/sum(f)
(Next we invoke the program of Exercise for the Reader 6.15 to randomly select the chromosome
numbers for the mating pool, according to Prop)
ind = GeneralFiniteSimulator(10,[1 2 ... N],Prop)
(Finally, we run through the indices of this vector to install the corresponding bit string chromosomes
as the rows of M)
FOR i = 1 TO N
     M(i,:) = P(ind(i),:) (i.e., set the ith row of M to be the ind(i)th row of P)
END (i FOR)
OUTPUT:   M
```

EFR 10.4: Here is the pseudocode for the mutation operator: `P = Mutation(C,pm)`

```
(Read off dimensions of C)
SET N = # of rows of C
```

```
SET s = # of columns of C
INITIALIZE output matrix: P = C
FOR i=1 TO N, FOR j=1 TO s
        IF rand < pm
            UPDATE P(i,j) = 1 - C(i,j)
        END IF
END (j FOR), END (i FOR)
OUTPUT:   P
```

EFR 10.5:   Here is the pseudocode for the mutation operator: `P = Inversion(C,pinv)`

(Read off dimensions of C*)*
```
SET N = # of rows of C
SET s = # of columns of C
INITIALIZE output matrix: P = C
FOR i=1 TO N
        IF rand < pinv
            SET a = floor((s-1)*rand + 1)
            SET b = floor((s-a)*rand + a + 1)
            UPDATE ath through bth entries of P(i,:)
                  to be these entries taken in reverse order
        END IF
END (i FOR)
OUTPUT:   P
```

EFR 10.7:   There are a total of 12 different such counterexamples (equaling 1.17% of the 1024 possible 5-vertex graphs). These can easily be found by running a brute-force program through all graphs, measuring the fitness of each, and counting those whose fitness = 10 (counterexamples).

Appendix C: Answers/Brief Solutions to Odd Numbered Exercises

CHAPTER 1: Section 1.1: **#1.** (a) False statement (MM 1926, JFK 1917) (b) Not a statement (c) True statement (d) Not a statement (e) False statement (f) False statement (Grant was the 18th president) **#3.** (a) True (b) False (c) True (d) False (e) False **#5.** (a) F or T = True (b) F and T = False (c) If F, then T = True (d) If T, then F = False (e) F if, and only if T = False **#7.** (a) 3, 4 (b) 3 (c) 3, 4 (d) 1, 2, 5 (e) 1, 4, 5 (f) 2, 3 (g) 1, 2, 3, 4, 5 **#9.** (a) If we go to a movie, then it will rain. Converse: If it will rain, then we will go to a movie. Contrapositive: If it will not rain, then we will not go to a movie. (b) If Yumi is going to the party, then I will go. Converse: If I go to the party, then Yumi will come. Contrapositive: If I do not go to the party, then Yumi will not go. (c) If I enter the tournament, then I can beat Norris this weekend. Converse: If I can beat Norris this weekend, then I will enter the tournament. Contrapositive: If I cannot beat Norris this weekend, then I will not enter the tournament. (d) If Tom makes the team, then he is able to run a mile in under six minutes. Converse: If Tom is able to run a mile in under six minutes, then he makes the team. Contrapostive: If Tom cannot run under a six minute mile, then he will not make the team. (e) If Carol gets an attractive job offer, then she will move to France. Converse: If Carol moves to France, then she got an attractive job offer. Contrapositive: If Carol does not move to France, then she did not get an attractive job offer. (f) If $\sum_{n=1}^{\infty} a_n$ converges, then the terms a_n tend to zero as n tends to infinity. Converse: If the terms a_n tend to zero as n tends to infinity, then $\sum_{n=1}^{\infty} a_n$ converges. Contrapositive: If $\sum_{n=1}^{\infty} a_n$ diverges (or does not converge), then the terms a_n do not tend to zero as n tends to infinity.

#11. (a)

P	~P → P	
T	F	T
F	T	F

(b)

P	P ∧ ~(P → P)		
T	F	F	T
F	T	F	T

(c) A tautology

P	Q	(P ∧ Q) → P	
T	T	T	T
T	F	F	T
F	T	F	T
F	F	F	T

(d)

P	Q	P ∨ (Q → P)	
T	T	T	T
T	F	T	T
F	T	F	F
F	F	T	T

(e)

P	Q	((P → Q) ↔ P) → ~Q)			
T	T	T	T	F	F
T	F	F	F	T	T
F	T	T	F	T	F
F	F	T	F	T	T

(f)

P	Q	((P ↔ Q) ∧ P) ⊕ Q		
T	T	T	T	F
T	F	F	F	F
F	T	F	F	T
F	F	T	F	F

(g)

P	Q	R	$(P \oplus Q) \to (Q \oplus R)$		
T	T	T	F	T	F
T	T	F	F	T	T
T	F	T	T	T	T
T	F	F	T	F	F
F	T	T	T	F	F
F	T	F	T	T	T
F	F	T	F	T	T
F	F	F	F	T	F

(h)

P	Q	R	$P \to (Q \leftrightarrow R)$	
T	T	T	T	T
T	T	F	F	F
T	F	T	F	F
T	F	F	T	T
F	T	T	T	T
F	T	F	T	F
F	F	T	T	F
F	F	F	T	T

(i)

P	Q	R	$(P \to Q) \to (Q \vee (R \leftrightarrow \sim P))$				
T	T	T	T	T	T	F	F
T	T	F	T	T	T	T	F
T	F	T	F	T	F	F	F
T	F	F	F	T	T	T	F
F	T	T	T	T	T	T	T
F	T	F	T	T	T	F	T
F	F	T	T	T	T	T	T
F	F	F	T	F	F	F	T

(j)

P	Q	R	$[P \to (R \wedge Q)] \;\leftrightarrow\; [\sim P \to (\sim R \vee Q)]$						
T	T	T	T	T	T	F	T	F	T
T	T	F	F	F	F	F	T	T	T
T	F	T	F	F	F	F	T	F	F
T	F	F	F	F	F	F	T	T	T
F	T	T	T	T	T	T	T	F	T
F	T	F	T	F	T	T	T	T	T
F	F	T	T	F	F	T	F	F	F
F	F	F	T	F	T	T	T	T	T

#13. (a) P unless $Q \equiv \sim P \to Q$. (b) If I do not go to the movies, then Diane will have called. Converse: If Diane calls, then I will not go to the movies. $\equiv$ Diane does not call unless I don't go to the movies. (This form is certainly awkward.) Contrapositive: If Diane does not call, then I will go to the movies. $\equiv$ Diane will have called unless I went to the movies. (Note: Do not worry to much about the form of the verb used in these constructions; there is some flexibility here since the logic does not concern itself with verb tenses.) **#15.** (a) The needed truth table is shown below:

P	Q	$P \vee Q$	$\sim (P \vee Q)$	$\sim P$	$\sim Q$	$\sim P \wedge \sim Q$
T	T	T	F	F	F	F
T	F	T	F	F	T	F
F	T	T	F	T	F	F
F	F	F	T	T	T	T

(b) The conjunction $\sim P \wedge \sim Q$ will be true (in one out of the four cases) when both $\sim P$ is true and $\sim Q$ is true. In other words, when both P and Q are false. Since any conjunction (of two logical variables) is true in three out of four cases, its negation will be true in exactly one out of four cases. With both P and Q false, the statement $\sim (P \vee Q)$ becomes $\sim (F \vee F) \equiv \sim F \equiv T$, thus $\sim (P \vee Q)$ has the same truth values as $\sim P \wedge \sim Q$.

#17. (i) (a) The one-step truth table of $Q \wedge P$ is readily seen to be identical to that for $P \wedge Q$ (Table 1.2). (b) Alternatively, any conjunction is true exactly when both parts are true, for either $P \wedge Q$ or $Q \wedge P$ this amounts to $P = T$ and $Q = T$, so the truth tables are identical. (ii) (a) The truth table shown below establishes the equivalence.   (b) Alternatively, using the fact that a conjunction of two

P	Q	R	$P \wedge Q$	$(P \wedge Q) \wedge R$	$Q \wedge R$	$P \wedge (Q \wedge R)$
T	T	T	T	T	T	T
T	T	F	T	F	F	F
T	F	T	F	F	F	F
T	F	F	F	F	F	F
F	T	T	F	F	T	F
F	T	F	F	F	F	F
F	F	T	F	F	F	F
F	F	F	F	F	F	F

statements is true precisely when each of the statements is true, we see that $(P \wedge Q) \wedge R =$ T exactly when $P \wedge Q = T$ and R $= T$. Using this fact again we get that $P \wedge Q = T$ exactly when $P = T$ and $Q = T$. In summary, the only row of the truth table for $(P \wedge Q) \wedge R$ that results in a true truth value is when all of P, Q, and $R = T$. A nearly identical argument shows that this is the only row of the truth table for $P \wedge (Q \wedge R)$ that makes this latter statement true. (iii) (a) The simple truth table for the double negation rule shown below establishes the equivalence $\sim (\sim P) \equiv P$. (b) Alternatively, $\sim (\sim P) = T$ exactly when $\sim P = F$, which happens exactly when P = T, so $\sim (\sim P) \equiv P$.

P	$\sim P$	$\sim (\sim P)$
T	F	T
F	T	F

(iv) (a) A truth table that establishes the exportation rule is shown below:

P	Q	R	$Q \to R$	$P \to (Q \to R)$	$P \wedge Q$	$(P \wedge Q) \to R$
T	T	T	T	T	T	T
T	T	F	F	F	T	F
T	F	T	T	T	F	T
T	F	F	T	T	F	T
F	T	T	T	T	F	T
F	T	F	F	T	F	T
F	F	T	T	T	F	T
F	F	F	T	T	F	T

(b)
Alternatively, working with the fact that implications are true "most of the time" (in three out of four cases), we separately determine what truth values make each of the two statements false. First, $P \to (Q \to R) = F$, if and only if $P = T$ and $Q \to R = F$. The latter equation forces $Q = T$ and $R = F$. Similarly, if we "solve the logical equation" $(P \wedge Q) \to R = F$, we get that $P \wedge Q = T$ and $R = F$. The former forces $P = Q = T$. Thus the values of P, Q, and R that render $P \to (Q \to R)$ false are the same as those that render $(P \wedge Q) \to R$ false, so the equivalence is proved.

#19. (a) The required truth table is shown below:

P	Q	R	$P \rightarrow Q$	$Q \rightarrow R$	$(P \rightarrow Q) \wedge (Q \rightarrow R)$	$P \rightarrow R$
T	T	T	T	T	T	T
T	T	F	T	F	F	F
T	F	T	F	T	F	T
T	F	F	F	T	F	F
F	T	T	T	T	T	T
F	T	F	T	F	F	T
F	F	T	T	T	T	T
F	F	F	T	T	T	T

To establish the implication $(P \rightarrow Q) \wedge (Q \rightarrow R) \Rightarrow P \rightarrow R$, we need only look at the rows of this truth table where the hypotheses (first part) $(P \rightarrow Q) \wedge (Q \rightarrow R)$ is true (these are the darker shaded rows), and check that the corresponding values of the conclusion (second part) $P \rightarrow R$ are also true. Since this is indeed the case, the implication is proved.   (b)   Assume that the hypothesis (i) $(P \rightarrow Q) \wedge (Q \rightarrow R)$ is true. We must prove that the conclusion $P \rightarrow R$ is also true. So assume that (ii) P is true. It suffices to show that R is also true (i.e., we assume P and must deduce R). By subtraction (Theorem 1.2 (b)) from (i) we get (iii) $P \rightarrow Q$.   From (ii) and (iii) and modus ponens (Theorem 1.2 (c)) we deduce (iv) Q. Another application of subtraction to (i) gives us (v) $Q \rightarrow R$, [1] which, when combined with (iv), another application of modus ponens gives R, as desired. **#21.** (a) The truth table needed to establish the implication $(P \vee Q) \wedge \sim P \Rightarrow Q$, is shown on the left. We need only check that when the hypothesis $(P \vee Q) \wedge \sim P$ is true, so is the conclusion Q. This is indeed the case (indicated by the shaded row of the truth table).

P	Q	$P \vee Q$	$\sim P$	$(P \vee Q) \wedge \sim P$	Q
T	T	T	F	F	T
T	F	T	F	F	F
F	T	T	T	T	T
F	F	F	T	F	F

(b) Here is an analytic proof: We assume that the hypothesis is true: (i) $(P \vee Q) \wedge \sim P$. Our task is to deduce Q (i.e., the truth of the conclusion). From (i) and subtraction (Theorem 1.2(b)) we deduce (ii) $P \vee Q$ and (iii) $\sim P$. We may rewrite the disjunction (ii) as an implication (iv) $\sim P \rightarrow Q$ (Theorem 1.1 II(a)). Applying modus ponens (Theorem 1.2 (c)) with (iii) and (iv) we deduce Q, as desired. **#23.** (In both proofs we use the names of each equivalence as given in Theorem 1.1.)
(a)

$P \leftrightarrow Q \equiv (P \rightarrow Q) \wedge (Q \rightarrow P)$ (Biconditional as Implications)
$\equiv (\sim P \vee Q) \wedge (\sim Q \vee P)$ (Implication as Disjunction)
$\equiv [\sim P \wedge (\sim Q \vee P)] \vee [Q \wedge (\sim Q \vee P)]$ (Distributivity)
$\equiv [(\sim P \wedge \sim Q) \vee (\sim P \wedge P)] \vee [(Q \wedge \sim Q) \vee (Q \wedge P)]$ (Distributivity)
$\equiv [(\sim P \wedge \sim Q) \vee \mathbf{F}] \vee [\mathbf{F} \vee (Q \wedge P)]$ (Contradictions)
$\equiv (\sim P \wedge \sim Q) \vee (Q \wedge P)$ (Identity Laws)
$\equiv (P \wedge Q) \vee (\sim P \wedge \sim Q)$ (Commutativity) □

---

[1] Technically, we also used commutativity (Theorem 1.1(a)) $P \wedge Q \equiv Q \wedge P$, in order to deduce Q from the subtraction implication $(P \wedge Q \Rightarrow P)$. In the future, we sometimes will use commutativity, associativity (Theorem 1.1(b)), and double negation (Theorem 1.1(e)) without explicit mention.

(b)

$(P \vee Q) \vee \sim R$

$$\equiv [\mathbf{T} \wedge (P \vee Q)] \vee \sim R \qquad \text{(Identity Laws)}$$
$$\equiv [(\sim P \vee P) \wedge (Q \vee P)] \vee \sim R \qquad \text{(Tautology, Commutativity)}$$
$$\equiv [(\sim P \wedge Q) \vee P] \vee \sim R \qquad \text{(Distributivity)}$$
$$\equiv [(\sim P \wedge Q) \vee \sim R] \vee P \qquad \text{(Commutativity)}$$
$$\equiv [\sim (P \vee \sim Q) \vee \sim R] \vee P \qquad \text{(De Morgan's law)}$$
$$\equiv \sim [(P \vee \sim Q) \wedge R] \vee P \qquad \text{(De Morgan's law)}$$
$$\equiv [(P \vee \sim Q) \wedge R] \rightarrow P \qquad \text{(Implication as Disjunction)} \;\; \square$$

#25. (a) This situation gives F $\rightarrow$ T, which is a true implication. The company probably did not intend for this to happen, but is does not logically contradict their claim. (b) The implication T $\rightarrow$ T would correspond to someone going to another dealer and paying more (too much) for a new Toyota. The logical truth of this statement corresponds to the intended truth of the claim. The implication T $\rightarrow$ F would correspond to someone going to another dealer and not paying more (too much) for a new Toyota. Logically, this implication is false and so would contradict the advertised claim. Finally, the implication F $\rightarrow$ F would correspond to someone coming (and buying) a car from Johnson Toyota and not paying too much (i.e., more than at other dealers). The logical truth of this statement agrees with the intended truth. (c) A phrase such as: "our prices are the lowest," would work. Such phrases are common (but not worth much without some sort of guarantee). **#27.** Italy = 1st, France = 2nd, Germany = 3rd **#29.** A is a liar, B is a truth teller. **#31.** (a) The two possible answers could not distinguish between three doors. (b) A simple strategy would be to use the idea of Part (b) of the solution of Example 1.6. First point to the first door and ask the guard: True or False? You are a liar or this door leads to freedom, but not both. As in the example, a true answer will always mean this door leads to freedom. If the guard answers false, then you know the freedom door must be one of the other two, so you can easily modify the Example 1.6 solution question to distinguish between the two remaining doors. **#35.** (a) Since $P \vee P \equiv P$, any grouping involving parentheses, a single logical variable P, and the $\vee$ symbol will always be equivalent (by repeatedly using the stated equivalence) to P. In particular, it will not be possible to represent the statement $\sim P$. (b) Repeated use of the identities $P \vee P \equiv P$ and $P \wedge P \equiv P$, allows us to reduce any logical statement built up with the single logical variable P, along with the operators $\wedge$ and/or $\vee$, along with balanced parentheses to prove such a statement is logically equivalent to P. It is therefore not possible to represent the statement $\sim P$ in this way. **#37.** *Sketch of proof:* First show that $P \uparrow P \equiv \sim P$ and $(P \uparrow P) \uparrow (Q \uparrow Q) \equiv P \vee Q$, and then use De Morgan's law (or Exercise 34(b)).

Section 1.2: **#1.** (a) $P(x) \wedge P(xy)$ (b) $P(x) \oplus P(y)$ (c) $(P(x) \wedge P(-y)) \rightarrow P(xy)$ (Since $y > 0$ if, and only if $-y < 0$.) Alternatively (but less elegantly), $(P(x) \wedge \sim P(y) \wedge (y \neq 0)) \rightarrow P(xy)$ (d) $(P(x) \wedge P(y)) \rightarrow P(-xy)$ (or $(P(x) \wedge P(y)) \rightarrow [\sim P(xy) \wedge (xy \neq 0)])$ **#3.** (a) $L(\text{Jane, Jimmy})$ (b) $H(\text{Jimmy, Jane})$ (c) $L(\text{Carla, Chuck}) \wedge L(\text{Chuck, Carla})$ (d) $\exists x \, H(x, \text{Dan})$ (e) $\forall x \, L(x, \text{Raymond})$ (f) $\forall x \sim H(x, \text{Susan})$ (Note: $H(x,y)$ is not the negation of $L(x,y)$. (Why?)) **#5.** (a) $G(\text{Helen, Tom})$ (b) $I(\text{Tom, Helen})$ (c) $\forall x \, [G(\text{Brad}, x) \wedge I(\text{Brad}, x)]$ (d) $\forall x \, [\sim G(\text{Esmerelda}, x) \wedge \sim I(\text{Esmerelda}, x)]$ (e) $\forall x \forall y \, [(x \neq y) \wedge G(x, y)) \rightarrow \exists z \, (G(z, x))]$ (Note that this statement involves three possibly different people, so we need three quantifiers.) (f) $\exists x \forall y \, [G(x, y) \rightarrow \sim G(y, x))]$ **#7.** (a) $\sim L(\text{Jane, Jimmy}) \equiv$ Jane does not love Jimmy. (b) $\sim H(\text{Jimmy, Jane}) \equiv$ Jimmy does not hate Jane. (c) (Using De Morgan's law) $\sim L(\text{Carla, Chuck}) \vee \sim L(\text{Chuck, Carla}) \equiv$ Either Carla does not love Chuck, or Chuck does not love Carla. Alternatively (and more elegantly) (At least) one of Carla and Chuck does not love the other. (d) $\forall x \sim H(x, \text{Dan}) \equiv$ No one hates Dan. (or Nobody hates Dan.) (e) $\exists x \sim L(x, \text{Raymond}) \equiv$ Somebody doesn't love Raymond. (f) $\exists x \, H(x, \text{Susan}) \equiv$ Somebody hates Susan. **#9.** (a) $\sim G(\text{Helen, Tom}) \equiv$ Helen doesn't gossip to Tom. (b) $\sim I(\text{Tom, Helen}) \equiv$ Tom listens to Helen. (c) (Using De Morgan's law) $\exists x \, [\sim G(\text{Brad}, x) \vee \sim I(\text{Brad}, x)] \equiv$ There is someone to whom either Brad does not gossip to or Brad listens to. (d) (Using De Morgan's law)

$\exists x\,[G(\text{Esmerelda},x) \vee I(\text{Esmerelda},x)]$ ≡ There is someone to whom either Esmerelda gossips to or does not listen to. (e) $\exists x \exists y \sim [(x \neq y) \wedge G(x,y)) \rightarrow \exists z\,(\sim I(z,x))]$. If we use Theorem 1.1 II(b) to negate the implication, this becomes $\exists x \exists y\,[(x \neq y) \wedge G(x,y)) \wedge \sim \exists z(\sim I(z,x))]$. Finally, since $\sim \exists z(\sim I(z,x)) \equiv \forall z(I(z,x))$, the whole negation becomes $\exists x \exists y\,[(x \neq y) \wedge G(x,y)) \wedge [\forall z(I(z,x))]]$. In English: There are two different people such that the first gossips to the second but no one listens to the first. (f) $\forall x \exists y \sim [G(x,y) \rightarrow \sim G(y,x)]$. As in Part (e), this can be rewritten as $\forall x \exists y\,[G(x,y) \wedge G(y,x)]$. In English: Everyone has another person with whom they exchange gossip.

#11. (a) True. Proof: Take $n = 1$ (or 0). (b) False. Counterexample: Take n to be any negative integer, n^3 will be negative, but n^2 will be positive. (c) True. *Proof:* For any integer n, $n \leq n^2$, so if n and m are any two integers, $n + m \leq n^2 + m^2$. (d) False. Counterexample: With $n = m = 1$, $(n+m)^2 \leq n^2 + m^2$ would say that $4 \leq 2$. **#13.** The truth values of the negations are the opposites of the truth values of the statements (see answers to Exercise 11). Here are the negations of each of the statements. (a) $\forall n\,(n^2 \neq n)$ (b) $\exists n\,(n^3 < n^2)$ (c) $\exists n \exists m (n + m > n^2 + m^2)$ (d) $\exists n \exists m ((n+m)^2 > n^2 + m^2)$ **#15.** (a) (i) $\exists t[P(\text{Wife},t)]$ (ii) Negation: $\forall t[\sim P(\text{Wife},t)] \equiv$ You can never please your wife. (b) (i) $\forall t[P(\text{Father},t)]$ (ii) Negation: $\exists t[\sim P(\text{Father},t)] \equiv$ Sometimes you can't please your father. Alternatively: You can't always please your father. (c) (i) $\forall t[\sim P(\text{Mother-in-law},t)]$ (ii) Negation: $\exists t[P(\text{Mother-in-law},t)] \equiv$ You can sometimes please your mother-in-law. (d) (i) $\exists x \forall t\,[P(x,t)]$ (ii) Negation: $\forall x \exists t\,[\sim P(x,t)] \equiv$ For any person, there is sometime that you will not be able to please him/her. (e) (i) $\forall x \exists t\,[P(x,t)]$ (ii) Negation: $\exists x \forall t\,[\sim P(x,t)] \equiv$ There are some people that you can never please. (f) (i) $\sim \forall x \forall t\,[P(x,t)] \equiv \exists x \exists t\,[\sim P(x,t)]$ (ii) Negation: $\forall x \forall t\,[P(x,t)] \equiv$ You can please all of the people all of the time. **#17.** (a) (i) For any two real numbers x and y, we have $xy = yx$. (ii) True; this is the commutative law of multiplication. (iii) Negation: $\exists x \exists y\,(xy \neq yx) \equiv$ There are two real numbers x, and y such that $xy \neq yx$. (b) (i) For any real number x, there exists a real number y such that $xy = x + y$. (ii) False. If we use $x = 1$, the equation becomes $y = 1 + y$, so there exists no y that solves this. (iii) Negation: $\exists x \forall y\,(xy \neq x + y) \equiv$ There is a real number x, such that for any real number y, we have $xy \neq x + y$. (c) (i) For any two real numbers x, and y, there exists a real number z such that $xyz = x + y + z$. (ii) False. If we use $x = 1$ and $y = 1$, the equation becomes $z = 2 + z$, so there exists no z that solves this. (iii) Negation: $\exists x \exists y \forall z\,(xyz \neq x + y + z) \equiv$ There exist a pair of real numbers x and y such that for any real number z, we have $xyz \neq x + y + z$. (d) (i) For any pair of real numbers a and b, there exists a real number x such that $ax + b = 0$. (ii) False. If we use $a = 0$ and $b = 1$, the equation becomes $1 = 0$, a contradiction. (iii) $\exists a \exists b \forall x\,(ax + b \neq 0) \equiv$ There exists a pair of real numbers a and b, such that for any real number x, we have $ax + b \neq 0$. (e) (i) For any three real numbers a, b, and c, for which $a \neq 0$, there exists a real number x for which $ax^2 + bx + c = 0$. (ii) False: If we use $a = c = 1$ and $b = 1$, the equation becomes $x^2 + 1 = 0$, which has no (real number) solution x. (iii) Negation: (Using Theorem 1.1 II(b) to negate the implication.) $\exists a \exists b \exists c\,(a \neq 0 \wedge \forall x(ax^2 + bx + c \neq 0)) \equiv$ There exist three real numbers a, b, and c, with $a \neq 0$, such that for any real number x, we have $ax^2 + bx + c \neq 0$. (f) (i) For any pair of real numbers a, and b, for which $a \neq 0$, there exists a real number x for which $ax + b = 0$. (ii) True: Take $x = -b/a$ (unique solution). (iii) Negation: (Again using Theorem 1.1 II(b)) $\exists a \exists b(a \neq 0 \wedge \forall x(ax + b \neq 0)) \equiv$ There exist a pair of real numbers a and b, with $a \neq 0$, such that for any real number x, we have $ax + b \neq 0$. **#21.** (a) This statement is true, but it takes a bit of work to see why. Here is a proof: In the equation $\dfrac{x}{z} + yz = 2x + 2y$, we cannot have $z = 0$, so it is equivalent to (multiply both sides by z): $x + yz^2 = 2(x + y)z$. Case 1: $y = 0$. The equation becomes

$x = 2xz$. This can always be solved for z: If $x = 0$, z can be taken to be any real number, while if $x \neq 0$, the unique value of z that will work is 1/2. Case 2: $y \neq 0$. Now we have a quadratic equation: $yz^2 - 2(x+y)z + x = 0$. This will have a real solution z if, and only if, the discriminant $b^2 - 4ac = 4(x+y)^2 - 4xy$ is nonnegative, i.e., $(x+y)^2 - xy \geq 0$. To prove that this is always the case (for any two real numbers x, and y), we may as well assume that $xy > 0$ (otherwise there is nothing to prove). We have $(x+y)^2 - xy = x^2 + 2xy + y^2 - xy = x^2 + xy + y^2 \geq x^2 - 2xy + y^2 = (x-y)^2 \geq 0$. The proof is complete. **#23.** (a) No hummingbirds are large (birds). (b) All pawnbrokers are honest people. **#25.** (a) The only books (in this library) that are recommended for reading are both bound and healthy in tone. (b) Here are the assumptions using reasonable notation: 1: $H \rightarrow C$, 2: $G \rightarrow S$, 3: $D \rightarrow A$, 4: $\sim Cv \vee P \equiv Cv \rightarrow P$ (since "unless" means "or"—see Exercise 13 of Section 1.1), 5: $C \rightarrow K$, 6: $T \rightarrow H$, 7: $K \rightarrow \sim S$, 8: $K \rightarrow Cv$, 9: $\sim T \rightarrow D$, 10: $P \rightarrow G$. We can piece together these statements (or their contrapositives) in the following order: 8, 4, 10, 2, 7, 5, 1, 6, 9, 3, to obtain the chain: $K \rightarrow Cv \rightarrow P \rightarrow G \rightarrow S \rightarrow \sim K \rightarrow \sim C \rightarrow \sim H \rightarrow \sim T \rightarrow D \rightarrow A$, which implies the following ultimate conclusion: $K \rightarrow A$. In (spoken) English, this becomes: I avoid all kangaroos.

Section 1.3: **#1.** (a) $\{0, 2, 3, 4, 6, 8, 9\}$ (b) $\{0, 6\}$ (c) $\{1, 3, 5, 7, 9\}$ (d) $\{2, 4, 8\}$ (e) $\{3, 9\}$ (f) $\{3, 9\}$ **#3.** (a) $\{a, b, c, d, e, i, j\}$ (b) $\{a\}$ (c) $\{f, g, h\}$ (d) $\{f, g, h\}$ (e) $\{f, g, h\}$ (f) $\{f, g, h\}$ **#5.** (a) the set of all even integers (b) $\{\pm\sqrt{2}\}$ (c) $\varnothing$ (the empty set) (d) $\{1\}$ **#7.** (a) $\{$all male basketball playing students$\}$ (b) $U = \{$all college students$\}$ (c) $\{$all female students who do not play basketball$\}$ (d) $\{$all female students who do not play basketball$\}$ **#9.** (a) $\sim C$ (b) $R \sim J$ (c) $\sim J \cap \sim C$ (d) $\sim C \cap \sim J \cap \sim R$

#11. (i) The four Venn diagrams are shown below:

(a) $(A \cup B) \cup C$
$= A \cup (B \cup C)$

(b) $\sim (A \cup B)$
$= \sim A \cap \sim B$

(c) $A \cup (A \cap B)$
$= A$

(d) $\sim (A \cup B \cup C)$
$= \sim A \cup \sim B \cup \sim C$

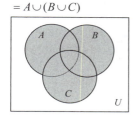
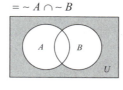
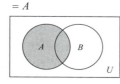
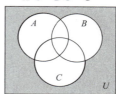

The analytic proofs of each part are rather simple consequences of the definitions and the corresponding rules of logic. Below, we will be using Parts I(b), (d) (and its extension) of Theorem 1.1. (a) (ii) $x \in (A \cup B) \cup C \Leftrightarrow x \in (A \cup B) \vee x \in C \Leftrightarrow (x \in A \vee x \in B) \vee x \in C \Leftrightarrow x \in A \vee (x \in B \vee x \in C)$ $\Leftrightarrow x \in A \vee x \in (B \cup C) \Leftrightarrow x \in A \cup (B \cup C)$. $\square$ (b) (ii) $x \in \sim (A \cup B) \Leftrightarrow x \notin A \cup B \Leftrightarrow \sim (x \in A \vee x \in B) \Leftrightarrow (x \notin A) \wedge (x \notin B) \Leftrightarrow x \in (\sim A \cap \sim B)$. $\square$ (c) (ii) Since $A \subseteq A$ and $A \cap B \subseteq A$, we certainly have $A \cup (A \cap B) \subseteq A$. On the other hand, since $A \subseteq A \cup S$, for any set S, we get the reverse inequality $A \subseteq A \cup (A \cap B)$. Thus we have proved that $A \cup (A \cap B) = A$. $\square$ (d) (ii) The analytic proof is basically the same as what was done in Part (b), except that we need to use the corresponding De Morgan law (of logic) for negating a disjunction of three propositions (such extensions were discussed in Section 1.1).

#13. (i) The two Venn diagrams for each part are shown below:

(a) $A \cap B$　　　　(b) $A \cap B \cap C$　　　　(c) $(A \sim B) \sim C$　　　　(d) $(A \sim B) \sim C$

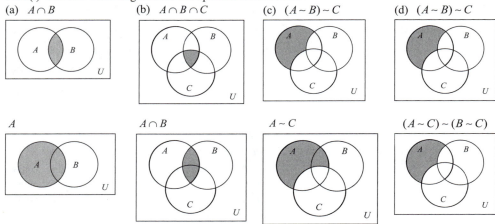

A　　　　　　$A \cap B$　　　　　　$A \sim C$　　　　　　$(A \sim C) \sim (B \sim C)$

From these Venn diagrams, the following (best possible) relationships are readily observed: [2]
(a)　$A \cap B \subseteq A$　　(b)　$A \cap B \cap C \subseteq A \cap B$　　(c)　$(A \sim B) \sim C \subseteq A \sim C$　　(d)　$(A \sim B) \sim C = (A \sim C) \sim (B \sim C)$

(ii) Here are the corresponding analytic proofs and counterexamples:
(a)　*Analytic proof:* If $x \in A \cap B$, this means $x \in A$ and $x \in B$, so, in particular $x \in A$.　(iii) Counterexample to show the reverse inclusion $A \subseteq A \cap B$ can fail: $A = \{1\}$ (or any nonempty set), $B = \varnothing$. (b) (ii) This inclusion can be easily proved analytically with minor changes to what was done in Part (a).　Alternatively, the inclusion $A \cap B \cap C \subseteq A \cap B$ follows from the one of Part (a) $A \cap B \subseteq A$ by replacing A with C and B with $A \cap B$ (and using commutativity and associativity of intersections—Parts (a) and (b) of Theorem 1.3). (iii) Counterexample to show the reverse inclusion $A \cap B \subseteq A \cap B \cap C$ can fail: $A = B = \{1\}$ (or any nonempty set), $C = \varnothing$.　(c) Analytic proof of this inclusion: If $x \in (A \sim B) \sim C$, this means that $x \in (A \sim B)$, but $x \notin C$. The first inclusion tells us that $x \in A$ (but $x \notin B$). In particular, the facts that $x \in A$ and $x \notin C$, tell us that $x \in A \sim C$.　(iii) Counterexample to show the reverse inclusion $A \sim C \subseteq (A \sim B) \sim C$ can fail: $A = B = \{1\}$ (or any nonempty set), $C = \varnothing$. (d) (ii) Analytic proof that $(A \sim B) \sim C = (A \sim C) \sim (B \sim C)$: We proceed to prove the two inclusions. $x \in (A \sim B) \wedge x \notin C \Rightarrow (x \in A \wedge x \notin B) \wedge x \notin C \Rightarrow (x \in A \wedge x \notin C) \wedge x \notin B$ $\Rightarrow (x \in A \sim C) \wedge (x \notin B \sim C) \Rightarrow x \in (A \sim C) \sim (B \sim C)$.　This proves the inclusion $(A \sim B) \sim C \subseteq (A \sim C) \sim (B \sim C)$.　　For　the　reverse　inclusion:　　$x \in (A \sim C) \sim (B \sim C) \Rightarrow$ $(x \in A \wedge x \notin C) \wedge (x \notin B \sim C) \Rightarrow (x \in A \wedge x \notin B) \wedge (x \notin C) \Rightarrow x \in (A \sim B) \sim C$.　With　the　reverse inclusion just now proved, the set equality is established. **15.** (a) Always True. Proof: Since the inclusion statement $x \notin B \rightarrow x \notin A$ is the contrapositive of the statement $x \in A \rightarrow x \in B$, it follows that $A \subseteq B$, if, and only if, $\sim B \subseteq \sim A$.　(b) Always True. Proof: Suppose that $x \in A \sim B$. Case 1: $x \in C$.　Then　$x \in A \cap C$, so (by definition of the symmetric difference) $x \notin A \Delta C$.　But $x \in (B \cup C) \sim (B \cap C) = B \Delta C = A \Delta C$ —a contradiction. This proves that $A \sim B = \varnothing$. By switching

---

[2] Note that the Venn diagrams also give us a recipe for constructing a counterexample to show that the reverse inclusion cannot hold in general (this is what we will have to do in Part (iii)); we simply have to give three sets A, B, and C, so that A and B have at least one common element that is not in C (i.e., an element in the shaded portion of the Venn diagram on the right that is not in the shaded portion of the Venn diagram on the right).

the roles of A and B, the same argument shows that $B \sim A = \varnothing$, and it follows that $A = B$. (c) Always True. *Proof:* If $x \in \bigcup_k A_k$, then $x \in A_k$ for some index k. But since

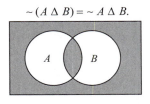

$\sim(A \Delta B) = \sim A \Delta B.$

$A_k \subseteq B_k$, we have also $x \in B_k$, and thus $x \in \bigcup_k B_k$. This proves the inclusion $\bigcup_k A_k \subseteq \bigcup_k B_k$. **17. (a)** The Venn diagram for both sets are shown on the left, and establishes the equality. **(b)** For the logical proof, we note that the definition of the symmetric difference states that for any two sets E and F, $x \in E \Delta F$ if, and only if, x lies $E \cup F$, but lies outside of $E \cap F$. This can be restated as saying that x lies in exactly one of the sets E and F. Thus, the condition $x \in \sim(E \Delta F)$ could be restated as either x lies in neither E nor F, or in both E and F. Viewing these observations as (expanded) definitions of the symmetric difference, we now proceed with a logic-based proof:

$x \in \sim A \Delta B$

$\Leftrightarrow x \in (\sim A \cup B) \wedge x \notin (\sim A \cap B)$	(Definition of Symmetric Difference)
$\Leftrightarrow [x \in \sim A \vee x \in B] \wedge \sim[x \in \sim A \wedge x \in B]$	(Definition of Union, Intersection)
$\Leftrightarrow [x \in \sim A \vee x \in B] \wedge [x \notin \sim A \vee x \notin B]$	(De Morgan's law)
$\Leftrightarrow [x \in \sim A \vee x \in B] \wedge [x \in A \vee x \in \sim B]$	(Definition of Set Complement)
$\Leftrightarrow (x \in \sim A \wedge [x \in A \vee x \in \sim B]) \vee$	
$\quad\quad (x \in B \wedge [x \in A \vee x \in \sim B])$	(Distributivity)
$\Leftrightarrow (x \in \sim A \wedge x \in A) \vee (x \in \sim A \wedge x \in \sim B) \vee$	
$\quad\quad (x \in B \wedge x \in A) \vee (x \in B \wedge x \in \sim B)$	(Distributivity)
$\Leftrightarrow \mathbf{F} \vee (x \in \sim A \wedge x \in \sim B) \vee$	
$\quad\quad (x \in B \wedge x \in A) \vee \mathbf{F}$	(Contradictions)
$\Leftrightarrow (x \in \sim A \wedge x \in \sim B) \vee (x \in B \wedge x \in A)$	(Identity Law)
$\Leftrightarrow (x \in \sim A \cap \sim B) \vee (x \in B \cap A)$	(Definition of Intersection)
$\Leftrightarrow x \in (\sim A \cap \sim B) \cup (B \cap A)$	(Definition of Union)
$\Leftrightarrow x \in \sim(A \Delta B)$	(Definition of Symmetric Difference) $\square$

#19. (a) True **(b)** False **(c)** False **(d)** True **(e)** True **(f)** False **#21. (a)** $2^8 = 256$ **(b)** 255 **(c)** $2^6 = 64$ **(d)** $2^5 = 32$ **(e)** 31 **#23. (a)** {(Red, Green), (Red, Yellow), (White, Green), (White, Yellow), (Blue, Green), (Blue, Yellow)} **(b)** {(Green, Red), (Green, White), (Green, Blue), (Yellow, Red), (Yellow, White), (Yellow, Blue)} **(c)** {(Green, Green, Green), (Green, Green, Yellow), (Green, Yellow, Green), (Green, Yellow, Yellow), (Yellow, Green, Green), (Yellow, Green, Yellow), (Yellow, Yellow, Green), (Yellow, Yellow, Yellow)} **(d)** Since A^2 has $3^2 = 9$ elements, it follows that $\mathscr{P}(A^2)$ has $2^9 = 512$ elements. Here are three elements in this power set: $\varnothing$ (always a member of any power set), {(Red, Red)}, {(White, Red), (Red, White)}. **#25. (a)** Sometimes True. All we need is one of the two factors A or B to be empty for their product $A \times B$ to be empty. For example, $\mathbb{Z} \times \varnothing = \varnothing$. **(b)** Sometimes True. If we take $A = \varnothing$ and $B = \{\varnothing\}$, then the assertion will be true if $C = \{\varnothing, \{\varnothing\}\} = \{A, B\}$, but false if $C = \{\{\varnothing\}\} = \{B\}$.

#31. (a) (c) A Venn diagram for five sets using five ellipses is shown at the right.

A five-set Venn Diagram.

CHAPTER 2:   Section 2.1:   **#1.**
(a) (i) $a = b$:
  $(-2,-2),\ (-1,-1),\ (1,1),\ (2,2)$
(ii) and (iii): see figure.
(b) (i) $2a = b$: $(-1,-2)$, $(1,2)$ (ii) and
(iii): see figure. (c) a is a factor of b:
  $(-2,-2),\ (-2,2),(-1,-2),\ (-1,-1),$
  $(-1,1),\ (-1,2),(1,-2),\ (1,-1),\ (1,1),\ (1,2),$
  $(2,-2),\ (2,2)$ (ii) and (iii): see figure.
(d) $a^2 = b^2$: $(\pm2,\pm2),\ (\pm1,\pm1)$ (ii) and
(iii): see figure.

R	-2	-1	1	2
-2	1	0	0	1
-1	0	1	1	0
1	0	1	1	0
2	1	0	0	1

Figure Exercise #1(d).

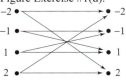

R	-2	-1	1	2
-2	1	0	0	0
-1	0	1	0	0
1	0	0	1	0
2	0	0	0	1

Figure Exercise #1(a).

R	-2	-1	1	2
-2	0	0	0	0
-1	1	0	0	0
1	0	0	0	1
2	0	0	0	0

Figure Exercise #1(b).

R	-2	-1	1	2
-2	1	0	0	1
-1	1	1	1	1
1	1	1	1	1
2	1	0	0	1

Figure Exercise #1(c).

#3. (a) The graph is the line $y = 5x$; see figure.   (b) To graph this relation, we note that $x^2 > y^2 \Leftrightarrow -|x| < y < |x|$. So the relation consists of the all points in the plane that lie (strictly) between the graphs of $y = |x|$ (above) and $y = -|x|$ (below). See figure.

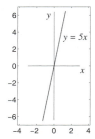

Figure for #3(b).

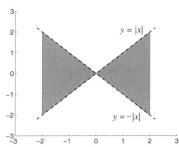

Figure for #3(b). Note: (0,0) is not in the region.

(c) Since $x^2 > |y| \Leftrightarrow -x^2 < y < x^2$, relation consists of all points in the plane that lie (strictly) between the graphs of $y = -x^2$ (below) and $y = x^2$ (above); see figure.

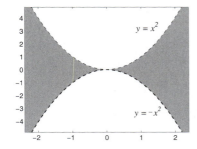

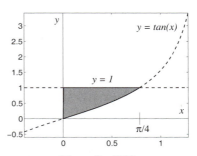

Figure for #3(c). Note: (0,0) is not in the region.   Figure for #3(d).

#5. (a) $R_1 \cup R_2 = \{(a,b) \in \mathbb{R}^2 : a \ge b\}$ (b) $R_1 \Delta R_2 = \{(a,b) \in \mathbb{R}^2 : a \ge b\}$ (c) $\sim (R_1 \cup R_3) =$
$\{(a,b) \in \mathbb{R}^2 : b^2 \le a \le b\}$ (d) $R_1 \cap R_3 = \{(a,b) \in \mathbb{R}^2 : b < a < b^2\}$ (e) $R_3 \sim R_1 =$
$\left\{(a,b) \in \mathbb{R}^2 : \begin{matrix} a < b^2, \text{ if } 0 \le b \le 1 \\ a \le b, \text{ otherwise} \end{matrix}\right\}$ (f) $R_1 \sim R_3 = \left\{(a,b) \in \mathbb{R}^2 : \begin{matrix} a \ge b, \text{ if } 0 \le b \le 1 \\ a > b^2, \text{ otherwise} \end{matrix}\right\}$ (The reader is
encouraged to draw graphs of the original three sets as well as for the above six related sets to check that the above descriptions are indeed valid.) **#7.** (a) $D = [0,\infty)$, $f(D) = [0,\infty)$ (b) $D = (-\infty,1)$,
$f(D) = (0,\infty)$ (c) $D = \mathbb{Z}^2$, $f(D) = \mathbb{Z}$ (d) $D = S \times T$, $f(D) = S$ **#9.** (a) $f(-5) = \sqrt{-1}$ is
undefined. $D = [-4,\infty)$ (b) $f(1)$ is undefined. $D = \{x \in \mathbb{R} : x \ne 1\}$ (c) $f(5) = \ln(0)$ is undefined.
$D = (-\infty,5)$ (d) $f(2) = \pm\sqrt{3}$, a function can have only one output for each input. $D = \{\pm 1\}$ **#11.** (a)
(i) one-to-one (ii) not onto (iii) (0,1], $\varnothing$ (b) (i) one-to-one (ii) onto (iii) (0,1], $(-\infty,0]$ (c) (i) one-
to-one (ii) onto (iii) [1,9], [2,3) (d) (i) one-to-one (ii) onto (iii) [−8,8], (−4,−1] (e) (i) not one-to-
one (ii) onto (iii) (−1,1], $\{\pi/2 + 2n\pi : n \in \mathbb{Z}\}$ (f) (i) not one-to-one (ii) not onto (iii) If $n = -m$, then
$h(n,m) = 2m^2$, and the set of these images is just the set of all doubles of square integers:
$\{2n^2 : n \in \mathbb{Z}\} = \{0,2,8,18,32,50,\cdots\}$, $\{(a,b) : a,b \in \{-1,0,1\}\}$ (g) (i) not one-to-one (ii) onto (iii) $\mathbb{Z}$,
$\{(a,b) \in \mathbb{Z} \times \mathbb{Z}_+ : a = 0, \text{ or } a = \pm b, \text{ or } a = \pm 2b\}$ **#13.** (a) bijection, $f^{-1}(x) = (x-5)/3$ (b) not a
bijection, but becomes one if co-domain is changed to $\mathbb{R}_+ = [0,\infty)$; formula for inverse function:
$\ln(x)$ (c) not a bijection, but becomes one if the co-domain is changed to the range of h, which is the
set $\{3k+2 : k \in \mathbb{Z}\}$; formula for inverse function: $(x-5)/3$ (d) not a bijection, cannot be made into
one by (only) restricting the co-domain since the function is not one-to-one **#15.** (a)
$(g \circ f)(x) = 2\cos x - 1$, $-3, 1$ (b) $(g \circ f)(x) = \sqrt{x^2}$ or $|x|$, $3, 4$ **#17.** (a) $\begin{pmatrix} 1 & 2 & 3 & 4 & 5 & 6 & 7 \\ 2 & 3 & 5 & 7 & 1 & 4 & 6 \end{pmatrix}$

(b) $\begin{pmatrix} 1 & 2 & 3 & 4 & 5 & 6 & 7 \\ 1 & 2 & 5 & 4 & 6 & 7 & 3 \end{pmatrix}$ (c) $\begin{pmatrix} 1 & 2 & 3 & 4 & 5 & 6 & 7 \\ 7 & 3 & 5 & 1 & 2 & 6 & 4 \end{pmatrix}$ (d) $\begin{pmatrix} 1 & 2 & 3 & 4 & 5 & 6 & 7 \\ 1 & 2 & 5 & 4 & 6 & 7 & 3 \end{pmatrix}$ **#19.** (a)
True. *Proof:* $(g \circ f)(x) = (g \circ f)(x') \Rightarrow g(f(x)) = g(f(x')) \Rightarrow$ (since g is 1-1) $f(x) = f(x')$
$\Rightarrow x = x'$ □ (b) True. *Proof:* $f(x) = f(x') \Rightarrow g(f(x)) = g(f(x'))$, i.e., $(g \circ f)(x) = (g \circ f)(x')$
$\Rightarrow$ (since $g \circ f$ is 1-1) $\Rightarrow x = x'$ □ (c) False. Here is one counterexample:
Take $g : \mathbb{R} \to \mathbb{R} :: g(x) = x^2$ and $f : \mathbb{R} \to \mathbb{R} :: f(x) = e^x$, then $g \circ f(x) = e^{2x}$ is 1-1, but g is not. **#21.**
(a) False. Counterexample: With $x = 1.6$, $\lfloor 2x \rfloor = 3 \ne 2 = 2\lfloor x \rfloor$. (b) False. Counterexample: With x
$= 0.5$, $\lfloor -x \rfloor = -1 \ne 0 = -\lfloor x \rfloor$. (c) True. *Proof:* For any real number x, there exists a unique integer

such that $k \le x < k+1$, and this implies $-(k+1) < -x \le -k$. The former inequality tells us that $\lceil x \rceil = k+1$, so that $-\lceil x \rceil = -(k+1)$. The latter inequality tells us that $\lfloor -x \rfloor = -(k+1)$, thus $\lfloor -x \rfloor = -\lceil x \rceil$, as desired. $\square$ **#23.** (a) True. In fact for any integer n, we have $\lfloor x+n \rfloor = \lfloor x \rfloor + n$. *Proof:* For any real number x, there exists a unique integer such that $k \le x < k+1$, and this implies $k+n \le x+n < k+n+1$. The former inequality tells us that $\lfloor x \rfloor = k$, and the latter tells us that $\lfloor x+n \rfloor = k+n$. Hence, $\lfloor x+n \rfloor = \lfloor x \rfloor + n$. $\square$ (b) True. *Proof:* Since $\lfloor x \rfloor$ is an integer, and the ceiling (or floor) of any integer is itself, we get that $\lceil \lfloor x \rfloor \rceil = \lfloor x \rfloor$. $\square$ (c) False. Counterexample: With $x=2$, $1/\lceil x \rceil = 1/2 \ne 1 = \lceil 1/x \rceil$. **#25.** (a) *Proof:* If $x \in f^{-1}(B_1)$, this means that $f(x) \in B_1$. But since $B_1 \subseteq B_2$, we certainly also have $f(x) \in B_2$, which means that $x \in f^{-1}(B_2)$. We have proved that $f^{-1}(B_1) \subseteq f^{-1}(B_2)$. $\square$ (b) *Proof:* If $y \in f(A_1)$, this means that there exists $x \in A_1$ such that $y = f(x)$. But since $A_1 \subseteq A_2$, we certainly also have $x \in A_2$, so that $y = f(x) \in f(A_2)$. This proves that $f(A_1) \subseteq f(A_2)$. $\square$ (c) *Proof:* If $x \in f^{-1}(\bigcap_k B_k)$, this means that $f(x) \in \bigcap_k B_k$. In particular, for any fixed index k, we have $f(x) \in B_k$, i.e., $x \in f^{-1}(B_k)$. Since this is true regardless of the index k, x must lie in the intersection of all of these sets: $x \in \bigcap_k f^{-1}(B_k)$, and result is established. $\square$ **#27.** (a) *Proof:* $x \in f^{-1}(B_1 \cup B_2) \Leftrightarrow f(x) \in B_1 \cup B_2 \Leftrightarrow [f(x) \in B_1] \vee [f(x) \in B_2] \Leftrightarrow [x \in f^{-1}(B_1)] \vee [x \in f^{-1}(B_2)] \Leftrightarrow x \in f^{-1}(B_1) \cup f^{-1}(B_2)$. $\square$ (b) *Proof:* $y \in f(A_1 \cup A_2) \Leftrightarrow y = f(x), \exists x \in A_1 \cup A_2 \Leftrightarrow [y = f(x), \exists x \in A_1] \vee [y = f(x), \exists x \in A_2] \Leftrightarrow y \in f(A_1) \cup f(A_2)$. $\square$ (c) *Proof:* $x \in f^{-1}(\bigcup_k B_k) \Leftrightarrow f(x) \in \bigcup_k B_k \Leftrightarrow \exists k [f(x) \in B_k] \Leftrightarrow \exists k [x \in f^{-1}(B_k)] \Leftrightarrow x \in \bigcup_k f^{-1}(B_k)$. $\square$ (d) *Proof:* $y \in f(\bigcup_k A_k) \Leftrightarrow \exists x \in \bigcup_k A_k [y = f(x)] \Leftrightarrow \exists k \exists x \in A_k [y = f(x)] \Leftrightarrow y \in \bigcup_k f(A_k)$. $\square$ **#29.** (a) *Proof:* Since identity functions are certainly one-to-one and onto, it follows (see the above proof in Exercise 19(b)) from $F \circ f = i_A$, that f is one-to-one. Similarly, from $f \circ F = i_B$, it follows that f must be onto; indeed, given $b \in B$, $b = i_B(b) = (f \circ F)(b) = f(F(b))$. Thus f is a bijection, so from Definition 2.5, its inverse function is (uniquely) specified by $f^{-1}(b) = f^{-1}(\{b\})$. Since $f(F(b)) = b$, we have $f^{-1}(\{b\}) = F(b)$, as desired. This completes the proof. $\square$ (b) *Proof:* We will use the criterion of Part (a) applied to the functions $\varphi \equiv g \circ f : A \to C$, $\Phi \equiv f^{-1} \circ g^{-1} : C \to A$. (We will also tacitly be using the fact that composition of functions is associative; see Exercise 28.) We have $\Phi \circ \varphi \equiv (f^{-1} \circ g^{-1}) \circ (g \circ f) = f^{-1} \circ (g^{-1} \circ g) \circ f = f^{-1} \circ (i_B \circ f) = f^{-1} \circ f = i_A$. A similar argument leads to $\varphi \circ \Phi \equiv i_B$. It follows from the result of Part (a) that $\varphi^{-1} = \Phi$, i.e., that $(g \circ f)^{-1} = f^{-1} \circ g^{-1}$. $\square$ **#31.** Each of the four parts is proved similarly, separating into four cases (depending on whether a and/or b is an integer). So we will give a proof only for Part (a). *Proof:* Let N denote the number of integers in the interval $[a, b]$. There are unique integers $k \ge n$ such that $k \le b < k+1$ and $n \le a < n+1$. *Case 1: a and b are both integers.* Thus $a = n$, $b = k$, and $N = k - n + 1 = \lfloor b \rfloor - \lceil a \rceil + 1$. *Case 2: a is an integer, b is not.* Thus $a = n$, $b > k$, and $N = k - n + 1 = \lfloor b \rfloor - \lceil a \rceil + 1$. *Case 3: b is an integer, a is not.* Thus $a > n$, $b = k$, and $N = k - n = k - (n+1) + 1 = \lfloor b \rfloor - \lceil a \rceil + 1$. *Case 4: Neither a nor b are integers.* Thus $a > n$, $b > k$ and $N = k - n = k - (n+1) + 1 = \lfloor b \rfloor - \lceil a \rceil + 1$. $\square$ **#33.** (a) *Proof:* We will use the criterion of Exercise 29(a). The cycles $f = (a_1 \ a_2 \ \cdots \ a_{k-1} \ a_k)$ and $g = (a_k \ a_{k-1} \ \cdots \ a_2 \ a_1)$ represent two permutations on the set $A = \{1, 2, \cdots, n\}$. It suffices to show that $f \circ g = i_A$ and $g \circ f = i_A$. Since the two cycles f and g are mirror images of one another, it suffices to show just one of these. We will show that $g \circ f = i_A$, i.e., that $(g \circ f)(j) = j$, for any integer $j \in A$. Since both of the permutations f

and g ignore (i.e., fix) elements outside the set $\{a_1, a_2, \cdots, a_k\}$, we may restrict our attention to showing $(g \circ f)(a_i) = a_i$. Indeed, $(g \circ f)(a_i) = g(f(a_i)) = g(a_{i+1}) = a_i$, as desired. (We have set $a_{k+1} = a_1$.)

□

Section 2.2: **#1.** (a) Not an equivalence relation; not symmetric. (b) Not an equivalence relation; neither reflexive nor symmetric. (c) Equivalence relation. (d) Not an equivalence relation. Need not be reflexive or transitive (although some might debate these claims). **#3.** (a) Equivalence relation. (b) Not an equivalence relation; neither reflexive nor transitive. (c) Not an equivalence relation; not transitive. (d) Equivalence relation. **#5.** (a) The fact that this is an equivalence relation follows directly from Proposition 2.3. $[\pi] = \{n\pi : n \in \mathbb{Z}\}$ (b) Since $a - b$ is even if, and only if 2 is a factor of $a - b$, this relation is the same as congruence modulo 2, which is an equivalence relation by Proposition 2.2. $[11] = \{\text{odd integers}\}$. (c) This can be verified axiom by axiom. More elegantly, we introduce the function $f : \{\text{four bit strings}\} \rightarrow \{0,1\}$, defined by

$$f(s) = \begin{cases} 0, & \text{if } s \text{ has an even number of ones} \\ 1, & \text{otherwise.} \end{cases}$$

Since $s \sim s' \Leftrightarrow f(s) = f(s')$, the fact that $\sim$ is an equivalence relation follows from Proposition 2.3. $[1011] = \{[0001], [0010], [0100], [1000], [0111], [1011], [1101], [1110]\}$. (d) Define $f : \{\text{Physical street addresses in the US}\} \rightarrow \{\text{Zip Codes}\} :: f(\text{street addresses}) = $ its zip code. Clearly, $a_1 R a_2 \Leftrightarrow f(a_1) = f(a_2)$, so Proposition 2.3 implies that R is an equivalence relation. $[x] = \{\text{all street addresses in the US with zip code 20500 (i.e., same zip code as the White House)}\}$. **#7.** (a) Let n, m, and k be integers. (i) Reflexivity: Since $n^2 = n^2$, we have $n \, R \, n$. (ii) Symmetry: If $n \, R \, m$, this means that $n^2 = m^2$, which is the same as $m^2 = n^2$, and this means $m \, R \, n$. (iii) Transitivity: If $n \, R \, m$ and $m \, R \, k$, these mean that $n^2 = m^2$ and $m^2 = k^2$. Combining these two equations gives $n^2 = k^2$, and this means $n \, R \, k$. □ (b) Simply apply Proposition 2.3 to the function $f : \mathbb{Z} \rightarrow \mathbb{Z} :: f(n) = n^2$. □ **#9.** (a) It is reflexive and symmetric, but not transitive. (b) For this part, assume throughout that x, y, and z are positive real numbers. (i) reflexivity: The condition xRx (substitute $y = x$) is $x^2 \geq x/x \, (=1)$. This fails if $x < 1$ so we don't have reflexivity. (ii) symmetry: The implication $xRy \Rightarrow yRx$ is equivalent to $xy \geq y/x \Rightarrow yx \geq x/y$. This can easily fail, for example if we take $x = 10$ and $y = 1/10$, the implication becomes $1 \geq 1/100 \Rightarrow 1 \geq 100$ (which is false) so we don't have symmetry. (iii) transitivity: The implication $[xRy \wedge yRz] \Rightarrow xRz$ is equivalent to $[xy \geq y/x \wedge yz \geq z/y] \Rightarrow xz \geq z/x$. This one we can prove as follows: $xz = xy \cdot \dfrac{z}{y} \geq \dfrac{y}{x} \cdot \dfrac{z}{y} \geq \dfrac{z}{x}$. (We only used the hypothesis $xy \geq y/x$.) So we do have transitivity. (c) This one is an equivalence relation. This can be easily seen by verifying each of the three axioms. Alternatively, it follows from Proposition 2.3, with the function $f : \mathbb{Z}^2 \rightarrow \mathbb{Z} :: f(n,m) = |n| + |m|$. For each nonnegative integer k, there is an equivalence class consisting of all ordered pairs of integers (n,m) such that $|n| + |m| = k$. For example, when $k = 0$, we get the single element equivalence class $\{(0,0)\}$. All positive integer values of k correspond to equivalence classes with more than one element in them. For example, $k = 3$ corresponds to the equivalence class $\{(\pm 1, \pm 2), (\pm 2, \pm 1), (0, \pm 3), (\pm 3, 0)\}$. **#11.** (a) Yes, for example the relation $R = \{(1,1), (2,2), (3,3), (1,2), (2,1), (2,3), (3,2)\}$ is clearly reflexive and symmetric, but not transitive since $1R2$, $2R3$, but $1\cancel{R}3$. (b) Yes, for example the relation $R = \{(1,1), (2,2), (3,3), (1,2)\}$ is clearly reflexive and transitive, but not symmetric. (c) Yes, for example $R = \varnothing$ (empty relation) vacuously satisfies the symmetry and transitivity axioms, but is definitely not reflexive. Many other nonempty examples exist too, e.g., $R = \{(1,1)\}$. **#13.** *Proof that $\sim$ is an equivalence relation on $\mathbb{Z}^2$:* Let $(n,m), (n',m')$, and $(\tilde{n}, \tilde{m})$ be elements of $\mathbb{Z}^2$. Reflexivity: If $m = 0$, then by (i) $(n,m) \sim (n,m)$. If $m \neq 0$, then since $n/m = n/m$, (ii) tells us that $(n,m) \sim (n,m)$. Symmetry: If $(n,m) \sim (n',m')$, then either (i) $m = m' = 0$, i.e., $m' = m = 0$, which gives $(n',m') \sim (n,m)$, or (ii) $n/m = n'/m'$, i.e., $n'/m' = n/m$, which gives

$(n', m') \sim (n, m)$.   Transitivity: Suppose that $(n, m) \sim (n', m')$ and $(n', m') \sim (\tilde{n}, \tilde{m})$. Case 1: $m = m' = 0$. Since $m' = 0$, the second equivalence $(n', m') \sim (\tilde{n}, \tilde{m})$ forces $\tilde{m} = 0$, and thus $(n, m) \sim (\tilde{n}, \tilde{m})$. Case 2: $n / m = n' / m'$.   In particular, this means that $m' \neq 0$, so the second equivalence $(n', m') \sim (\tilde{n}, \tilde{m})$ must mean that $n' / m' = \tilde{n} / \tilde{m}$, which, when combined with the previous equation gives $n / m = \tilde{n} / \tilde{m}$, meaning that $(n, m) \sim (\tilde{n}, \tilde{m})$. □ The equivalence class corresponding to an ordered pair of integers (n, m) with $m \neq 0$ consists of all ordered pairs (n', m'), whose ratio n' / m' is the same rational number n / m. The equivalence class of $(n, 0)$ is $\mathbb{Z} \times \{0\}$. Thus the equivalence classes with $m \neq 0$ are in a natural bijective correspondence with the set of rational numbers $\mathbb{Q}$. If we view the exceptional equivalence class with $m = 0$ as corresponding to infinity (i.e., dividing by 0), all of the equivalence classes are then in a natural correspondence with $\mathbb{Q} \cup \{\infty\}$. **#15.** Parts (a) and (b) are routine. (c) Apply Proposition 2.3 to the function $f : A \to \mathscr{P}$ :: $f(a) =$ the unique subset of the partition $\mathscr{P}$ that contains a. (d) For any equivalence relation R on a set A, by Theorem 2.1, the set of all equivalence classes forms a partition $\mathscr{P}$ of A. The resulting equivalence relation $\sim_{\mathscr{P}}$ of Part (a) is exactly the equivalence relation R. **#19.** (a) Partial order. (b) Not a partial order; not reflexive. (c) Partial order. (d) Partial order. **#21.** Hasse diagrams for each of the four Parts (in order, from left to right) of this problem are shown in the figure below:

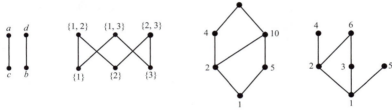

#23. Let (x, y), (x', y'), and $(\tilde{x}, \tilde{y})$ be any three points in the plane. (i) Reflexivity: Since $x \leq x$ and $y \leq y$, we certainly have $(x, y) \preceq (x, y)$. (ii) Antisymmetry: Suppose that $(x, y) \preceq (x', y')$ and that $(x', y') \preceq (x, y)$. Together these are equivalent to the double inequalities $x \leq x' \leq x$, and $y \leq y' \leq y$, which force $x = x'$ and $y = y'$, i.e., $(x, y) \preceq (x', y')$. (iii) Transitivity: Suppose that $(x, y) \preceq (x', y')$, and that $(x', y') \preceq (\tilde{x}, \tilde{y})$. Together these are equivalent to the double inequalities $x \leq x' \leq \tilde{x}$, and $y \leq y' \leq \tilde{y}$, which entails that $x \leq \tilde{x}$ and $y \leq \tilde{y}$, i.e., $(x, y) \preceq (\tilde{x}, \tilde{y})$. **#25.** This a routine verification that all of the axioms are inherited by any subset of a poset. **#27.** (a) Yes. (b) Not always. Counterexample: On $A = \{1, 2, 3\}$, both $R_1 = \{(1,1), (2,2), (1,2)\}$, $R_2 = \{(1,1), (2,2), (2,1)\}$ are partial orders but $R_1 \cup R_2$ is not (it is not anti-symmetric). **#29.** (a) The order preserving condition (of Definition 2.14) is automatically satisfied for any function $f : \{1,2\} \to \{1,2\}$, so all such functions are order preserving. There are four such functions: 1. $f(1) = f(2) = 1$, 2. $f(1) = f(2) = 2$, 3. $f(1) = 1, f(2) = 2$, 4. $f(1) = 2, f(2) = 1$. (b) For a function $f : \{1,2\} \to \{1,2,3\}$, to preserve the order relations $\leq$ on the domain and co-domain, we have to make sure that $f(1) \leq f(2)$. Of the eight possible functions $f : \{1,2\} \to \{1,2,3\}$: 1. $f(1) = f(2) = f(3) = 1$, 2. $f(1) = f(2) = f(3) = 2$, 3. $f(1) = 1, f(2) = f(3) = 2$, 4. $f(2) = 1, f(1) = f(3) = 2$, 5. $f(3) = 1, f(1) = f(2) = 2$, 6. $f(1) = f(2) = 1, f(3) = 2$, 7. $f(2) = 1, f(1) = f(3) = 2$, 8. $f(3) = 1, f(1) = f(2) = 2$; only those of 4, 5, 7, and 8 violate the order-preserving requirement, so those of 1, 2, 3, and 6 are the order preserving functions. (c) There are 16 possible functions $f : \{1,2\} \to \{1,2,3,4\}$. The order-preserving ones could be found by listing them all and checking each one. But it is perhaps more illuminating to separate into cases: Case 1: $f(1) = 1$. In this case, $f(2)$ can be assigned <u>any</u> of the four possible values and the function will be order preserving. Case 2: $f(1) = 2$. Then $f(2)$ must be 2, 3, or 4. Case 3: $f(1) = 3$. Then

$f(2)$ must be 3 or 4. Case 3: $f(1) = 4$. Then $f(2)$ must be 4. **#31.** There are seven five element four-level Hasse diagrams; they are shown below.

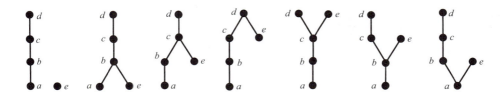

#23. (a) *Proof:* Suppose that a is a greatest element in a poset $(A, \preceq)$. To show that a is a maximal element, we need to show that if $x \in A$, $x \neq a$, then $a \npreceq x$. Indeed, if $a \preceq x$, then since we also have (since a is a greatest element) $x \preceq a$, anti-symmetry would imply $x = a$ —a contradiction! □ (b) *Proof:* Suppose that both a and b are greatest elements in a poset $(A, \preceq)$. The fact that a is a greatest element implies $b \preceq a$, and the fact that b is a greatest element implies $a \preceq b$. From these two relations, anti-symmetry implies that $a = b$, proving that greatest elements are unique. □ (c) The poset of the integers $(\mathbb{Z}, \leq)$ (endowed with its natural less than or equal to ordering) has no maximal (nor minimal) element. Proof that a finite poset $(A, \preceq)$ always has a maximal element: Let $(A, \preceq)$ be any finite poset with N elements in it, and suppose that there were no maximal elements. We will obtain a contradiction. If we negate the definition of an element $a \in A$ being maximal, we get $\exists x \in A \sim [x \neq a \Rightarrow a \npreceq x]$. Recalling that the negation of an implication $\sim [P \to Q] \equiv P \wedge \sim Q$ (Theorem 1.1 II(b)), the preceding can be expressed in the following more useful form: (*) $\exists x \in A [x \neq a \wedge a \preceq x]$. We use this fact to construct a sequence of $N + 1$ elements in $(A, \preceq)$ as follows: Let $a_1 \in A$ be any element. By (*) (since a_1 is not maximal), there exists an element $a_2 \in A$ with $a_2 \neq a_1 \wedge a_1 \preceq a_2$. Next, since (since a_2 is not maximal) (*) shows that there exists an element $a_3 \in A$ with $a_3 \neq a_2 \wedge a_2 \preceq a_3$. Notice that a_3 must also be different from a_1, since we know $a_1 \preceq a_2$ and $a_2 \preceq a_3$, so if $a_3 = a_1$ antisymmetry would imply $a_3 = a_2$, which we know is not true. In the same fashion (since a_3 is not maximal), we could find an element $a_4 \in A$, such that $a_4 \neq a_3 \wedge a_3 \preceq a_4$ and a_1, a_2, a_3, a_4 are all different. If we continue with this process, after $N + 1$ steps, we will have constructed a sequence of $N + 1$ different elements of A $a_1, a_2, a_3, \cdots, a_{N+1}$, but this is a contraction to the fact that A contains only $N + 1$ elements. (Note: This proof is actually a disguised version of a proof by mathematical induction; this proof technique will be developed in detail in Section 3.1.) **#35.** (Look at the Hasse diagrams for the Exercise 21.) (a) greatest: none, least: none, maximal: a, d, minimal: b, c. (b) greatest: none, least: none, maximal: $\{1,2\}$, $\{1,3\}$, $\{2,3\}$, minimal: $\{1\}, \{2\}, \{3\}$ (c) greatest: 20, least: 1, maximal: 20, minimal: 1 (d) greatest: none, least: 1, maximal: 4, 5, 6, minimal: 1 **#37.** (Look at the Hasse diagrams for the Exercise for the Reader 2.12.) greatest: 60, least: 1, maximal: 60, minimal: 1 **#39.** (a) True. $\hat{a}$ and $\hat{b}$ are the maximal elements of A and B respectively, if, and only if, $(\hat{a}, \hat{b})$ is a maximal element of C. (b) True.

Section 2.3: **#1.** (a) $(0 + \overline{1}) \cdot 1 = (0 + 0) \cdot 1 = 0 \cdot 1 = 0$ (b) $\overline{(0+1)} \cdot 1 = \overline{1} \cdot 1 = 0 \cdot 1 = 0$
(c) $\overline{\overline{\overline{(0+1)} \cdot (1+1)}} = \overline{\overline{\overline{1} \cdot \overline{1}}} = \overline{\overline{0 \cdot 0}} = \overline{\overline{0}} = 1$ (d) $0 + \overline{0} + \overline{\overline{0}} = 0 + 1 + 0 = 1$
#3. The results of each part are built up in the single table below. In cases where the Boolean function's column contains multiple values, the values in bold fonts are the values of the function.

x	y	$\bar{x}$	$\bar{y}$	$x\bar{y}$	$\bar{x}y$	(a) $x\bar{y}+\bar{x}y$	(b) $\overline{x\bar{y}}+\overline{\bar{x}y}$	(c) $\overline{x\bar{y}}+\overline{\bar{x}y}$	(d) $1+x\bar{y}+\bar{x}y$
1	1	0	0	0	0	0	1 1 1	0 0 0 : 1	1
1	0	0	1	1	0	1	0 1 1	1 1 0 : 0	1
0	1	1	0	0	1	1	1 1 0	0 1 1 : 0	1
0	0	1	1	0	0	0	1 1 1	0 0 0 : 1	1

#5. (a) Although the Boolean function $f(x,y,z)=x\bar{y}+\bar{x}y$ is defined as a function of three Boolean variables, the formula involves only the first two, so its values are given in Part (a) of the solution to Exercise 3. That is, the formula is independent of the value of the input value z.

NOTE: A table with eight rows giving the values of all possible combinations of the three Boolean variables could be easily written down from the two-variable table but there would be no point in doing this. By the same token, any Boolean expression can be viewed as a Boolean function in all of the involved Boolean variables, and any number of additional Boolean variables.

(b)

x	y	z	$\bar{x}$	$\bar{y}$	$\bar{z}$	$x\bar{y}z$	$\bar{x}y\bar{z}$	$\overline{\bar{x}y\bar{z}}$	$x\bar{y}z + \overline{\bar{x}y\bar{z}}$
1	1	1	0	0	0	0	0	1	1
1	1	0	0	0	1	0	0	1	1
1	0	1	0	1	0	1	0	1	1
1	0	0	0	1	1	0	0	1	1
0	1	1	1	0	0	0	0	1	1
0	1	0	1	0	1	0	1	0	0
0	0	1	1	1	0	0	0	1	1
0	0	0	1	1	1	0	0	1	1

(c)

x	y	z	$\bar{x}$	$\bar{y}$	$\bar{z}$	$x+\bar{y}+z$	$y\bar{z}$	$\bar{x}+y\bar{z}$	$\overline{(x+\bar{y}+z)(\bar{x}+y\bar{z})}$
1	1	1	0	0	0	1	0	0	0
1	1	0	0	0	1	1	1	1	0
1	0	1	0	1	0	1	0	0	0
1	0	0	0	1	1	1	0	0	0
0	1	1	1	0	0	1	0	1	0
0	1	0	1	0	1	0	1	1	1
0	0	1	1	1	0	1	0	1	0
0	0	0	1	1	1	1	0	1	0

(d)

w	x	y	z	$\bar{y}$	$w+\bar{y}$	xz	$w+\bar{y}+xz$
1	1	1	1	0	1	1	1
1	1	1	0	0	1	0	1
1	1	0	1	1	1	1	1
1	1	0	0	1	1	0	1
1	0	1	1	0	1	0	1
1	0	1	0	0	1	0	1
1	0	0	1	1	1	0	1
1	0	0	0	1	1	0	1
0	1	1	1	0	0	1	1
0	1	1	0	0	0	0	0

0	1	0	1	1	1	1	1
0	1	0	0	1	1	0	1
0	0	1	1	0	0	0	0
0	0	1	0	0	0	0	0
0	0	0	1	1	1	0	1
0	0	0	0	1	1	0	1

#7. (a) The following table establishes the identity: $x + \overline{(x\bar z)}y = x + y$.

x	y	z	$x+y$	$\bar z$	$x\bar z$	$\overline{(x\bar z)}$	$\overline{(x\bar z)}y$	$x+\overline{(x\bar z)}y$
1	1	1	1	0	0	1	1	1
1	1	0	1	1	1	0	0	1
1	0	1	1	0	0	1	0	1
1	0	0	1	1	1	0	0	1
0	1	1	1	0	0	1	1	1
0	1	0	1	1	0	1	1	1
0	0	1	0	0	0	1	0	0
0	0	0	0	1	0	1	0	0

(b) We now prove the identity by setting each side equal to zero and solving for the variables. First, $x+y=0 \Leftrightarrow x=y=0$. Next, $x+\overline{(x\bar z)}y=0 \Leftrightarrow x=0$ and $\overline{(x\bar z)}y=0$. The latter equality is equivalent to $x\bar z =1$ or $y=0$. Thus the right hand expression is zero if the left hand expression is zero. The only other possible way that the right hand expression could be zero would be if $x=0$, $y=1$, and $x\bar z = 1$, which could never happen since $x=0$ implies $x\bar z = 0$. This proves the two solution sets are identical, so the two Boolean expressions are equivalent.

(c) Below is a proof of the identity using Theorem 2.4. We freely use the commutative and associative laws of Boolean addition and multiplication (Theorem 2.4(a,b)).

$$
\begin{aligned}
x + \overline{(x\bar z)}y &= x + (\bar x + \bar{\bar z})\,y && \text{(De Morgan's Law—Theorem 2.4(d))}\\
&= x + (\bar x + z)\,y && \text{(Double Complementation—Theorem 2.4(e))}\\
&= x + xy + \bar x y + zy && \text{(Absorption—Theorem 2.4(f))}\\
&= x + 1\cdot y + zy && \text{(Unit Identity—Theorem 2.4(i))}\\
&= x + y + zy && \text{(Identity Law—Theorem 2.4(g))}\\
&= x + y && \text{(Absorption—Theorem 2.4(f))} \;\square
\end{aligned}
$$

#9. (a) The following table establishes the identity: $x\bar y + y\bar z + \bar x z = \bar x y + \bar y z + x\bar z$.

x	y	z	$\bar x$	$\bar y$	$\bar z$	$x\bar y$	$y\bar z$	$\bar x z$	$x\bar y + y\bar z + \bar x z$	$\bar x y$	$\bar y z$	$x\bar z$	$\bar x y + \bar y z + x\bar z$
1	1	1	0	0	0	0	0	0	0	0	0	0	0
1	1	0	0	0	1	0	1	0	1	0	0	1	1
1	0	1	0	1	0	1	0	0	1	0	1	0	1
1	0	0	0	1	1	1	0	0	1	0	0	1	1
0	1	1	1	0	0	0	0	1	1	1	0	0	1
0	1	0	1	0	1	0	1	0	1	1	0	0	1
0	0	1	1	1	0	0	0	1	1	0	1	0	1
0	0	0	1	1	1	0	0	0	0	0	0	0	0

(b) We now prove the identity by setting each side equal to zero and solving for the variables. First, $x\bar y + y\bar z + \bar x z = 0 \Leftrightarrow$ (1) ($x=0$ or $y=1$), (2) ($y=0$ or $z=1$), and (3) ($x=1$ or $z=0$). Similarly, $\bar x y + \bar y z + x\bar z = 0 \Leftrightarrow$ (4) ($x=1$ or $y=0$), (4) ($y=1$ or $z=0$), and (6) ($x=0$ or $z=1$).

To see these three sets of conditions lead to identical solutions, we separate into cases:

Case 1: $x=0$: (3) forces $z=0$ for the left expression, and then (2) forces $y=0$ for this expression. For the right hand expression, (4) forces $y=0$, and then (5) forces $z=0$.

Case 2: $x = 1$: (1) forces $y = 1$ for the left hand expression, and (2) then forces $z = 1$ for this expression. For the right hand expression, (6) forces $z = 1$, and then (5) forces $y = 1$.

#11. (a) (i) $(f \cdot g)(x, y, z) = (x\bar{y} + \bar{x}y)(x\bar{y}z + \overline{\bar{x}y\bar{z}})$　　We first simplify the more complicated second factor:

$$x\bar{y}z + \overline{\bar{x}y\bar{z}} = x\bar{y}z + x + y\bar{z} \qquad \text{(De Morgan's Law, Double Complementation—Theorem 2.4(d),(e))}$$
$$= x + y\bar{z} \qquad\qquad\quad \text{(Absorption—Theorem 2.4(f))}$$
$$= x + \bar{y} + z \qquad\qquad \text{(De Morgan's Law, Double Complementation—Theorem 2.4(d),(e))}$$

Substituting this into the above function expression, we may simplify it further by expanding it and freely using the zero identity law $x\bar{x} = 0$, and the dominance law $x0 = 0$, as follows

$$(f \cdot g)(x, y, z) = (x\bar{y} + \bar{x}y)(x + \bar{y} + z) = x\bar{y} + \bar{x} + z$$

(ii) $(f + \bar{h})(x, y, z) = x\bar{y} + \bar{x}y + \overline{(x + \bar{y} + z)(\bar{x} + y\bar{z})}$　　The last term may be simplified using Boolean identities from Theorem 2.4:

$$\overline{(x + \bar{y} + z)(\bar{x} + y\bar{z})}$$
$$= (x + \bar{y} + z) + \overline{(\bar{x} + y\bar{z})} \quad \text{(De Morgan's Law, Double Complementation—Theorem 2.4(d),(e))}$$
$$= x + \bar{y} + z + xy\bar{z} \qquad\quad \text{(De Morgan's Law, Double Complementation—Theorem 2.4(d),(e))}$$
$$= x + \bar{y} + z + x(\bar{y} + z) \quad \text{(De Morgan's Law, Double Complementation—Theorem 2.4(d),(e))}$$
$$= x + \bar{y} + z \qquad\qquad\quad \text{(Absorption—Theorem 2.4(f))}$$

Since another application of absorption lets us combine $x\bar{y} + x = x$, we arrive at the following simplified expression for the given function: $(f + \bar{h})(x, y, z) = x + \bar{x}y + \bar{y} + z$.

(b) The formulas for the Boolean functions f and k of Exercise 5 are already in sum-of-products form. For the other two functions we can read off the products of literals in the sum-of-products form by looking at the rows of the corresponding tables that were created for these functions in the solution of Exercise 5. The table for g has only one row that gives an output of 1: $x = 0, y = 1, z = 0$. Thus, the sums-of-products expansion of g consists of the single corresponding product: $g(x, y, z) = \bar{x}y\bar{z}$. The table for h is the opposite, all rows give an output of 1 except for the row: $x = 0, y = 1, z = 0$. Thus, the sums-of-products expansion of h consists of the sum of the remaining seven products of literals:
$$h(x, y, z) = xyz + xy\bar{z} + x\bar{y}z + x\bar{y}\,\bar{z} + \bar{x}yz + \bar{x}\,\bar{y}z + \bar{x}\,\bar{y}\,\bar{z}.$$

(c) $f^d(x, y, z) = [x\bar{y} + \bar{x}y]^d = (x + \bar{y})(\bar{x} + y)$, $g^d(x, y, z) = \left[x\bar{y}z + \overline{\bar{x}y\bar{z}}\right]^d = (x + \bar{y} + z)(\overline{\bar{x} + y + \bar{z}})$

$h^d(x, y, z) = \left[\overline{(x + \bar{y} + z)(\bar{x} + y\bar{z})}\right]^d = \overline{x}\bar{y}z + \bar{x}(y + \bar{z})$, $k^d(w, x, y, z) = [w + \bar{y} + xz]^d = w\bar{y}(x + z)$

#13. (a) $x[(x + \bar{z}) + y] = xy$

(b)

x	y	z	xy	$\bar{z}$	$x + \bar{z}$	$\overline{(x + \bar{z})}$	$\overline{(x + \bar{z})} + y$	$x[\overline{(x + \bar{z})} + y]$
1	1	1	1	0	1	0	1	1
1	1	0	1	1	1	0	1	1
1	0	1	0	0	1	0	0	0
1	0	0	0	1	1	0	0	0
0	1	1	0	0	0	1	1	0
0	1	0	0	1	1	0	1	0
0	0	1	0	0	0	1	1	0
0	0	0	0	1	1	0	0	0

(c) We now prove the identity by setting each side equal to one and solving for the variables. First, $xy = 1 \Leftrightarrow x = y = 1$. Next, $x[\overline{(x + \bar{z})} + y] = 1 \Leftrightarrow x = 1$ and $\overline{(x + \bar{z})} + y = 1$. The latter equality is equivalent to $x + \bar{z} = 0$ or $y = 1$. Thus the right hand expression is one if the left hand expression is one. The only other possible way that the right hand expression could be one would be if $x = 1, y = 0$, and $x + \bar{z} = 0$, which could never happen since $x = 1$ implies $x + \bar{z} = 1$. This proves the two solution sets are identical, so the two Boolean expressions are equivalent.

NOTE: This proof is just the "dual" of the proof of Exercise 7(b).

(d) We freely use the commutative and associative laws of Boolean addition and multiplication (Theorem 2.4(a,b)).

$$
\begin{aligned}
x[\overline{(x+\overline{z})}+y] &= x[(\overline{x}\,\overline{\overline{z}})+y] && \text{(De Morgan's Law—Theorem 2.4(d))}\\
&= x[(\overline{x}z)+y] && \text{(Double Complementation—Theorem 2.4(e))}\\
&= x(x+y)[(\overline{x}z)+y] && \text{(Absorption—Theorem 2.4(f))}\\
&= xx\overline{x}z + xxy + xy\overline{x}z + xxy && \text{(Distributivity—Theorem 2.4(c))}\\
&= xy + xy && \text{(Using } xx = x\text{—Theorem 2.4(h), and } x\overline{x}=0\text{—Thm. 2.4(i))}\\
&= xy && \text{(Idempotent Law—Theorem 2.4(h)) } \square
\end{aligned}
$$

#15. $\overline{x}y(x+\overline{y})$

#17. $\left[\overline{\overline{xyz}+\overline{y}}\right]\left[y+\overline{y}+\overline{z}\right]$

#19. (a)

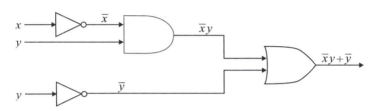

(b)

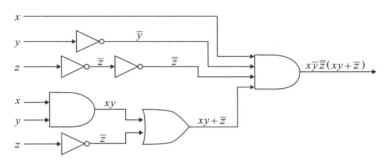

(c)

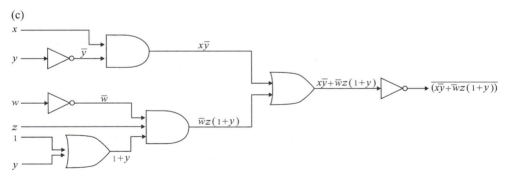

(d)

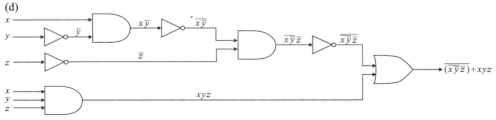

#21.   The Boolean function $f(x,y,z)$ for the circuit has inputs x, y, z, being the votes of the three voters (1 = yes, 0 = no). It is defined by $f(1,1,0) = f(1,0,1) = f(0,1,1) = f(1,1,1) = 1$ (pass), and it equals 0 (fail) at all other inputs. The sums-of-products expansion is thus given by:
$$f(x,y,z) = xy\bar{z} + x\bar{y}z + \bar{x}yz + xyz$$
This Boolean function is identical to that which was constructed in Example 2.17, so the circuit diagram has already been drawn in Figure 2.15.

#23.   The Boolean function $f(w,x,y,z)$ for the circuit has inputs w, x, y, z, being the votes of the four voters (1 = yes, 0 = no). In order to pass (f = 1) with three votes, since a yes vote of the first voter counts as two votes, either a yes vote from the first voter and at one from at least one other voter:
$$f(1,1,0,0) = f(1,0,1,0) = f(1,0,0,1) = f(1,1,1,0) = f(1,1,0,1) = f(1,0,1,1) = f(1,1,1,1) = 1,$$
or else yes votes from all but the first voter: $f(0,1,1,1) = 1$. The sums-of-products expansion is thus given by:
$$f(w,x,y,z) = wx\bar{y}\,\bar{z} + w\bar{x}y\bar{z} + w\bar{x}\,\bar{y}z + wxy\bar{z} + wx\bar{y}z + w\bar{x}yz + wxyz + \bar{w}xyz.$$
The corresponding circuit design is shown on the next page:

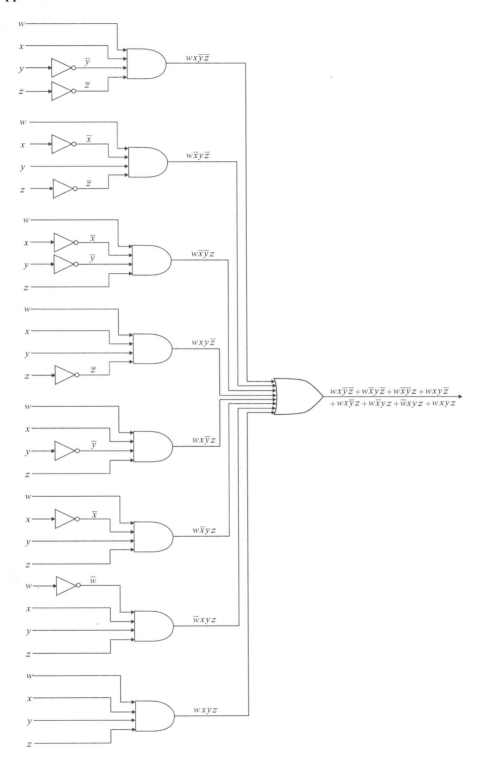

#25. Karnaugh maps along with the resulting simplified Boolean expression for each part are shown below. Note that in Parts (a) and (c) there is no real simplification, and the total simplifications of Parts (b) and (d) are identical.

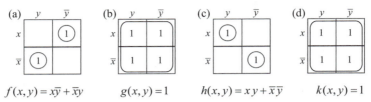

$$f(x,y) = x\bar{y} + \bar{x}y \qquad g(x,y) = 1 \qquad h(x,y) = xy + \bar{x}\bar{y} \qquad k(x,y) = 1$$

#27. (a) The Karnaugh map and (simplified) Boolean expression of Part (a) of the above solution of Exercise 25 work equally well for the same function when viewed as a function of three variables.
(b) The ones in the Karnaugh map can be read off from the table of values in the solution of Exercise 5. Below we illustrate two different ways to complete the Karnaugh map, along with the corresponding simplified Boolean expressions for each. We will usually proceed in only one way, but the reader should keep in mind that the resulting simplified Boolean expressions need not be unique.

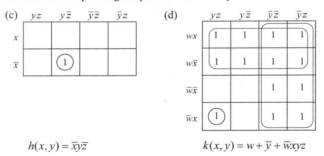

Method 1:   Method 2:

$$g(x,y) = x + \bar{y} + \bar{x}yz \qquad\qquad g(x,y) = x + z + \bar{x}\,\bar{y}\,\bar{z}$$

(c) and (d): The ones in the Karnaugh maps can be read off from the tables in the solution of Exercise 5. Their completion and the corresponding simplified Boolean expressions are shown below:

(c)

	yz	y$\bar{z}$	$\bar{y}\bar{z}$	$\bar{y}z$
x				
$\bar{x}$		1		

(d)

	yz	y$\bar{z}$	$\bar{y}\bar{z}$	$\bar{y}z$
wx	1	1	1	1
w$\bar{x}$	1	1	1	1
$\bar{w}\bar{x}$			1	1
$\bar{w}x$	1		1	1

$$h(x,y) = \bar{x}y\bar{z} \qquad\qquad k(x,y) = w + \bar{y} + \bar{w}xyz$$

#29. (a) The Boolean function of Figure 2.24(a) was found (in Exercise 15) to be $f(x,y) = \bar{x}y(x+\bar{y})$. This function is (easily) seen to be identically zero, for example: $f = 1 \Rightarrow x = 0,\ y = 1$, and $x + \bar{y} = 1$, but since the first two conditions imply $x + \bar{y} = 0$, it follows that f can never equal one, i.e., $f(x,y) = 0$. The Karnaugh map has all of its cells empty, and the circuit is empty. (Indeed, for a circuit that should always be "off" no device is needed.)

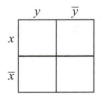

(b) The Boolean function of Figure 2.25(a) was found (in Exercise 17) to be $f(x,y,z) = \left[\overline{xyz + \bar{y}}\right]\left[y + \bar{y} + \bar{z}\right]$. To find the terms in the sum-of-products representation, a table could always be created; but here is a direct approach after a simplification using Theorem 12.4: Using the unit identity: $y + \bar{y} = 1$, the dominance law: $1 + \bar{z} = 1$, the identity law now shows that $f(x,y,z) = \overline{xyz + \bar{y}}$. Thus, $f(x,y,z) = 0 \Leftrightarrow xyz = 1$ and $y = 1 \Leftrightarrow x = y = z = 1$. Thus, xyz is the only product missing in the sum-of-products expansion, and so:

$$f(x,y,z) = xy\bar{z} + x\bar{y}z + \bar{x}yz + x\bar{y}\,\bar{z} + \bar{x}y\bar{z} + \bar{x}\,\bar{y}z + \bar{x}\,\bar{y}\,\bar{z}.$$

The Karnaugh map shown below shows that this Boolean function can be expressed using the following simpler formula: $f(x,y,z) = \overline{z} + \overline{y}z + \overline{x}yz$. The corresponding circuit design is shown to the right of the Karnaugh map.

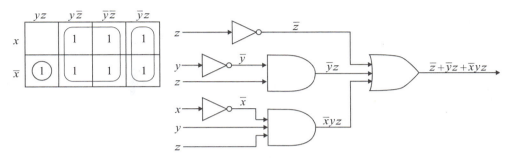

#31. (a) Either of the Karnaugh maps of our solution to Exercise 27(b) is readily modified to obtain the one shown at the right so as to incorporate all of the "don't care" conditions. The resulting Boolean expression $xy + x\overline{y}z$ serves as a Boolean function to match the given requirements.

	yz	$y\overline{z}$	$\overline{y}\,\overline{z}$	$\overline{y}z$
x	1	1	X	1
$\overline{x}$	X		X	X

(b) We determine the "1-cells" in an (ungrouped) Karnaugh map for the Boolean function of Exercise 19(c) (either by creating a table or via simplifying the expression and proceeding directly as in the solution of Exercise 29(b)). The result is shown on the left figure below. The right figure has the "don't care" conditions added and the groupings for the completion of the Karnaugh map. The resulting Boolean expression $w + \overline{w}\,\overline{z}$ will thus match the original Boolean function with the allowance of the "don't care" conditions.

	yz	$y\overline{z}$	$\overline{y}\,\overline{z}$	$\overline{y}z$
wx	1	1		
$w\overline{x}$	1	1	1	1
$\overline{w}\,\overline{x}$		1	1	
$\overline{w}x$		1	1	

$\rightarrow$

	yz	$y\overline{z}$	$\overline{y}\,\overline{z}$	$\overline{y}z$
wx	1	1	X	X
$w\overline{x}$	1	1	1	1
$\overline{w}\,\overline{x}$	X	1	X	X
$\overline{w}x$		1	1	

(c) The same approach used in Part (b) leads to the completion of the Karnaugh map as shown below:

	yz	$y\overline{z}$	$\overline{y}\,\overline{z}$	$\overline{y}z$
wx		1	1	1
$w\overline{x}$	1	1	1	1
$\overline{w}\,\overline{x}$	1	1	1	1
$\overline{w}x$		1	1	1

$\rightarrow$

	yz	$y\overline{z}$	$\overline{y}\,\overline{z}$	$\overline{y}z$
wx	X	1	1	1
$w\overline{x}$	1	1	1	1
$\overline{w}\,\overline{x}$	X	X	1	1
$\overline{w}x$		1	1	1

The resulting Boolean expression $w + \overline{w}\,\overline{y} + \overline{w}y\overline{z}$ will thus meet match the original Boolean function with the allowance of the "don't care" conditions.

#35. (a) $(x+y)(\overline{x}+\overline{y})$ (b) $x+\overline{y}+z$

(c) $(x+y+z)(x+y+\overline{z})(\overline{x}+y+z)(\overline{x}+\overline{y}+z)(\overline{x}+y+\overline{z})(x+\overline{y}+\overline{z})(\overline{x}+\overline{y}+\overline{z})$

(d) $(w+\overline{x}+\overline{y}+z)(w+x+\overline{y}+\overline{z})(w+x+\overline{y}+z)$

#37. (a) six

(b) Illustration of a six variable Karnaugh map diagram, along with the sets of six "adjacent" cells for a specific pair of cells: a corner cell (indicated by black-filled circle) and an interior cell (indicated by a gray-filled circle).

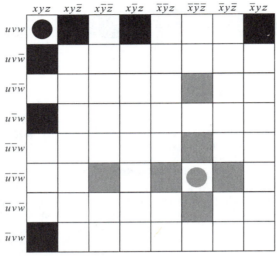

CHAPTER 3:

Section 3.1: **#1.** (a) Let $S(n)$ denote the statement $\sum_{i=1}^{n} 2^i = 2^{n+1} - 2$, for $n \in \mathbb{Z}_+$. 1. *Basis Step*:

$S(1)$ is obtained by substituting $n = 1$: $\sum_{i=1}^{1} 2^i = 2^{1+1} - 2$ or $2^1 = 2^2 - 2$, which is true. 2. *Inductive*

Step: Let $k \in \mathbb{Z}_+$. Our task is to assume that $S(k)$ is true, and use this (inductive hypothesis) do

deduce   that $S(k+1)$ is   true,   i.e.,   that $\sum_{i=1}^{k+1} 2^i = 2^{[k+1]+1} - 2$,   or,   in   simpler   terms:

$\sum_{i=1}^{k+1} 2^i = 2^{k+2} - 2$, (*). Breaking off the last term from the summation on the left, and invoking the

inductive hypothesis on the remaining terms, we obtain: $\sum_{i=1}^{k+1} 2^i = \sum_{i=1}^{k} 2^i + 2^{k+1} = 2^{k+1} - 2 + 2^{k+1}$.

But $2^{k+1} - 2 + 2^{k+1} = 2^{k+1} + 2^{k+1} - 2 = 2 \cdot 2^{k+1} - 2 = 2^{k+2} - 2$, which is the right side of (*), so the proof

is complete. (b) We can also prove the formula without induction by invoking the finite geometric

series formula (3) $\sum_{i=0}^{n} ar^i (\equiv a + ar + ar^2 + \cdots + ar^n) = (ar^{n+1} - a)/(r-1)$. To see this, we simply

rewrite $\sum_{i=1}^{n} 2^i$ as (make the index substitution $i \to i+1$) $\sum_{i=0}^{n-1} 2^{i+1} = \sum_{i=0}^{n-1} 2 \cdot 2^i$, and use (3) with

parameters $a \to 2, n \to n-1$, to get that $\sum_{i=0}^{n-1} 2 \cdot 2^i = (2 \cdot 2^{[n-1]+1} - 2)/(2-1) = (2^{n+1} - 2)/1 = 2^{n+1} - 2$,

as desired. □   **#3.**   For any positive integer n, we let $S(n)$ denote the equation

$3 + 7 + 11 + \cdots + (4n - 1) = n(2n + 1)$. 1. *Basis Step*: $S(1)$ is the equation $3 = 1 \cdot (2 \cdot 1 + 1)$, which is true.

(Note that when $n = 1$, $(4n - 1) = 3$, so the first term on the left of $S(1)$ is the last term.) 2. *Inductive*

Step: Let $k \in \mathbb{Z}_+$. Assume that $S(k)$ is true; we must deduce that $S(k+1)$ is true, i.e., that

$3 + 7 + 11 + \cdots + (4[k+1] - 1) = [k+1](2[k+1] + 1)$.   Note that the right side of this equals

$(k+1)(2k+3)$. To see that this equals the left side, we break off the last term, and apply the inductive

hypothesis (since the sum of the first k terms on the left of $S(k + 1)$ is just the left side of $S(k)$):

$$3 + 7 + 11 + \cdots + (4[k+1]-1) = 3 + 7 + 11 + \cdots + (4k+1) + (4[k+1]-1)$$
$$= k(2k+1) + (4[k+1]-1) \qquad \text{(By the inductive hypothesis)}$$
$$= k(2k+1) + (4k+3) = 2k^2 + 5k + 3 = (k+1)(2k+3). \ \square$$

#5. For any positive integer n, we let $S(n)$ denote the statement that the general triangle inequality $|a_1 + a_2 + \cdots + a_n| \le |a_1| + |a_2| + \cdots + |a_n|$ (with exactly n terms) is valid for any n real numbers $a_1, a_2, \cdots, a_n$. 1. *Basis Step:* $S(1)$ simply states that $|a_1| \le |a_1|$, for any real number a_1. This is trivially valid. 2. *Inductive Step:* Let $k \in \mathbb{Z}_+$. We assume that $S(k)$ is true, i.e., that the general triangle inequality is valid for k numbers, and use this (inductive hypothesis) to deduce that $S(k+1)$ is true, i.e., that the general triangle inequality is true for $k+1$ numbers. We will also make use of the (ordinary) triangle inequality (for two numbers) $|x+y| \le |x| + |y|$. (By the way, this fact easily follows from the fact that any sidelength of any triangle cannot exceed the sum of the other two sidelengths; alternatively, it can be easily proved analytically by considering the four cases depending on whether $x \ge 0$, and/or $y \ge 0$.) We have:

$$|a_1 + a_2 + \cdots + a_{k+1}| = |(a_1 + a_2 + \cdots + a_k) + a_{k+1}|$$
$$\le |a_1 + a_2 + \cdots + a_k| + |a_{k+1}| \qquad \text{(By the ordinary triangle inequality for two numbers)}$$
$$\le |a_1| + |a_2| + \cdots + |a_k| + |a_{k+1}| \ \text{(By the inductive hypothesis)} \ \square$$

#7. For any positive integer n, we let $S(n)$ denote the equation $\sum_{i=1}^{n} i^3 = [n(n+1)/2]^2$. 1. *Basis Step:* $S(1)$ is the equation $1^3 = [1 \cdot (1+1)/2]^2$, which is true. 2. *Inductive Step:* Let $k \in \mathbb{Z}_+$. Assume that $S(k)$ is true, we must deduce that $S(k+1)$ is true, i.e., that $\sum_{i=1}^{k+1} i^3 = [[k+1]([k+1]+1)/2]^2$. Note that the right side can be written more simply as $[(k+1)(k+2)/2]^2$. We proceed in the usual fashion (for this type of problem) by breaking off the first k terms on the left and applying the inductive hypothesis:

$$\sum_{i=1}^{k+1} i^3 = \sum_{i=1}^{k} i^3 + (k+1)^3$$
$$= [k(k+1)/2]^2 + (k+1)^3 \qquad \text{(By the inductive hypothesis)}$$
$$= (k+1)^2 \{[k/2]^2 + (k+1)\} = (k+1)^2 (1/2)^2 \{k^2 + 4(k+1)\}$$
$$= (k+1)^2 (1/2)^2 (k+2)^2 = [(k+1)(k+2)/2]^2 \ \square$$

#9. Let $S(n)$ denote the equation $1 \cdot 2 + 2 \cdot 3 + \cdots + n(n+1) = n(n+1)(n+2)/3$ $(n \in \mathbb{Z}_+)$. 1. *Basis Step:* $S(1)$ is the equation $1 \cdot 2 = 1 \cdot (1+1)(1+2)/3$, which is true. 2. *Inductive Step:* Let $k \in \mathbb{Z}_+$. We assume that $S(k)$ is true, and we must deduce that $S(k+1)$ is true, i.e., that $1 \cdot 2 + 2 \cdot 3 + \cdots + [k+1]([k+1]+1) = [k+1]([k+1]+1)([k+1]+2)/3$. Note that the right side can be written more simply as $(k+1)(k+2)(k+3)/3$.

$$1 \cdot 2 + 2 \cdot 3 + \cdots + [k+1]([k+1]+1) = 1 \cdot 2 + 2 \cdot 3 + \cdots k(k+1) +$$
$$\text{(By the inductive hypothesis)} \quad = k(k+1)(k+2)/3 + (k+1)(k+2)$$
$$= (k+1)(k+2)/3\{k+3\} = (k+1)(k+2)(k+3) \ \square$$

#11. Let $S(n)$ denote the equation $\sum_{i=1}^{n} (-1)^i i^2 = (-1)^n [n(n+1)/2]$ $(n \in \mathbb{Z}_+)$. 1. *Basis Step:* $S(1)$ is the equation $(-1)^1 \cdot 1^2 = (-1)^1 [1(1+1)/2]$, which is true. 2. *Inductive Step:* Let $k \in \mathbb{Z}_+$. We assume that $S(k)$ is true, and we must deduce that $S(k+1)$ is true, i.e., that $\sum_{i=1}^{k+1} (-1)^i i^2 = (-1)^{k+1}[(k+1)((k+1)+1)/2]$. Note that the right side can be written more simply as $(-1)^{k+1}[(k+1)(k+2)/2]$.

$$\sum_{i=1}^{k+1}(-1)^i i^2 = \sum_{i=1}^{k}(-1)^i i^2 + (-1)^{k+1}(k+1)^2$$
$$= (-1)^k[k(k+1)/2] + (-1)^{k+1}(k+1)^2 \qquad \text{(By the inductive hypothesis)}$$
$$= (-1)^{k+1}(k+1)\{-k/2 + (k+1)\} = (-1)^{k+1}(k+1)(k/2+1) = (-1)^{k+1}(k+1)(k+2)/2. \ \square$$

#13. (a) Two cents, and all integer postages greater than 3 cents. (b) We will prove using strong mathematical induction that any postage greater than 3 cents can be obtained using 2 cent and or 5 cent stamps. 1. *Basis Step*: 4 cents can be made using two 2 cent stamps. 2. *(Strong) Inductive Step*: Let k be a positive integer greater than 3, and assume that all postages from 4 cents to k cents can be obtained using 2 cent and/or 5 cent stamps. We need to show that a postage $k+1$ cents can be so obtained. Case 1: $k=4$, then $k+1=5$ can be trivially be achieved (with a single 5 cent stamp). Case 2: $k>4$, then $k+1-2 = k-1 > 3$, so by the (strong) inductive hypothesis, a postage of $k+1-2$ can be built using 2 cent and/or 5 cent stamps. Adding on a 2 cent stamp to this combination yields a desired representation of the postage $[k+1-2]+2 = k+1$ cents. $\square$

#15. Noninductive proof of (3): The expansion

$$(r-1)(a + ar + ar^2 + \cdots + ar^n) = \quad ar + ar^2 + ar^3 \cdots + ar^{n+1}$$
$$\underline{-a - ar - ar^2 - \cdots - ar^n}$$
$$-a \qquad\qquad\qquad + ar^{n+1}$$

gives $(r-1)(a + ar + ar^2 + \cdots + ar^n) = ar^{n+1} - a$, which becomes (3) when we divide both sides by $r-1$.

#17. Let $S(n)$ denote the equation $\sqrt{2 + \sqrt{2 + \sqrt{2 + \cdots + \sqrt{2}}}} = 2\cos\left(\dfrac{\pi}{2^{n+1}}\right)$ $(n \in \mathbb{Z}_+)$. It will be convenient (for the inductive step) to denote the left side of $S(n)$ as $L(n)$. 1. *Basis Step*: $S(1)$ is the equation $\sqrt{2} = 2\cos\left(\dfrac{\pi}{2^{1+1}}\right)$, which is true since (from basic trigonometry) $2\cos\left(\dfrac{\pi}{2^{1+1}}\right) = 2\cos\left(\dfrac{\pi}{4}\right) = 2\left(\dfrac{\sqrt{2}}{2}\right) = \sqrt{2}$. 2. *Inductive Step*: Let $k \in \mathbb{Z}_+$. We assume that $S(k)$ is true, and we must deduce that $S(k+1)$ is true, i.e., that $L(k+1) = 2\cos\left(\dfrac{\pi}{2^{[k+1]+1}}\right)$. We first note that $L(k+1) = \sqrt{2 + L(k)}$. If we apply the inductive hypothesis, i.e., that $L(k) = 2\cos\left(\dfrac{\pi}{2^{k+1}}\right)$, this equation becomes:

$$L(k+1) = \sqrt{2 + L(k)} = \sqrt{2 + 2\cos\left(\dfrac{\pi}{2^{k+1}}\right)} = 2\sqrt{\dfrac{1 + \cos\left(\dfrac{\pi}{2^{k+1}}\right)}{2}} = 2\sqrt{\dfrac{1 + \cos\left(2 \cdot \pi / 2^{k+2}\right)}{2}}.$$

If we apply the half angle formula $\cos(x) = \pm\sqrt{\dfrac{1 + \cos(2x)}{2}}$, the above equation becomes

$L(k+1) = 2\cos\left(\dfrac{\pi}{2^{k+2}}\right)$, which is what we needed to show. $\square$

#19. (a) Let $S(n)$ denote the equation $\sum_{i=1}^{n} i^3 = \left(\sum_{i=1}^{n} i\right)^2$. 1. *Basis Step*: $S(1)$ is the equation $1^3 = (1)^2$, which is true. 2. *Inductive Step*: Let $k \in \mathbb{Z}_+$. We assume that $S(k)$ is true, and we must deduce that $S(k+1)$ is true, i.e., that $\sum_{i=1}^{k+1} i^3 = \left(\sum_{i=1}^{k+1} i\right)^2$. We can rewrite the right side as follows:

$$\left(\sum_{i=1}^{k+1} i\right)^2 = \left(\sum_{i=1}^{k} i + (k+1)\right)^2$$

$$= \left(\sum_{i=1}^{k} i\right)^2 + 2\left(\sum_{i=1}^{k+1} i\right)(k+1) + (k+1)^2 \qquad \text{(using the identity } (x+y)^2 = x^2 + 2xy + y^2\text{)}$$

$$= \sum_{i=1}^{k} i^3 + 2\left(\sum_{i=1}^{k+1} i\right)(k+1) + (k+1)^2 \qquad \text{(using the inductive hypothesis).}$$

We need to see that this is equal to the left-hand side of $S(k+1)$, so it remains to show that $2\left(\sum_{i=1}^{k+1} i\right)(k+1) + (k+1)^2 = (k+1)^3.$ From (1), we get that $\left(\sum_{i=1}^{k+1} i\right) = k(k+1)/2$, hence $2\left(\sum_{i=1}^{k+1} i\right)(k+1) + (k+1)^2 = 2[(k(k+1)/2](k+1) + (k+1)^2 = (k+1)^2\{k+1\} = (k+1)^3$, as desired. $\square$

(b) Exercise 7 tells us that $\sum_{i=1}^{n} i^3 = [n(n+1)/2]^2 \quad (n \in \mathbb{Z}_+)$. On the other hand, formula (1) tells us that $\sum_{i=1}^{n} i = n(n+1)/2$. Squaring both sides of the latter equation gives $(\sum_{i=1}^{n} i)^2 = [n(n+1)/2]^2$, and if we combine this with the first equation we obtained, the result is $\sum_{i=1}^{n} i^3 = \left(\sum_{i=1}^{n} i\right)^2$, as desired.

$\square$

#21. For any positive integer n, we let $S(n)$ denote the statement that the general (set) distributive law $A \cap (B_1 \cup B_2 \cup \cdots \cup B_n) = (A \cap B_1) \cup (A \cap B_2) \cup \cdots \cup (A \cap B_n)$ is valid (with a set A, and exactly n sets $B_1, B_2, \cdots, B_n$). 1. *Basis Step*: $S(1)$ simply states that $A \cap (B_1) = (A \cap B_1)$, and this is trivially valid. 2. *Inductive Step*: Let $k \in \mathbb{Z}_+$. We assume that $S(k)$ is true, i.e., that the general k-set distributive law is valid, and use this (inductive hypothesis) do deduce that $S(k+1)$ is true, i.e., that the general $(k+1)$-set distributive law is valid. We will make use of the 2-set distributive law $A \cap (B \cup C) = (A \cap B) \cup (A \cap C)$, from Theorem 1.3(c). We have:

$$A \cap (B_1 \cup B_2 \cup \cdots \cup B_{k+1}) = A \cap ([B_1 \cup B_2 \cup \cdots \cup B_k] \cup B_{k+1})$$
$$= (A \cap [B_1 \cup B_2 \cup \cdots \cup B_k]) \cup (A \cap B_{k+1}) \qquad \text{(By the 2-set distributive law)}$$
$$= [(A \cap B_1) \cup (A \cap B_2) \cup \cdots \cup (A \cap B_n)] \cup (A \cap B_{k+1}) \text{ (By the inductive}$$
$$\text{hypothesis) } \square$$

(We have tacitly used the fact that general unions are associative; see Exercise for the Reader 3.4 and its solution.)

#23. The proof is nearly identical to the one we gave for Part (a) of Exercise for the Reader 3.4, except that unions are replaced by disjunctions, and the associative law for disjunctions of three logical statements (Theorem 1.1 I(b)) gets used in place of the corresponding rule for unions. We omit the details.

#25. For any positive integer n, we let $S(n)$ denote the statement that any n distinct lines in the plane that pass through a common point partition the plane into $2n$ regions. 1. *Basis Step*: $S(1)$ says simply that a single line partitions the plane into two regions. This is trivial. 2. *Inductive Step*: Let $k \in \mathbb{Z}_+$. We assume that $S(k)$ is true, and we must deduce that $S(k+1)$ is true. So suppose that we have $k+1$ lines drawn in the plane that pass through a common point. We need to show that these lines determine $2(k+1)$ regions. If we remove one of these lines (the dotted line in the right-hand side of the figure), then we have a configuration of k distinct lines in the plane that all pass through a common point (the left-hand side of the figure). By the inductive hypothesis, these k lines determine $2k$ planar regions.

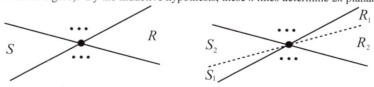

Now the deleted line must lie exactly two opposite regions in this configuration (these are the regions R and S in the left-hand side figure). When added, this $(k+1)$st line will in turn split each of R and S

into two new regions (these are the regions R_1, R_2 and S_1, S_2 in the right-hand side figure). Thus the k + 1 lines partition the plane into $2k - 2 + 4 = 2(k+1)$ regions, as we needed to show. □

#27. The proof will be accomplished by strong induction. For an integer $n > 1$, we let $S(n)$ denote the statement that an $n \times 6$ grid can be partitioned into L-shaped regions. The figure on the right shows that both $S(2)$ and $S(3)$ are true. Now, let $k > 2$ be an integer and assume that $S(1), S(2), \cdots, S(k)$ are all true. (This is the strong induction hypothesis.) We must show that $S(k + 1)$ is also true. Consider a $(k+1) \times 6$ grid. Using the validity of $S(2)$, we can tile the last two columns wih L-shaped regions.   This leaves a $(k-1) \times 6$ grid. Since $k - 1 \geq 2$, the inductive hypothesis tells us that this remaining portion can also be tiled with L-shaped regions. Thus the whole $(k+1) \times 6$ grid can be so-tiled. □

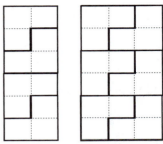

#29.   For an integer $n \geq 1$, we let $S(n)$ denote the statement that in a meeting with n people where everyone shakes everyone else's hand, there will be a total of $n(n-1)/2$ handshakes. 1. *Basis Step*: $S(1)$ says that the total number of handshakes if there are two people will be $1(1-1)/2 = 0$ handshakes; this is certainly true.   2. *Inductive Step*: Let $k \in \mathbb{Z}_+$. We assume that $S(k)$ is true, and we must deduce that $S(k+1)$ is true. So suppose that in a meeting with k people, there will be a total of $k(k-1)/2$ handshakes (this is the inductive hypothesis $S(k)$).   Now if one more person comes in (late); he/she will shake hands with each of the other k people. This will add k additional handshakes, bringing the total to $k(k-1)/2 + k = \{k(k-1) + 2k\}/2 = k(k-1+2)/2 = [k+1]([k+1]-1)/2$. This is what we needed to prove. □

#31. (Note: Even with the suggestion, this exercise is difficult.)   (a) We first prove the lemma: $n \geq \sqrt{n+1}$, whenever $n \geq 2$ is an integer. For $n \geq 2$, we let $S(n)$ be the statement that $n \geq \sqrt{n+1}$. 1. *Basis Step*: $S(2)$ says $2 \geq \sqrt{2+1}$ or $2 \geq \sqrt{3}$ (≈ 1.732), which is true. 2. *Inductive Step*: Let $k \in \mathbb{Z}_+$. We assume that $S(k)$ is true, and we must deduce that $S(k+1)$ is true, i.e., that $k + 1 \geq \sqrt{[k+1]+1}$.   The latter is equivalent to (squaring both sides) $(k+1)^2 \geq k + 2$, or (expanding) $k^2 + 2k + 1 \geq k + 2$. Now, by squaring both sides of the equation in the inductive hypothesis, we know that $k^2 \geq k + 1$, and if we subtract this from what we need to show, we have reduced the problem to showing that $2k + 1 \geq 1$, which is certainly true (since k is a positive integer). This proves the lemma. We are now ready to use induction to prove the main assertion: For a positive integer n, we let $T(n)$ denote the inequality $\left(1 - \dfrac{1}{\sqrt{2}}\right) \cdot \left(1 - \dfrac{1}{\sqrt{3}}\right) \cdots \cdot \left(1 - \dfrac{1}{\sqrt{n}}\right) < \dfrac{2}{n^2}$. 1. *Basis Step*: $T(2)$ is the inequality $\left(1 - \dfrac{1}{\sqrt{2}}\right) \cdot < \dfrac{2}{2^2}$.   This is easily seen to be true using a calculator. (Alternatively:

$$1 - \frac{1}{\sqrt{2}} = \frac{\sqrt{2}-1}{\sqrt{2}} = \frac{\sqrt{2}-1}{\sqrt{2}} \cdot \frac{\sqrt{2}+1}{\sqrt{2}+1} = \frac{1}{2+\sqrt{2}} < \frac{1}{2} = \frac{2^2}{2}.)$$ 2. *Inductive Step*: Let $k \geq 2$ is an integer. We assume that $T(k)$ is true, and we must deduce that $T(k+1)$ is true, i.e., that

$$\left(1 - \frac{1}{\sqrt{2}}\right) \cdot \left(1 - \frac{1}{\sqrt{3}}\right) \cdots \cdot \left(1 - \frac{1}{\sqrt{k+1}}\right) < \frac{2}{(k+1)^2}.$$ By the inductive hypothesis, the left side of this is less than (replace all but the last factor):

$$\frac{2}{k^2} \cdot \left(1 - \frac{1}{\sqrt{k+1}}\right) = \frac{2(\sqrt{k+1}-1)}{k^2\sqrt{k+1}} = \frac{2(\sqrt{k+1}-1)}{k^2\sqrt{k+1}} \cdot \frac{(\sqrt{k+1}+1)}{(\sqrt{k+1}+1)} = \frac{2k}{k^2(k+1+\sqrt{k+1})} = \frac{2}{k(k+1)+k\sqrt{k+1}}.$$

If we now invoke the lemma $(k \geq \sqrt{k+1})$, the last quantity on the right is less than or equal to:

$$\frac{2}{k(k+1)+\sqrt{k+1}\sqrt{k+1}}=\frac{2}{k(k+1)+(k+1)}=\frac{2}{(k+1)^2},$$

as desired. □

Section 3.2:

#1. (a) $a_0=1,\ a_1=2,\ a_2=4,\ a_3=8,\ a_4=16$ (b) $a_1=1,\ a_2=-1,\ a_3=-1,\ a_4=-3,\ a_5=-7$

(c) $a_1=1,\ a_2=2,\ a_3=2,\ a_4=8,\ a_5=22$

#3. (a) $a_0=1,\ a_1=3,\ a_2=7,\ a_3=15,\ a_4=31$ (b) $a_1=1,\ a_2=-1,\ a_3=-1,\ a_4=-1,\ a_5=-1$

(c) $a_1=1,\ a_2=2,\ a_3=2,\ a_4=16,\ a_5=56$

#5. (a) $\begin{cases}a_1=2\\a_n=a_{n-1}\ (n\ge2)\end{cases}$ (b) $\begin{cases}a_1=2\\a_n=a_{n-1}+2\,(n\ge2)\end{cases}$ (c) $\begin{cases}a_1=2\\a_n=2a_{n-1}\ (n\ge2)\end{cases}$

(d) $\begin{cases}a_1=2\\a_n=[(n-1)/n]a_{n-1}\ (n\ge2)\end{cases}$

#7. (a) $\begin{cases}a_1=11\\a_n=a_{n-1}+2\,(n\ge2)\end{cases}$ (b) $\begin{cases}a_1=2\\a_n=a_{n-1}+3\,(n\ge2)\end{cases}$ (c) $\begin{cases}a_1=5\\a_n=a_{n-1}+n-1\,(n\ge2)\end{cases}$

(d) $\begin{cases}a_1=6\\a_n=-na_{n-1}\ (n\ge2)\end{cases}$

#9. (a) $1,4,7,10,\dots\ \Rightarrow a_n=1+3(n-1)$. (b) $a_n=1+2+3+\cdots+n=n(n+1)/2$, by formula (1) of Example 3.1.

(c) $a_1=1,\ a_2=1.1+1,\ a_3=1.1(1.1+1)+1=(1.1)^2+(1.1)+1,\ a_4=1.1((1.1)^2+(1.1)+1)+1=$

$(1.1)^3+(1.1)^2+(1.1)+1$, and it follows that $a_n=\sum_{j=0}^{n-1}(1.1)^j=[(1.1)^n-1]/(1.1-1)=10[(1.1)^n-1]$,

where we have used the geometric series formula (3) (of Proposition 3.5).

#11. (a) Given that the current period balance is $A(n)$, the next period balance $A(n+1)$ is obtained by adding to $A(n)$ the interest earned over the period $=(r/k)A(n)$ and periodic payment PMT, and thus

$A(n+1)=A(n)+(r/k)A(n)+PMT=(1+r/k)A(n)+PMT$.

(b) Although this could be done using Theorem 3.8, we will do it by writing out the first few terms to discover a general pattern:

$A(1)=PMT$

$A(2)=(1+r/k)(PMT)+PMT$

$A(3)=(1+r/k)[(1+r/k)(PMT)+PMT]+PMT=(1+r/k)^2(PMT)+(1+r/k)(PMT)+PMT$

$A(4)=(1+r/k)[1+r/k)^2(PMT)+(1+r/k)(PMT)+PMT]+PMT$

$\quad=(1+r/k)^3(PMT)+(1+r/k)^2(PMT)+(1+r/k)(PMT)+PMT$

It now follows that (and this can be rigorously justified with an easy induction proof):

$A(n)=\sum_{j=0}^{n-1}(PMT)(1+r/k)^j$. If we apply the geometric series formula (3) (of Proposition 3.5), we

obtain: $A(n)=[(PMT)(1+r/k)^n-PMT]/(r/k)=PMT\cdot\dfrac{(1+r/k)^n-1}{r/k}$.

(c) $FV=\$1509.39$ (d) $FV=\$1,511,911.09$. Interest $=FV-n\cdot PMT=\$1,271,911.09$.

#13. Since there is only one way to climb a single stair, $a_1=1$. Two stairs can either be climbed at once or one stair at a time, so $a_2=2$. Now, if $n\ge3$, any scheme for climbing n stairs must begin with either climbing a single stair or two stairs. In the first case, the number of ways of climbing the remaining $n-1$ stairs is a_{n-1}. In the second case, the number of ways of climbing the remaining $n-2$ stairs is a_{n-2}. Since the cases are disjoint, it follows that $a_n=a_{n-1}+a_{n-2}$. Note that by comparing with the Fibonacci sequence of Example 3.10, we have $a_n=f_{n+1}$. We can use this along with the explicit formula (4) for the Fibonacci sequence to compute $a_{10} = f_{11} = 89$. (This could also be computed directly with the recursion formula.)

#15. (a) From the information given in the exercise, it follows that the sequences defined in the suggestion satisfy: $a_n = \alpha_n + \beta_n + \gamma_n$, and the following system of recursion formulas:

$$\begin{cases} \alpha_1 = 0, & \alpha_{n+1} = \gamma_n \ (n \geq 1) \\ \beta_1 = 1, & \beta_{n+1} = \alpha_n \ (n \geq 1) \\ \gamma_1 = 0, & \gamma_{n+1} = \beta_n + \gamma_n \ (n \geq 1) \end{cases}$$

(b) In the table below use these formulas to compute the first six terms of the sequence $\{a_n\}$:

n	α_n	β_n	γ_n	$a_n = \alpha_n + \beta_n + \gamma_n$
1	0	1	0	1
2	0	0	1	1
3	1	0	1	2
4	1	1	1	3
5	1	1	2	4
6	2	1	3	6
7	3	2	4	9

(c) From the formulas of Part (a), it follows that $a_{n+1} - a_n = \gamma_n$, and from the patterns seen in the table of Part (b), it becomes evident that $\gamma_n = n - 3$, for $n \geq 4$. Thus, $\{a_n\}_{n \geq 4}$ satisfies the recurrence relation: $\begin{cases} a_4 = 3 \\ a_{n+1} = a_n + n - 3 \ (n \geq 4). \end{cases}$

Hence, for $n \geq 5$, we may write:

$a_n = a_{n-1} + (n-4) = a_{n-2} + (n-5) + (n-4) = a_{n-3} + (n-6) + (n-5) + (n-4) =$
$\cdots = a_4 + 1 + 2 + \cdots + (n-4).$

Invoking formula (1) of Example 3.1 along with the initial value $a_4 = 3$ allows us to write

$a_n = 4 + (n-4)(n-3)/2$, which is valid for all $n \geq 4$.

#17. (a) Consider separate sequences as follows:

α_n = the number of rabbits of age 0 months at the end of month n.

β_n = the number of rabbits of age 1 month at the end of month n.

γ_n = the number of rabbits of age 2 months at the end of month n.

γ_n = the number of rabbits of age 3 months at the end of month n.

From the information given in the exercise, it follows that the sequences defined in the suggestion satisfy: $a_n = \alpha_n + \beta_n + \gamma_n + \delta_n$, and the following system of recursion formulas:

$$\begin{cases} \alpha_1 = 0, & \alpha_{n+1} = \gamma_n + \delta_n \ (n \geq 1) \\ \beta_1 = 1, & \beta_{n+1} = \alpha_n \ (n \geq 1) \\ \gamma_1 = 0, & \gamma_{n+1} = \beta_n \ (n \geq 1) \\ \delta_1 = 0, & \delta_{n+1} = \gamma_n \ (n \geq 1) \end{cases}$$

(b) In the table below use these formulas to compute the first six terms of the sequence $\{a_n\}$:

n	α_n	β_n	γ_n	δ_n	$a_n = \alpha_n + \beta_n + \gamma_n + \delta_n$
1	0	1	0	0	1
2	0	0	1	0	1
3	1	0	0	1	2
4	1	1	0	0	2
5	0	1	1	0	2
6	1	0	1	1	3
7	2	1	0	1	4

#19. (a) Clearly $k_1 = 2$. To compute k_3, note that the smaller disk needs first to be moved to peg 3 (so that the larger disk can be moved), and this takes 2 moves. We then can move the large disk to peg 2 (1 move). In order to move the large disk to peg 3, we need to move the small disk back to peg 1 (2 moves) and then move the large disk to peg 3 (1 move), and finally move the small disk back to peg 3 (2 moves). Thus we have shown that $k_3 = 2 + 1 + 2 + 1 + 2 \ (= 3 \cdot k_1 + 2) = 8$.

(b) The same reasoning that was used to compute k_3 in Part (a) may be applied to compute k_{n+1} to be $3k_n + 2$, simply replace each move of "the smaller disk" with "the n smaller disks." This produces the following recursive formula for the sequence: $\begin{cases} k_1 = 2 \\ k_{n+1} = 3k_n + 2 \ (n \geq 1). \end{cases}$

(c) Although this could be done using Theorem 3.8, we will do it by writing out the first few terms in a way that can be used to discover a general pattern:

$k_1 = 2$
$k_2 = 3[2] + 2$
$k_3 = 3[3(2) + 2] + 2 = 3^2 \cdot 2 + 3 \cdot 2 + 2$
$k_4 = 3[3^2 \cdot 2 + 3 \cdot 2 + 2] + 2 = 3^3 \cdot 2 + 3^2 \cdot 2 + 3 \cdot 2 + 2$

It now follows that (and this can be rigorously justified with an easy induction proof):
$k_n = 3^{n-1} \cdot 2 + \cdots + 3^2 \cdot 2 + 3 \cdot 2 + 2 = \sum_{j=0}^{n-1} 2 \cdot 3^j = [2 \cdot 3^n - 2]/[3-1] = 3^n - 1$, where in the third equality we have applied the geometric series formula (3) of Propostion 3.5.
(d) 2186

#21. Note that p_n is the number of (ordered) vectors whose entries are either 2, 3, or 5, and whose components add up to n.

(a) Vectors for 2¢: (2) $\Rightarrow p_2 = 1$. Vectors for 3¢: (3) $\Rightarrow p_3 = 1$. Vectors for 4¢: (2,2) $\Rightarrow p_4 = 1$. Vectors for 5¢: (2,3), (3,2), (5) $\Rightarrow p_5 = 3$.

(b) To obtain a recursive formula for p_n, we assume that $n \geq 6$, and separate into three cases of corresponding postage vectors depending on whether the first stamp (on the left) is 2¢, 3¢, or 5¢. Since the corresponding vectors in these cases form disjoint sets, it follows that $p_n = p_{n-2} + p_{n-3} + p_{n-5}$. This formula, together with the first four values of the sequence (and the obvious value $p_1 = 0$) found in Part (a) constitute a recursive definition of the sequence p_n.

(c) 14

#23. (a) We let $\mathscr{B}_n$ denote the set of length n binary strings that do not contain "000."

$b_3 = |\mathscr{B}_3| = |\{\text{length 3 bit strings}\} \sim \{000\}| = 2^3 - 1 = 7$.

For a length 4 bit string in $\mathscr{B}_4$, if the first bit is 1, the remaining three bits may be any string of $\mathscr{B}_3$ while if the first bit is 0, the remaining three bits may be any string of $\mathscr{B}_3$ except 001. Since these two cases are disjoint, it follows that $b_4 = |\mathscr{B}_4| = 7 + 6 = 13$.

To compute b_5, we separate the strings in $\mathscr{B}_5$ into three disjoint sets depending on which of the following three bit strings prefix the whole string: 1,   01,   001.   It follows that $b_5 = b_4 + b_3 + b_2 = 13 + 7 + 4 = 24$.

(c) The same argument used to compute b_5 yields the formula $b_n = b_{n-1} + b_{n-2} + b_{n-3} \ (n \geq 4)$. This formula, along with the three initial values $b_1 = 2, b_2 = 4, b_3 = 7$ constitute a recursive definition of the sequence b_n.

(d) The formula of Part (c) allows us to continue computing:
$b_6 = 44, \ b_7 = 81, \ b_8 = 149, \ b_9 = 274, \ b_{10} = 504$.

#25. $\text{OddFact}(n) = \begin{cases} 1, & \text{if } n = 0 \\ \text{OddFact}(n-1), & \text{if } n \geq 1, \ n \text{ even} \\ n \cdot \text{OddFact}(n-1), & \text{if } n \geq 1, \ n \text{ odd} \end{cases}$

#27. (a) The following recursive algorithm can be applied to find the maximum component of any vector of n real numbers $x_1, x_2, \cdots, x_n$ where $n > 1$.

$y = \text{MaxRecursive}(x_1, x_2, \cdots, x_n)$

IF $x_{n-1} > x_n$

    SET $z = x_{n-1}$

ELSE

    SET $z = x_n$

END IF

IF $n = 2$

    OUTPUT $y = z$, EXIT

ELSE

    SET $y = \text{MaxRecursive}(x_1, x_2, \cdots, x_{n-2}, z)$

END IF

(b) This recursive algorithm proceeds through $n-1$ iterations, each requiring a single comparison, so the total number of comparisons is $n-1$, the same as for the direct algorithm.

#29. (a) Characteristic Equation: $r - 2 = 0$. The general solution of the recurrence (by Theorem 3.7) is $a_n = x2^n$. Substituting the initial condition $a_1 = -5$ yields $-5 = 2x \Rightarrow x = -5/2$. Thus we have the explicit formula: $a_n = (-5/2)2^n = -5 \cdot 2^{n-1}$.

(b) Characteristic Equation: $r^2 - 2r - 1 = 0$. Since $r^2 - 2r - 1 = (r - 2)(r - 1)$, the roots are $r = 1, 2$, and the general solution of recurrence (by Theorem 3.6) is $a_n = x \cdot 1^n + y2^n = x + y2^n$. Substituting the initial conditions $a_1 = 3, a_2 = 7$ yields the system $\begin{cases} x + 2y = 3 \\ x + 4y = 7 \end{cases}$, which has solution $x = -1, y = 2$.   Thus we have the explicit formula: $a_n = -1 + 2 \cdot 2^n = 2^{n+1} - 1$.

(c) Characteristic Equation: $r^2 - r - 1 = 0$. This is the same as for the Fibonacci recursion, so the general solution is as in Example 3.10: $a_n = x\left(\dfrac{1 + \sqrt{5}}{2}\right)^n + y\left(\dfrac{1 - \sqrt{5}}{2}\right)^n$. Substituting the initial conditions $a_1 = 0, a_2 = 2$ yields the system $\begin{cases} x + y = 0 \\ [(1 + \sqrt{5})/2]x + [(1 + \sqrt{5})/2]y = 2 \end{cases}$, which has solution $x = 2/\sqrt{5}, y = -2/\sqrt{5}$.   Thus we have the explicit formula: $a_n = \dfrac{2}{\sqrt{5}}\left[\left(\dfrac{1 + \sqrt{5}}{2}\right)^n - \left(\dfrac{1 - \sqrt{5}}{2}\right)^n\right]$.

(d) Characteristic Equation: $r^2 - 6r + 5 = 0$. Since $r^2 - 6r + 5 = (r - 5)(r - 1)$, the roots are $r = 1, 5$, and the general solution of the recurrence (by Theorem 3.6) is $a_n = x \cdot 1^n + y2^n = x + y5^n$.   Since the sequence begins with the index $n = 7$, it is simpler if we rewrite this general solution as $a_n = x + y5^{n-7}$. (The difference in powers of 5 can be absorbed in the variable coefficient y.) Substituting the initial conditions $a_7 = 5, a_8 = 13$ yields the system $\begin{cases} x + y = 5 \\ x + 5y = 13 \end{cases}$, which has solution $x = 3, y = 2$.   Thus we have the explicit formula: $a_n = 3 + 2 \cdot 5^{n-7}$.

#31. (a) Characteristic Equation: $r^2 - 4r + 4 = 0$. Since $r^2 - 4r + 4 = (r - 2)^2$, there is a double root 2, and the general solution of the recurrence (by Theorem 3.6) is $a_n = x2^n + yn2^n$. Substituting the initial conditions $a_0 = 3$, $a_1 = 1$ yields the system $\begin{cases} x = 3 \\ 2x + 2y = 1 \end{cases}$, which has solution $x = 3, y = -5/2$.

Thus we have the explicit formula: $a_n = 3 \cdot 2^n - (5/2)n2^n = 3 \cdot 2^n - 5 \cdot n2^{n-1}$.

(b) Characteristic Equation: $r^2 - 2r + 1 = 0$. Since $r^2 - 2r + 1 = (r-1)^2$, there is a double root 1, and the general solution of the recurrence (by Theorem 3.6): $a_n = x \cdot 1^n + yn \cdot 1^n = x + yn$. Substituting the initial conditions $a_2 = 17$, $a_3 = 21$ yields the system $\begin{cases} x + 2y = 17 \\ x + 3y = 21 \end{cases}$, which has solution $x = 9, y = 4$.

Thus we have the explicit formula: $a_n = x + yn = 9 + 4y$.

#33. (a) Characteristic Equation: $r^4 - 8r^2 + 16 = 0$. Since $r^4 - 8r^2 + 16 = (r^2 - 4)^2 = (r-2)^2(r+2)^2$, there are two double roots ± 2, and the general solution of the recurrence (by Theorem 3.7) is $a_n = x \cdot 2^{n-1} + yn \cdot 2^{n-1} + z \cdot (-2)^{n-1} + wn \cdot (-2)^{n-1}$.

(b) Characteristic Equation: $r^4 - 10r^2 + 9 = 0$. Since $r^4 - 10r^2 + 9 = r^2(r^2 - 9) - (r^2 - 9) = (r^2 - 1)(r^2 - 9) = (r-1)(r+1)(r-3)(r+3)$, the roots are $\pm 1, \pm 3$, and the general solution of the recurrence (by Theorem 3.7) is $a_n = x \cdot 1^n + y(-1)^n + z3^n + w(-3)^n = x + y(-1)^n + z3^n + w(-3)^n$.

(c) Characteristic Equation: $r^3 - 3r^2 + 3r - 1 = 0$. Since $r^3 - 3r^2 + 3r - 1 = (r-1)^3$, there is a triple root 1, and the general solution of the recurrence (by Theorem 3.7) is $a_n = x \cdot 1^n + yn \cdot 1^n + zn^2 \cdot 1^n + wn^3 \cdot 1^n = x + yn + zn^2 + wn^3$.

(d)Characteristic Equation: $r^4 - 8r^3 + 24r^2 - 32r + 16 = 0$. Since $r^4 - 8r^3 + 24r^2 - 32r + 16 = (r-2)^4$, there is just one that has multiplicity 4, and the general solution of the recurrence (by Theorem 3.7) is $a_n = x \cdot 2^n + yn \cdot 2^n + zn^2 \cdot 2^n + wn^3 \cdot 2^n = (x + yn + zn^2 + wn^3) \cdot 2^n$.

#35. (a) The associated homogeneous recurrence relation is the same as for Exercise 29(a), for which the general solution was found to be $a_n^h = x2^n$. From Theorem 3.9, a particular solution of the inhomogeneous recurrence will have the form $a_n^p = y3^n$. Substituting the latter into the recurrence $a_n = 2a_{n-1} + 3^n$ produces $y3^n = 2y3^{n-1} + 3^n \Rightarrow 3y = 2y + 3 \Rightarrow y = 3$. Thus, from Theorem 3.8, the general solution of the given recurrence equation is $a_n^h + a_n^p = x2^n + 3^{n+1}$. Substituting the initial condition $a_2 = 19$ yields $4x + 27 = 19 \Rightarrow x = -2$. Thus, the explicit formula is $a_n = (-2)2^n + 3^{n+1} = 3^{n+1} - 2^{n+1}$.

(b) Just as in Part (a), the general solution of the associated homogeneous recurrence is $a_n^h = x2^n$. But this time since the inhomogeneity is a power of the root 2 of the characteristic equation, Theorem 3.9 tells us that a particular solution of the inhomogeneous recurrence will have the form $a_n^p = yn2^n$. Substituting the latter into the recurrence $a_n = 2a_{n-1} + 2^n$ produces $yn2^n = 2y(n-1)2^{n-1} + 2^n \Rightarrow ny = (n-1)y + 1 \Rightarrow y = 1$. Thus, from Theorem 3.8, the general solution of the given recurrence equation is $a_n^h + a_n^p = x2^n + n2^n = (x+n)2^n$. Substituting the initial condition $a_4 = 32$ yields $(x+4) \cdot 16 = 32 \Rightarrow x = -2$. Thus, the explicit formula is $a_n = (n-2)2^n$.

(c) The associated homogeneous recurrence relation is the same as for Exercise 31(a), for which the general solution was found to be $a_n^h = x \cdot 1^n + y2^n = x + y5^n$. From Theorem 3.9, a particular solution of the inhomogeneous recurrence will have the form $a_n^p = (zn + w)2^n$. Substituting the latter into the recurrence $a_n = 6a_{n-1} - 5a_{n-2} + n2^n$ produces:

$$(zn+w)2^n = 6(z(n-1)+w)2^{n-1} - 5(z(n-2)+w)2^{n-2} + n2^n$$
$$\Rightarrow 4zn+4w = 12z(n-1)+12w - 5(z(n-2)+w)+4n$$
$$\Rightarrow -3zn+(2z-3w) = 4n$$
$$\Rightarrow \begin{cases} -3z = 4 \\ 2z-3w = 0 \end{cases} \Rightarrow \begin{cases} z = -4/3 \\ w = -8/9. \end{cases}$$

(In the last line the system was obtained by equating coefficients of n and $1 = n^0$.)   Thus, $a_n^p = (zn+w)2^n = -(12n+8)2^n/9$, and from Theorem 3.8, the general solution of the given recurrence is $a_n^h + a_n^p = x + y5^n - (12n+8)2^n/9$.   Substituting the initial conditions $a_0 = 0$, $a_1 = 1$ yields the

system $\begin{cases} x+y = 8/9 \\ x+5y = 41/36 \end{cases}$, which has the solution $x = -1/4$, $y = 8/9$. Thus, the explicit formula for

the given sequence is $a_n = -1/4 + (41/36)\cdot 5^n - (12n+8)\cdot 2^n/9$.

(d) The associated homogeneous recurrence relation is the same as for Exercise 31(a), for which the general solution was found to be $a_n^h = x2^{n-2} + yn2^{n-2}$. (Once again, since the recursion starts at $n = 2$, it is convenient to subtract two from the powers in the exponentials; the multiplicative constants can absorb such changes.) From Theorem 3.9, and since the exponential base 2 in the inhomogeneity is a root of the characteristic equation, a particular solution of the inhomogeneous recurrence will have the form $a_n^p = n^2 w 2^n$. Substituting the latter into the recurrence $a_n = 4a_{n-1} - 4a_{n-2} + 2^{n-1}$ produces:

$$n^2 w 2^n = 4(n-1)^2 w 2^{n-1} - 4(n-2)^2 w 2^{n-2} + 2^{n-1}$$
$$\Rightarrow n^2 w 2^n = 4(n^2-2n+1)w 2^{n-1} - 4(n^2-4n+4)w 2^{n-2} + 2^{n-1}$$
$$\Rightarrow 4n^2 w = 4n^2 w - 8w + 2$$
$$\Rightarrow w = 1/4.$$

Thus, $a_n^p = n^2 2^{n-2}$, and from Theorem 3.8, the general solution of the given recurrence is $a_n^h + a_n^p = x2^{n-2} + yn2^{n-2} + n^2 2^{n-2}$. Substituting the initial conditions $a_2 = 1$, $a_3 = 6$ yields the system

$\begin{cases} x+2y = -3 \\ 2x+6y = -12 \end{cases}$, which has the solution $x = 3$, $y = -3$. Thus, the explicit formula for the given

sequence is $a_n = (3-3n+n^2)2^{n-2}$.

Section 3.3:

#1. (a) False (b) True (c) False (d) True (since $a = (-1)(-a)$). (e) False (only when $a = 0$ since by Definition 2.1, 0 cannot divide any integer, but if $a \neq 0$ the relation is true). (f) False

#3. Since $a \mid b$, there is an integer k such that $b = ak$, and since $b \mid a$, there is an integer ℓ such that $a = b\ell$. Substituting the first equation into the second gives: $a = (ak)\ell$, but since $a \neq 0$ (since it divides b; see Definition 2.1), we can divide both sides by a to obtain $1 = k\ell$, and this forces $\ell = \pm 1$. Thus $a = \ell b = \pm b$. □

#5. primes: (a), (e).

#7. (a) $24 = 2^3 \cdot 3$ (b) $88 = 2^3 \cdot 11$ (c) $675 = 3^3 \cdot 5^2$ (d) $6400 = 2^8 \cdot 5^2$ (e) $74{,}529 = 3^2 \cdot 7^2 \cdot 13^2$

(f) $183{,}495{,}637 = 13^3 \cdot 17^4$

#9. (a) $\pi(10^9) \rightarrow 10^9/\ln(10^9) \approx 48{,}254{,}942$

(b) $\pi(10^{10}) - \pi(10^9) \rightarrow 10^{10}/\ln(10^{10}) - 10^9/\ln(10^9) \approx 386{,}039{,}539$

(c) $\pi(10^{12}) - \pi(10^9) \rightarrow 10^{12}/\ln(10^{12}) - 10^9/\ln(10^9) \approx 36{,}142{,}951{,}883$

#11. (a) quotient: $q = 33$, remainder: $r = 1$

(b) quotient: $q = 21$, remainder: $r = 3$

(c) quotient: $q = -39$, remainder: $r = 1$

(d) quotient: $q = 67$, remainder: $r = 11$
(e) quotient: $q = 49$, remainder: $r = 3$
(f) quotient: $q = -44$, remainder: $r = 2$

#13. (a) 12 (b) 100 (c) 4 (d) 280 (e) 600 (f) 11025

#15. (a) $36 = 3 \cdot 12 + 0$ (Euclidean algorithm terminates after one step.) So $\gcd(36, 12) = 12$.

(b) $\begin{cases} 25 = 1 \cdot 20 + 5 \\ 20 = 4 \cdot 5 + 0 \end{cases} \Rightarrow \gcd(20, 25) = 5 \Rightarrow \text{lcm}(20, 25) = 20 \cdot 25 / 5 = 100$.

(c) $\begin{cases} 100 = 1 \cdot 56 + 44 \\ 56 = 1 \cdot 44 + 12 \\ 44 = 3 \cdot 12 + 8 \\ 12 = 1 \cdot 8 + 4 \\ 8 = 2 \cdot 4 + 0 \end{cases} \Rightarrow \gcd(100, 56) = 4$.

(d) $\begin{cases} 1400 = 2 \cdot 560 + 280 \\ 560 = 2 \cdot 280 + 0 \end{cases} \Rightarrow \gcd(560, 1400) = 280$.

(e) $\begin{cases} 120 = 2 \cdot 50 + 20 \\ 50 = 2 \cdot 20 + 10 \\ 20 = 2 \cdot 10 + 0 \end{cases} \Rightarrow \gcd(120, 50) = 10 \Rightarrow \text{lcm}(120, 50) = 120 \cdot 50 / 10 = 600$.

(f) $\begin{cases} 5788125 = 47 \cdot 121275 + 88200 \\ 121275 = 1 \cdot 88200 + 33075 \\ 88200 = 2 \cdot 33075 + 22050 \\ 33075 = 1 \cdot 22050 + 11025 \\ 22050 = 2 \cdot 11025 + 0 \end{cases} \Rightarrow \gcd(5788125, 121275) = 11025$.

#17. (a) Since one of the numbers is a factor of the other, we can do this by inspection: $12 = \gcd(36, 12) = 36 \cdot 0 + 12 \cdot 1$.

(b) Start with the equation $12 = 1 \cdot 8 + 4 \Rightarrow 4 = 12 - 1 \cdot 8$, from the Euclidean algorithm computation of #15(c), and working our way up the list to perform substitutions of the new remainders, we obtain:
$4 = 12 - 1 \cdot 8 = 12 - 1 \cdot (44 - 3 \cdot 12) = 4 \cdot 12 - 1 \cdot 44 \Rightarrow 4 = 4 \cdot 12 - 1 \cdot 44 =$
$4 \cdot (56 - 1 \cdot 44) - 1 \cdot 44 = 4 \cdot 56 - 5 \cdot 44 \Rightarrow 4 = 4 \cdot 56 - 5 \cdot 44 = 4 \cdot 56 - 5 \cdot (100 - 1 \cdot 56) = -5 \cdot 100 + 9 \cdot 56$.

(c) From the Euclidean algorithm computation of #15(d), since it terminated after two iterations, the first equation leads to the desired expression: $1400 = 2 \cdot 560 + 280 \Rightarrow 280 = -2 \cdot 560 + 1400$.

(e) Start with the equation $33075 = 1 \cdot 22050 + 11025 \Rightarrow 11025 = 33075 - 1 \cdot 22050$. From the Euclidean algorithm computation of #15(e), and working our way up the list to perform substitutions of the new remainders, we obtain: $11025 = 33075 - 1 \cdot 22050 = 33075 - 1 \cdot (88200 - 2 \cdot 33075) = 3 \cdot 33075$
$-1 \cdot 88200 \Rightarrow 11025 = 3 \cdot (121275 - 1 \cdot 88200) - 1 \cdot 88200 = 3 \cdot 121275 - 4 \cdot 88200 \Rightarrow 11025 = 3 \cdot 121275$
$-4 \cdot (5788125 - 47 \cdot 121275) = 191 \cdot 121275 - 4 \cdot 5788125$.

#19. (a) True (b) False (c) False (d) True (e) False (f) False

#21. (a) 14 (b) 8 (c) 16 (d) 7 (e) First notice that (working in $\mod 24$) $21^2 = 9$, so $21^4 = (21^2)^2 = 9^2 = 9$, and it follows that any power of 9, and hence any even power of 21 is 9.

#23. (a) 16 (b) 120 (c) 384 (d) 2400

#25. (a) Since n is even, we can write $n = 2^f m$, where m is odd and f is a positive integer. From Proposition 3.20, it follows that $\phi(n) = 2^{f-1}\phi(m)$ and $\phi(2n) = 2^f \phi(m)$, and from these two equations it follows that $\phi(2n) = 2\phi(n)$. (b) If n is odd, then it follows from Proposition 3.20 that $\phi(2n) = (2-1) \cdot 2^0 \cdot \phi(n) = \phi(n)$.

#27. (This can be done using Proposition 3.20.) (a) $\{2\}$ (b) $\{5, 8, 10, 12\}$ (c) $\varnothing$ (no solutions) (d) $\{13, 21, 26, 28, 36, 42\}$

#29. (a) 1 (b) 7 (c) 24 (d) 8

#31(a)	$a^k \pmod 6$	
	$k=1$	$k=2$
$a=1$	1	1
$a=5$	5	1

#31(b)	$a^k \pmod{12}$			
	$k=1$	$k=2$	$k=3$	$k=4$
$a=1$	1	1	1	1
$a=5$	5	1	5	1
$a=7$	7	1	7	1
$a=11$	11	1	11	1

From the table of modular powers, we see that $\operatorname{ord}_6(1) = \operatorname{ord}_{12}(1) = 1$. (Note: It is trivial that $\operatorname{ord}_n(1) = 1$ for any modulus n.) And the orders of all other elements are 2. Five is a primitive root (mod 6), but there are no primitive roots mod 12.

#33. (a) $\operatorname{ord}_{10}(3) = 4$

(b) $\operatorname{ord}_{21}(6)$ is undefined since 6 is not relatively prime to 21.

(c) $\operatorname{ord}_{304}(21) = 36$

#35. (a) (i) $12 = 2^2 \cdot 3$ is not of any of the forms listed in Theorem 3.23(1), so there are no primitive roots.
(b) (i) 13 is prime so by Theorem 3.23(1) there are $\phi(\phi(13)) = \phi(12) = 4$ primitive roots (mod 13). (ii) Since $\operatorname{ord}_{13}(2) = 12 = \phi(13)$, 2 is a primitive root mod 13.
(c) (i) Since $14 = 2 \cdot p$, with $p = 7$, Theorem 3.23(1) tells us that there are $\phi(\phi(14)) = \phi(6) = 2$ primitive roots mod 14. (ii) Since $\operatorname{ord}_{14}(3) = 6 = \phi(14)$, 3 is a primitive root mod 14.

#37. (a) (i) $25 = p^2$ with $p = 5$, so by Theorem 3.23(1) we know there are $\phi(\phi(25)) = \phi(20) = 8$ primitive roots mod 25. (ii) The possible mod 25 orders (for integers relatively prime to 25) must divide (by Proposition 3.22) $\phi(25) = 20 = 2^2 \cdot 5$, so can only be one of the following values: (1 only for $a = 1$), 2, 4, 5, 10, or 20 (in which case we have a primitive root). To find the smallest primitive root mod 25, go through all integers relatively prime to 25 starting with $a = 2$, and compute $a^2, a^4, a^5, a^{10} \pmod{25}$ until we find the smallest for which these powers are all different from 1, this will be the smallest primitive root. $2^2 \equiv 4, 2^4 \equiv 16, 2^5 \equiv 7, 2^{10} \equiv 24 \pmod{25}$. So we can stop: $g = 2$ is the smallest primitive root. (iii) By Theorem 3.23(2), if j is relatively prime to $\operatorname{ord}_{25}(g) = 20$, then g^j will also be a primitive root. Using $j = 3$, we get that $2^3 \equiv 8$ is another primitive root mod 25.
(b) (i) Since $39 = 3 \cdot 13$ is not of any of the forms listed in Theorem 3.23(1), there are no primitive roots mod 39. (ii) and (iii) are not applicable. (c) (i) Since 31 is prime, Theorem 3.23(1) tells us that there are $\phi(\phi(31)) = \phi(30) = 8$ primitive roots mod 31. (ii) The possible mod 31 orders (for integers relatively prime to 31) must divide (by Proposition 3.22) $\phi(31) = 30 = 2 \cdot 3 \cdot 5$, so can only be one of the following values: (1 only for $a = 1$), 2, 3, 5, 6, 10, 15, or 30 (in which case we have a primitive root). To find the smallest primitive root mod 31, go through all integers relatively prime to 31 starting with $a = 2$, and compute $a^2, a^3, a^5, a^6, a^{10}, a^{15} \pmod{25}$ until we find the smallest for which these powers are all different from 1, this will be the smallest primitive root. $2^2 \equiv 4, 2^3 \equiv 8, 2^5 \equiv 1 \pmod{31}$, so 2 does not work, and we move on to $a = 3$:

$$3^2 \equiv 9, 3^3 \equiv 27, 3^5 \equiv 26, 3^6 \equiv 16, 3^{10} \equiv 25, 3^{15} \equiv 30 \pmod{31}.$$

We can stop: $g = 3$ is the smallest primitive root. (iii) By Theorem 3.23(2), if j is relatively prime to $\operatorname{ord}_{25}(g) = 30$, then g^j will also be a primitive root. Using $j = 7$, we get that $3^7 \equiv 17$ is another primitive root mod 31.

(d) (i) Since $50 = 2 \cdot p^2$, with $p = 5$, Theorem 3.23(1) tells us that there are $\phi(\phi(50)) = \phi(20) = 8$ primitive roots mod 50. (ii) 27 can be found to be a primitive root of 50. (iii) By Theorem 3.23(2), if j is relatively prime to $\phi(50) = 20$, then G^j will also be a primitive root. Using $j = 3$, we get that $27^3 \equiv 33$ is another primitive root mod 50.

(e) (i) Since $52 = 2^2 \cdot 13$ is not of any of the forms listed in Theorem 3.23(1), there are no primitive roots mod 52. (ii) and (iii) are not applicable.

(f) (i) Since $961 = p^2$, with $p = 31$, Theorem 3.23(1) tells us that there are $\phi(\phi(31^2)) = \phi(930) = 240$ primitive roots mod 961. (ii) 3 can be found to be a primitive root of 961. (iii) By Theorem 3.23(2), if j is relatively prime to $\phi(961) = 930$, then G^j will also be a primitive root. Using $j = 7$, we get that $3^7 \equiv 265$ is another primitive root mod 961.

#39. (a) True. *Proof:* Since $a \mid b$ and $a \mid (b+1)$, Theorem 3.10(b) tells us that $a \mid [(b+1) - b]$, or $a \mid 1$, which forces $a = \pm 1$. □

(b) True. *Proof:* Since n is even, we can write $n = 2k$, for some integer k. Squaring gives $n^2 = 4k^2$, showing that $4 \mid n^2$. □

(c) True. *Proof:* Since a and b have the same parity, we have $a \equiv b \pmod 2$, or that $2 \mid a - b$. It follows that $2 \mid (a - b)(a + b) = a^2 - b^2$, showing that $a^2 - b^2$ is even. □ (Note: With a little more work, it can be shown that $4 \mid a^2 - b^2$.)

#41. (a) False: $4 \nmid 1 \cdot 2 \cdot 3$.

(b) True. *Proof:* First we factor $a^4 - a^2 = a^2(a^2 - 1) = a^2(a+1)(a-1)$. We separate into two cases: *Case 1:* a is even, i.e., $2 \mid a$. Then $4 \mid a^2$, so (by the factorization above) $4 \mid (a^4 - a^2)$. *Case 2:* a is odd, then $a + 1$ and $a - 1$ must both be even, i.e., $2 \mid a \pm 1$. Thus $4 = 2 \cdot 2 \mid (a+1)(a-1)$, and so (by the factorization above) $4 \mid (a^4 - a^2)$. □

(c) False. Counterexample: $4^9 - 3 = 11 \cdot 23831$.

#43. (a) Since $d \mid b$ and $d \mid c$ if, and only if $ad \mid ab$ and $ad \mid ac$, it follows that the set $\{ad : d$ is a common divisor of a and $b\}$ is exactly the set of common divisors of ab and ac. Therefore, gcd(ab, ac) is the greatest element of the set $\{ad : d$ is a common divisor of a and $b\}$, which is gcd(a,c).

(b) From Part (a), it follows that $d \gcd(b/d, c/d) = \gcd(d \cdot (b/d), d \cdot (c/d)) = \gcd(b,c) = d$, and this forces $\gcd(b/d, c/d) = 1$. This latter fact implies that the numerator and denominator of $\dfrac{b/d}{c/d}$ have no common factors, so this is the lowest terms representation of $\dfrac{b}{c}$. (This fact corresponds to the arithmetical fact that to put a fraction in lowest terms, we need to cancel out the greatest common factor of the numerator and denominator.)

(c) Use the Euclidean algorithm to obtain $d = \gcd(b,c)$, and then apply Part (b). For example (the Euclidean algorithm gives) gcd(1474,39463) = 67, so the lowest terms form of 1474/39463 is (1474/67)/(39463/67) = 22/589.

#45. If n had a nontrivial factorization $n = ab$, then the indicated identity would produce a nontrivial factorization of $2^{ab} - 1$. This proves that if n is composite, then so is $2^n - 1$, which is the contrapositive, and so logically equivalent to the indicated statement.

#47. *Proof:* If there were ever a repeated term in this sequence of powers $g, g^2, g^3, \cdots, g^{\phi(n)}$, say $g^i \equiv g^j \pmod n$, then Part (c) of Proposition 8.5 would tell us that $i \equiv j \pmod{\text{ord}_n(g)}$. Since g is a primitive root, $\text{ord}_n(g) = \phi(n)$, so this congruence is impossible if i and j are different integers in the

range $\{1, 2, \cdots, \phi(n)\}$. It follows that all of the modular powers $g, g^2, g^3, \cdots, g^{\phi(n)}$ are distinct mod n.
□

#49. All of the integers mod 223 for which order is defined (i.e., which are relatively prime to 223) are powers of the primitive root $g = 3$ of Part (a) (by Theorem 8.7). By Proposition 3.22, the order of any such element must divide $\phi(223) = 222 = 2 \cdot 3 \cdot 37$. Since 10 is not a factor there are no integers with order 10 (so answer to Part (d) is zero). But both 6 and 74 are factors. By the formula of Exercise 48(a) with a taken to be the primitive root 3: $\text{ord}_{223}(3^j) = \dfrac{\text{ord}_{223}(3)}{\gcd(j, \text{ord}_{223}(3))} = \dfrac{222}{\gcd(j, 222)}$. Thus, the mod 223 integers with order = 6 correspond to the modular powers 3^j $(1 \le j \le 222)$ that will make the latter fraction = 6, i.e., $\dfrac{222}{\gcd(j, 222)} = 6 \Rightarrow \gcd(j, 222) = 222/6 = 37$. The corresponding integers are clearly: $j = 37, 5 \cdot 37$ $(2 \cdot 37, 3 \cdot 37, 4 \cdot 37$ are out since they have other common factors with 222, and $7 \cdot 37$ is already too large). Thus there are 2 elements of order 6 (mod 223). Part (c) is done in the same fashion. Elements of order 74 must be of the form 3^j $(1 \le j \le 222)$ with $\dfrac{222}{\gcd(j, 222)} = 74 \Rightarrow \gcd(j, 222) = 222/74 = 3$. A computation shows that there are 36 such elements.

#51. Working mod 2, if $a \equiv 0$ or if $a \equiv 1$, the expression $a^{4n+1} - a$ is clearly congruent to zero, so $2 \mid (a^{4n+1} - a)$. In order to work mod 5, we factor: $a^{4n+1} - a = a(a^{4n} - 1) = a(a^{2n} + 1)(a^n + 1)(a^n - 1)$. If $a \equiv 0$, the first factor is zero, if $a \equiv \pm 1$, one of the last two factors is 0, and finally, if $a \equiv \pm 2$, then $a^2 \equiv 4 \equiv -1$, so $a^{4n} - 1 \equiv 0$. This proves that $5 \mid (a^{4n+1} - a)$, and when combined with $2 \mid (a^{4n+1} - a)$, produces $10 \mid (a^{4n+1} - a)$. This means that $a^{4n+1} \equiv a \pmod{10}$. □

Section 4.1:

Note: For simplicity of notation we will write all base b expansions as strings whenever $b \le 10$ or $b = 16$ (hex format).

#1. (a) (as in Table 4.2): 0, 1, 10, 11, 100, 101, 110, 111, 1000, 1001, 1010, 1011, 1100, 1101, 1110, 1111, 10000, 10001, 10010, 10011, 10100, 10101, 10110, 10111, 11000, 11001 (b) 0, 1, 2, 3, 4, 5, 6, 7, 10, 11, 12, 13, 14, 15, 16, 17, 20, 21, 22, 23, 24, 25, 26, 27, 30, 31 (c) see Table 6.2.

#3. (a) 1100100, 1100101, 1100110, 1100111, 1101000, 1101001, 1101010, 1101011, 1101100, 1101101, 1101110, 1101111, 1110000, 1110001, 1110010, 1110011, 1110100, 1110101, 1110110, 1110111, 1111000, 1111001, 1111010, 1111011, 1111100, 1111101 (b) 144, 145, 146, 147, 150, 151, 152, 153, 154, 155, 156, 157, 160, 161, 162, 163, 164, 165, 166, 167, 171, 172, 173, 174, 175 (c) 64, 65, 66, 67, 68, 69, 6A, 6B, 6C, 6D, 6E, 6F, 70, 71, 72, 73, 74, 75, 76, 77, 78, 79, 7A, 7B, 7C, 7D.

#5. (a) 42 (b) 14,043 (c) 11,259,375 (d) 703 (e) 4,886,735,530 (f) 17,343,427

#7. (a) $66 = 1000010$ (base 2) $= 102$ (base 8) $= 42$ (hex) (b) $237 = 11101101$ (base 2) $= 355$ (base 8) $=$ ED (hex) (c) $1925 = 11110000101$ (base 2) $= 3605$ (base 8) $= 785$ (hex) (d) $12587 =$ 11000100101011 (base 2) $= 30453$ (base 8) $= 312B$ (hex) (e) $28,000 = 110110101100000$ (base 2) $=$ 66540 (base 8) $= 6D60$ (hex) (f) $150,269 = 100100101011111101$ (base 2) $= 445375$ (base 8) $=$ 24AFD (hex).

#9. (a) $66 = 2110$ (base 3) $= 73$(base 9) $= [2\ 12]$ (base 27) (b) $237 = 22210$ (base 3) $= 283$ (base 9) $=$ [8 21] (base 27) (c) $1925 = 2122022$ (base 3) $= 2568$ (base 9) $= [2\ 17\ 8]$ (base 27) (d) $12,587 =$ 12202101 (base 3) $= 18235$ (base 9) $= [17\ 7\ 5]$ (base 27) (e) $28,000 = 11021020$ (base 3) $= 42361$ (base 9) $= [1\ 11\ 11\ 1]$ (base 27) (f) $150,269 = 21122010$ (base 3) $= 248115$ (base 9) $= [7\ 17\ 3\ 14]$ (base 27).

#11. (a) agent = 00000 00110 00100 01101 10011, met = 01100 00100 10011, liaison = '01011 01000 00000 01000 10010 01110 01101 (b) (i) help (ii) keller (iii) now

#13. (a) 11011; 12 + 15 = 27 (b) 11104; 2921 + 1755 = 4676 (c) 14464; 43,724 + 39,320 = 83,044 (d) [24 18]; 558 + 60 = 618

#15. (a) 1001100111; $428 + 187 = 615$ (b) 56100010; $11,979,065 + 112,338 = 12,091,400$ (c) A4444CB; $11,189,196 + 161,057,023 = 172,246,219$ (d) $[1 \quad 0 \quad 4 \quad 6 \quad 21]$; $355,233 + 38,063 = 393,296$

#17. (a) 0001 or 1; $12 - 11 = 1$ (b) 6062; $3721 - 599 = 3122$ (c) 1099CD; $11150028 - 10062079 = 1087949$ (d) $355,233 - 38,063 = 317,170$

#19. (a) 101010; $6 \times 7 = 42$ (b) 23177; $365 \times 27 = 9855$ (c)   679FEC; $2764 \times 2457 = 6,791,148$ (d) $[2 \quad 3 \quad 14 \quad 5]$; $558 \times 60 = 33,480$

#21. (a) 1100000111; $25 \times 31 = 775$ (b)   23413214; $2916 \times 1755 = 5,117,580$ (c)   66BE0D34; $43,724 \times 39,423 = 1,723,731,252$ (d) $[2 \quad 15 \quad 23 \quad 17 \quad 11 \quad 17]$.

#23. Each hexadecimal digit is equivalent to a string of 4 binary digits via Table 4.1. Since 16 is a power of two, the digit placements correspond so that the adjacent groups of four binary digits can be converted to their single hex equivalents, and the expansions will represent the same number. The binary expansion should initially be padded on the left with redundant zeros, so that the number of binary digits is a multiple of four. Here is a simple example: To convert the binary expansion: [10011] to hex, we first pad with three zeros on the left: [0001 0011]. Then we use Table 4.1 to look up the hex equivalents of the groups of four binary digits: [13] (hex). The reader should check that both expansions represent 19.

#25. Each corresponding pair of digits must be added with the current carry (in the formation of NewCar in Step 2), and this amounts to two additions for each of the n digits. Then, in computing s_i in the same step, we need to subtract a certain quantity from the number just computed, giving a total of n subtractions. Thus we have to do a total of $3n$ additions/subtractions. In case certain carries are zero, they need not be added, cutting down the number of additions by n. In total, thus there are between $2n$ and $3n$ additions/subtractions. (Actually, we can possibly cut this down a bit further if we avoid subtracting NewCar terms or single digits that are zero.)

#27. This follows from Theorem 4.1 with $b = 2$: Any positive integer W can be written as: $W = c_K 2^K + c_{K-1} 2^{K-1} + \cdots + c_1 2^1 + c_0$, where K is a nonnegative integer, and each coefficient c_i is either 0 or 1. If $W \le 2^{n-1}$, then we may assume in this expansion that $K < n$ (because 2^K would by itself be greater than W). This means that the weight W is the sum of all the weights 2^i whose coefficients c_i are 1.

#29. (a) (i) Since a lies in the range $-32 = -2^{6-1} \le a < 2^{6-1} = 32$, we can use a length 6 (or larger) two's complement representation. Since 17 is nonnegative, the first digit is 0. The remaining digits are given by the binary expansion of 17, which is (by Algorithm 4.1) 10001; so we have the two's complement representation $17 \sim [010001]$. (Note: it would also be correct to use any larger number of digits by tacking on additional zeros after the first (sign) digit, in this and any two's complement representation.) (ii) Since -22 lies in the range $-32 = -2^{6-1} \le a < 2^{6-1} = 32$, we can use length 6 (or larger) two's complement representation. Since -22 is negative, the first digit is 1, and the remaining 5 digits will be the binary expansion of $32 - |-22| = 10$, which (by Algorithm 4.1) is 01010. Thus we have the two's complement representation $-22 \sim [101010]$. (iii) Since -32 lies in the same range, we can use the same procedure to obtain the following two's complement representation: $-32 \sim [100000]$. (b) (i) Since the first digit is 1, the sign of the number a being represented is negative. The remaining five digits 10011 which are the binary representation of $16 + 2 + 1 = 19$, represents $32 - |a|$. Solving for $|a|$ gives $|a| = 13$, so the number being represented is -13. (ii) 12 (iii) 31

#31. (a) When a is nonnegative, the remainder of this division is just a, while if a is negative, then since $a = -|a|$, it follows that $a = (-1) \cdot 2^{n-1} + (2^{n-1} - |a|)$ so the remainder of the division of a by 2^{n-1}. In either case, it follows that the two's complement representation of a uses (after the first digit that determines the sign) precisely the binary expansion of this remainder as the remaining digits. (b) By the result of Part (a), here is a simple algorithm for converting an integer in the range $-2^{n-1} \le a < 2^{n-1}$, to its two's complement representation $a \sim [b_{n-1} \, b_{n-2} \cdots b_1 \, b_0]$:

Step 1: If $a < 0$, set $b_{n-1} = 1$, otherwise set $b_{n-1} = 0$.

Step 2: Compute the remainder r of the division of a by 2^{n-1}.

Step 3: Take $[b_{n-2}\ b_{n-3}\ \cdots b_1\ b_0]$ to be the binary expansion of r. (Padding with zeros on the left, as needed so there will be $n-1$ bits.)

(c) An algorithm for the reverse process can also be based on the result of Part (a). We assume that we are given an n bit two's complement representation $[b_{n-1}\ b_{n-2}\ \cdots b_1\ b_0]$: The following steps will compute the integer a that is represented.

Step 1: If $b_{n-1}=1$, set the sign of a to be negative, otherwise take it to be positive.

Step 2: Convert the binary expansion of $[b_{n-2}\ b_{n-3}\ \cdots b_1\ b_0]$ into an integer r. Then set $|a|=2^{n-1}-r$.

Step 3: Attach the sign found in Step 1 to the absolute value found in Step 2 to recover a.

Section 4.2:

#1. (a) True (b) False (c) False (d) True (e) False (f) False

#3. (a) 14 (b) 8 (c) 16 (d) 7 (e) First notice that (working in $\mathbb{Z}_{24}$) $21^2=9$, so $21^4=(21^2)^2=9^2=9$, and it follows that any power of 9, and hence any even power of 21 is 9. Since $223=2\cdot111+1$, we have: $21^{223}=21^{2\cdot111+1}=(21^2)^{111}\cdot21^1=(9)^{111}\cdot21=9\cdot21=21$.

#5.

$\mathbb{Z}_2$

+	0	1
0	0	1
1	1	0

$\mathbb{Z}_2$

×	0	1
0	0	0
1	0	1

$\mathbb{Z}_4$

+	0	1	2	3
0	0	1	2	3
1	1	2	3	0
2	2	3	0	1
3	3	0	1	2

$\mathbb{Z}_4$

×	0	1	2	3
0	0	0	0	0
1	0	1	2	3
2	0	2	0	2
3	0	3	2	1

In $\mathbb{Z}_2$ only 1 is invertible, while in $\mathbb{Z}_4$ the invertible elements are 1 and 3.

#7. The invertible elements in $\mathbb{Z}_8$ are $\{1, 3, 5, 7\}$ and each of these elements is its own inverse.

#9. All nonzero elements of $\mathbb{Z}_7$ are invertible; 2 and 4 are inverses, 3 and 5 are inverses, 1 and 6 are self-inverses.

#11. (a) 4 (b) 2 (c) 24 (d) 63

#13. Using the fact (see the solution to Exercise 7) that every odd number in $\mathbb{Z}_8$ is its own inverse, each of the given congruences can be solved:

(a) $3x\equiv5\ \Rightarrow\ x\equiv1\cdot x\equiv3^{-1}\cdot3x\equiv3^{-1}\cdot5\equiv3\cdot5\equiv15\equiv7\ (\bmod\,8)$

(b) $7x+2\equiv3\ \Rightarrow\ 7x\equiv3-2\equiv1\ \Rightarrow\ x\equiv(7^{-1}\cdot7\equiv7\cdot1\equiv)\ 7\ (\bmod\,8)$ (c.f. footnote to Example 4.11)

(c) $5x-2\equiv2\ \Rightarrow\ 5x\equiv2+2=4\ \Rightarrow\ x\equiv1\cdot x\equiv3^{-1}\cdot3x\equiv5^{-1}\cdot4\equiv5\cdot4\equiv20\equiv4\ (\bmod\,8)$

#15. (a) Initial Vectors: $U = [388, 1, 0]$, $V = [3, 0, 1]$. Since $V(1) = 3$ is positive, we update the vectors: $W = U - \text{floor}(U(1)/V(1))V = [1, 1, -129]$, $U = V = [3, 0, 1]$, $V = W = [1, 1, -129]$. Since $V(1) = 1$ is positive, we update the vectors: $W = U - \text{floor}(U(1)/V(1))V = [0, -3, 388]$, $U = V = [1, 1, -129]$, $V = W = [0, -3, 388]$. Since $V(1) = 0$, the algorithm terminates. From the last updated vector U we obtain: $d = 1$, $x = 1$, and $y = -129$; it can be readily checked that these numbers satisfy $d = 388x + 3y$. In light of Proposition 2.11, since $\gcd(388, 3) = 1$ and since $y = -129 \equiv 259\,(\bmod\,388)$, we have $3^{-1} = 259$ in $\mathbb{Z}_{388}$.

(b) Initial Vectors: $U = [388, 1, 0]$, $V = [55, 0, 1]$. Since $V(1) = 55$ is positive, we update the vectors: $W = U - \text{floor}(U(1)/V(1))V = [3, 1, -7]$, $U = V = [55, 0, 1]$, $V = W = [3, 1, -7]$. Since $V(1) = 3$ is positive, we update the vectors: $W = U - \text{floor}(U(1)/V(1))V = [1, -18, 127]$, $U = V = [3, 1, -7]$, $V = W = [1, -18, 127]$. Since $V(1) = 1$ is positive, we update the vectors: $W = U - \text{floor}(U(1)/V(1))V = [0, 55, -388]$, $U = V = [1, -18, 127]$, $V = W = [0, 55, -388]$. Since $V(1) = 0$, the algorithm terminates. From the last updated vector U we obtain: $d = 1$, $x = -18$, and $y = 127$; it can be readily checked that these numbers satisfy $d = 388x + 55y$. In light of Proposition 4.2, since $\gcd(388, 55) = 1$, we have $55^{-1} = y = 127$ in $\mathbb{Z}_{388}$.

(c) Computing as above we obtain: $149^{-1} = 125$ in $\mathbb{Z}_{388}$.   (d) Since (as the extended Euclidean algorithm will show) $\gcd(97, 388) = 97 > 1$, (by Proposition 4.2) 97 is not invertible in $\mathbb{Z}_{388}$.

#17. (a) Initial Vectors: $U = [1353, 1, 0]$, $V = [2, 0, 1]$. Since $V(1) = 2$ is positive, we update the vectors: $W = U - \text{floor}(U(1)/V(1))V = [1, 1, -676]$, $U = V = [2, 0, 1]$, $V = W = [1, 1, -676]$. Since $V(1) = 1$ is positive, we update the vectors: $W = U - \text{floor}(U(1)/V(1))V = [0, -2, 1353]$, $U = V = [1, 1, -676]$, $V = W = [0, -2, 1353]$. Since $V(1) = 0$, the algorithm terminates. From the last updated vector U we obtain: $d = 1$, $x = 1$, and $y = -676$; it can be readily checked that these numbers satisfy $d = 1353x + 2y$.   In light of Proposition 4.2, since $\gcd(1353, 2) = 1$ and since $y = -676 \equiv 677 \,(\text{mod}\,1353)$, we have $2^{-1} = 677$ in $\mathbb{Z}_{1353}$.

(b) Since (as the extended Euclidean algorithm will show) $\gcd(44, 1353) = 11 > 1$ (by Proposition 4.2), 44 is not invertible in $\mathbb{Z}_{1353}$. (c) The extended Euclidean algorithm terminates after six iterations to show that $d = 1353x + 886y$, where $d = 1$, $x = -203$, and $y = 310$; so it follows from Proposition 4.2 that $886^{-1} = 310$ in $\mathbb{Z}_{1353}$.   (d) The extended Euclidean algorithm terminates after nine iterations to show that $d = 1353x + 350y$, where $d = 1$, $x = 76$, and $y = -259$ $(\equiv 1094)$; so it follows from Proposition 4.2 that $350^{-1} = 1094$ in $\mathbb{Z}_{1353}$.

#19. Making use of the inverse information obtained in the Exercises 15 and 17, we can easily solve each of the indicated congruences:

(a) $3x \equiv 59 \;\Rightarrow\; x \equiv 1 \cdot x \equiv 3^{-1} \cdot 3x \equiv 3^{-1} \cdot 59 \equiv 259 \cdot 59 \equiv 149 \,(\text{mod}\,388)$

(b) $149x \equiv 225 \;\Rightarrow\; x \equiv 1 \cdot x \equiv 149^{-1} \cdot 149x \equiv 149^{-1} \cdot 225 \equiv 125 \cdot 225 \equiv 189 \,(\text{mod}\,388)$

(c) $2x \equiv 1225 \;\Rightarrow\; x \equiv 1 \cdot x \equiv 2^{-1} \cdot 2x \equiv 2^{-1} \cdot 1225 \equiv 677 \cdot 225 \equiv 789 \,(\text{mod}\,1353)$

(d) $886x \equiv 35 \;\Rightarrow\; x \equiv 1 \cdot x \equiv 886^{-1} \cdot 886x \equiv 886^{-1} \cdot 35 \equiv 310 \cdot 35 \equiv 26 \,(\text{mod}\,1353)$

#21. (a) Since $d = \gcd(3,18) = 3 \,|\, 6$, we can follow Algorithm 4.7 to obtain the $d = 3$ solutions of the congruence.
Step 1: We solve the modified congruence: $(3/3)y \equiv (6/3)\,(\text{mod}\,18/3)$, i.e., $y \equiv 2(\text{mod}\,6)$. Plainly the unique solution is 2 (mod 6).
Step 2: We may now list the $d = 3$ solutions of the original congruence: 2, $2 + 18/3$, $2 + 2 \cdot 18/3 = \{2, 8, 14\}$. (The reader may wish to check each of these.)
(b) Since $d = \gcd(15,51) = 3|21$, the congruence has 3 solutions. To find them, Algorithm 2.3 first has us solve the congruence $(15/3)y \equiv (21/3)(\text{mod}\,51/3)$, or $5y \equiv 7\,(\text{mod}\,17)$, which, when multiplied by $5^{-1} \equiv 7\,(\text{mod}\,17)$, yields the (unique) solution $y \equiv 15\,(\text{mod}\,17)$. The solutions of the original congruence are now $\{7, 7 + 51/3, 7 + 2(51/3)\} = \{15, 32, 49\} \,(\text{mod}\,51)$.
(c) Since $d = \gcd(8,28) = 4 \,|\, 12$, we can follow Algorithm 4.7 to obtain the $d = 4$ solutions of the congruence.
Step 1: We solve the modified congruence: $(8/4)y \equiv (12/4)\,(\text{mod}\,28/4)$, i.e., $2y \equiv 3(\text{mod}\,7)$. Since $2^{-1} = 4(\text{mod}\,7)$, we obtain $y \equiv 2^{-1} \cdot 2y \equiv 2^{-1} \cdot 3 \equiv 4 \cdot 3 \equiv 5(\text{mod}\,7)$.
Step 2: We may now list the $d = 4$ solutions of the original congruence: 5, $5 + 28/4$, $5 + 2 \cdot 28/4$, $5 + 3 \cdot 28/4 = \{5, 12, 19, 26\}$.
(d) Since $d = \gcd(8,28) = 4 \nmid 6$, we know that the congruence has no solution.

#23. (a) Since $d = \gcd(6,776) = 2\,|\,6$, we can follow Algorithm 4.7 to obtain the $d = 2$ solutions of the congruence. *Step 1:* We solve the modified congruence: $(6/2)y \equiv (28/2)\,(\text{mod}\,776/2)$, i.e., $3y \equiv 14(\text{mod}\,388)$. In Exercise #15(a) we computed $3^{-1} = 259\,(\text{mod}\,388)$, and we can use this to solve for y: $y \equiv 3^{-1} \cdot 14 \equiv 259 \cdot 14 \equiv 134\,(\text{mod}\,388)$. *Step 2:* We may now list the $d = 2$ solutions of the original congruence: 134, $134 + 776/2 = \{134, 522\}$.
(b) Since $d = \gcd(15,1940) = 5 \nmid 21$, we know that the congruence has no solution.
(c) Solution: $\{189, 577, 965, 1353\}$ (The answer to #15(c) will make this easy.)
(d) Solution: $\{747, 2100, 3453, 4806\}$ (The answer to #17(c) will make this easy.)
#25. (a) By the Chinese remainder theorem, the solution obtained will be unique modulo $N = 35$.

The inverse of $N/5 = 7$ (mod 5) is $e_1 = 3$, and the inverse of $N/7 = 5$ (mod 7) is $e_2 = 3$; thus, by (10) the solution is $x = 3 \cdot 3 \cdot 7 + 4 \cdot 3 \cdot 5 \equiv 18 \pmod{35}$. (b) By the Chinese remainder theorem, the solution obtained will be unique modulo $N = 165$. The inverse of $N/3 = 55 \equiv 1$ (mod 3) is $e_1 = 1$, the inverse of $N/5 = 33 \equiv 3$ (mod 5) is $e_2 = 2$, the inverse of $N/11 = 15 \equiv 4$ (mod 11) is $e_3 = 3$, thus, by (10) the solution is $x = 2 \cdot 1 \cdot 55 + 1 \cdot 2 \cdot 33 + 3 \cdot 3 \cdot 15 = 311 \equiv 146 \pmod{165}$. (c) Solution: $x \equiv 1886 \pmod{2730}$

#27. Letting x denote the number of coins, we have the following system:

$$\begin{cases} x \equiv 8 \pmod{15} \\ x - 8 \equiv 11 \pmod{14} \\ x - 8 - 11 \equiv 5 \pmod{13} \end{cases} \Rightarrow \begin{cases} x \equiv 8 \pmod{15} \\ x \equiv 5 \pmod{14} \\ x \equiv 11 \pmod{13}. \end{cases}$$

The Chinese remainder theorem is applicable and yields the following solution: $x \equiv 2273 \pmod{2730}$, and thus 2273 is the smallest possible number of coins.

#29. (a) 6, 1, 2 (b) Suppose that $x_1' x_2' \cdots x_{13}'$ differs from $x_1 x_2 \cdots x_{13}$ in exactly one digit. If the differing digits x_k', x_k shared an odd index, then since all odd index digits have coefficient 1 in (12), it would follow from (12) that the two odd digits would be equal (this could be seen by subtracting one version of (12) from the other) —a contradiction. So assume the differing digits have an even index. Subtracting (12) for $x_1 x_2 \cdots x_{13}$ from (12) for $x_1' x_2' \cdots x_{13}'$ would result (after removing common terms and simplifying) in the equation: $3x_k' \equiv 3x_k \pmod{10}$. But since 3 is invertible (mod 10), we could multiply both sides by the inverse to obtain $x_k' \equiv x_k \pmod{10}$, so in any case, we are led to a contradiction thus proving that two valid ISBN-13 numbers cannot differ in only a single digit. □

(c) We assume that $x_1 x_2 \cdots x_{13}$ was incorrectly typed as $y_1 y_2 \cdots y_{13}$, where each $y_i = x_i$, with exactly two adjacent permuted exceptions: $y_j = x_{j+1}$, $y_{j+1} = x_j$ with the two digits being different. Assume that both ISBNs checked with (12). Then, by subtracting the corresponding sides of the two equations (and simplifying), we would be left with $2(x_{i+1} - x_i) \equiv 0 \pmod{10}$. Since 2 is not invertible, we cannot conclude that $x_{i+1} \equiv x_i \pmod{10}$. More precisely, since $2(x_{i+1} - x_i) \equiv 0 \pmod{10}$ is equivalent to $10 \mid 2(x_{i+1} - x_i)$, which in turn (divide the divisibility by 2) is equivalent to $5 \mid (x_{i+1} - x_i)$. The error will not be detected if $x_{i+1} \equiv x_i \pmod{5}$. (But otherwise it would be detected.) (d) If $x_1 x_2 \cdots x_{13}$ violated (12), we could change just the check digit x_{13} so that (9) would hold. We could similarly easily change any of the other odd-indexed digits (that each have coefficient ± 1 in (12)) to assure that the modified sequence would satisfy (12).

#33. (a) We will prove that $n = \sum_{k=0}^{D} d_k \cdot 10^k$ is always congruent to

$(d_0 + d_2 + d_4 + \cdots - d_1 - d_3 - d_5 - \cdots) \pmod{11}$. (Note that this is more than we needed to show, since the indicated statement simply states that if one of these quantities is congruent to 0 (mod 11) then so is the other.) Collecting like terms in the difference, we obtain:

$$n - (d_0 + d_2 + d_4 + \cdots - d_1 - d_3 - d_5 - \cdots)$$
$$= [d_0 \cdot 10^0 + d_1 \cdot 10^1 + d_2 \cdot 10^2 + d_3 \cdot 10^3 + \cdots + d_D \cdot 10^D] - (d_0 + d_2 + d_4 + \cdots - d_1 - d_3 - d_5 - \cdots)$$
$$= [d_0 \cdot (10^0 - 1) + d_1 \cdot (10^1 + 1) + d_2 \cdot (10^2 - 1) + d_3 \cdot (10^3 + 1) + \cdots + d_D \cdot (10^D - (-1)^D)]$$

But since $10 \equiv -1 \pmod{11}$, it follows that $(10^{2k} - 1) \equiv ((-1)^{2k} - 1) \equiv 0$ and $(10^{2k+1} + 1) \equiv ((-1)^{2k+1} + 1) \equiv 0 \pmod{11}$, so that all of the terms in the last bracketed expression above must be congruent to zero (mod 11). □

(b) Collecting like terms in the difference, we obtain:

$$n - (d_0 + 10d_1 + 100d_2 - d_3 - 10d_4 - 100d_5 + d_6 + 10d_7 + 100d_8 - \cdots)$$
$$= [d_0 \cdot 10^0 + d_1 \cdot 10^1 + d_2 \cdot 10^2 + d_3 \cdot 10^3 + \cdots + d_D \cdot 10^D] - (d_0 + 10d_1 + 100d_2 - d_3 - 10d_4 - 100d_5$$
$$+ d_6 + 10d_7 + 100d_8 - \cdots)$$
$$= [d_0 \cdot (10^0 - 1) + d_1 \cdot (10^1 - 10) + d_2 \cdot (10^2 - 100) + d_3 \cdot (10^3 + 1) + d_4 \cdot (10^4 + 10) + d_5 \cdot (10^5 + 100)$$
$$d_6 \cdot (10^6 - 1) + d_7 \cdot (10^7 - 10) + d_8 \cdot (10^8 - 100) - \cdots]$$

Now each expression in parentheses in the last expression is (when the greatest common factor is 10 is factored out) of one of these two forms: $10^j \cdot (10^{6n} - 1)$ or $10^j \cdot (10^{6n+3} + 1)$, where $j = 0$, 1, or 2, and n is a nonnegative integer. But since $10^3 \equiv 3^3 \equiv -1 (\mathrm{mod}\, 7)$, it follows that $10^{6n} \equiv 1 (\mathrm{mod}\, 7)$, and thus each of these two expressions is congruent to zero (mod 7). The assertion now follows, just as it did in Part (a). □

#35. If $m > 1$ is even, then the sum $\sum_{k=1}^{m-1} k$ contains an odd number of terms. Apart from the middle term $m/2$, the remaining terms can be paired off as $m/2 + j$ and $m/2 - j$, as j runs from 1 to $m/2 - 1$. Each such pair adds up to $(m/2 + j) + (m/2 - j) = m \equiv 0\ (\mathrm{mod}\, m)$, so the entire sum must equal this remaining middle term: $m/2$. □

#37. (a) This follows directly from Proposition 4.5(b). (b) The statement of Part (a) fails if the assumption that a and b are relatively prime is omitted. Counterexample: $6 \equiv 2\ (\mathrm{mod}\, 4)$ and $6 \equiv 2\ (\mathrm{mod}\, 2)$, but $6 \not\equiv 2\ (\mathrm{mod}\, 2 \cdot 4)$.

#39. (a) This follows directly from Proposition 4.5(a), with $b = x^a - a$ (using the definition of a congruence). □ (b) From the result just proved, it follows that an integer a will have a square root x mod pq if, and only if x is a square root of a both mod p and mod q. From Exercise 38, this will happen if, and only if, a belongs to both of the sets: $\{0^2, 1^2, 2^2, \cdots, [(p-1)/2]^2\}\ (\mathrm{mod}\ p)$ and $\{0^2, 1^2, 2^2, \cdots, [(q-1)/2]^2\}\ (\mathrm{mod}\, q)$. Furthermore, since each such <u>nonzero</u> element (of either set) has two distinct square roots under the corresponding modulus and zero has only one square root, for each of the one, two, or four possibilities, the Chinese remainder theorem can be used to compute all of the distinct square roots. (c) (i) With $p = 5$ and $q = 7$, since $a = 9$ belongs to both sets: $\{0^2, 1^2, 2^2, \cdots, [(p-1)/2]^2\} = \{0, 1, 4\}\ (\mathrm{mod}\ p)$ and $\{0^2, 1^2, 2^2, \cdots, [(q-1)/2]^2\} = \{0, 1, 4, 2\}\ (\mathrm{mod}\, q)$, it follows that a does indeed have a square root, in fact four of them. Since $\sqrt{a} \equiv \sqrt{4} \equiv \pm 2 \equiv 2, 3 (\mathrm{mod}\, 5)$, and $\sqrt{a} \equiv \sqrt{2} \equiv \pm 3 \equiv 3, 4 (\mathrm{mod}\, 7)$, with any of the four combinations of these square roots, we can use the Chinese remainder theorem to obtain a square root of a mod pq. For example, using the smaller roots, the Chinese remainder theorem applied to the system $\begin{cases} x \equiv 2 (\mathrm{mod}\, 5) \\ x \equiv 3 (\mathrm{mod}\, 7) \end{cases}$ produces the unique solution $x = 17\ (\mathrm{mod}\, 35)$. The other three combinations produce the following three solutions: $x = 3$, 18, 32 (mod 35), and so $\sqrt{a} \equiv \sqrt{9} \equiv \{3, 17, 18, 32\}\ (\mathrm{mod}\, 35)$. (The reader may wish to verify that the squares of each of these numbers is congruent to 9 (mod 35).) (ii) $\sqrt{51} \equiv \{102, 391\}\ (\mathrm{mod}\, 493)$. (d) Part (a) would remain true (for the same reason as before). The result of Part (b) would be simplified somewhat since all integers have a unique square root mod 2. To see whether a has a square root mod $2q$, we need only check whether a belongs to $\{0^2, 1^2, 2^2, \cdots, [(q-1)/2]^2\}\ (\mathrm{mod}\, q)$. If a appears on the list as zero (i.e., $\sqrt{a} \equiv 0 (\mathrm{mod}\, q)$), then a will have a unique square root mod $2q$, obtained by using the Chinese remainder theorem to solve the system: $x \equiv \sqrt{a} (\mathrm{mod}\, 2), x \equiv 0 (\mathrm{mod}\, q)$, whereas if a appears on the list as a nonzero square, then a will have two square roots mod $2q$, each obtainable using the Chinese remainder theorem using one of the two square roots of a mod q. (e) 11 has no square roots mod 26 (since it has none mod 13), $\sqrt{68} \equiv \{5, 38\}\ (\mathrm{mod}\, 86)$.

#41. (i) By Exercise 61, $x^2 \equiv 46 (\mathrm{mod}\, 413)$ is equivalent to $\begin{cases} x^2 \equiv 46 \equiv 4 (\mathrm{mod}\, 7) \\ x^2 \equiv 46 (\mathrm{mod}\, 59) \end{cases}$. The first congruence plainly has two solutions $x \equiv \pm 2 (\mathrm{mod}\, 7)$. Proposition 4.6 tells us that the latter will have a solution if, and only if $46^{(59+1)/4} \equiv 46^{15}$ is a square root of 46 (mod 59). We work our way up to this power: $46^2 \equiv 51$, $46^4 \equiv 51^2 \equiv 5$, $46^8 \equiv 5^2 \equiv 25$, so

$46^{15} \equiv 46^8 \cdot 46^4 \cdot 46^2 \cdot 46 \equiv (25 \cdot 5) \cdot (51 \cdot 46) \equiv 7 \cdot 45 \equiv 20$. Since $20^2 \equiv 46$ (mod 59), we know there are two square roots of 46: $x \equiv \pm 20 \pmod{59}$. Applying the Chinese remainder theorem to the four possible square root combinations mod 7 and mod 59 produces the four square roots mod 413: $\sqrt{46} \equiv \{79, 138, 275, 334\} \pmod{413}$.

(ii) no such square roots.

(iii) $\sqrt{34} = \{6568, 8711, 14862, 17005\} \pmod{23573}$

Section 4.3:

#1. (a) 3×2 (b) 3 (c) $\begin{bmatrix} 10 & 0 \\ 9 & 3 \end{bmatrix}$ (d) $\begin{bmatrix} -8 & 7 & -1 \\ 2 & 4 & -5 \\ 13 & 13 & 3 \end{bmatrix}$ (e) undefined (f) $\begin{bmatrix} 19 & -5 & 8 \\ -4 & 4 & 16 \\ -2 & -5 & 21 \end{bmatrix}$

#3. (a) $\begin{bmatrix} 4 & 16 \\ 63 & 14 \end{bmatrix}$ (b) $\begin{bmatrix} 40 & -12 \\ -6 & -22 \end{bmatrix}$ (c) $A^2 = A \cdot A = \begin{bmatrix} -20 & -28 \\ 42 & 1 \end{bmatrix}$ (d) $\begin{bmatrix} 5 & 56 \\ 31 & 30 \\ -22 & -18 \end{bmatrix}$ (e) Undefined

(f) $\begin{bmatrix} -7 & -16 & 36 \\ -18 & -12 & 40 \\ 13 & -86 & 127 \end{bmatrix}$

#5. $A' = \begin{bmatrix} 2 & 6 \\ -4 & 5 \end{bmatrix}$, $C' = \begin{bmatrix} 9 & -2 & -5 \\ -4 & 0 & -6 \\ 3 & 7 & 6 \end{bmatrix}$, $D' = \begin{bmatrix} 2 & 7 & 5 \\ 6 & 4 & 6 \end{bmatrix}$

#7. $A^{-1} = \frac{1}{34}\begin{bmatrix} 5 & 4 \\ -6 & 2 \end{bmatrix}$, $B^{-1} = \frac{1}{28}\begin{bmatrix} 2 & 4 \\ 3 & -8 \end{bmatrix}$

#9. As pointed out in the text, multiplying an $n \times m$ matrix with an $m \times r$ matrix requires a grand total of $nr(2m-1)$ mathematical operations (additions and multiplications). Since A has size 100×2, B has size 2×100, computing the product AB will thus require $100 \cdot 100 \cdot (2 \cdot 2 - 1) = 30,000$ mathematical operations to produce a 100×100 matrix D. Then, since C has size 100×2, computing the product DC will require an additional $100 \cdot 2 \cdot (2 \cdot 100 - 1) = 39,800$ mathematical operations for a grand total of 69,800 mathematical operations in computing the triple product $(AB)C$. On the other hand, if we were to compute $A(BC)$, we would first compute BC with $2 \cdot 2 \cdot (2 \cdot 100 - 1) = 796$ mathematical operations to produce a 2×2 matrix E, and then compute the product AE with a total of $100 \cdot 2 \cdot (2 \cdot 2 - 1) = 600$ mathematical operations, yielding a grand total of only 1396 mathematical operations. Thus, the first method would have required 2850% times as many mathematical operations!

#11. (a) $\begin{bmatrix} 1 & 5 \\ 3 & 0 \end{bmatrix}$ (b) $\begin{bmatrix} 3 & 5 \\ 3 & 2 \end{bmatrix}$ (c) $\begin{bmatrix} 0 & 1 \\ 3 & 3 \end{bmatrix}$ (d) $\begin{bmatrix} 7 & 7 \\ 1 & 4 \end{bmatrix}$

#13. (a) $AB \equiv \begin{bmatrix} 1 & 5 \\ 3 & 0 \end{bmatrix}$, $BA \equiv \begin{bmatrix} 0 & 3 \\ 4 & 3 \end{bmatrix}$, $A^2 \equiv \begin{bmatrix} 0 & 2 \\ 2 & 1 \end{bmatrix}$, $CD \equiv \begin{bmatrix} 0 & 1 \\ 1 & 0 \\ 3 & 2 \end{bmatrix}$, DE is undefined,

$EC \equiv \begin{bmatrix} 3 & 4 & 1 \\ 2 & 3 & 0 \\ 3 & 4 & 2 \end{bmatrix}$ (b) $AB \equiv \begin{bmatrix} 4 & 6 \\ 3 & 4 \end{bmatrix}$, $BA \equiv \begin{bmatrix} 0 & 8 \\ 4 & 8 \end{bmatrix}$, $A^2 \equiv \begin{bmatrix} 0 & 2 \\ 2 & 1 \end{bmatrix}$, $CD \equiv \begin{bmatrix} 5 & 6 \\ 1 & 0 \\ 8 & 2 \end{bmatrix}$, DE is undefined,

$EC \equiv \begin{bmatrix} 3 & 4 & 6 \\ 2 & 8 & 0 \\ 3 & 4 & 7 \end{bmatrix}$ Note: Each of these answers should agree with the corresponding answers of

Exercise 3, after the latter are converted into mod 5 matrices.

#15. $\det(A) \equiv 7 \pmod 9$, $A^{-1} \equiv 7^{-1} \cdot \begin{bmatrix} 5 & -1 \\ -3 & 2 \end{bmatrix} \equiv 4 \cdot \begin{bmatrix} 5 & 8 \\ 6 & 2 \end{bmatrix} \equiv \begin{bmatrix} 2 & 5 \\ 6 & 8 \end{bmatrix} \pmod 9$; $\det(B) \equiv 3 \pmod 9$, since

3 is not relatively prime to 9, B is not invertible (mod 9).

#17. (a) 46 (b) 4 (c) 0 (d) $A^{-1} = \dfrac{1}{46} \begin{bmatrix} -1 & 23 & -19 \\ -15 & 23 & -9 \\ 14 & -46 & 36 \end{bmatrix}$,

(e) Since $4^{-1} \equiv 16 \pmod{21}$, $A^{-1} \equiv 4^{-1} \cdot \begin{bmatrix} -1 & 23 & -19 \\ -15 & 23 & -9 \\ 14 & -46 & 36 \end{bmatrix}$

$\equiv 16 \cdot \begin{bmatrix} 20 & 2 & 2 \\ 6 & 2 & 12 \\ 14 & 17 & 15 \end{bmatrix} \equiv \begin{bmatrix} 5 & 11 & 11 \\ 12 & 11 & 3 \\ 14 & 20 & 9 \end{bmatrix} \pmod{21}$

(f) Since $\det(A) \equiv 0$ is not invertible (mod 23) the matrix A is not invertible (mod 23).

#19. (a) (i) FKEEZNVRUFJHKNDRPMELUSROUQNA (ii) DR MHKEWZHPNQXTPHQR (iii) WDGIAQGYGJRTFLEURZAF (iv) FKYPLRVHYWTHQSAH (b) (i) Check into the Mayflower. (ii) Take out the ceiling panel by window. (iii) Assemble the unit. (iv) Bring and leave hotel immediately.

#21. (a) (i) AXEZAHBTFGJCTTERRIDMHTNGCBP (ii) LJJZHVHNATNYZUCYNL (iii) LWHYSQMYQAXEBDDNYBFBS (iv) AXEWDKRVZTMN MKQQQD (b) Note that

$A^{-1} \equiv \begin{bmatrix} 1 & 1 & 25 \\ 0 & 25 & 1 \\ 25 & 0 & 1 \end{bmatrix} \pmod{26}$. (i) Math is fun. (ii) Hill patented his cipher. (iii) The FBI keeps

files on all cryptographers. (iv) Celebrity gets delayed in cryptography.

#23. (a) The given information can be translated (using Table 4.2) into the following matrix equation: $A \begin{bmatrix} 19 & 4 & 13 \\ 22 & 14 & 1 \end{bmatrix} = \begin{bmatrix} 25 & 6 & 16 \\ 2 & 2 & 25 \end{bmatrix}$. If we restrict this equation to using only the first and third columns

of the partial plaintext and ciphertext matrices, the equation becomes $A \begin{bmatrix} 19 & 13 \\ 22 & 1 \end{bmatrix} = \begin{bmatrix} 25 & 16 \\ 2 & 25 \end{bmatrix}$, and

since $\det\left(\begin{bmatrix} 19 & 13 \\ 22 & 1 \end{bmatrix} \right) \equiv 19 \pmod{26}$, the latter matrix equation can be solved by right multiplying by

$\begin{bmatrix} 19 & 13 \\ 22 & 1 \end{bmatrix}^{-1} \equiv 19^{-1} \cdot \begin{bmatrix} 1 & -13 \\ -22 & 19 \end{bmatrix} \equiv 19^{-1} \cdot \begin{bmatrix} 1 & -13 \\ -22 & 19 \end{bmatrix} \equiv 11 \cdot \begin{bmatrix} 1 & 13 \\ 4 & 19 \end{bmatrix} \equiv \begin{bmatrix} 11 & 13 \\ 18 & 1 \end{bmatrix} \pmod{26}$. We point out

that the other three options for selecting two of the three columns in the matrix equation result in

noninvertible matrices. This yields $A \equiv \begin{bmatrix} 25 & 16 \\ 2 & 25 \end{bmatrix} \begin{bmatrix} 19 & 13 \\ 22 & 1 \end{bmatrix}^{-1} \equiv \begin{bmatrix} 25 & 16 \\ 2 & 25 \end{bmatrix} \begin{bmatrix} 11 & 13 \\ 18 & 1 \end{bmatrix} \equiv \begin{bmatrix} 17 & 3 \\ 4 & 25 \end{bmatrix}$. We

may now decrypt the ciphertext in the usual fashion, revealing the plaintext to be: "The revolt will be on Bastille Day." (b) Since the character correspondence represented by the plaintext portion "reou" occupies the 4th through the 7th characters, only the "eo" subportion fills an entire column of the plaintext integer matrix. Since all known plaintext columns begin with "e," whose integer equivalent is 4, it follows that the determinants of any of the 2×2 matrices formed by known plaintext columns will always be even, and hence not relatively prime to 26. Thus, none of these matrices can be inverted (mod 26), and so none of the corresponding partial encryption equations can be solved for the encryption matrix. (c) The given information can be translated (using Table 4.2) into the following matrix equation: $A \begin{bmatrix} 17 & 22 & 17 \\ 24 & 8 & 14 \\ 11 & 13 & 21 \end{bmatrix} = \begin{bmatrix} 1 & 24 & 5 \\ 20 & 21 & 10 \\ 2 & 24 & 24 \end{bmatrix}$. Since $\det\left(\begin{bmatrix} 17 & 22 & 17 \\ 24 & 8 & 14 \\ 11 & 13 & 21 \end{bmatrix} \right) \equiv 4 \pmod{26}$ is not

relatively prime to 26, it follows that this matrix cannot be inverted, so the matrix equation cannot be solved for A. The same problem occurs if we try to permute the columns of this matrix. Thus the given information is not sufficient to decrypt the plaintext.

#25. Any choice that would render the integer matrix (obtained using Table 4.2) to be invertible mod 26 would work. One such example, and a particularly convenient one at that, would be to use the strings: $abb\cdots bb$, $bab\cdots bb$, $bba\cdots bb$, $\cdots$, $bbb\cdots ab$, $bbb\cdots ba$.   This would result in the matrix equation $AI = C$, which directly gives the encoding matrix.

#27. *Proof:* We assume that the sizes of the matrices involved are compatible so that both sides of the identities are defined.   The (i,j) entry of $(A+B)C$ is the dot product of the ith row of $A+B$:

$$[a_{i1} + b_{i1} \quad a_{i2} + b_{i2} \quad \cdots \quad a_{im} + b_{im}] \quad \text{and} \quad \text{the} \quad j\text{th} \quad \text{column} \quad \text{of} \quad C: \begin{bmatrix} c_{1j} \\ c_{2j} \\ \vdots \\ c_{mj} \end{bmatrix}, \quad \text{and so equals}$$

$$\sum_{k=1}^{m}(a_{ik} + b_{kj})c_{kj} = \sum_{k=1}^{m} a_{ik}c_{kj} + \sum_{k=1}^{m} b_{ik}c_{kj}.$$ Since the last two sums give the (i,j) entries of AC and BC, respectively, the identity $(A+B)C = AC + BC$ is thus established. $\square$

#29. *Proof:* The (i,j) entry of $(A+B)'$ is the (j,i) of $A+B$, which is $a_{ji} + b_{ji}$. Since the first term is the (i,j) entry of A', and the second term is (i,j) entry of B', it follow that their sum is the (i,j) entry of $A' + B'$, proving that $(A+B)' = A' + B'$.

#31.   (a)   *Proof:* Let $C = B^{-1}A^{-1}$.   In order to show that $C = (AB)^{-1}$, we must show that $C(AB) = I$ and that $(AB)C = I$.   Using associativity of matrix multiplication we compute: $C(AB) = (B^{-1}A^{-1})(AB) = B^{-1}(A^{-1}A)B = B^{-1}(I)B = B^{-1}B = I$.   (We have used the definition of inverse for the matrices A and B, and the multiplicative identity property of the identity matrix that $IM = MI = M$.)   Similarly: $(AB)C = (AB)(B^{-1}A^{-1}) = A(B^{-1}B)A^{-1} = A(I)A^{-1} = AA^{-1} = I$. $\square$ (b)   *Proof:* Let $C = (A^{-1})^t$.   In order to show that $C = (A^t)^{-1}$, we must show that $CA^t = I$ and that $A^t C = I$.   This can be accomplished by repeatedly collapsing product of A and its inverse t times, in a similar fashion to what was done in the proof of Part (a).   For example:

$$CA^t = \underbrace{A^{-1}A^{-1}\cdots(A^{-1}}_{t \text{ factors}}\underbrace{A)A\cdots A}_{t \text{ factors}} = \underbrace{A^{-1}A^{-1}\cdots A^{-1}}_{t-1 \text{ factors}}I\underbrace{AA\cdots A}_{t-1 \text{ factors}} = \underbrace{A^{-1}A^{-1}\cdots A^{-1}}_{t-1 \text{ factors}}\underbrace{AA\cdots A}_{t-1 \text{ factors}} = \cdots A^{-1}A = I \quad \square$$

Section 4.4:

#1. (a) $x = .129 \times 10^5$ (b) $x = .12345 \times 10^{-2}$ (c) $x = .120 \times 10^2$ (d) $x = .54200 \times 10^{-3}$

#3.   (a)   1100   (b) $(3.1)^2 + (2.7)^2 = 9.6 + 7.2 = 16.$ (c) $(110/210)^2 = (.52)^2 = 0.27$ (d) $110^2/210^2 = 12000/44000 = 0.27$

#5. (a)   1110   (b) $(3.14)^2 + (2.71)^2 = 9.85 + 7.34 = 17.1.$ (c) $(111/210)^2 = (.528)^2 = 0.278$ (d) $111^2/210^2 = 12300/44100 = 0.278$

#7. (a) $11,060,000/0.000626 = 17,660,000,000$

(b) $12 + 111 + 1111 + 11110 + 111100 = 123 + 1111 + 11110 + 111100 = 1234 + 11110 + 111100 = 12340$ $+111100 = 123400$

(c) $111100 + 11110 + 1111 + 111 + 11 + 1 = 123300 + 1111 + 111 + 11 + 1 = 13330 + 111 + 11 + 1 = 133400 + 11 + 1 = 133400$ (d)

$42 \cdot 211 \cdot 2111 \cdot 21110 = 8862 \cdot 2111 \cdot 21110 = 19,590,000 \cdot 21110 = 413,500,000,000$

#9. (a)   For each positive integer N, we let S_N denote the corresponding finite partial sum defined by:

$S_N = \sum_{n=1}^{N} 1/n = 1 + 1/2 + 1/3 + \cdots + 1/N.$ We continue to compute $S_1, S_2, S_3, \cdots$ in 2-digit chopped arithmetic, until the terms (which are decreasing) no longer have any effect on the accumulated sum.
$S_1 = 1.0$, $S_2 = 1 + 1/2 = 1 + 0.50 = 1.5$, $S_3 = 1.5 + 1/3 = 1.5 + 0.33 = 1.8$, $S_4 = 1.8 + 1/4 = 1.8 + 0.25 = 2.0$,
$S_5 = 2.0 + 1/5 = 2.0 + 0.20 = 2.2$, $S_6 = 2.2 + 1/6 = 2.2 + 0.16 = 2.3$, $S_7 = 2.3 + 1/7 = 2.3 + 0.14 = 2.4$,
$S_8 = 2.4 + 1/8 = 2.4 + 0.12 = 2.5$, $S_9 = 2.5 + 1/9 = 2.5 + 0.11 = 2.6$, $S_{10} = 2.6 + 1/10 = 2.6 + 0.10 = 2.7.$

All subsequent terms will be at most $1/11 = .090$ (in floating point arithmetic) which would get chopped when added to the thus far accumulated sum of 2.7. Thus, in 2-digit floating point arithmetic, we have shown that $\sum_{n=1}^{\infty} 1/n = 2.7$.

(b) The idea was witnessed in Part (a) in 2-digit floating point arithmetic. Since the terms being added $(1/n)$ are decreasing to zero (as n gets larger and larger), and the accumulated sum will be at least 1, it follows that the terms will be chopped from the accumulated sum when n is sufficient large. For example, if $n > 10^{s-1}$, then $1/n < 10^{1-s} = 0.\underbrace{00 \cdots 0}_{s-2}1$, will get chopped when added to the accumulated sum, since the latter will be at least 1.

#11. If we work with a larger number of significant digits, then in adding positive numbers, less will get chopped so the resulting sums will be at least as large as corresponding sums computed using a smaller number of significant digits.

#13. Both of the inequalities can fail in floating point arithmetic due to changes in the true value of $\bar{x}$ when it is computed in floating point arithmetic. Here are two examples in $s = 2$ floating point arithmetic:

Example to show that the inequality $\min(x_1, x_2) \leq \mathrm{fl}(\bar{x})$ can fail:

Take $x_1 = x_2 = 0.99$. Then $\mathrm{fl}(\bar{x}) = \mathrm{fl}(1.98)/2 = 1.9/2 = 0.95 < \min(x_1, x_2)$.

Example to show that the inequality $\mathrm{fl}(\bar{x}) \leq \max(x_1, x_2)$ can fail:

Take $x_1 = x_2 = -0.99$. Then $\mathrm{fl}(\bar{x}) = \mathrm{fl}(-1.98)/2 = -1.9/2 = -0.95 > \max(x_1, x_2)$.

NOTE: If we are working with positive numbers only (in chopped arithmetic), then the inequality $\mathrm{fl}(\bar{x}) \leq \max(x_1, x_2)$ will be valid, since chopping can only make $\mathrm{fl}(\bar{x})$ less than (or equal to) $\bar{x}$. In rounded arithmetic, however, this inequality can fail for positive numbers; a similar example to what was given above can be constructed to justify this fact.

#15. Even if we bar overflows, counterexamples can still be concocted to show that multiplication is not associative in floating point arithmetic. Here is an example in 2-digit floating point arithmetic:

$$(0.11 \cdot 0.016) \cdot 500 = (0.0017) \cdot 500 = 0.85 \neq 0.88 = 0.11 \cdot (8) = 0.11 \cdot (0.016 \cdot 500).$$

#17. If we consider the equation $700x = 3$ in 2-digit ($s = 2$) floating point arithmetic with m sufficiently small and M sufficiently large (say $m = -3$, $M = 3$), there will be a total of 2 solutions: namely x = 0.0043, and $x = 0.0044$. If we work instead in 1-digit arithmetic, then the equation $7x = 0.8$ has no solutions since $7 \cdot 0.1 = 0.7$, and $7 \cdot 0.2 = 1.4 \to 1$.

Section 4.5:

#1. (a) Since the prime factorization of $p-1$ is $772 = 2^2 \cdot 193$, the only possibilities for $\mathrm{ord}_p(2)$ are (by Proposition 3.22(2)): 2, 4, 193, 386, and 772 (for a primitive root). Since $2^{193} \equiv 317$ and $2^{386} \equiv 772$, it follows that $\mathrm{ord}_p(2) = 772$, and thus $g = 2$ is a primitive root mod p. (b) $B = g^b \equiv 2^{603} \equiv 122 \pmod{773}$, $A = g^a \equiv 2^{333} \equiv 277 \pmod{773}$. (c) $K = B^a \equiv 122^{333} \equiv 75 \pmod{773}$, $K = A^b \equiv 277^{603} \equiv 75 \pmod{773}$.

#3. (a) $g = 307$ (Note: $\mathrm{ord}_p(301) = 97$, $\mathrm{ord}_p(303) = 388$, $\mathrm{ord}_p(305) = 97$) (b) $A = 1143$, $B = 124$ (c) $K = 245$

#5. (a) $C \equiv P^e \equiv 1234^{143} \equiv 4347 \pmod{n}$ (b) $C^d \equiv 4347^{47} \equiv 1234 \equiv P \pmod{n}$.

#7. (a) $C = 58684$ (b) $C^d \equiv 58684^{29401} \equiv 12345 \equiv P \pmod{n}$.

#9. (a) $\phi(n) = (p-1)(q-1) = 2376$, $\gcd(169, \phi(n)) = 1$, so $e = 169$ is a legitimate encryption exponent. We use the extended Euclidean Algorithm 4.6 to compute $d \equiv e^{-1} \equiv 2137 \pmod{\phi(n)}$, which is the decryption exponent. (b) $C = 1744$ (c) $C^d \equiv 1744^{2137} \equiv 1234 \equiv P \pmod{n}$.

#11. (a) Bob's public key is $B \equiv g^b \equiv 2^{603} \equiv 122 \pmod{773}$, and Alice's is $A \equiv g^a \equiv 2^{333} \equiv 277 \pmod{773}$. Alice computes $C \equiv B^a P \equiv 75 \cdot 321 \equiv 112 \pmod{773}$. Thus, the entire ciphertext would be $(A,C) = (277, 112)$.   (b) The decryption exponent is $p-1-b = 169$ and

$A^{169} \equiv 277^{169} \equiv 134 (\mathrm{mod}\,1231)$, and so $d_\kappa((A,C)) \equiv A^{p-1-b} C \equiv 134 \cdot 112 \equiv 321 \equiv P (\mathrm{mod}\,1231)$.

#13. (a) $g = 5055$ (b) $(A,C) = (3533, 2030)$

#15. (i) (a) The sequence is easily checked to be superincreasing, since $m = 175 > 171$ = the sum of the weights, and $\gcd(w, m) = 1$, the key is legitimate. (b) The public key is the mod m vector:

$w \cdot [a_1 \ a_2 \ \cdots a_6] \equiv 88 \cdot [3 \ 5 \ 8 \ 18 \ 36 \ 100] \equiv [89 \ 90 \ 92 \ 9 \ 18 \ 50] \,(\mathrm{mod}\,m)$.

(c) The ciphertext is $f_b([1,0,1,0,1,0]) = x_1 b_1 + x_2 b_2 + x_3 b_3 + x_4 b_4 + x_5 b_5 + x_6 b_6 = 1 \cdot 89 + 0 \cdot 90 + 1 \cdot 92$

$+ 0 \cdot 9 + 1 \cdot 18 + 0 \cdot 50 = 199 = s$.   (d) Using the extended Euclidean algorithm (Algorithm 4.6) we compute $w^{-1} \equiv 2$.   Since $w^{-1} \cdot s \equiv 2 \cdot 199 \equiv 48 (\mathrm{mod}\,m)$, the plaintext will be the solution of the superincreasing knapsack problem with weight vector $[a_1 \ a_2 \ a_3 \ a_4 \ a_5 \ a_6]$ and knapsack weight $s' = 17$. Algorithm 4.12 quickly produces the original plaintext.   (ii) (b) The public key is the mod m vector:

$w \cdot [a_1 \ a_2 \ \cdots a_6] \equiv 371 \cdot [18 \ 36 \ 100 \ 184 \ 360 \ 750] \equiv [878 \ 306 \ 850 \ 114 \ 160 \ 1300] \,(\mathrm{mod}\,m)$.   (c) $s = 1888$ (ciphertext).   (d)   Using the extended Euclidean algorithm (Algorithm 4.6) we compute $w^{-1} \equiv 981$.   Since $w^{-1} \cdot s \equiv 981 \cdot 1888 \equiv 478 (\mathrm{mod}\,m)$, the plaintext will be the solution of the superincreasing knapsack problem with weight vector $[a_1 \ a_2 \ a_3 \ a_4 \ a_5 \ a_6]$ and knapsack weight $s' = 17$. Algorithm 4.12 quickly produces the original plaintext. (iii) (b) The public key is the mod m vector: $w \cdot [a_1 \ a_2 \ \cdots a_6] \equiv 205 \cdot [5 \ 9 \ 18 \ 34 \ 72 \ 144] \equiv [167 \ 129 \ 258 \ 106 \ 174 \ 62] \,(\mathrm{mod}\,m)$.   (c) $s = 599$ (ciphertext).   (d)   Using the extended Euclidean algorithm we compute $w^{-1} \equiv 173$.   Since $w^{-1} \cdot s \equiv 173 \cdot 599 \equiv 95 (\mathrm{mod}\,m)$, the plaintext will be the solution of the superincreasing knapsack problem with weight vector $[a_1 \ a_2 \ a_3 \ a_4 \ a_5 \ a_6]$ and knapsack weight $s' = 17$. Algorithm 4.12 quickly produces the original plaintext.

#17. (a) The public key is the mod m vector:

$w \cdot [a_1 \ a_2 \ \cdots a_9] \equiv 365 \cdot [5 \ 9 \ 18 \ 34 \ 72 \ 144] \equiv [1095 \ 326 \ 287 \ 574 \ 1148 \ 524 \ 1204 \ 987 \ 932]$ $(\mathrm{mod}\,m)$.

(b) (i) 4831 (ii) 4666 (iii) 4025   (c) The main parameters needed for the decryptions are $w^{-1} \equiv 538$ (for each decryption), and (i) $w^{-1} \cdot s \equiv 1311 (\mathrm{mod}\,m)$ (ii) $w^{-1} \cdot s \equiv 982 (\mathrm{mod}\,m)$ (iii) $w^{-1} \cdot s \equiv 894 (\mathrm{mod}\,m)$.

#19. (a) If either C or $\tilde{C}$ were not relatively prime to n, then the corresponding gcd (which could be found with the Euclidean algorithm) would be one of the two (secret) prime factors of $n = pq$. Thus, the factorization of n could be found, and from this Eve could obtain the (secret) decryption exponent. She would thus be able to decrypt not only these two, but all future messages sent to either Bob or Ben (by anyone, not just Alice). (We point out that this case is typically quite rare.) Henceforth we assume that both C and $\tilde{C}$ are relatively prime to n.   Since $\gcd(e, \tilde{e}) = 1$, Eve could use the extended Euclidean algorithm (Algorithm 4.6) to determine integers x and y such that $ex + \tilde{e}y = 1$. From this, and the given ciphertext equations, we deduce that $C^x \tilde{C}^y \equiv (P^e)^x (P^{\tilde{e}})^y \equiv P^{ex + \tilde{e}y} \equiv P (\mathrm{mod}\,n)$.   From the equation $ex + \tilde{e}y = 1$, it follows that one of x or y is a positive integer, and the other is negative. Without loss of generality, we assume that $x < 0$. Now since $C^x \equiv (C^{-1})^{|x|}$, Eve only needs to first compute $C^{-1} (\mathrm{mod}\ n)$, using the extended Euclidean algorithm, and then raise the result and $\tilde{C}$ to the appropriate (positive) modular powers, and multiply them (mod n) to determine the plaintext P.   (b) The extended Euclidean Algorithm tells us that with $x = -587$ and $y = 38$, we have $ex + \tilde{e}y = 1$. Another application of the extended Euclidean algorithm tells us that $C^{-1} = 5179 (\mathrm{mod}\,n)$. Using fast modular exponentiation (Algorithm 4.5), we compute $C^x \equiv (C^{-1})^{587} \equiv 5179^{587} \equiv 1538 (\mathrm{mod}\,6887)$ and also $\tilde{C}^y \equiv 1902^{38} \equiv 2594$. The plaintext P will be the product of these two numbers (mod 6887): $P = 1538 \cdot 2594 \equiv 1999$. (As a check, the reader should verify that this number encrypts to C, and $\tilde{C}$, when the RSA encryption exponents e and $\tilde{e}$ are used, respectively. (c) As in Part (b), we now obtain $ex + \tilde{e}y = 1$, with $x = -578$, and $y = 2917$. Also, $C^{-1} = 405390(\bmod\, n)$. Hence $P \equiv C^x \tilde{C}^y = (C^{-1})^{578}(\tilde{C})^{2917} \equiv 856876 \cdot 766399 \equiv 888888(\bmod\, n)$.

#21. (a) $D_e : \mathbb{Z}_p \to \mathbb{Z}_p :: D_e(P) \equiv P^d (\bmod\, p)$, where d is the inverse of e (mod $p-1$). In the given setting, we compute $d \equiv 275^{-1} \equiv 11(\bmod\, p-1)$. Thus, $D_e(777) \equiv 777^{11} \equiv 600 \equiv P(\bmod\, p)$. (The reader may wish to verify that $E_e(P) \equiv 777$.) (b) This system is totally insecure. Eve could simply use the extended Euclidean algorithm to compute d (the private key) by inverting e (the public key) (mod $p-1$).

#23. Following the suggestion, Eve could use the extended Euclidean algorithm to compute x^{-1} (mod n). When Bob decrypts $\tilde{C} \equiv Cx^e \equiv P^e x^e \equiv (Px)^e (\bmod\, n)$, the result (that he gives to Eve) will be $\tilde{C}^d \equiv (Px)^{ed} \equiv Px(\bmod\, n)$, and Eve need only multiply the latter number by x^{-1} (mod n) to obtain the plaintext.

#25. (a) Since $g^{L_g(a)+L_g(b)} \equiv g^{L_g(a)} g^{L_g(b)} \equiv ab \equiv g^{L_g(ab)}(\bmod\, p)$, it follows that (see Proposition 3.22) $L_g(ab) \equiv L_g(a) + L_g(b)$ (mod $p-1$). (b) Since $g^{-L_g(a)} \equiv (g^{L_g(a)})^{(-1)} \equiv a^{-1}(\bmod\, p)$ $\equiv ab \equiv g^{L_g(ab)}$, it follows that $L_g(a^{-1}) = -L_g(a)$ (mod $p-1$). (c) Since $g^{kL_g(a)} \equiv (g^{L_g(a)})^k \equiv a^k(\bmod\, p)$, it follows that $L_g(a^k) = kL_g(a)(\bmod\, p-1)$.

(d) Since $h^{L_h(g) \cdot L_g(a)} \equiv (h^{L_h(g)})^{L_g(a)} \equiv (g)^{L_g(a)} \equiv a(\bmod\, p)$, it follows that $L_h(a) \equiv L_h(g) \cdot L_g(a)$ (mod $p-1$).

CHAPTER 5: Section 5.1:
#1. (a) $10 \cdot 26^3 \cdot 10^3 = 175,760,000$ (b) $10 \cdot 26 \cdot 25 \cdot 24 \cdot 9 \cdot 8 \cdot 7 = 78,624,000$ **#3.** $4 \cdot 4 = 16$ **#5.** (a) $5 \cdot 4 \cdot 3 \cdot 2 \cdot 1 = 120$ (b) $2 \cdot 1 \cdot 3 \cdot 2 \cdot 1 = 12$ (c) $2 \cdot 4 \cdot 3 \cdot 2 \cdot 1 = 48$ **#7.** (a) $12 \cdot 11 \cdot 10 = 1320$ (b) $1320 - 9 \cdot 8 \cdot 7 = 816$ (c) $7 \cdot 6 \cdot 5 + (10 \cdot 9 \cdot 8 - 7 \cdot 6 \cdot 5) + (9 \cdot 8 \cdot 7 - 7 \cdot 6 \cdot 5)$ $= 1014$ (d) $3 \cdot 2 \cdot 10 + 10 \cdot 9 \cdot 8 = 780$ **#9.** (a) $5^7 = 78,125$ (b) $2^3 \cdot 5^5 = 25,000$ (c) 0 **#11.** (a) $[(3 \cdot 2)/2] \cdot 4^2 = 48$ (b) $[(3 \cdot 2)/2] \cdot [(4 \cdot 3)/2]^2 = 216$ **#13.** (a) $62^5 + 62^6 + 62^7 = 3,579,330,974,624$ (b) $62^5 + 62^6 + 62^7 - (52^5 + 52^6 + 52^7) - 2 \cdot (36^5 + 36^6 + 36^7) + 2 \cdot (26^5 + 26^6 + 26^7) + (10^5 + 10^6 + 10^7)$ $= 2,386,621,947,840$ (c) $62^5 + 62^6 + 62^7 - [(52^5 + 52^6 + 52^7 + 2 \cdot (36^5 + 36^6 + 36^7)) -$ $+ 10^5 + 10^6 + 10^7)] = 2,386,621,947,840$ (d) $2^3 \cdot 5 \cdot 62^2 + 2^3 \cdot 6 \cdot 62^3 + 2^3 \cdot 7 \cdot 62^4 = 839,068,320$

#15. (a) 3257 (b) 5766 (c) For Part (a):

```
set count = 0;
for n =  1 TO n = 5999
    set a = n/3, b = n/5, c = n/7
    if floor(a)==a OR floor(b)==b OR floor(c)==c
        UPDATE count = count+1
    end
end
OUTPUT count
```

For the corresponding program for Part (b) change 5 to 11, and replace the three "OR"s with "AND"s. **#17.** (a) 3506 (b) The program is similar to the one for #15(a); just add one more letter d = n/11, and one more (OR) condition for d. **#19.** (a) 1598 (b) 830 (c) The program for Part (a) is similar to that of Parts (a) and (b) of Exercise 15. By first using the complement principle, the problem of Part (b) is reduced to one like in Part (a), and can be programmed accordingly. Alternatively, a program can be written directly as shown on the left:

```
set count = 0;
for n =  1 TO n = 3999
    set a = n/2, b = n/3, c = n/5, d = n/7, e = n/11
    if floor(a)~=a AND floor(b)~=b AND floor(c)~=c
            AND floor(d)~=d AND floor(c)~=c
        UPDATE count = count+1
    end
end
OUTPUT count
```

#21. (a) Rounded to the nearest inch, there are exactly 20 possible heights between 5-0 and 6-7 (these are the pigeons), therefore, in a group of 21 men in this height range, at least two must be the same height. (b) We use the generalized pigeonhole principle with the four suits serving as the pigeons. The smallest integer N for which $\lceil N/4 \rceil = 3$ is $N = 9$, and so this is the smallest number of cards to guarantee at least three will have the same suit. With only 8 cards, we could have 2 of each suit. **#23** (a) The pigeonholes are the nine smaller equilateral triangles (each having side length 1/3) without their vertices, as shown in the figure. If ten points are put in the interior of the big triangle, then at least two must lie in the same pigeonhole and therefore have distance between them less than 1/3. (b) If we take nine points to be the vertices of the smaller triangles that lie on the edges of the larger triangle, then any two of these will be at a distance of at least 1/3.

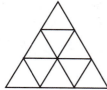

Section 5.2: **#1.** (a) CAT, CTA, ACT, ATC, TAC, TCA (b) AB, BA, AC, CA, AD, DA, AE, EA, BC, CB, BD, DB, BE, EB, CD, DC, CE, EC, DE, ED (c) {A,B}, {A,C}, {A,D}, {A,E}, {B,C}, {B,D}, {B,E}, {C,D}, {C,E}, {D,E} **#3.** (a) 6 (b) 1 (c) 132,600 (d) 15,504 (e) 5 (f) 1 **#5.** (a) $C(12,3) = 220$ (b) $\sum_{k=0}^{3} C(12,k) = 299$ **#7.** (a) $C(12,4) = 495$ (b) $P(12,7) = 3,991,680$ **#9.** (a) $C(8,5) = 56$ (b) $C(8,5) \cdot 3^5 = 13,608$ (c) $C(6,3) \cdot 3^5 + C(6,5) \cdot 3^5 = 6318$ (d) $C(6,3) \cdot 3^4 + C(6,5) \cdot 3^5 = 3078$ **#11.** (a) $C(26,2) = 325$ (b) $C(15,2) + C(11,2) = 160$ (c) $C(11,2) + 11 \cdot 15 = 220$ **#13.** (a) $C(39,4) + C(39,3) \cdot C(13,1) = 201,058$ (b) $13^4 = 28,561$ (c) $C(52,4) - C(13,1)^4$ (*all different suits*) $= 242,164$ **#15.** (a) $C(8,5) = 56$; $C(4,3) \cdot C(4,2) + C(4,4) \cdot C(4,1) = 24 + 4 = 28$ **#17.** (a) $6!2! = 1440$ (b) $5!3! = 720$ (c) $7!/3 = 1680$ (d) $5! = 120$ (e) $4! = 24$ **#19.** (a) 10,897,286,400 (b) $P(14,9) + P(15,9) = 2,542,700,160$ **#21.** (a) $x^7 + 7x^6z + 21x^5z^2 + 35x^4z^3 + 35x^3z^4 + 21x^2z^5 + 7xz^6 + z^7$ (b) $3125x^5 + 3125x^4y^3 + 1250x^3y^6 + 250x^2y^9 + 25xy^{12} + y^{15}$ (c) $\binom{5}{3}3^3(-4)^2 = 4320$ **#25.** (a) $216xyz^2 + 144xy^2z + 72x^2yz + x^4 + 16y^4 + 81z^4 + 8x^3y + 12x^3z + 24x^2y^2 + 54x^2z^2 + 32xy^3 + 108xz^3 + 96y^3z + 216y^2z^2 + 216yz^3$ (c) $\binom{20}{4,6,4,6}2^63^4 = 42,237,882,086,400$

#27.(a) $\binom{4}{2,2} = \binom{4}{2} = 6$ (b) $\binom{8}{3,2,2,1} = 1680$ (c) $\binom{9}{3,2,1,1,1,1} = 30,240$ (d) $\binom{10}{3,2,2,2,1} = 75,600$

#29. (a) $\binom{9}{3,2,4} = 1260$ (b) $3 \cdot 3 \cdot 3 - 1 = 26$ (can't have ggg) (c) $26 + 3^2 + 3 = 38$

#31. (a) Let x_A be the number of $500 increments placed in Fund A, and similarly for x_B, x_C. The number of permissible fund allocations is the number of nonnegative integer solutions of the equation $x_A + x_B + x_C = 30$; and by Theorem 5.10, this number is $\binom{30 + (3-1)}{3-1} = \binom{32}{2} = 496$.

(b) Here we seek the number of solutions of the integer equation above with the additional constraints that $x_A, x_C \geq 5$. If we define $y_A = x_A - 5, y_C = x_C - 5$, the problem is transferred to counting the number of nonnegative integer solutions of the equation $y_A + x_B + y_C = 20$. By Theorem 5.10, it follows that this number is $\binom{20 + (3-1)}{3-1} = \binom{22}{2} = 231$.

#33. Think of placing the 12 donuts in an arbitrary dozen into 8 bins, according to the type of donut. By Theorem 5.10, it follows that the number of dozens of donuts is given by $\binom{12 + (8-1)}{8-1} = \binom{19}{7} = 50,388$.

#35. (a) Let S be a set of n (distinct) objects. Any subset A of S containing k objects (i.e., a k-combination) naturally corresponds to a subset of S containing $n-k$ objects, namely its complement $\sim A = S \sim A$. Since this correspondence is one-to-one, it follws that the number of k-combinations, namely $\binom{n}{k}$, must equal the number of $(n-k)$-combinations, namely $\binom{n}{n-k}$. A noncombinatorial proof is simpler (but less revealing) due to the symmetry in the definition of the binomial coeffients:

$$\binom{n}{n-k} = \frac{n!}{(n-k)!(n-[n-k])!} = \frac{n!}{(n-k)!k!} = \frac{n!}{k!(n-k)!} = \binom{n}{k}.$$

(b) Answer: If n is even, $k = n/2$; if n is odd, $k = \lfloor n/2 \rfloor$ and $\lceil n/2 \rceil$.

Proof: Suppose that $0 < k \le n/2$. Since this implies that $n \ge 2k$, we obtain:

$$\binom{n}{k} \Big/ \binom{n}{k-1} = \frac{(k-1)!(n-k+1)!}{k!(n-k)!} = \frac{n-k+1}{k} \ge \frac{2k-k+1}{k} = \frac{k+1}{k} > 1.$$

This proves that that the left half of the binomial coefficients are an increasing sequence:

$$\binom{n}{0} < \binom{n}{1} < \binom{n}{2} < \cdots < \binom{n}{\lfloor n/2 \rfloor}.$$

From the symmetry result of Part (a), it follows that the corresponding right half of the binomial coefficient sequence is decreasing:

$$\binom{n}{\lfloor n/2 \rfloor + 1} > \binom{n}{\lfloor n/2 \rfloor + 2} > \binom{n}{\lfloor n/2 \rfloor + 3} > \cdots > \binom{n}{n}.$$

In case n is even, there is a single middle coefficient $\binom{n}{\lfloor n/2 \rfloor} = \binom{n}{n/2}$ that is larger than any other.

In case n is odd, there are two equal largest middle coefficients $\binom{n}{\lfloor n/2 \rfloor} = \binom{n}{\lfloor n/2 \rfloor + 1}$. $\square$

#37. (a) 0 (b) 6! (c) $C(6,2) \cdot 5!$ (d) $2^7 - 2$

#43. We consider the combinatorial problem of choosing a team of k people with a designated team leader from a group of n people. By the multiplication principle, the total number of ways to form such a team is
(the number of ways to choose a subset k people from n people) $\cdot$
(the number of ways of choosing a team leader from a team of k people)

$$= \binom{n}{k} \cdot k = k\binom{n}{k}.$$

This number can also be computed as
(the number of ways to choose a leader from a group of n people) $\cdot$
(the number of ways of choosing the $k-1$ non leaders from the remaining $n-1$ people)

$$= n\binom{n-1}{k-1}.$$

It follows that $k\binom{n}{k} = n\binom{n-1}{k-1}.$

Section 5.3: **#1.** (a) $1 - x + x^2 - x^3 + \cdots = \sum_{n=0}^{\infty} (-1)^n x^n$ (b) $6 + 4x^2 + 2x^4 + x^6$

#3. (a) $1 - x + x^2 - x^3 + \cdots - x^9 + x^{10} = \sum_{n=0}^{10} (-x)^n = \frac{1-(-x)^{11}}{1-(-x)} = \frac{1+x^{11}}{1+x}$. (By use of Theorem 3.5.)

(b) Using the binomial theorem, we may write:
$$C(10,0) + C(10,1)x + C(10,2)x^2 + \cdots + C(10,10)x^{10} = \sum_{n=0}^{10} C(10,n)x^n = \sum_{n=0}^{10} C(10,n)x^n 1^{10-n} = (x+1)^{10}.$$

(c) $20 - 40x + 80x^2 - 160x^3 + 320x^4 = 20\sum_{n=0}^{5} (-2x)^n = 20 \cdot \frac{1-(-2x)^5}{1-(-2x)} = 20 \cdot \frac{1+32x^5}{1+2x}$. (By use of

Theorem 3.5.)

(d) Using the binomial theorem, we may write:

$$C(5,0)x^2 + C(5,1)x^3 + C(5,2)x^4 + C(5,3)x^5 + C(5,4)x^6 + C(5,5)x^7$$
$$= x^2 \sum_{n=0}^{5} C(5,n)x^n = x^2 \sum_{n=0}^{5} C(5,n)x^n 1^{5-n} = x^2(x+1)^5.$$

#5. (a) $\sum_{n=0}^{\infty}(-x)^n \underset{\text{By (8)}}{=} \dfrac{1}{1-(-x)} = \dfrac{1}{1+x}$ **(b)** $\sum_{n=0}^{\infty}\dfrac{(-2)^n}{n!}x^n = \sum_{n=0}^{\infty}\dfrac{(-2x)^n}{n!} \underset{\text{By (9)}}{=} e^{2x}$

(c) $\sum_{n=0}^{\infty}(-x)^n + x - 5x \underset{\text{By (8)}}{=} \dfrac{1}{1-(-x)} + x - 5x = \dfrac{1}{1+x} - 4x$

(d) $\sum_{n=0}^{\infty}\dfrac{1}{(n+2)!}x^n = x^{-2}\sum_{n=0}^{\infty}\dfrac{x^{n+2}}{(n+2)!} = x^{-2}\sum_{n=2}^{\infty}\dfrac{x^n}{n!} \underset{\text{By (9)}}{=} \dfrac{e^x - 1 - x}{x^2}$

#7. (a) Using the binomial theorem: $x^3(x+5)^4 = x^3\sum_{n=0}^{4}C(4,n)x^n 5^{4-n} = \sum_{n=0}^{4}[C(4,n)5^{4-n}]x^{n+3}$, so

the sequence is: $a_n = C(4,n)5^{4-n}$ $(n=0,1,2,3,4)$.

(b) Using the binomial theorem:

$$(1-x)^3 - x^5 = C(3,0) + C(3,1)(-x) + C(3,2)(-x)^2 + C(3,3)(-x)^3 - x^5 = 1 - 3x + 3x^2 - x^3 - x^5,$$

so the sequence is: $a_0 = 1, \ a_1 = 3, \ a_2 = 3, \ a_3 = -1, a_5 = -1$.

(c) $1/(1+3x) = 1/(1-(-3x)) \underset{\text{By (8)}}{=} \sum_{n=0}^{\infty}(-3x)^n$, so the sequence is $a_n = (-3)^n$ $(n=0,1,2,\cdots)$

(d) From the expansion: $1/(1+x) - x/(1-2x) = 1/(1-(-x)) - x \cdot 1/(1-(2x)) \underset{\text{By (8)}}{=} \sum_{n=0}^{\infty}(-x)^n -$

$x\sum_{n=0}^{\infty}(-2x)^n = 1 + \sum_{n=0}^{\infty}[(-1)^n - (-2)^n]x^n$, the sequence is: $a_0 = 1, \ a_n = (-1)^n - (-2)^n$ $(n=1,2,3,\cdots)$.

(e) We clear out denominators in the partial fractions decomposition: $1/[(1+x)(1-2x)] = A/(1+x) + B/(1-2x)$, to obtain: $1 = A(1-2x) + B(1+x)$. Substituting $x = -1$ yields $A = 1/3$, and

substituting $x = 1/2$ yields $B = 2/3$. Next, $[1/(1+x) + 2/(1-2x)] \underset{\text{By (8)}}{=} (1/3)\sum_{n=0}^{\infty}(-x)^n +$

$(2/3)\sum_{n=0}^{\infty}(2x)^n = \sum_{n=0}^{\infty}[(-1)^n/2 + 2 \cdot 2^n/3]x^n$, so the sequence is $a_n = [(-1)^n + 2 \cdot 2^n]/3 (n=0,1,2,\cdots)$

(f) Using (9), we may write: $x^2(1+x)e^{-x} = (x^2 + x^3)\sum_{n=0}^{\infty}x^n/n! = \sum_{n=0}^{\infty}x^{n+2}/n! + \sum_{n=0}^{\infty}x^{n+3}/n! =$

$\sum_{n=2}^{\infty}x^n/(n-2)! + \sum_{n=3}^{\infty}x^n/(n-3)!$, so the sequence is $a_n = [1/(n-2)! + 1/(n-3)!]$ $(n=0,1,2,\cdots)$

#9. (a) 5   **(b)** 3   **(c)** −4   **(d)** 14

#11. (a) $\sum_{n=0}^{\infty}b_n x^n \triangleq x^3\sum_{n=0}^{\infty}a_n x^n = \sum_{n=0}^{\infty}a_n x^{n+3} = \sum_{n=3}^{\infty}a_{n-3}x^n \Rightarrow b_n = a_{n-3}$ (which tacitly implies

$b_0 = b_1 = b_2 = 0$, because of the convention that unassigned coefficients, like $a_{-1}, a_{-2}, \cdots$ are taken to

be zero).

(b) $\sum_{n=0}^{\infty}b_n x^n \triangleq (1-x)\sum_{n=0}^{\infty}a_n x^n = \sum_{n=0}^{\infty}a_n x^n - \sum_{n=0}^{\infty}a_n x^{n+1} = \sum_{n=0}^{\infty}[a_n - a_{n-1}]x^n$

$\Rightarrow b_n = a_n - a_{n-1}$ (which tacitly implies $b_0 = a_0$).

(c) $\sum_{n=0}^{\infty}b_n x^n \triangleq \sum_{n=0}^{\infty}a_n x^n \cdot 1/(1-x) \underset{\text{By (8)}}{=} \sum_{n=0}^{\infty}a_n x^n \cdot \sum_{n=0}^{\infty}x^n = \sum_{n=0}^{\infty}[\sum_{k=0}^{n}a_k]x^n$

$\Rightarrow b_n = \sum_{k=0}^{n}a_k.$

#13. (a) −56   **(b)** 0.00537075

#15. (a) With $a = -1/2$, (10) becomes $(1+x)^{1/2} = \sum_{n=0}^{\infty}\binom{-1/2}{n}x^n$. Using Definition 5.6, for $k > 1$ we

may write:

$$\binom{-1/2}{k} = \frac{(-1/2)(-1/2-1)(-1/2-2)\cdots(-1/2-k+1)}{k!} = \frac{(-1/2)(-3/2)(-5/2)\cdots([1-2k]/2)}{k!}$$

$$= \frac{(-1)^k}{2^k}\frac{(2k-1)(2k-3)\cdots 3\cdot 1}{k!}.$$

It follows that

$$(1+x)^{-1/2} = 1 - x/2 + 3x^2/8 - 5x^3/16 + 35x^4/128 - \cdots = \sum_{n=0}^{\infty}\frac{(-1)^n}{2^n}\frac{(2n-1)(2n-3)\cdots 3\cdot 1}{n!}x^n.$$

(b) (i) With $a = 7/2$, (10) becomes $(1+x)^{7/2} = \sum_{n=0}^{\infty}\binom{7/2}{n}x^n = 1 + \binom{7/2}{1}x + \binom{7/2}{2}x^2 + \binom{7/2}{3}x^3 + \cdots$

Since $\binom{7/2}{1} = \frac{7/2}{1!} = \frac{7}{2}, \binom{7/2}{2} = \frac{7/2\cdot 5/2}{2!} = \frac{35}{8}, \binom{7/2}{3} = \frac{7/2\cdot 5/2\cdot 3/2}{3!} = \frac{35}{16},$ we may write:

$(1+x)^{7/2} = 1 + 7x/2 + 35x^2/8 + 35x^3/16 + \cdots$

(b) (ii) Using (9) and the expansion obtained in Part (a), and then multiplying the two generating functions (according to Definition 5.5), we obtain:

$$e^x/\sqrt{1+x} = e^x\cdot(1+x)^{-1/2} = (1 + x + x^2/2! + x^3/3! + \cdots)\cdot(1 - x/2 + 3x^2/8 - 5x^3/16 + \cdots)$$
$$= 1 + x/2 + 3x^2/8 - x^3/48 + \cdots$$

#17. (a) Let $F(x) = \sum_{n=0}^{\infty}a_n x^n$. $a_n = 2a_{n-1} + 5 \ (n\geq 1) \Rightarrow \sum_{n=1}^{\infty}a_n x^n = 2\sum_{n=1}^{\infty}a_{n-1}x^n + 5\sum_{n=1}^{\infty}x^n$

$\Rightarrow F(x) - 3 = 2xF(x) + 5/(1-x) - 5 \Rightarrow F(x) = \dfrac{2x+3}{(1-2x)(1-x)} = \dfrac{A}{1-2x} + \dfrac{B}{1-x}.$ To find A and B in this

partial fractions expansion, we first clear out the denominators $2x+3 = A(1-x) + B(1-2x)$. Substituting $x = 1$ gives $B = -5$, and substituting $x = 1/2$ gives $A = 8$. Applying the expansion (8) we

obtain: $F(x) = \dfrac{8}{1-2x} - \dfrac{5}{1-x} = 8\sum_{n=0}^{\infty}(2x)^n - 5\sum_{n=0}^{\infty}x^n = \sum_{n=0}^{\infty}(2^{n+3}-5)x^n,$ and hence: $a_n = 2^{n+3} - 5.$

(b) Let $b_n = a_{n+2}(n\geq 0)$, so that $\begin{cases} b_0 = 1 \\ b_n = 3b_{n-1} - 1 \ (n\geq 1)\end{cases}.$ By the solution of Exercise for the Reader

5.21, we have $b_n = (3^n + 1)/2$, so it follows that $a_n = b_{n-2} = (3^{n-2}+1)/2 \ (n\geq 2).$

(c) Let $F(x) = \sum_{n=0}^{\infty}a_n x^n$. $a_n = 2a_{n-1} + 3n \ (n\geq 1) \Rightarrow \sum_{n=1}^{\infty}a_n x^n = 2\sum_{n=1}^{\infty}a_{n-1}x^n + 3\sum_{n=1}^{\infty}nx^n \Rightarrow$

$\sum_{n=1}^{\infty}a_n x^n = 2x\sum_{n=1}^{\infty}a_{n-1}x^{n-1} + 3x\sum_{n=1}^{\infty}nx^{n-1}.$ Using expansions (8) and (11), this translates

into: $F(x) - 1 = 2xF(x) + 3x/(1-x)^2 \Rightarrow F(x) = \dfrac{x^2+x+1}{(1-2x)(1-x)^2} = \dfrac{A}{1-2x} + \dfrac{B}{1-x} + \dfrac{C}{(1-x)^2}.$ To find $A, B,$

and C in this partial fractions expansion, we first clear out the denominators $x^2 + x + 1 = A(1-x)^2 + B(1-x)(1-2x) + C(1-2x)$. Substituting $x = 1$ gives $C = -3$, substituting $x = 1/2$ gives $A = 7$, and substituting $x = 0$ now gives $B = -3$. Applying the expansions (8) and (11) we

obtain: $F(x) = \dfrac{7}{1-2x} - \dfrac{3}{1-x} - \dfrac{3}{(1-x)^2} = 7\sum_{n=0}^{\infty}(2x)^n - 3\sum_{n=0}^{\infty}x^n - 3\sum_{n=0}^{\infty}(n+1)x^n,$ and hence:

$a_n = 7\cdot 2^n - 3(n+2).$

(d) Let $F(x) = \sum_{n=0}^{\infty}a_n x^n$. $a_n = 2a_{n-2} + 5 \ (n\geq 2) \Rightarrow \sum_{n=2}^{\infty}a_n x^n = 2\sum_{n=2}^{\infty}a_{n-2}x^n + 5\sum_{n=2}^{\infty}x^n =$

$2x^2\sum_{n=2}^{\infty}a_{n-2}x^{n-2} + 5\sum_{n=2}^{\infty}x^n.$ Using expansion (8), this translates into:

$F(x) - 1 - x = 2x^2F(x) + 5/(1-x) - 5 - 5x \Rightarrow F(x) = \dfrac{4x^2+1}{(1-2x^2)(1-x)} = \dfrac{A}{1-\sqrt{2}x} + \dfrac{B}{1+\sqrt{2}x} + \dfrac{C}{1-x}.$ To

find A, B, and C in this partial fractions expansion, we first clear out the denominators

$4x^2+1=A(1+\sqrt{2}x)(1-x)+B(1-\sqrt{2}x)(1-x)+C(1-2x^2)$.   Substituting x =   1, $x=1/\sqrt{2}$,

$x=-1/\sqrt{2}$, gives $C=-5, A=3/(2-\sqrt{2}), B=3/(2+\sqrt{2})$,   respectively. Applying the expansion (8)

we obtain: $F(x)=\dfrac{3/(2-\sqrt{2})}{1-\sqrt{2}x}+\dfrac{3/(2+\sqrt{2})}{1+\sqrt{2}x}-\dfrac{5}{1-x}=3/(2-\sqrt{2})\sum_{n=0}^{\infty}(\sqrt{2}x)^n$

$$+3/(2+\sqrt{2})\sum_{n=0}^{\infty}(-\sqrt{2}x)^n-5\sum_{n=0}^{\infty}x^n,$$

and hence:

$a_n=3(\sqrt{2})^n/(2-\sqrt{2})+3(-\sqrt{2})^n/(2+\sqrt{2})-5=(3/2)[2\sqrt{2}^n(1+(-1)^n)+\sqrt{2}^{n+1}(1-(-1)^n)]-5$.

(The reader is encouraged to check the correctness of this explicit formula by using the recursive one.)

(e) Let $F(x)=\sum_{n=0}^{\infty}a_nx^n$. $a_n=2a_{n-2}+a_{n-1}$ $(n\geq 2)\Rightarrow$ $\sum_{n=2}^{\infty}a_nx^n=2\sum_{n=2}^{\infty}a_{n-2}x^n+\sum_{n=2}^{\infty}a_{n-1}x^n=$

$2x^2\sum_{n=2}^{\infty}a_{n-2}x^{n-2}+x\sum_{n=2}^{\infty}a_{n-1}x^{n-1}\Rightarrow F(x)-1-2x=2x^2F(x)+x(F(x)-1)\Rightarrow F(x)=\dfrac{x+1}{(1-2x)(1+x)}$

$=\underset{\text{by (8)}}{\dfrac{1}{1-2x}=}\sum_{n=0}^{\infty}(2x)^n\Rightarrow a_n=2^n$.

#19. (a) We note that the recurrence relation $1=a_n+2a_{n-1}+3a_{n-2}+\cdots+na_1+(n+1)a_0$ remains valid

when $n=0$, and hence for all $n\geq 0$. Let $F(x)=\sum_{n=0}^{\infty}a_nx^n$. The recurrence relation implies that:

$$\sum_{n=0}^{\infty}x^n=\sum_{n=0}^{\infty}[a_n+2a_{n-1}+3a_{n-2}+\cdots+na_1+(n+1)a_0]x^n=\sum_{n=0}^{\infty}\left[\sum_{k=0}^{n}(k+1)a_{n-k}\right]x^n.$$

The expansion on the left is just (8), the generating function of $1/(1-x)$. By Definition 5.5 and by

(11), we see that the series on the right is just the product of the generating functions $1/(1-x)^2$ and

$F(x)$. Thus, the equation translates to: $\dfrac{1}{1-x}=\dfrac{F(x)}{(1-x)^2}$ $\Rightarrow F(x)=1-x$.   The sequence defined by

the recursion is thus the very simple sequence: $a_0=1, a_1=-1, a_n=0$ $(n\geq 2)$. (The reader may wish

to verify this directly.)

(b) We let $F(x)=\sum_{n=0}^{\infty}a_nx^n$. Unlike in Part (a), the recursion formula is not valid when $n=0$, it

gives: $\sum_{n=1}^{\infty}nx^n=\sum_{n=1}^{\infty}\left[\sum_{k=0}^{n}(k+1)a_{n-k}\right]x^n$.   The series on the left is $x\sum_{n=1}^{\infty}nx^{n-1}=$

$x\underset{\text{by (11)}}{\sum_{n=0}^{\infty}(n+1)x^n=}x/(1-x)^2$. The series on the right is the same as the corresponding series of Part

(a), less the zeroth term $\left[\sum_{k=0}^{0}(k+1)a_{0-k}\right]x^0=2$. We are thus led to the equation: $x/(1-x)^2$

$=F(x)/(1-x)^2-2\Rightarrow F(x)=x+2(1-x)^2=2-3x+2x^2$. The sequence defined by the recursion is

thus: $a_0=2, a_1=-3, a_2=2, a_n=0$ $(n\geq 3)$. (The reader may wish to verify this directly.)

#21. (a) (i) $(1+x+x^2+x^3+x^4+x^5+x^6+x^7)^3$ Seek coefficient of x^7.   (ii)   36

(b) (i) $(1+x+x^2+x^3+x^4+x^5+x^6+x^7)^2$ Seek coefficient of x^7.   (ii)   8

(c) (i) $(1+x+x^2+x^3+\cdots+x^{10})^3$ Seek coefficient of x^{10}.   (ii)   66

#23. (a) (i) $(1+x+x^2+x^3+x^4+x^5+x^6+x^7)\cdot(x+x^3+x^5+x^7)\cdot(x+x^2+x^3+x^4+x^5+x^6+x^7)$

Seek coefficient of x^7.   (ii)   12

(b) (i) $(x+x^3+x^5+x^7)\cdot(1+x+x^2+x^3+x^4+x^5)$ Seek coefficient of x^7.   (ii)   3

(c) (i) $(1+x+x^2+x^3+\cdots+x^{10})\cdot(x+x^2+x^3+\cdots+x^{10})\cdot(x^2+x^3+\cdots+x^{10})$ Seek coefficient of

x^{10}.   (ii)   36

#27. (a) For any positive integer k, the generating function for the number of parts of size k being used

in a given partition is: $P_k(x)=1+x^k+x^{2k}+x^{3k}+\cdots$, which by (8) with the substitution $x\mapsto x^k$ can be rewritten as $P_k(x) = 1/(1-x^k)$. The generating function for $p_m(n)$ will thus be the product m of these functions: $P_1(x) \cdot P_2(x) \cdots P_m(x) = 1/(1-x) \cdot 1/(1-x^2) \cdots 1/(1-x^m)$, as asserted.

Of course, to use this generating function to compute $p_m(n)$, we would work with the polynomial form: $P_1(x) \cdot P_2(x) \cdots P_m(x) = (1 + x + x^2 + x^3 + \cdots)(1 + x^2 + x^4 + \cdots) \cdots (1 + x^m + x^{2m} + \cdots)$, where in each parenthesized polynomial, only terms of degree at most n are listed. Thus, to compute $p(n)$, for $n = 4, 5, 6$, and 8, we need only look for the coefficients of x^4, x^5, x^6, x^8 in the expansion of:

$$(1 + x + x^2 + \cdots + x^8)(1 + x^2 + x^4 + x^6 + x^8)(1 + x^3 + x^6)(1 + x^4 + x^8)(1 + x^5)(1 + x^6)(1 + x^7)(1 + x^8).$$

Computing the indicated coefficients in this product leads us to: $p(4) = 5, p(5) = 7, p(6) = 11$, and $p(8) = 22$.

(b) The problem asks for the value of $p_5(15)$, and this will be the coefficient of x^{15} in $G_5(x)$. As explained above, we may work with the expansion:

$$(1 + x + x^2 \cdots + x^{15})(1 + x^2 + x^4 + \cdots + x^{14})(1 + x^3 + x^6 + \cdots + x^{15})(1 + x^4 + x^8 + x^{12})(1 + x^5 + x^{10} + x^{15}).$$

Computing the indicated coefficient in this product leads us to: $p_5(15) = 84$.

#29. (a) Since each positive integer k appears either exactly once or not at all in such a partition for n, the generating polynomial for the appearance of k is $1 + x^k$, and so the generating function of $p_D(n)$ is the product of these polynomials, over all positive integers k: $(1+x)(1+x^2)(1+x^3)\cdots$.

(b) To compute $p_D(n)$ for $n = 4, 5, 6, 7$, and 10, we need only look for the coefficients of $x^4, x^5, x^6, x^7, x^{10}$ in the expansion of:

$$(1+x)(1+x^2)(1+x^3)(1+x^4)(1+x^5)(1+x^6)(1+x^7)(1+x^8)(1+x^9)(1+x^{10}).$$

Computing the indicated coefficients in this product leads us to: $p_D(4) = 2, p_D(5) = 3, p_D(6) = 4, p_D(7) = 5$ and $p(10) = 10$.

#31. The generating function for the sequence $a_n =$ the number of ways to express n as a sum of distinct powers of 2 $(n \geq 1)$ is $F(x) = (1+x)(1+x^2)(1+x^4)(1+x^8)\cdots$. It suffices to show that $F(x) = 1 + x + x^2 + x^3 + \cdots \underset{\text{by (8)}}{=} 1/(1-x)$ (i.e, this implies $a_n = 1$ for all positive indices n), and we will accomplish this by showing that $(1-x)F(x) = 1$. We repeatedly apply the identity: $(x^k + 1)(x^k - 1) = x^{2k} - 1$:

$$\begin{aligned}(1-x)F(x) &= [(1-x)(1+x)](1+x^2)(1+x^4)(1+x^8)\cdots \\ &= [(1-x^2)(1+x^2)](1+x^4)(1+x^8)\cdots \\ &= [(1-x^4)(1+x^4)](1+x^8)\cdots \\ &\ \ \vdots \\ &= 1.\end{aligned}$$

More formally, by iteratively making the substitutions $(x^{2^j} + 1)(x^{2^j} - 1) = x^{2 \cdot 2^j} - 1$, the coefficient of any fixed positive power is shown to be zero.

#35. (b) 37,917

(c) $(x + x^2 + \cdots + x^9)(1 + x + x^2 + \cdots + x^9)^{k-1} = x(1-x^9)(1-x^{10})^{k-1}(1-x)^{-k}$

CHAPTER 6: Section 6.1:

NOTE: In some of these exercises, a particular scheme is suggested in the answer. This is simply meant to suggest one possible scheme; there may be others.

#1. (a) 1/2 (b) 1/4 (c) $2/13 \approx 0.1538$

#3. (a) $13 \cdot C(4,2) \cdot 12 \cdot 4 / C(52,3) \approx 0.1694$ (b) $1 - C(44,3)/C(52,3) \approx 0.4007$

(c) $4 \cdot C(13,3) / C(52,3) \approx 0.0518$

#5. (a) $1/13$ (b) $1/4$ (c) $13 \cdot C(4,2) / C(52,2) \approx 0.0588$ (d) $1 - C(48,2) / C(52,2) \approx 0.1493$

#7. (a) $13 \cdot 12 \cdot C(4,1) / C(52,5) \approx 2.401 \times 10^{-4}$ (i.e., about 1 in 4165)

(b) $C(13,2) \cdot C(4,2)^2 \cdot 44 / C(52,5) \approx 0.0475$

(c) A straight flush is completely determined by its starting denomination (there are 10 choices: A, 2, 3, ..., 10) and the suit (4 choices). $10 \cdot 4 / C(52,5) \approx 1.539 \times 10^{-5}$. Alternatively, if the royal straight flushes are removed from these, the probability would be $(10 \cdot 4 - 4) / C(52,5) \approx 1.385 \times 10^{-5}$.

#8. (a) $1 - C(71,5) / C(80,5) \approx 0.4584$ (b) $1 - C(71,10) / C(80,10) \approx 0.7196$

#9. (a) $C(538,20) / C(550,20) \approx 0.6382$ (b) $C(538,50) / C(550,50) \approx 0.3148$

#11. (a) $P(A \mid B) = P(A \cap B) / P(B) = 0.2 / 0.6 = 1/3$. $P(A \mid B) = P(A \cap B) / P(\sim B) = 0.2 / (1 - 0.6)$ $= 1/2$. (b) $P([\geq 3H] \mid H_1) = P([\geq 3H] \cap H_1) / P(H_1) = P(\{HHHH, HHHT, HHTH, HTHH\}) / 0.5$ $= 4/16 / 0.5 = 1/2 = 0.5$. $P([\geq 3H] \mid T_1) = P(\{HHHH\}) / 0.5 = 1/16 / 0.5 = 1/8 = 0.125$. (c) $P(S \mid D) =$ $P(S \cap D) / P(D) = 0.4 / 0.6 = 2/3 \approx 0.667$. $P(D \mid S) = P(S \cap D) / P(S) = 0.4 / 0.55 \approx 0.727$.

(d) Let $P_s \equiv P([\geq 1 \text{ five}] \mid [\text{Sum} = s]) = P([\geq 1 \text{ five}] \cap [\text{Sum} = s]) / P([\text{Sum} = s])$. By consulting Figure 6.3, we can count the outcomes in the numerator and denominator of this expression to obtain the following results: When $s < 6$, $P_s = 0$ (since the numerator is zero), $P_6 = 2/5$, $P_7 = 2/6 = 1/3$, $P_8 = 2/5$, $P_9 = 2/4 = 1/2$, $P_{10} = 1/3$, $P_{11} = 2/2 = 1$, $P_{12} = 0/1 = 0$.

#13. (a) $P(A \mid B) = P(A \cap B) / P(B) = 0.2 / 0.4 = 0.5 > 0.4 = P(A)$.

(b) $P(A \mid B) = P(A \cap B) / P(B) = 0.0005 / 0.2 = 0.0025 > 0.001 = P(A)$.

(c) $P(A \mid B) = P(A \cap B) / P(B)$. First, by counting outcomes, we get $P(A \cap B) = P(\{TTTT, TTTH, HTTT, THHH, HHHT\}) = 5 / 2^4 = 5/16$. Since the event A has one additional outcome ($HHHH$), we obtain $P(A) = 6/16 = 3/8 = 0.375$. Using complements, $P(B) = 1 - P(\{HHHH\}) = 1 - 1/16 = 15/16$. We may now obtain that $P(A \mid B) = P(A \cap B) / P(B) = (5/16) / (15/16) = 1/3 \approx 0.333$, which is less than $P(A)$.

(d) $P(A \mid B) = P(A \cap B) / P(B) = P(A \cap B) / P(B)$. The "at least" clauses in both events make it simpler to work with complementary probabilities. First: $P(\sim B) = P(\text{"no sixes"}) = 5^3 / 6^3 \approx 0.579$. Note also that $P(\sim A) = P(\text{"all different"}) = 6 \cdot 5 \cdot 4 / 6^3 \approx 0.556$. Using De Morgan's Law, and then the inclusion-exclusion principle, we obtain: $P(\sim [A \cap B]) = P(\sim A \cup \sim B) = P(\sim A) + P(\sim B) - P(\sim A \cap \sim B)$. The last of these probabilities is just $P(\text{"all different and no six"}) = 5 \cdot 4 \cdot 3 / 6^3 \approx 0.278$. Putting this together with our previously obtained complementary probabilities, we obtain $P(\sim [A \cap B]) \approx 0.556 + 0.579 - 0.278 \approx 0.857$. Finally, going back to the original conditional probability, we obtain: $P(A \mid B) = P(A \cap B) / P(B) = (1 - P(\sim [A \cap B])) / (1 - P(\sim B)) \approx$ $(1 - 0.857) / (1 - 0.579) \approx 0.34$ This is less than $P(A) = 1 - P(\sim A) \approx 1 - 0.556 \approx 0.444$.

#15. (a) $P(R_2) = P(R_2 \mid A)P(A) + P(R_2 \mid B)P(B) = (3/5)(1/2) + (8/11)(1/2) = 73/110 \approx 0.6636$

(b) $P(RR) = P(RR \mid A)P(A) + P(RR \mid B)P(B) = (3/5)(2/4)(1/2) + (8/11)(7/10)(1/2) \approx 0.4045$

(c) $P(A \mid RR) = P(RR \mid A)P(A) / P(RR) = (3/5)(1/2)(1/2) / P(RR) \approx 0.3708$ (d) $P([\geq 1R]) =$ $1 - P(BlBl) = 1 - [P(BlBl \mid A)P(A) + P(BlBl \mid B)P(B)] = 1 - [(2/5)(1/4)(1/2) + (3/11)(2/10)(1/2)] \approx 0.9227$

(e) $P(A \mid [\geq 1R]) = P([\geq 1R] \mid B)P(B) / P([\geq 1R]) = (1 - (3/11)(2/10))(1/2) / P([\geq 1R]) \approx 0.8724$

#16. (a) $P(R_2) = P(R_2 \mid A)P(A) + P(R_2 \mid B)P(B) = (3/5)(1/2) + (8/11)(1/2) \approx 0.6636$

(b) $P(RR) = P(RR \mid A)P(A) + P(RR \mid B)P(B) = (3/5)^2(1/2) + (8/11)^2(1/2) \approx 0.4445$

(c) $P(A \mid RR) = P(RR \mid A)P(A)/P(RR) = (3/5)^2(1/2)/P(RR) \approx 0.4050$ (d) $P([\geq 1R]) = 1 - P(BlBl) =$
$1 - [P(BlBl \mid A)P(A) + P(BlBl \mid B)P(B)] = 1 - [(2/5)^2(1/2) + (3/11)^2(1/2)] \approx 0.8828$

(e) $P(A \mid [\geq 1R]) = P([\geq 1R] \mid B)P(B)/P([\geq 1R]) = (1 - (3/11)^2)(1/2)/P([\geq 1R]) \approx 0.5242$

#17. (a) $P(H) = P(H \mid F)P(F) + P(H \mid D)P(D) + P(H \mid U)P(U) = (1/2)(1/2) + 1(1/4) + (8/10)(1/4)$
$= 7/10$ (b) $P(F \mid H) = P(F \cap H)/P(H) = P(F \mid H)P(F)/P(H) = (1/2)(1/2)/(7/10) = 5/14 \approx 0.3571$
(c) $P(D \mid H) = P(H \mid D)P(D)/P(H) = 1 \cdot (1/4)/(7/10) = 5/14 \approx 0.3571$ (d) 0

#18. (a) We let E denote the event that the sum of the money drawn is at least \$100.
$P(E) = P(E \mid A)P(A) + P(E \mid B)P(B) + P(E \mid C)P(C) = 0 + (6/7)(1/3) + 1 \cdot (1/3) = 13/21 \approx 0.6190$ For
the middle term, we computed $P(E \mid B) = P([2\ 50s] \mid B) + P([3\ 50s] \mid B) = C(5,2) \cdot C(2,1)/C(7,3) +$
$C(5,3)/C(7,3) = 30/35 = 6/7$. (b) $P(A \mid \sim E) = P(A \cap \sim E)/P(\sim E) = P(\sim E \mid A)P(A)/P(\sim E) =$
$1 \cdot (1/3)/(1 - 13/21) = 7/8 = 0.875$

#19. Let W denote the event that the drawn card is Card #3, and W_1 the event that the face of the card
drawn is white: $P(W \mid W_1) = P(W \cap W_1)/P(W_1) = (1/3)/(1/2) = 2/3$. Thus, it would be advantageous
for you to take this bet. Note: This is a famous problem. Many students incorrectly get 1/2 for the
answer.

#23. (a) $P(HHHH) = (1/2)^4 = 1/16$

(b) $1 - P(HHHH) = 1 - 1/16 = 15/16$ (c) $P(HHHH \mid H{*}{*}{*}) = P(HHHH \cap H{*}{*}{*})/P(H{*}{*}{*}) =$
$(1/2)^4/(1/2) = 1/8$ (d) $P(\sim HHHH \mid H{*}{*}{*}) = 1 - P(\sim HHHH \mid H{*}{*}{*}) = 1 - 1/8 = 7/8$

#25. (a) $\{T,\ HT,\ HHT,\ HHHT,\ \cdots\} \cup \{HHH \cdots\}$ (b) $E_n = \underbrace{HH \cdots H}_{n-1\ \text{times}} T$, $P(E_n) = 1/2^n$ (c) Since

disjoint event probabilities can be added, we have $P(\bigcup_{n=1}^{\infty} E_n) = \sum_{n=1}^{\infty} 1/2^n = (1/2)/(1 - 1/2) = 1$
(geometric series). The outcome $HHH \cdots$ (i.e., getting heads forever) lies outside $\bigcup_{n=1}^{\infty} E_n$, but this
event has zero probability of ever happening.

#27. If the distribution of tickets were indeed random and independent (on cars with parking
violations), the probability of those 16 Wednesday-only tickets would be $(1/5)^{16} \approx 6.55 \times 10^{-12}$. This
extremely small probability is good evidence against the independence hypothesis, and it would seem
that feeding parking meters only on Wednesdays would work well.

#29. (a) $C(n,2s) \cdot 2^{2s}/C(2n,2s)$ (b) $n \cdot C(n-1,2s-2) \cdot 2^{2s-2}/C(2n,2s)$

(c) $C(n,2) \cdot C(n-2,2s-4) \cdot 2^{2s-4}/C(2n,2s)$

#31. For each part, we let E_n denote the event in question with n people in the room. (a)
$P(E_n) = 1 - P(\overline{E_n}) = 1 - (364/365)^n$ (i) $n = 253$ (easily determined using a computer loop). (b)
$P(E_n) = 1 - P(\overline{E_n}) = 1 - (364/365)^n - n(1/365) \cdot (364/365)^{n-1}$ (i) $n = 613$ (easily determined using a
computer loop). (c) (*Difficult*) To compute $P(E_n)$ we use complementary probabilities:
$P(E_n) = 1 - P(D_1) - P(D_2)$, where $D_i = \{$maximum number of birthdays/day $= i\}$ (among a set of n
randomly chosen birthdays). In Example 7.4 (the ordinary birthday problem), we showed how to
compute $1 - P(D_1) \equiv B(n)$, so we have left to compute $P(D_2)$. Let D_2^j denote the number of
outcomes in D_2 for which there are exactly j pairs of common birthdays $(1 \leq j \leq \text{floor}(n/2))$. We
count D_2^j as follows:

$$|D_2^j| = \binom{n}{2j} \left[\frac{\binom{2j}{2} \cdot \binom{2j-2}{2} \cdots \binom{4}{2} \cdot \binom{2}{2}}{j!} \right] \prod_{k=1}^{n-j} (366-k) = \frac{n!}{2^j} \cdot \binom{365}{j} \cdot \binom{365-j}{n-2j}.$$

Thus, $|D_2| = \sum_{j=1}^{\text{floor}(n/2)} |D_2^j| = \sum_{j=1}^{\text{floor}(n/2)} \frac{n!}{2^j} \cdot \binom{365}{j} \cdot \binom{365-j}{n-2j}$, and so

$$P(D_2) = \frac{|D_2|}{365^n} = \frac{n!}{365^n} \sum_{j=1}^{\text{floor}(n/2)} \frac{1}{2^j} \cdot \binom{365}{j} \cdot \binom{365-j}{n-2j}.$$

(i) Putting this all together and using a computer loop, we find that $n = 88$ is the smallest number of people needed in the room for $P(E_n) > 0.5$.

Section 6.2:

#1. (a) (i) range$(X) = \{0,1,2,3,4\}$, (ii) $p(0) = P(TTTT) = 1/16$, $p(1) = P(\{HTTT,THTT,TTHT,TTTH\}) = 1/4$, $p(2) = C(4,2)/16 = 3/8$, $p(3) = 1/4$, $p(4) = 1/16$.

(iii) $F(x) = \begin{cases} 0, & \text{if } x < 0 \\ 1/16, & \text{if } 0 \le x < 1 \\ 5/16, & \text{if } 1 \le x < 2 \\ 11/16, & \text{if } 2 \le x < 3 \\ 15/16, & \text{if } 3 \le x < 4 \\ 1, & \text{if } x \ge 4 \end{cases}$

(iv) $\mu = \sum_{i=0}^{4} ip(i) = 0 \cdot \frac{1}{16} + 1 \cdot \frac{1}{4} + 2 \cdot \frac{3}{8} + 3 \cdot \frac{1}{4} + 4 \cdot \frac{1}{16} = 2$

(v) Using the definition:

$$\sigma^2 = E[(X-\mu)^2] = \sum_{i=0}^{4} (i-\mu)^2 p(i) = (0-2)^2 \cdot \frac{1}{16} + (1-2)^2 \cdot \frac{1}{4} + (2-2)^2 \cdot \frac{3}{8} + (3-2)^2 \cdot \frac{1}{4} + (4-2)^2 \cdot \frac{1}{16} = 1$$

(b) (i) range$(X) = \{0,1,2\}$, (ii) $p(0) = (6/7)^2 = 36/49$, $p(1) = 6/7 \cdot 1/7 + 1/7 \cdot 6/7 = 12/49$, $p(3) = (1/7)^2 = 1/49$

(iii) $F(x) = \begin{cases} 0, & \text{if } x < 0 \\ 36/49 & \text{if } 0 \le x < 1 \\ 48/49, & \text{if } 1 \le x < 2 \\ 1, & \text{if } x \ge 2 \end{cases}$

(iv) $\mu = \sum_{i=0}^{2} ip(i) = 0 \cdot \frac{36}{49} + 1 \cdot \frac{12}{49} + 2 \cdot \frac{1}{49} = \frac{14}{49}$

(v) Using the definition:

$$\sigma^2 = E[(X-\mu)^2] = \sum_{i=0}^{2} (i-\mu)^2 p(i) = \left(0-\frac{14}{49}\right)^2 \frac{36}{49} + \left(1-\frac{14}{49}\right)^2 \frac{12}{49} + \left(2-\frac{14}{49}\right)^2 \frac{1}{49} = \frac{12}{49}$$

#3. $X \sim \mathcal{B}(10,1/6)$ (a) $P(X=2) = C(10,2)(1/6)^2(5/6)^8 \approx 0.2907$

(b) $P(X \ge 2) = 1 - P(X < 2) = 1 - [C(10,0)(1/6)^0(5/6)^{10} + C(10,1)(1/6)^1(5/6)^9] \approx 0.5155$

(c) $C(10,0)(1/6)^0(5/6)^{10} \approx 0.1615$

#5. $X \sim \mathcal{B}(200,1/365)$ (a) $P(X=k) = C(200,k)(1/365)^k(364/364)^{200-k}$

(b) $P(X \le 5) = \sum_{k=0}^{5} C(200,k)(1/365)^k(364/364)^{200-k} \approx 0.999779$

#7. $p(1) = 11/36$, $p(2) = 9/36$, $p(3) = 7/36$, $p(4) = 5/36$, $p(5) = 3/36$, $p(6) = 1/36$

#9. $p(1) = 7/16$, $p(2) = 81/400$, $p(3) = 63/400$, $p(4) = 45/400$, $p(5) = 27/400$, $p(6) = 9/400$

#11. (a) Using the solution to #7,

$$E(X) = \sum_{i=1}^{6} i\,P(X=i) = 1\cdot 11/36 + 2\cdot 9/36 + 3\cdot 7/36 + 4\cdot 5/36 + 5\cdot 3/36 + 6\cdot 1/36 = 91/36 \approx 2.5277$$

(b) $E(X^2) = \sum_{i=1}^{6} i^2\,P(X=i) = 1\cdot 11/36 + 4\cdot 9/36 + 9\cdot 7/36 + 16\cdot 5/36 + 25\cdot 3/36 + 36\cdot 1/36 = 301/36$

so $\quad \text{Var}(X) = E(X^2) - E(X)^2 = 301/36 - (91/36)^2 \approx 2.2215.$ $\qquad$ (c) $\qquad E(R) = \sum_{i=1}^{6} i\,P(R=i) =$

$(1/6)(1+2+3+4+5+6) = 21/6 = 3.5$ so $E(R+X) = E(R) + E(X) = 3.5 + 91/36 \approx 6.0277$

(d) $E(R^2) = \sum_{i=1}^{6} i^2\,P(R=i) = (1/6)(1^2+2^2+3^2+4^2+5^2+6^2) = 91/6$ so $\text{Var}(R) = E(R^2) - E(R)^2 =$

$91/36 - (21/6)^2 \approx 2.9167,$ and thus $\text{Var}(R) + \text{Var}(X) \approx 5.1382$

(e) This part is the hardest one to compute. Basically we will be OK once we compute E(RX). (R and X are not independent, so we cannot simply set this equal to E(R)E(X).) Since the sample size is small enough, we can simply write out the outcomes of RX over each of the 36 possible outcomes in the sample space and then use these values to compute the expectation:

R	1	2	3	4	5	6
1	1	1	1	1	1	1
2	2	2	2	2	2	2
3	3	3	3	3	3	3
4	4	4	4	4	4	4
5	5	5	5	5	5	5
6	6	6	6	6	6	6

X	1	2	3	4	5	6
1	1	1	1	1	1	1
2	1	2	2	2	2	2
3	1	2	3	3	3	3
4	1	2	3	4	4	4
5	1	2	3	4	5	5
6	1	2	3	4	5	6

RX	1	2	3	4	5	6
1	1	1	1	1	1	1
2	2	4	4	4	4	4
3	3	6	9	9	9	9
4	4	8	12	16	16	16
5	5	10	15	20	25	25
6	6	12	18	24	30	36

(In each of these three tables, the top row bold-faced numbers indicate the outcome of the white die, and the left column bold-faced numbers are the outcomes of the red die. The 36 non-bold numbers inside are the corresponding outcomes of each of the three random variables indicated in the upper left corner of each table. Note the outcomes in the third table are simply the products of the corresponding entries in the first two tables.) From the last table, we can compute $\quad$ E(RX) $=$

$(1/36)[1\cdot 6 + 2 + 3 + 4\cdot 6 + 5 + 6\cdot 2 + 8 + 9\cdot 4 + 10 + 12\cdot 2 + 15 + 16\cdot 3 + 18 + 20 + 24 + 25\cdot 2 + 30 + 36] =$

$371/36.$ Now using previous computations, we obtain: $E([R+X]^2) = E(R^2 + 2RX + X^2) =$

$E(R^2) + 2E(RX) + E(X^2) = 91/6 + 2\cdot 371/36 + 301/36 \approx 44.1389.$ $\quad$ Thus, $\quad$ $\text{Var}([R+X]^2) =$

$E([R+X]^2) - E([R+X])^2 \approx 7.8057$ (Note that this is different than the answer to Part (d)!)

#13. $E[X] = \sum_{i=0}^{10} i C(32,i)\cdot C(468,10-i)/C(500,10) = 16/25 = 0.64$

#15. $X \sim \mathscr{B}(3, 0.3)$ so $E[X] = np = 3(0.3) = 0.9.$

#17. Only the pair in Part (a) is independent. For example,
$P(X=6, Y=1) = 0 \neq (1/36)\cdot(1/36) = P(X=6)\cdot P(Y=1)$ shows X and Y are not independent.

#19. Let N_H be the number of flips of a fair coin until we land a heads. Certainly N_H is at least one (to land a head, we need at least one flip). Since coin flips are independent, if we landed a tails on the first flip (which happens with probability 1/2), then the expected number of additional flips to land a head will also be $E(N_H)$. It thus follows that $E(N_H) = 1 + (1/2)E(N_H)$, from which it immediately follows that $E(N_H) = 2.$

#21. (a) Following the suggestion, note that $P(X_i = 1) = 1 - P(X_i = 0) = 1 - (4/5)^{10} \approx 0.8926$, thus

$E[X] = \sum_{i=1}^{5} E[X_i] \approx 5\cdot 0,8926 = 4.4631.$ $\qquad$ (b) $\qquad 5\cdot(1-(4/5)^N)$ $\qquad$ (c) $\qquad 1-(4/1004)^{10} +$

$4(1-(1003/1004)^{10}) \approx 1.0397$ (d) $1-(4/1004)^N + 4(1-(1003/1004)^N)$

#23. Letting X be the number of radioactive particles detected by the Geiger counter in one hour, we model X as using a Poisson random variable $Z \sim \mathscr{P}(8)$. More precisely, we set $X = 10^{-2}Z.$

(a) $\quad P(X \le 0.02) = P(Z \le 2) = e^{-8}8^0/0! + e^{-8}8^1/1! + e^{-8}8^2/2! \approx 0.0138 \ (=1.38\%).$

(b) $\quad P(X \le 0.05) = P(Z \le 2) + e^{-8}8^3/3! + e^{-8}8^4/4! + e^{-8}8^5/5! \approx 0.1912 \ (=19.12\%).$

(c) $\quad P(X > 0.1) = P(Z > 10) = 1 - P(Z \le 10)$

$= 1 - [P(Z \le 5) + e^{-8}8^6/6! + e^{-8}8^7/7! + e^{-8}8^8/8! + e^{-8}8^9/9! + e^{-8}8^{10}/10!] \approx 0.1841 \ (=18.41\%).$

#33. (a) Assuming $X \sim \mathcal{B}(n, p)$ and letting $q = 1 - p$, we may write,

$$E[X^2] = \sum_{k=0}^{n} k^2 P(X = k) = \sum_{k=0}^{n} (k^2 - k + k) \binom{n}{k} p^k q^{n-k}$$

$$= \sum_{k=0}^{n} (k^2 - k) \binom{n}{k} p^k q^{n-k} + \sum_{k=1}^{n} k \frac{n!}{k!(n-k)!} p^k q^{n-k} \triangleq A + B.$$

The sum B is simply $E[X]$, and this sum was analytically evaluated in the solution to Exercise for the Reader 6.21 to be np. We have left to evaluate the sum A:

$$A = \sum_{k=0}^{n} (k^2 - k) \binom{n}{k} p^k q^{n-k} = \sum_{k=2}^{n} k(k-1) \frac{n!}{k!(n-k)!} p^k q^{n-k}$$

$$= \sum_{k=2}^{n} \cancel{k} \, (k \cancel{-1}) \frac{n(n-1)(n-2)!}{\cancel{k} \, (k \cancel{-1})(k-2)!(n-k)!} p^k q^{n-k} = n(n-1) p^2 \sum_{k=2}^{n} \frac{(n-2)!}{(k-2)!(n-2-[k-2])!} p^{k-2} q^{n-2-(k-2)}$$

$$\underset{k \to k+2}{=} n(n-1) p^2 \sum_{k=0}^{n-2} \frac{(n-2)!}{k!(n-k)!} p^k q^{n-2-k} = n(n-1) p^2 \underbrace{\sum_{k=0}^{n-1} \binom{n-2}{k} p^k q^{n-2-k}}_{=(p+q)^{n-2}=1 \;\; (\text{Binomial Thm.})} = n(n-1) p^2.$$

Combining this sum with $B = np$, we obtain

$$E[X^2] = A + B = n(n-1) p^2 + np = np[(n-1)p + 1].$$

(b) Following the suggested strategy, we obtain:

$$Var(X) = E[X^2] - (E[X])^2 = np[(n-1)p + 1] - (np)^2 = np[(n-1)p + 1 + np] = np[1 - p] = npq.$$

CHAPTER 7: Section 7.1:

#1. (a) We provide details for each step of Algorithm 7.2.
Step 1: Initialize search interval endpoint indices: INITIALIZE: LeftIndex = 1, RightIndex = 16
Note that $t = 21$, and $n = 16$ (part of the inputs).

Step 2: Iteratively reduce search interval to a single number:
First Iteration: $m = $ floor([1+16]/2]) = 8 (Note that $a_8 = 68$)

UPDATE RightIndex = 8 (Since t is not greater than a_8.)

Second Iteration: $m = $ floor([1+8]/2]) = 4 (Note that $a_4 = 33$)

UPDATE RightIndex = 4 (Since t is not greater than a_4.)

Third Iteration: $m = $ floor([1+4]/2]) = 2 (Note that $a_2 = 23$)

UPDATE RightIndex = 2 (Since t is not greater than a_2.)

Third Iteration: $m = $ floor([1+2]/2]) = 1 (Note that $a_1 = 21$)

UPDATE RightIndex = 1 (Since t is not greater than a_9.)

OUTPUT $k = 1$ (Since $t = a_1$)

#5. (a) Input List: [86, 3, 33, 23, 77, 46, 38, 86, 36]

First Pass: 88 moves from front to end of list: → [3, 33, 23, 77, 46, 38, 86, 36, 86]

Second Pass: → [3, 23, 33, 77, 46, 38, 86, 36, 86] → [3, 23, 33, 46, 77, 38, 86, 36, 86]
→ [3, 23, 33, 46, 38, 77, 86, 36, 86] → [3, 23, 33, 46, 38, 77, 36, 86, 86]

Third Pass: → [3, 23, 33, 38, 46, 77, 36, 86, 86] → [3, 23, 33, 38, 46, 36, 77, 86, 86]

Fourth Pass: → [3, 23, 33, 38, 36, 46, 77, 86, 86]

Fifth Pass: → [3, 23, 33, 36, 38, 46, 77, 86, 86]

Sixth Pass: Done
#7. (a) Rather than working on one bracket list at a time, at each iteration we will break down any bracketed lists (with more than one element) that remain:
Input List: [86, 3, 33, 23, 77, 46, 38, 86, 36]

Method (i): Leftmost element selection
First Iteration: Select $x = 86$, and compare to form
[3, 33, 23, 77, 46, 38, 86] 86
Second Iteration: Select $x = 3$, and compare to form
3 [33, 23, 77, 46, 38, 86] 86
Third Iteration: Select $x = 33$, and compare to form
3 [23] 33 [77, 46, 38, 86] 86

Fourth Iteration: Select $x = 77$, and compare to form
3 23 33 [46, 38] 77 [86] 86

Fifth Iteration: Select $x = 46$, and compare to form
3 23 33 [38] 46 77 [86] 86
Ignoring brackets, we now have the sorted list.

Method (ii): Rightmost element selection
First Iteration: Select $x = 36$, and compare to form
[3, 33, 23, 36] 36 [86, 77, 46, 38, 86]
Second Iteration: Select $x = 36$ for the first bracket, $x = 86$ for the second bracket, and compare to form [3, 33, 23] 36 36 [86, 77, 46, 38] 86
Third Iteration: Select $x = 23$ for the first bracket, $x = 38$ for the second bracket, and compare to form [3] 33 [23] 36 36 [46] 38 [86, 77] 86
Fourth Iteration: Select $x = 77$, and compare to form
3 33 23 36 36 46 38 77 [86] 86
Ignoring brackets, we now have the sorted list.
#9. As in our solution to Example 7.4 we skip the last splitting step when the maximum list size is two.
(a) Input List: [86, 3, 33, 23, 77, 46, 38, 86, 36]
Split: [86, 3, 33, 23] [77, 46, 38, 86, 36]
Split: [86, 3] [33, 23] [77, 46] [38, 86, 36]
Split: [86] [3] [33] [23] [77] [46] [38] [86, 36]
Sort: [86] [3] [33] [23] [77] [46] [38] [36, 86]
Merge: [3, 86] [23, 33] [46, 77] [36, 38, 86]
Merge: [3, 23, 33, 86] [36, 38, 46, 77, 86]
Merge: [3, 23, 33, 36, 38, 46, 77, 86, 86]

#11. (a) It will always take the same number of iterations for RightIndex to equal LeftIndex and the binary search algorithm to halt. This number is $\lceil \log_2(n) \rceil$. This can be seen by first verifying it in the easier case in which n is a power of 2. At each iteration we have one comparison to make, thus the maximum and minimum number of comparisons is $\lceil \log_2(n) \rceil$.

(b) maximum = minimum = $n - 1$

#13. (a) $Q1 = 36$, $Q3 = 81.5$; $Q1 = 28$, $Q3 = 81.5$; $Q1 = 28$, $Q3 = 89$;
(b) We simply need to input the list of the appropriate indices of the corresponding sorted list $b_1 \le b_2 \le \cdots \le b_n$ for the quartiles. Using the definitions, we first deal with $Q1$.

Note that whether n is even or odd, $Q1$ is the median of $b_1, b_2, \cdots, b_{\text{floor}(n/2)}$. Set $k = \text{floor}(n/2)$.

If k is odd, then $Q1 = b_{\text{ceil}(k/2)}$, while if k is even then $Q1 = (b_{k/2} + b_{k/2+1})/2$. Algorithm 7.8 could thus be used by inputting the needed index (indices).

For $Q3$ it follows from the definitions that $Q3$ is the median of $b_{\text{ceil}(n/2)+1}, b_{k+3}, \cdots, b_n$. Set $\ell = \text{ceil}(n/2) + 1$, $m = n - \ell + 1$.

If m is odd, then $Q3 = b_{\ell + \text{ceil}(m/2) - 1}$, while if $n - \ell + 1$ is even then $Q3 = (b_{\ell + m/2 - 1} + b_{\ell + m/2})/2$.

#15. (a) Input List: [86, 3, 33, 23, 77, 46, 38, 86, 36]

Popping off the left most element from the input list, the sorted list is built up as follows:
[86] $\rightarrow$ [3, 86] $\rightarrow$ [3, 33, 86] $\rightarrow$ [3, 23, 33, 86] $\rightarrow$ [3, 23, 33, 77, 86] $\rightarrow$ [3, 23, 33, 46, 77, 86] $\rightarrow$
[3, 23, 33, 38, 46, 77, 86] $\rightarrow$ [3, 23, 33, 38, 46, 77, 86, 86] $\rightarrow$ [3, 23, 33, 36, 38, 46, 77, 86, 86] $\rightarrow$

(b) Since we are always popping numbers from the left of the input list the minimum number $n-1$ of comparisons (one for each iteration) would occur exactly if the list is a decreasing sequence of integers. Note that if the integers need not be distinct for this minimum number of comparisons.

Second Iteration: Select $x = 3$, and compare to form
3 [33, 23, 77, 46, 38, 86] 86
Third Iteration: Select $x = 33$, and compare to form
3 [23] 33 [77, 46, 38, 86] 86

Fourth Iteration: Select $x = 77$, and compare to form
3 23 33 [46, 38] 77 [86] 86

Fifth Iteration: Select $x = 46$, and compare to form
3 23 33 [38] 46 77 [86] 86
Ignoring brackets, we now have the sorted list.

#17. (a) The selection sort always requires $n(n-1)/2$ to sort a list of n elements.

(b) For each pass of the bubble sort algorithm, notice that any integer can move at most one place to the left (by contrast it can move all the way to the right). This means that if the smallest element appears at the right end of the list, a full $n-1$ passes will be needed requiring $n(n-1)/2$ comparisons.

(c) If at each iteration, the smallest/largest element of the lists are chosen for comparison, then the number of comparisons will be $n(n-1)/2$ (even if the input list was already sorted).

#19. We initialize a counter vector $(V_m, V_{m+1}, \cdots, V_M) = (0, 0, \cdots, 0)$. We then scan through each element a_j of the list, and increment the corresponding counter by 1: $V_{a_j} = V_{a_j} + 1$. Once this is done, we initialize an empty SortedList, then scan through the components of the counter vector, and for each V_i that is positive, we append V_i copies of the integer i to SortedList. The creation of the counting vector requires $M - m + 1$ assignments (zero for each component). Scanning the list to form the counting vector will require n additions. Thus the algorithm always requires $n + M - m + 1$ operations.

#21. The probability that t appears in any of the n locations of the list is the same, namely $1/n$. Thus,

$$E[X] = 1 \cdot P(X=1) + 2 \cdot P(X=2) + \cdots + n \cdot P(X=n) = \frac{1}{n}(1 + 2 + \cdots + n) = \frac{1}{\not{n}}\frac{\not{n}(n+1)}{2} = \frac{(n+1)}{2}.$$

(In the second-to-last equation, we used formula (1) of Section 3.1.)

Section 7.2:

#1. (a) True. (i) If $n \geq 150$, then $150 \leq 1 \cdot n$, which shows that $k = 150$ and $C = 1$ are witnesses to the big-O relationship $150 = O(n)$.

NOTE: We remind the reader that witness sets are never unique for any big-O estimate. For example, k = 1 and C = 150 is another set of witnesses for the big-O relationship $150 = O(n)$.

(ii) None of the theoretical results of this section give this basic big-O estimate. (Although it can be obtained using the result of Exercise for the Reader 7.13 and Proposition 7.2, this would amount to an overkill proof.)

(b) True. (i) If n is any positive integer, then $2n + 10 \leq 2n + 10n = 12n$. This shows that $k = 1$ and $C = 12$ are witnesses to the big-O relationship $2n + 10 = O(n)$. (ii) This also follows directly from the result of Exercise for the Reader 7.13 (or from the big-Theta estimate of Example 7.8, since a big-Theta estimate always implies the corresponding big-O estimate).

(c) False. (i) Given any positive constant C, if $n > C$, then $2n^3 + 10n > 2Cn^2 > Cn^2$. This proves that there can be no witnesses for the relationship $2n^3 + 10n = O(n^2)$, so it must be false.

(ii) This is also a direct consequence of Proposition 7.1(b).

(d)   True.   (i)   For any positive integer n, since $\log n < n$, and $\sqrt{n} \leq n$, it follows that $\log n \sqrt{n} < n \cdot n = n^2$, showing that $k = 1$ and $C = 1$ are witnesses to the big-O relationship $(\log n)\sqrt{n} = O(n^2)$.

(ii) This is also a direct consequence of the result of Example 7.7, with $s = 1, p = 1/2, p' = 2$.

(e) False. (i) If there were a constant C for which $n^n \le Cn!$, for all sufficiently large integers n, then

we would have $C \ge n^n / n! = \dfrac{n \cdot n \cdots n}{1 \cdot 2 \cdots n} = n \cdot \dfrac{n \cdots n}{2 \cdots n} > n$. But this is impossible since n can be

arbitrarily large. This proves that no witnesses can exist to support the (false) big-O relationship
$n^2 (1/2)^n = O(1)$.

(ii) None of the theoretical results of this section give this basic big-O estimate.

(f) True. (i) First note that if $n \ge 3$, then $10n < 27n \le n^3 \cdot n = n^4$. Since quotients of positive integers
get larger if their numerators are increased, or if their denominators are decreased, we may obtain the

following inequality if $n \ge 3$: $\dfrac{n^4 + 10n}{2n^2 + 4} < \dfrac{n^4 + n^4}{2n^2} = \dfrac{2n^4}{2n^2} = n^2$. This establishes $k = 3$ and $C = 1$ to be

witnesses to the big-O relationship $(n^4 + 10n)/(2n^2 + 4) = O(n^2)$.

(ii) None of the theoretical results of this section give this basic big-O estimate.

NOTE: In general it can be shown that if $p(n), q(n)$ are polynomials of degree D, D', respectively,

with positive leading coefficients, and $D > D'$, then $p(n)/q(n) = O(n^{D-D'})$. In fact, the corresponding
big-Theta estimate is true. See the solutions of Exercises 3(f) and 5(f) below for an idea towards a
general proof.

#3. (a) True. Since $\log n$ strictly increases (to infinity), as long as n is sufficiently large, we have
$150 \le \log n$. Note that for this inequality to hold, n would need to be at least 10^{150}.

(b) True. This follows directly from the result of Exercise for the Reader 7.13

(c) False. $2n^3 + 10n + 5 = \Theta(n^3)$, so this is not strictly dominated by n^3.

(d) True. Since $\log n < n$, and log is increasing for bases > 1, we may take logs on both sides and the
inequality will be preserved: $\log(\log n) \le \log n$. This gives the domination. The strict domination
follows from the fact that n strictly dominates $\log n$. Indeed, given $M > 0$, we know that
$n > M \log n$ for sufficiently large n. Since this remains to be true if n is replaced by any sufficiently
large real number x, it follows that (taking $x = \log n$) $\log n \ge M \log(\log n)$, provided that n is
sufficiently large.

(e) True. From Proposition 7.1(c) it follows that $n^2 \prec 1.5^n$, and since we trivially have $2^n \preceq 2^n$, it
follows from Proposition 7.2(2) that $n^2 2^n \prec 1.5^n \cdot 2^n = 3^n$.

(f) True. $\dfrac{n^2 + 4n}{2n + 3} = n \cdot \dfrac{n^2 + 4n}{2n + 3} = n \cdot \dfrac{(1/n)}{(1/n)} \dfrac{1 + 4/n}{2 + 3/n} = n \cdot \dfrac{1 + 4/n}{2 + 3/n}$. Now, if $n \ge 40$, then $4/n \le 1/10$, so

$\dfrac{1 + 4/n}{2 + 3/n} \le \dfrac{1.1}{2} = 0.55$. This proves that if $n \ge 40$, then $\dfrac{n^2 + 4n}{2n + 3} \le 0.55n$, i.e., $k = 40$, and $C = 0.55$ are

witnesses to the big-O relationship $(n^2 + 4n)/(2n + 3) = O(n)$.

#5. (a) (i) $O(n^4)$ (ii) Yes, this follows from the big-Theta estimate of Example 7.8. (iii) and (iv) have
the same answers as (i) and (ii), respectively.

(b) (i) $O(n)$ (ii) Yes. From Exercise for the Reader 7.15(a), it follows that $n \preceq 2n + 7 \log n$. From
Proposition 7.2(1) it follows that $2n + 7 \log n \preceq n$. These two dominations are together equivalent to the
big-Theta estimate $2n + 7 \log n \preceq \Theta(n)$. (iii) and (iv) have the same answers as (i) and (ii), respectively.

(c) (i) $O(n^2 \log n)$ (ii) Yes. Since the first term of $2n^2 \log n + 7(\log n)^2 n$ strictly dominates the
second (see Example 7.7) the big-Theta estimate follows. (iii) There is no best big-O estimate of the
form $O(n^a b^n)$, due to the presence of the log factor. (iv) N/A

(d) (i) $O(n^6)$ (ii) Yes. This follows (after expanding the polynomial) from the big-Theta estimate of
Example 7.8. (iii) and (iv) have the same answers as (i) and (ii), respectively.

(e) (i) $\ln 6^n + \sqrt{n}\log n = n\ln 6 + \sqrt{n}\log n = O(n)$ (ii) Yes. From Exercise for the Reader 7.15(a), it follows that $n \preceq 2n + 7\log n$. From Proposition 7.2(1) it follows that $2n + 7\log n \preceq n$. These two dominations are together equivalent to the big-Theta estimate $2n + 7\log n \preceq \Theta(n)$. (iii) and (iv) have the same answers as (i) and (ii), respectively.

(f) (i) $\dfrac{n + 4n^3}{n + 3} = n^2 \cdot \dfrac{4n + 1/n}{n + 3} = n^2 \cdot \dfrac{4 + 1/n^2}{1 + 3/n} = \Theta(n^2)$. (ii) Yes. This follows from the string of equations in (i) since in the second to last expression, all terms with powers of n in the denominator decrease to zero as n grows. (iii) and (iv) have the same answers as (i) and (ii), respectively.

#7. (a) We are given that $T(n) = Cn^2$, and $T(1000) = 0.1$, so it follows that $T(n) = 10^{-7}n^2$. Thus, $T(100,000) = 10^{-7}(10^5)^2 = 10^3$ seconds, or about 16 minutes. Also, $T(1\,\text{million}) = 10^{-7}(10^6)^2 = 10^5$ seconds, or about 28 hours.

(b) We are given that $T(n) = Cn\log n$, and $T(1000) = 0.1$, so it follows that $T(n) = \dfrac{n\log n}{3 \times 10^4}$. Thus, $T(100,000) = \dfrac{10^5 \log 10^5}{3 \times 10^4} \approx 17$ seconds, or about 16 minutes. Also, $T(1\,\text{million}) = \dfrac{10^6 \log 10^6}{3 \times 10^4} = 200$ seconds, or about 3 minutes.

#9. (a) False, since $n \succ 150$.

(b) True, the stronger big-Theta estimate is true by Example 7.8.

(c) True, expanding the left polynomial, Example 7.8 tells us it grows at the same rate as n^3, which by Proposition 7.1(b) (strictly) dominates n^2.

(d) False, since by the result of Example 7.7, $n^2 \succ n^{1/2}\log n$.

(e) True, since $\log n^3 = 3\log n = \Theta(\log n)$.

(f) True, since $\ln(n^4 + 10n) > \log(n^4) = 4\log n = \Theta(\log n)$. (Recall that a big-Theta estimate always implies, a big-Omega estimate.)

#11. The formula for the sum of a finite geometric series (see Proposition 3.5 of Section 3.1) tells us that

$$f(n) = 1 + r + r^2 + \cdots + r^n = \frac{r^{n+1} - 1}{r - 1}.$$

Case 1: $0 < r < 1$. In this case, as n grows larger and larger, r^{n+1} decays to zero, so the above expression shows that $f(n)$ will eventually remain as close as we wish to the positive number $\dfrac{0 - 1}{r - 1} = 1/(1 - r)$. This means that $f(n) = \Theta(1)$.

Case 2: $r = 1$. Since $r^k = 1$, for any exponent k, it follows that $f(n) = 1 + r + r^2 + \cdots + r^n = n + 1 = \Theta(n)$.

Case 3: $r > 1$. Working with the above expression, we may write $f(n) = \dfrac{r^{n+1} - 1}{r - 1} = r^n \cdot \dfrac{r - 1/r^n}{r - 1}$. As n grows larger and larger, $1/r^n$ decays to zero, so the expression $\dfrac{r - 1/r^n}{r - 1}$ will eventually remain as close as we wish to the positive number $\dfrac{r - 0}{r - 1} = r/(r - 1)$. Thus, $\dfrac{r - 1/r^n}{r - 1} = \Theta(1)$, and it follows from the result of Exercise for the Reader 7.15(b) that $f(n) = = r^n \cdot \dfrac{r - 1/r^n}{r - 1} = \Theta(r^n \cdot 1) = \Theta(r^n)$.

#13. (a) True. Let k, C be witnesses to the big-O relationship $f(n) = O(g(n))$, i.e., for $n \geq k$, we have $f(n) \leq Cg(n)$. Taking pth powers on both sides gives $f(n)^p \leq C^p g(n)^p$, which show that k, C^p are witnesses to the big-O relationship $f(n)^p = O(g(n)^p)$.

(b) True. Let k, C be witnesses to the big-O relationship $f(n) = O(g(n))$, i.e., for $n \geq k$, we have $f(n) \leq Cg(n)$. Taking logs on both sides gives $\log f(n) \leq \log[Cg(n)] = \log C + \log g(n) = \Theta(\log g(n))$.

(c) True. Using the product rule for logs, we may write:

$$\log n! = \log(1 \cdot 2 \cdots \cdot n) = \log(1) + \log(2) + \cdots + \log(n) \leq n \log n.$$

This shows that $k = 1$, $C = 1$ are witnesses to the big-O relationship $\log n! = O(n \log n)$.

#15. We assume, as in the explanation of the insertion sort algorithm, that numbers from the input list are always popped from the left side of the list.

(a) The best case occurs when the inputted list is sorted in decreasing order. In this case a total of $n - 1 = \Theta(n)$ comparisons is required.

(b) The worst case occurs ironically when the inputted list is already sorted (in increasing order). In this case the number of comparisons would be $1 + 2 + 3 + \cdots + (n-1) = n(n-1)/2 = \Theta(n^2)$.

(c) Let $[a_1, a_2, \cdots, a_n]$ be the inputted list. Consider the random variable X_i = the number of comparisons required when a_i gets inserted into $[b_1, b_2, \cdots, b_{i-1}] = \text{sort}([a_1, a_2, \cdots, a_{i-1}])$. Clearly $X_1 = 0, X_2 = 1$, so assume that $i > 2$.

If $a_i \leq b_1$, then $X_i = 1$, if $b_1 < a_i \leq b_2$, then $X_i = 2, \cdots$ if $b_{i-2} < a_i$, then $X_i = i - 1$. Since each of these possibilities is equally likely (with probability $1/(i-1)$), it follows that

$$E[X_i] = \frac{1}{i-1} \cdot 1 + \frac{1}{i-1} \cdot 2 + \cdots + \frac{1}{i-1} \cdot i = \frac{1}{i-1} \cdot \frac{i(i+1)}{2}.$$

Thus, using linearity of expectation, the total number of comparisons used in an insertion sort $X = \sum_{i=1}^{n} X_i$ has expectation given by:

$$E(X) = \sum_{i=1}^{n} E(X_i) = \sum_{i=1}^{n} E(X_i) = 1 + \sum_{i=3}^{n} \frac{i(i+1)}{2(i-1)} = \Theta\left(1 + \sum_{i=3}^{n} i\right) = \Theta(n^2).$$

The actual sum (before the first big-Theta estimate above) is approximately equal to $n(n+1)/4$, which is about half of the number of comparisons of the worst case, but as far as big-Theta estimates are concerned the average-case complexity is the same as the worst-case complexity.

NOTE: The situation is similar to what was found in Example 7.11 for the bubble sort.

#17. (a) At each iteration of the for loop of the iterative Fibonacci program, one addition is required, and three variable updates/assignments are also required. Thus a total of four computer/mathematical tasks is needed, which is $O(1)$. Since there are $n - 2 = O(n)$ iterations, it follows that the overall complexity of the algorithm is $O(n)$, and this is actually a big-Theta estimate. But since the input size is considered to be $\lfloor \log_2(n) \rfloor = \Theta(\log n)$, and $n = 2^{\log_2 n}$, it follows that the algorithm actually has exponential complexity.

(b) Let $g(n)$ = the number of computer/mathematical operations needed to compute the nth Fibonacci number f_n using the recursive algorithm `fibonacciv2` of Example 3.8. From the recursive formula of this algorithm, we obtain the following recursive formula for $g(n)$:

$$g(n) = \begin{cases} 1, & \text{if } n = 1, 2 \\ g(n-1) + g(n-2) + 1, & \text{if } n > 2. \end{cases}$$

(The "+ 1" term in the second line occurs from the addition.). This formula (except for the "+ 1" term in the second line) is identical to the defining formula for the Fibonacci sequence, and it readily follows that $f_n \leq g(n) \leq 2f_n$, so that $g(n) = \Theta(f_n)$. By formula (4) (of Section 3.2) for the Fibonacci numbers, it follows that $f_n = \Theta(([1+\sqrt{5}]/2)^n) = \Theta(([1+\sqrt{5}]/2)^{2^{\log_2(n)}})$. Thus, the recursive algorithm has a much higher (super-exponential) complexity than the exponential complexity of the iterative algorithm.

#19. To add two K digit base b strings using Algorithm 4.2, each digit in the sum comes from at most two additions (the digits, and the carry, if there is one), then a division and a subtraction (to compute the sum digit and the new carry). This gives an estimate of $K \cdot \Theta(1) = \Theta(K)$, i.e, we have shown linear complexity.

#21. To multiply two base b strings with K and L digits using Algorithm 4.2, the worst case would occur when both numbers have the same number K of digits. For each iteration of the outer for loop of the algorithm, the number of mathematical operations is at most a fixed constant times the number of digits in the second number, i.e., $\Theta(L)$. Thus the total complexity is $\Theta(L \cdot K) \le O(\max(K,L)^2)$. Thus we have quadratic complexity.

#27. As pointed out after the description of Algorithm 5.12 in the appendix to Section 5.3, the number of mathematical operations required to perform the algorithm is bounded above by $NW + Nq + 2N - 1 \le 3NW = O(NW)$. Since $W = 2^{\log_2 W}$, the complexity is exponential (in terms of the input), but it is still superior to that of the brute-force method.

#29. (a) It suffices to show that $f(n) \le 5n$, and we will do this using (strong) induction on n.
Basis Step: $n = 1, 2$: The inequality is trivially true.
Induction Step: We assume that $n > 2$, and that $f(k) \le 5k$ for all positive integers $k < n$. Since $\lfloor n/3 \rfloor \le n/3 < n$, it follows that $f(n) = 2f\left(\lfloor n/3 \rfloor\right) + n \le 2 \cdot 5 \cdot n/3 + n = 13n/3 < 5n$. This completes the proof.
(b) Yes. Since (from the recursion formula it follows that) $f(n)$ is always nonnegative, the equation $f(n) = 2f\left(\lfloor n/3 \rfloor\right) + n$ tells us that $f(n) \ge n = \Omega(n)$.

CHAPTER 8: Section 8.1:

#1. (a) $V(G_1) = \{u, v, x, w\}$; $V(G_2) = \{u, v, x, w, y\}$; $V(G_3) = \{u, v, x, w\}$; $V(G_4) = \{u, v, x, w, y\}$

(b) $E(G_1) = \{\{u, w\}, \{v, x\}\}$; $E(G_2) = \{\{u, y\}, \{v, y\}, \{x.y\}, \{w, y\}\}$; $E(G_3) = \{\{u, v\}, \{u, x\}, \{u.y\}, \{v, x\}\}$;

$E(G_4) = \{\{u, v\}, \{u, x\}, \{u.y\}, \{v, w\}, \{v, y\}, \{x, w\}, \{x, y\}, \{w, y\}\} \rightarrow$

(c) $\deg_{G_1}(u) = \deg_{G_1}(v) = \deg_{G_1}(x) = \deg_{G_1}(w) = 1$; $\deg_{G_2}(u) = \deg_{G_2}(v) = \deg_{G_2}(x) = \deg_{G_2}(w) = 1$,

$\deg_{G_2}(y) = 4$; $\deg_{G_3}(u) = 3$, $\deg_{G_3}(v) = \deg_{G_3}(x) = 2$, $\deg_{G_3}(w) = 1$; $\deg_{G_4}(u) = \deg_{G_4}(v) =$

$= \deg_{G_4}(x) = \deg_{G_4}(w) = 3$, $\deg_{G_4}(y) = 4$

#3. (a) $V(G_1) = \{u, v, x, w\}$, $V(G_2) = \{u, v, x, w, y\}$, $V(G_3) = \{u, v, x, w\}$, $V(G_4) = \{u, v, x, w, y\}$

(b)
$E(G_1) = \{\{u, u\}, \{u, w\}, \{v, v\}, \{v, x\}_1, \{v, x\}_2\}$; $E(G_2) = \{\{u, y\}, \{v, y\}_1, \{v, y\}_2, \{w.y\}_1, \{w, y\}_2, \{w, y\}_3\}$;

$E(G_3) = \{\{u, v\}, \{u, x\}, \{u, w\}_1, \{u, w\}_2, \{u, w\}_3, \{v, x\}_1, \{v, x\}_2, \{v, x\}_3, \{w, w\}_1, \{w, w\}_2\}$

$E(G_4) = \{\{u, v\}_1, \{u, v\}_2, \{u, x\}_1, \{u, x\}_2, \{u.y\}, \{v, w\}_1, \{v, w\}_2, \{v, y\}, \{x, w\}_1, \{x, w\}_2, \{x, y\}, \{w, y\}\}$

(c) $\deg_{G_1}(u) = 2$, $\deg_{G_1}(v) = 4$, $\deg_{G_1}(x) = 2$, $\deg_{G_1}(w) = 1$; $\deg_{G_2}(u) = 1$, $\deg_{G_2}(v) = 2$, $\deg_{G_2}(x) = 0$,

$\deg_{G_2}(w) = 3$, $\deg_{G_2}(y) = 6$; $\deg_{G_3}(u) = 5$, $\deg_{G_3}(v) = \deg_{G_3}(x) = 4$, $\deg_{G_3}(w) = 7$; $\deg_{G_4}(u) =$

$= \deg_{G_4}(v) = \deg_{G_4}(x) = \deg_{G_4}(w) = 5$, $\deg_{G_4}(y) = 4$

#5.

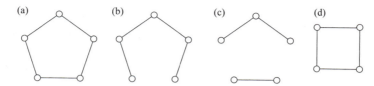

#7. For each part, except for Part (b), we apply Algorithm 8.1.
Part (a): Initial Sequence: (4, 4, 3, 3, 2, 1)
Iteration #1: (i) (3, 2, 2, 1, 1), (ii) (3, 2, 2, 1, 1)
Iteration #2: (i) (1, 1, 0, 1), (ii) (1, 1, 1, 0)
Iteration #3: (i) (0, 1, 0), (ii) (1, 0, 0)
Iteration #4: (i) (−1, 0), (ii) (0, −1) ⇒ Not graphic

Part (b): In an n-vertex simple graph, the largest possible degree is $n − 1$, so the sequence is not graphic.
Part (c): Initial Sequence: (5, 5, 3, 3, 3, 3)
Iteration #1: (i) (4, 2, 2, 2, 2), (ii) (4, 2, 2, 2, 2)
Iteration #2: (i) (1, 1, 1, 1), (ii) (1, 1, 1, 1)
Iteration #3: (i) (0, 1, 1), (ii) (1, 1, 0)
Iteration #4: (i) (0, 0), (ii) (0, 0) ⇒ Graphic

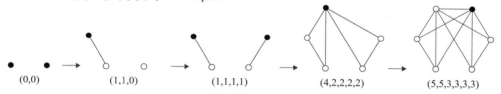

(0,0) (1,1,0) (1,1,1,1) (4,2,2,2,2) (5,5,3,3,3,3)

Part (d): Initial Sequence: (4, 4, 4, 4, 1, 1)
Iteration #1: (i) (3, 3, 3, 0, 1), (ii) (3, 3, 3, 1, 0)
Iteration #2: (i) (2, 2, 0, 0), (ii) (2, 2, 0, 0)
Iteration #3: (i) (1, -1, 0), (ii) (1, 0, -1) ⇒ Not graphic

#9. Part (a): No. This would correspond to a graph in which an odd number of vertices (seven) each had odd degree (three), which is impossible by Corollary 8.2.
Part (b): If we place the n people equally spaced around a circle, and have everyone shake hands with their two neighbors and with the person who is diametrically across the circle, then everyone shakes hands with three people. Here is a more formal graph construction: Write $n = 2k$, define a graph G by: $V = \{1, 2, \cdots, 2k\}$,

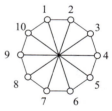

$$E = \{(1,2),(2,3),\cdots,(2k-1,2k),(2k,1),(1,k+1),(2,k+2),\cdots,(k,2k)\}.$$

The graph on the right illustrates the case when there are $n = 10$ people.
#11. By Corollary 8.2, the degree sequence of Part (b) cannot be realized with a general graph. Since all of the others have an even number of vertices of odd degree, they can be realized by general graphs; the figures below show constructions using the given procedure:

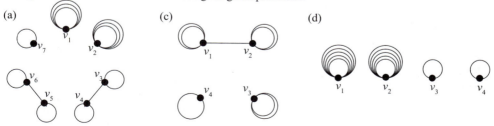

#13. Always True: The same bipartition (intersected with the vertices of the subgraph) will serve as a bipartition for the subgraph.

#15. *Proof:* The degree sequence of a simple graph with n vertices will be of the form $(d_1, d_2, \cdots, d_n)$, where $n-1 \geq d_1 \geq d_2 \geq \cdots \geq d_n \geq 0$. If these n degrees were all different, this would force $d_1 = n-1$, and $d_n = 0$. But this situation is not possible since it would mean on one hand that one vertex v_n had no neighbors, and another v_1 was a neighbor to all other vertices (including v_n) – a contradiction. $\square$

#17. P_n is always a subgraph of C_n, and $K_{n,n}$, but not conversely. C_n is a subgraph of $K_{n,n}$ if, and only if n is even. *Proof:* By the solution to Exercise 13, a subgraph of a bipartite graph must be bipartite, and since by Example 8.4 C_n is bipartite if, and only if n is even, the "only if" part follows. To see why the "if" part is true, suppose that n is even, and write $n = 2k$. Using the notation of Defintion 8.6, consider the following vertex, edge sequence of $K_{n,n}$:

$$u_1, v_2, u_3, v_4 \cdots, u_{2k-1}, v_{2k}, u_1$$

Each adjacent pair of vertices determines an edge belonging to $E(K_{n,n})$, so the subgraph consisting of these edges and the above vertices is identical (after a relabelling of the vertices) to C_n. $\square$

#19. Not isomorphic: In the right side graph, the two vertices u, v of degree 2 are neighbors, while in the left side graph, the two vertices f, b of degree 2 are not adjacent.

#21. Isomorphic: The mapping: $a \mapsto v, b \mapsto w, c \mapsto u, d \mapsto y, f \mapsto x$ is an isomorphism.

#23. G and H are isomorphic, J and K are isomorphic, but the first two are not isomorphic to the latter two. *Proof:* All of the graphs are 4-regular (regular graphs tend to be the most difficult to check for isomorphisms since the degrees are all equal). The corresponding complement graphs will thus all be 3-regular and will be less complicated to analyze; they are shown in the figure below:

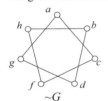 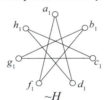 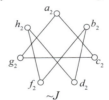 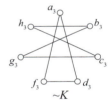

If we begin at any vertex of $\sim G$ and follow along the edges (in either direction), we will proceed through all of the edges and vertices exactly once and then return to the vertex we started. The same is easily seen to hold true of $\sim H$ and so it follows that $\sim G \cong \sim H \cong C_7$. It follows (from Exercise for the Reader 8.11) that $G \cong H$. A specific isomorphism can be obtained by identifying the vertices in the order they arise in any two such cycle constructions. For example: $a \mapsto a_1, c \mapsto d_1, f \mapsto h_1, h \mapsto c_1, b \mapsto g_1, d \mapsto b_1, g \mapsto f_1$. In the same fashion that the first two graphs were seen to be single copies of C_7, each of the latter two graphs are easily seen to consist of separate copies of C_3, C_4. A specific isomorphism can be obtained by first following vertices around the C_3 portion, and then around the C_4 portion. For example: $a_2 \mapsto a_3, c_2 \mapsto d_3, g_2 \mapsto f_3, b_2 \mapsto b_3, d_2 \mapsto g_3, h_2 \mapsto c_3, f_2 \mapsto h_3$. This is also an isomorphism between J and K.

#25. No two of the four graphs are isomorphic. *Proof:* G and H contain no copies of K_3, while J contains two copies, and K contains four copies. Thus, J and K are not isomorphic to any other graphs in the group. H is easily seen to be bipartite (e.g., color the vertices a_1, c_1, f_1, h_1 black and the remaining vertices white) but G cannot be bipartite since it contains a copy of a cycle of odd length: e.g., the sequence of vertices a, b, c, d, i, a and the edges between correspond to a subgraph copy of C_5. Since odd length cycles are not bipartite by Example 8.4, G thus contains a non-bipartite subgraph, so it follows (see Exercise 13) that G also is not bipartite. Thus $G \not\cong H$. $\square$

#27. $(n-1-d_1, n-1-d_2, \cdots, n-1-d_n)$

#29. (a) Adjacency matrices: Although it is not necessary, we label the rows/columns to make the adjacency matrices easier to read.

$$A(G_1) = \begin{array}{c} \\ u \\ v \\ w \\ x \end{array}\begin{array}{cccc} u & v & w & x \\ \left[\begin{array}{cccc} 0 & 0 & 1 & 0 \\ 0 & 0 & 0 & 1 \\ 1 & 0 & 0 & 0 \\ 0 & 1 & 0 & 0 \end{array}\right] \end{array}, \quad A(G_2) = \begin{array}{c} \\ u \\ v \\ w \\ x \\ y \end{array}\begin{array}{ccccc} u & v & w & x & y \\ \left[\begin{array}{ccccc} 0 & 0 & 0 & 0 & 1 \\ 0 & 0 & 0 & 0 & 1 \\ 0 & 0 & 0 & 0 & 1 \\ 0 & 0 & 0 & 0 & 1 \\ 1 & 1 & 1 & 1 & 0 \end{array}\right]\end{array}, \quad A(G_3) = \begin{array}{c} \\ u \\ v \\ w \\ x \end{array}\begin{array}{cccc} u & v & w & x \\ \left[\begin{array}{cccc} 0 & 1 & 1 & 1 \\ 1 & 0 & 0 & 1 \\ 1 & 0 & 0 & 0 \\ 1 & 1 & 0 & 0 \end{array}\right]\end{array},$$

$$A(G_4) = \begin{array}{c} \\ u \\ v \\ w \\ x \\ y \end{array}\begin{array}{ccccc} u & v & w & x & y \\ \left[\begin{array}{ccccc} 0 & 1 & 0 & 1 & 1 \\ 1 & 0 & 1 & 0 & 1 \\ 0 & 1 & 0 & 1 & 1 \\ 1 & 0 & 1 & 0 & 1 \\ 1 & 1 & 1 & 1 & 0 \end{array}\right]\end{array}$$

(b) Incidence matrices: Although it is not necessary, we label the rows/columns to make the incidence matrices easier to read.

(i) The lexicographic ordering of the two edges of G_1 is: $e_1 = \{u,w\}$, $e_2 = \{v,x\}$ and the corresponding incidence matrix is:

$$B(G_1) = \begin{array}{c} \\ u \\ v \\ w \\ x \end{array}\begin{array}{cc} e_1 & e_2 \\ \left[\begin{array}{cc} 1 & 0 \\ 0 & 1 \\ 1 & 0 \\ 0 & 1 \end{array}\right]\end{array}.$$

(ii) The lexicographic ordering of the four edges of G_2 is: $e_1 = \{u,y\}$, $e_2 = \{v,y\}$, $e_3 = \{w,y\}$, $e_4 = \{x,y\}$ and the corresponding incidence matrix is:

$$B(G_2) = \begin{array}{c} \\ u \\ v \\ w \\ x \\ y \end{array}\begin{array}{cccc} e_1 & e_2 & e_3 & e_4 \\ \left[\begin{array}{cccc} 1 & 0 & 0 & 0 \\ 0 & 1 & 0 & 0 \\ 0 & 0 & 1 & 0 \\ 0 & 0 & 0 & 1 \\ 1 & 1 & 1 & 1 \end{array}\right]\end{array}.$$

(iii) The lexicographic ordering of the four edges of G_3 is: $e_1 = \{u,v\}$, $e_2 = \{u,w\}$, $e_3 = \{u,x\}$, $e_4 = \{v,x\}$ and the corresponding incidence matrix is:

$$B(G_3) = \begin{array}{c} \\ u \\ v \\ w \\ x \end{array}\begin{array}{cccc} e_1 & e_2 & e_3 & e_4 \\ \left[\begin{array}{cccc} 1 & 1 & 1 & 0 \\ 1 & 0 & 0 & 1 \\ 0 & 1 & 0 & 0 \\ 0 & 0 & 1 & 1 \end{array}\right]\end{array}.$$

(iv) The lexicographic ordering of the eight edges of G_4 is: $e_1 = \{u,v\}$, $e_2 = \{u,x\}$, $e_3 = \{u,y\}$, $e_4 = \{v,w\}$, $e_5 = \{v,y\}$, $e_6 = \{w,x\}$, $e_7 = \{w,y\}$, $e_8 = \{x,y\}$ and the corresponding incidence matrix is:

$$B(G_4) = \begin{array}{c} \\ u \\ v \\ w \\ x \\ y \end{array}\begin{array}{cccccccc} e_1 & e_2 & e_3 & e_4 & e_5 & e_6 & e_7 & e_8 \\ \left[\begin{array}{cccccccc} 1 & 1 & 1 & 0 & 0 & 0 & 0 & 0 \\ 1 & 0 & 0 & 1 & 1 & 0 & 0 & 0 \\ 0 & 0 & 0 & 1 & 0 & 1 & 1 & 0 \\ 0 & 1 & 0 & 0 & 0 & 1 & 0 & 1 \\ 0 & 0 & 1 & 0 & 1 & 0 & 1 & 1 \end{array}\right]\end{array}.$$

(c) Adjacency lists:

G_1		G_2		G_3		G_4	
Vertex	**Neighbors**	**Vertex**	**Neighbors**	**Vertex**	**Neighbors**	**Vertex**	**Neighbors**
u	w	u	y	u	v, w, x	u	v, x, y
v	x	v	y	v	u, x	v	u, w, y
w	u	w	y	w	u	w	v, x, y
x	v	x	y	x	u, v	x	u, w, y
		y	u, v, w, x			y	u, v, w, x

(d) Edge lists:

G_1	
u	w
v	x

G_2	
u	y
v	y
w	y
x	y

G_3	
u	v
u	w
u	x
v	x

G_4	
u	v
u	x
u	y
v	w
v	y
w	x
w	y
x	y

#31. (a) Adjacency matrices:

$$A(G_1) = \begin{array}{c} \\ u \\ v \\ w \\ x \end{array}\begin{array}{cccc} u & v & w & x \\ \begin{bmatrix} 1 & 0 & 1 & 0 \\ 0 & 1 & 0 & 2 \\ 1 & 0 & 0 & 0 \\ 0 & 2 & 0 & 0 \end{bmatrix} \end{array}, \quad A(G_2) = \begin{array}{c} \\ u \\ v \\ w \\ x \\ y \end{array}\begin{array}{ccccc} u & v & w & x & y \\ \begin{bmatrix} 0 & 0 & 0 & 0 & 1 \\ 0 & 0 & 0 & 0 & 2 \\ 0 & 0 & 0 & 0 & 3 \\ 0 & 0 & 0 & 0 & 0 \\ 1 & 2 & 3 & 0 & 0 \end{bmatrix} \end{array}, \quad A(G_3) = \begin{array}{c} \\ u \\ v \\ w \\ x \end{array}\begin{array}{cccc} u & v & w & x \\ \begin{bmatrix} 0 & 1 & 3 & 1 \\ 1 & 0 & 0 & 3 \\ 3 & 0 & 2 & 0 \\ 1 & 3 & 0 & 0 \end{bmatrix} \end{array},$$

$$A(G_4) = \begin{array}{c} \\ u \\ v \\ w \\ x \\ y \end{array}\begin{array}{ccccc} u & v & w & x & y \\ \begin{bmatrix} 0 & 2 & 0 & 2 & 1 \\ 2 & 0 & 2 & 0 & 1 \\ 0 & 2 & 0 & 2 & 1 \\ 2 & 0 & 2 & 0 & 1 \\ 1 & 1 & 1 & 1 & 0 \end{bmatrix} \end{array}$$

(b) Incidence matrices:

(i)   The lexicographic ordering of the five edges of G_1 is: $e_1 = \{u,u\}, e_2 = \{u,w\}, e_3 = \{v,v\}$,

$e_4 = \{v,x\}_1, e_5 = \{v,x\}_2$ and the corresponding incidence matrix is: $B(G_1) = \begin{array}{c} \\ u \\ v \\ w \\ x \end{array}\begin{array}{ccccc} e_1 & e_2 & e_3 & e_4 & e_5 \\ \begin{bmatrix} 2 & 1 & 0 & 0 & 0 \\ 0 & 0 & 2 & 1 & 1 \\ 0 & 1 & 0 & 0 & 0 \\ 0 & 0 & 0 & 1 & 1 \end{bmatrix} \end{array}.$

(ii) The lexicographic ordering of the six edges of G_2 is: $e_1 = \{u,y\}, e_2 = \{v,y\}_1, e_3 = \{v,y\}_2$,

$e_4 = \{w,y\}_1, e_5 = \{w,y\}_2, e_6 = \{w,y\}_3$ and the corresponding incidence matrix is:

$$B(G_2) = \begin{array}{c} \\ u \\ v \\ w \\ x \\ y \end{array}\begin{array}{cccccc} e_1 & e_2 & e_3 & e_4 & e_5 & e_6 \\ \begin{bmatrix} 1 & 0 & 0 & 0 & 0 & 0 \\ 0 & 1 & 1 & 0 & 0 & 0 \\ 0 & 0 & 0 & 1 & 1 & 1 \\ 0 & 0 & 0 & 0 & 0 & 0 \\ 1 & 1 & 1 & 1 & 1 & 1 \end{bmatrix} \end{array}.$$

(iii) The lexicographic ordering of the 10 edges of G_3 is: $e_1 = \{u,v\}, e_2 = \{u,w\}_1, e_3 = \{u,w\}_2$,

$e_4 = \{u,w\}_3, e_5 = \{u,x\}, e_6 = \{v,x\}_1, e_7 = \{v,x\}_2, e_8 = \{v,x\}_3, e_9 = \{w,w\}_1, e_{10} = \{w,w\}_2$ and   the corresponding incidence matrix is:

$$B(G_3) = \begin{array}{c} \\ u \\ v \\ w \\ x \end{array}\begin{array}{cccccccccc} e_1 & e_2 & e_3 & e_4 & e_5 & e_6 & e_7 & e_8 & e_9 & e_{10} \\ \begin{bmatrix} 1 & 1 & 1 & 1 & 1 & 0 & 0 & 0 & 0 & 0 \\ 1 & 0 & 0 & 0 & 0 & 1 & 1 & 1 & 0 & 0 \\ 0 & 1 & 1 & 1 & 0 & 0 & 0 & 0 & 2 & 2 \\ 0 & 0 & 0 & 0 & 1 & 1 & 1 & 1 & 0 & 0 \end{bmatrix} \end{array}.$$

(iv) The lexicographic ordering of the 12 edges of G_4 is: $e_1 = \{u,v\}_1, e_2 = \{u,v\}_2, e_3 = \{u,x\}_1,$ $e_4 = \{u,x\}_2, e_5 = \{u,y\}, e_6 = \{v,w\}_1, e_7 = \{v,w\}_2, e_8 = \{v,y\}, e_9 = \{w,x\}_1, e_{10} = \{w,x\}_2, e_{11} = \{w,y\}, e_{12} = \{x,y\}$

and the corresponding incidence matrix is:

$$B(G_4) = \begin{array}{c} \\ u \\ v \\ w \\ x \\ y \end{array}\begin{array}{c} \begin{array}{cccccccccccc} e_1 & e_2 & e_3 & e_4 & e_5 & e_6 & e_7 & e_8 & e_9 & e_{10} & e_{11} & e_{12} \end{array} \\ \left[\begin{array}{cccccccccccc} 1 & 1 & 1 & 1 & 1 & 0 & 0 & 0 & 0 & 0 & 0 & 0 \\ 1 & 1 & 0 & 0 & 0 & 1 & 1 & 1 & 0 & 0 & 0 & 0 \\ 0 & 0 & 0 & 0 & 0 & 1 & 1 & 0 & 1 & 1 & 1 & 0 \\ 0 & 1 & 0 & 0 & 0 & 1 & 1 & 0 & 1 & 1 & 1 & 0 \\ 0 & 0 & 0 & 0 & 1 & 0 & 0 & 1 & 0 & 0 & 1 & 1 \end{array}\right] \end{array}.$$

#33. (a) $V(\vec{G}_1) = \{u,v,x,w\}, V(\vec{G}_2) = \{u,v,x,w,y\}, V(\vec{G}_3) = \{u,v,x,w\}, V(\vec{G}_4) = \{u,v,x,w,y\}$

(b)

$E(\vec{G}_1) = \{(v,x),(w,u)\}; E(\vec{G}_2) = \{(u,x),(u,y),(v,u),(v,y),(w,v),(w,y),(x,w),(x,y)\}; E(\vec{G}_3) = \{(v,y),$

$(w,y),(y,v),(y,w)_1,(y,w)_2\}; E(\vec{G}_4) = \{(u,w),(v,u),(v,x)_1,(v,x)_2,(w,u)_1,(w,u)_2,(w,w)_1,(w,w)_2,(x,u),(x,v)\}$

(c)

$\deg^+_{\vec{G}_1}(u) = 0, \deg^-_{\vec{G}_1}(u) = 1, \deg^+_{\vec{G}_1}(v) = 1, \deg^-_{\vec{G}_1}(v) = 0, \deg^+_{\vec{G}_1}(w) = 1, \deg^-_{\vec{G}_1}(w) = 0, \deg^+_{\vec{G}_1}(x) = 0, \deg^-_{\vec{G}_1}(x) = 1;$

$\deg^+_{\vec{G}_2}(u) = 2, \deg^-_{\vec{G}_2}(u) = 1, \deg^+_{\vec{G}_2}(v) = 2, \deg^-_{\vec{G}_2}(v) = 1, \deg^+_{\vec{G}_2}(w) = 2, \deg^-_{\vec{G}_2}(w) = 1, \deg^+_{\vec{G}_2}(x) = 2, \deg^-_{\vec{G}_2}(x) = 1,$

$\deg^+_{\vec{G}_2}(y) = 0, \deg^-_{\vec{G}_2}(y) = 4;\ \deg^+_{\vec{G}_3}(u) = 0, \deg^-_{\vec{G}_3}(u) = 1, \deg^+_{\vec{G}_3}(v) = 1, \deg^-_{\vec{G}_3}(v) = 1,\ \deg^+_{\vec{G}_3}(w) = 1,$

$\deg^-_{\vec{G}_3}(w) = 2, \deg^+_{\vec{G}_1}(x) = 0,\ \deg^-_{\vec{G}_1}(x) = 0;\ \deg^+_{\vec{G}_4}(u) = 1,\ \deg^-_{\vec{G}_4}(u) = 4, \deg^+_{\vec{G}_4}(v) = 3, \deg^-_{\vec{G}_4}(v) = 1,$

$\deg^+_{\vec{G}_4}(w) = 4, \deg^-_{\vec{G}_4}(w) = 3, \deg^+_{\vec{G}_4}(x) = 2, \deg^-_{\vec{G}_4}(x) = 2$

#35. (a) Adjacency matrices:

$$A(\vec{G}_1) = \begin{array}{c} \\ u \\ v \\ w \\ x \end{array}\begin{array}{c} \begin{array}{cccc} u & v & w & x \end{array} \\ \left[\begin{array}{cccc} 0 & 0 & 0 & 0 \\ 0 & 0 & 0 & 1 \\ 1 & 0 & 0 & 0 \\ 0 & 0 & 0 & 0 \end{array}\right] \end{array},\quad A(\vec{G}_2) = \begin{array}{c} \\ u \\ v \\ w \\ x \\ y \end{array}\begin{array}{c} \begin{array}{ccccc} u & v & w & x & y \end{array} \\ \left[\begin{array}{ccccc} 0 & 0 & 0 & 1 & 1 \\ 1 & 0 & 0 & 0 & 1 \\ 0 & 1 & 0 & 0 & 1 \\ 0 & 0 & 1 & 0 & 1 \\ 0 & 0 & 0 & 0 & 0 \end{array}\right] \end{array},\quad A(\vec{G}_3) = \begin{array}{c} \\ u \\ v \\ w \\ x \\ y \end{array}\begin{array}{c} \begin{array}{ccccc} u & v & w & x & y \end{array} \\ \left[\begin{array}{ccccc} 0 & 0 & 0 & 0 & 0 \\ 0 & 0 & 0 & 0 & 1 \\ 0 & 0 & 0 & 0 & 1 \\ 0 & 0 & 0 & 0 & 0 \\ 1 & 1 & 2 & 0 & 0 \end{array}\right] \end{array},$$

$$A(\vec{G}_4) = \begin{array}{c} \\ u \\ v \\ w \\ x \end{array}\begin{array}{c} \begin{array}{cccc} u & v & w & x \end{array} \\ \left[\begin{array}{cccc} 0 & 0 & 1 & 0 \\ 1 & 0 & 0 & 2 \\ 2 & 0 & 2 & 0 \\ 1 & 1 & 0 & 0 \end{array}\right] \end{array}$$

(b) $\vec{G}_1$: Edges: $e_1 = (v,x), e_2 = (w,u)$ Incidence matrix: $B = \begin{array}{c} \\ u \\ v \\ w \\ x \end{array}\begin{array}{c} \begin{array}{cc} e_1 & e_2 \end{array} \\ \left[\begin{array}{cc} 1 & 0 \\ 0 & -1 \\ -1 & 0 \\ 0 & 1 \end{array}\right] \end{array}.$

$\vec{G}_2$: Edges: $e_1 = (u,x), e_2 = (u,y), e_3 = (v,u), e_4 = (v,y), e_5 = (w,v), e_6 = (w,y), e_7 = (x,w), e_8 = (x,y)$

Incidence matrix: $B = \begin{array}{c} \\ u \\ v \\ w \\ x \\ y \end{array}\begin{array}{c} \begin{array}{cccccccc} e_1 & e_2 & e_3 & e_4 & e_5 & e_6 & e_7 & e_8 \end{array} \\ \left[\begin{array}{cccccccc} -1 & -1 & 1 & 0 & 0 & 0 & 0 & 0 \\ 0 & 0 & -1 & -1 & 1 & 0 & 0 & 0 \\ 0 & 0 & 0 & 1 & -1 & -1 & 1 & 0 \\ 1 & 0 & 0 & 0 & 0 & 0 & -1 & -1 \\ 0 & 1 & 0 & 1 & 0 & 1 & 0 & 1 \end{array}\right] \end{array}$

$\vec{G}_3$: Edges: $e_1 = (v,y), e_2 = (w,y), e_3 = (y,u), e_4 = (y,v), e_5 = (y,w)_1, e_6 = (y,w)_2$ Incidence matrix:

$$B = \begin{array}{c} \\ u \\ v \\ w \\ x \\ y \end{array}\begin{array}{c} \begin{array}{cccccc} e_1 & e_2 & e_3 & e_4 & e_5 & e_6 \end{array} \\ \left[\begin{array}{cccccc} 0 & 0 & 1 & 0 & 0 & 0 \\ -1 & 0 & 0 & 1 & 0 & 0 \\ 0 & -1 & 0 & 0 & 1 & 1 \\ 0 & 0 & 0 & 0 & 0 & 0 \\ 1 & 1 & -1 & -1 & -1 & -1 \end{array}\right] \end{array}$$

$\vec{G}_4$: Edges: $e_1 = (u,w), e_2 = (v,u), e_3 = (v,x)_1, e_4 = (v,x)_2, e_5 = (w,u)_1, e_6 = (w,u)_2, e_7 = (w,w)_1,$

$e_8 = (w,w)_2, e_9 = (x,u), e_{10} = (x,v)$ Incidence matrix:

$$B = \begin{array}{c c} & \begin{array}{c c c c c c c c c c} e_1 & e_2 & e_3 & e_4 & e_5 & e_6 & e_7 & e_8 & e_9 & e_{10} \end{array} \\ \begin{array}{c} u \\ v \\ w \\ x \end{array} & \left[\begin{array}{c c c c c c c c c c} -1 & 1 & 0 & 0 & 1 & 1 & 0 & 0 & 1 & 0 \\ 0 & -1 & -1 & -1 & 0 & 0 & 0 & 0 & 0 & 1 \\ 1 & 0 & 0 & 0 & -1 & -1 & 2 & 2 & 0 & 0 \\ 0 & 0 & 1 & 1 & 0 & 0 & 0 & 0 & -1 & -1 \end{array}\right] \end{array}$$

(c) Adjacency lists:

$\vec{G}_1$	
Vertex	**Head-Neighbors**
v	x
w	u

$\vec{G}_2$	
Vertex	**Head-Neighbors**
u	x, y
v	u, y
w	v, y
x	w, y

(d) Edge lists:

$\vec{G}_1$	
Tail	**Head**
v	x
w	u

$\vec{G}_2$	
Tail	**Head**
u	x
u	y
v	u
v	y
w	v
w	y
x	w
x	y

#37. (a) The digraphs are not isomorphic. Here is one reason: On the left digraph, there is an edge going from the unique vertex (d) of total degree 4 to the unique vertex (f) of total degree 3, but the edge between the corresponding vertices (w,z) on the right digraph goes in the opposite direction.
(b) The underlying graphs are isomorphic. The following vertex bijection gives an isomorphism:
$$a \mapsto x, b \mapsto s, c \mapsto u, d \mapsto w, f \mapsto z, g \mapsto y, i \mapsto v$$

#39. (a) The vertex f_1 in $\vec{H}$ has outdegree 2 and indegree 0; the other digraphs have no such vertex, so $\vec{H}$ is not isomorphic to either $\vec{G}, \vec{K}$. But $\vec{G} \cong \vec{K}$; the vertex mapping $a \mapsto b_2, b \mapsto a_2, d \mapsto c_2,$

$c \mapsto d_2, f \mapsto f_2$ is a (digraph) isomorphism

(b) The underlying graphs are identical (if the subscripts are removed from the vertex labels) and hence isomorphic.

#41. *Proof that isomorphism is an equivalence relation on the set of general graphs:*
We assume that $G = (V, E, \Phi)$, $G' = (V', E', \Phi')$, and $G'' = (V'', E'', \Phi'')$ be general graphs.

(i) *Reflexivity*: The identity functions $i_V : V \to V :: i_V(v) = v, i_E : E \to E :: i_E(e) = e$ satisfies the condition $i_E(e) = \{i_V(u), i_V(v)\}$, and so is an isomorphism, thus $G \cong G$.

(ii) *Symmetry*: Assume that $G \cong G'$. This means there exists an isomorphism, i.e., a pair of *bijective* functions: $f : V \to V'$, $g : E \to E'$, such that $e = \{u, v\} \in E \implies g(e) = \{f(u), f(v)\} \in E'$. It follows that

$g^{-1}(g(e)) = \{u,v\} = \{f^{-1}(f(u)), f^{-1}(f(v))\} \in E,$ and since the edges $\{g(e): e \in E\}$ comprise all of the edges of E', this shows that $G' \cong G$.

(iii) *Transitivity*: Assume that $G \cong G'$, and that $G' \cong G''$. This means there exists pairs of bijections: $f: V \to V'$, $g: E \to E'$, such that $e = \{u,v\} \in E \Rightarrow g(e) = \{f(u), f(v)\} \in E'$, and $f': V' \to V''$, $g': E' \to E''$, such that $e' = \{u',v'\} \in E' \Rightarrow g'(e') = \{f'(u'), f'(v')\} \in E''$.

If we successively apply these two implications, we arrive at the following:

$$e = \{u,v\} \in E \Rightarrow (g' \circ g)(e) = \{(f' \circ f)(u), (f' \circ f)(v)\} \in E''.$$

The pair of (bijective) compositions $f' \circ f: V \to V''$, $g' \circ g: E \to E''$, thus provide the isomorphism $G \cong G''$. □

#43. (a) General digraph isomorphisms: $\vec{K} \cong \vec{M}$; no other pairs are isomorphic.

Proof: To show that $\vec{K} \cong \vec{M}$, we start by defining the following vertex bijection $f: V(\vec{K}) \to V(\vec{M})$: $u_2 \mapsto x_3, v_2 \mapsto u_3, w_2 \mapsto v_3, x_2 \mapsto w_3$. (Note: This vertex mapping is necessary in order for an isomorphism to exist. For example, the unique directed edge (w_2, x_2) in $\vec{K}$ going from one of the two vertices with a double loop to the other, must correspond to the directed edge (v_3, w_3) in $\vec{M}$, and this forces the vertex correspondences: $w_2 \mapsto v_3, x_2 \mapsto w_3$.) The compatibility condition for the edge function $g: E(\vec{K}) \to E(\vec{M}) :: g(u,v) = (f(u), f(v))$ will now uniquely define it; it will complete the desired isomorphism provided that edges in the range of g coincide exactly with those of $E(\vec{M})$. It is readily checked that this is indeed the case.

The following observation shows that there can be no other general directed graph isomorphisms between any other pairs:

The two distinguished vertices having self-loops are neighbors that are directly connected by two oppositely directed edges in $\vec{G}$; neighbors that are directly connected by a single directed edge in each of $\vec{K}, \vec{M}$, and are not neighbors in $\vec{G}$. □

(b) General (underlying) graph isomorphisms: $K \cong M$, no other pairs are isomorphic.

Proof: Any digraph isomorphism always specializes to a general graph isomorphism by simply ignoring the restriction of preserving edge directions. The observation that ruled out any other general digraph isomorphisms also works to rule out the corresponding underlying graph isomorphisms. □

#45. Always true: The edge compatibility condition for simple directed graphs is more restrictive than that for underlying graphs (which ignores the preservation of directions), so any directed graph isomorphism will also be simple graph isomorphism of the underlying graphs.

#47. Suppose that $H_1 \cong H$ is a subgraph of G_1. Consider the subgraph H_2 of G_2 defined by $V(H_2) = \{f(v): v \in V(H_1)\}$, $E(H_2) = \{\{f(v), f(w)\}: \{v,w\} \in E(H_1)\}$. Since f preserves adjacency and is injective, each of the latter edges belongs to $E(G_2)$, and this shows that $H_2 \cong H$ is a subgraph of G_2.

#49. *Proof:* Let $n = |V(G)|$. Since $|E(\sim G)| = C(n,2) - |E(G)|$, it follows that if $G \cong \sim G$, then (since $|E(\sim G)| = |E(G)|$) $C(n,2) = 2|E(G)|$, i.e., the binomial coefficient $C(n,2)$ is even. If we write $n = 4k + j$, where k and j are nonnegative integers with $j = 0, 1, 2, 3$, then $C(n,2) = n(n-1)/2 = (4k+j)(4k+j-1)/2 = (4\ell + j(j-1))/2 = 2\ell + j(j-1)/2$. The parity is thus determined by the parity of $j(j-1)/2$. As j runs through its possible values: $j = 0, 1, 2, 3$, the corresponding values for $j(j-1)/2$ are $0, 0, 1, 3$. Thus, in order for $C(n,2)$ to be even, we must have that $j = 0$ or 1, which is equivalent to $n = 4k + j \equiv 0$ or $1 \pmod 4$. □

#51. *Proof:* Let $D = \max_{v \in V} \deg(v)$. From the handshaking theorem $\sum_{v \in V} \deg(v) = 2|E(G)|$, it follows that $nD = \sum_{v \in V} D \geq \sum_{v \in V} \deg(v) = 2|E(G)| = 2m \Rightarrow D \geq 2m/n$. □

#53. (a) *Proof:* Pick any person vertex $v \in V$.

Case 1: $\deg_G(v) \geq 3$. Let x, y, z be neighbors in G of v. If any two of these three neighbors are themselves neighbors in G, then these two neighbors together with v along with all edges between any two of the three will form a triangle in G, i.e., a subgraph of G that is isomorphic to K_3. On the other hand, if no two of x, y, z are neighbors in G, then all of them will be neighbors in $\sim G$, so that $\sim G$ will contain a copy of K_3.

Case 2: $\deg_G(v) \leq 2$. In this case $\deg_{\sim G}(v) = 5 - \deg_G(v) \geq 3$, so the argument of Case 1 can be modified by reversing the roles of G and $\sim G$ to again show that one of these two graphs must contain a copy of K_3. $\square$

(b) This social proposition follows directly from the result of Part (a), if we define the graph G by letting the vertices be the six people at the party and joining two vertices by an edge if, and only if, the people they represent are acquaintances.

(c) The figure on the right makes it clear that neither $G = C_5$ nor its complement (which is isomorphic to C_5) contains a copy of K_3.

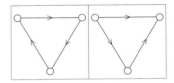

#55. (a) The two digraph isomorphism classes are shown on the right.

(b) If multi-edges are allowed, an additional six digraph isomorphism classes arise, and are shown below (bringing the total to eight).

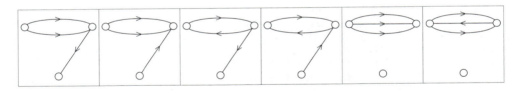

#57. (a) Suppose that f, g are both automorphisms of G, and let u, v be two vertices in G. We have:

u, v are adjacent $\Leftrightarrow$ $f(u), f(v)$ are adjacent $\Leftrightarrow$ $g(f(u)), g(f(v)) = (g \circ f)(u), (g \circ f)(v)$ are adjacent

Thus, the composition $g \circ f$ is also a bijection $V \to V$ that preserves adjacency and non-adjacency, and so is also an automorphism.

(b) Suppose that f is an automorphisms of G, and let u, v be two vertices in G, and write $u = f(w), v = f(x)$. We then can translate the equivalence:

$$w, x \text{ are adjacent} \Leftrightarrow f(w), f(x) = u, v \text{ are adjacent}$$

into:

$$u, v \text{ are adjacent} \Leftrightarrow f^{-1}(u), f^{-1}(v) = w, x \text{ are adjacent}$$

This is the defining condition for f^{-1} to preserve adjacency and nonadjacency; hence this vertex bijection is also an automorphism.

#59. (a) (i) maximal independent sets for first graph: $\{a, d\}$; for second graph: $\{u, v\}$

(ii) indepence numbers for both graphs are 2

(iii) cliques in the first graph: $\{a, b, c\}$, $\{b, c, d\}$; for second graph: $\{u, w, x\}$, $\{v, w, x\}$

(iv) clique numbers for both graphs are 2

(b) (i) $\{a, h\}$, $\{b, i\}$, $\{c, f\}$, $\{d, g\}$, $\{a, c, g, i\}$, $\{b, d, f, h\}$ (ii) 4

(iii) 12 cliques are all pairs of adjacent vertices (iv) 2

(c) (i) $\{a_1, d_1\}, \{b_1, h_1\}, \{c_1, g_1\}, \{f_1, i_1\}, \{a_1, c_1, f_1, h_1\}, \{b_1, i_1, d_1, g_1\}$ (ii) 4

(iii) 12 cliques are all pairs of adjacent vertices (iv) 2

(d) (i) $\{a_2, c_2, g_2\}, \{a_2, d_2, g_2\}, \{a_2, d_2, h_2\}, \{b_2, d_2, h_2\}, \{b_2, f_2, h_2\}, \{b_2, f_2, i_2\}, \{c_2, f_2, i_2\}, \{c_2, g_2, i_2\}$

(ii) 3 (iii) cliques are all pairs of adjacent vertices (iv) 2

#61. (a) $\alpha(P_n) = \lceil n/2 \rceil$ *Proof:* A maximum sized independent set can be formed by selecting alternate vertices starting with the vertex on one end. If n is odd, then the vertex at the other end will be selected and so $(n+1)/2$ vertices will be in the maximum independent set, while if n is even, the last vertex will not be selected and $n/2$ will be the independence number. In summary:

$$\alpha(P_n) = \begin{cases} (n+1)/2, & \text{if } n \text{ is odd} \\ n/2, & \text{if } n \text{ is even} \end{cases} = \lceil n/2 \rceil. \quad \square$$

(b) $\alpha(C_n) = \lfloor n/2 \rfloor$ *Proof:* A maximum sized independent set can be formed by selecting alternate vertices (starting with any vertex in the cycle). If n is odd, the final vertex selected will have two vertices between it and the first selected, so $(n-1)/2$ vertices will be in the maximum independent set, while if n is even, then exactly $n/2$ vertices will be selected. In summary:

$$\alpha(C_n) = \begin{cases} (n-1)/2, & \text{if } n \text{ is odd} \\ n/2, & \text{if } n \text{ is even} \end{cases} = \lfloor n/2 \rfloor. \quad \square$$

(c) $\alpha(K_n) = 1$

(d) $\alpha(\text{Petersen Graph}) = 4$ *Proof:* The Petersen graph has two C_5 subgraphs with its five inside vertices and its five outside vertices. Thus, by Part (b), any independent set can contain at most two inside vertices and at most two outside vertices, and this means $\alpha \leq 4$. On the other hand, the vertex set $\{a, j, f, h\}$ (see Figure 8.14) is clearly an independent set, and so $\alpha = 4$. $\square$

(e) $\alpha(Q_n) = 2^{n-1}$ *Proof:* By Exercise for the Reader 8.3(b), Q_n is bipartite, and a bipartition can be formed by coloring half of its 2^n vertices black and the other half white. In any bipartite graph, either of the two sets of vertices of a single color forms an independent set, and thus $\alpha(Q_n) \geq 2^{n-1}$. We prove the reverse inequality $\alpha(Q_n) \leq 2^{n-1}$ by induction on n. The basis step is trivial, since $Q_1 = P_2$, and (either trivially or by Part (a)) $\alpha(P_2) = 1$. For the inductive step, we assume that $\alpha(Q_n) \leq 2^{n-1}$, and we must infer that $\alpha(Q_{n+1}) \leq 2^{[n+1]-1} = 2^n$. We will assume that Q_{n+1} has an independent set S with more than 2^n vertices, and seek to find a contradiction. By the recursive contruction of the hypercubes given in the section, Q_{n+1} contains two disjoint copies of Q_n, and so (by the generalized pigeonhole principle) one of these copies must contain at least $2^n/2 + 1 = 2^{n-1} + 1$ vertices of S. But these vertices would then give rise to an independent set of Q_n with more than 2^{n-1} elements—a contradiction to the inductive hypothesis. $\square$

#63. Both parts readily follow from the observation that a set S of vertices in a simple graph is independent if, and only if, it forms a clique in the complement graph.

Section 8.2:

#1. (a) (i) yes (ii) 3 (iii) yes (iv) no;   (b) (i) no; (c) (i) yes (ii) 5 (iii) yes (iv) no;
(d) (i) yes (ii) 4 (iii) no (iv) yes

#3. (a) (i) no; (b) (i) yes (ii) 5 (iii) no (iv) no; (c) (i) yes (ii) 7 (iii) yes (iv) yes; (d) (i) no

#5. The graph G is not connected; it has four components that correspond to the following four pairs of vertices and whose edge sets are the single edges that join them: $\{s, x\}, \{t, w\}, \{u, z\}, \{v, y\}$.

#7. The digraph $\vec{G}$ is weakly connected; this can be seen, for example, from the fact that the vertex sequence $<v, u, w, z, y, z, v>$ forms a cycle in the underlying graph that passes through every vertex.

The digraph $\vec{G}$ is not connected, since, for example the vertex u has no edges that are directed towards it, so no other vertex can be joined to u via a (directed) path in the digraph.

The digraph $\vec{H}$ is weakly connected; this can be seen, for example, from the fact that the vertex sequence $<v,z,x,u,v,y,w,t,v>$ forms a circuit in the underlying graph that passes through every vertex. The digraph $\vec{H}$ is not connected, since, for example the vertex y has no edges that are directed towards it, so no other vertex can be joined to y via a (directed) path in the digraph.

#9. (a) dist$(t,u)=1$, dist$(v,z)=3$, dist$(x,w)=2$; (b) diam$(G)=3$, rad$(G)=2$;

(c) ecc$(t)=$ ecc$(v)=3$; (d) Central vertices are u and y (they have minimal eccentricity 2).

#11.   (a)   The component of any actor/actress X in the actor/movie graph is the set of all actors/actresses Y who made a movie (in the movie database) for which there is a sequence of movies: $M_1, M_2, \cdots, M_k$ (in the movie database) such that X appeared in movie M_1, Y appeared in movie M_k, and for any index $i<k$, the movies M_i, M_{i+1} share at least one common actor/actress.

(b) This would be possible if there were a movie in the database all of whose actors/actresses did not appear in any other movie in the database.

#13. (a) diam$(P_n)=n-1$, rad$(P_n)=$ ceil$((n-1)/2)$,

Central vertices: single middle vertex if n is odd or two middle vertices if n is even. In the notation of Definition 8.4 (and Figure 8.5), the central vertex is $v_{(n+1)/2}$ if n is odd, and the central vertices are $v_{n/2}, v_{n/2+1}$ if n is even.

(b) Since for the Petersen graph the radius $= 2 =$ the diameter, all vertices are central vertices.

(c) In the complete graph K_n, $n \geq 2$, the radius $= 1 =$ the diameter, so all vertices are central vertices. In K_1, since there is only one vertex, the farthest away vertex is itself (with distance zero), so the radius $= 0$, and again all vertices are central.

#15. (a) $A = \begin{bmatrix} 0 & 1 & 1 \\ 1 & 0 & 0 \\ 0 & 1 & 0 \end{bmatrix}$

(b) The theorem tells us that the $(2,3)$ entry of the matrix $A + A^2 + A^3$ equals the number of routes to go from city 2 (Baltimore) to city 3 (Rockville) using 1, 2, or 3 flights. There are 0 direct flights, 1 two flight route (via Washington), and 0 three flight routes; the total is thus 1.

(c) Since all entries of $I + A + A^2 = \begin{bmatrix} 2 & 2 & 1 \\ 1 & 2 & 1 \\ 1 & 1 & 1 \end{bmatrix}$ are positive, it follows that the diameter of the network is 2.

(d) $A = \begin{bmatrix} 0 & 0 & 1 & 1 & 0 \\ 1 & 0 & 1 & 0 & 0 \\ 1 & 0 & 0 & 0 & 1 \\ 0 & 0 & 1 & 0 & 1 \\ 0 & 1 & 1 & 0 & 0 \end{bmatrix}$

(e) As in Part (b), from the figure we see that there is 1 direct flight from Rome to Paris (4 to 3), 1 two-flight route (via Madrid), and 3 three-flight routes (via Madrid and Frankfurt, and also these two: $4 \rightarrow 3 \rightarrow 1 \rightarrow 3$, $4 \rightarrow 3 \rightarrow 5 \rightarrow 3$), so the $(4,3)$ entry of $A + A^2 + A^3$ equals $1+1+3 = 5$.

(f) $I + A + A^2 = \begin{bmatrix} 2 & 0 & 2 & 1 & 2 \\ 2 & 1 & 2 & 1 & 1 \\ 1 & 1 & 3 & 1 & 1 \\ 1 & 1 & 2 & 1 & 2 \\ 2 & 1 & 2 & 0 & 2 \end{bmatrix}$, and $I + A + A^2 + A^3 = \begin{bmatrix} 3 & 2 & 5 & 2 & 3 \\ 3 & 2 & 5 & 2 & 3 \\ 4 & 1 & 5 & 1 & 4 \\ 3 & 2 & 5 & 2 & 3 \\ 3 & 2 & 5 & 2 & 3 \end{bmatrix}$, so it follows that

the diameter of the network is 3.

#17. (a)

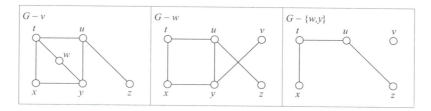

(b)

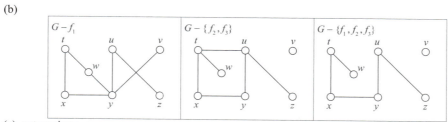

(c) cut vertices: u, y

(d) bridges: $\{u,z\}$, $\{v,y\}$

(e) 1 (any bridge is an edge cut)

#19. P_2 has arbitrarily long paths (just go back and forth between the vertices any number of times).

#21. (a) 3; P_3 (b) 2 (the other isomorphism class has degree sequence (3,1,1,1)). (c) $n-1$; P_n

#23. (a) *Proof:* $\text{rad}(G) = \min_{v \in V(G)} \{\text{ecc}(v)\} \leq \max_{w \in V(G)} \{\text{dist}(v,w)\} = \text{diam}(G)$. To prove the reverse inequality, fix a central vertex v, and let u, w be any vertices. Let $P(u,v)$ be a shortest path from u to v, and $P(v,w)$ be a shortest path from v to w. Then each of these paths has length less than $\text{rad}(G) = \text{ecc}(v)$. Since the juxtaposition of these two paths is a path P from u to w, it follows that

$$\text{dist}_G(u,w) \leq \text{length}(P) = \text{length}(P(u,v)) + \text{length}(P(v,w)) \leq \text{rad}(G) + \text{rad}(G) = 2\text{rad}(G). \quad \text{Since the}$$

vertices u,w were arbitrary, it follows that $\text{diam}(G) \leq 2\text{rad}(G)$. $\square$

(b) If a graph is not connected, then $\text{diam}(G) = \infty = \text{rad}(G)$, so the inequalities remain trivially valid (all sides are infinite).

#25. (a) $n-2, n, 1$ (b) 0, 0, 3 (c) Except when $n = 2$, K_n has no vertex cuts or bridges. K_2 has a bridge (its only edge) but no vertex cuts. As was shown in the solution to Exercise for the Reader 8.24, the size of a minimum edge cut of K_n, is $n-2$ if $n \geq 2$. Since K_1 has no edges, a minimum edge cut for K_1 is undefined.

#27. (a) n

(b) If either n or $m = 1$, then $K_{n,m}$ contains no cycles. Otherwise, the largest value of k for which $K_{n,m}$ contains a copy of C_k is $2\min(n,m)$. Such a cycle can be formed by starting at the "first" black vertex, assuming there are at least as many black vertices as there are white vertices. We alternate back and forth between subsequent white and black vertices until we reach the last white vertex. Then we take an edge back to the original black vertex.

(c) Answer 9: Reason: Using the labeling of Figure 8.14, here is a 9-cycle in the Petersen graph $<a,d,k,h,e,b,j,g,c,a>$. Showing that the Petersen graph does not contain a 10-cycle it quite a bit more difficult. A justification of this fact may be found in the solution to Exercise for the Reader 9.5 (in this solution, the word Hamiltonian means that a graph has a cycle that passes through all of its vertices).

#29. All three of the given graphs are 3-regular 8-vertex bipartite graphs. Each contains a 9-cycle subgraph, examples of which are shown below:

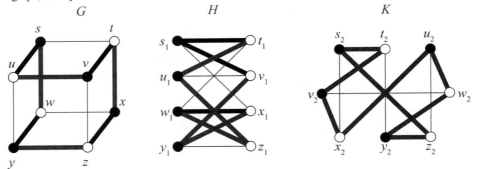

$$G \qquad\qquad H \qquad\qquad K$$

Let us use the above cycles to define vertex bijections. If we begin the vertex matching by starting all cycles at the s-vertices, we obtain the following bijection for the first two graphs:

$$f : V(G) \to V(H) :: s \mapsto s_1, u \mapsto v_1, v \mapsto y_1, t \mapsto x_1, x \mapsto w_1, z \mapsto z_1, y \mapsto u_1, w \mapsto t_1.$$

By construction, this vertex mapping preserves all edge adjacencies of the corresponding edges for each cycle, so to show this map is an isomorphism, we need only check that it preserves the remaining four edge adjacency relations. Indeed $\{f(s), f(t)\} = \{s_1, x_1\}$, $\{f(u), f(y)\} = \{v_1, u_1\}$, $\{f(v), f(z)\} = \{y_1, z_1\}$, $\{f(w), f(x)\} = \{t_1, w_1\}$ are precisely the non-highlighted edges of H. This proves that f is an isomorphism, so that $G \cong H$. In a similar fashion, the induced vertex mapping:

$$g : V(G) \to V(K) :: s \mapsto s_2, u \mapsto t_2, v \mapsto v_2, t \mapsto x_2, x \mapsto u_2, z \mapsto w_2, y \mapsto y_2, w \mapsto z_2,$$

is readily shown to be an isomorphism. Thus, all three graphs are isomorphic.

#31. We begin by observing that the Petersen graph contains a copy of P_{10}; one such copy is illustrated below (the bold edges on the left). Notice also that we have used the disjoint subset labeling of Figure 8.53. We will also use the structural property that the five "outer vertices" of P lie on a 5-cycle, as do the five "inner vertices" and these two 5-cycles are disjoint in P. In G the five vertices on top and the five on bottom lie also on disjoint 5-cycles. We will try to label the vertices of G using the disjoint subset scheme as we proceed to build a 10-path. We begin by labeling y_1 by 23 (the start of the cycle), and then label v_1 as 14 (second point of the cycle) to correspond to the labeling in P. The cycle then moves to one more point x_1 on the "lower 5-cycle" of G, which we label as 35, and then should move to t_1 (in the upper 5-cycle), and we label it as 24 since the path in P also switches 5-cycles here. The next vertex 15 in P is adjacent to 23, so the next vertex in our path in G should be u_1, which we label as 15. From here, the next vertices visited in G should be (in this order): r_1, v_1 (adjacent to the already visited 35 in both graphs), w_1, s_1, q_1, and these get the corresponding labels: 34, 12, 45, 13, 25. With the labels all shown in the middle figure below, it is clear that they satisfy the disjointness/adjacency property that characterizes the Petersen graph. Thus the labels determine an isomorphism between P and G. A similar argument using the path and labels drawn in the graph H below shows that we have an isomorphism between P and H.

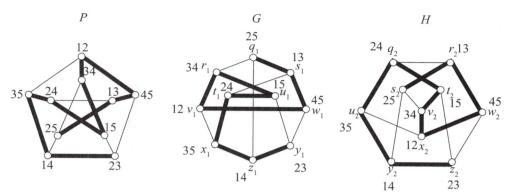

#33. Answer: 8; Explanation: One of the components will be either K_1 or K_2. In the first case the other component will be a connected four vertex graph, of which there are six (see the solution to EFR 8.10). In the second case the other component will be a connected three vertex graph, for which there are two: P_3, K_3.

#35. *Proof:* For a connected bipartite graph, there are exactly two black/white bipartition vertex colorings, i.e., the color reversals of one another. If a graph G has $C(G)$ components, since either of these two colorings can be chosen for each component, it follows from the multiplication principle that the number of different such colorings is $2^{C(G)}$. □

#37. *Proof:* ($\Rightarrow$) If G is bipartite, let $V(G) = X \cup Y$ be a bipartition. So both X and Y are independent sets. Any subgraph H of G will have at least half of its vertices either belonging to X or to Y, and this set will be independent in H.

($\Leftarrow$) We assume that every subgraph of G has an independent set consisting of at least half of its vertices. For a contradiction, we also assume that G is not bipartite. By Theorem 8.7, G must contain a copy of a cycle C_n of odd length (i.e., n is odd). But the size of a maximum independent set of C_n is floor($n/2$), which is less than $n/2$ when n is odd. So this odd length cycle subgraph $H = C_n$ does not have in independent set consisting of at least half of its vertices—a contradiction. □

#39. *Proof:* Suppose that G is a simple graph that is not connected. We must show that the complement $\sim G$ is connected. Let C be a nontrivial component of G, so that $V(C)$ is a proper subset of $V(G)$. Since $E(G)$ can contain no edge connecting any vertex in $V(C)$ with any vertex in $\sim V(C)$, it follows that $E(\sim G)$ contains all such edges. It follows that any two vertices can be joined by a path in $\sim G$ of length at most two: They are connected by an edge if they lie in different sets of $V(C), \sim V(C)$, while they can be joined by a path of length two if they lie in the same one of these sets (take an edge from one vertex to an third vertex in the other set, and then take the edge from the third vertex back to the other vertex). □

#41. (a) For a simple graph G on n vertices with adjacency matrix A, it follows from Theorem 8.6 and the reasoning that led to Algorithm 8.2 that G is connected if, and only if, all of the entries of the matrix $I + A + A^2 + \cdots + A^{n-1}$ are nonzero. It furthermore follows in general that the component of the ith vertex of G will simply be the column numbers corresponding to the nonzero entries of this matrix. The following algorithm thus follows:

Input: The n by n adjacency matrix A of a graph or a directed graph with $n > 2$ vertices, with respect to some ordering of the vertices. Output: A length n vector Comp whose entries come from the set $\{1,2,\ldots,k\}$, where k is the number of components of G, and Comp(i) = Comp(j) if, and only if, vertex i and vertex j are in the same component of G.

Step 1: Compute the matrix $C = I + A + A^2 + \cdots + A^{n-1}$, INITIALIZE Comp to the length n vector of 0s, compNum = 1, and index ind = 1.

Step 2: (*Iterative Step*)
WHILE ind $< n$
        IF Comp(ind) = 0

Find the set S of all column indices for which the ind row of the matrix C are nonzero, and set Comp(i) = comNum, for

each $i \in S$.

Update compNum = compNum + 1

END IF

UPDATE ind = ind + 1

END WHILE

(b) Comp = [1, 1, 1, 2, 2, 2]   (c) Comp = [1, 2, 1, 2, 1, 2]

#43. (a) From Theorem 8.6, it follows that b_{ii} is the number of length two paths from vertex i to itself. Such a path can only be achieved (since we are dealing with a simple graph) by taking an edge to a neighboring vertex, and then taking the same edge back. So it follows that $b_{ii} = \deg(i)$.

(b) From Theorem 8.6, it follows that c_{ii} is the number of length two paths from vertex i to itself. Any such path must constitute a 3-cycle in G: Take an edge from vertex i to a second vertex j, then take an edge from j to a third vertex k, and take an edge from k back to i. Since there are no other length three paths from vertex i to itself and c_{ii} counts the oppositely directed cycles $<i, j, k, i>$, $<i, k, j, i>$ separately, the result follows.

Section 8.3:

#1. (a) forest but not a tree (b) tree (and thus also a forest) (c) not a forest (and thus not a tree) (d) not a forest (and thus not a tree)

#3. (a) 99 (b) 101 (c) $100 - 4 = 96$ (d) $100 + 10 = 110$ (e) No, there would have to be at least $110 - 4 = 106$ edges (see EFR 8.27).

#5. (a) For all values of n, P_n is a tree. (b) K_n is a tree only when $n = 1, 2$ (for larger values of n it always has cycles).

#7. (a) There are six isomorphically distinct trees on six vertices; drawings of each of these are shown below:

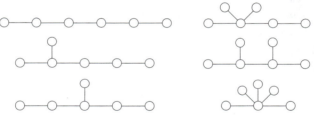

(b) The five nonisomorphic four-vertex forests are shown below, listed in order of decreasing numbers of components.

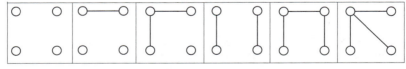

#9. (a) e, f, g, h, i, j (b) d (c) e (d) i, d, r (e) f, g, m, n, s, t (f) 4 (g) 7 (h) r, a, b, d, e, g, i, m

#11.

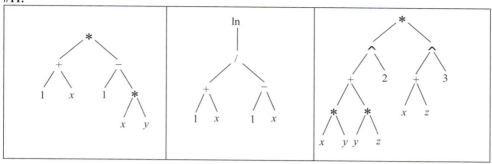

#13. (a) Since tournament trees are arranged to be balanced, trees with the leaves being the number of teams, it follows from Corollary 8.12 that the number of rounds of the tournament is $h = \text{ceil}(\log_2(25)) = 5$.

(b)   The number of byes will be $2^h - \ell = 32 - 25 = 7$.   (b)   The number of games = the number of internal vertices = $i = \ell - 1 = 24$.

#15. (a) From Theorem 8.11, a complete 5-ary tree of depth h has $(5^{h+1} - 1)/4$ vertices. This quantity first exceeds 163 when $h = 5$, so it follows that it would take at least five days for the gossip to spread to everyone.

(b) From the information given, we can conclude the following estimates for the number $P(n)$ of new people that learn the gossip on day n: $P(0) = 1,\ P(1) = 5,\ P(2) \geq 5 \cdot 2,\ P(3) \geq 5 \cdot 2 \cdot 2 = 5 \cdot 2^2,$

$P(4) \geq 5 \cdot 2^2 \cdot 2 = 5 \cdot 2^3, \cdots$ It follows that the total number of people who know the gossip after n days

is $\sum_{k=0}^{n} P(k) \geq 1 + \sum_{k=0}^{n-1} 5 \cdot 2^k = 1 + 5 \dfrac{2^n - 1}{2 - 1} = 1 + 5(2^n - 1),$ where we used the formula for the sum of a

geometric series. This quantity first exceeds 163 when $n = 6$, so it follows that it would take at most six days for the gossip to spread to everyone.

#17. (a) PreOrder $= [32,\ 14,\ 4, 12,\ 27,\ 20,\ 54,\ 47,\ 36,\ 52,\ 88]$; InOrder $= [12,\ 4,\ 14,\ 20,\ 27,$ 32, 36, 47, 52, 54, 88];

PostOrder $= [12,\ 4,\ 20,\ 27,\ 14,\ 36,\ 52,\ 47,\ 88,\ 54,\ 32]$

(b)   PreOrder $= [r, a, e, k, \ell, b, f, g, m, s, t, n, c, d, h, i, o, p, q, j]$;   InOrder $=$

$[k, \ell, e, a, r, f, s, t, m, g, n, b, c, h, d, o, i, p, q, j]$;

PostOrder $= [k, \ell, e, a, f, s, t, m, n, g, b, c, h, o, p, q, i, j, d, r]$

#19.

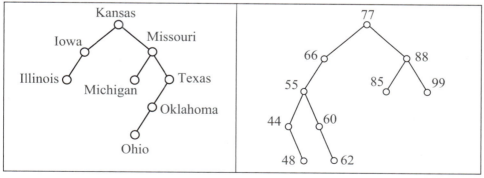

#21.

Vertex	Parent
r	NULL
a	*r*
b	*r*
c	*r*
d	*r*
e	*a*
f	*b*
g	*b*
h	*d*
i	*d*
j	*d*
k	*e*
ℓ	*e*
m	*g*
n	*g*
o	*i*
p	*i*
q	*i*
s	*m*
t	*m*

Vertex	Child 1	Child 2	Child 3	Child 4
r	*a*	*b*	*c*	*d*
a	*e*	NULL	NULL	NULL
b	*f*	*g*	NULL	NULL
c	NULL	NULL	NULL	NULL
d	*h*	*i*	*j*	NULL
e	*k*	*ℓ*	NULL	NULL
f	NULL	NULL	NULL	NULL
g	*m*	*n*	NULL	NULL
h	NULL	NULL	NULL	NULL
i	*o*	*p*	*q*	NULL
j	NULL	NULL	NULL	NULL
k	NULL	NULL	NULL	NULL
ℓ	NULL	NULL	NULL	NULL
m	*s*	*t*	NULL	NULL
Rest	NULL	NULL	NULL	NULL

#23. (a) The folder in which the file is stored.
(b) Two files/folders that lie in the same folder.
(c) A file/folder belonging to the folder.
(d) The nested sequence of folders containing the file/folder (corresponding to the folder path from the root).
(e) The contents of a folder.
(f) The number of folders needed to navigate down from the root directory to reach the file/folder.
(g) The maximum number of nested folders in which any file is stored.
(h) The nonempty folders.

#25. (a) 2 (b) A strategy that uses at most two weighings is on the right:

#27. (a) 16, 3 (b) A strategy that uses at most three weighings is below:

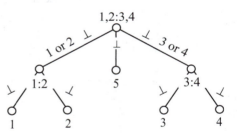

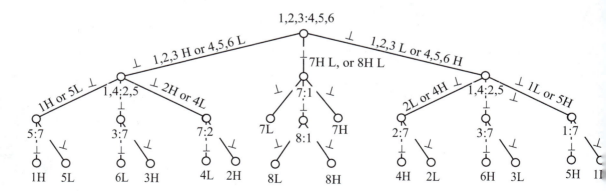

#29. (a) (i) $4 - 6 = -2$ (ii) $2*4 + 6 = 14$ (iii) $\sqrt{\dfrac{13+12}{6^2}} = 5/6$.

(b) (i) 3^y (ii) $(x - y) + z$ (iii) $\ln\left(\dfrac{xy}{1 + x^2}\right)$

(c) (i) $* + 1\ x - 1 * x\ y$ (ii) $\ln / + 1\ x - 1\ x$ (iii) $*\ ^\wedge + * x\ y * y\ z\ 2\ ^\wedge + x\ z\ 3$

(d) Same answers as (c).

(e) *Proof:* We use induction on the height h of a tree representing an algebraic expression.

Basis Step: $h = 0$: In this case, the representing tree consists of a root vertex alone, so must just be a number or a variable. The preorder will be a one-element vector with this same number/variable as its only component, so will be the same as the Polish notation of the expression.

Inductive Step: Assume that the result is valid for algebraic expression trees of height h, and consider an algebraic expression tree T of height $h + 1$. Now T will have either one or two child trees, depending on whether its root r represents a unitary or binary operator, respectively. The preorder Algorithm 8.4, outputs the root r first followed by the preorder traversals of these one or two trees, from left to right. By the induction hypotheses, these latter preorders coincide with the prefix notation of their corresponding algebraic expressions, and hence so does the preorder of T. $\square$

#31. *Proof:* By virtue of their definition, trees have no cycles. In particular, they have no cycles of odd length, so by Theorem 8.7, it follows that all trees are bipartite. $\square$

#33. (a) *Proof:* Assume that T is a tree. By definition, trees are connected, so any two vertices can always be joined by a path. Suppose that there were two different paths P, Q from a vertex u to another vertex v. By Proposition 8.5, we may assume that both of these paths are simple with no repeated vertices. Let w be the first vertex at which these paths split up, and let x be the first vertex after w at which they rejoin. Then the path $P(w,x)$ followed by $Q(x,w)$ would constitute a cycle in T, contradicting the fact that trees have no cycles. $\square$

(b) *Proof:* Assume that T is a simple graph satisfying (4). The existence of paths between any pair of vertices is equivalent to connectedness. We have left to show that T has no cycles. But any cycle C of T would give rise to multiple paths between vertices, e.g., if u and v are adjacent vertices of C, then the edge between them, and path obtained by following C in the opposite direction would be two different paths. Since this is not possible by (4), it follows that T can have no cycles. $\square$

(c) *Proof:* Assume that T is a simple graph satisfying (2). In order to show (5), we must show that every edge of T is a bridge. This amounts to showing that an n vertex graph with (at most) $n - 2$ edges cannot be connected, and this is a special case of the result of Exercise for the Reader 8.27.

(d) *Proof:* Assume that T is a simple graph satisfying (5). In order to show (2), we must show that T has exactly $n - 1$ edges. It follows from the result of Exercise for the Reader 8.27 that T must have at least $n - 1$ edges. We will prove by induction on the number n of vertices of T that if T is connected and every edge is a bridge, than T has at most $n - 1$ edges.

Basis Step: Since any simple graph T satisfying the hypothesis must have at least three vertices, the basis step begins the induction with $n = 3$. A simple graph with three edges and vertices must be K_3, which has no bridges.

Induction Step: We assume the statement is true for all simple graphs with n or fewer vertices, where $n \geq 3$ is an integer, and we suppose that T is a simple connected graph on $n + 1$ vertices such that every edge is a bridge. For a contradiction, we will assume that T has at least $n + 1$ edges. If we take any edge e of T, since all edges are bridges, it follows that the vertex deletion graph will consist of two connected components T_1, T_2 whose numbers of vertices $n_1 = |V(T_1)|$, $n_2 = |V(T_2)|$ satisfy $n_1 + n_2 = n + 1$, and whose numbers of edges $m_1 = |E(T_1)|$, $m_2 = |E(T_2)|$ satisfy $m_1 + m_2 \geq n$. It follows that we must have either $m_1 \geq n_1$, or $m_2 \geq n_2$, which would lead to a contradiction with the inductive hypothesis for the corresponding graph T_1 or T_2. $\square$

#35. (a) *Proof:* Assume that G is a connected simple graph with n edges and n vertices. By Theorem 8.10, G has too many edges to be a tree, so it must have a cycle C. Since removing any edge e of C from G will not destroy the connectivity of G, it follows that $G - e$ is a tree. If G had another cycle $D \neq C$, then some edge f of D does not belong to C, some edge e of C does not belong to D, and the preceding argument shows that $G - \{e, f\}$ is connected. But this simple graph has n vertices and only $n - 2$ edges, so we have a contradiction with the result of Exercise for the Reader 8.27. □

(b) Not necessarily: Counterexample: The graph G obtained by starting with the 4-cycle C_4 and adding one edge e between a pair of opposite vertices (say vertex #1 and vertex #3). This graph has four vertices, five edges but has exactly three cycles: the 4-cycle C_4, and the two 3-cycles that share the edge e, and contain different third vertices.

#37. If T is a tree on n vertices, then T has $n - 1$ edges, so the complement graph $\sim T$ will have $C(n, 2) - [n - 1] = n(n-1)/2 - (n-1)$ edges. In order for $\sim T$ to be a tree, this number must equal $n - 1$, which is equivalent to the equation $n^2 - 5n + 4 = (n-4)(n-1) = 0$. Thus the only possibilities for n are 1 and 4. When $n = 1$, there is only one simple graph: the graph with a single vertex and no edges, and it is its own complement. When $n = 4$, there are two trees: P_4, and the tree with degree sequence $(3,1,1,1)$. The former is its own complement, but the complement of the latter is K_3, which is not a tree.

#39. *Proof:* Following the suggestion and letting n_i denote the number of vertices of degree i, the equation $n_1 + n_2 + n_3 = n$, simply breaks down the numbers of vertices of each degree with the total number of vertices. The equation $n_1 + 2n_2 + 3n_3 = 2(n-1)$ is simply the handshaking theorem, the number on the left being the sum of all vertex degrees, and the number on the right being twice the number of edges. Combining these two equations yields:

$$n_1 + 2n_2 + 3n_3 = 2(n_1 + n_2 + n_3 - 1) = 2n_1 + 2n_2 + 2n_3 - 2 \Rightarrow n_3 = n_1 - 2.$$

Substituting the last equation into $n_1 + n_2 + n_3 = n$, yields $n_2 + 2n_3 + 2 = n \Rightarrow 2n_3 \leq n - 2 \Rightarrow n_3 \leq \lfloor (n-2)/2 \rfloor$, as asserted. □

#41. *Proof:* We proceed by induction on the number n of vertices.

Basis Step: When $n = 1$, the only tree is a single isolated vertex with no edges and hence with degree sequence (0). The equation $\sum_{i=1}^{n} d_i = 2n - 2$ becomes $d_1 = 0$.

Induction Step: We assume the statement is true for some positive integer n, and must show that it remains true for $n + 1$. Suppose that $(d_1, d_2, \cdots, d_n, d_{n+1})$ is the degree sequence of a tree T. We may assume that the degrees are listed in decreasing order. We know from Theorem 8.8 that T has at least two leaves. Let u be a leaf of T and consider the vertex deletion graph $T - u$. This graph is still a tree, and has degree sequence $(\tilde{d}_1, \tilde{d}_2, \cdots, \tilde{d}_n)$, where all but one of these degrees satisfies $\tilde{d}_i = d_i$, and the exception (corresponding to the neighbor of u) satisfies $\tilde{d}_i = d_i - 1$. By the inductive hypothesis, we know that $\sum_{i=1}^{n} \tilde{d}_i = 2n - 2$. From the last equation, it follows that $\sum_{i=1}^{n+1} d_i = \sum_{i=1}^{n} \tilde{d}_i + 2 = 2n - 2 + 2 = 2(n+1) - 2$, as desired. For the converse, we assume that the sequence of positive integers $(d_1, d_2, \cdots, d_n, d_{n+1})$ satisfies $\sum_{i=1}^{n+1} d_i = 2(n+1) - 2$. We may assume that the degrees are listed in decreasing order. It follows that $d_{n+1} = 1$, and for some index j, $d_j \geq 2$. We define a sequence $(\tilde{d}_1, \tilde{d}_2, \cdots, \tilde{d}_n)$, by setting $\tilde{d}_i = \begin{cases} d_i, & \text{if } i \neq j \\ d_j - 1, & \text{if } i = j \end{cases}$. Since $\sum_{i=1}^{n} \tilde{d}_i = 2n - 2$, it follows from the induction hypothesis that there exists a tree $\tilde{T}$ with degree sequence $(\tilde{d}_1, \tilde{d}_2, \cdots, \tilde{d}_n)$. By adding one new vertex with an edge connected to the jth vertex of $\tilde{T}$ we arrive at a tree with degree sequence $(d_1, d_2, \cdots, d_n, d_{n+1})$. □

CHAPTER 9: Section 9.1:

#1. (a) All vertices of both graphs have even degree so by Theorem 9.1 both are Eulerian and (b) have an Euler path.

#3. Since G has four vertices of odd degree it follows from Theorem 9.1 that G is neither Eulerian nor has an Euler path. Since all vertices of H have even degree, Theorem 9.1 tells us that H is Eulerian and has an Euler path by Theorem 9.1.

#5. (a) P_n, $n \geq 2$ is never Eulerian but always has an Euler path.

(b) K_n, $n \geq 2$ will be Eulerian and will have an Euler path when n is odd. When $n = 2$, it has (only) an Euler path, and for larger even values of n K_n is not Eulerian and does not have an Euler path.

#7. For Exercise #1: Euler tour for G: $< a,b,c,d,a,e,f,g,c,h,i,j,k,h,g,k,e,j,a >$

Euler tour for H: $< a,b,c,f,c,f,g,b,g,h,a >$

For Exercise #3: Euler tour for H: $< a,h,a,b,g,c,f,d,e,f,g,h,b,c,d,e,c,f,b,g,a >$

#9. $\vec{G}$ is neither Eulerian nor has an Euler path since it has four vertices of odd degree (Theorem 9.2). $\vec{H}$ is Eulerian (and has an Euler path) by Theorem 9.2. Euler tour for $\vec{H}$:

$< a,e,a,b,e,b,c,e,c,d,e,d,a >$

#11.

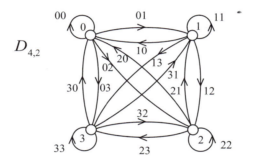

Euler tour in $D_{4,2}$: $< 0,0,1,1,2,2,3,3,2,1,3,1,0,2,0,3,0 >$

(4,2)-de Bruijn sequence: $< 0,0,1,1,2,2,3,3,2,1,3,1,0,2,0,3 >$

#13. G is Hamiltonian; Hamilton tour: $< a,g,b,c,d,e,f,a >$

H is Hamiltonian; Hamilton tour: $< a,b,c,d,e,f,a >$

#15. G is Hamiltonian; Hamilton tour: $< a,b,c,d,e,f,g,a >$

H is Hamiltonian. Here are the details of a construction of a Hamilton tour using Proposition 9.8: Any Hamilton tour C would have to contain the edges *ef*, *ed*, *gf*, *gh*. Since *f* has two used incident edges, we may delete the edges *fj*, *af*.   Since C cannot contain any smaller cycles; we must have *ji* $\in C$, and exactly one of *dj*, *jh* $\in C$. If *dj* $\in C$, We can delete *di*, so *ib* $\in C$, and thus we can delete *hb*. Thus vertex *h* has only one edge of C adjacent to it—a contradiction. But if *jh* $\in C$, then we can delete *hb*, and since *cb* can also be deleted (lest C contain the 3-cycle $< a,b,c,a >$), we must have *ib* $\in C$. So we may delete *di*, and conclude *dc* $\in C$. We now have a Hamilton tour: $< a,b,i,j,h,g,f,e,d,c,a >$

#17. We shall prove the following result (asked for in Part (d)); from it, the results of Parts (a) through (c) will follow.

Proposition: If n and m are positive integers, the graph $R_{m,n}$ is Hamiltonian if, and only if at least one of m or n is odd.

Proof: Part I: (*Sufficiency*)   Since   clearly

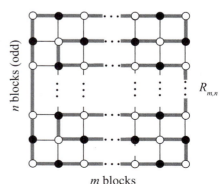

$R_{n,m} \cong R_{m,n}$, we may assume that n is even. A generic Hamilton cycle can now be constructed as shown in the figure.

Part II: (*Necessity*) The "checkerboard" coloring of the vertices of $R_{m,n}$ (see figure) shows it to be bipartite. The number of vertices of $R_{m,n}$ is $(m+1)(n+1)$. Thus, when both n and m are even, $R_{m,n}$ is a bipartite graph with an odd number of vertices, so it cannot be Hamiltonian (since a Hamilton cycle would be a cycle of odd length, and bipartite graphs cannot have these). □

(e) The 6 by 6 grid of students problem is represented by $R_{5,5}$, and a rearrangement would correspond to moving students along a Hamilton path, so by the proposition such a move is possible, but is not possible for a 7 by 7 grid.

#19. Since all of these graphs were shown to be Hamiltonian, all of the Hamilton tours give rise to Hamilton paths, for example, by deleting the last vertex.

#21. The Petersen graph can be thought of as two copies of C_5 (the inside vertices and the outside pentagon vertices) whose vertices are joined by five edges (but not preserving the order of the cycles). In light of this symmetry, it suffices to show that a single vertex deletion graph is Hamiltonian.   Such a Hamilton tour is shown in the figure on the right on the vertex deletion graph obtained from deleting the uppermost vertex on the standard drawing of the Petersen graph.

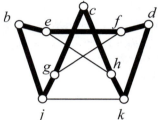

#23. (a)

(b) K_4 is Hamiltonian since it contains a copy of C_4; but is not Eulerian since every vertex has degree 3 (Theorem 9.1)

(c) C_3 (d) P_3 (or any tree)

#25. If v is any cut vertex, just apply Proposition 9.7 with $S = \{v\}$.

#27. Fleury's Algorithm gives the following Euler path: $< a,b,a,c,a,d,c,b,e^*,f,a,f^*,d,f^*,e^*,c^* >$.
The asterisks indicate the locations where bridges had to be chosen.

#31.   Theorem 8.7 tells us that bipartite graphs cannot contain cycles of odd length.   Since a Hamiltonian graph contains a cycle passing through all of its vertices, it follows that the number of vertices must be even.

Section 9.2:

#1. (a) $\{b,d\},\{b,g\},\{a,f\},\{c,f\}$ (b) $\{a,f\}$; $\{b,d\},\{b,g\},\{a,f\},\{a,d\},\{a,g\}$

(c) $\{a,d\},\{a,g\},\{b,f\},\{c,d\},\{c,g\}$ (d) $\{b,f\}$; $\{a,d\},\{a,g\},\{b,d\},\{b,g\},\{c,d\},\{c,g\}$

#3. The Prim spanning trees for Parts (a), (b), (c) are shown in the figures below with bold edges. Note that all of these trees are the same, but the discovery numbers are different.

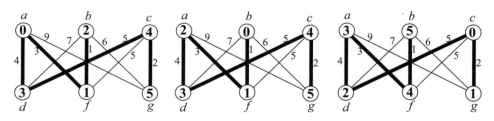

#5. The Prim spanning trees for Parts (a), (b), (c) are shown in the figures below with bold edges. Note that all of these trees are the same, but the discovery numbers are different.

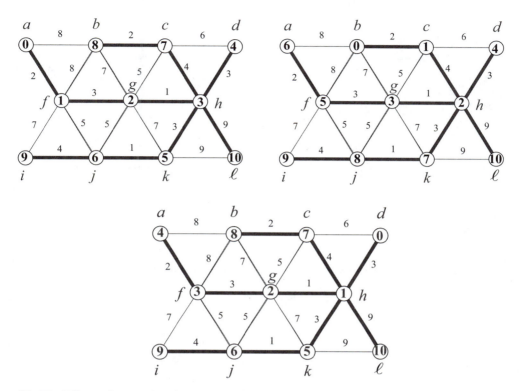

#7. The Dijkstra distance trees for Parts (a), (b), (c) are shown in the figure below with bold edges. Discovery numbers are indicated in bold in the vertex circles, distances to the starting vertex are indicated in square brackets, and the minimum distance tree's edges are drawn with thick edges with directions going from parent to child.

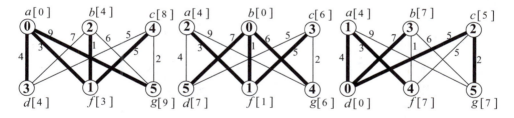

#9. (a) through (c): The Dijkstra distance trees for Parts (a), (b), (c) are shown in the figures below with bold edges. Discovery numbers are indicated in bold in the vertex circles, distances to the starting vertex are indicated in square brackets, and the minimum distance tree's edges are drawn with thick edges with directions going from parent to child.

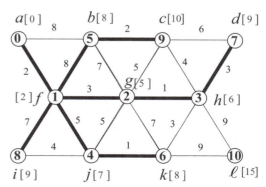

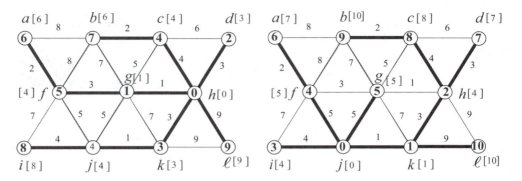

(d) The result of the directed version of Dijkstra's algorithm with starting vertex *a* is described in the following figure:

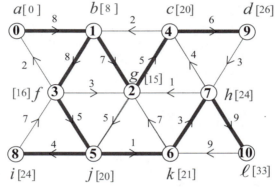

#11. For each part, the figures below show the resulting depth-first search tree (left) and the breadth-first search tree (right), with root (starting) vertex *a*. Discovery numbers are indicated in bold and tree edges are drawn with thick lines.

(a)

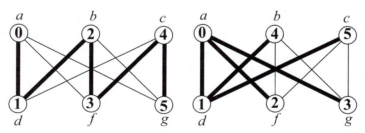

(b)

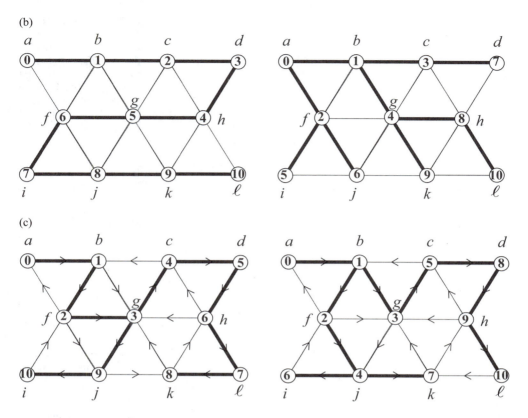

(c)

#13. (a) One such graph drawing is shown on the right.

(b) The brute-force method (which should be done with a computer) shows that the minimum possible distance of a tour is 18 miles. There are four tours that realize this minimum: <1,6,5,4,2,3,1>, <1,3,6,5,4,2,1>, and the corresponding reverse order tours.

(c) The nearest insertion heuristic will construct the tour: <1,3,6,5,4,2,1> which is one of the exact solutions (of minimum distance 18) of the TSP (see the answer to Part (b)). So the heuristic solution's excess driving distance is 0% over that of the optimal solution.

(d) Applying the cheapest insertion heuristic (Algorithm 9.10(c)) starting with vertex 1 (Home) produces the following tour <1,6,5,4,3,2,1> that has a distance of 24 miles; which is 1/3 =33 1/3% more than the distance of an optimal tour.

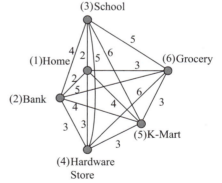

#15. (a) The problem is to find the shorted path in the weighted digraph of Figure 9.41 from vertex #0 to vertex #5 that has minimum length. We first need to use the data given in the problem to find the edge weights. From the way edges are priced it is clear that the weight of any edge in such a maintenance graph depends only on its length. We go through a few sample edge weight calculations to help clarify this.

Edge (0,1): This corresponds to trading in the truck after the first year. The total cost of this (from time zero to time one year) are $600 maintenance + $20,000 purchase price of new truck − $16,000 trade-in = $4600.

Edge (1,2): This edge corresponds to trading in, at the end of the second year, the truck that was bought new at the end of the first year. This one-year-old truck would have had $600 maintenance costs and Otis would need to spend $20,000 to buy the new truck at time 1 year, but get back $16,000 from trading in his one-year-old truck at time two years. So the net cost of this edge would be $4600 (the same as the edge (0,1)).

Edge (2,5): This edge corresponds to trading in, at the end of the fifth year, the truck that was bought new at the end of the second year. Over the three years, the maintenance would costs would be $600 for the first year of driving, $1500 for the second year, and $2000 for the third year. Since this edge terminates at vertex #5, we take the original purchase cost $20,000 of the truck, less the final trade in value (of the three-year-old truck) $11,000. So the net cost of this edge would be $4100 maintenance plus $9000 (original purchase less final trade in), or $13,100.

The weights of all (directed) edges are shown in the figure below:

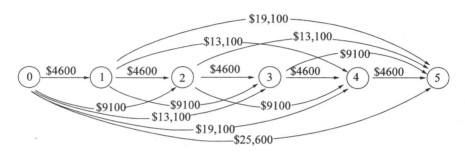

If we apply Dijkstra's algorithm we will easily find the a shortest path from vertex #0 to vertex #5 is <0, 2, 5>, which yields a minimum cost of $22,000.

(b)   In general, the vertex set of such a maintenance digraph will be $V = \{0,1,2,\cdots,n\}$, and the corresponding set of directed edges will be $E = \{(i, j) \mid 0 \le i < j \le n\}$. It follows that the number of edges is just the number of 2-combinations of a set of $n+1$ objects, i.e., $C(n+1,2) = n(n+1)/2$. For example, in case $n = 5$, this gives $C(6,2) = 15$, which nicely checks with Figure 9.41.

#17.   Following the suggestion, we work with the edge-weighted graph whose weights are $\tilde{W}(i, j) = -\log_{10} W(i, j)$. If $< v_0, v_1, \cdots, v_n >$ is any path in the network, then it will fail only if one of its links fails. By independence, the probability that the path does not fail is the product of the probabilities that each of its links does not fail, which is $W(v_0, v_1) \cdot W(v_1, v_2) \cdot \cdots \cdot W(v_{n-1}, v_n)$. Thus we seek to find a path from vertex d to vertex c that will minimize this (failure) probability. Since $0 < W(v_0, v_1) \cdot W(v_1, v_2) \cdot \cdots \cdot W(v_{n-1}, v_n) < 1$, and the function $t \mapsto -\log_{10}(t)$ is positive and decreasing for $t \in (0,1)$, minimizing a product $W(v_0, v_1) \cdot W(v_1, v_2) \cdot \cdots \cdot W(v_{n-1}, v_n)$ is equivalent to maximizing

$$-\log_{10}[W(v_0, v_1) \cdot W(v_1, v_2) \cdot \cdots \cdot W(v_{n-1}, v_n)]$$
$$= -\log_{10}(W(v_0, v_1)) - \log_{10}(W(v_1, v_2)) - \cdots - \log_{10}(W(v_{n-1}, v_n))$$
$$= \tilde{W}(v_0, v_1) + \tilde{W}(v_1, v_2) + \cdots + \tilde{W}(v_{n-1}, v_n).$$

We have thus shown that a most reliable path in the original communication network from vertex d to vertex c will simply be a shortest path in the corresponding edge-weighted network with weights $\tilde{W}(i, j)$. The figure on the right shows the shortest path (thickened edges) from vertex d to vertex c resulting from applying Dijkstra's algorithm with starting vertex d.   The probability that this

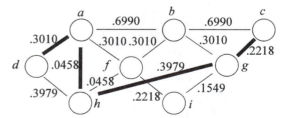

path does not fail in the original network is obtained by multiplying the corresponding edge weights in the original graph: $0.5 \cdot 0.9 \cdot 0.4 \cdot 0.6 = 0.108$.

#19.   (a)   Change the "Next Edge Selection" scheme to the following: A boundary edge from BdyEdges of maximum weight is selected. Proposition 9.9 can be correspondingly modified and proved (the reader should do this).

(b) The maximum spanning trees produced by the algorithm along with the vertex discovery numbers are shown below (with thickened edges).
(i)

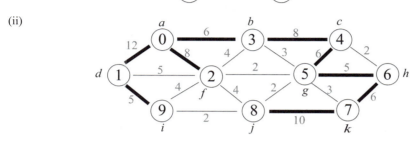

(ii)

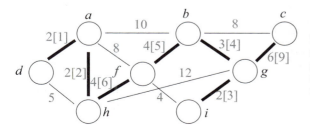

#21. The figure below shows a very simple example of how Prim's algorithm can fail for digraphs. If we start the algorithm at vertex a, it produces the tree with the single directed edge (a,b), which is not a minimum spanning tree since the tree with the oppositely directed edge is 90% lighter.

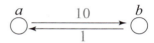

#23. (a) The table below gives the edge-data sorted by weight. After the table we give the results of the iterations of Kruskal's algorithm, and then provide an illustration of the resulting minimum spanning tree

i	1	2	3	4	5	6	7	8	9	10	11	12	13
e_i	$\{a,d\}$	$\{a,h\}$	$\{g,i\}$	$\{b,g\}$	$\{b,f\}$	$\{f,h\}$	$\{f,i\}$	$\{d,h\}$	$\{c,g\}$	$\{a,f\}$	$\{b,c\}$	$\{a,b\}$	$\{g,h\}$
w_i	2	2	2	3	4	4	4	5	6	8	8	10	12

Step 1: n=8
$\mathcal{E} = \{a,d\}$ index=2

Step 2, 1st iteration, $\mathcal{E} = \{\{a,d\},\{a,h\}\}$ index=3

　　　　2nd iteration, $\mathcal{E} = \{\{a,d\},\{a,h\},\{g,i\}\}$ index=4

　　　　3rd iteration, $\mathcal{E} = \{\{a,d\},\{a,h\},\{g,i\},\{b,g\}\}$ index=5

　　　　4th iteration, $\mathcal{E} = \{\{a,d\},\{a,h\},\{g,i\},\{b,g\},\{b,f\}\}$ index=6

　　　　5th iteration, $\mathcal{E} = \{\{a,d\},\{a,h\},\{g,i\},\{b,g\},\{b,f\},\{f,h\}\}$ index=7

　　　　6th iteration, $\mathcal{E} = \{\{a,d\},\{a,h\},\{g,i\},\{b,g\},\{b,f\},\{f,h\},\{c,g\}\}$ index=8

The minimal spanning tree is indicated by thickened edges. The numbers in square brackets next to the edge weights give the order numbers when the corresponding edge was added to the tree.

For Parts (b) and (c) the figures below

show the minimum spanning trees that were produced by Kruskal's algorithm.

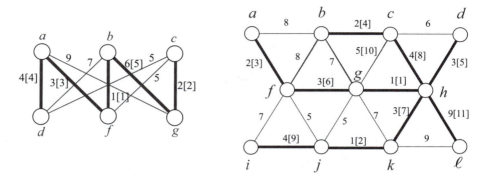

#25. Let e_0 be the inputted edge (that must be included in the spanning tree), set $\mathscr{E} = \{e_0\}$, and let the remaining edges (if any) be denoted by $e_1, e_2, \cdots, e_m$. Modify Step 1 of Kruskal's Algorithm 9.11 to the following: If there are no remaining edges, output $T = \mathscr{E}$ and exit the algorithm. Otherwise, we begin by sorting the edges $e_1, e_2, \cdots, e_m$ in such a way that the corresponding weights of these edges $w_1, w_2, \cdots, w_m$ satisfy: $w_1 \le w_2 \le \cdots \le w_m$. Update $\mathscr{E} = \{e_0, e_1\}$. If $m = 1$, output $T = \mathscr{E}$ and exit the algorithm. Otherwise, initialize Index = 2, Set $G_{\mathscr{E}}$ to be the subgraph of G determined by the edges of $\mathscr{E}$ and their endpoints.

#29. Let the vertices of the inputted graph be $v_1, v_2, \cdots, v_n$. The following algorithm allows components to be discovered very efficiently using depth-first searches.

INITIALIZE: CompNumber = 1; COMP = a length n vector of zeros.
(*Once the algorithm is complete, the ith component of vector COMP will be the positive integer corresponding to the component number of vertex v_i.*)

WHILE min(COMP) = 0 (*vertices remain whose component numbers have not yet been determined*)
  SET Index = the smallest index for which COMP(Index) = 0.
  Apply the depth-first search algorithm starting with vertex v_{Index}.

  SET Disc = the vertices discovered in this search (including v_{Index}).

  (*The vertices of Disc will be those reachable from v_{Index} and hence be the component $C(v_{\text{Index}})$.*)

  UPDATE the components of COMP to be CompNumber at the indices corresponding to Disc
  UPDATE CompNumber = CompNumber + 1
END WHILE
This program allows components to be discovered very efficiently using depth-first searches.

#31. *Proof:* *Step 1:* We assume that v is a cut vertex of the connected graph G. This means (Definition 8.28) that the graph $H \triangleq G - v$ has at least two components. Let u, w be two vertices of H that belong to different components of H. Any path in G joining u to w must pass through the vertex v, since if it did not, all of its vertices would belong to H, and this would mean u, w would belong to the same component of H.

Step 2: To prove the converse, we assume the existence of two vertices u, w of G, such that any path that joins them must pass through the vertex v. As in Step 1, it follows that when viewed as vertices of the graph $H \triangleq G - v$, u, w are not reachable from one another in H, and hence must lie in different components of H. Thus $H \triangleq G - v$, so v is a cut vertex of G. □

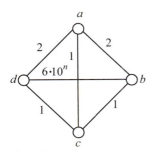

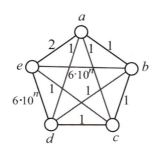

#35. (a) The graph presented in Example 9.18 gives the idea for the 4-vertex graph example shown on the left below. If we start at vertex a and apply the nearest neighbor heuristic, we obtain the following tour: $<a, c, b, d, a>$, which has weight $6 \cdot 10^n + 4$. This is more than 10^n times greater than 6, the length of the optimal TSP tour. (b) If we apply the nearest insertion heuristic to the 5-vertex TSP problem shown on the right side of the figure, starting at vertex a, the construction proceeds as follows: $<a>$, $<a, b, a>$, $<a, b, c, a>$, $<a, b, c, d, a>$. At this point, no matter where vertex e gets added, an edge of cost $6 \cdot 10^n$ will get added. Thus the cost of the tour will be $6 \cdot 10^n + 4$, which is more than 10^n times greater than 6, the length of the optimal TSP tour. Although we made selections (in cases of ties) using lexicographic order, it is clear that the same example would work regardless of which priority scheme is used.

Section 9.3:

#1. (a) Not admissible: The inflow to vertex b is 6, but the outflow is 8.

(b) Admissible; value of flow is 16.

#3. (a) Capacity of $(\{s\}, \{a,b,c,t\}) = 30$, capacity of $(\{s,a\}, \{b,c,t\}) = 42$, capacity of $(\{s,a,b,c\}, \{t\})$ $= 36$.

(b) Capacity of $(\{s\}, \sim \{s\}) = 30$, capacity of $(\{s,a,e,d\}, \{b,c,d,t\}) = 54$, capacity of $(\{s,e\}, \sim \{s,e\})$ $= 42$.

#5. (a) Value of maximum flow $= 30$; corresponding minimum cut is $(\{s\}, \sim \{s\})$.

(b) Value of maximum flow $= 26$; corresponding minimum cut is $(\sim \{t\}, \{t\})$.

#7. (a) *First iteration labels*: (from the initial flow) Label$(a) = (s, +, 14)$, Label$(b) = (s, +, 6)$, Label(c) $= (a, +, 8)$, Label$(t) = (a, +, 6)$. The resulting augmented flow is shown in the figure; it has value 16. *Second iteration labels*: (from the first iteration flow) Label$(a) = (s, +, 8)$, Label$(b) = (s, +, 6)$, Label(c) $= (a, +, 8)$, Label$(t) = (b, +, 6)$. The resulting augmented flow is shown in the figure; it has value 22. *Third iteration labels*: (from the second iteration flow) Label$(a) = (s, +, 8)$, Label$(b) = (a, +, 8)$, Label$(c) = (a, +, 8)$, Label$(t) = (b, +, 4)$. The resulting augmented flow is shown in the figure; it has value 26.

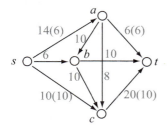

Iteration #1 Flow

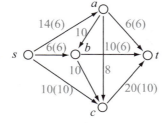

Iteration #2 Flow

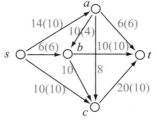

Iteration #3 Flow

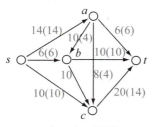

Iteration #4 Flow

Fourth iteration labels: (from the third iteration flow) Label$(a) =$ $(s, +, 4)$, Label$(b) = (a, +, 4)$, Label$(c) = (a, +, 4)$, Label$(t) = (c, +,$ $4)$. The resulting augmented flow is shown in the figure; it has value 30.

In the fifth iteration, apart from the vertex s (which always gets labeled) nothing else gets labeled, but it was quite clear that the iteration #4 flow was maximum since all edges from s are flowing at capacity.

(b) *First iteration labels*: (from the initial flow) Label(a) = (s, +, 14), Label(b) = (s, +, 2), Label(c) = (s, +, 10), Label(t) = (a, +, 6). The resulting augmented flow is shown in the figure; it has value 10.
Second iteration labels: (from the first iteration flow) Label(a) = (s, +, 8), Label(b) = (s, +, 2), Label(c) = (s, +, 10), Label(t) = (b, +, 2). The resulting augmented flow is shown in the figure; it has value 12.
Third iteration labels: (from the second iteration flow) Label(a) = (s, +, 8), Label(c) = (s, +, 10); Label(b) = (a, +, 8), Label(t) = (c, +, 10). The resulting augmented flow is shown in the figure; it has value 22.
Fourth iteration labels: (from the third iteration flow) Label(a) = (s, +, 8), Label(b) = (a, +, 8), Label(c) = (a, +, 8); Label(t) = (b, +, 4). The resulting augmented flow is shown in the figure; it has value 26.
Fifth iteration labels: (from the fourth iteration flow) Label(a) = (s, +, 4), Label(b) = (a, +, 4), Label(c) = (a, +, 4); Label(t) = (c, +, 4). The resulting augmented flow is shown in the figure; it has value 30.
The sixth iteration detects the maximum flow.

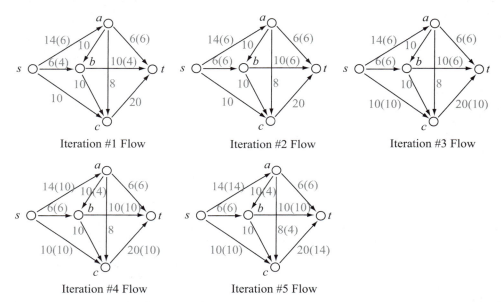

Iteration #1 Flow   Iteration #2 Flow   Iteration #3 Flow

Iteration #4 Flow   Iteration #5 Flow

(c) The iterations look very much like those of Part (b), except that in iteration #2 the flow gets augmented by 6 rather than 2.

(d) *First iteration labels*: (from the initial flow) Label(c) = (s, +, 20); Label(d) = (c, +, 18), Label(e) = (c, +, 8); Label(t) = (d, +, 10). The resulting augmented flow is shown in the figure; it has value 20.
Second iteration labels: (from the first iteration flow) Label(c) = (s, +, 10); Label(d) = (c, +, 8), Label(e) = (c, +, 8); Label(a) = (e, −, 5); Label(b) = (e, +, 6); Label(t) = (e, +, 1). The resulting augmented flow is shown in the figure; it has value 21.

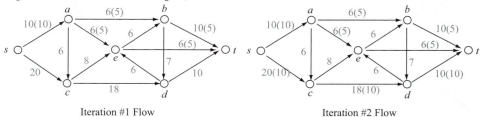

Iteration #1 Flow   Iteration #2 Flow

Third iteration labels: (from the second iteration flow) Label(*c*) = (*s*, +, 9); Label(*d*) = (*c*, +, 7),

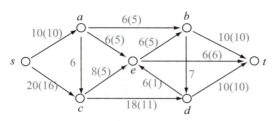

Label(*e*) = (*c*, +, 8); Label(*a*) = (*e*, −, 5), Label(*b*) = (*e*, +, 6), Label(*t*) = (*b*, +, 5). The resulting augmented flow is shown in the figure; it has value 26.

Fourth iteration labels: (from the third iteration flow) Label(*c*) = (*s*, +, 4); Label(*d*) = (*c*, +, 4), Label(*e*) = (*c*, +, 4); Label(*a*) = (*e*, −, 5), Label(*b*) = (*e*, +, 1). Since *t* does not get labeled, the fourth iteration detects the maximum flow = 26 (third iteration flow).

Iteration #3 Flow

(e) It takes six iterations for the maximum flow of 26 to be found (and detected).
(f) It takes six iterations for the maximum flow of 26 to be found (and detected).

#9. (a) *First iteration labels:* (from the initial flow) Label(*a*) = (*b*, +, 3), Label(*b*) = (*s*, +, 9), Label(*c*) = (*b*, +, 8), Label(*d*) = (*b*, +, 6), Label(*e*) = (*s*, +, 6), Label(*f*) = (*e*, +, 6), Label(*g*) = (*s*, +, 20), Label(*t*) = (*c*, +, 8). Corresponding flow value = 18.
Second iteration labels: (from the first iteration flow—a drawing of which should be helpful) Label(*a*) = (*b*, +, 1), Label(*b*) = (*s*, +, 1), Label(*c*) = (*a*, +, 1), Label(*d*) = (*b*, +, 1), Label(*e*) = (*s*, +, 6), Label(*f*) = (*e*, +, 6), Label(*g*) = (*s*, +, 20), Label(*t*) = (*d*, +, 1). Corresponding flow value = 19.
Third iteration labels: (from the second iteration flow—a drawing of which should be helpful) Label(*b*) = (*d*, −, 1), Label(*c*) = (*d*, +, 6), Label(*d*) = (*e*, +, 6), Label(*e*) = (*s*, +, 6), Label(*f*) = (*e*, +, 6), Label(*g*) = (*s*, +, 20), Label(*t*) = (*d*, +, 5). Corresponding flow value = 24.
Note: If we continue, the maximum flow of 44 will be detected in the 9th iteration. The algorithm will also produce the following minimum cut: $(\{s,d,e,f,g\},\{a,b,c,t\})$.

(b) *First iteration labels:* (from the initial flow) Label(*a*) = (*s*, +, 10), Label(*b*) = (*s*, +, 9), Label(*c*) = (*a*, +, 2), Label(*e*) = (*s*, +, 6), Label(*g*) = (*s*, +, 15), Label(*t*) = (*a*, +, 10). Corresponding flow value = 15.
Second iteration labels: (from the first iteration flow—a drawing of which should be helpful) Label(*a*) = (*b*, +, 3), Label(*b*) = (*s*, +, 9), Label(*c*) = (*b*, +, 8), Label(*d*) = (*b*, +, 6), Label(*e*) = (*s*, +, 6), Label(*f*) = (*e*, +, 6), Label(*g*) = (*s*, +, 15), Label(*t*) = (*c*, +, 8). Corresponding flow value = 23.
Third iteration labels: (from the second iteration flow—a drawing of which should be helpful) Label(*a*) = (*b*, +, 1), Label(*b*) = (*s*, +, 1), Label(*c*) = (*a*, +, 1), Label(*d*) = (*b*, +, 1), Label(*e*) = (*s*, +, 6), Label(*f*) = (*e*, +, 6), Label(*g*) = (*s*, +, 15), Label(*t*) = (*d*, +, 1). Corresponding flow value = 24.
Note: If we continue, the maximum flow of 44 will be detected in the 10th iteration. The algorithm will also produce the following minimum cut: $(\{s,d,e,f,g\},\{a,b,c,t\})$.

(c) *First iteration labels:* (from the initial flow) Label(*a*) = (*b*, +, 3), Label(*b*) = (*s*, +, 9), Label(*c*) = (*b*, +, 8), Label(*d*) = (*b*, +, 6), Label(*e*) = (*s*, +, 6), Label(*f*) = (*e*, +, 6), Label(*g*) = (*s*, +, 20), Label(*t*) = (*c*, +, 8). Corresponding flow value = 18.
Second and third iterations now same as first and second iterations of Part (a).
Note: If we continue, the maximum flow of 44 will be detected in the 10th iteration. The algorithm will also produce the following minimum cut: $(\{s,d,e,f,g\},\{a,b,c,t\})$.

(d) *First iteration labels:* (from the initial flow) Label(*a*) = (*c*, +, 4), Label(*b*) = (*c*, +, 4), Label(*c*) = (*s*, +, 4), Label(*d*) = (*s*, +, 7), Label(*f*) = (*d*, +, 4), Label(*g*) = (*s*, +, 4), Label(*h*) = (*s*, +, 3), Label(*i*) = (*g*, +, 4), Label(*t*) = (*h*, +, 3). Corresponding flow value = 7.
Second iteration labels: (from the first iteration flow—a drawing of which should be helpful) Label(*a*) = (*c*, +, 4), Label(*b*) = (*c*, +, 4), Label(*c*) = (*s*, +, 4), Label(*d*) = (*s*, +, 7), Label(*f*) = (*d*, +, 4), Label(*g*) = (*s*, +, 4), Label(*h*) = (*d*, +, 4), Label(*i*) = (*g*, +, 4), Label(*t*) = (*b*, +, 4). Corresponding flow value = 11.
Third iteration labels: (from the second iteration flow—a drawing of which should be helpful) Label(*a*) = (*c*, +, 3), Label(*c*) = (*s*, +, 3), Label(*d*) = (*s*, +, 7), Label(*f*) = (*d*, +, 4), Label(*g*) = (*s*, +, 4), Label(*h*) = (*d*, +, 4), Label(*i*) = (*g*, +, 4), Label(*t*) = (*f*, +, 4). Corresponding flow value = 15.
Note: If we continue, the maximum flow of 21 will be detected in the 8th iteration. The algorithm will also produce the following minimum cut: $(\sim\{t\},\{t\})$.

(e) *First iteration labels*: (from the initial flow) Label(a) = (s, +, 4), Label(b) = (a, +, 2), Label(c) = (s, +, 1), Label(d) = (s, +, 7), Label(g) = (s, +, 4), Label(h) = (s, +, 3), Label(t) = (a, +, 4).   Corresponding flow value = 7.

Second iteration labels: (from the first iteration flow—a drawing of which should be helpful) Label(a) = (c, +, 1), Label(b) = (c, +, 1), Label(c) = (s, +, 1), Label(d) = (s, +, 7), Label(f) = (d, +, 4), Label(g) = (s, +, 4), Label(h) = (s, +, 3), Label(i) = (g, +, 4), Label(t) = (h, +, 3).   Corresponding flow value = 10.

Third iteration labels: (from the second iteration flow—a drawing of which should be helpful) Label(a) = (c, +, 1), Label(b) = (c, +, 1), Label(c) = (s, +, 1), Label(d) = (s, +, 7), Label(f) = (d, +, 4), Label(g) = (s, +, 4), Label(h) = (d, +, 4), Label(i) = (g, +, 4), Label(t) = (b, +, 1).   Corresponding flow value = 11.

Note: If we continue, the maximum flow of 21 will be detected in the 9th iteration. The algorithm will also produce the following minimum cut: ($\sim \{t\}, \{t\}$).

(f) *First iteration labels*: (from the initial flow) Label(a) = (s, +, 4), Label(b) = (a, +, 2), Label(c) = (s, +, 1), Label(d) = (s, +, 7), Label(g) = (s, +, 4), Label(h) = (s, +, 3), Label(t) = (a, +, 4).   Corresponding flow value = 4.

Second and third iterations now same as first and second iterations of Part (a).

Note: If we continue, the maximum flow of 21 will be detected in the 9th iteration. The algorithm will also produce the following minimum cut: ($\sim \{t\}, \{t\}$).

#11. If a network has several different parallel edges directed from one vertex to another, their effect on the flow network would be equivalent to a single edge whose capacity is the sum of those of all such parallel edges. Although the Ford–Fulkerson theorem and algorithm would still be valid for such networks, it would entail additional work (a larger number of edges), and any applications could be realized by this edge consolidation process. The Ford–Fulkerson theorem and algorithm can be extended to the setting networks with oppositely directed edges connecting pairs of vertices (i.e., undirected networks). This extension allows one to model undirected networks or partially undirected networks, such as the "counting edge disjoint paths" application that was described in Exercise for the Reader 9.22.

#13. (a) Let v be an internal vertex in a flow network with a flow constraint cap(v).   In case this capacity is at least as large as the sum of capacities of edges coming into v, or the sum of the capacities of all edges going out of v, then no modifications are necessary. If this is not the case, we can modify the network as follows: introduce an edge from v to a new vertex $\tilde{v}$ with capacity cap(v), transplant all edges going out of v to instead go out of this new vertex. The process is illustrated in the figure below:

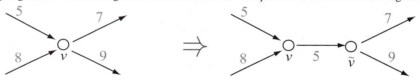

Vertex v Flow Limit = 5

#15. If we order the vertices as follows: s, Job(1), ..., Job(5), T(0,1), ..., T(6,8), t, then the first iteration of Step 3 of the algorithm gives a flow with value 1; Job(1) is assigned to T(0,1). The second iteration gives a flow value of 2; in addition to the assignment in the first iteration, Job(1) is assigned to T(1,2). The third iteration gives a flow value of 2; in addition to the assignment in the second iteration, Job(1) is assigned to T(2,4) (for the remaining 1 unit of time to complete it).

Note: A maximum flow of 15 is computed in the 15th iteration. This shows that two machines will suffice to solve this instance of the machine shop scheduling problem.

#17. The first iteration produces the path: $< s, b, t >$. The second iteration produces the additional disjoint path $< s, h, t >$. The third iteration produces the additional disjoint path $< s, i, t >$. The fourth iteration produces the additional disjoint path $< s, a, f, t >$. The fifth iteration produces the additional disjoint path $< s, g, j, t >$. The sixth iteration produces the additional disjoint path $< s, d, f, c, t >$. The seventh iteration certifies that the flow is a maximum flow (so the number of disjoint paths from s to t is 6). It also shows that $(\{s\}, \sim \{s\})$ is a corresponding minimum cut of capacity 6.

#19. (a) Here is a maximum matching for G: $\{\{a, f\}, \{c, d\}, \{g, h\}\}$.

Here is a maximum matching for H: $\{\{a, g\}, \{b, g\}, \{c, k\}, \{d, i\}, \{f, h\}\}$.

(b) Here is a maximal matching of G that is not a maximum matching: $\{\{a, g\}, \{b, f\}\}$.

Here is a maximal matching of H that is not a maximum matching: $\{\{a,g\}, \{c,i\}, \{d,g\}, \{f,h\}\}$.

#21. The flow network is illustrated in the figure below (all edges have capacity 1).

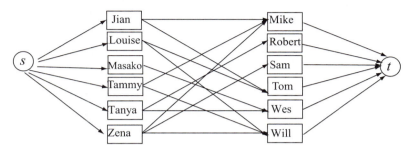

Typically you can save a lot of time by finding a good initial flow (of course, the problem is small enough to do by trial and error, but this is not the intention of the problem. Here is what happens with the algorithm if we begin with the zero flow:

First iteration: Jian and Mike are married. Second iteration: Louise and Tom are married.
Third iteration: Masako and Wes are married. Fourth iteration: Tammy and Will are married.
Fifth iteration: Tanya and Robert are married. Sixth iteration: Zena and Sam are married.
Seventh iteration: Maximum flow is confirmed.

#23. We will relate an instance of such a game of solitaire to a marriage problem involving 13 women $w_1, w_2, \cdots, w_{13}$ and 13 men $m_1, m_2, \cdots, m_{13}$. A woman w_i is marriage compatible with a man m_j if, and only if the ith column of the cards contains a card having the jth denomination, where the 13 denominations are ranked in some order (e.g., A,2,3,4,5,6,7,8,9,10,J,K,Q). Notice that a win of the game of solitaire corresponds to marrying off all of the women and men according to this compatibility definition. Hall's marriage theorem thus tells us that this will be possible if, and only if for any set of k women, the collective number of suitable men has at least k members. But this means that for any k columns, the number of different denominations of cards which appear must be at least k. Since each column has four cards, and there are exactly four cards of each denomination, the result immediately follows.

References

[Ahl-79] Ahlfors, Lars Valerian, *Complex Analysis, Third Edition*, McGraw-Hill, New York (1979)

[AhMaOr-93] Ahuja, Ravindra K., Thomas L. Magnanti, and James B. Orlin, *Network Flows*, Prentice-Hall, Upper Saddle River, NJ (1993)

[AMOR-95] Ahuja, Ravindra K., Thomas L. Magnanti, James B. Orlin, and M. R. Reddy, *Applications of network optimization*, pp. 1–88 in *Network Models*, edited by M. Ball, T. Magnanti, C. Monman, and G. Nemhauser, North-Holland, (1995)

[AlSp-00] Alon, Noga, and Joel H. Spencer, *The Probabilistic Method, Second Edition*, John Wiley & Sons, New York (2000)

[ABCC-06] Applegate, David L., Robert E. Bixby, Vašek Chvátal, and William J. Cook, *The Traveling Salesman Problem*, Princeton University Press, Princeton, NJ (2006)

[ALMSS-92] Arora, Sanjeev, Carsten Lund, Rajeev Motwani, Madhu Sudan, and Mario Szegdy, *Proof, verification, and intractability of approximation problems*, Proceedings of the 33rd IEEE Symposium on the Foundations of Computer Science, pp. 3–22, IEEE Press, Pittsburgh, PA (1992)

[AsDD-99] Ash, Robert B., and Catherine A. Doléans-Dade, *Probability & Measure Theory, Second Edition*, Elsevier, Amsterdam (1999)

[BaSh-96] Bach, Eric, and Jeffrey Shallit, *Algorithmic Number Theory, Vol. 1: Efficient Algorithms*, The MIT Press, Cambridge, MA (1996)

[BePi-82] Beker, Henry J., and Fred C. Piper, *Cipher Systems: The Protection of Communications*, John Wiley & Sons, New York (1982)

[BFLL-00] Bilbao, J. M., J. R. Fernández, A. Jiménez Losada, and J. J. López, *Generating functions for computing power indices efficiently*, Sociedad de Estadística e Investigatión Operativa, vol. 8, no. 2, pp. 191–213 (2000)

[Bil-85] Billingsley, Partick, *Probability and Measure, Second Edition*, John Wiley & Sons, New York (1985)

[BoDoTh-99] Bonabeau, Eric, Marco Dorigo, and Guy Theraulaz, *Swarm Intelligence: From Natural to Artificial Systems*, Oxford University Press, Oxford, UK (1999)

[Bra-08] Brass, Peter, *Advanced Data Structures*, Cambridge University Press, Cambridge, UK (2008)

[Bru-04] Brualdi, Richard A., *Introductory Combinatorics, Fourth Edition*, Prentice-Hall, Hoboken, NJ (2004)

[Bur-05] Burton, David M., *Elementary Number Theory, 6th Edition*, McGraw-Hill, New York (2005)

[CaWi-93] Campbell, Keith W., and Michael J. Wiener, *DES is not a group*, Proceedings of CRYPTO '93, Advances in Cryptography, published by Springer-Verlag (Lecture Notes in Computer Science, vol. 1976), pp. 30–43

[ChKaTo-99]   Chandra, Barun, Howard Karloff, and Craig Tovey,   *New results on the old k-opt algorithm for the traveling salesman problem*,   SIAM Journal on Computing, vol. 28(6), pp. 1998–2029 (1999)

[Coh-93]   Cohen, Henri,   *A Course in Computational Number Theory*,   Springer-Verlag, New York (1993)

[Coo-71]   Cook, Stephen A.,   *The complexity of theorem proving procedures*,   Proceedings of the 3rd Annual ACM Symposium on the Theory of Computing,   pp. 151–158 (1971)

[Chu-79]   Chung, Kai Lai.,   *Elementary Probability Theory with Stochastic Processes*,   Springer-Verlag, New York (1979)

[CoWi-90]   Coppersmith, Don, and Shmuel Winograd,   *Matrix multiplication via arithmetic progressions*,   Journal of Symbolic Computation,   vol. 9, no. 3, pp. 251–280 (1990)

[CLRS-01]   Cormen, Thomas H., Charles E. Leisserson, Ronald L. Rivest, and Clifford Stein, *Introduction to Algorithms, 2nd Edition*,   McGraw-Hill, New York (2001)

[DaLaPo-93]   Damgård, Ivan, Peter Landrock, and Carl Pommerance,   *Average case error estimates for the strong probable prime test*,   Mathematics of Computation, vol. 61(203), pp. 177–194 (1993)

[DaFuJo-54]   Dantzig, George B., Delbert Ray Fulkerson, and Selmer Johnson,   *Solution of a large scale traveling salesman problem*,   Technical Report P-510, RAND Corporation, Santa Monica, CA (1954)

[DPV-08]   Dasgupta, Sanjoy, Christos Papadimitriou, and Umesh Vazirani,   *Algorithms,*   McGraw-Hill, New York (2008)

[DiHe-76]   Diffie, B. Whitfield, and Martin E. Hellman,   *New directions in cryptography*,   IEEE Transactions on Information Theory, vol. 22, pp. 644–654 (1976)

[DiSz-78]   Dixon, L. C. W., and G. P. Szego,   *The Optimization Problem: An Introduction, Towards Global Optimization II*, North Holland Publishing, New York (1978)

[EdKa-72]   Edmonds, Jack, and Richard M. Karp,   *Theoretical Improvements in Algorithmic Efficiency for Network Flow Problems*, Journal of the Association of Computing Machinery, vol. 19 (2) (April) pp. 248–264 (1972)

[Edw-89a]   Edwards, Anthony W. F.,   *Venn diagrams for many sets*, Bulletin of the International Statistical Institute, 47th Session, Paris (1989). Contributed papers, Book 1, pp. 311–312

[Edw-89b]   Edwards, Anthony W. F.,   *Venn diagrams for many sets*, New Scientist, 7 (January 1989) pp. 51–56

[End-77] Enderton, Herbert B.,   *Elements of Set Theory*, Academic Press, New York (1977)

[ErSz-35]   Erdös, Paul, and George Szekeres,   *A combinatorial problem in geometry*, Compositio Mathematica, vol 2, pp. 464–470 (1935)

[GaJo-79]   Garey, Michael R., and David S. Johnson, *Computers and Intractability, A Guide to the Theory of NP-Completeness*, WH Freeman and Company, New York (1979)

[GlLa-97] Glover, Fred, and Manuel Laguna, *Tabu Search*, Springer-Verlag, New York (1997)

[Gol-89]   Goldberg, David E., *Genetic Algorithms in Search, Optimization, and Machine Learning*, Addison Wesley, Reading, MA (1989)

[GoPr-71] Goldstein, A. A., and I. F. Price, *On descent from local minima*, Math. Computation, vol 25, pp. 115–177 (1971)

[GoVL-83] Golub, Gene, H., and Charles F. Van Loan, *Matrix Computations*, The Johns Hopkins University Press, Baltimore (1983)

[GrGl-55] Greenwood, Robert E., and Andrew M. Gleason, *Combinatorial Relations and Chromatic Graphs*, pp. 1–7, Canadian Journal of Mathematics. vol. 7 (1955)

[GrHa-98] Gross, Donald, and Carl M. Harris, *Fundamentals of Queueing Theory, Third Edition*, John Wiley & Sons, New York (1998)

[GrYe-06] Gross, Jonathan L., and Jay Yellen, *Graph Theory and its Applications, Second Edition*, CRC Press, Boca Raton, FL (2006)

[Hak-61] Hakimi, S. L., *On the realizability of a set of integers as degrees of the vertices of a graph*, SIAM Journal of Applied Mathematics, 10, pp. 496–506 (1961)

[HaWr-80] Hardy, Godfrey H., and Edward M. Wright, *An Introduction to the Theory of Numbers*, Oxford University Press, Oxford, UK (1980)

[Hav-55] Havel, V., *A remark on the existence of finite graphs* (Czech), Časopis Pěst. Mat., vol. 80, pp. 477–480 (1955)

[HiLi-05] Hillier, Frederick S., and Gerald J. Lieberman, *Introduction to Operations Research, Eighth Edition*, McGraw-Hill, Boston, MA (2005)

[HiSt-01] Hinsley, F. H., and Alan Stripp (Editors), *Codebreakers: The Inside Story of Bletchley Park*, Oxford University Press, Oxford, UK (2001)

[HoKu-71] Hoffman, Kenneth, and Ray Kunze, *Linear Algebra*, Prentice-Hall, Englewood Cliffs, NJ (1971)

[Hun-96] Hungerford, Thomas W., *Abstract Algebra, An Introduction, Second Edition*, Brooks Cole, Pacific Grove, CA (1996)

[JüReRi-95] Jünger, M., G. Reinelt, and G. Rinaldi, *The traveling salesman problem*, in *Network Models*, edited by M. Ball, T. Magnanti, C. Monma, and G. Nemhauser, North-Holland, pp. 225–330 (1995)

[Kah-96] Kahn, David, *The Codebreakers: The Comprehensive History of Secret Communication from Ancient Times to the Internet*, Scribner Press, Princeton NJ (1996)

[Kar-53] Karnaugh, Maurice, *The Map Method for Synthesis of Combinational Logic Circuits*, Trans. AIEE. part I, 72(9):593–599, November (1953)

[KaLaMo-07] Kaye, Phillip, Raymond Laflamme, and Michele Mosca, An Introduction to Quantum Computing, Oxford University Press, New York (2007)

[KePfPi-04] Kellerer, Hans, Ulrich Pferschy, and David Pisinger, *Knapsack Problems*, Scribner Press, Princeton NJ (1996), Springer-Verlag, New York (2004)

[Ker-64] Kéry, Gerzson, *On a theorem of Ramsey*, Matematikai Lapok. Bolyai Janos Matematikai Tarsulat 15 (1964) 204-224

[Kle-00] Kleiner, Israel, *From Fermat to Wiles: Fermat's Last Theorem Becomes a Theorem*, Elem. Math. 55 (2000) 19–37

[Knu-97] Knuth, Donald E., *The Art of Computer Programming* (three volume set), Addison Wesley, Hoboken, NJ (1997)

[Kob-94] Koblitz, Neal, *A Course in Number Theory and Cryptography, Second Edition*, Springer-Verlag, New York (1994)

[KoHi-99] Kolman, Bernard, and David R. Hill, *Elementary Linear Algebra, Seventh Edition*, Prentice-Hall, Upper Saddle River, NJ (1999)

[LLRS-85] Lawler, Eugene L., Jan Karel Lenstra, A. H. G. Rinnooy Kan, and David B. Shmoys, *The Traveling Salesman Problem: A Guided Tour of Combinatorial Optimization*, John Wiley & Sons Inc., NewYork (1985)

[LiKe-73] Lin, Shen, and Brian W. Kernighan, *An effective heuristic algorithm for the traveling salesman problem*, Operations Research, vol. 21, pp. 498–516 (1973)

[LiXi-04], Ling, San, and Chaoping Xing, *Coding Theory: A First Course*, Cambridge University Press, Cambridge, UK (2004)

[MeOoVa-96]   Menezes, Alfred, Paul van Oorschot, and Scott Vanstone, *Handbook of Applied Cryptography*, Chapman Hall/CRC Press, Boca Raton, FL (1996)

[MeHe-78] Merkle, Ralph C., and Martin E. Hellman, *Hiding information and signatures in knapsack trapdoors*, IEEE Transactions on Information Theory, vol. IT-24, pp. 525–530 (1978)

[MeHe-81] Merkle, Ralph C., and Martin E. Hellman, *On the security of multiple encryption*, Communications of the ACM, vol. 24, pp. 465–467 (1981)

[Mil-76] Miller, Gary L., *Riemann's hypothesis and tests for primality*, Journal of Computer System Sciences, vol. 13, no. 3 pp. 300–317 (1976)

[Mit-98] Mitchell, Melanie, *An Introduction to Genetic Algorithms*, MIT Press, Cambridge, MA (1998)

[MiMäTo-03] Miettinen, Kaisa, Mäkelä, Marko M., and Jari Toivanen, *Numerical comparison of some penalty-based constraint handling techniques in genetic algorithms*, Journal of Global Optimization, vol. 27, pp. 427–446 (2003)

[Mol-03] Mollin, Richard A., *RSA and Public Key Cryptography*, Chapman & Hall/CRC, Boca Raton, FL (2003)

[Mol-05] Mollin, Richard A., *Codes, The Guide to Secrecy from Ancient to Modern Times*, Chapman & Hall/CRC, Boca Raton, FL (2005)

[Mon-05] Montgomery, Douglas C., *Design and Analysis of Experiments*, John Wiley & Sons Inc., NewYork (2005)

[Moo-05] Moon, Todd, K., *Error Correction Coding: Mathematical Methods and Algorithms*, John Wiley & Sons, Hoboken, NJ (2005)

[Mun-75] Munkres, James R., *Topology*, Prentice-Hall, Upper Saddle River, NJ (1975)

[MyMo-02] Myers, Raymond H., and Douglas C. Montgomery, *Response Surface Methodology, Process and Product Optimization Using Designed Experiments*, 2nd ed., John Wiley & Sons Inc., NewYork (2002)

[Nag-01] Nagell, Trygve, *Introduction to Number Theory*, *2nd Edition*, American Mathematical Society, Providence, RI (2001)

[NYYY-07] Nguyen, Hung Dinh, Ikuo Yoshihara, Kunihito Yamamori, Moritoshi Yasunaga, *Implementation of an effective hybrid GA for large-scale traveling salesman problems*, IEEE Transactions on Systems, Man, and Cybernetics, Part B, vol. 37(1), pp. 92–99 (2007)

[Poo-05] Poole, David, *Linear Algebra: A Modern Introduction*, Brooks Cole Publishing, Pacific Grove, CA (2005)

[Pri-57] Prim, Robert C., *Shortest connection networks and some generalizations*, Bell System Technical Journal, vol. 36, 1389–1401 (1957)

[PBSh-94] Prügel-Bennett, Adam, and Jonathan L. Shapiro, *An analysis of genetic algorithms using statistical mechanics*, Physical Review Letters vol. 72 no. 9 (1994) 1305–1309

[PBSh-97] Prügel-Bennett, Adam, and Jonathan L. Shapiro, *The dynamics of a genetic algorithm for simple randomising systems*, Physica D vol. 104 (1997) 75–114

[Rab-76] Rabin, Michael O., *Probabilistic algorithm for testing primality*, Journal of Number Theory, vol. 12, no. 1 pp. 128–138 (1980)

[RiPaLi-89] Richardson, John M., Palmer, Mark R., Liepins, Gunnar E., and Michael R. Hilliard, *Some guidelines for genetic algorithms with penalty functions*, Proceedings of the Third International Conference on Genetic Algorithms (J. D. Schaffer, editor), Morgan Kaufmann Publishers, San Mateo, CA (1989)

[RiShAd-78] Rivest, Ronald L., Adi Shamir, and Leonard Adleman, *A method for obtaining digital signatures and public-key cryptosystems*, Communications of the A.C.M., vol. 21, pp. 120–126 (1978)

[Rob-05] Robinson, Sara, *Toward an optimal algorithm for matrix multiplication*, SIAM News, vol. 38, no. 9, published by the Society of Industrial and Applied Math.(SIAM), Philadelphia, PA (2005)

[Ros-02] Ross, Sheldon M., *A First Course in Probability, Sixth Edition*, Prentice-Hall, Upper Saddle River, NJ (2002)

[Ros-06] Ross, Sheldon M., *Simulation, Fourth Edition*, Academic Press, New York (2006)

[RoStLe-77] Rosenkrantz, Daniel J., Richard E. Stearns, and Philip M. Lewis II, *An analysis of several heuristics for the traveling salesman problem*, SIAM Journal of Computation, Vol 6 (No. 3), pp. 563–581 (1977)

[Roy-88] Royden, Halsey, *Real Analysis, Third Edition*, Prentice-Hall, Upper Saddle River, NJ (1988)

[Sha-82] Shamir, Adi, *A polynomial-time algorithm for breaking the Merkel-Hellman cryptosystem*, Proceedings of the 23rd Annual Symposium on Foundations of Computer Science (Chicago, IL, 1982), published by the IEEE, New York, pp. 145–152 (1982)

[SmTa-93] Smith, Alice E., and David M. Tate, *Genetic optimization using a penalty function*, Proceedings of the 5th International Conference on Genetic Algorithms (S. Forrest, editor) Morgan Kaufmann Publishers, San Mateo, CA (1993)

[Sta-05] Stanoyevitch, Alexander, *Introduction to MATLAB® with Numerical Preliminaries*, John Wiley & Sons, Hoboken, NJ (2005)

[Sta-10] Stanoyevitch, Alexander, *Introduction to Cryptography with Mathematical Foundations and Computer Implementations*, Chapman & Hall/CRC, Boca Raton, FL (2010)

[Sti-06]   Stinson, R. Douglas, *Cryptography, Theory and Practice, Third Edition*, Chapman & Hall/CRC, Boca Raton, FL (2006)

[Str-69]   Strassen, Volker, *Gaussian elimination is not optimal*, Numerishe Mathematik, vol. 13, pp. 354-356 (1969)

[Ton-04]   Toni, A., *Lower bounds on zero-one matrices*, Linear Algebra and its Applications, vol. 376 pp. 275–282 (2004)

[Tve-80]   Tverberg, Helge, *A proof of the Jordan curve theorem*, Bulletin of the London Mathematical Society, vol. 12, no. 1, pp.34–38 (1980)

[TrWa-06]   Trappe, Wade, and Lawrence C. Washington, *An Introduction to Cryptography with Coding Theory, Second Edition*, Prentice-Hall, Upper Saddle River, NJ (2006)

[Vau-01]   Vaudenay, Serge, *Cryptanalysis of the Chor-Rivest cryptosystem*, Journal of Cryptology, vol. 14, no. 2, pp. 87–100 (2001)

[Ven-80]   Venn, John, *On the diagrammatic and mechanical representation of propositions and reasonings*, The London, Edinburgh, and Dublin Philosophical Magazine and Journal of Science, 9 pp.1–18. (1880)

[VoLi-91]   Vose, Michael, and Gunar E Liepins, *Punctuated equilibria in genetic search*, Complex Systems, vol. 5 (1991) 31–44

[WHLZ-06]   Wang, Yuping, Lixia Han, Yinghua Li, and Shuguang Zhao, *A new encoding based genetic algorithm for the traveling salesman problem*, Engineering Optimization, Vol. 38, No. 1, pp. 1–13 (2006)

[Wil-90]   Wilf, Herbert S., *Generatingfunctionology*, Academic Press, San Diego, CA (1990)

[Wil-98]   Williams, Hugh C., *Edouard Lucas and Primality Testing*, John Wiley & Sons, Hoboken, NJ (1998)

[Yak-08]   Yakuba, Vyacheslav, *Evaluation of Banzhaf index with restrictions on coalitions formation*, Mathematical and Computer Modelling, Vol. 48, no. 9–10, pp. 1602–1610 (2008)

Index of Theorems, Propositions, Lemmas, and Corollaries

Index of Algorithms

Index

A

Absorption, 10, 46, 93
Additive identity, 212
Adelman, Leonard, 278, 287
Al-Khwarizmi, Abu Muhammad, 449–450
Algorithms
    addition, with base b expansions, 194–196
    ant colony optimization algorithms, 741–744
    Banzhaf indices for weighted voting systems, for computing, 372–373
    big-O notation. *see* Big-O notations
    binary search. *see* Binary search algorithms
    bubble sort algorithms. *see* Bubble sort algorithms
    complexity theory of, 311, 480–482
    definition of, 132, 449
    deterministic. *see* Deterministic algorithms
    Dijkstra's algorithm, 653–657, 659, 660–661, 663
    Euclidian algorithm. *see* Euclidian algorithm
    fast modular exponentiation, 228–230
    Fermat's primality test, 180, 183
    Fleury's algorithm, 622–623, 644, 645
    Ford–Fulkerson maximum flow. *see* Ford-Fulkerson maximum flow
    genetic. *see* Genetic algorithms
    hill climbing algorithms. *see* Hill climbing algorithms
    k-opt local search algorithm for, 737, 738–739
    linear search. *see* Linear search algorithms
    merge sort algorithms. *see* Merge sort algorithms
    multiplication, with base b expansions, 199–202
    optimization algorithms. *see* Optimization algorithms
    ordered tree traversal algorithms. *see* Ordered tree traversal algorithms
    polynomial time, in, 487
    Prim's algorithm, 651–653, 654
    quick sort algorithms. *see* Quick sort algorithms
    randomized. *see* Randomized algorithms
    recursive graphic sequence determination algorithms, 507, 508–509
    RSA public key cryptosystem algorithm. *see* RSA cryptosystems
    searching/sorting, use in, 450, 571
    selection sort algorithms. *see* Selection sort algorithms
    simulated annealing algorithms, 744–745
    speed of, 470–472
    stochastic. *see* Randomized algorithms
    subtraction, with base b expansions, 196–199

Continued from inside front cover

$\det(A)$	The determinant of the square matrix A
A'	The determinant of the matrix A
A^n	The nth power of the square matrix A; $A^0 = I, = A \cdot A \cdots A \,(n \text{ times})$, if $n \in \mathbb{Z}_+$
$\mathrm{adj}(A)$	The classical adjoint of the square matrix A
$\mathrm{fl}(x)$	The floating point representation of the real number x

Counting and Probability

$|S|$	The cardinality of the set S = the number of elements in S
$P(n,k)$	The number of permuations of n objects taken k at a time
$C(n,k)$	The number of combinations of n objects taken k at a time
$\dbinom{n}{k}$	The binomial coefficient; $= C(n,k)$
$\dbinom{n}{n_1, n_2, \cdots, n_k}$	The binomial coefficient; $= \dfrac{n}{n_1! n_2! \cdots n_k!}$ (where $n_1 + n_2 + \cdots + n_k = n$)
$P(E)$	The probability that event E occurs
$P(E \mid F)$	The conditional probability of event E given that event F has occurred
$p_X(x)$	The probability distribution function (pdf) or a random variable X; $= P(X = x)$
$F_X(x)$	The cumulative distribution function (cdf) or a random variable X; $= P(X \le x)$
$X \sim \mathscr{B}(n, p)$	X is a binomial random variable with parameters n, p
$X \sim \mathscr{U}(a, b)$	X is a random variable that is uniformly distributed over the interval $[a,b]$
$E[X]$ or μ	The expectation (or expected value) of a random variable X
$\mathrm{Var}[X]$ or σ^2	The expectation (or expected value) of a random variable X
$1_A(x)$	The indicator function of the set A; $1_A(x) = 1$, if $x \in A$, otherwise $1_A(x) = 0$
$\mathrm{Cov}(X,Y)$	The covariance of the pair of random variables X, Y

Graphs

$V(G)$	The set of vertices of a graph G
$E(G)$	The set of edges of a graph G
$\deg(v)$ or $\deg_G(v)$	The degree of the vertex v (in the graph G)
K_n	The complete graph on n vertices
P_n	The n-path graph
C_n	The n-cycle graph $(n \ge 3)$
Q_n	The n-hypercube graph
$K_{n,m}$	The complete bipartite graph
$\sim G$	The complement graph of the graph G
$A(G)$	The adjacency matrix of the graph G
$B(G)$	The incidence matrix of the graph G
$\deg^-(v)$ or $\deg_{\vec{G}}^-(v)$	The indegree of the vertex v (in the directed graph $\vec{G}$)